Current Protocols
Essential Laboratory Techniques

EDITED BY

Sean R. Gallagher
UVP, LLC
Upland, California

Emily A. Wiley
Joint Science Department
Claremont McKenna, Pitzer, and Scripps Colleges
Claremont, California

Published by John Wiley & Sons, Inc.

Cover Image: ©Jason Reed/Digital Vision/Getty Images

Published by John Wiley & Sons, Inc., Hoboken, New Jersey.
Published simultaneously in Canada.

Library of Congress Cataloging-in-Publication Data

Current protocols essential laboratory techniques / edited by Sean R. Gallagher, Emily A. Wiley.
p. ; cm.
Includes bibliographical references and index.
ISBN 978-0-470-08993-4 (cloth)
1. Medical laboratories. 2. Diagnosis, Laboratory I. Gallagher, Sean R. II. Wiley, Emily A.
[DNLM: 1. Laboratory Techniques and Procedures. 2. Clinical Laboratory Techniques. 3. Clinical Protocols. QY 25 C976 2008]
RB36.C87 2008
610.28′4—dc22 2007037114
CIP

Printed in the United States of America

10 9 8 7 6 5 4 3 2 1

Contents

10 Enzymatic Reactions

Appendices

Index

Foreword

Advances in molecular biology and genomic methodologies in the past quarter century have unified as well as revolutionized the biological and medical sciences. This has led many thinkers to suggest that we have entered into a golden age of biological research. Recombinant DNA technology, invented in the early 1970s, was built upon the tools that bacterial geneticists developed after World War II. DNA sequencing, the polymerase chain reaction (PCR), high-throughput genomics, and bioinformatics further revolutionized the arsenal of techniques available to modern biological scientists. These methods of modern biological research are the threads that keep the somewhat disparate biological fields, as diverse as ecology and biophysics, sewn together. The acquisition of these diverse and sophisticated laboratory techniques is essential for the success of almost all modern research, but the breadth of biological methods that life science researchers are now expected to master is almost overwhelming. How to keep abreast of the continually increasing storehouse of molecular methods that includes protocols from bacteriology, genetics, molecular biology, cell biology, biochemistry, protein chemistry, biophysics, and bioinformatics?

Laboratory methods books are a good place to start. The *Current Protocols* series, of which *Current Protocols Essential Laboratory Techniques* is a part, is the most comprehensive set of published biological protocols. First published in 1987, *Current Protocols* has been a source of the latest methods in a variety of biological disciplines, with separate titles in Molecular Biology, Cell Biology, Immunology, Microbiology, Protein Science, Nucleic Acid Chemistry, and Bioinformatics, to name a few. *Current Protocols in Molecular Biology* was the first title in the series and is constantly being updated both in print and online. The updating feature has allowed all of the *Current Protocols* family to keep current by the addition of new protocols and the modification of old ones. Many molecular biology techniques are not only highly sophisticated but change rapidly. This is especially true in recent years as the rate of development of highly sophisticated high-throughput methods has rapidly increased.

Since first published in 1987, *Current Protocols in Molecular Biology* (CPMB) has been a rich source of many "basic" laboratory techniques. The latest edition of CPMB, now online, contains over 1,200 protocols, and the entire *Current Protocols* series contains a staggering 10,000 protocols and continues to grow.

But, CPMB, as well as the other *Current Protocols* titles, assumes a relatively sophisticated grounding in a variety of basic laboratory techniques. The fundamental concept of *Current Protocols Essential Laboratory Techniques* is to provide a more basic level of understanding by explaining in depth the principles that underlie key protocols, principles that are often taken for granted even by those who have been performing the techniques for years. This will be helpful not only for students or technicians who are just beginning their training in the laboratory, but also for experienced researchers who have learned techniques through "lab lore," but have never learned the theory behind the technical steps. *Current Protocols Essential Laboratory Techniques* is also designed to complement the general descriptions of laboratory techniques that students typically encounter in college and graduate-level biology courses. Thus, *Current Protocols Essential Laboratory Techniques* includes chapters on weighing and measuring volumes, spectroscopy, reagent preparation, growing bacterial and animal cells, centrifugation, purification and measurement of concentrations of nucleic acids, electrophoresis, protein and nucleic acid blotting, microscopy, and molecular enzymology, including the polymerase chain reaction (PCR). Importantly, *Current Protocols Essential Laboratory Techniques* also includes information not found in other *Current Protocol* titles, including appendices on how to keep a lab notebook and how to prepare and present a poster and PowerPoint presentation.

By providing training in and understanding of the theory behind basic techniques, *Current Protocols Essential Laboratory Techniques* eases the transition to the use of more sophisticated techniques and allows beginning researchers to adopt a much more sophisticated approach to experimental design and troubleshooting, two areas that are particularly challenging to beginners who may not fully understand the underlying principles of the techniques that they are using. The philosophy underlying *Current Protocols Essential Laboratory Techniques*, and other *Current Protocol* titles, is that it is not sufficient simply to master the steps of a protocol, but that it is also critical to understand how and why a technique works, when to use a particular technique, what kind of information a technique can and cannot provide, and what the critical parameters are in making the technique work. The goal is to provide a theoretical and practical foundation so that researchers do not simply use techniques by rote but obtain a level of understanding sufficient to develop new techniques on their own. This is particularly useful as more and more commercial molecular biology "kits" are becoming available to carry out a large variety of techniques. While kits often save time, they obviate the need to understand how a technique works, making it difficult to troubleshoot if something goes wrong or to perform the technique in a more efficient or more cost-effective way.

Preparing reliable but relatively simple protocols with clearly defined parameters and troubleshooting sections is a tremendous amount of work. *Current Protocols Essential Laboratory Techniques* reflects the extensive expertise and experience of the editors, Sean R. Gallagher and Emily A. Wiley, in technology development and in teaching biology laboratory courses at the Claremont Colleges, respectively. A good protocol is useless, however, if it is not described clearly with step-by-step instructions without ambiguities that lead down false paths. One of the most important features of the *Current Protocols* series is extensive editing and strict adherence to the proven *Current Protocols* "style." Protocols that work the vast majority of time with minimum expenditure of time and effort are extraordinarily valuable, collected assiduously, and highly prized by laboratory researchers.

We welcome you, the user of this manual, both newcomers to molecular biology and old hands who want to brush up on the basics. It is you who will be inventing the next generation of molecular biology protocols that will form the basis of the next revolution in this fast-moving field. Happy experimenting!

Fred Ausubel
Department of Genetics
Harvard Medical School and
Department of Molecular Biology
Massachusetts General Hospital

Preface

Current Protocols Essential Laboratory Techniques (CPET) is a fundamentally new type of laboratory manual. Although written for those new to life science, it will appeal to an unusually wide range of scientists—advanced undergraduates, graduate students, and professors alike—as a reference book and a "how to" lab bench manual for research. With its breadth and focus on techniques and applications spanning from PCR to high-resolution digital imaging, complete with detailed step-by-step instructions, CPET will be an indispensable resource for the laboratory worker.

Historically, scientific researchers have learned basic techniques by reading protocols or observing others at the bench. Often, protocol steps and their nuances are "lab lore" that have been passed down through generations of post-docs and graduate students. Through this process, the theoretical reasoning behind various procedural steps is often lost. Most molecular techniques manuals provide detailed descriptions of technical protocols, but little in the way of theoretical explanations. Yet, a more complete understanding of technical theory enables researchers to design the most appropriate and cost-effective experiments, to interpret data more accurately, and to solve problems more efficiently. A primary aim of CPET is to elucidate the fundamental chemical and physical principles underlying techniques central to molecular research. This information is especially useful in today's world where reagents for many techniques are packaged and sold in kits, with minimal technical explanation, complicating troubleshooting and data interpretation.

This manual explains the most common, or "essential," bench techniques and basic laboratory skills. A complete set of references, including the seminal papers for a technique, are also provided for better understanding of the historical basis of a protocol. It is easy to forget the origin and original authors of a procedure even though those early publications have a wealth of troubleshooting information that should be consulted prior to experimentation.

Although it is written in a style accessible for undergraduate and graduate students, even the seasoned experimenter will find a large number of useful explanations within, many related to making decisions for efficient and cost-effective laboratory setup. It was the editors' goal to compile information that would enable beginning researchers to more quickly become independent at the bench. In addition to covering broadly transferable skills such as reagent preparation and weight and volume measurement, other essential training—including keeping experimental records, proper manipulation of digital data images for publication, and effective presentation of research—is addressed. This manual thus serves as a companion for more advanced technical manuals, such as the other titles in the *Current Protocols* series, and is a basic standard reference for any molecular laboratory. As this manual covers technical skills taught in many undergraduate molecular and cell biology classrooms, it may also be used as a text for these courses. The editors expect that it will not only support early research experiences but will continue to be a key reference throughout a student's subsequent years of training and practice.

HOW TO USE THIS MANUAL

Organization

Chapters and appendices

Subjects are organized by chapters, which are composed of units. Page numbering reflects this modular arrangement: for example, page 10.3.7 refers to Chapter 10 (Enzymatic Reactions), Unit 3 (Real-Time PCR), page 7.

In addition to the material provided in the chapters, several appendices are included to provide the reader with the tools he or she needs to analyze, record, and disseminate the results of experiments.

Units

Units provide either step-by-step protocols or detailed overviews, depending on the subject matter being presented. In the case of units presenting protocols, each unit is built along the same general format to allow the reader to easily navigate the provided information.

The first part of each unit presenting protocols includes a detailed discussion of the theory behind the techniques presented within ("Overview and Principles"), important choices to make before the experiment is undertaken ("Strategic Planning"), and specific safety precautions ("Safety Considerations").

This is followed by a section entitled "Protocols." Each protocol begins with a brief introduction to the technique, followed by a list of materials in the order they are used (reagents and equipment being treated separately), and finally by the protocol steps, which are often supported by italicized annotations providing additional information. See below for an explanation of the different types of protocols and how they interrelate.

Recipes specific to the techniques described in the unit are provided under the head "Reagents and Solutions," while recipes for more general reagents can be found in *UNIT 3.3*.

NOTE: Unless otherwise stated, deionized, distilled water should be used in all protocols in this manual, and in the preparation of all reagents and solutions. Protocols requiring aseptic technique (see *UNIT 4.1*) are indicated.

Guidance towards understanding the product of the experiment ("Understanding Results") and solving common problems ("Troubleshooting") is provided next. When applicable, this may be followed by a list of resources where the reader can find details about variations of the technique that are beyond the scope of this basic skills manual.

Sections providing resources where the researcher can find more information—Literature Cited, Key References, and Internet Resources—are provided in the last part of the unit.

References

While the Editors have attempted to follow a logical progression in arranging the methodology described in this manual, real experiments often have a unique order of techniques that must be executed. Furthermore, while every effort has been made to provide the basic skill set needed to work in the modern life science laboratory, it is impossible to include every technique and variation the researcher may need. To address these issues, throughout the book readers are referred to related techniques described in other units within this manual, in other titles in the *Current Protocols* series, in other literature, and on Web sites.

Protocols

Many units in this manual contain groups of protocols, each presented with a series of steps. A *basic protocol* is presented first in each unit and is generally the recommended or most universally applicable approach. Additional basic protocols may be included where appropriate. *Alternate protocols* are provided where different equipment or reagents can be employed to achieve similar ends, where the starting material requires a variation in approach, or where requirements for the end product differ from those in the basic protocol. *Support protocols* describe additional steps that are required to perform the basic or alternate protocols; these steps are separated from the core protocol because they might be applicable to more than one technique or because they are performed in a time frame separate from the protocol steps which they support.

Commercial Suppliers

Throughout the manual, commercial suppliers of chemicals, biological materials, and equipment are recommended. In some cases, the noted brand has been found to be of superior quality, or is the only suitable product available in the marketplace. In other cases, the experience of the author of that protocol is limited to that brand. In the latter situation, recommendations are offered as an aid to the novice experimenter in obtaining the tools of the trade. Experienced investigators are therefore encouraged to experiment with substituting their own favorite brands.

Safety Considerations

Anyone carrying out these protocols may encounter the following hazardous or potentially hazardous materials: (1) pathogenic and infectious biological agents, (2) recombinant DNA, (3) radioactive substances, and (4) toxic chemicals and carcinogenic or teratogenic reagents. Most governments regulate the use of these materials; it is essential that they be used in strict accordance with local and national regulations.

APPENDIX 1 provides information relating to general laboratory safety and should be considered required reading before performing any experiment. Safety Considerations are provided for units describing protocols, and additional cautionary notes are included throughout the manual. However, it must be emphasized that users must proceed with the prudence and caution associated with good laboratory practice. Radioactive substances must of course be used only under the supervision of licensed users, following guidelines of the appropriate regulatory body, e.g., the Nuclear Regulatory Commission (NRC). See *UNIT 2.3* for more detail.

ACKNOWLEDGMENTS

A project of this scope is only possible with the assistance of a great number of people. We are greatly indebted to the staff at John Wiley & Sons, Inc., who helped get this project going and have continued to support it. We are especially grateful to Tom Downey and Virginia Chanda for their editorial efforts, and to Scott Holmes, Tom Cannon, Jr., Marianne Huntley, Susan Lieberman, Maria Monte, Sylvia Muñoz de Hombre, Allen Ranz, Erica Renzo, and Joseph White for their skillful assistance. We would also like to acknowledge David Sadava who graciously provided helpful advice during the early stages of this projects.

We extend our thanks to the contributors for sharing their knowledge; without them this manual would not be possible. We thank members of our laboratories and our colleagues throughout the world for their ongoing support. Our warm thanks also go out to our families for their encouragement and understanding.

Sean R. Gallagher and Emily A. Wiley

Contributors

Jennifer A. Armstrong
Joint Science Department
Claremont McKenna, Pitzer, and Scripps Colleges
Claremont, California

David J. Asai
Harvey Mudd College
Claremont, California

Donna Bozyczko-Coyne
Cephalon, Inc.
Frazer, Pennsylvania

Steve Bursik
The Salk Institute
La Jolla, California

Tomasz Bykowski
University of Kentucky College of Medicine
Lexington, Kentucky

Bulbul Chakravarti
Proteomics Center
Keck Graduate Institute of Applied Life Sciences
Claremont, California

Deb N. Chakravarti
Proteomics Center
Keck Graduate Institute of Applied Life Sciences
Claremont, California

Stephen Chang
Keck Graduate Institute of Applied Life Sciences
Claremont, California

Buena Chui
GE Healthcare
Piscataway, New Jersey

Eric S. Cole
St. Olaf College
Northfield, Minnesota

Thomas Davis
Joint Science Department
Claremont McKenna, Pitzer, and Scripps Colleges
Claremont, California

Bruce Dorsey
Cephalon, Inc.
Frazer, Pennsylvania

Dennis H. Dowhan
Diamantina Institute for Cancer Immunology and Metabolic Medicine
University of Queensland
Woolloongabba, Queensland
Australia

Steven Fenster
Ashland University
Ashland, Ohio

Dean Fraga
College of Wooster
Wooster, Ohio

Sean R. Gallagher
UVP, LLC
Upland, California

Michael Guzy
Ohaus Corporation
Pine Brook, New Jersey

Matthew Hall
Joint Science Department
Claremont McKenna, Pitzer, and Scripps Colleges
Claremont, California

Johanna Hardin
Pomona College
Claremont, California

Karl A. Haushalter
Harvey Mudd College
Claremont, California

Laura L. Mays Hoopes
Pomona College
Claremont, California

Nick Huang
Joint Science Department
Claremont McKenna, Pitzer, and Scripps Colleges
Claremont, California

John Kloke
Pomona College
Claremont, California

Christine D. Kuslich
Molecular Profiling Institute
Phoenix, Arizona

Scott Larsen
Cephalon, Inc.
Frazer, Pennsylvania

George Lunn
Baltimore, Maryland

Buddhadeb Mallik
Proteomics Center
Keck Graduate Institute of Applied Life Sciences
Claremont, California

Glenn M. Manthey
City of Hope
Beckman Research Institute
Duarte, California

Stephen D. Mastrian
Translational Genomics Research Institute
Phoenix, Arizona

Scott Medberry
Agilent Technologies, Inc.
Palo Alto, California

Jill Meisenhelder
The Salk Institute
La Jolla, California

Tea Meulia
Ohio Agricultural Research and Development Center
Wooster, Ohio

Butch Moomaw
Hamamatsu Photonic Systems
Spring Branch, Texas

Rob Morris
Ocean Optics, Inc.
Dunedin, Florida

Maria Cristina Negritto
Pomona College
Claremont, California

Benson Ngo
Joint Science Department
Claremont McKenna, Pitzer, and Scripps Colleges
Claremont, California

Sean P. Riley
University of Kentucky College of Medicine
Lexington, Kentucky

Cindy Santangelo
Worthington Biochemical
Lakewood, New Jersey

Joachim Sasse
Shriners Hospital for Crippled Children
Tampa, Florida

Joseph R. Schulz
Occidental College
Los Angeles, California

Jerry Sedgewick
University of Minnesota
Minneapolis, Minnesota

Christine Sjolander
Keck Graduate Institute of Applied Life Sciences
Claremont, California

David Skrincosky
Worthington Biochemical
Lakewood, New Jersey

Brian Stevenson
University of Kentucky College of Medicine
Lexington, Kentucky

Warren Strober
National Institute of Allergy and Infectious Diseases
Bethesda, Maryland

Sharon Torigoe
Joint Science Department
Claremont McKenna, Pitzer, and Scripps Colleges
Claremont, California

Jeffrey W. Touchman
Translational Genomics Research Institute
Phoenix, Arizona

Emily A.Wiley
Joint Science Department
Claremont McKenna. Pitzer, and Scripps College
Claremont, California

Michael Williams
Cephalon, Inc.
Frazer, Pennsylvania

Michael E. Woodman
University of Kentucky College of Medicine
Lexington, Kentucky

Carl T. Yamashiro
Arizona State University
Tempe, Arizona

Andrew Zanella
Joint Science Department
Claremont McKenna, Pitzer, and Scripps Colleges
Claremont, California

Common Conversion Factors

INTRODUCTION

Presented here is a brief overview of some of the more common units of measure used in the life sciences. Table 1 describes the prefixes indicating powers of ten for SI units (International System of Units). Table 2 provides conversions for units of volume. Table 3 provides some temperatures commonly encountered in the life science laboratory in equivalent Celsius and Fahrenheit degrees. Table 4 lists some of the more common conversion factors for units of measure, grouped categorically.

Table 1 Powers of Ten Prefixes for SI Units

Prefix	Factor	Abbreviation
Atto	10^{-18}	a
Femto	10^{-15}	f
Pico	10^{-12}	p
Nano	10^{-9}	n
Micro	10^{-6}	μ
Milli	10^{-3}	m
Centi	10^{-2}	c
Deci	10^{-1}	d
Deca	10^{1}	da
Hecto	10^{2}	h
Kilo	10^{3}	k
Myria	10^{4}	my
Mega	10^{6}	M
Giga	10^{9}	G
Tera	10^{12}	T
Peta	10^{15}	P
Exa	10^{18}	E

CONVERTING UNITS OF VOLUME

Conversion of units of volume is a fundamental part of life science research, and it often presents difficulties for the novice. For these reasons, the volume conversions are presented in Table 2, rather than including them in a larger table that addresses other conversion units.

CONVERTING TEMPERATURES

Table 3 provides conversions between degrees Celsius and Fahrenheit for some temperatures that are commonly used in the laboratory. Other conversions may be made using the equations below.

Celsius temperatures are converted to Fahrenheit temperatures by multiplying the Celsius figure by 9, dividing by 5, and adding 32; or by multiplying the Celsius figure by 1.8 and adding 32:

$$°F = (9/5)(°C) + 32 = 1.8(°C) + 32$$

Table 2 Units of Volume Conversion Chart

To convert:	Into:	Multiply by:
Cubic centimeters (cm^3)	Cubic feet (ft^3)	3.531×10^{-5}
	Cubic inches ($in.^3$)	6.102×10^{-2}
	Cubic meters (m^3)	10^{-6}
	Cubic yards	1.308×10^{-6}
	Gallons, U.S. liquid	2.642×10^{-4}
	Liters	10^{-3}
	Pints, U.S. liquid	2.113×10^{-3}
	Quarts, U.S. liquid	1.057×10^{-3}
Liters	Bushels, U.S. dry	2.838×10^{-2}
	Cubic centimeters (cm^3)	10^3
	Cubic feet (ft^3)	3.531×10^{-2}
	Cubic inches ($in.^3$)	61.02
	Cubic meters (m^3)	10^{-3}
	Cubic yards	1.308×10^{-3}
	Gallons, U.S. liquid	0.2642
	Gallons, imperial	0.21997
	Kiloliter (kl)	10^{-3}
	Pints, U.S. liquid	2.113
	Quarts, U.S. liquid	1.057
Microliters (μl)	Liters	10^{-6}
Milliliters (ml)	Liters	10^{-3}
Ounces, fluid	Cubic inches ($in.^3$)	1.805
	Liters	2.957×10^{-2}
Quarts, dry	Cubic inches ($in.^3$)	67.20
Quarts, liquid	Cubic centimeters (cm^3)	946.4
	Cubic feet (ft^3)	3.342×10^{-2}
	Cubic inches ($in.^3$)	57.75
	Cubic meters (m^3)	9.464×10^{-4}
	Cubic yards	1.238×10^{-3}
	Gallons	0.25
	Liters	0.9463

Fahrenheit temperatures are converted to Celsius temperatures by subtracting 32 from the Fahrenheit figure, multiplying by 5, and dividing by 9; or by subtracting 32 from the Celsius figure and dividing by 1.8:

$$°C = (5/9)(°F - 32) = (°F - 32)/1.8$$

To convert to the Kelvin scale (absolute temperature), add 273.15 to the temperature in degrees Celsius:

$$K = °C + 273.15 = [(°F - 32)/1.8] + 273.15$$

Table 3 Commonly Encountered Temperatures

Degrees Celsius (°C)	Degrees Fahrenheit (°F)
−80	−112.0
−20	−4.0
−4	30.8
0	32.0
4	39.2
20	68.0
25	77.0
30	86.0
37	98.6
72	161.6
100	212.0

By SI convention, temperatures expressed on the Kelvin scale are known as "kelvins" rather than "degrees Kelvin," and thus do not take a degree symbol.

A NOTE ABOUT WRITING UNITS OF MEASURE

By SI convention, unit names should *not* be treated as proper nouns when written in text and should thus be written with the initial letter lowercase—e.g., meter, newton, pascal—unless starting a sentence, included in a title, or in similar situations dictated by style. The exception is "Celsius," which is treated as a proper noun.

Unit symbols should likewise be written lowercase, unless derived from a proper noun, in which case the first letter is capitalized—e.g., m for meter, N for newton, Pa for pascal. The one exception is the interchangeable uppercase and lowercase L for liter, which is provided to prevent misunderstanding in circumstances where the number one and the lowercase letter "l" might be confused. Refer to *http://www.bipm.org/en/si/si_brochure* for more information.

INTERNET RESOURCES

http://www.bipm.org/en/home

Homepage of the Bureau International des Poids et Measures, the body charged with ensuring world-wide uniformity of measurements and their traceability to the International System of Units (SI).

http://www.bipm.org/en/si/si_brochure

The electronic version of the SI brochure. This document contains the definitive definitions of units of the SI system, as well as information about their conversion and proper usage.

http://www.nist.gov

Homepage of the U.S. National Institute of Standards and Technology (NIST), a nonregulatory federal agency within the U.S. Department of Commerce. NIST's mission is to promote U.S. innovation and industrial competitiveness by advancing measurement science, standards, and technology in ways that enhance economic security and improve the quality of life. In addition to detailed information about the use and conversion of units of measure, NIST is the source of reference standards within the United States.

http://ts.nist.gov/WeightsAndMeasures/Publications/upload/h4402_appenc.pdf

The General Tables of Units of Measure, provided by NIST. An excellent source of conversion factors appropriate for the needs of the average user as well as users requiring conversion factors with a large number of decimal places.

Table 4 starts on the following page.

Table 4 Units of Measurement Conversion Chart[a]

To convert:	Into:	Multiply by:
Angle		
Degrees (°) of angle	Minutes (min)	60.0
	Quadrants, of angle	1.111×10^{-2}
	Radians (rad)	1.745×10^{-2}
	Seconds (sec)	3.6×10^{4}
Quadrants of angle	Degrees (°)	90.0
	Minutes (min)	5.4×10^{3}
	Radians (rad)	1.571
	Seconds (sec)	3.24×10^{5}
Radians (rad)	Degrees (°)	57.30
	Minutes (min)	3,438
	Quadrants	0.6366
	Seconds (sec)	2.063×10^{5}
Concentration[b]		
Milligrams per liter (mg/liter)	Parts per million (ppm)	1.0
Distance		
Centimeters (cm)	Feet (ft)	3.281×10^{-2}
	Inches (in.)	0.3937
	Kilometers (km)	10^{-5}
	Meters (m)	10^{-2}
	Miles	6.214×10^{-6}
	Millimeters (mm)	10.0
	Mils	393.7
	Yards	1.094×10^{-2}
Inches (in.)	Centimeters (cm)	2.540
	Feet (ft)	8.333×10^{-2}
	Meters (m)	2.540×10^{-2}
	Miles	1.578×10^{-5}
	Millimeters (mm)	25.40
	Yards	2.778×10^{-2}
Kilometers (km)	Centimeters (cm)	10^{5}
	Feet (ft)	3,281
	Inches (in.)	3.937×10^{4}
	Meters (m)	10^{3}
	Miles	0.6214
	Yards	1,094
Millimeters (mm)	Centimeters (cm)	0.1
	Feet (ft)	3.281×10^{-3}

continued

Table 4 Units of Measurement Conversion Chart[a], *continued*

To convert:	Into:	Multiply by:
	Inches (in.)	3.937×10^{-2}
	Kilometers (km)	10^{-6}
	Meters (m)	10^{-3}
	Miles	6.214×10^{-7}
Electric current and charge		
Amperes per square centimeter (amp/cm^2)	Amperes per square inch (amp/$in.^2$)	6.452
	Amperes per square meter (amp/m^2)	10^4
Amperes per square inch (amp/$in.^2$)	Amperes per square centimeter (amp/cm^2)	0.1550
	Amperes per square meter (amp/m^2)	1.55×10^3
Ampere-hours (amp-hr)	Coulombs (C)	3.6×10^3
	Faradays	3.731×10^{-2}
Coulombs (C)	Faradays	1.036×10^{-5}
Coulombs per square centimeter (C/cm^2)	Coulombs per square inch (C/$in.^2$)	64.52
	Coulombs per square meter (C/m^2)	10^4
Coulombs per square inch (C/$in.^2$)	Coulombs per square centimeter (C/cm^2)	0.1550
	Coulombs per square meter (C/m^2)	1.55×10^3
Faradays	Ampere-hours (amp-hr)	26.80
	Coulombs (C)	9.649×10^{-4}
Energy		
British thermal units (Btu)	Ergs	1.0550×10^{10}
	Gram-calories (g-cal)	252.0
	Horsepower-hours (hp-hr)	3.931×10^4
	Joules (J)	1,054.8
	Kilogram-calories (kg-cal)	0.2520
	Kilogram-meters (kg-m)	107.5
	Kilowatt-hours (kW-hr)	2.928×10^{-4}
British thermal unit per minute (Btu/min)	Foot-pounds per second (ft-lb/sec)	12.96
	Horsepower (hp)	2.356×10^{-2}
	Watts (W)	17.57
Foot-pounds per minute (ft-lb/min)	British thermal units per minute (Btu/min)	1.286×10^{-3}
	Foot-pounds per second (ft-lb/sec)	1.667×10^{-2}
	Horsepower (hp)	3.030×10^{-5}
	Kilogram-calories per minute (kg-cal/min)	3.24×10^{-4}

continued

Table 4 Units of Measurement Conversion Chart[a], *continued*

To convert:	Into:	Multiply by:
	Kilowatts (kW)	2.260×10^{-5}
Horsepower (hp)	Horsepower, metric	1.014
Joules (J)	British thermal units (Btu)	9.480×10^{-4}
	Ergs	10^{7}
	Foot-pounds (ft-lb)	0.7376
	Kilogram-calories (kg-cal)	2.389×10^{-4}
	Kilogram-meters (kg-m)	0.1020
	Newton-meter (N-m)	1
	Watt-hours (W-hr)	2.778×10^{-4}
Kilowatts (kW)	British thermal units per minute (Btu/min)	56.92
	Foot-pounds per minute (ft-lb/min)	4.426×10^{4}
	Horsepower (hp)	1.341
	Kilogram-calories per minute (kg-cal/min)	14.34
Watts (W)	British thermal units per hour (Btu/hr)	3.413
	British thermal units per min (Btu/min)	5.688×10^{-2}
	Ergs per second (ergs/sec)	10^{7}
Force		
Dynes (dyn)	Joules per centimeter (J/cm)	10^{-7}
	Joules per meter (J/m) or newtons (N)	10^{-5}
	Kilograms (kg)	1.020×10^{-6}
	Pounds (lb)	2.248×10^{-6}
Newtons (N)	Dynes (dyn)	10^{5}
	Kilograms, force (kg)	0.10197162
	Pounds, force (lb)	4.6246×10^{-2}
Pounds, force (lb)	Newtons (N)	21.6237
Mass		
Grams (g)	Decigrams (dg)	10
	Decagrams (dag)	0.1
	Dynes (dyn)	980.7
	Grains	15.43
	Hectograms (hg)	10^{-2}
	Kilograms (kg)	10^{-3}
	Micrograms (μg)	10^{6}
	Milligrams (mg)	10^{3}
	Ounces, avoirdupois (oz)	3.527×10^{-2}

continued

Table 4 Units of Measurement Conversion Chart[a], *continued*

To convert:	Into:	Multiply by:
	Ounces, troy	3.215×10^{-2}
	Pounds (lb)	2.205×10^{-3}
Micrograms (μg)	Grams (g)	10^{-6}
Milligrams (mg)	Grams (g)	10^{-3}
Ounces, avoirdupois	Drams	16.0
	Grains	437.5
	Grams (g)	28.349527
	Pounds (lb)	6.25×10^{-2}
	Ounces, troy	0.9115
	Tons, metric	2.835×10^{-5}
Ounces, troy	Grains	480.0
	Grams (g)	31.103481
	Ounces, avoirdupois (oz)	1.09714
	Pounds, troy	8.333×10^{-2}
Pressure		
Atmospheres (atm)	Bar	1.01325
	Millimeters of mercury (mmHg) or torr	760
	Tons per square foot (tons/ft^2)	1.058
Bar	Atmospheres (atm)	0.9869
	Dynes per square centimeter (dyn/cm^2)	10^6
	Kilograms per square meter (kg/m^2)	1.020×10^4
	Pounds per square foot (lb/ft^2)	2,089
	Pounds per square inch (lb/in.^{2} or psi)	14.50
Inches of mercury (in. Hg)	Atmospheres (atm)	3.342×10^{-2}
	Kilogram per square centimeter (kg/cm^2)	3.453×10^{-2}
	Kilograms per square meter (kg/m^2)	345.3
	Pounds per square foot (lb/ft^2)	70.73
	Pounds per square inch (lb/in.^{2} or psi)	0.4912
Millimeters of mercury (mmHg) or torr	Atmospheres (atm)	1.316×10^{-3}
	Kilograms per square meter (kg/m^2)	136.0
	Pounds per square foot (lb/ft^2)	27.85
	Pounds per square inch (lb/in.^{2} or psi)	0.1934
Pascal (P)	Newton per square meter (N/m^2)	1

continued

Table 4 Units of Measurement Conversion Chart[a], *continued*

To convert:	Into:	Multiply by:
Pounds per square foot (lb/ft^2)	Atmospheres (atm)	4.725×10^{-4}
	Inches of mercury (in. Hg)	1.414×10^{-2}
	Kilograms per square meter (kg/m^2)	4.882
	Pounds per square inch (lb/in.^{2} or psi)	6.944×10^{-3}
Pounds per square inch (lb/in.^{2} or psi)	Atmospheres (atm)	6.804×10^{-2}
	Inches of mercury (in. Hg)	2.036
	Kilograms per square meter (kg/m^2)	703.1
	Pounds per square foot (lb/ft^2)	144.0
	Bar	6.8966×10^{-2}
Torr (*see* millimeter of mercury)		
Resistance		
Ohms (Ω)	Megaohms (MΩ)	10^6
	Microhms (μΩ)	10^{-6}
Time		
Days	Hours (hr)	24.0
	Minutes (min)	1.44×10^3
	Seconds (sec)	8.64×10^4
Velocity		
Centimeters per second (cm/sec)	Feet per minute (ft/min)	1.1969
	Feet per second (ft/sec)	3.281×10^{-2}
	Kilometers per hour (km/hr)	3.6×10^{-2}
	Meters per minute (m/min)	0.6
	Miles per hour (miles/hr)	2.237×10^{-2}
	Miles per minute (miles/min)	3.728×10^{-4}

[a]See Table 2 for conversion of units of volume.

[b]Refer to *UNIT 3.1* for a detailed description of different means for describing concentration.

Combining Techniques to Answer Molecular Questions

INTRODUCTION

This manual is a collection of basic techniques central to the study of nucleic acids, proteins, and whole-cell/subcellular structures. The following is an overview of how the basic techniques described in this manual relate to each other, and describes sequences of techniques that are commonly used to answer questions about proteins and nucleic acids. Flowcharts are provided to orient the novice researcher in the use of fundamental molecular techniques, and provide perspective regarding the technical units in this manual.

NUCLEIC ACIDS

Listed below are common questions about nucleic acids and techniques used to answer them. Also refer to Figure 1.

Genomic and Plasmid DNA Analyses

Does a particular genomic locus or region of plasmid DNA contain a sequence of interest? Where does it reside?

Techniques:
Restriction enzyme digestion (*UNIT 10.1*)
Agarose gel electrophoresis (*UNIT 7.2*)
Southern blot (*UNITS 8.1 & 8.2*)

How many genomic loci contain a particular sequence of interest, or how many copies of that sequence does a genome contain?

Technique:
Southern blot (*UNITS 8.1 & 8.2*)

What is the sequence of a specific DNA fragment?

Technique:
DNA sequencing (*UNIT 10.4*)

Gene Expression (Transcription) Analyses

What is the size of a specific gene transcript?

Technique:
Northern blot (*UNITS 8.1 & 8.2*)

Is a gene of interest expressed (transcribed)?

Technique:
Northern blot (*UNITS 8.1 & 8.2*)

Is transcription of a gene altered (increased or decreased) under different conditions?

Technique:
Northern blot (*UNITS 8.1 & 8.2*)
Real-time PCR (for more quantitative comparison; *UNIT 10.3*)

What is the relative abundance of mRNAs made from a specific gene compared to that from other genes?

Technique:
Real-time PCR (*UNIT 10.3*)

What are characteristics (abundance and size) of rRNAs or tRNAs?

Technique:
Northern blot (*UNITS 8.1 & 8.2*)
Real-time PCR (*UNIT 10.2*)

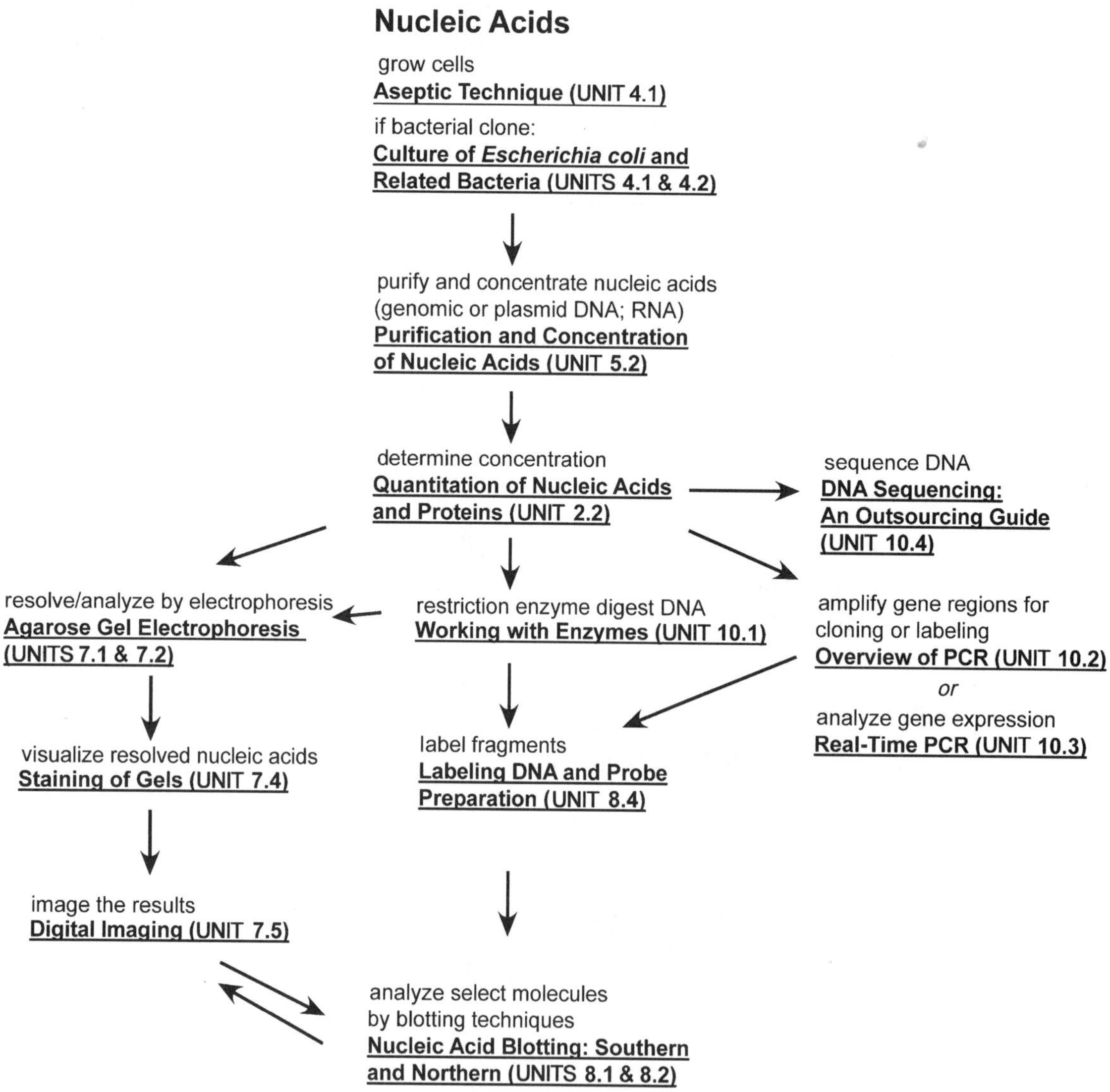

Nucleic Acids

grow cells
Aseptic Technique (UNIT 4.1)
if bacterial clone:
Culture of *Escherichia coli* and Related Bacteria (UNITS 4.1 & 4.2)

↓

purify and concentrate nucleic acids (genomic or plasmid DNA; RNA)
Purification and Concentration of Nucleic Acids (UNIT 5.2)

↓

determine concentration
Quantitation of Nucleic Acids and Proteins (UNIT 2.2)

sequence DNA
DNA Sequencing: An Outsourcing Guide (UNIT 10.4)

resolve/analyze by electrophoresis
Agarose Gel Electrophoresis (UNITS 7.1 & 7.2)

restriction enzyme digest DNA
Working with Enzymes (UNIT 10.1)

amplify gene regions for cloning or labeling
Overview of PCR (UNIT 10.2)
or
analyze gene expression
Real-Time PCR (UNIT 10.3)

visualize resolved nucleic acids
Staining of Gels (UNIT 7.4)

label fragments
Labeling DNA and Probe Preparation (UNIT 8.4)

image the results
Digital Imaging (UNIT 7.5)

analyze select molecules by blotting techniques
Nucleic Acid Blotting: Southern and Northern (UNITS 8.1 & 8.2)

Figure 1 Flowchart for answering questions related to nucleic acids.

For all of the above techniques, nucleic acids (RNA or DNA, genomic or plasmid) must first be isolated and concentrated from cells (*UNIT 5.2*). The concentration of the nucleic acid preparation must then be determined (*UNIT 2.2*). The preparation can then be analyzed by gel electrophoresis (*UNITS 7.1 & 7.2*) with or without prior restriction enzyme digestion (*UNIT 10.1*), depending on the experiment. DNA preparations can also be used in other enzymatic reactions, including PCR (*UNITS 10.2*) to amplify specific regions for cloning, sequencing (*UNIT 10.4*), or labeling for various experimental applications including the generation of probes for Southern or northern blotting (*UNIT 8.4*). Plasmid preparations (*UNIT 4.2*) can be used directly to obtain sequence information.

PROTEINS

Listed below are common questions about proteins and common techniques used to answer them. Also refer to Figure 2.

In which cellular structures or organelles do specific proteins reside?

Techniques:
Cell fractionation (*UNIT 5.1*)
Immunoblotting (*UNITS 8.1 & 8.3*)
Immunofluorescence (*UNIT 9.2*)

What is the molecular mass of a specific protein? Is it post-translationally modified?

Technique:
Immunoblotting (*UNITS 8.1 & 8.3*)

How pure is a particular protein preparation?

Techniques:
SDS-PAGE (*UNIT 7.3*)
Staining gels (*UNIT 7.4*)
Immunoblotting (if necessary; *UNITS 8.1 & 8.3*)

How does one isolate and analyze a particular protein?

Techniques:
Chromatography (*UNITS 6.1 & 6.2*)
Analysis by SDS-PAGE (*UNIT 7.3*)
Gel staining (*UNIT 7.4*) *or* immunoblotting (*UNIT 8.3*)

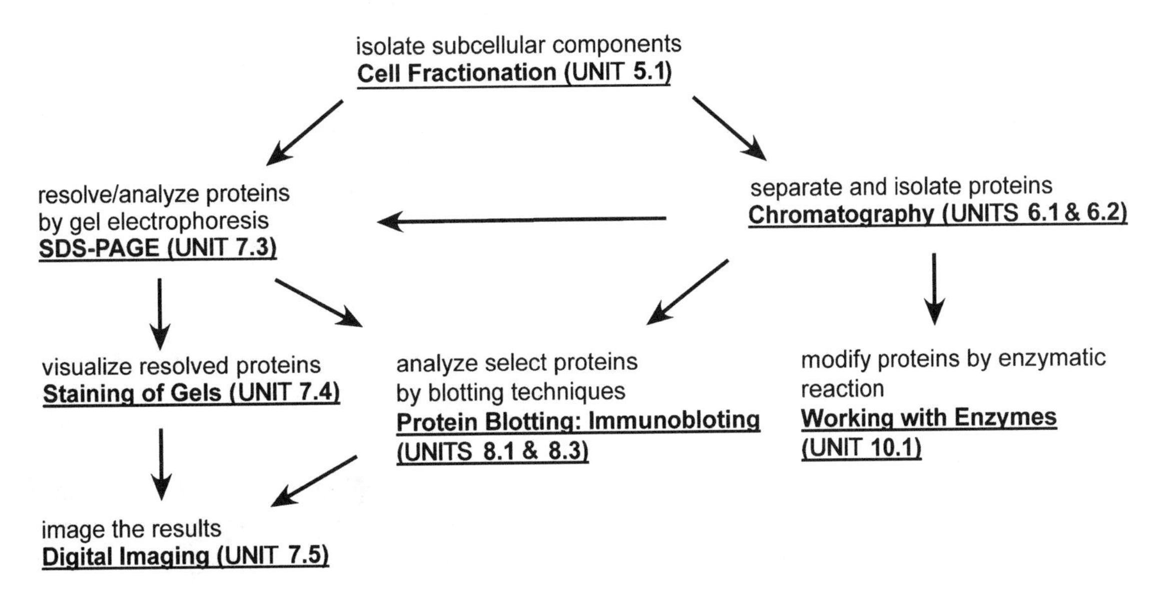

Figure 2 Flowchart for answering questions related to proteins.

For many experiments, the concentration of protein in the sample must first be quantified (*UNIT 2.2*). For example, this is often done prior to performing SDS-PAGE and/or an immunoblot to ensure equal loading of different protein samples for comparison. To determine the localization of specific proteins, cells can first be lysed and fractionated by centrifugation (*UNIT 5.1*), followed by immunoblotting of the proteins (*UNIT 8.3*) from fractions containing specific cell substructures.

A chromatography step would further resolve proteins from the various fractions (*UNITS 6.1 & 6.2*). Alternatively, localization of specific proteins to distinct cellular structures can be done using the immunofluorescence technique (*UNIT 9.2*).

WHOLE CELLS AND SUBCELLULAR STRUCTURES

This manual also includes techniques for studying whole cell structures. These include cell fractionation by centrifugation (*UNIT 5.1*), cell imaging by conventional light microscopy (*UNIT 9.1*), and imaging by fluorescence microscopy (*UNIT 9.2*). Refer to Figure 3.

These techniques can be used to answer questions such as:

Does cell morphology change under different treatment conditions?

Does cell behavior change under different treatment conditions?

Do genetically altered cell lines display morphological phenotypes?

In which cellular substructures does an endogenous or altered protein reside?

Whole Cells and Subcellular Structures

image whole cells and structures by light microscopy
Conventional Light Microscopy (UNIT 9.1)

localize proteins to subcellular structures
Immunofluorescence Microscopy (UNIT 9.2)

fractionate and collect subcellular structures
Centrifugation and Cell Fractionation (UNIT 5.1)

Figure 3 Techniques used to answer questions about cellular and subcellular structure.

Conventional light microscopy can be used to image most cell organelles and structures by using the appropriate microscopy technique. Common variations and their applications are described in *UNIT 9.1*. Fluorescence microscopy is used to image specific organelles with fluorescent dyes, or to study the localization of specific proteins (*UNIT 9.2*).

GENERAL

For any experiment performed, it is essential to keep thorough records in the form of a laboratory notebook. *APPENDIX 2* outlines the best practices for organizing and recording experimental details to optimize their usefulness and completeness.

Results from many techniques in this manual require digital imaging for documentation in a laboratory notebook and for publication. *APPENDIX 3A* and *APPENDIX 3B* present important ethical and practical considerations for capturing, manipulating, and storing digital images, as well as guidelines for preparing them for publication.

Some experimental results will require statistical analyses. *APPENDIX 4* provides guidelines for selecting and using appropriate statistical tests in the life sciences.

Chapter 1

Volume/Weight Measurement

Volume Measurement

Thomas Davis[1] and Andrew Zanella[1]

[1]Joint Science Department, Claremont McKenna, Pitzer, and Scripps Colleges, Claremont, California

OVERVIEW

Volume

There are two primary operations in volumetric measurement depending on the goals of the experimenter. The first is to measure out and then deliver a known volume of liquid to a container, solvent, or solution. In transferring a known volume (aliquot) of liquid from one container to another, one of the many types of volumetric apparatus, which are described below, is normally used. The desired accuracy and magnitude of the volume will dictate the choice of particular type of pipet or other apparatus. Refer to Table 1.1.1 for a list of manufacturers and distributors of volumetric apparatus.

The second goal involves preparing a solution of known volume containing a known concentration of solute or combination of solutes. The accuracy and precision of these measurements will be related to requirements of the nature of the experiment, as well as the quantity of the final sample. For relatively large volumes, a volumetric flask is employed for preparation of such a solution, whereas microscale solutions are sometimes prepared in microcentrifuge tubes.

The basic unit of volume in scientific laboratories is the liter, also expressed as the cubic decimeter (dm^3). More commonly, for smaller volumes, the milliliter (ml) and the microliter (μl) are used. The milliliter is equivalent to the cubic centimeter (cm^3 or cc).

Table 1.1.1 Manufacturers and Distributors

Company	URL
Representative list of volumetric apparatus manufacturers	
Biohit	*http://www.biohit.com/view/products.asp?document_id=276&cat_id=276*
Corning Glass Company	*http://www.corning.com*
Eppendorf[a]	—
Finnpipette	*http://www.thermo.com*
Gilson	*http://www.gilson.com*
Kimble Glass Company	*http://www.kimble.com*
Rainin	*http://www.rainin.com*
Laboratory supply companies[b]	
Fisher Scientific	*http://fishersci.com/*
VWR International	*http://vwr.com/*

[a]Generally sold through authorized laboratory supply distributors such as Fisher and VWR (see Laboratory Supply Companies in this table).
[b]Both distributors carry many brands and types of volumetric apparatus.

Current Protocols Essential Laboratory Techniques 1.1.1-1.1.15

The density of water is close to 1.000 g/ml, and at 20°C one milliliter of water has a mass of 0.99923 g.

Solutions

A solution is a combination of one or more solutes and a solvent such as water or an organic liquid. Solutes can be either solids or liquids, including more concentrated solutions which are then diluted to the appropriate concentrations. The preparation of solutions of known concentrations requires apparatus that is calibrated to high accuracy. Preparation of such solutions requires an array of volumetric apparatus ranging from micropipettors to volumetric flasks.

Temperature Effects

Considerations of temperature are important because the volumes of the liquids and of the containers themselves change with temperature, although most laboratory operations are performed near room temperatures between 20° and 25°C. The change in volume of the solvent may need to be considered if working at temperatures significantly different from 20°C, where most volumetric apparatus is calibrated, as for example in a cold room. The variation of the density and the volume of water with temperature (Table 1.1.2) shows that these variations are small but can become significant. For data regarding organic solvents, the *CRC Handbook of Chemistry and Physics* or a similar reference source should be consulted.

The volume (V) of the container itself also is affected by changes in temperature because of the material's coefficient of expansion, which is a factor multiplied by the given volume and the temperature difference to obtain the change in volume at a temperature different from a standard temperature. For example, the volume of borosilicate glass at a temperature (t) other than 20°C is given by:

$$V_t = V_{20}\,[1 + \alpha\,(t - 20)]$$

where α is the coefficient of cubic expansion (0.000010/°C; Hughes, 1959). For containers made of other materials, such as various types of plasticware, the supplier's information sheets or Web site should be consulted to determine if the effects are significant.

Table 1.1.2 Variation of the Density and Volume of Water with Temperature[a]

Temperature (°C)	Density (g/ml)	Volume (ml/g)
0	0.99987	1.00013
5	0.99999	1.00001
10	0.99973	1.00027
15	0.99913	1.00087
20	0.99823	1.00177
25	0.99707	1.00294
30	0.99567	1.00435
35	0.99406	1.00598
40	0.99224	1.00782
45	0.99025	1.00985
50	0.98807	1.01207

[a]Adapted from CRC Handbook of Chemistry and Physics (Lide, 1996-1997).

Accuracy and Precision

In order to ensure the accuracy of the solution and transferred volumes, high-quality apparatus must be employed. Thus, a "Class A" volumetric apparatus is recommended for all advanced experimental operations. (The "A" classification is the highest based upon NIST standards for accuracy—see below.) If it is desirable that at least three significant figures be obtained in the measurements, then volumetric equipment of the appropriate tolerance must be utilized. For example, if a 3.00 ml aliquot of solution is transferred, then a pipet with an accuracy rating of 3.00 ± 0.01 ml would be suitable for that operation, as opposed to 3.0 ± 0.1. Likewise, the precision or reproducibility of the measurement must be within the limits desired for the specific experiment. If multiple trials of the same kind are carried out, the calibration (see below) of the apparatus, as well as the skill in the way it is actually used, determines whether the resulting solution or volume transferred is of the desired precision.

Calibration of Volumetric Apparatus

Normally, the experimenter relies upon the specifications provided by the manufacturer attesting to the accuracy of the specified volume of the apparatus. The catalog from the manufacturer or a general scientific supply company lists the tolerances for each specific type of volumetric apparatus. The tolerance describes the maximum deviation from the nominal volume of a given item, and the relative tolerance usually decreases as the items increase in capacity. The tolerances are based upon accuracy standards set by the American Society for Testing and Materials (ASTM), and certificates of traceability to standard equipment of that exact type are provided by the National Institute for Standards and Technology (NIST), a U.S. federal government agency. For pipettors, the ISO 8655 Standard for accuracy is generally followed by the manufacturers, and these specifications are usually indicated in the catalog description. The International Organization of Standards (ISO) develops voluntary technical standards used worldwide.

In cases where there is a concern that the incorrect volume is being measured or delivered, there are methods to check the actual volume obtained, which involve weighing the volume of water contained in a vessel or transferred from a particular type of pipet. This technique requires measurements on an analytical balance which can be read to 0.1 mg (0.0001 g; see *UNIT 1.2*). The conversion factors shown in Table 1.1.3, based on data from Hughes (1959), can be applied to the measured mass of the water at the temperature of the measurement in order to correct the volume to the temperature of calibration, normally 20°C. This volume is then compared to the nominal volume of the apparatus to see if a significant correction needs to be applied.

In terms of precision, the skill of the experimenter may be the determining factor on how reproducible each replicate measure is. Therefore, if necessary, the standard deviation of a series of repeated measurements, e.g., transferring 1.00 ml of solution, can be determined by the same weighing method just described.

Glassware and Plasticware

The apparatus used in volumetric applications is generally made of glass or different types of plastic. The glass is most often borosilicate glass, commonly called Pyrex, a brand name of Corning Glass, also manufactured by Kimble Glass (Kimex brand), and sold as generic house brands by scientific laboratory supply companies. Borosilicate has a very low coefficient of expansion and thus can endure rapid changes in temperature without cracking. Soda-lime glass ("soft glass"), which melts at a lower temperature than borosilicate and is not as resistant to temperature changes, is less often used. Some types of soft glass apparatus, particularly serological pipets, are produced for one-time use. Soft glass is also used for drinking glasses, whereas borosilicate glass is used for measuring cups, baking dishes, and coffee carafes.

Table 1.1.3 Correction to Mass Measurement of Water to Determine the True Volume of a Nominally 10.0000 ml Borosilicate Glass Container[a]

Temperature of mass measurement (°C)	Value to add to the observed mass of water delivered from the pipet to obtain the actual volume (ml)
15	0.0200
16	0.0214
17	0.0229
18	0.0246
19	0.0263
20	0.0282
21	0.0302
22	0.0322
23	0.0344
24	0.0367
25	0.0391
26	0.0415
27	0.0441
28	0.0468
29	0.0495
30	0.0583

[a]Adapted from Hughes (1959).

Plastic apparatus is manufactured from a variety of different organic polymers including polyethylene, polypropylene, polystyrene, polycarbonate, polymethylpentene, and polytetrafluoroethylene (Teflon). Most often, plasticware pipets are used for one-time applications and then disposed of. Disposable micropipet tips are usually made from polypropylene, including autoclavable tips. The plastic polymers are generally chemically inert or resistant, but the manufacturer's specifications (which come with the apparatus or are available on the company's Web site) should be consulted before using strong acid, strong base, or organic solvents. (For example, see the Corning Web site on chemical compatibility under Internet Resources at the end of this unit.) Durability under autoclaving as well as biocompatibility with microorganisms should also be checked. (See Corning Web site on physical properties under Internet Resources.)

Safety Precautions

As with carrying out experiments in a typical biology laboratory, normal safety procedures need to be followed when using volumetric apparatus (see *APPENDIX 1*). Eye protection (such as goggles), impermeable gloves, and proper laboratory clothing should be worn. Experiments with materials which produce noxious odors or toxic gases should always be carried out in a certified fume hood. Special circumstances may require other safety devices, such as face shields or masks, especially when dealing with corrosive chemicals or toxic agents. Pipets should never be filled by mouth. Cracked or jagged glassware should be discarded and disposed of properly in a glass waste container. Used Pasteur pipets should be disposed in a glass waste container or in a "sharps" container. Any residues involving biohazards or radioactive materials should be disposed according to standard guidelines for handling them.

MICROPIPETTORS

There are many choices in micropipettors, and an example of both a single-channel and multichannel model are shown in Figure 1.1.1. The commonly available sizes range from 0.5 μl to 20 ml (sizes 1000 μl and larger are often designated as pipettors). There are fixed-volume and adjustable-volume micropipettors. There are air-displacement as well as positive-displacement models, as well as both single- and multichannel styles. There are also repetitive micropipettors. Most brands of manual micropipettors are now offering low-force models to decrease the force required, so as to minimize repetitive stress injuries (RSI) and operator fatigue. Micropipettors are also available in electronic as well as manual models. Determine what level of precision and accuracy you need, what functions you want and need, and what meets your ergonomic needs. Also make sure that the micropipettor you are getting fits your hand and that the placement of controls is convenient for you.

Manual Micropipettors

Fixed-volume micropipettors are convenient when a specific volume must be pipetted frequently. Because they are not adjustable they decrease the risk of the volume being incorrectly set either because of being set for something else, or because of parallax (in which viewing from the side will shift the apparent position of the marking). They are also less prone to being damaged by someone trying to set them above their maximum setting (especially a consideration in a teaching laboratory).

Adjustable micropipettors are convenient because they can be used for any volume in their usable range. In general, larger micropipettors have larger permissible systematic errors than smaller micropipettors, so accuracy is generally improved by selecting a micropipettor where the desired volume is near the top of the range. When setting the volume, it is necessary to view the markings straight on to avoid parallax. Also, when adjusting the volume, it is generally advised to slowly dial down to the volume and then to let the micropipettor rest a minute before using (see the instruction manual for your specific model).

Positive-displacement micropipettors work by having a disposable piston incorporated in the tip that goes all the way to the end of the tip. (Fig. 1.1.2) This avoids an air-to-liquid interface, and the piston wipes the sample out of the tip. Positive displacement models are especially suited for

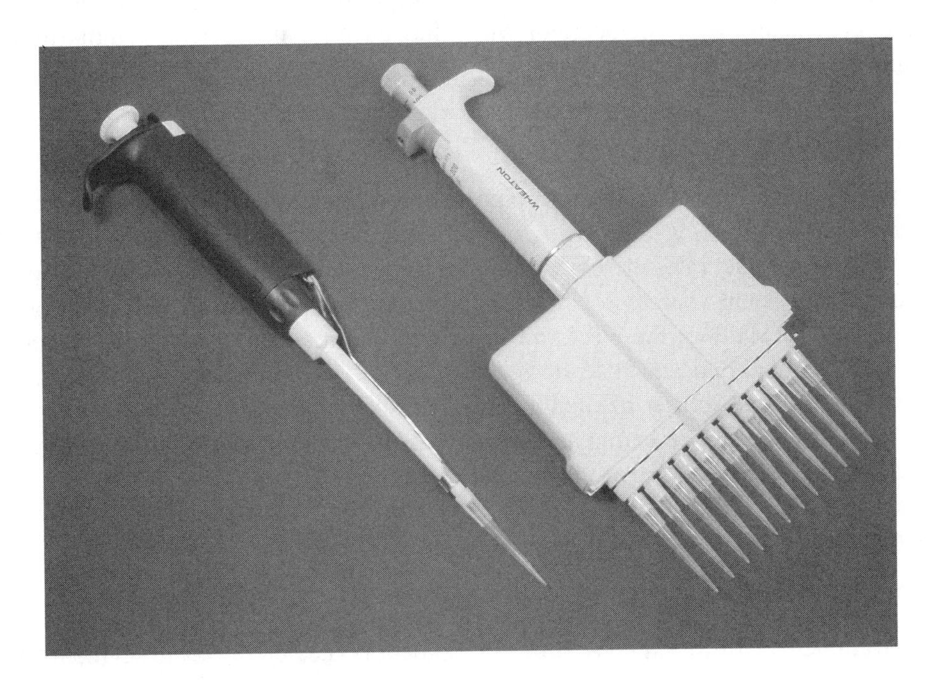

Figure 1.1.1 Micropipettors—single and multichannel.

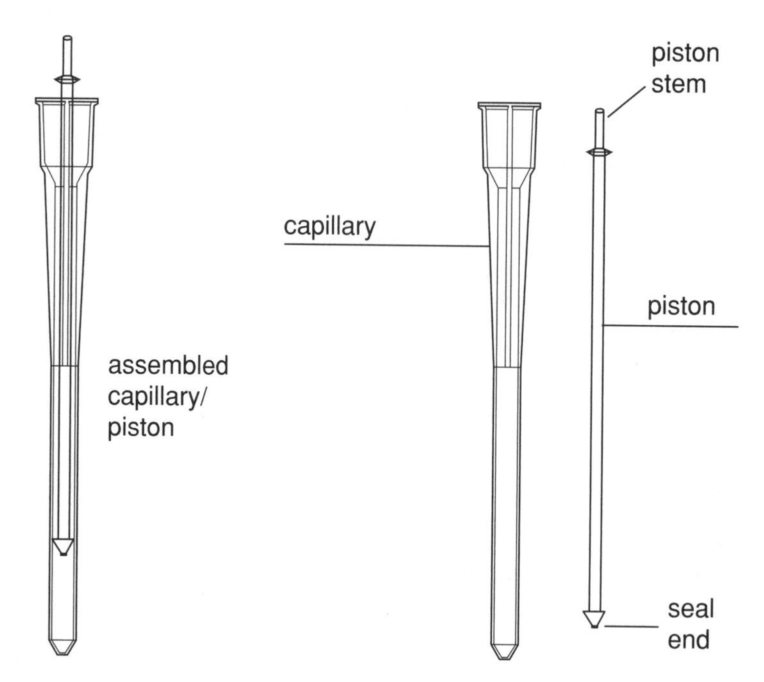

Figure 1.1.2 Sketch of a positive displacement disposable capillary with piston (illustration used with the permission of Rainin Instruments LLC).

problem liquids (liquids that are viscous, volatile, dense, or have high surface tension.) Because there is not a head space above the liquid there is also no risk of aerosol cross-contamination, and because the tips and pistons are ejected, cross-contamination for DNA amplification and similar experiments is also minimized.

Multichannel micropipettors (Fig. 1.1.1, right) are designed for working with multiwelled plates and racked tips. A whole row (or pair of rows) can be pipetted at once, thereby decreasing the time required as well as minimizing the chance that a given well might be missed. Multichannel micropipettors are available in 8-, 12-, 16-, and 24-channel styles.

Repetitive micropipettors draw up a volume and multiple small samples are dispensed. The electronic versions generally warn the operator when they have an insufficient quantity to deliver the next aliquot.

Electronic Pipettors

The electronic models have a stepper motor controlled by an electronic unit. Therefore they generally require little force for pipetting, decreasing operator fatigue, and the possibility of RSI. They also aspirate and dispense at set rates thereby decreasing the variance between samples. (See the Biohit Web site under additional sources for more information on specifics on how electronic models decrease variability.) Many of the electronically controlled models have various preprogrammed modes so they can add, add and mix, do serial dilutions, and do repetitive samples. Some can interface with computers, so user-defined modes are also available. Some models also remind the operator when periodic service is due and some assist in tip selection. The electronic models do come at an increased price relative to the manual versions.

General Concerns and Use of Micropipettors

Instructions for individual micropipettors will vary but in general the following should hold.

1. Never over ratchet the adjustable volume micropipettors (i.e., do not try to set them for volumes outside their range).

2. Do not invert the micropipettors; avoid getting fluid up in the barrel.
3. Use of appropriate-sized tips (without excess head volume) improves accuracy.
4. Volume measurements are most accurate when the micropipettor and the solution are at the same temperature.
5. Release the plunger slowly to increase the accuracy (within reason; too slow a release can also lead to inaccuracy). Also be sure to allow sufficient time for the sample to drain (especially with viscous solutions).
6. In forward pipetting, you will be blowing out the tip by pressing the plunger to the second stop. In reverse pipetting you will not be blowing out the tip. If you are pipetting in a blow out mode using a small volume (generally $\leq$10 μl) you may wish to rinse out the tip by aspirating and dispensing the receiving liquid several times. You will then want to get a new tip to pipet your next sample.
7. Reverse pipetting is a technique used with air-displacement micropipettors when working with viscous or foamy solutions.
    a. Push the plunger into the 2nd stop.
    b. Place the tip into the solution.
    c. Slowly draw up the solution. This will give you a volume larger than the set volume.
    d. Dispense the set volume by slowly depressing the plunger to the 1st stop. Do not blow out.
8. Most instruction manuals call for prerinsing the tips two times with the solution to be pipetted. (If pipetting at other than ambient temperatures generally do not prerinse and use a new tip each time you pipet. This keeps the tip from getting progressively further from ambient as you pipet more samples and the pipetted volume from continually changing due to temperature.)
9. Tips should be below the surface (generally 2 to 3 mm for the smaller volumes, somewhat deeper for the larger volume micropipettors) before attempting to aspirate.
10. Aspiration should be done with the micropipettor in a vertical (or near vertical) position. Dispensing should be done with the micropipettor at a 45° angle. While dispensing, the tip should be touching the side of the receiving vessel.
11. Filtered tips can be used to minimize aerosol contamination (see *UNIT 10.2*).
12. Use care when pipetting for an extended period of time. Your hand will warm the micropipettor and change the volume dispensed due to the expansion of components. Breaks should be taken both to allow the pipettor to cool and to minimize operator fatigue.
13. Increased care needs to be used when pipetting solvents, acids, and other compounds that could damage your micropipettor. Check your instruction manual for chemical compatibility. Some solvents that do not damage the material of your micropipettor may still require that you more frequently service your micropipettor in terms of replaced seals.
14. Avoid using your micropipettor with chemicals that will harm it and be sure to see your instruction manual for instructions on cleaning your micropipettor if any solutions do get up into the barrel of your micropipettor.
15. Your micropipettor will function best within certain temperature and humidity ranges. See your instruction manual.
16. Note that the manufacturer's claims regarding accuracy and precision are done with their tips and that in general they do not claim that such accuracy will be reached with other tips.

17. Note that micropipettors are calibrated with deionized water. To pipet solutions that vary greatly in density or behavior from water may require an adjustment in set volume to achieve the volume desired. If the density of the solution is known this correction can be determined by finding the set volume that gives the desired mass of solution.

18. Dispose of used tips appropriately especially when biohazard and radioactivity concerns need to be addressed.

Storage and Care

Pipettors should be stored in a near vertical position. Pipettors with any solution in them should not be laid on their side. See your instruction manual for proper methods/solvents for cleaning your pipettor.

Calibration and servicing

Most manufacturers have mail-in pipettor servicing and there are companies which provide manufacturer approved onsite calibration and servicing. Depending on your accuracy levels you may want to check that they are ISO 17025 or equivalent accredited. Semiannual servicing is the norm with these services but other intervals are available. Note that your pipettors need to be clean and decontaminated before sending them in for service or calibration.

Some pipettors have user-replaceable parts and the instruction manuals give directions on calibration. To accurately calibrate your pipettor you need to give it a few hours to come to thermal equilibrium with the room where it is to be tested. You will need an analytical balance (accurate to 0.0001 g for volumes ≥50 μl, 0.00001 g for smaller volumes) and appropriate glassware for handling the water samples. You will also need to know the barometric pressure, humidity, and temperature of the room.

PIPETS

There are a wide variety of pipets which can be used to transfer and dispense volumes of differing orders of magnitudes and accuracy. Some of the more common ones are summarized in Table 1.1.4 and are described below.

Pasteur and Transfer Pipets

Pasteur pipets are generally made of borosilicate glass or soft glass and hold about 1 to 2 ml of liquid and are similar to eye droppers. They are intended to be disposable and therefore should not be reused for dispensing different reagents. The liquid is drawn into the pipet by means of either a

Table 1.1.4 Summary of Pipet Types

Type	Max. volume	Materials	Comments
Pasteur	1 to 2 ml	Borosilicate or soft glass	Disposable
Transfer (Beral)	0.3 to 23 ml	Polyethylene	Disposable
Volumetric	1 to 100 ml	Borosilicate	TD–drain; TC–blow out
Serological	0.1 to 50 ml	Borosilicate or polyethylene	TD–blow out
Mohr	1.0 to 50 ml	Borosilicate	TD–to mark
Micropipettors	1 to 1000 μl	Disposable polypropylene tips	
Pipettors	1 to 20 ml	Disposable polypropylene tips	

small red rubber bulb or a latex bulb. They should be rinsed and placed in a glass waste container or in a "sharps" container for disposal.

Transfer pipets (or Beral pipets) are usually made from transparent polyethylene and are also disposable. The size varies up to about 20 ml, and some types yield a specified number of drops per ml, which depend upon the capacity and bore of the pipet. They are also available presterilized and with graduations. The accuracy of these pipets is of a lower standard compared to those described below.

Measuring Pipets

Two common types of measuring (graduated) pipets can be used to dispense different volumes of liquid from the same pipet. Biologists often use serological pipets (Fig. 1.1.3), which are graduated along their entire length. When drained completely these need to have the residual liquid in the tip blown out and are designated "TC and blow out." Disposable serological pipets are made of plastic (usually polystyrene) which can be also be presterilized, but glass disposable serological pipets are also available.

With the Mohr style of pipet (Fig. 1.1.3) the graduations end well above the tip (similar to a buret) at the specified maximum volume of the pipet. There are no graduations beyond this mark, so that in measuring out a known volume the liquid cannot be drained past that point. These pipets are normally made of borosilicate glass.

Volumetric Pipets

Volumetric pipets (Fig. 1.1.4) are also called transfer pipets and are usually designed "to deliver" (TD) a specified volume of liquid by filling to a calibration line and then letting the pipet drain

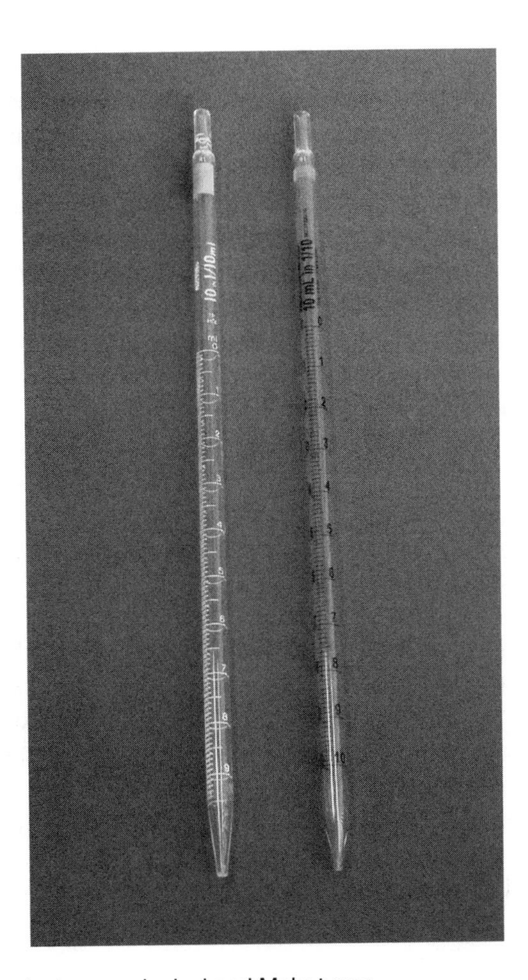

Figure 1.1.3 Measuring pipets—serological and Mohr types.

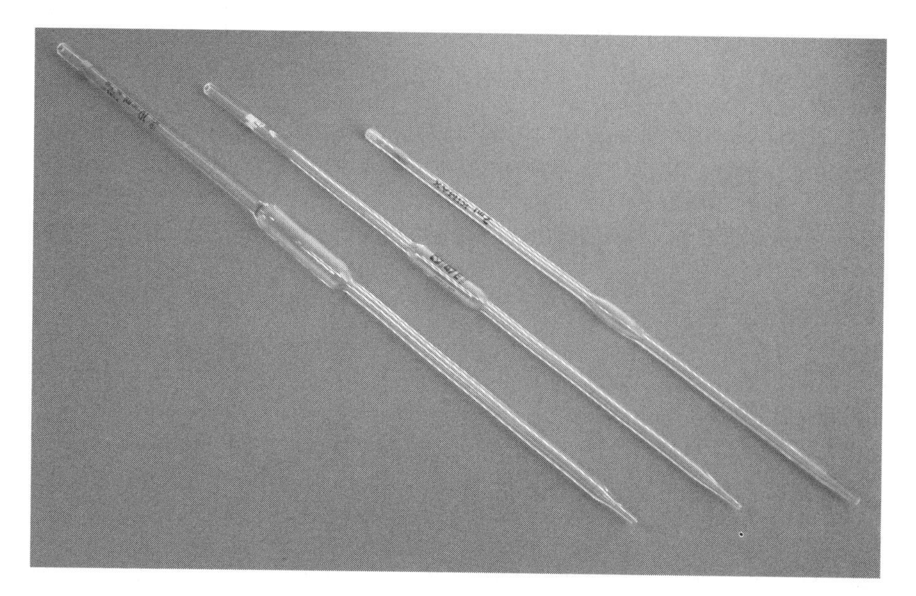

Figure 1.1.4 Typical volumetric pipets.

while touching the tip to the wall of the receptacle. Each pipet size has a color code and a designated "flow time" to allow delivery of the nominal volume of the liquid. There are also volumetric pipets which are designated "to contain" (TC) in which the remaining drop of liquid needs to be blown out. Furthermore, some incorporate both styles by having two "fill lines" on them, depending on which mode of use is preferred.

Filling the Pipet

In filling a pipet, a partial vacuum must be created inside the pipet so that liquid can fill the tube by being forced upward by atmospheric pressure. The user should be careful that the pipet tip is not touching the bottom of the container, which could prevent liquid from entering the pipet. The liquid should be drawn up above the calibration mark, and then some liquid is drained out of the pipet (into a waste container) until the meniscus coincides with the mark. Any excess liquid should be wiped off the tip (unless sterile conditions need to be maintained), and then the liquid can be transferred into the desired container. To make sure that it drains freely, the tip of the pipet should touch the side of the container. If the pipet is a TC volumetric pipet or a serological pipet, the tiny volume of liquid left in the pipet needs to be forced out. The pipet tip should not be placed under the surface of any solvent in the receiving container.

Pipet Bulbs and Fillers

There are a wide variety of devices which can be attached to the top of the pipets in order to draw up liquid. Most science supply catalogs carry a large number of these products, so only several of the more common ones are mentioned here.

The traditional way to draw up liquid is to use a hand-held rubber bulb which fits directly over the pipet's top or uses a plastic adapter to form a seal with the pipet. The bulb is evacuated by squeezing before being fitted onto the pipet. This can be awkward to use since it requires detaching the bulb and holding a finger over the top of the pipet in order to control the process of measuring out or transferring liquid. More convenient and common in the biology laboratory is the 3-valve rubber bulb, which can be used with one hand and does not need to be detached from the pipet when dispensing liquid (Fig. 1.1.5). Pressing the top valve helps to evacuate the bulb, a second valve is used to draw up the liquid, and the remaining valve introduces air to force liquid out of the pipet in a well-controlled manner. There is also a more chemical-resistant silicone version of this style with glass valves.

Figure 1.1.5 Three-valve pipet bulb.

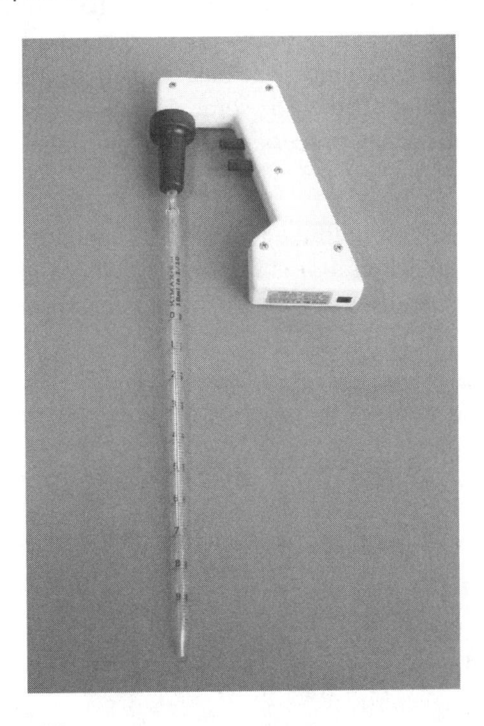

Figure 1.1.6 Electronic pipet filler.

Besides bulbs, there are numerous mechanical plastic pipet fillers which can be operated with one hand. Electronic fillers containing motors which run on line power or on batteries, usually rechargeable, are also available. Some of these are also programmable for repetitive operations. (Fig. 1.1.6). As the pipet filler becomes more elaborate, its cost also increases.

CAUTION: Liquids should never be drawn up into a pipet by mouth. Pipets which have broken or jagged tops or tips should be discarded.

VOLUMETRIC CONTAINERS

Beakers and Erlenmeyer Flasks

Beakers and Erlenmeyer (conical) flasks are usually made of borosilicate glass to withstand heating and cooling extremes, but plastic versions are available. They are primarily intended to serve as containers and indicate the volumes of the reagents only roughly. Therefore, they should not be used to measure liquids quantitatively when accurate concentrations are needed.

Volumetric Flasks

Volumetric flasks are staple items in a laboratory where solutions of known concentrations of reagents, including buffers and media, need to be prepared. They are normally made of borosilicate glass and are calibrated by the manufacturers; Class A flasks should be used when accurate concentrations are required. They range in size from one milliliter up to several liters, depending upon the volume of reagent required. Normally, a volumetric flask (Fig. 1.1.7) is designed to contain (TC) a well-defined volume by filling to the calibration mark on the neck with solvent and mixing well. The solute can be either a more concentrated stock solution of a reagent, or a solid. An aliquot of a stock solution can be introduced by using a pipet of the desired capacity. Solids may be weighed directly in the flask, if the balance has the weight capacity to do so. It is advisable to add the solid through a funnel. A solid may also be transferred after weighing and dissolving it in another container, such as a beaker, by pouring the solution through a funnel into the flask and rinsing the container several times with the solvent to fill the flask about half full. The flask is then shaken to mix well, more solvent is added to slightly below the mark, and a Pasteur pipet is used to add solvent until its meniscus is even with the mark. The flask is then inverted and shaken several times while pressing the stopper on securely. If there is some decrease in volume due to mixing, more drops of solvent are added and the mixing process is repeated until the solution's meniscus is even with the mark (see Chapter 3).

Various materials are used for stoppers, with the two most common being ground glass (standard taper) and a polyethylene plug. The chief aim is to provide a very tight seal against leakage and evaporation, as well as to be chemically inert. Glass stoppers can pose a problem with solutions

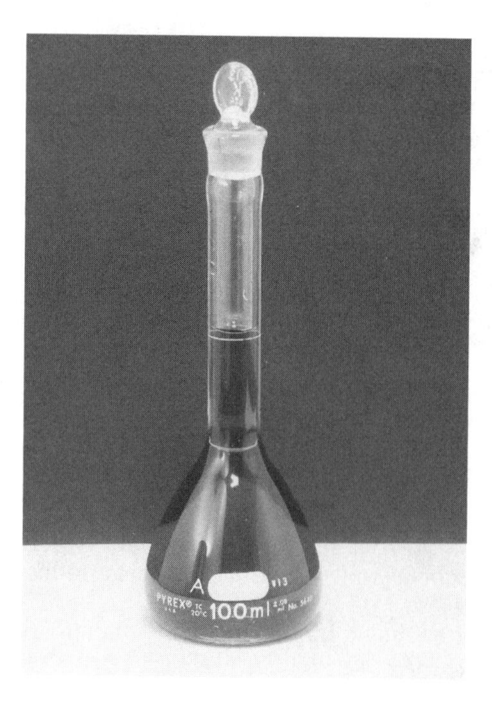

Figure 1.1.7 Volumetric flask.

of bases because they may "freeze" to the neck of the flask, so a plastic stopper is recommended for basic solutions. Volumetric flasks should not be used for long-term storage of solution, since bottles are much less expensive for that purpose.

Volumetric flasks are available in nonglass, most commonly made of polypropylene or polymethylpentene. These are light-weight, autoclavable, and nonbreakable; however one must be a aware of any solvent incompatibility as described by the manufacturer if a nonaqueous solvent or strong acid or base is to be used.

BURETS AND GRADUATED CYLINDERS

Burets can be employed to dispense somewhat larger volumes (up to 50 ml normally) with readings to two decimal places. A buret may be more convenient than a measuring pipet in some situations where variable amounts of a liquid are dispensed. As with other measurements, the meniscus of the liquid is used to determine the reading on the markings on the buret tube and estimating between the graduation lines to obtain the value in the last decimal place. The initial reading is subtracted from the final one to determine the volume obtained. Burets now often include tapered Teflon stopcocks to avoid the need for greasing a glass stopcock, which eliminates the potential for grease to contaminate the liquid in the buret and for stopcocks to "freeze."

A graduated cylinder, on the other hand, can usually be read only to the first decimal place, thereby providing two or three significant figures depending on the magnitude of the volume. They are not proper substitutes for pipets or volumetric flasks, but they are useful for containing or measuring out approximate volumes of a liquid.

CLEANING VOLUMETRIC APPARATUS

Washing

The normal procedure for cleaning reusable volumetric glassware is similar to that for other glassware. Firstly, glassware should be cleaned as soon as possible after use. Letting it sit while the solvent evaporates and forms deposits on the surface makes the task of cleaning more difficult. If the apparatus cannot be immediately thoroughly cleaned, then rinsing with tap water and deionized water will facilitate the cleaning process later.

Glassware should be washed in a hot tap water solution of laboratory detergent, rinsed thoroughly (at least three times) with hot or warm tap water, and then rinsed at least three times with deionized water before draining or drying. Also, detergent should be used sparingly to avoid the need to rinse many times (e.g., the maker of Alconox recommends a 1% w/v solution). The effectiveness of the cleaning process is judged by noting if the rinse water forms beads or streaks on the glass surface, which indicates that the surface has not been thoroughly cleaned. The water should form a uniform layer on the glass.

Some microorganism cultures may be especially sensitive to detergents, so extra rinsing is recommended in such cases. Under some circumstances, ultrapure water may be required for experiments. Therefore, rinsing with that quality of water will be necessary before drying.

When grease sticks on the glass, an organic solvent such as acetone or hexanes soaked on a cotton swab may be used to remove the impurity, followed by rinsing with more of the organic solvent. A high grade of solvent, such as reagent grade, is preferred to avoid contamination by any solvent impurities (*CAUTION*: This procedure should be performed in a well-ventilated fume hood, also see *APPENDIX 1*). Persistent stains can sometimes be removed by soaking in dilute (1 M) nitric acid overnight, followed by thorough rinsing with deionized water.

In the past, dirty glassware was often soaked in chromic acid solution, a strong oxidizing agent which also removes organics including grease. However, because of the toxicity of chromium(VI), which is a potential carcinogen, as well as the concentrated sulfuric acid used in the solution, this method is not recommended, particularly in the case of undergraduate research laboratories. Instead, repeated and diligent cleaning as described above should be sufficient for almost all glassware.

Pipets can be cleaned by drawing up the detergent solution with a bulb and then likewise rinsed with tap water followed by deionized water. However, if pipets have been freshly used, they may be cleaned by drawing the particular solvent up into the pipet past the mark and draining several times. They can then be left to drain more thoroughly on a rack. The same procedure holds for volumetric flasks, which do not ordinarily need to be dried if the same solvent is to be used in them for making solutions.

If a large number of reusable pipets routinely need to be cleaned, then investment in a pipet washing system which can be connected to tap water and to deionized water should be considered. Manufacturers of different systems provide the details of their cleaning procedures.

For cleaning plasticware, although most types are compatible with ordinary cleaning methods, the manufacturer's specifications should be consulted especially regarding the temperature of the wash water used. This also holds for the effects of organic cleaning agents and strong acids and bases.

Drying

In general, volumetric glassware can be dried in an oven at 105° to 110°C for an hour or more to ensure complete removal of water. If organic solvents have been used to rinse the glassware, it should be left in a fume hood to allow evaporation to occur. If necessary, the glassware may then be put in the oven for a shorter time to remove any traces of water. In the past there was concern about oven drying affecting the volumes of volumetric flasks or pipets, but a published study suggests that this is not a significant problem (Burfield and Hefter, 1987). If volumetric flasks are going to be used again with the same solvent, often water, then simply draining them on a rack will remove most of the excess water. For plastic apparatus, draining and air drying should be adequate in most circumstances.

SAFETY NOTE: Do not apply full vacuum to volumetric flasks, since they may shatter or collapse. Removing traces of volatile liquids by applying partial vacuum through a narrow pipet should be done very carefully. Do not attempt to remove residual water this way, since it is normally unnecessary.

Sterilization

When biological agents are present in experiments, they need to be removed from any reusable volumetric glassware (or plasticware) as part of the cleaning process. A variety of techniques, including rinsing with microcidal reagents (e.g., bleach or alcohol), thermal methods (e.g., autoclaving at 15 min at 121°C at 15 to 20 psi), and irradiation by UV light can be used, depending upon the specific organism and type of apparatus. Consult *UNIT 4.1* for more information.

LITERATURE CITED

Burfield, J. and Hefter, G. 1987. Oven drying of volumetric glassware. *J. Chem. Educ.* 64:1054.

Hughes, J.C. 1959. Testing of glass volumetric apparatus. *National Bureau of Standards Circular* 602.

Lide, D. (Ed.) 1996-1997. Variation in the density of water with temperature. *CRC Handbook of Chemistry and Physics* 77:6-10, 8-10.

INTERNET RESOURCES

Sites for further information are:

http://www.biohit.com/view/products.asp?document_id=1102&cat_id=276

http://www.rainin.com/lit.asp

Chemical compatibility and other properties of plastics:

http://www.corning.com/Lifesciences/technical_information/techDocs/chemcompplast.asp

http://www.corning.com/Lifesciences/technical_information/techDocs/prpertyplast.asp

UNIT 1.2

Weight Measurement

Michael Guzy[1]
[1]Ohaus Corporation, Pine Brook, New Jersey

OVERVIEW AND PRINCIPLES

The balance, or scale, is one of the most frequently used pieces of laboratory equipment. Although the terms "balance" and "scale" have become interchangeable in everyday use, there is a technical difference, as will be outlined in the next section.

Most balances are designed to be simple to operate. However, since accurate weighing is essential for achieving accurate experimental results, it is important to learn to use and calibrate a laboratory balance correctly. There are a wide variety of laboratory balances available (see Table 1.2.1 for a list of suppliers), with an ever-increasing range of options. This unit presents an overview of the different types of balances commonly used in laboratory settings, and reviews selection criteria and proper operating and maintenance techniques.

What's Being Measured: Mass versus Weight

When using a balance or scale to weigh an object, one is typically attempting to calculate the mass of the object. In scientific and technical terminology, "mass" refers to the amount of matter in an object (Scale Manufacturers Association, 1981). The units of measure for mass include grams (g) and kilograms (kg) in the Systeme Internationale (SI), or metric system, and ounces (oz) and pounds (lb) in the avoirdupois, or English system (Barry, 1995; see Table 1.2.2). In contrast, "weight" refers to the gravitational force exerted on an object (Scale Manufacturers Association, 1981). Units of measure of weight include newtons (N) and pound-force (lbf; Barry, 1995). Since the earth's gravitational force is comparable at different points on its surface, then, if the mass of an object is constant, its weight will be quite similar at all points on earth. There will, however, be slight variations. The difference between mass and weight is most apparent beyond the earth's gravitational field. If the same object is weighed on the moon, its mass will remain constant but its weight will be $\sim\frac{1}{6}$ its weight on earth.

It is important to understand these definitions for two very practical reasons. First of all, they contrast with practical and commercial usage (Barry, 1995). Generally, when a person tells somebody his weight in pounds or kilograms, he is really describing his mass. Since this unit refers to the use of balances and scales in a laboratory setting, it will use the term "mass." Certain situations may

Table 1.2.1 Suppliers of Laboratory Balances

Ohaus	*http://www.ohaus.com*
Mettler Toledo	*http://www.mt.com*
Sartorius	*http://www.sartorius.accurate-scale.com/*
AND Weighing	*http://www.andweighing.com/*
IWT (Intelligent Weighing Technology)	*http://intelligentwt.com/*
Shimadzu	*http://www.shimadzu.com/*
Acculab	*http://www.acculab.com/*

Current Protocols Essential Laboratory Techniques 1.2.1-1.2.11

Table 1.2.2 Differences Between Mass and Weight[a,b]

	Mass	Weight
Definition	The quantity of material in a body	The force by which a mass is attracted to the center of the earth by gravity
Units, Systeme Internationale	milligram (mg) gram (g) kilogram (kg)	dyne (dyn) newton (N) kilonewton (kN) kilogram-force (kgf)
Units, avoirdupois	ounce (oz) pound (lb) ton	ounce-force (ozf) pound-force (lbf) poundal ton-force

[a]Sources: Scale Manufacturers Association (1981); Barry (1995).
[b]Also see the section on conversion factors at the beginning of this volume.

require one to determine the "weight" of something in kilograms, pounds, or other units of mass. In these cases, "weight" is being used as a nontechnical term for mass.

Another practical difference between "mass" and "weight" is that mechanical balances directly measure mass, by comparing the mass of the object to a reference. Therefore, whether an object is "weighed" on earth or on the moon, a two-pan balance or a triple-beam balance (Fig. 1.2.1) will provide the same value for its mass. In contrast, an electronic scale will measure the force at which an object pushes down on the weighing platform, as measured by internal measurement devices, known as load cells. Therefore, they are directly measuring weight, not mass. Electronic scales would provide different values for the "mass" of an object on earth and on the moon. Yet, electronic scales are routinely used in the laboratory for calculating mass. Although they directly measure weight, the value is then divided by the gravitational constant in order to calculate the mass (see Crowell, 2006). Therefore, the value displayed represents the mass of the object

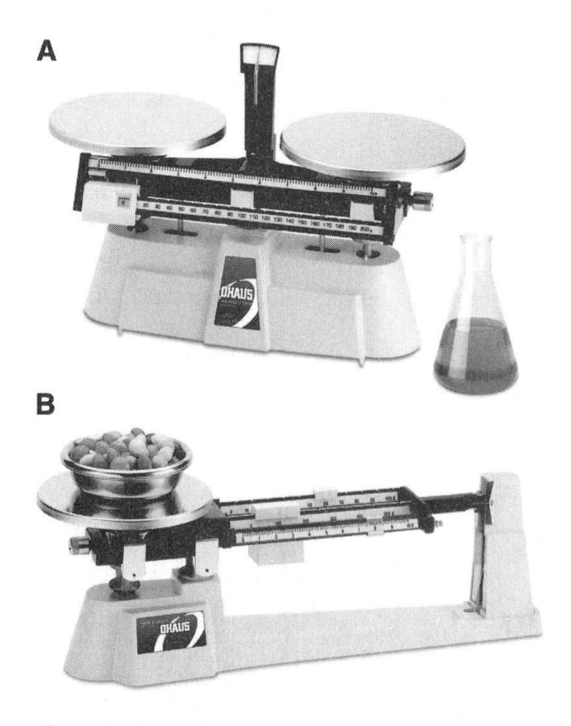

Figure 1.2.1 Examples of mechanical balances. (**A**) Ohaus Harvard Trip Balance. (**B**) Ohaus Triple Beam Balance.

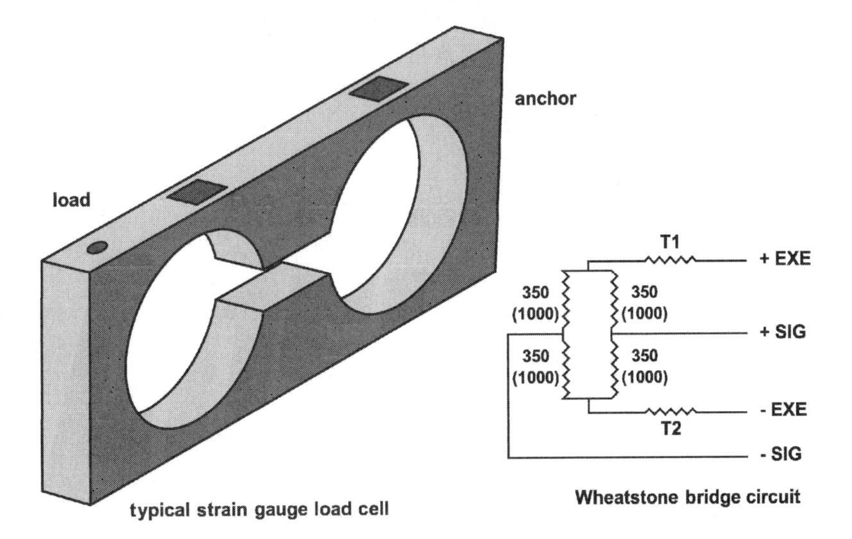

Figure 1.2.2 Drawing of a typical load cell, showing its operating principle. Abbreviations: SIG, signal; EXE, excitation.

being weighed. However, since the gravitational force does vary slightly at different locations on the earth's surface, each scale needs to be calibrated for its specific location. A more detailed discussion of the difference between balances and scales can be found elsewhere (see "weighing scale" at *http://en.wikipedia.org*). Despite this difference, the terms "balance" and "scale" are often used interchangeably in general usage.

How do load cells measure force?

Inside each load cell is a lever with electronic strain gauges attached. Pressure on the lever will create deflection of the strain gauge and change the electrical resistance. The strain gauges are sensitive to very small changes in pressure. Changing the resistance will, in turn, create changes in the electrical output of the system. Typically, a load cell will have four strain gauges attached, which will amplify the output sufficiently to have a detectable signal. Placing too much mass on the balance pan or adding mass too roughly or abruptly may permanently deform the lever and strain gauges, which are extremely delicate and precisely calibrated components. The operating principle behind a load cell is illustrated in Figure 1.2.2.

STRATEGIC PLANNING

When selecting the proper balance for your measurements, it is important to be familiar with both the various types of weighing instruments available and your specific experimental needs. Table 1.2.1 presents a list of balance suppliers.

Types of Scales

Mechanical balances

The most commonly used laboratory weighing instruments come in two main forms: mechanical and electronic (see Figs. 1.2.1 and 1.2.3). The classical mechanical balance is best represented by the classic "scales of justice": two hanging pans balanced on a fulcrum. If the mass of an object placed in one pan is equal to the reference object of known mass placed in the second pan, then the two pans will balance. This same basic principle has been updated, resulting in the more precise and easy-to-read mechanical balances often used in laboratory, industrial, or educational settings. A two-pan balance, such as the one shown in Figure 1.2.1A, is typically used to determine the difference in mass between two objects. To use this type of balance for comparative weighing, place each object in one of the pans and move the sliding weights so that the pans are in balance.

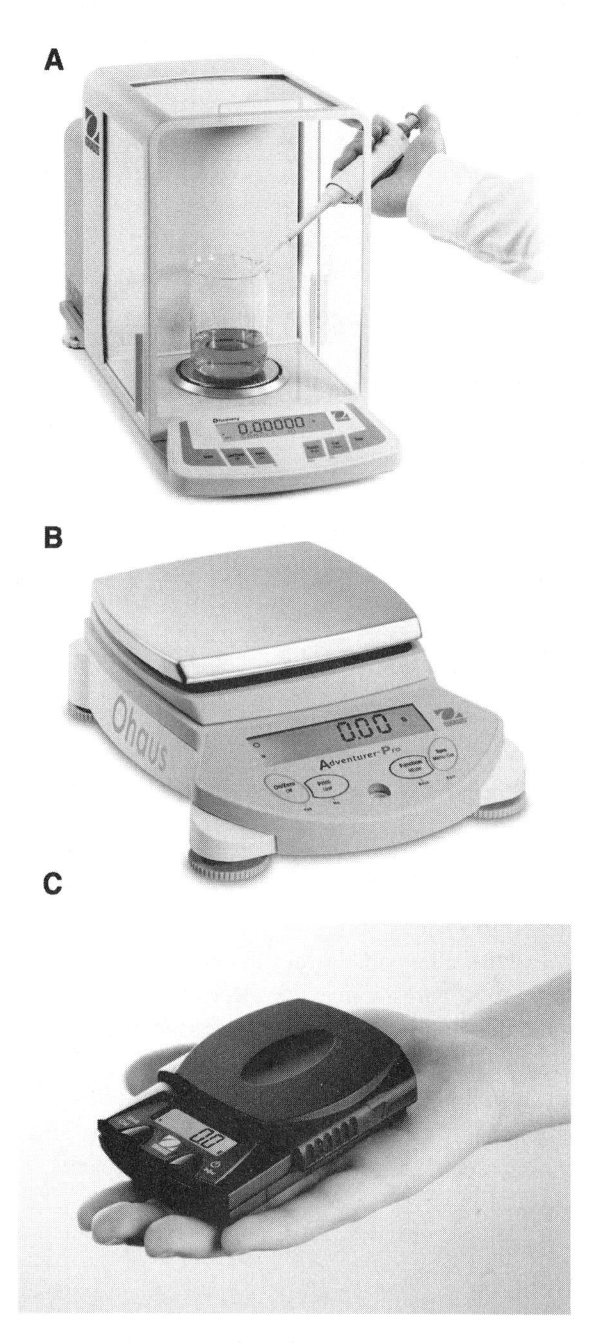

Figure 1.2.3 Examples of electronic scales. (**A**) Ohaus Discovery. (**B**) Ohaus Adventurer Pro. (**C**) Ohaus pocket scale.

The position of the sliding weights will indicate the difference in mass. Similarly, the balance can be used to calculate the mass of a single object. If only one pan is used, the positions of the sliding weights will indicate the mass of the unknown object.

A mechanical triple-beam balance is shown in Figure 1.2.1B. This is an example of a single-pan mechanical balance. The position of the sliding weights must be adjusted so that the beams are in balance. The triple-beam balances used in laboratories are quite similar in operation to upright scales used in doctors' offices.

Electronic balances

Electronic balances are commonly used in science and industrial laboratories. There is tremendous variety in the types of electronic balances available today; samples of three different types are shown in Figure 1.2.3. Electronic balances have several advantages over mechanical balances. They can provide greater accuracy and sensitivity than mechanical balances. The most sensitive balances, known as analytical balances, can measure to the nearest 0.00001 g (Fig. 1.2.3A). This sensitivity is due to the extreme sensitivity of the load cell to very small changes in tension. A second advantage of electronic balances is that most are equipped with a computer interface, so that data can be entered directly into a software program. Electronic balances may also perform simple calculations, including averaging, calibrating pipet volume (also see *UNIT 1.1*), and counting the number of items of comparable mass.

Features Available in Electronic Balances

When choosing an electronic balance, it is important to determine which features and specifications are most important for the types of measurements one will be making. First, one needs to consider what the balance will be used for. How precise do your measurements need to be? Some balances provide measurements to the nearest 0.01 mg (or 0.00001 g), others to the nearest 0.1 g. How heavy are the things that need to be measured? Some balances accurately measure in the milligram range, others in the kilogram range. Does the balance need to do certain computations automatically? The user's manual will include a specifications table summarizing all of the features of the balance, allowing one to easily compare the properties of different balances. An example of a specifications sheet for a typical analytical balance is shown in Figure 1.2.4. Some of the more important features to consider, which vary among the different makes and models of scales, are:

1. **Capacity**–maximum mass that can be placed on the balance. Balances also have a minimum capacity, or minimum mass that the balance can accurately measure.
2. **Readability (or precision)**–value of the smallest division which the balance displays, expressed as the number of decimal points to which a reading is expressed on the display. For example, a reading of 0.005 g has a readability of three decimal places, or 0.001 g. This is a more sensitive reading than 0.05 g, which has a readability of two decimal places.
3. **Repeatability (or reproducibility)**–ability of the balance to consistently display the same value when an object is placed on the balance more than once.

Furthermore, certain balances will include some of the following computational features:

4. **Animal/dynamic weighing**–calculates the mass of a moving object, such as a laboratory animal.
5. **Density**–calculation of the density of a solid or liquid. The object must be first weighed in the air, and then in a liquid.
6. **Parts counting**–the ability to determine the number of pieces placed on the weighing pan. The average mass of a piece is first entered by the user.
7. **Pipet calibration**–checks the accuracy and precision values of pipets by weight analysis.
8. **Statistics**–comparing a number of samples and examining the relative deviation of the samples along with other statistical data.
9. **Totalization**–measures the cumulative weight of objects.

Certain other features that are less easily quantified should also be considered when selecting a balance. These include sensitivity, robustness, and portability.

Discovery Semi-Micro and Analytical Balances

Specifications

Model	DV114C	DV214C	DV314C	DV215CD
Capacity (g)	110	210	310	81*/210
Readability (mg)	0.1	0.1	0.1	0.01*/0.1
Repeatability (Std.dev.) (mg)	0.1	0.1	0.2	0.02*/0.1
Linearity (mg)	±0.2		±0.5	±0.03*/0.2
Weighing Units	Milligram, Gram, Carat, Ounce, Ounce Troy, Grain, Pennyweight, Momme, Hong Kong Tael, Singapore Tael, Taiwan Tael, Custom			
Application Modes	Weighing, Parts Counting with Automatic Sample Recalibration, Percent Weighing, Checkweighing, Animal/Dynamic Weighing, Gross/Net/Tare Weighing, Totalization, High Point, Density, Statistics, Pipette Calibration			
Features	Selectable Environmental Filters, RS232 Interface with full GLP/GMP Protocol, Easy to Clean Stainless Steel Platform, Windring and Glass & Steel Construction, Protective In-Use Cover, Integral Weigh Below Hook, Easy to Use Keypad, Up-Front Level Indicator			
Tare Range	To Capacity by Subtraction			
Stabilization Time (s)	4		8	12/5
Sensitivity Drift (10-30°C)	±2ppm/°C			
Operating Temp Range	10° to 40°C			
Calibration	Automatic internal, Pushbutton internal, Manual external			
Power Requirements	External Adapter, 100-120VAC 150mA, 220-240VAC 100mA, 50/60 Hz Plug configuration for US, Euro, UK, Japan & Australian			
Display Type	2-line Alphanumeric Backlit LCD Display			
Display Size (in/cm)	4 x 1 / 10 x 2.5			
Pan Size (Diameter) (in/cm)	3.5/9			
Height above pan (in/cm)	9.5/24			
Dimension WxHxD (in/cm)	7.9 x 11.8 x 18 / 20 x 30 x 45.7			
Net Weight (lb/kg)	22.5/10.2			

*FineRange™

Other Standard Features and Equipment
Stability indicator, mechanical and software overload/underload protection, AC adapter, user selectable span calibration points, auto tare, auto shut-off, user selectable printing options, user selectable communications settings, user selectable data print options, user definable project and user ID's, software reset menu, software lockout menu

Approvals
CE, FCC, $_{C}UL_{US}$, $_{C}CSA_{US}$, NRTL, OIML

Accessories .... Ohaus Number
Impact printer, 42 column SF42
RS232 cable, SF42 printer 80500571
TAL Software WinWedge SW12W
SF42 printer paper (5 pack) 12101511
SF42 printer ribbon 12101512
RS232 cable, IBM 9 pin 80500525
RS232 cable, IBM 25 pin 80500524
Density determination kit 77402-00
Storage Cover 9773-79
Security device 77401-00
100g Calibration Mass, ASTM Class 1 49015-11
200g Calibration Mass, ASTM Class 1 49025-11

Industry Leading Quality and Support
All Ohaus Discovery balances are manufactured under an ISO 9001:2000 Registered Quality Management System. Our rugged construction and stringent quality control have been hallmarks of all Ohaus products for nearly a century.

Ohaus Corporation
* 19A Chapin Road
P.O. Box 2033
Pine Brook, NJ 07058 USA
Tel: 800.672.7722
973.377.9000
Fax: 973.593.0359
www.ohaus.com

With offices throughout Europe, Asia, and Latin America

* **ISO 9001: 2000**
Registered Quality Management System

Figure 1.2.4 Example of a specifications sheet provided by the manufacturer. This is for the Ohaus Discovery series of analytical balances. Highlighted terms are defined in Strategic Planning, Features Available in Electronic Balances.

10. **Sensitivity**–another term for readability; however, it also includes vulnerability to interference from drafts and other external sources.

11. **Robustness**–how well a balance will stand up to use without being damaged.

12. **Portability**–ability of a balance to be moved from place to place. An example of a pocket-sized electronic scale is shown in Figure 1.2.3C.

13. **Accuracy**–how well a balance displays the correct results, determined by its ability to display a value that matches a standard known mass.

One feature commonly found in analytical balances, which require a high degree of precision and accuracy, is the presence of draft shields to protect the weighing pan from air currents and other environmental interference. An example is shown in Figure 1.2.3A.

Information about these important features, and their inclusion on specific balance models, is available from the manufacturer's product specifications sheet.

Laboratory Setup

Typically, a biology laboratory will have more than one electronic balance, since no one balance is appropriate for all weighing situations. A common setup is to have two balances: a top-loading balance with a relatively high capacity (typically ~600 g) and precision of 0.1 g, and an analytical balance with a capacity of 110 g and a precision of 0.0001 g. If the laboratory routinely needs to weigh samples heavier than 600 g, an additional balance may be required. Industrial laboratories frequently require repetitive measurements and automatic computerized record keeping. The laboratory procedure may require certain computations, such as the need for a quality control laboratory to conduct statistical analysis or parts counting. Therefore, it is necessary to select a balance with those computational features and program it for the appropriate task. If a certain procedure is performed frequently, the laboratory director may choose to dedicate a balance for one specific task.

Certain laboratory supplies are required when using a balance. The substance being weighed must be placed on a weighing vessel. For small quantities of powder reagents, a creased piece of nonabsorbent weighing paper can be used. Weighing boats are available for larger quantities of solid reagents or for liquid reagents (Fig. 1.2.5). These generally are made of disposable, nonabsorbent virgin plastic, which resists static electricity. Weighing vessels with an antistatic surface are important for weighing extremely fine powders. Special small spatulas are available for removing reagents from their containers. Weighing paper, weighing boats, and spatulas are available from most general laboratory supply companies.

Operation and Maintenance of a Laboratory Balance

There are numerous makes and models of laboratory balances available today. The general operation of mechanical balances was described above, in the Mechanical Balances section. Because each electronic balance is operated in a different fashion, it may be necessary to consult the manufacturer's instruction manual to learn how to operate your particular unit. Each unit will have its own protocol for weighing an object, setting the tare value (to cancel out the mass of the container), and performing more complex computational procedures. In general, manufacturers strive to make the user interface clear and straightforward, so that operating the scale can be easily mastered. There are, however, several things to know about setting up and maintaining a balance, in order to obtain the most accurate measurements possible.

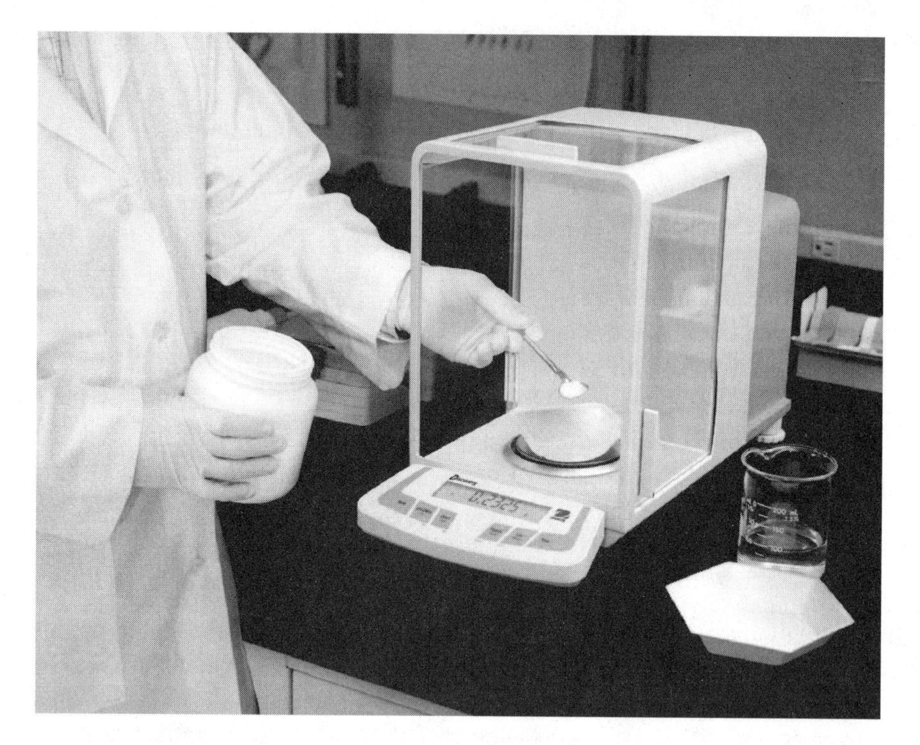

Figure 1.2.5 Weighing boats are available for larger quantities of solid reagents or for liquid reagents.

Installation

Select a location that is a smooth, level, steady surface. Changes in temperature, air currents, humidity, and vibrations all affect the performance of the balance. Therefore, select a location free from these environmental factors. Do not install the balance near open windows or doors, causing drafts or rapid temperature changes, near air conditioning or heat sources, or near magnetic fields or equipment that generate magnetic fields. If unsure as to whether a magnetic device affects the performance of a balance, compare the performance of the balance when the magnetic device is on and when it is off. In general, analytical balances should not be used near magnetic stir plates when the stir plates are on, because both the magnetic field and the vibrations of the stir plate may interfere with the performance of the balance. Be sure that the balance is level, adjusting it as necessary. The balance must be on a stable surface that does not wobble or vibrate. This may require the use of a specialized weigh table–either heavy granite or one with active antivibrational features. Since surface vibration can cause unacceptable variances in balance performance, it is best to always select an antivibration weigh table when precision measurements are required.

Calibration

It is necessary to calibrate your balance when it is first installed and anytime that it is moved to a new location. This will ensure accurate results at that particular location. As described above, electronic balances need to be calibrated for the specific location in order to adjust for slight differences in the local gravitational force. Furthermore, balances must be calibrated to adjust for any minor changes that occurred when it was moved. Therefore, both mechanical and electronic balances must be calibrated whenever they are installed in a new location. The accuracy of the balance should be verified periodically by weighing a reference mass. The balance can then be recalibrated if necessary. It is important that the reference mass be certified to be accurate and maintain its accuracy over time and in different environments. The National Institute of Standards and Technology (NIST; *http://www.NIST.gov*) is the federal agency in the United States that establishes standards for use in US industry. NIST standardized masses can be purchased from most suppliers of laboratory balances (see Table 1.2.1).

Mechanical balances are calibrated by setting the sliding weights to zero when the pans are empty and adjusting a calibration knob until the beam is balanced. Electronic balances will have a calibration program to be followed and may require the use of reference masses.

Calibration frequency depends on what the balance is to be used for. For applications that involve measuring exact dosage, the balance should be calibrated prior to each use. Other applications may require less frequent calibration. For example, if the application requires mixing two or more substances in a specific proportion, then the exact calibration is less critical to the outcome, since any calibration variation will apply equally and proportionally to all substances within the mixture. All laboratory balances should be calibrated at least once a year. Most calibrations can be done by laboratory personnel, using NIST standardized reference masses. It may also be necessary to have your balance serviced and calibrated periodically by a professional service technician who will open the balance, clean the internal mechanisms, reassemble, and recalibrate the balance. The cleanliness of the environment and the balance will determine how frequently your balance will need to be cleaned and calibrated professionally. Contact a service technician whenever a spill occurs in which materials may have entered the internal compartment.

For precision measurements, it is recommended that the technician use a check weight at regular intervals to gauge proper calibration. The technician can specify an acceptable tolerance range, within which the balance need not be calibrated. In addition, it is recommended that the technician record a written history of tolerance readings for a specific balance. This calibration can be done by the user using traceable test weights, or they can have a service person come in periodically and perform the calibration and issue a certificate. Calibration has to be done on site.

The manufacturer can provide a list of certified calibration services in the user's area.

Maintenance

Most laboratory balances require very little maintenance. Keep the balance and its platform clean and free from foreign material. If necessary, use a soft damp cloth and mild detergent. Never allow liquid to enter the balance. Contact a service technician to clean and recalibrate your balance, if a spill occurs. Check the accuracy of the balance periodically with reference weights, and calibrate as necessary. Keep calibration masses in a safe, dry place. It is recommended that the balance be "locked" when not in use. Please refer to the manufacturer's instruction manual for more detail.

Improper use or maintenance of the balance can destroy the ability of the balance to measure mass accurately. Placing too large a mass on the platform or dropping items on the balance too roughly can deform the strain gauges in the load cell. Similarly, the strain gauges can be deformed if the balance is moved without being locked properly. Liquids that seep under the balance pan can interfere with the proper operation of the electronic components of the load cell. Although analytical balances are generally easy to use, they are also highly sensitive instruments that must be used and maintained with care in order to remain accurate.

SAFETY CONSIDERATIONS

Laboratory balances in themselves pose little risk to the technician. The appropriate safety procedures will be determined by the reagents or other objects being weighed. For example, gloves, mask, and safety goggles are commonly used when handling laboratory reagents (see *APPENDIX 1*).

PROTOCOLS

The following protocols apply to most top-loaded precision balances (readability of 0.1 g) and analytical balances (readability of 0.1 to 0.01 mg). The user interfaces of different balances vary greatly; therefore, it is necessary to consult the manual for the balance being used in order to understand specifically how to operate it. The following protocols are for manually recording mass; many balances provide computer interface for recording mass automatically.

Basic Protocol 1: Measuring Mass Using a Top-Loading Balance

1. Turn on balance and wait for display to read 0.0 g.
2. Place weighing vessel on the balance pan (e.g., creased weighing paper, weigh boat)
3. Press tare button so that the display reads 0.0 g.
4. Gently add the substance being weighed to the weighing vessel.
5. Record mass.
6. Remove weighed sample.
7. Clean spills off balance with brush or absorbent laboratory tissue. Discard any disposable weighing vessel.

Basic Protocol 2: Measuring Mass Using an Analytical Balance

1. Turn on balance and wait for display to read 0.0000 g.
2. Check the level indicator and do not lean on table while weighing.
3. Place weighing vessel on the balance pan (e.g., creased weighing paper, weigh boat).
4. Close the sliding doors and wait for stability light indicator, indicating that the weight is stable.
5. Press tare button so that the display reads 0.0000 g.
6. Gently add the substance being weighed to the weighing vessel.
7. Close the sliding door.
8. Wait for stability light indicator before recording mass.
9. Remove weighed sample.

   If using the same vessel for multiple measurements, do not remove with your bare hands, since fingerprints can add weight. Use tongs, tissue, or other device.

10. Clean spills off balance with brush or absorbent laboratory tissue. Discard any disposable weighing vessel.

UNDERSTANDING RESULTS

The output of the balance will be the mass of the object in grams, kilograms, pounds, or whatever unit the scale has been set to display. As discussed above, although electronic laboratory scales directly measure weight, the weight will be converted internally to the mass of the object, by dividing by the gravitational constant (Crowell, 2006).

TROUBLESHOOTING

If one is experiencing trouble using their laboratory balance, or receiving an error message, it may be necessary to consult the manufacturer's instruction manual. There are numerous manufacturers of laboratory scales, and each manufacturer produces an ever-increasing number of models. Therefore, it is not feasible to provide examples here of all the different possible error messages. The manufacturer's instruction manual, however, will provide a clear explanation for any error message that may be received. Some errors, such as unstable readings, may require changes in the laboratory environment, including minimizing temperature fluctuations or eliminating vibrations.

1

Other problems, such as damaged load cells or liquid seeping into the internal mechanisms, will require professional servicing.

Laboratory balances are generally easy to use and require little maintenance. However, in order to provide accurate laboratory results, the balance must be used correctly and one must select a balance with the capacity, precision, and reliability that is needed. Also, one should select a quality instrument from a reputable manufacturer that can stand up to years of use. In addition to weighing objects, electronic scales can perform simple computations, calibrations, and statistical analyses. The information in this unit should help with selecting the balance that will best meet your laboratory needs. This information should also provide enough background to help one use the laboratory balance correctly.

LITERATURE CITED

Barry, T. 1995. NIST Special Publication 811: Guide for the Use of the International System of Units (SI). National Institute of Standards and Technology (NIST), Gaithersburg, Md. Available at *http://physics.nist.gov/Pubs/pdf.html.*

Crowell, B. 2006. Light and Matter: Newtonian Physics. Edition 2.3 Fullerton Calif. Available at *http://www.lightandmatter.com/area1book1.html*

Scale Manufacturers Association. 1981. Terms and definitions for the weighing industry, 4th ed. Scale Manufacturers Association, Naples, Fla.

INTERNET RESOURCES

http://www.dartmouth.edu/~chemlab/

Dartmouth University Chemlab Web site, which provides information and hands-on instruction on the proper operation, use, care, cleaning, and maintenance of both analytical and precision balances.

Chapter 2

Concentration Measurement

UNIT 2.1

Spectrophotometry

Rob Morris[1]

[1]Ocean Optics, Inc., Dunedin, Florida

2

OVERVIEW AND PRINCIPLES

"Let there be light!" was the very first order of creation. It is the light from the Big Bang that remains as evidence of the very first moments of the beginning. As light encounters matter, it can be changed. It is this interaction between light and matter that is the very basis of our ability to sense our universe, to observe the physical world, and to construct our theories and models of how the universe works.

The human eye is an exquisite sensory organ that provides information about the intensity and spectral distribution of light. Our vision is limited, however, in terms of the spectral range we can perceive, and in using such sensory information to make quantitative measurements (also see *APPENDIX 3B*).

Spectroscopy Origins

In 1666, Isaac Newton discovered that white light from the sun could be split into different colors if it passed through a wedge-shaped piece of glass called a prism. This device was called a spectroscope, or literally, a device "to see" (from the Greek *spektra*) the colors. In 1800, Herschel and Ritter discovered infrared and ultraviolet light that was invisible to the eye but still refracted by the prism. In 1814, Joseph Fraunhofer discovered dark lines in the spectra of the sun. In 1859, Kirchoff discovered that each element emitted and absorbed light at specific wavelengths, laying the groundwork for quantitative spectroscopy. Around the same time, Balmer found that the wavelengths of the line spectra of atomic hydrogen could be described with simple mathematical formulae based on a series of integer values (the Balmer series). In 1913, Neils Bohr explained the Balmer series as resulting from transitions of electrons from one orbital state to another. Bohr's work laid the groundwork for modern quantum theory and our understanding of atomic and molecular structure.

In 1873, James Clerk Maxwell discovered that light was related to magnetism and electricity. His explanation for the operation of a simple electrical experiment revealed that electromagnetic energy could propagate through space, and that the same equations that describe the Laws of Faraday, Gauss, and Coulomb predict that the speed of this radiation is the same as the speed of light.

Spectroscopy Techniques

In fact, light is just the small fraction of the vast electromagnetic spectrum that can be detected by our eyes. Electromagnetic radiation ranges from very high-energy, short-wavelength gamma rays, to X-rays, vacuum ultraviolet, ultraviolet, visible, near-infrared, infrared, microwaves, and radio waves. For each band, there exists specialized technology to resolve and detect this radiation, and to perform spectroscopic measurements.

When light travels through a vacuum, its speed is a constant, notated with the symbol c. The value of c is $\sim 1 \times 10^8$ m sec^{-1} for light traveling through space. However, when light travels through matter, it slows down. The ratio of the speed of light in space to the speed of light in the substance is called the refractive index of the substance. If the speed of light is slower in a substance, the substance is characterized as more optically dense, and its refractive index higher, than another substance. So, glass is more optically dense than air, and air is more optically dense than a vacuum.

Today, light is used in a wide variety of measurement techniques. However, all these techniques are based on inferring the properties of matter from one of these light-matter interactions:

Reflection
Refraction
Elastic scattering
Absorbance
Inelastic scattering
Emission

2

Reflection

The boundary between one substance and another is called the interface. For example, the boundary between glass and water in a fish tank is an interface. When light rays encounter an interface, the light energy may pass into the new material and/or it may reflect back into the original material (also see *UNIT 9.1*).

The energy that bounces off the interface is said to be reflected. If the interface is flat, the reflection will be at an angle that is the same but opposite to the normal (perpendicular) line drawn to the surface. This kind of reflection is called specular (mirror-like). How much energy is reflected depends on the refractive index of the two substances and the angle at which the light ray strikes the surface. Reflection is at a minimum when the ray strikes at a normal angle (perpendicular to the surface). It increases as the angle increases until it reaches 100% at the critical angle.

Refraction

The light that passes into the new material will change speed. If the light ray is perpendicular to the interface, the light will enter the material at the same angle. If the ray is at another angle, then it will change direction, or refract, when it enters the new material. If the new material is more optically dense, it will bend toward the normal. If the new material is less optically dense, it will bend away from the normal. This is the same phenomena observed by Newton as light passed from air into the denser glass of his prism.

A refractometer is an instrument that measures this angle and relates it to the concentration of solids dissolved in liquids. Refractive index of solutions is one of the colligative properties; the others are vapor pressure, freezing-point depression, boiling-point elevation, and osmotic pressure, as described by Jennings (1999). The refractive index value is related primarily to the numbers of solute particles per unit volume.

Elastic scattering

Light rays travel through a perfect vacuum in a perfectly straight path, at the same speed, forever. If a light ray traveled through a substance in a perfect straight line, then that substance would be perfectly transparent. If the light ray encounters a particle—for example, a water droplet in the air—then it may be elastically scattered. Elastic scattering means that the direction of the light is changed, but the wavelength is the same. Refraction and reflection usually refer to light encountering flat, well-defined interfaces. Elastic scattering describes what happens when the surface is rough or the interface is between a bulk substance and a very small particle that is suspended in it.

Turbidity

A solution with suspended particles is called turbid, and instruments that measure the amount of scattering by the suspended particles are called turbidometers. One type of turbidometer that is widely used to measure ground water and drinking water is called a nephelometer. The instrument measures light that is scattered by suspended particles by placing a detector at 90° to an illumination beam. The light may be white—i.e., broadband—or restricted to NIR wavelength ranges to avoid errors caused by colored particles. Typically, turbidity increases as the concentration of the particles increases. A glass of pure, filtered water will be nearly perfectly transparent, and have zero turbidity. Drinking water will have suspended particles (perhaps bacteria or bacterial spores) and have a measurable turbidity that is used as a quality parameter. Pond or river water may be very turbid, especially if, for example, it includes soil particles from run-off.

Absorbance

Absorbance is the capture of light energy by a molecule when it encounters a photon. The energy may be re-emitted as light or converted to heat. The amount of energy absorbed depends on the wavelength of the light. This property is used extensively to measure the concentration of absorbing molecules in a sample; the technique is called absorbance spectrophotometry. The absorbance of the light occurs when the frequency or wavelength of light is matched to the frequency of molecular vibrations or to the energy-level differences of electrons as they shift energies. Generally, electronic interactions require high-energy, shorter-wavelength light. Vibrational energy tends to be at longer wavelengths. Electronic interactions account for most of the absorbance features of compounds observed in the ultraviolet (UV) and visible (vis) ranges. Near-infrared and infrared spectra are due mostly to vibrational phenomena.

Inelastic scattering

In 1928, the Indian scientist Sir Chandrasekhara Venkata Raman discovered that photons also can be scattered inelastically by molecular bonds *(http://en.wikipedia.org/wiki/Raman_scattering).* Inelastic scattering changes both the direction and the wavelength of the light. This phenomenon is used in Raman spectroscopy to identify the molecular bonds in samples. Monochromatic laser light illuminates the sample and a sensitive spectrometer detects the faint Raman-shifted light at higher or lower wavelengths. The shift is diagnostic of a certain bond frequency and the technique is used to provide fingerprints of compounds.

Emission, fluorescence, and spectrofluorimetry

Substances may emit light or luminesce. Typically, chemical, nuclear, mechanical, electrical, or optical energy provides excitation or stimulation, and this energy is converted to photons that are emitted. Fluorescence is luminescence in which a high-energy (short-wavelength) photon is absorbed by a molecule. The excited molecule then emits a lower-energy (longer-wavelength) photon. The lifetime of the excited state varies considerably and is dependent on the particular molecule and its surrounding environment. Molecules with very long lifetimes are often called phosphorescent. Shorter-lifetime molecules are called fluorescent. Most fluorophores are excited by ultraviolet light and emit in the visible.

Fluorescence can be used to quantify the concentration of substances. A simple instrument designed for this purpose is the fluorometer. Excitation and detection wavelengths are controlled using transmission filters and quantification is accomplished by comparing the fluorescent intensity of an unknown to that of a standard solution. A more sophisticated approach is to measure fluorescence intensity as a function of excitation or emission wavelengths, or both. This device is called a spectrofluorometer.

COMPONENTS OF A SPECTROPHOTOMETER

Spectrophotometric instruments have been used in laboratories, educational settings, and industrial environments for more than 70 years. There are many designs and hundreds of models, but they all operate on a few basic principles.

Some spectrophotometers have the components hidden from view—i.e., they are presented as a "black box" or appliance designed specifically for one kind of measurement. One of the most common examples of this kind of system is the SPECTRONIC 20+ (*http://www.thermo.com/com/cda/product/detail/1,1055,12100,00.html*). Introduced over 40 years ago, this rugged and reliable device provides optical absorbance measurements of liquids at one wavelength, which is selected by the user. It can be set to wavelengths in the visible portion of the spectra—i.e., from 400 nm (blue) to 700 nm (red).

Other systems are made from modular parts. For example, Ocean Optics and other manufacturers offer product lines designed to use optical fibers to connect the various components into functional systems. In modular systems, the components are easy to see, and are selected and specified by the user.

Traditional spectrophotometer designs, like the SPECTRONIC 20+, are usually pre-dispersive. This means that white light from the source is spread into spectra, and one wavelength at a time is selected to go through the sample and to the detector. One advantage of this architecture is that the sample is exposed to much less light energy than in post-dispersive designs. The main disadvantage is that only one wavelength at a time can be analyzed, and spectra must be assembled from data points taken at different times.

The invention of micro-electronic detector arrays has enabled a new architecture. In post-dispersive systems, broadband (white) light is projected through the sample and spread into a spectra that strikes an array of detectors simultaneously (Fig. 2.1.1). In the Ocean Optics USB4000 Spectrometer (*http://www.oceanoptics.com/products/usb4000.asp*), for example, the detector strikes a line of 3648 detectors. Each detector is only 14-μm wide, and effectively each detector sees only one wavelength of light. With detector arrays, entire spectra can be acquired in milliseconds, and all the wavelengths are acquired during the same time interval.

Figure 2.1.1 shows the standard components of an absorbance spectrophotometer, including the optical fiber that transmits light through the system and the computer hardware that provides the user interface in most modern spectrophotometer systems. The following paragraphs will briefly describe each of the components of the system.

Light Sources

Absorbance spectrophotometry requires broadband light sources that have very stable power output. For ultraviolet work, deuterium gas discharge bulbs are used. Xenon bulbs can also provide broadband UV but are not as stable as deuterium bulbs. Tungsten filament incandescent lamps are the standard choice for visible through near-infrared wavelengths. The stability of the light sources

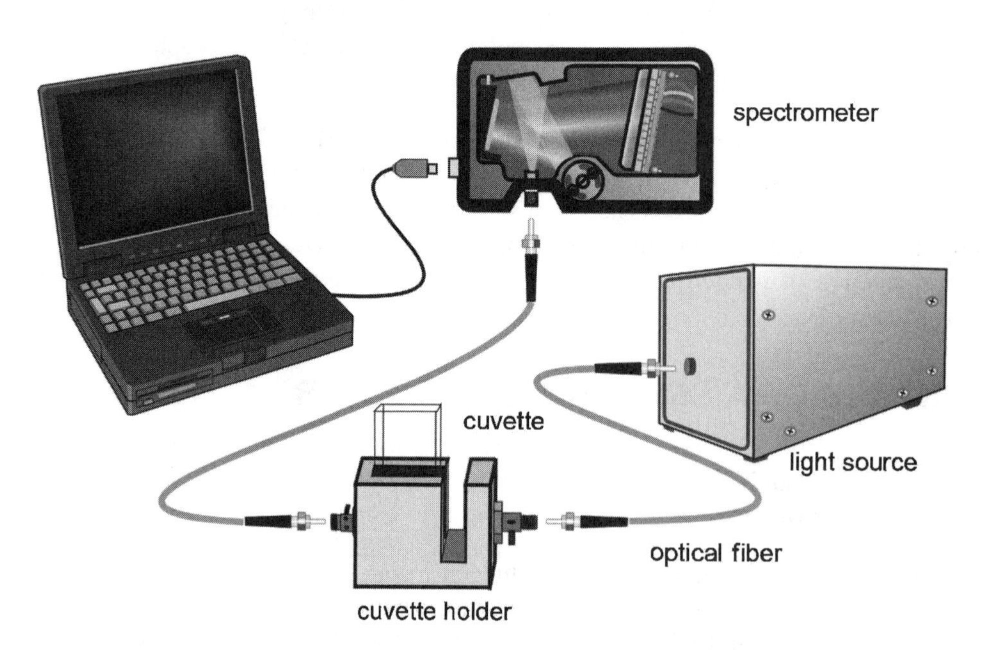

Figure 2.1.1 Typical post-dispersive spectrophotometer with key components.

Table 2.1.1 Spectrophotometric Light Sources

Type	Intended use	Typical wavelength range	Output	Measurement suitability
Deuterium-tungsten Halogen	Illumination or excitation	200-2000 nm	Continuous	Absorbance, reflectance, fluorescence, transmission
Deuterium	Illumination or excitation	200-400 nm	Continuous	Absorbance, reflectance, fluorescence, transmission
Xenon	Illumination or excitation	200-750 nm	Pulsed or continuous	Absorbance, reflectance, fluorescence, transmission
LEDs	Excitation	Various wavelengths from UV-visible	Pulsed or continuous	Fluorescence
Tungsten halogen	Illumination	360-2000 nm	Continuous	Absorbance, reflectance, transmission
Mercury argon	Calibration (wavelength)	253-1700 nm	Continuous	As standard for spectrometer wavelength calibration
Argon	Calibration (wavelength)	696-1704 nm	Continuous	As standard for spectrometer wavelength calibration
Calibrated deuterium-tungsten halogen	Calibration (radiometric)	200-1050 nm	Continuous	As radiometric standard for absolute irradiance measurements
Calibrated tungsten halogen	Calibration (radiometric)	300-1050 nm	Continuous	As radiometric standard for absolute irradiance measurements

over time is of paramount importance, as measurements are based on comparing samples with standards measured at different times. Table 2.1.1 lists some common spectrophotometric light sources and their uses.

Sampling Optics

Absorbance is defined as the attenuation of light in a parallel beam traveling perpendicular to a plane parallel slab of sample with a known thickness. This arrangement is provided by the sampling optics that manage the light beam, the sample holder and, in the case of liquids or gases, the sample container or cuvette (Fig. 2.1.2).

Wavelength Selector or Discriminator (Filters, Gratings)

Absorbance varies with wavelength; the wavelength discriminator provides the means to select or discriminate across wavelengths. The simplest wavelength discriminator is an optical pass band filter. This filter passes one wavelength (or one wavelength band) while blocking the other regions of the spectra. The user may insert the appropriate filter into the instrument, or may turn a wheel to select one of several filters mounted on the wheel. The term colorimeter is often used to describe filter-based systems.

Spectrophotometers usually use a diffraction grating for wavelength discrimination. Diffraction gratings are similar to prisms in that white light is spread into a spectrum by redirecting light at angles that are wavelength-dependent. A spectroscope is an optical system consisting of a small

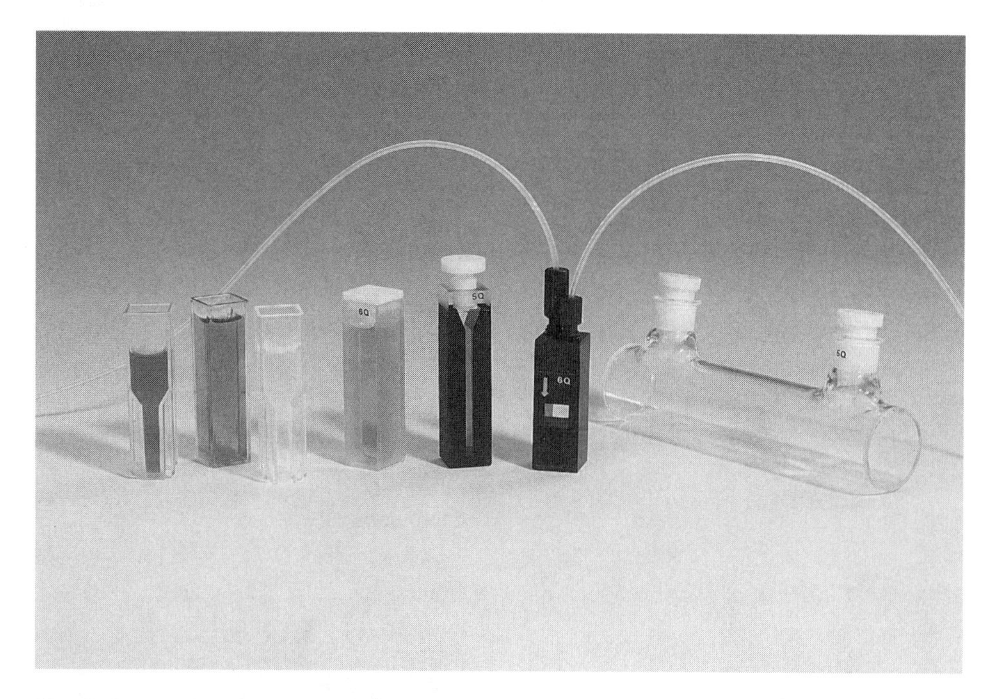

Figure 2.1.2 Collection of commonly used cuvettes (sample cells), differentiated by shape, volume, dimensions, and pathlength. Test tubes and spectropipettors are other types of sample containers.

aperture or slit, mirrors or lenses for directing the light from the slit onto the grating, and mirrors or lenses for re-imaging the light onto an exit aperture.

Detector

The light that passes through the sample and the wavelength discriminator must be detected and quantified. In the early days, this was done using photographic film. Modern instruments use electro-optical detectors and electronics to detect, amplify, and digitize the signal. Silicon photodiodes are used to detect ultraviolet, visible, and shortwave near-infrared light (Fig. 2.1.3). Different detector materials are required for longer wavelengths. Indium gallium arsenide is used for near-infrared. Mercury cadmium telluride is used for infrared. Detectors can be fabricated as single devices for measuring one wavelength at a time, or in arrays, to measure many wavelengths at a time. A spectroscope with a detector is called a spectrometer (Fig. 2.1.4).

Signal conditioner and processor

The electrical signal from the detector must be amplified, conditioned, and captured. In older instruments, this was accomplished entirely with analog circuitry. Modern devices convert the analog signal to numbers and use digital logic to accomplish the same tasks. In modern devices, the circuitry and the detector are often fabricated into the same device. Thus, a CCD array includes a set of photodiodes, amplifiers, digitizers, shift registers, and digital logic, all on a single chip.

User interface

The conditioned signal from the detector must be captured by the user, and stored, graphed, analyzed, and interpreted. In addition, the user must be able to control the instrument settings to maximize the quality of the measurement. Collectively, these functions are embodied as a user interface, which may involve dials, buttons, a display, and software.

Figure 2.1.3 Silicon linear CCD array (Toshiba TCD1304AP).

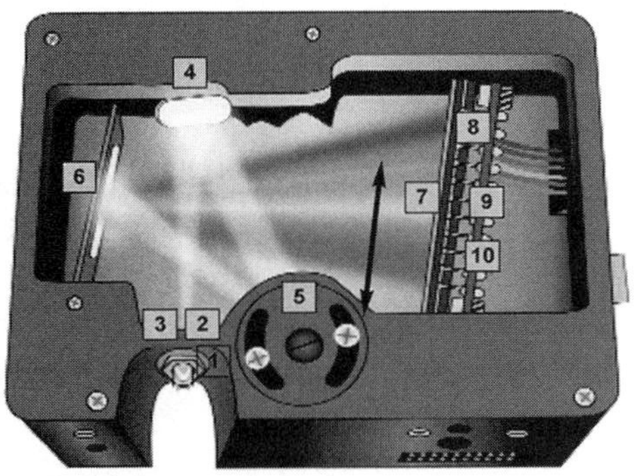

1 **SMA 905 Connector**
Light from a fiber enters the optical bench through an SMA 905 Connector. The SMA 905 bulkhead provides a precise locus for the end of the optical fiber, fixed slit, absorbance filter and fiber clad mode aperture.

2 **Fixed Entrance Slit**
Light passes through an installed slit, which acts as the entrance aperture. Slits come in various widths from 5 μm to 200 μm. The slit is fixed in the SMA 905 bulkhead to sit against the end of a fiber.

3 **Longpass Absorbing Filter**
An absorbance filter can be installed between the slit and the clad mode aperture in the SMA 905 bulkhead. The filter is used to block second- and third-order effects or to balance color.

4 **Collimating Mirror**
A collimating mirror is matched to the 0.22 numerical aperture of an optical fiber. Light reflects from this mirror, as a collimated beam, toward the grating. Some manufacturers also offer UV-absorbing mirrors.

5 **Grating & Wavelength Range**
Gratings are typically installed on a platform that is rotated to select the starting wavelength.

6 **Focusing Mirror**
This mirror focuses first-order spectra on the detector plane. Collimating and focusing mirrors are designed to ensure the highest reflectance and the lowest stray light possible.

7 **Detector Collection Lens**
This cylindrical lens is fixed to the detector to focus the light from the tall slit onto the shorter detector elements. It increases light-collection efficiency.

8 **Detector**
In photodiode or CCD linear detector arrays, each pixel responds to the wavelength of light that strikes it. Electronics bring the complete spectrum to the software.

9 **Variable Longpass Order-sorting Filter**
Filters can be installed to precisely block second- and third-order light from reaching specific detector elements.

10 **UV Detector Window**
A quartz detector window designed to allow light to transmit <340 nm.

Figure 2.1.4 Light moves through the optical bench of a spectrometer (spectroscope + detector) with an asymmetrical crossed Czerny-Turner design. The wavelength discriminator is the diffraction grating.

HOW A SPECTROPHOTOMETER WORKS

Quantitative Absorbance Spectrophotometry

Absorbance spectrophotometry can be used as a qualitative tool to identify or "fingerprint" substances, and as a quantitative tool to measure the concentration of a colored substance (chromophore) in a transparent solvent. In some situations, the absolute absorbance, or extinction coefficient, of the unknown substance is desired. In other situations, the concentration is related empirically to standard solutions of known concentration. In either case, the derivation of absorbance is called the Beer-Lambert Law (Brown et al., 1997).

Beer's Law

The Beer-Lambert Law, more commonly known as Beer's Law, states that the optical absorbance of a chromophore in a transparent solvent varies linearly with both the sample cell pathlength and the chromophore concentration (see Equ. 2.1.3). Beer's Law is the simple solution to the more general description of Maxwell's far field equations describing the interaction of light with matter. Beer's

Law is valid only for infinitely dilute solutions, but in practice Beer's Law is accurate enough for a range of chromophores, solvents, and concentrations, and is a widely used relationship in quantitative spectroscopy.

2

Absorbance and transmittance

Absorbance is measured in a spectrophotometer by passing a collimated beam of light at wavelength λ through a plane parallel slab of material that is normal to the beam. For liquids, the sample is held in an optically flat, transparent container called a cuvette (Fig. 2.1.2). The light energy incident on the sample (I_0) is more intense than the light that makes it through the sample (I) because some of the energy is absorbed by the molecules in the sample.

Transmission (T), expressed as a percentage, is the ratio:

$$T = 100\% \times I/I_0$$

Equation 2.1.1

Transmission ranges from 0% for a perfectly opaque sample to 100% for a perfectly transparent sample. If there are absorbing molecules in the optical path, the transmission will be <100%. The attenuation will increase as the number of molecules in the path increases. The number of molecules that interact with the light will depend on the pathlength (l) and the concentration of the molecules (c). The attenuation will also depend on the ability of the molecule to absorb light at that wavelength, expressed as the extinction coefficient or molar absorptivity ε.

$$T = I/I_0 = e^{-\varepsilon l c}$$

Equation 2.1.2

Absorbance(A_λ) is the negative $\log_{10}$ of transmission. This mathematical transformation is used to make the relationship between concentration and absorbance linear:

$$A_\lambda = -\log(I/I_0) = \varepsilon c l$$

Equation 2.1.3

Absorbance, which is a dimensionless number, ranges from 0 for perfectly transparent samples ($T = 100\%$) to infinity for perfectly opaque samples ($T = 0\%$). The value of the extinction coefficient ε depends on the units of c. Typically, c is in units of molarity (moles liter^{-1}), l is measured in centimeters, and ε has units of absorbance cm^{-1} liters mole^{-1}. Thus, the extinction coefficient is a measure of the ability of a substance to absorb or extinguish light. Although the absorbance scale goes from zero to infinity, the useful range is much less than that. Most instruments operate from zero to ~3 or 4 absorbance units. The combination of random errors that scale with light intensity, and sensitivity of analysis that scales with extinction coefficient, results in an optimum region for making absorbance measurements between 0.5 and 1 (see Fig. 2.1.5).

Because absorbance is linear, absorbance is also additive. For example, if cuvette no. 1 has an absorbance of 0.5 and cuvette no. 2 has an absorbance of 0.3, then putting both cuvettes into the light path in a series would yield an absorbance of 0.8. If two substances are present in the same sample, then the total absorbance will equal the sum of the individual absorbances.

Absorbance spectrophotometry and dark current

The photodiode detectors in spectrophotometers operate with a bias voltage across the silicon junction. As photons are absorbed, they create a current that is amplified and measured by the

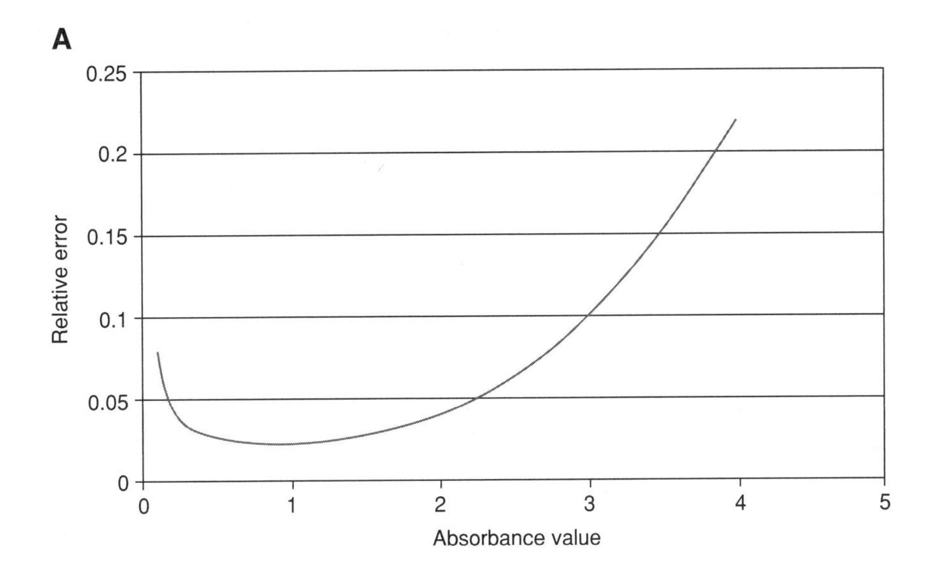

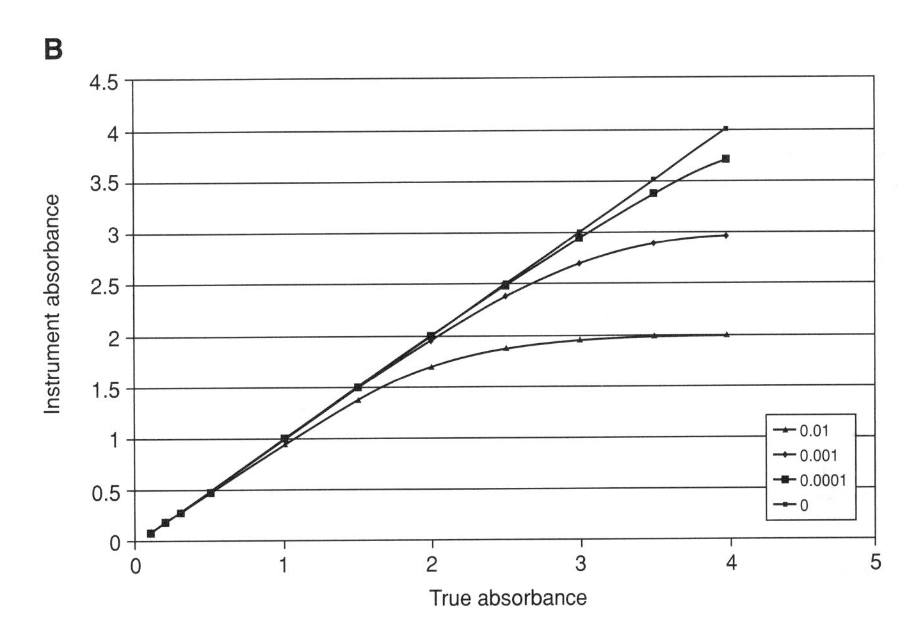

Figure 2.1.5 Absorbance error. (**A**) The combination of random errors that scale with light intensity and sensitivity of analysis that scales with extinction coefficient result in an optimum region for making absorbance measurements between 0.5 and 1. (**B**) All instruments have stray light, which affects absorbance linearity. However, it is possible to measure the stray light and remove it from experimental consideration.

electronics circuitry. However, random thermal events can also cause a current, even in the absence of light. This leakage current, or "dark current" must be accounted for. In practice, this background reading is recorded and subtracted from the other measurements as a "dark" reading, or by setting 0% transmission looking at an opaque sample.

Reference Sampling and Calibration

In an absorbance experiment, light is attenuated not only by the chromophore being studied, but also by reflections from the four interfaces: air to cuvette, cuvette to sample, sample to cuvette, and cuvette to air. In addition, light is absorbed by the solvent and by fingerprints and dust on the cuvette surface, and is affected by various other factors. These phenomena can be quantified separately, but are often removed in total by defining I_0 as the light passing through a sample "blank," also known as a "baseline" or "reference" sample. The definition of the blank sample thus becomes an important qualifier of the measurement.

For example, we might measure the absorbance of the dye thymol blue dissolved in water. The appropriate blank would be a cuvette filled with water. If the dye was dissolved in a buffer solution, the appropriate blank would be a cuvette filled with buffer solution. The terminology "absorbance of (sample) versus (blank)" is often used to define the meaning of the measurement.

You can account for the absorbance of a blank solution either explicitly (by measuring it and subtracting it from the sample absorbances) or automatically (by setting the instrument to start at 100% T, or 0 absorbance). This is much like setting the tare on a balance (*UNIT 1.2*). In older instruments, it was common to set the blank solution to zero by adjusting a potentiometer until the needle on a meter read "0." A variation in analog instrument design is the double-beam spectrophotometer. These devices split the light beam into two beams, and direct one beam through the sample cuvette and the other through the reference or blank cuvette. Optical choppers and phase-locked loop amplifiers allow one detector to measure both beams and report the difference between them. In modern digital instruments, the blank value is simply stored in memory and subtracted during the calculation stage.

Beer's Law Limitations

Standard curves

In an ideal experiment, Beer's Law as expressed in Equation 2.1.3 could be used to calculate the concentration of substance from its measured absorbance and the cuvette pathlength alone. However, many factors can affect the validity of Beer's Law, so it is much more common to perform a calibration of the assay. The calibration is performed by measuring the absorbance of a series of solutions with known concentrations of the chromophore. A graph of the results—concentration versus measured absorbance—is called the standard curve. The standards are prepared and measured as much as possible in exactly the same way as the unknown samples will be treated, so that the "calibration" will remove errors in the experiment, the equipment, and the batch of reagents (such as cuvettes of unknown pathlength). If a particular procedure is shown to be linear—i.e., follows Beer's Law—then a single-point standard can be used with the blank to establish the slope of the calibration. If the data is somewhat non-linear, then a series of standards and a higher-order polynomial fit may be required.

Linearity

In essence, all calibrations will be non-linear because of the non-linear relationship between concentration and the optical signal, and because of limitations of the instrument. The effects of signal to noise on measurements are shown in Figure 2.1.5A. The detector fundamentally measures the numbers of photons striking the detector during a given period, called the integration time. The arrival of the photons is random, and the random variation in how many photons strike the detector is equal to the square root of the number of photons. If a sample has a high absorbance, the number of photons reaching the detector is low and the signal to noise is relatively poor. At an absorbance of 4, for example, only 1 in 10,000 photons reaches the detector. At low absorbance values, there are plenty of photons and higher signal to noise. However, the relative absorbance error is higher because the denominator is smaller. This is the same as saying the test is relatively insensitive to changes in absorbance (concentration). There is an ideal region, from ~0.5 to 1.0 absorbance, where there is maximum sensitivity to changes in absorbance and reasonable signal to noise. It is customary to modify the procedures to ensure that samples are measured in this range, by adjusting the concentration of the samples through dilution, or by choosing cuvettes with different pathlengths, or both.

Stray light

Stray light refers to any light that strikes the detector that is either the wrong wavelength, or did not travel through the sample. Stray light can arise from many sources, including ambient light that leaks into the instrument, light that bypasses the sample (e.g., wave-guiding of light through a cuvette wall), higher orders of light from the diffraction grating, and scattering of optical surfaces in the spectrometer. Instruments are designed to minimize the effects, but there is a physical limit imposed by the zero-order scattering of the grating, and the stray light of any instrument is both measurable and quantifiable.

2

Stray light (s) always appears as "extra" signal. It appears in both the sample measurement (I) and the reference (I_0) measurement in the absorbance equation:

$$A_\lambda = -\log\,[(I + s)/(I_0 + s)]$$

Equation 2.1.4

When the absorbance equals zero, such as setting the instrument to read zero with the blank solution, the stray light terms cancel out. As absorbance increases, I decreases and stray light begins to reduce the absorbance value. Stray light is usually expressed as a percentage of the reference value. The effects on absorbance linearity are shown in Figure 2.1.5B. A good spectrometer will offer stray light as low as 0.001. As shown in Figure 2.1.5, absorbance readings in the 0.5 to 1.0 range are not significantly affected. However, all instruments have stray light, and at higher absorbance readings, the effects become important.

Stray light is a property of the spectrometer, and of the light source used in the instrument. It is possible to measure stray light and remove it from consideration. One technique is to use as the "dark" a sample that is known to have a high concentration of the species of interest. For example, if a concentrated sample is known to have an absorbance near 5, then the light reaching the detector would be reduced by a factor of 10,000. Any light above this level could be considered stray light. This sample could be used to take the dark reading, if an answer at only one wavelength is desired. In instruments like the Ocean Optics USB4000 Spectrometer, this value can be entered into the software and then scaled and subtracted from all other readings.

Many spectrophotometric assays rely on chemical equilibrium between the species of interest and color-forming reagents. The equilibrium conditions of these chemical species can often affect linearity as well. Most assays are designed to be relatively linear over a certain range by using high concentrations of color-forming reagents, such that their concentration is not appreciably affected by reaction with the species being assayed.

Beer's Law also assumes that there is a single absorbing species or chromophore at a given wavelength. This is difficult to ascertain using single-wavelength instruments. Full spectral analysis is useful for observing more complex systems. Figure 2.1.6 shows the absorbance spectra of the pH indicator dye bromocresol green dissolved in sol-gel at a variety of pH levels. The indicator dye appears blue when it is dissolved in high pH solutions, yellowish when it is dissolved in low pH solutions, and various shades of green in intermediate pH solutions. This system is in equilibrium between H^+ and the indicator dye molecule (abbreviated as A^-):

$$HA = H^+ + A^-$$

Equation 2.1.5

The spectrum of the protonated form, HA, is shown for the trace labeled pH 1. At this pH virtually all the molecules are in the HA form (as the equilibrium is shifted toward the left). The absorbance curve peaks ~460 nm, and diminishes toward zero at ~600 nm.

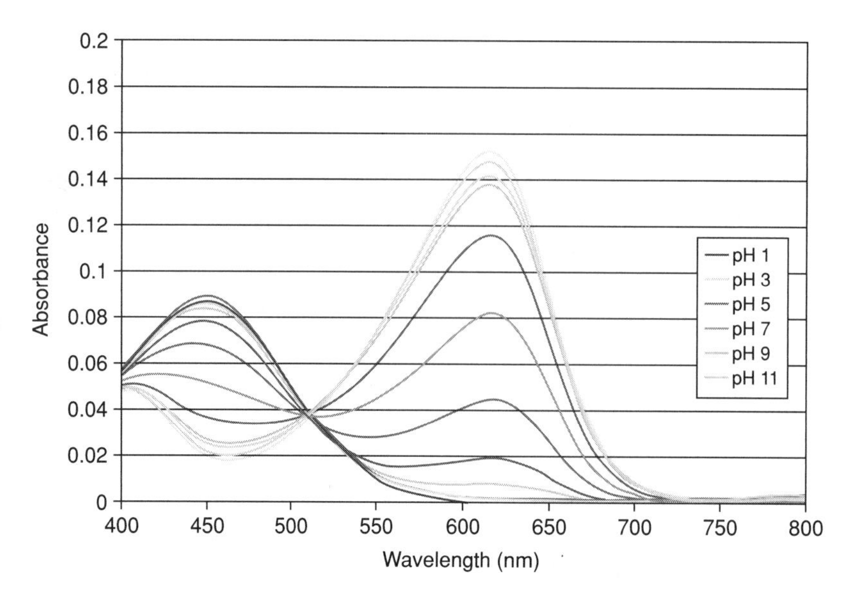

Figure 2.1.6 A pH indicator dye such as bromocresol green will appear as different colors at different pH levels. Full spectral analysis, such as that provided by a spectrometer, is required to discern such complex differences.

The un-protonated form of the dye (A^-) has an absorbance spectrum that peaks ~620 nm and diminishes toward zero at 725 nm.

The A^- form of the dye is easily measured using Beer's Law at 620 nm, because it is the only species that absorbs at that wavelength. Absorbance is generally best measured at the peak of the absorbance curve, because this is where the assay is most sensitive to changes in concentration. In contrast, measurements at 460 nm will reflect the contributions of both A^- and HA molecules, and a simple Beer's Law analysis will not work.

STRATEGIC PLANNING

Regardless of the sources of the deviations from Beer's Law, their effects are best determined by analyzing a series of samples in which the concentration of a single component is being varied. The absorbance of each is plotted against the concentration and the result is called a standard curve. If the system follows Beer's Law, the plot is linear. If the curve is non-linear, it can still be used by fitting a polynomial function to the points (see *APPENDIX 4* for more information on regression analysis).

Absorbance spectrophotometry can be used to measure the concentration of a chromophore in a clear solvent. The term chromophore literally means "colored," i.e., absorbs in the visible, but the term also extends to substances that absorb light in the UV, NIR, and IR regions. Clear solvent refers to a liquid that can dissolve the chromophore, but which has little or no absorbance in the wavelength range of interest.

Many biological molecules exhibit strong UV, VIS, and NIR spectra that make them amenable to direct absorption spectroscopy. Some examples are shown in Figure 2.1.7. Other molecules are not strong chromophores on their own, but they can react with specific reagents to form colored complexes. A good example of this from Figure 2.1.6 is bromocresol green as a color reagent for the assay of H^+ ions. H^+ ions by themselves have negligible absorbance anywhere in the optical portion of the electromagnetic spectrum. To form a colored substance that can be measured spectrophotometrically requires the reaction of H^+ with the bromocresol green anion.

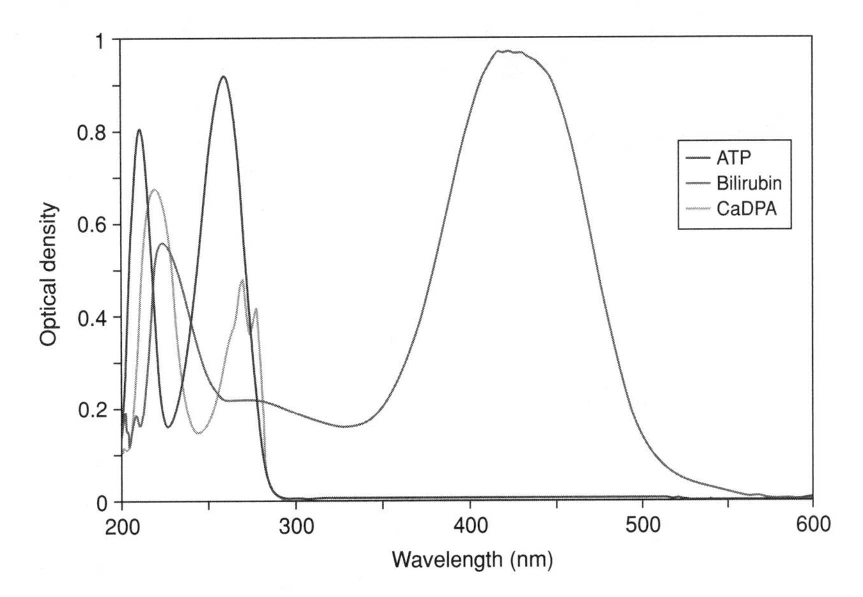

Figure 2.1.7 Biological molecules such as adenosine 5′-triphosphate (ATP), bilirubin, and calcium dipicolinate (CaDPA) have strong absorbance response.

Standard Curve

For direct spectroscopy, the standard curve is developed by preparing a series of solutions in which the concentration of the chromophore varies. The exact procedure depends on the choice of volumetric devices, the solubility of the chromophore, and the volume required for analysis. Typically, the volume of the stock solution (Vs) is varied, and the final volume of the diluted sample (Vd) is kept the same (by adding more or less solvent). The concentration is determined as:

$$Cd = (Vs \times Cs)/Vd$$

Equation 2.1.6

where (Cs) is the concentration of the stock solution and (Cd) is the concentration of the diluted stock solution.

Stock solutions

Accuracy of the stock solutions depends on the accuracy of the volumetric devices (*UNIT 1.1*), the volumes used, and the preparation of the stock solution (*UNIT 3.1*). This technique can be used only if the chromophore can be obtained in pure form and is stable once diluted. Many biological molecules are sold in frozen or lyophilized (freeze-dried) state to prevent degradation. Others are stable as solids kept at room temperature. In some cases, commercially prepared standard solutions are available from manufacturers such as GFS Chemicals, Inc. (*http://www.in-spec.com/products/standards.php*).

Stock solutions may be prepared gravimetrically or volumetrically (see Chapter 3). The chromophore standard substance is weighed and diluted to a known volume, or added to a known weight of solvent. If the solid form of the chromophore is hygroscopic, it will be necessary to dry it in a desiccator prior to weighing. If the manufacturer lists the purity of the substance, it is necessary to include this factor in the calculation of the final concentration. Many biological molecules are polymers with a variety of possible combinations of monomers. For example, there is not a standard for protein in general. It is necessary to specify the source of the protein, such as bovine serum albumin, as well.

Extinction coefficients for proteins can be estimated by adding the contributions of the individual amino acids that comprise the protein (Pace et al., 1995). The absorbance of a protein at 280 nm is due almost entirely to the UV absorbance of the side chains on the tryptophan, tyrosine, and cystine amino acid monomers in the polymeric chain. If the numbers of each of these units are known for a particular protein, then the molar extinction coefficient at 280 nm can be estimated as no. Trp (5500) + no. Tyr (1490) + no. Cys (125). If pure samples of proteins are available, and the molecular weight is known, the extinction coefficients can be measured using the technique presented below. Some values of ε (M^{-1} cm^{-1}) for some common proteins are:

bovine serum albumin	43,890
insulin	5510
luciferase A	53,745

The example used in the following sections is both newsworthy and typical; it is the assay of dipicolinic acid (DPA). DPA is a unique constituent of the cortex of bacterial endospores of the genera *Bacillus* and *Clostridium.* As shown in Figure 2.1.8, DPA has a broad absorbance in the UV region of the spectrum, featuring absorbance peaks at 220, 270, and 277 nm. It is thought that the UV absorbance provides the spore's DNA protection from UV radiation, making the spores highly resistant. Extraction and measurement of DPA is a technique used by the U.S. Department of Homeland Defense and first responders to rapidly detect *Bacillus anthracis*, the spores that cause the lethal disease anthrax, in suspicious powders (Samuels et al., 2003).

The DPA assay is a good example of direct spectroscopy, in that color-forming reagents are not required. DPA is readily available in pure form, which means that standards can be easily prepared. In principle, this technique can be used with any substance. The practical limitations are (1) the substance must absorb light at a wavelength in the range of the available equipment, (2) the substance must be dissolved in a solvent that is transparent at the same wavelength, and (3) there must not be other substances in the sample that also absorb at the same wavelength.

Preparation

The desired concentration of the stock solution and the final dilution depends on the extinction coefficient. These can be obtained in the literature for many substances of interest. If this information is not available, a semi-quantitative 10× serial dilution of the stock solution can provide a quick test

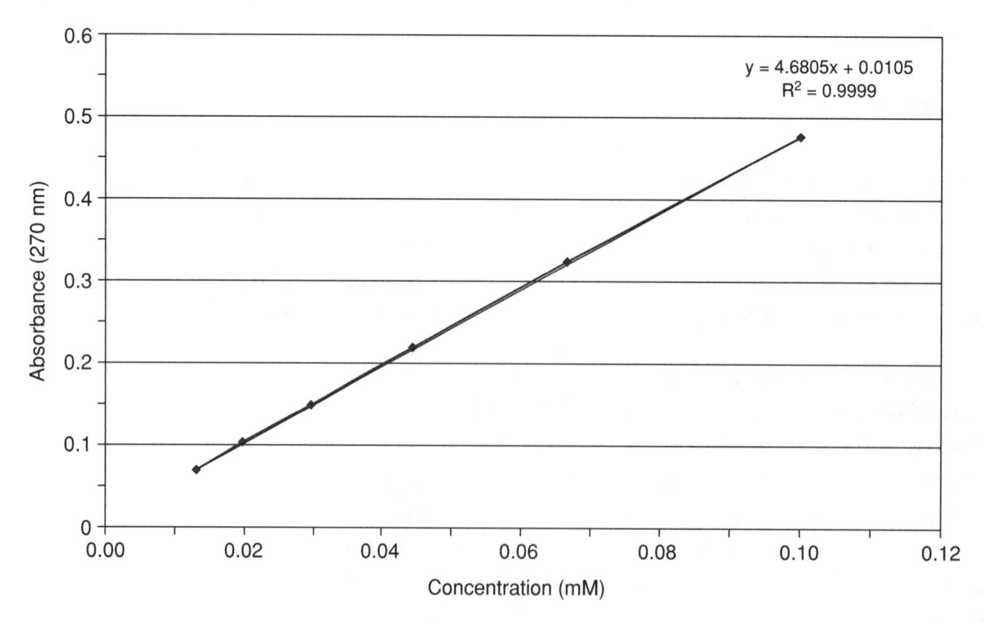

Figure 2.1.8 Dipicolinic acid (DPA), a chemical unique to spores, has broad absorbance in the UV region and is believed to protect the DNA in spores from UV radiation.

to establish the likely range. In the serial dilution, the stock solution is diluted 10:1, that resulting solution is diluted 10:1, and so forth until a sample is obtained with an absorbance ~1.

In our DPA example, the extinction coefficient at 270 nm is given as ~5000 M cm^{-1}. An absorbance of ~1 in a 1-cm cuvette is desired. Equation 2.1.3 is rearranged as follows to solve for concentration yields:

$$c = A_\lambda / \varepsilon l = (1)/(5000\ \text{M cm}^{-1})(1\ \text{cm}) = 2 \times 10^{-4}\ \text{M}$$

Equation 2.1.7

To create a range of solutions to bracket 2×10^{-4} M, make dilutions in the range of 10:1, so that the stock solution should be $\sim 2 \times 10^{-3}$ M. Weigh $\sim 2 \times 10^{-4}$ mol of DPA into a 100-ml volumetric flask.

The literature search also mentions the best choice of solvent. DPA will dissolve in alkaline solutions far more easily than in water. It exists in the endospores in a complex state with Ca^{2+} ions. The recommended solvent is 10 mM NaOH and 5 mM $CaCl_2$.

Cuvettes

There are several strategies for using cuvettes. The best choice depends on the desired level of accuracy, the number of samples to be analyzed, and any time constraints imposed by labile (unstable) samples, lack of temperature control, or impatience.

There is a tolerance associated with the pathlength of a cuvette. A set of 1-cm cuvettes with three significant figure precision would have to be within ±0.0001 cm dimensional tolerance, or about 40 thousandths of an inch. It is possible to buy premium "matched" quartz or glass cuvettes, where their absolute dimensions are within 1/1000th of each other. Disposable cuvettes, which are molded, are not likely to meet such specifications. Cuvette pathlengths can also be measured optically. One cuvette is chosen as the master blank (reference), and the absorbance of the other cuvettes filled with blank solution is recorded and subtracted from subsequent samples.

It is also possible to use a single cuvette for all the measurements. The advantage is that pathlength will ultimately cancel out when the standard curve is used to analyze sample readings obtained in the same cuvette. The trade-off is that the cuvette must be thoroughly rinsed with the next sample being analyzed to eliminate cross-over contamination.

An additional source of error comes in handling and positioning the cuvette. It is recommended that the cuvettes always be used in the same orientation. Most cuvettes have index marks as a guide. It is also important not to touch the optical surfaces. Oils from your skin, particles from wiping tissues, etc. can affect the readings. An alternative to handling the cuvette is to use a flow-through cuvette (Fig. 2.1.2). Here the fluids are pumped into the cuvette through tubing ports. Flow-through cells are more likely to suffer from cross-over contamination and also are subject to trapping air bubbles.

For open-top square cuvettes, perhaps the most effective approach is to use a slender transfer or Pasteur pipet to add and remove fluids. The tiny tip allows for suction of fluid from the corners, minimizing the carry-over volume. The typical procedure is to rinse the cuvette with the next sample to be analyzed at least three times. If the residual fluid is <10% of the wash fluid (it is more likely to be ≤1%), the carry-over is reduced to $1/1000^{th}$. It is important that the pipets also be washed with the sample, and that they are not allowed to touch or scratch the inside optical surfaces.

SAFETY CONSIDERATIONS

When working with chemicals and solvents, consult the Material Safety Data Sheets provided by the manufacturer. (Also, MSDS information for many items is available through an Internet search.) In most instances, the use of safety goggles and gloves is recommended. DPA stock solution is caustic. Wear goggles, avoid contact with skin, and flush immediately with water if contact with skin occurs. Refer to *APPENDIX 1* for more information.

Also, when working with lamps that emit UV radiation, it is recommended to wear UV-resistant safety goggles and to shutter or block the light path of the source when disconnected from a fiber optic. Many such lamps have built-in safety shutters for such occurrences.

2

PROTOCOLS

A standard absorbance spectrophotometry method is described here for preparing a standard curve for DPA. Preparing standard curves is a common spectroscopy method, as described in the literature by Ebbing (1987) and others.

Basic Protocol 1: Preparation of a Standard Curve for Dipicolinic Acid (DPA)

This protocol is the first of two protocols to determine extinction spectra for DPA. Included are steps for preparing the solvent, stock, and standard solutions.

Materials

Sodium hydroxide (NaOH)
Calcium chloride ($CaCl_2$)
Dipicolinic acid (DPA, Sigma cat. no. D-0759)

1-liter volumetric flask(s) with stopper
Weighing paper or weighing boat
Balance (*UNIT 1.2*)
Pipets

NOTE: There are dozens of suppliers of DPA, including Sigma, Spectrum Chemicals and Laboratory Products, and Alfa Aesar. Labware is available from a number of suppliers. In particular, Ocean Optics is a good source for cuvettes, cells, and reference materials for spectroscopy.

Prepare DPA solvent solution

1. Prepare 1 liter of an aqueous solution of 10 mM NaOH and 5 mM $CaCl_2$ as follows:

   a. Add 0.4 g of NaOH to a 1-liter volumetric flask. Carefully add water (the dissolution is exothermic) until the flask is about half full.

   b. Add 0.556 g $CaCl_2$ to the flask and swirl to dissolve.

   c. Bring the flask to 1 liter with water and mix thoroughly.

   d. Stopper the volumetric flask, invert, and shake. Repeat this procedure ten times.

   This is the solvent for the stock DPA solutions. It is also the diluent (dilutant) for the standard solutions; the wash solution for the cuvettes and the volumetric glassware; and the solvent used for analyzing samples. Its exact composition is not critical, but it must be homogeneous.

Prepare DPA stock solution

2. Prepare quantitatively 100 ml of $\sim 2 \times 10^{-3}$ M dipicolinic acid in 10 mM NaOH/5 mM $CaCl_2$ solvent.

   DPA is $C_7H_5NO_4$ *with a molecular weight of 167.12 g* mol^{-1}. *The target weight for the stock solution is:*

   2×10^{-4} mol DPA $\times$ 167.12 g mol^{-1} = $\sim$0.03 g DPA

The sample can be weighed using any of the established quantitative techniques (UNIT 1.2). Use weighing paper or boat to hold the DPA. Tare the balance (set to zero) with the weigh boat in place. Add ~0.03 g of DPA. Close the balance doors and obtain an accurate weight of boat plus DPA (W_1). Carefully transfer the DPA to a 100-ml volumetric flask. It is not necessary to transfer all of the solid DPA out of the boat, but it is necessary to ensure that all DPA that leaves the boat is captured in the flask. Return the boat to the balance, close the doors and obtain the final boat weight (W2). The amount transferred to the flask = $W_1 - W_2$. Use this precise weight to calculate the actual concentration of the stock solution. The final precision depends on the precision of the balance and the volumetric glassware. Precision to three significant figures is generally possible (UNITS 1.1 & 1.2).

Use a transfer pipet and the solvent to carefully wash any particles of DPA down the side of the flask neck. Add solvent, swirl to dissolve, and bring up the solution to the final volume. Again, a homogeneous solution is critical. Invert and mix ten times.

For best practices, volumetric glassware should be used at the specified temperature. A temperature-regulated water bath is used to bring the volumetric flask and its contents to temperature before the final addition of solvent is made. This can be avoided by using gravimetric units (e.g., molality). As shown in Equation 2.1.6, temperature-induced errors in volumetric glassware used in dilutions will tend to cancel out.

The stock solution is used to prepare diluted standard solutions. It is critical that concentration does not change. Always transfer the solution to a secondary container before pipetting. Never return any solution to the original flask, and always keep the flask stoppered to prevent evaporation.

Prepare standard solutions

3. Prepare a series of solutions that range in concentration by at least one order of magnitude.

If the highest sample has an absorbance of ~1, the lowest should be ~0.1. The blank solution (solvent only) will be an additional point for calculating the Beer's Law plot.

It is very easy to do the dilution calculations and regression analysis in spreadsheet software. Table 2.1.2 shows the calculations for six solutions.

a. Use volumetric pipets or an adjustable pipettor with disposable tips to transfer the volume of stock solution indicated in Table 2.1.2 into the appropriate 100-ml volumetric flask. Make sure the stock solution is mixed well before using it.

There is some degree of skill required to obtain optimum precision with volumetric devices (see UNIT 1.1). It is recommended to practice pipetting with water. Check for precision and accuracy by pipetting the water into a container on an analytical or top-loading balance. The theoretical weight of the water that is dispensed can be calculated from the density of the water, and the nominal volume of the pipetting device. Precision can be determined from the variance in replicate pipettings. It is likely that different devices in your laboratory may have different systematic errors. Labeling the devices and applying correction factors can improve overall laboratory results.

Table 2.1.2 Dilution Calculations for Standard Solutions

	Volume (ml)			Concentration (M)	
Sample	Stock solution	Solvent	Total	Stock solution	Final
d1	10	90	100	1.010×10^{-3}	1.010×10^{-4}
d2	7	93	100	1.010×10^{-3}	7.070×10^{-5}
d3	4	96	100	1.010×10^{-3}	4.040×10^{-5}
d4	3	97	100	1.010×10^{-3}	3.030×10^{-5}
d5	2	98	100	1.010×10^{-3}	2.020×10^{-5}
d6	1	99	100	1.010×10^{-3}	1.010×10^{-5}

b. Add the 10 mM NaOH/5 mM $CaCl_2$ solvent to the flasks to bring them up to desired volume. Invert and mix ten times.

Most solutions appear well mixed to the eye when in fact they are still measurably inhomogeneous. Solutions that sit for a long time may form density gradients and regions of variable concentration. A good rule of thumb is to mix solutions until you are sure they are well mixed, then mix some more.

2

Basic Protocol 2: Determination of a Standard Curve for Dipicolinic Acid (DPA) and Measurement of DPA Concentration in Unknown Samples

This protocol is designed to determine extinction spectra for DPA. Included are steps for optimizing the spectrophotometer and using Beer's Law to calculate extinction data.

Materials

Reference solution
Sample solutions

Spectrophotometer
Light source
Spectroscopy software
Optical fiber
Sample compartment (cuvette holder)
1-cm cuvettes

NOTE: There are hundreds of models of spectrophotometers (a good directory of spectrometer companies is available at *http://edirectory.spectroscopymag.com*). Details of their use will vary. To optimize results, consult the technical documentation provided by the manufacturer. The following lists some general guidelines on spectrophotometer operation, as well as details for use of an Ocean Optics USB4000 Spectrometer equipped with a DH-2000 Deuterium Tungsten Halogen Light Source, fiber optics, and a 1-cm pathlength CUV-UV cuvette holder (sample compartment).

Prepare spectrophotometer setup

1. Turn on the instrument and turn on the light sources. Allow the equipment to warm up for at least 1 hr.

   Electronics and light sources generate heat, and the equipment must reach thermal equilibrium with the surroundings before it reaches a stable temperature. The USB4000 Spectrometer is powered by the USB cable plugged into a PC.

2. Start the software (SpectraSuite) and confirm that spectra are being acquired.

   Most PC-based instruments are controlled by the software—i.e., the electronics must be turned on through software. It is a good idea to make sure the instrument is fully functional and then let it warm up to a stable operating temperature.

3. Optimize instrument settings using a wavelength of 270 nm as described in the Support Protocol.

4. Measure the dark spectra.

   In general, it is desirable to block the light rather than turn it off so as not to affect the thermal equilibrium of the system. Some instruments have built-in shutters, some use opaque cuvettes. The CUV-UV sample chamber has a slot for accepting filters or for inserting a light block. The cuvette cover is designed to fit into this slot, and can be pushed from side to side to open the light path or block it. Storing the dark reading varies from instrument to instrument. In the USB4000, the dark reading is stored in software by pressing the dark light bulb icon (or using the software menu system function, SAVE DARK).

   The dark reading includes the true dark current of the detector and the baseline condition of the amplifiers. Any change in operating conditions such as integration period or smoothing, or changes in temperature will affect this value. It is best to re-take dark measurements often.

5. Acquire a reference spectrum.

   The reference spectrum is that of the cuvette filled with solvent and no chromophore. It is the zero-concentration sample. In some instruments, you put the reference sample into the cuvette holder and then turn a knob or push a button to set absorbance to zero (or transmission to 100%). In dual beam designs, you put this cuvette into the reference slot and leave it there for the duration of the experiment. In the USB4000, you simply store this spectrum.

6. Insert the cuvette with solvent into the cuvette holder. Replace the cover.

   Cuvette holders are designed to help you place the cuvette in a reproducible position. It is a good idea to practice your technique, and to quantify the "positioning error" by repeating this process several times and observing the change in signal detected by the spectrometer. There will be a reference surface, which the cuvette presses against, and springs to push the cuvette to that position. It is also important to orient the cuvette the same way each time it is inserted.

7. Click on the lit light bulb icon (or in the software menu system, choose SAVE REFERENCE), then switch to absorbance view by clicking on the A icon.

   You should now observe a spectrum (or a value for single-wavelength devices) that is near zero, and perhaps randomly bouncing above and below zero over time.

8. Insert the cuvette filled with sample into the cuvette holder. Alternatively, using a transfer pipette to suction the old sample, rinse and suction three times with the new sample, and fill with the new sample.

   Rinsing is based in diluting residual fluid at one concentration with a large volume of fluid at the new concentration. The dilution factor that is required depends on the difference in concentration between the old fluid and the new fluid. The rinsing is most effective when the residual volume is minimized and the rinse volume is maximized. The resulting dilution factor is multiplied if the step is repeated. A good rule of thumb is to rinse three times.

9. Record the absorbance at 270 nm.

   The spectra of the six DPA standard solutions are shown in Figure 2.1.9. The spectra are essentially parallel—i.e., absorbance at every wavelength is proportional to the concentration. The data grows noisy at wavelengths <220 nm, due to low light levels. The near-zero absorbance at wavelengths >300 nm indicates that the experiment went smoothly. Problems with particles in the sample, or interference from fingerprints, bubbles, and other factors, would show up as a change in baseline. The shapes of the spectra match what is expected for DPA, again a good indication that the data is valid.

   The absorbance of the standard solutions at 270 nm can be recorded manually and input into a spreadsheet to calculate the Beer's Law relationship. If using SpectraSuite software, this can be done automatically by using the Beer's Law wizard. Most computer-driven spectrophotometers have a built-in Beer's Law function. In the example using DPA, 100 ml of each standard was prepared. This allows for replicate measurements. If replicate measurements are desired, be sure to replicate as many of the steps as possible. Measure the solutions in reverse or random order the second and third times through. If running unknowns at the same time, analyze the standard solutions at the beginning and end of the procedure.

10. Calculate Beer's Law equation by entering the absorbance values at 270 nm into a spreadsheet table.

    By convention, tables are constructed with one variable per column and multiple records in rows. In this format, it is easy to generate x,y plots and to perform regression analysis on the data. From an experimental point of view, concentration is the independent variable and the response (absorbance) is the dependent variable (see Table 2.1.3).

11. Make a graphical plot of absorbance versus concentration. In Microsoft Excel, highlight the concentration and absorbance columns and use Insert, Graph, x,y.

    As shown in Figure 2.1.10, the points fall close to a straight line. This confirms that Beer's Law was followed over this range. The deviations from the straight line appear to be random and are therefore random experimental error.

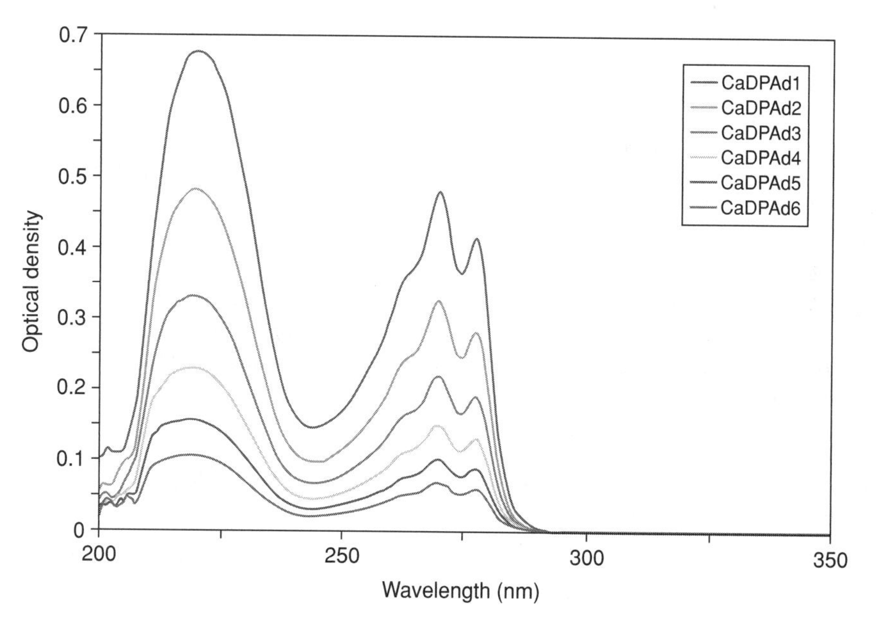

Figure 2.1.9 Spectra of six dipicolinic acid (DPA) standard solutions. The absorbance at every wavelength is proportional to the concentration of the solution.

Table 2.1.3 Concentration and Absorbance Values for Six Dipicolinic Acid Solutions

Solution no.	Concentration (M)	Absorbance
d1	1.010×10^{-4}	0.477
d2	7.070×10^{-5}	0.324
d3	4.040×10^{-5}	0.220
d4	3.030×10^{-5}	0.150
d5	2.020×10^{-5}	0.103
d6	1.010×10^{-5}	0.070
Blank	0	0.000

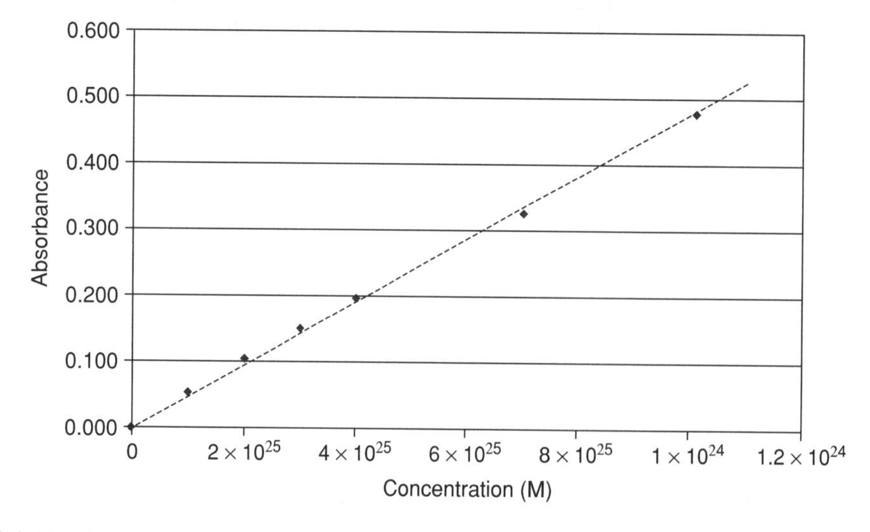

Figure 2.1.10 Absorbance of DPA at 270 nm. The points in a plot of absorbance versus concentration for a chromophore such as the DPA shown here should be nearly linear. Deviations from a straight line may indicate random experimental error.

12. Use linear regression (see *APPENDIX 4*) to calculate the best straight line that minimizes the residual errors.

Regression analysis is available in graphing calculators, or Microsoft Excel, or in SpectraSuite as part of the Beer's Law wizard. In Microsoft Excel, use TOOLS, DATA ANALYSIS, REGRESSION. Enter the x range (concentration), y range (absorbance, output range, a starting cell to put the answers into, and whatever options you like. The output is the equation of the form:

$$y = mx + b$$

Equation 2.1.8

where y is equal to absorbance, x is equal to concentration, m is the slope and b is the intercept. The output from Microsoft Excel will look like Table 2.1.4.

13. Determine the degree to which the data resembles a straight line as determined by the r^2 value, with 1 being perfect (*APPENDIX 4*).

The slope is "x variable 1," i.e., 4640. This is equal to εl in Equation 2.1.3. Since the pathlength is 1 cm, the extinction coefficient or molar absorptivity for DPA is 4640 Mcm^{-1}. The intercept should ideally go through zero, because the blank is set to be zero. Here the deviation is small, 0.005, and represents the effects of random error on the fitting of the line. In some assays, this value can be significant, and might represent a physical or chemical situation such as differences in cuvettes, differences in composition between samples and blank solutions, frost on windows, spectrometer or lamp drift, or the appearance of suspended particles in samples. Therefore, for DPA, the resulting Beer's Law plot is:

$$A = 4640(c) + 0.005$$

Equation 2.1.9

Table 2.1.4 Linear Regression Analysis of Six DPA Solutions

Summary output		
Regression statistics		
Multiple R	0.999493258	
R Square	0.998986772	
Adjusted R Square	0.998784127	
Standard Error	0.005775623	
Observations	7	
ANOVA		
	Df	SS
Regression	1	0.164444868
Residual	5	0.000166789
Total	6	0.164611657
	Coefficients	Standard error
Intercept	0.00515916	0.003375668
X Variable 1	4640.578943	66.09381375

Equation 2.1.9 is transformed to calculate the concentrations of unknowns:

$$c = (A - 0.005)/4640$$

Equation 2.1.10

Analyze unknown samples

Samples of suspicious powders can be analyzed by extracting the DPA with the solvent prepared in step 1 (10 mM NaOH/5 mM $CaCl_2$). If there are spores present, the DPA will dissolve into the solvent. The process may be aided by heat, sonication, or grinding. Anthrax spores are of course quite hazardous. There are other species of bacillus spores that are safe to handle and that can be purchased as powders or suspensions from biological supply companies (e.g., Microbiologics, Inc.). *Bacillus thuringiensis* spores are found in the horticultural dusting product Dipel. The powders can be added to the solvent using a cotton swab wet with the solvent. Suspensions can be added by pipet.

14. For measuring unknown samples, follow the procedure in steps 1 through 9 for the standard solutions.

    It is usual to work with both the samples and the standards during the same session.

15. Record the absorbance of the sample. If it is higher than the highest value for the standard solutions, or if it is >2.5 absorbance units, make a dilution using the solvent prepared in step 1. The ideal absorbance value should be <1. Record the dilution factor (f) for the sample, which is calculated as the ratio of final volume (V_f) to initial volume (V_i)

$$f = V_f/V_i$$

Equation 2.1.11

16. Use Equation 2.1.10 to calculate the concentration of the sample. If the sample was diluted, multiply the answer by the value of f calculated in Equation 2.1.11.

Support Protocol: Optimizing Spectrophotometer Settings

Wavelength

For single-wavelength devices, set the wavelength to 270 nm by turning the wavelength dial. For spectrophotometers such as the USB4000, complete spectra will be acquired. However, you can choose to record the specific value at 270 nm by moving the cursor to that position.

Integration period

Use a cuvette filled with the solvent only when setting up the operating conditions. Insert the cuvette into the cuvette holder, and open the light path by sliding the cuvette cover to the open position. Choose the optimum integration period. In the USB4000, this is the integration period that gives the highest signal without going off-scale.

The integration period is analogous to the shutter speed on a camera. To obtain more signal (light), make the integration period longer. To obtain less signal, make the period shorter.

Smoothing

Smoothing is setting the optical bandwidth (in nanometers) over which signals will be accumulated or averaged. In general, better signal to noise is obtained when the bandwidth is wider (because more photons are sampled). Most compounds dissolved in solvents exhibit broad absorption bands, and bandwidths as wide as 10 nm are more than sufficient. If the compound has narrow spectral features, or if there are multiple components, narrower bandwidths are recommended. A general guide, if uncertain, is to use 1 nm. In many instruments, this bandwidth is a fixed feature of the design. In some instruments, this bandwidth is changed by adjusting the slit width. In the USB4000, this is accomplished by a mathematical manipulation called a boxcar average. For every pixel in the detector array, the boxcar function reports the average of that pixel and a number of adjacent pixels. In SpectraSuite software, set the boxcar average to 3. This will smooth the data over each pixel and the 3 adjacent pixels to the one side and 3 to the other side.

Signal averaging

This involves setting the number of scans over which a signal will be accumulated. Signal averaging increases signal to noise, at the expense of longer assay time. If samples are stable, then the signal averaging can be increased as desired to achieve the desired results. In the USB4000, select the number of scans to average. Use a minimum of 10 scans and as many as desired.

UNDERSTANDING RESULTS

A standard curve is a basic tool for determining the concentration of a substance. So, as Beer's Law demonstrates, the amount of light that is absorbed by the substance is proportional to the concentration. The more molecules that are "packed" into the sample cell in a spectrophotometer setup, the more likely it is that a molecule will interact with the light if it can. Furthermore, we expect that if we use a smaller sample cell, there will be fewer molecules in the path of the light, and the absorption should therefore be smaller than if we used a larger cell. So the amount of absorbed light should also be proportional to the pathlength of the light through the sample. Finally, we might also expect that some molecules are better at absorbing light than others; the probability of a species absorbing at a given wavelength of light might not be the same as that of a different species. All of these factors are accounted for in the Beer-Lambert Law.

Beer's Law is valid for the ideal case of an infinitely dilute solution of infinitely small particles. In practice, it is accurate enough for a considerable range of chromophores, solvents, and concentrations; it is the most widely used relationship in quantitative spectroscopy.

TROUBLESHOOTING

Detector Saturation

In some absorbance measurements, more light may reach the detector of the spectrophotometer than can be detected successfully. This saturation effect can interfere with the reference measurement and appears as a graph that is off-scale. Saturation can be avoided by attenuating the signal before it reaches the detector, through the use of filters or smaller fiber optics, or by adjusting the integration period of the detector to a shorter interval, so that the detector "sees" fewer photons during the measurement.

Drift

Lamp drift, changes in temperature, and (possibly) changes in samples (depending on the chemistry) will show up as a drift away from zero absorbance in the blank solution. If fluid is spilled on the cuvette, a fingerprint is left on the cuvette, or something else has happened that you suspect has affected the readings, simply take a new dark and a new reference measurement. The reference can be retaken at any time, and should be checked throughout the assay. The spectrum you observe will be "quiet" (not jagged) over most of the UV region and will get noisier at lower wavelengths. This is a consequence of less signal to noise as the detector sensitivity and UV intensity drops off, approaching 200 nm.

The spectra of a series of samples with differing concentrations of only the chromophore will be parallel—i.e., Beer's Law will be followed at all wavelengths. Deviations from this pattern suggest additional chemistry such as other species, changing equilibrium conditions, and other factors.

If there are regions of the spectra with zero absorbance, the values here can be used to detect drift among samples. In many cases, the values at non-absorbing regions can be used to correct the readings at other wavelengths.

Random errors in the procedure are best quantified by measuring samples and standards in replicate. Systematic errors are best observed by analyzing the standard curve. Non-linearity and a line that does not go through the origin are indications of problems.

Despite all care, answers obtained using different instruments, equipment, reagents, and technicians will still vary. In situations where these differences are important, such as in clinical chemistry, an additional layer of quantification is employed. Samples (in clinical chemistry, blood serum) are pooled, blended, and distributed to a set of laboratories participating in the program. The pooled substance is called a control. All the participants report their analysis of the control and compare their results. The "normal" ranges for each laboratory are adjusted to account for their deviation from the mean value obtained by all the laboratories.

VARIATIONS

There are several ratiometric variations used to extend the utility of spectrophotometric analysis. For example, reaction kinetics can be studied by measuring the change in absorbance, caused by the disappearance of a reactant or the appearance of a product in a chemical reaction. This is quite commonly used to measure enzymes in biological systems. The answers are expressed as enzyme activity, or international units (IU, defined as the rate of change in concentration of reactant or product; $1\ IU = 1\ M^{-1}\ sec^{-1}$). It is necessary to carefully control the sample temperature (to within $\pm 0.1°C$) during the test. One benefit of this technique is that it is not necessary to make a careful reference measurement, as only the change in absorbance is required.

Using multiple wavelengths can sometimes add to the accuracy of the method. For example, the ratio of the absorbance readings at the absorbance maxima of the alkaline and acidic forms of the pH indicator in Figure 2.1.6 is a function only of pH and does not depend on the concentration of the dye.

If the samples have a non-absorbing region (e.g., at 700 nm in Fig. 2.1.6), then this value can be taken as the baseline. Any drift or cuvette-to-cuvette differences can be removed from the data by using the absorbance difference (between the absorbance maxima and the non-absorbing wavelength).

Some samples have two or more chromophores with absorbance regions that overlap. For example, proteins absorb at 280 nm, while DNA and RNA bases absorb most strongly at 260 nm. By measuring at both wavelengths, the contributions of both components can be estimated.

LITERATURE CITED

Brown, L., LeMay, H.E., and Bursten, B.E. 1997. General Chemistry: The Central Science. pp. 921-924. Prentice Hall Englewood Cliffs, N.J.

Ebbing, D. 1987. Experiments in General Chemistry. Houghton Mifflin Company, Boston.

Jennings, T.A. 1999. Lyophilization: Introduction and Basic Principles. Taylor & Francis CRC Press, Oxford.

Pace, C.N., Vajdos, F., Fee, L., Grimsley, G., and Gray, T. 1995. How to measure and predict the molar absorption coefficient of a protein. *Protein Science* 4:2411-2423.

Samuels, A.C., DeLucia, F.C., Jr., McNesby, K.L., and Miziolek, A.J. 2003. Laser-induced breakdown spectroscopy of bacterial spores, molds, pollens, and protein: Initial studies of discrimination potential. *Applied Optics* 42:6205-6209.

UNIT 2.2

Quantitation of Nucleic Acids and Proteins

Sean R. Gallagher[1]
[1]UVP, LLC, Upland, California

2

OVERVIEW AND PRINCIPLES

Nucleic Acid Quantification

Reliable quantitation of nanogram and microgram amounts of DNA and RNA in solution is essential to researchers in the life sciences. In deciding what method of nucleic acid measurement is appropriate, three issues are critical: specificity, sensitivity, and interfering substances. Properties of the four assays described in this section are listed in Table 2.2.1.

UV absorbance

The traditional method for determining the amount of DNA in solution is by measuring absorbance at 260 nm (see Basic Protocol 1). The UV absorption of DNA and RNA (with a maximum at ~260 nm) is due to the purine and pyrimidine bases of the nucleotide components of the DNA (*UNIT 5.2*; Voet et al., 1963). Relatively large amounts of DNA are required to get accurate readings—for example, 500 ng/ml DNA is equivalent to only 0.01 A_{260} units. Furthermore, the method cannot discriminate between RNA and DNA, and UV-absorbing contaminants such as protein will cause discrepancies. However, because many potential contaminants of DNA and RNA preparations absorb in the UV range at different peak values, absorption spectroscopy is a reliable method for assessing both the purity of a preparation and the quantity of DNA or RNA present using a ratio of OD_{260}/OD_{280}. Absorption spectroscopy does have serious limitations, but recent improvements in the technique include the use of basic buffers for accurate measurements of RNA (Wilfinger et al., 1997). UV absorption of RNA varies with pH due to the pH-dependent ionization of the pyrimidine and purine bases. Measurements of the OD at 260 and 280 nm are less variable and more reliable at pH 8.0 (see *UNIT 5.2*).

Hoechst 33258

An assay using Hoechst 33258 dye (Alternate Protocol 1) is specific for DNA (i.e., it does not measure RNA). This assay is in common use for rapid measurement of low quantities of DNA, with a detection limit of ~1 ng DNA. Concentrations of DNA in both crude cell lysates and purified preparations can be quantified (Labarca and Paigen, 1980). Because the assay accurately quantifies a broad range of DNA concentrations—from 10 ng/ml to 15 µg/ml—it is useful for the measurement of both small and large amounts of DNA (e.g., in verifying DNA concentrations prior

Table 2.2.1 Sensitivity of Absorbance and Fluorescence Spectrophotometric Assays for DNA and RNA

	UV	Fluorescence methods		
	Absorbance (A_{260})	H33258	Ethidium bromide	PicoGreen
DNA	1-50 µg/ml	0.01-15 µg/ml	0.1-10 µg/ml	25 pg/ml-1000 ng/ml
RNA	1-40 µg/ml	Not applicable	0.2-10 µg/ml	Minimal sensitivity
Ratio of signal (DNA/RNA)	0.8	400	2.2	>100

to performing electrophoretic separations and Southern blots). The Hoechst 33258 assay is also useful for measuring products of the polymerase chain reaction (PCR) synthesis.

Hoechst 33258 is nonintercalating and binds to the minor groove of the DNA, with a marked preference for AT sequences (Portugal and Waring, 1988). The binding to the minor grove has been reviewed by Neidle (2001) and is thought to be dependent upon a combination of structural preferences (e.g., the minor groove with a series of contiguous AT base pairs is more narrow) and electronic interactions between the Hoechst 22358 and DNA. Hoechst 33258, like other minor groove binding ligands, is positively charged, likely contributing to a preference for the negative potential of a stretch of AT base pairs. Upon binding to the nonpolar minor groove of the double helix DNA, the fluorescence characteristics of Hoechst 33258 change dramatically, showing a large increase in emission at ~458 nm. The fluorochrome 4′,6-diamidino-2-phenylindole (DAPI; Daxhelet et al., 1989) has similar characteristics to H33258 and binds to the minor groove as well. DAPI is also appropriate for DNA quantitation, although it is not as commonly used as Hoechst 33258. DAPI is excited with a peak at 344 nm. Emission is detected at ~466 nm, similar to Hoechst 33258.

Ethidium bromide

Ethidium bromide is best known for routine staining of electrophoretically separated DNA and RNA (e.g., see *UNIT 7.2*), but it can also be used to quantify both DNA and RNA in solution (Le Pecq, 1971). Unlike Hoechst 33258, ethidium bromide is a flat, planar molecule that inserts (intercalates) directly between base pairs and across the dsDNA. Ethidium needs double stranded structures to bind DNA and RNA and is much less dependent upon the % GC of the sample; its fluorescence is not significantly impaired by high GC content. The ethidium bromide assay (Alternate Protocol 2), with excitation at 546 nm, is ~20-fold less sensitive than the Hoechst 33258 assay.

PicoGreen

In addition to the advantages mentioned in the protocol itself (see Alternate Protocol 3), the PicoGreen assay protocol also minimizes the fluorescence contribution of single-stranded DNA (ssDNA) and RNA, in contrast to the Hoechst 33258-based method. The Hoechst 33258-based method does exhibit a large fluorescence enhancement from ssDNA when the assay is carried out in the recommended high-salt buffer and significant fluorescence signal from RNA when it is carried out in TE alone (10 mM Tris·Cl/1 mM EDTA, pH 7.5, with no NaCl added). Using the PicoGreen double-stranded DNA (dsDNA) quantitation reagent as described in Alternate Protocol 3, researchers can quantitate dsDNA in the presence of equimolar concentrations of ssDNA and RNA with minimal effect on the quantitation results.

Protein Quantification

The basic principles of the protein assays described in this unit are given in Table 2.2.2. The methods are sensitive to both the amino acid composition of protein and the sample buffer components that carry over into the assay when taking the measurement (Table 2.2.3 and Fig. 2.2.1). Depending upon the protein, one assay may be more appropriate than another. For example, if the protein has an unusually high content of arginine, the Coomassie blue assay is not recommended. For proteins with unknown characteristics or for complex samples, the Lowry or the BCA assay is recommended.

Gel-Based Quantification

Electrophoretically separated proteins and nucleic acid bands can be quantitated by comparing stained nucleic acids or proteins to stained standards of known amounts separated on the same gel. With many staining procedures, the intensity of the stain is dependent at least in part on the base pair composition of the nucleic acid or the amino acid composition of a specific protein. Ideally, the

Table 2.2.2 Protein Quantitation Methods

Method	Detection range	Detection wavelength	Comments	Reference
Ultraviolet spectroscopy	20-3000 μg	280 nm	Measures UV absorption of proteins, primarily tyrosine and tryptophan residues; variation in amino acid content affects absorbance	*CP Protein Science Unit 3.1* (Grimsley and Pace, 2003)
OD_{205}/OD_{280}	1-100 μg	205 and 280 nm	Measures peptide bond (205 nm) and tyrosine and tryptophan (280 nm) absorption	Scopes (1974)
OD_{260}/OD_{280}	20-3000 μg	260 and 280 nm	Correction for nucleic acid in the sample based on 260 nm value; protein concentration read at 280 nm (maximum absorption of tyrosine and tryptophan)	Warburg and Christian (1941) as modified by Layne (1957)
Lowry assay (Folin reagent)	2-100 μg	750 nm	Reduction of Folin reagent (phosphomolybdic-tungstic acid) by biuret reaction (reduced copper protein complex); tyrosine and tryptophan key contributors to color development; variation due to amino acid composition of proteins	Lowry et al. (1951); Peterson (1979); Sapan et al. (1999); *CP Protein Science Unit 3.4* (Olson and Markwell, 2007)
Bicinchoninic acid (BCA)	0.2-50 μg	562 nm	Folin reagent of Lowry assay replaced by bicinchoninic acid to detect reduced copper; less interference by nonionic detergents; variation similar to Lowry	Smith et al. (1985); Sapan et al. (1999); *CP Protein Science Unit 3.4* (Olson and Markwell, 2007)
Bradford Assay (Coomassie blue G250)	0.2-20 μg	595 nm	Positively ionized dye binds to proteins (a variety of mechanisms); variation possibly due in part to basic amino acid (lysine and arginine) composition of proteins	Bradford (1976); Sapan et al. (1999); *CP Protein Science Unit 3.4* (Olson and Markwell, 2007)

[a]Range data from Stoscheck (1990).

standard should have the same percent GC, or it should be the identical protein of a known amount quantitated by an alternate technique such as spectrophotometry or the Lowry protein assay.

Gel-based protein and nucleic acid quantitation is straightforward. The stain intensity of the sample band is used to estimate the sample band amount. By carefully using visual inspection of the gel, estimates of the amount of protein or nucleic acid in the sample band can also be made. However, the error on these measurements is typically quite high. A more quantitative approach involves creating a standard curve of stain intensity versus amount of protein by measuring the stain intensities of the different amounts of a protein standard on the gel, typically via CCD digital imaging and analysis (see UNIT 7.5).

Table 2.2.3 General Guide to Concentration Limits of Interfering Chemicals in Protein Assays[a]

Substance	Concentration limits in four types of assays				
	Enhanced copper[b]	BCA[c]	Dye[d]	UV[e]	
				280 nm	205 nm
Acids and bases					
HCl		0.1 M	0.1 M	>1 M	0.5 M
NaOH		0.1 M	0.1 M	>1 M	25 mM
PCA	<1.25%	<1%		10%	1 M
TCA	<1.25%	<1%		10%	<1%
Buffers					
Acetate		0.2 M	0.6 M	0.1 M	10 mM
Ammonium sulfate	>28 mM	20%	1 M	>50%	9%
Borate		10 mM			>100 mM
Citrate	2.5 mM	<1 mM	50 mM	5%	<10 mM
Glycine	2.5 mM	1 M	0.1 M	1 M	5 mM
HEPES	2.5 µM	100 µM	100 mM		<20 mM
Phosphate	250 mM	250 mM	2 M	1 M	50 mM
Tris	250 mM	0.1 M	2 M	0.5 M	40 mM
Detergents					
Brij 35		1%		1%	1%
CHAPS		1%		10%	<0.1%
Deoxycholate	625 µg/ml		0.25%	0.30%	0.1%
Digitonin				10%	
Lubrol PX		1%		10%	
Octylglucoside		1%		10%	
SDS	1.25%	1%	0.10%	0.10%	0.10%
Triton X-100	0.25%	1%	0.10%	0.02%	<0.01%
Triton X-100(R)				>10%	2%
Tween 20	0.10%	1%		0.30%	0.1%
Reductants					
Dithiothreitol	50 µM	<1 mM	1 M	3 mM	0.1 mM
2-mercaptoethanol	1.8 µM	<1%	1 M	10 mM	<10 mM
Miscellaneous					
DNA/RNA	0.2 mg	0.1 mg	0.25 mg	1 µg	
DMSO	>6.2%	5%		20%	<10%
EDTA	125 µM	10 mM	0.1 M	30 mM	0.2 mM
Glycerol	25%	10%	100%	40%	5%
KCl	30 mM	<10 mM	1 M	100 mM	50 mM

continued

Table 2.2.3 General Guide to Concentration Limits of Interfering Chemicals in Protein Assays[a], *continued*

Substance	Concentration limits in four types of assays				
	Enhanced copper[b]	BCA[c]	Dye[d]		UV[e]
NaCl	1.75 M	1 M	5 M	>1 M	0.6 M
Sucrose	50 mM	40%	1 M	2 M	0.5 M
Urea	>200 mM	3 M	6 M	>1 M	<0.1 M

[a]This table is a general guide. Figures preceded by (<) or (>) symbols indicate that the tolerable limit for the chemical is unknown but is, respectively, less than or greater than the amount shown. Blank spaces indicate that data were unavailable. Reproduced with permission from Stoscheck (1990). Originally published in *CP Protein Science Unit 3.4* (Olson and Markwell, 2007).
[b]Concentration of the chemical allowed in the final assay volume (Lowry assay; see Basic Protocol 2 and Alternate Protocol 4).
[c]Concentration of the chemical in a 25-µl sample prior to addition to the assay (see Basic Protocol 3).
[d]Concentration of the chemical in a 50-µl sample prior to addition to the assay (Coomassie blue assay; see Basic Protocol 4).
[e]Concentration of the chemical that does not produce an absorbance of 0.5 over water (see Basic Protocol 5).
Abbreviations: CHAPS, 3-[(3-cholamidopropyl)dimethylammonio]-1-propanesulfonate; DMSO, dimethyl sulfoxide; EDTA, ethylenediamine tetraacetic acid; HEPES, 4-(2-hydroxyethyl)-1-piperazineethanesulfonic acid; PCA, perchloric acid; SDS, sodium dodecyl sulfate; R, reduced; TCA, trichloroacetic acid.

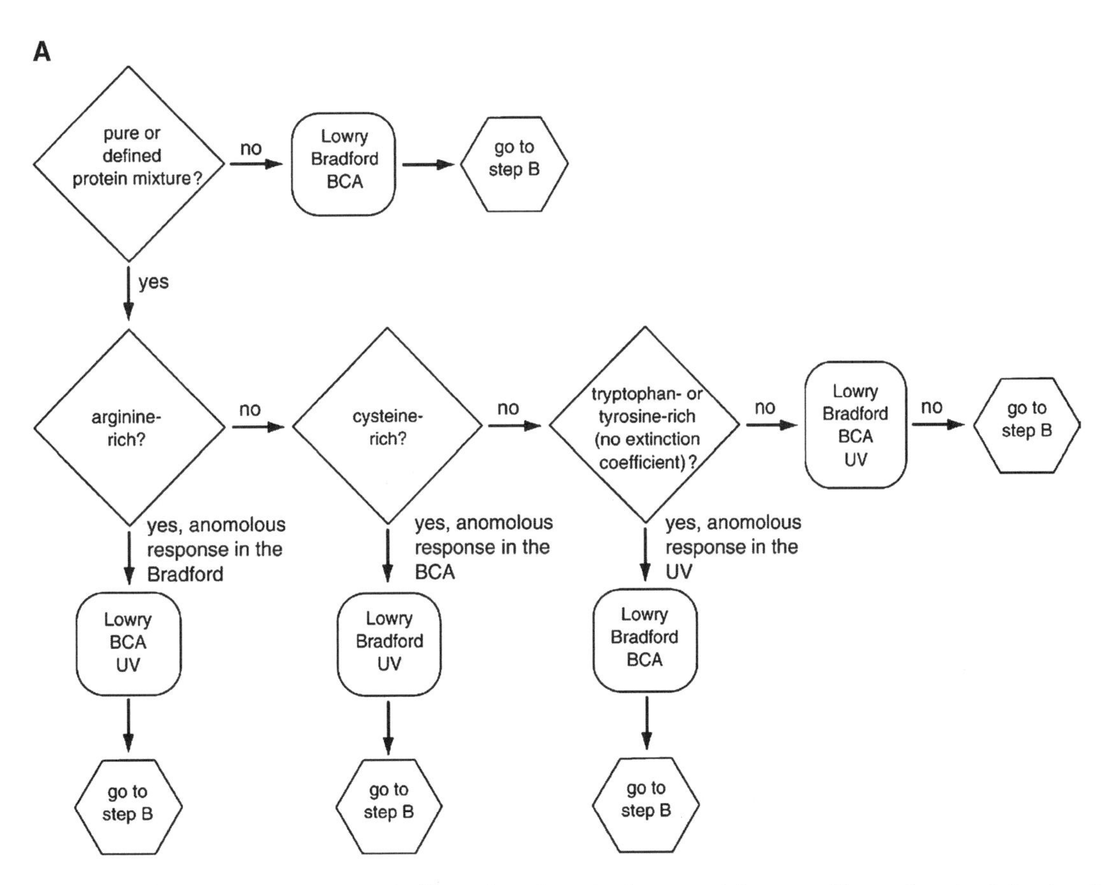

Figure 2.2.1 *(continues on next page)* Flow chart for selecting a protein assay. To use the chart, begin with the steps in panel **A** to find the protein assays that are most compatible with the sample. Then proceed to the steps in panel **B** to obtain the list of assays compatible with the buffer system. Compare the results from each step to find the most compatible assay for the sample. Refer to the text for the assay and Table 2.2.3 to confirm assay compatibility before using a protein assay strategy. Abbreviations: 2-ME, 2-mercaptoethanol; DTT, dithiothreitol; SDS, sodium dodecyl sulfate. Originally published in *CP Protein Science Unit 3.4*; Olson and Markwell (2007).

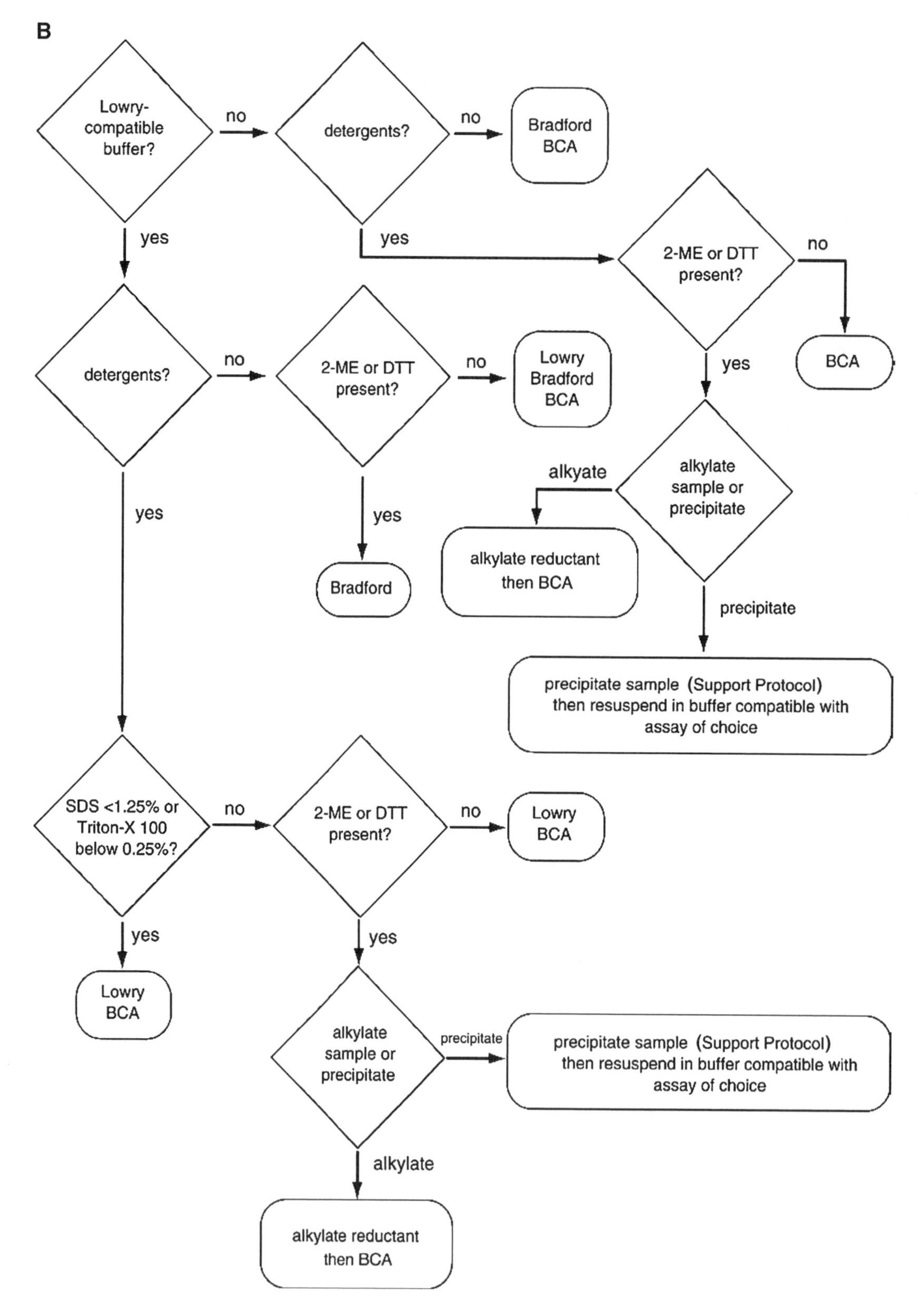

Figure 2.2.1 *(continued)*

Two types of digital imaging–based quantitation are typically performed: For the first approach, the peak intensity of each band is calculated and used to create a standard curve. A generally more accurate approach involves adding the intensity of all of the pixels of the standard and sample bands. This is referred to as the volume or total intensity of the band. Both approaches require background subtraction for the best result. Volume-based approaches are potentially more accurate when sample overloading causes band spreading to occur. The choice of method depends on which approach gives a more reliable standard curve with the standard protein or nucleic acid used for calibration.

STRATEGIC PLANNING

Assay Choice

Numerous assays are available for both protein and nucleic acid quantitation. Accuracy, ease of use, sensitivity, and lack of interference from common assay buffers or other reagents are all important considerations when deciding which assay to use. The purity of the sample is also a consideration. A crude sample runs a higher risk of containing interfering compounds compared to a purified protein or nucleic acid. If the protein or nucleic acid sample is reasonably pure, the simplest approach to quantitation is the use of absorption spectroscopy. While not as sensitive as fluorescent techniques and prone to interference by contaminants, UV spectroscopy can be performed in very small volumes, using micro-volume cuvettes, and is nondestructive (*CP Molecular Biology Appendix 3D*; Gallagher and Desjardin, 2006). This is by far the most popular approach for nucleic acids; for greater sensitivity, colorimetric or fluorescent assays are required. Generally, colorimetric assays are used for protein quantitation, although even greater sensitivity is possible with fluorescent protein assays. For crude samples, the H33258 assay for dsDNA and the Peterson modification of the Lowry assay for proteins are useful.

Linear Range

To be accurate, the sample readings need to be within the linear range of the assay. Using known amounts of a standard is a simple way to confirm the accuracy and linearity of the assay (e.g., *UNIT 2.1*). In addition, the high and low value of a standard should bracket the value of the sample so that the sample falls within the range set by the standards. If a sample is outside either the linear range or the range set by the standards, simply using more or less material is a straightforward way to keep the sample amount within the linear range.

The linear range of an assay is dependent on numerous factors, including the assay chemistry and the instrument range and sensitivity. Tables 2.2.1 and 2.2.2 list the typical detection ranges of the assays described in this unit. Both standards and samples should be run in separate assay tubes at least in duplicate but ideally triplicate so the average and standard deviation (variability) can be determined (see *APPENDIX 4*). In order to produce a realistic standard curve, at least three standards should be used: low, high, and middle range.

Cuvettes

Care should be taken when handling sample cuvettes in all spectrophotometric procedures. Fluorometers use cuvettes with four optically clear faces, because exciting and emitting light enter and leave the cuvette through directly adjacent sides. Thus, fluorometric cuvettes should be held by the upper edges only. In contrast, transmission spectrophotometers use cuvettes with two opposite optical windows, with the sides frosted for easy handling. It is important to check that the optical faces of cuvettes are free of fingerprints and scratches. In addition, for accurate absorbance readings, spectrophotometer cuvettes must be perfectly matched.

Cuvettes should be filled so that the meniscus bottom is 3 to 4 mm above the beam of light passing through the cuvette. Placing a white card in the light path will give a good indication of the correct volume and positioning of cuvettes. Disposable plastic cuvettes are also available that transmit in the ultraviolet for UV analysis of protein and nucleic acid (e.g., CVD-UV1S, Ocean Optics), but caution is advised because plastic cuvettes tend to have higher variability; they should not be reused because they are easily damaged.

Quantifying Nucleic Acids

Considerations when performing the Hoechst 33258 assay

Sensitivity of the Hoechst 33258 fluorescence assay decreases with nuclease degradation, increasing GC content, or denaturation of DNA (Labarca and Paigen, 1980; Stout and Becker, 1982). Increased temperature of the assay solution and ethidium bromide contamination also decrease the Hoechst 33258 signal. Sodium dodecyl sulfate (>0.01% final concentration) also interferes with accurate readings (Cesarone et al., 1979). The pH of the assay solution is critical to sensitivity and should be pH ~7.4 (Labarca and Paigen, 1980; Stout and Becker, 1982). At a pH <6.0 or >8.0, the background becomes much higher and there is a concomitant loss of fluorescence enhancement.

High-quality dsDNA is recommended, although single-stranded genomic DNA also works well with the Hoechst assay. However, with very small fragments of DNA, the Hoechst 33258 dye binds to dsDNA only. Thus, the assay will not work with single-stranded oligomers. Linear and circular DNA give approximately the same levels of fluorescence (Daxhelet et al., 1989). When preparing DNA standards, an attempt should be made to equalize the GC content of the standard DNA and that of the sample DNA. In most situations, salmon sperm or calf thymus DNA is suitable. An extensive list of estimated GC content for various organisms is available (Marmur and Doty, 1962). Eukaryotic cells vary somewhat in GC content but are generally in the range of 39% to 46%. Within this range, the fluorescence per microgram of DNA does not vary substantially. In contrast, the GC content of prokaryotes can vary from 26% to 77%, causing considerable variation in the fluorescence signal. In these situations, the sample DNA should first be quantitated via transmission spectroscopy and compared to a readily available standard (e.g., calf thymus DNA). Future measurements would then use calf thymus as a standard, but with a correction factor for difference in fluorescence yield between the two DNA types. For further troubleshooting, see Van Lancker and Gheyssens (1986), where the effects of interfering substances on the Hoechst 33258 assay and several other assays are compared.

Considerations when performing the ethidium bromide assay

In the ethidium bromide assay, ssDNA gives approximately half the signal of double-stranded calf thymus DNA. Ribosomal RNA also gives about half the fluorescent signal of dsDNA, and RNase and DNase both severely decrease the signal. Closed circular DNA also binds less ethidium bromide than nicked or linear DNA. Further critical parameters of the ethidium bromide assay are described by Le Pecq (1971).

Considerations when performing the PicoGreen assay

With PicoGreen, dsDNA can be quantitated in the presence of equimolar concentrations of single-stranded nucleic acids with minimal interference. Table 2.2.4 shows the concentrations of RNA or ssDNA that, for a given dsDNA concentration, result in less than a 10% change in the signal intensity using the PicoGreen assay protocol. Fluorescence due to PicoGreen reagent binding to RNA at high concentrations can be eliminated by treating the sample with DNase-free RNase. The use of RNase A/RNase T1 with S1 nuclease will eliminate all single-stranded nucleic acids and ensure that the entire sample fluorescence is due to dsDNA.

Time considerations

The three assays described can be performed in a short period of time. In a well planned series of assays, 50 samples can be prepared and read comfortably in 1 hr. To save time (although some error might be introduced) DNA samples can be sequentially added to the same cuvette containing

Table 2.2.4 Sensitivity of the PicoGreen dsDNA Assay for Quantitating dsDNA in the Presence of Single-Stranded Nucleic Acids

dsDNA concentration	RNA concentration[a]	RNA/dsDNA ratio	ssDNA concentration[b]	ssDNA/dsDNA ratio
1 μg/ml	10 μg/ml	10	300 ng/ml	0.3
500 ng/ml	500 ng/ml	1	50 ng/ml	0.1
10 ng/ml	100 ng/ml	10	30 ng/ml	3
5 ng/ml	50 ng/ml	10	15 ng/l	3
100 pg/ml	1 ng/ml	10	1 ng/ml	10
50 pg/ml	500 pg/ml	10	500 pg/ml	10

[a]The upper concentration limit of RNA that results in no more than a 10% increase in the sample's signal intensity for the indicated concentration of dsDNA.

[b]The upper concentration limit of ssDNA that results in no more than a 10% increase in the sample's signal intensity for the indicated concentration of dsDNA.

Abbreviations: dsDNA, double-stranded DNA; ssDNA, single-stranded DNA.

2

working dye solution. The increase in fluorescence with each sample is noted and subtracted from the previous reading to give relative fluorescence or concentration of the new sample, eliminating the need to change solutions for each sample. Be certain that the final amount of DNA does not exceed the linear portion of the assay (see Table 2.2.1 and *UNIT 2.1*).

Protein Measurement

For both colorimetric and ultraviolet spectrophotometry–based quantitation methods, there is considerable variation in color intensity or absorption value per microgram protein, depending on the protein measured (e.g., see Sapan et al., 1999; Peterson, 1983; *CP Protein Science Unit 4.3*, Olson and Markwell, 2007). For the colorimetric assays, this variation is due to the reactivity of the reagents with specific amino acid groups and varies with amino acid composition. For the UV absorption assays, the extinction coefficient of a protein varies with the amino acid composition and the wavelengths used. For example, while tryptophan and tyrosine are the main contributors to absorption at 280 nm, numerous other amino acids absorb at wavelengths of <240 nm, and the peptide bond absorbs below 210 nm. Using 25 μg BSA as a standard set to 1.00 and comparing 14 other proteins relative to BSA, Peterson (1983) found the average for the modified Folin phenol (Peterson, 1977) to be 1.18 ± 0.27; Coomassie blue, (Bradford, 1976) 0.94 ± 0.94; 280 nm, 1.81 ± 1.19; 224 to 236 nm (Groves et al., 1968), 1.08 ± 0.43; and 205 (Scopes, 1974), 1.06 ± 0.09. For a protein concentration of 1 mg per milliliter, the range of values varies significantly within each assay: UV absorption at 280 nm showed the most dramatic range (.02 to 4.19), while UV absorption at 205 nm ranged from 0.91 to 1.13. The BCA assay gives a response similar to Lowry for a range of proteins (Smith et al., 1985).

The standard protein used for calibration is a critical decision. Ideally, a purified protein identical to the same protein being analyzed should serve as the standard. Furthermore, the standard protein composition and amount should be confirmed by the use of quantitative amino acid analysis on the sample (*CP Protein Science Unit 3.2;* Dunn, 1995). For an unknown protein or a complex mixture, BSA is typically used as a standard protein.

Each assay is also sensitive to potential interference from various cellular and buffer components that carry over during purification (Layne, 1957; Peterson, 1977, 1979, 1983; Stoscheck, 1990; *CP Protein Science Unit 3.4*; Olson and Markwell, 2007). The type and level of interference by detergents, salts, buffer components, etc. varies depending upon the assay (Table 2.2.3). For example, the Coomassie blue dye and UV spectroscopy assays are sensitive to detergent interference by disodium dodecyl sulfate (SDS) while the BCA and the Folin phenol assays can handle 10-fold higher concentrations of SDS. Detergents likely compete for the dye and contribute to an artificially high signal unrelated to the protein present. As another example, in contrast to the Lowry and the BCA assay, the Coomassie blue dye assay is relatively insensitive to the reductants dithiothreitol and 2-mercaptoethanol.

In practice, the simplest approach to dealing with interference is to keep the final concentration of the reagent in the assay below the critical amount that leads to incorrect results. This can be readily tested by assaying a range of protein concentrations in the presence of the potential interfering reagent. A reagent blank, which contains no protein but all the assay components, is also included to monitor any effect on background. It is a good idea to test a range of concentrations of the interfering reagent against a range of protein concentrations to give a complete picture of possible effects on the assay. If needed, the protein can be precipitated to remove the interfering reagent and then suspended into the protein assay reagent (Support Protocol).

PROTOCOLS: NUCLEIC ACID QUANTIFICATION

In addition to methods for traditional absorbance measurements at 260 nm (see Basic Protocol 1), procedures for three more sensitive fluorescence techniques are presented below (see Alternate Protocols 1 to 3). These four protocols allow nucleic acid measurement over a range of 25 pg/ml to 5 to 10 ng/ml (see Table 2.2.1).

Absorbance measurements (A_{260}) are straightforward as long as any contribution from contaminants and the buffer components are taken into account. Fluorescence assays are less prone to interference than absorbance measurements and are also simple to perform. For both methods, a reading from a reagent blank (containing no nucleic acids) is taken prior to adding the nucleic acid. In instruments where the readout can be set to indicate concentration, a known concentration is used for calibration and subsequent readings are expressed as µg/ml, ng/ml, or pg/ml of nucleic acids.

Basic Protocol 1: Detection of Nucleic Acids Using Absorption Spectroscopy

Absorption of the sample is measured at several different wavelengths to assess purity and concentration of nucleic acids. While absorbance readings cannot discriminate between DNA and RNA, A_{260} measurements are quantitative for relatively pure nucleic acid preparations in microgram quantities. The A_{260}/A_{280} ratio can be used as an indicator of nucleic acid purity. Ratios of 1.8 to 1.9 (for DNA) and 1.9 to 2.0 (for RNA) indicate highly purified preparations. An A_{260}/A_{280} value can indicate contamination by proteins, which have a peak absorption at 280 nm. Absorbance at 325 nm indicates particulates in the solution or dirty cuvettes, and contaminants containing peptide bonds or aromatic moieties (e.g., protein and phenol) absorb at 230 nm.

This protocol is designed for a single-beam ultraviolet to visible range (UV/visble) spectrophotometer. If available, a double-beam spectrophotometer will simplify the measurements because it will automatically compare the cuvette holding the sample solution to a reference cuvette that contains the blank. In addition, more sophisticated double-beam instruments will scan various wavelengths and report the results automatically (see *UNIT 2.1*).

Materials

1× TNE buffer (see recipe)
DNA sample to be quantitated in 1× TNE buffer (see recipe)

Matched quartz semi-micro spectrophotometer cuvettes (1-cm path length)
Single- or dual-beam spectrophotometer with ultraviolet to visible light source

NOTE: Sample cuvettes should be rinsed and drained of any liquid to eliminate carryover prior to adding the subsequent sample.

For single-beam spectrophotometers

1a. Pipet 1.0 ml of 1× TNE buffer into a quartz cuvette.

2a. Place the cuvette in a single-beam spectrophotometer, take a reading at 325 nm and zero the instrument.

Always hold the cuvettes by the corners only and wipe clean with optical tissue between readings, if necessary (see UNIT 2.1*).*

3a. Remove blank cuvette and insert the cuvette containing the nucleic acid sample or standard suspended in the same solution as the blank and take a reading.

It is important that the sample be suspended in the same solution as used for the blank.

Variations between cuvettes (typically minimal) can be checked by measuring both cuvettes with blank solution. Any additional background can be subtracted.

4a. Repeat this process at 280, 260, and 230 nm for each sample before moving on to the next.

For dual-beam spectrophotometers

1b. Pipet 1.0 ml of 1× TNE buffer into two quartz cuvettes.

2b. Place the cuvettes in a dual-beam spectrophotometer, read at 325 nm and zero the instrument. Use this blank solution as the reference.

3b. Remove the cuvette from the sample position and replace the blank solution with the nucleic acid sample suspended in the same solution as the blank. Insert the cuvette into the spectrophotometer and take a reading.

4b. Repeat readings at 280, 260, and 230 nm for each sample before moving on to the next.

It is important that the nucleic acid be suspended in the same solution as used for the blank.

More sophisticated double-beam instruments will initially scan various wavelengths for each sample, and this step may not be required.

5. Determine the concentration of nucleic acid present, using the A_{260} reading in conjunction with one of the following equations:

$$\textit{Single-stranded DNA: } \text{pmol/µl} = \frac{A_{260}}{10 \times S}$$

$$\textit{Single-stranded DNA: } \text{µg/ml} = \frac{A_{260}}{0.027}$$

$$\textit{Double-stranded DNA: } \text{pmol/µl} = \frac{A_{260}}{13.2 \times S}$$

$$\textit{Double-stranded DNA: } \text{µg/ml} = \frac{A_{260}}{0.020}$$

$$\textit{Single-stranded RNA: } \text{µg/ml} = \frac{A_{260}}{0.025}$$

$$\textit{Oligonucleotide: } \text{pmol/µl} = A_{260} \times \frac{100}{1.5\,N_A + 0.71\,N_C + 1.20\,N_G + 0.84\,N_T}$$

where S represents the size of the DNA in kilobases and N is the number or residues of base A, G, C, or T.

For dsDNA or ssDNA and single-stranded RNA (ssRNA): These equations assume a 1-cm path length spectrophotometer cuvette and neutral pH.

The calculations are based on the Lambert-Beer law, $A = ECl$, where A is the absorbance at a particular wavelength, C is the concentration of DNA, l is the path length of the spectrophotometer cuvette (typically 1 cm), and E is the extinction coefficient.

For solution concentrations given in mol/liter and a cuvette of 1-cm path length, E is the molar extinction coefficient and has units of $M^{-1} cm^{-1}$. If concentration units of μg/ml are used, then E is the specific absorption coefficient and has units of $(\mu g/ml)^{-1} cm^{-1}$. The values of E used here are as follows: ssDNA, 0.027 $(\mu g/ml)^{-1} cm^{-1}$; dsDNA, 0.020 $(\mu g/ml)^{-1} cm^{-1}$; ssRNA, 0.025 $(\mu g/ml)^{-1} cm^{-1}$. Using these calculations, an A_{260} of 1.0 indicates 50 μg/ml dsDNA, ~37 μg/ml ssDNA, or ~40 μg/ml ssRNA (adapted from Applied Biosystems, 1987; also see UNIT 2.1*).*

For oligonucleotides: Concentrations are calculated in the more convenient units of pmol/μl. The base composition of the oligonucleotide has significant effects on absorbance, because the total absorbance is the sum of the individual contributions of each base (Table 2.2.5). These equations require S and N, which are known for synthesized oligonucleotides.

6. Estimate the purity of the nucleic acid sample using the A_{260}/A_{280} ratio and readings at A_{230} and A_{325}.

Ratios of 1.8 to 2.0 and 1.9 to 2.0 indicate highly purified preparations of DNA and RNA, respectively. Contaminants that absorb at 280 nm (e.g., protein) will lower this ratio.

Proteins in general have A_{280} readings considerably lower than nucleic acids on an equivalent weight basis. Thus, even a small increase in the A_{280} relative to A_{260} (i.e., a lowering of the A_{260}/A_{280} ratio) can indicate severe protein contamination.

Table 2.2.5 Molar Extinction Coefficients (ε) of DNA Bases at 260 nm

Base	$\varepsilon_{260\,nm}$[a] $(M^{-1}cm^{-1})$
Adenine	15,200
Cytosine	7,050
Guanosine	12,010
Thymine	8,400

[a]Wallace and Miyada (1987). Detailed spectrophotometric properties of nucleoside triphosphates are listed in *CP Molecular Biology Unit 3.4* (Struhl, 1993).

Table 2.2.6 Spectrophotometric Measurements of Purified DNA[a,b]

Wavelength (nm)	Absorbance[c]
325	0.01
280	0.28
260	0.56
230	0.30

[a]Typical absorbance readings of highly purified calf thymus DNA suspended in 1× TNE buffer.
[b]The concentration of DNA was nominally 25 μg/ml, determined by the supplier. The concentration of DNA calculated from the A_{260} value using the appropriate equation in Basic Protocol 1 was 28 μg/ml.
[c]Note that A_{260}/A_{280} is 2.0, typical of highly purified DNA.

Other commonly used buffer components absorb strongly at 260 nm and can cause interference if present in high enough concentrations. EDTA, for example, should not be present at >10 mM.

Absorbance at 230 nm reflects contamination of the sample by phenol or urea.

Absorbance at 325 nm suggests contamination by particulates and dirty cuvettes. Light scatter at 325 nm can be magnified 5-fold at 260 nm (K. Hardy, pers. comm.).

Typical values at the four wavelengths for a highly purified preparation are shown in Table 2.2.6.

Alternate Protocol 1: DNA Detection Using the DNA-Binding Fluorochrome Hoechst 33258

Use of fluorometry to measure DNA concentration has gained popularity because it is simple and much more sensitive than spectrophotometric measurements. The Hoechst 33258 fluorochrome is specific for nanogram amounts of DNA, has little affinity for RNA, and works equally well with either whole-cell homogenates or purified preparations of DNA. However, the fluorochrome is sensitive to changes in DNA composition, with preferential binding to AT-rich regions. A fluorometer capable of generating an excitation wavelength of 365 nm and detecting an emission wavelength of 460 nm is required for this assay.

Additional Materials (also see Basic Protocol 1)

Hoechst 33258 assay solution (working solution; see recipe)
DNA standards (e.g., lambda, calf thymus; Bio-Rad, Invitrogen)

Dedicated filter fluorometer (e.g., DQ 300, Hoefer; TBS-380, Turner BioSystems) *or* scanning fluorescence spectrophotometer (RF-5301 PC, Shimadzu; F-2500, Hitachi)
Fluorometric square glass cuvettes *or* disposable acrylic cuvettes (Sarstedt)
Teflon stir rod

1. Prepare the scanning fluorescence spectrophotometer by setting the excitation wavelength to 365 nm and the emission wavelength to 460 nm.

   The dedicated filter fluorometer has fixed wavelengths at 365 and 460 nm and does not need adjustment.

2. Pipet 2.0 ml Hoechst 33258 assay solution into the cuvette and place in the sample chamber.

3. Take a reading without DNA and note the readings in relative fluorescence units. Use this measurement as a background reading.

   Follow the manufacturer's instructions. Some instruments can store readings and create a standard curve automatically.

4. With the cuvette still in the sample chamber, add 2 µl DNA standard to the blank Hoechst 33258 assay solution.

5. Mix with a Teflon stir rod or by capping and inverting the cuvette.

6. Read the emission in relative fluorescence units or set the concentration readout equal to the final DNA concentration.

   Small-bore tips designed for loading sequencing gels minimize errors of pipetting small volumes. Pre-rinse tips with sample and make sure no liquid remains outside the tip after drawing up the sample.

7. Repeat measurements with remaining DNA standards using fresh assay solution (take background zero reading and zero instrument, if necessary).

   If necessary, the DNA standards should be quantitated by A_{260} measurement (Basic Protocol 1) before being used here.

   Read samples in duplicate or triplicate, with a blank reading taken each time. Unusual or unstable blank readings indicate a dirty cuvette or particulate material in the solution, respectively.

8. Repeat steps 3 to 5 with unknown samples.

   A dye concentration of 0.1 μg/ml is adequate for final DNA concentrations up to ~500 ng/ml. Increasing the working dye concentration to 1 μg/ml Hoechst 33258 will extend the assay's range to 15 μg/ml DNA, but will limit sensitivity at low concentrations (5 to 10 ng/ml). Sample volumes of ≤10 μl can be added to the 2.0-ml aliquot of Hoechst 33258 assay solution.

9. Plot the DNA standard data using a simple linear regression (see *UNIT 2.1* and *APPENDIX 4*) and estimate the DNA concentration of the unknown.

   Some instruments will read out concentration directly, once calibrated.

2

Alternate Protocol 2: DNA and RNA Detection with Ethidium Bromide Fluorescence

In contrast to the fluorochrome Hoechst 33258, ethidium bromide is relatively unaffected by differences in the base composition of DNA. Ethidium bromide is not as sensitive as Hoechst 33258 and, although capable of detecting nanogram levels of DNA, will also bind to RNA. In preparations of DNA with minimal RNA contamination or with DNA samples having an unusually high guanine and cytosine (GC) content (where the Hoechst 33258 signal can be quite low), ethidium bromide offers a relatively sensitive alternative to the more popular Hoechst 33258 DNA assay. A fluorometer capable of generating an excitation wavelength of 302 or 546 nm and detecting an emission wavelength of 590 nm is required for this assay.

Additional Materials (also see Basic Protocol 1)

Ethidium bromide assay solution (see recipe)

CAUTION: Please follow established safety procedures for handling DNA binding dyes such ethidium bromide (*APPENDIX 1*).

1. Set the fluorometer excitation wavelength to 302 nm or 546 nm and emission wavelength to 590 nm.

   The excitation wavelength of this assay can be either in the UV range (~302 nm) using a quartz cuvette or in the visible range (546 nm) using a glass cuvette. In both cases, the emission wavelength is 590 nm.

2. Pipet 2.0 ml ethidium bromide assay solution into the cuvette and place in sample chamber.

3. Take a reading without DNA and note the readings in relative fluorescence units. Use this measurement as a background reading.

   Follow the manufacturer's instruction. Some instruments can store readings and create a standard curve automatically.

4. Measure standards and samples and determine nucleic acid content as described in Alternate Protocol 1, steps 4 to 9.

   A dye concentration of 5 μg/ml in the ethidium bromide assay solution is appropriate for final DNA concentrations up to 1000 ng/ml. 10 μg/ml ethidium bromide in the ethidium bromide assay solution will extend the assay's range to 10 μg/ml DNA, but is only used for DNA concentrations >1 μg/ml. Sample volumes of up to 10 μl can be added to the 2.0-ml aliquot of ethidium bromide assay solution.

Alternate Protocol 3: DNA Detection Using PicoGreen dsDNA Quantitation Reagent

PicoGreen dsDNA quantitation reagent enables researchers to quantitate as little as 25 pg/ml of dsDNA (50 pg dsDNA in a 2-ml assay volume) using a standard spectrofluorometer with fluorescein excitation and emission wavelength capability. This sensitivity exceeds that achieved with the Hoechst 33258-based assay (Alternate Protocol 1) by at least 400-fold. Using a fluorescence microplate reader, it is possible to detect as little as 250 pg/ml dsDNA (50 pg in a 200-μl assay volume).

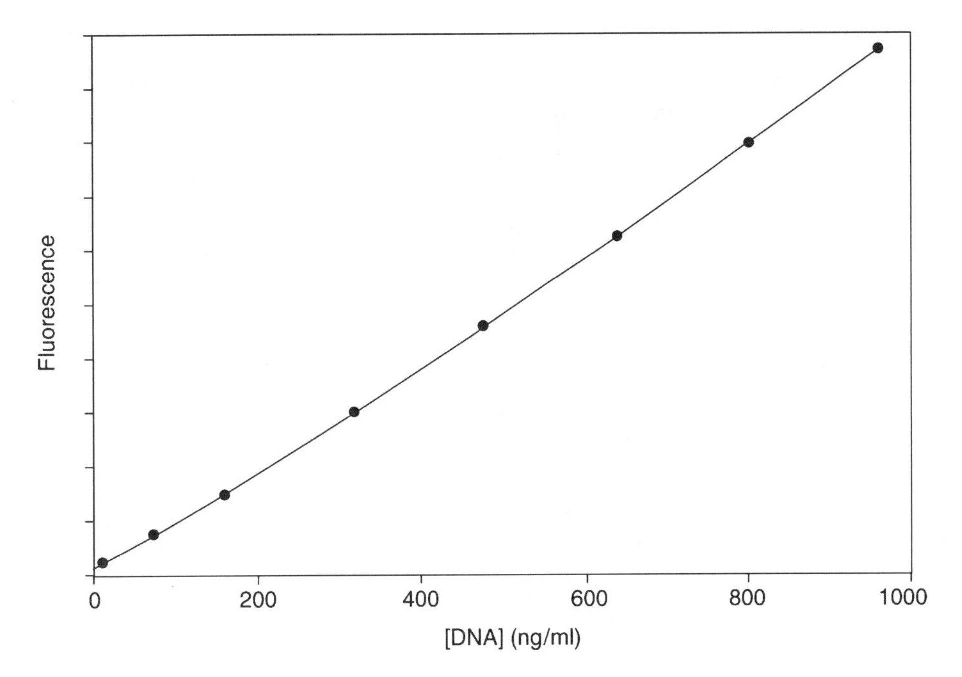

Figure 2.2.2 Dynamic range and sensitivity of the PicoGreen dsDNA quantitation assay.

The standard PicoGreen assay protocol is also simpler than that for Hoechst 33258 because a single concentration of the PicoGreen reagent allows detection over the full dynamic range of the assay. In order to achieve more than two orders of magnitude in dynamic range with Hoechst-based assays, two different dye concentrations are recommended. In contrast, the linear detection range of the PicoGreen assay in a standard fluorometer extends over more than four orders of magnitude in DNA concentration—from 25 pg/ml to 1000 ng/ml—with a single dye concentration (Fig. 2.2.2). Linearity is maintained in the presence of several compounds that commonly contaminate nucleic acid preparations, including salts, urea, ethanol, chloroform, detergents, proteins, and agarose.

The PicoGreen reagent is available as a reagent or as part of a kit. In both cases sufficient reagent is supplied for 200 assays using an assay volume of 2 ml according to the protocol below. Note that the assay volume is dependent on the instrument used to measure fluorescence; with a microplate reader and a 96-well microplate, the assay volume is reduced to 200 μl and 2000 assays are possible. The PicoGreen reagent supplied in the kits is exactly the same as the reagent sold separately.

CAUTION: No data are available addressing the mutagenicity or toxicity of PicoGreen dsDNA quantitation reagent. Because this reagent binds to nucleic acids, it should be treated as a potential mutagen and handled with appropriate care. The DMSO stock solution should be handled with particular caution because DMSO is known to facilitate the entry of organic molecules into tissues. It is strongly recommended that double gloves be used when handling the DMSO stock solution. As with all nucleic acid dye reagents, solutions of PicoGreen reagent should be poured through activated charcoal before disposal. The charcoal must then be incinerated to destroy the dye.

Additional Materials (also see Basic Protocol 1)

PicoGreen dsDNA quantitation kit (Invitrogen) containing:
- PicoGreen dsDNA quantitation reagent (Component A), 1 ml solution in DMSO (store up to 6 months at −20°C, protected from light)
- 20× TE (component B), 25 ml of 200 mM Tris·Cl/20 mM EDTA, pH 7.5: store up to 6 months at 4°C (preferred) or up to 6 months at −20°C
- Lambda DNA standard (component C), 1 ml of 100 μg/ml in TE: store up to 6 months at 4°C (preferred) or up to 6 months at −20°C

DNase-free water (e.g., Millipore), sterile

Spectrofluorometer or fluorescence microplate reader

2

Prepare the reagent

1. On the day of the experiment, allow the concentrated DMSO solution of PicoGreen reagent to warm to room temperature before opening the vial.

2. Prepare a 1× TE working solution by diluting the concentrated buffer 20-fold with sterile, distilled, DNase-free water.

    Because the PicoGreen dye is an extremely sensitive detection reagent for dsDNA, it is imperative that the TE solution be free of contaminating nucleic acids. The 20× TE buffer included in the PicoGreen dsDNA Quantitation Kits is certified to be nucleic acid-free and DNase-free.

3. To prepare enough working solution to assay 20 samples in a 2-ml final volume, add 100 μl PicoGreen dsDNA quantitation reagent to 19.9 ml TE (a 200-fold dilution).

    The author recommends preparing this solution in a plastic container rather than glass because the reagent may adsorb to glass surfaces.

    Protect the working solution from light by covering it with foil or placing it in the dark because the PicoGreen reagent is susceptible to photodegradation.

    For best results, this solution should be used within a few hours of its preparation.

Establish the DNA standard curve

4. Prepare a 2 μg/ml stock solution of dsDNA in 1× TE by diluting the lambda DNA standard in the PicoGreen kit 50-fold in 1× TE to make the 2 μg/ml working solution, (e.g., 30 μl of the DNA standard mixed with 1.47 ml of TE).

    For a standard curve, the author commonly uses bacteriophage lambda DNA (provided with kit) or calf thymus DNA (e.g., Sigma, Worthington Biochemicals), although any purified dsDNA preparation may be used.

    It is sometimes preferable to prepare the standard curve with DNA similar to the type being assayed (e.g., long or short linear DNA fragments when quantitating similar-sized restriction fragments or plasmid when quantitating plasmid DNA). However, most linear dsDNA molecules yield approximately equivalent signals, regardless of fragment length.

    Results have shown that the PicoGreen assay remains linear in the presence of several compounds that commonly contaminate nucleic acid preparations, although the signal intensity may be affected (see Table 2.2.7). Thus, to serve as an effective control, the dsDNA solution used to prepare the standard curve should be treated the same way as the experimental samples and should contain similar levels of such compounds.

5. Determine the DNA concentration on the basis of absorbance at 260 nm (A_{260}) in a cuvette with a 1-cm path length (see Basic Protocol 1).

    An A_{260} of 0.04 corresponds to 2 μg/ml dsDNA solution.

6a. *For high-range standard curve:* Create a five-point standard curve from 1 ng/ml to 1 μg/ml by combining the 2 μg/ml stock prepared in step 2 with 1× TE, in disposable cuvettes (or in plastic test tubes for transfer to quartz cuvettes), according to Table 2.2.8.

6b. *For low-range standard curve:* Prepare a 40-fold dilution of the 2 μg/ml DNA solution to yield a 50 ng/ml DNA stock solution. Create a five-point standard curve from 25 pg/ml to 25 ng/ml by combining this 50 ng/ml stock with 1× TE in disposable cuvettes (or in plastic test tubes for transfer to quartz cuvettes), according to Table 2.2.8.

    To create the low-range standard curve, adjust the fluorometer gain to accommodate the lower fluorescence signals.

7. Add 1.0 ml of the aqueous working solution of PicoGreen reagent (prepared in step 1) to each cuvette. Mix well and incubate for 2 to 5 min at room temperature, protected from light.

8. After incubation, measure the sample fluorescence using a spectrofluorometer or fluorescence microplate reader and standard fluorescein wavelengths (excitation ~480 nm, emission

Table 2.2.7 Effects of Several Contaminants of Nucleic Acid Preparations on Signal Intensity of the PicoGreen dsDNA Quantitation Assay

Contaminant	Concentration	Signal change[a]
Salts		
Ammonium acetate	50 mM	3% decrease
Magnesium chloride	50 mM	33% decrease
Sodium acetate	30 mM	3% increase
Sodium chloride	200 mM	30% decrease
Zinc chloride	5 mM	8% decrease
Urea	2 M	9% increase
Organic solvents		
Chloroform	2%	14% increase
Ethanol	10%	12% increase
Phenol	0.1%	13% increase
Detergents		
Sodium dodecyl sulfate	0.01%	1% decrease
Triton X-100	0.1%	7% increase
Proteins		
Bovine serum albumin	2%	16% decrease
IgG	0.1%	19% increase
Other compounds		
Agarose	0.1%	4% increase
Polyethylene glycol	2%	8% increase

[a]The compounds were incubated at the indicated concentrations with the PicoGreen reagent in the presence of 500 ng/ml calf thymus DNA. All samples were assayed in a final volume of 200 µl in 96-well microplates using a CytoFluor microplate reader (PerSeptive Biosystems). Samples were excited at 485 nm, and fluorescence intensity was measured at 520 nm.

~520 nm). Set the instrument's gain so that the sample containing the highest DNA concentration yields a fluorescence intensity near the fluorometer's maximum (to ensure that the sample readings remain in the detection range of the fluorometer). Keep the time for fluorescence measurement constant for all samples (to minimize photobleaching effects).

9. Subtract the fluorescence value of the reagent blank from that of each of the samples. Use corrected data to perform a regression analysis and generate a standard curve of fluorescence versus DNA concentration (see Fig. 2.2.2).

Analyze samples

10. Add 1.0 ml of the aqueous working solution of the PicoGreen reagent (prepared in step 1) to each sample.

11. Incubate 2 to 5 min at room temperature, protected from light.

12. Measure fluorescence of the sample using instrument parameters that correspond to those used for standard curve (see steps 4 to 9). Keep the time for fluorescence measurement constant for all samples (to minimize photobleaching effects).

13. Subtract the fluorescence value of the reagent blank from that of each sample. Determine DNA concentration of the sample from the standard curve.

14. If desired, repeat the assay using a different dilution of the sample to confirm results.

2

Table 2.2.8 Preparing a Standard Curve with the PicoGreen Reagent (Alternate Protocol 3)

Volume (μl) DNA stock	Volume (μl) 1× TE	Volume (μl) PicoGreen working solution	Final DNA concentration in PicoGreen assay
High-range (2 μg/ml DNA stock)			
1000	0	1000	1 μg/ml
100	900	1000	100 ng/ml
10	990	1000	10 ng/ml
1	999	1000	1 ng/ml
0	1000	1000	0
Low-range (50 ng/ml DNA stock)			
1000	0	1000	25 ng/ml
100	900	1000	2.5 ng/ml
10	990	1000	250 pg/ml
1	999	1000	25 pg/ml
0	1000	1000	0

PROTOCOLS: PROTEIN QUANTIFICATION

The most common procedure for protein quantitation is based on the original Lowry assay. However, as Figure 2.2.1 illustrates, there are many reasons to choose one assay over another. For example, all assays are prone in varying degrees to amino acid composition errors and interference from assay solution components. The simplest approach, although relatively insensitive, is UV spectroscopy. More sensitive assays include ones that use Coomassie blue binding (Bradford), bicinchoninic acid (BCA), and the Lowry reaction. For detection ranges of these assays see Table 2.2.2.

DTT or 2-ME will interfere with some protein assays (e.g., BCA), but are present in some electrophoresis buffers. Creating an S-carboxymethyl derivative of the DTT or 2-ME through alkylation with iodoacetamide (*CP Protein Science Unit 3.4*; Olson and Markwell, 2007) prevents the interference.

Basic Protocol 2: Lowry Protein Assay

The following protein assay is a modification and simplification of the method of Lowry et al. (1951) as described by Peterson (1977). Variations on this include a micro method (Alternate Protocol 4) and the use of a DOC-TCA protein precipitation step to concentrate proteins and remove interfering substances from samples (Support Protocol). There are also a number of commercially available reagents for the Lowry assay (e.g., Sigma or Pierce).

Materials

0.5 mg/ml standard protein stock solution (see recipe)
Sample protein
Copper-tartrate-carbonate (CTC; see recipe)
Lowry reagents A and B (see recipe)

5-ml test tubes
Spectrophotometer with visible light source

1. Prepare protein standards in duplicate in distilled water as follows:

Protein (μg)	Standard (μl)	Water (μl)
0	0	1000
25	50	950
50	100	900
100	200	800.

Protein standards should be diluted to cover the expected range of the sample proteins. Typically, 0, 25, 50 and 100 μg of standard protein are used to create the standard curve.

2. Adjust the sample volume to 1 ml with distilled water.
3. To each standard and sample, add 1 ml reagent A, mix, and let stand 10 min at room temperature.
4. Add 0.5 ml reagent B to each tube from step 3 and immediately mix by flicking the test tube or vortexing. Let stand 30 min at room temperature.
5. Measure absorbance at 750 nm of the blank, standards, and samples (see Basic Protocol 1, steps 2a,b and 3a,b).
6. Plot the protein standard data using a simple linear regression (see *UNIT 2.1* and *APPENDIX 4*) and estimate the protein concentration of the unknown.

For extended concentrations (to 200 μg), the curve is nonlinear but can be accurately plotted on a log/log plot.

Alternate Protocol 4: Lowry Protein Assay, Reduced Volume

This protocol allows for measuring up to 20 μg of protein and an assay volume of 0.5 ml. Double the amounts for a final volume of 1.0 ml.

Materials

See Basic Protocol 2

1. Prepare protein standards in distilled water as follows:

Protein (μg)	Standard (μl)	Water (μl)
0	0	200
25	50	150
50	100	100
100	200	0.

Protein standards should be diluted to cover the expected range of the sample proteins. Typically, 0, 5, 10, and 100 μg of standard protein are used to create the standard curve.

2. Adjust the sample volume to 0.2 ml with distilled water.
3. Add 0.2 ml reagent A, mix, and let stand 10 min at room temperature.
4. Add 0.1 ml reagent B to each tube from step 3 and immediately mix by flicking of the test tube or vortexing. Let stand 30 min at room temperature.
5. Measure absorbance at 750 nm of the blank, standards, and samples (see Basic Protocol 1, steps 2a,b and 3a,b).
6. Plot the protein standard data using a simple linear regression and estimate the protein concentration of the unknown (*UNIT 2.1* and *APPENDIX 4*).

For extended concentrations (to 20 μg), the curve is nonlinear but can be accurately plotted on a log/log plot (Peterson, 1977).

Support Protocol: Deoxycholate-Trichloroacetic Acid (DOC-TCA) Sample Precipitation for Removal of Interfering Compounds and Sample Concentration

This precipitation protocol is ideally suited for microcentrifuge tubes and final volumes of 1 ml. Dilute protein samples can be effectively concentrated, and interfering compounds (e.g., detergents) can be removed via TCA precipitation of a deoxycholate/protein mixture.

Materials

Protein sample (dilute or with interfering compounds)
0.15% (w/v) deoxycholate (DOC)
72% (w/v) trichloroacetic acid (TCA)

1.5-ml microcentrifuge tubes
Filter paper

2

1. In a 1.5-ml microcentrifuge tube, add 0.1 ml of 0.15% DOC to 1 ml of sample. Cap, briefly mix, and let stand 10 min at room temperature.

2. Uncap, add 0.1 ml of 72% TCA, recap, and briefly mix. Centrifuge 5 min at 3000 × *g*, room temperature.

3. Gently decant the supernatant and drain any remaining drops onto filter paper. Carefully aspirate any remaining liquid from the microcentrifuge tube.

4a. *To perform the standard Lowry assay:* Proceed with Basic Protocol 2, adding reagent A to the microcentrifuge tube and then transferring the dissolved protein to a test tube to complete the assay.

4b. *To perform the reduced volume Lowry assay:* Proceed with Alternate Protocol 4, adding reagent A directly to the microcentrifuge tube (immediately dissolving the precipitated protein).

Basic Protocol 3: BCA Protein Assay

The bicinchoninic acid (BCA) assay is based on the original publication by Smith et al. (1985). In comparison to the Lowry method, the BCA assay replaces the Folin-Ciocalteau reagent with BCA to detect the Cu^{1+} product of the biuret reaction. The assay described in this protocol is for a 2-ml reaction. Reagents are commercially available premixed for immediate use (e.g., Sigma or Pierce).

Materials

0.5 mg/ml standard protein stock solution (see recipe)
Protein sample
Bicinchoninic acid (BCA) standard working reagent (SWR; see recipe or commercially available, e.g., Pierce)

5-ml test tubes
37°C or 60°C water bath
Spectrophotometer with visible light source

1. Add 100 µl sample to 2 ml SWR and mix.

   The assay can be scaled appropriately to accommodate a range of volumes keeping the ratio of 1 volume of sample to 20 volumes of the SWR.

2. Incubate 2 hr at room temperature, 30 min at 37°C, or 30 min at 60°C.

   To maximize the assay sensitivity and minimize color yield differences among various standard proteins, the reaction is optimal at 60°C, although for routine work any of the above conditions is adequate.

3. Measure absorbance at 562 nm of the blank, standards, and samples (see Basic Protocol 1, steps 2a,b and 3a,b).

4. Plot the protein standard data using a simple linear regression (*UNIT 2.1* and *APPENDIX 4*) and estimate the protein concentration of the unknown.

Basic Protocol 4: Coomassie Blue Protein Assay (Bradford Assay)

Like Coomassie blue staining of proteins in gels, the following assay is based on the binding of Coomassie blue to protein (Bradford, 1976). However, in this case the proteins are in solution, and the dye is Coomassie blue G250. The method described below is designed for a final volume of 1 ml and accurately measures from 1 to 20 μg of protein.

Materials

Sample protein
0.5 mg/ml standard protein stock solution (see recipe)
1 M sodium hydroxide
Coomassie protein reagent (see recipe or available commercially, e.g., Pierce)

Disposable plastic cuvettes or test tubes
Spectrophotometer with visible light source

1. Pipet 0.1 ml sample or standard protein dilution into a disposable plastic cuvette or test tube. Turn on the spectrophotometer and allow it to warm up for five minutes prior to use or according to manufacturer's recommendations.
2. Add 100 μl of 1M sodium hydroxide and mix.

   The addition of sodium hydroxide prevents sample precipitation. Although this step is optional (because not all protein samples will form a precipitate upon addition of the Coomassie protein reagent), it is recommended for routine assays.
3. Add 1.0 m of the Coomassie protein reagent to the sample and mix.
4. Incubate 2 min at room temperature.
5. Measure absorbance at 595 nm of the blank, standards, and samples (see Basic Protocol 1, steps 2a,b and 3a,b).
6. Plot the protein standard data using a simple linear regression (*UNIT 2.1* and *APPENDIX 4*) and estimate the protein concentration of the unknown.

Basic Protocol 5: UV Spectrophotometry

Ultraviolet spectrometry has long been used to quantitate purified proteins in solution, typically by measuring the sample absorbance at 205 nm (Scopes, 1974) and 224 to 236 nm (Groves et al., 1968) as well as by calculating the A_{260}/A_{280} ratio (Warburg and Christian, 1941; Layne, 1957). The main difficulty with protein quantitation via absorption spectroscopy is that many potential contaminants of protein samples also absorb in the UV spectrum, and the actual optical density of a protein solution is highly dependent on the amino acid composition of the protein. Other factors such as protein conformation and pH also affect the reading. Measurement at 260 nm is also beneficial to help correct, via the formula below, for UV absorbing nucleic acids that might contaminate the sample and create an artificially high reading with protein (Alternate Protocol 5).

Materials

Sample protein dissolved in compatible buffer (see Table 2.2.3)
0 to 3 mg/ml standard protein stock solution in compatible buffer (see Table 2.2.3)
Protein standard

Spectrophotometer with UV light source
UV transparent quartz or methacrylate cuvettes

1. Turn on the spectrophotometer and warm up as recommended by the manufacturer.

   Typically, this requires <15 minutes.

2. Adjust the spectrophotometer to 280 nm and zero the instrument with a distilled water blank to confirm a clean baseline and spectrophotometer performance.
3. Replace the distilled water blank with the same buffer used to dissolve the standard and sample, and note the A_{280} of the blank.

   If the contribution of absorbance is too high from the blank then the readings of the sample will be limited in dynamic range.

4. Zero the spectrophotometer against the buffer blank.

   For dual beam instruments, the blank cuvette will simply contain the buffer while the sample cuvette will contain the protein of interest.

5. Add the protein standard to the sample cuvette and note the A_{280} nm reading.
6. Construct a standard curve and calculate, via linear regression (*UNIT 2.1* and *APPENDIX 4*), the line parameters and correlation coefficient.

   This equation will be used to calculate the amount of protein in the unknown. $A_{280} = (\text{slope} \times \text{protein amount}) + y \text{ intercept}$.

7. Remove the standard and add the sample. Note the A_{280} reading. From the equation in the step 6 annotation, calculate the amount of protein: protein amount = $(A_{280} - y \text{ intercept})/\text{slope}$

Alternate Protocol 5: Protein Quantitation with UV Spectroscopy and Correction for Like-Acid Contamination

When measuring the concentration of protein by optical density at 280 nm, the contribution of nucleic acids can be significant. While proteins maximally absorb at 280 nm due to the aromatic amino acids (primarily tryptophan and tyrosine), and nucleic acids maximally absorb at 260 nm, a significant contribution to the measurement at 280 nm can come from nucleic acids. Numerous strategies have been proposed to deal with the contribution from nucleic acids to the overall absorption when measuring protein concentration. The most straightforward implementation was originally described by Warburg and Christian (1941) and modified by Layne (1957). This simply consists of measuring OD at 260 nm and 280 nm and apply the values in the following formula:

$$\text{Protein concentration (mg/ml)} = 1.55\ OD_{280} - 0.76\ OD_{260}$$

PROTOCOL COMMON TO NUCLEIC ACIDS AND PROTEINS

Basic Protocol 6: Gel Based Quantitation of Protein and Nucleic Acids

Electrophoretic separation, in addition to indicating purity and size information, provides an excellent way to quantitate the amount of a particular band of protein or nucleic acid in a sample. In brief, the separated protein or nucleic acid is stained, and the intensity of the stained unknown is compared to a standard of known amount also separated on the same gel. Typically the standards and samples are separated in duplicate to give adequate data points for averaging and estimating error. As in the in-solution assays described in the previous protocols, the choice of the standard is important. Ideally, the standard should be similar or identical to the sample being analyzed, and it should be independently quantitated using another technique (e.g., UV spectrophotometry). Ready-to-use standards from commercial suppliers greatly simplify this process, although the range of standard proteins and nucleic acids is limited.

Materials

Stained gel containing protein or nucleic acid samples and quantitated protein or nucleic acid standards

Protein electrophoresis standards (e.g., Pierce, Sigma Aldrich, or Fermentas); prequantitated protein standards for solution assays are acceptable

Nucleic acids electrophoresis standards (e.g., NEB, Sigma Aldrich, or Fermentas)

CCD imaging system with image analysis software (*UNIT 7.5*)

1. Capture the stained gel image of either white light (e.g., Coomassie blue) or fluorescently stained (e.g., ethidium bromide) gel using a CCD imaging system (see *UNITS 7.4 & 7.5*).
2. Using the image analysis software, perform a lane-based or area-based analysis (see Fig. 2.2.3).
3. Correct for background using one of many options for changing the baseline in the lane-based analysis or by selecting an empty area of the gel and designating that area as background (Fig. 2.2.3; also see *UNIT 7.5*).

   This process varies greatly depending on the software used; the software manual should be consulted to understand the effects of background subtraction. At its most basic level, the background subtraction should only remove the signal contributed by a portion of the gel which contains no sample-based signal. It is useful to process the data with and without background subtraction to identify any loss of real signal due to the subtraction of the background.

If using analysis software to estimate sample in bands

4a. Calibrate the intensity of the standard band to the amount (see Fig. 2.2.3) of standard.

5a. Use this value to generate estimates of the amount present in the sample bands with the analysis software.

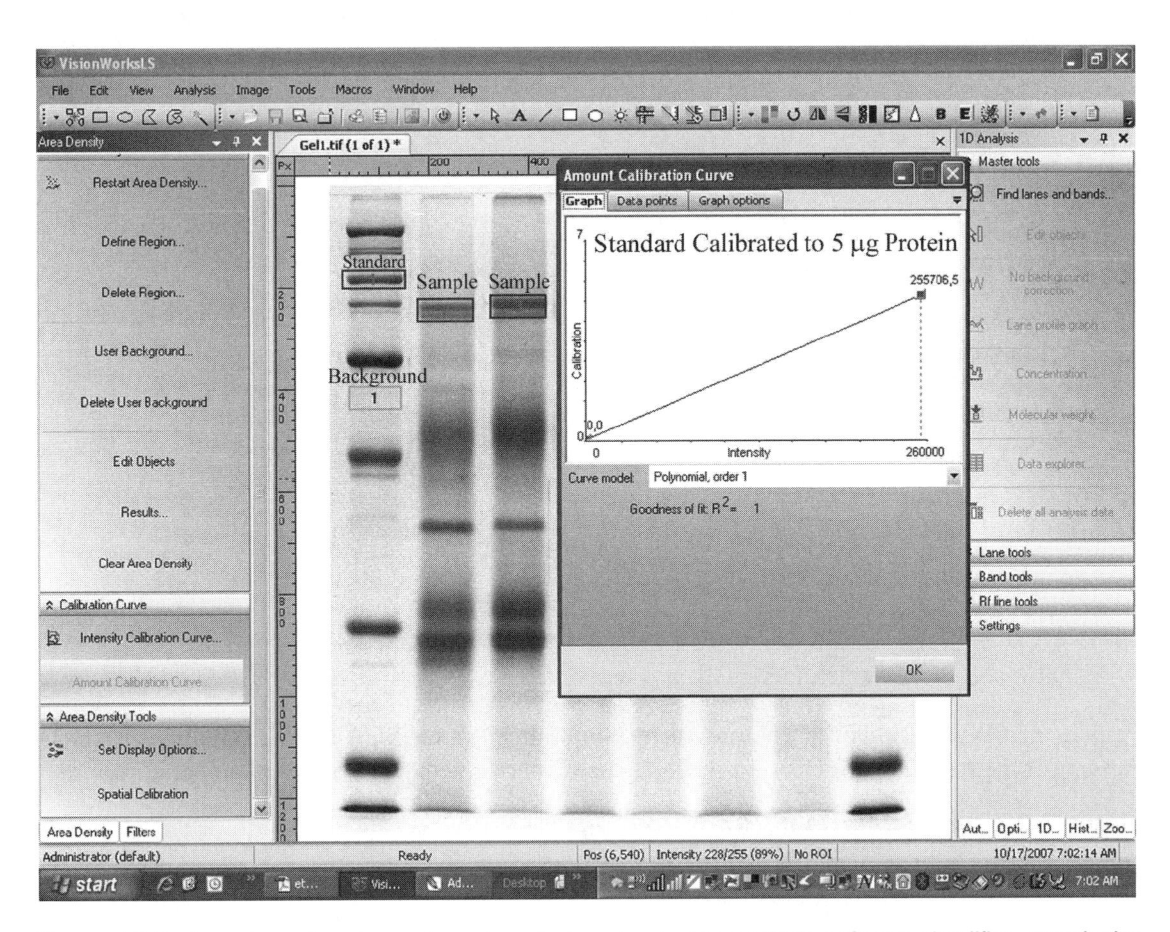

Figure 2.2.3 Lane-based and area-based analysis using image analysis software simplifies quantitative analysis and also enables rapid data storage for later retrieval. Background correction is important for accurate quantitation. However, the background is frequently inconsistent across the gel so picking the correct value for background can be a challenge. The simplest approach is to select an area of the gel containing no sample. Alternatively, automated background correction routines can average a ring of pixels around the band of interest to obtain a background value. Calibration routines vary depending upon the software used. Typically the known standard amount (e.g., µg or ng protein or nucleic acid) is equated to an intensity value determined from the standard. Standards representing a range of amounts are used to construct a calibration curve. Once the intensity value is calculated for the unknown it is compared to the standard curve to derive an estimate of the amount of protein or nucleic acid in the unknown band.

If using standard graphing programs to estimate sample in bands

4b. Export the band intensity data to scientific graphing software (e.g., SigmaPlot or Excel) and use linear regression to calculate the standard curve.

5b. Use the linear regression curve to determine the amount of protein or nucleic acid in the sample band.

Briefly, linear regression analysis plots the band intensity on the y axis and the amount of the standard on the x axis and gives an estimate of the linear correlation or relationship between intensity and amount. Ideally, duplicate standard lanes would be run so that duplicate values can be averaged. This allows a more realistic estimate of the amount and also gives an estimate of the error in the measurements.

For the standards, the parameters of the standard line are generated by the program or the calculator by simply inputting the amount (x axis) and intensity (background corrected, y axis) pairs. The equation of the line is $y = mx + b$, where m is the slope and b is the y intercept.

For quantitation of sample protein or nucleic acids separated by electrophoresis, rearrange the equation: $x = (y - b)/m$ or amount in the stained band = (intensity of the band− the y intercept)/slope (see UNIT 2.1 *and* APPENDIX 4*).*

The correlation coefficient (r) is also routinely generated and is a measure of strength of the relationship, with +1 a perfect positive correlation. In some cases, it is necessary to use a limited range of the curve to be on a linear portion so a reasonable r value is obtained. Typical standard curves should be >0.90.

REAGENTS AND SOLUTIONS

Use deionized, distilled water in all recipes and protocol steps. For common stock solutions, see **UNIT 3.3**.

Bicinchoninic acid (BCA) standard working reagent (SWR)

BCA Reagent A
1% (w/v) bicinchoninic acid (BCA-Na_2)
2% (w/v) Na_2CO_3 H_2O
0.16% (w/v) Na_2 tartrate
0.4% (w/v) NaOH
0.95% (w/v) $NaHCO_3$
Adjust pH to 11.25 with 50% NaOH if needed

BCA Reagent B
4% (w/v) $CuSO_4{\cdot}5H_2O$

Reagents A and B are stable indefinitely at room temperature.

Prepare the SWR weekly or just before use by combining 100 vol BCA Reagent A with 2 vol BCA Reagent B in quantities sufficient for the assay being performed. Store at room temperature.

Coomassie protein reagent (Bradford reagent)

100 mg Coomassie brilliant blue G-250 (0.01% w/v)
50 ml 95% ethanol (4.7% w/v)
100 ml 85% phosphoric acid (8.5% w/v)

Add 100 mg of Coomassie brilliant blue G-250 to 50 ml of 95% ethanol and mix. Slowly add 100 ml of 85% phosphoric acid and dilute with distilled water to a final volume of 1 liter. Filter through Whatman no. 1 filter paper. Store up to 1 month at room temperature or 3 months at 4°C.

Final concentrations are given in parentheses.

Premixed commercial reagents also available from a variety of suppliers (e.g., Bio-Rad, Pierce, or Sigma).

Copper-tartrate-carbonate (CTC)

0.2 g copper sulfate, pentahydrate (0.1% w/v)
0.4 g potassium tartrate (0.2% w/v)
100 ml 20% sodium carbonate (10% w/v)

Mix the copper sulfate and potassium tartrate together in 100 ml H_2O. With stirring, add 100 ml of 20% sodium carbonate solution to the copper sulfate/potassium tartrate solution. Store indefinitely at 4°C or up to 1 month at room temperature.

Final concentrations are given in parentheses.

2

Ethidium bromide assay solution

Add 10 ml of 10× TNE buffer (see recipe) to 89.5 ml water. Filter through a 0.45-μm filter to remove particulates, then add 0.5 ml of 1 mg/ml ethidium bromide. Store at 4°C, protected from light.

CAUTION: *Ethidium bromide is hazardous; wear gloves and use appropriate care in handling, storage, and disposal. Also see* APPENDIX 1.

Add the dye after filtering because ethidium bromide will bind to most filtration membranes.

Hoechst 33258 assay solutions

Stock solution: Dissolve 1 mg/ml Hoechst 33258 (Invitrogen) in H_2O. Store up to ~6 months at 4°C, protected from light.

Working solution: Add 10 ml of 10× TNE buffer (see recipe) to 90 ml water. Filter through a 0.45-μm filter, then add 10 μl of 1 mg/ml Hoechst 33258 stock solution.

CAUTION: *Hoechst 33258 is hazardous; use appropriate care in handling, storage, and disposal.*

Hoechst 33258 is a fluorochrome dye with a molecular weight of 624 and a molar extinction coefficient of $4.2 \times 10^4\ M^{-1} cm^{-1}$ *at 338 nm.*

The dye is added after filtering because it will bind to most filtration membranes.

Lowry reagent

Lowry reagent A: For a 100 ml stock solution, mix 25 ml each of CTC (see recipe), 10% SDS, 0.8 N NaOH, and water. Alternatively, for smaller or larger volumes, simply mix equal amounts of the listed stock reagents. Store up to 2 weeks at room temperature.

Use a high-purity grade of SDS to minimize background.

Lowry reagent B: For a 60 ml stock solution, mix 10 ml Folin-Ciocalteu phenol reagent (e.g., Pierce, Sigma, or prepare as described by Peterson, 1977) and 50 ml of water. Alternatively for smaller or larger volumes, simply mix one part stock reagent to five parts water. Store up to several months at room temperature in an amber bottle.

Standard protein stock solution, 0.5 mg/ml

Typically, BSA is used for standardization. Other proteins can be used (e.g., IgG) because, like all of the colorimetric and fluorescent protein assays, the Lowry assay does not respond identically to all proteins (Peterson, 1983; Sapan et al., 1999). Standards are also commercially available (e.g., Invitrogen, Pierce, Sigma).

TNE buffer, 10×

100 mM Tris base
10 mM EDTA
2.0 M NaCl
Adjust pH to 7.4 with concentrated HCl
Store up to 6 months at 4°C.
Dilute with water to desired concentration, as necessary.

Trichloroacetic acid (TCA), 72% (w/v)

TCA is hygroscopic and very caustic and will damage clothes and skin. It is frequently easier to use a new bottle of solid TCA and add water to this rather than try to measure TCA on a balance. Carefully add water so that you do not dilute past 72% for the given weight in the bottle. If an identical empty bottle is available, the volume can be checked visually on the empty bottle prior to a final measurement of the TCA solution with a graduated cylinder.

UNDERSTANDING RESULTS

Nucleic Acid Quantification

The detection limit of absorption spectroscopy will depend on the sensitivity of the spectrophotometer and any UV-absorbing contaminants that might be present. The lower limit is generally ~0.5 to 1 μg nucleic acid. Typical values for a highly purified sample of DNA and RNA are shown in Table 2.2.1.

For the Hoechst 33258, ethidium bromide, and PicoGreen assays, a plot of relative fluorescence units or estimated concentration (y axis) versus actual concentration (x axis) typically produces a linear regression with a correlation coefficient (r^2) of 0.98 to 0.99 (Fig. 2.2.4). Table 2.2.1 provides a comparison of the sensitivities and specificities of the three assays.

Protein Quantification

Basic features of the four assays described in this unit are listed in Table 2.2.2. Numerous factors contribute to accurate protein quantitation. For both colorimetric and the ultraviolet spectrophotometry–based quantitation methods, there is considerable variation in color intensity or absorption value depending on the protein measured (e.g., see Peterson, 1983; Sapan et al., 1999; *CP Protein Science Units 3.1* and *3.4*, Grimsley and Pace, 2003 and Olson and Markwell, 2007, respectively). Using BSA as a standard normalized to 1.00 and comparing 14 other proteins at the same concentration as

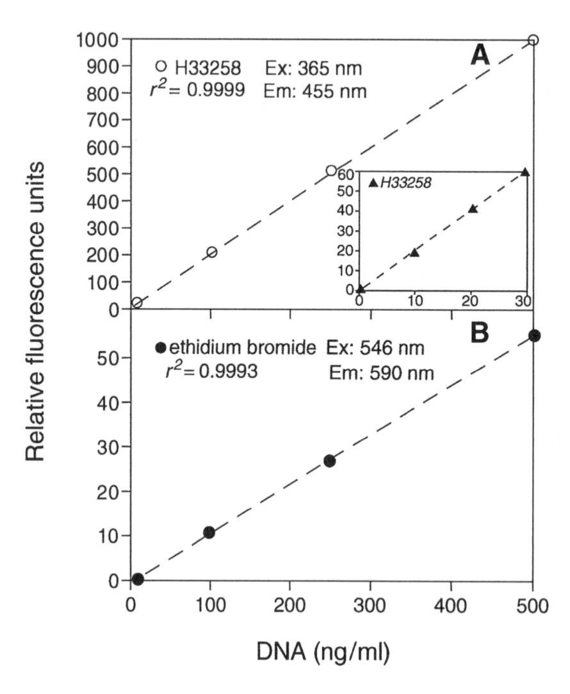

Figure 2.2.4 Fluorochrome Hoechst 33258 (H33258) (**A**) and ethidium bromide (**B**) DNA concentration standard curves. (Originally published in *CP Molecular Biology Appendix 3D*; Gallagher and Desjardins, 2006).

BSA, Peterson (1983) found the average predicted amount relative to BSA when using the following assays to be:

Modified Folin phenol (Peterson, 1977): 1.18 ± 0.27
Coomassie blue (Bradford, 1976): 0.94 ± 0.94
280 nm: 1.81 ± 1.19
224 to 236 nm (Groves et al., 1968): 1.08 ± 0.43
205 nm (Scopes, 1974): 1.06 ± 0.09.

For the same concentration of 1 mg/ml, the range of values for a particular protein in distilled water varies significantly within each assay. Note that UV absorption at 280 nm showing the most dramatic range (0.02 to 4.19), while UV absorption at 205 nm ranged from 0.91 to 1.13. See Table 2.2.2 for causes of such variability.

TROUBLESHOOTING

Each assay is also sensitive to potential interference from various cellular components that carry over during purification and deliberately added components of the buffer (Layne, 1957; Peterson, 1977, 1979, 1983; Stoscheck, 1990). The type and level of interference by detergents, salts, and buffer components, and other contaminants varies depending upon the assay (Table 2.2.3). For example, the Coomassie blue dye and UV spectroscopy assays are sensitive to detergent interference (SDS), while the BCA and the Folin phenol assays can handle 10-fold higher concentrations of SDS. Detergents form micelles with hydrophobic interiors that compete for the dye and contribute to an artificially high signal unrelated to the protein present. However, the Coomassie blue dye assay is relatively insensitive to the reductants dithiothreitol and 2-mercaptoethanol.

In practice, the simplest approach to dealing with interference is to keep below the concentration that leads to incorrect results. This can be readily tested by simply testing a range of protein concentrations, including a blank, which contains no protein but all the assay components, in the presence of the potential interfering reagent. It is a good idea to test a range of concentrations of the interfering reagent against a range of protein concentrations to give a complete picture of possible effects on the assay. If needed, the protein can be precipitated to remove the interfering reagent and then suspended into the protein assay reagent.

Drift and unstable readings can be minimized through use of filtered solutions and constant temperature. The most common issue is the presence of particulates that float in and out of the light beam leading to jumps in the signal. Poorly mixed solutions can have a similar effect. Adding a cold solution to the cuvette will cause condensation on the glass surfaces that can interfere with light transmission and create unstable readings. Fluorescence is also temperature sensitive and is more intense with colder samples; the intensity will drop off as the solution warms. Keeping the solution at room temperature should alleviate the temperature effects, while filtering the assay solution through 0.45-μm filters before adding the dye (since most filters will bind dye) will remove particulates. Dilute sample solutions should give stable readings, assuming the instrument is stable, but the signal level can be near the limits of the detection of the instrument, making determination of values and ratios difficult. Be sure to use samples with enough material to be accurately detected. Consult the instrument specifications to understand the limits of detection.

VARIATIONS

Protein Quantification

10-fold to 100-fold improvements in sensitivity can be achieved with fluorescence-based protein quantitation (Jones et al., 2003). While these assays require the use of a fluorometers, the speed and sensitivity for fluorescence-based assays is very useful in high-throughput settings and when working with limited amounts of protein.

LITERATURE CITED

Applied Biosystems. 1987. User Bulletin Issue 11, Model No. 370. Applied Biosystems, Foster City, Calif.

Bradford, M.M. 1976. A rapid and sensitive method for the quantitation of microgram quantities of protein utilizing the principle of protein-dye binding. *Anal. Biochem.* 72:248-254.

Cesarone, C.F., Bolognesi, C., and Santi, L. 1979. Improved microfluorometric DNA determination in biological material using 33258 Hoechst. *Anal. Biochem.* 100:188-197.

Daxhelet, G.A., Coene, M.M., Hoet, P.P., and Cocito, C.G. 1989. Spectrofluorometry of dyes with DNAs of different base composition and conformation. *Anal. Biochem.* 179:401-403.

Dunn, B. 1995. Quantitative amino acid analysis. *Curr. Protoc. Protein Sci* . 0:3.2.1-3.2.3.

Gallagher, S.R. and Desjardins, P.R. 2006. Quantitation of DNA and RNA with absorption and fluorescence spectroscopy. *Curr. Protoc. Mol. Biol.* 76:A.3D.1-A.3D.21.

Grimsley, G.R. and Pace, C.N. 2003. Spectrophotometric determination of protein concentration. *Curr. Protoc. Protein Sci.* 33:3.1.1-3.1.9.

Groves, W.E., Davis, F.C., Jr., and Sells, B.H. 1968. Spectrophotometric determination of microgram quantities of protein without nucleic acid interference. *Anal. Biochem.* 22:195-210.

Jones, L.J., Haugland, R.P., and Singer, V.L. 2003. Development and characterization of the NanoOrange protein quantitation assay: A fluorescence-based assay of proteins in solution. *BioTechniques* 34:850-858.

Labarca, C. and Paigen, K. 1980. A simple, rapid, and sensitive DNA assay procedure. *Anal. Biochem.* 102:344-352.

Layne, E. 1957. Spectrophotometric and turbidimetric methods for measuring proteins. *Methods Enzymol.* 3:447-454.

Le Pecq, J.-B. 1971. Use of ethidium bromide for separation and determination of nucleic acids of various conformational forms and measurement of their associated enzymes. *In* Methods of Biochemical Analysis, Vol. 20 (D. Glick, ed.) pp. 41-86. John Wiley & Sons, New York.

Lowry, O.H., Rosebrough, N.J., Farr, A.L., and Randall, R.J. 1951. Protein measurement with the Folin phenol reagent. *J. Biol. Chem.* 193:265-275.

Marmur, J. and Doty, P. 1962. Determination of the base composition of deoxyribonucleic acid from its thermal denaturation temperature. *J. Molec. Biol.* 5:109-118.

Neidle, S. DNA minor-groove recognition by small molecules. 2001. *Nat. Prod. Rep.* 18:291-309.

Olson, B.J.S.C. and Markwell, J. 2007. Assays for determination of protein concentration. *Curr. Protoc. Protein Sci.* 48:3.4.1-3.4.29.

Peterson, G.L. 1977. A simplification of the protein assay method of Lowry et al. which is more generally applicable. *Anal. Biochem.* 83:346-356.

Peterson, G.L. 1979. Review of the Folin phenol protein quantitation method of Lowry, Rosebrough, Farr and Randall. *Anal. Biochem.* 100:201-220.

Peterson, G.L. 1983. Determination of total protein. *Methods Enzymol.* 91:95-119.

Portugal, J. and Waring, M.J. 1988. Assignment of DNA binding sites for 4′,6-diamidine-2-phenylindole and bisbenzimide (Hoechst 33258): A comparative footprinting study. *Biochem. Biophys. Acta* 949:158-168.

Sapan, C.V., Lundblad, R.L., and Price, N.C. 1999. Colorimetric protein assay techniques. *Biotechnol. Appl. Biochem.* 29:99-108.

Scopes, R.K. 1974. Measurement of protein by spectrophotometry at 205 nm. *Anal. Biochem.* 59:277-282.

Smith, P.K., Krohn, R.I., Hermanson, G.T., Mallia, A.K., Gartner, F.H., Provenzano, M.D., Fujimoto, E.K., Goeke, N.M., Olson, B.J., and Klenk, D.C. 1985. Measurement of protein using bicinchoninic acid. *Anal. Biochem.* 150:76-85.

Stoscheck, C.M. 1990. Quantitation of protein. *Methods Enzymol.* 182:50-68.

Stout, D.L. and Becker, F.F. 1982. Fluorometric quantitation of single-stranded DNA: A method applicable to the technique of alkaline elution. *Anal. Biochem.* 127:302-307.

Struhl, K. 1993. Reagents and radioisotopes used to manipulate nucleic acids. *Curr. Protoc. Mol. Biol.* 9:3.4.1-3.4.11.

Van Lancker, M. and Gheyssens, L.C. 1986. A comparison of four frequently used assays for quantitative determination of DNA. *Anal. Lett.* 19:615-623.

Voet, D., Gratzer, W.B., Cox, R.A., and Doty, P. 1963. Absorption spectra of nucleotides, polynucleotides, and nucleic acids in the far ultraviolet. *Biopolymers* 1:193–208.

Wallace, R.B. and Miyada, C.G. 1987. Oligonucleotide probes for the screening of recombinant DNA libraries. *In* Methods of Enzymology, Vol. 152: Guide to Molecular Cloning Techniques (S.L. Berger and A.R. Kimmel, eds.) pp. 432-442. Academic Press, San Diego.

Warburg, O. and Christian, W. 1941. *Biochem. Z.* 310-384.

Wilfinger, W.W., Mackey, K., and Chomczynski, P. 1997. Effect of pH and ionic strength on the spectrophotometric assessment of nucleic acid purity. *Biotechniques* 22:474-476.

KEY REFERENCES

Labarca and Paigen, 1980. See above.
Contains a detailed description of the Hoechst 33258 fluorometric DNA assay.

Peterson, 1997. See above.
Describes in detail the Lowry assay used in this chapter.

Sapan et al., 1999. See above.
An excellent overview of the used and limits of protein assays.

2

UNIT 2.3

Radiation Safety and Measurement

Jill Meisenhelder[1] and Steve Bursik[1]
[1]The Salk Institute, La Jolla, California

2

INTRODUCTION

The use of radioisotopes to label specific molecules in a defined way has greatly furthered the discovery and dissection of biochemical pathways. The development of methods to inexpensively synthesize such tagged biological compounds on an industrial scale has enabled them to be used routinely in laboratory protocols, including many detailed in this manual. Although most of these protocols involve the use of only small, microcurie (μCi) amounts of radioactivity, some (particularly those describing the metabolic labeling of proteins or nucleic acids within cells) can require activity amounts on the order of tens of millicuries (mCi). In all cases where radioisotopes are used, depending on the quantity and nature of the isotope, certain precautions must be taken to ensure the safety of the scientist. This unit outlines a few such considerations relevant to the isotopes most frequently used in biological research, which are presented briefly in Table 2.3.1.

In designing safe protocols for the use of radioactivity, the importance of common sense, based on an understanding of the general principles of radioactive decay and the importance of continuous monitoring with a hand-held radiation monitor (e.g., Geiger counter), cannot be overemphasized. In addition, it is also critical to take into account the rules, regulations, and limitations imposed by each specific institution. These are usually not optional considerations: an institution's license to use radioactivity normally depends on strict adherence to such rules.

BACKGROUND INFORMATION

The Radioactive Decay Process

As anyone who has taken a basic chemistry course will remember, each element is characterized by its atomic number (Z), defined as the number of protons in the nucleus of that atom. Z is therefore unique to each element and determines the identification and chemistry of that particular element. Isotopes of a given element exist because some atoms of each element, while by definition having the same number of protons, have a different number of neutrons and therefore a different atomic weight (A), the atomic mass number which indicates the total number of nucleons (protons + neutrons). It should be noted that, generally, the number of electrons outside the nucleus remains the same for all isotopes of a given element, so all isotopes of a given element are equivalent with respect to their chemical reactivity.

Radioactive decay is defined as the spontaneous change in the structure of an atom accompanied by the emission of energy. This often results in the change of an atom of one element into an atom of a totally different element, a process termed nuclear transmutation, as the number of protons (or neutrons) in the atom changes after decay. The energy released can be particulate, i.e., alpha (α) and beta (β) particles, or nonparticulate, i.e., gamma (γ) and X-rays. These are the primary types of radiation encountered in biological research. Emitted radiation is usually measured in units of keV (kilo-electron volts) or MeV (mega-electron volts).

An α particle is essentially the nucleus of a helium atom, or two protons plus two neutrons. They are relatively large, heavy particles that move relatively slowly (compared to a β particle having the same amount of energy). Containing two protons, the α particle has a positive charge of 2+.

Table 2.3.1 Physical Characteristics of Commonly Used Radionuclides[a]

Isotope	Half-life	Emission	Energy, max (MeV)[b]	Range in air (max)	Specific activity 100% pure (Ci/mg)	Decay product	Typical use in laboratory
^{3}H[c]	12.43 years	β-	0.0186	0.42 cm	9.6	$^{3}_{2}He$	Cell proliferation assays (^{3}H-thymidine); tagging cellular proteins (^{3}H-amino acids)
^{14}C[c]	5370 years	β-	0.156	21.8 cm	4.4 mCi/mg	$^{14}_{7}N$	Tagging cellular proteins (^{14}C-amino acids)
^{32}P[d]	14.3 days	β-	1.71	610 cm	287	$^{32}_{16}S$	Probes for northern and Southern blots (α-dNTPs); in vitro kinase reactions (γ-ATP); metabolic labeling of cellular proteins (ortho-^{32}P)
^{33}P[c]	25.4 days	β-	0.249	49 cm	156	$^{33}_{16}S$	Probes for northern and Southern blots (α-dNTPs)
^{35}S[c]	87.4 days	β-	0.167	24.4 cm	43	$^{35}_{17}Cl$	Metabolic labeling of cellular proteins or in vitro translations (^{35}S-methionine/cysteine)
^{125}I[e]	60 days	γ and X rays	0.027–0.035	0.02 mm lead half-value layer	17	$^{125}_{52}Te$	Labeling of cellular surface proteins or of purified proteins in vitro (free ^{125}I); immunoblot detection reagent (^{125}I-Protein A)

[a] Table compiled based on information in Lederer et al. (1967), Shleien (1987), and the Princeton Radiation Safety Manual (*http://web.princeton.edu/sites/ehs, see radioisotope fact sheets*).

[b] A 100-W light bulb burning for 1 hr uses 2.2×10^{18} MeV. The energy of visible light is 1.8 to 3.1 eV.

[c] Shielding is not needed for activity amounts typically used in the laboratory.

[d] Recommended shielding for ^{32}P is clear acrylic plastic (up to 1 cm thick for mCi amounts).

[e] Recommended shielding for ^{125}I is lead foil or (for mCi amounts) a leaded acrylic workstation.

With a relatively high electronic charge, it only travels short distances before it readily interacts with some other atom via coulombic forces. Emitted α particles have discrete energies and are emitted from isotopes having high mass nuclei (atomic number $Z > 82$; e.g., thorium or uranium); such isotopes are not commonly used in biological research except for specific applications in electron microscopy and X-ray diffraction studies.

In contrast to α particles, β particles are light, high-speed, singly charged particles. Negatively charged β particles are essentially electrons of nuclear origin emitted when an intranuclear neutron changes to a proton with the attendant release of a neutrino (which is very weakly interacting). Release of a β particle thus changes the atomic number and elemental status of the isotope. β radiation is emitted across a spectrum of different energies, the most energetic being stated as E_{max}. The average energy can be approximated as $E_{ave} = 1/3\ E_{max}$. An example is ^{32}P, with an E_{max} of 1.71 MeV and an average energy of 0.57 MeV. The neutrino is emitted with an energy equal to the difference between the actual emitted β energy and E_{max}. Most of the β emitters used in the biology laboratory are "pure β emitters," meaning there is no other radiation emitted besides the β and neutrino.

γ radiation exhibits both particle and wave properties and its wavelength falls within the range of X-rays. Physically, there is no difference between γ and X-ray radiation. The only difference is their physical origin: γ radiation is defined as that originating from an atomic nucleus, and X-ray radiation as that originating from the electron cloud surrounding the nucleus. Unlike β particle emission, the emission of γ radiation by itself produces an isotopic change rather than an elemental one; however, the resultant nuclei may be unstable and decay further, possibly releasing β particles. γ radiation is emitted with a discrete energy.

Isotopic decay may involve a chain or sequence of events rather than just a single decay. Resultant daughter products may also be radioactive (unstable) and thus pose a hazard to workers.

Following their emission, α and β radiation travel varying distances at varying speeds, depending on their initial energy and the atomic number of the material through which they are moving. The distance they actually travel before interacting with the electrons or nuclei of another atom is termed their range and is a defined value for each kind of material. This range is usually expressed as a maximum for each type of particle and material. The energy and type of particulate radiation released (and therefore its potential range) dictates what type of shielding, if any, is necessary for protection against the radiation generated by the decay of the isotope. As discussed previously, α radiation, even with a high amount of energy, will not penetrate very far into common materials. In fact, a sheet of paper can block most α radiation, and therefore this radiation is not considered an external hazard, the major concern being internal exposure. Theoretically, γ and X-ray radiation can travel forever and do not have a specific range. Practically however, there is a measurable decrease in the intensity of γ radiation as it penetrates greater thicknesses of material. This decrease is usually expressed as the half-value layer (HVL) or tenth-value layer (TVL) of a specific material (e.g., lead) and is used to calculate the needed thickness of shielding.

When high-energy β particles released during the decay of ^{32}P encounter the nuclei of atoms with a high atomic number, a coulombic interaction occurs. The β particle decelerates and loses energy in the form of X-rays. Such X-rays are termed bremsstrahlung radiation (German for "braking radiation"); they are detectable using most radiation survey meters, especially those designed specifically for the detection of γ or X-rays. The amount of bremsstrahlung radiation produced is directly proportional to both the energy of the incident β particle and the atomic number, Z, of the absorber. Hence, much more bremsstrahlung is produced with ^{32}P β radiation than ^{35}S β radiation (E_{max} for ^{35}S is 0.167 MeV) when incident on the same material, and with a higher-Z material such as lead, the ^{32}P β radiation would produce more bremsstrahlung than when incident on plastic.

α, β, and γ emissions all have the potential, upon encountering an atom, to interact with and ionize the atom. Thus, these three types of emissions are called ionizing radiation. The formation of such ions may result in the perturbation of biochemical processes: therein lies the health concern associated with radioactivity!

Units Used to Measure Radioactivity

There are several measurable properties used to describe the amount and physical effects of ionizing radiation: activity, exposure, dose, and dose equivalent.

The first is termed activity. Activity is the amount of radioactive material in a sample and is measured in units of curies (Ci) or becquerels (Bq). A curie by definition is that amount of radioactive material that will produce 3.7×10^{10} disintegrations (ions) per second. This was originally determined as the number of disintegrations that occur during the radioactive decay of 1 g of radium-226. The becquerel is defined as 1 disintegration per second. In biological research, the curie is a very large unit and the becquerel a very small unit, so prefixes are added to these roots to express the activity amounts between these two extremes which are commonly used, i.e., microcurie (μCi) and megabecquerel (MBq).

Exposure is defined as that amount of ionization (measured as electrical charge) produced in a particular volume of air at standard temperature and pressure (STP) by the passage of γ or X-ray radiation. This unit is **only** defined for γ or X-ray radiation. The exposure unit is the *X* unit, which is slowly supplanting the roentgen (R) for expressing this property. One *X* unit is 1 coulomb per kilogram of air. This property can be measured directly with ionization chamber-based radiation survey meters.

Another property is the deposition of energy and is termed dose. Dose is defined as the amount of deposited radiation energy per mass unit of absorber. A deposition of 100 ergs in 1 g of material is equivalent to 1 rad. There are 100 rads in 1 gray, another common unit of dose. Dose is important, but only considers physical factors.

The property of dose equivalent is most correctly used to describe the potential for damage to an irradiated individual. The unit for dose equivalent is the rem. The number of rems is obtained by multiplying the number of rads by a "quality factor." The quality factor is based on the type of ionizing radiation delivering the dose. For β particles and γ or X-rays this factor is 1 and therefore rems and rads are equal for β radiation. In contrast, the quality factor associated with α particles is 20, so a dose of 1 rad due to α particles would be recorded as 20 rem. Thus, dose equivalent is based on both physical and biological factors. The dose equivalent is the most meaningful quantity used for radiation protection purposes and is the unit used to record dosimeter badge readings. There are 100 rems in a sievert, the unit of dose equivalent used in most countries other than the U.S.

Measuring Exposure to Ionizing Radiation

The radiation dose received by materials (cells, scientists, etc.) near a radioactive source depends not only on the specific type and energy of the radiation absorbed, but also on the subject's distance from the source, the existence of any intervening layers of attenuating material (shielding, clothing, etc.), and the length of time spent in the vicinity of the radiation source. To best measure doses to personnel, everyone working with or in close proximity to radioactive sources should wear the appropriate type of radiation dosimeter badge (in addition to using a portable radiation monitor that can give an immediate indication of the presence of radiation). This is normally a requirement (not an option) for compliance with an institution's radioactive materials license. Such badges are usually furnished by the radiation safety department, collected at regular intervals, and sent for processing to a contracted company. Most institutions use either TLDs (thermoluminescent dosimeters) or OSLDs (optically stimulated luminescent dosimeters). Both devices use crystals of calcium fluoride, lithium fluoride, or aluminum oxide, the electron structure of these crystalline materials being altered following exposure to ionizing radiation. For processing, stimulation by either heat (for TLD's) or laser light (for OSLD's) will cause the crystals to luminesce, the intensity of which is directly proportional to the dose of ionizing radiation that has been absorbed. The results are then compared with known controls to determine the actual dose equivalent. Different types of badges are sensitive to different types of radiation. Usually your Safety Office will determine what badge is best to wear while working in your particular facility. Workers should be trained to always wear their dosimeter badge on the outside of their laboratory coat, chest high, facing toward their work, as the results will read low if the badge has any layers of material (e.g., laboratory coats, jackets, and pocket material) covering it. Pregnant women are required to be educated and given the option to wear a dosimeter to better monitor the dose equivalent to their developing fetus. When working with >1 mCi of high-energy β emitters (^{32}P) or with any γ/X-ray emitting isotope (^{125}I) there is most likely an institutional license requirement for researchers to wear a ring badge to measure dose to the unshielded (though gloved!) fingers and hands (extremities). The limit for "acceptable" exposure to the extremities is ten times more than the limit for the whole body. However, the authors have found that the dose equivalent recorded with dosimeter rings is often significant as compared to the action limit for extremities set by the institution when working with multi-millicuries of ^{32}P. U.S. federal annual dose limits are as follows: 5,000 mrem for the whole body (trunk and upper extremities); 50,000 mrem for the skin, extremities, and thyroid; and 15,000 mrem for the lens of the eye.

2

What is known about the risks to humans after exposure to low levels of radiation (i.e., levels that would be received when briefly handling small amounts, μCi or mCi, of radioactivity)? Unfortunately, the nature of the problem precludes the ability of scientists to perform controlled studies to make this determination, and a complete consensus has not been reached. However, most experts think the use of a linear model extrapolated down from the quantitatively determined effects of high doses of ionizing radiation is a conservative means to determine the risk per rem. These studies make use of data obtained from studies of personnel exposures received during radiation accidents, medical treatments, occupational exposures, and the largest study group–survivors of the atomic bombs that were dropped on Hiroshima and Nagasaki during WWII. Genetic risks to subsequent generations are estimated using data from animal experiments and the families of atomic bomb survivors. Each form of extrapolation is subject to caveats, and given that predictions based on such extrapolations cannot be perfect, most health and safety personnel aim for radiation exposure levels to be ALARA or "as low as reasonably achievable." In short, scientists believe that there would not be any biological effect in an individual exposed to less than the dose limits during each year of their professional life (40 years used). An extensive discussion of both the studies and the statistics on which the federal annual dose limits are based, which is updated on a regular basis, may be found in the Biological Effects of Ionizing Radiation series (BRER, 2006; available online in an open book form at *http://www.nap.edu/books/030909156X).*

MINIMIZING EXPOSURE

Minimizing exposure to ionizing radiation can be accomplished by adjusting several parameters of the exposure: minimizing the number and duration of exposures, increasing the distance between the researcher and the source, and using appropriate shielding between the researcher and the source.

Time is of the Essence

When designing any experiment using radioactivity, every effort should be made to limit the time spent directly handling vials or tubes containing radioactive material or working in close proximity to radioactive materials. This includes minimizing the chance of error so as not to have to unnecessarily repeat an experiment. Work at a comfortable pace–not dawdling, but also not so fast as to cause spillage or error! Have everything needed for the experiment ready at hand before the radioactivity is introduced into the work area. This includes materials, equipment, and a thorough knowledge of the procedure being performed. The more time spent familiarizing yourself with your protocol, the more smoothly the work will go.

Keep Your Distance

When possible, experiments involving radioactivity should be performed in an area separate from the rest of the laboratory. Many institutions require that such work be performed in a designated "hot lab"; however, if many people in the laboratory routinely use radioisotopes, it is less than feasible to move them all into what is usually a smaller space. No matter where an individual is working, it is his or her responsibility to monitor the work area and ensure his or her own safety and the safety of those working nearby. To protect bystanders, remember that the intensity of radiation from a small source (moving through air) falls off in proportion to the square of the distance. Thus, if standing 1 foot away from a source for 5 min would result in an exposure of 45 mrem, standing 3 feet away for the same amount of time would result in an exposure $(1/3)^2$ of 45 mrem, or about 5 mrem. This factor is also relevant when considering the storage of large amounts of radioactivity, particularly ^{125}I or ^{32}P, as sometimes radiation cannot be completely attenuated.

2

Shielding: The Great Wall

When handling radioactive samples, it may be necessary to work behind shielding. When shielding is properly used, it will successfully minimize researcher (and neighbor) exposure to radioactive materials. However, when used improperly it can lead to worker fatigue, awkward movements, and a higher chance of spillage. Set up your work station in advance to make sure you are comfortable with the physical layout of your equipment and shielding. If feasible, start with small amounts of activity. After the worker has gained valuable experience and is comfortable working with these activity amounts, the transition to using shielding for higher activity amounts will be easier to make. And keep in mind—when using mCi amounts of ^{32}P, shielding will always be needed to minimize worker exposure.

As mentioned before, the energy of the particle(s) released during the decay of an isotope determines what type of shielding, if any, is appropriate. β particles released during the decay of ^{14}C and ^{35}S possess roughly ten times the energy of those released when ^{3}H decays. However, all three of these isotopes emit β particles of relatively low energy which do not travel very far in air and cannot penetrate through solid surfaces. Therefore, these isotopes are not considered an external hazard and shielding barriers are not necessary when working with them. The major health hazard from these isotopes is internal exposure, which could occur through their accidental ingestion, inhalation, absorption through the skin, or introduction through the skin by a wound.

β particles released during the decay of ^{32}P may have a 10-fold higher energy than those released from ^{14}C and may pose a very real concern to workers, especially if multi-mCi amounts of activity are being handled. (One potential biological effect is the induction of cataracts in the unshielded eye; however, the threshold dose for cataracts is on the order of 500 rem, which could only be delivered in a laboratory setting when using mCi amounts of ^{32}P with inadequate shielding over many years.) As explained previously, the fact that these high-energy β particles can potentially generate significant amounts of bremsstrahlung radiation is the reason that low-*Z* (atomic number) materials are used as the primary layer of shielding for ^{32}P β radiation. Water, glass, and plastic are suitable low-*Z* materials (as opposed to lead). Obviously water is unsuitable as a shielding layer for work on the bench, although it does a reasonable job when samples are incubating in a water bath. Shields made from a thickness of glass sufficient to stop these particles would be extremely heavy and cumbersome (as well as dangerous if dropped). Fortunately, plastic or acrylic materials variously called Plexiglas, Perspex, or Lucite are available for shielding against ^{32}P β radiation. Shields, sample storage boxes, waste container boxes, and sample racks constructed of various thicknesses of Plexiglas are necessary equipment in laboratories where ^{32}P is used. A thickness of plastic or acrylic material at least 0.64- to 1.0-cm thick is adequate for shielding up to 5 mCi of ^{32}P. To shield against bremsstrahlung radiation when using higher activity amounts of ^{32}P, it is necessary to add a layer of high-*Z* material (such as 0.38 to 2.29 mm lead) to the outside of the Plexiglas shield that is opposite the radioactive source (Klein et al., 1990).

γ/X-rays released during the decay of ^{125}I will easily penetrate the plastic materials used to shield the β particles from ^{32}P; this radiation must be reduced in intensity with a high-Z material, such as lead. Lead foil of varying thicknesses (0.76 to 1.0 mm) can be purchased in rolls and can be cut and molded to cover any container, or taped to a Plexiglas shield (used in this instance for support). Obviously, this latter arrangement has the disadvantage that it is impossible to see what one is doing through the shield. For routine shielding of manipulations involving mCi activities of ^{125}I, it is useful to purchase a lead-impregnated, transparent, Plexiglas shield (which can be very heavy as well as relatively expensive). When deciding how thick is "thick enough," consult the half-value layer (HVL) measurement for each type of shielding material and γ/X-ray energy. The HVL for ^{125}I is 0.02 mm of lead. Thus, a small thickness of lead is sufficient to shield most activity amounts of ^{125}I used in the laboratory. When using prelabeled kits, shielding is not needed, as the activity used in these kits is typically very small ($<$10 μCi).

GENERAL PRECAUTIONS

Before going on to a discussion of specific precautions to be taken with individual isotopes, a short list of general precautions to be taken with all isotopes seems pertinent:

1. **Know the rules.** Be sure that each individual is authorized to perform an approved procedure using a particular isotope and activity amount in the approved work area.
2. **Don the appropriate apparel.** Whenever working at the laboratory bench, it is good safety practice to wear a laboratory coat for protection; when using radioactivity, wearing a laboratory coat is imperative. Disposable paper/synthetic coats of various styles are commercially available: at $4 each these may be conveniently thrown out if contaminated with radioactivity during an experiment, rather than held for decay as might be preferable with cloth coats costing ~$30 each. As an alternative, disposable sleeves can be purchased and worn over the usual cloth coat. Other necessary accessories include radiation dosimeter badges, protective eyewear, shoes that cover the top of the foot, and two pairs of gloves. When one of the outer gloves becomes contaminated, it is easy to slowly peel it off, replace if necessary with another glove, and continue working with only minor interruption. Removal of contaminated gloves should always be performed over a bench or waste container so that microdroplets of contamination do not fall on the floor and get tracked about!
3. **Protect the work area as well as the workers.** Laboratory bench tops and the bases of any shields should be covered with a disposable, preferably absorbent, layered paper sheet. Blue absorbent pads ("hospital diapers") work quite well.
4. **Use appropriately designated equipment.** It is very convenient, where use justifies the expense, to have a few adjustable pipettors dedicated and labeled for use with each particular isotope. Likewise, it is good practice to use only certain labeled centrifuges and microcentrifuge rotors for radioactive samples so that all of the rotors in the laboratory do not become contaminated. Although such equipment should be cleaned after each use, complete decontamination is often not possible. A few pipettors or a single microcentrifuge can easily be stored (and used) behind appropriate shielding. Contamination of the insides and tip ends of pipettors can be greatly reduced by using tips supplied with internal aerosol barriers such as those used for PCR reactions. To prevent contamination of the outside of the pipettor barrel, simply wrap the hand-grip in Parafilm, which can be discarded later.
5. **Know where to dispose of radioactive waste, both liquid and solid.** Most institutions require that radioactive waste be segregated by isotope and physical form. This is done not only so that appropriate shielding can be placed around waste containers, but so that some waste can be allowed to decay prior to disposal through normal (nonradioactive) trash methods. With a decreasing number of radioactive waste disposal facilities able or willing to accept radioactive waste for burial (and a concomitant increase in dumping charges from those that still do), this

practice of on-site decay can save an institution thousands of dollars a year in disposal charges. With this in mind, the volume of radioactive waste present at your institution can be minimized by surveying all items before placing them into the radioactive waste. This will result in lower costs associated with the disposal of radioactive materials.

6. **Label your label!** It is only common courtesy (as well as common sense) to alert coworkers to the existence of anything and everything radioactive that is left where they may come in contact with it! A simple piece of tape affixed to the sample box with the investigator's name, the activity amount and type of isotope, and the date written on it should suffice. Yellow hazard tape printed with the international symbol for radioactivity is commercially available in a variety of widths for this purpose. Similarly, all equipment used with radioactive material should be labeled.

7. **Monitor for radioactive contamination early and often.** It is imperative that each laboratory authorized to use radioactive material have access to the use of a portable radiation survey meter. Some meters are more suited for β detection and some are more suited for γ detection. Keep the appropriate survey meter nearby. Switch it on before touching anything on your laboratory bench to avoid contaminating the switch on the meter. Always check the batteries before using a survey meter! Use a monitor with an adequate detector efficiency (β detector for ^{35}S and ^{32}P; γ detector for ^{125}I) before beginning, during, and after all procedures. The more frequently fingers, hands, relevant equipment, and your work area are monitored, the more quickly a spill or glove contamination will be detected. Timely detection will keep both the potential spill area and the cleanup time to a minimum. While it is tempting to cover the monitor's detector tube with Parafilm to protect it from contamination, remember that this will prevent the detection of low-energy β radiation from ^{35}S and ^{14}C! Because the low-energy β emitter ^{3}H cannot be detected at all using these monitors, obtaining wipe samples of the bench and equipment and subsequently counting them with a liquid scintillation counter (LSC) is necessary to ensure that contamination of the work area did not occur.

8. **Clean up contamination as soon as possible after discovery!** If contamination is discovered, it is to everyone's (yours and your neighbors') benefit that it be cleaned up as soon as possible. This will prevent the inadvertent spread of contamination to other areas (and people) in the laboratory. To wait over a weekend until Monday, or even one evening, to clean up contamination can invite disaster. A relatively small spill can turn into a large mess if tracked about. All things considered, it is imperative to take the time to clean up a spill immediately.

SPECIFIC PRECAUTIONS

The following sections describe precautions to be taken when working with individual isotopes in specific forms. Although the sections dealing with ^{35}S- or ^{32}P-labeling of proteins in intact cells are presented in terms of mammalian cells, most of the instructions are also pertinent (with minimal and obvious modifications) to the labeling of proteins in other cells (e.g., bacterial, insect, plant, etc.).

Working with ^{3}H

Tritiated compounds used in the biochemical laboratory include ^{3}H-thymidine (used in cell proliferation assays) and ^{3}H-amino acids (used to label newly synthesized proteins). As discussed above, the β radiation resultant from the decay of tritium is of such weak energy that no type of shielding is necessary to protect the scientist during the experiment. In fact, these β particles cannot even be detected using a typical hand-held Geiger-Müller (GM) monitor. Therefore, to avoid accidental ingestion or absorption through unintentional/unrealized contact, it is imperative that the researcher perform wipe tests of the experimental area and equipment used, to determine if any contamination exists. These wipe tests most often consist of both random and specific (most-likely candidate) swipes of surfaces with a paper filter that is subsequently counted, with fluor, in a liquid scintillation counter to determine if any ^{3}H is/was present.

Working with ^{14}C/^{35}S

As discussed above, the β radiation generated during ^{14}C and ^{35}S decay is not strong enough to make additional shielding necessary. The risk associated with both these isotopes then comes primarily through their ingestion and subsequent concentration in various target organs, depending on the compound to which the radioisotope is attached. Although willful ingestion of either seems unlikely, accidental or unknowing ingestion may occur. Like ^{3}H, ^{14}C is used to label amino acids and various reagents used in assays, as well as molecular weight standards for protein gels. The half-life of ^{14}C is 5730 years, so don't spill this!

2

Working with ^{35}S

Using ^{35}S to label cellular proteins and proteins translated in vitro

As reported several years ago (Meisenhelder and Hunter, 1988), ^{35}S-labeled methionine and cysteine, which are routinely used to label proteins synthesized in intact cells and by in vitro translation, break down chemically to generate a volatile radioactive component. Because this breakdown occurs independently of cellular metabolism, the radioactive component is generated to the same extent in stock vials as in cell culture dishes. The process seems to be promoted by freezing and thawing ^{35}S-labeled materials. The exact identity of this component is not known, although it is probably SO_2 or CH_3SH. What is known is that it dissolves readily in water and is absorbed by activated charcoal or copper.

The amount of this volatile radioactive component released, despite stabilizers added by the manufacturers, is about 1/8000 of the total radioactivity present. The amount of this radioactivity that a scientist is likely to inhale while using these compounds is presumably even smaller. Nevertheless, such a component can potentially contaminate a wide area because of its volatility, and would tend to concentrate in target organs. Thus, it is advisable to thaw vials of ^{35}S-labeled amino acids in a controlled area such as a mini-hood equipped with a charcoal filter. This charcoal filter will become quite contaminated and should be changed every few months. If such an area is not available, the stock vial should be thawed using a needle attached to a charcoal-packed syringe to vent and trap the volatile compound.

Anyone who has ever added ^{35}S-labeled amino acids to dishes of cells for even short periods knows that the incubator(s) used for such labeling may quickly become highly contaminated with ^{35}S. Such contamination is not limited to the dish itself, nor to the shelf on which the dish was placed. Rather, the radioactive component's solubility in water allows it to circulate throughout the moist atmosphere of the incubator and contaminate all of the inside surfaces of the incubator. For this reason, in laboratories where such metabolic labeling is routine, it is highly convenient to designate one incubator to be used solely for working with ^{35}S-labeled samples. Such an incubator can be fitted with a large honeycomb-style filter the size of the incubator shelf, made of pressed, activated charcoal. These filters are available from local air-quality-control companies. Such a filter will quickly become contaminated with radioactivity and should therefore be monitored and changed as necessary (e.g., every three months if the incubator is used several times a week). The water used to humidify the incubator will also become quite "hot" (contaminated with radioactivity); keeping the water in a shallow glass pan on the bottom of the incubator makes it easy to change after every use, thus preventing contamination from accumulating. Even with the charcoal filter and water as absorbents, the shelves, fan, and inner glass door of the incubator will become contaminated, as will the tray on which the cells are carried and incubated. Routine wipe tests and cleaning when necessary will help to minimize potential spread of this contamination.

If such work is done infrequently or there is not a "spare" incubator, dishes of cells can be placed in a box during incubation. This box should be made of plastic, which is generally more easily decontaminated than metal. Along with the dishes of cells, a small sachet made of activated charcoal wrapped loosely in tissue (Kimwipes work well) should be placed in the box. If the box is sealed, it will obviously need to be gassed with the correct mixture of CO_2; otherwise, small holes can be incorporated into the box design to allow equilibration with the incubator's atmosphere. In either case, the incubator used for the labeling should be carefully monitored for radioactivity after each experiment.

Working with ^{32}P

μCi amounts of ^{32}P

The amount of ^{32}P-labeled nucleotide used to label nucleic acid probes for northern or Southern blotting is typically under 250 μCi, and the amount of [γ ^{32} P]ATP used for in vitro phosphorylation of proteins does not usually exceed 50 μCi for a single kinase reaction (or several hundred μCi per experiment). However, handling even these small amounts, given the time spent on such experiments, can result in a measurable dose equivalent if proper shielding is not used. With no intervening shielding, the dose rate 1 cm away from 1 mCi ^{32}P is 200,000 mrads/hr; the local dose rate to basal cells resulting from a skin contamination of 1 μCi/cm^2 is 9200 mrads/hr (Shleien, 1987). Such a skin contamination could be easily obtained though careless pipetting and the resultant creation of an aerosol of radioactive microdroplets, because the concentration of a typical stock solution of labeled nucleotide may be 10 μCi/μl.

For proper protection during these types of experiments, besides the usual personal attire (glasses, gloves, coat, closed-toe shoes, and ring and lapel dosimeter badges) it is necessary to use some form of Plexiglas shield (see Fig. 2.3.1A) between the body and the samples. Check the level of radiation

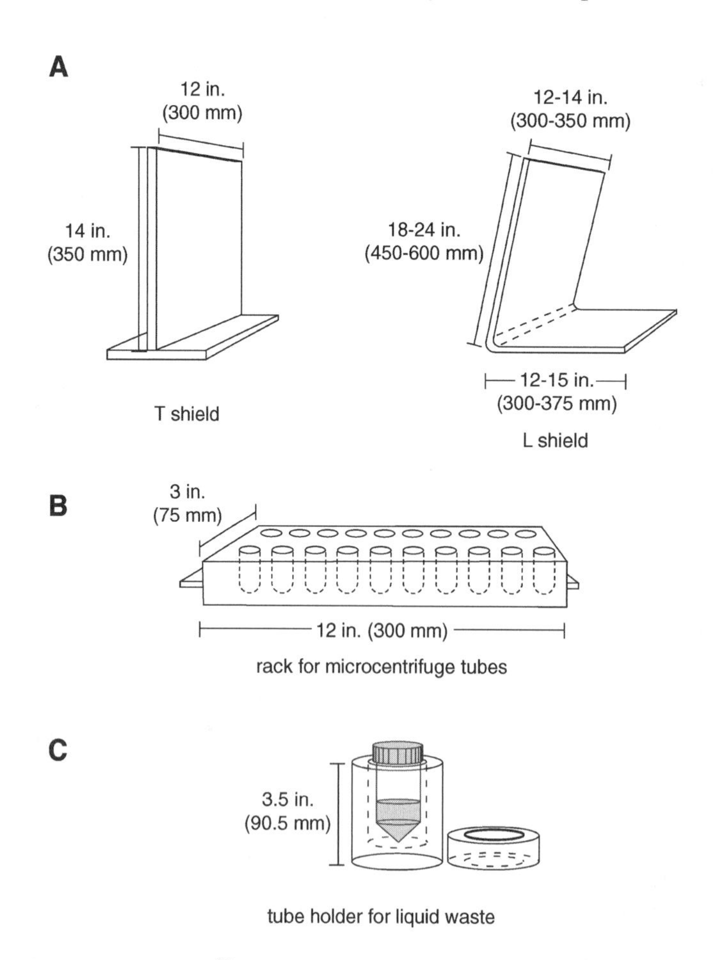

Figure 2.3.1 Plexiglas shielding for ^{32}P. (**A**) Two portable shields (L and T design) made of 0.5-in. (12.5-mm) Plexiglas. Either can be used to directly shield the scientist from the radioactivity being used. Turned on its side, the L-shaped shield can be used to construct two sides of a cage around a temporary work area, providing shielding for workers directly across from or to the sides of the person working with ^{32}P. (**B**) Tube rack for samples in microcentrifuge tubes. (**C**) Tube holder for liquid waste collection.

coming through the outside of the shield with a portable monitor to ensure that the thickness of the Plexiglas is adequate. Hands can be shielded from some exposure by placing the sample tubes in a solid Plexiglas rack, (see Fig. 2.3.1B) which is also useful for transporting samples from the bench to a centrifuge or water bath.

These experiments often include an incubation step performed at a specific temperature, usually in a water bath. Although the water surrounding the tubes or hybridization bags will effectively stop β radiation, shielding should be added over the top of the tubes where there is no water (e.g., using a simple flat piece of Plexiglas). If the frequency of usage justifies the expense, an entire lid for the water bath can be constructed from Plexiglas. When hybridization reactions are performed in bags, care should be taken to monitor (and shield) the apparatus used to heat-seal the bags. It is also important to ensure that the water in the bath does not become contaminated by leakage from the hybridization bags.

The waste generated during the experiments should also be shielded. It is convenient to have a temporary, satellite waste container right on the bench. Discard pipet tips and other solid waste into a Plexiglas box lined with a plastic bag and placed behind the shield. This bag can then be emptied into the appropriate shielded laboratory waste container when the experiment is done. Liquid waste can be pipetted into a disposable tube set in a stable rack behind the shield (see Fig. 2.3.1C).

When radiolabeled probes or proteins must be gel-purified, it may be necessary to shield the gel apparatus during electrophoresis if the samples are particularly hot. Be advised that the electrophoresis buffer is likely to become very radioactive if the unincorporated label is allowed to run off the bottom of the gel; check with radiation safety personnel for instructions on how to dispose of such buffer. It is also prudent to check the gel plates with a radiation survey meter after the electrophoresis is completed since they can become contaminated as well.

mCi amounts of ^{32}P

In order to study protein phosphorylation in intact mammalian cells, cells in tissue culture dishes are incubated in phosphate-free medium with ^{32}P-labeled orthophosphate for a period of several hours or overnight to label the proteins. The amount of ^{32}P used in such labels can be substantial. Be sure to calculate and pipet the actual activity, rather than just partitioning the label into equal volume aliquots. Cells are normally incubated in 1 to 2 mCi of ^{32}P/ml labeling medium; for each 6-cm dish of cells, 2.5 to 5 mCi ^{32}P may be used. When this figure is multiplied by the number of dishes necessary per sample, and the number of different samples in each experiment, it is clear that the amount of ^{32}P used in one experiment can easily reach 25 mCi or more. Because so much radioactivity is used in the initial labeling phase of such experiments, it is necessary for a researcher to take extra precautions in order to adequately shield him or herself and coworkers.

When adding ^{32}P label to dishes of cells, it is important to work as rapidly and as smoothly as possible. An important contribution to the speed of these manipulations is to have everything that will be needed at hand before even introducing the label into the work area. Prepare the work area in advance, arranging shielding and covering the bench with blue diapers. Set out all necessary items, including pipettors and tips needed, a portable detection monitor, extra gloves, and a cell house (see Fig. 2.3.2A).

Research using this much radioactivity should be done behind a Plexiglas shield at least 1 cm thick; the addition of a layer of lead (at least 2.3 mm) to the outside lower section of this shield to stop bremsstrahlung radiation is also needed. If one shield can be dedicated to this purpose at a specific location, a sheet of lead can be permanently screwed to the Plexiglas (as shown in Fig. 2.3.2B) However, this lead makes the shield extremely heavy and therefore less than portable. If space constraints do not permit the existence of such a permanent labeling station, a layer or two of thick lead foil can be taped temporarily to the outside of the Plexiglas shield.

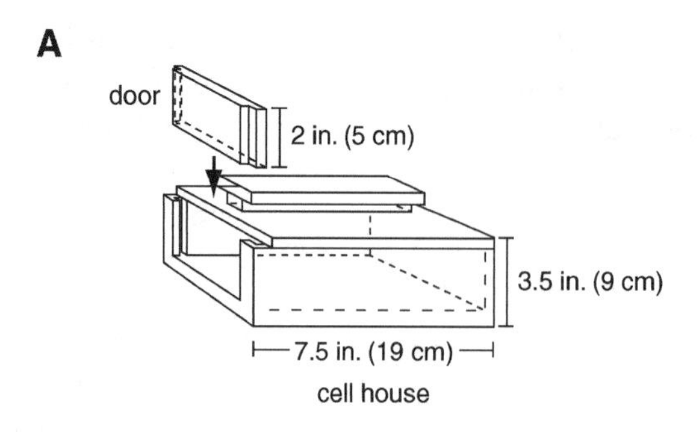

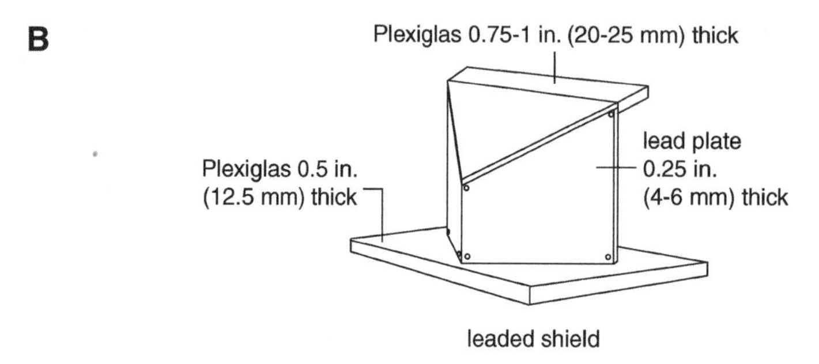

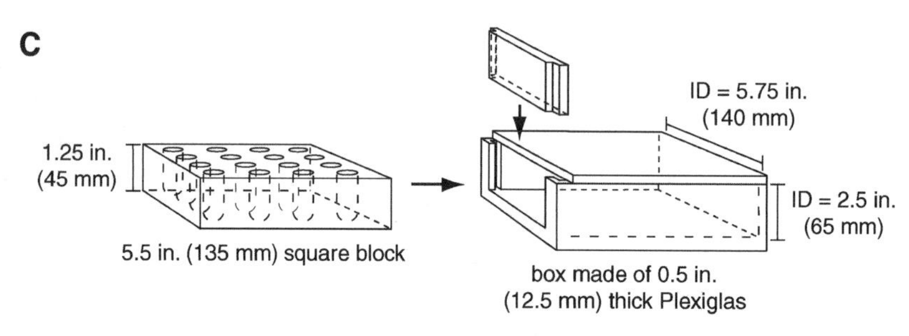

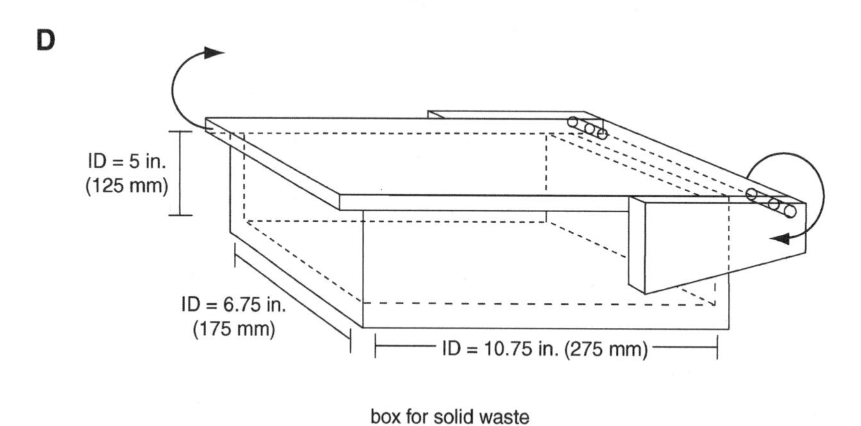

Figure 2.3.2 (**A**) Box for cell incubation (a "cell house"). (**B**) Stationary leaded shield. (**C**) Sample storage rack and box made of 0.5-in. Plexiglas. (**D**) Box for solid waste collection made of 0.5-in. Plexiglas. ID, interior dimension.

Again, each worker should take care to shield not only him or herself, but also bystanders on all sides. Handling of label should be done away from the central laboratory, if possible, to take maximum advantage of distance as an additional means of dose reduction. It is also advisable not to perform such experiments in a tissue culture room or any other room that is designed for a purpose vital to the whole laboratory. An accident involving this much ^{32}P would seriously inconvenience future work in the area, if not make it altogether uninhabitable! If care is taken to minimize the amount of time the dish of cells is open when adding the label, use of a controlled air hood to prevent fungal or bacterial contamination of the cells should not be necessary.

In the course of doing experiments to determine which hand receives the most exposure during such cell labeling procedures, extremity exposure was shown to vary as much as ten-fold depending on which finger the dosimeter ring was worn on, with the index finger of the left hand receiving the most exposure for a right-handed person (Bursik et al., 1999). As would be expected, the most exposure is received as the worker adds label to the dishes of cells and as the cells are lysed (see below). In order to mitigate this extremity exposure, Plexiglas dish covers (Fig. 2.3.3) can be used to shield each individual dish: the tissue culture dish fits snugly into the bottom Plexiglas piece while the top Plexiglas piece is joined to the top of the tissue culture dish using tape so that the two lids together can be handled as one unit. Tissue culture dishes of cells are fitted/taped into the Plexiglas dish covers immediately before adding the ^{32}P. As the top and bottom pieces of the dish covers do not form a seal, the medium can equilibrate with the CO_2 of the incubator for proper pH adjustment. Use of such dish covers reduces extremity exposure by 8- to 10-fold, despite the stream of radiation that passes through the small crack between the top and bottom.

Once the label has been added to the dishes of cells (and whether or not one is using the dish covers discussed above), they will also need to be shielded for transport to and from the incubator and other work areas. Plexiglas boxes that are open at one end (for insertion of the dishes) and have a handle on top (for safe carrying) make ideal "cell houses" (see Fig. 2.3.2A). A Plexiglas door that slides into grooves at the open end is important to prevent dishes from sliding out if the box is tilted at all during transport. If this door is only two-thirds the height of the house wall, the open slot thus created will allow equilibration of the CO_2 level within the house with that in the incubator. Obviously, this slot will also allow a substantial stream of radiation to pass out of the cell house, so the house should be carried and placed in the incubator with its door facing away from the worker (and others)! Use of the Plexiglas dish covers adds considerable bulk to the dishes of cells, and larger cell houses designed with handles on their sides and a hinged lid are more easily handled (see Fig. 2.3.3).

Following incubation with label and any treatments or other experimental manipulations, the cells are usually lysed in some type of detergent buffer. It is during this lysis procedure that a worker's

Figure 2.3.3 Use of Plexiglas dish shields for ^{32}P reduces extremity exposure.

hands will receive their greatest exposure to radiation, because it is necessary to handle pipettors directly over open cell dishes for a period of several minutes. It is therefore very important to streamline this procedure and use shielding whenever possible. If the cell lysates must be made at 4°C, as required by most protocols, working on a bench in a cold room is preferable to placing the dishes on a slippery bed of ice. In either case, make the lysate using the same sort of shielding (with lead if necessary) that was used when initially adding the label. Using disposable transfer pipets, remove the labeling medium and any solution used to rinse unincorporated radioactivity from the cells into a small tube held in a solid Plexiglas holder (shown in Fig. 2.3.1C). The contents of this tube can later be poured into the appropriate liquid waste receptacle. If possible, it is a good practice to keep this high-specific-activity ^{32}P liquid waste separate from the lower-activity waste generated in other procedures so that it can be removed from the laboratory in a specially designed shielded box as soon as possible following the experiment. The Radiation Safety Officer should be asked to remove this high-activity waste as soon as possible. If it is necessary to store it in the laboratory for any time, the shielding for the waste container should also include a layer of lead.

The solid waste generated in the lysis part of these experiments (pipet tips, disposable pipets, cell scrapers, and dishes) is very hot and should be placed immediately into a shielded container to avoid further exposure to the hands. A Plexiglas box similar in design to that in Figure 2.3.2D is convenient; placed to the side of the shield and lined with a plastic bag, it will safely hold all radioactive waste during the experiment, and is light enough to be easily carried to the main laboratory waste container, where the plastic bag (and its contents) can be dumped after the experiment is completed. If the lid of the box protrudes an inch or so over the front wall, it can be lifted using the back of a hand, thus decreasing the possibility of spreading contamination with hot gloves.

When scraping the cell lysates from the dishes, it is good practice to add them to microcentrifuge tubes that are shielded in a solid Plexiglas rack; this will help to further reduce the exposure to which the hands are subjected. At this point, the lysates are usually centrifuged at high speed (10,000 × *g*) to clear them of unsolubilized cell material. Use screw-cap tubes for this clarification step, as these will contain the labeled lysate more securely than flip-top tubes, which may open during centrifugation. No matter what type of tube is used, the rotor of the centrifuge often becomes contaminated, most probably due to tiny drops of lysate (aerosol) initially present on the rim of the tubes that are spun off during centrifugation. It is important to monitor and clean out the centrifuge after each use so contamination does not accumulate.

The amount of ^{32}P taken up by cells during the incubation period varies considerably, depending on the growth state of the culture as well as on the cell type and its sensitivity to radiation. This makes it difficult to predict the percentage of the radioactivity initially added to the cells that is incorporated into the cell lysate; however, this figure probably does not exceed 10%. Thus, the amount of radioactivity being handled decreases dramatically after lysis, making effective shielding much simpler. However, at least ten times more radioactivity is still involved compared to other sorts of experiments! It is easy to determine if the shielding is adequate–just use both β and γ survey meters to measure the radiation passing through the shielding. As a rule of thumb, if the "pancake" GM meter reads more than 5000 cpm, additional shielding is needed. Again, be sure that people working nearby (including those across the bench) are also adequately shielded. It is sometimes necessary to construct a sort of cage of Plexiglas shields around the ice bucket that contains the lysates.

At the end of the day or the experiment, it may be necessary to store radioactive samples; in some experiments, it may be desirable to save the cell lysates. These very hot samples are best stored in tubes placed in solid Plexiglas racks that can then be put into Plexiglas boxes (see Fig. 2.3.2C). Such boxes may be of similar construction to the cell houses described above, but they should have a door that completely covers the opening. Be sure to check for γ radiation coming through these layers and add lead outside the box if necessary.

Working with ^{33}P

Using ^{33}P-labeled nucleotides to label nucleic acid probes or proteins

Several of the major companies that manufacture radiolabeled biological molecules also sell nucleotides labeled with ^{33}P (both α- and γ-structural forms). ^{33}P offers a clear advantage over ^{32}P with respect to ease of handling, because the maximum energy of the emitted β radiation is between that of ^{35}S and ^{32}P and does not require the shielding needed for ^{32}P. In fact, the β radiation emitted (E_{max}= 0.248 MeV) can barely penetrate through two pairs of gloves and the outer dead layer of skin, so the external exposure hazard associated with even millicurie amounts of ^{33}P is minimal (as reported in the DuPont NEN product brochure). Gel bands visualized on autoradiographs of ^{33}P-labeled compounds are sharper than bands labeled with ^{32}P because the lower-energy β radiation does not have the scatter associated with the higher-energy β radiation emitted by ^{32}P. The half-life of ^{33}P is also longer (25 days compared to 14 days for ^{32}P). Despite its higher cost, these features have led some researchers to choose ^{33}P-labeled nucleotides for use in experiments such as band/gel shift assays where discrimination of closely-spaced gel bands is important.

As alluded to above, additional shielding when using this isotope will probably not be needed. The best way to make this determination is to monitor the source using a β-sensitive radiation survey meter. If counts can be detected, add a layer of Plexiglas as described previously for ^{32}P.

Working with ^{125}I

Using ^{125}I to detect immune complexes (immunoblots)

^{125}I that is covalently attached to a molecule such as staphylococcal protein A is not volatile and therefore is much less hazardous than the unbound or free form. Most institutions do not insist that work with bound ^{125}I be performed in a hood, but shielding of the γ radiation may still be necessary. Lead is a good high-*Z* material used to shield these γ rays; its drawback is its opacity. Commercially available shields for ^{125}I are made of lead-loaded Plexiglas; although heavy, these have the advantage of being transparent. Alternatively, a piece of lead foil may be taped to a structural support, although this arrangement does not provide shielding for the head as a worker peers over the lead!

Incubations of the membrane or blot with the [^{125}I] protein A solution and subsequent washes are usually done on a shaker. For shielding during these steps, a piece of lead foil may simply be wrapped around the container. Solutions of ^{125}I can be conveniently stored for repeated use in a rack placed in a lead box.

Using ^{125}I to label proteins or peptides in vitro

Any experiments that call for the use of free, unbound ^{125}I should be done behind a shield in a hood that exhausts the air through a charcoal filter (which absorbs the volatile iodine). Most institutions require that such experiments be done in a special hot laboratory to which access is limited. Since ingested, absorbed, or inhaled iodine is concentrated in the thyroid, a portable γ monitor should be used to scan the thyroid (neck area) at least 24 hr after completing each experiment. This procedure is called a bioassay and is a requirement of the institution's radioactive materials license. In the experience of the authors, another very common means of incurring an internal deposit of ^{125}I during this procedure is by the spread of surface contamination with subsequent ingestion or absorption through the skin.

MEASURING RADIOACTIVITY

There are two main purposes for counting radioactive materials in the laboratory. The first is to detect the presence and quantity of radioactive materials present in the laboratory as surface contamination or as an inadequately shielded source of radiation for the purposes of safety and regulatory compliance. The second is to determine the presence and/or quantity of the amount of radioactive label for the purpose of experimental analysis.

Safety and Regulatory Compliance

Regulations governing the use of radioactive materials in the laboratory address the concerns associated with exposure to radiation present as surface contamination or as an external radiation field (i.e., exposure at a distance, typically from an inadequately shielded radioactive source). Surface contamination is simply the unintentional presence of radioactive material and occurs by the transfer of actual radioactive material from one surface to another. To determine the presence of external surface contamination, two different methodologies are available to the researcher: the use of hand-held (portable) detectors and/or liquid scintillation counters.

Hand-held detectors

The first method is to use a portable survey meter, i.e., either a meter with a Geiger-Muller (GM) detector or a portable scintillation detector. The choice depends upon what isotope is being used. GM instruments can be used for detecting the presence of the low-energy β emitters ^{14}C, ^{35}S, and ^{33}P with an efficiency of detection being about 3% to 5%. For detecting the high-energy β emitter ^{32}P, the efficiency of a GM instrument is ~30%. As stated previously, a layer of Parafilm, commonly used to prevent contamination of the GM tube in some laboratories, actually prevents the meter from detecting the low-energy β emitters ^{14}C, ^{35}S, and ^{33}P, and must be removed prior to use.

The GM detector can also be used to determine the presence of ^{125}I contamination, although with much less efficiency (<1%). Because of this fact, if very small activity amounts of ^{125}I are being used (such as packaged in prelabeled kits with <10 μCi), it is best to use a portable scintillation detector, the efficiency of which is ~20% for ^{125}I.

Keep in mind that ^{3}H (tritium) β radiation has minimal penetration power and cannot be detected using either a GM or scintillation detector. The liquid scintillation counter (see below) is the only readily available means to count ^{3}H, either present on a wipe sample or as an analytical sample.

Unless the meter survey is performed properly, one can actually overlook contamination by either moving the meter probe too fast and/or by holding the probe out too far above the surface being monitored. This is especially true for the low-energy β emitters, ^{14}C, ^{35}S, and ^{33}P. The survey meter probe (GM or scintillation detector) should be moved slowly, about 1 to 2 cm/sec, and about 1 cm above the surface of concern. Any response of the meter higher than background must be considered contamination and cleaned up as such. The advantage of a GM survey is the greater area one can cover when doing the survey. The downside, though, is that low-level contamination can be overlooked.

GM detectors: How they operate

A GM detector works on the principle of the Townsend avalanche. A Townsend avalanche occurs when a single gas molecule inside a high-voltage electric field becomes ionized, i.e., loses an electron by the passage of ionizing radiation through the gas inside the GM tube. This single electron then acquires enough kinetic energy traveling across the electric potential to initiate subsequent ionizations in the gas, i.e., a cascade, throughout the gas volume. This creates a large migration of free electrons to a positively charged anode, which are then used to generate an electrical pulse in the electrical circuit. The avalanche in the tube is usually stopped by including a quench gas in the tube, which absorbs enough of the free electron energy to terminate the discharge. [This is a simplified explanation; there is more, complicated physics involved that is not presented here (Knoll, 1979).] In a GM tube, the voltage in the tube can be 600 to 2000 V, and is supplied by batteries and a voltage regulator in the detector body. The output of the device is read on a dial, either with or without a speaker output. The GM detector tube itself has a very thin window over one end to allow the passage of ionizing radiation into the tube to initiate the avalanche. The tube will implode if the window is accidentally punctured, because the pressure inside the tube is lower than atmospheric pressure.

In essence, a single ionizing event inside the GM detector is amplified so it can be detected. As stated previously, the GM counter is not 100% efficient at detecting all isotopes. This is because a threshold β energy is required to penetrate the thin window. The β energy from ^{3}H is below this energy threshold.

Liquid scintillation counters

The second method used for determining the presence of radioactive surface contamination is a wipe test with subsequent counting in a liquid scintillation counter. A dry Whatman filter paper (no. 2 works well) is used to wipe a flat surface, using moderate pressure, in the shape of a large "S" ~6 in. tall. The wipe is then put into a liquid scintillation vial, scintillation fluid (cocktail) is added, and the sample counted in a scintillation counter. Any counts greater than background will be indicated on the output of the counter. This method, while not in "real time," has the advantage of quantifying the contamination present on the wipe sample, which is not easy to do with a GM survey meter. Except for tritium, isotopes counted with this method can be detected with >90% efficiency. (Counts can be lost due to an effect termed quenching, which is a loss of low-energy events resulting in a shifting of the entire energy spectrum to lower energies. This results in a lower counting efficiency as compared to unquenched samples). Also, this method can detect **any** isotope used in the laboratory, and can be employed to find contamination in microcentrifuges and other tight spots where a survey meter cannot fit. However, the down side is that unless one wipes the surface pressing directly on the contamination, the contamination will be overlooked and a false negative can occur.

The major advantage of a liquid scintillation counter (LSC) over other radiation detection equipment in the laboratory is its ability to detect and count low-energy β radiation with good efficiency. As stated earlier, it is the only instrument that can detect tritium (^{3}H). Another advantage is its ability to discriminate between high- and low-energy β radiation. Indeed, samples that contain more than one isotope can be counted and the proportion of each separate isotope determined. The device is quite complex, and its operation depends on many principles of electronics, electromechanics, and physics.

Vital to liquid scintillation counting is the use of a scintillation fluid. A scintillation fluid is a specially formulated liquid comprised of two kinds of molecules: a great many solvent molecules and, in much lower quantity, scintillant molecules. Together these molecules serve the purpose of transforming the energy of the emitted radiation inside a sample to an energy level that falls in the sensitivity range of a photomultiplier tube (PMT). The radiation energy in the sample is first transferred to the solvent, which has an energy level above that of the scintillant and the PMT. The energy is then transferred to the scintillant, which is then stimulated to a higher energy level. As the scintillant returns to its ground state, it releases energy or light that lies within the sensitivity of the PMT. The light enters the PMT and initiates a cascade down a string of dynodes, where the PMT converts a relatively small light signal at one end into a large light pulse at the other. This pulse is then processed electrically and counted.

The fact that the sample is in intimate contact with the fluid, usually dissolved or in solution, accounts for the greater sensitivity of this device over that of the GM detector. The important fact is that higher β energies in the sample will stimulate greater numbers of solvent molecules in the first place, leading to a larger number of stimulated scintillant molecules. This in turn results in a more intense event at the start of the dynode string and a greater pulse at the other end. This explains how these detectors can discriminate β radiation of different energies.

In terms of liquid scintillation cocktails (fluors), keep in mind that some chemicals, notably solvents historically used in LSC cocktail formulations, are very difficult and expensive to dispose of as radioactive waste, as some are classified as "mixed wastes." It is best to try to avoid these solvents. Some institutions have restrictions on their use, so the Radiation Safety Officer should be contacted for more direction. The best cocktails to use are environmentally friendly cocktails.

External Radiation Fields: Exposure at a Distance

Radiation present in the form of an external radiation field (exposure at a distance) is also possible in the laboratory. This occurs most often when inadequately shielded mCi activities of ^{32}P are present. In this scenario, high local dose rates are possible at near distances and, because of the long range of ^{32}P β radiation in air, dose rates even a few feet from the source can be substantial. This radiation field **cannot** be measured directly with a GM type of survey instrument, although a rough estimate of the dose rate can be made. Using the very common, large-area "pancake" type of GM probe, a meter response to ^{32}P of 2000 to 3000 cpm indicates an approximate dose rate of 1 mrem per hr. This is a good level to aim for when determining the adequacy of any added shielding used to minimize the dose rate from mCi amount ^{32}P sources. Again, this is only a rough estimate, and Radiation Safety Department personnel should be contacted to provide more accurate determinations of external dose rates. Ion chamber type instruments are available that can measure actual dose or exposure rates more closely than do GM survey instruments.

Obviously, the best way to perform a complete laboratory survey is to use a combination of the two types of surveys (direct meter survey and wipe test survey) in such a way as to increase the probability of finding surface contamination and/or any inadequately shielded radioactive sources. One can never do too many surveys, and the more experience one gains, the more successful the researcher will become at locating and cleaning radioactive contamination in the laboratory.

Experimental Analysis

Due to technological advances with imaging machines and computer software, there is now an increasing number of ways to determine the amount of radioactivity present in experimental samples. However, if one wants to know the actual activity (dpm or cpm) of an isotope present, the most straightforward way is still to use a scintillation counter to count the material. While this is often the final stage of an experiment, in that the samples have been prepared with this quantitation being the goal, sometimes this measurement is necessary at several points in the experiment to monitor recovery of the sample. Since low-energy β emitters (^{3}H, ^{14}C, or ^{35}S) require the use of a scintillation fluor, the portion of sample that is counted is lost in terms of further analysis. While liquid scintillation counting is more accurate, ^{32}P samples offer the option of using Cerenkov counting, which preserves the entire sample for future types of analysis. Cerenkov counting requires a β E_{max} of at least 0.7 MeV, so ^{32}P samples can be counted with good efficiency using this method. For Cerenkov counting, the sample is simply put into an empty scintillation vial and counted; the counts appearing in the wide window (^{3}H plus ^{14}C plus ^{32}P) are proportional to the amount of ^{32}P present. Keep in mind that Cerenkov counting of a dry sample (a gel band or protein pellet, for example) cannot be directly compared with that of a sample counted with liquid scintillation counting because the liquid in the sample will act to quench some of the radioactivity. For ^{125}I samples a γ scintillation counter is used; samples can either be dry or in liquid form, and no fluor is necessary. Liquid scintillation can also be used to count ^{125}I samples.

The advent of phosphor screen technology has enabled researchers to both image and quantitate gels, blots, or plates by exposing them to a screen and then enabling the instrument to read the screen. The screens can detect ^{35}S, ^{32}P, and ^{125}I with roughly 5- to 10-fold higher sensitivity than film (depending on the isotope) and with a detection range of sample variability over 5 orders of magnitude (as opposed to 2 orders for film). The scientist then uses computer software to select areas of the resultant image and "extract" the amount of radioactivity therein. The numbers thus generated are of arbitrary units, but can be used for comparative purposes, the caveats being that the extracted areas must be from the same screen, which itself must be undamaged so that its sensitivity is uniform. If standards of known radioactive content are used to expose the screen alongside the sample, the arbitrary units generated can then be translated into real cpm. Several companies now make such instruments, with Fuji, Bio-Rad, and Molecular Dynamics (now GE Healthcare) being among the best known. The cost of these instruments is still very high; for this reason many institutions purchase one for use in a core facility. The ability to "cut and count" on a computer rather than on real samples is highly convenient, and the larger range of sensitivity of the phosphor screen eliminates the need for multiple exposures, as would be necessary when using film and a densitometer to quantify the signal present.

RESPONDING TO SPILLS

Despite the best intentions and utmost caution, accidents happen! Accidents involving spills of radioactive materials are particularly insidious because they can be virtually undetectable if a monitor is not present and turned on. For this reason it is best to foster a community spirit in any laboratory where radioisotopes are routinely used—specifically, a sense of cooperation that extends from shielding with others in mind to helping each other clean up after accidents occur.

The specific measures to be taken following a spill of radioactive materials naturally depend on the type and activity amount of isotope involved, associated chemical or biological hazards, and the physical parameters of the spill (i.e., where and onto what the isotope was spilled). However, there are several immediate steps that should be taken following any spill.

1. **Alert coworkers that there has been a spill**. This will give them the opportunity to protect themselves if need be and to help prepare to clean up as well. Notify the Radiation Safety Office.
2. **Restrict movement to and away from the site of the spill.** This ensures that radioactive contamination is not spread around the laboratory. It is especially important to address those individuals who may have come into contact with the radioactive materials.
3. **Perform a meter survey on any individuals** who may have come in contact with the radioactive material, including their exposed skin, protective clothing, and street clothing. If someone's skin is contaminated, first use a portable monitor to identify the specific areas of contamination. Then put that part of the body under room temperature running water in a sink. Wash the affected area with gentle soap and a soft sponge or washcloth. Try to restrict cleaning to the contaminated area only, so as not to spread the contamination to other parts of the body. Dry the area, survey, and repeat as necessary or as directed by Radiation Safety personnel. A shower room may even be needed. Contaminated clothing can be carefully removed, placed in a plastic bag, and given to Radiation Safety personnel. Contaminated strands of hair can be washed or cut (a new hairstyle may even be in order).
4. **Perform an area survey.** Use an appropriate survey meter and work in from the supposed outer limits of the spill towards the center. With a nonpermanent marker, outline the actual hot spots where counts are detected. Do not neglect to survey the sides of walls, cabinets, and equipment close to the spill area, as radioactive materials may have splashed up onto these surfaces.
5. **Clean the area.** When attempting to clean any contaminated equipment, floors, benches, etc., begin by soaking up any visible radioactive liquid with paper towels and promptly disposing of the towels in a plastic bag taped to the side of the laboratory bench. Apply a small amount of decontamination solution to the marked "hot spots" and let it set for a few minutes. Then, using three to four paper towels, wipe an area from the outer edge to the center of the spill with one swipe of the towels. Then immediately dispose of the paper towels. Repeat this movement with new paper towels each time, working your way around the spill. Do not reuse any paper towels. This method will minimize the chance of spreading contamination to an even greater area. Continue this procedure until all the marked areas have been cleaned.
6. **Perform another radiation survey.** Survey the entire area to make sure contamination has not been overlooked.

There is quite a range of commercially available foams and sprays made specifically to clean radioactive contamination. A dilute solution of phosphoric acid works well to pick up ^{32}P. Decontamination of centrifuge rotors can be tricky, as their anodized surfaces are sensitive to many detergents; check with the rotor manufacturer for appropriate cleansing solutions. Many surfaces (particularly metals) prove resistant to even herculean cleaning efforts; in these instances the best that can be done is to remove all contamination possible and then shield whatever remains until the radioactivity decays to minimal levels.

CONCLUSION

When working with radioisotopes, it is best to plan ahead and then plan ahead some more. The more thoroughly familiar a researcher becomes with all aspects of their work–by being proactive and addressing all questions and concerns beforehand (the science, the equipment, and the technique)–the more successful researchers will be at providing for their own safety and the safety of those working around them.

ACKNOWLEDGEMENTS

Many of the procedures and precautions described here have evolved (and are evolving still) over the years and through the millicuries in the authors' department (currently Molecular and Cell Biology) at the Salk Institute. The authors are indebted to those from whom they have learned about the safe use of radioactivity in the laboratory. Most of the designs for the shields and other safety equipment shown in the figures were created at the Salk Institute in collaboration with Dave Clarkin, Mario Tengco, and Steve Berry. Safety equipment of similar design is available from several commercial vendors, including CBS Scientific and Research Products International.

LITERATURE CITED

Board on Radiation Effects Research (BRER). 2006. Health risks from exposure to low levels of ionizing radiation: BEIR VII Phase 2. BRER, National Research Council, The National Academies Press, Washington, D.C. Available online at *http://www.nap.edu/books/030909156X/html.*

Bursik, S., Meisenhelder, J., and Spahn, G. 1999. Characterization and minimization of extremity doses during ^{32}P metabolic cell labeling. *Health Phys.* 77:595-600.

Klein, R., Reginatto, M., Party, E., and Gershey, E. 1990. Practical radiation shielding for biomedical research. *Radiat. Prot. Manage.* 7:30-37.

Knoll, G. 1979. Radiation Detection and Measurement. John Wiley & Sons, New York.

Lederer, C.M., Hollander, J.M., and Perlman, I. (eds.) 1967. Table of Radioisotopes. 6th ed. John Wiley & Sons, New York.

Meisenhelder, J. and Hunter, T. 1988. Radioactive protein-labeling techniques. *Nature* 335:120.

Shleien, B. 1987. Radiation Safety Manual for Users of Radioisotopes in Research and Academic Institutions. Nucleon Lectern Associates, Olney, Md.

KEY REFERENCES

Faires, R.A. and Boswell, G. 1981. Radioisotope Laboratory Techniques. Butterworths, Boston.

Contains useful information on techniques mentioned in this unit.

INTERNET RESOURCES

http://web.princeton.edu/sites/ehs

Princeton University Environmental Health and Safety Web site containing radioisotope fact sheets.

Chapter 3

Reagent Preparation

UNIT 3.1

Reagent Preparation: Theoretical and Practical Discussions

Deb N. Chakravarti,[1] Bulbul Chakravarti,[1] and Buddhadeb Mallik[1]

[1]Proteomics Center, Keck Graduate Institute of Applied Life Sciences, Claremont, California

REAGENT PREPARATION

Background and Precautions

In general, the experiments described in this title utilize aqueous solutions. The standard practice is to use either distilled water or deionized water to prepare most reagent solutions. Many of these reagents are adequately buffered for maintaining specific hydrogen ion concentration measured by the pH of the solution. Reagents are often sterilized to avoid microbial contamination (UNIT 4.1) or suitable chemical agents are added to stop microbial growth.

3

General Guidelines for Preparation of Reagents

For preparation of reagents, such as solutions and buffers, the following general guidelines should be considered.

Protective clothing

Gloves and safety glasses should be worn as much as possible. The use of gloves not only protects the user from exposure to the reagents but also protects the reagents from unwanted contamination from the hands of the user.

Clean glass- and plasticware

Any reagent preparation should be undertaken using clean glass- or plasticware as necessary. The cleanliness of glass- and plasticware plays a very important role in the success of the experiment being carried out, since the contamination of reagents through use of dirty or improperly cleaned glass- and plasticware can lead to erroneous results. Disposable plasticware, such as microcentrifuge tubes, 15- and 50-ml conical tubes, plastic syringes, etc., as well as disposable glassware, like culture tubes, should be of good quality, since these are used directly without washing. However, residual contaminants like plasticizer in microcentrifuge tubes have been known to cause problems in certain kinds of analyses. Plasticware made from polypropylene or polyethylene are generally suitable for use with acids, bases, and aqueous solutions. However, when using organic chemicals, information on the property of that particular plastic material and its resistance to such chemicals should be considered. Plasticware should not, in general, be treated with oxidizing agents like dichromate since these tend to produce free carboxyl groups on the plastic surface. Glassware can be cleaned with a suitable detergent such as Alconox (1:100 dilution), PCC-54 (2% to 20% solution), and RBS-35 (2% to 20% solution). If it is necessary to remove any remaining traces of organic materials from glassware, soaking them in chromic acid (made by mixing sodium or potassium dichromate and sulfuric acid) is very effective. To prepare chromic acid, dissolve 5 g of sodium or potassium dichromate in 5 ml of water in a 250-ml beaker, then slowly add 100 ml of concentrated sulfuric acid with constant stirring. The temperature will rise to 70° to 80°C. Allow the mixture to cool to room temperature and store in a glass-stoppered bottle. Chromic acid prepared this way can be reused. Over time, the color of the solution changes from reddish-brown to green; if this occurs, it should be discarded and a fresh chromic acid solution should be prepared. Like any other acidic solution, care should be taken to follow proper guidelines for disposal of this solution. Similar efficient cleaning can be obtained with Nochromix (Godax Laboratories) laboratory glass cleaning reagent. After washing with tap water, all glassware should be rinsed with deionized water and dried. Glass may be considered to be clean when water spreads evenly over the surface and does not form drops or patches. It is recommended that volume measuring apparatus, like volumetric flasks, measuring cylinders, etc., should not be heated during drying because this might affect subsequent accuracy of measurement. Very dilute protein solutions are often adsorbed to glass and some enzymes are known to be inactivated under such conditions. One way to overcome this is to add excess of a carrier protein like bovine serum albumin (BSA) to the dilute protein solution. An alternate approach is to siliconize the glass surface with a suitable agent like dimethyldichlorosilane ($(CH_3)_2SiCl_2$, DMDCS). Usually, glassware is treated with a 10% solution of DMDCS in toluene, hexane, or methylene chloride followed by a methanol rinse and blow drying with nitrogen or air. This attaches silane molecules to create an inert surface. For specific procedures, see *UNIT 3.3* in this title, and *CP Molecular Biology Appendix 3B* (Seed, 2003).

NOTE: Also refer to *UNIT 1.1* for information on glass- and plasticware.

3

High-purity reagents and safe handling of chemicals and biochemicals

In general, high-grade purity reagents, such as acids, bases, salts, biochemicals, etc., should be used. Reagents are available in several defined grades depending on purity. If possible, and whenever available, ACS-grade reagents should be obtained. ACS-grade signifies reagents having very high purity that meets or exceeds the purity standards set by the American Chemical Society (ACS) Committee on Analytical Reagents, which defines the specifications for most chemicals used in analytical testing and is contained in the current edition of *Reagent Chemicals, 10th edition*. Grades of reagents like analytical reagent (AR), guaranteed reagent (GR), ultra-pure reagent, molecular biology grade reagent, etc., are also available that usually have high purity levels similar to the ACS grade. Chemically pure (CP), lab-grade, or purified-grade reagents can also be used. Reagents that meet or exceed requirements such as US Pharmacopeia, known as USP grade, may also be used for most laboratory purposes. Chemicals that meet or exceed purity requirements of different national Pharmacopeias, such as US Pharmacopeia, British Pharmacopoeia, European Pharmacopeia, Japanese Pharmacopoeia, or Indian Pharmacopoeia, are acceptable for food, drug, or medicinal use. Similarly, Food Chemicals Codex (FCC) is a compendium of standards that define quality and safety of chemicals used for food additives. Technical-grade reagents used for commercial and industrial purposes may not be pure enough for many laboratory applications. Where no appropriate specification exists for a particular chemical, the one with the highest level of purity available should be used. Individual manufacturers often have their specific grades of reagents. For example, Mallinckrodt Baker provides a line of reagents known as GenAR that have been developed specifically for use in biotechnology and genetic research. For such proprietary grades, individual manufacturers' specifications should be consulted. When comparing the same reagent of different grades from different manufacturers, the certificate of analysis or specifications usually available in the label information providing impurities present should be carefully considered.

For general laboratory chemicals that are inexpensive, amounts necessary to perform the experiment several times should be obtained. For expensive chemicals, appropriate decisions should be made taking into considerations that some experiments may fail or would need to be repeated. Most reagents used for experiments described in this manual may not have an expiration date; however, if there is one, it should be noted and the reagent should be discarded after the date. Whether or not there is an expiration date, the date on which a particular chemical was received or opened for the first time should always be noted on the label with an indelible marker. The label of each chemical usually indicates if storage conditions other than room temperature such as storage at 4°C, −20°C, −70°C, etc., are necessary. Some chemicals may require storage in the absence of light. They usually come in amber-colored bottles. It is also advisable to check the actual lot analysis of the specific bottle of chemical received to see that it does not significantly deviate from the specifications required. All of the recipes in this manual are required to specify storage conditions and shelf life.

For each chemical, a very important item to check is the material safety data sheet (MSDS). An MSDS provides users with the proper procedures for handling a particular substance and includes information such as physical data (melting point, boiling point, flash point, etc.), toxicity, health effects, first aid, reactivity, storage, disposal, necessary protective equipment, and procedures to deal with accidents involving spills or leakages. MSDSs can be obtained from several places. Each laboratory should keep track of all the MSDSs that come with each chemical. If unavailable, they can be obtained from the customer service department of the distributor or manufacturer of the chemical. MSDSs are also available on the Web. Some useful websites that provide MSDSs are *http://www.ilpi.com/msds*, *http://www.msdssearch.com*, etc. Many institutions buy proprietary software or subscription services for MSDS. In the MSDS for each chemical, it is important to read section 3 (hazards identification), section 4 (first aid measures), section 5 (fire fighting measures), section 6 (accidental release measures), section 7 (handling and storage), section 8 (exposure controls, personal protection), section 9 (physical and chemical properties), section 10 (stability and reactivity), section 11 (toxicological information), and section 13 (disposal considerations).

General information about some reagents and their safe handling are summarized below. Refer to *APPENDIX 1* for more information on general laboratory safety.

Acids

Since the dilution of concentrated acid with water is exothermic, concentrated acid should slowly be added to water with gentle stirring when preparing dilutions. Water should never be added to the acid. Preparation of acids should be carried out in a chemical fume hood and eye protection should always be worn. Acids and bases should always be stored in separate cabinets.

Acrylamide

Acrylamide is absorbed through the skin and has strong neurotoxic properties. One should wear a mask and gloves when weighing acrylamide and methylenebisacrylamide. Appropriate gloves should also be worn when handling solutions containing these compounds. Polyacrylamide gels are presumed to be nontoxic; however, the possibility exists that small amounts of the monomer (i.e., monomeric) acrylamide may be present in these gels, necessitating careful handling. Cheaper grades of acrylamide and bisacrylamide should be avoided since they frequently contain metal ion impurities. When solutions containing acrylamide and bisacrylamide are stored, they are slowly converted to acrylic acid and bisacrylic acid, respectively.

NOTE: Refer to *UNIT 7.3* for futher discussion of acrylamide and bisacrylamide.

Ammonia

Ammonia vapors form inside any closed storage container containing aqueous ammonia solution, which is also known as ammonium hydroxide. This container should always be opened carefully to permit the venting of any ammonia vapor. To facilitate this, bottles of ammonia should be chilled and opened inside a chemical fume hood, particularly in warm weather. Eye protection must be worn. Ammonia vapor is irritating to the eyes while the liquid will cause burns. First aid measures involve immediate flushing with copious amounts of water.

EDTA

The disodium salt of ethylenediamine tetraacetic acid (EDTA) dissolves in water only when the pH of the solution is adjusted to 8.0 by the addition of NaOH solution.

Ethidium bromide

Ethidium bromide is a mutagenic agent. Wear a mask and gloves when weighing it.

Hygroscopic agents

Some chemicals such as magnesium chloride ($MgCl_2$) and some peptides are extremely hygroscopic (i.e., readily absorb moisture from the atmosphere) in nature. Buy smaller amounts of these chemicals; once opened, bottles should not be stored for a long period of time; however, such chemicals may be stored in a desiccator over suitable desiccant.

Phenol

Phenol is highly corrosive, toxic, and can cause severe burns. Gloves, protective clothing, and safety glasses should be worn when phenol is handled. It is highly recommended to restrict phenol use to a chemical hood. If phenol is in contact with the skin, it should be washed immediately with a large volume of water or polyethylene glycol.

Sodium dodecyl sulfate (SDS)

SDS is used frequently for the preparation of reagents for polyacrylamide gel electrophoresis of proteins. One should wear a mask when weighing SDS.

Tris

Many inferior quality preparations of Tris in strong aqueous solutions such as 1 M may have a yellow color. These should be discarded and good quality Tris should be obtained. Tris binds metal ions; thus, the use of Tris buffers should be avoided when the presence of metal ions is critical. Diethyl pyrocarbonate (DEPC) is commonly used at 0.1% (v/v) concentration to inactivate RNases for the preparation of RNase-free solutions. However, DEPC reacts with primary amines and cannot be applied directly to Tris buffers. For preparing RNase-free Tris buffers, the water used should first be treated with DEPC followed by the dissolution of Tris. Since the pK_a of Tris at 25°C is 8.1, Tris-based buffers (such as Tris·Cl) can be used to maintain pH in the range of 8.1 ±1 pH unit, i.e., from pH 7.1 to 9.1. For Tris buffers, pH increases ~0.03 unit per 1°C decrease in temperature, and decreases 0.03 to 0.05 unit per ten-fold dilution, according to Sigma-Aldrich.

An exercise on choice and information on chemicals

Fisher Scientific (ThermoFisher Scientific) is a supplier of laboratory chemicals. Go to the website of the company *http://www.fishersci.com* (or access US Fisher Scientific location through *http://www.thermofisher.com*). In the Product Search field, type "sodium chloride" and hit the Go button. Go through the list of the first 25 items to recognize different grades, such as ACS, USP, DNAse/RNAse/Protease-Free, etc., as well as different physical forms available, such as crystalline, granular, or aqueous solutions. Click on any one of the items such as Sodium Chloride (Cryst./Certified ACS), Fisher Chemical. For most laboratory purposes, the Poly Bottle (1 kg) will be sufficient. First click on the MSDS button to see the MSDS of the product and check the sections mentioned above. Now click on the Specification tab to view the product specifications. Note the comment that the Actual Lot Analysis is reported on the label. Click on the Certificate of Analysis button where links to the Certificate of Analysis of recent lots of this product are found.

Use of high purity water

IMPORTANT NOTE: The recipes provided throughout this manual assume the use of deionized or distilled water unless otherwise stated.

Deionized water (DI-H_2O)

Most of the reagents that are prepared for experiments described in this title are aqueous solutions; therefore, knowledge of the quality of water used is essential. All reagents should be prepared in deionized (DI) or ultra-pure DI water or distilled or double-distilled water (ddH_2O). Usually water contains two different kinds of impurities, ionic such as inorganic salts, and non-ionic such as organic substances. The ionic purity of water is usually determined by measuring the electrical resistance (or conductance). As ionic impurities are removed, the water becomes less conducting and the electrical resistance of DI water increases to a value as high as 18.2 MΩ cm at room temperature. Most laboratories contain a house deionized water supply, available via a specially marked tap. This water is purified by passing tap water through an ion exchange resin cartridge (also see *UNIT 6.1*) containing both cation and anion exchangers, usually as a mixed bed containing both, as well as through an activated charcoal cartridge for removal of organic impurities. Although the deionized water obtained from the house supply is sufficiently pure for many biochemical studies, further purification of this water is often carried out to produce ultrapure DI water, usually with 18.2 MΩ cm resistance and <10 ppb total organic content (TOC), using a polishing unit such as the Milli-Q (Millipore Corporation) or the Barnstead Nanopure (Barnstead International) units. These units contain cation and anion exchangers as well as activated charcoal cartridges for the further removal of organic materials, and may also contain an ultrafiltration cartridge to ensure very low nuclease and pyrogen levels, as well as ultraviolet lamps for oxidation of organic molecules and lowering bacterial levels within the system. The purified water is finally dispatched through a 0.22-μm membrane filter, which essentially sterilizes the water. Continuous or intermittent water circulation maintains water purity when the system is not in use. Using suitable polishing systems, it is possible to obtain water with extremely low ionic, organic, pyrogen, and nuclease contaminations. Such water is ideal for polymerase chain reaction (PCR; *UNIT 10.2*), two-dimensional (2-D) gel electrophoresis (*UNIT 7.3*), and DNA sequence analysis (*UNIT 10.4*). This water can also be used for preparing buffers and reagents for mammalian cell culture. It is advisable to obtain ultrapure DI water on an as-needed-basis. It should not be stored in glass or plastic containers, which leach contaminants at very low concentrations (Gabler et al., 1983)

Distilled water (ddH₂O)

In contrast to DI water, distilled water is produced by boiling water in an all-glass apparatus followed by the condensation and collection of steam. For some biochemical applications requiring exceptionally high-purity water, distilled water is subjected to redistillation to obtain double-distilled water. If really high-quality distilled water is required, water must be distilled from acid permanganate. For this purpose, 1 g of potassium permanganate ($KMnO_4$) and 1 ml of phosphoric acid (H_3PO_4) are added to 1 liter of water prior to distillation. The hot permanganate oxidizes contaminating organic substances and the distilled water has very low UV absorption and fluorescence. In theory, the pH of distilled or DI water should be 7.0 (also see *UNIT 3.2*). However, in practice, distilled or DI water has a slightly acidic pH ~5.0 due to the presence of CO_2 absorbed from the air. Distilled or DI water purchased in plastic containers from stores generally contain leached-out contaminants.

Detecting substandard water

Use of substandard quality distilled or DI water may be sometimes evident from careful visual inspection of reagents prepared with it. For example, when preparing Laemmli gel running buffer by diluting Tris glycine and SDS (*UNIT 7.3*), the appearance of white flocculent precipitate is indicative of presence of metal ions in the water.

ACCURACY OF WEIGHING AND PIPETTING

Volume and weight measurements are described in *UNITS 1.1 & 1.2*, respectively. Accuracy in both of these is extremely important in preparation of reagents.

USE OF CALIBRATED pH METERS

Proper use of a pH meter is discussed in *UNIT 3.2*.

Biochemical reactions require strict control of pH. For example, enzymes have specific pH optima where they are most active (also see *UNIT 10.1*). On the other hand, a protein may be denatured and lose biological activity when exposed to specific acidic or alkaline conditions. Also, when exposed to a solution at the isoelectric point of the protein, it will precipitate. Unless otherwise stated, the adjustment of pH of a solution is usually carried out with 1 N HCl or 1 N NaOH. Accordingly, accurate measurement of the pH of the reagents prepared is extremely important. The pH of any reagent solution is measured with a pH meter that consists of a thin-walled glass electrode probe and an analog or digital display to show the pH value. However, for accurate determination of the pH of a solution, the pH meter must first be calibrated with reference standard buffer solutions.

AVOIDING CHEMICAL AND MICROBIAL CONTAMINATION OF REAGENTS

Proper use of aseptic technique is described in *UNIT 4.1*.

Contamination can be avoided by storing reagents in a properly cleaned container. Fingers should not come in contact with reagents or the surface of the container of the reagents. Fingertips are a potent source of contamination such as fats, oils, amino acids, nucleases (see *UNIT 8.2*), keratins, etc. Reagent containers can be sealed against vapor of carbon dioxide, ammonia, or hydrochloric acid using Parafilm. Whenever possible, all buffers and solutions prepared should be sterilized to ensure absence of microbial contaminations (see *UNIT 4.1*). Most inorganic salt solutions such as NaCl, $CaCl_2$, $MgCl_2$, and buffers such as sodium phosphate buffer containing a mixture of mono- and di-sodium hydrogen phosphate can be sterilized by autoclaving. Both heat-labile as well as heat-stable reagents can be sterilized by filtration using a 0.22-µm filter. Another common practice is to use a bacteriostatic agent, such as 0.02% (w/v) sodium azide (NaN_3). However, the activities of several enzymes, such as the activity of membrane adenosine triphosphatase (ATPase) of *Escherichia coli*, are inhibited by low concentrations of azide ions. Care should be taken to note that sodium azide is poisonous; thus, inhalation of the powder should be avoided and caution should be exercised when handling solutions containing NaN_3. It should not be used for experiments with live organisms. Dithiothreitol (DTT), β-mercaptoethanol, or solutions containing these chemicals should not be autoclaved. Solutions containing 10% (w/v) or higher SDS concentrations do not need sterilization. Bacterial growth can be avoided for many aqueous solutions by storing in a refrigerator or freezer or by addition of a drop of chloroform to the reagent. Growth is often inhibited in more concentrated solution. It is prudent practice not to pour unused reagent back into the original container unless the reagent is really expensive.

PREPARING REAGENT OR BUFFER SOLUTIONS

A solution is a homogeneous mixture of two or more substances. In a solution, at least one substance, the solute, is dissolved in another substance, the solvent, which is usually a liquid. In general, the solvent is the major component of a solution.

The amount of a solute present in a specific volume of a solution is a measure of its concentration, which can be expressed in different ways as described below. Qualitatively speaking, one solution may be more concentrated than another or vice versa or a solution may be saturated with the solute at a particular temperature. Concentration is commonly expressed as: weight per unit volume (e.g., g/liter, mg/ml, µg/µl), percent composition (w/w, w/v, v/v), parts per million (ppm) or parts per billion (ppb), normality (N), molarity (M), molality (m), and mole fraction. In the following section, each of the concentration terms will be discussed separately.

Weight per Unit Volume

In this system, concentration is expressed as the weight of a solute present in a specific volume of the solution. Depending on the amount, concentration is generally expressed as g/liter, mg/ml, or μg/μl. Although 1 g/liter = 1 mg/ml = 1 μg/μl, the amount of reagent prepared or the quantity used often dictates the unit in which the concentration is expressed. For example, if 1 liter of a 1 g/liter solution is prepared and 75 ml of it is used in an experiment, the concentration of the solution will most likely be denoted as 1 g/liter. If 100 ml of the same solution is prepared and 5 ml of that is used in an experiment, then the concentration of the solution will most likely be indicated as 1 mg/ml. If 10 ml of the same solution is prepared and 5 μl is used in an experiment, the concentration will invariably be expressed as 1 μg/μl. In fact, in most molecular biology experiments, like enzyme assays, microliter amounts of reagents are frequently used and the concentration is thus designated as μg/μl. The solubility of a particular solute in a specific solvent is expressed similarly. The term solubility signifies the maximum amount of a solute that can be dissolved in a given volume of a specific solvent at a particular temperature. The MSDS of a chemical usually mentions the solubility in section 9 (physical and chemical properties).

Percent Composition

Percent composition can be expressed either as weight percent (w/w %), weight/volume percent (w/v %), or volume percent (v/v %).

Weight/weight (w/w)

Weight percent is defined as the ratio of the weight of a solute to the total weight of the solution (not solvent) multiplied by 100. If 150 g of an aqueous NaCl solution contains 25 g of solid NaCl, then the percent composition of the solution is $(25/150) \times 100$ or 16.7% (w/w).

Weight/volume (w/v)

Weight/volume percent is defined as the ratio of the weight of a solute to the total volume of the solution (not solvent) multiplied by 100. If 150 ml of an aqueous NaCl solution contains 25 g of solid NaCl, then the percent composition of the solution is $(25/150) \times 100$ or 16.7% (w/v). Although the units of the numerator and the denominator are different, this is a very convenient way to express the concentration of a solution.

Volume/volume (v/v)

Volume percent is defined as the ratio of the volume of a liquid solute to the total volume of the solution (not solvent) multiplied by 100. When 70 ml of ethanol is diluted with water to a total volume of 100 ml, the concentration of ethanol in the solution is 70% (v/v).

Since the specific gravity of water is ~1 over a wide range of temperatures, for the aqueous solution of a solid chemical, the numerical value of the concentration expressed as w/w and w/v are almost identical. However, for almost all nonaqueous solvents, the specific gravity is not 1, and thus the numerical value of the concentration in terms of w/w and w/v will be different. For the bacteriostatic agent NaN_3 above, the concentration in aqueous solution is expressed as w/v, since NaN_3 is weighed and dissolved in the required amount of water. On the other hand, the concentration of a highly viscous liquid like glycerol in an aqueous solution can also be expressed in w/v units, as in the following example. Since glycerol is very viscous, it is not easily measured as small volumes (e.g., 1 ml) using a pipet, it is far more accurate to weigh the required volume of glycerol and add the requisite amount of water to it. In some other cases when relatively large amounts of a solution is prepared (e.g., 500 ml) where glycerol is a major liquid component, the concentration of glycerol may be expressed in v/v units as well. An example of this is one of the stock buffers used in 2-D gel electrophoresis for reduction and alkylation of immobilized pH gradient strips following isoelectric focusing: 100 mM Tris·Cl (pH 8.0)/6 M urea/30% (v/v) glycerol/2% (w/v) SDS.

Parts per Million (ppm) and Parts per Billion (ppb)

For very dilute solutions, the above concentrations can be expressed as parts per million parts, which is known as parts per million or ppm. So the NaCl solution mentioned above will have a concentration of (25/150) × 10^6 or 166666.7 ppm. If 1 ml of this solution is diluted with 999 ml of water, which represents 1000-fold dilution, the concentration of this solution will be 166.7 ppm. For even more dilute solutions, the concentration can be expressed as parts per billion, known as ppb. In general, trace impurities in any chemical are expressed in ppm or ppb.

Normality (N)

Normality (N) is defined as the gram equivalent weight of a solute per liter of the solvent. The concentration of acids and alkalis are usually expressed in this unit. The gram equivalent weight is the molecular weight divided by the number of H^+ or OH^- ions released from one molecule of the acid or the base, respectively, in solution, or in the case of salts, the oxidation state of the metal ions. For example, the gram molecular weight of sulfuric acid (H_2SO_4) is 98 g and each molecule of the acid releases two protons (H^+ ions) in solution. So, 1 liter of 1 N H_2SO_4 solution contains 49 g of sulfuric acid. The oxidation state of copper in copper sulfate ($CuSO_4$, mol. wt. 159.6 g) is 2. So, 1 liter of 1 N $CuSO_4$ solution contains 79.8 g of anhydrous copper sulfate.

3

The following is an example that shows how the normality of a commercially available acid, such as HCl can be calculated from the specifications on the label. Hydrochloric acid is prepared by dissolving HCl gas in water. The concentrated grade of HCl is usually produced in solution up to a concentration of 38% (v/v), which has a specific gravity of 1.19. The molecular weight of HCl is the sum of the atomic weights of individual elements in the stoichiometry in which they are present:

1 mole of hydrogen atom = 1.0 g mol^{-1}
1 mole of chlorine atom = 35.46 g mol^{-1}

Thus, the molecular weight of HCl is 36.46 g.

Since the concentration of HCl is 38% (v/v):

100 ml of this concentrated HCl solution contains 38 ml HCl.
Thus, 1 liter of this concentrated HCl solution contains 380 ml HCl.

Since mass = volume × density (i.e., specific gravity):

The weight of 380 ml HCl is 380 × 1.19 g = 452.2 g

Thus, 1 liter of the solution contains 452.2 g HCl, from which the concentration of HCl is calculated as described above:

$$\frac{452.2}{36.46} = 12.4(N)$$

Thus, this particular lot of concentrated HCl containing 38% (v/v) of HCl with a specific gravity of 1.19 is equivalent to a 12.4 N solution of HCl. However, the percentage composition and the specific gravity may be slightly different for different batches of concentrated HCl and in each case the normality can be calculated using the above example.

Molarity (M)

Molarity (M) is a very widely used unit of concentration. It is the number of moles present in 1 liter of solution. For NaOH (mol. wt. 40), 1 liter of 1 M aqueous solution contains 40 g of NaOH. Since one molecule of NaOH produces one hydroxyl ion (OH^-), a 1 M solution is also a 1 N solution. On the other hand, 1 liter of 1 M H_2SO_4 solution contains 98 g of H_2SO_4 while 1 liter of 1 N H_2SO_4 solution contains 49 g of H_2SO_4.

For proteins and peptides, the weight in micrograms and the corresponding number of moles is used very frequently. This can be calculated using the following equation(s):

$$\frac{\text{wt. } (\mu\text{g})}{\text{mol. wt.}} \times 10^3 \text{ (nanomole)}$$

$$\frac{\text{wt. } (\mu\text{g})}{\text{mol. wt.}} \times 10^6 \text{ (picomole)}$$

Molality (m)

Molality (m) expresses the number of moles per 1000 g, i.e., per kilogram (kg) of solvent. Thus, molality depends on the density of the solvent. At 25°C, 1 liter of water weighs ~1 kg and molarity and molality become equal. However, the density of most of the common organic solvents is different from 1 and, thus, molarity and molality will be different. Furthermore, molarity refers to the volume of the solution, which is temperature-dependent, whereas molality is not. However, molal solutions are not usually used in biochemical experiments.

3

Mole Fraction (X)

Mole fraction (X) represents the fraction of moles of a specific solute component present in a multi-component solution. In a four-component solution having n_1, n_2, n_3, and n_4 number of moles of components 1, 2, 3, and 4, respectively, the mole fraction X_1 of component 1 is

$$X_1 = \frac{n_1}{n_1 + n_2 + n_3 + n_4}$$

It should be mentioned that the sum of all mole fractions in a solution is always equal to 1.

Preparing and Diluting Stock Solutions

It is common laboratory practice to prepare a stock solution of a reagent as long as this solution is stable over a period of time, when stored properly. The storage may be at room temperature or 4°C. When the reagent is required, small amounts of the stock solution are diluted, usually with distilled or deionized water, to obtain the working concentration. Stock solutions reduce the variability between a number of similar experiments. The preparation of the stock solution usually involves accurate weighing of a relatively large amount of the solute, which is a time-consuming process; but needs to be done only once for the preparation of the stock solution. If V_1 and S_1 represent the volume and the concentration, respectively, of the stock solution, and V_2 and S_2 represent the final volume and the concentration, respectively, of the diluted solution, then:

$$V_1 S_1 = V_2 S_2$$

The units of volume and concentration should be the same on both sides of the equation. Consider a stock solution of 1 M NaCl and a 150 mM working dilution. The molecular weight of NaCl is 58.44 g. To prepare a 1 M stock solution, 58.44 g of NaCl must be accurately weighed out and then transferred into a 1-liter volumetric flask or measuring cylinder. About 250 ml of distilled or deionized water should be initially added and the contents gently swirled to dissolve as much of the NaCl as possible. More water should then be gradually added until most of the NaCl dissolves. At this point the volume should be made up to the 1-liter mark so that the lower meniscus of the liquid just touches the mark of the volumetric flask or measuring cylinder. A magnetic stir bar should be introduced and the solution stirred on a magnetic stirrer for ~10 min so that the solution becomes homogeneous. Subsequently, the solution should be transferred into a suitable glass or plastic container and stored. It should be noted that the volume of the solution should be made up to the required final volume, i.e., 1 liter, but 1 liter of water should not be added directly to the NaCl. Theoretically, the total amount of water in the final solution will be a little less than 1 liter because the solute, i.e., NaCl, also contributes to the total volume. Although the volume of the solute may be very small in the dilute solution, it is significant in a concentrated solution.

Preparation of 100 ml of 150 mM (i.e., 0.15 M) working dilution from a 1 M stock solution can be calculated as follows:

$$V_1 S_1 = V_2 S_2$$

or

$$V_1 = \frac{V_2 S_2}{S_1}$$

3

So,

$$V_1 = \frac{100 \times 0.15}{1} = 15$$

Thus, 15 ml of the 1 M NaCl stock solution should be pipetted out and the volume made up to 100 ml with distilled or deionized water using a volumetric flask or a measuring cylinder.

MAKING BUFFER SOLUTIONS

Background Theory of Buffer Action

Aqueous solutions of partly neutralized weak acids (such as acetic acid) or weak bases (such as Tris) resist change of H^+ ion concentration (i.e., pH) if a small amount of a strong acid or a strong base is added to it. The property of such a solution to resist changes in hydrogen ion concentration was described by Fernbach and Hubert (1900) as buffering. For example, the theoretical pH of distilled water is 7.0. Although freshly prepared distilled water will have a pH very close to 7.0, most water samples will have a slightly acidic pH, i.e., <7.0. As mentioned above, this is due to absorption of CO_2 from the air to form bicarbonate ions. When 1 ml of 1 M HCl or 1 ml of 1 M NaOH is added to 1 liter of pure water, the pH shifts to 3.0 or 11.0, respectively. However, if a 1-liter solution containing a mixture of 390 ml of 0.1 M sodium phosphate monobasic (NaH_2PO_4) and 610 ml of 0.1 M sodium phosphate dibasic (Na_2HPO_4) is prepared, this solution will have a pH of $\sim$7.0. When 1 ml of 1 M HCl or 1 ml of 1 M NaOH is added to this solution, the pH will remain unchanged at 7.0. This solution is thus known as 0.1 M sodium phosphate buffer solution. This particular solution will act as a buffer in the pH range 5.8 to 8.0. If a pH of 8.5 must be maintained, a Tris·Cl buffer solution with a pH range of 7.2 to 9.0 should be used. Solutions having this type of buffering property can be prepared from pairs of chemicals such as: glycine-HCl (pH range 2.2 to 3.6), citric acid–sodium citrate (pH range 3.0 to 6.2), citric acid–dibasic sodium phosphate (pH range 2.6 to 7.0), acetic acid–sodium acetate (pH range 3.6 to 5.6), sodium carbonate–sodium bicarbonate (pH range 9.2 to 10.6), etc. Biochemical reactions require strict control of pH. For example, enzymes have specific pH optima where they are most active. On the other hand, a protein may be denatured and lose biological activity when exposed to specific acidic or alkaline conditions. Also, when exposed to a solution at the isoelectric point of the protein, it will precipitate. Physiological fluids maintain a tightly regulated pH under normal condition. For example, the pH of human blood is maintained between 7.35 and 7.45. The principal buffers in blood are bicarbonate, proteins, and phosphates. Therefore, when performing experiments on biological and biochemical systems in the laboratory, control of pH is a primary requirement.

It is obvious from the above discussion that buffer solutions are made of a weak acid and its conjugate base or a weak base and its conjugate acid. Consider a weak acid, HA, in aqueous solution. By virtue of its weak dissociation property, which distinguishes these from the strong acids or strong bases, the solution consist of undissociated HA (the weak acid) and A^- (the conjugate base), and at equilibrium:

$$HA + H_2O \rightleftharpoons H_3O^+ + A^-$$

The equilibrium constant (also called the dissociation constant), K_a, for this dissociation is generally very low (H_3O^+ is replaced with H^+ in the following equation to represent hydrogen ions):

$$K_a = \frac{[H^+][A^-]}{[HA]}$$

According to Le Chatelier's principle, if this equilibrium process is perturbed by the addition of a similar component, the system will quickly restore the preexisting equilibrium process. If a small amount of strong base is added to the system, the protons will consume the base. On the other hand, if a small amount of strong acid is added to the system, the conjugate base will consume the extra protons. As a result, the respective concentrations of H^+ and A^- will change. At that point, further dissociation of HA will take place to keep the K_a the same and as a result the pH of the solution will remain unchanged.

3

By taking the negative logarithm of both sides of the above equation, the expression for equilibrium constant can be rearranged as follows.

$$\mathrm{pH} = \mathrm{pK_a} + \log\frac{[A^-]}{[HA]}$$

In this equation, $\mathrm{pK_a}$ is equal to $-\log K_a$. This is commonly known as the Henderson-Hasselbalch equation. The capacity of a buffer to resist changes in pH is maximum when A^- and HA are present in equal amounts and the pH becomes equal to $\mathrm{pK_a}$. A weak acid and its conjugate base or a weak base and its conjugate acid can effectively buffer when the pH is within the ± 1 range of $\mathrm{pK_a}$.

The effect of temperature on the pH of a buffer solution depends on the particular buffer being used. The pH of Tris buffer is significantly affected by temperature, whereas the pH of phosphate buffer is less dependent on variation of temperature. Since the pH of a buffer solution can change with dilution, whenever a buffer solution is diluted, the final pH should always be checked.

How to Prepare a Buffer Solution

From the Henderson-Hasselbalch equation, it is possible to determine the relative amounts of weak acids or weak bases and their respective conjugate counterparts to prepare a buffer solution at a certain pH. For most common buffers used in biochemical experiments, these values are usually obtained from buffer preparation tables such as those found in *UNIT 3.3* and in many biochemical handbooks. Two widely used collections are that by Gomori (1955) and Stoll and Blanchard (1990). Another very convenient source is Calbiochem's *Buffers: A Guide for the Preparation and Use of Buffers in Biological Systems*, which is available online (*http://www.emdbiosciences.com/docs/docs/LIT/Buffers_CB0052_E.pdf*). Detailed knowledge about buffer solutions using biochemical studies is available (Good and Izawa, 1972; Blanchard, 1984).

An example of the preparation of a phosphate buffer stock solution (pH range is 5.8 to 8.0) is provided in Table 3.3.5.

Example: Preparing a phosphate buffer stock solution

The molecular weight of sodium phosphate monobasic monohydrate ($NaH_2PO_4.H_2O$) is 138.0 g (23 + 2 + 31 + 64 + 18). Note the importance of including the single molecule of water of crystallization in the calculation. It should be noted that sodium phosphate monobasic is also available as NaH_2PO_4 anhydrous (i.e., without any water of crystallization; molecular weight = 23 + 2 + 31 + 64 = 120.0 g) as well as the dihydrate form $NaH_2PO_4.2H_2O$ (molecular weight = 23 + 2 + 31 + 64 + 36 = 156.0 g). Many such anhydrous forms of the same salt dissolve slowly in water, while the hydrated forms dissolve rapidly.

To prepare 1 liter of 100 mM stock monobasic sodium phosphate solution, dissolve 13.8 g of $NaH_2PO_4.H_2O$ (solid) into distilled or deionized water. Adjust the final volume to 1 liter with water.

The molecular weight of sodium phosphate dibasic heptahydrate ($NaH_2PO_4.7H_2O$) is 268.0 g (46 + 1 + 31 + 64 + 126). Note the importance of including the seven molecules of water of crystallization in the calculation. Dibasic sodium phosphate is available in an anhydrous as well as dihydrate form.

3

To prepare 1 liter of 100 mM stock dibasic sodium phosphate solution, dissolve 26.8 g of $NaH_2PO_4.7H_2O$ into distilled or deionized water. Adjust the final volume to 1 liter.

These two stock solutions can be mixed in different proportions as indicated in Table 3.3.5 to obtain buffer solutions of different pH. This follows from the Henderson-Hasselbalch equation described above. To prepare 100 ml of 100 mM sodium phosphate buffer (pH 7.0), 39 ml of the 100 mM stock monobasic sodium phosphate solution is mixed with 61 ml of 100 mM stock dibasic sodium phosphate solution. The pH should be checked and adjusted to 7.0 with 1 N phosphoric acid (H_3PO_4) or 1 N NaOH as required. If 100 ml of the above mixture is diluted with distilled or deionized water to 200 ml, the resulting buffer solution will have a concentration of 50 mM. Calculation of this concentration is based on the equation $V_1 \times S_1 = V_2 \times S_2$ described above. In practice, during dilution, the volume is kept slightly below the final volume required followed by adjustment of pH. Subsequently, the volume is adjusted to the required final volume and the pH is checked again.

Example of preparation of multi-component buffer solution

The elution buffer used in *UNIT 6.2* for column chromatography is 10 mM Tris·Cl/150 mM NaCl (pH 7.5). To prepare this, two stock solutions of 1 M Tris·Cl (pH 7.5) and 1 M NaCl must first be prepared.

Preparation of 1 liter of 1 M Tris·Cl (pH 7.5) stock solution

Using a chemical balance, weigh 121 g of Tris base (mol. wt. 121.0) and dissolve in ~850 ml distilled or deionized water in a beaker. Adjust the pH to 7.5 with 1 N HCl using a calibrated pH meter containing an electrode compatible for Tris buffer. Transfer the solution to a 1-liter measuring cylinder and adjust the volume to 980 ml with distilled or deionized water. Stir on a magnetic stirrer using a magnetic stir bar. Check the pH. If necessary, readjust the pH. At this point, remove the measuring cylinder from the magnetic stirrer and, using a magnetic stir bar retriever, remove the stir bar from the solution. Carefully adjust the final volume to the 1-liter mark with distilled or deionized water. Stir the solution again and recheck the pH. Transfer the solution to a clean bottle. If necessary, sterilize the solution by filtration using a 0.22-μm filter and store up to 1 year at room temperature as long as aseptic conditions are used to transfer the reagent.

Prepare 1 liter of 1 M NaCl stock solution

This has been described earlier in this unit (see Preparing and Diluting Stock Solutions).

Prepare 1 liter of 10 mM Tris·Cl/150 mM NaCl (pH 7.5)

Using a pipet, transfer 10 ml of the Tris·Cl stock solution into a 1-liter beaker. Using a 250-ml measuring cylinder, add 150 ml of the NaCl stock solution to the beaker. Using the volume markings on the side of the beaker as a guide, add distilled or deionized water to ~950 ml. The solution is stirred using a magnetic stirrer and the pH checked as mentioned above and adjusted to 7.5 if necessary. The contents are then transferred to a 1-liter measuring cylinder, the contents of the beaker rinsed with distilled or deionized water, and the volume made up to 1 liter. The pH is rechecked. The solution is transferred to a clean bottle. If necessary, the solution may be sterilized by filtration using a 0.22-μm filter and stored.

LITERATURE CITED

Blanchard, J.S. 1984. Buffers for enzymes. *Methods Enzymol.* 104:404-414.

Fernbach, A. and Hubert, L. 1900. De l'influence des phosphates et de quelques autres matières sur la diastase protèolitique du malt. *Compt. Rend.* 131:293-295.

Gabler, R., Hegde, R., and Hughes, D. 1983. Degradation of high purity water on storage. *J. Liq. Chromatogr.* 6:2565-2570.

Gomori, G. 1955. Preparation of buffers for use in enzyme studies. *Methods Enzymol.* 1:138-146.

Good, N.E. and Izawa, S. 1972. Hydrogen ion buffers. *Methods Enzymol.* 24:53-68.

Seed, B. 2003. Silanizing glassware. *Curr. Protoc. Mol. Biol.* 28:A.3B.1-A.3B.2.

Stoll, V.S. and Blanchard, J.S. 1990. Buffers: Principles and practice. *Methods Enzymol.* 182:24-38.

UNIT 3.2

Measurement of pH

Emily A. Wiley[1] and Deb N. Chakravarti[2]

[1]Joint Science Department, Claremont McKenna, Pitzer, and Scripps Colleges, Claremont, California
[2]Proteomics Center, Keck Graduate Institute of Applied Life Sciences, Claremont, California

3

INTRODUCTION

The pH is a measure of the acidity or alkalinity of a solution. These properties can have dramatic effects on the structure and function of biological molecules. For example, pH affects folding of polypeptide chains, and the biological activity of most enzymes is optimal within a narrow pH range (also see UNIT 10.1). For this reason, the correct pH of biological solutions is critical for the success of many different biomolecular techniques. The pH value reflects the relative quantity of hydrogen ions $[H^+]$ in a solution. Since the values of $[H^+]$ for most solutions are very small and difficult to compare, a more practical quantity of measure, that of pH, was formulated by Danish biochemist Soren Sorenson in 1909. The pH value of a solution is the negative log of its hydrogen ion activity (α), which in turn is the product of hydrogen ion concentration $[H^+]$ and the activity coefficient of hydrogen (Y_{H+}) at that concentration.

$$\text{pH} = -\log \alpha = -\log Y_{H+} [H^+]$$

In pure water and dilute solutions, such as most biological buffers, Y_{H+} is given a value of 1, and thus (α) can be considered to be the same as the H^+ concentration. In this case, pH is defined as the negative log of the hydrogen ion concentration, and uses the following equation:

$$\text{pH} = -\log [H^+]$$

According to this equation, the greater the concentration of H^+, the lower the pH. The pH value indicates the degree of acidity or alkalinity relative to the ionization of water. Pure water dissociates to yield equimolar amounts (10^{-7} M) of $[H^+]$ and $[OH^-]$ at 25°C. This means that the pH of pure water is 7, a value that is considered to be neutral pH, meaning that there is no excess acidity or basicity. (See UNIT 3.1 for a discussion of why, in practice, the pH of water deviates from 7.)

$$\text{pH}_{\text{water}} = -\log [H^+] = -\log 10^{-7} = 7$$

The pH scale ranges from 0 to 14. Solutions with a higher $[H^+]$ than water (pH <7) are acidic; solutions with a lower $[H^+]$ than water (pH >7) are basic, or alkaline.

MEASURING pH

There are three basic components in a standard pH measurement system: a pH electrode, a pH meter, and a temperature compensation element. The electrode measures the potential of solutions with unknown $[H^+]$ using a sensing half-cell relative to a known reference potential in a reference half-cell. Sensing and reference half-cells must be used together to complete the pH circuit (Fig. 3.2.1). The pH meter is an electronic meter that converts the voltage difference between the two half-cells to pH values. Most electrodes made today are combination electrodes that contain both the reference and sensing half-cells in the same body.

Reference half-cells contain a conductor surrounded by a solution with known $[H^+]$. The potential between the conductor and the known solution is constant, providing a stable reference potential. *Sensing half-cells* (measuring half-cells) contain a nonconducting tube sealed to a conductive glass

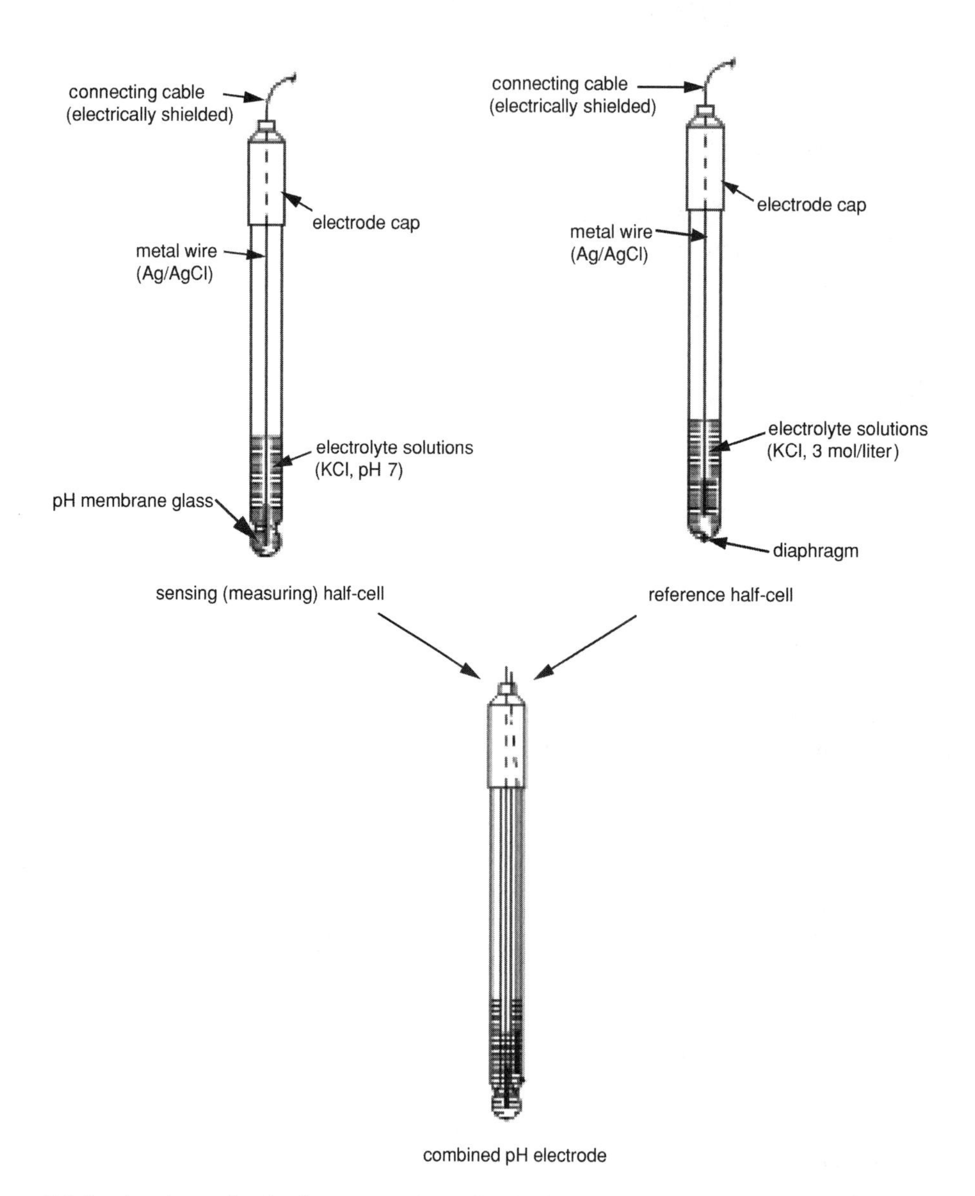

Figure 3.2.1 A schematic of reference and sensing half-cells in pH electrodes. The two half-cells are commonly combined into one electrode body making a combination electrode. Reproduced with kind permission from Metrohm Group.

membrane. Similar to the reference half-cell, it also contains a conductor surrounded by a solution with known $[H^+]$, which maintains constant voltages on the inner surface of the glass membrane.

When the pH electrode is immersed in the solution to be measured, a potential is established on the surface of the sensing glass membrane. If the solution is neutral, the sum of the voltages on the inner surface of the glass membrane and on the sensing conductor approximately equals the voltage on the outer surface of the glass membrane and the reference half-cell. This results in a total potential difference of 0 mV and a pH value of 7. If a known standard buffer of pH 7 is measured, the electrode should ideally read 0 mV. This is referred to as the "zero point" of the electrode.

In acidic or alkaline solutions, the voltage on the outer membrane surface changes proportionally to changes in $[H^+]$. The pH meter detects the change in potential and determines $[H^+]$ of the unknown by the Nernst equation:

$$E = E^\circ + RT/nF \times \ln\,(\text{unknown }[H^+]/\text{reference }[H^+]);\text{ or}$$
$$E = E^\circ + 2.3RT/nF \times \log\,(\text{unknown }[H^+]/\text{reference }[H^+])$$

where:

E = total potential difference (measured in mV)
E° = reference potential
R = gas constant (8.3145 J × K^{-1} × mol^{-1})
T = temperature in Kelvin
n = number of electrons transferred (moles)
F = Faraday's constant (electrical charge of 1 mol of electrons: 96,485 J × V^{-1} × mol^{-1})
$[H^+]$ = hydrogen ion concentration.

In practice, it is rare that an electrode reliably reproduces the theoretical zero point. The zero point may actually vary from −30 mV to +30 mV, and as an electrode ages, the zero point tends to drift upward overall. Due to this inherent inconsistency, the exact zero point must be established prior to taking pH measurements, a process called electrode calibration, which is described later in this unit.

3

Temperature Effects and Compensation

The Nernst equation shows that temperature has a proportional relationship to E (the total potential difference). Since the pH measured by the pH meter is directly proportional to E (in mV), temperature has a direct influence on the pH value. The electrode voltage changes linearly in response to changes in pH, and the temperature of the solution affects the slope of the linear relationship. The slope increases with increasing solution temperature, and decreases with decreasing temperature. All temperature-dependent slope lines intersect at the zero point: pH 7. If an electrode is calibrated at 25°C then it should be capable of reliable linear measurement throughout the entire pH range at 25°C. However, if the electrode is then used at a different temperature, the electrode slope will change (as dictated by the Nernst equation). One pH unit corresponds to 59.16 mV at 25°C. This is the standard voltage and temperature that all calibrations are compared to. The electrode voltage decreases to 54.20 mV/pH unit at 0°C and increases to 74.03 mV/pH unit at 100°C (Table 3.2.1).

Since pH values are temperature dependent, pH measuring systems require some way to compensate for temperature in order to produce standardized pH values. Both pH meters and pH controllers with manual temperature compensation require the user to enter the temperature. Many modern systems have automatic temperature compensation (ATC), and they receive continuous temperature

Table 3.2.1 Dependency of Nernst Potential U_N on Temperature[a]

Temperature (°C)	Slope U_N (mV)	Temperature (°C)	Slope U_N (mV)
0	54.20	50	64.12
5	55.19	55	65.11
10	56.18	60	66.10
15	57.17	65	67.09
20	58.16	70	68.08
25	59.16	75	69.07
30	60.15	80	70.07
35	61.14	85	71.06
37	61.54	90	72.05
40	62.13	95	73.04
45	63.12	100	74.03

[a]Taken from "Basics of Potentiometry," *Metrosensor Electrodes, Theoretical Background* (*http://www.metrohm.com*) with permission from Metrohm Group.

information from a temperature sensing element typically located on the pH electrode. These systems, which automatically correct the pH value based on the temperature of the solution, are generally considered to be more practical for most pH measurements in the laboratory. Ideally, temperature and pH should be measured at the same location in the solution. In modern pH electrodes, the temperature sensor is often located within the electrode in the immediate vicinity of the glass membrane. Some sensors may be located outside the membrane. These external sensors usually require more care when handling and cleaning.

Laboratory pH Meters

A pH meter is an electronic voltmeter capable of detecting and reporting millivolt changes from the pH electrode. pH meters range in cost and complexity. They are available as inexpensive pen-like devices or as complex and expensive instruments that have computer interfaces and several inputs from pH sensor electrodes, reference electrodes, and temperature sensors. An example of a typical meter for solution measurements in life science laboratories is shown in Fig. 3.2.2. Modern meters have temperature compensation capabilities (manual or automatic) to correct for changes in slope that result from changes in temperature. Current technology has allowed for other useful features on modern meters, including calculation of slope and % efficiency, and automatic recognition of calibration buffers. Specialty meters for use in atypical applications, such as harsh environments, are also available. Inexpensive pocket pH meters that automatically compensate for temperature can now be purchased from most manufacturers.

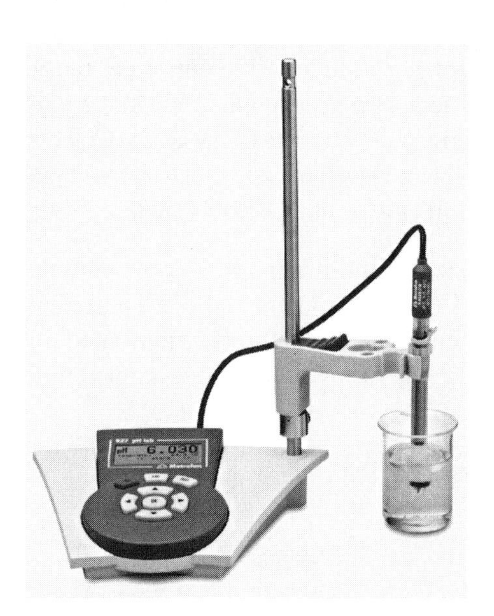

Figure 3.2.2 An example of a typical pH meter used for solution measurements in the molecular life science laboratory. Used with permission from Metrohm Group (*http://www.metrohm.com*).

Laboratory pH Electrodes

Electrodes are made to be used in the field or in the laboratory. This discussion will focus on laboratory electrodes, which manufacturers generally classify into the following categories: general purpose, fermentation, ion-selective, or specialty.

Fermentation electrodes are sturdy, autoclavable, and can be used continuously for several days. Ion-selective electrodes determine the concentration of specific ions and utilize solid state, gas sensing, and liquid membrane electrodes. The majority of this discussion will focus on general-purpose electrodes, the kind commonly found in most life science laboratories and used extensively for the preparation of solutions. A typical electrode for these purposes is shown in Figure 3.2.3.

It is critical that the correct type of electrode be selected. A very high proportion of pH measurement problems are caused by the use of an electrode that is inappropriate for the solution being measured.

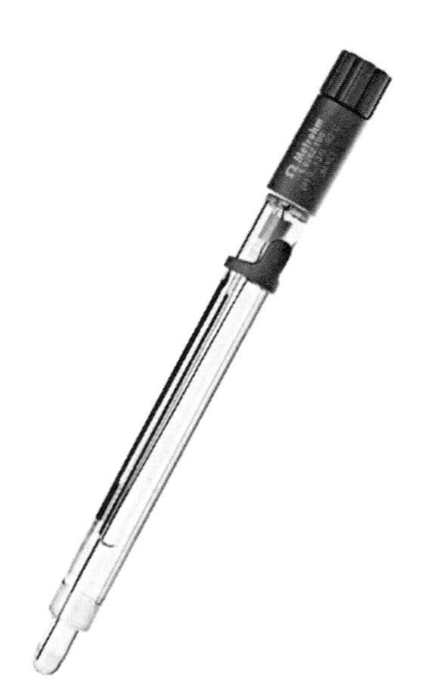

Figure 3.2.3 An example of a typical electrode used for pH measurement of solutions prepared in the molecular life science laboratory. Used with permission from Metrohm Group (*http://www.metrohm.com*).

There are several choices that must be made when selecting an appropriate electrode. Decisions most relevant to the life science laboratory are the following:

1. *Type of sensor material.* Sensors are usually glass membranes, but the type of glass can vary depending on the requirements of the specific application.
2. *Type of internal conductor.* Most common is silver/silver chloride, but other materials may be chosen if one is measuring solutions containing molecules that react with silver.
3. *Refillable or sealed body.* This determines how the reference solution is maintained.
4. *Material of electrode body.* The choices are usually glass or epoxy.
5. *Type of reference junction (the junction between sensing and reference half-cells).* These can be single or double junctions made usually of glass, but can also be made of other materials for special applications.

Type of Sensor Material

Glass sensors

Glass is the most common material used for pH sensing electrodes in the life science laboratory, where aqueous solutions are most commonly being measured. When immersed in an aqueous solution, a thin hydrated layer forms on the silicate mesh of the glass surface of the sensing membrane. Glass that is covered by the layer of hydration is softened, allowing ions to penetrate and alter the electrochemical properties of the glass. The hydrated layer is thus essential for pH measurement. The type of glass used is usually lithium silicate and it is optimized so that generally only protons can penetrate. The selectivity of glass for protons is reflected in its selectivity coefficient, a value that can vary between different electrodes. The values typically range from 10^{-1} to 10^{-15}, lower values representing better selectivity. The most problematic ions that can interfere with pH measurements are sodium and potassium ions, which are also able to penetrate the glass membrane to a certain degree. This is more of an issue in solutions with high concentrations of these ions as they interfere with accurate pH readings at relatively low $[H^+]$ (high pH). Error caused by sodium and potassium ions is referred to as "alkali error." Detailed information about selectivity of the glass surface should be available from the manufacturer.

The glass membranes can be constructed out of different types of glass for use in a variety of conditions. Different glasses can expand the temperature range, or even reduce alkali error at high pH. The standard general purpose glass used is effective over the entire pH range (0 to 14) and at

Table 3.2.2 Overview of Different Electrode Membrane Glasses Used by Metrohm[a]

Application	U glass (green)	T glass (blue)	M glass (colorless)	Aquatrode glass (colorless)
pH range	0–14	0–14	0–14	0–13
Temperature range				
Continuous	0°–80°C	0°–80°C	0°–60°C	0°–80°C
Short-term	0°–100°C			
Membrane surface	Electrodes with large membrane surfaces	Electrodes with medium to large membrane surfaces (mini-electrodes)	Electrodes with small membrane surfaces (microelectrodes)	Electrodes with large surfaces
Special features	For strongly alkaline solutions, long-term measurements and measurements at high temperatures	Measurements in non-aqueous sample solutions	Measurements in small-volume samples	Responds very quickly, so particularly suitable for measurements in ion-deficient or weakly buffered solutions
Membrane resistance (MN)	150–500	40–600	300–700	80–200

[a]Taken from "Basics of Potentiometry," *Metrosensor Electrodes, Theoretical Background* (*http://www.metrohm.com*), with kind permission of Metrohm Group.

temperatures up to 100°C. Amber glass has lower alkali error and is also effective over the entire pH range and at temperatures up to 110°C. Blue glass is effective over a pH range of 0 to 13, but can withstand temperatures up to 110°C. As an example, Table 3.2.2 provides descriptions of different kinds of glasses used by one electrode manufacturer.

Solid-state metal sensors

Solid-state metal sensors are also available. Generally, these are built around ion selective field effect transistors (ISFET). They are typically much smaller and sturdier than glass sensing electrodes and can be wiped clean for dry storage, and thus are more practical for many kinds of field work. However, they are generally not as accurate as glass electrodes, and they cannot be used with standard pH meters unless a special interface is purchased.

ISFET electrodes measure the flow of current between two semiconductor elements generated by an electrostatic field resulting from protonation of an oxide gate. The degree of protonation depends on the proton concentration of the solution, $[H^+]$. The resulting current flowing through the transistor is measured and converted to a pH value.

Type of Internal Conductor

Silver/silver chloride

Ag/AgCl is the most common internal conducting element. It is suitable for most applications in the life science laboratory and has a temperature limit of 80°C. These are often called "general-purpose" electrodes, and can be used with solutions containing at least 5% water and that do not contain substances that react with silver, such as Tris.

Calomel

Hg/Hg_2Cl_2 is another popular option that should be used for solutions containing organics, heavy metals, proteins, or Tris, which may react with silver and/or clog the reference junction (the junction between the reference and measuring half cells). Calomel electrodes are often referred to as SCEs, for standard calomel electrodes.

Refillable or Sealed Body and Type of Refilling Solution

The reference conductor is immersed in a chloride reference solution, most commonly 3 M or saturated potassium chloride (KCl), in part because it has a small diffusion potential. In refillable electrodes, the internal solution slowly flows to the outside through the reference junction. As the internal solution is lost, it must be replaced through a small fill hole near the top of the electrode. This type of electrode is economical and durable. Sealed electrodes are even more durable and require little to no maintenance (no refilling); however, they must be replaced as the reference solution is lost over time.

The type of reference solution can also vary for certain applications. The choices are summarized in Table 3.2.3.

Body Material

The two most common choices for body material are epoxy or glass. Epoxy body electrodes are sturdy and impact-resistant, but should not be used at high temperatures or for inorganic solutions. Glass-body electrodes are more fragile, but can be used at high temperatures and with highly corrosive solvents.

Type of Reference Junction

In combination electrodes (both reference and measuring electrode housed in the same body), the reference junction allows H^+ to pass freely between the reference and sensing half-cells to complete the electrical circuit. The most common type of junction, the single junction, is economical and suitable for general-purpose applications. Double junctions are available and recommended for use with solutions that contain sulfides, heavy metals, or Tris buffers to prevent contamination of the reference cell. The double junction contains an additional chamber between the reference

Table 3.2.3 Alternatives to the Standard Reference Electrolyte [c(KCl) = 3 mol/liter][a]

Medium	Problem with the standard electrolyte	Alternative electrolyte
Silver ions	Reaction with Cl^- with precipitation of AgCl: slow response	KNO_3 saturated
Non-aqueous	Precipitation of KCl, solutions and electrolyte immiscible: unsteady signal	2 mol/liter LiCl in ethanol or LiCl saturated in ethanol
Ion-deficient water	Contamination of the medium by salt: drift	KCl solution of lower concentration
Proteins/polypeptides	Precipitation of the proteins with KCl and AgCl: zero point shift/reduced slope	Idrolyte[b]
Semi-solid substances	Contamination of diaphragm: zero-point shift/slow response	Solid electrolyte in combination with pinhole diaphragm
Surfactants (proteins)	Adsorption on diaphragm: zero-point shift/reduced slope	Porolyte[c]

[a]Taken from "Basics of Potentiometry," *Metrosensor Electrodes, Theoretical Background* (*http://www.metrohm.com*) with permission from Metrohm Group.

[b]Idrolyte is a glycerol-based electrolyte whose chloride ion activity corresponds to that of a KCl solution with c(KCl) = 3 mol/liter. This means that the latter can also be readily replaced by Idrolyte. Idrolyte is excellent for use with solutions containing proteins and aqueous solutions with an organic fraction.

[c]Porolyte is a KCl solution that has been gelled by polymerization and is used in electrodes with a capillary diaphragm (Porotrode).

electrode and external solution. Solution must diffuse through both junctions before reaching the reference electrode, thus minimizing the likelihood of reference contamination. For this reason, double junction electrodes usually have a longer lifespan but they are more expensive.

Most reference junctions are made of H^+-permeable glass. Porous Teflon junctions are also available for use with solutions that may clog glass junctions, such as protein-rich or oily solutions.

pH ELECTRODE CARE

Tips for General Handling

Electrodes are extremely fragile. Always handle carefully and avoid bumping the bottom edge that contains the sensor bulb.

It is recommended that the electrode have a protective plastic cap to protect the glass membrane from damage during use. These are sold by most manufacturers.

3

Glass electrodes must always be immersed in aqueous solution to keep the glass membrane hydrated. Use the solution recommended by the manufacturer, or a neutral solution of 3 M KCl for storage. If accidentally dried, it should be soaked for 24 hr in storage solution before use.

Electrodes should be rinsed with deionized or distilled water between measurements. Excess water can be removed by lightly blotting with lint-free tissue paper. Never rub or wipe the surface of the glass membrane, as static electricity generated can cause inaccurate readings and scratches will permanently disable the electrode.

Do not allow the electrode to dry out internally. Keep it filled with the appropriate filling solution recommended by the manufacturer (usually 3 to 4 M KCl solution).

Do not immerse the electrode in dehydrating solutions such as ethanol, or solutions that can dissolve glass such as hydrofluoric acid or concentrated alkalis.

If using the electrode with solutions of substances that can clog the junction or stick to the surface of the glass membrane, clean the electrode as soon as possible after use (see Cleaning).

Storage and Refilling

The electrode must always be kept moist. The best solution for storage is a 4 M KCl solution, pH 7. Alternatively, the electrode may be stored in a buffer solution at pH 4 or 7 (do not use Tris). Avoid storage in deionized or distilled water as ions will leach out of the glass membrane and disable the electrode. Also avoid long-term storage in solutions with high K^+ or Na^+ ions as they will penetrate the glass over time and cause an increase in electrode response time as the H^+ ions replace the other ions in the hydrated layer. To be safe, special storage solutions available from the manufacturer should be used. A protective rubber boot that fits over the end with the glass membrane is usually supplied with the electrode. The bulb may be used for long-term storage by filling it with enough storage solution to cover the glass bulb.

Electrodes should contain the appropriate amount of filling fluid—filled to, but not past, the refill hole. Care should be taken to add the fill solution when its level within the electrode is 2 cm or more below the fill hole. The filling hole should be closed for storage, but the filling hole plug should be removed, i.e., opened during calibration and pH measurement to allow for flow through the reference junction. The filling solution is typically a saturated KCl solution (3 to 4 M KCl). The presence of a few crystals of KCl indicates that the solution is indeed saturated.

Conditioning

Prior to using the electrode for the first time, it must be conditioned. First, rinse with distilled or deionized water. Soak the electrode for 20 min in 4 M KCl, or a buffer at pH 4 or 7. After soaking, rinse with distilled or deionized water before measuring pH.

Cleaning

The electrode should be cleaned when one of the following occurs: the calibration slope falls outside 0.975 and 1.025, a slow pH response time, or the pH meter indicates that the electrode is not working properly.

Cleaning steps

1. A mixed acid cleaner is recommended. It is made by combining concentrated H_2SO_4 and HNO_3 at a 2:1 ratio.
2. Immerse the end on the electrode (~1 cm) into the acid cleaner for the appropriate amount of time:

   For daily cleaning: 5 min
   After prolonged storage: 15 min
   After use in oily or viscous materials: 12 to 24 hr.

3. Thoroughly rinse the electrode in deionized water.
4. Soak in filling solution or 4 M KCl for 10 min before use.

3

Table 3.2.4 Possible Sources of Errors and Solutions for pH Glass Electrodes[a]

Source of error	Effects	Cleaning	Alternatives
High pH value and high alkali content	Increased alkali error: pH too low		Use of electrodes with amber glass
High temperatures	Rapid rise in membrane resistance by aging: increased polarizability and drift		Use of electrodes with green glass. Careful treatment with etching salt and HCl
Measurements at low temperatures	High membrane resistance: polarization effects		Use of electrodes with blue glass and Idrolyte as reference electrolyte
Dry storage	Zero point drift	Store in water overnight	Storage in reference electrolyte
Reaction of a solution component with the glass	Slow response, zero point shift, slope reduction	Reactivation by treatment with etching salt/HCl	Try other types of glass
Non-aqueous media	Reduced sensitivity	Store in water	Blue glass/non-aqueous electrolyte
Deposition of solids on membrane surface	Slow response, zero point shift, slope reduction	Solvent or strong acid or reactivation by treatment with etching salt/HCl	
Deposition of proteins on membrane surface	Slow response, zero-point shift, slope reduction	5% pepsin in 0.1 mol/liter HCl	

[a]Taken from "Basics of Potentiometry," *Metrosensor Electrodes, Theoretical Background* (*http://www.metrohm.com*) with permission from Metrohm Group.

Alternative method using HCl

For general cleaning, soak electrode in 0.1 M HCl for 30 min.

To clean off proteins, soak in 1% pepsin in 0.1 M HCl for 15 min.

To clean off grease (oil), rinse electrode with a detergent solution.

Following all of the above soaking steps, the electrode should be soaked in filling solution for 1 hr.

If the above methods do not improve electrode performance, then try soaking it for 4 to 8 hr in 1 M HCl, rinse, and then soak in buffer, pH 7 for 1 hr.

Common Problems and Solutions

Table 3.2.4 summarizes common electrode problems and solutions.

MEASUREMENT OF pH

Measuring with pH Meter

Calibration of the pH Meter

For accurate determination of the pH of a solution, the pH meter connected to a specific pH electrode must first be calibrated with reference standard buffer solutions (RSBSs). These ready-to-use solutions have very precise pH at room temperature. The pH values, which are accurate to two decimal places, are supplied on the label. For accurate calibration of a pH meter, three different RSBSs, one at acidic pH (such as 4.0 ±0.01), one at neutral pH (such as 7.0 ±0.01), and one at basic pH (such as 10.0 ±0.01) are used. It is recommended to obtain buffer solutions that the manufacturer claims to be traceable to a recognized standard such as the U.S. National Institute of Standards and Technology (NIST) standard reference materials. In older model pH meters, the gain and offset settings needed repeated adjustments as the electrode was alternately placed in two RSBSs (two-point calibration; see below) until accurate readings were obtained in both solutions. This procedure is completely automated in modern instruments and usually requires single immersion of the electrode in each RSBS. The calibration process correlates the voltage produced by the electrode with the pH scale. A general guideline on calibration of a pH meter is described below.

A pH meter is usually calibrated by the two-point calibration method, which involves calibration of the meter with two different RSBSs with known pH and then checking the meter against a pH 7.00 RSBS solution to validate the calibration. The pH meter should be calibrated at least once on the day it is used. If the final pH of the solution is between pH 7.0 and 10.0, then the RSBSs of pH 7.0 and 10.0 should be used for two-point calibration. On the other hand if the final pH is between pH 4.0 and 7.0, then the RSBSs of pH 7.0 and 4.0 should be used. In between the following steps for calibration, the electrode should be rinsed with deionized or distilled water observing the precautions mentioned in the section on Tips for General Handling of pH Electrodes. If the pH meter does not have an ATC, the temperature dial on the pH meter should be adjusted to room temperature as long as that represents the temperature of the RSBS used.

For two-point calibration, the tip of the electrode is first placed into the pH 7.0 RSBS and the necessary button (usually "Cal" or "Calibration") is pressed. The pH meter will show the pH. Following washing, the tip of the electrode is then placed in the second RSBS (either pH 4.0 or 10.0) and the necessary button is pressed again. The pH meter will display the pH. The pH meter will also show the electrode slope value. If the slope is <0.95 or >1.05, one should consult the troubleshooting guide for the specific pH meter: usually the electrode requires cleaning. For three-point calibration, all three RSBSs (i.e., pH 4.0, 7.0, and 10.0) are used instead of the two used above. It should be noted that the automated calibration described here uses preset values for the RSBSs, which should be available in the manual for the pH meter. These values should always be checked against the pH of the standards being used. If they are different, the preset values should be modified accordingly.

Calibration against one pH reference buffer is usually sufficient to ensure accurate pH measurement; however, two- and three-point calibrations are used for the most reliable results and will result in a pH calibration curve that is informative about electrode function (Fig. 3.2.4).

Measurement of pH

Following calibration, the electrode is rinsed with deionized or distilled water and the tip is immersed in the solution of unknown pH. The necessary button on the pH meter, usually Read, is pressed. The pH reading should stabilize within a short time and the determined pH value is recorded. The electrode is then removed, rinsed again, and stored as stated before. For measurement of pH of solutions not at room temperature, care, as previously stated, should be taken.

3

Adjustment of pH

The pH of a solution can be adjusted with the addition of acid or base solution. If adjustments are made, it is best to keep the solution stirring constantly while the acid or base is added, but care should be taken to ensure that the stir bars do not come in contact the electrode. After each addition of acid or base, the electrode should be allowed to stabilize before the next addition is made.

Measuring with pH Strips

A rough estimation of the pH of a solution can be made with pH strips (or pH papers). These consist of paper (or fabric for better accuracy) that has been impregnated with pH indicators: weak acids and bases that change color depending on their protonation state. In most cases, the pH must change by 2 units to result in a complete color change. To measure pH, immerse the strip of paper in the solution and then wait several seconds for any color change. The colors on the strip are then compared to the color scale, usually printed on the box.

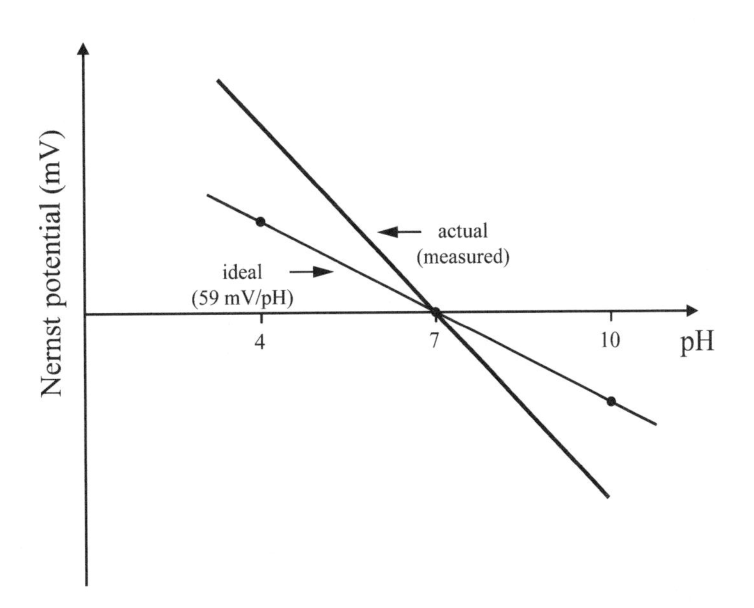

Figure 3.2.4 An example of the ideal (theoretical) slope of 59 mV/pH versus the actual measured slope calculated from calibration standards. The difference is portrayed as the fraction of the ideal. For example, if the actual slope shown in this figure is 61 mV/pH, then the fraction would be 1.03. Adapted from “Basics of Potentiometry,” *Metrosensor Electrodes, Theoretical Background* (*http://www.metrohm.com*) with permission from Metrohm Group.

Strips are available to test the entire pH range (0 to 14) or to test narrower windows. Most strips have a shelf life of at least 2 years if stored in the original box and out of direct sunlight. The strips are generally effective with many different kinds of solutions, but deeply colored solutions may interfere with the color readout.

INTERNET RESOURCES

http://www.pH-measurement.com/pdf.shtml
This document provides a broad range of comprehensive, in-depth explanations on instrumentation, equipment choices, and pH measurement principles.

http://www.pH-meter.info
This site provides excellent explanations about concepts related to pH measurement.

http://www.coleparmer.com/techinfo/techinfo.asp?htmlfile=pHMeasurement.htm
This site provides comprehensive information related to pH instrumentation and basic measurement theory.

UNIT 3.3

Recipes for Commonly Encountered Reagents

INTRODUCTION

This unit describes the preparation of buffers and reagents commonly encountered in the life science laboratory. Prior to preparing any reagent, please review *UNIT 3.1* and *APPENDIX 1*.

When preparing solutions, use deionized or distilled water and reagents of the highest available grade (see *UNIT 3.1*). Sterilization (see *UNIT 4.1*)—by filtration through a 0.22-µm filter or by autoclaving—is recommended for most solutions stored at room temperature and is essential for cell culture applications (e.g., *UNIT 4.2*). Where storage conditions are not specified, store up to 6 months at room temperature. Discard any reagent that shows evidence of contamination, precipitation, or discoloration. Where indicated, solutions may be frozen in aliquots of sizes appropriate to their use. Refreezing thawed solutions is generally discouraged.

CAUTION: Handle strong acids and bases with care. See *APPENDIX 1* for more information concerning the use of hazardous chemicals.

RECIPES

Acid, concentrated stock solutions

See Table 3.3.1.

Acid precipitation solution

1 M HCl (Table 3.3.1)
0.1 M sodium pyrophosphate

Nucleic acids can also be precipitated with a 10% (w/v) solution of trichloroacetic acid (TCA); however, this recipe is cheaper, easier to prepare, and just as efficacious.

Table 3.3.1 Molarities and Specific Gravities of Concentrated Acids and Bases[a]

Acid/base	Molecular weight	% by weight	Molarity (approximate)	Specific gravity	1 M solution (ml/liter)
Acetic acid (glacial)[b]	60.05	99.6	17.4	1.05	57.5
Ammonium hydroxide	35.0	28	14.8	0.90	67.6
Formic acid[b]	46.03	90	23.6	1.205	42.4
		98	25.9	1.22	38.5
Hydrochloric acid	36.46	36	11.6	1.18	85.9
Nitric acid	63.01	70	15.7	1.42	63.7
Perchloric acid	100.46	60	9.2	1.54	108.8
		72	12.2	1.70	82.1
Phosphoric acid[b]	98.00	85	14.7	1.70	67.8
Sulfuric acid	98.07	98	18.3	1.835	54.5

[a]*CAUTION:* Handle strong acids and bases carefully (see *APPENDIX 1*).
[b]Also see Table 3.3.3.

Ammonium acetate, 10 M

Dissolve 385.4 g ammonium acetate in 150 ml H_2O
Add H_2O to 500 ml

Ammonium hydroxide, concentrated stock solution

See Table 3.3.1.

Ammonium sulfate, saturated

76 g ammonium sulfate
100 ml H_2O
Heat with stirring to just below boiling point
Let stand overnight at room temperature

3

ATP, 100 mM

1 g ATP (adenosine triphosphate)
12 ml H_2O
Adjust pH to 7.0 with 4 M NaOH
Adjust volume to 16.7 ml with H_2O
Store in aliquots indefinitely at −20°C

Base, concentrated stock solutions

See Table 3.3.1.

BCIP, 5% (w/v)

Dissolve 0.5 g 5-bromo-4-chloro-3-indolyl phosphate disodium salt (stored at −20°C) in 10 ml of 100% dimethylformamide (DMF). Store wrapped in aluminum foil up to 6 months at 4°C.

The BCIP may not dissolve completely. Vortex the solution immediately before use and pipet with a wide-mouth pipet tip.

Discard the solution if it turns pinkish.

BSA (bovine serum albumin), 10% (w/v)

Dissolve 10 g BSA (e.g., Sigma) in 100 ml H_2O. Filter sterilize using a low-protein-binding 0.22-µm filter. Store indefinitely at 4°C.

Lower-concentration stock solutions (e.g., 1%), which are useful for various applications, can be made by diluting 10% stock appropriately with sterile water. See UNIT 3.1 for a discussion of stock solutions.

BSA is available in various forms that differ in fraction of origin, preparation, purity, pH, and cost; the most commonly used is fraction V. Use the form that is appropriate for the application; this may need to be optimized empirically.

$CaCl_2$, 1 M

147 g $CaCl_2 \cdot 2H_2O$
H_2O to 1 liter

Denhardt's solution, 100×

10 g Ficoll 400
10 g polyvinylpyrrolidone
10 g BSA (Pentax Fraction V; Miles Laboratories)
H_2O to 500 ml
Filter sterilize and store in 25-ml aliquots at −20°C

DEPC (diethylpyrocarbonate)–treated solutions

Refer to *UNITS 5.2 & 8.2*.

DNase I, RNase-free (1 mg/ml)

Prepare a solution of 0.1 M iodoacetic acid plus 0.15 M sodium acetate and adjust pH to 5.3. Filter sterilize. Add sterile solution to lyophilized RNase-free DNase I (e.g., Worthington) to give a final concentration of 1 mg/ml. Heat 40 min at 55°C and then cool. Add 1 M $CaCl_2$ (see recipe) to a final concentration of 5 mM. Store at −80°C in small aliquots.

dNTPs: dATP, dTTP, dCTP, and dGTP

Concentrated stocks: Purchase deoxyribonucleoside triphosphates (dNTPs) from a commercial supplier either as ready-made 100 mM solutions (the preferred form for shipping and storage) or in lyophilized form. If purchased lyophilized, dissolve dNTPs in deionized H_2O to an expected concentration of 30 mM, then adjust to pH 7.0 with 1 M NaOH (to prevent acid-catalyzed hydrolysis). Determine the actual concentration of each dNTP by UV spectrophotometry at 260 nm (*UNIT 2.1*), referring to the extinction coefficients given in Table 3.3.2.

Table 3.3.2 Molar Extinction Coefficients of DNA Bases

Base	Molar extinction coefficient (ε)[a]
Adenine	15,200
Cytosine	7050
Guanosine	12,010
Thymine	8400

[a]1 M solution measured at 260 nm; see Wallace and Miyada (1987).

Working solutions: Prepare working solutions of desired concentration (commonly 2 mM) for each dNTP by diluting concentrated stocks appropriately. Remember that the molarity of the 3dNTP and 4dNTP mixes refers to the concentration of *each* precursor present in the solution.

4dNTP mixes: For use in various molecular biology applications, prepare mixed dNTP solutions containing equimolar amounts of all four DNA precursors, e.g.,

2 mM 4dNTP mix: 2 mM each dATP, dTTP, dCTP, dGTP
1.25 mM 4dNTP mix: 1.25 mM each of dATP, dTTP, dCTP, dGTP.

3dNTP mixes: For use in radioactive labeling procedures (see *UNIT 8.4*), prepare similar stocks lacking one particular dNTP but containing equimolar amounts of the remaining three precursors, e.g., 2 mM 3dNTP mix (minus dATP): 2 mM each of dTTP, dCTP, dGTP.

Store dNTPs and dNTP mixtures as aliquots up to 1 year at −20°C.

Also refer to UNIT 10.2.

DPBS (Dulbecco's phosphate-buffered saline)

8.00 g NaCl (0.137 M)
0.20 g KCl (2.7 mM)
0.20 g KH_2PO_4 (1.1 mM)
0.10 g $MgCl_2 \cdot 6H_2O$ (0.5 mM)
2.16 g $Na_2HPO_4 \cdot 7H_2O$ (8.1 mM)
0.10 g anhydrous $CaCl_2$ (0.9 mM)
H_2O to 1 liter

continued

There are different formulations commonly referred to as "PBS." Be sure that the formulation you are using is appropriate to the task at hand. See recipes for phosphate-buffered saline (PBS) and phosphate-buffered saline containing potassium (KPBS) for examples of alternative formulations.

DPBS may be made or purchased without Ca^{2+} and Mg^{2+} (CMF-DPBS). These components are optional and usually have no effect on an experiment; in a few cases, however, their presence may be detrimental. Prior to use, determine if the presence or absence of these components is optimal.

DTT (dithiothreitol), 1 M

Dissolve 1.55 g DTT in 10 ml water
Filter sterilize
Store in aliquots at −20°C

EDTA (ethylenediaminetetraacetic acid), 0.5 M (pH 8.0)

Dissolve 186.1 g disodium EDTA dihydrate in 700 ml water. Adjust pH to 8.0 with 10 M NaOH (~50 ml; add slowly). Add water to 1 liter and filter sterilize.

Begin titrating before the sample is completely dissolved. EDTA, even in the disodium salt form, is difficult to dissolve at this concentration unless the pH is increased to between 7 and 8.

3

Ethidium bromide staining solution

Refer to UNITS 2.2 & 7.2.

Formamide loading buffer, 2×

Prepare in deionized formamide:
0.05% (w/v) bromphenol blue
0.05% (w/v) xylene cyanol FF
20 mM EDTA
Do not sterilize
Store up to 6 months at −20°C

HCl, 1 M

Mix in the following order:
913.8 ml H_2O
86.2 ml concentrated HCl (Table 3.3.1)

CAUTION: *Always handle concentrated acids carefully.*

KCl, 1 M

74.6 g KCl
H_2O to 1 liter

2-ME (2-mercaptoethanol), 50 mM

Prepare 1 M stock:
0.5 ml 14.3 M 2-ME
6.6 ml H_2O
Prepare 50 mM stock:
5 ml 1 M 2-ME
95 ml H_2O
Store up to 1 week at 4°C or up to 6 months at −20°C

Dilute to 50 µM (final) for use in media.

$MgCl_2$, 1 M

20.3 g $MgCl_2 \cdot 6H_2O$
H_2O to 100 ml

$MgSO_4$, 1 M

24.6 g $MgSO_4 \cdot 7H_2O$
H_2O to 100 ml

MOPS buffer

0.2 M MOPS [3-(*N*-morpholino)-propanesulfonic acid], pH 7.0 (Table 3.3.3)
0.5 M sodium acetate
0.01 M EDTA
Store in the dark and discard if it turns yellow

Refer to UNIT 7.2 *for MOPS running buffer.*

Table 3.3.3 pK_a Values and Molecular Weights for Components of Some Common Biological Buffers[a]

Name	Chemical formula or IUPAC name	pK_{a1}	Useful pH range	MW (g/mol)
Phosphoric acid[b]	H_3PO_4	2.12 (pK_{a1})	–	98.00
Citric acid[c]	$C_6H_8O_7(H_3Cit)$	3.06 (pK_{a1})	–	192.1
Formic acid[b]	HCOOH	3.75	–	46.03
Succinic acid	$C_4H_6O_4$	4.19 (pK_{a1})	–	118.1
Citric acid[c]	$C_6H_7O_7^-(H_2Cit^-)$	4.74 (pK_{a2})	–	
Acetic acid[b]	CH_3COOH	4.75	–	60.05
Citric acid[c]	$C_6H_6O_7^-(HCit^{2-})$	5.40 (pK_{a3})	–	
Succinic acid	$C_4H_5O_4^-$	5.57 (pK_{a2})	–	
MES	2-(*N*-morpholino)ethanesulfonic acid	6.15	5.5-6.7	195.2
Bis-Tris	bis(2-hydroxyethyl)imino tris(hydroxymethyl)methane	6.50	5.8-7.2	209.2
ADA	*N*-(2-acetamido)-2-iminodiacetic acid	6.60	6.0-7.2	190.2
PIPES	Piperazine-*N*,*N′*-bis(2-ethanesulfonic acid)	6.80	6.1-7.5	302.4
ACES	*N*-(carbamoylmethyl)-2-amino-ethanesulfonic acid	6.80	6.1-7.5	182.2
Imidazole	1,3-diaza-2,4-cyclopentadiene	7.00	–	68.08
Diethylmalonic acid	$C_7H_{12}O_4$	7.20	–	160.2
MOPS	3-(*N*-morpholino)propanesulfonic acid	7.20	6.5-7.9	209.3
Sodium phosphate, monobasic	NaH_2PO_4	7.21 (pK_{a2})	–	120.0
Potassium phosphate, monobasic	KH_2PO_4	7.21 (pK_{a2})	–	136.1
TES	*N*-tris(hydroxymethyl)methyl-2-aminoethanesulfonic acid	7.40	6.8-8.2	229.3
HEPES	*N*-(2-hydroxyethyl)piperazine-*N′*-(2-ethanesulfonic acid)	7.55	6.8-8.2	238.3
HEPPSO	*N*-(2-hydroxyethyl)piperazine-*N′*-(2-hydroxypropanesulfonic acid)	7.80	7.1-8.5	268.3
Glycinamide HCl	$C_2H_6N_2O \cdot HCl$	8.10	7.4-8.8	110.6

continued

Table 3.3.3 pK_a Values and Molecular Weights for Components of Some Common Biological Buffers[a], *continued*

Name	Chemical formula or IUPAC name	pK_{a1}	Useful pH range	MW (g/mol)
Tricine	*N*-tris(hydroxymethyl)methylglycine	8.15	7.4-8.8	179.2
Glycylglycine	$C_4H_8N_2O_3$	8.20	7.5-8.9	132.1
Tris	Tris(hydroxymethyl)aminomethane	8.30	7.0-9.0	121.1
Bicine	*N,N*-bis(2-hydroxyethyl)glycine	8.35	7.6-9.0	163.2
Boric acid	H_3BO_3	9.24	–	61.83
CHES	2-(*N*-cyclohexylamino)ethane-sulfonic acid	9.50	8.6-10.0	207.3
CAPS	3-(cyclohexylamino)-1-propane-sulfonic acid	10.40	9.7-11.1	221.3
Sodium phosphate, dibasic	$Na_2H_3PO_4$	12.32 (pK_{a3})	–	142.0
Potassium phosphate, dibasic	$K_2H_3PO_4$	12.32 (pK_{a3})	–	174.2

[a]Some data reproduced from Buffers: A Guide for the Preparation and Use of Buffers in Biological Systems (Mohan, 1997) with permission of Calbiochem.
[b]Also see Table 3.3.1.
[c]Available as a variety of salts, e.g., ammonium, lithium, sodium.

NaCl, 5 M

292 g NaCl
H_2O to 1 liter

NaOH, 10 M

Dissolve 400 g NaOH in 450 ml H_2O
Add H_2O to 1 liter

PCR amplification buffer

Refer to *UNIT 10.2*.

Phenol, water-saturated

Refer to *UNIT 5.2*.

Phenol/chloroform/isoamyl alcohol, 25:24:1 (v/v/v)

Refer to *UNIT 5.2*.

Phosphate-buffered saline (PBS)

0.23 g NaH_2PO_4 (anhydrous; 1.9 mM)
1.15 g Na_2HPO_4 (anhydrous; 8.1 mM)
9.00 g NaCl (154 mM)
Add H_2O to 900 ml
Adjust to desired pH (7.2 to 7.4) using 1 M NaOH or 1 M HCl (Table 3.3.1)
Add H_2O to 1 liter

There are different formulations commonly referred to as "PBS." Be sure that the formulation you are using is appropriate to the task at hand. See recipes for Dulbecco's phosphate-buffered saline (PBS) and phosphate-buffered saline containing potassium (KPBS) for examples of alternative formulations.

Phosphate-buffered saline containing potassium (KPBS)

8.00 g NaCl (0.137 M)
0.20 g KCl (2.7 mM)
0.24 g KH_2PO_4 (1.4 mM)
1.44 g Na_2HPO_4 (0.01 M)
H_2O to 1 liter

There are different formulations commonly referred to as "PBS." Be sure that the formulation you are using is appropriate to the task at hand. See recipes for Dulbecco's phosphate-buffered saline (PBS) and phosphate-buffered saline (PBS) for examples of alternative formulations.

PMSF (phenylmethylsulfonyl fluoride), 100 mM

Dissolve 0.174 g PMSF in 10 ml of 100% ethanol, isopropanol, or methanol. Store in aliquots up to 2 years at −20°C.

CAUTION: *Phenylmethylsulfonyl fluoride is toxic.*

Make fresh dilutions from the alcohol stock for each use, because the half-life of PMSF in aqueous solution is <30 min at room temperature and a few hours on ice.

If PMSF is being added to a solution without detergent, the solution should be stirred vigorously during PMSF addition because PMSF has a tendency to form an insoluble precipitate in aqueous solution.

3

Potassium acetate buffer, 0.1 M

Solution A: 11.55 ml glacial acetic acid (see Tables 3.3.1 and 3.3.3) per liter (0.2 M) in water
Solution B: 19.6 g potassium acetate ($KC_2H_3O_2$) per liter (0.2 M) in water

Referring to Table 3.3.4 for desired pH, mix the indicated volumes of solutions A and B, then dilute with water to 100 ml. Filter sterilize if necessary. Store up to 3 months at room temperature.

This may be made as a 5- or 10-fold concentrate by scaling up the amount of sodium acetate in the same volume. Acetate buffers show concentration-dependent pH changes, so check the pH by diluting an aliquot of concentrate to the final concentration.

To prepare buffers with pH intermediate between the points listed in Table 3.3.4, prepare the closest higher pH, then titrate with solution A.

Table 3.3.4 Preparation of 0.1 M Sodium and Potassium Acetate Buffers[a,b]

Desired pH	Solution A (ml)	Solution B (ml)
3.6	46.3	3.7
3.8	44.0	6.0
4.0	41.0	9.0
4.2	36.8	13.2
4.4	30.5	19.5
4.6	25.5	24.5
4.8	20.0	30.0
5.0	14.8	35.2
5.2	10.5	39.5
5.4	8.8	41.2
5.6	4.8	45.2

[a]Adapted with permission of CRC (1975).
[b]See Tables 3.3.1 and 3.3.3 for more information regarding acetic acid.

Potassium phosphate buffer, 0.1 M

Solution A: 27.2 g KH_2PO_4 per liter (0.2 M final) in water

Solution B: 34.8 g K_2HPO_4 per liter (0.2 M final) in water

Referring to Table 3.3.5 for desired pH, mix the indicated volumes of solutions A and B, then dilute with water to 200 ml. Filter sterilize if necessary. Store up to 3 months at room temperature.

This buffer may be made as a 5- or 10-fold concentrate simply by scaling up the amount of potassium phosphate in the same final volume. Phosphate buffers show concentration-dependent changes in pH, so check the pH of the concentrate by diluting an aliquot to the final concentration.

To prepare buffers with pH intermediate between the points listed in Table 3.3.5, prepare the closest higher pH, then titrate with solution A.

Table 3.3.5 Preparation of 0.1 M Sodium and Potassium Phosphate Buffers[a,b]

Desired pH	Solution A (ml)	Solution B (ml)
5.7	93.5	6.5
5.8	92.0	8.0
5.9	90.0	10.0
6.0	87.7	12.3
6.1	85.0	15.0
6.2	81.5	18.5
6.3	77.5	22.5
6.4	73.5	26.5
6.5	68.5	31.5
6.6	62.5	37.5
6.7	56.5	43.5
6.8	51.0	49.0
6.9	45.0	55.0
7.0	39.0	61.0
7.1	33.0	67.0
7.2	28.0	72.0
7.3	23.0	77.0
7.4	19.0	81.0
7.5	16.0	84.0
7.6	13.0	87.0
7.7	10.5	90.5
7.8	8.5	91.5
7.9	7.0	93.0
8.0	5.3	94.7

[a]Adapted by permission of CRC (1975).
[b]Also see Table 3.3.3 for more information.

RNase A stock solution, DNase-free (2 mg/ml)

Dissolve RNase A (e.g., Sigma) in DEPC-treated H_2O (see UNIT 8.2) to 2 mg/ml. Boil 10 min in a 100°C water bath. Store up to 1 year at 4°C.

The activity of the enzyme varies from lot to lot; therefore, prepare several 10-ml aliquots of each dilution to facilitate standardization.

Saline, 0.9% (w/v)

9 g NaCl (154 mM final)
H_2O to 1 liter

SDS, 20% (w/v)

Dissolve 20 g sodium dodecyl sulfate (SDS) in H_2O to 100 ml total volume with stirring. Pass through a 0.45-μm filter to remove particulates.

It may be necessary to heat the solution slightly to fully dissolve the powder.

SDS sample buffer

See Table 3.3.6.

Table 3.3.6 Preparation of SDS Sample Buffer

Ingredient	2×	4×	Final conc. in 1× buffer
0.5 M Tris·Cl, pH 6.8[a]	2.5 ml	5.0 ml	62.5 mM
SDS	0.4 g	0.8 g	2% (w/v)
Glycerol	2.0 ml	4.0 ml	10% (v/v)
Bromphenol blue	20 mg	40 mg	0.1% (w/v)
2-mercaptoethanol[a,b,c]	400 μl	800 μl	~300 mM
H_2O	to 10 ml	to 10 ml	—

[a]See recipe.
[b]Alternatively, dithiothreitol (DTT; see recipe), at a final concentration of 100 mM, can be substituted for 2-mercaptoethanol.
[c]Add just before use.

Silanized glassware

For smaller items: In a well vented fume hood, place glassware or plasticware (e.g., tubes, tips) in a dedicated vacuum desiccator with an evaporating dish containing 1 ml dichlorodimethylsilane. Apply a vacuum with an aspirator and allow ~50% of the liquid to evaporate (several minutes). Turn off the aspirator and allow the items to remain under vacuum for 30 min. Remove the lid and allow fumes to vent into the hood for ~30 min. Rinse with water or autoclave, if desired.

For larger items (e.g., glass plates for denaturing polyacrylamide sequencing gels): Silanize items that do not fit in a desiccator by briefly rinsing with or soaking in a solution of ~5% dichlorodimethylsilane in a volatile organic solvent (e.g., chloroform, heptane). Remove the organic solvent by evaporation, allowing deposition of dichlorodimethylsilane.

CAUTION: *Dichlorodimethylsilane vapors are toxic and highly flammable. Always use a fume hood.*

IMPORTANT NOTE: *Do not leave the desiccator attached to the vacuum pump. This will suck away the silane, minimizing deposition and damaging the pump.*

Refer to UNIT 3.1 for a discussion of when silanization might be useful.

Sodium acetate, 3 M

Dissolve 408 g sodium acetate trihydrate ($NaC_2H_3O_2 \cdot 3H_2O$) in 800 ml H_2O
Adjust pH to 4.8, 5.0, or 5.2 (as desired) with 3 M acetic acid
Add H_2O to 1 liter
Filter sterilize

Sodium acetate buffer, 0.1 M

Solution A: 11.55 ml glacial acetic acid (see Tables 3.3.1 and 3.3.3) per liter (0.2 M) in water
Solution B: 27.2 g sodium acetate ($NaC_2H_3O_2 \cdot 3H_2O$) per liter (0.2 M) in water

Referring to Table 3.3.4 for desired pH, mix the indicated volumes of solutions A and B, then dilute with water to 100 ml. Filter sterilize if necessary. Store up to 3 months at room temperature.

This may be made as a 5- or 10-fold concentrate by scaling up the amount of sodium acetate in the same volume. Acetate buffers show concentration-dependent pH changes, so check the pH by diluting an aliquot of concentrate to the final concentration.

To prepare buffers with pH intermediate between the points listed in Table 3.3.4, prepare the closest higher pH, then titrate with solution A.

Sodium phosphate buffer, 0.1 M

Solution A: 27.6 g $NaH_2PO_4 \cdot H_2O$ per liter (0.2 M) in water
Solution B: 53.65 g $Na_2HPO_4 \cdot 7H_2O$ per liter (0.2 M) in water

Referring to Table 3.3.5 for desired pH, mix the indicated volumes of solutions A and B, then dilute with water to 200 ml. Filter sterilize if necessary. Store up to 3 months at room temperature.

This buffer may be made as a 5- or 10-fold concentrate by scaling up the amount of sodium phosphate in the same final volume. Phosphate buffers show concentration-dependent changes in pH, so check the pH by diluting an aliquot of the concentrate to the final concentration.

To prepare buffers with pH intermediate between the points listed in Table 3.3.5, prepare the closest higher pH, then titrate with solution A.

SSC (sodium chloride/sodium citrate), 20×

Dissolve the following in 900 ml H_2O:
175 g NaCl (3 M final)
88 g trisodium citrate dihydrate (0.3 M final)
Adjust pH to 7.0 with 1 M HCl (Table 3.3.1)
Adjust volume to 1 liter
Filter sterilize

SSPE (sodium chloride/sodium phosphate/EDTA), 20×

175.2 g NaCl
27.6 g $NaH_2PO_4 \cdot H_2O$
7.4 g disodium EDTA
800 ml H_2O
Adjust pH to 7.4 with 6 M NaOH, then bring volume to 1 liter with H_2O
Filter sterilize

The final sodium concentration of 20× SSPE is 3.2 M.

T4 DNA ligase buffer, 10×

500 mM Tris·Cl, pH 7.6 (see recipe).
100 mM $MgCl_2$
10 mM DTT (see recipe)

continued

10 mM ATP (see recipe)
250 μg/ml BSA
Store up to 6 months in aliquots at −20°C

TAE buffer

Refer to *UNIT 7.2*.

TBE buffer

Refer to *UNIT 7.2*.

TBS (Tris-buffered saline)

100 mM Tris·Cl, pH 7.5 (see recipe)
0.9% (w/v) NaCl
Store up to several months at 4°C

TCA (trichloroacetic acid), 100% (w/v)

500 g TCA
227 ml H_2O

TE (Tris/EDTA) buffer

10 mM Tris·Cl, pH 7.4, 7.5, or 8.0 (or other pH; see recipe)
1 mM EDTA, pH 8.0 (see recipe)

TEA (triethanolamine) solution

50 mM triethanolamine, pH ∼11.5
0.1% (v/v) Triton X-100
0.15 M NaCl

Add Triton X-100 from a 10% stock (see recipe).

Tris·Cl, 1 M

Dissolve 121 g Tris base in 800 ml H_2O
Adjust to desired pH with concentrated HCl (Table 3.3.1)
Adjust volume to 1 liter with H_2O
Filter sterilize if necessary
Store up to 6 months at 4°C or room temperature

IMPORTANT NOTE: *The pH of Tris buffers changes significantly with temperature, decreasing approximately 0.028 pH units per 1°C. Tris-buffered solutions should be adjusted to the desired pH at the temperature at which they will be used. Because of the pK_a, Tris should not be used as a buffer below pH ∼7.2 or above pH ∼9.0 (see Table 3.3.3).*

Approximately 70 ml HCl is needed to achieve a pH 7.4 solution and ∼42 ml for a solution that is pH 8.0.

Tris buffer cannot be used with DEPC (see UNITS 5.2 & 8.2).

Triton X-100, 10% (w/v)

1 g Triton X-100
H_2O to 10 ml
Stir to dissolve
Pass through a 0.45-μm filter to remove particulates
Store protected from light

TTBS (Tween 20/TBS)

Dissolve 0.1% (w/v) polyoxyethylenesorbitan monolaurate (Tween 20) in TBS (see recipe). Store up to several months at 4°C.

Urea loading buffer, 2×

5 mg bromphenol blue (0.05% w/v)
5 mg (w/v) xylene cyanol FF (0.05% w/v)
4.8 g urea (8 M)
186 mg EDTA (50 mM)
H_2O to 10 ml
Do not sterilize

LITERATURE CITED

Chemical Rubber Company (CRC). 1975. CRC Handbook of Biochemistry and Molecular Biology, Physical and Chemical Data, 3rd ed., Vol. 1. CRC Press, Boca Raton, Fla.

Mohan, C. (ed.) 1997. Buffers: A Guide for the Preparation and Use of Buffers in Biological Systems. Calbiochem, San Diego.

Wallace, R.B. and Miyada, C.G. 1987. Oligonucleotide probes for the screening of recombinant DNA libraries. *In* Methods of Enzymology, Vol. 152: Guide to Molecular Cloning Techniques (S.L. Berger and A.R Kimmel, eds.) pp. 432-442. Academic Press, San Diego.

Chapter 4

Cell Culture Techniques

UNIT 4.1

Aseptic Technique

Tomasz Bykowski[1] **and Brian Stevenson**[1]
[1]University of Kentucky College of Medicine, Lexington, Kentucky

OVERVIEW AND PRINCIPLES

Aseptic technique is a set of routine measures that are taken to prevent cultures, sterile media stocks, and other solutions from contamination by unwanted microorganisms (i.e., sepsis). While such actions are sometimes called "sterile technique," that terminology is appropriate only in reference to preventing introduction of any organisms to laboratory or medical equipment and reagents (such those as used for surgery). Since the goal of a biologist is to grow microorganisms or eukaryotic cells without introduction of extraneous organisms, aseptic techniques are crucial for accurate and meaningful experimentation.

One should keep in mind that a completely sterile working environment does not exist. However, there are a number of simple, common sense procedures that will reduce the risk of culture contaminations. Examples of aseptic technique are cleaning and disinfecting laboratory surfaces prior to use, limiting the duration that cultures or media are uncapped and exposed to the air, keeping petri dishes closed whenever possible, effectively sterilizing inoculating loops and other equipment that comes into contact with cultures or media, and avoiding breathing on cultures or sterile instruments. These precautions will become second nature after practical laboratory experience.

The Bunsen Burner

Probably the easiest way to create a relatively sterile environment on the laboratory bench is by using a simple gas-powered burner. This common piece of equipment burns a continuous stream of a flammable gas (usually natural gas [methane]), based upon a design made almost 150 years ago by the German chemist Robert Wilhelm Bunsen (1811–1899; Fig. 4.1.1). A major purpose of the open flame in aseptic technique is to create a cone of hot air above and around the laboratory bench to reduce the viability of organisms on suspended dust particles. The ability of the Bunsen burner flame to heat things very quickly also makes it an ideal choice to sterilize inoculating loops, warm glass bottle necks, or ignite alcohol on culture spreaders.

The Laminar Flow Unit

A laminar flow unit (or hood) is a sophisticated appliance that can further help protect reagents and biological cultures from contamination. Used correctly, it provides the work space with clean, ultra-filtered air. It also keeps room air from entering the work area, and both suspends and removes airborne contaminants introduced into the work area by personnel.

The most important part of a laminar flow hood is a high-efficiency bacteria-retentive filter called HEPA (high-efficiency particulate air). A certified HEPA filter must capture a minimum of 99.97% of dust, pollen, mold, bacteria, and any airborne particles with a size of >0.3 µm at a flow of 85 liters per min (liters/min). The first HEPA filters were developed in the 1940s by the United States Atomic Energy Commission as part of the Manhattan Project (the development of the atomic bomb) to provide an efficient, effective way to filter radioactive particulate contaminants. HEPA filter technology was declassified after World War II, allowing extensive research and commercial use.

Current Protocols Essential Laboratory Techniques 4.1.1-4.1.11

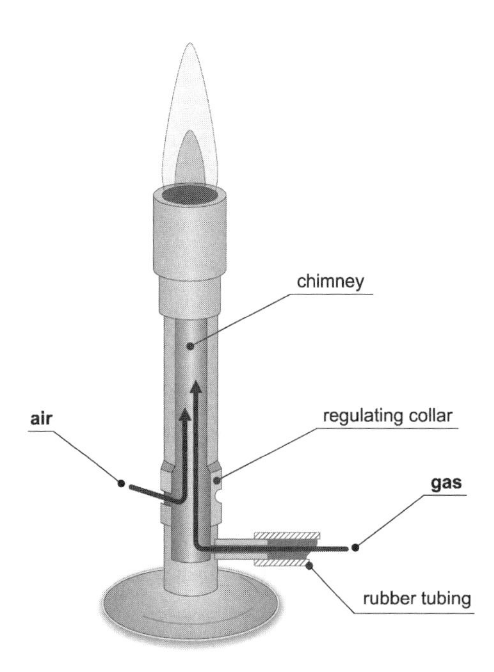

Figure 4.1.1 The Bunsen burner. Gas burner consists of a vertical metal tube through which a narrow jet of natural gas is directed. Air is drawn in via air holes located near the stand. The gas/air mixture burns above the upper opening. The metal collar can be turned to cover or partly cover the air holes, which allows regulation of the amount of air sucked in and hence the heat and shape of the flame.

STRATEGIC PLANNING

Contamination Risk Assessment and Organizing Space

Localizing potential sources of biological contamination is an important part of arranging space when a room is used as a laboratory. Benches for aseptic work should be organized away from air conditioning vents, cooling fans, and open windows or blowers from heating or refrigerating systems. Containers used for the disposal of biological material may contain large numbers of viable microorganisms, even when emptied. Some contaminants, such as sporulating bacteria, yeasts or fungi, are very difficult to eradicate from the working environment. Waste should be kept very well covered or, if feasible, in a different room, especially if using particularly harmful or sturdy microorganisms.

General Cleanliness

In principle, keeping high standards of cleanliness has to be a universally accepted norm in biological laboratories. People should wear laboratory coats that are worn only in the laboratory. Researchers should wash their hands often during the course of the day, especially prior to any and all handling of biological cultures, media, or sterile supplies. Dust and stains must be removed regularly, and spills cleaned and decontaminated immediately. Bench tops and shelves should be washed immediately before all uses with 10% bleach (sodium hypochlorite). Organic disinfectants such as 70% ethanol are generally less effective than bleach solutions, as ethanol may evaporate too quickly to effectively sterilize surfaces. Outerwear may be covered with dirt or dust, and so should be kept in a cloakroom away from the work space. These simple rules of cleanliness will eliminate the bulk of potential contaminations from the work area.

Sterilizing Equipment and Reagents as a Prelude to Effective Aseptic Work

Starting aseptic work with material that is contaminated is a path to nowhere. Many reagents are supplied by their manufacturer as ready to use for cultures, while many others require sterilization (i.e., the complete destruction or elimination of all viable organisms) before use. Others do not have to be sterilized at all (such as ethanol, phenol, or concentrated detergents).

A wide variety of sterilization methods have been developed, based on heat and high pressure, liquid or gas chemicals, radiation, or physical removal of microorganisms. Although only a few, straightforward protocols are usually applied in a general research laboratory, all should be carried out very carefully, because there are no degrees of sterilization: an object is either sterile or it is not.

Filter sterilization

Reagents such as antibiotics, drugs, sugars, amino acids, vitamins, and complex media which are either flammable or would be denatured by heat are usually filter-sterilized. A range of syringe-based or bottle-top filters of different pore sizes are commercially available. These filters physically remove (exclude) living organisms, spores, etc. Most living organisms are retained by a filter with a 0.45-μm average pore size. Such filters are often used as prefilters, since they will also remove particulates from liquids that would rapidly clog smaller filters with a smaller pore size. However, because some bacteria readily pass through 0.45-μm pores, a 0.22-μm filter should be used to better ensure sterilization of fluids. Be aware that some bacteria, such as *Leptospira* species, may pass through even a 0.22-μm filter, although such organisms are often very fastidious and do not normally present a contamination problem. Filtering procedures are usually carried out in a laminar flow unit or on a bench equipped with a Bunsen burner (see below) to avoid recontamination of sterilized material.

Autoclaving

CAUTION: Items containing solvents, volatile or corrosive chemicals (e.g., phenol, trichloroacetic acid, ether, chloroform etc.), or any radioactive isotopes **cannot** be autoclaved under any circumstances.

4

Autoclaving (using steam under pressure) is a rapid method suitable for sterilizing almost anything except heat-labile substances. A temperature higher than the boiling point of water (super-heating of liquids) can be achieved inside the autoclave because the system is under pressure. Typical autoclaving conditions of 121°C (250°F) for 15 to 30 min at 103 kPa (15 psi) are sufficient to kill all forms of life, including bacterial endospores. However, the time of the sterilization cycle should be modified according to the amount and type of loaded items, as some objects have a slower surface heating or steam penetration and should therefore be autoclaved longer, e.g., a large volume of medium (>1 liter), bulky bags of biohazard waste, or a ≥1-liter beaker full of microcentrifuge tubes with aluminum foil covering. Most substances cannot be autoclaved for too long, so when in doubt, it is generally better to extend an autoclaving time.

General principles of loading an autoclave can be summarized as follows:

1. Polypropylene and polycarbonate can be autoclaved, but polyethylene and high-density polyethylene cannot. Initials imprinted on bottom of containers can help in identification of the material type (PP = polypropylene, PC = polycarbonate, PE = polyethylene, HDPE = high-density polyethylene).

2. The autoclave must not be overcrowded; the steam needs to be able to penetrate everywhere. All materials must be placed in an autoclavable tray (polypropylene plastic or stainless steel tubs) with sides at least 10-cm high to catch spills and eliminate possible damage to the machine. Dry items must be wrapped in aluminum foil to prevent them from recontamination after the procedure.

3. Liquids should be sterilized separately, each in a vessel at least twice the volume of the liquid to be autoclaved to avoid boil-overs and spilling; heat-resistant borosilicate glass (Pyrex) vessels are preferred. To prevent bottle bottoms from breaking, fill the tub with 3 to 5 cm water. Bottles must be loosely capped to avoid shattering during pressurization. For screw-cap containers, close the lid hand-tight and then loosen the lid at least one-half turn to avoid pressure-differential implosions/explosions. Be aware that large bottles with narrow necks can simulate sealed containers if filled with too much liquid.

To optimize autoclaving of common laboratory goods two general types of exhaust cycles have been designed:

1. The "Gravity" or "Fast Exhaust" cycle is for all dry items such as glassware, tips, flasks, tubes, etc. The chamber is charged with steam, and set pressure and temperature conditions are held for a desired period of time. At the end of the cycle, an exhaust valve opens quickly and the chamber rapidly returns to atmospheric pressure. Drying time may also be added to the cycle.
2. "Liquid" or "Slow Exhaust" cycle is used for liquids to prevent explosive boil-overs. Pressure is slowly released which allows the fluids (super-heated) to cool gradually. This type of cycle should also be used when mixed loads (i.e., liquids and dry goods) are to be processed. Be aware that liquids may be superheated even when the pressure in the autoclave has returned to atmospheric, and liquids may boil over or explode. For safety, leave the autoclave door open wide for 10 to 20 min to allow adequate cooling. A face shield and full body covering (such as long-sleeved laboratory coat) is essential when handling liquids in the autoclave.

Sterilized goods can be marked using a special indicator tape which contains a chemical that changes color when the appropriate autoclaving conditions have been met. Some types of packaging have built-in indicators on them.

Oven sterilization

Selected goods such as glassware, metal, and other objects that will not melt at temperatures between 121°C and 170°C can be sterilized with dry heat (in a hot air oven). Because the heat takes much longer to be transferred to the organism, routinely a temperature of 160°C is maintained for at least 2 hr, or 170°C for 1 hr. The procedure certainly cannot be applied for liquids and rubber or any plastic objects. However, the method has some advantages. For instance, it can be used on powders and other heat-stable items that are adversely affected by steam, and it does not cause rusting of steel objects. It is also a method of choice for the treatment of glassware used for work with ribonucleic acid (RNA), since the process of baking not only kills any remaining organisms but also inactivates any residual RNA-degrading enzymes (RNases; see *UNITS 5.2 & 8.2*).

General considerations

Note, that a standard microwave oven (such as a household appliance), although capable of bringing water to boiling does not provide a useful method for sterilizing laboratory goods and should only be used for warming up the liquids.

Sterilized solutions and other reagents will need to cool down before use, which takes time and should be performed well before the actual experimental work begins.

It is best for each researcher in the laboratory to keep his/her own stocks of reagents (such as media, buffers, salts, glycerol, etc.) to prevent accidental contamination by a coworker. In this manner, each worker is solely responsible for the condition of his/her reagents and for impacts on experimental results. This can be facilitated by preparing media in several small bottles (aliquots), one for each researcher, with only one container per worker opened at a time.

Although growth media are more likely to produce noticeable biological contamination (by turning color, developing clumps, or becoming turbid), most other liquids used in the laboratory can also support the growth of organisms. Check solutions before using by gentle swirling and looking for presence of cloudy material.

Useful Materials to Have on Hand

A list of simple materials that might be useful in aseptic work and that should be kept "on hand" is presented in Table 4.1.1.

Table 4.1.1 Common Laboratory Items Useful for Aseptic Work

Material	Common use
Aluminum foil	Widely used in laboratories for packing goods for autoclaving. It protects them from recontamination and facilitates storage. Simple lids can be made from it for beakers, flasks, or bottles. Wrapping boxes of disposable tips, centrifuge tubes, bundles of glass pipets, or spatulas with aluminum foil not only protects but also helps to keep them organized by size or type.
Disposable rubber gloves	Should be worn at all times during aseptic work. Usually made of latex rubber, PVC (polyvinyl chloride), or nitrile (*APPENDIX 1*). Recently a range of mild to severe allergic reactions from exposure to natural rubber latex has been widely recognized. It is recommended to provide the laboratory with both powder-free low-protein latex disposable gloves and nonlatex gloves. After removing latex gloves hands should be washed with a mild soap and dried thoroughly. Particular caution should be used when gloves are worn in immediate proximity of an open flame (e.g., Bunsen burner) since rubber may catch fire or melt easily.
Disposable wipes (lint-free; e.g., Kimwipes)	Since they do not leave unwanted particles behind, considered as very useful for wiping bench tops, inside of the laminar flow hood, or any other object during cleaning and disinfection
70% ethanol or other Industrial Methylated Spirits (IMS)	Simple, flammable disinfectants used for cleaning items and surfaces from microorganisms and stains. A 70% water solution evaporates slower than the pure alcohol and therefore stays longer on surfaces and has a better disinfecting capability. Also used for dipping culture spreaders. See Safety Considerations (*APPENDIX 1*) for usage instructions.
Eye protection and thermal gloves	Equipment for extra protection should be kept within reach and used during procedures such as opening an autoclave
Flint-on-steel striker	Used to quickly ignite Bunsen burner when needed
10% household bleach (sodium hypochlorite)	Chemical more effective for disinfecting surfaces than quickly-evaporating ethanol. Should not be stored for longer than 1 month at room temperature, since sodium hypochlorite will degrade.
Laboratory coat	Clean, long coat of the right size should be used to protect the researcher against toxic substances and biohazard material as well as samples from contamination from human skin and apparel
Spray or squirt plastic bottles	Very useful in dispensing ethanol, water, and other liquids for disinfection and cleaning of large surfaces and places difficult to reach with a wipe

SAFETY CONSIDERATIONS

Autoclaving

An autoclave produces steam which can cause burns and requires a lot of caution in handling. The most important moment is opening the machine after completion of the cycle. Consider some basic rules that should be applied while opening an autoclave:

1. Wear laboratory coat, eye protection, loose fitting thermal (insulating) gloves, and closed-toe shoes.

2. Check that the chamber pressure is zero.

3. Stand behind the door when opening and use it as a shield. Slowly open a crack and wait for steam to be released. Beware of rush of steam. Open door fully to remove tray.

4. After the slow exhaust cycle, open autoclave door and allow liquids to cool for 10 to 20 min before removing. Do not jolt hot bottles or move them, especially if the liquid is bubbling or boiling.

Working with Open Flame

Burners should only be on for as long as needed for a particular procedure. The blue flame of a hot Bunsen burner can be difficult to see. **Never** leave an open flame unattended, not even for "a few seconds."

Since burns are among the most common laboratory accidents, making sure that face, clothing, and hair are not above or near the opening of the burner tube is a crucial safety feature. Temperatures in the hottest region of the burner flame approach 1500°C and objects will heat extremely quickly. Any heated object, glass or metal, needs time to cool down before it can be safely touched.

It is also absolutely necessary to remove flammable and combustible materials from the vicinity of the flame. For instance, nitrocellulose membranes, commonly used for blotting techniques, are extremely flammable.

Ethanol used for dipping culture spreaders is an exception to the rule regarding flammables, but only small volumes ($\leq$20 ml) in glass beakers should be used at a time. An accidental alcohol fire in a beaker can be quickly extinguished simply by covering the top of the beaker with a piece of aluminum foil. Keeping the alcohol beaker covered with such a piece of foil when not in use will help reduce accidental fires and will slow evaporation.

A fire extinguisher should always be kept nearby.

The tubing connecting the Bunsen burner to the laboratory gas supply should be checked regularly. Latex rubber tubing tends to harden and crack after exposure to air, making it a poor tubing choice. Silicone tubing, such as Tygon, is generally more durable.

Ultraviolet Light

Many laminar flow hoods are equipped with ultraviolet (UV) lights, which can help reduce contamination by inducing DNA damage to potential contaminants. Such light is also harmful to laboratory workers. Never look directly at a UV light, as this can cause eye damage. The UV light can also cause skin burns ("sun burns") to anyone in the room, so UV lights should always be turned off whenever the room is occupied.

TECHNIQUES FOR MAINTAINING ASEPTIC CONDITIONS

Usually, there is no need to equip the laboratory with expensive paraphernalia just for culture of hardy bacteria such as *Escherichia coli* (*UNIT 4.2*), since that bacteria's rapid growth rate generally allows it to out-compete any contaminants: a typical laboratory bench supplied with a Bunsen burner is sufficient. A cone of heat produced by the burning gas has been used in laboratories worldwide since the 19th century. It is straightforward and does not require considerable financial outlays. However, a laminar flow hood is the method of choice when a more rigid aseptic technique is required. The rich media used for eukaryotic cell cultures and many microorganisms, coupled with their slow growth rates, necessitates high levels of security from contamination. For some organisms, work must always be performed inside a laminar flow unit to meet biosafety standards and to protect personnel against potentially harmful microorganisms or viruses (*APPENDIX 1*).

Any aseptic work should be completed as quickly as is comfortable to minimize the risk of contamination. Consider that working in an aseptic manner may take longer than when being less cautious. Reserve extra time to avoid being rushed, which may result in spilling or breaking important samples, dishes, or solutions.

Bench-Top Aseptic Work Using a Bunsen Burner

Before starting work locate the gas and air controls on the burner. They can be in different places on different models. In some cases it is necessary to use the laboratory gas valve to control the gas flow rate.

Once a Bunsen burner has been adjusted for optimal air and gas flow, the settings can be left in place, making future usage more efficient.

1. Close the burner's air control. If the burner has its own gas control valve, shut and then open it, allowing only a small amount of gas to get into the burner tube.

2. To ignite the gas using a flint-on-steel striker, hold the striker 2 to 3 cm above and slightly to the side of the burner tube top. Squeeze and release striker repeatedly until a spark ignites the flame. If using a match, light it and slowly bring up from the bottom along the side of the burner tube until the flame ignites.

   If a match is stuck into the middle of the gas stream, the high-pressure gas stream often blows out the flame before the burner can light.

   The initial flame is low on oxygen, bushy orange in appearance, and not very hot (Fig. 4.1.2A and B).

3. Open the air-control vent until the central blue cone of the flame forms and a slight buzzing sound is audible (Fig. 4.1.2C). If the flame is too small, gradually increase both the supply of gas and air flow to the burner.

   The hottest part of the flame is just above the central blue cone.

4. When finished using the burner, turn off the laboratory gas valve.

   CAUTION: *Do not leave a lit burner unattended.*

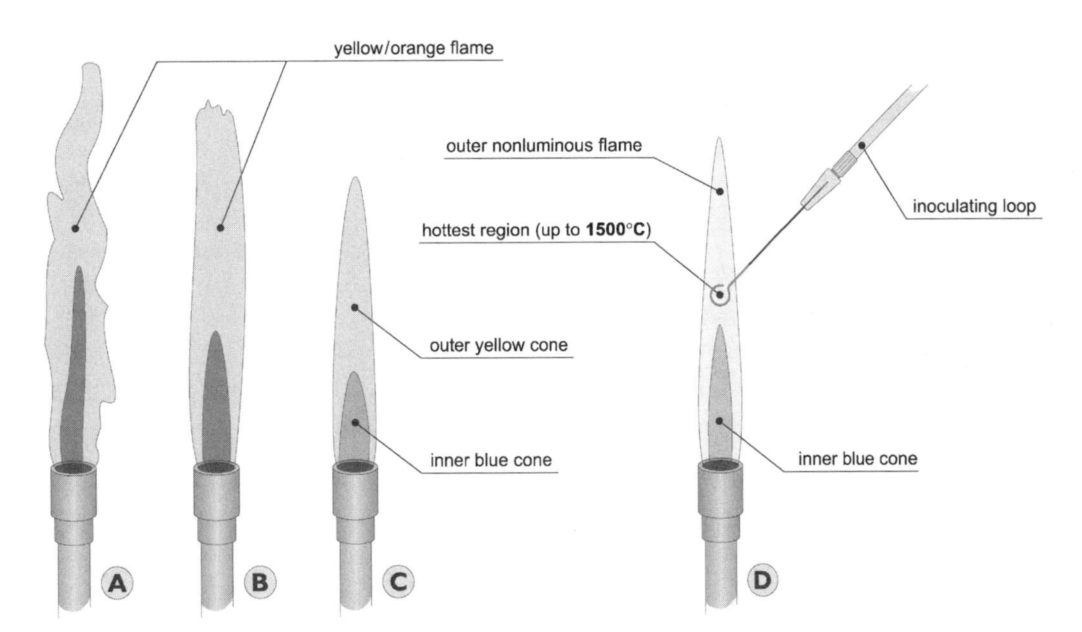

Figure 4.1.2 Adjusting the Bunsen burner and heating an inoculation loop. A lighted Bunsen burner with its air holes closed produces a cool, yellow-orange flame (**A**), and is not particularly useful for heating objects. The more the air holes are open, the more blue and fierce the flame becomes (**B** and **C**). A properly adjusted Bunsen burner gives a blue flame containing two cones, an outer (pale blue) and an inner (deeper blue) (**D**). There is no combustion inside the central cone. Temperatures in the part of the flame just above the central blue cone approach 1500°C. See text for more details.

Handling and Pipetting Liquids

Try to open, pipet, and aliquot any growth media, cultures, or sterile reagents inside the cone of heat that is created above and around the flame of the burner. Bottles and test tubes should be arranged near the burner but not directly within the flame. Brief (1 to 2 sec) flaming of lips of tubes and flasks should only be done when caps are removed during transfer of liquids and cultures. This is especially important when the content is to be poured from a container (e.g., pouring petri dishes or dividing large volumes of liquid into aliquots). The purpose of flaming is actually not to sterilize, but to warm the opening and create air convection currents up and away from the opening. This "canopy" of warm, rising air helps to prevent the entrance of dust particles, and flaming should be performed immediately upon opening and just before closing tubes and bottles. Lips of disposable plastic containers and containers of flammable solutions should never be flamed. Autoclaved beakers containing sterile wooden toothpicks do not have to be flamed.

NOTE: The following instructions assume the experimenter is right-handed. In the case of a left-handed individual, switch the hands in the instructions from right to left and vice versa, e.g., when manipulating liquids, hold the container in your right hand and the instrument in your left.

Manipulating vessels containing liquids

Most manipulations of cultures or sterile reagents in tubes, bottles, or flasks should be performed as follows.

4

1. Loosen closures (i.e., lids, caps, etc.) of all containers prior to any manipulation.

   This will ensure that a procedure will not have to be stopped midway.

2. Hold the container in your left hand at a 45° angle so that dust cannot fall in when open.
3. Hold the instrument to be used for manipulation (e.g., inoculating loop, pipet, needle, toothpick, etc.) in the right hand.
4. Grasp the container closure using the little finger of your right hand, and lift from the container.

   Do not set the container closure down. Doing so increases the risk of it becoming contaminated from the bench top. Remember that even a cleaned bench top is not sterile.

5. While continuing to hold the closure with the little finger, lightly flame the container opening.
6. Quickly manipulate the instrument into the container, then withdraw.
7. Replace container closure immediately.

Transferring liquids from one test tube to another

8. Hold both tubes in the left hand.
9. Remove closures from both tubes with little finger of right hand.
10. Flame lips of both tubes
11. Take culture, reagent, etc., from first tube.
12. Add culture, reagent, etc., to second tube.
13. Replace both closures.

Storing and Using Pipets

Cans of sterile glass pipets should be kept in horizontal positions, to reduce introduction of dust and other airborne contaminants. Pipets should be removed in a manner that prevents the tips from contacting any potentially contaminated surfaces: remove pipets by inserting as little of fingers as possible, and as the pipet is withdrawn, hold it away from any surface to prevent its contamination (including the ends of other pipets, since they have come into contact with your hands, and are therefore considered nonsterile).

Individually wrapped, presterilized plastic pipets are an alternative to glass pipets. These are generally inexpensive when purchased in small quantities but expensive over the long term. To use, open from one end, then peel the wrapping backward, allowing removal of the pipet without contact to the outer surface. Remember that the outside of the wrappings are considered contaminated. Bulk-wrapped, presterilized pipets are somewhat less expensive, but it is often difficult to maintain sterility of the large quantities of pipets in each package.

Using an Inoculating Loop

An inoculating loop (*UNIT 4.2*) is a common laboratory tool used for transferring small, live biological samples (such as bacteria or yeast) from one culture medium (such as a plate or flask) to the other.

Before each use, an inoculating loop must be sterilized using a burner, as follows:

1. Place the junction between the loop wire and the handle just above the inner blue cone (hottest point) until the wire turns red.
2. Slowly draw the wire through the blue flame, making sure that every part of the wire is heated to glowing red. Heat the loop tip last.
3. Cool the loop by making contact with another sterile surface, such as an unused section of an agar plate.

   Do not blow on the loop or wave it in the air to cool it. Such actions will contaminate it.
4. Use the loop immediately. Do not put the loop down or touch it to a nonsterile surface before using.
5. Flame loop again immediately after use before setting it down.

Using Petri Dishes ("Plates")

Before pouring medium into sterile petri dishes, first remove the plates from their container (plastic sleeve, etc.), and arrange on the laboratory bench convenient to the burner. Flame the lip of the container of liquefied medium immediately before pouring medium into plates, and again after each 5 to 10 plates are poured. Remove petri dish lids only when needed for pouring the medium, and close immediately after. When the lid is removed, it should be held over the plate as a shield and never placed on the bench top.

For all manipulations of cultures in petri dishes, lids should only be lifted for as short a time as possible. Lids are never to be placed on the bench top. Do not walk around the room with an open plate. As with all other media, do not breathe on open plates.

Working in the Laminar Flow Unit

An additional layer of security against contamination is the use of a laminar flow hood. Although several modifications of the two major flow hood design types (horizontal and vertical) are commercially available, major design concepts are similar (Fig. 4.1.3). Room air is taken into the unit and passed through a prefilter to remove gross contaminants (e.g., lint, dust, etc.). The air is subsequently compressed and channeled through the HEPA filter that removes nearly all of the bacteria from the air. Purified air flows out over the entire work surface in parallel layers at a uniform velocity with no disruption between the layers. A cluttered hood or other poor techniques can easily overcome the desired airflow and reverse currents, potentially introducing contaminants into the work area.

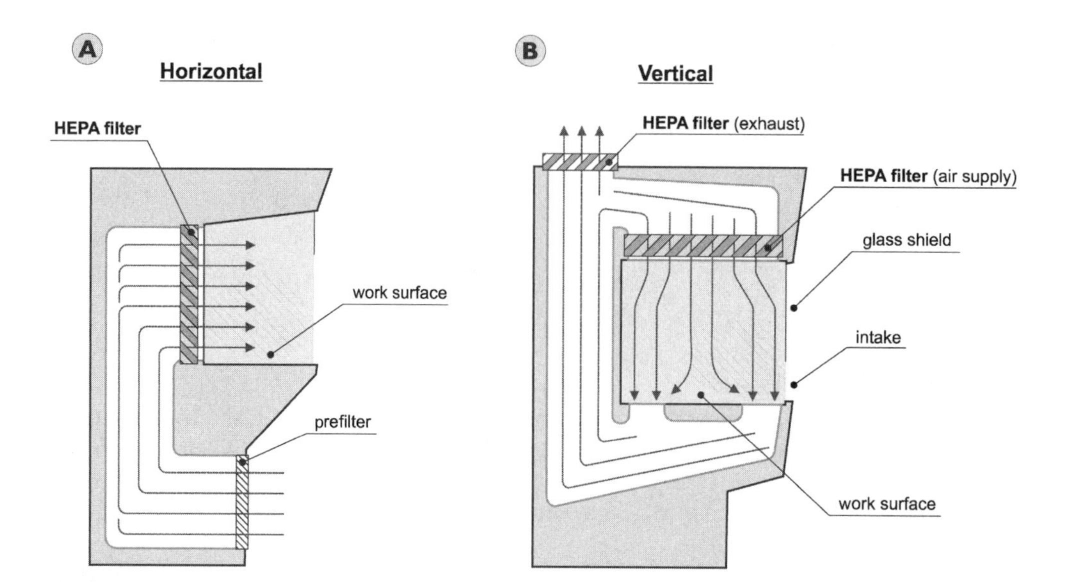

Figure 4.1.3 Laminar flow hoods. (**A**) Horizontal and (**B**) vertical are the two major types of laminar flow units. See text for details.

4

Thoroughly read through your laminar flow unit supplier's manual for detailed information on the operation of a particular appliance, especially with regard to the position of air ducts and the track of laminar flow.

Cleaning, disinfecting, and arranging the space inside hood (before use)

Before starting work or re-entering the hood, remove jewelry from the hands and wrists, tie long hair back, button up your laboratory coat and roll the sleeves up. Wash your hands and forearms and don fresh disposable gloves. Exterior of gloves must be sprayed with disinfectant and then hands rubbed together before putting them inside the hood. Keep hands within the cleaned area of the hood as much as possible without touching hair, face, or clothing. Hand cleanliness is reduced each time bottles and other nonsterile items are touched, so the disinfecting procedure should be repeated periodically.

To maintain efficient air flow, the laminar flow hood should remain turned on 24 hr a day. If turned off for any reason, it should be on for at least 30 min before use. The hood should be thoroughly cleaned before any use. Spray the working surface liberally with 70% alcohol solution (ethanol or other Industrial Methylated Spirit) and swab immediately with lint-free wipes.

Any object that is introduced into the laminar hood environment, even glassware or plastic containers marked as "sterile," should be disinfected by liberally spraying the outside with alcohol. Always remove outer pouches and wraps from disposable equipment (such as pipets, loops, spreaders, and cell scrapers) at the edge of the work area just before the sterile contents are pulled into the hood. To help reduce introduction of contaminants from outside the hood, always keep a set of clean pipettors inside for exclusive use.

Leave a wide clear space in the center of the hood (not just the front edge) on which to work. Arrange the area to have easy access to all of it without having to reach over one item to get another (especially over an open bottle or flask). Reaching over items increases chances of knocking things over, or transferring contamination through the workplace.

Never place large objects near the back of the hood nor clutter the area. Not only do these objects contaminate everything downstream, they also disrupt the laminar flow pattern of air which normally suspends the contaminants and removes them from the area.

A common mistake is to keep waste, old cultures and media, empty boxes, notebooks, manuals, protocols, pencils, and other unnecessary items inside the hood. Such objects should never be kept inside the laminar flow unit.

Working in a laminar flow hood

Perform all work at a distance of no less than 6 in. from the front edge of the work surface. At a lesser distance, laminar flow air begins to mix with the outside air and contamination is possible.

To ensure the proper air flow, maintain a direct path between the filter and the area inside the hood where the manipulations are being performed (nothing should touch the filter). Undesired turbulence may also be produced by coughing or sneezing into the hood, quick movements, rotating, talking, etc. To help keep air disturbances to a minimum, the hood should be located outside of the stream of traffic in the laboratory.

Remember that a laminar flow hood is *not* a completely sterile environment. All the rules of aseptic work on the bench top apply to work in a hood. Minimize the time bottles or plates are open, avoid touching extraneous surfaces with pipets, etc.

In case of any spill, clean the work surface immediately with sterile, distilled water and follow by spraying liberally with alcohol. Use side to side motion starting at the back of the hood and then working forward.

If you fully comply with all the above instructions, chances for the introduction of sepsis into the cultures or reagents are low. Never become so engrossed in your work that you forget these basic rules.

Finishing work in the laminar flow hood

When finished with using the hood, remove all unnecessary items and then clean all surfaces by liberally spraying with 70% alcohol. Most laminar flow hoods are equipped with a UV lamp, which helps kill introduced microorganisms by irradiation. Since UV light is also harmful to humans, do not switch on the light when anyone is working in the laboratory. Turn on the UV light at the end of the work day to help keep down levels of potential contaminants.

IMPORTANT NOTE: Before switching on the UV light, remove all important samples of living organisms to a sheltered location, so they are not affected by the harmful conditions.

TROUBLESHOOTING

Most lapses in aseptic technique will become apparent by contamination of cultures or supposedly sterile media and reagents.

A simple method to check the quality of your aseptic technique is to perform manipulations of media, etc., without intentionally adding bacteria. Incubate the media, etc., overnight at 37°C, then observe for signs of contaminant growth (e.g., turbidity). A small amount of contamination is not always evident until the media is incubated at an appropriate temperature or atmospheric conditions.

Some bacterial cultures may grow briefly, then lyse. This may be due to bacteriophage contamination as a result of poor aseptic technique. If this is suspected, thoroughly clean all laboratory benches and equipment with disinfectants, and strengthen use of aseptic techniques.

Many contaminations occur through the poorly designed or defective air handling systems, possibly transferring microorganisms from neighboring laboratories. Your facility's physical plant management should be consulted if you suspect such possibilities, as your laboratory's air handling system may need to be re-engineered.

UNIT 4.2

Culture of *Escherichia coli* and Related Bacteria

Sean P. Riley,[1] Michael E. Woodman,[1] and Brian Stevenson[1]
[1]University of Kentucky College of Medicine, Lexington, Kentucky

OVERVIEW AND PRINCIPLES

This unit focuses on the culture and maintenance of the bacterium *Escherichia coli* and other related bacterial species. *E. coli* is the world's most widely understood microorganism and is an extremely important model organism in many fields of research, including fields outside the study of microbiology. Methods for genetic manipulation are well defined and straightforward to perform in *E. coli*, and when combined with the rapid growth rate of the bacterium, *E. coli* can be used as "factories" to produce large quantities of DNA, proteins, or metabolites. A well-known example of genetic manipulation was performed by Eli Lilly in 1982, in which the human insulin gene was inserted into *E. coli*, permitting simple isolation of treatment-grade insulin from the bacteria (Wright, 1986).

E. coli is one of the many resident bacteria that colonize the human intestinal tract; as such, they are considered part of our resident microflora. Theodor Escherich, a pediatrician and bacteriologist, discovered the bacteria in 1886, and the genus was named in his honor in 1919. The bacterium plays a vital role in human digestion, as it is a source of vitamins B12 and K, but there are three major situations where *E. coli* can cause disease: (1) when bacteria contaminate the urinary tract, (2) when it enters the abdomen through a perforation of the intestinal tract, or (3) if a person consumes a toxigenic strain.

The most commonly used *E. coli* in life science laboratories are derivatives of the strain K12. Derivatives of this strain have been maintained in laboratories for many years, and a wide variety of genetically manipulated strains have been made. The genome sequence of *E. coli* strain K12 is available from the Institute for Genomic Research (*http://www.tigr.org*).

While *E. coli* growth requirements are well defined, bacteria, in general, have unbelievably diverse metabolic capabilities, requiring the use of various media and culture conditions for each species. The efforts of over a century of research have resulted in the ability to culture many bacteria, especially for those species of importance to human health. Yet, only ~1% of all bacterial species can be cultured in laboratories. In most cases, too little is known about those noncultivable microorganisms to create an appropriate medium for growth. Researchers who plan to cultivate bacterial species other than *E. coli* are advised to consult references specific to the proposed organism, such as those found in *Current Protocols in Microbiology*.

Media that contain ingredients for which all components are not precisely defined are called complex media. For example, LB (see recipe below) is a complex medium, because it contains all the components of yeast cells. Generally, these media are used for routine culture, because they provide a rich environment for growth where the bacteria are not under stress to synthesize cellular components from scratch. The other type of growth medium is defined medium. In this case, all ingredients are simple chemicals that are added at specific concentrations. This type of medium is used most often for metabolic studies, and in some cases metabolic genes are used as selective agents to differentiate strains. An example of this type of medium is M9, which is described later in the unit.

SAFETY CONSIDERATIONS

Laboratory strains of *E. coli* such as those used for genetic manipulation and protein production are generally harmless to humans, but caution must be exercised. Many different strains of the species exist, some of which can potentially cause disease. Also, if the bacteria are being used to produce proteins or metabolites, that can potentially be toxic. Individuals should wash their hands immediately upon completion of work with bacteria. You should never eat, drink, smoke, or touch your eyes in the presence of bacteria. All cultures, stocks, and other wastes must be decontaminated by an approved method before disposal. Materials to be decontaminated outside the immediate laboratory are to be placed in a durable, leak-proof container and closed for transport from the laboratory.

Many culture techniques involve the use of a burner flame and/or ethanol (*UNIT 4.1*). Appropriate care must be taken to avoid burns. Note that ethanol is very flammable and burns with a clear flame.

CAUTION: E. coli is a Biosafety Level 1 (BSL-1) pathogen, with some pathogenic human isolates requiring BSL-2 precautions. Follow all appropriate guidelines and regulations for the use and handling of pathogenic microorganisms. See *APPENDIX 1, CP Microbiology Unit 1A.1* (Coico and Lunn, 2005), and other pertinent resources (e.g., *http://www.absa.org*) for more information.

COMMONLY USED TOOLS

4

Inoculating Loop

The inoculating loop is used for the transfer of liquid or solid bacterial cultures. It consists, simply, of a handle with a thin protruding wire that is molded into a circle at the end. Many varieties of loops, made of nickel-chromium or platinum wire, are available from many suppliers. The wire may be permanently attached to a handle or replaceable, and may be obtained as either a straight wire or a twisted loop. The loop shape allows efficient transfer of liquids, which form a film over the loop. Standardized, volumetric loops are commercially available for reproducible transfers of specific volumes of liquid. Both looped and straight wires are suitable for transferring bacteria from solid media.

Construction: A loop can be easily made on the end of a straight wire by bending it around the tip of a sharp pencil or needle-nose pliers, and should be 2 to 3 mm in diameter. The loop must be complete, with the tip of the wire just touching the opposite side of the loop, to allow formation of liquid film for transfer. The wire should be 4 to 5 cm in length (Fig. 4.2.1).

The inoculating loop must be sterilized immediately before and immediately after use. To sterilize, first place the end of the wire closest to the handle in the blue (hottest) part of a burner flame, until the wire glows red. Then slowly draw the wire through the flame, ensuring that the entire length of the wire glows red. The loop can be cooled quickly by gently touching it to an unused part of an agar plate or in liquid medium. Do not blow on a hot loop or wave it in the air to cool it off.

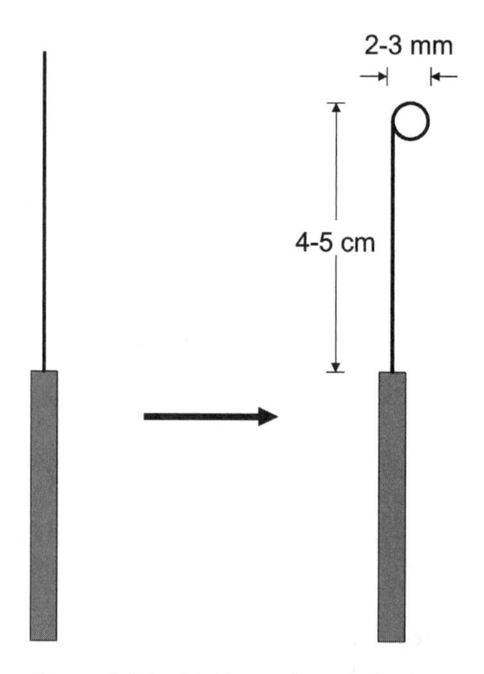

Figure 4.2.1 Making an inoculating loop.

Disposable, plastic loops are also available, although they are expensive over the long term and create a large amount of waste.

Toothpicks

Autoclaved toothpicks can be used for transfer of cultures as an inexpensive and quick alternative, since they do not need to be flamed like a loop. Toothpicks should be placed into a 100-ml beaker with the narrowest end at the bottom of the beaker, covered with aluminum foil, and autoclaved. The sterile container should be turned on its side during removal of individual toothpicks, to prevent contaminants from falling into the open beaker.

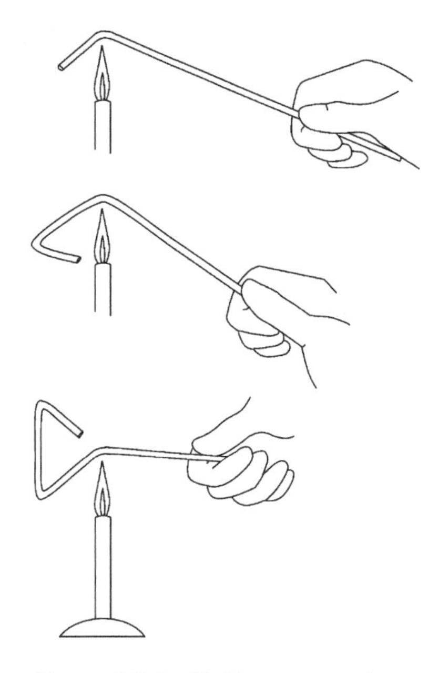

Figure 4.2.2 Making a spreader.

Culture Spreader

A spreader is used to evenly distribute bacterial cells from a liquid suspension over the surface of a plate. This will yield either a "lawn" of bacteria on nonselective medium, or isolated colonies on selective medium. A spreader can be easily made by heating and bending a piece of a 4- to 5-mm diameter glass rod, as illustrated in Figure 4.2.2. In a pinch, a much less durable spreader can be quickly made from a Pasteur pipet. For safety in handling, melt away sharp ends of rod as a first step. A pair of needle-nose pliers or similar tool is useful for forming bends in glass.

Before each use, sterilize the spreader by immersing the triangular end in a beaker of 95% ethanol, passing the spreader quickly through a flame, then allowing the flame to burn out. Be careful not to let the burning ethanol drip into the beaker or onto flammable objects. Sterilize after use by immersing in the ethanol.

Presterilized, disposable, plastic culture spreaders are also available from a variety of sources. As with plastic inoculating loops, plastic spreaders are expensive over a long period of time and generate considerable waste.

REAGENTS

Use deionized, distilled water in all recipes and protocol steps, with the possible exception of media for *E. coli* (see Commonly Used Bacterial Media). For common stock solutions, see *UNIT 3.3*. All solutions and equipment that comes into contact with living cells must be sterile and aseptic technique (*UNIT 4.1*) must be used.

CULTURING BACTERIA ON SOLID MEDIA

Many bacteria, including *E. coli*, possess the ability to form discrete colonies on solid media. Each single colony is typically the result of outgrowth from a single bacterium, and since *E. coli* divides asexually, all cells in a colony are genetically identical, and are defined as "clones." Since each colony is a clone and each separate colony may be different from others on a plate, only well-separated single colonies should be used for further analyses. Studies of mixed populations will yield inconclusive results. The section below describes common techniques for growing bacteria on solid media.

Spreading a Culture on a Plate

E. coli and many other species of bacteria can be efficiently spread across the surface of an agar plate using a culture spreader (described above).

1. Place a small volume of culture (50 to 500 μl) in the middle of the agar surface.
2. Sterilize the spreader by immersing the hooked end into 95% ethanol in a small beaker, then burn off the ethanol by passing through a Bunsen burner flame.

   CAUTION: *Be careful not to let the burning ethanol drip into the beaker or onto flammable objects.*

3. Cool the spreader by touching it lightly to the unused agar surface, then spread the bacterial culture uniformly around the plate.

   Rotating "lazy Susan" devices, which spin the plate, are readily available from many sources, and can aid uniform spreading of cultures.

4. Return spreader to ethanol.
5. Allow liquid to dry such that there are no drops visible on the surface.
6. Incubate plate with agar facing down at appropriate temperature (see Incubation below).

Streaking a Culture on a Plate

For *E. coli* and many other species of bacteria, clonal populations can be derived by streak plating. Each successive streak dilutes the previous streak, such that isolated colonies (clones) can be obtained (Fig. 4.2.3).

NOTE: Some bacteria, such as *Proteus* species are very motile, and will not form isolated colonies on solid media but will instead swarm across the entire surface. Other bacterial species will not grow on the surface of solid media, and thus cannot be cultured by this procedure.

1. First, streak a loopful of liquid or solid culture across a small (~2- to 3-cm long) area near one side of an agar plate.
2. Flame the loop to sterilize.
3. Touch the loop to an unoccupied area of the plate to cool.
4. Lightly drag the loop **once** through the first streak, then continue to drag in a zigzag manner over a section of the agar surface. Be careful not to cross previous lines.
5. Flame the loop again, and cool by touching agar surface.
6. Again, drag the loop **once** through the previous streak, then continue to make a second streak.
7. Repeat for a total of four to five streaks.
8. Incubate plate with agar facing down at appropriate temperature.

   Alternatively, this procedure can be performed with toothpicks. In this case, it is important to use a new toothpick at every step where loop flaming is needed.

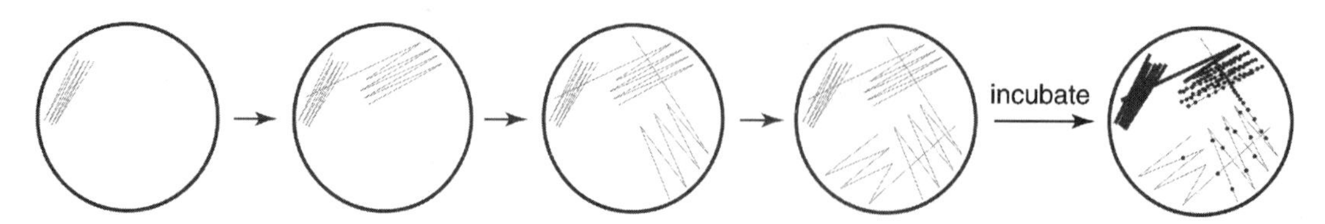

Figure 4.2.3 Proper technique for streaking bacteria on solid medium.

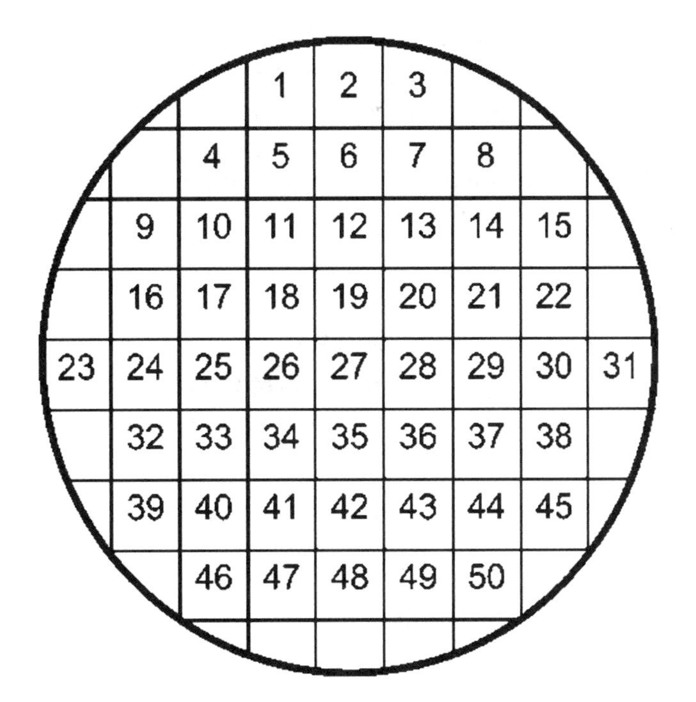

Figure 4.2.4 Template for organization of Petri plate.

Organization of Clones

Following certain manipulations of *E. coli*, such as transformation or mutagenesis, mixed populations are generated. Individual colonies (clones) should be transferred to new plates to continue growth and encourage good organization of clones. Given that it is impractical to streak each clone on an individual plate, a template is provided for arranging 50 clones on a single plate (Fig. 4.2.4). Photocopy the template (four or more copies can be arranged on a single sheet of paper), cut out the gridded circle, then attach with tape to the bottom of a 100-mm Petri dish with the ink face against the plastic. Use sterile toothpicks or a loop to individually pick a portion of each colony and spread it within one of the grid squares. Since all cells from the original colony are identical, it is not necessary to re-streak for isolated colonies. Incubate the plate overnight as described below, and the next morning you will have 50 well-separated and organized clones. Retain the library plate as a reference until all analyses of the clones are completed.

Incubation

For *E. coli* and its relatives, Petri dishes are incubated with the agar side at the top. The reason for this is that condensation often forms in dishes during incubation. If the agar side is at the bottom, water may drip from the lid onto the agar surface, spreading bacteria around the plate, thereby preventing isolation of individual colonies. The temperature at which *E. coli* is typically grown is 37°C, because this is similar to normal human body temperature.

GROWING BACTERIA IN LIQUID CULTURE

By inoculating a single clone into liquid medium, you can produce many grams of identical bacteria. Some variables must be taken into account for appropriate growth. The following sections describe techniques for proper growth in liquid media.

Oxygen Environment

Growth of *E. coli* is generally enhanced by the presence of oxygen, as the Krebs cycle and oxidative phosphorylation require oxygen for complete catabolism of complex carbon sources to carbon dioxide. Liquid cultures should be oxygenated by shaking during incubation (usually ~200 rpm). Special Erlenmeyer flasks with baffles on the bottom are ideal for cultures of ≥50 ml, as the baffles perturb liquid flow and help dissolve air into the medium. Capped test tubes work well for volumes <10 ml. A variety of tube sizes are available from commercial sources, as are metal or plastic caps or foam plugs for tubes and Erlenmeyer flasks. For all containers, caps should be loosely attached, or vented caps or other enclosures used. To maximize exposure of liquid to air, do not fill culture containers more than $\frac{1}{3}$ full.

Inoculation and Growth of Small Batch Cultures

A small culture (<10 ml) can be efficiently grown overnight from a single colony in LB. A single colony inoculated into a larger volume (e.g., 100 ml) of a very rich medium such as TB or SD will also grow to stationary phase overnight.

1. Follow the instructions above for proper sterilization of the inoculating loop.
2. With a cooled loop or sterile toothpick, scoop a small portion of a single isolated colony.
3. Insert the loop/toothpick into the liquid medium and shake lightly to dislodge the bacteria. Alternatively, touch the bacteria on the loop to the glass just above the liquid.

   The inoculum will enter the liquid once it is put on a shaker. It is not necessary to see the bacteria in the liquid medium, because even a few cells will grow to a high final density after incubation.

4. Place the culture in a 37°C shaking (~200 rpm) water bath or 37°C dry incubator and grow overnight.

   Water baths must contain water to approximately the same level as the broth culture to maintain incubation temperature. However, be certain the water level is not too high as to splash water into the tops of the culture flasks/tubes.

Growth of Large-Volume Cultures

E. coli requires a minimal starting density for efficient growth and that density depends on the culture medium. A culture from a colony will only reach stationary phase in ~10 ml for LB. Larger cultures (e.g., 1 liter) will necessitate two successive nights of culture. The colony is first grown in 10 ml of LB for the first night, and then this "starter" culture is added to the larger volume for growth over the second night.

Monitoring Growth in Liquid Medium

When measuring the population growth in a bacterial culture, a defined progression of the culture can be observed. These phases of growth are commonly referred to as the following: lag, exponential/logarithmic, stationary, and death (Fig. 4.2.5).

Lag phase is characterized by no net growth. During this phase, the bacteria are adjusting to the fresh medium. Typically, the lag phase is shorter when the bacteria are transferred from an exponential phase culture into the same medium that has been prewarmed to culture temperature. Once the bacteria adapt during lag phase, they enter the exponential growth phase. This phase is characterized by increased binary fission and a rapidly increasing number of cells. During exponential phase, *E. coli* can reach its maximum doubling time of ~20 min if grown at optimal conditions. After a period of rapid growth, bacteria will enter stationary phase, after they have exhausted available nutrients, limited oxygen, or accumulated toxic byproducts. Death will occur after varying periods of time depending on the growth media and specific bacterial species or strains. Death occurs because of depletion of cellular energy or production of autolytic enzymes. The phases of growth can be determined by measuring cell number in a culture using several techniques, followed by generation of a simple growth curve as described below.

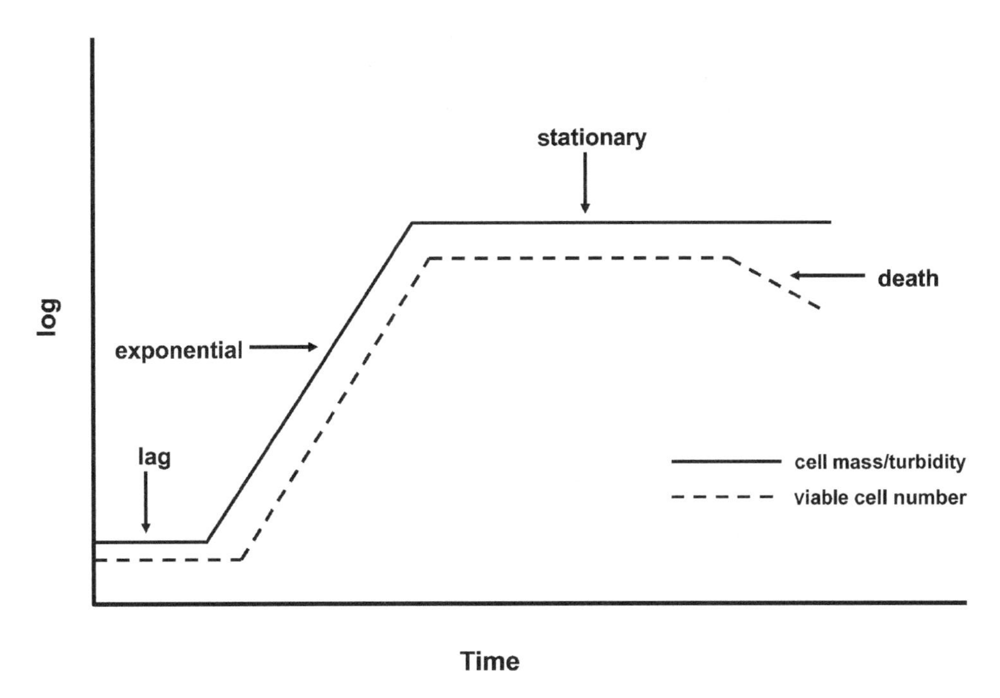

Figure 4.2.5 Bacterial growth curve.

Optical density determination

A simple, quick, and commonly used method to determine growth is to measure the turbidity of the bacterial culture using a spectrophotometer (*UNIT 2.1*). As the number of cells in a solution increases, the solution becomes increasingly turbid. However, this technique is sensitive to all mass present in the culture, so both living and dead cells contribute to the turbidity, as shown in Figure 4.2.5. Not only is the turbidity influenced by cell number, it is also dependent on the shape and composition of the bacteria. Therefore, this technique is specific to each bacterial species and cannot be compared between species. Different strains of the same species of bacteria may also have different sizes, so be aware of this possible inaccuracy when using this method for determining cell density. Bacteria may also change cell size or shape at different stages of growth and in different growth media, which could introduce inaccuracies to the cell count. Absorbance or optical density (OD) is measured at an appropriate wavelength, e.g., 600 nm is often used for *E. coli*. When measuring growth of a culture, the term optical density (OD) is normally used to more correctly represent the light scattering (versus absorbance) that is occurring.

Zero the spectrophotometer with a blank sample containing unused, sterile culture medium. The final OD of the sample should range between 0.1 and 1.0 for the most accurate readings. When cell densities are too high ($OD > 1.0$), artificially low OD readings will occur because the light is re-scattered and directed towards the phototube. If a culture is at too high a density for an accurate reading, it should be diluted with sterile medium to obtain a spectrophotometer reading within the optimal range.

Generalized estimates can be used to get an idea of the cell number based upon the OD reading. An OD reading of 0.1 is $\sim 10^7$ cells/ml, and the culture will appear mildly turbid. This OD-to-bacteria/ml ratio can be extrapolated for any OD reading obtained, e.g., an OD of 1.0 would roughly equal 10^8 bacteria/ml. Barely perceptible turbidity in the culture would have an OD reading of ~ 0.01 (below lower limit of detection for this method) and equal $\sim 10^6$ bacteria/ml. It is important to remember that these are estimations. To obtain an accurate cell number using a spectrophotometer, a standard curve should be generated, where cell number is determined by using another method such as a counting chamber or viable plate counts and plotted as a function of OD. After the standard curve is made, the OD of a culture can be measured and the cell number obtained from the curve. Again, it is important to remember that not only do different bacterial species have different sizes and shapes but also different strains of the same bacterial species have different average sizes, so the ratio of OD to bacterial density for one strain may not be appropriate for a second strain. Therefore, a new standard curve should be made not only for each bacterial species but also for different strains of the same bacterium.

Klett meters are inexpensive, simple spectrophotometers designed to measure the growth of cultures. Specialized culture flasks, consisting of an Erlenmeyer flask with a test tube jutting at an angle, which can be used to directly measure OD without opening the flask, may be purchased or can be made in a glasswork shop.

Hemacytometer

A hemacytometer, Petroff-Hausser counting chamber, or other similar cell counter can be used with a microscope to determine culture density. Such counters consist of a glass slide with a fine grid etched into the glass, and a cover slide that is suspended above the first. The two slides hold a specific volume of liquid, which enters by capillary action as shown in Figure 4.2.6.

To count bacterial numbers in a liquid culture using a hemacytometer, first place the coverslip onto the slide. The volume that fits between the coverslip and slide is fixed, but it is important to make sure the entire counting area under the coverslip is filled with liquid. A 10-μl aliquot of undiluted or diluted culture (depending on the bacteria density in the culture) is typically added to each side of the hemacytometer under the coverslip. As the liquid is added slowly from the edge, capillary action will draw the liquid across the entire counting area. If the bacteria are flowing rapidly over the counting surface when you first look through the microscope, let the hemacytometer sit for a few minutes; currents in the chamber will subside and allow for easier and more accurate counting. Count the bacteria in the five squares illustrated in Figure 4.2.6. If you count both sides of the chamber, then the counts should be very similar and you can average them together if desired. To determine the number of bacteria per milliliter, multiply the number of cells counted on each side or the average for both sides by the dilution (if used) by cell depth by the area. Based on the hemacytometer depicted in Figure 4.2.6 to determine the number of bacteria per milliter, use the following equation:

$$\text{number of cells counted} \times \text{dilution (if used)} \times 50{,}000$$

where $50{,}000 = 10$ (cell depth 0.1 mm) $\times$ 5000 (5 squares counted $\times$ 1000 mm^3) (Hauser Scientific). Note that 1000 mm^3 = 1 ml.

It is important to remember that volumes and mathematical ratios vary between units, so be sure to thoroughly read manufacturer's instructions of the chamber before use. This method is advantageous because it can be performed very easily and quickly. However, a disadvantage to this method, as with measuring turbidity described above, is distinguishing living cells from dead cells.

Viable plate count

Another common method to determine cell number is the viable plate count. This method involves making serial dilutions of the culture of interest to obtain enough colonies to ensure an accurate measurement, but not so many that colonies are too close to each other to count individual colonies. Figure 4.2.7 provides a demonstrative example of a serial dilution of a culture.

A good target number of colonies to ensure accurate bacterial counts on each standard 90-mm plate is between 30 and 300 colonies. To obtain individual colonies, it is important to spread the liquid evenly onto the plate using the technique described above and allow it to absorb into the agar, preferably using a volume of culture of $\leq$0.1 ml. This will prevent excess liquid on the surface from causing colonies to run together. Incubate the plate overnight at 37°C with the lid side down.

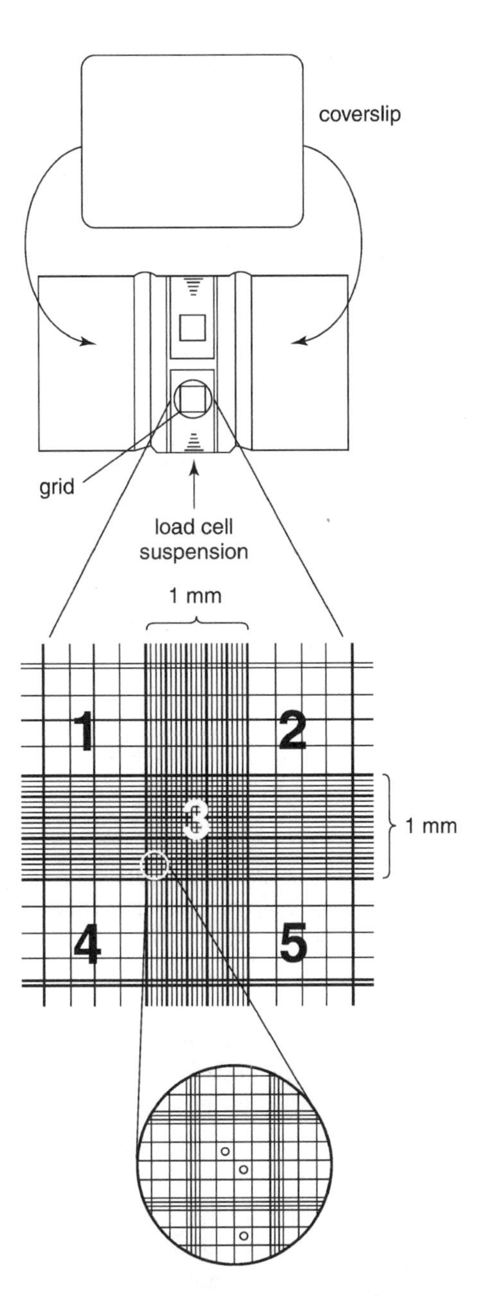

Figure 4.2.6 Example illustrating a gridded counting chamber, a hemacytometer slide (Improved Neubauer), and a coverslip. The coverslip is applied to the slide and the cell suspension is added to the counting chamber using a mechanical pipettor or a Pasteur pipet. Each counting chamber has a 3 × 3–mm grid (enlarged). The four corner squares (1, 2, 4, and 5) and the central square (3) are counted on each side of the hemacytometer (numbers added).

By counting each colony the next day, the total number of colony forming units (CFU) on the plate is determined. To remember which colonies have been counted, use a marking pen to check off each colony on the underside of the plate as it is counted. The term CFU/ml is typically used instead of bacteria/ml for the results since one colony may not equal one bacterium. By multiplying this count by the total dilution of the solution, it is possible to find the total number of CFU/ml in the original sample.

Based on the sample serial dilution shown in Figure 4.2.7, the CFUs per milliliter can be calculated from the colonies that grew on the plates. In this example, the best plate to use for the calculation would be the one with the middle dilution. By counting the number of colonies on the plate and multiplying by the final dilution, the CFU per milliliter can be calculated. For example, the second plate in Figure 4.2.7 contains 50 colonies. The liquid culture was twice diluted 10-fold (1 ml into

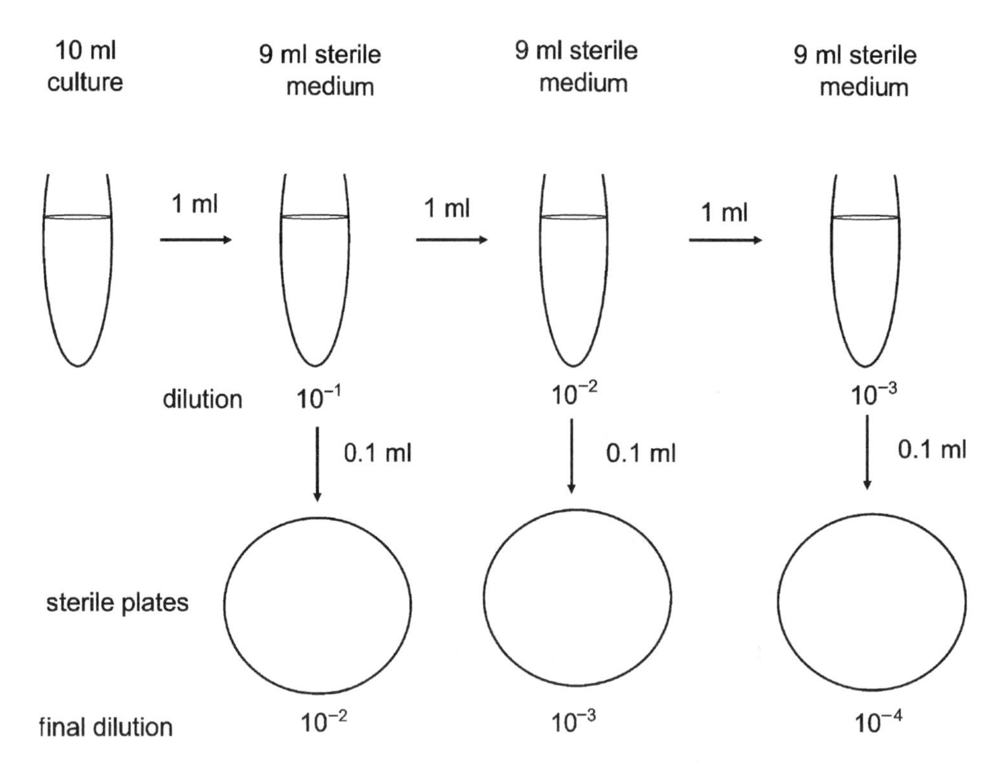

Figure 4.2.7 Serial dilution.

9 ml), then spread 0.1 ml onto the plate, making the final dilution on that plate $1{:}10^{-3}$ of the initial culture. Therefore, the original liquid culture was 10^3 or 1000 times more concentrated than the plated sample. Thus, the original culture contained 50×10^3 or 5.0×10^4 CFU/ml.

This technique for counting bacteria can be easily performed in any microbiology laboratory using readily available supplies, and has the major advantage of determining the number of living cells in a culture compared to the total number of living and dead cells. One disadvantage of the viable plate count is the assumption that each colony arises from one cell. A single colony may not represent a single bacterium because if multiple bacteria aggregate together, a single colony will form, although it actually arose from multiple bacteria. For this reason, it is especially important to spread a small volume of the liquid culture thoroughly over the plate, and to perform multiple dilutions. High numbers of spread bacteria increase the risk of multiple bacteria forming single colonies, as well as making it difficult to discriminate individual colonies on a crowded plate. Another drawback to this method is that it requires an overnight incubation of the plates. Therefore, if the bacterial concentration of the culture is needed immediately, one of the techniques described above should be performed instead.

It is important that care is taken in performing the dilutions and plating to avoid errors. To ensure the dilutions and plating are accurate, measurements of the bacterial counts from two separate dilutions should provide similar bacterial concentrations. The growth rate of different strains of *E. coli* can vary significantly, so the incubation time must be varied accordingly depending on the bacteria being studied. Along with growth medium and temperature, the plasmid content and other genetic manipulations made to the bacteria can have a profound effect on the doubling time.

ANALYSIS AND PURIFICATION OF PLASMID DNAs

A wide variety of bacteria have been found to contain plasmids, relatively small DNAs that replicate independently of the bacterial chromosome. Some naturally occurring plasmids are as small as 2 or 3 kb, while others may be several hundred kb in size. Some naturally occurring plasmids carry genes that can benefit the bacterium, such as by conferring resistance to antibiotics, by producing a toxin or other pathogenesis-related factors, or by permitting utilization of novel nutrients. Other natural plasmids do not appear to provide any benefits to the bacterium. In general, plasmids can be thought of as parasites or commensals of the host bacterium, replicating their DNA at the expense of the bacterium while possibly providing a benefit in return.

Many molecular biology techniques take advantage of plasmids that have been genetically modified for a wide variety of laboratory uses. Researchers can readily move plasmids from one bacterium to another, making plasmids ideal tools for genetic manipulation. DNAs can be cut and ligated in the test tube to form novel plasmids that can then be put into a bacterium, which will in turn replicate the new plasmid to form limitless identical copies. Many plasmids have been derived that are useful for producing recombinant proteins: a gene of interest is simply inserted into a specific site in the plasmid, the plasmid construct is then introduced into bacteria, and finally the bacteria are used to produce large quantities of a specific protein.

An important caveat to the use of plasmids is that many are extremely host specific. Many plasmids replicate using their host bacterium's DNA polymerase and other components of its DNA replication machinery. Those plasmids generally are unable to interact with the replication proteins of other bacteria, and therefore cannot replicate in any bacterium other than one specific host species. Please note that most recombinant plasmids used as cloning vectors, recombinant protein "factories," etc. are derived from plasmids that will replicate only in *E. coli* and closely related species.

In this section, methods for introduction of plasmids into *E. coli* (a process called "transformation"), confirming successful transformation, and re-isolation of plasmids from *E. coli* are described. The described methods are optimized for *E. coli*, and may not be applicable to all bacterial species. For details on transformation of other species, including description of recombinant plasmids that will replicate in those species, refer to a more specific guidebook such as *Current Protocols in Microbiology*.

Transformation of *E. coli*

The ability of a bacterium to take up exogenous naked DNA is called "competence." *E. coli* is not naturally competent, but artificial competence can be induced by a relatively simple chemical procedure that makes membranes more permeable for DNA to enter cells. An alternative method, electroporation, uses a pulse of electricity to force DNA through a suspension of bacteria, with the result that DNA is physically driven into a proportion of the bacteria.

A significant advantage to use of chemically competent bacteria is that transformation is largely unaffected by ions in the DNA solution, and they can be used to directly transform DNA ligation reactions. Electroporation is highly sensitive to ions, and small amounts of salts can cause the applied electric field to arc, thereby killing all the bacteria and ruining the experiment. Salts in DNA ligation reactions must be removed prior to electroporation. Electroporation also requires a specialized, relatively expensive piece of equipment, making this technique unsuitable for many laboratories. However, electroporation is a very quick method for introducing DNA into *E. coli*, and it is much simpler to prepare "electrocompetent" than chemically competent bacteria.

Many strains of both chemically competent and "electrocompetent" *E. coli* are available from a variety of commercial suppliers, but at prices that will be prohibitive to many researchers. In addition, some researchers may need to transform specific strains of *E. coli* that are not commercially available, such as metabolic or methylation-defective mutants. Procedures for preparation of both chemically competent and "electrocompetent" *E. coli* follow.

$CaCl_2$-Induced Competence and Transformation with Plasmid DNA

The following protocol has been successfully used in the authors' laboratory for many years with a wide variety of *E. coli* strains and plasmids, and is based upon the original method of Mandel and Higa (1970). Most commercially available chemically competent *E. coli* are prepared in similar manners, and the manufacturers' recommended transformation procedures generally mirror steps 11 to 18.

Materials

LB broth (see Commonly Used Bacterial Media)
E. coli strain
$CaCl_2$/Tris buffer: 50 mM $CaCl_2$/10 mM Tris·Cl, pH 8.0 (sterilize by filtering through a 0.22-µm filter; store indefinitely at 4°C), ice cold
LB plates with appropriate antibiotic(s) (see Commonly Used Bacterial Media)

15-ml culture tubes
37°C incubator
Spectrophotometer (*UNIT 2.1*)
Refrigerated centrifuge (*UNIT 5.1*)
Vortex
37° to 42°C heating block or water bath

1. Inoculate 5 to 10 ml LB broth with desired *E. coli* strain, and culture overnight at 37°C with aeration.

2. Dilute an aliquot of the overnight culture into fresh LB broth at a ratio of ~1:100. Cultivate at 37°C with aeration until the culture reaches an A_{600} of ~0.5 (~5×10^7 bacteria/ml), where the bacteria are in mid-exponential phase, and a large proportion are of optimal health.

   For efficiency, inoculate a volume of broth appropriate for the amount of competent bacteria desired. The final volume of competent cells will be 1/10 the starting culture volume. For example, if 300 µl competent cells are required, inoculate a 3-ml broth culture. Small volume cultures are best grown in capped test tubes that fit into the researcher's centrifuge, such that cultures can be grown, centrifuged, treated with $CaCl_2$, recentrifuged, and retreated all in the same tube.

3. Chill on ice 10 min.

4. Centrifuge 5 min at 4000 × *g*, 4°C.

   Bacteria should form a dense pellet at this point.

5. Aseptically decant the supernatant.

6. Add 1/2 volume sterile, ice-cold $CaCl_2$/Tris buffer and suspend by vortexing. Place on ice for 15 min.

7. Centrifuge 5 min at 4000 × *g*, 4°C.

   The initial incubation in $CaCl_2$ will cause changes to the bacterial membranes. At this stage, the bacterial pellet should appear very diffuse. These E. coli are relatively fragile, and should be handled delicately.

8. Aseptically decant the supernatant.

9. Add 1/10 volume sterile, ice-cold $CaCl_2$/Tris buffer. Gently suspend bacteria (swirl or rock the tube by hand). Do not vortex.

10. Dispense 50 to 300 µl bacterial suspension per tube (each aliquot will be sufficient for one transformation).

    Bacteria may be transformed with plasmid DNA immediately. Further incubation on ice or at 4°C overnight generally increases the ability of E. coli to take up plasmid DNA. Do not hold freshly made competent bacteria for longer than 24 hr, or competence levels diminish significantly.

    Alternatively, competent bacteria may be frozen at −80°C. The length of time at which frozen bacteria will remain viable and competent can vary between strains, and therefore should be determined empirically.

11. Add DNA to bacterial suspension.

In a suspension of competent cells, only a small proportion are actually competent. For that reason, the efficiency of transformation decreases as more DNA is added (i.e., fewer transformed bacteria will be obtained per microgram plasmid added). A quantity amounting to 1 or 2 ng plasmid is usually sufficient. It is important, however, to minimize the volume of the introduced DNA, to avoid overly diluting the competent cells. A 1:5 ratio of DNA to cells gives good results.

The optimal amount of DNA used in each transformation will depend upon the nature of the plasmid DNA. Note that efficiency of transformation declines as the concentration of DNA increases. For example, 1 μl of a plasmid preparation may yield 100 colonies, whereas 10 μl of the same preparation may yield 600 colonies. If one is transforming E. coli with a pure solution of a previously made plasmid, then all resulting colonies will contain the same plasmid. Since only a few such colonies will be required (actually, only one colony), transformation with a small amount of plasmid is most desirable. However, if constructing a novel plasmid through recombinant DNA technology, then only a portion of the transformed bacteria will actually contain the desired plasmid. In such cases, it is best to use the entire ligation reaction for transformation. Even though the efficiency of transformation will be reduced, the total number of colonies obtained will be maximized, enhancing the likelihood of obtaining the desired recombinant plasmid.

12. Incubate bacterial suspension/plasmid on ice for 30 min.

$CaCl_2$ interacts with DNA and increases its resistance to bacterial DNases, thereby increasing the efficiency of transformation.

13. Transfer the bacterial suspension/plasmid to a 37° to 42°C heating block or water bath and incubate for 2 min.

The warmer temperature causes the DNA to adhere to bacterial membranes. A range of 37° to 42°C can be used without significantly affecting transformation efficiency. Do not use a dry air incubator, as it is important that the temperature change be rapid and complete.

14. Return to ice for 1 to 2 min.

15. Add fresh culture broth to tube, at a ratio of 5 volumes broth per volume bacterial suspension (i.e., add 250 μl culture broth to a 50-μl bacterial suspension).

A broth with high osmolarity, such as SOC is best, as bacteria are still very fragile at this stage. A larger proportion of bacteria may lyse by osmotic shock when using a lower osmolarity medium such as LB.

16. Incubate (1-hr incubation is sufficient) at 37°C to allow bacteria to recover from the $CaCl_2$ treatment, and to permit transformed bacteria to express the plasmid-encoded selectable marker.

Some researchers prefer to incubate the bacteria in a still incubator/heat block, since shaking may damage the fragile bacteria and reduce transformation efficiency. Others prefer to shake the newly transformed bacteria, to increase aeration and thereby aid recovery. The gains and losses of each method generally cancel each other out, so either method can be used.

17. Transfer aliquots of transformed bacteria to plates of solid culture medium, and spread across entire plate surface (e.g., solidified LB plates with appropriate antibiotic(s)).

Recombinant plasmids contain at least one gene encoding resistance to an antibiotic. Only bacteria that were transformed with the introduced plasmid will be resistant to that antibiotic, and will be able to grow and form colonies.

The volume of bacteria plated depends upon the number of colonies needed per experiment. If transformation is done with a previously made, pure plasmid, then all resulting bacterial colonies will contain the same plasmid, and therefore a small number of colonies is all that is required. In such a case, a single plate with 10 to 50 μl of the transformed culture will be more that adequate. If one is constructing a novel plasmid by recombinant DNA technology, then only a portion of the transformed bacteria will actually contain the desired plasmid. In such cases, the entire transformation culture should be plated at 50 to 500 μl per plate on as many plates as necessary.

18. Incubate plates overnight at 37°C.

Some commonly used plasmids confer resistance to beta-lactam antibiotics (such as penicillin, ampicillin, or carbenicillin). The encoded beta-lactamase enzyme degrades those antibiotics. If such a plasmid is used, care should be taken not to let resultant colonies grow too large. Transformed bacteria will secrete beta-lactamase into the medium, reducing the antibiotic concentration around the colony, thereby permitting other, nontransformed bacteria to also grow and form colonies. These "satellite" colonies can rapidly grow to large sizes, obscuring the colony of truly transformed bacteria.

19. If colonies need to be screened for a specific plasmid (such as if a novel recombinant plasmid is being constructed), it is most efficient to transfer colonies to a new plate(s) to which a numbered grid has been fastened to the back—see Organization of Clones.

Electroporation of *E. coli* with Plasmid DNA

Due to the high voltages used during the electroporation procedure, it is essential that bacteria be suspended in a medium with low conductivity. Bacteria are washed extensively in salt-free solution. It is also essential that the plasmid solution be free of salts. Salts should be removed from DNA solutions, including ligation reaction, by ethanol precipitation *(UNIT 5.2)* followed by resuspension in water or a low-salt buffer such as TE buffer.

The following protocol is based upon the method of Dower et al. (1988). Commercially available electrocompetent *E. coli* are prepared essentially the same way.

4

Materials

LB broth (see Commonly Used Bacterial Media)
E. coli strain
10% (v/v) glycerol (filter sterilize using a 0.22-μm filter and store indefinitely at 4°C), ice cold
DNA in water or TE buffer
SOC broth (see Commonly Used Bacterial Media)
Solid culture medium plates with appropriate antibiotic(s) (e.g., LB medium with antiobiotics; see Commonly Used Bacterial Media)

37°C incubator
Spectrophotometer (*UNIT 2.1*)
Refrigerated centrifuge (*UNIT 5.1*)
Vortex
1.5-ml polypropylene tubes
0.2-cm gap electroporation cuvette
Electroporator
17 × 100–mm capped test tube

1. Inoculate 5 to10 ml LB broth with desired *E. coli* strain, and culture overnight at 37°C with aeration.
2. Dilute an aliquot of the overnight culture into fresh LB broth at a ratio of ~1:100. Cultivate at 37°C with aeration until the culture reaches an A_{600} of ~0.5 (~5×10^7 bacteria/ml), where the bacteria are in mid-exponential phase, and a large proportion are of optimal health.
3. Chill on ice for 10 min.
4. Centrifuge 5 min at 4000 × *g*, 4°C.
5. Aseptically decant the supernatant.
6. Add 1/2 volume sterile, ice-cold 10% (v/v) glycerol and suspend by vortexing.
7. Centrifuge 5 min at 4000 × *g*, 4°C.

8. Add 1/2 volume sterile, ice-cold 10% (v/v) glycerol and suspend by vortexing.
9. Centrifuge 5 min at 4000 × *g*, 4°C.
10. Add ~1/500 volume sterile, ice-cold 10% (v/v) glycerol and suspend by vortexing.

    Bacteria from a 1-liter culture will be resuspended in a final volume of ~2 to 3 ml.

    Bacteria at this point can be dispensed into aliquots (~40-μl aliquots) and stored almost indefinitely at −70°C or colder.

11. In a sterile 1.5-ml polypropylene tube, mix 40 μl bacterial suspension with 1 to 5 μl DNA (in either water or a low-salt buffer such as TE buffer).
12. Transfer bacteria and DNA to a pre-chilled, 0.2-cm gap electroporation cuvette. Ensure bacteria are evenly dispersed along the bottom of the cuvette by tapping it lightly two to three times on a table top.
13. Set electroporator to 25 μF, 2.5 kV, and 200 Ω. Apply one pulse.
14. Immediately add 1 ml SOC broth to cuvette.
15. Transfer to a sterile 17 × 100–mm capped test tube.
16. Incubate 1 hr at 37°C with shaking.
17. Transfer aliquots of transformed bacteria to plates of solid culture medium, and spread across entire plate surface (e.g., solid LB medium with appropriate antibiotic(s)).

    Recombinant plasmids contain at least one gene encoding resistance to an antibiotic. Only bacteria that were transformed with the introduced plasmid will be resistant to that antibiotic, and will be able to grow and form colonies.

    The volume of bacteria plated depends upon the number of colonies needed per experiment. If transforming with a previously made, pure plasmid, then all resulting bacterial colonies will contain the same plasmid, and so a small number of colonies is all that is required. In such a case, a single plate with 10 to 50 μl of the transformed culture will be more than adequate. If one is constructing a novel plasmid by recombinant DNA technology, then only a portion of the transformed bacteria will actually contain the desired plasmid. In such cases, the entire transformation culture, with 50 to 500 μl per plate, should be plated on as many plates as necessary.

18. Incubate plates overnight at 37°C.

    Some commonly used plasmids confer resistance to beta-lactam antibiotics (such as penicillin, ampicillin, or carbenicillin). The encoded beta-lactamase enzyme degrades those antibiotics. If such as plasmid is used, care should be taken not to let resultant colonies grow too large. Transformed bacteria will secrete beta-lactamase into the medium, reducing the antibiotic concentration around the colony, thereby permitting other, non-transformed bacteria to also grow and form colonies. These "satellite" colonies can rapidly grow to large sizes, obscuring the colony of truly transformed bacteria.

19. If colonies need to be screened for a specific plasmid (e.g., if constructing a novel recombinant plasmid), it is most efficient to transfer colonies to a new plate(s) to the back of which a numbered grid has been fastened—see Organization of Clones.

Screening Bacteria for Plasmid DNA (Colony PCR)

Following transformation of *E. coli* with a pure preparation of a previously constructed plasmid, it is generally safe to conclude that all resulting bacterial colonies contain the plasmid of interest. But if the researcher had attempted construction of a new plasmid via ligation of DNA fragments, then used the ligation reaction mix to transform *E. coli*, not every colony may contain the desired plasmid. Some colonies may contain plasmid DNA that is not assembled correctly. The researcher therefore needs to screen through colonies to find one that contains the desired plasmid.

A simple PCR-based method allows simultaneous screening of multiple colonies with very little effort, and can be completed in only a few hours. Colony PCR takes advantage of the fact that the desired plasmid contains a unique arrangement of DNA bases. PCR of the desired plasmid using oligonucleotide primers complementary to that plasmid's sequence will yield a distinctive amplicon, whereas all other DNAs will either yield a product of different size, or no product at all.

A major advantage of colony PCR is that this procedure does not require DNA purification. A small aliquot of whole bacteria is added to a PCR mix (specific oligonucleotide primers, heat-stable DNA polymerase, dNTPs, and buffer), then subjected to PCR. The initial PCR step, heating to 94°C, is sufficient to lyse all bacteria and liberate their DNA contents (*UNIT 10.2*).

Oftentimes, the researcher will be cloning a distinct DNA fragment into the multiple cloning site of a previously constructed plasmid vector. Many widely used cloning vectors contain multiple cloning sites flanked by conserved DNA sequences. In many plasmids, these conserved sequences are based on DNAs originally found in bacteriophage M13, and are generally referred to as "M13-forward" and "M13-reverse," with one sequence located on either side of the multiple cloning site. Some cloning vectors contain sequences based on bacteriophage T7, while others may contain both M13- and T7-derived sequences. The researcher should examine the cloning vector's DNA sequence to determine whether or not any of these conserved DNA sequences are present. Complete sequences of many commonly used cloning vectors are available from databases such as GenBank (*http://www.ncbi.nlm.nih.gov/Genbank*). Commercially obtained cloning vector plasmids usually include such data on product information sheets. Many vendors provide this information at no cost through the World-Wide Web, therefore, product information sheets may be downloaded by the researcher even without purchasing from that supplier.

One can take advantage of conserved vector sequences to minimize the expense of PCR screening. For example, the commonly used vector pCR2.1 (Invitrogen) contains an M13-forward sequence on one side of the multiple cloning site/TA cloning site, and an M13-reverse sequence on the other side. Thus, any pCR2.1-based plasmid will yield an amplicon from PCR with M13-forward and -reverse primers. Vectors without any insert will yield a very small product (<100 bp), vectors containing the desired DNA fragment will yield an appropriately sized amplicon, and vectors with incorrect inserts will generally yield amplicons of different sizes. An added advantage to the use of such primers is that a PCR product will be obtained regardless of whether or not each plasmid contains an insert, which serves as a reaction control. Researchers can purchase their standard PCR-screening oligonucleotides in large quantities, which cut costs, and then use those oligonucleotides for all screenings.

This procedure has many applications in addition to screening transformed *E. coli*. Any DNA sequence in a bacterium can be detected and/or isolated regardless of its location. The authors' laboratory has had excellent success amplifying specific genetic loci from the chromosomes of intact *E. coli*, the spirochete *Borrelia burgdorferi*, and other species of bacteria.

Materials

Oligonucleotide primers
DNA polymerase, heat-stable
dNTPs
PCR buffer *(UNITS 3.3 & 10.2)*

PCR tubes
Sterile, new toothpicks
PCR thermal cylcer (e.g., Perkin-Elmer 9700)

This procedure works best when using a multiwell PCR thermal cycler with a heated lid, such as the Perkin-Elmer 9700. Such a device allows efficient PCR analyses of volumes of ≤10 µl.

4

1. Prepare a "master mix" of specific oligonucleotide primers, heat-stable DNA polymerase, dNTPs, and buffer (*UNIT 10.2*).

   The volume of master mix prepared will depend upon the number of reactions and volume of each reaction. For example, if performing 30 reactions of 10 μl each, then 300 μl master mix will be required. It is best to include 50 to 100 μl extra master mix, to account for potential pipeting errors. Also see the discussion of stock solutions in UNIT 3.1.

2. Dispense aliquots of master mix into PCR tubes.

3. Using a sterile, new toothpick, remove a small amount of a bacterial colony to be tested.

   The amount of bacteria should be just barely visible to the unaided eye: too little, and PCR amplification will not be efficient; too much, and bacterial components can interfere with PCR. The researcher will learn from experience how much E. coli is desirable.

4. Perform PCR in a thermal cycler using conditions appropriate for the DNA being tested.

   For production of a detectable amplicon, 25 cycles is sufficient. A rule of thumb for PCR is to allow a 1-min extension time for every 1 kb DNA being amplified.

5. Mix 5 to 10 μl of each reaction mixture with 5 μl agarose gel electrophoresis loading buffer (*UNIT 7.2*).

   For efficiency, loading buffer may be added directly to the reaction tube. However, if cloning or direct sequencing of the PCR product is desired, then an aliquot of each reaction should be removed from the reaction tube and mixed with loading buffer.

6. Separate DNA on a 0.7% (w/v) agarose gel (*UNIT 7.2*). Visualize DNA by staining with ethidium bromide and UV light.

Purification of Plasmid DNA from *E. coli*

Due to their small sizes relative to chromosomal DNA, plasmids can be efficiently purified using methods that cause chromosomal DNA to form insoluble pellets but retain solubility of small DNAs. This can be achieved by DNA denaturation followed by rapid renaturation. Two rapid methods are described that use either alkaline pH or heat as the denaturant. The alkaline lysis procedure takes longer but yields cleaner DNA, while the boiling lysis procedure takes less time (<30 min). Both procedures are generally adequate in producing DNA for restriction endonuclease digestion (*UNIT 10.1*), however, boiling may not be suitable for DNA sequencing (*UNIT 10.4*; phenol/chloroform extraction and treatment with RNase A are mandatory for DNA intended for sequencing). Note that the RNA, which contaminates DNA prepared by these rapid methods, makes it impossible to determine DNA concentration by absorption spectrophotometry (*UNIT 2.2*).

Small-Scale Plasmid Isolation by Alkaline Lysis

This procedure, based on the method of Birmboim and Doly (1979) denatures DNA by alkaline pH in the presence of SDS. Rapid neutralization at cold temperature results in renaturation of small plasmids, while larger DNA, such as the *E. coli* chromosome, forms a tangled, insoluble clump that is readily removed following centrifugation.

Commercial plasmid purification kits, such as those available from Qiagen, also use this method. Qiagen buffers 1, 2, and 3 are similar to the corresponding buffers 1, 2, and 3 found in the following protocol. Many such kits contain RNase in buffer 1, to remove contaminating RNAs. The following procedure can be similarly modified, if desired, by addition of RNase A to buffer 1 at a final concentration of 100 μg/ml.

Materials

LB broth (see Commonly Used Bacterial Media)
E. coli strain

Buffer 1: 50 mM glucose, 10 mM EDTA, 25 mM Tris·Cl, pH 8.0 (store 3 to 6 months at 4°C); immediately before use, add lysozyme to stock solution to a final concentration of 2 mg/ml
Buffer 2: 0.2 N NaOH, 1% (w/v) sodium dodecylsulfate (store ≤1 month at room temperature)
Buffer 3: 3 M sodium acetate, pH 4.8; adjust pH as necessary with glacial acetic acid
0.1 M sodium acetate/ 0.05 M Tris·Cl, pH 8.0
70% and 95% ethanol
TE buffer *(UNIT 3.3)*

1.5-ml microcentrifuge tubes
Refrigerated microcentrifuge

1. Inoculate 2 ml LB broth with desired *E. coli* strain, and culture overnight at 37°C with aeration.
2. Transfer ~1.4 ml overnight culture to a 1.5-ml microcentrifuge tube. Cap and centrifuge 15 sec at 15,000 × *g*, room temperature.
3. Remove all the supernatant using a pipet.
4. Completely resuspend cells in 100 µl buffer 1, by vortexing.
5. Add 200 µl buffer 2, mix by gently inverting four to five times.

   Do not vortex.
6. Place on ice for 5 to 10 min.
7. Add 150 µl buffer 3, mix by gently inverting four to five times.

   Do not vortex.
8. Place on ice for 30 to 60 min.
9. Pellet chromosomal DNA, proteins, and other cellular debris by centrifuging 5 min at 15,000 × *g*, 4°C.
10. Transfer 400 µl of supernatant into a clean 1.5-ml microcentrifuge tube.
11. Add 1 ml of 95% ethanol and incubate 15 to 30 min at −20°C.
12. Pellet DNA by centrifuging 15 min at ≥20,000 × *g*, 4°C.

    DNA should be visible as a white pellet at the bottom of the tube.
13. Remove supernatant.
14. Resuspend DNA in 100 µl of 0.1 M sodium acetate/0.05 M Tris·Cl, pH 8.0.
15. Add 200 µl of 95% ethanol and incubate 15 to 30 min at −20°C.
16. Pellet nucleic acids by centrifuging 15 min at ≥20000 × *g*, 4°C and remove supernatant.
17. Rinse tube with ~500 µl of 70% ethanol. Remove all traces of ethanol with a pipet.
18. Dry for 2 to 3 min with cap open at room temperature.

    Do not overdry, as that can make it more difficult to redissolve the nucleic acids.
19. Resuspend nucleic acids in 50 to 100 µl TE buffer.

    Note that product contains both plasmid DNA and contaminating RNAs.

Small-Scale Plasmid Isolation by Boiling Lysis

This rapid alternative takes only ~30 min to complete, but generally yields DNA that is less pure than can be obtained by alkaline lysis. The procedure is based on that of Holmes and Quigley (1981).

Materials

LB broth (see Commonly Used Bacterial Media)
E. coli strain
STET, per 100 ml:
  1 ml 1 M Tris·Cl, pH 8.0 (*UNIT 3.3*)
  10 ml 0.5 M EDTA, pH 8.0 (*UNIT 3.3*)
  500 µl Triton X-100
  8 g sucrose
  Add water to a final volume of 100 ml
Lysozyme solution: 10 mg lysozyme/ml of 10 mM Tris·Cl, pH 8.0, prepare fresh daily
Isopropanol
70% ethanol
TE buffer (*UNIT 3.3*)

1.5-ml microcentrifuge tubes
Vortexer
Boiling water bath
Sterile toothpicks

1. Inoculate 2 ml LB broth with desired *E. coli* strain, and culture overnight at 37°C with aeration.
2. Transfer ~1.4 ml overnight culture to a 1.5-ml microcentrifuge tube. Cap and centrifuge 2 min at ~5000 × *g*, room temperature.
3. Remove all the supernatant using a pipet.
4. Add 300 µl STET and vortex until pellet just disappears.
5. Add 30 µl fresh lysozyme solution and vortex ~5 sec.
6. Place tube in boiling water bath for 1 min.

   NOTE: *Open tube lids before boiling. Otherwise, the heated liquid inside may force caps open during boiling, potentially spilling the contents.*

   Begin heating water bath at the start of the isolation procedure, so that it is already at a boil when you reach step 6. For safety, keep a close watch on the water bath during all heating stages.

7. Centrifuge 5 min at ≥20,000 × *g*, room temperature.
8. Remove pellet (chromosome and other cellular debris) with a clean, sterile toothpick or micropipet tip. Discard pellet.
9. Add 330 µl isopropanol. Mix thoroughly by vortexing.
10. Pellet nucleic acids by centrifuging 15 min at ≥20,000 × *g*, 4°C.
11. Rinse tube with ~500 µl of 70% ethanol. Remove all traces of ethanol with a pipettor.
12. Dry for 2 to 3 min with cap open at room temperature.

    Do not overdry.

13. Resuspend nucleic acids in 20 to 50 µl TE buffer.

    Note that product contains both plasmid DNA and contaminating RNAs.

Additional Purification of Plasmid DNA

Contaminating RNAs can be removed from plasmid preparations by addition of RNase A to a final concentration of 100 mg/ml, followed by incubation for 1 hr at 37°C. Stock solutions of RNase A may be prepared ahead of time in TE buffer, and stored for at least 1 month at −20°C. Be certain that RNase A is labeled as being free of DNases.

Contaminating proteins (including RNase added as above) can be removed by phenol extraction using the following protocol.

CAUTION: Phenol can cause chemical burns to skin and other tissues. Wear gloves when handling phenol solutions. Do not inhale.

Materials

Nucleic acid solution
TE buffer (*UNIT 3.3*)
3 M sodium acetate, pH 7.0 (*UNITS 3.1 & 5.2*)
Phenol (previously saturated with 10 mM Tris·Cl, pH 7.5)
95:5 (v/v) chloroform/isoamyl alcohol
70% and 90% ethanol

1.5-ml microcentrifuge tubes
Vortexer

1. Increase volume of nucleic acid solution to 450 µl by addition of TE buffer.

   Larger volumes should be aliquoted accordingly.

2. Add 50 µl of 3 M sodium acetate, pH 7.0.
3. Add ~1 ml phenol (previously saturated with 10 mM Tris·Cl, pH 7.5). Vortex vigorously.
4. Centrifuge at ≥20,000 × *g* for 5 min at room temperature.
5. Transfer aqueous (upper) phase to a clean microcentrifuge tube. Do not transfer any of the lower (phenol) phase. If necessary, leave a little of the aqueous phase behind in the original tube.
6. Add ~1 ml 95:5 (v/v) chloroform/isoamyl alcohol. Vortex vigorously.
7. Centrifuge 5 min at ≥20,000 × *g*, room temperature.
8. Transfer aqueous (upper) phase to a clean 1.5-ml microcentrifuge tube. As above, avoid transferring any of the lower phase, and leave a little of the aqueous phase in the original if necessary.
9. Repeat chloroform/isoamyl alcohol extraction one additional time.
10. Add 2 vol of 95% ethanol to aqueous phase. Incubate 15 to 30 min at −20°C.
11. Pellet nucleic acids by centrifuging 15 min at ≥20,000 × *g*, 4°C. Remove supernatant.
12. Rinse tube with ~500 µl of 70% ethanol. Remove all traces of ethanol with a pipet.
13. Dry for 2 to 3 min with cap open at room temperature.

    Do not overdry.

14. Resuspend nucleic acids in the original volume of TE buffer.

STORAGE OF CULTURES

It is extremely important to have an adequate method for long-term storage of bacteria, especially if genetic manipulations have been performed. *E. coli* on Petri plates are generally viable for up to 2 weeks at 4°C if the plate is sealed around the edges with Parafilm. Bacteria should not be left in liquid cultures longer than it takes them to reach high density.

It is important in many cases that the bacteria do not reach stationary phase, because energy salvaging processes performed by starving bacteria can affect the stability of recombinant DNA.

The protocol for long-term storage of *E. coli* is simple, but requires either a −80°C freezer or a liquid nitrogen container. The day before freezing, inoculate the bacteria into 5 to 10 ml rich media (TB/LB; see recipe) with appropriate antibiotics, and incubate at 37°C with shaking. Add 50% (v/v) glycerol to the overnight culture at a ratio of 1:1 (25% final glycerol). Glycerol is a cryoprotectant (antifreeze), which prevents damage of the bacteria from the shear forces of ice formation. Vortex this mixture and transfer to one or more cryogenic vials (two or more vials for precious strains). The size and type of cryogenic vials are determined by individual laboratories based on availability of freezer space. Rapidly freeze the cryogenic vials and contents by placing in a dry-ice/ethanol bath or in liquid nitrogen, then place in a −80°C freezer. Alternatively, vials can be placed directly in the −80°C freezer. The cultures are now stably frozen and can be kept indefinitely.

To resurrect bacteria, scrape the surface of the frozen stock with a sterile wooden probe or other suitable tool, then transfer to growth medium. Return stock to freezer immediately. Do not allow stocks to thaw, as freeze-thaw cycles significantly reduce viability.

COMMONLY USED BACTERIAL MEDIA

Following are several common media used to culture *E. coli*, which are also suitable for cultivation of some other species. These media are generally sterilized by autoclaving, although certain media or supplements, including antibiotics, cannot be autoclaved and must instead be sterilized by passage through a 0.22-μm filter (*UNIT 4.1*).

CAUTION: To prevent explosions, loosen caps of bottles prior to autoclaving.

For certain media, components must be sterilized independently, and then combined later. Certain mixtures of salts or other compounds can yield toxic byproducts when subjected to autoclaving conditions. Other components, such as sugars, are generally filter sterilized to prevent burning (caramelization) during autoclaving. Antibiotics are usually degraded by the heat of the autoclave.

For most bacteria, deionized or distilled water is used to make culture media. For less fastidious bacterial species such as *E. coli*, tap water is satisfactory or even preferable for making many culture media, as the trace minerals in such water can be beneficial.

Sterilized liquid media can be stored indefinitely at room temperature.

Pouring Plates

Petri plates are the most common containers for storage of solid medium. The dish is partially filled with warm liquid agar along with a particular mix of nutrients, salts, and amino acids and, when required, antibiotics. The solidifying medium, 1.5% (w/v) agar, is added prior to autoclaving. It is easiest to dissolve all other ingredients except those that are heat labile (e.g., certain antibiotics) as per instructions, and then add the agar immediately before autoclaving. The autoclaving process will melt the solidifying agent. After autoclaving, swirl the medium gently to evenly disperse ingredients. If solid medium is be used for plates, first cool the medium to ∼50°C in a water bath, as this will prevent melting of plastic Petri dishes, facilitate handling of the container flask, and reduce condensation in the solidified plates. After the medium has cooled to 50°C, filter-sterilized heat-labile components should be added to the liquid and swirled. Pre-warming such components to ∼50°C will avoid over-cooling and premature solidification of the agar-containing solution.

Flaming the lip of the flask between plate pours can help prevent introduction of contaminants. Pour 20 to 25 ml per 100-mm diameter Petri dish. Swirl plates if necessary to cover entire dish. Bubbles on medium surface can be removed by quickly passing a burner flame over the surface prior to solidification.

Solidified plates should be left for 1 to 2 days at room temperature to allow evaporation of excess moisture, which will reduce condensation during later culture incubations. Alternatively, solidified plates can be dried by incubating for ~30 min at 37°C with lids ajar. A perfect plate for growth of *E. coli* has a slightly rippled surface. The rippled surface is a good indicator that there is no excess moisture on the surface of the medium. If there is excess moisture on the surface of the plate, cells will be able to move through that fluid and will create a smear, rather than individual colonies. Plates can be stored almost indefinitely at 4°C if wrapped in plastic to prevent dehydration. Plates containing antibiotics should not be stored for longer than 1 month at 4°C.

If not used immediately for pouring plates, solid media can be stored indefinitely at room temperature in tightly-capped bottles. To use, loosen cap, melt by re-autoclaving or microwaving, then proceed as above.

LB (Luria-Bertani) medium

Sometimes also referred to as Luria broth or L-broth, this is a good general-purpose medium for culturing *E. coli*. The tryptone and yeast extract provide all nutrients needed, and NaCl provides a proper osmotic pressure for growth. This medium is commonly used for solid medium by adding a solidifying agent (e.g., agar).

10 g tryptone
5 g yeast extract
5 g NaCl
Add H_2O to a volume of 1 liter

4

Some researchers adjust the pH to ~7 by slowly adding 1 N NaOH, but this is not necessary.

Terrific broth (TB)

E. coli grows rapidly in this very rich medium. A small inoculum will grow to stationary phase overnight in 100 ml TB, making it a good choice for growing bacteria for plasmid preparations. The medium is made from two individually autoclaved components that are combined when cooled, because a precipitate will form if autoclaved together.

12 g tryptone
24 g yeast extract
4 ml glycerol

Add H_2O to a volume of 900 ml. Dispense 90-ml aliquots into screw-cap bottles and autoclave. Before use, add 10 ml TB-potassium salts (see below) to each bottle.

TB-potassium salts

125.5 g K_2HPO_4
23 g KH_2PO_4
Add H_2O to a volume of 1 liter. Dispense into 100-ml aliquots and autoclave.

Super broth (SB)

This is an extremely rich medium often used for production of proteins within the bacteria. Components needed for protein production are provided in excess. This medium can also be used for overnight growth of *E. coli* to high cell density.

32 g trypone
20 g yeast extract
5 g NaCl
Add H_2O to 1 liter
Autoclave

SOC

Transformed bacteria are very fragile and require a similar osmolarity inside and outside of the cell to prevent swelling. This rich medium has a high osmolarity, and is ideal for the recovery step following transformation of *E. coli* and similar bacteria.

20 g tryptone
5 g yeast extract
10 ml 1 M NaCl
2.5 ml 1 M KCl

Add H_2O to 980 ml, autoclave, and after cooling to at least 45°C, add the following:

10 ml 2 M $MgCl_2$ (filter sterilized)
20 ml 20% (w/v) glucose (filter sterilized)

M9 minimal (5× concentrated stock)

30 g Na_2HPO_4
15 g KH_2PO_4
5 g NH_4Cl
2.5 g NaCl
0.015 g $CaCl_2$ (optional)

Add H_2O to 1 liter. This 5× concentrated medium can be stored for many months at 4°C. Add ~50 ml chloroform to stock solution as a preservative (this separates into an organic layer at the bottom of the bottle, which should not be included in final preparation).

Before use, dilute 1:5 in water and sterilize by autoclaving. After cooling to <50°C, add the following sterile solutions:

1 ml 1 M $MgSO_4$-$7H_2O$ (filter sterilized)
10 ml 20% carbon source (filter sterilized)

The type of carbon source used for growth is dependent on experimental design and strain requirements. The most common sugar used for casual growth and maintenance is glucose.

Glycerol, 50% (v/v)

Use to freeze bacterial cultures for long-term storage. Glycerol is too viscous to pipet, but a 50:50 mix of glycerol and water is readily pipetted.

50 ml glycerol
50 ml water

Pour glycerol, then water into a 100-ml graduated cylinder, seal with Parafilm, then shake vigorously to mix thoroughly. Pass through a 0.22-μm filter to sterilize.

Antibiotic Supplements for Media

The antibiotics described below and in Table 4.2.1 are commonly added to culture media for growth of *E. coli* to prevent outgrowth of bacteria that do not contain the antibiotic resistance gene in those strains or clones of interest and to ensure plasmid stability in the bacteria.

To prevent their destruction by excess heat, antibiotics should be added only after media have cooled to <50°C. If solid media are being prepared, autoclaved media should be thoroughly cooled in a 50°C water bath, antibiotic(s) added, then plates poured.

Antibiotics should be sterilized by passage through 0.22-μm filters. Antibiotics dissolved in alcohols do not require additional sterilization.

Table 4.2.1 Antibiotic Stock Solutions for Use with *E. coli* Cultures

Antibiotic	Stock concentration	Working concentration for *E. coli*	Mechanism of action[a]
Ampicillin (Na salt) (carbenicillin is more stable, but the two are interchangeable)	10 mg/ml in H_2O	50 μg/ml	β-lactam antibiotic that inhibits cell wall synthesis through inactivation of transpeptidases on bacterial membrane. Bacteriocidal.
Carbenicillin	10 mg/ml in H_2O	50 μg/ml	Same mechanism as described above for ampicillin
Chloramphenicol	6 mg/ml in 100% ethanol	30 μg/ml	Inhibits translation on the 50S ribosomal subunit of bacterial ribosomes by blocking transpeptidation and elongation of the amino acid chain. Bacteriostatic.
Erythromycin	10 mg/ml in H_2O	50 μg/ml	Macrolide antibiotic that binds to the 50S subunit of bacterial ribosomes and inhibits translation by blocking transpeptidation. Bacteriostatic.
Gentamicin	3 mg/ml in H_2O	15 μg/ml	Aminoglycoside antibiotic that inhibits translation by binding to the 30S ribosomal subunit of bacterial ribosomes. Bacteriocidal.
Kanamycin	10 mg/ml in H_2O	50 μg/ml	Inhibits translation by binding 70S ribosomal subunit preventing translocation and eliciting miscoding errors. Bacteriocidal
Rifampicin	20 mg/ml in either methanol or DMSO	100 μg/ml	Inhibits initiation of RNA transcription by binding RNA polymerase. Bacteriocidal. Light-sensitive.
Spectinomycin	20 mg/ml in H_2O	100 μg/ml	Aminoglycoside antibiotic that inhibits translation by binding to the 30S ribosomal subunit of bacterial ribosomes. Bacteriocidal.
Streptomycin	10 mg/ml in H_2O	50 μg/ml	Aminoglycoside antibiotic that inhibits translation by binding to the 30S ribosomal subunit of bacterial ribosomes. Bacteriocidal.
Tetracycline-HCl	2.5 mg/ml in 50% ethanol	12.5 μg/ml	Inhibits translation by binding to the 30S ribosomal subunit. Bacteriostatic. Light-sensitive.

[a]Bacteriocidal: kill bacteria; bacteriostatic: inhibit growth or reproduction of bacteria.

It saves time and effort to prepare concentrated stock solutions of each antibiotic. Table 4.2.1 provides the concentration of the antibiotic stock solutions as well as the final concentration of the antibiotic after addition to media. Additionally, the mechanism of action for each antibiotic is described. For simplicity, Table 4.2.2 can be used to quickly determine the volume of antibiotic stock solution to add to a specific culture volume to obtain the desired final concentration. For example, if you wanted to start a 50-ml culture, you would go down the culture volume column to 50 ml, then across to the volume of antibiotic to add, in this case, 250 μl. Concentrated solutions should be kept no more than 1 month at 4°C or 6 months at −20°C.

Table 4.2.2 Volumes of Stock Solutions from Table 4.2.1 to be Added to Culture Media to Obtain Correct Final Antibiotic Concentrations

Culture volume	Add this volume of antibiotic stock
1 ml	5 µl
2 ml	10 µl
3 ml	15 µl
5 ml	25 µl
10 ml	50 µl
20 ml	100 µl
25 ml	125 µl
50 ml	250 µl
100 ml	500 µl
200 ml	1 ml
500 ml	2.5 ml
1 liter	5 ml

Media containing antibiotics should be kept for no more than 1 month. Protect media containing light-sensitive antibiotics, such as tetracycline and rifampicin, from light by wrapping in aluminum foil and/or storing in a dark place.

4

LITERATURE CITED

Birnboim, H.C. and Doly, J. 1979. A rapid alkaline extraction procedure for screening recombinant plasmid DNA. *Nucl. Acids Res.* 7:1513-1523.

Coico, R. and Lunn, G. 2005. Biosafety: Guidelines for working with pathogenic and infectious microorganisms. *Curr. Prot. Microbiol.* 0:1A.1.1-1A.1.8.

Dower, W.J., Miller, J.F., and Ragsdale, C.W. 1988. High efficiency transformation of *E. coli* by high voltage electroporation. *Nucl. Acids Res.* 16:6127-6145.

Holmes, D.S. and Quigley, M. 1981. A rapid boiling method for the preparation of bacterial plasmids. *Anal. Biochem.* 114:193-197.

Mandel, M. and Higa, A. 1970. Calcium-dependent bacteriophage DNA infection. *J. Mol. Biol.* 53:159-162.

Wright, S. 1986. Recombinant DNA technology and its social transformation, 1972-1982. *Osiris* 2:303-360.

Chapter 5

Sample Preparation

UNIT 5.1

Centrifugation

Sean R. Gallagher[1]

[1]UVP, LLC, Upland, California

INTRODUCTION

Centrifuges use a spinning rotor (designed to hold sample tubes) to apply centrifugal force to a sample. Centrifuges are commonplace for a variety of applications in life science laboratories, and they vary from low-speed, uncooled benchtop models to ultracentrifuges capable of generating more than 100,000 × g (see Table 5.1.1). Models designed to generate relatively low centrifugal force are useful for quickly pelleting a protein or nucleic acid precipitate, or for performing affinity purification by forcing the sample suspension through an affinity matrix in a small centrifuge tube (e.g., see *UNIT 5.2*). High-speed centrifuges are used (under varying conditions) mainly for generation of whole-cell pellets (e.g., see *UNIT 4.2*), removal of cellular debris following homogenization, or differential centrifugation of larger organelles such as nuclei, mitochondria, and chloroplasts. Ultracentrifuges (see Griffith, 1986; Fig. 5.1.1), capable of generating extremely high centrifugal force, are used in purifying smaller organelles and membranes. Analytical ultracentrifugation is a

Table 5.1.1 Typical Applications of Centrifugation

Application	Microcentrifuge	High-speed centrifuge	Analytical ultracentrifuge[a]	Ultracentrifuge
Upper limit, rpm[b]	14,000	30,000	50,000	130,000
Upper limit, RCF[b]	18,000	100,000	250,000	1,000,000
Cell pelleting and recovery (see *UNIT 4.2*)	√	√		
CsCl nucleic acid purification				√
Density gradient (subcellular) fractionation of organelles		√		√
Differential (subcellular) fractionation of organelles (mitochondria, nuclei)	√	√		
Differential (subcellular) fractionation: organelles (plasma membrane, Golgi, ER)		√		√
DNA/RNA size fractionation				√
Organic extraction and precipitation protein and nucleic acid (see *UNIT 5.2*)	√	√		
Protein size, shape, and subunit structure			√	
Spin columns: filtration, desalting, and affinity column purification (see *UNIT 5.2* and Chapter 6)	√	√		

[a]See Scott et al. (2006) and *CP Immunology Unit 18.8* (Schuck and Braswell, 2000).

[b]Maximum allowable depends on the rotor (see Tables 5.1.3 to 5.1.6) and tubes or bottles being used.

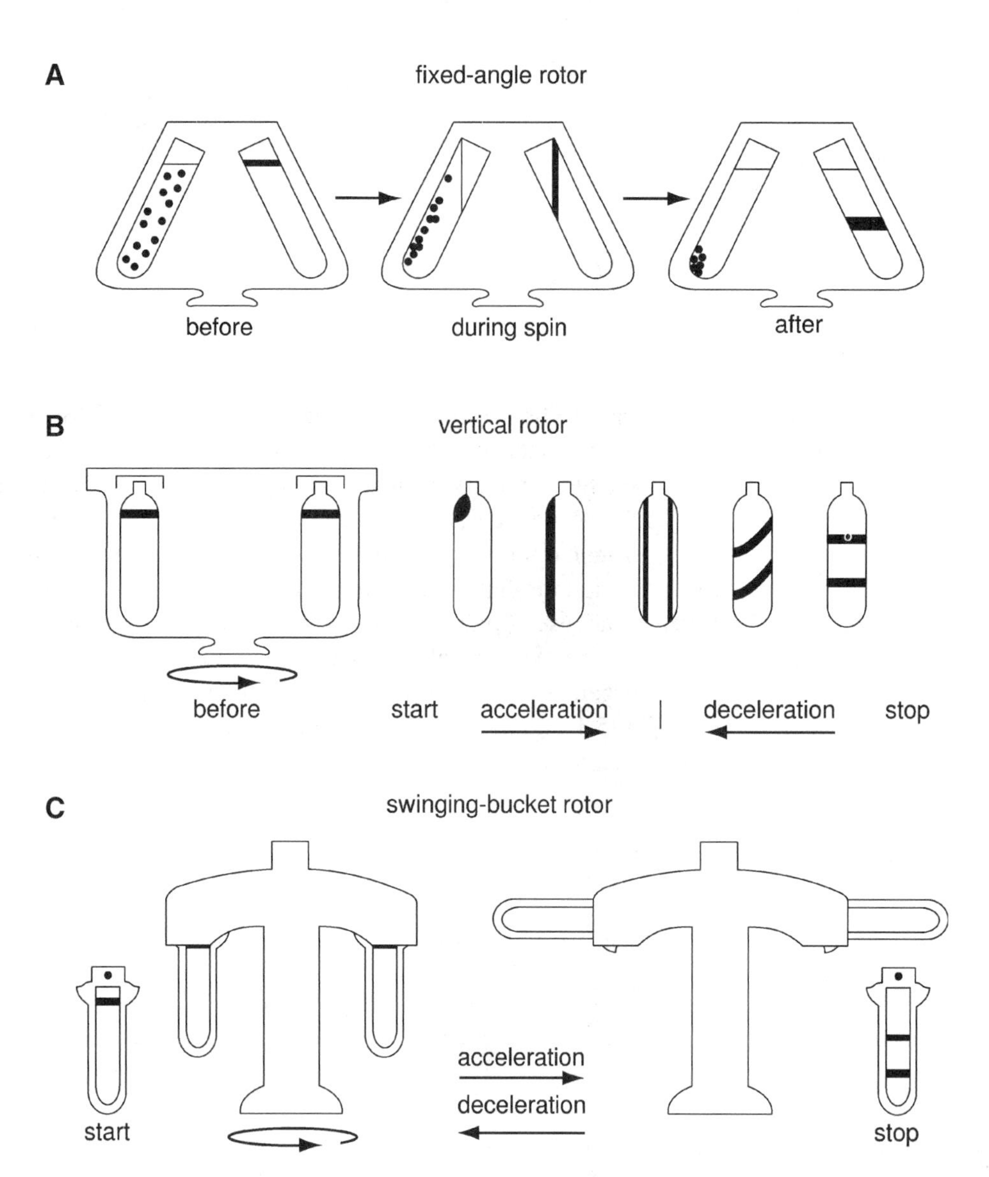

Figure 5.1.1 Rotors used in ultracentrifuges. (**A**) Fixed-angle rotor. In each profile, the tube on the left shows the formation of a pellet from a uniform suspension, and the tube on the right shows sedimentation of a band layered at the top above a higher-density fluid. Sedimentation is radial (particles move toward the outer wall of the tube and then down this wall to the pellet). Note how the band in the right tube re-orients during the centrifugation and then again at the end of the run. (**B**) Vertical rotor. The rotor profile shows tubes with sample layered above a higher-density solution before a run. Note that during the run (successive tube profiles shown at right) the layer re-orients along the inner wall of the tube, and then the bands resolve vertically. During deceleration the bands re-orient to a horizontal position. (**C**) Swinging-bucket rotor. Note that the tube on the left in the upper profile has a sample layered at the top. The buckets are mounted vertically on the rotor. During the run, the buckets reorient to the horizontal position, and the bands separate along the length of the tube. Because sedimentation is radial, the particulates in the bands are more concentrated at the walls facing and away from the viewer than in the center of the tube. Toward the end of deceleration, the buckets re-orient to the vertical position. Originally published in *CP Cell Biology Unit 3.1* (Castle, 1998).

highly specialized application of ultracentrifugation used in protein conformation analysis (Scott et al., 2006).

Cell fractionation methods illustrate the importance of centrifugation in the life sciences. For a typical cell fractionation study, the tissue is homogenized (e.g., with a mortar and pestle) at 4°C in buffer to break open the cells. The homogenate is usually strained to remove large pieces of tissue and then subjected to a set of differential centrifugations. A low-speed centrifugation removes smaller debris. The supernatant is subjected to high-speed centrifugation, which results

Table 5.1.2 Preparing Organelle Fractions from Mammalian Tissues and Cells[a]

Fraction	Procedure	Reference
Organelles		
Nucleus	Centrifugation through high-density sucrose	Blobel and Potter (1966)
Endoplasmic reticulum	Discontinuous sucrose gradient	Adelman et al. (1973)
Golgi complex	Sucrose gradient, continuous and discontinuous	Bergeron et al. (1982)
Secretion granules		
Endocrine	Metrizamide gradient	Loh et al. (1984)
Exocrine	Discontinuous sucrose gradient	Cameron and Castle (1984)
Synaptic vesicles	Chromatography on controlled-pore glass	Carlson et al. (1978); Huttner et al. (1983)
Plasma membrane	Discontinuous sucrose gradient	Hubbard et al. (1983); *CP Cell Biology Unit 3.2* (Tuma and Hubbard, 1999)
Endosomes	Free-flow electrophoresis	Marsh et al. (1987)
	Density shift with sucrose gradient	Beaumelle et al. (1990)
Lysosomes	Metrizamide gradient	Wattiaux et al. (1978)
Mitochondria	Velocity sedimentation in sucrose	Schnaitman and Greenawalt (1968)
Peroxisomes	Sucrose gradient, discontinuous and continuous	Leighton et al. (1968)

[a]Originally published in *CP Cell Biology Unit 3.1* (Castle, 1998).

in a pellet of relatively large cell organelles, e.g., nuclei and mitochondria. The supernatant from this centrifugation contains smaller organelles such as Golgi and membrane fragments of other organelles, including endoplasmic reticulum (ER), plasma membrane, mitochondria, and lysosomes and other vacuoles. Ultracentrifugation is then used to collect this material in a pellet. The pellet is suspended in buffer and then centrifuged, again using an ultracentrifuge, through a sucrose gradient (see Differential centrifugation, below) to separate and enrich specific organelles based on size and density (see Table 5.1.2).

DEFINITIONS

Given the wide range of applications in which centrifugation is used, various types of centrifugation are defined below to give the reader an overview of some of the more common types. For additional information, see more detailed references that describe the theory and practical use of centrifugation, e.g., Griffith (1986) and *CP Cell Biology Unit 3.1* (Castle, 1998).

Preparative centrifugation. Centrifugation is used to purify and concentrate cells and subcellular components in sufficient quantity for further study.

Analytical centrifugation. Centrifugation is used to monitor the properties of a sedimenting particle (e.g., a protein complex). Key applications include protein conformation and subunit structure analysis.

Differential centrifugation. The force generated by centrifugation will cause particles such as organelles to form a pellet at the bottom of the centrifuge tube. How quickly this happens depends on the particle size, shape, and density. The size and density differences among the various organelles provide a way to optimize separation. By using a stepped series of increasing centrifugation speeds (and thus *g* force), a concentrated pellet enriched for a particular organelle (e.g., nuclei, plasma membrane) can be obtained. Although the pellets are still mixtures of several cellular components, this is an important enrichment step prior to further purification.

Density gradient centrifugation. Centrifugation is used to force a sample containing a mixture of different organelles through a linear gradient of increasing density. The typical implementation uses a sucrose gradient from a high concentration (i.e., high density) of sucrose at the bottom of the centrifugation tube to a lower concentration (lower density) at the top of the centrifugation tube (see Fig. 5.1.2B). Although these gradients can be prepared by hand using manual gradient makers, automated gradient makers (e.g., Teledyne ISCO programmable gradient system) are more precise and greatly simplify preparation of sucrose gradients in the lab. They handle the mixing and delivery automatically via preset programs for a variety of different gradient profiles. In addition to sucrose, a variety of other reagents (e.g., see *CP Cell Biology Unit 3.1*; Castle, 1998) are used for creating linear gradients.

The particle mass and density of the organelles affect their sedimentation rates through the gradient. Organelle densities range from ~1.06 to 1.2 g/cm^3 (a span easily covered by a sucrose gradient). For example, a sucrose solution that is 10% to 50% by weight has a density range of 1.03 to 1.23 g/cm^3. If the gradient is shallow, the sedimentation rate is the key factor in separating one population of organelles from another. By forming a gradient that spans the full range of densities in the sample, organelles with characteristic densities will band together (form a highly enriched disk) in the gradient where the gradient is the same density as the organelle (see bands in Fig. 5.1.1). Typically, a high-density step at the bottom of the gradient prevents a pellet from forming and material from being lost from the gradient.

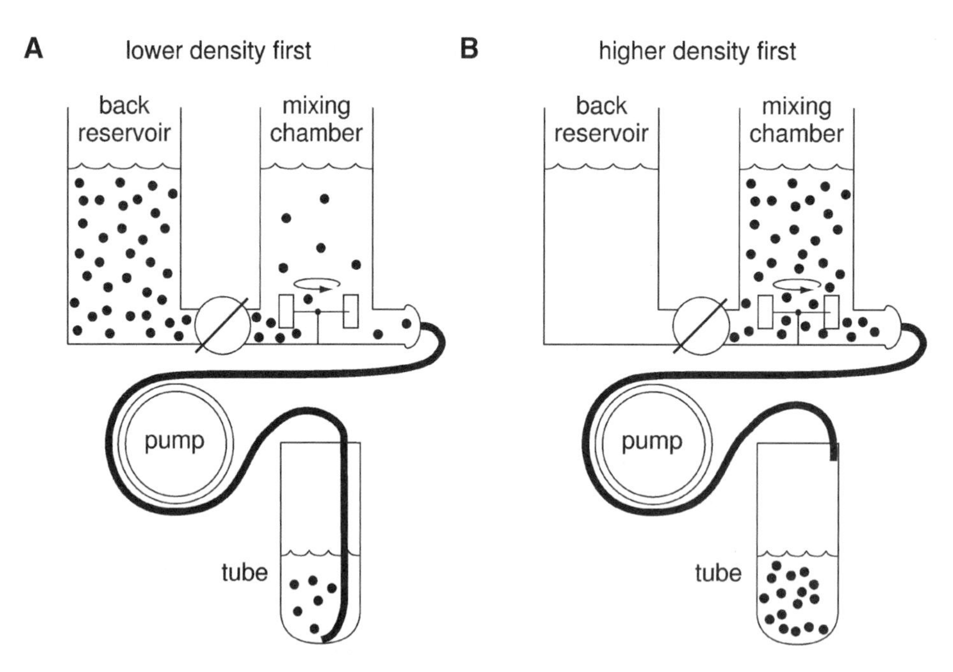

Figure 5.1.2 Linear gradient-forming device showing the back reservoir connected to the mixing chamber via a channel with a stopcock. A delivery tube leads from the mixing chamber to a peristaltic pump and then to the centrifuge tube. (**A**) Configuration used when the lower-density solution is delivered first; the lower-density solution is in the mixing chamber and the outlet is at the bottom of the centrifuge tube. (**B**) Configuration used when the higher-density solution is delivered first; the higher-density solution is in the mixing chamber and the outlet tube is always above the surface of the liquid entering the tube. Originally published in *CP Cell Biology Unit 3.1* (Castle, 1998).

Sedimentation coefficient (S). The sedimentation rate or velocity divided by centrifugal force, in Svedbergs (S; equivalent to 10^{-13} seconds). Estimates of sedimentation coefficients are available in the literature and can also be determined with an ultracentrifuge (e.g., see Griffith 1986). Large organelles such as mitochondria can have sedimentation coefficients >10,000S while individual proteins are typically <10S. The 30S and 50S designations for ribosomal subunits are another application of this unit.

K factor (k). A value specific to the rotor design, used to estimate the centrifugation time needed to pellet a particle of a known sedimentation coefficient. For a rotor running at maximum speed, the k factor gives the pelleting time (t) for a given sedimentation coefficient (e.g., at 20°C in water; $S_{20,w}$):

$$t = k/S_{20,\,W}$$

In addition, the k factor is used to compare the efficiency of rotors with different designs (a lower k factor means faster pelleting).

ROTORS

Centrifugation runs usually specify a relative centrifugal force (RCF) expressed as some number times the force of gravity ($\times\ g$), corresponding to a speed in revolutions per minute (rpm) for a particular centrifuge and rotor model. Available equipment will vary from laboratory to laboratory, and the investigator must be able to adapt specifications for centrifugation conditions to various centrifuges and rotors.

The relationship between RCF and rpm is given by the following equation:

$$\text{RCF} = 1.12r\ (\text{rpm}/1000)^2$$

where r is the rotating radius between the particle being centrifuged and the axis of rotation. In most cases, an accurate conversion from speed to relative centrifugal force (or vice versa) can be obtained using the maximum value of r (r_{max}), which is equal to the distance between the axis of rotation and the bottom of the centrifuge tube as it sits in the well or bucket of the rotor.

5

Tables 5.1.3 to 5.1.6 (at the end of this unit) provide r_{max} values for commonly used rotors manufactured by Du Pont (Sorvall) and Beckman. There are situations (e.g., where an adapter is used to fit a smaller tube into a larger rotor well) where the listed r_{max} will not accurately represent the effective rotating radius and must be modified. In such cases, the manual for the rotor and/or the manufacturer's Web site should be consulted to obtain the appropriate value.

As an alternative to using the tables or equation above, the nomograms in Figures 5.1.3 and 5.1.4 make it possible to determine the RCF where rpm and r_{max} are known, or rpm where RCF and r_{max} are known. This is done by aligning a ruler across the two known values and reading the unknown value at the point where the ruler crosses the remaining column. Figure 5.1.3 should be used for centrifuge runs <21,000 rpm, while Figure 5.1.4 should be used for faster centrifugations.

CAUTION: Do not exceed maximum rotor speed! For Beckman ultracentrifuges, the maximum speed for each rotor is denoted by its name, e.g., the maximum speed for the Beckman VTi 80 rotor is 80,000 rpm. This speed refers only to centrifugation of solutions below a particular allowed density, which differs among rotors (see the user manual). For centrifugation of high-density solutions, a reduced rotor maximum speed is calculated:

$$\text{reduced rpm} = \text{rpm}_{max}(A/B)^{1/2}$$

where A is the allowed density and B is the density of a more concentrated solution. $A = 1.7$ g/ml for several vertical rotors (including VTi 80 and VTi 50), and 1.2 g/ml for several swinging-bucket rotors (including SW 55 Ti, 28, 28.1, 40 Ti, 50.1). For more dense gradients (e.g., using heavy

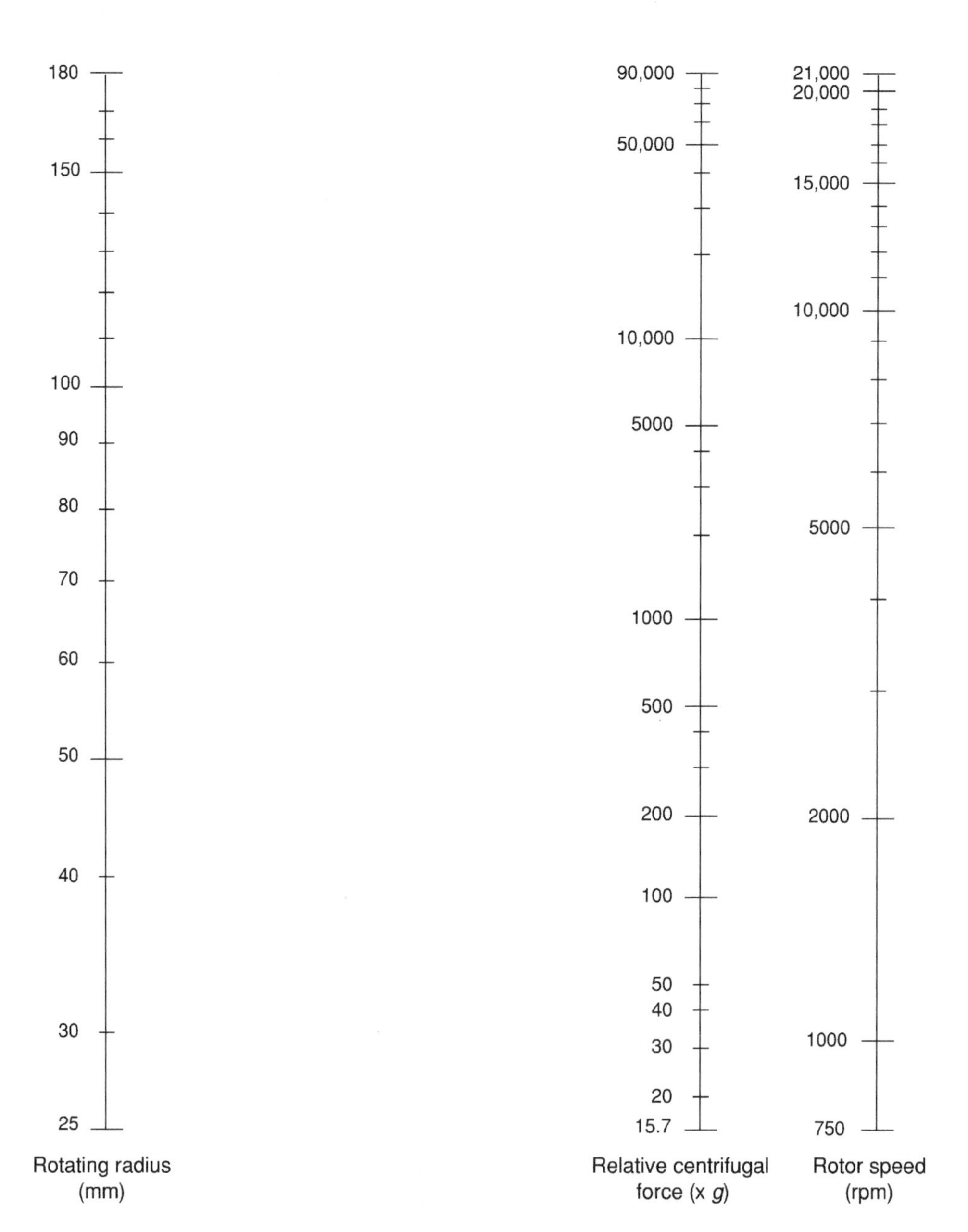

Figure 5.1.3 Nomogram for conversion of relative centrifugal force to rotor speed in low-speed (<21,000 rpm) centrifuge runs. To determine an unknown value in a given column, align a ruler through known values in the other two columns. The desired value is found at the intersection of the ruler with the column of interest. For faster centrifugations, use Figure 5.1.4. A more precise conversion can be obtained using the equation at the beginning of this unit. See Tables 5.1.3 to 5.1.6 for rotating radii of commonly used rotors. Originally published in *CP Human Genetics Appendix 2B* (Haines et al., 2000).

salts such as CsCl) and particularly at low temperatures, the increased concentration of salt at the bottom of the tube can precipitate, exceeding the operating tolerance of the rotor and leading to rotor failure. In such cases, the maximum rpm should be reduced to prevent precipitation (see user manual for guidelines).

NOTE: In this manual, for microcentrifuges built to the Eppendorf standard, a shortened style of reference including only the speed (in rpm) is used. All of these instruments have approximately the same rotating radius; hence, the same speed will yield approximately the same RCF value from machine to machine. Microcentrifugations may also be described as at "top speed" or "maximum speed," meaning 12,000 to 14,000 rpm, which is the maximum speed for all Eppendorf-type microcentrifuges.

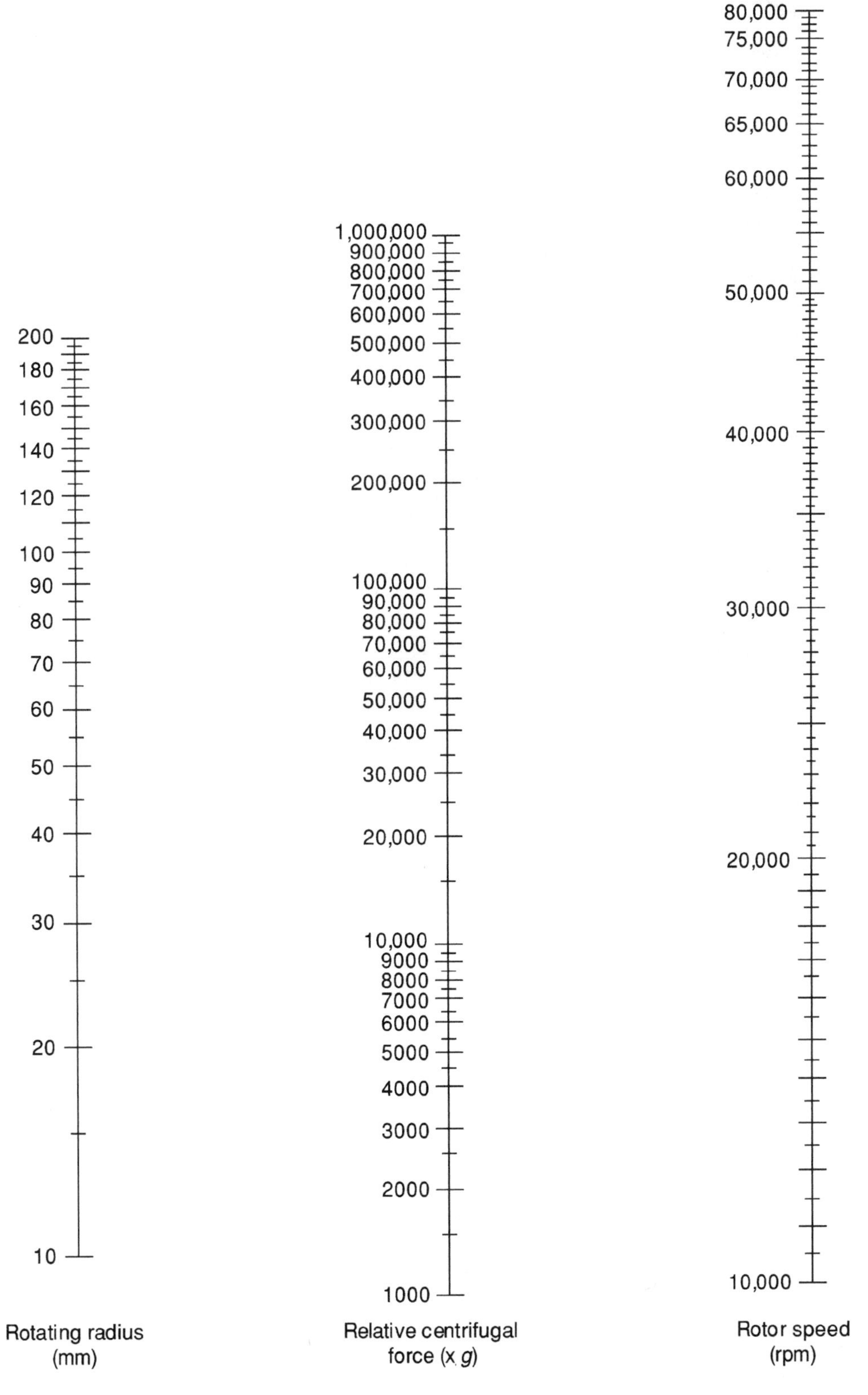

Figure 5.1.4 Nomogram for conversion of relative centrifugal force to rotor speed in high-speed (>21,000 rpm) centrifuge runs. For lower-speed centrifugations and instructions for using the nomogram, see Figure 5.1.3. A more precise conversion can be obtained using the equation described in the text. Originally published in *CP Human Genetics Appendix 2B* (Haines et al., 2000).

5

LITERATURE CITED

Adelman, M.R., Blobel, G., and Sabatini, D.D. 1973. An improved cell fractionation procedure for the preparation of rat liver membrane-bound ribosomes. *J. Cell Biol.* 56:191-205.

Beaumelle, B.D., Gibson, A., and Hopkins, C.R. 1990. Isolation and preliminary characterization of the major membrane boundaries of the endocytic pathway in lymphocytes. *J. Cell Biol.* 111:1811-1823.

Bergeron, J.J.M., Rachubinski, R.A., Sikstrom, R.A., Posner, B.I., and Paiement, J. 1982. Galactose transfer to endogenous acceptors within Golgi fractions of rat liver. *J. Cell Biol.* 92:139-146.

Blobel, G. and Potter, V.R. 1966. Nuclei from rat liver: Isolation method that combines purity with high yield. *Science* 154:1662-1665.

Cameron, R.S. and Castle, J.D. 1984. Isolation and compositional analysis of secretion granules and their membrane subfraction from the rat parotid gland. *J. Memb. Biol.* 79:127-144.

Carlson, S.S., Wagner, J.A., and Kelly, R.B. 1978. Purification of synaptic vesicles from elasmobranch electric organ and the use of biophysical criteria to demonstrate their purity. *Biochemistry* 17:1188-1199.

Castle, J.D. 1998. Overview of cell fractionation. *Curr. Protoc. Cell Biol.* 0:3.1.1-3.1.9.

Griffith, O.M. 1986. Techniques of Preparative Zonal and Continuous Flow Ultracentrifugation. Beckman SBU8. 94088.

Haines, J.L., Korf, B.R., Morton, C.C., Seidman, C.E., Seidman, J.G., and Smith, D.R. (eds.) 2000. Current Protocols in Human Genetics, pp. A.2B.1-A.2B.5. John Wiley & Sons, New York.

Hubbard, A.L., Wall, D.A., and Ma, A.K. 1983. Isolation of rat hepatocyte plasma membranes: I. Presence of the three major domains. *J. Cell Biol.* 96:217-229.

Huttner, W.B., Schiebler, W., Greengard, P., and DeCamilli, P. 1983. Synapsin I (Protein I), a nerve terminal–specific phosphoprotein: III. Its association with synaptic vesicles studied in a highly purified synaptic vesicle preparation. *J. Cell Biol.* 96:1374-1388.

Leighton, F., Poole, B., Beaufay, H., Baudhuin, P., Coffey, J.W., Fowler, S., and DeDuve, C. 1968. The large-scale separation of peroxisomes, mitochondria, and lysosomes from the livers of rats injected with Triton WR-1339. *J. Cell Biol.* 37:482-513.

Loh, Y.P., Tam, W.W.H., and Russell, J.T. 1984. Measurement of pH and membrane potential in secretory vesicles isolated from bovine pituitary intermediate lobe. *J. Biol. Chem.* 259:8238-8245.

Marsh, M., Schmid, S., Kern, H., Harms, E., Male, P., Mellman, I., and Helenius, A. 1987. Rapid analytical and preparative isolation of functional endosomes by free flow electrophoresis. *J. Cell Biol.* 104:875-886.

Schnaitman, C. and Greenawalt, J.W. 1968. Enzymatic properties of the inner and outer membranes of rat liver mitochondria. *J. Cell Biol.* 38:158-175.

Scott, D.J., Harding, S.E., and Rowe, A.J. (eds.) 2006. Analytical Ultracentrifugation: Techniques and Methods. RSC Publishing, Cambridge.

Schuck, P. and Braswell, E.H. 2000. Measuring protein-protein interactions by equilibrium sedimentation. *Curr. Protoc. Immunol.* 40:18.8.1-18.8.22.

Tuma, P.L. and Hubbard, A.L. 1999. Isolation of rat hepatocyte plasma membrane sheets and plasma membrane domains. *Curr. Protoc. Cell Biol.* 2:3.2.1-3.2.16.

Wattiaux, R., Wattiaux-DeConnick, S., Ronveaux-Dupal, M.-F., and Dubois, F. 1978. Isolation of rat liver lysosomes by isopycnic centrifugation in a metrizamide gradient. *J. Cell Biol.* 78:349-368.

Tables 5.1.3 through 5.1.6 appear starting on the next page.

Table 5.1.3 Sorvall Superspeed and Supraspeed Rotors

Model no.	Rotor type	Axis of rotation	r_{min} (mm)	r_{max} (mm)	k factor	Tube slots	Capacity (ml)	Max RCF (g)
Superspeed rotors								
F-20/MICRO	Fixed angle, aluminum	45°	85.7	115.1	187	32	1.5	51,427
GS-3	Fixed angle, aluminum	20°	39.4	151.3	4203	6	500	13,689
GSA	Fixed angle, aluminum	28°	NS	145.6	2023	6	250	27,485
HB-6	Swinging bucket, body/titanium buckets	90°	45.0	146.3	1765	6	50	27,617
HS-4	Swinging bucket, aluminum	90°	72.2	172.3	3912	4	250	10,826
SA-300	Fixed angle, aluminum	34°	23.5	96.7	573	6	50	67,509
SA-512[a]	Fixed angle, aluminum	23°	47.0 63.7	98.4 115.2	492 394	32	16	41,794 48,930
SA-600	Fixed angle, aluminum	34°	55.2	129.6	747	12	50	41,837
SA-800	Fixed angle, aluminum	20°	41.2	104.5	560	8	100	49,054
SE-12	Fixed angle, aluminum	40°	38.1	93.3	335	12	14	70,450
SH-MT[b]	Fixed angle, aluminum	90°	NS	NS	NS	60	1.8	NS
SH-3000	Swinging bucket, aluminum	90°	91.1	185.4	8138	4	750	4,575
SH-80	Swinging bucket, aluminum body/titanium buckets	90°	54.0	101.6	400	8	10	45,395
SLA-1000	Fixed angle, aluminum	23°	NS	117.7	1725	4	250	35,793
SLA-1500	Fixed angle, aluminum	23°	36.6	135.9	1475	6	250	34,155
SLA-3000	Fixed angle, aluminum	20°	39.4	151.3	2364	6	500	24,336
SLA-600TC	Fixed angle, aluminum	34°	79.9	147.0	913	12	50	27,750
SLC-1500	Fixed angle, composite	23°	37.4	137.0	1676	6	250	29,994
SLC-3000	Fixed angle, composite	23°	41.2	157.6	3394	6	500	17,604

continued

Table 5.1.3 Sorvall Superspeed and Supraspeed Rotors, *continued*

Model no.	Rotor type	Axis of rotation	r_{min} (mm)	r_{max} (mm)	k factor	Tube slots	Capacity (ml)	Max RCF (g)
SLC-4000	Fixed angle, composite	22°	25.5	167.8	5114	4	1000	15,182
SLC-6000	Fixed angle, composite	20°	54.2	195.9	3549	6	1000	15,810
SM-24[a]	Fixed angle, aluminum	28°	34.1 53.8	91.0 110.7	591 434	24	16	42,717 51,965
SS-34	Fixed angle, aluminum	34°	32.7	107.0	714	8	50	50,228
SS-34/KSB	Continuous flow, aluminum	NS	32.7	107.0	750	8	50	47,808
SV-288	Vertical, aluminum	0°	64.7	90.2	210	8	36	40,301
SV-80	Vertical, aluminum	0°	88.4	101.6	98	16	5	40,969
TZ-28	Zonal, titanium	NS	36.2	95.3	612	1	1350	42,580
TZ-28/GK	Continuous flow, titanium	NS	36.2	95.3	678	NS	1350	38,428
Supraspeed rotors								
F-16/250	Fixed angle, aluminum	23°	36.7	136.0	1295	6	250	38,889
F-28/13	Fixed angle, aluminum	23.5°	73.4	114.3	143	16	12.5	100,096
F-28/36	Fixed angle, aluminum	23.5°	61.8	114.3	198	12	36	100,096
F-28/50	Fixed angle, titanium	34°	40.4	114.7	33	10	50	100,446
S-20/17	Swinging bucket, aluminum rotor/titanium buckets	90°	64.5	165.9	598	6	17	74,124
S-20/20	Swinging bucket, aluminum rotor/titanium buckets	90°	72.1	129.3	369	6	20	57,771
S-20/36	Swinging bucket, aluminum rotor/titanium buckets	90°	72.1	161.0	508	6	36	71,935

[a]Rotor contains two rows of slots, numbers are for inner and outer rows respectively.

[b]r_{min}, r_{max}, k factor, and maximum RCF vary with tube.

Abbreviation: NS, not specified.

Table 5.1.4 Sorvall Ultraspeed Rotors

Model no.	Rotor type	Axis of rotation	r_{min} (mm)	r_{avg} (mm)	r_{max} (mm)	*k* factor	Tube slots	Capacity (ml)	Max RCF (*g*)
A-1256	Fixed angle, aluminum	24°	40.6	61.30	82.0	56.7	12	12.5	287,239
A-621	Fixed angle, aluminum	23°	36.7	86.35	136.0	751	6	250	66,993
A-641	Fixed angle, aluminum	22.5°	33.9	66.60	99.3	162	6	100	186,453
A-841	Fixed angle, aluminum	23.5°	38.4	64.70	91.0	130	8	36	170,869
AH-629 (17 ml)	Swinging bucket, aluminum body/titanium buckets	90°	64.5	115.20	165.9	284	6	17	155,846
AH-629 (20 ml)	Swinging bucket, aluminum body/titanium buckets	90°	72.1	100.70	129.3	176	6	20	121,464
AH-629 (36 ml)	Swinging bucket, aluminum body/titanium buckets	90°	72.1	116.55	161.0	242	6	36	151,243
AH-650	Swinging bucket, aluminum body/titanium buckets	90°	62.8	84.40	106.0	53.0	6	5	296,005
StepSaver 50 V39[a]	Vertical, titanium body	0°	60.8	73.70	86.6	35.8	8	39	241,831
StepSaver 65 V13[a]	Vertical, titanium body	0°	68.5	76.70	84.9	12.9	8	13.5	400,671
StepSaver 70 V6[a]	Vertical, titanium body	0°	72.1	78.75	85.4	8.7	8	6	467,420
Surespin 630 (17 ml)	Swinging bucket, titanium rotor and buckets	90°	64.0	115.0	166.0	268	6	17	166,880
Surespin 630 (36 ml)	Swinging bucket, titanium rotor and buckets	90°	64.0	115.0	166.0	219.0	6	36	166,880
T-1250	Fixed angle, titanium	24°	54.7	81.25	107.8	68.7	12	36	301,032

continued

Table 5.1.4 Sorvall Ultraspeed Rotors, *continued*

Model no.	Rotor type	Axis of rotation	r_{min} (mm)	r_{avg} (mm)	r_{max} (mm)	k factor	Tube slots	Capacity (ml)	Max RCF (g)
T-1270	Fixed angle, titanium	24°	40.6	61.30	82.0	36.3	12	12.5	448,811
T-647.5	Fixed angle, titanium	22.5°	33.9	66.60	99.3	121.0	6	100	250,259
T-865	Fixed angle, titanium	23.5°	38.4	64.70	91.0	51.7	8	36	429,459
T-865.1	Fixed angle, titanium	23.5°	46.2	66.65	87.1	38.0	8	12.5	411,053
T-875	Fixed angle, titanium	23.5°	46.2	66.65	87.1	28.5	8	12.5	547,260
T-880	Fixed angle, titanium	25.5°	41.5	62.90	84.3	28.0	8	12.5	602,644
T-890	Fixed angle, titanium	25°	34.2	55.35	76.5	25.1	8	12.5	692,149
T-8100	Fixed angle, titanium	26°	35.6	53.7	71.8	17.7	8	6.5	802,006
TFT-45.6[b]	Fixed angle, titanium	20°	60.7 76.0	75.50 90.80	90.3 105.6	49.6 41.1	40	6	204,252 238,859
TFT-80.2	Fixed angle, titanium	25°	33.9	47.00	60.1	22.6	12	2	429,643
TFT-80.4	Fixed angle, titanium	25°	34.1	49.80	65.5	25.8	10	4.4	468,246
TH-641	Swinging bucket, titanium rotor and buckets	90°	71.9	112.55	153.2	114	6	13.2	287,660
TH-660	Swinging bucket, titanium rotor and buckets	90°	64.6	93.05	121.5	44.4	6	4.4	488,576
TV-1665	Vertical, titanium	0°	74.6	81.20	87.8	9.8	16	6	414,357
TV-860	Vertical, titanium	0°	59.2	71.95	84.7	25.2	8	35	340,596
TV-865	Vertical, titanium	0°	71.7	78.30	84.9	10.1	8	6	400,671
TV-865B	Vertical, titanium	0°	59.2	71.95	84.7	21.4	8	18.5	399,727
TZ-28	Zonal, titanium	NS	36.2	65.75	95.3	312	1	1350	83,457

[a]Multiple rotor systems are available. Numbers given assume that the rotor system with the largest capacity is being used.
[b]Rotor contains two rows of slots, numbers are for inner and outer rows respectively.
Abbreviation: NS, not specified.

Table 5.1.5 Beckman High-Speed Rotors

Model no.	Rotor type	Axis of rotation (deg)	r_{min} (mm)	r_{avg} (mm)	r_{max} (mm)	k factor	Tube slots	Capacity (ml)	Max RCF (g)
C0650	Fixed angle, aluminum	25°	32	62	92	2680	6	50	10,400
C1015	Fixed angle, aluminum	25°	38	65.5	93	2270	10	15	10,397
F0485	Fixed angle, aluminum	30°	21	55.5	90	920	4	85	40,248
F0603	Fixed angle, aluminum	30°	23	50.5	78	454	6	38.5	59,860
F0650	Fixed angle, aluminum	25°	21	52.5	84	795	6	50	41,400
F0685	Fixed angle, aluminum	25°	25	61	97	1428	6	85	26,320
F0850	Fixed angle, aluminum	25°	33	63.5	94	973	8	50	28,611
F1202	Fixed angle, aluminum	45°	31	47.5	64	204	12	1.85	64,396
F2402H	Fixed angle, aluminum	45°	50	65	80	185	24	1.8	62,084
F3602[a]	Fixed angle, aluminum	45°	47	62	77	260	36	1.8	41,666
			57	72	87	224			47,618
F1010	Fixed angle, aluminum	35°	20	48	76	500	10	10	57,438
FX301.5	Fixed angle, material	45°	67	84	100	519	30	2.2	19,515[b]
FX6100	Fixed angle, aluminum	25°	98	66.5	35	NS	6	100	11,400

continued

5

Table 5.1.5 Beckman High-Speed Rotors, *continued*

Model no.	Rotor type	Axis of rotation (deg)	r_{min} (mm)	r_{avg} (mm)	r_{max} (mm)	k factor	Tube slots	Capacity (ml)	Max RCF (g)
H6002	Horizontal, aluminum	90°	35	64.5	94	NA	60	1.8	12,400
S0410	Swinging bucket, aluminum	90°	26	61	96	3310	4	10	10,733
S2096	Swinging bucket, aluminum	90°	70	NS	110	NS	2	NA[c]	1107
S5700	Swinging bucket, aluminum	90°	138.6	153.4	168.5	NS	10	NA[c]	6130
SX241.5	Swinging bucket, aluminum	90°	35	54	74	NA	24	2.2	19,515[d]
SX4250	Swinging bucket, aluminum	90°	71	101	172	NS	4	250	3398[e]
TA-10–250	Fixed angle, aluminum	25°	35	86	137	3450	6	250	15,300
TA-14–50	Fixed angle, aluminum	25°	33.0	64.5	96	1380	8	50	21,100
TA-15–1.5	Fixed angle, aluminum	45°	67.0	83.5	100	428	30	1.5	25,160
TS-5.1–500	Swinging bucket, aluminum	90°	87.0	138.5	190	7600	4	500	5500

[a]Rotor contains two rows of slots, numbers are for inner and outer rows respectively.
[b]Maximum RCF is 21,920 in a refrigerated centrifuge.
[c]Holds microtiter plates.
[d]Maximum RCF is 16,220 in a refrigerated centrifuge.
[e]Maximum RCF is 3901 in a refrigerated centrifuge.
Abbreviations: NS, not specified; NA, not applicable.

Table 5.1.6 Beckman Ultraspeed Centrifuges[a]

Model no.	Rotor type	Axis of rotation	r_{min} (mm)	r_{avg} (mm)	r_{max} (mm)	*k* factor	Slots	Capacity (ml)	Max RCF (*g*)
NVT 100	Near vertical, titanium	8°	48.3	57.6	67.0	8	8	5.1	750,000
NVT 65	Near vertical, titanium	7.5°	59.5	72.2	84.9	21	8	13.5	402,000
NVT 65.2	Near vertical, titanium	8.5°	68.8	78.4	87.9	15	16	5.1	416,000
NVT 90	Near vertical, titanium	8°	52.4	61.8	71.1	10	8	5.1	645,000
SW 28.1	Swinging bucket, aluminum head/titanium buckets	90°	72.9	122.1	171.3	276	6	18	150,000
SW 32 Ti	Swinging bucket, titanium	90°	66.8	109.7	152.5	204	6	38.5	175,000
SW 32.1	Swinging bucket, titanium	90°	64.4	113.6	162.8	228	6	17	187,000
SW 40 Ti	Swinging bucket, titanium head and buckets	90°	66.7	112.7	158.8	137	6	14	285,000
SW 41 Ti	Swinging bucket, titanium head and buckets	90°	67.4	110.2	153.1	124	6	13.2	288,000
SW 55 Ti	Swinging bucket, titanium head and buckets	90°	60.8	84.6	108.5	48	6	5	368,000
SW 60 Ti	Swinging bucket, titanium head and buckets	90°	63.1	91.7	120.3	45	6	4	485,000
SW28	Swinging bucket, aluminum head/titanium buckets	90°	75.3	118.2	161.0	246	6	39	141,000

continued

5

Table 5.1.6 Beckman Ultraspeed Centrifuges[a], *continued*

Model no.	Rotor type	Axis of rotation	r_{min} (mm)	r_{avg} (mm)	r_{max} (mm)	k factor	Slots	Capacity (ml)	Max RCF (g)
Type 19	Fixed angle, aluminum	25°	34.4	83.9	133.4	951	6	250	53,900
Type 25[b]	Fixed angle, aluminum	25°	NS	NS	100.4 116.3 132.1	84 71 62	100	1	70,300 81,400 92,500
Type 42.2 Ti	Fixed angle, titanium	30°	NS[c]	NS	113.0	9	72	0.23	223,000
Type 45 Ti	Fixed angle, titanium	24°	35.9	69.8	103.8	133	6	94	235,000
Type 50.2 Ti	Fixed angle, titanium	24°	54.4	81.2	107.9	69	12	39	302,000
Type 50.4 Ti[d]	Fixed angle, titanium	20°	80.8	96.1	111.5	39 33	44	6.5	270,000 312,000
Type 70 Ti	Fixed angle, titanium	23°	39.5	65.7	91.9	44	8	39	504,000
Type 70.1 Ti	Fixed angle, titanium	24°	40.5	61.2	82.0	36	12	13.5	450,000
Type 90 Ti	Fixed angle, titanium	25°	34.2	55.4	76.5	25	8	13.5	694,000
Type100 Ti	Fixed angle, titanium	26°	39.5	55.5	71.6	15	8	6.8	802,000
VTi 50	Vertical, titanium	0°	60.8	73.7	86.6	36	8	39	242,000
VTi 65.1	Vertical, titanium	0°	68.5	76.7	84.9	13	8	13.5	402,000
VTi 65.2	Vertical, titanium	0°	74.7	81.3	87.9	10	16	5.1	416,000
VTi 90	Vertical, titanium	0°	57.9	64.5	71.1	6	8	5.1	645,000

[a]Beckman centrifuges carry a safety classification A to T. Be sure to match the safety rating of the centrifuge (located on a decal on the machine) and rotor. Rotors currently in production are classified as H, R, or S.

[b]Rotor contains three rows of slots, and multiple numbers are for inner, middle, and outer rows, respectively.

[c]$r_{meniscus}$ = 104 mm. This rotor is for microcentrifuge tubes. Because it is specific for processing microsamples, an $r_{meniscus}$ rather than r_{min} is given.

[d]Rotor contains two rows of slots, and k factor and max RFC numbers are for inner and outer rows, respectively. Radii apply to the outer row.

Abbreviation: NS, not specified.

UNIT 5.2

Purification and Concentration of Nucleic Acids

Dennis H. Dowhan[1]

[1]Diamantina Institute for Cancer, Immunology and Metabolic Medicine, University of Queensland, Woolloongabba, Queensland, Australia

OVERVIEW AND PRINCIPLES

NOTE: The material in this unit was adapted from earlier work published in *CP Molecular Biology* by Kingston et al. (1996; RNA) and Moore and Dowhan (2002; DNA).

Making Cellular Lysate

This unit outlines the basic principles for manipulating solutions of DNA or RNA through commonly used purification and concentration procedures. The protocols and techniques are useful when proteins or solute molecules need to be removed from nucleic acids in aqueous solutions or when solutions containing nucleic acids need to be concentrated for reasons of convenience. Removal of contaminating proteins, carbohydrates, lipids, and other biological molecules from nucleic acids is usually required before DNA or RNA can be utilized in subsequent enzymatic manipulations or analytical analysis.

Nucleic acids that are routinely used for experimentation and analysis in molecular biology are primarily isolated from tissue, mammalian cell cultures, or bacteria. Eukaryotic and prokaryotic cells need to be broken apart and the cellular contents released to make crude cellular lysates. This is done by using a variety of ionic detergents and/or protease treatments to break open the cells. Common lysis methods have been developed for specific procedures. For preparation of crude lysates from bacteria, the alkaline lysis method (Alternate Protocol 1 and UNIT 4.2), which uses sodium dodecyl sulfate (SDS) to denature bacterial proteins (Birnboim and Doly, 1979), or treatment with lysozyme, Triton (a nonionic detergent), and heat (Holmes and Quigley, 1981) are common methods. For animal organs and tissue, the isolated tissues are usually snap frozen in liquid nitrogen and crushed prior to lysis; this allows the tissues to be broken into digestible-sized pieces and the individual cells can be disrupted to release their cellular contents. DNA is a fairly robust molecule; however, RNA undergoes alkaline hydrolysis when in solution at alkaline pH and the presence of ribonucleases (RNases) can rapidly degrade RNA; therefore, particular methods are required for preparation of RNA from eukaryotic and prokaryotic cells. When lysing cells to isolate RNA, all solutions should be treated with diethylpyrocarbonate (DEPC), which inactivates RNases by covalent modification. The chemical guanidine thiocyanate is a strong protein denaturant and inhibits RNase activity; therefore, it is commonly used to disrupt cells or tissues in RNA extraction procedures. Specific protocols relating to the cellular lysis of tissue, mammalian cells, or bacteria for collection of nucleic acids can be found in *CP Molecular Biology Unit 1.6* (Engebrecht et al., 1991), *Unit 1.7* (Heilig et al., 1998), *Unit 1.8* (Seidman et al., 1997), *Unit 2.2* (Strauss, 1998), *Unit 2.3* (Richards et al., 1994), and *Unit 2.4* (Wilson, 1997). After the cells have been lysed to produce crude cellular lysate, the nucleic acids can then be isolated and purified from the nuclear and cellular proteins, lipids, and other biological molecules. In general, carbohydrate and polysaccharide molecules are not major contaminants when isolating nucleic acids from bacteria or most eukaryotic cells, including mammalian cells, but may be a problem when trying to isolate nucleic acids from plants. When isolating nucleic acids from plants, carbohydrate and polysaccharide molecules can be removed using silica membrane spin columns (Alternate Protocol 1), cesium chloride:ethidium bromide density gradient centrifugation, or anion-exchange chromatography instead of phenol/chloroform extraction; these methods can be found in *CP Molecular Biology Unit 1.6* (Engebrecht et al., 1991), *Unit 1.7* (Heilig et al., 1998), *Unit 1.8* (Seidman et al., 1997), *Unit 2.2* (Strauss, 1998), *Unit 2.3* (Richards et al., 1994), and *Unit 2.4* (Wilson, 1997).

DNA Purification

Once a cellular lysate is obtained, it is often necessary to purify or concentrate solutions of nucleic acids (DNA or RNA) prior to further enzymatic manipulations or analytical studies. Proteins and lipids are a common contaminant and should be removed from nucleic acids prior to experimentation. The most commonly used method for deproteinizing DNA and RNA is extraction with phenol (see Basic Protocol 1), which efficiently denatures proteins and probably dissolves denatured protein (Kirby, 1957). The purpose of the phenol is to act as a protein solvent to disassociate nucleic acids from protein and, in conjunction with salt solutions, isolate nucleic acids from nuclear proteins bound to nucleic acids. The ability of phenol to do this is due to its ability to denature proteins and lipids but not nucleic acids. Phenol is itself hydrophobic and is therefore able to solvate or form interactions with the hydrophobic interiors of proteins and with the long hydrocarbon chains in lipids. DNA on the other hand, due to its sugar-phosphate backbone, is very hydrophilic. For this reason, DNA will partition to the aqueous phase and is not soluble in phenol.

The chemical structure of phenol is shown in Figure 5.2.1. Chloroform is also a useful protein denaturant with somewhat different properties–it stabilizes the rather unstable boundary between an aqueous phase and a pure phenol layer. The phenol/chloroform mixture reduces the amount of aqueous solution retained in the organic phase (compared to a pure phenol phase), maximizing the yield (Penman, 1966; Palmiter, 1974). Isoamyl alcohol, which is often added to the mix, prevents foaming of the mixture upon vortexing and aids in the separation of the organic and aqueous phases (Marmur, 1961). Denatured protein forms a layer at the interface between the aqueous and organic phases and is thus isolated from the bulk of the DNA in the aqueous layer. This procedure is rapid, inexpensive, and easy to perform.

For DNA, silica membrane spin columns—modified from principles originally described in Vogelstein and Gillespie (1979)—provide a simple, nontoxic method for removing DNA from contaminating impurities (see Alternate Protocol 1). In the presence of high chaotropic salt concentrations, DNA binds to a silica membrane inside a spin column. The resulting precipitate is washed to remove NaI (or other chaotropic salts like guanidine HCl or sodium perchlorate), along with impurities from the original sample, and subsequent suspension in water or TE buffer causes dissociation (elution) of the DNA from the silica membrane. Because fewer manipulations are required, this method is faster and easier to perform than organic-based extraction methods. However, the yields may be somewhat lower, generally ranging from 50% to 75% of the starting material. The procedure seems to work best with DNA fragments between 0.5 to 50 kb in size. The recovery for DNA smaller than 500 bp may be reduced, as some shorter fragments may bind tightly and irreversibly to the silica membrane. Plasmids >50 kb will not elute efficiently from silica membrane spin columns; therefore, when isolating plasmids >50 kb or genomic DNA, cesium chloride:ethidium bromide density gradient centrifugation or anion-exchange chromatography techniques should be used.

RNA Purification

A significant consideration when purifying and concentrating RNA in comparison to DNA is the fact that RNA is less stable than DNA, and extra care must be taken to prevent RNA degradation. The reason for the reduced stability of RNA is that the RNA ribose sugar has two hydroxyl (OH) groups attached to the pentose ring, whereas DNA has only one (Fig. 5.2.2A). RNA is less stable than DNA because the extra hydroxyl group on the pentose ring makes RNA more prone to hydrolysis from nuclease enzymes called ribonucleases, which are capable of cleaving the phosphodiester bonds, which covalently link the ribonucleotide subunits of RNA, specifically those linked to pyrimidine bases such as uracil. Ribonucleases do not hydrolyze or cleave DNA, because DNA lacks the second hydroxyl group that is essential for the formation of cyclic

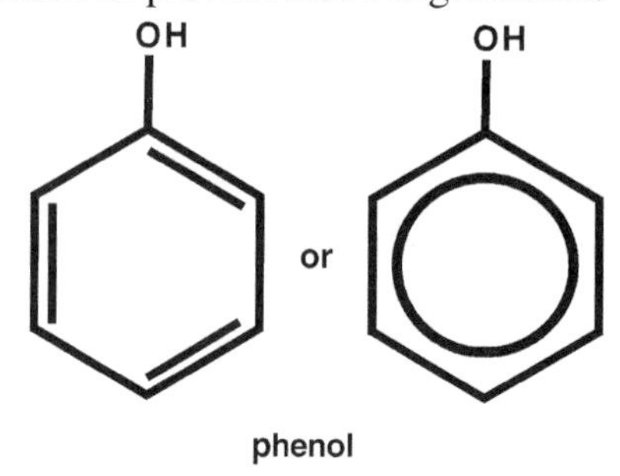

Figure 5.2.1 The chemical structure of phenol.

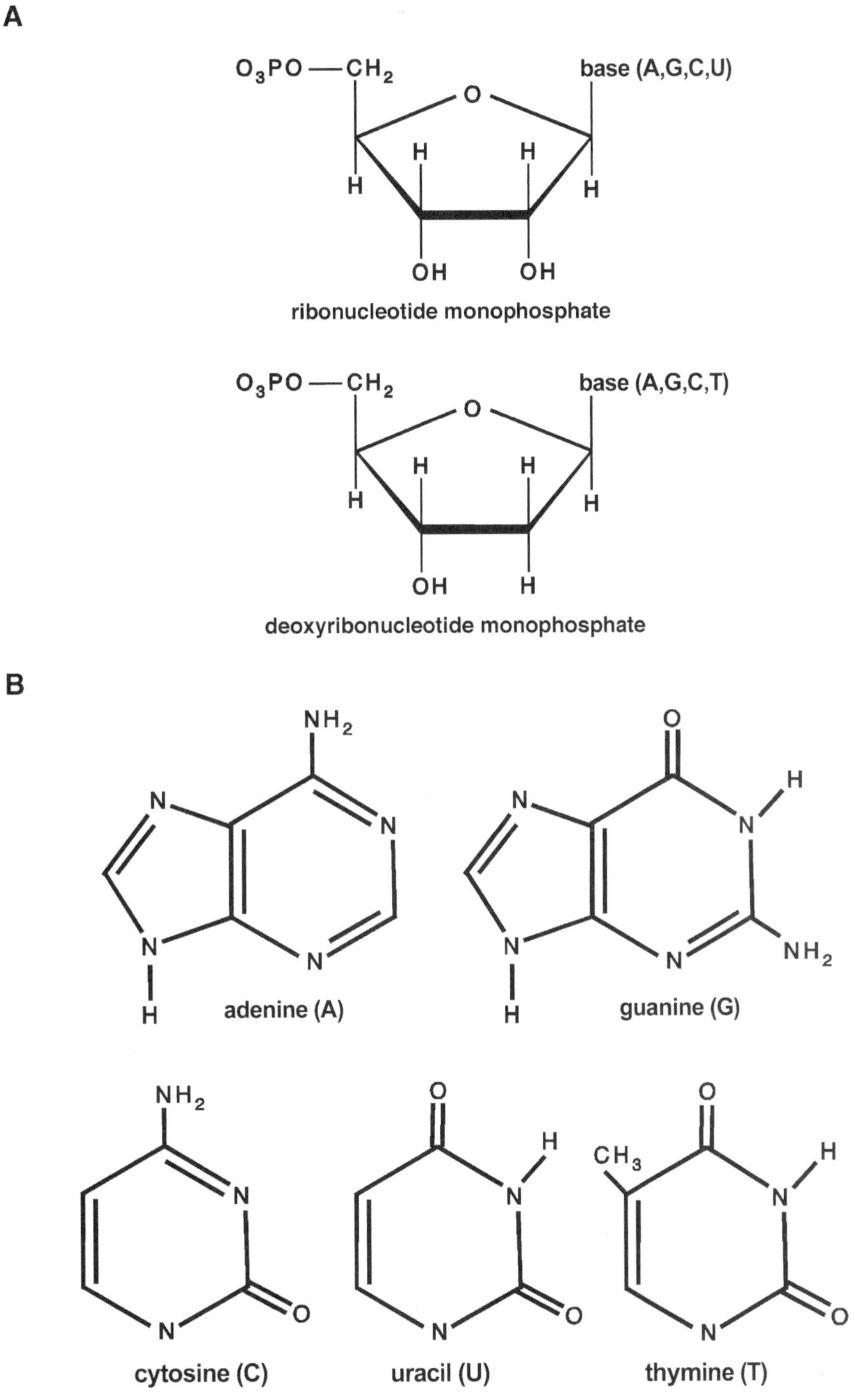

A

ribonucleotide monophosphate

deoxyribonucleotide monophosphate

B

adenine (A)

guanine (G)

cytosine (C)

uracil (U)

thymine (T)

Figure 5.2.2 (**A**) The chemical structure of the ribose and deoxyribose sugars that are a constituent of RNA and DNA, respectively. (**B**) The pyrimidine nucleobases found in nucleic acids are cytosine (C), thymine (T), and uracil (U). The purine nucleobases found in nucleic acids are adenine (A) and guanine (G). The chemical structure of each of the nucleobases is shown.

intermediates required for the cleavage process. The chemical structure of the ribose and deoxyribose sugars, along with the pyrimidine and purine bases found in RNA and DNA, are illustrated in Figures 5.2.2A and 5.2.2B.

The additional hydroxyl group on the pentose ring of the RNA molecule also causes RNA molecules to be susceptible to alkaline hydrolysis. If placed in an aqueous solution (H_2O) with a pH >7, the phosphodiester bond between two ribonucleotides can be broken by alkaline hydrolysis. Thus, RNA should not be placed in highly basic solutions unless the intent is to break these bonds.

The major source of failure in any attempt to purify RNA is contamination by ribonucleases (RNases). RNases are very stable enzymes and generally require no cofactors to function. Therefore, a small amount of RNase in an RNA preparation will create a real problem. Any water or salt solutions used in RNA preparation should be treated with the chemical DEPC. This chemical inactivates ribonucleases by covalent modification. Solutions containing Tris cannot be effectively treated with DEPC because Tris reacts with DEPC to inactivate it. Labware used in the preparation of RNA should be treated to remove residual RNase activity. Many RNases will not be fully inactivated by autoclaving. Glassware can be baked at 300°C for 4 hr. Certain kinds of plasticware (e.g., some conical centrifuge tubes and pipets) can be rinsed with chloroform to inactivate RNase. When done carefully, this rinse with chloroform is an effective treatment. Keep in mind, however, that many plastics (e.g., gel boxes) will melt when treated with chloroform. Plasticware straight out of the package is generally free from contamination and can be used as is. Direct contact with skin is also a major source of contaminating RNase, and clean gloves should be used at all times. Refer to *UNIT 8.2* for more information.

Guanidine thiocyanate is one of the most effective protein denaturants known. The use of guanidine to lyse cells was originally developed to allow purification of RNA from cells high in endogenous ribonucleases (Cox, 1968; Ullrich et al., 1977; Chirgwin et al., 1979). Guanidine thiocyanate acts as a protein denaturant by chemical denaturation. Chemical denaturants like guanidine thiocyanate or urea are capable of disrupting strong secondary bonds, like hydrogen bonds, that normally hold proteins together. The guanidinium cation and the thiocyanate anion are chaotropic agents that have ability to disrupt the regular hydrogen bond structures in water, and this directly affects protein secondary structure and increases the water solubility of the protein. The chemical structure of guanidine thiocyanate is shown in Figure 5.2.3.

The single-step method of RNA isolation described in the basic RNA protocol (Basic Protocol 2) is based on the ability of RNA to remain water soluble in a solution containing 4 M guanidine thiocyanate, pH 4, in the presence of a phenol/chloroform organic phase. Under such acidic conditions, most proteins and small fragments of DNA (50 bases to 10 kb) will be found in the organic phase while larger fragments of DNA and some proteins remain in the interphase. The fragmentation of DNA during homogenization helps to remove DNA from the water phase. The phase partitioning of DNA and RNA in phenol extractions is pH dependent, as RNA will partition to the aqueous phase at low pH (<4.5) and both DNA and RNA will partition to the aqueous phase at pH 7.5 to 8.0, although the physical basis for this phenomenon is currently not fully understood.

Since its introduction (Chomczynski and Sacchi, 1987), the single-step method has become widely used for isolating RNA from a large number of samples. In addition, the procedure permits recovery of total RNA from small quantities of tissue or cells, making it suitable for gene expression studies whenever the quantity of tissue or cells available is limited. The protocol presented here is an updated version of the original method that further shortens the time for RNA isolation. All commercial application of the method is restricted by a U.S. patent (Chomczynski, 1989).

There are several commercial kits for total RNA isolation utilizing guanidine-based methods, the majority based on the single-step method. They can be divided into two groups. The first, exemplified by the RNA Isolation Kit from Stratagene, includes kits containing denaturing solution, water-saturated phenol, and sodium acetate buffer prepared according to the single-step protocol described here (see Basic Protocol 2). The use of these kits saves the time needed to make components of the single-step method, but at a substantially higher price. The second group of kits is based on a commercial version of the single-step method combining denaturing solution, phenol, and buffer in a single monophase solution. These kits offer an improved yield and shorter RNA isolation time (Chomczynski and Mackey, 1995). In this second group, the authors have tested and can recommend the following kits: Isogen (Nippon Gene), RNA-Stat 60

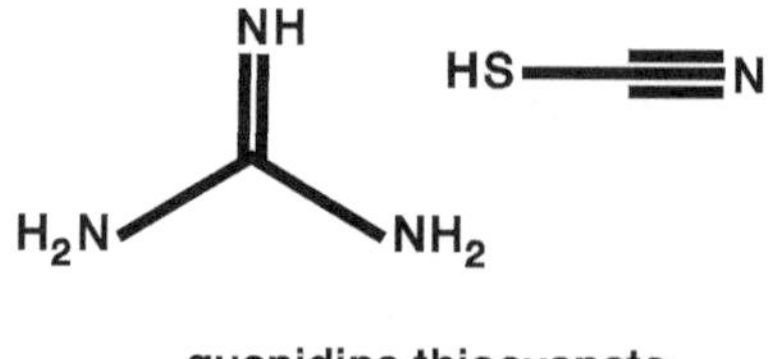

NH
HS≡N
H_2N NH_2
guanidine thiocyanate

Figure 5.2.3 The chemical structure of guanidine thiocyanate.

(Tel-Test), RNAzol B (Cinna Scientific), Tri-Pure Isolation Reagent (Boehringer Mannheim), TRI Reagent (Molecular Research Center), TRIzol Reagent (Invitrogen), and TRI reagent (Sigma). All the kits in the second group, except RNAzol B, allow simultaneous isolation of DNA and proteins from a sample used for RNA isolation.

Nucleic Acid Precipitation

Ethanol precipitation is useful for concentrating both DNA and RNA solutions, and for removing residual phenol and chloroform from the deproteinized aqueous solution. It is also useful for providing DNA that is relatively free of solute molecules when buffer conditions need to be changed. Ethanol and isopropanol are both commonly used for the precipitation of nucleic acids. Isopropanol is less volatile than ethanol and takes longer to remove by evaporation. It should be noted that some salts are less soluble in isopropanol (compared to ethanol) and will be precipitated along with nucleic acids. Extra washings may therefore be necessary to eliminate these contaminating salts. Isopropanol has the advantage that the volume required for nucleic acid precipitation is half that of the given volume of ethanol.

In the presence of relatively high (0.1 to 0.5 M) concentrations of monovalent cations, ethanol induces a structural transition in nucleic acid molecules, which causes them to aggregate and precipitate from solution (Eickbush and Moudrianakis, 1978). However, most salts and small organic molecules are soluble in 70% ethanol; thus, ethanol precipitation and washing of the pellet in 70% ethanol will effectively desalt DNA. Although sodium chloride, sodium acetate, and ammonium acetate are each capable of inducing precipitation, it is more difficult to remove sodium chloride due to its lower solubility in 70% ethanol. Additionally, lithium chloride is used for precipitation of RNA, as it has the benefit of being unable to effectively precipitate carbohydrate, protein, or DNA (Barlow et al., 1963). A list of salts and their uses in precipitation of nucleic acids is listed in Table 5.2.1.

Table 5.2.1 Salt Solutions Commonly Used in Nucleic Acid Precipitation

Salt	Stock concentration	Working concentration	Specific use	Comments
Sodium chloride (NaCl)	3 M	0.3 M	Precipitation of DNA or RNA	It can be difficult to remove sodium chloride due to its lower solubility in 70% ethanol compared to sodium acetate If the solution contains SDS, sodium chloride is used at a final concentration of 0.2 M, as the SDS will remain soluble in the 70% ethanol and thus will remain in the supernatant
Sodium acetate ($C_2H_3NaO_2$)	3 M	0.3 M	Precipitation of DNA or RNA	Higher solubility in 70% ethanol than sodium chloride
Ammonium acetate ($C_2H_7NO_2$)	7.5 M	2.5 M	Precipitation of DNA or RNA	If not removed, ammonium acetate will inhibit polynucleotide kinase phosphorylation of DNA Ammonium ions do not precipitate free nucleotides in solution
Lithium chloride (LiCl)	8 M	0.8 M	RNA precipitation	LiCl does not precipitate DNA, protein, or carbohydrate

5

STRATEGIC PLANNING

DNA

Approximately 90 min should be allowed for carrying out steps 1 through 9 of the basic DNA protocol (Basic Protocol 1) on twelve DNA samples in polypropylene microcentrifuge tubes. Nucleic acids should not be left in the presence of phenol, but can be left to stand indefinitely precipitated in alcohol or dried after precipitation. When processing nucleic acids, minimize exposure to acids, bases, or solvents, including DNA denaturing solutions used in DNA isolation procedures. The silica membrane spin column protocol (Alternate Protocol 1) can be performed on twelve samples in 15 to 20 min.

RNA

The isolation of total RNA by the single-step method (Basic Protocol 2) can be completed in <4 hr. The procedure can be interrupted at one of the isopropanol precipitations or at the ethanol wash steps. RNA samples can be stored at −20°C overnight or at −80°C for up to one month if the procedure is interrupted at these steps. Avoid keeping samples in denaturing solution for >30 min. As with any RNA preparative procedure, care must be taken to ensure that solutions are free of ribonuclease. Solutions that come into contact with the RNA after adding the guanidine solution are all treated with DEPC, with the exception of the TES solution (Tris inactivates DEPC). To remove RNase enzymatic activity by DEPC treatment, add DEPC to a final concentration of 0.1% and incubate 3 to 4 hr, followed by autoclaving for 45 min to hydrolyze and inactivate any remaining DEPC. Most investigators wear gloves at all times when working with RNA solutions, as hands/skin are a likely source of ribonuclease contamination (also see *UNIT 8.2*).

SAFETY CONSIDERATIONS

When working with chemicals used in the purification and concentration of nucleic acids, basic laboratory safety precautions should be followed. Chemicals such as phenol, chloroform, and DEPC are toxic and harmful if swallowed, inhaled, or absorbed through the skin. Safety precautions include wearing appropriate personal protective equipment such as laboratory coat, safety goggles, and gloves. Avoid contact of material or chemicals with skin or eyes, and use in an area with adequate ventilation.

Probably one of the more dangerous chemicals used in routine molecular biology is phenol. Phenol is toxic to humans and can burn the skin and other tissue that it comes into contact with. Phenol can act as an anesthetic and burning off the skin or tissues may not be initially felt. For these reasons, care must be taken at all times using phenol-containing products.

If in doubt concerning the hazards of any particular chemical, refer to the material data safety sheet (MSDS) provided with each reagent. See *APPENDIX 1* for more information.

PROTOCOLS

Basic Protocol 1: Phenol Extraction and Ethanol Precipitation of DNA

This protocol describes the most commonly used method of purifying and concentrating DNA preparations. The DNA solution is first extracted with a phenol/chloroform/isoamyl alcohol mixture to remove protein contaminants, and then precipitated with 100% ethanol. The DNA is pelleted after the precipitation step, washed with 70% ethanol to remove salts and small organic molecules, and resuspended in buffer at a concentration suitable for further experimentation.

The oxidation products of phenol can damage nucleic acids: only molecular-biology-grade phenol should be used. Typical oxidation products of phenol like para-benzoquinone can form a complex or "adduct" with the deoxycytidine, deoxyadenosine, and deoxyguanosine bases to cause damage to the DNA. If your phenol turns a brown or pink color, this means it is extensively oxidized and should not be used.

For complete deproteinization, extractions should be repeated until no protein precipitate remains at the aqueous/organic interface. The protein precipitate will look like a white fluffy precipitate between the aqueous and organic phases.

In general, alcohol precipitation of nucleic acids requires the presence of at least 0.1 M monovalent cation in the starting aqueous solution. Precipitation of nucleic acids at low concentrations requires cooling to low temperatures to give good recovery. Precipitation of nucleic acids at high concentrations ($\geq$0.25 mg/ml after addition of ethanol) is very rapid at room temperature. Formation of a visible precipitate after adding alcohol and mixing well indicates complete precipitation, and no chilling or further incubation is needed.

Materials

1 mg/ml DNA to be purified (e.g., see *UNIT 4.2*)
25:24:1 (v/v/v) phenol/chloroform/isoamyl alcohol
3 M sodium acetate, pH 5.2 (*UNIT 3.3*)
100% ethanol, ice cold
70% ethanol, room temperature
TE buffer, pH 8.0 (*UNIT 3.3*)

1.5 ml-polypropylene microcentrifuge tubes
Microcentrifuge
200-µl pipettor
Desiccator or Speedvac evaporator (Savant)

Purify using phenol

1. Add an equal volume of phenol/chloroform/isoamyl alcohol to the DNA solution to be purified in a 1.5-ml microcentrifuge tube.

   DNA solutions containing $\leq$0.5 M monovalent cations can be used. Extracting volumes $\leq$100 µl is difficult; small volumes should be diluted to obtain a volume that is easy to work with.

   High salt concentrations can cause the inversion of the aqueous and organic phases. If this happens, the organic phase can be identified by its yellow color.

2. Vortex vigorously for 10 sec and microcentrifuge 15 sec at maximum speed, room temperature.

   Phases should be well separated. If the DNA solution is viscous or contains a large amount of protein, it should be microcentrifuged longer (1 to 2 min).

   If purifying high-molecular-weight DNA, like genomic DNA, vortex gently, as vigorous vortexing may cause shearing of high-molecular-weight DNA.

3. Carefully remove the top (aqueous) phase containing the DNA using a 200-µl pipettor and transfer to a new tube. If a white precipitate is present at the aqueous/organic interface, repeat steps 1 to 3.

   If starting with a small amount of DNA (<1 µg), recovery can be improved by re-extracting the organic phase with 100 µl TE buffer, pH 8.0. This aqueous phase can be pooled with that from the first extraction.

Add salt (cation)

4. Add 1/10 vol 3 M sodium acetate, pH 5.2, to the recovered aqueous solution that contains the DNA. Mix by vortexing briefly or by flicking the tube several times with a finger.

   If the solution contains a high concentration of NaCl or sodium acetate (0.3 to 0.5 M) prior to the phenol extraction step, then no additional salt should be added. It is advisable to make appropriate dilutions to keep NaCl and sodium acetate concentrations below 0.5 M. For high concentrations of DNA (>50 to 100 μg/ml), precipitation is essentially instantaneous at room temperature.

To prevent carryover of residual phenol, the aqueous phase can be re-extracted with 24:1 (v/v) chloroform/isoamyl alcohol. However, this should not be necessary if the final pellet is washed well with 70% ethanol, or if an additional ethanol precipitation step is included.

Precipitate with ethanol

5. Add 2 to 2.5 vol (calculated *after* salt addition) ice-cold 100% ethanol. Mix by vortexing (or by gentle inversion for high-molecular-weight DNA) and place in crushed dry ice for 5 min or longer.

   This precipitation step can also be done in a −70°C freezer for 15 min or longer, or in a −20°C freezer for at least 30 min. A slurry of dry ice and ethanol may also be used, but tube labels are less often lost when only crushed dry ice is used.

6. Microcentrifuge 5 min at maximum speed and remove the supernatant.

   For large pellets, the supernatant can simply be poured off. For small pellets (<1 μg), aspirate off the ethanol supernatant with a pipetting device such as a Pasteur pipet or mechanical pipettor. This is best accomplished by drawing off liquid from the side of the tube opposite that against which the DNA precipitate was pelleted. Start at the top and move downward as the liquid level drops.

7. Add 1 ml room-temperature 70% ethanol. Invert the tube several times and microcentrifuge as in step 6.

   If the DNA molecules being precipitated are very small (<200 bases), use 95% ethanol at this step.

8. Remove the supernatant as in step 6. Dry the pellet in a desiccator under vacuum or in a Speedvac evaporator.

   The DNA pellet will not stick well to the walls of the tube after the 70% ethanol wash and care must be taken to avoid aspirating the pellet out of the tube.

9. Dissolve the dry pellet in an appropriate volume of water if it is going to be used for further enzymatic manipulations requiring specific buffers. Dissolve in TE buffer, pH 8.0, if it is going to be stored indefinitely.

   DNA pellets will not dissolve well in high-salt buffers. To facilitate resuspension, the DNA concentration of the final solution should be kept at <1 mg/ml.

   If DNA is resuspended in a volume of TE buffer or water to yield a DNA concentration of <1 mg/ml, small quantities (<25 μg) of precipitated plasmids or restriction fragments should dissolve quickly upon gentle vortexing or flicking of the tube. However, larger quantities of DNA may require vortexing and brief heating (5 min at 65°C) to resuspend. High-molecular-weight genomic DNA may require one to several days to dissolve and should be shaken gently (not vortexed) to avoid shearing, particularly if it is to be used for cosmid cloning or other applications requiring high-molecular-weight DNA. Gentle shaking on a rotating platform or a rocking apparatus is recommended.

Alternate Protocol 1: Purification of Plasmid DNA Using Silica Membrane Spin Columns

The use of glass beads or silica gel particles has become a popular method for isolating DNA. The evolution of this principle has resulted in the introduction of silica membrane spin columns. The proposed molecular basis for silica-DNA interaction is that silanol groups (SiOH) on the surface of silica functions as a hydrogen donor molecule that can form hydrogen bonds with another molecule acting as a hydrogen acceptor. DNA contains exocyclic nitrogens and carbonyl oxygens on the purine/pyrimidine bases, along with oxygen atoms in the phosphate backbone of the DNA that can function as hydrogen receptors and therefore bind to the silica by the formation of hydrogen bonds (Mao et al., 1994). The basic principle of silica gel solid support spin columns is fairly simple. DNA is bound to the silica membrane spin column in the presence of a high concentration of chaotropic salt, contaminants are washed away, and the DNA is then eluted from the silica membrane in water or a low-salt buffer.

This method can be used for isolation of plasmid DNA as small as 100 bp and up to 50 kb in size. Spin column kits are available from different companies for specific DNA isolation applications, such as isolation and purification of genomic DNA. Protocols for the large-scale isolation of plasmid DNA (>10 μg) can be found in *CP Molecular Biology Unit 1.7* (Heilig et al., 1998). This alternate protocol is designed for when a 1- to 5-ml bacterial culture is used as the starting material for small scale purification of plasmid DNA (minipreps).

A critical parameter in the purification of DNA using silica membrane spin columns is the pH of the chaotropic salt solution. For efficient binding to the silica membrane, the salt solution should be at pH 6.5. Many different companies sell silica membrane spin columns as kits supplied with all necessary reagents and buffers. One variable is the type of chaotropic salt solution supplied with the kit (NaI, guanidine HCl, or guanidine isothiocyanate). For the most part, these salt solutions are only a variation on the same theme and may be interchangeable between kits or columns from different vendors.

The advantages of using silica membrane spin columns for DNA purification is the fact that the silica is bound to a solid support, the method is quick and convenient, and can produce a high yield of pure DNA. Silica membrane spin columns are available from many companies, including Qiagen, Promega, Invitrogen, and Novagen, as kits including the columns and all appropriate buffers necessary for DNA purification.

NOTE: Refer to Chapter 6 for more information about chromatography.

Materials

1×10^9 cell/ml bacterial culture containing DNA of interest *or* 0.1 to 1 mg/ml DNA to be purified (5 to 10 μg DNA total); see *UNIT 4.2*
6 M sodium iodide (NaI) solution (filter through filter paper, store up to 3 months in the dark at 4°C)
Resuspension buffer (see recipe)
Lysis solution: 0.2 M NaOH/1.0% (w/v) SDS (store indefinitely at room temperature)
Neutralization/binding solution (see recipe)
Wash buffer (see recipe)
TE buffer, pH 8.5; (*UNIT 3.3*) *or* nuclease-free H_2O

Silica membrane spin columns (e.g., Qiagen, Promega, Invitrogen, Novagen)
1.5-ml microcentrifuge tubes

5

Prepare DNA solution–plasmid miniprep

1. Harvest 1 to 5 ml bacterial culture by centrifuging 1 min at 10,000 × *g*, room temperature, then discard the supernatant.

    To clean a DNA solution not generated from bacterial lysates, add 3 vol of 6 M NaI solution to DNA in a 1.5-ml microcentrifuge tube, mix, and proceed to step 6. The NaI is a chaotropic salt and functions to put the DNA in a hydrophobic environment. Under the hydrophobic conditions, the silica membrane of the spin columns become a suitable binding partner for the nucleic acids.

2. Add 250 μl cell resuspension buffer, resuspend the cell pellet by vortexing or pipetting, and transfer to a 1.5-ml microcentrifuge tube.

3. Add 250 μl lysis solution and mix by inversion (do not vortex). Allow lysis to proceed 3 to 5 min; do not allow lysis reaction to proceed for more than 5 min.

    In steps 2 and 3, the cell resuspension buffer and lysis solution rupture the cell wall and cell membrane of the bacteria by alkaline lysis. The lysis reaction should not be allowed to proceed for more than 5 min, as exposing the plasmid DNA to the NaOH in the lysis solution may begin to denature the DNA. The SDS in the lysis buffer solubilizes the phospholipids and proteins of the cell membrane, causing the release of the cellular contents. The NaOH denatures proteins as well as genomic and plasmid DNA. The short lysis time allows for the minimal exposure of plasmid DNA to denaturing conditions and the release of plasmid DNA from the cell, but not the release of genomic DNA, which remains bound to the cell wall.

4. Add 350 μl neutralization/binding solution and mix by inversion.

   The neutralization/binding solution is used to neutralize the bacterial lysate and increase the salt concentration. The addition of the neutralization/binding solution precipitates cellular proteins, linear (genomic) DNA, and SDS into a fluffy white precipitate, which is removed in step 5. Upon the addition of the neutralization/binding solution, the linear (genomic) DNA precipitates, presumably due to interstrand reassociation of denatured, linear DNA molecules at multiple sites. This results in the formation of a large insoluble DNA network that precipitates out of solution with the protein-SDS complexes. The plasmid DNA renatures in part due to its smaller and covalently closed conformation and remains in solution to be applied to the silica membrane spin column.

5. Microcentrifuge 10 min at maximum speed (a white pellet will form in the bottom of the tube).

Purify DNA on silica membrane spin column

6. Apply the supernatant to the silica membrane spin column.

7. Microcentrifuge spin column inside its collection tube for 1 min at maximum speed. Remove the spin column from the collection tube and discard the flowthrough. Reinsert the spin column into the collection tube.

8. Wash the spin column by adding 750 μl of wash buffer and microcentrifuging 1 min at maximum speed. Remove the spin column from the collection tube and discard the flowthrough.

   The plasmid DNA binds to the silica membrane when in the presence of high salt, while the RNA and proteins do not. The wash step helps to remove any unbound RNA and protein while reducing the salt concentration.

9. Reinsert the spin column in the collection tube and microcentrifuge for an additional 1 min to remove any residual wash buffer (ethanol) from the column membrane.

   Failure to remove all the ethanol will inhibit the alkaline TE buffer from increasing the pH of the membrane, which is required for DNA to be released from the membrane and eluted.

10. Transfer the spin column to a 1.5-ml microcentrifuge tube and add 75 to 100 μl nuclease-free water or TE buffer, pH 8.5, to the center of the membrane. Let stand for 2 to 10 min, then microcentrifuge 1 min at maximum speed.

11. Collect DNA and store at 4°C or −20°C until use.

Support Protocol 1: Precipitation of DNA Using Isopropanol

Equal volumes of isopropanol and DNA solution are used in precipitation. Note that the isopropanol volume is half that of the given volume of ethanol in precipitations. This allows precipitation from a large starting volume (e.g., 0.7 ml) in a single microcentrifuge tube. Isopropanol is less volatile than ethanol and takes longer to remove by evaporation. Some salts are less soluble in isopropanol (compared to ethanol) and will be precipitated along with nucleic acids. Extra washings may be necessary to eliminate these contaminating salts.

Support Protocol 2: Concentration of DNA Using Butanol

It is generally inconvenient to handle large volumes or dilute solutions of DNA. Water molecules (but not DNA or solute molecules) can be removed from aqueous solutions by extraction with *sec*-butanol (2-butanol). This procedure is useful for reducing volumes or concentrating dilute solutions before proceeding with Basic Protocol 1.

Additional Materials (also see Basic Protocol 1)

sec-Butanol
25:24:1 (v/v/v) phenol/chloroform/isoamyl alcohol
Polypropylene tube

1. Add an equal volume of *sec*-butanol to the sample and mix well by vortexing or by gentle inversion (if the DNA is of high molecular weight). Perform extraction in a polypropylene tube, as butanol will damage polystyrene.

2. Centrifuge 5 min at 1200 × *g* (2500 rpm), room temperature, or in a microcentrifuge for 10 sec at maximum speed.

3. Remove and discard the upper (*sec*-butanol) phase.

4. Repeat steps 1 to 3 until the desired volume of aqueous solution is obtained.

5. Extract the lower, aqueous phase with 25:24:1 phenol/chloroform/isoamyl alcohol and ethanol precipitate (see Basic Protocol 1, steps 1 to 9).

   Addition of too much sec-butanol can result in complete loss of the water phase into the sec-butanol layer. If this happens, add 1/2 vol water back to the sec-butanol, mix well, and spin. The DNA can be recovered in this new aqueous phase and can be further concentrated with smaller amounts of sec-butanol.

   The salt concentration will increase in direct proportion to the volume decrease. The DNA can be precipitated with ethanol to readjust the buffer conditions.

Basic Protocol 2: Single-Step RNA Isolation from Cultured Cells or Tissues

The following procedure describes a single-step isolation method employing liquid-phase separation to selectively extract total RNA from tissues and cultured cells. Cultured cells or tissues are homogenized in a denaturing solution containing 4 M guanidine thiocyanate. The homogenate is mixed sequentially with 2 M sodium acetate (pH 4.0), phenol, and finally chloroform/isoamyl alcohol or bromochloropropane. The resulting mixture is centrifuged, yielding an upper aqueous phase containing total RNA. In this single-step extraction, the total RNA is separated from proteins and DNA that remain in the interphase and organic phase. Following isopropanol precipitation, the RNA pellet is redissolved in denaturing solution (containing 4 M guanidine thiocyanate), reprecipitated with isopropanol, and washed with 75% ethanol.

5

Several other popular protocols for the purification of RNA use cesium chloride centrifugation. These procedures, detailed in *CP Molecular Biology Unit 4.2* (Kingston et al., 1996), take advantage of the fact that RNA can be separated from DNA and protein by virtue of its greater density. These methods have received widespread use because they require very few manipulations. This increases the chance of producing intact RNA and reduces hands-on time for the experimenter. The disadvantage is that these methods require an ultracentrifuge and rotor, which generally limits the number of samples that can easily be processed simultaneously. These protocols should be used when very high-quality RNA from a limited number of samples is required.

Materials

Denaturing solution (see recipe)
2 M sodium acetate, pH 4 (see recipe)
Phenol, water-saturated (see recipe)
49:1 (v/v) chloroform/isoamyl alcohol or bromochloropropane
100% isopropanol
75% ethanol (prepared with DEPC-treated water)
DEPC-treated water or freshly deionized formamide (see recipes)

Glass Teflon homogenizer
5-ml polypropylene centrifuge tube
Sorvall SS-34 rotor (or equivalent; *UNIT 5.1*)
1.5-ml polypropylene microcentrifuge tubes

Homogenize cells

1a. *For tissue:* Add 1 ml denaturing solution per 100 mg tissue and homogenize with a few strokes in a glass Teflon homogenizer.

1b. *For cultured cells:* Either centrifuge suspension cells and discard supernatant, or remove the culture medium from cells grown in monolayer cultures. Add 1 ml denaturing solution per 10^7 cells and pass the lysate through a pipet seven to ten times.

Do not wash cells with saline. Cells grown in monolayer cultures can be lysed directly in the culture dish or flask.

The procedure can be carried out in sterile, disposable, round-bottom polypropylene tubes with caps; no additional treatment of the tubes is necessary. Before using, test if the tubes can withstand centrifugation at 10,000 × g with the mixture of denaturing solution and phenol/chloroform.

2. Transfer the homogenate into a 5-ml polypropylene tube. Add 0.1 ml of 2 M sodium acetate, pH 4.0, and mix thoroughly by inversion. Add 1 ml water-saturated phenol, mix thoroughly, and add 0.2 ml of 49:1 chloroform/isoamyl alcohol or bromochloropropane. Mix thoroughly and incubate the suspension 15 min at 0° to 4°C.

   Make sure that caps are tightly closed when mixing. The volumes used are per 1 ml denaturing solution.

   Bromochloropropane is less toxic than chloroform and its use for phase separation decreases the possibility of contaminating RNA with DNA (Chomczynski and Mackey, 1995).

3. Centrifuge 20 min at 10,000 × *g* (9000 rpm in SS-34 rotor), 4°C. Transfer the upper aqueous phase to a clean tube.

   The upper aqueous phase contains the RNA, whereas the DNA and proteins are in the interphase and lower organic phase. The volume of the aqueous phase is ∼1 ml, equal to the initial volume of denaturing solution.

Isolate RNA

4. Precipitate the RNA by adding 1 ml (1 vol) of 100% isopropanol. Incubate the samples 30 min at −20°C. Centrifuge 10 min at 10,000 × *g*, 4°C, and discard supernatant.

   For isolation of RNA from tissues with a high glycogen content (e.g., liver), a modification of the single-step method is recommended to diminish glycogen contamination (Puissant and Houdebine, 1990). Following this isopropanol precipitation, wash out glycogen from the RNA pellet by vortexing in 4 M LiCl. The LiCl precipitation can precipitate RNA, but does not efficiently precipitate proteins or DNA, so the 4 M LiCl precipitation will help to remove proteins or DNA. Sediment the insoluble RNA 10 min at 5000 × g. Dissolve the pellet in denaturing solution and follow the remainder of the protocol.

5. Dissolve the RNA pellet in 0.3 ml denaturing solution and transfer into a 1.5-ml microcentrifuge tube.

6. Precipitate the RNA with 0.3 ml (1 vol) of 100% isopropanol for 30 min at −20°C. Centrifuge 10 min at 10,000 × *g*, 4°C, and discard supernatant.

7. Resuspend the RNA pellet in 75% ethanol, vortex, and incubate 10 to 15 min at room temperature to dissolve residual amounts of guanidine contaminating the pellet.

8. Centrifuge 5 min at 10,000 × *g*, 4°C, and discard supernatant. Dry the RNA pellet in a vacuum for 5 min.

   Do not let the RNA pellet dry completely, as this greatly decreases its solubility. Avoid drying the pellet by centrifugation under vacuum, as over drying the RNA will make it hard to solubilize. Drying is not necessary for solubilization of RNA in formamide.

9. Dissolve the RNA pellet in 100 to 200 μl of DEPC-treated water or freshly deionized formamide by passing the solution a few times through a pipet tip. Store RNA dissolved in water at −70°C and RNA dissolved in formamide at either −20° or −70°C.

If the RNA is to be run on a formaldehyde-agarose gel for northern blotting (UNIT 8.2), resuspending the RNA in formamide will allow a larger amount of RNA to be run on the gel rather than diluting the RNA in DEPC-treated water first. For all other applications, resuspending the RNA in DEPC-treated water can be done to avoid an extra precipitation step to remove the to avoid Incubate 10 to 15 min at 55° to 60°C.

RNA dissolved in formamide is protected from degradation by RNase and can be used directly for formaldehyde-agarose gel electrophoresis in northern blotting (Chomczynski, 1992; UNIT 8.2). However, before use in real time PCR (UNIT 10.3) RNA should be precipitated from formamide by adding 4 vol ethanol and centrifuging 5 min at 10,000 × g.

Quantitate RNA

10. Quantitate RNA by diluting 5 µl in 1 ml alkaline water and reading the A_{260} and A_{280} (see UNIT 2.2).

Water used for spectrophotometric measurement of RNA should have pH >7.5. Acidic pH affects the UV absorption spectrum of RNA and significantly decreases its A_{260}/A_{280} ratio (Willfinger et al., 1997). Typically, distilled water has pH >6 see (UNIT 3.1). Adjust water to a slightly alkaline pH by adding concentrated Na_2HPO_4 solution to a final concentration of 1 mM.

Alternate Protocol 2: Purification and Concentration of RNA and Dilute Solutions of DNA

The following adaptations to the purification procedure (see Basic Protocol 2) are used if RNA or dilute solutions of DNA are to be purified.

Purify and concentrate RNA

The procedure outlined in Basic Protocol 1 is identical for purification of RNA, except that 2.5 vol ethanol should be used routinely for the precipitation (step 5). It is essential that all water used directly or in buffers be treated with DEPC (see recipe) to inactivate RNase.

Dilute solutions of DNA

When DNA solutions are dilute (<10 µg/ml) or when <1 µg of DNA is present, the ratio of ethanol to aqueous volume should be increased to 3:1 and the time on dry ice (step 5) extended to 30 min. Microcentrifugation should be carried out for 15 min in a cold room to ensure the recovery of DNA from these solutions.

Nanogram quantities of labeled or unlabeled DNA can be efficiently precipitated by the use of carrier nucleic acid. A convenient method is to add 10 µg of commercially available tRNA from *E. coli*, yeast, or bovine liver to the desired DNA sample. The DNA will be co-precipitated with the tRNA. The carrier tRNA will not interfere with most enzymatic reactions, but will be phosphorylated efficiently by polynucleotide kinase and should not be added if this enzyme will be used in subsequent radiolabeling reactions.

Recovery of small quantities of short DNA fragments (<50 bp) and oligonucleotides can be enhanced by adding magnesium chloride to a concentration of <10 mM before adding ethanol (step 4). However, DNA precipitated from solutions containing >10 mM magnesium or phosphate ions is often difficult to redissolve and such solutions should be diluted prior to ethanol precipitation.

DNA in large aqueous volumes (>0.4 to 10 ml)

Larger volumes can be accommodated by simply scaling up the amounts used in the basic protocols or by using butanol concentration (see Support Protocol 2). For the phenol extraction (see Basic Protocol 1, steps 1 through 3), tightly capped 15- or 50-ml polypropylene tubes should be used, since polystyrene tubes cannot withstand the phenol/chloroform mixture. Centrifugation steps should be performed for 5 min at speeds not exceeding 1200 × *g* (2500 rpm), room temperature. The ethanol precipitate (Basic Protocol 1, step 6) should be centrifuged in thick-walled Corning glass test tubes (15- or 30-ml capacity) for 15 min in fixed-angle rotors at 8000 × *g* (10,000 rpm), 4°C.

REAGENTS AND SOLUTIONS

Use deionized, distilled water in all recipes and protocol steps. For common stock solutions, see UNIT 3.3.

Denaturing solution

Stock solution: Mix 293 ml water, 17.6 ml of 0.75 M sodium citrate, pH 7.0, and 26.4 ml of 10% (w/v) *N*-lauroylsarcosine (Sarkosyl). Add 250 g guanidine thiocyanate and stir at 60° to 65°C to dissolve. Store up to 3 months at room temperature.

Working solution: Add 0.35 ml 2-mercaptoethanol (2-ME) to 50 ml of stock solution. Store up to 1 month at room temperature.

Final concentrations are 4 M guanidine thiocyanate, 25 mM sodium citrate, 0.5% Sarkosyl, and 0.1 M 2-ME.

Diethylpyrocarbonate (DEPC)-treated solutions

Add 0.2 ml DEPC to 100 ml of the solution to be treated. Shake vigorously to get the DEPC into solution. Autoclave the solution to inactivate the remaining DEPC.

Many investigators keep the solutions they use for RNA work separate to ensure that "dirty" pipets do not go into them.

Formamide

Prepare freshly deionized formamide by stirring with 1 g AG 501-X8 ion-exchange resin (Bio-Rad) per 10 ml formamide for 30 min and filter at room temperature. Alternatively, use commercially available, stabilized, ultrapure formamide (Formazol, Molecular Research Center).

Neutralization/binding solution

Add 477.65 g guanidine HCl and 49.09 g potassium acetate to 500 ml water and stir to dissolve. Adjust pH to ~4.2 with acetic acid, dilute solution to 1 liter with water, and filter sterilize. Store indefinitely at 4°C.

5

Phenol, water-saturated

Dissolve 100 g phenol crystals in water at 60° to 65°C. Aspirate the upper water phase and store up to 1 month at 4°C.

Do not use buffered phenol in place of water-saturated phenol. Water-saturated phenol is acidic and will partition RNA into the aqueous phase, while buffered phenol that has a neutral or slightly alkaline pH will partition DNA to the interface or phenol phase.

Resuspension buffer

50 mM Tris·Cl, pH 8.0 (*UNIT 3.3*)
10 mM EDTA
100 mg RNase A
Store indefinitely at 4°C

Sodium acetate, 2 M

Add 16.42 g sodium acetate (anhydrous) to 40 ml water and 35 ml glacial acetic acid. Adjust solution to pH 4 with glacial acetic acid and dilute to 100 ml final with water (solution is 2 M with respect to sodium ions). Store up to 1 year at room temperature.

Wash buffer

1 part 10 mM Tris·Cl, pH 7.5 (*UNIT 3.3*)
1 part 100 mM NaCl
4 parts 100% ethanol (final 80%)
Store indefinitely at room temperature

UNDERSTANDING RESULTS

These procedures should result in virtually complete removal of proteins and quantitative recovery of nucleic acids. Recovery of nucleic acids can be quantified by several methods, the most common is by obtaining the optical density (OD) of the DNA/RNA solution by UV absorbance spectroscopy at 260 nm (*UNIT 2.2*). DNA quantitation has traditionally also been done by running DNA samples on agarose gels followed by ethidium bromide staining (*UNITS 2.2 & 7.2*). The new DNA can then be compared to a sample of DNA of known concentration. Newer methods utilizing fluorescent dyes that undergo fluorescence enhancement upon binding to dsDNA and the use of a fluorometer to measure the change in fluorescence are becoming more common for high-throughput applications. For a comprehensive explanation of different nucleic acid quantification methods, see *UNIT 2.2*.

It is important to remember that sequential extractions or precipitations require care and attention to detail to prevent accumulation of small losses at each step. It is particularly important to carefully recover the aqueous phase and re-extract the organic phase to ensure full recovery of small amounts of DNA from phenol/chloroform extractions. The yield of nucleic acids resulting from the silica membrane spin column procedure can be similarly improved (to >80% recovery) by subjecting supernatants to an additional binding step and increasing the amount of elution buffer to 100 μl or more.

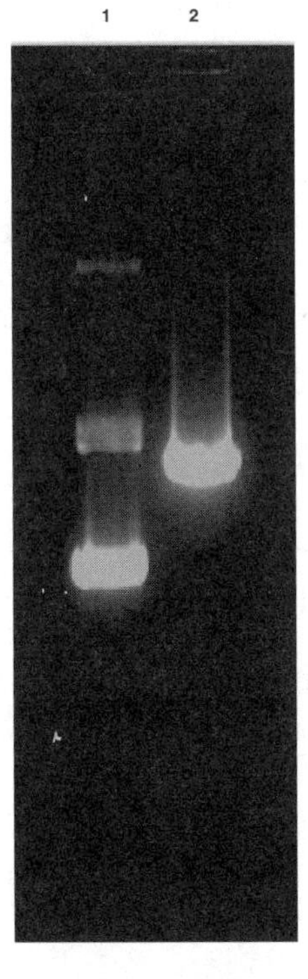

Figure 5.2.4 Plasmid DNA. The plasmid pUC19 was isolated from a 3 ml bacterial culture using Alternate Protocol 1. The DNA was electrophoresed in a 0.8% agarose gel containing ethidium bromide and viewed under UV light. Lane 1 contains 500 ng of uncut pUC19 DNA, the majority of the DNA is in the supercoiled configuration (lower band), with some nicked and open circular DNA comprising the less intense higher bands of DNA on the gel. Lane 2 contains 500 ng of pUC19 plasmid DNA cut with the *Eco*R1 restriction enzyme, forming a single linear band on the gel.

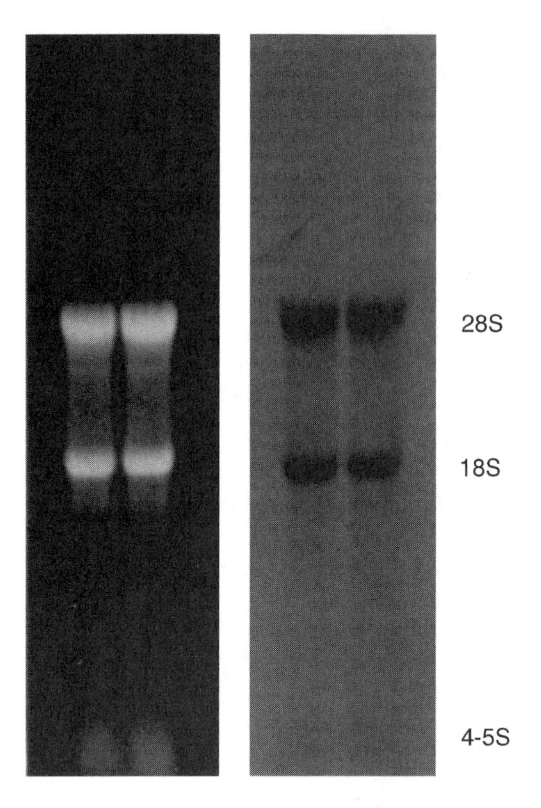

Figure 5.2.5 Rat liver RNA (5 μg) isolated using Basic Protocol 2 was electrophoresed in a formaldehyde 1% agarose gel containing ethidium bromide (left), transferred to a hybridization membrane, and stained with methylene blue stain (Molecular Research Center; Herrin and Schmidt, 1988, right). Methylene blue stain is an aqueous solution of methylene blue designed for quantitative and qualitative examination of RNA on hybridization membranes in northern blotting. Shown are 28S (4.7 kb) and 18S (1.9 kb) ribosomal RNAs, and 4S to 5S (0.10 to 0.15 kb) RNA containing mix of tRNA and 5S ribosomal RNA. Originally published in *CP Molecular Biology Unit 4.2* (Kingston et al., 1996).

The electrophoretic pattern of plasmid DNA isolated by Alternate Protocol 1, is exemplified in Figure 5.2.4, which shows the results of agarose gel electrophoresis of pUC19 plasmid DNA. In a high-quality plasmid DNA isolation, the majority of plasmid DNA will be in the supercoiled form, and only a small percentage (<10% to 20%) in the nicked and open circular plasmid conformation.

RNA

The single-step method yields the whole spectrum of RNA molecules, including small (4S to 5S) RNAs. The amount of isolated RNA depends on the tissue used for isolation. Typically, 100 to 150 μg of total RNA is isolated from 100 mg of muscle tissue and up to 800 μg is isolated from 100 mg of liver. The yield of total RNA from 10^7 cultured cells ranges from 50 to 80 μg for fibroblasts and lymphocytes and 100 to 120 μg for epithelial cells. The A_{260}/A_{280} ratio of DNA or RNA is an indication of purity and is based on the extinction coefficients of nucleic acids at 260 nm and 280 nm. Pure DNA and RNA preparations have expected A_{260}/A_{280} ratios of ≥1.8 (Glasel, 1995). If the isolated RNA has an A_{260}/A_{280} ratio <1.8, there may still be a significant amount of contaminants (protein) in the RNA solution. The RNA should be added to 1 ml denaturing solution and Basic Protocol 2 (Single-Step RNA Isolation from Cultured Cells or Tissues) should be repeated.

The electrophoretic pattern of RNA isolated by the single-step method is exemplified in Figure 5.2.5, which shows the results of formaldehyde-agarose gel electrophoresis of rat liver RNA.

TROUBLESHOOTING

Preparation of Crude Lysate

This is a reliable procedure and one in which few irretrievable disasters can occur, but is very important for successful isolation of plasmid DNA. It is important to make sure the SDS in the SDS/NaOH lysis solution is actually in solution. If stored at 4°C, the SDS may fall out of solution. Simply warm the lysis solution to get the SDS to go back into solution prior to use. An important parameter of this procedure is the incubation time when using the SDS/NaOH lysis solution. Leaving the DNA in the SDS/NaOH lysis solution longer than recommended (3 to 5 min) can result in increased levels of plasmid DNA that do not renature and will result in increased levels of open circular DNA (see Fig. 5.2.4), reducing the yield of supercoiled plasmid DNA. One potential problem is failure of chromosomal DNA and proteins to precipitate after addition of potassium acetate solution. The cause of this is probably improper pH of the potassium acetate solution. If precipitation fails to occur because the pH is incorrect, the preparation can be saved by adding concentrated formic or acetic acid dropwise to the solution. Mix after each addition until the viscosity decreases, which will happen suddenly. A precipitate will then appear and the alkaline lysis procedure can then be completed normally.

DNA and RNA Purification

The oxidation products of phenol can damage nucleic acids and only molecular-biology-grade or redistilled phenol should be used. For complete deproteinization of DNA or RNA, extractions should be repeated until no protein precipitate remains at the aqueous/organic interface when performing phenol extractions. In organic extractions involving DNA or RNA, low yields may be the result of loss of nucleic acid at the interface and into the organic phase, this problem is minimized by back-extracting the organic phase to minimize the loss of any nucleic acids. An additional extraction with 24:1 chloroform/isoamyl alcohol can be performed to remove any residual phenol from the purified DNA or RNA, which may interfere with absorption spectroscopy quantitation or subsequent enzymatic reactions.

In the RNA protocol, care must be taken to ensure that solutions are free of ribonuclease. Solutions that come into contact with the RNA after adding the guanidine solution are all treated with DEPC, with the exception of the TES solution (Tris inactivates DEPC). Most investigators wear gloves at all times when working with RNA solutions, as hands are a likely source of ribonuclease contamination (see *UNIT 8.2* for further discussion of RNase contamination). Low yields may result from failing to allow sufficient time for resuspension of the RNA pellet after centrifugation. This pellet is not readily soluble, and sufficient time and vortexing should be allowed to dissolve it. There are two important points to consider in the isolation and purification of RNA. First, fresh tissue is preferable for RNA isolation. Alternatively, tissue should be frozen immediately in liquid nitrogen and stored at −70°C. In the latter case, tissue should be pulverized in liquid nitrogen and homogenized, using a Polytron or Waring blender, in denaturing solution without thawing. Second, it is important not to let the final RNA pellet dry completely, as that will greatly decrease its solubility. This is critical in all RNA isolation methods. Partially solubilized RNA has an A_{260}/A_{280} ratio <1.6. Solubility of RNA can be improved by heating at 55°C to 60°C with gentle vortexing or by passing the RNA solution through a pipet tip.

Nucleic acid precipitation is a straightforward procedure. The most important parameter is that one is using an adequate cation concentration and appropriate amount of ethanol or isopropanol. Once precipitated and resuspended in water or TE buffer, if the isolated DNA fails to cut with restriction endonucleases, the most common cause is inadequate washing of the pellets after the ethanol precipitation step. Precipitating the DNA a second time with ethanol, or washing the pellets from the first precipitation with 70% ethanol, will usually clean up the DNA enough for restriction enzyme cutting.

Once the DNA or RNA has been obtained, the A_{260}/A_{280} ratio should be obtained by absorption spectroscopy to test the purity of the recovered nucleic acids. If the A_{260}/A_{280} ratio indicates the samples have significant levels of protein contaminants (a ratio <1.8), the samples need to be processed again. For DNA, the samples can be subjected either to Basic Protocol 1 (Phenol Extraction and Ethanol Precipitation of DNA) or Alternate Protocol 1 (Purification of Plasmid DNA Using Silica Membrane Spin Columns). RNA should be added to 1 ml denaturing solution and Basic Protocol 2 (Single-Step RNA Isolation from Cultured Cells or Tissues) should be repeated.

LITERATURE CITED

Barlow, J.J., Mathias, A.P., Williamson, R., and Gammack, D.B. 1963. A simple method for the quantitative isolation of undegraded high molecular weight ribonucleic acid. *Biochem. Biophys. Res. Commun.* 13:61-66.

Birnboim, H.C. and Doly, J. 1979. A rapid alkaline extraction procedure for screening recombinant plasmid DNA. *Nucleic Acids Res.* 7:1513-23.

Chirgwin, J.J., Przbyla, A.E., MacDonald, R.J., and Rutter, W.J. 1979. Isolation of biologically active ribonucleic acid from sources enriched in ribonuclease. *Biochemistry* 18:5294.

Chomczynski, P. 1989. Product and process for isolating RNA. U.S. Patent #4,843,155.

Chomczynski, P. 1992. Solubilization in formamide protects RNA from degradation. *Nucl. Acids Res.* 20:3791-3792.

Chomczynski, P. and Mackey, K. 1995. Substitution of chloroform by bromochloropropane in the single-step method of RNA isolation. *Anal. Biochem.* 225:163-164.

Chomczynski, P. and Sacchi, N. 1987. Single-step method of RNA isolation by acid guanidine thiocyanate-phenol-chloroform extraction. *Anal. Biochem.* 162:156-159.

Cox, R.A. 1968. The use of guanidine chloride in the isolation of nucleic acids. *Methods Enzymol.* 12:120-129.

Eickbush, T.H. and Moudrianakis, E.N. 1978. The compaction of DNA helices into either continuous supercoils or folded-fiber rods and toroids. *Cell* 13:295-306.

Engebrecht, J., Brent, R., and Kaderbhai, M.A. 1991. Minipreps of plasmid DNA. *Curr. Protoc. Mol. Biol.* 15:1.6.1-1.6.10.

Glasel, J.A. 1995. Validity of nucleic acid purities monitored by A260/A280 absorbance ratios. *Biotechniques* 18:62-63.

Heilig, J.S., Elbing, K.L., and Brent, R. 1998. Large-scale preparation of plasmid DNA. *Curr. Protoc. Mol. Biol.* 41:1.7.1-1.7.16.

Herrin, D.L. and Schmidt, G.W. 1988. Rapid, reversible staining of northern blots prior to hybridization. *Biotechniques* 6:196-200.

Holmes, D.S. and Quigley, M. 1981. A rapid boiling method for the preparation of bacterial plasmids. *Anal. Biochem.* 114:193-197.

Kingston, R.E., Chomczynski, P., and Sacchi, N. 1996. Guanidine methods for total RNA preparation. *Curr. Protoc. Mol. Biol.* 36:4.2.1-4.2.9.

Kirby, K.S. 1957. A new method for the isolation of deoxyribonucleic acids: Evidence on the nature of bonds between deoxyribonucleic acid and protein. *Biochem. J.* 66:495-504.

Mao, Y., Daniel, L.N., Whittaker, N., and Saffiotti, U. 1994. DNA binding to crystalline silica characterized by Fourier-transform infrared spectroscopy. *Environ. Health Perspect.* 10:165-71.

Marmur, J. 1961. A procedure for the isolation of deoxyribonucleic acid from microorganisms. *J. Mol. Biol.* 3:208-218.

Moore, D. and Dowhan, D. 2002. Manipulation of DNA. *Curr. Protoc. Mol. Biol.* 59:2.1.1-2.1.10.

Palmiter, R.D. 1974. Magnesium precipitation of ribonucleoprotein complexes. Expedient techniques for the isolation of undegraded polysomes and messenger ribonucleic acid. *Biochemistry* 13:3606-3615.

Penman, S. 1966. RNA metabolism in the HeLa cell nucleus. *J. Mol. Biol.* 17:117-130.

Puissant, C. and Houdebine, L.M. 1990. An improvement of the single-step method of RNA isolation by acid guanidine thiocyanate-phenolchloroform extraction. *Biotechniques* 8:148-149.

Richards, E., Reichardt, M., and Rogers, S. 1994. Preparation of genomic DNA from plant tissue. *Curr. Protoc. Mol. Biol.* 27:2.3.1-2.3.7.

Seidman, C.E., Struhl, K., Sheen, J., and Jessen, T. 1997. Introduction of plasmid DNA into cells. *Curr. Protoc. Mol. Biol.* 37:1.8.1-1.8.10.

Strauss, W.M. 1998. Preparation of genomic DNA from mammalian tissue. *Curr. Protoc. Mol. Biol.* 42:2.2.1-2.2.3.

Ullrich, A., Shine, J., Chirgwin, J., Pictet, R., Tischer, E., Rutter, W.J., Vogelstein, B., and Gillespie, D. 1979. Preparative and analytical purification of DNA from agarose. *Proc. Nat. Acad. Sci. U.S.A.* 76:615-619.

Vogelstein, B. and Gillespie, D. 1979. Preparative and analytical purification of DNA from agarose. *Proc. Nat. Acad. Sci. U.S.A.* 76:615-619.

Wilfinger, W.W., Mackey, K., and Chomczynski, P. 1997. Effect of pH and ionic strength on the spectrophotometric assessment of nucleic acid purity. *Biotechniques* 22:474-476.

Wilson, K. 1997. Preparation of genomic DNA from bacteria. *Curr. Protoc. Mol. Biol.* 27:2.4.1-2.4.5.

Chapter 6

Chromatography

UNIT 6.1

Overview of Chromatography

Buddhadeb Mallik,[1] Bulbul Chakravarti,[1] and Deb N. Chakravarti[1]

[1]Proteomics Center, Keck Graduate Institute of Applied Life Sciences, Claremont, California

OVERVIEW

The term chromatography encompasses a group of methods for analysis or preparative separation of mixtures of chemical or biological molecules based on their differential elution from a stationary phase by a mobile phase, without the application of any external electric field (such as in electrophoresis; see *UNIT 7.1*). In general, this is due to differential distribution of those molecules between the stationary and the mobile phase. In a typical chromatography method, the mixture to be separated is usually applied to a solid stationary phase, followed by elution with a mobile phase of constant or varying composition, which carries the components of the mixture through the stationary phase. Based on specific interactions among the components, the stationary phase, and the mobile phase, some molecules of a specific type will have a higher tendency to move with the mobile phase, while others would like to remain with the stationary phase. The former will thus move through the system at a faster rate than the latter, and in effect will be separated from each other. As a part of the process, the separated components are detected, collected, and quantified as necessary. When the mobile phase is in a gaseous state, the chromatographic method is known as gas chromatography (GC), and when it is in a liquid state, the method is known as liquid chromatography (LC). In this unit, the authors will concentrate their discussion on LC only.

Mikhail Tswett, a Russian botanist, invented the chromatographic technique around the year 1900 with the separation of plant leaf photosynthetic pigments using a column of calcium carbonate as the stationary phase and established the presence of several chemical compounds. He coined the term “chromatography” from the Greek words “chroma” meaning “color” and “graphein” meaning “to write” since he obtained colored (chroma) bands (graphein) as he washed the pigments down the column. Chromatography now includes such methods for the separation of the components of a mixture of chemicals or biomolecules for preparation of pure components or for quantitative analysis. The components do not need to be colored, however; suitable methods for detection of separated components, such as optical absorbance, conductivity, or mass spectrometry should be available. The 1952 Nobel Prize in chemistry was awarded to Archer John Porter Martin and Richard Laurence Millington Synge for their invention of partition chromatography. Partition chromatography resulted from the combination of two techniques, that of chromatography and countercurrent solvent extraction. In liquid partition chromatography, separation is based mainly on differences between the solubilities of the sample components in the mobile and stationary phases. Beginning around that time, chromatographic techniques have undergone rapid development, as specialized methods have been developed for separation of much larger molecules, such as peptides and proteins. It is interesting to note that in a more recent method like reversed-phase high-performance liquid chromatography (RP-HPLC), small molecules may be retained on the stationary phase by partitioning, while polypeptides are generally retained by adsorption.

Application of chromatography for purification of proteins originated during the early nineteen fifties. Low-molecular-weight proteins cytochrome *c* (Paléus and Neilands, 1950) and ribonuclease (Hirs et al., 1951) were the first molecules of their size to be purified by ion exchange chromatography using the synthetic polymethacrylic acid resin Amberlite IRC-50 as the stationary phase and aqueous salt solutions as the mobile phase. However, the primary drawbacks of such resins as matrices for chromatography of proteins were their hydrophobicity and low porosity. With the introduction of ion exchangers prepared from cellulose (Peterson and Sober, 1956), hydroxyapatite as a highly selective adsorbent (Hjerten et al., 1956), and cross-linked dextran (Sephadex; Porath and Flodin, 1959), polyacrylamide (Hjerten and Mosbach, 1962), and agarose (Hjerten, 1962) for gel filtration, the modern era of protein chromatography began. Researchers started using many of these methods in combination to purify a large number of enzymes and other proteins. Introduction of the cyanogen bromide (CNBr) activation method by Axén et al. (1967) led to rapid advances in affinity chromatography techniques, which are among the most powerful techniques for purification of proteins. CNBr reacts with hydroxyl groups on agarose to form CNBr-activated agarose–containing imidocarbonate groups to which proteins, nucleic acids, or other biopolymers can be attached through their primary amino groups or similar nucleophilic groups using mild conditions. Subsequently, a number of techniques such as hydrophobic interaction chromatography, metal chelate chromatography, and chromatofocusing, as well as a variety of chromatographic media, have been developed for analysis and purification of proteins. This is due to the fact that proteins show an enormous variety of structures and functions, and thus any one or a single set of methods may not be generally applicable in the case of different proteins. In general, the major advantages of chromatography are the relative simplicity of the process, ease of automation, retention of biological properties of the protein molecules, and easy scalability from purification of several micrograms to kilogram quantities of material without significant effect on the degree of purification. Methods developed in the laboratory using columns with a few milliliters of stationary phase can be transferred to industrial scale columns of several hundred liters of volume.

Since the late 1970s, remarkable advances in speed, quality, and automation in the field of protein purification and analysis have been obtained with the introduction of the techniques of high-performance liquid chromatography (HPLC). Although the basic principles of separation, such as gel filtration or ion exchange, remain the same as in conventional LC, HPLC uses novel stationary phase materials that allow elution with mobile phases using moderate to high pressure. One of the major differences between the conventional LC, often referred to as "open column" liquid chromatography, and HPLC is the application of pressure in the elution process. Pressure makes the elution process quicker; however, operation under higher pressures required development of various specialized instrument components and software, including special injectors for sample application, automated devices for unattended sample application (auto samplers), pumps for delivery of mobile phases, modifications of the physical parameters of the stationary phase materials, and software for automated purification or analysis of a series of samples, as well as data collection, etc.

In summary, in LC and HPLC, the solution of a sample (usually a mixture) is applied to an immobile *stationary phase*. LC and HPLC techniques, in general, use *columns* that are tubes packed with the stationary phase. The *mobile phase* is then allowed to flow (or forced as in the case of HPLC) through the immobile and immiscible stationary phase. The sample is transported through the column by such continuous addition of the mobile phase and the process is called *elution*. LC or HPLC can be a preparative, as well as an analytical technique.

THEORY OF CHROMATOGRAPHY

The theory of chromatography will be discussed here with particular reference to LC, in which the stationary phase is a solid matrix and the mobile phase is a liquid. Details about the general terminology and quantitative expressions described in this unit can be found in Willard et al. (1974), Janson and Jönsson (1989), Lindsay (1992), and Wilson and Walker (1994). LC is most commonly carried out as column chromatography where the stationary phase is packed in a tubular column and the mobile phase (eluent) is passed through the column using gravity flow or a pump. The sample, which contains several components, is introduced into one end of the column followed by elution with the mobile phase in the direction towards the other end. Different components move at different rates through the column depending on their partition or distribution coefficients between the mobile and the stationary phases as explained below. The separated components are collected at the other end of the column and detected, as well as recorded as a plot of time or the volume

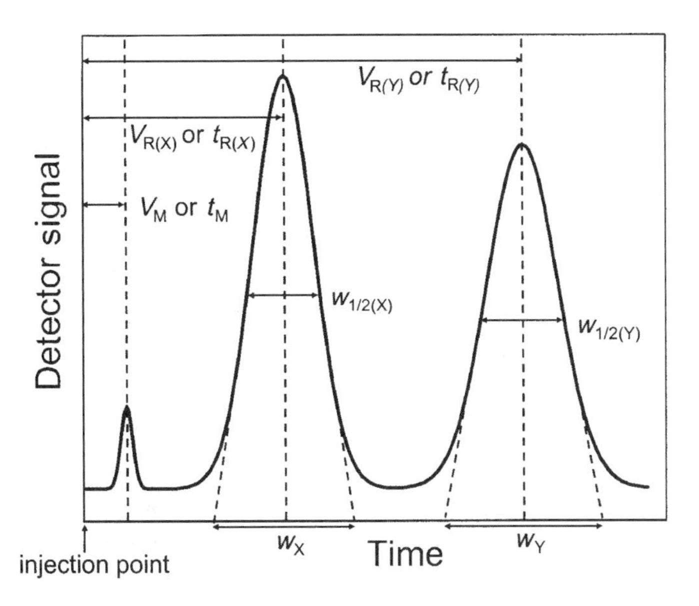

Figure 6.1.1 Schematic representation of a chromatogram showing various chromatographic parameters. Definition of each parameter is given in the text.

of the mobile phase versus concentration of the components. Such a plot is called a chromatogram (Fig. 6.1.1). Ideally, each component is represented in the chromatogram by a Gaussian distribution curve having a maximum along the concentration axis (i.e., the detector signal). LC is widely used for separation of biomolecules like proteins, nucleic acids, carbohydrates, etc. Gas chromatography is used for the separation of more volatile compounds, such as methyl esters of fatty acids, in which the mobile phase is a gas.

Partition or Distribution Coefficient

The principle of the chromatographic separation lies in the fact that following application of the sample when the mobile phase starts running through the column, the components in the sample distribute between the stationary and the mobile phase. Considering the simple case that the sample consists of two components **X** and **Y**, each of them will distribute between the mobile phase (mp) and the stationary phase (sp): $X_{mp} \rightleftharpoons X_{sp}$ and $Y_{mp} \rightleftharpoons Y_{sp}$.

At equilibrium, the equilibrium constant for the distribution of a solute component between the stationary phase and the mobile phase is known as the equilibrium partition coefficient or equilibrium distribution coefficient of that component specific for those pair of phases and at a specific temperature:

$$K_X = \frac{[X]_{sp}}{[X]_{mp}} \text{ and } K_Y = \frac{[Y]_{sp}}{[Y]_{mp}}$$

Equation 6.1.1

where $[X]_{sp}$ and $[Y]_{sp}$ are the concentrations of solutes **X** and **Y** in the stationary phase and $[X]_{mp}$ and $[Y]_{mp}$ are the concentrations of **X** and **Y** in the mobile phase. It is obvious from Equation 6.1.1 that the higher the value of K of a component, the higher will be the fraction of that component in the stationary phase. If $K_Y > K_X$, $[X]_{mp}$ will be higher than $[Y]_{mp,}$ as a result of which **X** will be eluted faster than **Y**.

The mechanism by which a solute particle is associated with the stationary phase is, however, different in different forms of chromatography. For example, the solute could be adsorbed on the surface of the solid stationary phase by van der Waal's forces (e.g., the interaction between the plant pigments and calcium carbonate in Mikhail Tswett's experiment) or through ionic interactions (e.g., separation of charged protein molecules using ion-exchange chromatography; see the Ion-Exchange Chromatography section later in this unit). It should be pointed out that the general theory of chromatography was developed on the basis of partitioning of small molecules between the stationary phase and the mobile phase, and thus may not be applicable for peptides and proteins. For example, separation of small molecules by RP-HPLC involves their continuous partitioning between the hydrophobic stationary phase and the mobile phase. On the other hand, peptides and proteins, due to their large size, are unable to do so. They adsorb to the hydrophobic stationary phase and remain adsorbed until the precise critical concentration of the organic modifier in the mobile phase is applied to cause desorption.

Elution Parameters

The quality of a chromatographic separation can be expressed by a number of parameters such that the optimization of these parameters individually or collectively would enhance the performance of the method. The relationships among such chromatographic parameters is described below. A representative chromatogram along with different notations used for these parameters is shown in Figure 6.1.1.

Retention time and retention volume of the mobile phase

In Figure 6.1.1, the first or the smallest peak denotes elution of a solute which is not retained in the column and moves through the column at the same speed as that of the mobile phase. This will be the case where the interaction of the solute with the stationary phase is extremely unfavorable, i.e., partition coefficient K equals to zero, since the concentration of the component in the stationary phase will be zero. The retention time corresponding to the maximum concentration (represented by the maximum of the Gaussian peak) of this solute is t_M. The retention time depends on the length of the column, as well as the flow rate of the eluting mobile phase. The corresponding volume of the mobile phase V_M is known as the void volume or dead space of the column and is equal to the volume of the mobile phase around and within the stationary phase particles in the column. Retention times are particularly of great value in analytical applications of chromatography, for example, if initial experiments show that a particular component is eluted at 5 min 45 sec in a defined chromatographic setup, then a peak eluting at the same time for similar sample can always be identified with that component.

6

Retention time and retention volume of sample components

The volume of the mobile phase required to elute a solute to its maximum concentration is called the ***retention volume***, V_R, and the corresponding time is known as the ***retention time***, t_R. If the flow rate of the mobile phase is v_M (in volume unit) per unit time, the following equation will measure the retention volume of component **X** (Fig. 6.1.1):

$$V_{R(X)} = v_M \times t_{R(X)}$$

Equation 6.1.2

Relationship between retention parameters and the partition coefficient

It is obvious from Figure 6.1.1 that at $t_{R(X)}$ half of the solute has eluted. The amount of component **X** in the corresponding retention volume is given by:

$$V_{R(X)} \times [X]_{mp}$$

Equation 6.1.3

The remaining half of component **X** is, however, still distributed between the stationary phase and the mobile phase around it within the column, i.e., V_M. Thus the amount of component **X** in the mobile phase is $V_M \times [X]_{mp}$ and in the stationary phase is $V_S \times [X]_{sp}$, where V_S is the volume of the stationary phase. The sum of these two quantities is equal to the quantity in Equation 6.1.3, as mentioned above. Therefore:

$$V_{R(X)} \times [X]_{mp} = V_M \times [X]_{mp} + V_S \times [X]_{sp}$$

Equation 6.1.4

or

$$(V_{R(X)} - V_M) = V_S \frac{[X]_{sp}}{[X]_{mp}}$$

Equation 6.1.5

As defined earlier, the quantity (Equ. 6.1.1) is the partition coefficient of component **X**. Thus we have:

$$(V_{R(X)} - V_M) = K_X V_S$$

Equation 6.1.6

Retention factor and selectivity factor

The ***retention factor*** or ***retention ratio*** (k) and the ***selectivity factor*** (α) are two important parameters often used to describe the migration of a solute in a column, as well as to determine the efficiency of a chromatographic column. The retention factor of component **X**, k_X is defined as:

$$k_X = \frac{V_{R(X)} - V_M}{V_M}$$

Equation 6.1.7

When k_X is equal to zero, i.e., $V_{R(X)}=V_M$, there will be no separation. The parameter k_X can also be written in a different form as follows (see Equ. 6.1.2):

$$k_X = \frac{t_{R(X)} - t_M}{t_M}$$

Equation 6.1.8

When k_X is very low, i.e., $t_{R(X)}$ is very close to t_M, a clear separation may not be obtained. Similarly, when k_X is very high, the elution time will be very high, which may not be advantageous for a process chromatography. For optimum separation k should be in the range between 1 and 10.

A more quantitative relationship between t_R and the length of the column (L) and the mobile phase velocity (v_M) can be derived from the preceding equation (Equ. 6.1.8):

$$k_X = \frac{t_{R(X)}}{t_M} - 1$$

Equation 6.1.9

or

$$1 + k_X = \frac{t_{R(X)}}{t_M}$$

Equation 6.1.10

Since $t_M \times v_M$ is equal to L:

$$t_{R(X)} = \frac{L}{v_M}(1 + k_X)$$

Equation 6.1.11

Thus, $t_{R(X)}$ is directly proportional to L and inversely proportional to v_M.

The quality of the chromatographic separation of two solutes is judged by their positions in the chromatogram and the efficiency with which they are separated. This can be quantitatively expressed in terms of properties of both the stationary phase and the mobile phase. The selectivity factor (α) is defined as the ratio of the retention factors of the respective components:

$$\alpha = \frac{k_Y}{k_X} = \frac{t_{R(Y)} - t_M}{t_{R(X)} - t_M}$$

Equation 6.1.12

The selectivity factor α is defined in such a way that it is always ≥ 1.0. However, if $\alpha = 1$, no separation is obtained. A value of α which may be slightly >1.0 (such as 1.02) can indicate a meaningful separation in some processes.

Resolution

The disadvantage with the selectivity factor α defined above is that it does not account for widths of the peaks, which are extremely important to assess the quality of separation. A quantity, R_S, has been defined to represent the resolution of two solutes in terms of their retention times (or retention volumes) and average peak widths (see Fig. 6.1.1). Thus resolution depends on two opposing forces: differential migration and zone spreading (see band broadening below).

$$R_S = \frac{t_{R(Y)} - t_{R(X)}}{0.5(w_Y + w_X)} = \frac{V_{R(Y)} - V_{R(X)}}{0.5(w_Y + w_X)}$$

Equation 6.1.13

To obtain the peak width, tangents are drawn at two inflection points of the Gaussian curve and then extended to the time (volume) axis as shown by dotted lines in the leading and trailing edges of the last two eluting peaks in Figure 6.1.1. The distance between two intersections at the baseline of a specific peak represents the corresponding peak width (denoted by w_X and w_Y in Fig. 6.1.1). Because of the Gaussian shape of the peaks, $R_S = 1.5$ is taken as standard for the two peaks to be clearly resolved (100% resolution) at the baseline (Fig. 6.1.2A). If $R_S = 1.0$, the resolution is ~98% complete, which corresponds to a 2% overlap of peak areas (Fig. 6.1.2B). The value of R_S determines the purity of two recovered components. A value of $R_S = 0.8$ is considered to be the lowest minimum for qualitative analysis. A good resolution, among other different operating

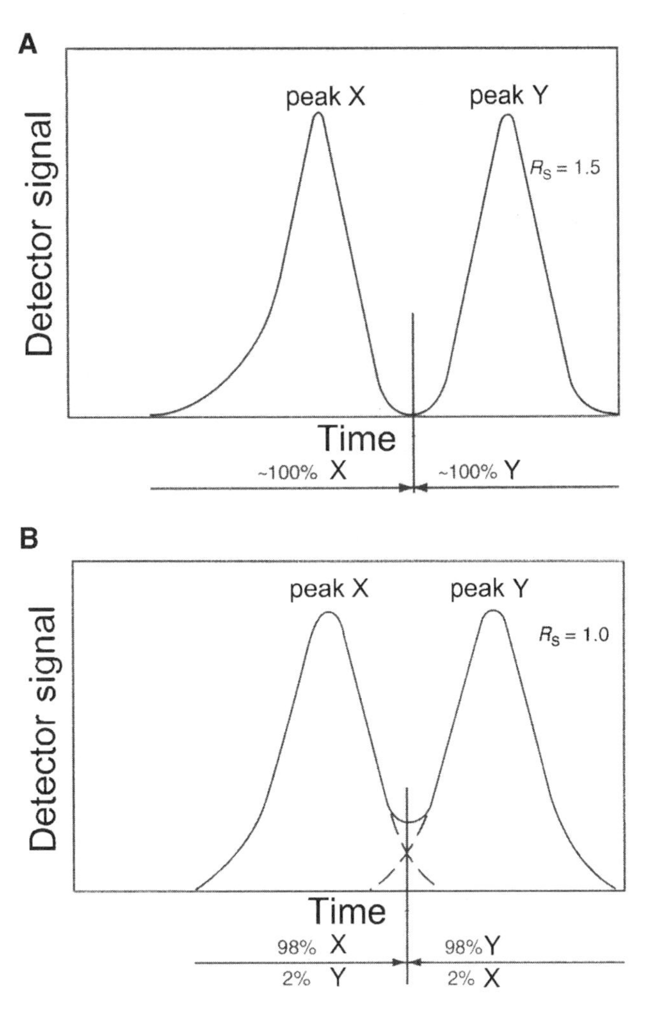

Figure 6.1.2 Measurement of chromatographic peak resolution. (**A**) Peaks **X** and **Y** are 100% separated. (**B**) Peaks **X** and **Y** are overlapped to some extent.

conditions such as temperature, flow rates, etc., is directly related to the property of the column itself.

Another form of resolution, as shown below, can be easily derived from the above equations. This incorporates a new column parameter, the number of theoretical plates in a column (discussed in Theoretical Plate Model of Chromatography, below):

$$R_S = \frac{N^{1/2}}{4}\left(\frac{\alpha - 1}{\alpha}\right)\left(\frac{\bar{k}}{1+\bar{k}}\right)$$

Equation 6.1.14

where $\bar{k}$ is the average capacity factor for two species and N is the number of theoretical plates. This equation is often used to optimize the column performance. The way the term N is optimized is discussed in the following section (Theoretical Plate Model of Chromatography, below). The retention factor can be improved by changing the mobile phase composition. The part of the selectivity factor is rather complex and can be optimized mostly by trial and error method. The most probable targets to optimize this part are mobile phase composition, column temperature, and stationary phase composition. The complexity of optimization of resolution increases with increasing number of components in the sample.

Theoretical Plate Model of Chromatography

A.J.P. Martin and R.L.M. Synge, the inventors of partition chromatography, developed the theoretical plate model of chromatography (Martin and Synge, 1941). To describe the widths of chromatographic peaks, this model hypothesizes that the chromatographic column consists of a large number of separate adjacent layers called theoretical plates (Fig. 6.1.3), and an instantaneous equilibrium distribution of solute molecules between the stationary and the mobile phases occurs in each plate. Sample components move down the column through the transfer of equilibrated mobile phase from one plate to the next. It should be pointed out that such plates do not actually exist, but they help us to understand how the separation process works inside the column. The theoretical plate model of column chromatography introduces two hypothetical parameters, ***plate number*** and ***plate height***, that are shown to dictate the column efficiency and consequently the quality of separation. The total number of theoretical plates in a column is denoted by N. It is hypothesized that with higher values of N, the efficiency of the column will usually be better. However, it will be described later that when the column is made longer, the width of the individual peaks also increase. This phenomenon is called ***band broadening***. Minimal band broadening leads to sharp symmetrical peaks that give rise to optimal separation. In addition, as the column is made longer, the elution time is also increased, which may not be desirable. Another parameter is the plate height (H) or the ***height equivalent to a theoretical plate*** (HETP). If the length of the column is L and the total number of plates is N, then:

$$\text{HETP} = \frac{L}{N}$$

Equation 6.1.15

The smaller the plate height, the better the resolution. The Gaussian distribution of a peak originates from the fact that the molecules of an analyte travel different distances along the column. The width of the peak is thus directly related to the standard deviation σ of the distance or that of the retention time. Statistical methods can be applied to calculate these standard deviations and relate them to the theoretical plate count N. Since the chromatographic parameters are more easily accessible than these standard deviations, the following equation relating N in terms of retention time and peak

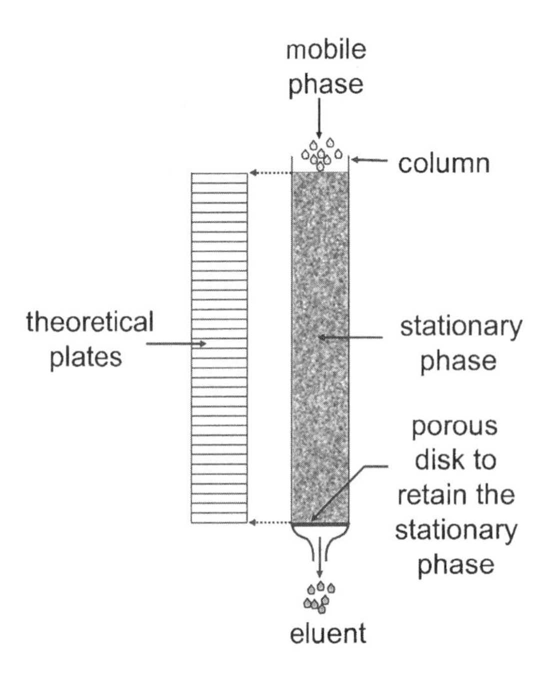

Figure 6.1.3 Schematic representation of the plate theory of chromatography.

width can be derived:

$$N = 16\left(\frac{V_R}{w}\right)^2 = 16\left(\frac{t_R}{w}\right)^2$$

Equation 6.1.16

Since the peak width and the plate number in the above equation (Equ. 6.1.16) follow an inverse relationship, narrow peak width requires higher plate numbers to increase the efficiency of the separation. Obviously a large plate number means an efficient column. In reality, however, $w_{1/2}$, the peak width at the half height of the Gaussian peak (Fig. 6.1.1), is more accurately measured than w. This is because the baseline width calculation is frequently associated with asymmetry at the leading or trailing edges of a peak. A more accurate representation of N in terms of $w_{1/2}$ can similarly be derived as follows:

$$N = 5.54\left(\frac{V_R}{w_{1/2}}\right)^2 = 5.54\left(\frac{t_R}{w_{1/2}}\right)^2$$

Equation 6.1.17

One of the purposes of theoretical plate model is to characterize chromatographic performance. Although retention time and peak width are solute-specific and the above equations imply different N for different solutes for the same column, it is assumed that approximately the same plate number can be observed for different solute peaks. Therefore, N can be fairly treated as the property of the column and is used to optimize column efficiency.

The Rate Theory of Chromatography

The theoretical plate model assumes that the equilibration time of a sample component between stationary and mobile phases within a plate is negligibly small. However, while the components are inside the column, the equilibration is not instantaneous. The solute takes finite time for equilibration. With longer time of equilibration, the dispersion of the solute becomes more active. This leads to band broadening, which plate theory is unable to explain. A more realistic relationship between HETP and the velocity of the mobile phase that takes into account the band broadening is offered by the van Deemter equation:

$$\text{HETP} = A + \frac{B}{v_M} + Cv_M$$

Equation 6.1.18

The above equation describes three important reasons of band broadening and thus defines column efficiency. The contributions of three important factors such as eddy diffusion (A), longitudinal diffusion (B), and the resistance to mass transfer (C) are described below (Fig. 6.1.4).

Eddy diffusion (A): A represents the contribution from eddy diffusion, which is independent of the mobile phase velocity. Different molecules of a particular component might take different paths to travel along the column, which causes band broadening (Fig. 6.1.4A). Therefore A depends on the diameter of the particles that makes the stationary phase. Finer particles reduce the nonuniformity of the stationary phase, leading to more uniform travel distances, which consequently reduce the eddy diffusion and increase the plate numbers. Since eddy diffusion is independent of v_M, the efficiency of a column cannot be improved further from a preset threshold value defined by the quality of packing by changing v_M.

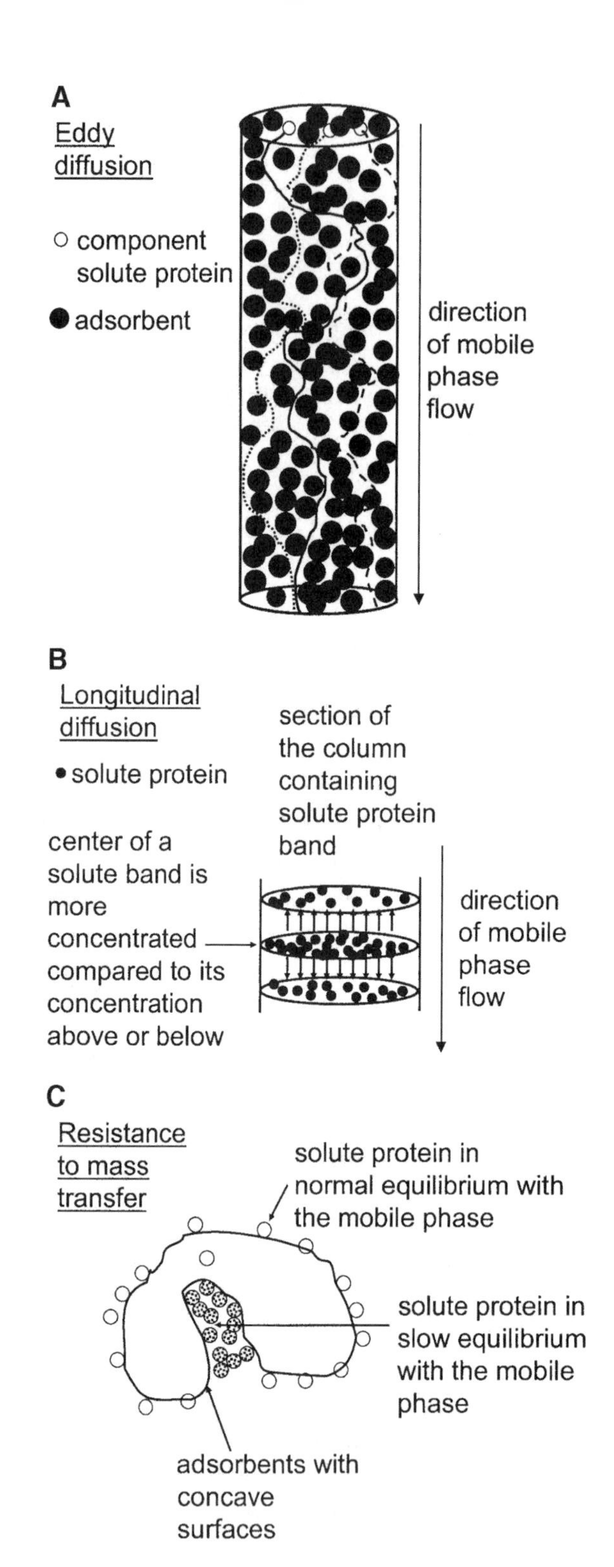

Figure 6.1.4 Schematic representations of three major band-broadening effects: (**A**) eddy diffusion, (**B**) longitudinal diffusion, and (**C**) resistance to mass transfer.

Longitudinal diffusion (B): B represents the longitudinal diffusion that originates from the concentration difference of a solute at the edges and the center of a band of sample (Fig. 6.1.4B). The longitudinal diffusion is a spontaneous process that tries to increase the system's entropy (the second law of thermodynamics). So it cannot be controlled one hundred percent but can be reduced by increasing v_M. The concentration is always greater at the center of a zone that could diffuse to the less concentrated regions ahead and behind that zone (Fig. 6.1.4B). Therefore, some of the molecules take longer to elute when the diffusion occurs opposite to the direction of the flow of the mobile phase, and some molecules take less time when the diffusion occurs in the same direction as that of the mobile phase and elute earlier than the average molecules. This term is directly proportional to the diffusion coefficient of the solute and inversely proportional to the mobile phase velocity. Because the diffusion is very slow in liquid, the term B is less important, provided the solutes do not stay too long inside the column, which is the case when the mobile phase velocity is high.

Resistance to mass transfer (C): C represents the resistance to mass transfer between two phases (Fig. 6.1.4C). Depending on the affinity to the stationary phase, a solute molecule either moves slower or faster along the column. Some molecules may become strongly adsorbed at the stationary phase by diffusing into the porous structures, and consequently the process of desorption for those molecules is slower than for those which are adsorbed at the surface. If the mobile phase velocity increases, those molecules, which are weakly adsorbed on the stationary phase surface, will elute faster. Consequently, there is a difference in elution times between the molecules that have entered the pores of the stationary phase and those that have not entered the pores, and the band broadening becomes more pronounced. Similar to term A, the term C also depends on the diameter of the stationary phase particle. Thus the term C depends on both the mobile phase flow rate and the stationary phase particle size.

Different terms of the van Deemter equation discussed above are plotted as HETP versus v_M separately in Figures 6.1.5A and 6.1.5B. The sum of the three plots gives the idea of the minimum plate height and the optimum mobile phase velocity at which the column efficiency is maximum.

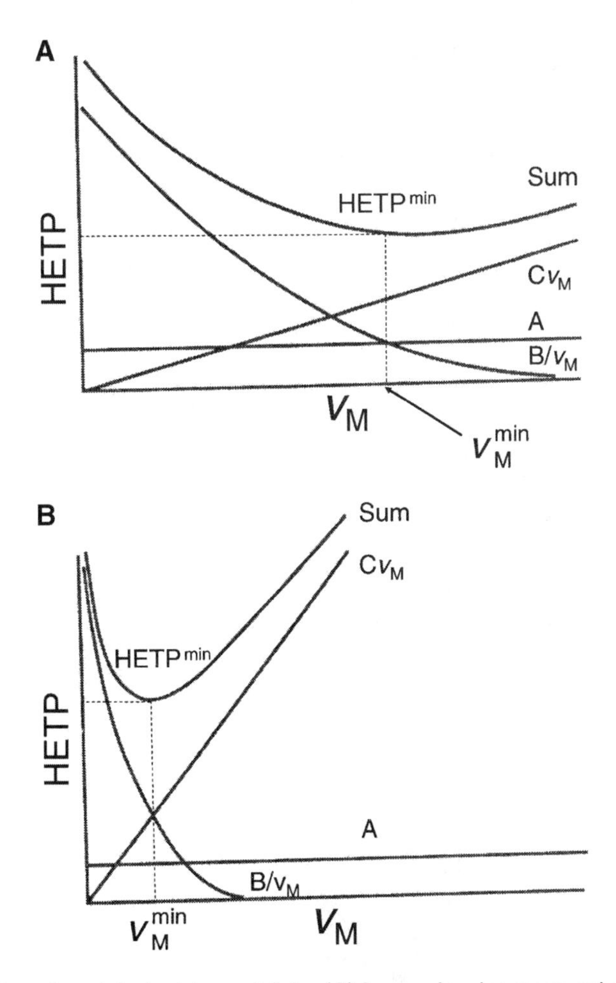

Figure 6.1.5 van Deemter plots (not to scale) for (**A**) low-molecular-mass solutes and (**B**) high-molecular-mass solutes, such as proteins. The definition and the effects of each term are given in the text. Adapted from Scopes (1987) with kind permission from Springer-Verlag.

This means that van Deemter equation is a practical guide to optimize the mobile phase velocity and the stationary phase property such as particle diameter.

However, in some practical processes the estimated optimum mobile phase velocity turns out to be very low. For example, optimal mobile phase velocity obtained for high-molecular-weight solutes is much lower than that for the low-molecular-weight solutes (Fig. 6.1.5B). Figure 6.1.6 compares overall HETP versus v_M plots for a small peptide and two large proteins. The mass transfer effects of large proteins have dramatically steep slopes. Moreover, none of the three molecules predicts any minimum v_M, which means that longitudinal diffusion is negligible in LC. A reduced mobile phase velocity,

$$v_{\mathrm{M}}^{red} = \frac{v_{\mathrm{M}} . d_{\mathrm{p}}}{D}$$

Equation 6.1.19

where d_p is the stationary phase particle diameter and D is the diffusion coefficient of the solute in the mobile phase, is recommended in such cases to take into account the effects of mass transfer and longitudinal parameters. This in turn can be used to decide a suitable mobile phase velocity. Giddings (1965) has offered a very similar but more complete expression. This expression includes more parameters that cause band broadening, in addition to the factors that are influenced by the velocity of the mobile phase (as described by the van Deemter equation):

$$\sigma = \sqrt{L\left(\frac{2D}{v_{\mathrm{M}}} + 2R(1-R)v_{\mathrm{M}}t_{\mathrm{d}} + 2\lambda d_{\mathrm{p}}\right)}$$

Equation 6.1.20

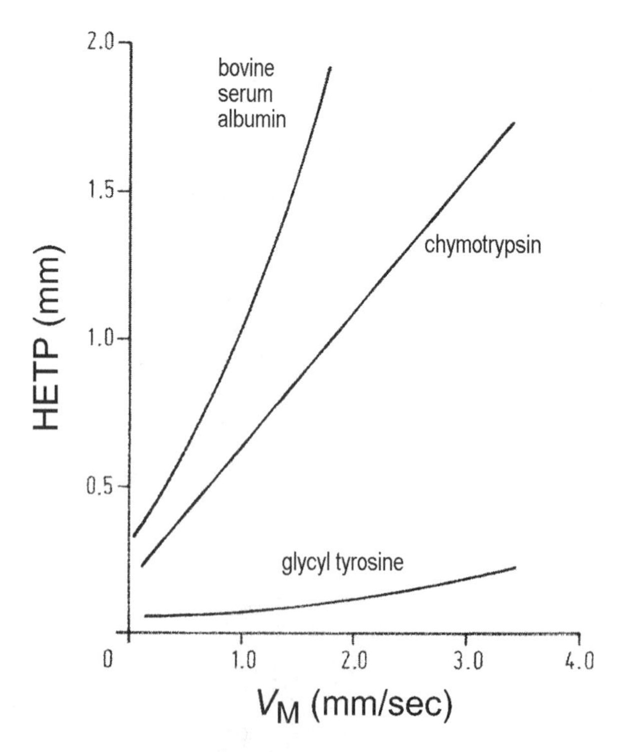

Figure 6.1.6 Real data presenting van Deemter plots of low-molecular-mass peptide and high-molecular-mass proteins. Adapted from Regnier and Gooding (1980), with kind permission from Elsevier, Ltd.

The various terms are: σ, standard deviation in retention time (4σ is approximately equal to the baseline peak width for a symmetrical Gaussian peak); D, the diffusion coefficient of the solute in the mobile phase; R, the ratio of zone velocity to v_M; t_d, average time spent by a solute between adsorption and desorption; λ, the geometric factor; and d_p, the stationary phase particle diameter. The terms L and v_M have been defined before. Similar to van Deemter equation as explained before, eddy diffusion, longitudinal diffusion, and resistance to mass transfer are the principal components of this equation as well.

Deviations in Case of Protein Chromatography

The plate theory is not very useful for chromatography of protein molecules. In specific cases where the composition of the mobile phase changes with time or gradient elution is used, variable adsorption or partition thermodynamics in different regions of the column become dominant. For protein separation, either of the above two methods is frequently used. Consequently, the plate theory becomes irrelevant. Even when isocratic elution is followed, some other reasons exist that do not comply with the basic assumptions of the plate theory. The primary reason is that the plate theory is based on the assumption that a linear adsorption isotherm exists between the solutes in the mobile phase and the solutes in the stationary phase. Moreover, each solute is assumed to bind only one adsorbent site in a reversible fashion. However, in reality, the adsorption isotherms of proteins are primarily highly nonlinear or curved. Proteins are composed of residues having a range of electrical polarity. Consequently, the interactions between protein molecules and the adsorbent particles are expected to be much stronger and involve multiple sites at the same time. The situation can be dealt from a biochemist's point of view such that in a column the protein-adsorbent interactions consist of a large number of association-dissociation kinetics instead of one-to-one adsorption-desorption processes that exist for simple molecules. Mathematically, this is a very complex situation to deal with. However, if some approximations are used, a working kinetic model of protein chromatography can be developed. In the following paragraph, the conceptual difference between the protein adsorption isotherms and that of small molecules is discussed.

Let us consider (Scopes, 1987) that multiple dissociation kinetics that exist between a solute protein and adsorbents can be approximated by a single "apparent" dissociation constant (K_P). Therefore, at equilibrium:

$$K_p = \frac{mp}{q}$$

Equation 6.1.21

where m is the concentration of free binding sites in adsorbent, p is the concentration of free protein in solution, and q is the concentration of bound protein. If total concentration of the adsorption sites and the proteins are m_t and p_t respectively, then K_P can be expressed in terms of a modified partition function ϕ as follows:

$$p_t^2\phi - \phi(m_t + p_t + K_p) + m_t = 0$$

Equation 6.1.22

where

$$\phi = \frac{q}{p_t}$$

Equation 6.1.23

It is obvious that ϕ lies between 0 and 1. It is known that the partition function is an important parameter in the determination of individual peak positions of components in a chromatogram. An

6

estimate of ϕ can be obtained from the knowledge of K_P, m_t, and p_t. The above relationship between ϕ and K_P is important for optimizing column efficiency. The expression of K_P in terms of column parameters above suffers from some other serious issues that have not been taken into account. First, some accessible adsorbent sites delay partition equilibrium in such a way that a sizeable amount of protein remains in the column during rapid elution. Second, larger proteins may block access to adsorbent sites that are further buried. Such unsaturated adsorption pattern in practice would result in faster adsorption kinetics than that predicted from the theoretical model. Third, inhomogeneity of adsorbents could cause a wrong estimate of the total binding sites of the adsorbent (m_t), which is calculated from the average protein binding per unit volume. This, in turn, calculates a higher value of ϕ. Despite having many such drawbacks, the above adsorption kinetics is still a valuable model for protein chromatography.

COMMON TYPES OF CHROMATOGRAPHY FOR PURIFICATION OF PROTEINS AND UNDERLYING PRINCIPLES

Three commonly used chromatographic techniques for purification of proteins will be discussed in this unit. These include: size exclusion chromatography, ion exchange chromatography, and affinity chromatography.

Size Exclusion Chromatography

In size exclusion chromatography (SEC), molecules are separated according to their size and shape. The method is also known as gel filtration chromatography, molecular sieve chromatography, or gel permeation chromatography (GPC). The gels that constitute the "stationary phase" usually consist of beads of a hydrated sponge-like material containing pores through which molecules over a specific range of size and shape can permeate. Molecules that are too large to pass through the pores are excluded from the solvent present inside the gel beads and are eluted first. Larger molecules pass through the column in a smaller eluent volume, i.e., more rapidly than the smaller molecules that enter and pass through the solvent present within the pores of the gel (Fig. 6.1.7). The minimum

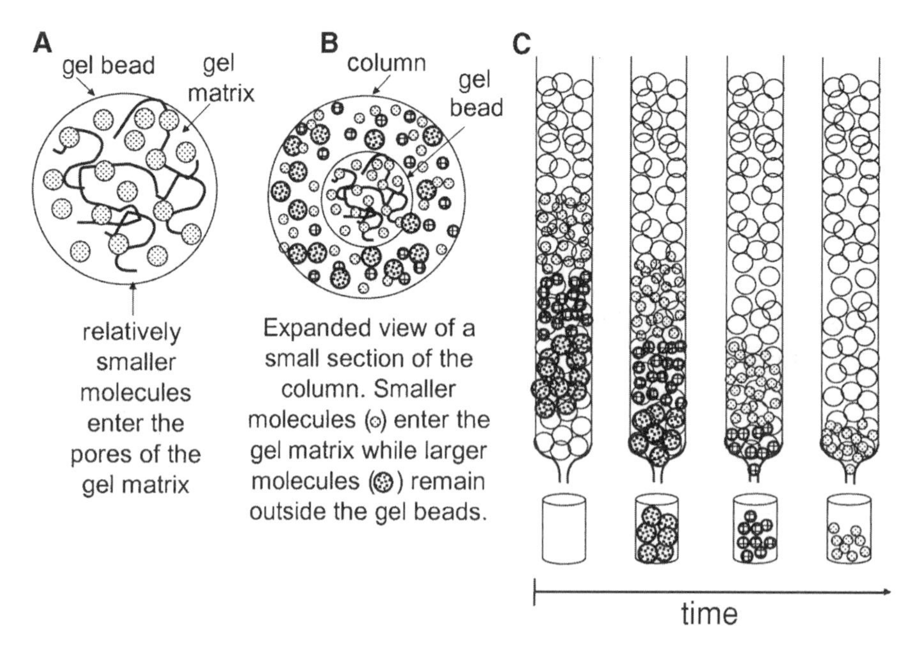

Figure 6.1.7 Schematic representations of the principle of gel filtration chromatography: (**A**) enlarged view of a hypothetical gel bead, its polymeric matrix network, and the access of relatively smaller molecules inside the bead; (**B**) a small section of the column representing a gel bead and its surrounding; (**C**) a three-component solute mixture is separated stepwise based on the sizes (larger molecules are separated first followed by the smaller molecules).

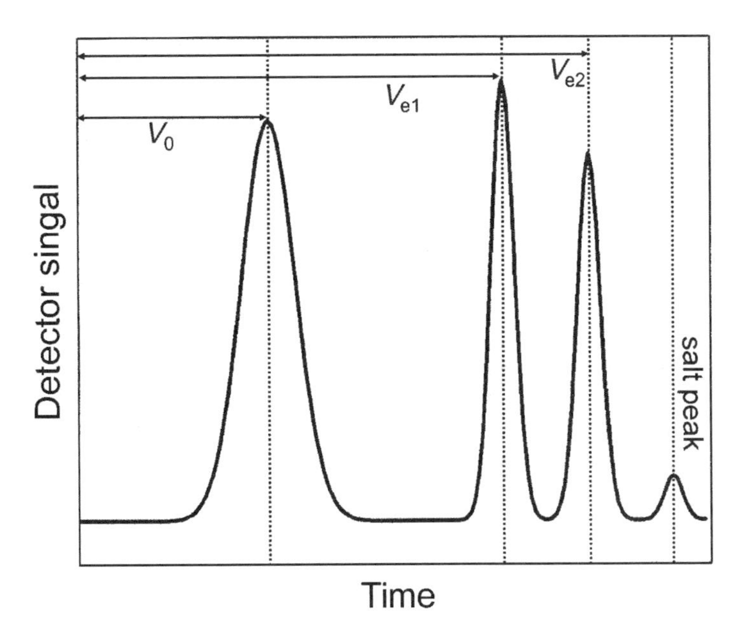

Figure 6.1.8 Schematic representation of a theoretical chromatogram of gel filtration chromatography. Definitions of parameters have been provided in the text.

molecular mass of the sample molecule unable to penetrate the pores of a given gel is known as the *exclusion limit* of the specific gel. The exclusion limit, to some extent, is a function of the shape of the sample molecule as well.

Let us consider that:

V_x = volume occupied by the gel beads

V_o = volume of the solvent space surrounding the gel beads or the void volume of the column

V_t = total volume of the gel beads in the column or the total bed volume of the column

Then $V_t = V_x + V_o$

The value of V_t can be determined from the internal diameter of the column and the height of the gel bed in the column. In general, V_o is one-third of V_t. For accurate determination of V_t and V_o, see UNIT 6.2. The elution volume of a molecule, V_e, from the column can also be determined (Fig. 6.1.8). For a particular molecule on a given gel, V_e/V_o, which is the relative elution volume, is independent of the size of the particular column used. The plot of the relative elution volume versus the logarithm of molecular masses for a variety of proteins on a Sephadex G-200 column at pH 7.5 is shown (Fig. 6.1.9). The G-types of Sephadex (such as 10, 15, 25, 50, 75, 100, 150, and 200) differ in their degree of cross-linking, and thus in their degree of swelling and fractionation range of molecules in terms of molecular weight. Some commonly used gel materials used in size exclusion chromatography in conventional LC and their protein fractionation range are shown in Table 6.1.1.

Ion-Exchange Chromatography

The principle of ion exchange chromatography is based on reversible electrostatic attraction of a charged molecule, for example a protein, to a solid matrix which contains covalently attached side groups of opposite charge. Proteins are charged molecules and each protein has a net charge of zero at its characteristic isoelectric point or pI (Fig. 6.1.10). A protein carries a net positive charge when present in a solution more acidic than its pI and a net negative charge when present in a solution more alkaline than its pI. In ion exchange chromatography, charged protein molecules in the

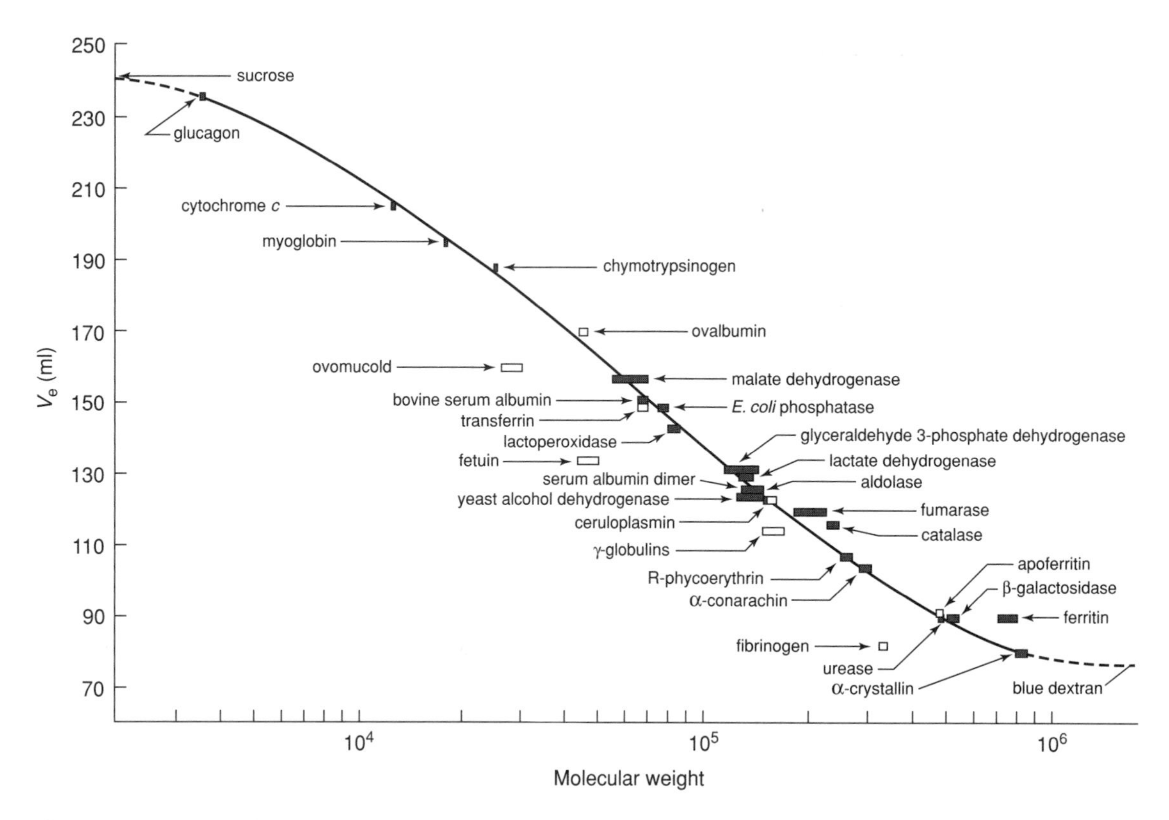

Figure 6.1.9 Use of gel filtration chromatography to determine the molecular mass of proteins. The plot demonstrates the relative elution volume versus the logarithm of molecular mass for a variety of proteins on a Sephadex G-200 column at pH 7.5. Adapted from Andrews (1965) with kind permission from The Biochemical Society.

Table 6.1.1 List of Commonly Used Gel Filtration Materials

Name[a]	Type	Fractionation range (Da)[b]
Sephadex G-10	Dextran	$<7 \times 10^2$
Sephadex G-25	Dextran	1×10^2 to 5×10^3
Sephadex G-50	Dextran	5×10^2 to 1×10^4
Sephadex G-100	Dextran	4×10^3 to 1.05×10^5
Bio-Gel P-2	Polyacrylamide	1×10^2 to 1.8×10^3
Bio-Gel P-6	Polyacrylamide	1×10^3 to 6×10^3
Bio-Gel P-10	Polyacrylamide	1.5×10^3 to 2×10^4
Bio-Gel P-30	Polyacrylamide	2.4×10^3 to 4×10^4
Bio-Gel P-100	Polyacrylamide	5×10^3 to 1×10^5
Sepharose 4B	Agarose	6×10^4 to 2×10^7
Sepharose 6B	Agarose	1×10^4 to 4×10^6
Sepharose CL-2B	Agarose	7×10^4 to 4×10^7
Sepharose CL-4B	Agarose	6×10^4 to 2×10^7

[a]Sephadex and Sepharose gels are products of GE Healthcare Bio-Sciences; Bio-Gel gels are products of Bio-Rad Laboratories.
[b]The ranges shown are usually applicable for peptides and proteins.

6

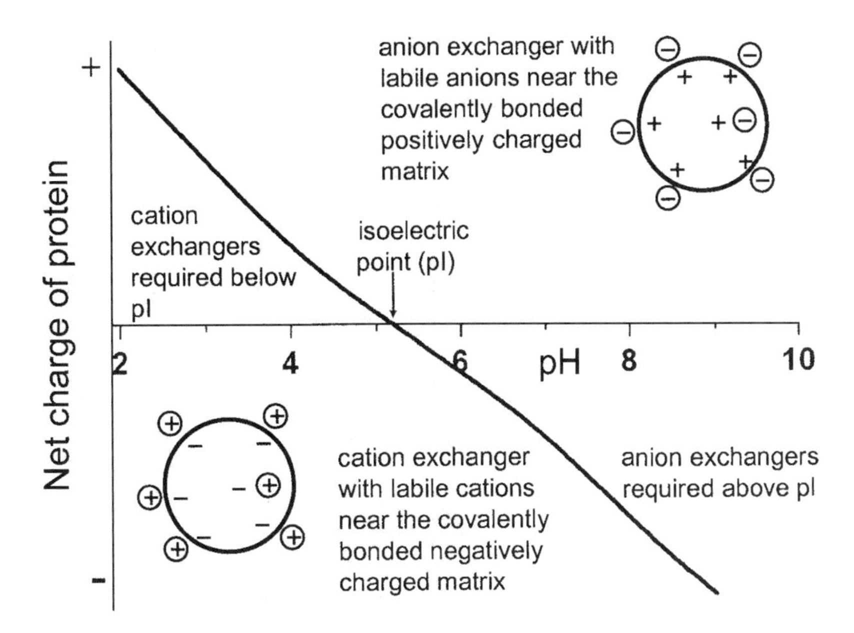

Figure 6.1.10 Selection of ion exchangers in ion-exchange chromatography. Net residual positive charge requires cation exchanger whereas net residual negative charge requires anion exchanger.

sample are allowed to bind through ionic interactions to an ion exchanger at a specific pH and salt concentration. This is usually followed by differential elution of proteins bound to the ion exchanger using increasing salt concentration or pH, or a combination of both. Each component protein in the sample is eluted at a specific salt concentration or pH that destabilizes the ionic interaction.

There are two different kinds of ion exchangers: *anion exchangers* and *cation exchangers* (Fig. 6.1.10). Anion exchangers contain covalently attached positively charged groups that bind anionic proteins, i.e., proteins with a net negative charge. On the other hand, cation exchangers contain covalently attached negatively charged groups that bind cationic proteins, i.e., proteins with a net positive charge. In addition, ion exchangers are also classified as strong or weak. Strong ion exchangers remain ionized over a wide range of pH, whereas weak ion exchangers are ionized within a narrow pH range. The principle of ion exchange chromatography using an anion exchanger is schematically shown in Figure 6.1.11.

For use in ion-exchange chromatography an ideal support matrix should be inert, structurally rigid to facilitate scale-up, and operate at high flow rates, as well as be highly porous. When packed in large columns, soft matrices often pack closely together, i.e., become compact, even under their own weight, which may lead to reduction of mobile phase flow rates. Porous gels allow proteins to enter their internal matrix structure, which is desirable since it increases the binding capacity of the column. Commonly used support matrices include agarose (Sepharose), cellulose, dextran (Sephadex), silica, various organic polymers (such as Fractogel, a cross-linked polymethacrylate), cross-linked agarose (Sepharose CL), etc. Recently tentacle-type ion exchangers have been introduced that consist of hydrophilic support beads to which linear charged flexible polymers, the tentacles, are attached. This provides a large amount of sterically accessible ligands for the binding of biomolecules without any steric hindrance. Biomolecules in the sample are thus much more tightly bound during the separation process.

Elution of sample components from ion exchangers is often carried out by a gradual increase of salt concentration in appropriate buffer. This can be achieved by:

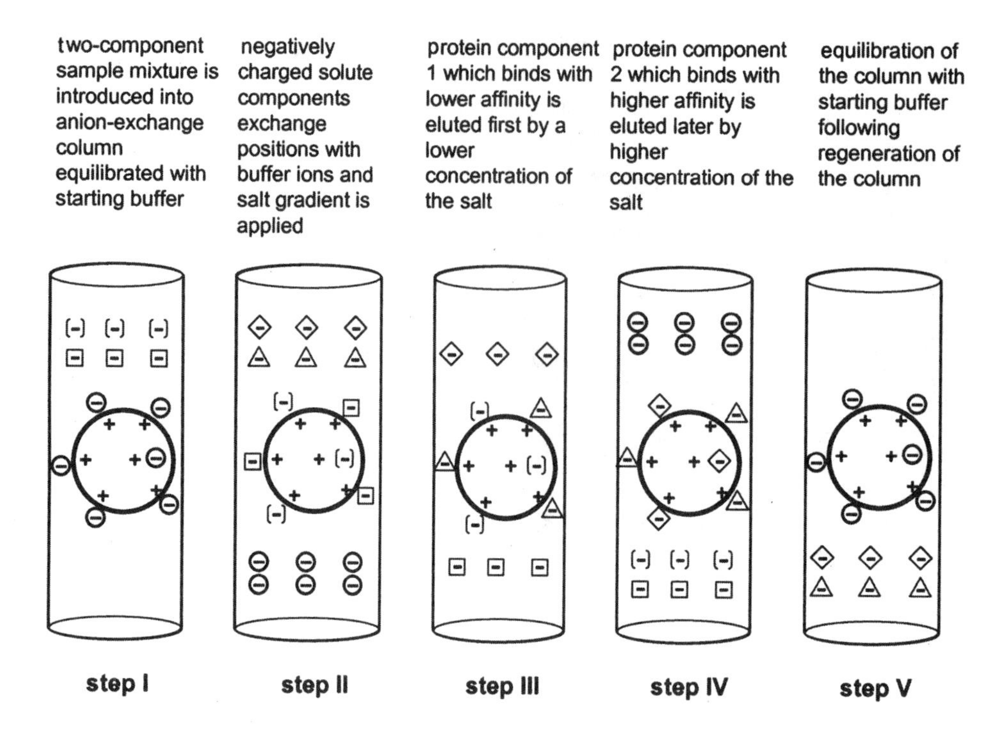

Key to symbols used:

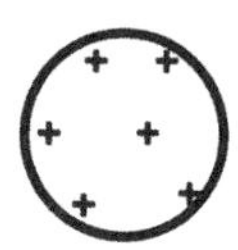

covalently bonded positive ions present in the stationary phase of an anion exchanger

⊖ starting buffer ions

⊟ protein component 1 in the sample which binds to the stationary phase with low affinity and is eluted by salt ions at lower concentration

[-] protein component 2 in the sample which binds to the stationary phase with high affinity and is eluted by salt ions at higher concentration

△ salt ions at lower concentration

◇ salt ions at higher concentration

Figure 6.1.11 Schematic representation of the principle of ion-exchange chromatography. An anion exchange column has been considered as an example. The column is preequilibrated with the starting buffer. As a result, the starting buffer ions (represented by a negative charge enclosed within a circle) are now the counter-ions of the covalently bonded positive ions on the stationary phase (represented by the large circle). Step I: A two-component protein mixture (Protein component 1 represented by a negative charge enclosed within a square and Protein component 2 represented by a negative charge enclosed within a square bracket) is introduced into the column. Step II: Protein molecules replace the labile buffer ions and become bound to the stationary phase. At this point, the mobile phase with salt gradient is applied to the column. This is shown by salt ions at lower concentration (represented by a negative charge enclosed within a triangle) which is followed by salt ions at higher concentration (represented by a negative charge enclosed within a diamond). Step III: Salt ions at a lower concentration replace the proteins bound with lower affinity (Protein component 1) leading to their elution from the column. Step IV: The protein bound with higher affinity (Protein component 2) is eluted when higher salt concentration is reached. After completion of elution of the proteins from the column, it is regenerated by washing with very high salt concentration to remove any remaining bound sample components or other impurities (step not shown). Subsequently the column is washed and equilibrated with the starting buffer. Step V: The column is now ready for next batch of separation.

a. Step elution: In this method, sample components are eluted using successive steps of higher concentration of a salt (such as NaCl) in an appropriate buffer solution. For example, following loading the sample on an anion exchanger, the column may be eluted with, say 50 ml of the buffer solution 20 mM Tris·Cl (pH 8.0) containing 25 mM NaCl. In step elution this will be followed by successive elutions with 50 ml each of the same 20 mM Tris·Cl (pH 8.0) buffer solution containing 50 mM, 100 mM, 150 mM, 200 mM, and 250 mM NaCl, in order.

b. Gradient elution: In this method sample components are most commonly eluted using a continuous linear gradient from lower to higher concentration of a salt (such as NaCl) in an appropriate buffer solution. For example, instead of being eluted with successive steps of 50 ml each of six different salt concentrations as described in (a) above, the column will be eluted with a continuous linear gradient from 20 mM Tris·Cl (pH 8.0)/25 mM NaCl to 20 mM Tris·Cl (pH 8.0)/250 mM NaCl in a total volume of 50 ml × 6 = 300 ml. Methods for generating linear concentration gradients are described in *UNIT 6.2*.

Figure 6.1.12 shows a graphical representation of the example used above. Some commonly used ion exchangers are summarized in Table 6.1.2.

Affinity chromatography

The principle of affinity chromatography is based on specific and reversible binding of a protein [**P**] to an immobilized ligand [**L**]. The specific ligands may be:

Substrate analog for an enzyme protein.

Inhibitor of an enzyme protein.

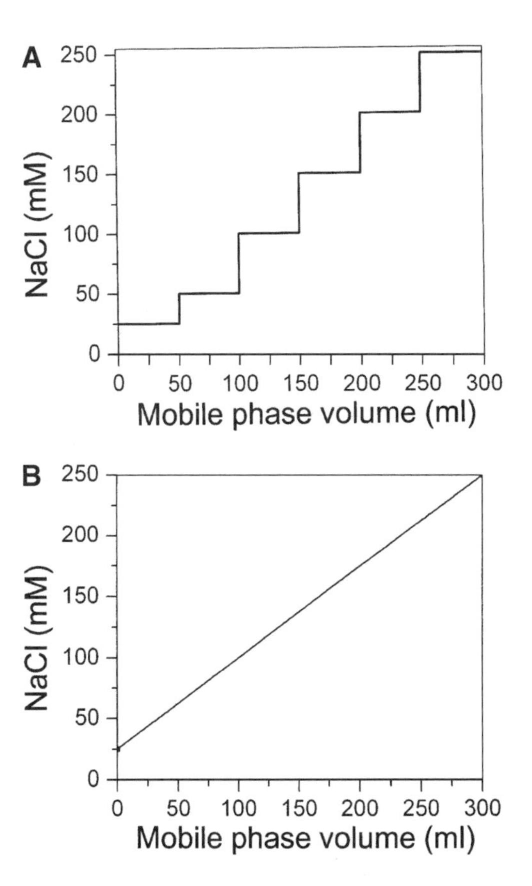

Figure 6.1.12 Plots of two different elution modes which are used commonly: (**A**) step elution and (**B**) gradient elution.

Table 6.1.2 List of Commonly Used Ion Exchangers

Name[a]	Type	Ionizable group
DEAE-Cellulose	Basic	$-CH_2CH_2N(C_2H_5)_2$
CM-Cellulose	Acidic	$-CH_2COOH$
P-Cellulose	Strongly and weakly acidic	$-OPO_3H_2$
DEAE-Sephadex	Basic cross-linked dextran gel	$-CH_2CH_2N(C_2H_5)_2$
CM-Sephadex	Acidic cross-linked dextran gel	$-CH_2COOH$
Bio-Gel CM 100	Acidic cross-linked polyacrylamide gel	$-CH_2COOH$

[a]Sephadex gels are products of GE Healthcare Bio-Sciences; Bio-Gel gels are products of Bio-Rad Laboratories. The abbreviations are: DEAE: Diethylaminoethyl; CM: Carboxymethyl; P: Phosphate.

Specific binding protein. For example, Protein A is a cell wall component produced by several strains of the bacteria *Staphylococcus aureus*. This protein binds specifically to the Fc region of antibody molecules, particularly to those of immunoglobulin type G (IgG). Immobilized Protein A is routinely used for purification of IgG from the serum of many different species as well as monoclonal IgG produced by hybridoma cultures.

Antibody specific for the protein. This method is known as *immunoaffinity chromatography*. Purified polyclonal antibodies having a high specificity for the protein of interest or purified monoclonal antibodies raised against the particular protein are extremely useful ligands for immunoaffinity purification of the protein.

Immobilized metal ions. This method is specifically known as *immobilized metal affinity chromatography (IMAC)*.

Synthetic dye molecule. This method is specifically known as *dye affinity chromatography*.

The steps involved in affinity chromatography are as follows and are shown in Figure 6.1.13:

1. *Preparation of the affinity purification matrix by immobilization or coupling of the ligand to a suitable support matrix.* A number of support matrices, such as agarose (Sepharose), cellulose, silica, various organic polymers, cross-linked agarose (Sepharose CL), etc., are available for immobilization of ligands for affinity chromatography. The support matrix should be physically and chemically stable, should have sufficient rigidity to allow high mobile phase flow rates, should allow chemical derivatization for immobilization of ligands but at the same time be inert enough to refrain from contributing to non-specific binding of proteins, should be inexpensive, and should have a high degree of porosity—with pores sufficiently large to allow free entry of protein molecules to internally bound ligands that would increase the binding capacity of the column. A large number of chemical methods are available for immobilization of the ligands to the matrix. The choice is primarily dictated by the available chemical groups on the support matrix and that in the ligand.

2. *Adsorption or binding of the specific sample components to the affinity matrix.* The sample contained in an appropriate buffer solution is applied in general to a column containing the specific affinity chromatography matrix. During this the specific protein in the sample binds to the immobilized ligand in the affinity matrix and all other proteins (impurities) are eluted out.

3. *Elution of the bound protein from the affinity matrix.* The bound protein is subsequently eluted from the affinity matrix column using any one of a variety of methods that does not affect the activity of the protein. These include elution with acidic buffers, buffers containing chaotropes such as 6 M guanidine hydrochloride, or 4 to 8 M urea, etc.

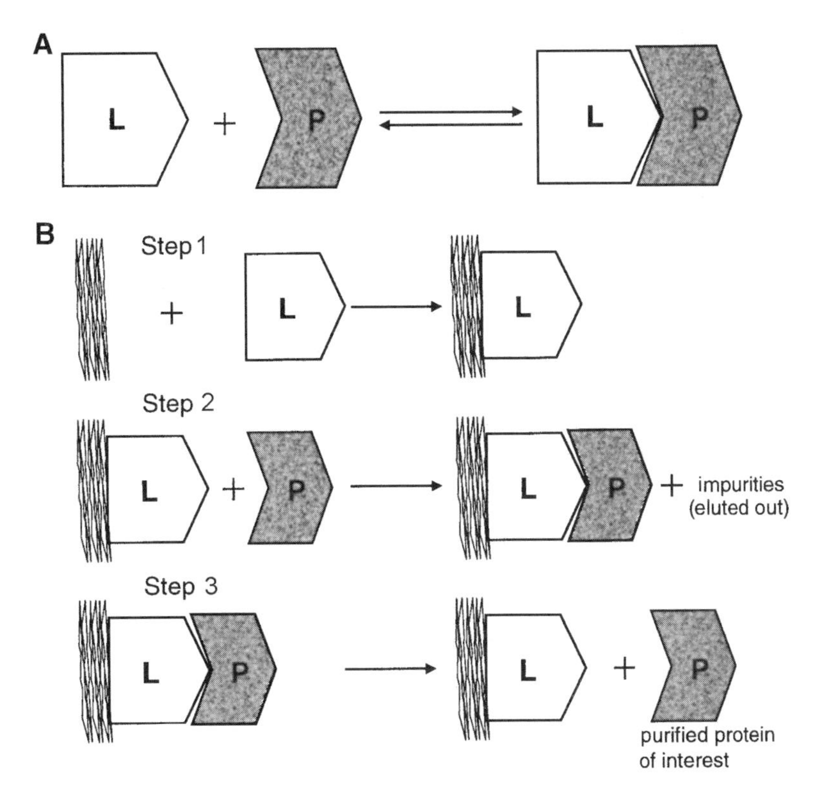

Figure 6.1.13 (**A**) Specific and reversible binding of a protein (P) to a ligand (L); (**B**) Three main stages for purification of proteins using affinity chromatography. In Step 1, the desired ligand is immobilized to a stable support matrix. In Step 2, the protein sample along with the associated impurities is introduced into the column. The protein molecules are adsorbed by the ligand whereas the impurities are eluted out. In Step 3, the adsorbed protein molecules are eluted and the desired purification is achieved in a single step.

In essence, whereas mainly ionic interactions are involved in ion-exchange chromatography, ionic, as well many other types of noncovalent interactions, may be involved in affinity chromatography. The elution buffer destabilizes these interactions.

Affinity chromatography in general allows a high degree of purification and yield to be obtained in a single step. This may not usually be matched by other chromatographic methods for purification of proteins. Although affinity chromatography is an ideal technique for research purposes, for commercial production scale purification of proteins it has some drawbacks. Biospecific ligands may be quite expensive, the stability of the ligands may be limited, methods for coupling ligands to the matrix may involve complex chemistries that may require hazardous or costly chemicals, or leaching of coupled ligands may reduce the capacity of the column and may require further steps for purification of traces of ligands from the purified protein product.

LITERATURE CITED

Andrews, P. 1965. The gel-filtration behaviour of proteins related to their molecular weights over a wide range. *Biochem. J.* 96:595-606.

Axén, R., Porath, J., and Ernback, S. 1967. Chemical coupling of peptides and proteins to polysaccharides by means of cyanogen halides. *Nature* 214:1302-1304.

Giddings, J.C. 1965. Dynamics of Chromatography. Part 1. Principles and Theory, Mercel Dekker Inc., New York.

Hirs, C.H.W., Stein, W.H., and Moore, S. 1951. Chromatography of proteins. Ribonuclease. *J. Am. Chem. Soc.* 73:1893.

Hjerten, S. 1962. Chromatographic separation according to size of macromolecules and cell particles on columns of agarose suspensions. *Arch. Biochem. Biophys.* 99:466-475.

Hjerten, S. and Mosbach, R. 1962. Molecular-sieve chromatography of proteins on columns of cross-linked polyacrylamide. *Anal. Biochem.* 3:109-118.

Hjerten, S., Levin, O., and Tiselius, A. 1956. Protein chromatography on calcium phosphate columns. *Arch. Biochem. Biophys.* 65:132-155.

Janson, J.C. and Jönsson, J.Å. 1989. Introduction to chromatography. *In* Protein Purification (J.C. Janson and L. Ryden, eds.) pp. 35-62. VCH Publishers, Inc., New York.

Lindsay, S. 1992. High Performance Liquid Chromatography. 2nd ed. John Wiley & Sons Ltd., London.

Martin, A.J.P. and Synge, R.L.M. 1941. A new form of chromatography employing two liquid phases: A theory of chromatography. 2. Application to the micro-determination of the higher monoamino-acids in proteins. *Biochem. J.* 35:1358-1368.

Paléus, S. and Neilands, J.B. 1950. Preparation of cytochrome *c* with the aid of ion exchange resin. *Acta. Chem. Scand.* 4:1024-1030.

Peterson, E.A. and Sober, H.A. 1956. Chromatography of proteins. 1. Cellulose ion-exchange adsorbents. *J. Am.Chem. Soc.* 78:751-755.

Porath, J. and Flodin, P. 1959. Gel filtration: A method for desalting and group separation. *Nature* 183:1675-1679.

Regnier, F.E. and Gooding, K.M. 1980. High-performance liquid-chromatography of proteins. *Anal. Biochem.* 103:1-25.

Scopes, R.K. 1982. Protein Purification: Principles and Practice. Springer-Verlag, New York.

Willard, H.H., Merritt, L.L., and Dean, J.A. 1974. Instrumental methods of analysis. 5th ed. D. Van Nostrand Company, New York.

Wilson, K. and Walker, J. 1994. Principles and techniques of practical biochemistry. 4th ed. Cambridge University Press, Cambridge.

UNIT 6.2

Column Chromatography

Bulbul Chakravarti,[1] Buddhadeb Mallik,[1] and Deb N. Chakravarti[1]
[1]Proteomics Center, Keck Graduate Institute of Applied Life Sciences, Claremont, California

OVERVIEW

Development of chromatographic methods for the separation and purification of biomolecules like proteins from natural or recombinant sources has been at the forefront of the progress made during the last forty to fifty years in molecular biology and biotechnology. The most common form in which chromatography, particularly chromatography of proteins, is performed in the laboratory uses tubular columns in which the stationary phase is packed (Fig. 6.2.1). In contradistinction to column chromatography, a batch process involves addition of solid stationary phase to the liquid sample followed by removal of liquids eluted under appropriate conditions by decantation, centrifugation (*UNIT 5.1*), or filtration. Although more complicated to set up than the batch process, usually column chromatography leads to higher resolution and is thus more efficient in terms of purification of proteins. Column chromatography occupies a central role in many successful protein purification methods developed and practiced in modern life science laboratories. In column chromatography, one end of the column contains a meshed or perforated disc, called the frit (Fig. 6.2.1), which retains the stationary phase within the column while allowing liquids to elute through it. Thus, a chromatographic column can be envisioned as a continuous filtration device. The liquid eluted from the column, the eluent, is channeled through a stopcock followed by flexible plastic

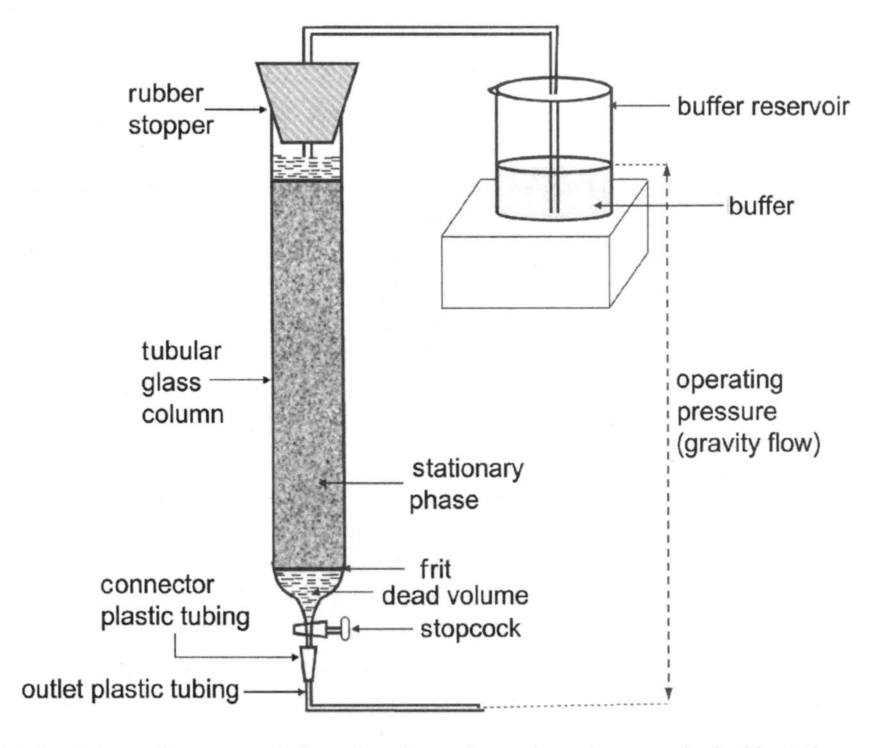

Figure 6.2.1 Schematic representation of a chromatography column packed with stationary phase. In this typical example of a gravity flow system for elution of the column, the eluent buffer reservoir is placed at a higher level than the column outlet. The flow rate in such systems can be regulated by changing the operating pressure, which is dependent on the difference of height between the level of the eluent in the buffer reservoir and the outlet of the column. Different parts of the column and necessary associated accessories are shown as well.

tubing (Fig. 6.2.1) for collection of separated fractions using an automated fraction collector. The protein content in these fractions can be monitored either online with a flow-through detector system placed between the column and the fraction collector or offline using a spectrophotometer. Although specific features such as sample application or conditions for elution may be different, this general set up is applicable for any type of laboratory-based column chromatographic technique, such as size exclusion chromatography, ion exchange chromatography, affinity chromatography, etc. for purification of proteins. (See *UNIT 6.1* for a discussion of the principles underlying these types of chromatography.) Availability of improved stationary phases for chromatography with high efficiency and selectivity, as well as automated instruments, have greatly aided the field of protein purification, which interestingly is still considered by many to be more of an art than science.

In column chromatography, elution of the proteins retained by the stationary phase with the mobile phase can be achieved by the application of either high pressure or low pressure. In general, liquid chromatography carried out at pressures <5 bar (1 bar = 14.5 pounds per square inch) is considered as low pressure, that between 6 and 50 bar as medium pressure, and in excess of 50 bar as high pressure (Wilson and Walker, 1994). However, this unit will focus on column chromatography carried out with low pressure or gravity flow for performing three different chromatographic techniques most commonly used for separation of proteins, size exclusion chromatography (separation based on size and shape of the molecules), ion exchange chromatography (separation based on net charge of the molecules), and affinity chromatography (separation based on specific affinity for immobilized ligands, inhibitors, receptors, antibodies, etc.). The chromatographic conditions for each of these separation methods are quite different. If an affinity chromatography method having great specificity for a specific protein is available, it has the ability to produce, in a single step, a highly purified preparation of that protein from a mixture of proteins contained in a crude biological extract. However, in general, several different chromatographic methods, one followed by the other, are usually necessary to obtain a protein in a highly purified form from such extracts. At the end of each of these steps, the partially purified protein of interest is detected in the separated column chromatography fractions, using specific methods of detection, such as enzyme activity; the active fractions are pooled, then concentrated if necessary, followed by application of another column chromatography method. It is a common practice to use affinity chromatography or ion-exchange chromatography during the initial phases of purification of a protein, while gel filtration is used later in the process. For example, human serum has a very high protein content of ~60 to 80 mg/ml and contains hundreds of proteins. A rather minor constituent of human serum is a protein known as human complement protein C6, which is normally present at a concentration of ~11 μg/ml. To purify C6 from human serum, three different column chromatography steps are used. The first step involves an immunoadsorbent affinity chromatography step, in which antibodies that specifically bind to human C6 have been immobilized on Sepharose CL-4B using cyanogen bromide activation (see *UNIT 6.1*). Human serum is passed through this column of anti-C6-Sepharose CL-4B followed by washing with the appropriate buffer to remove nonspecifically adsorbed proteins. C6 fractions eluted from the column are pooled and purified further using DEAE-Sephacel ion exchange chromatography. For this step, a linear gradient of 0 to 0.4 M NaCl in appropriate buffer is used. Finally, the pooled C6 fractions from the ion exchange chromatography step are purified by size-exclusion chromatography using Sephacryl S-200 (superfine) to obtain C6 that is essentially homogeneous. The characterization of homogeneity of a protein is usually carried out by sodium dodecyl sulfate-polyacrylamide gel electrophoresis (SDS-PAGE; *UNIT 7.3*). Absence of heterogeneity is the generally accepted criterion of homogeneity. Demonstration of homogeneity by SDS-PAGE is used very widely. The above process for purification of human complement protein C6 has been described in details by Chakravarti and Muller-Eberhard (1988); SDS-PAGE analysis of purified C6 showing homogeneity, i.e. absence of other protein contaminants, is shown in Figure 6.2.2.

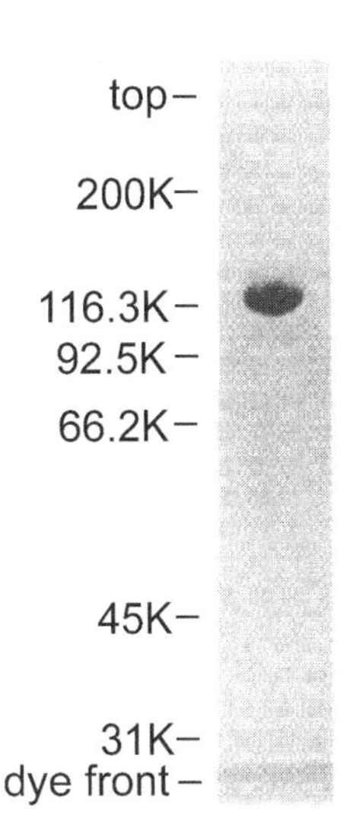

Figure 6.2.2 SDS-PAGE (7.5%, w/v, acrylamide; *UNIT 7.3*) analysis of purified human complement protein C6 showing homogeneity, i.e., absence of other protein contaminants. The positions of molecular weight markers run in an adjacent lane of the same gel are shown. Reproduced with permission from Chakravarti and Muller-Eberhard (1988).

CHOICE OF CHROMATOGRAPHY COLUMN

Columns used in low pressure or gravity flow chromatography are usually made of glass or plastic material. The column contains a frit close to its base, just before the eluent exits the column. The purpose and the specification of the frit are described below. Some very precise column chromatography methods require that the column be maintained at a constant temperature during the separation. This is achieved by using water-jacketed columns in which water at the desired temperature is circulated using a thermostatically controlled water bath. However, for separation of proteins, it is far more common to carry out the separation in the cold, usually at a temperature of 4°C to 8°C. Elution of the column in the cold usually protects the proteins from proteolysis with proteolytic enzymes present in the sample and also helps retention of biological activity. This is usually carried out by performing column chromatography in a temperature-controlled cold room or in a refrigerated cold cabinet. The latter should have a glass door through which the process can be visually inspected, and should also have necessary electrical outlets inside the cabinet.

Empty chromatography columns can be obtained from several different manufacturers. Bio-Rad Laboratories (*http://www.bio-rad.com*) offers a full line of chromatography columns and accessories. Bio-Rad's Econo-Columns are made of borosilicate glass. These columns are available in lengths from 5 to 170 cm, with inner diameters from 0.5 to 5 cm, and, having capacities from 1 to 600 ml, can withstand operating pressures up to 1 bar and can be sanitized with NaOH. Bio-Rad also offers disposable low-pressure chromatography columns made of polypropylene. Even flow of eluent through the column and low dead volumes (see below), such as $<0.1\%$ of the total column volume, minimize broadening of eluted peaks of the components.

From Equation 6.1.14, it follows that resolution (R_s) is directly proportional to the square root of the number of theoretical plates (N) which in turn is directly proportional to the length of the column (L). Thus separation of two proteins in gel filtration increases as the square root of column length. Thus columns used for gel filtration should be long. For analytical purposes, a column with an internal diameter of ~2.5 cm is widely used. In case of wide columns, dilution of the eluting peaks is a problem. However, if size exclusion is used for desalting or buffer exchange, the column should be short and wide rather than long and narrow. This allows higher flow rate and hence speedy elution of the proteins. Similar considerations apply for ion-exchange and affinity column chromatography.

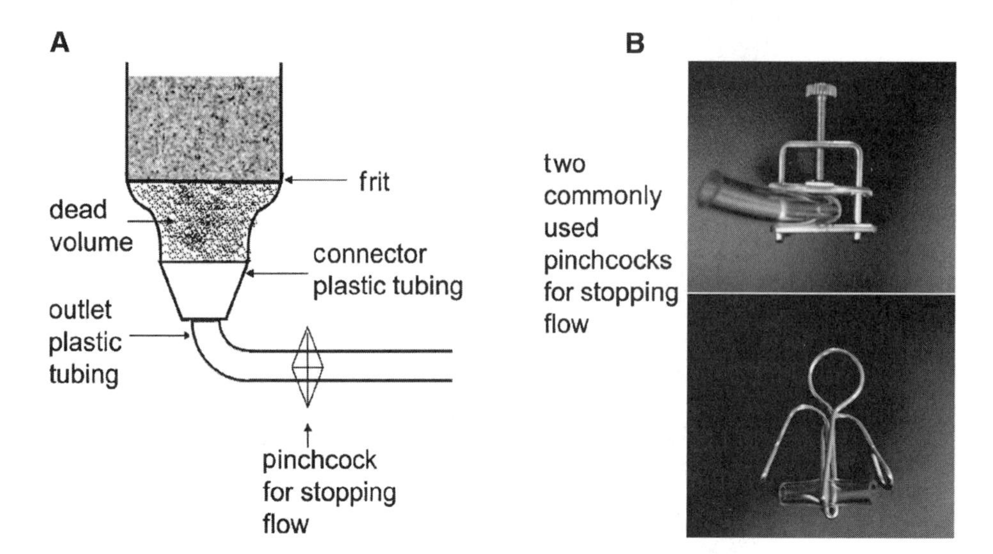

Figure 6.2.3 (**A**) Enlarged view of the bottom part of a chromatography column showing dead volume, connector plastic tubing, outlet plastic tubing, and the commonly used pinchcock for stopping flow. The connector plastic tubing is used to attach the outlet plastic tubing to the column. In practice, the connector plastic tubing may not always be necessary to attach the outlet plastic tubing to the column outlet. (**B**) Two commonly used pinchcocks for stopping the flow of the eluent from the column are shown.

For example, in the case of purification of human complement protein C6 mentioned above, starting from 1 liter blood plasma from which serum was prepared, the size of the columns were as follows:

Anti-C6-Sepharose CL-4B affinity: 4-cm internal diameter × 20-cm length
DEAE-Sephacel ion exchange: 5.7-cm internal diameter × 65-cm length
Sephacryl S-200 (superfine): 2.8-cm internal diameter × 110-cm length

As evident from the above, the size of a chromatography column is defined by two parameters: the length and the internal diameter. It is important to know the precise internal diameter of the column for calculation of bed volume of the stationary phase and also for determination of the linear flow rate of the column (see section on Resolution as a Function of Flow Rate). To determine the internal diameter of the column, the easiest way is to close the stopcock at the outlet of the column (Fig. 6.2.1) and fill up the column with water up to a certain height, followed by measuring the height of the water in the column using a measuring tape, and then opening the stopcock and collecting the water in a measuring cylinder. The internal diameter of the column can be calculated using the following equation:

$$V = \pi r^2 h$$

where V is the volume of the water in the column and h is the height of the water in the column. The internal diameter ($2r$) is determined by multiplying the internal radius (r) of the column by 2. Some chromatography columns may not have a stopcock. In that case, flow of liquid can be turned on or off using a pinchcock on the tubing coming out of the column (Fig. 6.2.3).

FRITS, STOPCOCKS, TUBING, AND OTHER ACCESSORIES

During passage of the mobile phase, the actual chromatographic separation takes place within the column containing the stationary phase. However, a number of accessories are necessary to retain the stationary phase material within the column and connect the column to other parts of the chromatographic system. The design and choice of some of these accessories are very important to maintain the optimum efficiency of the column. The materials that come in contact with the eluent should be noncorrosive and compatible with eluent buffers used.

Frits

Frits, also known as bed support in the case of column chromatography, allow retention of stationary phase in the column. In a chromatography column, the frit should be located as close to the base of the column as possible. This is necessary to minimize the so called 'dead volume' (Fig. 6.2.1) below the frit where mixing of the separated components can take place. Frits are porous filters usually made of glass fibers or granules or of porous polymeric material. Frits in disc form, made from borosilicate or quartz glass fused to the internal wall of one end of a glass column, are frequently used in column chromatography. Nylon nets with appropriate mesh size are also used for this purpose (Fig. 6.2.4). In some column systems, a top and a bottom frit is used (Fig. 6.2.4). These columns can be eluted in the reverse direction, i.e. from the bottom to the top as well. If a column with a frit is not available, a glass wool plug can be used to retain the stationary phase. Bio-Rad's Econo-Columns contain frits with a 28-μm pore size that are capable of retaining fine particles of the stationary phase, while the disposable columns have a 20-μm inert, hydrophilic frit. Many of GE Healthcare's columns contain nylon nets for retention of the stationary phase with 10-μm mesh size in XK columns and 10 or 80-μm mesh for K 9 columns. Reusable glass columns for chromatography available from Pierce Biotechnology (*http://www.piercenet.com*) contain a polyethylene porous disc with 10-μm pore size.

Stopcocks

Glass columns used for purification of proteins sometime contain a stopcock (Fig. 6.2.1) which can be opened or closed to turn on and off the flow of the eluent. The disadvantages of ground

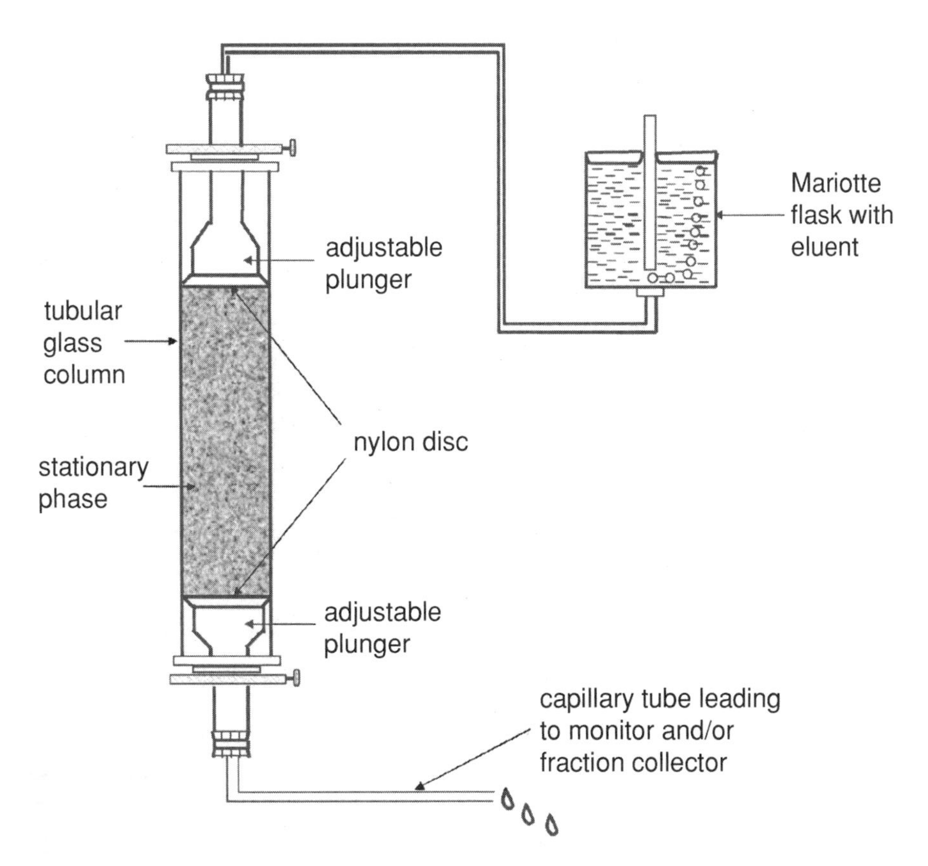

Figure 6.2.4 Chromatography column system in which the stationary phase is packed in between the top and the bottom frits (e.g., nylon discs) is shown. The plungers are adjustable column end pieces that support the stationary phase. This column can also be eluted in the reverse direction, i.e. from the bottom to the top. Although this is a gravity flow system, instead of an open buffer reservoir (such as a beaker in Fig. 6.2.1), a Mariotte flask is used to keep the operating pressure constant. When an open buffer reservoir is used, with the elution of the column the operating pressure will drop due to the lowering of the level of the eluent in the reservoir. Operation under constant pressure allows elution of the column at a steady uniform rate.

glass stopcocks are that they may be a potential source for leakage and may contain lubricant that may be a source of contamination. In this respect plastic stopcocks used in some plastic columns tend to be far better. Many columns like Bio-Rad's Econo-Columns and Pierce's Reusable Glass Columns come with an end cap to stop flow. For these columns, during storage of the column, the tubing through which eluent exits the column is removed and replaced with the end cap. However, a common and useful technique to stop flow of eluent in a chromatography column is to use a pinchcock on the plastic tubing coming out of the column (Fig. 6.2.3).

Tubing

Tubing with appropriate internal diameter, usually made of polyethylene or polypropylene is attached to the column through which the eluent flows to the online detector (usually UV monitor) or the fraction collection system (Figs. 6.2.1 and 6.2.4). The plastic tubing is usually attached to the end of the column through one or more pieces of tubing that act as a connector (Fig. 6.2.3).

BUFFER SYSTEMS

Composition

The composition of the elution buffer does not usually influence the resolution obtained in size-exclusion chromatography. However, for size-exclusion chromatography of proteins, which are charged molecules, the eluent is often a Tris·Cl or sodium phosphate buffer to control the pH (see Chapter 3). The ionic strength of the buffer (10 to 50 mM), usually maintained with 20 to 50 mM NaCl, is often used to minimize any possible ionic interaction between the proteins and the gel matrix. If the protein is going to be lyophilized, a volatile buffer such as ammonium acetate or ammonium bicarbonate is used. During lyophilization, the components of such buffer are removed under vacuum. It should be noted that the pH and the ionic strength of the buffer as well as presence of denaturing compounds and detergents in the elution buffer can cause conformational changes as well as dissociation of proteins into subunits. This must be taken into account for the choice of the gel as well as interpretation of results.

Eluting the Column

The sample containing components to be separated is uniformly applied to the top of the stationary phase bed and separated by passing a suitable eluent (mobile phase) through the stationary phase (see *UNIT 6.1*). The eluent is kept in a buffer reservoir and it is important that the eluent flow through the column at a stable uniform rate. For low-pressure column chromatography, this can be done either under gravity flow or using a peristaltic pump. In gravity flow systems the buffer reservoir is placed at a level higher than the column outlet (Fig. 6.2.1). In many cases the reservoir is actually placed at a level above the top of the column. For such gravity feed, the flow rate can be regulated by varying the operating pressure, which is dependent on the difference of height between the level of the eluent in the buffer reservoir and the outlet of the column. If an open buffer reservoir such as a conical flask or a beaker is used, the flow rate will change over time during the course of the experiment as the eluent will pass through the column, since the operating pressure will drop due to lowering of the level of the eluent in the reservoir. A simple device like the Mariotte flask can be used to keep the operating pressure constant in the gravity flow process (Fig. 6.2.4). The Mariotte flask is an ingenious but at the same time very simple device for obtaining a constant flow of liquid which is unaffected by the concomitant lowering of the level of the liquid in the reservoir (McCarthy, 1934). This can be explained as shown in Figure 6.2.5, in which a stoppered reservoir is supplied with an air inlet and a siphon tube. When water leaves the reservoir due to siphon action, air will enter the reservoir through the air inlet tube and the pressure at the bottom end of the air inlet tube will be identical to the atmospheric pressure. If it were greater, air would not enter the reservoir; also it cannot be less since it is in contact with the atmosphere. If the end of the siphon tube within the reservoir is at the same depth, it will always receive liquid at atmospheric pressure and will deliver a flow under a constant head pressure equal to h (Fig. 6.2.5), which will

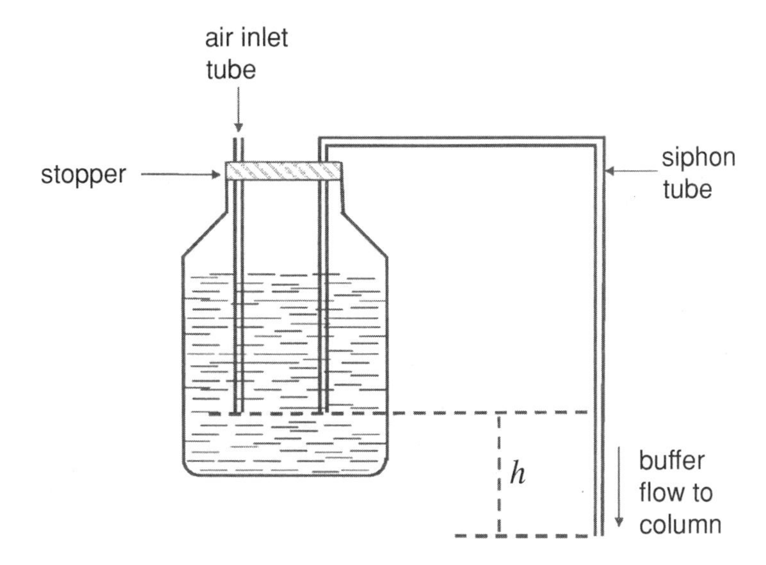

Figure 6.2.5 Mariotte flask for obtaining a constant flow of liquid which is unaffected by lowering of the liquid level in the reservoir. The ends of the air inlet tube and the siphon tube within the reservoir are at the same level. In the arrangement shown, liquid will deliver under a constant head pressure equal to *h*, which will not be affected by the change of level of liquid within the reservoir. Refer to Figure 6.2.4 to see how this flask is attached to the column apparatus.

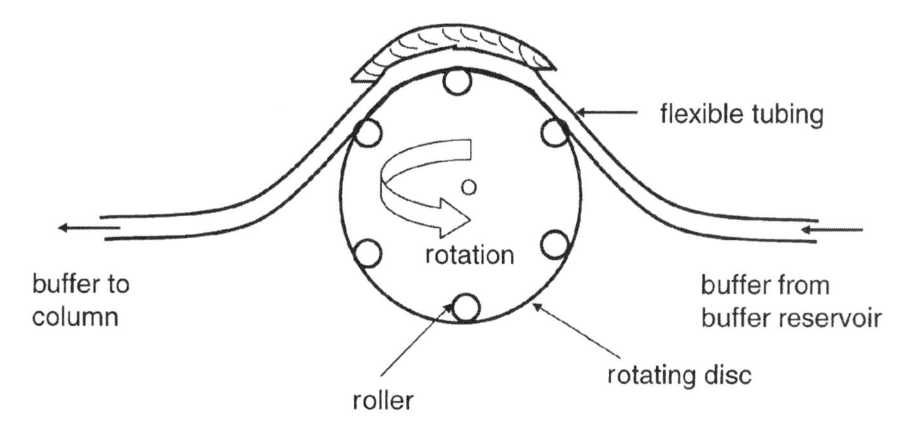

Figure 6.2.6 The principle of a peristaltic pump for elution of a column at a constant flow rate. The rollers on the disc come in contact with the flexible tubing and compress them. Because of this, when the disc is rotated, the liquid inside the tubing moves in the same direction the rollers move. Rotation of the disc at a specified speed thus leads to delivery of liquid at a constant flow rate.

not be affected by the change of the level of liquid within the reservoir. A similar principle is used in the device shown in Figure 6.2.4 which is attached to the column. Alternatively, a peristaltic pump (Fig. 6.2.6) can be used to pump the eluent through the column at a constant flow rate. These pumps are commonly used in low-pressure systems (0.1 MPa). Excellent animations which show how a peristaltic pump works is available online (*http://www.coleparmer.ca/techinfo/techinfo.asp?htmlfile=HowMflexPumpHeadsWork.htm&ID=339* or *http://www.animatedsoftware.com/pumpglos/peristal.htm*). An adjustable constant low flow rate can be achieved using a peristaltic pump. Usually the peristaltic pump uses silicone, PVC, or fluoro-rubber tubing. However, whether using gravity flow system or a peristaltic pump care should be taken to ensure that the operating pressure applied does not exceed that allowed for the particular column material. At higher operating pressures than that recommended, many of the stationary phase materials used in low-pressure column chromatography tend to break, down and that slows down the flow rate of the eluent through the column.

Elution Using Single (Isocratic) Versus Multiple Buffers

Elution using a single eluent is referred to as isocratic elution. Isocratic elution is used in size exclusion and affinity chromatography. While the elution in size exclusion chromatography is carried out entirely with a single buffer composition, affinity chromatography uses different buffers—one for binding of the protein to the column, one for washing out nonspecifically bound proteins, and one for elution of the specifically bound material. However, for ion-exchange chromatography, it is necessary to change the salt concentration and/or pH of the eluent. Such elution is known as, gradient elution. As explained in *UNIT 6.1* a this change can be either stepwise (step gradient) or continuous. For continuous gradient elution, two different buffer solutions have to be mixed in correct proportion before they enter the column. The commonly used buffer reservoir for linear gradient elution used for ion-exchange column chromatography is known as a gradient mixer (Fig. 6.2.7), and is different from that used for isocratic elution. Such gradient mixers are available commercially. It consists of two separate cylindrical chambers of identical internal diameter, a reservoir (for the limit buffer), and a mixing chamber (for the start buffer), both of which are at the same level. The chambers are connected through an interconnecting stopcock. A second stopcock regulates the flow of eluent from the mixing chamber and is connected to the column through tubing. The mixing chamber is placed on a magnetic stirrer and the contents are continually mixed with a stir bar. The interconnected reservoir and the mixing chamber can be considered as a U-tube. As the eluent leaves the mixing chamber and flows to the column, an equal amount of buffer from the reservoir enters the mixing chamber so that the heights of the liquids in both chambers remain the same. For generating a 300 ml linear NaCl gradient from buffer A (20 mM Tris·Cl/25 mM NaCl; pH 8.0) to buffer B (20 mM Tris·Cl/250 mM NaCl; pH 8.0) as described graphically in Figure 6.1.12 of *UNIT 6.1*, 150 ml of buffer B is placed in the reservoir and the same amount of buffer A is placed in the mixing chamber with the interconnecting stopcock in the off position. The mixing chamber is placed on a magnetic stirrer and the outlet from the mixing chamber is connected to the column. The magnetic stirrer is turned on, followed by opening the stopcock connecting the reservoir and the mixing chambers. The relative diameter of the two chambers determines whether the gradient is linear, convex, or concave. A three-channel peristaltic pump can be used to generate precise gradients as well.

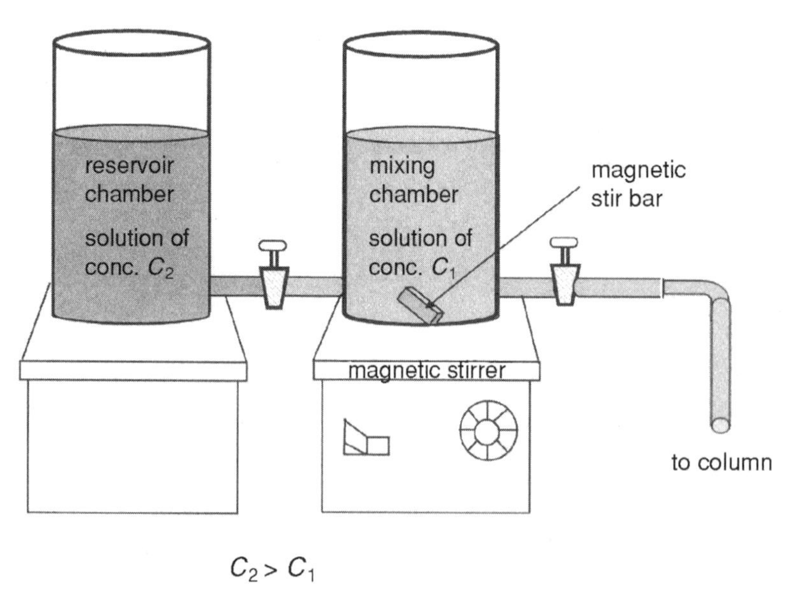

Figure 6.2.7 Commonly used gradient mixture for generating a continuous linear gradient, such as that shown in Figure 6.1.12 B, from a buffer solution of lower salt concentration (C_1) to a buffer solution with a higher salt concentration (C_2).

Resolution as a Function of Flow Rate

Resolution in size exclusion chromatography usually decreases with increasing eluent flow rate through the column. Maximum resolution is generally obtained when a long column is used with a low eluent flow rate. For size exclusion chromatography using a long column, if peaks are well separated at a low flow rate, experiments may be performed to determine if flow rate can be increased, if a shorter column can be used, or more sample can be applied to the column.

For column chromatography used for purifying a protein, application of a higher flow rate to obtain rapid separation is quite often more desirable than the corresponding loss of resolution. However, the gel matrix used as the stationary phase must be able to tolerate the higher operating pressure that is required to produce the increased flow rate. Approximate maximum flow rates and pressures for a specific gel type are available from the manufacturer and should be given due consideration.

For measuring flow rate, the column is allowed to run over a fixed interval of time, and the eluted buffer is collected and measured using a measuring cylinder. The interval of time is chosen such that a sufficient amount of liquid, say 5 to 10 ml, is collected that can be accurately measured. Flow rate for a low-pressure column is usually expressed in ml/hr. Another way to express the flow rate is to divide this value with the cross-sectional area of the specific column. This linear flow rate, i.e. the flow rate per unit cross-sectional area of the column is expressed in cm/hr. In the case of purification of human complement protein C6, the flow rate used to run the DEAE-Sephacel column was 60 ml/hr while that for the Sephacryl S-200 (superfine) column was 30 ml/hr.

PLANNING AND RUNNING THE COLUMN CHROMATOGRAPHY EXPERIMENT

First, a decision has to be taken on what specific type of column chromatographic separation technique, such as size exclusion or ion exchange or affinity, etc., will be used. Usually affinity chromatography can be used with a very crude sample when a suitable specific affinity matrix is available. It can generate a highly purified protein component in a single step. Ion exchange chromatography is suitable for purification of proteins from a crude sample by applying pH or salt gradient as explained in *UNIT 6.1*. On the other hand, size exclusion chromatography is more applicable for further purification of products already purified by affinity or ion-exchange chromatography. Thus it is common practice to apply affinity or ion-exchange chromatography as an initial chromatographic step for the purification of a crude protein sample.

Optimum design of any column chromatography step will depend on the type of proteins (e.g., net charge and molecular weight) that will be separated. The following give some general guidelines regarding setting up a column chromatography experiment. The ultimate goal of any one of these is to achieve optimum resolution of the protein components of interest while allowing maximum recovery of protein as well as maximum biological activity with the least amount of sample dilution, which can be carried out as quickly as possible.

Choice of Gel Type

Size exclusion chromatography

For size exclusion chromatography where desalting or buffer exchange (see below) is required, there is generally considerable difference in the molecular weight of the proteins and the salts. For this, a gel should be chosen where the proteins will elute in the void or exclusion volume of the column and low-molecular-weight salts should elute near the total volume of the column. Usually, Sephadex G-25 is an excellent choice for this purpose. The numbering system of Sephadex gels is based on their water regain values (gram water/gram dry gel); for example, this value is 2.5 g/gram for Sephadex G-25, 5.0 g/gram for Sephadex G-50, 7.5 g/gram for Sephadex G-75, 10 g/gram for Sephadex G-100, etc. When a number of proteins with molecular weight range of close proximity need to be separated, one should choose the gel matrix carefully, taking into account the separation range of the gels to be used. Since the molecular weight selectivity of different gels overlap (see Table 6.1.1 of *UNIT 6.1*), more than one gel type may be suitable for a particular separation. In this case, it is advisable to use the gel where the protein of interest will elute earlier.

Ion-exchange chromatography

For ion-exchange chromatography, the relationship of the net charge of a protein as a function of pH and the pH ranges in which the protein is bound to anion or cation exchangers has been described in *UNIT 6.1* (Fig. 6.1.10). In general, if the protein of interest is most stable below its pI, a cation exchanger is used and if the protein is most stable above its pI, an anion exchanger is used. If the protein is stable over a wide range of pH, either a cation exchanger or anion exchanger can be used successfully. The pH of the starting buffer should be at least one pH unit above or below the pI of the protein to allow proper binding to the ion-exchanger.

If the pI of the protein of interest is not known, a simple method to choose the starting pH for binding of the protein to the ion-exchanger can be carried out in a test tube. For this, a series of test tubes containing an equal amount of the ion-exchanger are taken and equilibrated to a different pH by washing with excess of the start buffer at different pH. This can be in a range of pH 5 to 9 for anion exchangers and pH 4 to 8 for cation exchangers, with 0.5 pH unit difference. A known constant amount of sample is then added to each tube and mixed for 10 min. After the gel settles, the supernatant is tested for the presence of the proteins, e.g. by SDS-PAGE (*UNIT 7.3*). It is preferable to use a pH where the substance should bind to the ion-exchanger but close to the pH where it is released from the ion-exchanger, i.e. detected in the supernatant. Choice of too low or too high pH for binding may result in the use of buffer with high salt concentration to elute the protein from the gel.

A similar test as described above can be performed to determine the optimum ionic strength (e.g. NaCl concentration) that allows binding and elution of the protein from the gel. In most instances, the conditions are chosen so that the protein of interest binds to the ion-exchange matrix. However, conditions can also be chosen where impurities bind to the ion-exchange column and the protein of interest elutes without binding. A well-known example is the purification of IgG antibodies on DEAE-ion exchanger at around pH 8.0.

Preparation of the Stationary Phase

Many stationary phase materials are available as a liquid suspension. For those that are supplied as a dry solid, manufacturer's instructions for swelling the solid to a gel with appropriate buffer should be followed.

Column Packing

Column packing is a very important step for successful separation in any type of column chromatography. For packing a column, excess dissolved air should be removed (i.e. degassed under vacuum) from the buffer or gel slurry to be used for column packing. Usually a 20% to 50% slurry of the hydrated gel support in the elution or start buffer is used and the column, buffer, and gel slurry should be equilibrated to room temperature before packing of the column. Fine particles of the gel may be removed by decantation of the buffer from the top of the gel slurry after allowing it to settle. The empty column should be clamped into an upright position and the outlet should be fitted with tubing and closed as described above. The column should be partially filled with the buffer. Packing of the column should be carried out by pouring the slurry of the stationary phase gently down a glass rod into the mobile phase. Care should be taken to avoid air bubble formation. The gel should be allowed to settle in the column for at least 30 min. The outlet of the column should be opened to allow the buffer to drain from the column until it reaches close to the top of the gel bed. Further amounts of gel slurry may be added if necessary until the desired height of the bed has been obtained. The gel bed should not be allowed to become dry or get filled up with air bubbles. The packed stationary phase bed of the column should be checked for uniformity of packing as well as absence of trapped air bubbles with transmitted light from a lamp held behind the column. The packed bed should be eluted with at least two bed volumes of elution buffer in order to allow the system to reach equilibration and to stabilize the bed.

Sample Size

For size exclusion chromatography, the sample volume should be within 1% to 5% of the bed volume of the column for optimum resolution. However, for desalting and buffer exchange (see below), up to 30% of the column bed volume can be used to minimize sample dilution.

Sample Application and Elution

The most common practice for sample application is to open the top of the column and allow eluent to drain to the top of the stationary phase bed. The sample solution is then applied to the bed with a pipet and allowed to drain into the stationary phase. When the entire sample has been applied, the top of the column is washed with the elution buffer (or the starting buffer) and the column is connected for elution. Care should be taken that during sample application and elution, the column should never go dry.

Storage

At the completion of the run, ion exchange columns may be regenerated for reuse according to the manufacturer's instructions. All columns should be appropriately stored with buffers containing appropriate antimicrobial agents.

DETECTORS AND RECORDERS

In general, the eluent continuously flows through a chromatography column in order to separate a mixture of various protein molecules. Usually, a detector is connected at the outlet of the column for optically detecting each separated component and a chart recorder is connected to the detector. Detection is usually based on UV absorbance (*UNIT 2.2*). Each component detected in the detector is electrically measured in a photoelectric tube, or the like, to produce electric indication, which is transmitted to the recorder to be recorded continuously as a graph on a chart paper in the recorder. Although a computer can be used to record the detector signal, many chromatographers still like the chart recorder in which a pen continuously plots the absorbance registered by the detector on a chart paper that rolls out of the recorder. In a chart recorder with two pens, the second pen can be used to record the absorbance at a second wavelength, if the detector has that capability or to mark changes in fractions collected if the fraction collector has that ability. The advantage of the chart recorder over many computer-based systems is that the recording of absorbance takes place in real time without any lag for data processing and the chart can be manually marked with a pen or marker to make notation in real time.

6

Detection of Protein Content in Column Effluent

The protein content in the column effluent can be monitored either offline, usually with appropriate spectrophotometric procedures, or online with a flow-through optical (UV) detection system (*UNITS 2.1 & 2.2*).

The most common procedure for offline detection of protein content in eluted fractions is to determine the absorbance at 280 nm in a UV spectrophotometer using the column buffer as the blank. For column chromatography involving multi-step buffer elution, the starting buffer or the buffer with which the column is equilibrated is used as the blank. Proteins absorb strongly at 280 nm due to the presence of aromatic amino acid residues, tryptophan, tyrosine, and phenylalanine. The absorbance at 280 nm of a 1 mg/ml pure protein solution contained in a quartz cuvette with 1-cm light path is usually between 0.5 and 1.5. Because of this, for an unknown protein solution, this value is often approximated to be 1.0. Thus the numerical value of the absorbance of the fractions determined at 280 nm using a cuvette with 1-cm light path can be regarded as the concentration of the protein in mg/ml. The advantage of this procedure is that additional time required for protein estimation is not required, and, for proteins present in limited amounts, no loss of protein is incurred. However, estimation of protein content by Lowry-Folin Ciocalteau assay (*UNIT 2.2*) to determine protein content is carried out as well.

FRACTION COLLECTORS

In column chromatography, separated components elute from the column at different times. Collection of these components as separate fractions is thus a critical part of column chromatography. In a very small scale column chromatography experiment, particularly involving affinity chromatography, it might be possible to collect eluate manually. However, collection of eluted fractions over a long period of time or from a liquid chromatography experiment performed in the cold warrants the use of a fully automatic fraction collection machine, known as the fraction collector. A very interesting description of an early prototype machine is available (Stein and Moore, 1948). The fraction collector is generally designed to collect the eluate from the chromatographic column in successive fractions of known volume. The fractions are usually collected in tubes held in appropriate test tube racks that are part of the instrument. In the drop counting mode, each drop of eluate falling from the end of the tubing exiting the column intercepts a light beam which is focused upon a light-sensing device, such as a phototube, in the drop head of the machine. After the preset number of drops has been registered, the drop counter resets itself to zero and advances to the next fraction collection tube. One possible disadvantage of the drop count method is that drop size may change in some cases when the composition of the eluent changes, such as during gradient elution. Fraction collection based on a preset period of time is also an option available in these machines. If the flow rate is not uniform, this mode will collect fractions of different volumes. There is a bias towards fraction collection based on time, since it is considered to be the more robust of the two methods of fraction collection. Protein purification methods are often carried out in the cold and there is a chance of occasional fogging that may lead to malfunctioning of the drop counting mechanism. Advanced fraction collectors are able to collect eluting peaks, such as by peak level that uses a percentage of full scale signal or by peak slope that determines the extent of baseline change to start and stop fraction collection.

POOLING FRACTIONS

In addition to protein content, if an assay for the biological activity of the protein of interest is available, it should be carried out with the fractions. It may not be necessary to assay each fraction for biological activity. Testing of activity at fixed intervals, say every second or every fifth fraction, depending on the number of fractions collected and the ease of the assay, should be carried out. In addition, samples taken at a fixed interval should be subjected to SDS-PAGE analysis (*UNIT 7.3*) to determine the purity of the protein in those fractions. For each step of purification, the protein content and activity data should be recorded in a table that will have all the relevant information, such as information on the total volume of the sample (ml), protein concentration (mg/ml), total protein content (mg), biological activity (units/ml), total activity (units), specific activity (units/mg protein), yield of total activity (%), and purification factor (fold in increase of specific activity). Usually, with subsequent steps of purification, the total protein content and the total biological activity will decrease, whereas the specific activity and the purification factor will increase along with the degree of purity as adjudged by SDS-PAGE analysis. In each step of purification, based on the protein profile of the fractions, the activity, and purity, a choice is made to pool fractions for the next step of purification. The idea is to maximize the yield of the specific protein while minimizing the amounts of impurities.

A schematic diagram showing a column chromatographic process is shown in Figure 6.2.8.

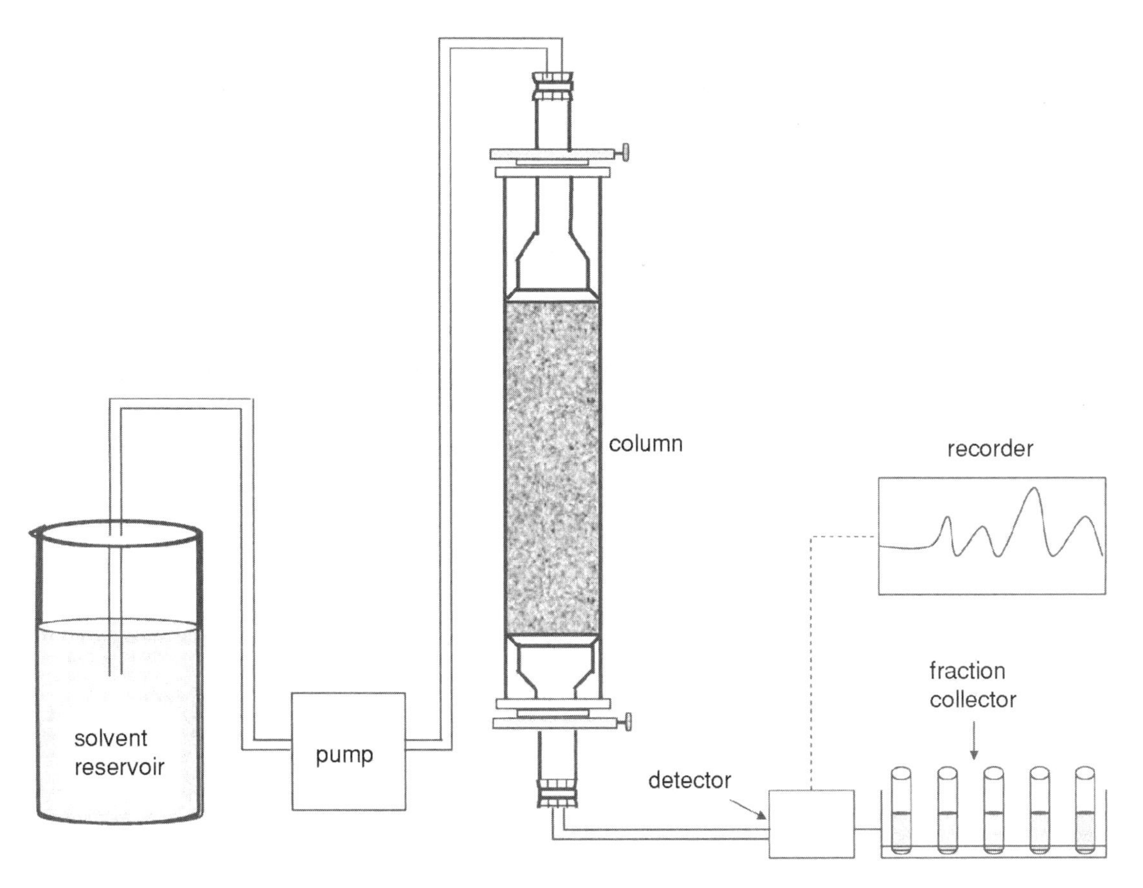

Figure 6.2.8 Schematic flow diagram showing a low-pressure column chromatography system using a peristaltic pump, online detector, chart recorder, and fraction collector.

PREPACKED COLUMNS AND KITS

For beginners in research, it is often advisable to use commercially available prepacked columns or chromatography kits. Prepacked columns are often available and are optimized for a specific separation. Kits for a specific separation are generally supplied with proprietary reagents that have been developed for optimum separation and yield. These save considerable amounts of time required for optimization. Examples of commonly used commercially available packed columns and column chromatography kits are given below.

6

Disposable PD-10 Desalting Columns

These columns packed with Sephadex G-25 medium are available from GE Healthcare. PD-10 columns are used for fast buffer exchange and desalting of protein solutions with minimal loss of protein. Let us consider a protein present in solution in Buffer B1 (such as 10 mM Tris·Cl/150 mM NaCl; pH 7.2) that needs to be transferred to a solution in Buffer B2 (such as 10 mM Tris·Cl; pH 7.2). During buffer exchange, the buffer solution containing the protein is changed from Buffer B1 to Buffer B2. One common approach to achieve this in the laboratory is to perform dialysis. In dialysis, the protein solution is contained in a dialysis bag made of a semi-permeable membrane with pores that allow small molecules like buffer ions and salts to pass through it while larger molecules like proteins are retained. The bag in this case will be placed in a solution of Buffer B2 and allowed to equilibrate. After a couple of hours, fresh Buffer B2 is added and the process is repeated few times. After the final buffer change, the sample is collected from the dialysis bag. Comparatively much faster buffer exchange or desalting can be achieved with the PD-10 columns by following the manufacturer's instructions.

Purification of Antibody Using Protein A or Protein A/G

Reagents for affinity purification of antibodies are available from several vendors. These are very useful for a beginner to gain experience in protein purification.

LITERATURE CITED

Chakravarti, D.N. and Muller-Eberhard, H.J. 1988. Biochemical characterization of the human complement protein C6. Association with alpha-thrombin-like enzyme and absence of serine protease activity in cytolytically active C6. *J. Biol. Chem.* 263:18306-18312.

McCarthy, E.L. 1934. Mariotte's bottle. *Science* 80:100.

Stein, W.H. and Moore, S. 1948. Chromatography of amino acids on starch columns: Separation of phenylalanine, leucine, isoleucine, methionine, tyrosine and valine. *J. Biol. Chem.* 176:337-365.

Wilson, K. and Walker, J. 1994. Principles and Techniques of Practical Biochemistry, 4th ed. Cambridge University Press, Cambridge.

Chapter 7

Electrophoresis

UNIT 7.1

Overview of Electrophoresis

Sean R. Gallagher[1]

[1]UVP, LLC, Upland, California

INTRODUCTION

Electrophoresis is a gateway technique, fundamental to modern life science research, that opens up a remarkable range of analysis options for protein and nucleic acids (Tables 7.1.1 and 7.1.2; Gallagher, 1999a). To illustrate the importance of electrophoresis, one only needs to review the latest journals of life science research such as *Science* and *Nature* to see many panels of data and images of electrophoretically separated proteins and DNA. The technique is applied to purification of DNA and protein, and the determination of their presence or absence, size, structure, amount, and modifications. Uses include the determination of sample purity, mutations in DNA, gene expression at both the mRNA and the protein levels, detection of post-translational modifications to proteins, and identification of antibodies to disease organisms indicating an infection. While there are many variations of electrophoresis methods, including isoelectric focusing (IEF) and two-dimensional SDS-polyacrylamide gel electrophoresis (SDS-PAGE), the techniques outlined in the following sections focus on the standard procedures used in day-to-day research in the lab. The more specialized techniques are cross-referenced to appropriate manuals for more details.

GENERAL PRINCIPLES

Electrophoresis is the movement of a charged molecule under the influence of an electric field (see Table 7.1.3). In a typical configuration, the sample is loaded near the negative electrode of the gel tank. When power is applied, the negatively charged molecules (e.g., nucleic acid- or SDS-coated proteins) will migrate to the positive electrode. Fortunately, under standard pH and separation conditions, nucleic acids have a negative charge due to their phosphate backbones. Nucleic acid electrophoresis is the first step in Southern and northern blotting (see *UNIT 8.2*), where the separated nucleic acids are transferred and bound to nylon membranes for subsequent analysis using specific DNA and RNA probes to determine the presence and amount of a particular sequence of DNA or RNA.

The situation with proteins is more complex. Proteins are polyampholytes composed of chains of amino acids, many of which have a variety of ionizable acidic and basic groups. Depending on the pH of the aqueous buffer, a protein will have either a net positive, neutral (a special case referred to as the isoelectric point), or negative charge. The native charge on the protein is exploited for a variety of separation techniques, including isoelectric focusing (IEF), which resolves proteins according their isoelectric point and is the first dimension of two-dimensional SDS-PAGE—see *CP Molecular Biology Unit 10.3* (Adams, 1992) and *CP Molecular Biology Unit 10.4* (Adams and Gallagher, 2004). In the most common protein separation technique, SDS-PAGE (*UNIT 7.3*; Laemmli, 1970), the negatively charged detergent sodium dodecyl sulfate (SDS) is used to both solubilize and impart a negative charge so the proteins migrate to the positive electrode.

In order to accurately determine the size of a protein molecule, the contribution from the native charge must be eliminated. With SDS-PAGE, the intrinsic charge is swamped out by the addition of the negatively charged detergent SDS to solubilize the protein sample. SDS typically binds at 1.4 grams of detergent per gram of protein (Reynolds and Tanford, 1970), resulting in a constant negative charge-to-mass ratio, allowing size-dependent separations regardless of the original native charge on the protein. With the continued and expanded interest in the protein complement in a cell and how the expressed proteins interact within cells and tissues, the ability to analyze for complexity, amount, and modifications of proteins continues to be of the highest importance to life

Current Protocols Essential Laboratory Techniques 7.1.1-7.1.6

Table 7.1.1 Key Applications of Protein Electrophoresis[a]

Technique	Separation principle	Application	Reference
Native PAGE	Native charge, size, shape	Purification, size of native protein and protein complex, isoenzyme analysis	*CP Molecular Biology Unit 10.2B* (Gallagher, 1999b)
SDS-PAGE	Size dependent; SDS imparts negative charge to proteins, giving a constant charge-to-mass ratio	Size estimation, purity, purification, subunit composition, protein expression and turnover, post translational modifications	*CP Molecular Biology Unit 10.2A* (Gallagher, 2006)
IEF	Intrinsic charge (both native and denatured proteins)	Purification, purity check, isoelectric point analysis, isoenzyme detection	*CP Molecular Biology Unit 10.3* (Adams, 1992) and *Unit 10.4* (Adams and Gallagher, 2004)
2-D SDS-PAGE	Isoelectric point in the first dimension and size in the second	Protein expression, purification, post-translational analysis, proteomics, MS analysis of separated protein	*CP Molecular Biology Unit 10.3* (Adams, 1992) and *Unit 10.4* (Adams and Gallagher, 2004)

[a]Table adapted from Gallagher (1999a).
Abbreviations: SDS-PAGE, sodium dodecyl sulfate polyacrylamide gel electrophoresis; IEF, isoelectric focusing; 2-D, 2-dimensional.

Table 7.1.2 Key Applications of Nucleic Acid Electrophoresis[a]

Technique	Separation principle	Application	Reference
Large fragment DNA and RNA separation	Size; due to the negatively charged phosphate backbone, the nucleic acids have a constant charge-to-mass ratio	Purification, determination of purity and size, genetic analysis, and preparation of DNA probes	*CP Molecular Biology Unit 2.7* (Chory and Pollard, 1999) and *Unit 2.8* (Smith and Nelson, 2004)
Small fragment DNA and RNA separation	Size; see above	Purification, determination of purity and size, genetic analysis, and preparation of DNA probes	*CP Molecular Biology Unit 2.5A* (Voytas, 2000) and *Unit 2.5B* (Finney, 2000)
DNA sequencing	Size of single-stranded oligonucleotide; denaturing gels and high temperature keep DNA single stranded	DNA sequence determination; mutation and polymorphism analysis	*CP Molecular Biology* Chapter 7 (Ausubel et al., 2007)

[a]Table adapted from Gallagher (1999a).

science research. The simplicity of SDS-PAGE enables scientists to get a quick snapshot of the proteins from within contexts ranging from subcellular organelles to tissue to the whole organism. SDS-PAGE is also the second step in two-dimensional SDS-PAGE, a technique for high-resolution separation of hundreds of proteins on a single gel, and it is the first step in immunoblotting (*UNIT 8.3*), where the electrophroretically separated proteins are transferred onto a membrane for analysis using protein-specific antibodies.

Table 7.1.3 Important Formulas Used in Electrophoresis and Their Definitions[a]

Formula	Definition
QE	The driving force of the charged protein or DNA is a product of the charge Q and the voltage or potential gradient E in the electrophoresis gel.
$F = 6\pi r v \eta$	F is the resistance to the movement of protein or DNA. r represents the molecule radius, v is the velocity in solution, and η is the viscosity of the liquid surrounding the molecule.
$V = IR$	Ohm's Law: useful for understanding how to use power supplies and electrophoresis separation equipment for optimum separation. V is voltage in volts, I is current in amps (A), and R is resistance in Ohms.
$I = V/R$	Typically, SDS-PAGE is performed at a constant current (I) of 10 to 30 milliamps (mA) per gel.
$P = VI$ $P = I^2R$	P is power in watts. Excess power generates heat that will cause distortion and overall loss of resolution. Typically, SDS-PAGE electrophoresis is performed at <5 watts per gel.
R_f = Distance migrated by band/distance migrated by reference marker	Relative mobility (R_f) is a convenient way to compare gels and sample migrations from approximate molecular weight relative to the tracking dye.
$\%T$ = g total monomer (acrylamide + bisacrylamide) /100 ml final gel solution	$\%T$ (compared to % acrylamide) is a more accurate way (together with $\%C$) to express the exact composition of the gel.
$\%C$ = g bisacrylamide/g (acrylamide + bisacrylamide) ×100	$\%C$ is an important value to report so the exact composition of the gel can be reproduced.

[a]Table adapted from Gallagher (1999a).

Historically, electrophoresis has been performed both in free solution and a variety of stabilizing matrices, including starch, agarose, and polyacrylamide. For routine work in the lab, there are two matrices typically used for electrophoresis: acrylamide and agarose. The choice of one or the other depends, in general, on the porosity required. Acrylamide is made up of acrylamide monomers periodically cross-linked with bisacrylamide to form a porous sieving network. Acrylamide gels are useful for relatively small fragments of DNA and most proteins. The larger pore structure of agarose allows separation of larger fragments of DNA typically used in molecular biology techniques.

The format of separation is also largely dictated by the matrix used. Agarose gels are prepared by simply heating the powdered agarose (a purified galactan hydrocolloid from seaweed; *UNIT 7.2*) with the electrophoresis buffer to dissolve the agarose. This is followed by pouring the molten agarose into a gel tray to solidify and form the rectangular gel, which can then be used in DNA and RNA separation. Typically the gels are poured, loaded with samples, and run horizontally for the sake of simplicity.

In contrast, acrylamide polymerization is inhibited by oxygen, requiring that the gels be prepared in the absence of oxygen, sealed between two glass plates. In this case, the gel is cast, loaded with samples, and run in a vertical format. The original electrophoresis units consisted of individual vertical glass tubes containing gels of polymerized acrylamide. Each tube was loaded with a single sample and then carefully extruded from the glass tube for staining and visualization. The use of slab gels to separate and analyze multiple samples simultaneously greatly simplifies lane-to-lane comparison while improving throughput.

Equations describing the movement of a charged protein or DNA molecule through a gel (Table 7.1.3) in an electric field show that a number of factors affect the migration. These include the molecule size, shape, and charge, as well as solution viscosity, temperature, and voltage gradient. Consideration of these equations can provide some general guidelines: for a given charge-to-mass ratio, larger molecules migrate more slowly. If the voltage is increased, then the migration speed will increase, but with the effect of adding additional heat to the separation. Without proper cooling, the increased temperature will lead to localized heating affects causing lane and band distortions that limit resolution. The shape of the molecule has an important impact on the speed and separation resolution. A compact molecule will have a higher mobility than the same molecule in the extended form, as illustrated in Figure 7.1.1. Even though all have the same sequence (and thus size), the DNA plasmid pBR322 migrates at a different rate, depending on whether it is the closed, circular, or supercoiled form I; nicked or open circular form II; or straight linear form III. The supercoiled DNA has a smaller hydrodynamic radius and thus interacts less with the surrounding medium and moves more quickly through the gel.

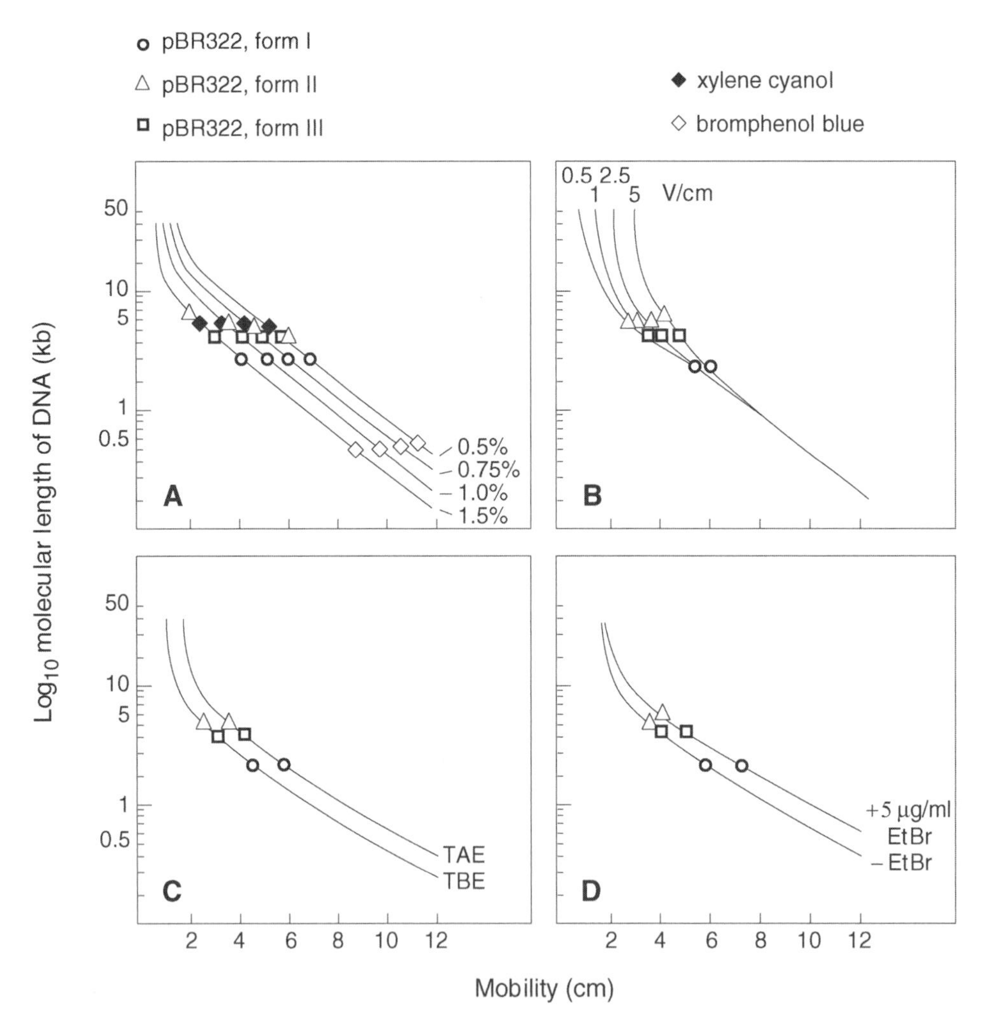

Figure 7.1.1 Effects of (**A**) agarose concentration, (**B**) applied voltage, (**C**) electrophoresis buffer, and (**D**) ethidium bromide on migration of DNA molecules through agarose gels. Note that the mobility of the DNA and the two tracking dyes (xylene cyanol and bromphenol blue) varies with agarose concentration (panel A). In addition, the mobility of the same DNA molecule will vary depending on the conformation. Form I is a compact supercoiled plasmid, and moves more quickly through the gel compared to the relaxed or nicked circular plasmid form II or the linear form III. The apparent molecular sizes of the tracking dyes range from 800 to 4000 bp (xylene cyanol) and 100 to 300 bp (bromphenol blue) and are used for relative mobility (R_f) calculation and visual inspection of the agarose gel during separation. Unless otherwise indicated, all electrophoresis runs were performed using an applied voltage of 1 V/cm, TAE buffer, a 16-hr run time, and a 1% agarose gel. Published previously in *CP Molecular Biology Unit 2.5A* (Voytas, 2000).

VISUALIZATION OF RESOLVED MOLECULES

To visualize the separated proteins or nucleic acid, the gel is typically immersed in a staining solution, allowing the stain to diffuse into the gel and bind to the protein or nucleic acid (*UNIT 7.4*). Coomassie Blue, for example, will diffuse and bind to the proteins, staining them blue under white light. Other stains, such as SYPRO Orange (proteins) and SYBR Green (DNA), also bind after diffusion but are visualized using fluorescence. The bound dyes have an excitation peak in the ultraviolet, and by placing the gel on an ultraviolet light box or transilluminator, the bands are visualized by the emitted fluorescence. Typically, the resulting fluorescence is in the visible spectrum, although some fluorescent stains extend out into the infrared. Another approach is to preload the gel with stain prior to electrophoresis; this method is commonly used with ethidium bromide staining in DNA separations.

Immunoblotting is a special technique relying on two electrophoresis steps and similar visualization equipment (*UNIT 8.3*). The process of immunoblotting involves, first, protein separation by SDS-PAGE, followed by electrophoretic transfer of the separated proteins to a nitrocellulose or polyvinylidine difluoride (PVDF) membrane. The transferred proteins form an exact replica of the gel, except that proteins now sit on the surface of the membrane, making them easily accessible to specific antibody probes. Visualization is accomplished with chromogenic, fluorescent, or chemiluminescent reactions that are visualized from overhead. The availability of off-the-shelf chemiluminescence chemistry and low-light digital cameras have made chemiluminescence much more quantitative. Similar kits and chemistries are available for both Southern (DNA) and northern (RNA) blotting (*UNIT 8.2*).

IMAGING RESOLVED MOLECULES

The intensity of the staining color or emission (fluorescent or chemiluminescent) can be quantitated, and is linear over a wide range of protein or nucleic acid amounts. For many years, documentation was performed either with Polaroid or negative film, followed by a traditional silver-based printing process. The current use of digital image capture by scientific charge-coupled device (CCD)-based cameras or laser scanners has greatly simplified and lowered the cost of image acquisition, storage, display, printing, and analysis (*UNIT 7.5*). A wide range of digital imaging options are available depending on the application. For simple documentation a system of low-cost uncooled cameras is relatively straightforward to use. For documenting results that have a wide dynamic range or that require long exposures, cooled cameras are generally used to reduce the CCD noise during longer exposures. For applications such as chemiluminescent imaging of immunoblots, cooled cameras are a requirement.

However, with digital imaging comes a new level of responsibility because the images can be readily modified with a number of general photography-based imaging software programs (e.g., Photoshop; see *APPENDIX 3*). This includes being able to make regions of an image brighter or darker as well as cutting and pasting part of one image into another. For visual display and emphasis in teaching for example, one might want to enhance or cut out a region of the gel. However, for quantitative analysis and illustration of a result in a scientific publication, any manipulations must be made according to the journal's guidelines, with the original unaltered image kept as support should questions arise. In any case, the process of digital image capture is no different than any method that would be reported in a publication. Details on how the image was captured and the resolution, bit depth, aperture, exposure time, and any post-acquisition manipulations that occurred need to be reported. The current version of Photoshop (CS3 extended) has additional features for nondestructive quantitative analysis as well as for selectively enhancing the visual display of a photograph.

Twenty years ago, if a researcher wanted to perform protein analysis via electrophoresis, the gels had to be prepared in the lab, usually on the day they were to be used. Stocks of buffers had to be prepared and maintained by each individual lab and, indeed, even appropriate standard proteins frequently had to be mixed and blended by the individual laboratories to include for electrophoresis for calibration. Proteins were stained with the relatively insensitive Coomassie blue over a 2-day process that first immersed the gel in the staining solution and then destained the gel to clear the background and visualize the bands. Although this is still the approach taken by many laboratories, commercially available precast gels, premade electrode solutions, and rapid staining kits for both protein and nucleic acid separations have simplified this process to the point where electrophoretic techniques are ubiquitous in the modern life science laboratory.

LITERATURE CITED

Adams, L.D. 1992. Two-dimensional gel electrophoresis using the ISO-DALT system. *Curr. Protoc. Mol. Biol.* 20:10.3.1-10.3.12.

Adams, L.D. and Gallagher, S. 2004. Two-dimensional gel electrophoresis. *Curr. Protoc. Mol. Biol.* 67:10.4.1-10.4.23.

Ausubel, F.M., Brent, R., Kingston, R.E., Moore, D.D., Seidman, J.G., Smith, J.A., and Struhl, K. (eds.) 2007. Current Protocols in Molecular Biology. Chapter 7. John Wiley & Sons, Hoboken, N.J.

Chory, J. and Pollard, J.D. Jr. 1999. Separation of small DNA fragments by conventional gel electrophoresis. *Curr. Protoc. Mol. Biol.* 47:2.7.1-2.7.8.

Finney, M. 2000. Pulsed-field gel electrophoresis. *Curr Protoc. Mol. Biol.* 51:2.5B.1-2.5B.9.

Gallagher, S.R. 1999a. Gel electrophoresis of proteins and nucleic acids. *In* The Encyclopedia of Bioprocess Technology: Fermentation, Biocatalysis, and Bioseparation (M.C. Flickinger and S.W. Drew, eds.) pp. 900-909. John Wiley & Sons, New York.

Gallagher, S.R. 1999b. One-dimensional electrophoresis using nondenaturing conditions. *Curr Protoc. Mol. Biol.* 47:10.2B.1-10.2B.11.

Gallagher, S.R. 2006. One-dimensional SDS gel electrophoresis of proteins. *Curr. Protoc. Mol. Biol.* 75:10.2A.1-10.2A.37.

Laemmli, U.K. 1970. Cleavage of structural proteins during the assembly of the head of bacteriophage T4. *Nature* 227:680-685.

Reynolds, J.A. and Tanford, C. 1970. The gross conformation of protein-sodium dodecyl sulfate complexes. *J. Biol. Chem.* 245:5161-5165.

Smith, A. and Nelson, R.J. 2004. Capillary electrophoresis of DNA. *Curr. Protoc. Mol. Biol.* 68:2.8.1-2.8.17.

Voytas, D. 2000. Agarose gel electrophoresis. *Curr. Protoc. Mol. Biol.* 51:2.5A.1-2.5A.9.

UNIT 7.2

Agarose Gel Electrophoresis

Jennifer A. Armstrong[1] **and Joseph R. Schulz**[2]

[1]Joint Science Department, Claremont McKenna, Pitzer, and Scripps Colleges, Claremont, California
[2]Occidental College, Los Angeles, California

OVERVIEW AND PRINCIPLES

Nucleic acid separation by agarose gel electrophoresis is one of the most commonly used techniques in molecular biology. It is the method of choice for analyzing PCR products (*UNIT 10.2*) as well as for analyzing plasmid DNA digested by restriction enzymes (*UNIT 10.1*). Agarose gel electrophoresis is used to purify DNA restriction enzyme fragments and PCR products for cloning purposes, and is also the starting point for the identification of specific DNA fragments by Southern blotting or RNA fragments by northern blotting (*UNIT 8.2*). Given the central importance of agarose gel electrophoresis, understanding the basic principles underlying this technique is critical.

Electrophoresis

Electrophoresis separates charged molecules migrating in an electric field (see *UNIT 7.1*). A buffer with pH 8 to 8.3 is included in the electrophoresis chamber to counter pH changes and allow for the passage of current. As a result of this passage of current and consequent electrolysis of water at the platinum electrodes, hydrogen gas is formed at the cathode (negative side) and oxygen gas is formed at the anode (positive side). DNA and RNA have regularly repeating phosphodiester linkages that carry negative charges near neutral pH (Fig. 7.2.1). These macromolecules migrate to the anode of the electrophoresis chamber under an applied constant voltage. In the presence of an agarose (or polyacrylamide) matrix, the nucleic acid polymers will electrophorese according to their sizes, with the smaller molecules sieving faster through the pores of the matrix. A voltage range of 1 to 10 V per cm gel length (V/cm) is used, depending on the size of the nucleic acid macromolecule. A voltage that is too high results in poor resolution and overheating, while a voltage that is too low results in diffusion of smaller-sized nucleic acids.

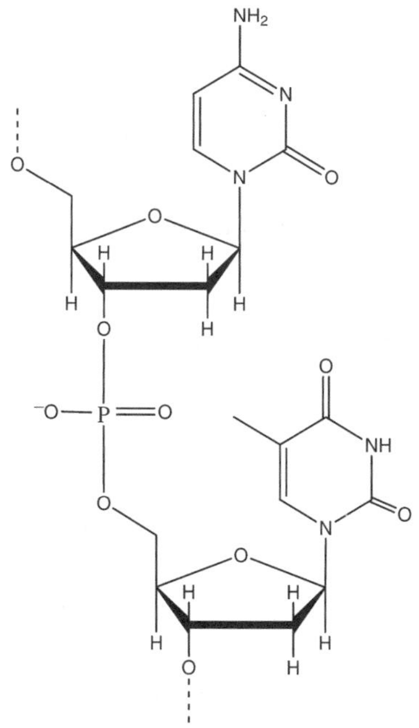

Figure 7.2.1 A negatively charged phosphate group connects nucleosides through phosphodiester linkages in nucleic acid polymers.

Agarose Gels

Agarose is composed of long polymers of mainly uncharged repeating disaccharides (Fig. 7.2.2). An agarose gel is formed when a solution of agarose polymers is first heated and then allowed to cool. As the agarose cools below ~40°C, hydrogen bonds form to create the gel matrix. Agarose is the gelling component of agar (agar-agar), an extract from red algae (mainly *Gelidium* and *Gracilaria* species). Agar is used in food manufacture as a stabilizer and a gelling and thickening agent, and is widely used in microbiology as a culture medium due to its resistance to bacterial enzymatic digestion. However, agar (though present in most molecular biology labs) is unsuitable for gel electrophoresis because every agar contains a small percentage of agarose polysaccharide chains with attached

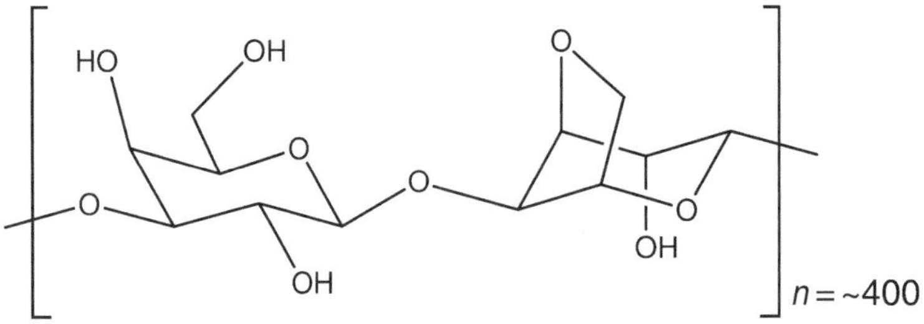

Figure 7.2.2 Chemical structure of the repeating D-galactose, 3,6-anhydro L-galactose disaccharide of agarose. Hydrogens bound to carbons are omitted for clarity.

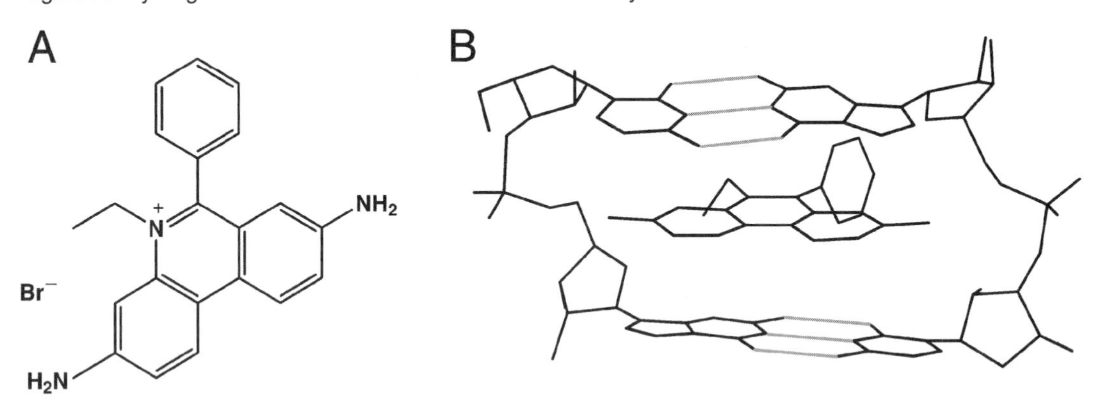

Figure 7.2.3 (**A**) Planar structure of ethidium bromide and (**B**) stick-model representation of ethidium bromide intercalated into a C,G base-pair stack. Hydrogen bonding of the base pairs is indicated by gray lines. The bromide ion is the counterion for the positively charged ethidium molecule; the nitrogen of ethidium bromide has a positive charge. In the electric field, the positively charged ethidium will migrate toward the negative pole, and it is that molecule which is seen under UV light.

negatively charged sulfate groups. The positively charged counterions to these sulfate groups cause problems in electrophoresis, since they migrate toward the negative pole of the gel along with the coincident osmotic flow of water. This phenomenon, referred to as electroendosmosis (EEO), interferes with the electrophoretic migration of DNA and RNA molecules in the gel (which migrate towards the positive pole). While commercial agarose preparations contain only a small percentage of polysaccharides with attached sulfates, high-quality (molecular-biology grade, also labeled low EEO) agarose should be purchased to reduce the effects of electroendosmosis. Molecular-biology grade agarose has the added benefit of being certified free of DNA- and RNA-destroying nucleases. Specialty agarose preparations are also available such as NuSieve 3:1 for small DNA fragments—50 basepairs (bp) to 1 kilobasepair (kb)—or SeaKem Gold for large DNA fragments—1 to 1000 kb. Both of these agarose preparations are available from Lonza.

Visualizing DNA and RNA

Nucleic acids on their own are not visible but can be visualized by staining agarose gels with various DNA-binding fluorescent dyes. One of the most common is ethidium bromide (Fig. 7.2.3A). This aromatic, largely planar compound interacts with DNA by intercalating among the stacked base pairs (Fig. 7.2.3B), and with RNA by intercalating between the RNA bases, thereby concentrating the dye on the molecules and stimulating increased fluorescence emission at 590 nm when excited with UV light (302 nm).

STRATEGIC PLANNING

When performing DNA agarose gel electrophoresis, one must choose (1) the size of the gel, (2) the percentage of agarose, (3) the running buffer, (4) the voltage at which to run the gel, and (5) the method used for staining the DNA. Each of these decisions can affect the overall time required.

When running an RNA gel for a northern blot, many of the same decisions must be made (gel size, percentage agarose, voltage, choice of stain).

Gel Size

Minigels (~8 cm long) are sufficient for many applications, including analysis of plasmid digestions (*UNIT 4.2*) and PCR products (*UNIT 10.2*). Larger gels may be used for Southern blotting (see *UNIT 8.2*) of digested genomic DNA. A larger gel will allow greater resolution of similarly sized DNA fragments. Gel rigs vary depending on the manufacturer. Become familiar with your gel rig before beginning the protocol.

Agarose Percentage

The percentage of agarose is chosen based on the size of DNA one wishes to clearly separate. Prepare a higher-percentage gel to create smaller pores (~100 nm) to resolve small DNA fragments, and a lower-percentage gel to create larger pores (~300 nm) to resolve larger DNA fragments. Table 7.2.1 provides guidelines for choosing the percentage agarose depending on the size range to be separated. For example, a common agarose concentration is 0.8% because it allows separation of DNA fragments ranging from ~0.7 to 11 kb. You will also choose a DNA size standard appropriate for the expected size range of your samples (load according to manufacturer's instructions). A DNA size standard that covers a range of 0.1 to 12 kb is Invitrogen's 1 kb Plus DNA ladder, and a number of manufacturers offer similar DNA size standards.

Table 7.2.1 Choosing the Percentage of Agarose for DNA Gels[a]

Effective range of resolution (linear double-stranded DNA)	Percentage agarose
0.2 to 3 kb	1.5
0.4 to 7 kb	1.2
0.5 to 10 kb	1.0
0.8 to 12 kb	0.7
1.0 to 30 kb	0.5

[a]Adapted from *CP Molecular Biology Unit 2.5A* (Voytas, 2000).

Running Buffer

For DNA gels, one may use Tris-acetate-EDTA (TAE) or Tris-borate-EDTA (TBE) running buffer. TAE is commonly used since it is easy to make and store. (TBE may form precipitates upon storage.) TBE has a superior buffering capacity, as the pK_a of Tris (8.08) is closer to that of boric acid (9.24) than it is to the pK_a of acetic acid (4.76, also see *UNIT 3.3*). TBE is not necessarily always best for large DNA fragments (Miura et al., 1999). Smaller DNA fragments (<1 kb) may be better resolved in TBE gels, since TBE interacts with agarose more tightly, effectively forming smaller pores. TAE should be used if the DNA is to be purified from the gel, since the interaction between TBE and agarose can reduce the amount of DNA recovered (Ogden and Adams, 1987). While many labs make their own solutions for agarose gel electrophoresis (see Reagents and Solutions), premade stock solutions of running buffers and gel loading buffers are available from a variety of manufacturers (including Bio-Rad and Genesee Scientific).

Voltage

As a constant electric field is desired, agarose gels are run at a constant voltage. Most agarose gels can be run at 10 V/cm. An 8-cm gel can therefore be run at 80 V. The resolution of large DNA (above 5 kb) is improved if the gel is run more slowly (1 to 5 V/cm); this is often the case when the gel will be used for a Southern blot (*UNIT 8.2*).

Stains

Ethidium bromide (also see discussion in Overview and Principles) is most commonly used to stain DNA and RNA agarose gels. For DNA gels, one can choose to stain the gel after the run is complete (as described in the Basic Protocol 1) or to include ethidium bromide in the gel. Adding ethidium bromide directly to the DNA agarose gel to a final concentration of 0.5 μg/ml will allow staining of DNA during the run and visualization immediately following (Sharp et al., 1973; Borst, 2005). However, as discussed in Understanding Results, this technique can result in poor staining of small DNA fragments (Fig. 7.2.7). To prevent this, one can include ethidium bromide at 0.5 μg/ml in the running buffer. However, doing so tends to generate a greater volume of ethidium bromide liquid waste. See the discussion in Understanding Results to determine which staining method is more appropriate for your purposes. Refer to the discussion in Variations for alternative staining methods using SYBR Gold, SYBR Green I and II, and Gelstar.

Time Considerations

DNA agarose gel electrophoresis

An 8-cm-long minigel will require a total of 2.5 to 3 hr to prepare, load, run, stain, and photograph. A large agarose gel (25 cm long) can require >8 hr. Large gels are often run overnight at lower voltage to save working time. If small DNA fragments (<1 kb) are being analyzed, a short gel run is desired to prevent significant diffusion; longer DNA fragments (>5 kb) will require more run time to allow for adequate separation of these more slowly migrating molecules.

The time required for several of the steps in Basic Protocol 1 for DNA electrophoresis can be adjusted, thereby allowing the researcher to space agarose gel electrophoresis among other laboratory activities. For example, once the gel solidifies, it does not have to be loaded immediately. Running buffer should be poured over the gel to prevent it from drying out, and it is best to use a gel within 1 day. The gel running time is flexible; the voltage can be reduced slightly to increase the running time. The ethidium bromide staining and destaining times can be increased up to 1 hr each. However, the DNA will begin to diffuse immediately after the run is complete, so it is desirable to keep the staining and destaining times as short as possible. Staining requires enough time (∼30 min) to allow diffusion of the dye into the gel matrix. This is followed by destaining (∼30 min), to allow any dye that is not intercalated into the DNA to be washed out. A total of ∼1 hr is thus needed for staining and destaining.

Denaturing RNA agarose gel electrophoresis

Running a denaturing formaldehyde RNA gel (Basic Protocol 2) will take ∼6 hr. The inclusion of formaldehyde in denaturing RNA gel electrophoresis limits the flexibility in the timing of preparation and running of the gel.

SAFETY CONSIDERATIONS

The power supplies used for electrophoresis are capable of delivering several hundred milliamps (mA) of current and should be considered potentially dangerous, especially in combination with the buffered solution in the electrophoresis chamber. Care should be taken to apply voltage to the apparatus only after plugging in the electrical leads to the power supply and electrophoresis chamber, to reduce any possibility of electrocution.

Ethidium bromide used in staining DNA in agarose gels is not, on its own, mutagenic, but can be converted by liver enzymes into mutagenic forms (McCann et al., 1975). It is therefore appropriate to handle both the agarose gels and solutions containing ethidium bromide with care. Appropriate lab wear is recommended (lab coat, gloves, closed-toed shoes, and safety goggles). Solutions containing ethidium bromide can be stored in waste containers for future absorption on activated charcoal (Extractor EtBr System from Whatman) or for chemical decontamination (see discussion in *CP Molecular Biology Appendix 1H*; Lunn and Lawler, 2002). Contact your department's safety officer to determine the most appropriate method for disposing of ethidium-containing agarose gels and solutions.

Gloves should be used for all steps in agarose gel electrophoresis, as the pipettors may be contaminated with ethidium bromide or other chemicals. Use of gloves is also critical to protect your DNA and RNA samples from degradation by nucleases secreted by your hands.

Formaldehyde and formamide, used in denaturing RNA gel electrophoresis, must be handled with gloves and safety goggles in a fume hood. Contact your department's safety officer to determine the most appropriate method for disposing of formaldehyde-containing agarose gels and solutions.

PROTOCOLS

Basic Protocol 1: DNA Agarose Gel Electrophoresis

A slab agarose gel allows separation of double-stranded DNA based on size, and is most useful for resolving DNA fragments ranging from 0.1 to 12 kb. The technique is broken down into several steps, which are summarized in Table 7.2.2. Two variables to consider in the preparation of DNA samples for agarose gels are mass and volume. The smallest amount of DNA easily visualized on ethidium bromide–stained agarose gels is ~20 ng. The volume that can be loaded in a single well is determined by the size of the comb (both length and width of each tooth), and the thickness of the gel. A 10-tooth comb to be used in a minigel typically holds up to 25 μl, but smaller volumes are easier to load and less prone to artifacts from spillover into the next well. Gel combs vary depending on the manufacturer; see your instruction manual for exact volumes recommended for use with your comb.

This protocol is written with DNA samples in mind; however one could just as easily run RNA samples on these gels. In the absence of the denaturant formaldehyde, the RNA will form secondary structures and will not run strictly according to its size. Use of formaldehyde gels for RNA separation is described in Basic Protocol 2. However, running RNA on nondenaturing agarose gels can quickly assess the abundance of RNA samples, for example, determining whether in vitro transcription reactions have yielded a sufficient quantity of product. When working with RNA samples, one must use RNase-free solutions and equipment to avoid degradation; refer to Chapter 3 and UNIT 8.2.

Table 7.2.2 Time Requirements for Performing DNA Agarose Gel Electrophoresis

Procedure	Time requirement
Prepare and pour the gel	45 min to 1 hr
Prepare and load DNA samples	10 min
Run the gel	45 min to overnight
Stain and photograph the gel	1 hr

Materials

Running buffer: 1× TAE running buffer (see recipe for 50×) *or* 0.5× TBE running buffer (see recipe for 5×)
Agarose, molecular-biology grade
6× loading buffer (see recipe)
DNA samples to be run
DNA size standard (often called DNA ladder or DNA marker)
Ethidium bromide working stain solution (see recipe)

Erlenmeyer flask of appropriate size for preparation of agarose gel
Autoclave gloves
Horizontal gel electrophoresis system, including a buffer chamber, gel tray, comb
Electrophoresis power supply
Containers for staining and destaining gels (Rubbermaid-type work well)
Platform shaker or orbital shaker
UV light box
Gel documentation system: Polaroid-film based or CCD digital–camera based

Prepare and pour the agarose gel

1. Prepare sufficient running buffer to fill the buffer chamber and for casting the gel.

   For a typical minigel, it is sufficient to make 300 to 400 ml. Either 1× TAE or 0.5× TBE running buffer can be used.

   Some labs prefer one buffer over the other. In the authors' experience, TBE gels typically run cooler, and can therefore be run at higher voltage. If DNA is to be recovered from the gel, TAE is recommended, since TBE interacts with agarose.

2. In a flask, add agarose to the appropriate amount of running buffer (Fig. 7.2.4A).

   A typical minigel will be 45 ml in volume. To create a 0.8% agarose gel, add 0.36 g agarose to 45 ml of buffer. The typical thickness for an agarose gel is 0.6 to 0.8 cm.

3. Microwave the agarose to melt. Begin on medium power and watch the flask carefully (agarose easily boils over). Stop the microwave periodically (i.e., every 20 to 30 sec) and, using autoclave gloves or other heat-resistant gloves, gently swirl to mix.

   CAUTION: *Overheated solutions can boil over and cause severe burns by soaking through fabric gloves.*

   The agarose is completely dissolved when you no longer see transparent flecks floating in the clear solution (Fig. 7.2.4B).

4. Let the gel solution sit on the bench until it is cool enough to handle without a protective glove. Monitor every ~5 min, since the exact cooling time depends on the gel volume.

   The gel will be ready to pour when the temperature drops to ~55° to 60°C, but it is unnecessary to measure the exact temperature. If the gel is poured when the agarose is too hot (>60°C) the Plexiglas gel-casting platform can be damaged. If you are adding ethidium bromide directly to the gel (described below), add it before the gel solution has cooled, and gently swirl to mix. While the gel is cooling, avoid swirling or shaking the solution, as this will introduce unnecessary bubbles. If using a stain other than ethidium bromide, follow supplier's recommendations (also see Variations).

5. Set up the gel tray, insert the comb, and pour the gel (Figure 7.2.4C).

   Some types of gel electrophoresis systems require that the gel tray be taped on the sides; the tape is removed after the gel sets. Other systems have a rubber gasket in the sides of the gel tray. In these systems the tray is placed in the buffer chamber 90° with respect to the usual running orientation (or in a casting tray), and the gel is then poured (Fig. 7.2.4C). After the gel sets, the tray is lifted out, rotated 90°, and replaced in the buffer chamber.

7

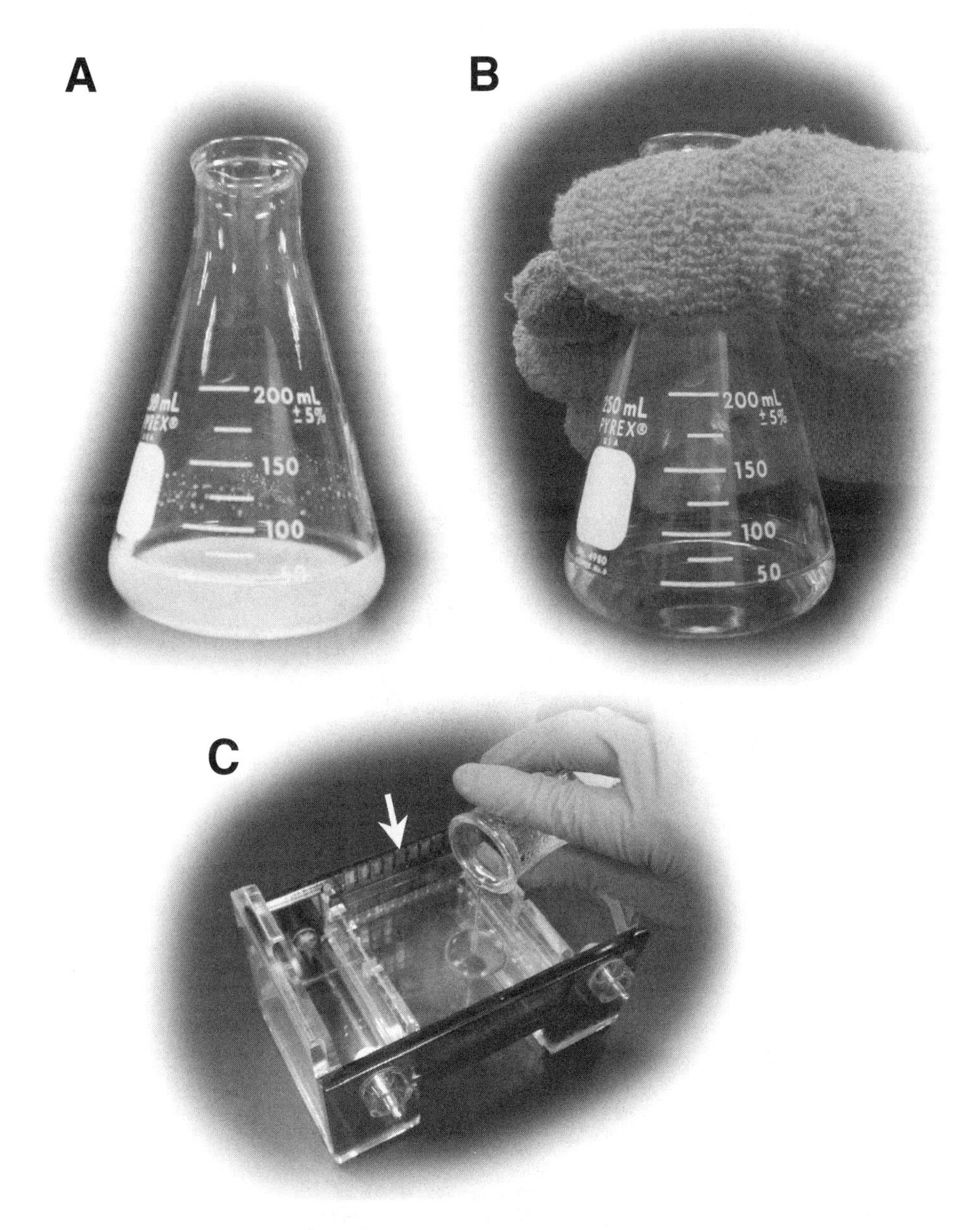

Figure 7.2.4 Pouring the agarose gel. (**A**) Addition of agarose to 1× TAE running buffer. (**B**) After dissolving the agarose in a microwave, the gel solution is clear, with no transparent specks of agarose evident. (**C**) Once the gel solution has cooled to allow handling (55° to 60°C), it can be poured. For the gel rig pictured, the gel tray is placed in the buffer chamber 90° with respect to the usual running orientation, and the gel is poured. Rubber gaskets in the sides of the gel tray prevent leaking. Note the presence of the comb (arrow). This particular comb is double sided, with one set of teeth thicker than the other to allow a choice of narrow or thick wells.

Prepare DNA samples and load the agarose gel

6. Add sufficient 6× loading buffer to each of your DNA samples for a final concentration of 1× (i.e., 1 µl of 6× sample loading buffer for every 5 µl of DNA sample).

   The loading buffer contains glycerol to increase the density of the DNA solution such that when loaded into the well, it will sink to the bottom of the well rather than float away in the buffer. In addition, loading buffer contains dye(s) to facilitate gel loading and track migration through the gel. Most laboratories use the standard dyes bromphenol blue (dark blue) and the slower-migrating xylene cyanol (light blue). Bromphenol blue migrates with an apparent size of 600 bp, while xylene cyanol migrates with an apparent size of 5 kb. Both dyes are visible in pictures of gels and can interfere with the visualization of DNA. Orange G is a useful dye that migrates further than bromphenol blue (further than even small DNA fragments of 150 bp) and is less visible in photographs.

7. Once the gel has set (~30 min), remove the comb (Fig. 7.2.5A).

If the comb is removed before the gel has set, the wells can tear. When fully set, the gel looks milky and translucent without any spots. Pull the comb straight up in a slow, smooth motion, using two hands. A comb pulled sideways will tear the wells.

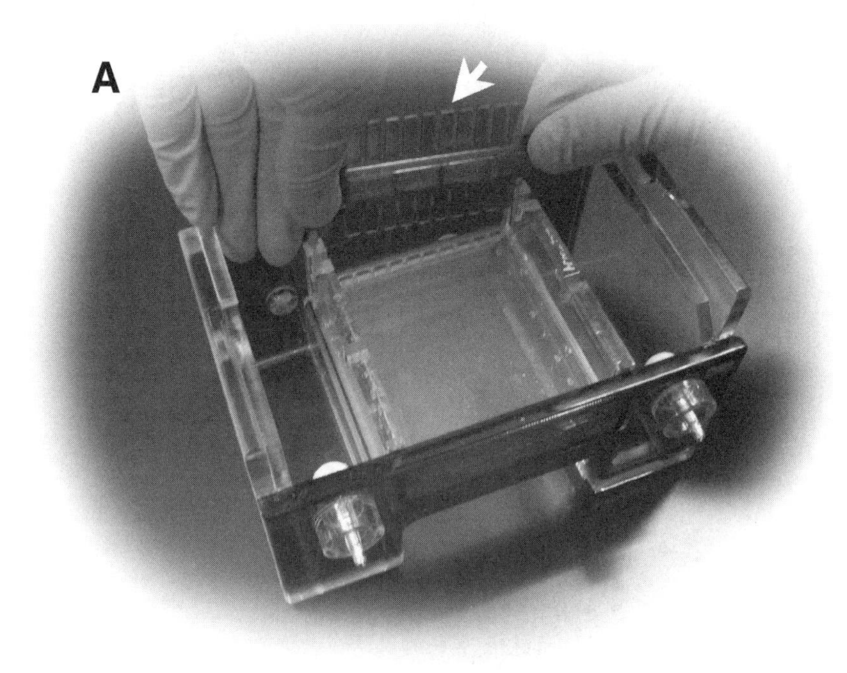

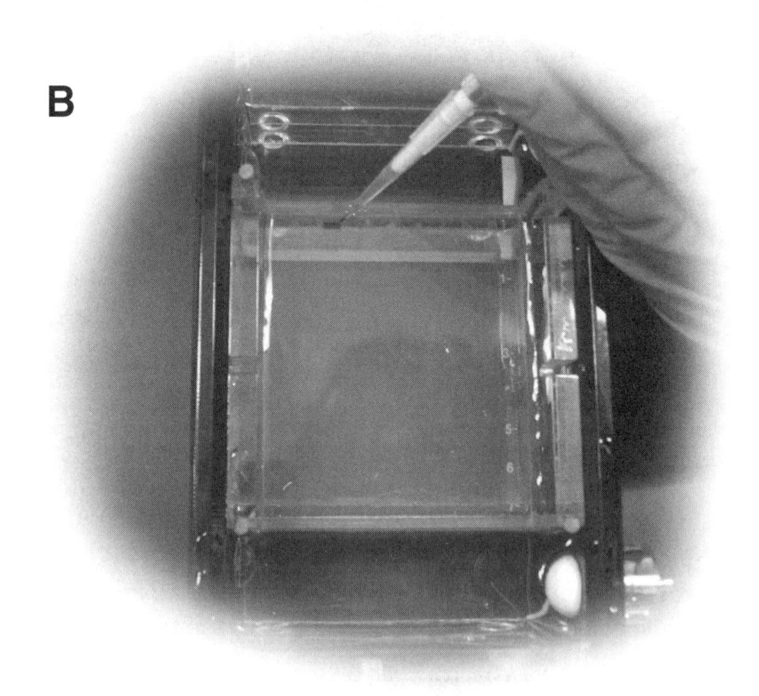

Figure 7.2.5 Loading the agarose gel. (**A**) The gel has fully set when it looks translucent, but not clear in any spots; this usually takes ~30 min. Pull the comb (arrow) straight up in a slow, smooth motion, using two hands. In the gel rig pictured, the tray will be lifted out, rotated 90°, and replaced in the buffer chamber. (**B**) When loading the gel, stabilize the micropipettor with your other hand. Avoid tearing or puncturing the well with the pipet tip.

8. Position the gel in the buffer chamber and add a sufficient volume of running buffer (1× TAE or 0.5× TBE) to fill the buffer chamber and cover the gel by ~0.5 cm.

   The gel rig should be close to the power supply before you begin loading, to avoid having to move the loaded gel.

9. Load the DNA size standard, following manufacturers instructions to determine the correct amount, and then load your DNA samples in the remaining wells (Fig. 7.2.5B).

   Avoid bubbles in the ends of the micropipet tips, as they will be ejected in front of your sample, then rise through the sample and disrupt it. With your free hand, brace your arm to stabilize the pipettor. Pipet slowly and steadily so that the sample falls down into the well without spilling over into the next well (Fig. 7.2.5B). Avoid touching the micropipet tip to the bottom of the well, as the bottom of the well is easily punctured. Likewise, avoid pressing the tip into the side of the well.

   A DNA size standard should be included on every gel.

Run the agarose gel

10. Place the lid on the buffer chamber and plug the leads into the power supply (Fig. 7.2.6A). Orient the lid such that the black power supply lead (the cathode, negative lead) is behind the DNA lanes, with the DNA running towards the red power supply lead (the anode, positive lead) as illustrated in Figure 7.2.6B.

    By convention, DNA gels are viewed with the wells at the top (see Fig. 7.2.7). It therefore makes sense to orient the buffer chamber such that the wells are at the top (as seen looking down at the gel), and load the gel in this way. If this orientation is used, the rhyme "black in back" can be used to remember the correct orientation of power supply leads. Do not reposition the gel after the samples are loaded, as the samples are easily dislodged. If the gel is placed such that the red lead (typically positive) is behind the DNA, the power leads may be plugged into the power supply backwards to generate the correctly oriented electric field—i.e., to allow DNA to migrate toward the anode (positive).

11. Run the gel at the desired voltage until the bromphenol blue dye has migrated two-thirds the length of the gel (this dye comigrates with DNA ~600 bp in size).

    Most agarose gels can be run at 10 V/cm. The average minigel can therefore be run at 80 to 100 V, and the average run will take from 45 min to 1 hr. If large DNA (above 5 kb) is to be separated, greater resolution will result if the gel is run more slowly (1 to 5 V/cm). A gel containing digested genomic DNA to be used for a Southern blot (UNIT 8.2) should be run at lower voltage for better resolution of a large size range of DNA fragments. On the other hand, small DNA fragments (<500 bp) should be run at 10 V/cm to prevent fuzzy bands caused by diffusion.

Stain the agarose gel with ethidium bromide, image, and photograph

12. In a plastic container, incubate the gel in sufficient ethidium bromide working stain solution to cover the gel. Shake on a platform or orbital shaker for ~30 min at room temperature.

    CAUTION: *This solution should be disposed of as hazardous waste. Contact your safety officer for your institution's procedure for disposal of liquid ethidium bromide waste. Detailed protocols for disposal of ethidium bromide are provided in CP Molecular Biology Appendix 1H (Lunn and Lawler, 2002). It is not advisable to add bleach to ethidium bromide solutions, as the combination forms carcinogenic compounds.*

13. Destain the gel in water at room temperature on a shaker for ~30 min.

    The more often the water is changed, the more quickly the gel will destain. Small DNA will diffuse quickly and should be destained for <1 hr; larger DNA (>1 kb) can be destained up to 2 hr.

    Although the water used to destain will contain a minimal amount of ethidium bromide, this is often not considered enough to qualify as hazardous waste. Consult your safety officer for institutional guidelines.

A

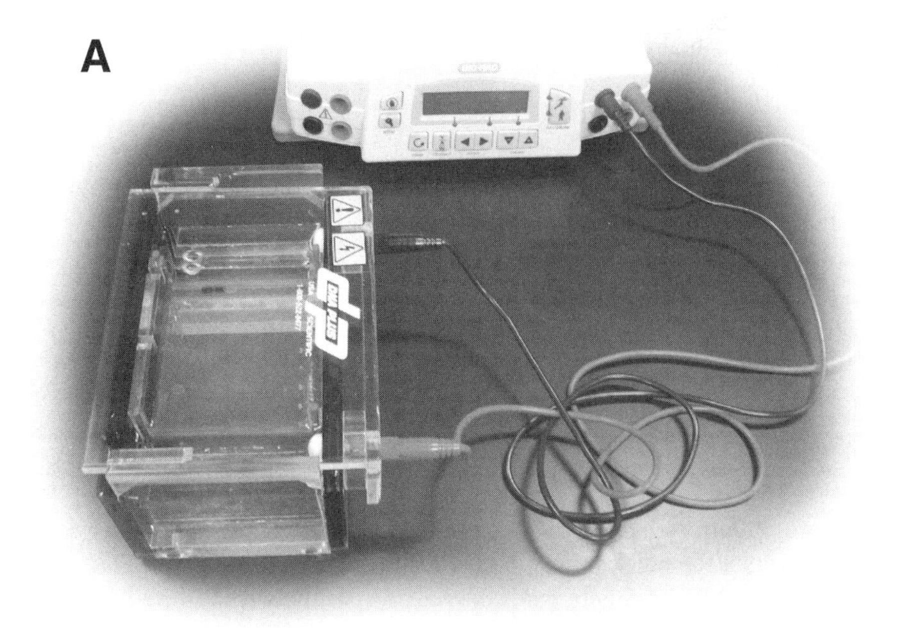

B

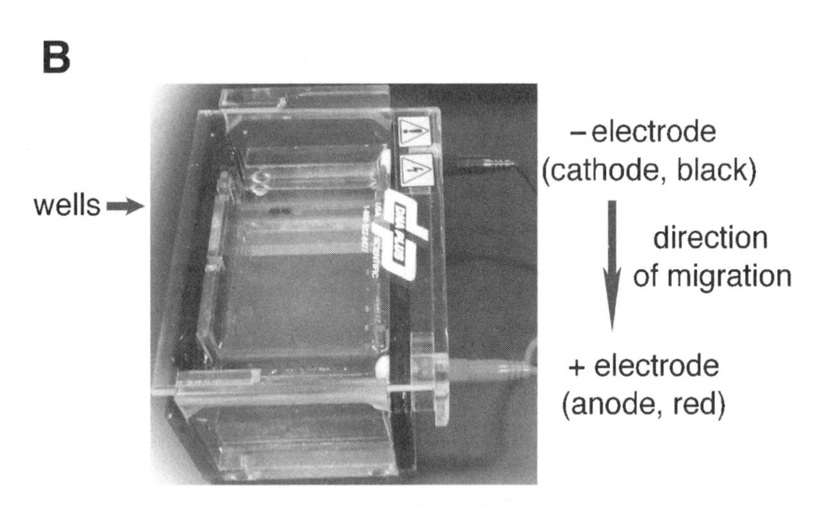

Figure 7.2.6 Running the agarose gel. (**A**) Place the lid on the buffer chamber and connect the leads to the power supply. (**B**) Orient the lid such that the black power supply lead (the cathode, negative lead) is behind the DNA lanes, with the DNA running towards the red power supply lead (the anode, positive lead). This gel rig has the leads built into the cover.

14. Using a spatula or gloved hands, carefully transfer the gel to a UV light box. Photograph using a digital gel documentation device (*UNIT 7.5*) or Polaroid camera with hood-like attachment designed for photographing agarose gels.

    CAUTION: *Avoid exposure of skin and eyes to UV light. Use a face shield or safety goggles. Avoid touching any surface with gloves that are possibly contaminated with ethidium bromide. Especially prone to contamination are the computer keyboard and mouse of the gel documentation system. Contact your safety officer for your institution's procedure for disposal of ethidium bromide–stained gels.*

    If the DNA is to be isolated from the gel and used for cloning, limit the UV exposure time to avoid mutations. If possible, use a long wavelength UV lamp.

    A red filter (Wratten #22) or interference filter with peak transmission of 590 nm allows for ideal contrast of ethidium bromide–stained gels.

Basic Protocol 2: Denaturing RNA Agarose Gel Electrophoresis

This protocol should be used together with those in *UNIT 8.2*, which provides instructions for blotting the RNA gel. The protocol is similar in many respects to DNA agarose gel electrophoresis. The largest difference is the inclusion of formaldehyde in the RNA samples and in the agarose gel. Formaldehyde is necessary to denature the RNA and prevent the formation of regions of double-stranded RNA (which would affect the migration of the RNA). The inclusion of formaldehyde also necessitates that the gel be prepared and run in a fume hood.

NOTE: All solutions used for RNA agarose gel electrophoresis should be prepared with DEPC-treated sterile deionized or ultrapure water as described in *UNIT 8.2*.

Materials

Agarose, molecular biology grade
10× and 1× MOPS running buffer (see recipe)
37% (12.3 M) formaldehyde, pH >4.0
RNase-free H_2O (*UNIT 8.2*)
Formamide
Formaldehyde loading buffer (see recipe)
RNA molecular weight ladder and/or RNA of known size
0.5 M ammonium acetate
0.5 μg/ml ethidium bromide in 0.5 M ammonium acetate (prepare from 10 mg/ml ethidium bromide stock)

Erlenmeyer flask of appropriate size for preparation of agarose gel
Autoclave gloves
55°C water bath or heat block
Horizontal gel electrophoresis system, including a buffer chamber, gel tray, and comb
Electrophoresis power supply
RNase-free containers for staining and destaining gels (Rubbermaid-type work well; see *UNIT 8.2* for elimination of RNase contamination)
Platform or orbital shaker
UV box
Gel documentation system: Polaroid-film based or CCD digital digital–camera based

Prepare and pour the RNA denaturing agarose gel

The following instructions will make 100 ml of a 1% agarose gel, which is ideal for separating RNA 0.5 kb to 10 kb in size. The gel concentration can be increased to 2% for smaller RNA molecules, and decreased to 0.7% for larger RNA molecules. The gel volume can be scaled up or down depending on the size of the gel to be prepared.

1. In a flask, add 1.0 g agarose to 72 ml of water. Heat the agarose-containing solution in a microwave oven to melt the agarose. Begin on medium power, and watch the flask carefully (agarose solutions easily boil over). Stop the microwave periodically (i.e., every 20 to 30 sec) and, using autoclave gloves or other heat-resistant gloves, gently swirl to mix.

   CAUTION: *Overheated solutions can boil over and cause severe burns by soaking through fabric gloves.*

   The agarose is completely dissolved when you no longer see transparent flecks floating in the clear solution (Fig. 7.2.4B).

2. Allow the agarose solution to cool to ∼60°C.

3. Working in a fume hood, add 10 ml of 10× MOPS running buffer and 18 ml of 37% formaldehyde (final concentrations are 1× MOPS buffer and 6.66% formaldehyde).

7

CAUTION: *Formaldehyde is toxic. It can be absorbed directly through the skin or the vapors can be inhaled. All work with formaldehyde must be done in a fume hood while wearing gloves and safety goggles.*

4. Set up the gel tray, insert the comb, and pour the gel in the fume hood (Fig. 7.2.4C).

   IMPORTANT NOTE: *For RNA gels, the gel tray, comb, and gel rig must be treated to inactivate RNases (UNIT 8.2). If RNA gels are frequently run, it is helpful to dedicate specific equipment for RNA work.*

   Some types of gel electrophoresis systems require that the gel tray be taped on the sides; the tape is removed after the gel sets. Other systems have a rubber gasket in the sides of the gel tray. In these systems, the tray is placed in the buffer chamber 90° with respect to the usual running orientation (or in a casting tray), and the gel is then poured (Fig. 7.2.4C). After the gel sets, the tray is lifted out, rotated 90°, and replaced in the buffer chamber.

Prepare RNA samples and load the agarose gel

5. Add RNase-free water to each of your RNA samples for a final volume of 5.5 μl, with each sample containing the appropriate quantity of RNA to be loaded per lane of the gel.

   If running total RNA, 10 to 50 μg should be loaded per lane; if you have already selected for poly(A)$^+$ mRNA, load less (0.5 to 10 μg per lane). Refer to CP Molecular Biology Unit 4.5 (Kingston, 1993), for isolation of poly(A)$^+$ mRNA.

   If the volume of any single RNA sample to be loaded on the gel is >5.5 μl, see UNIT 5.2 for instructions on how to concentrate your RNA. Alternatively, use a larger comb to allow for a greater volume to be loaded. 5.5 μl of RNA will be loaded as a final volume of 30 μl once all necessary additions (see steps 6 to 8) have been made.

6. Working in a fume hood, to each sample add:

   2.5 μl 10× MOPS running buffer
   4.5 μl 37% formaldehyde
   12.5 μl formamide.

   CAUTION: *Formamide is a teratogen, and should only be handled with gloves and safety goggles in a fume hood.*

   This recipe for sample preparation will generate a final volume of 30 μl for each RNA sample. The volumes can be adjusted for a larger or smaller total volume (depending upon the size of the comb used).

7. Mix by vortexing. Microcentrifuge briefly to collect the samples at the bottoms of the tubes and incubate 5 min at 55°C in a water bath or heat block.

8. Working in a fume hood, add 5 μl formaldehyde loading buffer to each RNA sample, vortex briefly, and pulse centrifuge.

9. Once the RNA agarose gel has set (~30 min), remove the comb (Fig. 7.2.5A).

   If the comb is removed before the gel has set, the wells can tear. When fully set, the gel looks milky and translucent without any spots.

   Pull the comb straight up in a slow, smooth motion, using two hands: a comb pulled sideways will tear the wells.

10. Assemble the buffer chamber in the fume hood and position the gel in the buffer chamber. Add a sufficient volume of 1× MOPS running buffer to fill the buffer chamber and cover the gel by ~0.5 cm.

    The gel rig should be close to the power supply before you begin loading to avoid having to move the loaded gel.

11. Load the RNA samples (load the entire 30 μl). In outside lanes that can be cut off and stained, you may include an RNA molecular weight ladder, RNA of known size, and/or a duplicate lane of one or more of your RNA samples.

Avoid bubbles in the ends of the micropipet tips, as they will be ejected in front of your sample, then rise through the sample and disrupt it. With your free hand, brace your arm to stabilize the pipettor. Pipet slowly and steadily so the sample falls down into the well without spilling over into the next well (Fig. 7.2.5B). Avoid touching the micropipet tip to the bottom of the well, as the bottom of the well is easily punctured. Likewise, avoid pressing the tip into the side of the well.

Running the denaturing RNA gel

12. Place the lid on the buffer chamber and plug the leads into the power supply (Fig. 7.2.6A). Orient the lid such that the black power supply lead (the cathode, negative lead) is behind the RNA lanes, with the RNA running towards the red power supply lead (the anode, positive lead) as illustrated in Figure 7.2.6B.

By convention, RNA gels and DNA gels are viewed with the wells at the top (see Fig. 7.2.7). It therefore makes sense to orient the buffer chamber such that the wells are at the top (as seen looking down at the gel), and load the gel in this way. If this orientation is used, the rhyme "black in back" can be used to remember the correct orientation of power supply leads. Do not reposition the gel after the samples are loaded, as the samples are easily dislodged. If the gel is placed such that the red lead (typically positive) is behind the RNA, the power leads may be plugged into the power supply backwards to generate the correctly oriented electric field—i.e., to allow RNA to migrate toward the anode (positive).

13. Run the gel at 5 V/cm until the bromphenol blue dye has migrated between one-half and two-thirds the length of the gel (~3 hr).

One should not reduce the voltage and run the gel longer than 5 hr without increasing the amount of formaldehyde—see CP Molecular Biology Unit 4.9 (Brown et al., 2004).

Staining the RNA agarose gel with ethidium bromide

14. With a new razor blade, carefully cut off the lanes to be stained (see step 11). In an RNase-free plastic or glass container, incubate the gel slice in 0.5 M ammonium acetate for 40 min at room temperature, changing the solution halfway through the incubation period.

These lanes will not be transferred to the membrane during the northern blot.

RNA that is denatured by formaldehyde does not stain with ethidium bromide. Ammonium acetate is used to remove the formaldehyde.

15. Pour off the solution. Place the gel slice in enough 0.5 M ammonium acetate containing 0.5 μg/ml ethidium bromide to cover the gel.

16. Incubate at room temperature on a shaker for ~40 min. Dispose of the solution as hazardous waste.

Contact your safety officer for your institution's procedure for disposal of liquid ethidium bromide waste. Detailed protocols for disposal of ethidium bromide are provided in CP Molecular Biology Appendix 1H (Lunn and Lawler, 2002). Is not advisable to add bleach to ethidium bromide solutions, as the combination forms carcinogenic compounds.

17. If desired, destain the gel slice in 0.5 M ammonium acetate at room temperature on a shaker for up to 60 min.

Although the solution used to destain will contain a minimal amount of ethidium bromide, this is often not considered enough to qualify as hazardous waste. Consult your safety officer for institutional guidelines.

18. Using a spatula or gloved hands, carefully transfer the gel to a UV light box. Photograph using a digital gel documentation device (UNIT 7.5) or Polaroid camera with hood-like attachment designed for photographing agarose gels.

It is important to include a ruler placed alongside the gel so that band positions can later be identified on the northern blot membrane. A red filter (Wratten #22) or interference filter with peak transmission of 590 nm allows for ideal contrast of ethidium bromide–stained gels. Contact your safety officer for your institution's procedure for disposal of ethidium bromide–stained gels.

19. Prepare the RNA gel for transfer by northern blot (UNIT 8.2).

REAGENTS AND SOLUTIONS

Use deionized, distilled water in all recipes and protocol steps. For common stock solutions, see **UNIT 3.3**.

Ethidium bromide working stain solution

Prepare a stock solution of 10 mg/ml ethidium bromide in water. Store up to 1 year at 4°C, protected from light. Prior to use, prepare a fresh working solution of 0.5 to 1.5 μg/ml ethidium bromide in water. Discard used solution in an appropriate waste container.

CAUTION: *Weigh ethidium bromide carefully in a fume hood while wearing gloves and goggles. Refer to CP Molecular Biology Appendix 1H (Lunn and Lawler; 2002) for additional information.*

Premade liquid stock solutions of ethidium bromide are commercially available (e.g., Bio-Rad).

Formaldehyde loading buffer

1 mM EDTA, pH 8.0 (UNIT 3.3)
0.25% (w/v) bromphenol blue
0.25% (w/v) xylene cyanol FF
50% (w/v) glycerol

To avoid growth of microorganisms, pass through a 0.45-μm filter to sterilize. Store in small (0.5-ml) aliquots up to 1 year (if unopened) at 4°C. Once opened, aliquots can be used up to 6 months.

Loading buffer, 6×

0.25% (w/v) bromphenol blue
0.25% (w/v) xylene cyanol FF (optional)
30% (w/v) glycerol

To avoid growth of microorganisms, pass through a 0.45-μm filter to sterilize. Store in small (0.5-ml) aliquots up to 1 year (if unopened) at 4°C. Once opened, aliquots can be used up to 6 months.

Bromphenol blue and xylene cyanol FF can be substituted with 0.25% (w/v) orange G (e.g., Sigma-Aldrich).

MOPS running buffer (10×)

0.4 M 3-(*N*-morpholino)-propanesulfonic acid (MOPS), pH 7.0
0.1 M sodium acetate (UNIT 3.3)
10 mM EDTA (add from 0.5 M EDTA, pH 8.0; UNIT 3.3)
Store up to 3 months at 4°C
Dilute to 1× with H_2O as needed

TAE buffer, 50×

242 g Tris base
57.1 ml glacial acetic acid
100 ml 0.5 M EDTA, pH 8.0 (UNIT 3.3)
Bring volume up to 1 liter with H_2O
Store up to 1 year at room temperature
Dilute to 1× with H_2O as needed

CAUTION: *Add glacial acetic acid only after the majority of water has been added.*

TBE buffer, 5×

54 g Tris base
27.5 g boric acid
20 ml 0.5 M EDTA, pH 8.0 (*UNIT 3.3*)
Bring volume to 1 liter with H_2O
Store up to 1 year at room temperature
Dilute to 0.5× with H_2O as needed

UNDERSTANDING RESULTS

This section is devoted to understanding and interpreting DNA gels.

Linear DNA Runs According to Its Size

Agarose gel electrophoresis can be used to determine the size of linear DNA fragments (Takahashi et al., 1969). The gels shown in Figure 7.2.7 are actual size; therefore, the DNA bands in the molecular weight standards can be measured and used to reproduce the standard curve shown in Figure 7.2.8. Figure 7.2.8 illustrates the relationship between the distance traveled by DNA through an agarose gel and the log of the number of base pairs (bp) in the DNA. The two gels in Figure 7.2.7 contain identical DNA samples. The gel on the left was stained with ethidium bromide after the run was complete, while the gel on the right was prepared with 0.5 µg/ml ethidium bromide and was visualized immediately after the run. Ethidium bromide was not included in the running buffer, as discussed below.

The DNA run on these gels is a 5.7-kb plasmid vector containing a small DNA insert. The plasmid in lanes 2 and 6 was linearized with a restriction enzyme (*UNIT 10.1*). Restriction enzymes were used to cut the insert out of the vector in lanes 3 and 7. The uncut plasmid (circular) is in lanes 4 and 8.

To determine the approximate size of a linear DNA molecule, generate a standard curve by plotting the log of the base pairs of each of the bands in the DNA size standard against the distance traveled by each band (Fig. 7.2.8; Serwer, 1989). The distance traveled is relative to an arbitrary line, usually drawn through the center of the wells, and measured to the center of each DNA band. One limitation on the accuracy of determining the size of a DNA fragment is measuring the distance migrated; a difference of <1 mm can greatly affect the determined size of the DNA fragment.

Once a standard curve is generated, one can estimate the size of a DNA fragment by measuring the distance traveled and then determining the size in base pairs using the standard curve. If you choose to use an image capture and analysis program (*UNIT 7.5*) or graphing program—e.g., Microsoft Excel, Igor (WaveMetrics), or Sigma Plot (Sigma)—you can use the equation for the best-fit line to solve for the size of the DNA. For example, the equation for the linear portion of the standard curve for the gel in the left panel of Figure 7.2.7 is $y = 4.7019 - 0.32178x$, where y is the log of the number of base pairs and x is the relative migration in centimeters. On the left gel panel in Figure 7.2.7, the distance from the well to the furthest-migrating DNA band (marked with an arrow) is 5.7 cm. On the standard curve, 5.7 cm corresponds to 2.868, the antilog of which is 738 bp.

Note that while both large and small DNA fragments are resolved on the 0.8% gel in Figure 7.2.7, DNA fragments >6 kb and <0.65 kb are outside the linear range of the standard curve (Fig. 7.2.8). Linear range refers to the points of the DNA standard whose distance of migration is correlated to the log of their number of basepairs by a linear relationship. To accurately determine the sizes of smaller DNAs, use a higher-percentage agarose gel. To determine the sizes of larger DNAs, use a lower-percentage agarose gel. One can also use the closest-migrating bands in the DNA standards to determine the size of a DNA fragment (Elder and Southern, 1983). Estimates are often made by visually comparing the DNA band to the closest migrating bands in the molecular weight standards. In Figure 7.2.7, the insert band (marked by white arrow) migrates between the 0.65- and 0.8-kb marker bands.

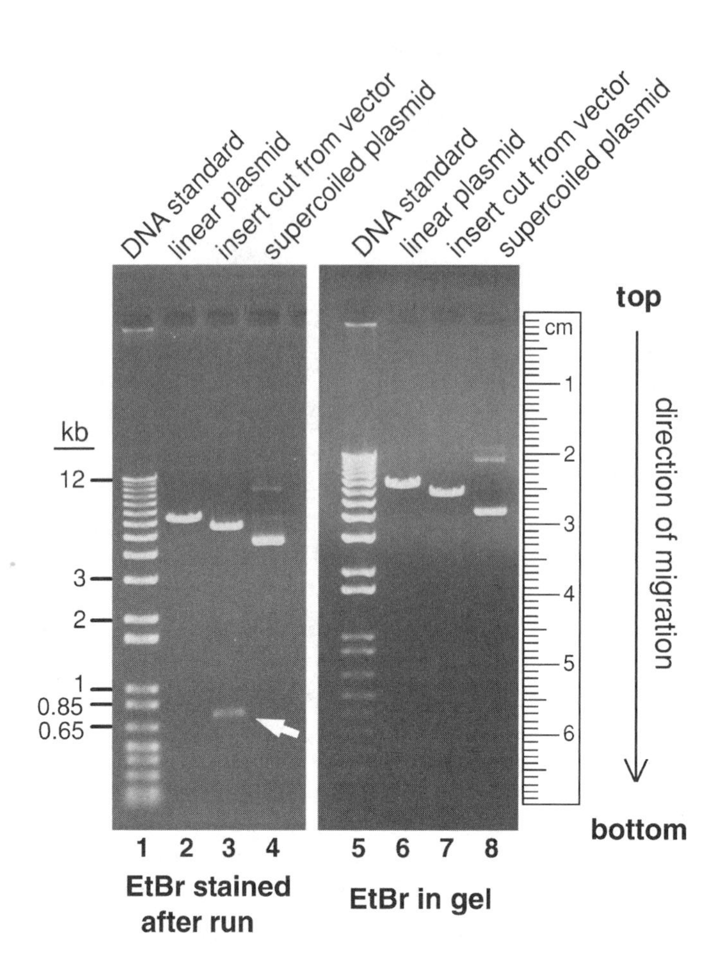

Figure 7.2.7 Linear DNA migrates according to its size. Shown are two 0.8% agarose gels loaded with identical DNA samples and run at 100 V until the bromphenol blue was three-quarters down the gel (~50 min). 500 ng of 1 Kb Plus DNA ladder (Invitrogen) was loaded into lanes 1 and 5; 300 ng of DNA was loaded into lanes 2 to 4 and 6 to 8. Lanes 2 and 6 contain the linear plasmid (6.4 kb); lanes 3 and 7 contain two DNA fragments, the vector (5.7 kb), and the DNA insert (738 bp); lanes 4 and 8 contain the uncut plasmid (supercoiled). Note that, although the supercoiled plasmid is 6.4 kb, it migrates significantly further than 6.4-kb linear DNA (for example compare lane 4 to lane 2). The slower-migrating band in lanes 4 and 8 is relaxed (nicked) circular plasmid. The gel on the left was stained with ethidium bromide after the gel run was complete. Ethidium bromide (EtBr) was added to the gel on the right, allowing the DNA to be stained during the run. Note the insert DNA band of 738 bp (white arrows) is more difficult to detect in the gel on the right. The gels are shown at actual size.

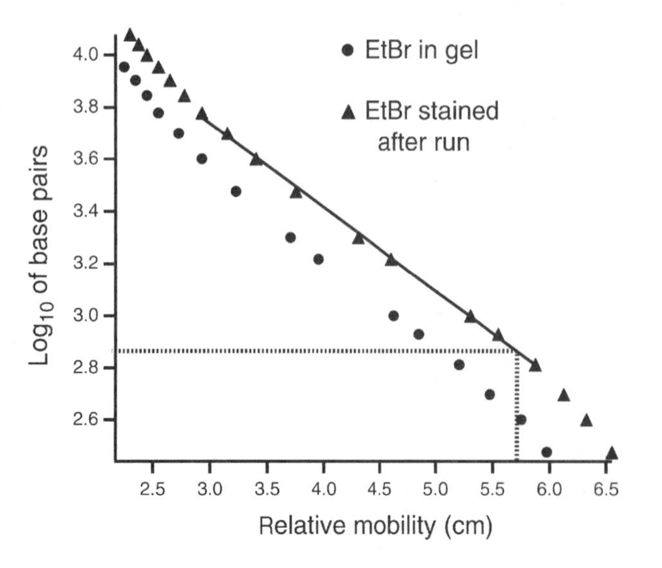

Figure 7.2.8 Standard curves of DNA size standards in Figure 7.2.7. Note that inclusion of ethidium bromide (EtBr) in the gel results in a reduction in the distance migrated by DNA. This is evidenced by the shift to the left of the standard curve. The standard curve can be used to estimate the size of other DNA bands on the same gel. The dashed line at 5.7 cm corresponds to the insert DNA (white arrow in Figure 7.2.7).

Staining Method Can Affect the Resolution and Detection Limits of the Gel

Figure 7.2.7 shows two gels with identical DNA samples that were run identical distances (as judged by the migration of the bromphenol blue marker, not evident in the figure). The gel on the right was run in the presence of 0.5 μg/ml ethidium bromide, while the gel on the left was stained with ethidium bromide after the run.

Note several consequences of including ethidium bromide in the gel:

1. The gel can be visualized immediately after the run is complete.
2. Low amounts of small DNA fragments may be difficult to see. Ethidium bromide is positively charged (Fig. 7.2.3A); ethidium bromide not bound to DNA will therefore migrate in the opposite direction to the DNA. If 0.5 μg/ml ethidium bromide is not included in the running buffer, smaller-sized DNA will be in the presence of a lower concentration of ethidium bromide and will therefore be more difficult to detect. While the 738-bp fragment is evident in the gel stained after the run (arrow), it is nearly undetectable in the gel that contained ethidium bromide (Fig. 7.2.7). There is ~35 ng of DNA in this band, which is near the detection limits of ethidium bromide–stained agarose gels.
3. DNA shows reduced migration in gels containing ethidium bromide. Intercalation of ethidium bromide into DNA can increase the DNA length by up to 27% (Freifelder, 1971), and the positive charges of ethidium bromide partially neutralize the negative charges of DNA. Together, these effects reduce the migration rates of the DNA and affect the resolution of large DNA fragments. For example, the large bands in the DNA standard are fully resolved in the gel on the left (no ethidium bromide added), but are not fully resolved in the gel on the right (ethidium bromide added). This difference in migration rate and resolution can be seen in the standard curves of the two gels in Figure 7.2.8. The points corresponding to the standards run on the agarose gel containing ethidium bromide are shifted to the left, reflecting a slower rate of migration. Furthermore, the points corresponding to the three largest bands of the DNA standard were not resolved on the gel containing ethidium bromide. Since they could not be measured, they are not present on the standard curve.

Different Forms of DNA Migrate Differently

Under most conditions, circular, supercoiled DNA runs faster than linear DNA of the same size (Fig. 7.2.7; compare linear plasmid to supercoiled plasmid). As a result, one cannot use average linear DNA size standards to determine the size of a supercoiled plasmid. A few manufacturers do offer supercoiled DNA markers, but it is most common to digest a plasmid with a restriction enzyme that cuts once, in order to determine its size.

TROUBLESHOOTING

Table 7.2.3 lists several common problems encountered when running DNA agarose gels, their potential causes, and suggested solutions.

VARIATIONS

Precast Gels

Precast agarose gels are available from several sources, e.g., Cambrex, Sigma-Aldrich, and Invitrogen. These precast agarose gels often contain ethidium bromide or other DNA stain and are ready to use out of the package. Some can be used in a standard electrophoresis buffer chamber, while others require their own specialized units. These specialized units may include a UV light, allowing visualization of the DNA as the gel runs. Given the relative ease with which an agarose gel can be prepared, precast gels are not currently cost-effective for routine use.

Table 7.2.3 Troubleshooting DNA Agarose Gels

Problem	Potential causes	Suggested solutions
Wells collapse when comb removed	Agarose gel not completely set	Allow more time for gel to set before removing comb (particularly important for low-percentage gels)
DNA sample leaks out the bottom of the well	Bottom of well punctured by micropipettor tip	Put dark piece of paper under gel rig to better visualize well when loading. Do not place tip so far down into the well.
DNA sample leaks into neighboring well and/or DNA bands from one lane inappropriately appear in another	Too much DNA sample loaded	Load less sample
	Agarose walls between wells cracked	Keep comb perfectly vertical during removal
	DNA sample loaded too quickly	Load DNA sample more slowly
Gel melts during run	Buffer omitted from gel preparation (water used instead)	Include buffer when preparing gel
	Buffer capacity depleted due to long gel run or low gel box capacity	Recirculate buffer using a peristaltic pump or stop the gel run and recirculate manually
	Voltage too high	Reduce voltage
High ethidium bromide background on gel	Gel not fully destained	Destain gel for at least same amount of time as was stained
DNA bands fuzzy	DNA diffusion due to running gel at very low voltage for long period of time	Run gel at higher voltage for less time
	DNA diffusion due to too much time elapsed between end of gel run and photography	Reduce time for staining and destaining gel
Poor resolution of DNA fragments	DNA too large or small to be properly resolved	Alter agarose concentration (see Table 7.2.1)
Bands not rectangular in shape	Well deformed by tip when loaded sample	Avoid pressing tip into side of well
	Comb was removed before gel set	Allow gel to completely set before removing comb
DNA bands are smeared	Sample degraded	Prepare new sample
	Too much DNA loaded	Load less DNA
	High salt in sample	Remove salt with a commercially available spin column or by ethanol precipitation and washing in 70% ethanol (*UNIT 5.2*)
	Agarose not completely melted	Make sure gel is completely mixed and agarose is completely dissolved
	Gel partially solidified before it was poured	Pour gel before agarose solidifies
	Gel run at too-high voltage	Run gel at lower voltage

continued

Table 7.2.3 Troubleshooting DNA Agarose Gels, *continued*

Problem	Potential causes	Suggested solutions
Gel is blank	Polarity incorrect and DNA ran backwards	Run DNA towards positive electrode (anode)
	DNA not stained	Stain gel for adequate amount of time

Isolating DNA from Agarose

There are many methods available for the recovery of fractionated DNA from agarose gels for subsequent cloning, sequencing, or other applications. These methods include electroelution, isolation on NA-45 paper (available from Whatman), and recovery from low-melting point agarose (available from Cambrex). While beyond the scope of this manual, these protocols are thoroughly described in *CP Molecular Biology Unit 2.6* (Moore et al., 2002). In addition to these classical methods, several kits are routinely used, including the spin column–based QIAquick Gel Extraction Kit (Qiagen), the glass-milk-based GeneClean kits (MP Biomedicals), and the Quantum Prep Freeze 'N Squeeze DNA gel extraction spin column (Bio-Rad). Follow the detailed manufacturers' instructions.

Alternative DNA and RNA Stains

There are several alternatives to ethidium bromide that may allow detection of smaller amounts of DNA and RNA.

SYBR Gold (Invitrogen) has an excitation maximum at 495 nm, with a smaller excitation peak at 300 nm, and an emission peak at 537 nm. The dye is reported to detect as little as 25 pg of DNA or RNA.

SYBR Green I (Invitrogen) has an excitation maximum at 497 nm, with smaller excitation peaks at 290 nm and 380 nm, and a peak emission at 520 nm. The sensitivity is given as 25 to 100 times more sensitive than ethidium bromide for double-stranded DNA. Note that this dye does not detect RNA.

SYBR Green II (Invitrogen) has an excitation maximum at 497 nm, and a peak emission at 513 nm. The sensitivity is given as 5 to 10 times more sensitive than ethidium bromide in RNA formaldehyde agarose gels. It can detect 15 ng of RNA.

Gelstar (Cambrex) has an excitation maximum at 493 nm, with a smaller excitation peak at 300 nm and a peak emission at 527 nm. Gelstar's sensitivity is listed as 25 to 100 times greater than ethidium bromide. It is reported to detect as little as 20 pg of DNA and 10 ng of RNA (Cambrex, 2007).

Procedures for the above dyes require staining in buffered solutions (such as running buffer) for 15 to 40 min. However, no destaining is required. These dyes require specific photographic filters for optimal sensitivity; however, all of them can be imaged directly using a UV transilluminator. No data are available on mutagenicity or toxicity of these dyes. Treat them as potential mutagens and treat waste solutions in a similar manner to those containing ethidium bromide. The dyes listed above all come as stock solutions in DMSO (dimethylsulfoxide). DMSO facilitates the passage of dissolved solutes into exposed tissues, requiring special care in handling the dye stock solutions.

Variations in Separation of Nucleic Acids by Electrophoresis

Pulsed-field gel electrophoresis: An alternative approach for resolving DNA fragments in the 10 to 2000 kb size range; see *CP Molecular Biology Unit 2.5B* (Finney, 2000).

Denaturing polyacrylamide gels: A useful alternative to agarose gels (Ogden and Adams, 1987) for the separation of short DNA and RNA fragments as well as for DNA sequencing; single base length differences in nucleic acid polymers can be distinguished using this approach; see *CP Molecular Biology Unit 7.6* (Slatko and Albright, 1991).

Capillary DNA electrophoresis: An alternative approach for the rapid separation and detection of small amounts of DNA; see *CP Molecular Biology Unit 2.8* (Smith and Nelson, 2004).

LITERATURE CITED

Borst, P. 2005. Ethidium DNA agarose gel electrophoresis: How it started. *IUBMB Life* 57:745-747.

Brown, T., Mackey, K., and Du, T. 2004. Analysis of RNA by northern and slot-blot hybridization. *Curr. Protoc. Mol. Biol.* 67:4.9.1-4.9.19.

Cambrex. 2007. The Sourcebook: A Handbook for Gel Electrophoresis. Cambrex Corp., Rockland, Me.

Elder, J.K. and Southern, E.M. 1983. Measurement of DNA length by gel electrophoresis II: Comparison of methods for relating mobility to fragment length. *Anal. Biochem.* 128:227-231.

Finney, M. 2000. Pulsed-field gel electrophoresis. *Curr. Protoc. Mol. Biol.* 51:2.5B.1-2.5B.9.

Freifelder, D. 1971. Electron microscopic study of the ethidium bromide–DNA complex. *J. Mol. Biol.* 60:401-403.

Kingston, R. 1993. Preparation of poly(A)$^+$ RNA. *Curr. Protoc. Mol. Biol.* 21:4.5.1-4.5.3.

Lunn, G. and Lawler, G. 2002. Safe use of hazardous chemicals. *Curr. Protoc. Mol. Biol.* 58:A.1H.1-A.1H.33.

McCann, J., Choi, E., Yamasaki, E., and Ames, B.N. 1975. Detection of carcinogens as mutagens in the Salmonella/microsome test: Assay of 300 chemicals. *Proc. Natl. Acad. Sci. U.S.A.* 72:5135-5139.

Miura, Y., Wake, H., and Kato, T. 1999. TBE, or not TBE, that is the question: Beneficial usage of tris-borate for obtaining a higher resolution of small DNA fragments by agarose gel electrophoresis. *Nagoya Med. J.* 43:1-6.

Moore, D., Dowhan, D., Chory, J., and Ribaudo, R.K. 2002. Isolation and purification of large DNA fragments from agarose gels. *Curr. Protoc. Mol. Biol.* 59:2.6.1-2.6.12.

Ogden, R.C. and Adams, D.A. 1987. Electrophoresis in agarose and acrylamide gels. *Methods Enzymol.* 152:61-87.

Serwer, P. 1989. Sieving of double-stranded DNA during agarose gel electrophoresis. *Electrophoresis* 10:327-331.

Sharp, P.A., Sugden, B., and Sambrook, J. 1973. Detection of two restriction endonuclease activities in *Haemophilus parainfluenzae* using analytical agarose–ethidium bromide electrophoresis. *Biochemistry* 12:3055-3063.

Slatko, B.E. and Albright, L.M. 1991. Denaturing gel electrophoresis for sequencing. *Curr. Protoc. Mol. Biol.* 19:7.6.1-7.6.13.

Smith, A. and Nelson, R.J. 2004. Capillary electrophoresis of DNA. *Curr. Protoc. Mol. Biol.* 68:2.8.1-2.8.17.

Takahashi, M., Ogino, T., and Baba, K. 1969. Estimation of relative molecular length of DNA by electrophoresis in agarose gel. *Biochim. Biophys. Acta.* 174:183-187.

Voytas, D. 2000. Agarose gel electrophoresis. *Curr. Protoc. Mol. Biol.* 51:2.5A.1-2.5A.9.

SDS-Polyacrylamide Gel Electrophoresis (SDS-PAGE)

Sean R. Gallagher[1]
[1]UVP, LLC, Upland, California

OVERVIEW AND PRINCIPLES

Although electrophoresis has been studied for over two centuries, Arne Tiselius (University of Uppsala, Sweden) put the technique on the map with his Ph.D. thesis in 1930, which demonstrated moving zones of serum proteins in a special, square-shaped, free-solution electrophoresis cell. His original work showed for the first time that protein components in serum could be separated, giving the now familiar designations for α, β, and γ globulin. Arne Tiselius was awarded the Nobel Prize in Chemistry in 1948, "... for his research on electrophoresis and adsorption analysis, especially for his discoveries concerning the complex nature of the serum proteins ..." (*http://nobelprize.org/chemistry/laureates/1948/index.html*).

Electrophoresis is used to separate complex mixtures of proteins (e.g., from cells, subcellular fractions, column fractions, or immunoprecipitates) to investigate subunit compositions and to verify homogeneity of protein samples. It can also serve to purify proteins for use in further applications. In polyacrylamide gel electrophoresis (PAGE), proteins migrate in response to an electrical field through pores in the gel matrix, which consists of polymers of cross-linked acrylamide; pore size is determined by acrylamide concentration. The combination of gel pore size and protein charge, size, and shape determines the migration rate of the protein.

Electrophoresis

Electrophoresis is the movement of a charged particle, including large molecules such as DNA and proteins, in a liquid medium under the influence of an electric field. The driving force (QE) on the charged molecule is a product of the charge (Q) and the electric field (E) across the separation gel. Larger proteins move more slowly, as do proteins with a lower net charge. Furthermore, if a protein is in native (compact) form, it will migrate more quickly than the same protein in a fully denatured and extended form, where it experiences more frictional resistance with the surrounding medium.

Resistance in electrophoresis is defined as $f = 6\pi r v \eta$, where f is the resistance of the medium to electrophoretic movement, r the radius of the protein (assumed to be a sphere), v the electrophoretic velocity, and η the viscosity of the fluid. The two equations indicate that protein charge, size, shape, solution viscosity, and applied voltage are key factors influencing electrophoretic separation.

The most widely used method for gel electrophoresis is the discontinuous system described by Laemmli (1970), where a negatively charged detergent, sodium dodecyl sulfate (SDS), is used to denature proteins and give a constant negative charge-to-mass ratio, overwhelming any native charge of the protein and enabling separation largely based on size. Discontinuous SDS-PAGE uses three buffers (i.e., phases) for protein separation: (1) the same Tris glycine buffer for both electrodes, (2) a Tris·Cl buffer, pH 6.8, for the stacking gel, and (3) another Tris·Cl buffer, pH 8.8, for the resolving gel. These composition and pH discontinuities (or phases) act in concert to give exquisite resolution of separated proteins. Because the discontinuous gel system concentrates the proteins in each sample into narrow bands, the applied sample may be more dilute than that used for continuous electrophoresis.

Current Protocols Essential Laboratory Techniques 7.3.1-7.3.25

Continuous electrophoresis systems, where the same buffer is used in the tank and in making the gel, have been popular in the past because of their versatility and simplicity. The typical phosphate system is based on that of Weber et al. (1972). Although unable to produce the high-resolution separations of the discontinuous SDS-PAGE procedures, continuous SDS-PAGE, with only one basic buffer and no stacking gel (see Discontinuous Electrophoresis), uses fewer solutions. Additionally, artifacts are less likely to occur in continuous systems. Pepsin, for example, migrates anomalously on Laemmli-based discontinuous SDS-PAGE but has the expected mobility after electrophoresis in the phosphate-based continuous system. This is also true of cross-linked proteins.

Discontinuous Electrophoresis

While the theory of discontinuous or multiphasic electrophoresis is fairly complex (e.g., see Ornstein, 1964; Jovin, 1973), the application is straightforward for SDS-PAGE. The use of leading and trailing ions bracket and concentrate the proteins in the stacking gel when a potential (voltage) is applied. In this discontinuous system the sample first passes through a stacking gel, which has large pores and is made with a buffer having a lower pH (6.8), compared to the resolving gel buffer (pH 8.8). The stacking gel buffer contains chloride ions (called the leading ions) with an electrophoretic mobility greater than the mobility of the proteins in the sample. The electrophoresis buffer contains glycine ions (called the trailing ions) with an electrophoretic mobility less than the mobility of the proteins in the sample. The pH in the stacking gel ensures that the mobility of the glycine ions is less than that of the proteins. The net result is that the faster-migrating chloride ions leave a zone of lower conductivity between themselves and the migrating protein. The glycine provides a high electric field strength "stop" that forces the protein stack to stay in the zone between the leading and trailing ions. The higher voltage gradient in this zone allows the proteins to move faster and to "stack" between the leading and trailing ions. Under these conditions, even dilute solutions of proteins are concentrated and staged into thin zones prior to separation in the resolving gel. After leaving the stacking gel, the protein enters the separating gel. The separating gel has a smaller pore size, a higher salt concentration, and higher pH compared to the stacking gel. Under these conditions in the separating gel, the glycine ions migrate past the proteins, and the proteins separate according to either molecular size in a denaturing gel (containing SDS), or molecular shape, size, and charge in a nondenaturing gel (see Variations).

One-Dimensional Gel Electrophoresis

One-dimensional gel electrophoresis under denaturing conditions (i.e., in the presence of 0.1% SDS; Fig. 7.3.1A,B) separates proteins based on molecular size as they move through a polyacrylamide gel matrix toward the anode. The polyacrylamide gel is cast as the separating gel (sometimes called resolving or running gel), topped by the stacking gel, and secured in an electrophoresis apparatus. Proteins are denatured and solubilized by heating in the presence of 0.1% SDS (a negatively charged detergent) and a low-molecular-weight thiol such as dithiothreitol (DTT; Fig. 7.3.1C) or 2-mercaptoethanol (2-ME; Fig. 7.3.1D) to reduce disulfide bonds, ensuring that only single polypeptides are resolved. Most proteins bind SDS in a constant-weight ratio of 1.4 g SDS/g protein, leading to a constant charge-to-mass ratio and identical charge densities for the denatured proteins. The larger the protein, the longer the polypeptide chain, and the more SDS bound. The constant charge-to-mass ratio means that the protein migrates through the electrophoresis gel based on size—the larger the protein, the more it interacts with the gel pores and the more slowly it migrates. Thus, the SDS-protein complexes migrate in the polyacrylamide gel according to size, not charge. After sample proteins are denatured, an aliquot of the protein solution is applied to a gel lane and the individual proteins are separated electrophoretically.

Most proteins are resolved on polyacrylamide gels containing from 5% to 15% acrylamide and 0.2% to 0.5% bisacrylamide. Typically, the ratio of bisacrylamide to acrylamide is 1:37.5. Specialty applications such as separation of large proteins may require a larger bisacrylamide to acrylamide ratio to create a firmer gel at low acrylamide concentrations. The relationship between the relative

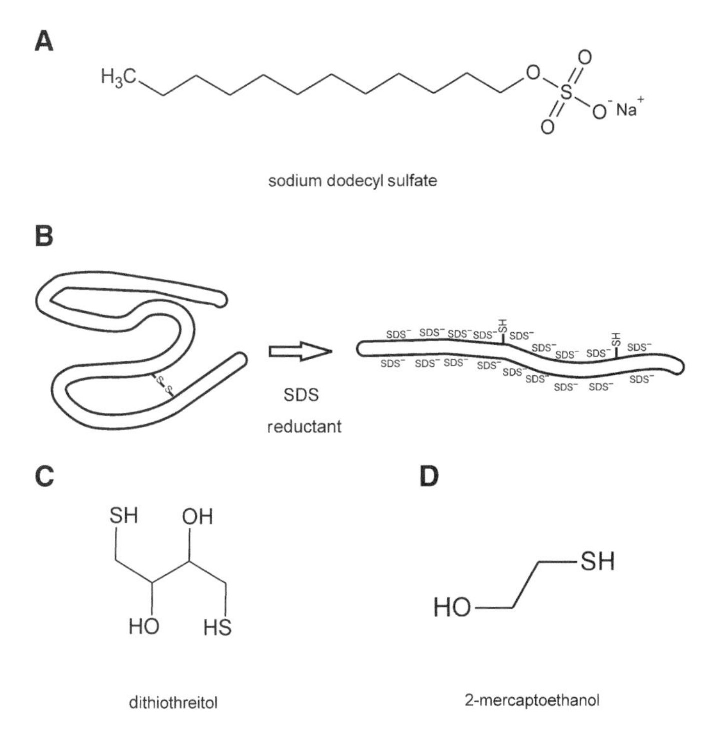

Figure 7.3.1 (**A**) Structure of sodium dodecyl sulfate. (**B**) SDS coating a denatured protein chain illustrating the constant binding of SDS per unit length of protein. Dithiothreitol and 2-mercaptoethanol, reductants used to break disulfide bonds in proteins so they are fully denatured, are shown in **C** and **D**, respectively.

mobility and log molecular weight is linear over these ranges. With the use of plots like those shown in Figure 7.3.2, the molecular weight of an unknown protein (or its subunits) may be determined by comparing it to known protein standards (Table 7.3.1). Note that gradient gels have a wide molecular weight–resolution range, while single-concentration gels have a more limited linear separation range, but the proteins are farther apart, simplifying excision and analysis. In general, all of the procedures in this unit are suitable without modification for radiolabeled and biotinylated proteins.

Polymerization of Polyacrylamide Gels

Polyacrylamide gels form through the copolymerization of monomeric acrylamide to form the long chains and cross-linking of those chains by bisacrylamide monomer (*N,N′*-methylenebisacrylamide; see Fig. 7.3.3). The polymerization reaction is initiated by the addition of ammonium persulfate, and the reaction is accelerated by *N,N,N′,N′*-tetramethylethylenediamine (TEMED), which catalyzes the formation of free radicals from ammonium persulfate. Because oxygen inhibits the polymerization process, deaerating the gel solution before the polymerization catalysts are added will speed polymerization; hence, the procedures presented in this unit include the use of a side-arm flask to facilitate deaeration. An additional factor is temperature. If the temperature is too low, then polymerization will proceed slowly or not at all. Keeping the persulfate concentration and temperature and deaeration conditions the same will help ensure consistent day-to-day polymerization.

Ohm's Law and Electrophoresis

Gel rig configuration

Understanding how a gel apparatus is connected to the power supply requires a basic understanding of Ohm's law: voltage (V) = current (I) × resistance (R). A gel can be viewed as a resistor and the power supply as the voltage and current source. Most power supplies deliver constant current or constant voltage. Some will also deliver constant power: power = voltage × current, or $VI = I^2R$. The discussion below focuses on constant current because this is the most common mode in vertical SDS-PAGE.

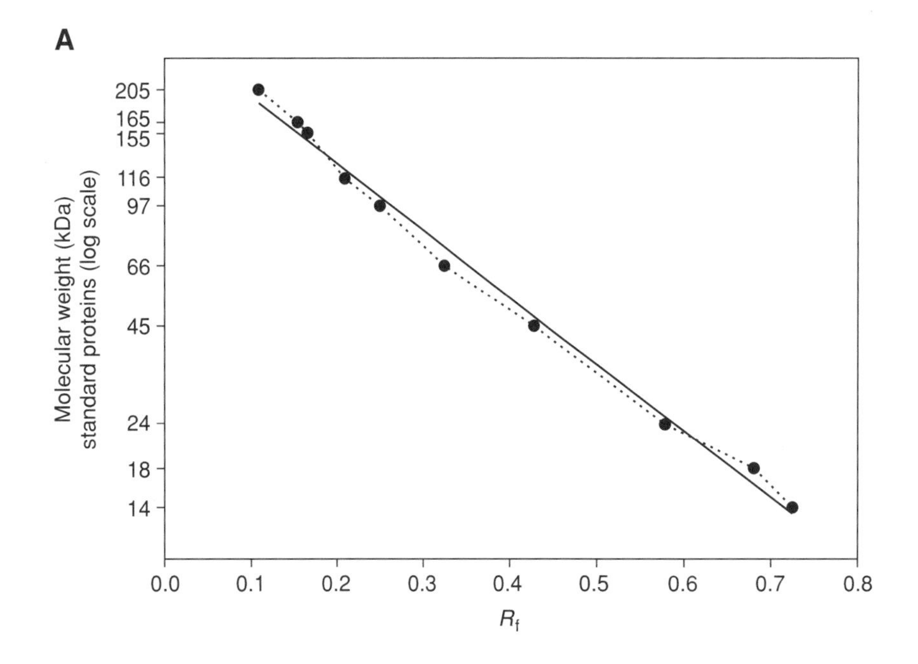

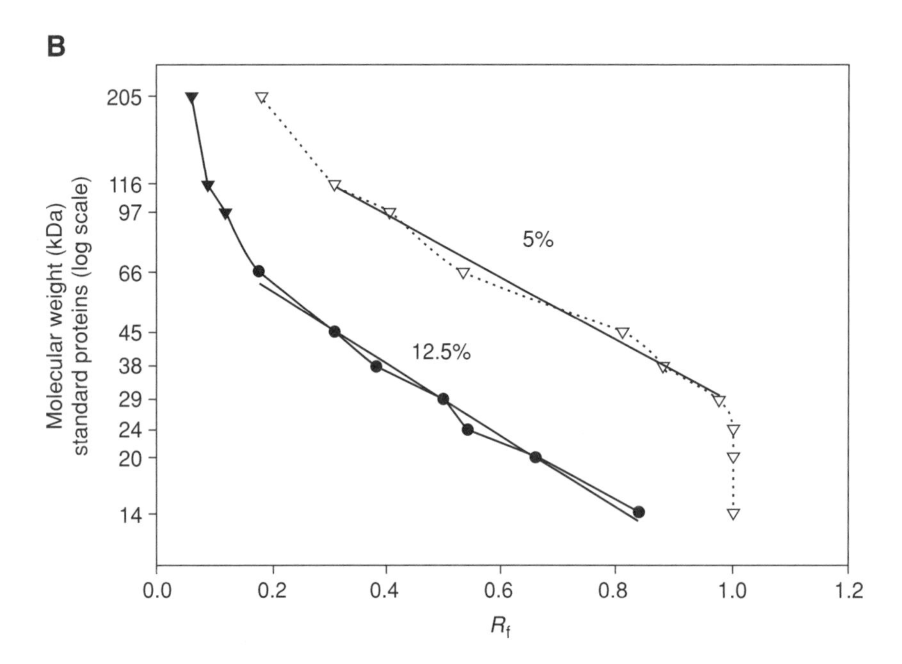

Figure 7.3.2 Standard curves for (**A**) a gradient 5% to 20% gel and (**B**) two single-concentration gels. Protein standards were separated via SDS-PAGE and visualized by staining with the protein-specific stain Coomassie Blue (*UNIT 7.4*). Their positions were measured relative to the dye front to give the relative mobility (R_f). Note the single-concentration gels have a more limited range of linearity for molecular-weight measurements than the gradient gel. The standard curves permit the calculation of the molecular weight of an unknown by simply using the R_f of the unknown to predict the molecular weight.

Components of Power

$$V = IR$$
$$I = V/R$$
$$P = IV$$
$$P = I \times (IR)$$
$$P = I^2R$$

Table 7.3.1 Molecular Weights of Protein Standards for Polyacrylamide Gel Electrophoresis[a]

Protein	Molecular weight (Da)
Cytochrome *c*	11,700
α-lactalbumin	14,200
Lysozyme (hen egg white)	14,300
Myoglobin (sperm whale)	16,800
β-lactoglobulin	18,400
Trypsin inhibitor (soybean)	20,100
Trypsinogen, PMSF treated	24,000
Carbonic anhydrase (bovine erythrocytes)	29,000
Glyceraldehyde-3-phosphate dehydrogenase (rabbit muscle)	36,000
Lactate dehydrogenase (porcine heart)	36,000
Aldolase	40,000
Ovalbumin	45,000
Catalase	57,000
Bovine serum albumin	66,000
Phosphorylase *b* (rabbit muscle)	97,400
β-galactosidase	116,000
RNA polymerase (*E. coli*)	160,000
Myosin, heavy chain (rabbit muscle)	205,000

[a]Protein standards are commercially available in kits (e.g., Bio-Rad, GE/Amersham, Invitrogen).

Most modern commercial equipment is color-coded so that the red or positive terminal of the power supply can simply be connected to the red lead of the gel apparatus, which goes to the lower buffer chamber. The black lead is connected to the black or negative terminal and goes to the upper buffer chamber. This configuration is designed to work with vertical slab gel electrophoreses in which negatively charged proteins or nucleic acids move to the positive electrode in the lower buffer chamber (an anionic system).

Movement of ions in an electric field

Anionic systems

When a single gel is attached to a power supply, the negative charges flow from the negative cathode (black) terminal into the upper buffer chamber, through the gel, and into the lower buffer chamber. The lower buffer chamber is connected to the positive anode (red) terminal to complete the circuit. Thus, negatively charged molecules (e.g., SDS-coated proteins and nucleic acids) move from the negative cathode attached to the upper buffer chamber toward the positive anode attached to the lower chamber. SDS-PAGE is an anionic system because of the negatively charged SDS. SDS is highly soluble in water, compatible with the commonly used Tris and phosphate buffer systems, and a potent protein denaturant that binds to the protein through a combination of electrostatic and hydrophobic interactions (Fig. 7.3.1).

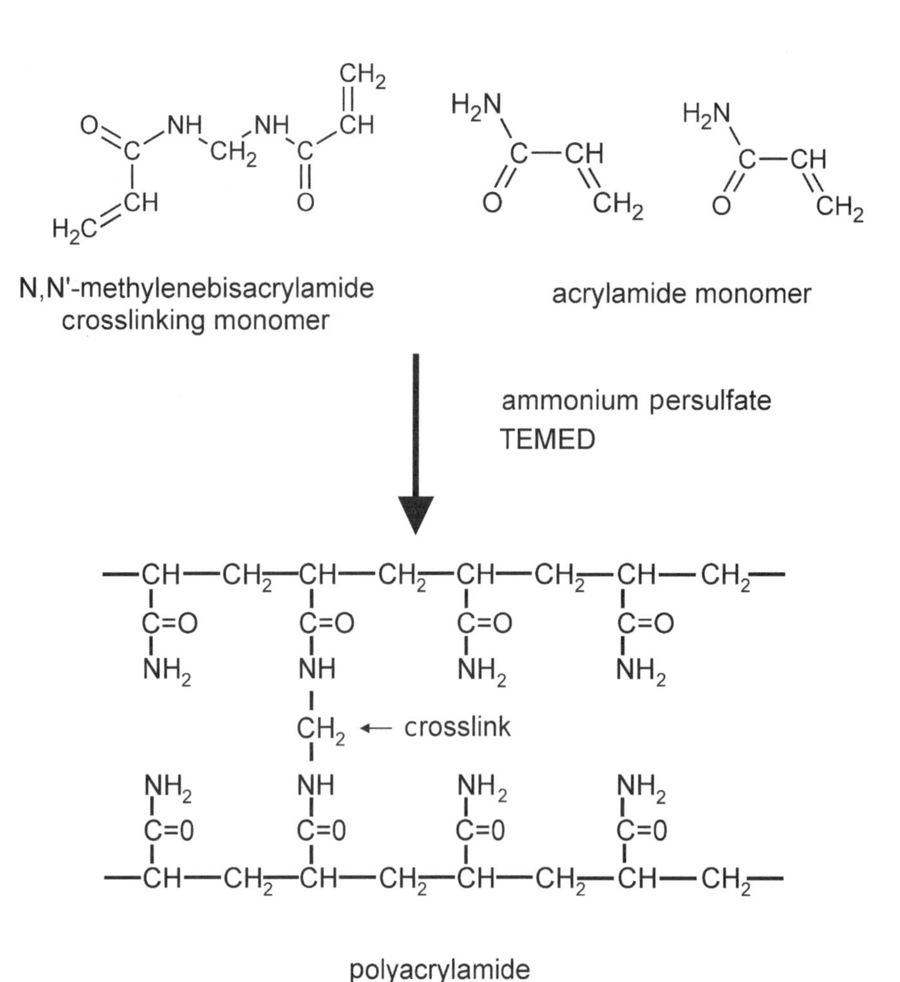

Figure 7.3.3 Free radical chain polymerization of acrylamide and *N,N′*-methylenebisacrylamide. Polymerization is initiated via free radicals generated by a redox pair, ammonium persulfate and *N,N,N′,N′*-tetramethylethylenediamine (TEMED). Polymerizing the acrylamide monomer by itself produces long linear chains of polymer in the form of a thick liquid with a syrup-like consistency. By adding the *N,N′*-methylenebisacrylamide, a cross-link with four potential branch points for the linear chain is copolymerized with the acrylamide to yield a mesh or gel. Note that the reaction is exothermic, and the outside of the gel plate will be warm during polymerization. The mesh creates pores of varying size ranges, depending on the concentration of acrylamide and *N,N′*-methylenebisacrylamide, to produce the sieving gel that is used for protein separation.

Cationic systems

Occasionally, proteins are separated in cationic systems. In these gels, the proteins are positively charged because of the very low pH of the gel buffers (e.g., acetic acid/urea gels for histone separations) or the presence of a cationic detergent (e.g., cetyltrimethylammonium bromide, CTAB). Proteins move toward the negative electrode (cathode) in cationic gel systems, and the polarity is reversed compared to SDS-PAGE: the red lead from the lower buffer chamber is attached to the black outlet of the power supply, and the black lead from the upper buffer chamber is attached to the red outlet of the power supply.

Change of resistance in the Laemmli system under constant current

Most SDS-PAGE separations are performed under constant current. (Consult instructions from the manufacturer to set the power supply for constant current operation.) The resistance of the gel will increase during SDS-PAGE in the standard Laemmli system. If the current is constant, then the voltage will increase during the run as the resistance increases.

Effect of multiple gels on Ohm's law

Power supplies usually have more than one pair of outlets. The pairs are connected in parallel with one another internally. If more than one gel is connected directly to the outlets of a power supply, then these gels are connected in parallel. In a parallel circuit, the voltage is the same across each gel: in other words, if the power supply reads 100 V, then each gel has 100 V across its electrodes. The total current, however, is the sum of the individual currents going through each gel; therefore, under constant current it is necessary to increase the current for each additional gel that is connected to the power supply. Two identical gels require double the current to achieve the same starting voltages and electrophoresis separation times.

Relationship between current and gel thickness

Gel thickness affects the above relationships. A 1.5-mm gel can be thought of as consisting of two 0.75-mm-thick gels run in parallel. Because currents are additive in parallel circuits, a 0.75-mm gel will require half the current of the 1.5-mm gel to achieve the same starting voltage and separation time. If gel thickness is doubled, then the current must also be doubled. There are limits to the amount of current that can be applied, however. Thicker gels require more current, generating more heat that must be dissipated. Unless temperature control is available in the gel unit, a thick gel should be run more slowly than a thin gel.

STRATEGIC PLANNING

When planning your experiment, you will need to choose an appropriate gel composition (acrylamide concentration and acrylamide:bisacrylamide ratio), gel length, and gel thickness. You have the option of making the gels yourself (Support Protocol 1) or purchasing them precast from a number of suppliers (Table 7.3.2).

Gel Composition (Single Concentration versus Gradient)

To achieve the maximum separation between two proteins of a specific size, an optimized one-dimensional gel is best. The most appropriate acrylamide concentration to use depends on the size range over which you require the best resolution (Fig. 7.3.2). In general, higher acrylamide concentrations produce gels with smaller pore sizes and provide better resolution of smaller polypeptides, whereas lower acrylamide concentrations produce larger pores and better resolution of larger polypeptides. However, to visualize the widest range of protein sizes on a single lane with sharp bands, gradient gels are recommended (see Variations).

Gel Size

While gel sizes range from mini (8 × 10–cm) to large (25 × 25–cm) format, minigels have become widely popular because of their commercial availability in precast form, shorter run time, and need for smaller quantities of reagents. Minigels are useful for routine one-dimensional separation and

Table 7.3.2 Precast Gels Available from Selected Suppliers

	Format		Application				Instrument compatibility			
	Mini	Large	Native	SDS	Peptide	2-D[a]	Bio	Cambrex	Hoefer	Invitrogen
Bio-Rad	X	X	X	X	X	X	X			
Cambrex	X		X	X	X	X	X	X	X	X
Jule	X	X	X	X	X	X	X	X	X	X
Invitrogen	X		X	X	X	X	X	X	X	X

[a]Two-dimensional analysis (see Variations).

low-resolution two-dimensional separations. Large-format gels, on the other hand, are useful for achieving maximum separation between two proteins in one- and two-dimensional separations (see Variations) or for loading more material on the gel for preparative work or because of a dilute sample. Generally, in any electrophoretic separation, the use of larger gels results in better resolution.

Minigels: Good Things Come in Small Packages

Minigels are generally considered to be in the ~8 × 10–cm size range, although there is considerable variation in exact size. Every technique that is used on larger systems can be translated with little difficulty into the minigel format. This includes standard and gradient SDS-PAGE and separations for immunoblotting and peptide sequencing. Two-dimensional SDS-PAGE electrophoresis also adapts well, but here the limitation of separation area becomes apparent; for high-resolution separations, large-format gels are required. Gradient minigels are popular due to the combination of separation range and resolution (Matsudaira and Burgess, 1978). They are particularly useful for separation of proteins prior to peptide sequencing.

Gel Thickness

Determining which thickness of gel to use depends on several factors. For the highest resolution, gels <1 mm thick should be used because fixation and staining reagents penetrate the gel more quickly, minimizing any loss of proteins during these steps. In addition, thin gels show little front to back distortion, which can lower the observed resolution of the gel. However, thinner gels hold less protein before overloading (see Troubleshooting, Vertical Streaking) and also are limited in the volume that can be loaded. Lastly, thinner gels are more prone to tearing and damage during post-electrophoresis processing. In contrast, thicker gels are easier to handle and hold more material, making them useful for dilute samples or larger loading volumes.

Gel Sources and Gel Storage

Self-made gels

While precast gels offer a wide assortment of sizes and applications, cost for gels start at $8 (excluding shipping), which can be prohibitive. Typical self-cast gels cost about $0.50 in materials and are available as soon as the polymerization is complete (1 to 2 hr). Although stock solutions must be maintained, simple tracking of the age of the solutions and monitoring of the performance by including standards in gel runs is generally enough to ensure consistent quality.

Self-made gels should be used within 2 days if stored at 4°C.

Precast gels

Precast gels for commonly used vertical minigel and standard-sized SDS-PAGE apparatuses are available from several manufacturers (Table 7.3.2). When using precast gels, pay strict attention to the shelf life, which may vary from 3 to 12 months depending on the chemistry used. However, manufacturers generally overrate the shelf life, and the sooner the gels are used, the better. When reasonably fresh, precast gels provide excellent resolution, comparable to a typical gel cast in the laboratory.

Protein Amount and Concentration

Assuming a complex mixture, the protein of interest should be present in an amount of 0.2 to 1 µg if the gel will be stained by Coomassie blue (*UNIT 7.4*). Typically, 25 to 50 µg of a complex protein mixture in a total volume of <20 µl is loaded onto a 0.75-mm-thick slab minigel. Silver and fluorescent staining techniques require 10- to 100-fold less material loaded on the gel.

SAFETY CONSIDERATIONS

Many researchers are poorly informed concerning the electrical parameters of running a gel. It is important to note that the voltages and currents used during electrophoresis are dangerous and potentially lethal. Thus, safety should be an overriding concern. A working knowledge of electricity is an asset in determining what conditions to use and in troubleshooting the electrophoretic separation, if necessary. For example, an unusually high or low voltage for a given current (milliampere) might indicate an improperly made buffer or an electrical leak in the chamber.

Connecting Leads

Never remove or insert high-voltage leads unless the voltage on the power supply is turned down to zero and the power supply is turned off. Always grasp high-voltage leads one at a time with one hand only; never insert or remove high-voltage leads with both hands. This can shunt potentially lethal electricity through the chest and heart should electrical contact be made between a hand and a bare wire. On older or homemade instruments, the banana plugs may not be shielded and can still be connected to the power supply at the same time they make contact with a hand. Carefully inspect all cables and connections and replace frayed or exposed wires immediately.

Working with Power Supplies

Always start with the power supply turned off and the power supply controls turned all the way down to zero. Then hook up the gel apparatus following these basic general steps.

1. Connect the red high-voltage lead to the red outlet and the black high-voltage lead to the black outlet.
2. Turn the power supply on with the controls set at zero and the high-voltage leads connected.
3. Turn up the voltage, current, or power to the desired level.

Reverse the process when the power supply is turned off: i.e., to disconnect the gel, turn the power supply down to zero, wait for the meters to read zero, turn off the power supply, and then disconnect the gel apparatus one lead at a time.

CAUTION: If the gel is first disconnected and then the power supply turned off, a considerable amount of electrical charge is stored internally. The charge will stay in the power supply over a long time. This will discharge through the outlets even though the power supply is turned off and can deliver an electrical shock.

PROTOCOLS

The standard Laemmli method described below is used for discontinuous gel electrophoresis under denaturing conditions, i.e., in the presence of sodium dodecyl sulfate (SDS; Fig. 7.3.1). The popular minigel format (e.g., 7.3 × 8.3 cm) can be adapted for full-size gels (e.g., 14 × 14 cm). Although minigels provide rapid separation they give lower resolution than a full-size gel.

Descriptions of variations of the standard Laemmli method and their applications are provided at the end of the unit.

Basic Protocol: Denaturing (SDS) Discontinuous Gel Electrophoresis: The Laemmli Gel Method

This protocol is designed for a vertical slab gel with a maximum size of 0.75 mm × 10 cm × 10 cm. For thicker gels or large format gels, the volumes of stacking and separating gels (Support Protocol 1), and the operating current must be adjusted.

Materials

Protein sample to be analyzed: 1 to 50 μg in <20 μl (depending on sample complexity) when staining with Coomassie blue; 10- to 100-fold less protein when silver staining (0.01 to 5 μg in <20 μl)
2× or 6× SDS sample buffer (see recipes)
1× SDS sample buffer: dilute from 2× or 6× stock
Protein molecular-weight-standards mixture (Table 7.3.1 and Fig. 7.3.2)
Polyacrylamide gel, purchased precast (e.g., Table 7.3.2) or self-cast (Support Protocol 1)
1× SDS electrophoresis buffer (see recipe)

100°C and 56°C water bath
Screw-cap microcentrifuge tube
100°C heating block
Electrophoresis apparatus (see Table 7.3.2), small format with 100-mA capability constant-current power supply (allowing running of two gels simultaneously), including clamps, glass plates, casting stand, and buffer chambers (e.g., XCell SureLock Mini-Cell, Invitrogen; Mini-Protean 3, Bio-Rad; or SE 250 10-cm unit, Hoefer)
Absorbent paper

CAUTION: The voltages and currents used during electrophoresis are dangerous and potentially lethal. It is extremely important to read the section entitled Safety Considerations before performing any electrophoresis.

Prepare the sample

1a. *For aqueous samples:* Dilute a portion of the protein sample to be analyzed 1:1 (v/v) with 2× SDS sample buffer. Heat 3 to 5 min at 100°C in a sealed screw-cap microcentrifuge tube. To achieve the highest resolution possible, use the following precautions:

Prior to adding the sample buffer, keep samples at 0°C.

Add the SDS sample buffer (room temperature) directly to the 0°C sample (still on ice) in a screw-top microcentrifuge tube.

Cap the tube to prevent evaporation, mix by vortexing, and transfer directly to a 100°C water bath for 3 to 5 min.

Do not *leave the sample in SDS sample buffer at room temperature without first heating to 100°C to inactivate proteases. Endogenous proteases are very active in SDS sample buffer and will cause severe degradation of the sample proteins after even a few minutes at room temperature. To test for possible proteases, mix the sample with SDS sample buffer without heating and leave 1 to 3 hr at room temperature. A loss of high-molecular-weight bands and a general smearing of the banding pattern indicate a protease problem. Once heated, the samples can sit at room temperature for the time it takes to load samples.*

1b. *For dilute protein solutions:* Treat as in step 1a, except add 5:1 (v/v) protein solution/6× SDS sample buffer to increase the amount of protein loaded.

Proteins can also be concentrated by precipitation in acetone, ethanol, or trichloroacetic acid (TCA), but losses will occur. Refer to CP Protein Science Unit 1.1 (Scopes, 1995) for a thorough overview of protein purification.

1c. *For immunoprecipitates:* Dissolve in 1× SDS sample buffer 1 hr at 56°C prior to boiling in the 100°C water bath.

1d. *For protein pellets:* Dissolve in 50 to 100 μl 1× SDS sample buffer at room temperature and heat in a heating block 3 to 5 min at 100°C.

Heating blocks provide precise temperature control and accommodate a variety of tube sizes.

2. Dissolve the protein molecular-weight standards mixture in 1× SDS sample buffer according to supplier's instructions; use these standards as a control.

3. Microcentrifuge samples 5 min at maximum speed to remove any particulates.

   Particulates in the sample can create streaks in the stained gel lane.

Load the gel

4a. *For gels cast in-house:* Carefully remove the Teflon comb without tearing the edges of the polyacrylamide wells and rinse wells with 1× SDS electrophoresis buffer. Tip the gel onto absorbent paper to remove the rinse solution.

   Wells with torn arms should not be loaded with sample due to potential for leakage of the sample into adjacent lanes. Instead add sample buffer to maintain electrical consistency across the gel.

   The rinse removes unpolymerized monomer; otherwise, the monomer will continue to polymerize after the comb is removed, creating uneven wells that will interfere with sample loading and subsequent separation.

4b. *For precast gels:* Place the precast gel on the electrophoresis unit and then remove the comb as in step 4a just prior to applying the sample.

   Refer to manufacturer's instructions for specific directions.

5. Using a Pasteur pipet, fill the wells with 1× SDS electrophoresis buffer.

   If well walls are not upright loading the gel can be difficult. Any wells laying flat or distorted at an angle can be manipulated upright with a flat-tipped needle attached to a syringe.

6. Fill the lower buffer chamber with the recommended amount of 1× SDS electrophoresis buffer.

   Alternatively the upper buffer chamber can be filled first; this may be useful for detecting buffer leaks from this chamber.

7. Place sandwich attached to upper buffer chamber into lower buffer chamber.

8. Partially fill the upper buffer chamber with 1× SDS electrophoresis buffer so that the sample wells of the stacking gel are filled with buffer.

   Completely filling the chamber with the electrophoresis buffer makes it difficult to see the top of the wells during the sample loading process. Partially filling the chamber (just enough to fill the wells) minimizes disturbance to the loaded sample when adding the rest of the electrophoresis buffer to fill the chamber.

   Bubbles should float free of the wells. If necessary, use a pipet to gently inject electrode buffer into the well to dislodge bubbles.

9. Monitor the upper buffer chamber for leaks and, if necessary, reassemble the unit.

   A slow leak in the upper buffer chamber may cause arcing around the upper electrode, damaging the upper buffer chamber.

10. Using a standard disposable 200-µl pipet tip, load the protein sample(s) into one or more wells by carefully depositing the sample through the electrophoresis buffer and letting it fall as a thin layer at the bottom of the well.

   The samples will layer on the bottom of the wells because the glycerol in the sample buffer gives the solution a greater density than the electrophoresis buffer. The bromphenol blue in the sample buffer makes sample application easy to follow visually.

   A 25- or 100-µl syringe with a flat-tipped needle will enable filling the well by carefully underlayering the sample as a thin layer at the bottom, giving better resolution and control over inadvertent mixing of the sample while loading. However, because of convenience, disposable tips are more popular.

Preparing the samples at approximately the same concentration and loading an equal volume to each well will ensure that all lanes start with the same width and the proteins run evenly. If unequal volumes of sample buffer are added to wells, the lane with the larger volume will spread during electrophoresis and constrict the adjacent lanes, causing distortions (see Troubleshooting).

For an 0.8-cm-wide well, 25 to 50 μg total protein in <20 μl is recommended for a complex mixture when staining with Coomassie blue, while 1 to 10 μg total protein is needed for samples containing one or a few proteins. If silver staining is used, 10- to 100-fold less protein can be applied (0.01 to 5 μg in <20 μl, depending on sample complexity).

11. Load at least one well with molecular-weight standards. Add an equal volume of 1× SDS sample buffer to any empty wells to prevent spreading of sample to adjoining lanes.

12. Fill the remainder of the upper buffer chamber with additional 1× SDS electrophoresis buffer so that the upper platinum electrode is completely covered. Do this slowly so that samples are not swept into adjacent wells.

Run the gel

13. Connect the power supply to the cell. For a minigel that is 0.75 mm thick, run the gel at 10 mA constant current until the bromphenol blue tracking dye enters the separating gel, and then increase the current to 15 mA.

CAUTION: *Refer to Safety Considerations prior to connecting or disconnecting the apparatus to or from the power supply.*

For gels of other sizes, increase the current to 0.8 mA per cm^2. In general, the current should be doubled for thicker gels.

Electrophoresis is normally performed at 15°C to 20°C, with the temperature held constant using a circulating water bath. For air-cooled electrophoresis units, lower currents and thus longer run times are recommended.

For a standard minigel, 125 V per 0.75- to 1.0-mm-thick gel will be complete in <90 min.

If the level of buffer in the upper chamber decreases, a leak has occurred.

14. After the bromphenol blue tracking dye (670 Da) has reached the bottom of the separating gel, disconnect the power supply.

Bromphenol blue simply runs as a dye front, and not at a particular molecular weight. It serves as the relative mobility (R_f) marker for mobility calculations.

Disassemble the gel

15. Discard upper electrode buffer and remove the upper buffer chamber with the attached gel sandwich.

16. Orient the gel so that the order of the sample wells is known. Remove the sandwich from the upper buffer chamber and lay onto a sheet of absorbent paper or paper towels.

17. Carefully slide one of the spacers halfway from the edge of the sandwich along its entire length. Use the exposed spacer as a lever to pry open the glass plate, exposing the gel.

Disassembly of precast gels varies according to the manufacturer. For all gels, carefully monitor to which plate the gel is adhering and carefully remove the other plate by letting the gel fall on the adhered plate.

18. Cut a small triangle off one corner of the gel so the lane orientation is not lost during staining and drying manipulations. Carefully remove the gel from the plate by sliding gloved fingers under the gel and lifting it off.

The gel can be left on the plate for staining if it is damaged or difficult to remove (e.g., a low-percentage gel). The gel will typically float off the plate in the process. The gel can be stained with Coomassie blue or silver (see UNIT 7.4*), or proteins can be electroeluted, electroblotted onto a polyvinylidene difluoride (PVDF) membrane for subsequent staining or sequence analysis, or transferred to a* *membrane for immunoblotting (see UNIT 8.3). If the proteins are radiolabeled, they can be detected by autoradiography.*

Once proteins have been detected, molecular mass can be determined as described in Support Protocol 2.

Support Protocol 1: Casting a Gel for Use in Denaturing Discontinuous Electrophoresis

Separating (or resolving) gels are prepared and poured individually, which requires a total time of 2 to 3 hr. Separation of proteins in a small-gel format is becoming increasingly popular for applications that range from isolating material for peptide sequencing to performing routine protein separations. The unique combination of speed and high resolution is the foremost advantage of small gels. Additionally, small gels are easily adapted to single-concentration, gradient, and two-dimensional SDS-PAGE procedures. The minigel procedures described below are adaptations of larger gel systems.

Materials

Detergent, laboratory quality (e.g., Alconox or RBS-35; Pierce)
30% acrylamide/0.8% bisacrylamide solution (see recipe), room temperature
1× and 4× Tris·Cl (pH 8.8)/SDS (see recipe), room temperature
10% ammonium persulfate (APS), prepare fresh
N,N,N′,N′-tetramethylethylenediamine (TEMED)
Water-saturated isobutyl alcohol (see recipe)

Glass plates (part of electrophoresis apparatus; see Basic Protocol):
Laboratory marker (e.g., Sharpie)
0.75-, 1.0-, or 1.5-mm Teflon spacers
Casting stand
25-ml Erlenmeyer side-arm flask with solid rubber stopper
Vacuum pump with cold trap
Teflon comb (same thickness as spacers) with 1, 3, 5, 10, 15, or 20 teeth

Prepare the glass-plate sandwich

1. Clean the unassembled glass plates using a laboratory-quality detergent.

   These aqueous-based solutions are compatible with silver and Coomassie blue staining procedures (UNIT 7.4). Standard liquid dish soap is not acceptable due to the nonionic detergents and brighteners typically found in household detergents.

2. Assemble the glass-plate sandwich of the electrophoresis apparatus according to manufacturer's instructions, using two clean glass plates and two spacers. Mark the location of the bottom of the sample well on the outside of the gel plate with a laboratory marker.

   The mark will wash of during cleaning.

3. Lock the sandwich in the casting stand.

Prepare the separating gel

4. In a 25-ml Erlenmeyer side-arm flask, mix 30% acrylamide/0.8% bisacrylamide solution, 4× Tris·Cl (pH 8.8)/SDS, and water according to the desired final concentration (Table 7.3.3).

   All reagents must be warmed to room temperature before use because low temperatures (<20°C) will also slow or prevent polymerization.

   The desired percentage of acrylamide in the separating gel depends on the molecular size of the protein being separated (see Table 7.3.4).

Table 7.3.3 Recipes for Polyacrylamide Separating Gels[a]

	Final acrylamide concentration in separating gel[d]									
Stock solution[b,c]	5%	6%	7%	7.5%	8%	9%	10%	12%	13%	15%
30% acrylamide/0.8% bisacrylamide (see recipe)	2.50	3.00	3.50	3.75	4.00	4.50	5.00	6.00	6.50	7.50
4× Tris·Cl (pH 8.8)/SDS (see recipe)	3.75	3.75	3.75	3.75	3.75	3.75	3.75	3.75	3.75	3.75
Water	8.75	8.25	7.75	7.50	7.25	6.75	6.25	5.25	4.75	3.75

[a]The recipes produce a final volume of 15 ml which is adequate for a gel of dimensions 0.75 mm × 14 cm × 14 cm. The recipes are based on the SDS (denaturing) discontinuous buffer system of Laemmli (1970).

[b]All reagents and solutions must be prepared with Milli-Q-purified water or equivalent. Volumes for solutions used to make gels are in milliliters.

[c]For best results, prepare fresh.

[d]The desired percentage of acrylamide in the separating gel depends on the molecular size of the protein being separated. See Table 7.3.4.

Table 7.3.4 Separation Size Range for Proteins in Acrylamide

Acrylamide	Size range (Da)[a]
5%	25,000–200,000 (36,000–200,000)
10%	14,000–200,000 (14,000–66,000)
15%	14,000–66,000 (14,000–36,000)
20%	10,000–45,000 (10,000–45,000)
5%–20% gradient	10,000–300,000 (10,000–300,000)

[a]The linear range of separation is indicated in parentheses.

5. Stopper the flask and connect the side arm to a vacuum pump with cold trap. Degas under vacuum ~5 min.

   Oxygen inhibits polymerization, and the gels are cast between two glass plates in part to minimize this inhibition, once the gels are poured and overlay applied. Normal percentage gels (i.e., >10%) will typically polymerize without the degassing step. However, for lower percentage gels (<10%), deaerating under vacuum is important for achieving strong and consistent polymerization. Labs doing a variety of gel work will typically have a degassing step in their process to maintain day-to-day consistency in polymerization.

6. Add 50 µl of 10% APS and 10 µl TEMED. Swirl gently to mix.

   Use the solution immediately; otherwise it will polymerize in the flask.

7

Pour the separating gel

7. Using a Pasteur pipet or syringe, smoothly apply the separating gel solution to the sandwich along an edge of one of the spacers (to prevent trapping bubbles) until the height of the solution between the glass plates is 1 cm below the mark denoting the bottom of the sample well.

   A typical minigel will utilize a 7-cm separating gel.

   Sample volumes <10 µl do not require a stacking gel. In this case, cast the separating gel as you normally would, but extend the resolving gel into the comb to form the well. The proteins are then separated under the same conditions as used when a stacking gel is present. No isobutyl alcohol overlay is needed in this case (see step 8). Although this protocol works well with single-concentration gels, a gradient gel is recommended for maximum resolution (see Variations).

8. Using another Pasteur pipet, slowly cover the top of the gel with a layer (~1 cm thick) of water-saturated isobutyl alcohol by gently pipetting the solution against the edge of one and then the other spacer. Be careful not to disturb the gel surface.

   The overlay provides a barrier to oxygen, which inhibits polymerization, and allows a flat interface to form during gel formation.

   The isobutyl alcohol overlay should not be left on the gel longer than 2 hr. If longer storage is required (up to 2 days), remove the isobutyl alcohol and overlay with 1× Tris·Cl (pH 8.8)/0.1% SDS.

   Water-saturated n-butyl alcohol is less prone to form droplets that crash into the gel interface when adding the overlay. However, if carefully applied, a 1× resolving gel buffer or water overlay can be used. If water is used, then extra care is required to evenly overlay without creating drops that fall into the gel and create an uneven interface.

9. Allow the gel to polymerize 30 to 60 min at room temperature.

   Polymerization should be optimized so it starts within 10 min as judged by the sharp optical discontinuity at the interface between the overlay and the gel. If it polymerizes too fast, the resulting gel will have an uneven interface or wavy polymerization lines leading to distortion of the protein bands and poor separation. To correct fast polymerization, the amount of APS should be reduced by one-third to one-half. If the gels polymerize too slowly or fail to polymerize all the way to the top, mix a new gel solution and use fresh APS or increase the amount of APS by one-third to one-half.

   A sharp optical discontinuity (a line) at the overlay/gel interface will be visible upon polymerization. Failure to form a firm gel usually indicates a problem with the APS, TEMED, or both. Ammonium persulfate solution should be made fresh before use and the dry chemical should "crackle" when added to the water, indicating low moisture content. If this effect is not seen, fresh APS should be purchased. Purchase TEMED in small bottles so that, if necessary, a new, previously unopened source can be tried.

Prepare the stacking gel

10. Within 1 day of use, pour off the layer of water-saturated isobutyl alcohol or buffer and rinse with 1× Tris·Cl (pH 8.8)/SDS. Tip the gel cassette at an angle to drain the overlay off the gel. Wick the last remaining liquid off onto a sheet of blotter paper or laboratory tissue.

    The stacking gel should be poured within 1 day of use; otherwise, there will be a gradual diffusion-driven mixing of buffers between the two gels, which will cause a loss of resolution. Regardless of the concentrations used in the separating gel, the concentrations of acrylamide/bisacrylamide in the stacking gel are constant.

    Residual isobutyl alcohol can reduce resolution of the protein bands; therefore, it must be completely removed.

11. In a 25-ml side-arm flask, mix 0.65 ml of 30% acrylamide/0.8% bisacrylamide, 1.25 ml of 4× Tris·Cl (pH 6.8)/SDS, and 3.05 ml water.

12. Stopper the flask and connect to a vacuum pump with cold trap. Degas under vacuum 10 to 15 min.

13. Add 25 µl of 10% APS and 5 µl TEMED. Swirl gently to mix.

    Use the solution immediately; otherwise it will polymerize in the flask.

Pour the stacking gel

14. Using a Pasteur pipet, slowly allow the stacking gel solution to trickle into the center of the sandwich along an edge of one of the spacers until the height of the solution in the sandwich is ~1 cm from the top of the plates.

    Be careful not to introduce air bubbles into the stacking gel.

15. Insert a Teflon comb with desired number of teeth into the layer of stacking gel solution. If necessary, add additional stacking gel to fill the spaces in the comb completely.

    Again, be careful not to trap air bubbles in the tooth edges of the comb; they will cause small circular depressions in the well after polymerization that will lead to distortion in the protein bands during separation. If there are bubbles in the gel or on the surface of the gel, inserting the comb from one edge at an angle and sweeping from one edge to another will simultaneously insert the comb and push the bubbles to one side and out of the way of the gels.

16. Allow the stacking gel solution to polymerize 30 to 45 min at room temperature.

    A sharp optical discontinuity will be visible around wells upon polymerization. The completed gel can be stored up to 1 day at 4°C.

Support Protocol 2: Calculating Molecular Mass

Determining the molecular mass of an unknown protein or nucleic acid fragment is straightforward given the use of calibration standards in the same gel. Typically, a set of protein standards are separated along with protein sample containing the unknown. The standards are used to create a standard curve of mobility versus size or molecular mass.

Although digital image analysis has greatly simplified calculating the mass of an unknown protein separated by electrophoresis (Fig. 7.3.4), manual assessments of molecular mass are useful and employ the same basic calculations.

Materials

Gel containing separated proteins (Basic Protocol)
Molecular mass/R_f acetate overlay calculator (Fig. 7.3.5).
Calculator or analysis program for performing linear regression

1. Place the molecular mass/R_f acetate overlay calculator (Fig. 7.3.5) on the gel. Align the top and bottom of the overlay with the top of the gel and the dye front, respectively, to get a read out of relative mobility (R_f).

2. Calculate the relative mobility using following formula:

    R_f = distance migrated by protein/distance migrated by marker.

    Typically, a tracking dye such as bromphenol blue is added to the sample prior to loading on the gel (e.g., see recipe for SDS sample buffer). The tracking dye moves ahead of the proteins and serves as the relative mobility marker. Day-to-day variations in separation conditions due to temperature

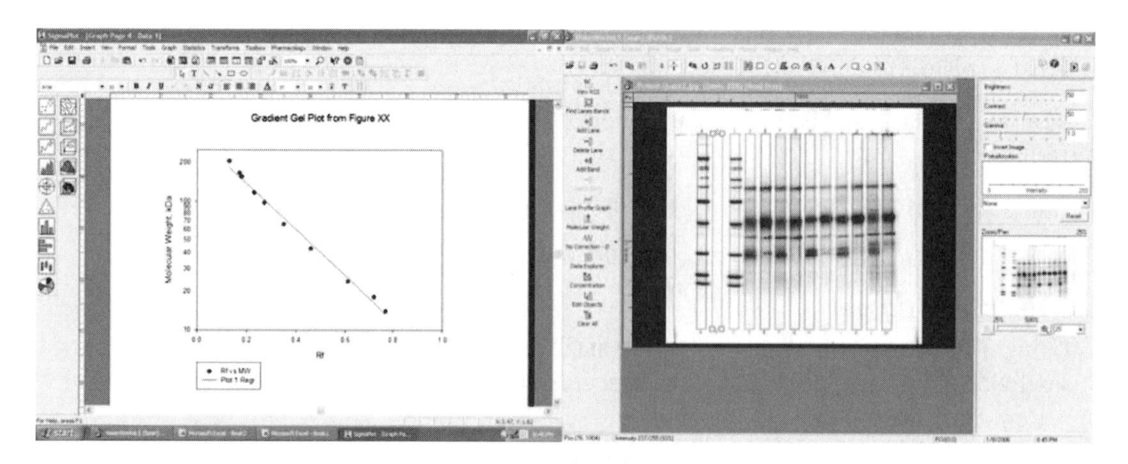

Figure 7.3.4 Computerized analysis of separated proteins. Acquiring the image of the stained protein gel with a digital camera greatly simplifies subsequent analysis. The protein bands are automatically identified, assigned an R_f, and given an estimated molecular mass via comparison to the standard curve generated from the separated standard proteins. The software illustrated in this figure is Vision Works LS (UVP).

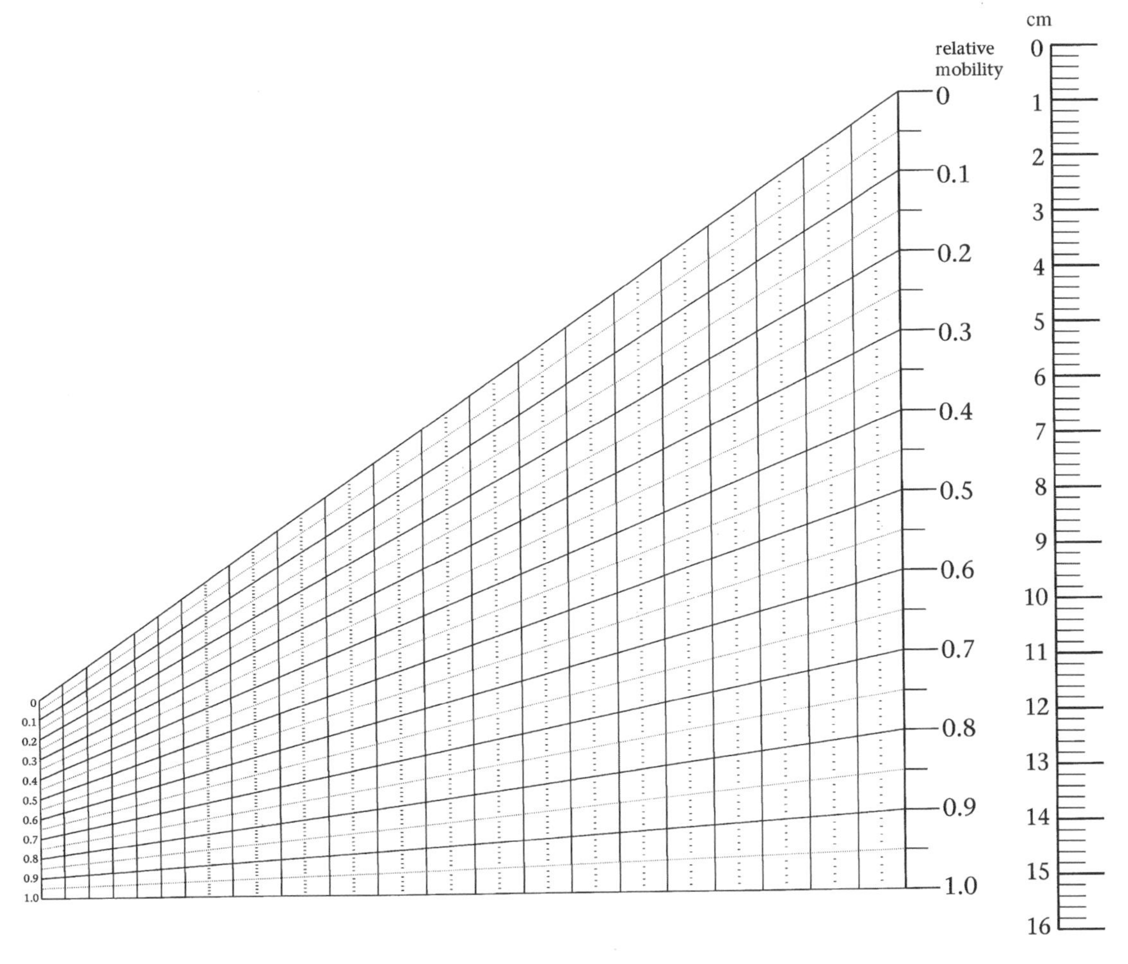

Figure 7.3.5 Example of a relative mobility (R_f) calculator. This sheet can be copied to transparency film using a paper copier and used as an overlay on the gel. When the transparency is placed on top of the gel, so that the top of the gel aligns with the top of the calculator and the dye front lines up with the bottom of the calculator, the R_f can be read directly off the overlay. Note that the calculator accommodates a range of gel lengths. The overlay should be copied at a 1:1 ratio so that the centimeter scale remains accurate. However, as long as the overlay can fit the top and bottom of the gel, the R_f numbers will be accurate.

fluctuations, slight differences in pH or buffer concentration, or differences in the sample conductivity will lead to variations in distance migrated by the proteins. To simplify comparison among separate gels, the band positions must be normalized to common standards, typically the bromphenol blue dye front. In situations where the dye front is not obvious (e.g., gradient gels or where the dye front was run off the gel), a standard protein near the bottom of the gel can serve as the marker. These "relative front" or R_f markers greatly simplify the comparison of samples run on different gels.

3. Plot log protein mass (y axis) versus relative mobility on the x axis of the standards.
4. Perform linear regression using a calculator or analysis program.
5. Use the linear regression equation ($y = mx + b$) to estimate the mass of the unknown:

$$\text{Log molecular weight} = (\text{slope})(\text{mobility of the unknown}) + y\text{-intercept}$$

Support Protocol 3: Recrystallizing SDS

High-purity SDS is available from several suppliers, but for some sensitive applications (e.g., protein sequencing) recrystallization is useful. Commercially available electrophoresis-grade SDS is usually of sufficient purity for most applications; however, if proteins will be electroeluted or electroblotted for protein sequence analysis, it may be desirable to crystallize the SDS twice from ethanol/water (Hunkapiller et al., 1983).

Materials

SDS
100% ethanol, room temperature
Water, 55°C
Activated charcoal, Norit 1 (Sigma)
100% reagent-grade ethanol, −20°C

55°C water bath
Buchner funnel and Whatman no. 5 paper
Coarse-frit (porosity A) sintered-glass funnel
Vacuum source
Desiccator containing charged phosphorous pentoxide (P_2O_5)
Dark bottle

1. Add 100 g SDS to 450 ml 100% ethanol and heat to 55°C. While stirring, gradually add 50 to 75 ml hot (~55°C) water until all the SDS dissolves.
2. Add 10 g activated charcoal to solution.
3. After 10 min, filter solution through Whatman no. 5 paper on a Buchner funnel to remove the charcoal. Chill the filtrate 24 hr at 4°C and 24 hr at −20°C.
4. Collect crystalline SDS on a coarse-frit (porosity A) sintered-glass funnel and wash with 800 ml −20°C ethanol (reagent grade).
5. Repeat the crystallization without adding activated charcoal.
6. Dry recrystallized SDS under vacuum overnight at room temperature.
7. Store up to 1 year in a dark bottle in a desiccator over P_2O_5.

REAGENTS AND SOLUTIONS

Use deionized, distilled water in all recipes and protocol steps. For common stock solutions, see **UNIT 3.3**.

Acrylamide (30%)/bisacrylamide (0.8%)

Mix 30.0 g acrylamide and 0.8 g *N,N′*-methylenebisacrylamide (bisacrylamide) with water in a total volume of 100 ml. Pass the solution through a 0.45-μm filter and store up to 30 days at 4°C in the dark. Discard after 30 days because acrylamide gradually hydrolyzes to acrylic acid and ammonia.

CAUTION: *Acrylamide monomer is a neurotoxin. A mask should be worn when weighing acrylamide powder. Gloves should be worn while handling the solution, and like all solutions, it should not be pipetted by mouth.*

The 2× crystallized grades of acrylamide and bisacrylamide are recommended.

SDS electrophoresis buffer, 5×

15.1 g Tris base (0.125 M final)
72.0 g glycine (0.96 M final)
5.0 g SDS (0.5% final), recrystallization optional (see Support Protocol 3)
H_2O to 1000 ml

Dilute to 1× or 2× with water as appropriate.

Do not adjust the pH of the stock solution, as the solution is pH 8.3 when diluted. Store at 0° to 4°C until use (up to 1 month).

If an electrophoretically separated protein will be electroeluted or electroblotted for sequence analysis, the highest-purity reagents available should be used. If necessary, SDS can be purified by recrystallization following the procedure given in Support Protocol 3.

SDS sample buffer, 2×

25 ml 4× Tris·Cl (pH 6.8)/SDS (see recipe)
20 ml glycerol (20% final)
4 g SDS (4% final; recrystallization optional)
2 ml 2-mercaptoethanol (2-ME; 0.2% final concentration) *or* 3.1 g dithiothreitol (DTT; 0.2 M final concentration)
1 mg bromphenol blue (0.001% final concentration)
Add H_2O to 100 ml and mix
Dispense into 1-ml aliquots and store up to 6 months at −70°C

To avoid reducing proteins to subunits (if desired), omit 2-ME or DTT (reducing agent) and add 10 mM iodoacetamide to prevent disulfide interchange.

SDS sample buffer, 6×

7 ml 4× Tris·Cl (pH 6.8)/SDS (see recipe)
3.0 ml glycerol (30% final concentration)
1 g SDS (10% final concentration)
0.93 g DTT (0.6 M final concentration)
1.2 mg bromphenol blue (0.012% final concentration)
Add H_2O to 10 ml, if necessary
Dispense into 0.5-ml aliquots and store up to 6 months at −70°C

Recrystallization of the SDS (Support Protocol 3) is optional.

Tris·Cl, 4× (pH 6.8)/SDS

Dissolve 6.05 g Tris base in 40 ml water. Adjust to pH 6.8 with 1 N HCl. Add water to 100 ml total volume. Pass the solution through a 0.45-μm filter, add 0.4 g SDS, and store up to 1 month at 4°C.

Final concentrations are 0.5 M Tris·Cl and 0.4% (w/v) SDS.

Tris·Cl, 4× (pH 8.8)/SDS

Dissolve 91 g Tris base in 300 ml water. Adjust to pH 8.8 with 1 N HCl. Adjust the volume to 500 ml with water. Pass the solution through a 0.45-μm filter, add 2 g SDS, and store up to 1 month at 4°C.

Final concentrations are 1.5 M Tris·Cl and 0.4% (w/v) SDS.

Water-saturated isobutyl alcohol

Prepare by shaking isobutyl alcohol and water in a glass vial. The butanol phase sits on top of the water. Prepare enough for 1 month (e.g., 25 ml). Store up to 1 month at room temperature.

UNDERSTANDING RESULTS

Separating proteins based on size alone

Polyacrylamide gel electrophoresis performed under denaturing and reducing conditions should resolve any two proteins of nonidentical size. Resolution of proteins in the presence of SDS is a function of gel concentration and the molecular mass of the proteins being separated. Under nondenaturing conditions, the biological activity of a protein will be maintained.

Looking for disulfide bonds

Comparison of reducing and nonreducing denaturing gels can also provide valuable information about the number of disulfide cross-linked subunits in a protein complex (Fig. 7.3.6). If the subunits are held together by disulfide linkages, the protein will separate in denaturing gels as a complex or as smaller-sized subunits under nonreducing or reducing conditions, respectively. However, proteins separated on nonreducing denaturing gels appear more diffuse and exhibit less overall resolution than those separated on reducing gels.

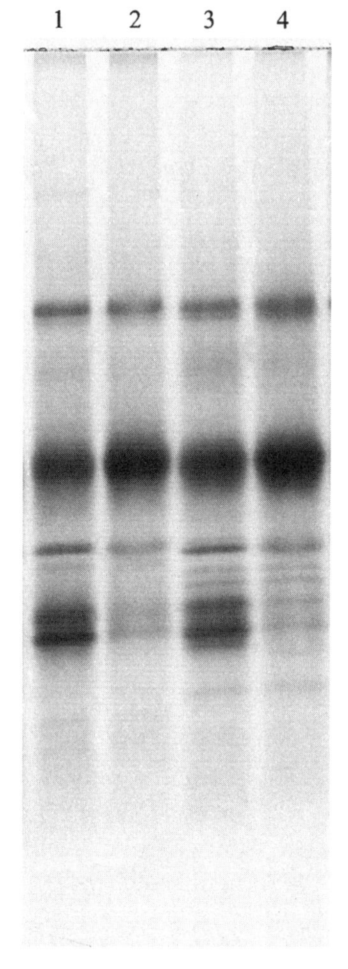

Figure 7.3.6 Enriched plasma membrane fractions from corn were separated by denaturing gradient SDS-PAGE. Lanes 1 and 3 contain denatured and reduced proteins, while 2 and 4 are the same samples which, while denatured, have not been reduced. Note that reducing the proteins generated heavier bands of lower molecular-weight proteins, indicating the presence of a larger multisubunit complex of proteins in the unreduced state.

TROUBLESHOOTING

Heat Transfer (Smiling)

Uneven heating of the gel causes differential migration of proteins, with the outer lanes moving more slowly than the center lanes (called smiling; Fig. 7.3.7). Increased heat transfer eliminates smiling and can be achieved by filling the lower buffer chamber with buffer all the way to the level of the sample wells and stirring the lower buffer with a magnetic stirrer. This helps to maintain an optimal constant temperature between 10°C to 20°C. Alternatively, decrease the heat load by running at a lower current.

Diffuse Protein Bands

If the tracking dye band is diffuse, prepare fresh buffer and acrylamide monomer stocks. If the protein bands are diffuse, increase the current by 25% to 50% to complete the run more quickly and minimize band diffusion, use a higher percentage of acrylamide, or try a gradient gel. Lengthy separations using gradient gels generally produce good results (see Fig. 7.3.8). Check for possible proteolytic degradation that may cause loss of high-molecular-weight bands and create a smeared banding pattern.

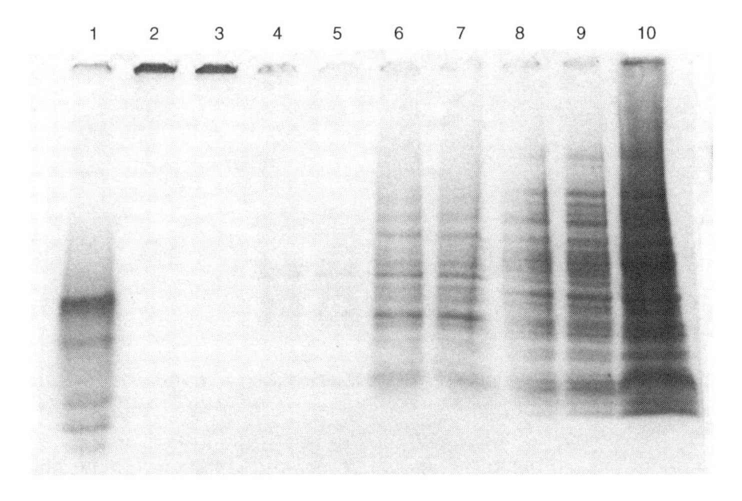

Figure 7.3.7 An example of a gel with uneven separation and staining. For optimal results, each sample lane in a gel should be loaded with the same amount of either sample or blank sample buffer so that the electrical characteristic of each lane are consistent. This prevents lane spreading (lane 10). Uneven staining can be minimized by gentle shaking in a large volume of staining solution during the staining process. Loss of high molecular-weight proteins and a general "fuzziness" in separation (lane 1) can result from the use of old or expired gels and buffers, or by protease activity in the sample.

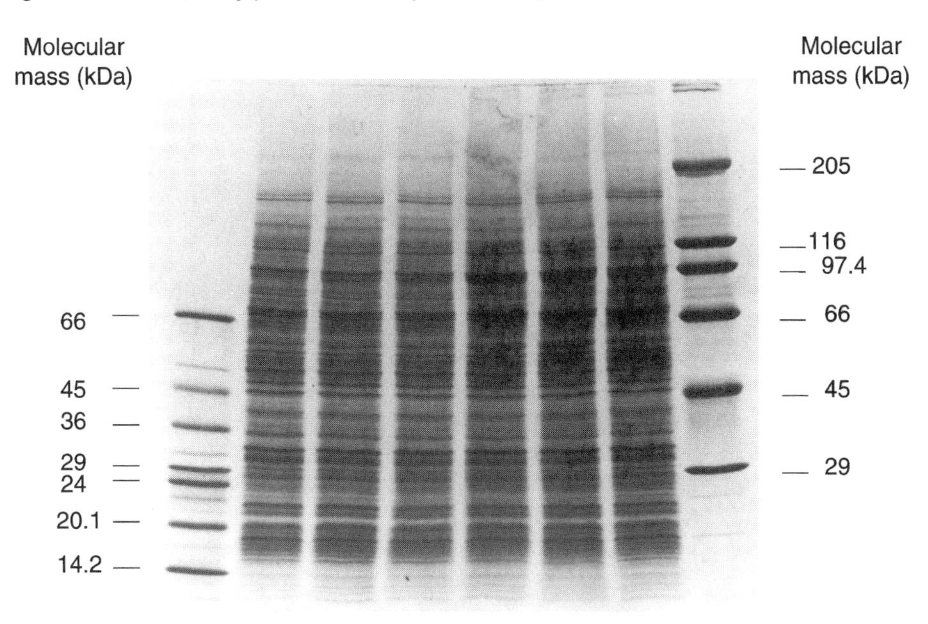

Figure 7.3.8 Separation of membrane proteins by 5.1% to 20.5% T polyacrylamide gradient SDS-PAGE. (%T = g acrylamide + g bisacrylamide/100 ml × 100) Approximately 30 μl of 1μ SDS sample buffer containing 30 μg of Alaskan pea (*Pisum sativum*) membrane proteins were loaded in wells of a 14 × 14–cm, 0.75-mm-thick gel. Standard proteins were included in the lanes 1 and 10. The gel was run 15 hr at 4 mA.

Vertical Streaking

If there is vertical streaking of protein bands (Fig. 7.3.9), decrease the amount of sample loaded on the gel, further purify the protein of interest to reduce the amount of contaminating protein applied to the gel, or reduce the current by 25%. Vertical streaking of protein bands may also result from precipitation, which can sometimes be eliminated by centrifuging the sample or by reducing the percentage of acrylamide in the gel.

Protein Runs Faster than Expected

Proteins can migrate faster or slower than their actual molecular weight would indicate. Abnormal migration is usually associated with a high proportion of basic or charged amino acids (Takano

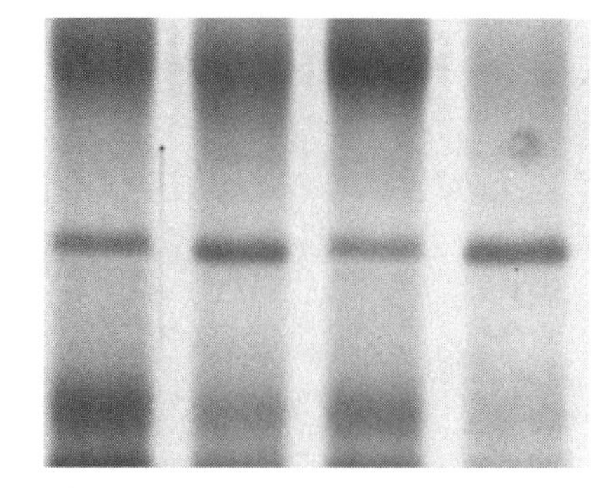

Figure 7.3.9 An example of vertical streaking. The vertical streaks of stained material are caused by skin flakes polymerized into the gel during gel preparation. When the dye front passes through during electrophoresis, the proteins in the skin flake solubilize and electrophoresis, causing the tail of stained material. Filtering solutions and blowing out dust from the gel sandwich before polymerization will minimize these issues. The circular stain in the far right lane was caused by another source of protein: a small insect that dropped onto the gel during handling.

et al., 1988). Other problems can occur during isolation and preparation of the protein sample for electrophoresis. Proteolysis of proteins by endogenous proteases during cell fractionation can cause subtle band splitting and smearing in the resulting electrophoretogram (electrophoresis pattern). Many endogenous proteases are very active in SDS sample buffers and will rapidly degrade the sample; thus, heating the samples to 70°C to 100°C for 3 min before loading them onto the gel is recommended.

Heating Artifacts and Proteolysis

In some cases, heating to 100°C in sample buffer will cause selective aggregation of proteins, creating a smeared layer of Coomassie blue-stained material at the top of the gel (Gallagher and Leonard, 1987; Fig. 7.3.10). To avoid heating artifacts and also to prevent proteolysis, the use of specific protease inhibitors during protein isolation and/or lower heating temperatures (70°C to 80°C) have been effective (Dhugga et al., 1988).

Although continuous gels suffer from poor band sharpness, they are less prone to artifacts caused by aggregation and protein cross-linking. If streaking or aggregation appears to be a problem with the Laemmli system, then the same sample should be subjected to continuous SDS-PAGE to see if the problem is intrinsic to the Laemmli gel or the sample.

Lateral Spreading

If the protein bands spread laterally from gel lanes, the time between applying the sample and running the gel should be reduced in order to decrease the diffusion of sample out of the wells. Alternatively, the acrylamide percentage should be increased in the stacking gels from 4% to 4.5% or 5% acrylamide, or the operating current should be increased by 25% to decrease diffusion in the stacking gel. Use caution when adding 1× SDS electrophoresis buffer to the upper buffer chamber. Samples can get swept into adjacent wells and onto the top of the well arm.

Distorted Protein Bands

If the protein bands are uneven, the stacking gel may not have been adequately polymerized. This can be corrected by deaerating the stacking gel solution thoroughly or by increasing the ammonium persulfate and TEMED concentrations by one-third to one-half. Another cause of distorted bands is salt in the protein sample, which can be removed by dialysis, gel filtration, or precipitation. Skewed protein bands can be caused by an uneven interface between the stacking and separating gels, which can be corrected by starting over and being careful not to disturb the separating gel while overlaying with isobutyl alcohol.

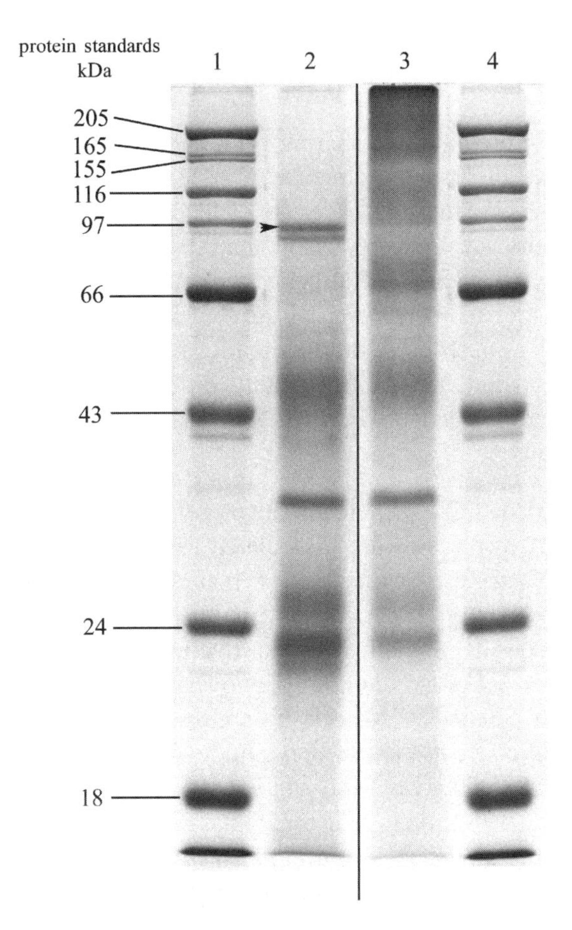

Figure 7.3.10 An example of the effect of heating-related aggregation and endogenous proteases on the pattern of separated proteins. Lane 2 contains purified membrane proteins that were not heated in SDS sample buffer. The standard samples in lanes 1 and 4 were heated, as was the membrane sample in lane 3. Note that heating caused aggregation of the membrane protein so that the proteins collected at the top of the separating gel. Unheated sample proteins (lane 2) show the major band of the ATPase ion transport catalytic subunit (arrow) otherwise lost using typical sample preparation and heating protocols. Heating the sample in SDS is typically used to denature the sample and inactivate proteases that are active in the SDS sample buffer; however, this can lead to aggregation of membrane proteins. Eliminating heating and using protease inhibitors instead can prevent heat-induced aggregation. Protease inhibitors added to the SDS sample buffer gave a slightly higher molecular weight for several proteins in the sample, indicating degradation by endogenous proteases (not shown). The center line indicates two portions of the same gel that were cut and pasted together for ease of comparison.

Improper Run Times

If a run takes too long, the buffers may be too concentrated or the operating current too low. If the run is too short, the buffers may be too dilute or the operating current too high.

Double Bands

If double bands are observed, the protein may be partially oxidized or partially degraded. Oxidation can be minimized by increasing the 2-mercaptoethanol concentration in the sample buffer or by preparing a fresh protein sample. If fewer bands than expected are observed and there is a heavy protein band at the dye front, increase the acrylamide percentage in the gel. A very heavy band at the dye front indicates proteins too small to be resolved by the acrylamide concentration used. The dye front can include degraded proteins.

VARIATIONS

Gradient Gels

Gradient gels provide superior protein-band sharpness and resolve a larger size range of proteins, making them ideal for most types of experiments in spite of being more difficult to prepare. Molecular-weight calculations are simplified because of the extended linear relationship between size and protein position in the gel. Increased band sharpness of both high- and low-molecular-weight proteins on the same gel greatly simplifies survey experiments such as gene expression studies where the characteristics of the responsive protein are not known. Furthermore, the increased resolution dramatically improves autoradiographic analysis. Preparation of gradient gels is straightforward and preparation times are not significantly different; however, practice with gradient solutions containing dye is recommended. Gradient gels can be stored for several days at 0°C to 4°C before casting the stacking gel.

The Basic Protocol relies on denaturing proteins in the presence of SDS and 2-ME or DTT. Under these conditions, the subunits of proteins are dissociated and their biological activities are lost. A true estimate of a protein's molecular size can be made by comparing the relative mobility of the unknown protein to proteins in a calibration mixture (Fig. 7.3.2). Gradient gels simplify molecular-weight determinations by producing a linear relationship between log molecular weight of the protein and log total percent of acrylamide (%T) over a much wider size range than single-concentration gels. Although percent acrylamide monomer is a more common measure of gel concentration, %T is the total concentration of acrylamide and bisacrylamide monomers (%T = g acrylamide + g bisacrylamide/100 ml × 100) and is used for molecular-weight calculations in gradient gels. The %T for a stained protein is estimated assuming that the acrylamide gradient is linear. For example, proteins in the gel shown in Figure 7.3.8 were separated in a 5.1% to 20.5% T acrylamide gradient. The %T of the point halfway through the resolving gel is 12.5% T. Simply

Table 7.3.5 Key Applications of Protein Electrophoresis[a]

Technique	Separation principle	Application	For more information
Native PAGE	Native charge, size, and shape	Purification, determination of native protein size, and identification of protein complexes	*CP Molecular Biology Unit 10.2B* (Gallagher, 1999)
1-D SDS-PAGE	Size dependent: SDS imparts a negative charge to proteins, giving a constant charge to mass ratio	Size estimation, purity check, purification, subunit composition, protein expression	*CP Molecular Biology Unit 10.2A* (Gallagher, 2006)
Peptide SDS-PAGE	Size dependent, as in 1-D SDS-PAGE	Small protein separation	*CP Molecular Biology Unit 10.2A* (Gallagher, 2006); Schagger and von Jagow (1987); Okajima et al. (1993)
IEF	Intrinsic charge with both native and denatured proteins	Purification, purity check, isoenzyme analysis	*CP Protein Science Unit 10.2* (Ploegh, 1995)
2-D SDS-PAGE	Isoelectric point in the first dimension, size in the second	Protein expression, purification, posttranslational analysis, proteomics	*CP Molecular Biology Unit 10.4* (Adams and Gallagher, 2004)

[a]Abbreviations: 1-D, one dimensional; 2-D, two dimensional; IEF, isoelectric focusing; PAGE, polyacrylamide gel electrophoresis; SDS, sodium dodecyl sulfate.

plotting log molecular mass versus distance moved into the gel (or R_f) also produces a relatively linear standard curve over a fairly wide size range.

Other Gel Systems

Table 7.3.5 provides other commonly used gel systems and a reference where a more thorough treatment can be found.

LITERATURE CITED

Adams, L.D. and Gallagher, S. 2004. Two-dimensional gel electrophoresis. *Curr. Protoc. Mol. Biol.* 67:10.4.1-10.4.23.

Dhugga, K.S., Waines, J.G., and Leonard, R.T. 1988. Correlated induction of nitrate uptake and membrane polypeptides in corn roots. *Plant Physiol.* 87:120-125.

Gallagher, S.R. and Leonard, R.T. 1987. Electrophoretic characterization of a detergent-treated plasma membrane fraction from corn roots. *Plant Physiol.* 83:265-271.

Gallagher, S.R. 1999. One-dimensional electrophoresis using nondenaturing conditions. *Curr. Protoc. Mol. Biol.* 47:10.2B.1-10.2B.11.

Gallagher, S.R. 2006. One-dimensional SDS gel electrophoresis of proteins. *Curr. Protoc. Mol. Biol.* 75:10.2A.1-10.2A.37.

Hunkapiller, M.W., Lujan, E., Ostrander, F., and Hood, L.E. 1983. Isolation of microgram quantities of proteins from polyacrylamide gels for amino acid sequence analysis. *Methods Enzymol.* 91:227-236.

Jovin, T.M. 1973. I. Steady-state moving-boundary systems formed by different electrolyte combinations. *Biochemistry* 12:871-879.

Laemmli, U.K. 1970. Cleavage of structural proteins during the assembly of the head of bacteriophage T4. *Nature* 227:680-685.

Matsudaira, P.T. and Burgess, D.R. 1978. SDS microslab linear gradient polyacrylamide gel electrophoresis. *Anal. Biochem.* 87:386-396.

Okajima, T., Tanabe, T., and Yasuda, T. 1993. Nonurea sodium dodecyl sulfate-polyacrylamide gel electrophoresis with high-molarity buffers for the separation of proteins and peptides. *Anal. Biochem.* 211:293-300.

Ornstein, L. 1964. Disc electrophoresis. I. Background and theory. *Ann. N.Y. Acad. Sci.* 121:321-349.

Ploegh, H.L. 1995. One-dimensional isoelectric focusing of proteins in slab gels. *Curr. Protoc. Prot. Sci.* 0:10.2.1-10.2.8.

Schagger, H. and von Jagow, G. 1987. Tricine-sodium dodecyl sulfate-polyacrylamide gel electrophoresis for the separation of proteins in the range from 1 to 100 kDa. *Anal. Biochem.* 166:368-379.

Scopes, R.K. 1995. Overview of protein purification and characterization. *Curr. Protoc. Protein Sci.* 0:1.1.1-1.1.6.

Takano, E., Maki, M., Mori, H., Hatanaka, N., Marti, T., Titani, K., Kannagi, R., Ooi, T., and Murachi, T. 1988. Pig heart calpastatin: Identification of repetitive domain structures and anomalous behavior in polyacrylamide gel electrophoresis. *Biochemistry* 27:1964-1972.

Weber, K., Pringle, J.R., and Osborn, M. 1972. Measurement of molecular weights by electrophoresis on SDS-acrylamide gel. *Methods Enzymol.* 26:3-27.

KEY REFERENCE

Hames, B.D. and Rickwood, D. (eds.) 1990. Gel Electrophoresis of Proteins: A Practical Approach 2nd Edition. Oxford University Press, New York.

An excellent book describing gel electrophoresis of proteins.

UNIT 7.4

Staining Proteins in Gels

Sean R. Gallagher[1] and Joachim Sasse[2]
[1]UVP, LLC, Upland, California
[2]Shriners Hospital for Crippled Children, Tampa, Florida

OVERVIEW AND PRINCIPLES

The location of a protein in a polyacrylamide gel is typically determined by either Coomassie blue staining (Basic Protocol 1) or silver staining (Basic Protocol 2). The former is easier and more rapid; however, silver staining methods are considerably more sensitive and thus can be used to detect smaller amounts of protein. Fluorescent staining has become a popular alternative to the traditional staining procedures, mainly because it is more sensitive than Coomassie staining and is often as sensitive as silver staining. Therefore, this unit also includes a protocol describing SYPRO Orange or SYPRO Red staining of proteins in SDS-polyacrylamide gels (Basic Protocol 3).

Coomassie blue remains the method of choice for simple, low-cost protein visualization. The reasons for this are many: the dark blue proteins can be visually inspected, the stain is quantitative, and protocol variations produce fast and very sensitive results (also see Westermeier, 2006). In addition, Coomassie blue can be analyzed digitally with both desktop scanners and CCD imaging systems (see *UNIT 7.5* for more discussion) and is compatible with mass spectrometry. Coomassie brilliant blue binds nonspecifically to proteins (Wilson, 1983). It is thought to work as an anion in acidic staining solutions, interacting with the NH_3^+ groups associated with the separated protein (Fazekas de St. Groth et al., 1963). Because the dye does not bind to the polyacrylamide gel, proteins will be detected as blue bands surrounded by clear gel zones.

Silver staining relies on differential reduction of silver ions, which is the basis for photographic processes. It is performed via two common methods: reduction of silver diamine complexes in basic solution (e.g., Oakley, 1980) and silver ion reduction in acidic solution (e.g., Morrissey, 1981; Merril et al., 1984). The use of silver nitrate with acidic development (Morrissey, 1981) is the method of choice for simple, fast visualization of gels <1 mm thick and is presented in Basic Protocol 2. The mechanism of silver staining has been reviewed in detail (Merril, 1990; Rabilloud, 1990). Briefly, silver ions or complexes bind to proteins through interaction with several amino acids, including cysteine, lysine, aspartate, glutamate, and histidine (Rabilloud, 1990). Selective reduction of the silver interacting with the protein leads to the visualization. The dependence on certain amino acids leads to composition-dependent staining, meaning that not all proteins will show the same intensity of staining for a given amount of silver stain. Silver staining permits the detection of polypeptides in gels at more than 100× lower concentrations than Coomassie brilliant blue (i.e., femtomole levels of protein). The silver staining protocol is a very rapid method originally described by Bloom et al. (1987).

Fluorescent protein gel stains provide a number of advantages over conventional colorimetric stains. The SYPRO Orange and Red protein gel stains described in this unit can detect 1 to 2 ng protein per minigel band, more sensitive than Coomassie brilliant blue staining and as sensitive as many silver staining techniques. In addition, staining is complete in <1 hr. After electrophoresis, the gel is simply stained, rinsed, and photographed; no separate fixation or destaining step is required, and there is no fear of overstaining the gel (Steinberg et al., 1996a,b, 1997). In addition, stained proteins can be visualized using a standard 300-nm UV transilluminator, blue light transilluminator, or a laser scanner. Because the dyes interact with the SDS coat around all proteins in the gel, they give more consistent staining between different types of proteins compared to Coomassie or silver staining. Furthermore, the dyes detect a variety of proteins down to ~6500 Da without staining nucleic acid or lipopolysaccharide contaminants that are sometimes found in protein preparations derived from cell or tissue extracts.

STRATEGIC PLANNING

Staining Considerations

Silver staining

The high sensitivity of the silver staining technique renders it susceptible to impurities and staining artifacts. It is mandatory that the polyacrylamide gels and all staining solutions be prepared from high-quality reagents in order to avoid staining artifacts. Especially important is the use of high-quality water (glass-distilled or deionized, carbon-filtered). The glassware used for gel polymerization and plastic containers should be cleaned thoroughly, and gels should be handled with vinyl, powder-free gloves.

Fluorescence staining

If immunoblotting (*UNIT 8.3*) and other blotting techniques are to be performed, staining the gel with SYPRO Tangerine protein gel stain (which does not require organic solvents or acetic acid fixation) or staining the blot directly with SYPRO Ruby protein blot stain is recommended.

Changing the recommended concentration of fluorescent stain will not result in better detection. Diluting fluorescent stain below the recommended concentration will result in reduced staining sensitivity. Using higher staining concentrations than recommended will result in increased background and quenching of the fluorescence from dye molecules crowded around the proteins.

SYPRO Red and Orange stains cannot be used to prestain protein samples for SDS gels. Loading solutions contain so much SDS that the stain is sequestered by the excess SDS, effectively lowering the dye available to bind to and stain the proteins.

The SDS front at the bottom of the gel (where protein are not fractionated and run near an R_f of 1.0) stains very heavily with SYPRO stains. Unless the proteins of interest co-migrate with the SDS front, it is advantageous to run the SDS front off the gel. Colored stains and marker dyes, as well as commercially prestained protein markers, interfere with SYPRO dye staining and will quench fluorescence.

Time Considerations

Coomassie blue staining requires 8 to 12 hr after the proteins have been separated. Fixation may be extended for several days before staining with Coomassie blue. Detection of protein bands by rapid Coomassie blue staining (Alternate Protocol 1) requires ≤90 min for running a minigel (30 to 60 min), fixing the gel (10 min), and placing it in staining solution (5 to 10 min); however, additional time may be necessary for larger gels.

Silver staining requires ~5 hr after the proteins have been separated. Fixation may be extended up to several weeks before silver staining. Separated proteins stained using the rapid silver staining method (Alternate Protocol 2) can be visualized in ~35 min.

The staining time for SYPRO dyes is 10 to 60 min, depending on the thickness and percentage of acrylamide in the gel. For 1-mm-thick, 15% polyacrylamide gels, the fluorescence signal is typically optimal after 40 to 60 min of staining.

PROTOCOLS

Basic Protocol 1: Coomassie Blue Staining

Detection of protein bands in a gel by Coomassie blue staining depends on nonspecific binding of a dye (Coomassie brilliant blue R) to proteins. The detection limit is 0.3 to 1 μg/protein band. In this procedure, proteins separated in a polyacrylamide gel are precipitated using a fixing solution containing methanol and acetic acid. The location of the precipitated proteins is then detected using Coomassie blue (which turns the entire gel blue). After destaining, the blue protein bands appear against a clear background. The gel can then be stored in acetic acid or water, photographed, or dried to maintain a permanent record. A rapid staining procedure using Coomassie blue is provided in Alternate Protocol 1.

Materials

Polyacrylamide gel with separated proteins (see *UNIT 7.3*)
Fixing solution for Coomassie blue and silver staining (see recipe)
Coomassie blue staining solution (see recipe)
Methanol/acetic acid destaining solution (see recipe)
7% (v/v) aqueous acetic acid

Plastic container (slightly larger than the gel)
Orbital shaker or rocking platform
Heat-sealable plastic bags
Camera (optional)
Whatman 3MM filter paper (optional)
Plastic wrap (optional)
Gel dryer (optional)

1. Place the polyacrylamide gel in a plastic container and cover with three to five gel volumes of fixing solution. Agitate gently 2 hr at room temperature on an orbital shaker or rocking platform.

   If agitation is too rapid, the gel may break apart. Use fixing solution only once.

2. Pour out the fixing solution. Cover the gel with Coomassie blue staining solution and gently agitate 4 hr at room temperature.

   Use staining solution only once.

3. Pour out the staining solution. Rinse the gel briefly with ~50 ml fixing solution.

   Do not store gels in fixing solution because the protein bands will eventually lose the bound dye and disappear.

4. Pour out the fixing solution. Cover the gel with methanol/acetic acid destaining solution for 2 hr and agitate gently 4 hr at room temperature.

5. Pour out the destaining solution. Add fresh methanol/acetic acid destaining solution and continue destaining until blue bands and a clear background are obtained.

   Destaining time depends on the number of destaining solution changes, volume of destaining solution used, agitation, and temperature. Destaining can usually be completed in 12 to 24 hr.

6. Store the gel in 7% aqueous acetic acid or water up to 1 year at 4°C in a heat-sealed plastic bag.

7. *Optional:* Photograph the gel.

8. *Optional:* Dry the stained gel to maintain a permanent gel record. Place the gel on two sheets of Whatman 3MM filter paper and cover the top with plastic wrap. Dry in a conventional gel dryer 1 to 2 hr at ~80°C (or according to the manufacturer's instructions). Store indefinitely once dried.

Alternate Protocol 1: Rapid Coomassie Blue Staining

Protein bands stained using this protocol can be detected within 5 to 10 min after adding rapid Coomassie staining solution. Because the Coomassie blue concentration is lower than that used in Basic Protocol 1, the gel background never stains very darkly and the bands can be seen even while the gel remains in the staining solution. Another difference is that isopropanol is substituted for methanol in the fixing solution. This method is faster but slightly less sensitive than the standard procedure (Basic Protocol 1).

Additional Materials (also see Basic Protocol 1)

Isopropanol fixing solution (see recipe)
Rapid Coomassie blue staining solution (see recipe)
10% (v/v) acetic acid

1. Place the polyacrylamide gel in a plastic or glass container. Cover the gel with three to five gel volumes isopropanol fixing solution and shake gently at room temperature. For a 0.7-mm-thick gel, shake 10 to 15 min; for a 1.5-mm-thick gel, shake 30 to 60 min.

2. Pour out the fixing solution. Cover the gel with rapid Coomassie blue staining solution and shake gently until the desired intensity is reached (2 hr to overnight) at room temperature.

   Bands will become visible even in the staining solution within 5 to 30 min, depending on gel thickness. The gel background will never stain very darkly.

3. Pour out the staining solution. Cover the gel with 10% acetic acid to destain, shaking gently ≥2 hr at room temperature, until a clear background is obtained.

4. If necessary, pour out the 10% acetic acid and add more. Continue destaining until clear background is obtained.

   It is usually unnecessary to add more destaining solution.

5. Store the gel up to 6 months at 4°C in 7% acetic acid or water or wrapped in plastic wrap.

6. *Optional:* Photograph or dry the gel (see Basic Protocol 1, steps 7 and 8).

Basic Protocol 2: Nonammoniacal Silver Staining

This nonammoniacal silver staining procedure uses more stable solutions and detects certain proteins (e.g., cytochrome *c*, insulin B) not stained using older silver staining methods (Morrissey, 1981). This method is more sensitive than the rapid silver staining procedure provided in Alternate Protocol 2, and it ensures the detection of very small proteins through glutaraldehyde fixation, which crosslinks the proteins in the gel.

Materials

Polyacrylamide gel with separated proteins (*UNIT 7.3*)
Fixing solution for Coomassie blue and silver staining (see recipe)
Methanol/acetic acid destaining solution (see recipe)
10% (v/v) glutaraldehyde: prepare fresh from 50% stock (Kodak)
5 μg/ml dithiothreitol (DTT; *UNIT 3.3*)
0.1% (w/v) silver nitrate solution: prepare fresh before use
Carbonate developing solution (see recipe)
2.3 M citric acid
0.03% (w/v) sodium carbonate (optional)

Glass or polyethylene container (slightly bigger than the gel)
Orbital shaker or rocking platform
Camera (optional)

Gel dryer (optional)
Heat-sealable plastic bags (optional)

NOTE: Wear gloves at all times to avoid fingerprint contamination.

1. Place a polyacrylamide gel in a glass or polyethylene container and add 100 ml fixing solution for Coomassie blue and silver staining. Agitate gently 30 min at room temperature on an orbital shaker.

   Times and volumes given in the steps are appropriate for all gel sizes.

2. Pour out the fixing solution. Immerse the gel in methanol/acetic acid destaining solution and agitate gently 30 min at room temperature.

3. Pour out the destaining solution. Cover the gel with 50 ml of 10% glutaraldehyde and agitate gently 10 min at room temperature in a fume hood.

   CAUTION: *Wear gloves and work only in a fume hood.*

   Fixing in glutaraldehyde ensures that the gel will not bleach (and will probably last longer) and is important for retention and detection of very small proteins.

4. Pour out the glutaraldehyde. Wash the gel *thoroughly* in several changes of water (or in running water) for 2 hr to ensure low background levels.

5. Pour out the water. Soak the gel in 100 ml of 5 μg/ml DTT for 30 min at room temperature.

   Treating proteins with DTT results in more reproducible silver staining.

6. Pour out the DTT. Without rinsing, add 100 ml 0.1% silver nitrate solution and agitate gently for 30 min at room temperature.

7. Pour out the silver nitrate. Quickly wash the gel once with a small amount of water and then twice with a small amount of carbonate developing solution.

8. Soak the gel in 100 ml carbonate developing solution and agitate gently until the desired level of staining is achieved.

9. Stop the staining by adding 5 ml 2.3 M citric acid per 100 ml carbonate developing solution for 10 min and agitate slowly.

   Pay attention to the volumes of carbonate and citric acid solutions (steps 8 and 9). These must be balanced carefully to bring the pH to neutrality. If the pH is too high, the reaction will not stop; if the pH is too low, the gel will bleach.

10. Pour off the solution. Wash the gel several times in water, agitating gently 30 min at room temperature.

11. *Optional:* Photograph the gel.

12. *Optional:* Dry the well rinsed gel under low heat and store indefinitely.

13. *Optional:* Soak the gel 10 min in 0.03% sodium carbonate and store up to 6 months at 4°C in plastic wrap or a heat-sealed plastic bag.

Alternate Protocol 2: Rapid Silver Staining

This protocol (Bloom et al., 1987) is based upon the nonammoniacal silver staining method in Basic Protocol 2. It is rapid and gives low background but may not be quite as sensitive in detecting very small proteins because there is no glutaraldehyde fixation. This may be an advantage because special handling is not required to avoid glutaraldehyde's irritant effects.

Additional Materials (also see Basic Protocol 2)

Formaldehyde fixing solution (see recipe)
0.2 g/liter sodium thiosulfate ($Na_2S_2O_3$)
Thiosulfate developing solution (see recipe)
Drying solution (see recipe)
Dialysis membrane soaked in 50% methanol
Glass plates

1. Place the polyacrylamide gel in a plastic container and add 50 ml formaldehyde fixing solution. Agitate gently 10 min at room temperature on an orbital shaker or rocking platform.

   Times indicated are flexible and are appropriate for a 0.75-mm × 5.5-cm × 8-cm, 12.5% acrylamide slab gel. Each gel is placed in an 8 × 14–cm plastic container.

2. Pour out the fixing solution. Wash the gel twice with water, 5 min for each wash, with gentle agitation.

3. Pour out the water. Soak the gel 1 min in 50 ml of 0.2 g/liter $Na_2S_2O_3$, agitating gently.

4. Pour out the $Na_2S_2O_3$. Wash the gel twice with water, 20 sec for each wash.

5. Pour out the water. Soak the gel 10 min in 50 ml of 0.1% silver nitrate solution, agitating gently.

6. Pour out the silver nitrate. Quickly wash the gel with water and then with a small volume of thiosulfate developing solution.

7. Soak the gel in 50 ml fresh thiosulfate developing solution and agitate gently until band intensities are adequate (~1 min).

   Development continues a little after stopping (next step), so do not overdevelop here.

8. Add 5 ml of 2.3 M citric acid per 100 ml thiosulfate developing solution and agitate slowly for 10 min.

   Pay attention to the volumes of thiosulfate and citric acid solutions (steps 7 and 8). These must be balanced carefully to bring the pH to neutrality. If the pH is high, the reaction will not stop; if the pH is too low, the gel will bleach.

9. Pour off the solution. Wash the gel in water, agitating slowly for 10 min.

10. Pour off the water. Soak the gel 10 min in 50 ml drying solution.

11. Sandwich the gel between two pieces of wet dialysis membrane on a glass plate. Clamp edges of the plate with notebook clamps and dry overnight at room temperature.

Basic Protocol 3: Fluorescent Staining Using SYPRO Orange or Red

Fluorescent dyes have a number of advantages over traditional protein stains. SYPRO Orange and Red protein gel stains can detect 1 to 2 ng protein/band. This is more sensitive than Coomassie brilliant blue staining and as sensitive as many silver staining techniques. Staining is straightforward, with less hands-on time than typical silver staining protocols, and is complete in <1 hr. Stained proteins can be visualized using a standard 300-nm UV transilluminator, blue light transilluminator, or a laser scanner. Although the protocol below is limited to one-dimensional SDS-PAGE, alternative kits for native, isoelectric focusing (IEF), and two-dimensional SDS-PAGE applications are available from the supplier (Molecular Probes). Coomassie Fluor Orange protein gel stain, a premixed fluorescent stain, is also available from Molecular Probes.

The staining properties of the two SYPRO dyes are similar, and both are equally suitable for use in most procedures. The SYPRO Orange gel stain is slightly brighter, whereas the SYPRO Red gel stain has somewhat lower background fluorescence. For those using a laser-excited gel scanner, the authors recommend the SYPRO Orange stain for argon-laser-based instruments and the SYPRO Red stain for instruments that employ green He-Ne or Nd/YAG (neodymium/yttrium-aluminum-garnet) lasers. Both dyes are efficiently excited by UV or broad-band illumination and (with the proper filters) work nicely with CCD camera archiving systems. The Support Protocol describes photography of fluorescently stained proteins.

Materials

Protein sample for analysis
SYPRO Orange or Red fluorescent staining solution (see recipe)
7.5% (v/v) acetic acid
0.1% (v/v) Tween 20

Additional reagents and equipment for performing one-dimensional SDS-PAGE (*UNIT 7.3*)

1. Prepare and separate proteins using SDS-PAGE (*UNIT 7.3*), but use 0.05% SDS in the running buffer instead of the usual 0.1% SDS.

   Gels run in 0.1% SDS show the same sensitivity for staining as those run with lower SDS concentrations, but require either more time in staining solution or a 10-min rinse in water before staining to reduce the background fluorescence that is produced by dye interaction with SDS. Gels run in 0.05% SDS show no change in the migration of proteins and can be photographed sooner because they require less time in the staining solution to clear the SDS from the gel. Gels run in SDS concentrations <0.05% or in old running buffer exhibit poor resolution of bands and other problems, so it is essential that the SDS stock solution used to prepare the running buffer be fresh and at the proper SDS concentration.

   Do not fix the proteins in the gel with methanol-containing solutions. Methanol removes the SDS coat from proteins, strongly reducing the signal from SYPRO Orange or Red stains.

2. Pour SYPRO Orange or Red fluorescent staining solution into a small plastic dish. For one or two standard-size minigels, use ~50 ml staining solution. For larger gels, use between 500 and 750 ml staining solution.

   Staining dishes should be cleaned and rinsed well before use because detergent will interfere with staining.

   Alternatively, use Coomassie Fluor Orange protein gel stain, which is available as a premixed staining solution.

3. Place the gel into the staining solution. Cover the container with aluminum foil to protect the dye from bright light.

   The staining solution may be reused up to four times. However, due to reduced sensitivity, the use of fresh staining solution is recommended.

   Several alternate methods of staining can be used. (1) Gels may be stained in sealable plastic bags. However, it is still important to use the proper amount of staining solution. (2) For low-percentage gels and for very small proteins, increasing the staining solution from 7.5% to 10% acetic acid will result in better retention of the protein in the gel, without compromising sensitivity. (3) Protein gel stains can be dissolved 5000-fold into the cathode (top) running buffer to stain proteins as the gel runs. The dye moves through the gel with the SDS front, so that all sizes of protein are stained. Staining does not influence relative migration of proteins through the gel. This method results in poorer protein staining than the standard post-staining method, and requires the same amount of time because the gel must be destained for 15 to 40 min in 7.5% acetic acid to reduce background fluorescence.

4. Gently agitate at room temperature for 10 to 60 min.

   The staining time depends on the thickness and percentage of the gel. For 1-mm-thick, 15% polyacrylamide gels, the signal is typically optimal after 40 to 60 min staining. Once the optimal signal is achieved, additional staining time (several hours to overnight) does not enhance or degrade the signal. Gels can be left in stain for up to a week with only a small loss in sensitivity; the detection limits under these conditions are ~2 to 4 ng/band.

5. Rinse briefly (<1 min) with 7.5% acetic acid.

   This brief rinse removes excess stain from the gel surface to reduce background fluorescence on the surface of the transilluminator or gel scanner (see Support Protocol).

6. Store the gel in staining solution, protected from light.

   The signal decreases somewhat after several days, but (depending on the amount of protein in the bands) gels may retain a usable signal for many weeks. Gels may be left in staining solution overnight without losing sensitivity. However, fixation in acetic acid is relatively mild, so, for low-percentage gels or very small proteins, photographs should be taken as soon as possible after staining, before the proteins begin to diffuse.

   Gels may be dried between cellophane membrane backing sheets (Bio-Rad), although there is sometimes a slight decrease in sensitivity. If the gels are dried onto paper, the light will scatter and the sensitivity will decrease. Other types of plastic sheets are not typically transparent to UV light. Store dried gels in the dark to prevent photobleaching.

7. Destain the gel by incubating overnight in 0.1% Tween 20.

   Alternatively, incubating in several changes of 7.5% acetic acid will eventually remove all of the stain. Incubating in methanol will strip off dye and SDS, but will also precipitate proteins.

8. Photograph the gel (see Support Protocol).

Support Protocol: Photography of SYPRO Orange or Red Fluorescently Stained Gels

Photographing the fluorescently stained gel is essential to obtaining high sensitivity in detecting the protein bands. The camera's integrating effect can make visible bands that are not visible to the eye. Place the fluorescently labeled gel directly on a standard 300-nm UV transilluminator or a blue-light transilluminator (e.g., Clare Chemical, Invitrogen, or UVP). It is important to clean the surface of the transilluminator after each use with deionized water and a soft cloth (e.g., cheesecloth). Otherwise, fluorescent dyes (e.g., SYPRO stains, SYBR stains, and ethidium bromide) will accumulate on the glass surface and cause high background fluorescence. Plastic wraps, such as Saran Wrap, should not be used, because they fluoresce naturally and will fluoresce even more when exposed to SYPRO Orange or Red stain. This results in a large background signal, making it impossible to achieve good sensitivity. Pharmacia PhastGels have a polyester backing material (Gel bond) that is not only highly autofluorescent, but also binds the SYPRO Orange and Red protein gel stains, producing additional background fluorescence. Consequently, the plastic backing should be removed before trying to visualize bands (Pharmacia markets a PhastGel backing remover).

The use of a photographic camera or charge-coupled device (CCD; see *UNIT 7.5*) camera and the appropriate filters is essential to obtaining the greatest sensitivity. The instrument's integrating capability can make visible bands that cannot be detected by eye. Using a CCD camera, images are best obtained by digitizing at ~1024 × 1024 pixels resolution with 12- or 16-bit grayscale levels per pixel. The camera manufacturer should be consulted for recommendations on filter sets to use. A CCD camera-based image analysis system can gather quantitative information that will allow comparison of fluorescence intensities between different bands or spots. For those using a laser-excited gel scanner, the SYPRO Orange stain is recommended for argon laser-based instruments, and the SYPRO Red stain is recommended for instruments that employ green He-Ne or Nd/YAG lasers.

REAGENTS AND SOLUTIONS

*Use high-quality deionized, distilled water (≥18 MΩ) in all recipes and protocol steps. For common stock solutions, see **UNIT 3.3**.*

Carbonate developing solution

0.5 ml 37% formaldehyde per liter of solution
3% (w/v) sodium carbonate
Prepare fresh before use

Coomassie blue staining solution

50% (v/v) methanol
0.05% (w/v) Coomassie brilliant blue R-250 (Bio-Rad or Pierce)
10% (v/v) acetic acid
40% H_2O

Dissolve Coomassie brilliant blue R-250 in methanol before adding acetic acid and water. Store for up to 6 months at room temperature.

If precipitate is observed following prolonged storage, filter to obtain a homogeneous solution.

Drying solution

10% (v/v) ethanol
4% (v/v) glycerol
86% H_2O
Store up to ~1 month at room temperature

Fixing solution for Coomassie blue and silver staining

50% (v/v) methanol
10% (v/v) acetic acid
40% H_2O
Store up to ~1 month at room temperature

Formaldehyde fixing solution

40% (v/v) methanol
0.5 ml 37% formaldehyde per liter solution
60% H_2O
Store up to ~1 month at room temperature

Isopropanol fixing solution

25% (v/v) isopropanol
10% (v/v) acetic acid
65% H_2O
Store indefinitely at room temperature

Methanol/acetic acid destaining solution

5% (v/v) methanol
7% (v/v) acetic acid
88% H_2O
Store up to ~1 month at room temperature

Rapid Coomassie blue staining solution

10% (v/v) acetic acid
0.006% (w/v) Coomassie brilliant blue G-250 (Bio-Rad)
90% H_2O
Store indefinitely at room temperature

SYPRO Orange or Red fluorescent staining solution

Allow the stock vial of SYPRO Orange or Red protein gel stain (Molecular Probes) to warm to room temperature. If particles of dye are present, dissolve by briefly sonicating the tube or vortexing it vigorously after warming. Briefly microcentrifuge to deposit the dimethyl sulfoxide (DMSO) solution at the bottom of the vial. Dilute stock 1:5000 (v/v) in 7.5% (v/v) acetic acid

continued

and mix vigorously. Store in very clean, detergent-free glass or plastic bottles, protected from light, at 4°C (stable ≥3 months). Store the stock solutions 6 months to 1 year at room temperature, 4°C, or −20°C, protected from light.

SYPRO Orange: 300 and 470 nm excitation, 570 nm emission.

SYPRO Red: 300 and 550 nm excitation, 630 nm emission.

Thiosulfate developing solution

3% (w/v) sodium carbonate
0.0004% (w/v) sodium thiosulfate
Store indefinitely at room temperature

Add 0.5 ml 37% formaldehyde per liter solution immediately before use.

UNDERSTANDING RESULTS

The sensitivity of Coomassie blue gel staining is 0.3 to 1 µg/protein band; the sensitivity of silver staining is 2 to 5 ng/protein band. The sensitivity of both stains varies in an unpredictable manner with the protein being stained.

For fluorescent dyes, detection limits are typically ~500 ng protein/band in room light, ~50 ng protein/band with 300-nm transillumination, and ~1 to 2 ng protein/band in a photograph taken with Polaroid 667 black-and-white print film. The authors achieve detection limits of 1 to 2 ng/band using a Fotodyne Foto/UV 450 ultraviolet transilluminator, which has six 15-watt bulbs that provide peak illumination at 312 nm. When using weaker illumination sources, exposures must be correspondingly longer. Although the authors' detection limits are 1 to 2 ng/band for most proteins, it should be emphasized that bands containing 5 to 10 ng/protein are more readily detected. Bands containing less than 5 to 10 ng protein require longer exposures and sharp bands for good visualization. Longer exposures can result in higher background.

TROUBLESHOOTING

Uneven Staining

With all staining procedures, the gel should remain fully hydrated and submerged in fixation and staining solutions with continuous gentle agitation using a reciprocating or rotary shaker. This prevents uneven staining due to uneven exposure to reagents. If dust or particulates are present in the solution, agitation prevents the debris from settling on the gel and showing up as spots. For prolonged diffusion staining using Coomassie blue, the tray should be covered to prevent evaporation, which can concentrate the staining solution, cause precipitation of the dye, and lead to poorly controlled staining and speckled or high backgrounds.

Poor Sensitivity

Due to different staining properties of proteins, dual staining procedures can reveal proteins with one procedure that the other has not visualized. SYPRO-stained gels can be restained with either Coomassie brilliant blue or silver stain procedures. In fact, for some silver staining methods, prestaining with SYPRO dyes actually increases the rate of staining and the sensitivity for detection.

To fluorescently stain gels that have previously been stained with Coomassie, the Coomassie stain must be completely removed because it will quench the fluorescence of SYPRO dyes. Soaking the gel in either 30% methanol or 7.5% acetic acid with several changes of destaining solution is effective in removing Coomassie stain. Once the Coomassie has been removed, the gel should be incubated in 0.05% SDS for 30 min before staining with the SYPRO stain as usual.

Highly colored prosthetic groups (e.g., heme) that remain bound in native gels will quench the fluorescence of the SYPRO Orange and Red stains.

Triton X-100 at ≥0.1% interferes with SYPRO dye staining. If Triton X-100 is used in the gel, the authors recommend soaking the gel in two to three changes of buffer to be sure the Triton X-100 is diluted out, and then incubating the gel in 0.05% SDS for 30 min before staining as usual.

Odd Marks

Odd marks on stained gels can be caused by several factors. If the gel is squeezed, a mark appears that stains heavily with the SYPRO dyes. This is probably due to a localized high concentration of SDS that has difficulty diffusing out. Glove powder can also give background markings, so rinsing or washing gloves is recommended prior to handling gels. Staining with the SYPRO Orange dye occasionally results in gels with scattered fluorescent speckles.

VARIATIONS

Westermeier (2006) describes variations on the basic Coomassie blue staining. These include staining the gel at 50°C as well as several variation of colloidal blue stains, including "Blue Silver" with sensitivity rivaling silver staining (Candiano et al., 2004).

The SYPRO Ruby protein gel stain is recommended for immunoblot total protein staining and two dimensional SDS-PAGE. It has two excitation maxima (at ~280 nm and ~450 nm) and an emission maximum near 610 nm. Proteins stained with the dye (see recipe) can be visualized using a 300-nm UV transilluminator, a blue-light transilluminator, or a laser scanner. The stain has exceptional photostability, allowing long exposure times for maximum sensitivity.

LITERATURE CITED

Bloom, H., Beier, H., and Gross, H.S. 1987. Improved silver staining of plant proteins, RNA and DNA in polyacrylamide gels. *Electrophoresis* 8:93-99.

Candiano, G., Bruschi, M., Musante, L., Santucci, L., Ghiggen, G.M., Carnerolla, B., Orecchia, P., Zardi, L., and Righetti, P.G. 2004. Blue silver: A very sensitive colloidal Coomasie G-250 staining for proteome analysis. *Electrophoresis* 25:1327-1333.

Fazeks de St. Groth S., Webster, R.G., and Datyner, A. 1963. Two new staining procedures for quantitative estimation of proteins on electrophoretic strips. *Biochim. Biophys. Acta.* 71:337-391.

Merril, C.R. 1990. Silver staining of proteins and DNA. *Nature* 343:779-780.

Merril, C.R., Goldman, D., and Van Keuren, M.L. 1984. Gel protein stains: Silver stain. *Methods Enzymol.* 104:441-447.

Morrissey, J.H. 1981. Silver stain for proteins in polyacrylamide gels: A modified procedure with enhanced uniform sensitivity. *Anal. Biochem.* 117:307-310.

Oakley, B.R., Kirsch, D.R., and Morris, N.R. 1980. A simplified ultrasensitive silver stain for detecting proteins in polyacrylamide gels. *Anal. Biochem.* 105:361-363.

Rabilloud, T. 1990. Mechanisms of protein silver staining in polyacrylamide gels: A 10-year synthesis. *Electrophoresis* 11:785-794.

Steinberg, T.H., Haugland, R.P., and Singer, V.L. 1996a. Applications of SYPRO orange and SYPRO red protein gel stains. *Anal. Biochem.* 239:238-245.

Steinberg, T.H., Jones, L.J., Haugland, R.P., and Singer, V.L. 1996b. SYPRO orange and SYPRO red protein gel stains: One-step fluorescent staining of denaturing gels for detection of nanogram levels of protein. *Anal. Biochem.* 239:223-237.

Steinberg, T.H., White, H.M., and Singer, V.L. 1997. Optimal filter combinations for photographing SYPRO orange or SYPRO red dye-stained gels. *Anal. Biochem.* 248:168-172.

Westermeir, R. 2006. Sensitive, quantitative, and fast modifications for Coomassie blue staining of polyacrylamide gels. *Proteomics* 6:61-64.

Wilson, C.M. 1983. Staining of proteins on gels: Comparison of dyes and procedures. *Methods Enzymol.* 91:236-247.

Overview of Digital Electrophoresis Analysis

Butch Moomaw,[1] Scott Medberry,[2] and Sean R. Gallagher[3]
[1]Hamamatsu Photonic Systems, Spring Branch, Texas
[2]Agilent Technologies, Inc., Palo Alto, California
[3]UVP, LLC, Upland, California

INTRODUCTION

Gel electrophoresis has become a ubiquitous method for separating biomolecules. This prominence is the result of several factors, including the robustness, speed, and potential throughput of the technique. The results of this method are traditionally documented using silver-halide-based photography followed by manual interpretation. While this remains an excellent method for qualitative documentation of single-gel results, digital capture offers a number of significant advantages when documentation requires quantitation and sophisticated analysis. Digital images of gel electropherograms can be obtained rapidly using an image-capture device, and the images can be easily manipulated using image analysis software.

NOTE: This unit provides information on capturing a digital image. For digital manipulation, the reader is referred to APPENDICES 3A & 3B.

REASONS FOR DIGITAL DOCUMENTATION AND ANALYSIS

There are several reasons to consider digital documentation and analysis of electrophoresis results. These justifications can usually be categorized into issues of ease of handling, accuracy, reproducibility, and cost.

Ease of Data Storage and Retrieval

A major advantage of the digital revolution has been in storage and retrieval of information. Storage in notebooks and filing cabinets previously meant that searching for specific data or experiments was a tedious manual process. Due to an increased reliance on image-based data, the much higher data density of high-throughput microtiter formats and array-based experiments, the need for data tracking and security, and shared data access, electronic notebooks are becoming more popular (see APPENDIX 2). With digital information, modern search engines can quickly find specific information in a fraction of the time usually required for a manual search. Making backup copies of nondigital data can be difficult, expensive, and time consuming since it requires copying, retyping, or photographic reproduction. Copies of digital data can be generated more easily and at reduced costs.

Manipulation of information is also easier when it is in a digital format. While the cut-and-paste analogy comes from physical documentation, it takes on a new perspective when applied digitally. Electrophoresis images can be resized, cropped, and inserted into reports. Data can be passed to spreadsheets and statistical packages for analysis and later insertion into notebooks and reports. These reports can be distributed via the Internet to colleagues throughout the world. A single individual can do all this in a few hours.

Digital analysis also provides an easier method for handling the data when comparing large numbers of results or large numbers of separate experiments. Research that requires comparing the banding patterns on 1000 gels containing 50 lanes each can be an undertaking of heroic proportions if the analysis is performed manually. Database software can dramatically speed the analysis and handle the more mundane tasks, leaving the researcher free to interpret the data.

Accuracy

The human eye is an extremely versatile measuring instrument. It can handle light intensities covering a range of nearly nine orders of magnitude and is sensitive to a fairly wide spectrum of light (Russ, 1995). Yet the eye cannot accurately and reproducibly quantitate density and patterns, nor can it deal with large numbers of bands or spots. Accuracy of measurement is a primary reason for using digital analysis on electrophoretically separated proteins and nucleic acids. Two categories of accuracy are key to digital analysis: positional accuracy, which is important for mobility determinations such as molecular weight, and quantitative accuracy.

Positional accuracy is based on both resolution of the recording medium and measuring accuracy. Silver-halide-based recording has a theoretical resolution based on ~2000 imaging elements (silver grains) per inch. Measurement traditionally occurs using a ruler, with an accuracy of ~20 to 40 elements (50 to 100 elements per inch). In comparison, typical digital systems have 200 to 600 picture elements (pixels) per inch. The advantage that digital systems have lies in their measuring accuracy, which can occur at the level of a single imaging element.

Quantitative accuracy is also an issue. The amount of material represented by a band or spot is difficult to determine accurately from an image of a gel unless it is a digital image. On a digital image, the amount present is directly correlated with the derived volume of the band or spot—the volume is calculated using the intensity values of the pixels within the object.

Reproducibility

Any technique or measurement is only as good as its ability to be faithfully replicated. With software-defined routines, measurements are performed in the same manner every time. Allowing the computer to do repetitive tasks and complicated calculations minimizes the chance for individual errors. This does not imply that such measurements are correct, just that they are reproducible: an incorrect routine or algorithm can also invalidate data.

Cost

A consideration when evaluating any laboratory method is cost. Digital electrophoresis analysis equipment can be expensive. In many cases however, it offers the only method for achieving acceptable analysis performance. In other cases, equal performance can be achieved using silver-halide technology. However, traditional photography can also be expensive when the costs of consumable supplies, such as film and developers and other expensive requirements, such as developing tanks and dark rooms, are included. Often, digital methods can be a good choice when all costs are considered.

KEY TERMS FOR IMAGING

There are several specialized terms encountered during digital image analysis. The most commonly encountered are contrast, brightness, gamma, saturation, resolution, and dynamic range. They describe controls on how the light detectors report a range of light intensities. Below is a brief description of each. Please note that in order to retain the full data from the image, both camera capture and post-capture adjustment need optimization. Many software packages currently available use presets and automation to provide starting points.

Contrast

Contrast describes the slope of the light intensity response curve. An increase in the contrast increases the slope of the curve. The result is a more detailed display over a narrowed range of intensities with less detail in the remaining portions of the intensity range. This is depicted in Figure 7.5.1A and 7.5.1B, where a normal, unadjusted image and a contrast-adjusted image are displayed, respectively. The contrast was increased on midrange intensity values in Figure 7.5.1B to highlight band intensity differences at the expense of background information. Images with a narrow range of informative intensities can benefit from increasing the contrast since it effectively increases the scale and improves detection of minor differences in intensity. Contrast settings should be lowered if information is being lost outside of the contrast range. For example, in Figure 7.5.1B, loss of background information between peaks indicates that this adjusted image should not be used for quantitation.

Brightness

While brightness can have many different definitions, only one is considered here. Brightness shifts the light intensity response curve without changing its slope as is shown in Figure 7.5.1C. Another name for brightness is black level since it is commonly used to control the number of black picture elements (pixels) in an image. Incorrect brightness levels can lead to either high background and potential image saturation or, as is illustrated in Figure 7.5.1C, to loss of background information entirely and partial loss of band information.

Gamma

Nonlinear corrections are often applied to images to compensate for how the eye perceives changes in intensity, how display devices reproduce images, or both. The most common correction is an exponential one with the exponent in the equation termed the gamma. A typical gamma value is 0.45 to 0.50 for camera-based systems and is illustrated in Figure 7.5.1D. This is a compromise value that compensates for the 2.2 to 2.5 gamma present in most video monitors and the print dynamics of most printers. Since it is a nonlinear correction, special care must be taken if quantitation is desired. Unless otherwise directed by the manufacturer, gamma values other than 1.0 should be avoided when quantitating. More information on gamma correction can be found on Poynton's Gamma FAQ (*http://www.poynton.com/GammaFAQ.html*).

Dynamic Range

Dynamic range describes the breadth of intensity values detectable by a system and is usually expressed in logarithmic terms such as orders of magnitude, decades, or optical density (OD) units. A large dynamic range is important when trying to quantitate over a wide range of concentrations. The most accurate quantitation occurs in the linear part of the dynamic range, which is usually not the complete dynamic range of the system. An additional consideration is the dynamic range of the visualization method. Many popular visualization methods have linear dynamic ranges of 1 to 2.5 orders of magnitude. An imaging system with greater dynamic range analyzing the results of such a visualization method will not improve the dynamic range.

Saturation

Saturation occurs when a detector or visualization method receives input levels beyond the maximum end of the dynamic range. This results in a loss of detail and quantitative information from those data points that are saturated. For fluorescent and chemiluminescent samples, reduction in the sampling time or increasing the binning (albeit at a loss of resolution) can correct saturation problems. Optical density-based visualization techniques can also generate saturated images, as is illustrated in Figure 7.5.1E; this can sometimes be avoided with longer sampling times or increased detection source

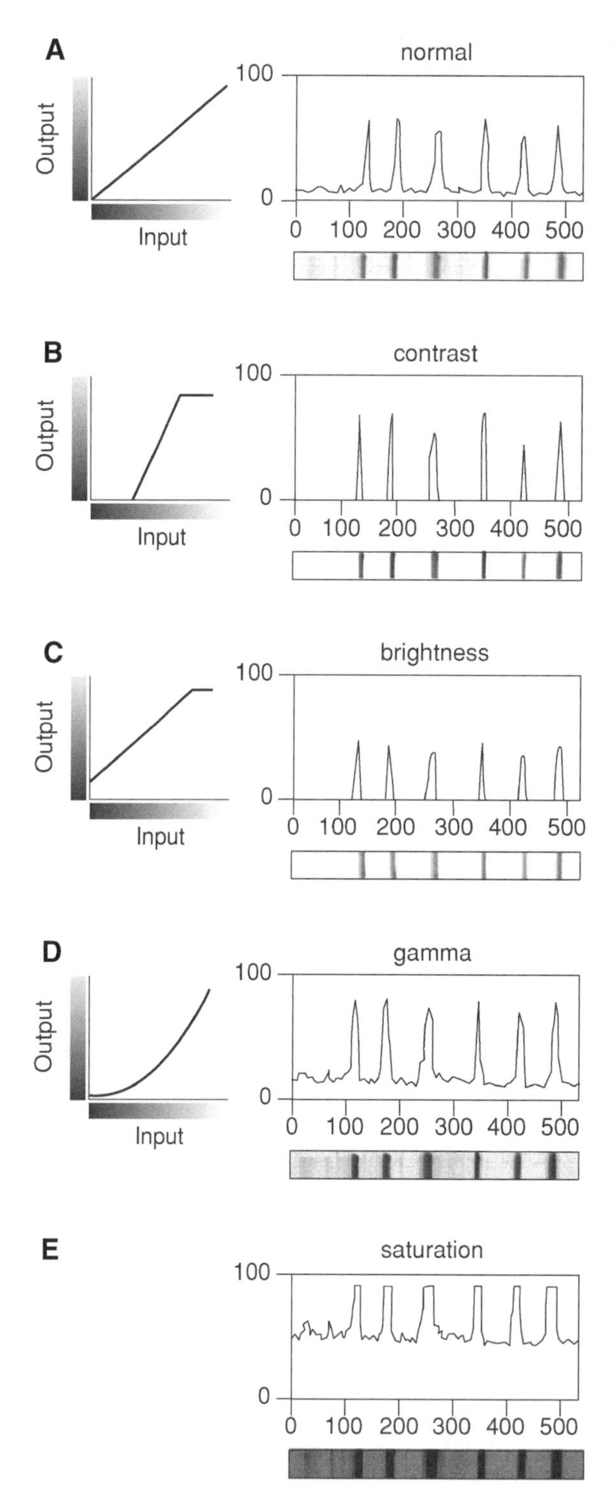

Figure 7.5.1 Examples of how altering image capture settings affects the image and the analysis. The graph on the left displays the light intensity response curve used for image capture, while the image and resulting lane profile on the right display how the setting affects the image. The lane profile displays pixel position versus normalized pixel intensity. (**A**) In this case, the output has not been altered, giving a straight line with a slope of 1 on the response curve. (**B**) The image acquisition was adjusted to increase the contrast of the displayed image. Although useful for images with a narrow range of informative intensity values, increasing the contrast can lead to a loss of low and high values. (**C**) Decreasing the brightness reduces peak values but also leads to a loss of the weak bands and original background. (**D**) Gamma adjusts raw data to appear more visually accurate. Note that this leads to a loss of fidelity between the adjusted image and the original. (**E**) Saturation indicates that the detector is reporting its maximum value or that the dynamic range for the visualization method has been exceeded. Originally published in *CP Molecular Biology Unit 10.5* (Medberry et al., 2004).

intensities. More often, it will be necessary to perform another electrophoresis with more dilute samples or to alter the visualization process to generate a less optically dense material.

Resolution

Resolution is the ability of a system to distinguish between two closely placed or similar objects. Three types of resolution are important for analysis—spatial resolution, intensity resolution, and technique-dependent resolution.

Spatial resolution is the ability to detect two closely placed objects in one-, two-, or three-dimensional space. It is most accurately described as the closest distance at which two objects can be placed and still be detected as separate objects. In practice, it is often defined nominally in terms of the number of detectors per unit area such as dots per inch (dpi) or the number of detectors present in total or in each dimension such as 512 × 512 (262,144 total detectors). Actual resolution is less than half the nominal resolution due to the need for two detectors for every resoluble object (one for the object and one for the separation space) and the effects of optical resolution. Figure 7.5.2 demonstrates how spatial resolution can affect detection of objects. The 42-μm resolution image allows detection of closely spaced bands, the 168-μm resolution image detects fewer bands, and the 840-μm resolution image detects only major bands. For instruments with online detection systems, a pseudo spatial resolution is often reported in units of time from the start of the separation or the time interval between two objects crossing the detection path.

Intensity resolution is the ability to identify small changes in intensity. It is a function of both the dynamic range of a detector and the number of potential values that detector can report. Greater dynamic range decreases the intensity resolution of a given detector. The number of potential values a detector reports is described by its bit depth. An 8-bit detector can report 256 (2^8) different possible values, while a 12-bit detector can report 4,096 (2^{12}) values, and a 16-bit detector can report 65,536 (2^{16}) values. The higher the bit depth, the greater the intensity resolution.

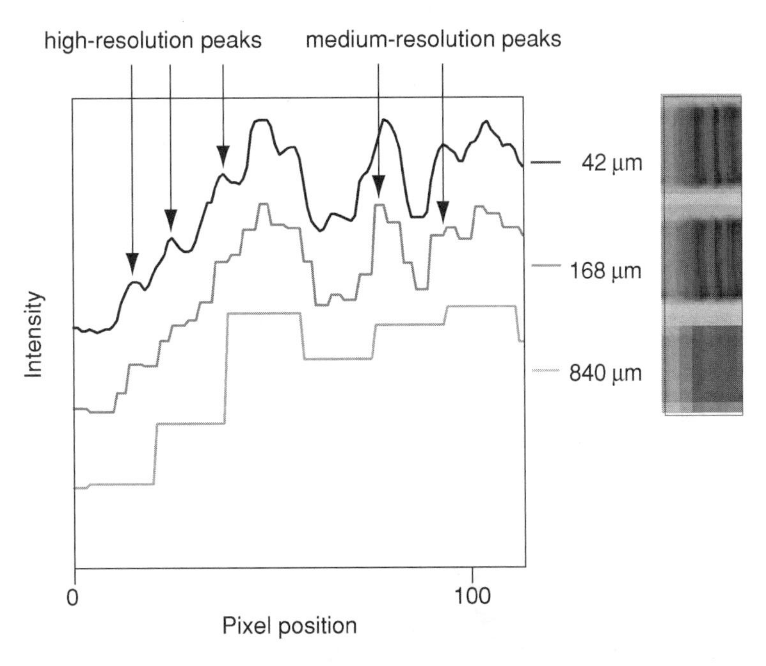

Figure 7.5.2 The effect of spatial resolution on the ability to detect closely spaced objects. Whole-cell protein lysates from *E. coli* were separated using SDS-PAGE and visualized with Coomassie blue staining. An image was captured at 42 μm (600 dpi), 168 μm (150 dpi), and 840 μm (30 dpi) from a segment of the lane, and a lane profile was generated for each image. The lane profiles have offset intensities to allow for comparison. Only major bands can be detected with the low-resolution image (840 μm); at higher resolutions more bands are detectable. Originally published in *CP Molecular Biology Unit 10.5* (Medberry et al., 2004).

Technique-dependent resolution directly affects the spatial and intensity resolution. Electrophoretic separation techniques that generate overlapping objects or that have object separation distances shorter than the spatial resolution will fail to provide reliable data. Many factors, including the amount of sample loaded, gel pore size, buffer constituents, and electrophoresis field strengths, can dramatically affect separation and resolution of biomolecules. Likewise, detection methods that can only generate a small range of discrete intensity values will not benefit from systems with improved intensity resolution.

IMAGE CAPTURE

Systems

Several components make up a digital imaging system (Fig. 7.5.3). These include: (1) a light source for illuminating the sample, either for nonfluorescent white-light imaging or for fluorescent excitation of a sample stained with a fluorescent dye; (2) a filter on the lens to act as a contrast or signal-to-noise (S/N) enhancer for nonfluorescent applications, or an emission wavelength isolation filter for fluorescent detection applications; (3) a fixed or zoom lens; (4) the CCD camera for acquiring the image; and (5) the computer and software for analysis. Commercially available systems are generally packaged to accomplish a specific range of applications, from simple documentation to low-light bioluminescence-based analysis. To provide basic and low-cost documentation of a fluorescently stained DNA or protein sample, a system limited to short exposures (<30 sec) producing a file and a printout for the laboratory notebook is ideal. For experiments using chemiluminescence or bioluminescence, exposures may run 1 hr or more, requiring that the imaging cabinet be completely light-tight and the camera CCD thermoelectrically cooled to minimize noise during long exposure (Tables 7.5.2 and 7.5.3).

Filters used for digital imaging have a number of functions depending on the application (Table 7.5.1). For nonfluorescent imaging, colored glass or wavelength-isolation filters are typically used to enhance the contrast and S/N. For example, a Coomassie blue–stained protein gel shows enhanced contrast over background illumination with use of an orange filter. This is very useful for acquiring

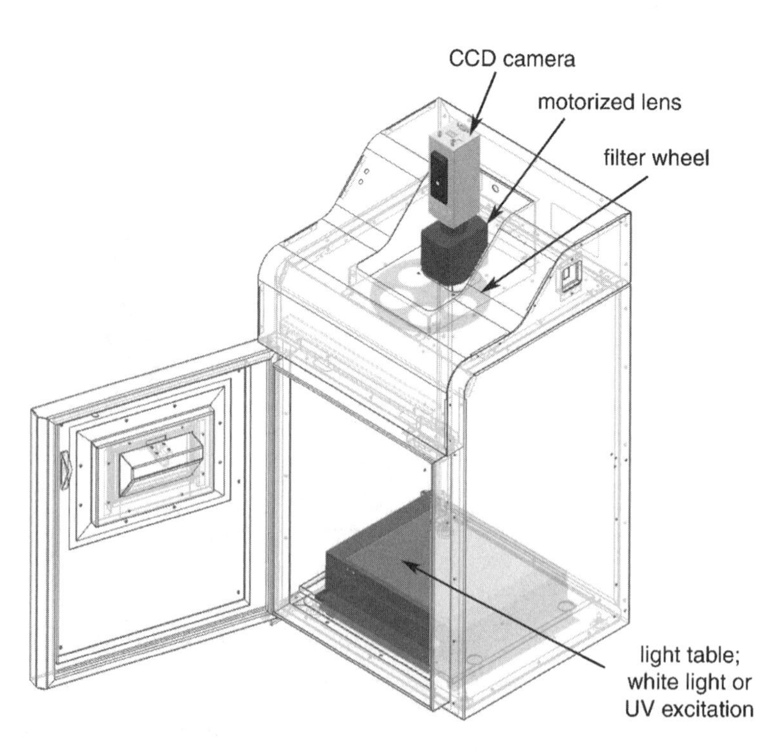

Figure 7.5.3 Typical light-tight desktop system for UV/visible colorimetric, chemiluminescent/bioluminescent, and fluorescent imaging. Originally published in *CP Molecular Biology Unit 10.5* (Medberry et al., 2004).

Table 7.5.1 Filters for Digital Imaging Applications

Application	Desired effect	Typical filters[a]
Optical protein stains (e.g., Coomassie Blue)	Contrast enhancement	Colored glass
		Broad band-pass
		Short-pass
		Long-pass
Fluorescent nucleic acid and protein stains, green fluorescent protein	Signal to noise (S/N) enhancement	Interference filters
	Excitation and emission wavelength isolation	Broad and short band-pass
		Short-pass
		Long pass

[a]Suppliers: Edmund Scientific, Omega Optical, Semrock, Schott Glass.

faintly stained bands on a gel. With fluorescent applications, the lens filter acts to isolate and detect the emission wavelength from the fluorescent sample or gel. These filters are typically very accurate band-pass interference filters.

Devices

Capturing digital images involves a detection beam or source, a sensor for that beam or source, and some method of assembling a two-dimensional image from the data generated. Most systems use a light source for detection. The light wavelengths used range from ultraviolet (UV) to infrared (IR) and can be broad spectrum or narrow wavelength. Broad-spectrum detection is more versatile since it can often be used for more than one detection wavelength. A typical UV light table will emit a broad spectrum of excitation light that peaks at 302 nm, useful for excitation of many fluorescent stains. However, when compared to narrow-wavelength sources such as lasers, broad-spectrum detection suffers from reduced sensitivity and reduced dynamic range. While many types of light sensors have been used, including charge-coupled devices (CCDs), charge-injection devices (CIDs), and photon multiplier tubes (PMTs), technology advances in CCDs have led to their dominance. CCDs are semiconductor imaging devices that convert photons into charge. This charge is then read and converted into a digital format via an analog-to-digital converter (ADC).

The method of image assembly depends on the light source and detector geometry. One method is to capture the image all at once using a two-dimensionally arrayed CCD detector similar to the detectors found in digital and video cameras (see Advances in CCD Technology). Typically, a camera-type sensor is paired with a light source that evenly illuminates the sample. This same sensor is often used with fluorescent and chemiluminescent detection methods as its ability to detect light continuously over the entire sample reduces image-capture times. Another method of image assembly is to capture the image a line at a time. This typically involves a linearly arrayed CCD scanning slowly across the sample in conjunction with the detection beam of light. The data from each line are then compiled into a composite image. Spatial resolution in this method can be significantly better on large-format samples compared to the resolution of a camera-based system. This method is also advantageous when OD-based detection is used, since the more focused light beam is usually of higher intensity and can penetrate denser material. A third method of image assembly is to use a point light source and single-element detector on each point on a sample. The image is then compiled from each point sampled. This method is slower than the others but can offer extremely high resolution and sensitivity. A fourth commonly encountered method is that of generating a pseudoimage of electrophoresis results through the use of a finish line type of detection system. This is composed of a light source positioned at the bottom of the gel (i.e., the end opposite the site of sample loading) and light detectors positioned next to each lane to detect the transmitted light or emitted fluorescence. A lane trace is generated using time on the x axis and light intensity on the y axis. The pseudoimage is then generated from this data (Sutherland et al., 1987).

Capture Process

Sample separation and calibration

Prior to image capture, electrophoretic separation and any visualization steps are performed. To calibrate the separation process, standards are usually run at the edges of the gel and often at internal positions. If quantitation of specific proteins or nucleic acids is to be performed, a dilution series of standards with similar properties to the experimental samples should also be included. After separation, the protein or nucleic acid is visualized, if necessary. Visualization can include binding of a fluorochrome or chromophore (e.g., Coomassie blue), precipitation of metal ions (e.g., copper, silver, or gold), enzymatic reactions, and exposure of film or phosphor screens to radiant sources. These methods can be grouped based on the type of detection into optical density, fluorescence, chemiluminescence, and radioactivity. The suitability of popular detection devices with these methods is described in Table 7.5.2. Once visualization has occurred, image capture consists of the following steps: previewing the image while adjusting capture parameters, capturing the image, and saving the image for later analysis.

Preview

During the preview process, capture parameters are optimized for data content and for ease and rapidity of later processing steps. Typically, the first step is to place the sample so that when the image is captured, the rectangular edges of the gel are horizontal and vertical on the monitor and any lanes are either horizontal or vertical. Since band and spot detection will be much easier if the image is properly oriented, this eliminates the need to later rotate the image digitally. Image rotation is time consuming and can result in spatial linearity errors (a change in the size and shape of objects in the image) resulting from rectangular image-capture device geometries. The next step for camera-based systems is to adjust magnification and to focus the sample image. For thicker samples, it might be necessary to reduce the aperture on camera-based systems to get a sufficient depth-of-field to focus the entire sample. Often at this point image imperfections—e.g., dust, liquid, or other foreign objects that will detract from later analysis—are detected, and they need to be removed. Next, image intensity is set. Within the area of interest on the image, band or spot peaks should have values less than the maximum saturated value, and the background should have nonzero values. This is usually

Table 7.5.2 Compatibility of Popular Image Capture Devices with Common Visualization Methods[a]

	Image capture device				
Visualization method	Silver-halide photography	CCD camera	Desktop scanner	Storage phosphor	Fluorescent scanner
Optical density[b]	+	+	++	−	−
Fluorescence	+	+	−	−	++
Chemiluminescence	+	++	−	±	−
Radioactivity	+	−	−	++	−

[a]The device with the highest sensitivity and greatest dynamic range for a visualization method is marked with a ++, other devices that can detect this visualization method are indicated with a +, and devices that are not suitable for a visualization method are indicated with a −. A ± indicates that only some devices of this type can be used with this visualization method.
[b]Optical density methods include Coomassie blue staining.

accomplished through adjustment of the light-source intensity or the sensor signal integration. If the device allows precapture optimization of other parameters such as spatial resolution, contrast, brightness, gain, or gamma, they are adjusted next. Note that this only applies to controls that affect the response of the sensor or processing of the image prior to a data reduction step and not to controls that affect the image at later stages. The latter process can enhance visualization of specific features, but is best left to adjustments in look-up tables (LUTs) in later analysis steps rather than during image capture since there is a risk of data loss during post-acquisition image processing. LUTs are indexed palettes or tables where each index value corresponds to color or gray-scale intensity values present in an image. Many image analysis programs alter LUTs instead of image values directly since it is faster and does not change the original image data.

Capture

Once all the capture parameters are optimized, the image-capture process is initiated. This might take less than a second for images captured with camera-based detectors and up to hours for scanning single-point detectors. When the image has been captured, it should be carefully examined for content. It should fully capture the area of interest and the parameters should have been set so that all necessary information is detectable. Furthermore, it should be in a form that will allow for easy analysis. Extra time spent on optimizing the capture parameters will often result in a reduction in total image analysis time and an increase in data quality. When the best possible image has been captured, it often contains information outside the area of interest. While this is unlikely to cause problems with later analysis, it is often advantageous to crop the image so that the only portion containing the area of interest is saved. This reduces the amount of disk space necessary to store the image, and the image usually will load and analyze faster with the analysis software.

File save and archive

The last step in image capture is to save the image. Several options are available at this point, including choosing the location at which to save the image, what file type or format to use, and whether to use some form of compression.

The location where the image is saved is not as trivial a question as the image might need to be transferred to another computer at some point. File sizes can easily exceed 30 megabytes on high-resolution images. Readily available hard drives now exceed terabyte capacity. Due to the robust backup and high reliability of centralized servers, data should be backed up and stored on such servers. If the computer used to help capture the image is connected to a network, the image files can easily be transferred this way or potentially saved on a central server. Alternatively, several types of high-capacity removable media are available (e.g., removable hard drives, DVDs, CDs, and compact Flash). This usually requires the installation of additional hardware onto two or more computers but does make backing up data easier if a server-based or Internet-based backup is not available.

Data compression and file formats

Since image files can be very large, compression techniques are sometimes used to reduce disk space requirements. Compression algorithms use several methods, typically by replacing frequent or repetitive values or patterns with smaller reference values and by replacing pixel values with the smaller difference values describing the change in adjacent pixels. When the file is later decompressed, the compressed values are then replaced with the original information. Not all images compress equally, with simple images containing mostly repetitive motifs compressible by $\geq$90%, while complex images will benefit much less from compression. Because compression is a much slower method of saving files and not every file will benefit from it, compression is not used to save all files. Several different forms of compression are available but are separable into two main classes, lossless and lossy. Lossless methods faithfully and completely restore the image when it is decompressed (no loss of data) but offer only moderate file compression. Compression values range from ~10% to 90%, depending on the image. Examples of lossless compression include Huffman coding (Huffman, 1952), RLE (Run Length Encoding), and LZW (Lempel, Ziv, and Welch; Welch, 1984). In comparison, lossy methods such as JPEG (Joint Photographic Experts Group), MPEG (Moving Picture Experts Group), or fractal compression schemes can reach compression values of ≥98% (Russ, 1995). The trade-off is that not all information from the original file is recovered during decompression. Lossy compression is sometimes necessary for applications with extremely large image files such as real time video capture, but it usually represents an unacceptable loss of data if used with electrophoresis image capture.

Digital Image Formats

Many different file types have been developed to store digital images. Some of these file types are proprietary or are hardware specific. For example, PICT is a Macintosh format and BMP is a PC-compatible format (see descriptions of both systems, below). Each file type has its own structure. Some types do not allow compression, for others it is optional, and for some it is mandatory. They vary in the types of images they support, particularly in the number of colors or gray levels. Below is a brief description of a few of the more prevalent file types.

Tagged-Image File Format (TIFF)

TIFF is one of the most commonly used formats. It is particularly versatile since it is an open format that can be modified for specific applications. One reason for its versatility is the ability to attach or tag data to the image. The tags can include information such as optical density calibration, resolution, experimenter, date of capture, and any other data that the application software supports. TIFF images can be monochrome, 4-, 8-, or 16-bit gray scale, or one of many color-image formats. Compression is optional, with LZW, RLE, and JPEG often supported (Russ, 1995). Since TIFF is supported by both Macintosh and PC computers, it is a good choice for multiple-platform environments. The versatility of TIFF can also be a weakness. Since there are many different tagging schemes and since not all programs support all possible compression and color schemes, it is sometimes not possible for one program to access the information in a TIFF file generated by a different program.

Graphics Interchange Format (GIF)

GIF is a file format that is widely encountered on the Internet due to its compactness and standardization. Its compactness is attributable to a mandatory modified LZW compression. Another feature of GIF is the use of a LUT to index the values in the image. One interesting ability of GIF is that it supports storing multiple images within a single file. This can offer some advantages for applications such as time-lapse image capture. A GIF image can contain no more than 256 individual colors or gray levels and therefore does not support intensity resolutions higher than 8 bit. In addition, since the image is implemented as a LUT, it also is not a true gray-scale image. Due to these limitations and others, alternative formats such as PNG have been developed to replace GIF.

Joint Photographic Experts Group (JPEG)

JPEG is a family of file formats with both lossy and lossless compression strategies (*http://www.jpeg.org*). The simplest implementation is a lossy strategy that does result in compression artifacts. Alternative approaches to JPEG include user-adjustable compression to maintain a predefined quality. For image storage, ideally an uncompressed storage approach should be used. If compression is required for reducing files size, be sure to save, reopen, and then reanalyze the image to make sure critical data features are not lost.

Picture (PICT)

PICT is a file format and graphics metafile language (it contains commands that can be played back to recreate an image) designed for the Apple Macintosh. It can contain both bitmap images and vector-based objects such as polygons and fonts. It supports a ≤256-gray-level LUT, and monochrome images can be RLE compressed. Because it only offers a 256-gray-level LUT, it has the same weaknesses that GIF does with true gray-scale and high-intensity-resolution images. In addition, any vector objects in the image are difficult to translate on a PC since they are designed to be interpreted by Macintosh QuickDraw routines.

Bitmap (BMP)

BMP is the native bitmap file format present on Windows-based PCs. It supports 2-, 16-, 256-, or 16-million level images. With images of ≤256 gray levels, it implements a LUT, while the highest-resolution image is implemented directly. RLE compression is optional for 16- and 256-gray-level images. Since compression is prohibited on 16 million-gray-level images and there is no intermediate level supported beyond 256 levels, BMP is not a good choice for images with high-intensity resolution requirements.

ADVANCES IN CCD TECHNOLOGY

Currently there is a growing emphasis on using charge-coupled devices (CCDs) to capture the intensity or spectral data produced in electrophoretic assays. A CCD offers great flexibility in data capture, since it provides both location and intensity information at the same time. Once the data are captured, they must be understood and interpreted. The most difficult component in this scheme is the experimental design necessary to meet the requirements of this type of detector, or the choice of the appropriate detector for a particular experimental design. Then, it is necessary to know the influence of the detector on the data set. To make good choices in the experimental setup and proper interpretation of the data, a thorough and intimate knowledge of the detector is essential.

CCD detectors operate on a simple principle. When a photon of acceptable wavelength interacts with the CCD substrate (a silicon crystal lattice chip), the energy of the photon generates an electron and a "hole pair." This effect produces a negative charge in that region of lattice, and the accumulation of these electrons and hole pairs is directly proportional to the number of photons that successfully interact with the silicon lattice. Short-wavelength photons interact near the surface of the lattice, and progressively longer wavelengths interact at progressively deeper depths of the lattice.

A CCD substrate is electronically segregated into discrete regions called pixels by creating electrical currents and gates at the surface of a silicon wafer. This is the same process used in other semiconductor manufacturing processes to produce memory chips and other microelectronic devices. Historically, the CCD was developed at AT&T Bell Labs in 1969 out of an effort to create new semiconductor memory technology, and is fascinating reading (Janesick, 2001). When appropriate voltages and timing signals are passed through the microcircuits, a CCD chip becomes divided into pixels and is able to convert incoming photons into an electron signal, create a two-dimensional map of this signal (Fig. 7.5.4), and then transfer this signal to an amplifier. Each pixel in the CCD has an X and Y value, measured in microns, plus a depth in the substrate. These values determine a volume of the pixel and how many electron hole pairs can be generated and segregated in each pixel. This capacity is referred to as the full-well capacity or saturation level of a pixel.

When the CCD chip is used to create a camera, the amplified signal is passed to a monitor or recording device for viewing or documentation. If the camera is designated as a digital camera, the analog signal is passed to an analog-to-digital converter (ADC) before being sent to a computer. A key point here is to remember that all CCDs are analog devices. Cameras may be analog or digital depending on how the camera is designed, but the CCD is always analog.

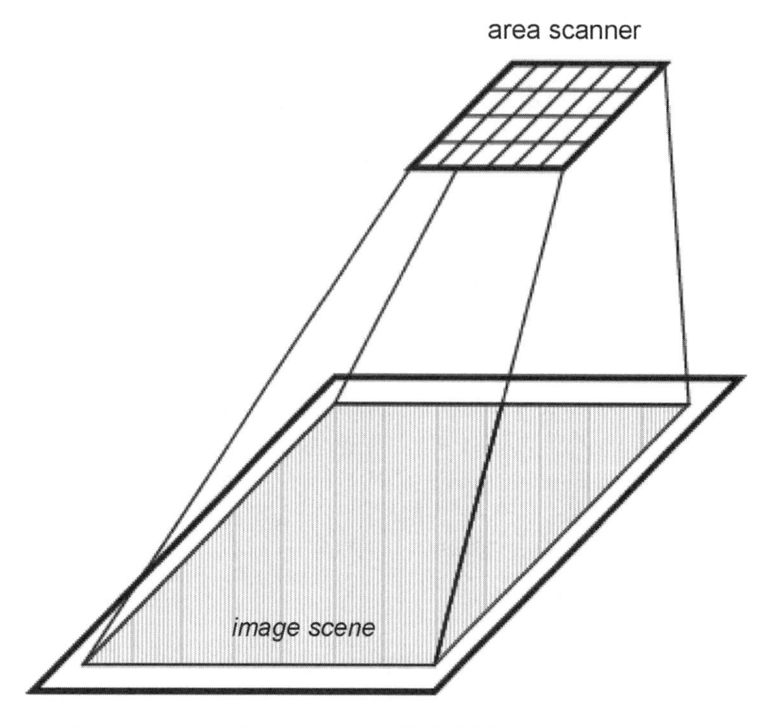

Figure 7.5.4 Two-dimensional area image capture. With CCD area arrays the sample image is captured with a single exposure. The advantages are many: standards can be included with the image at the same time the sample image is captured, no moving parts are needed as with scanners, and the captures are very fast, typically well under 1 sec.

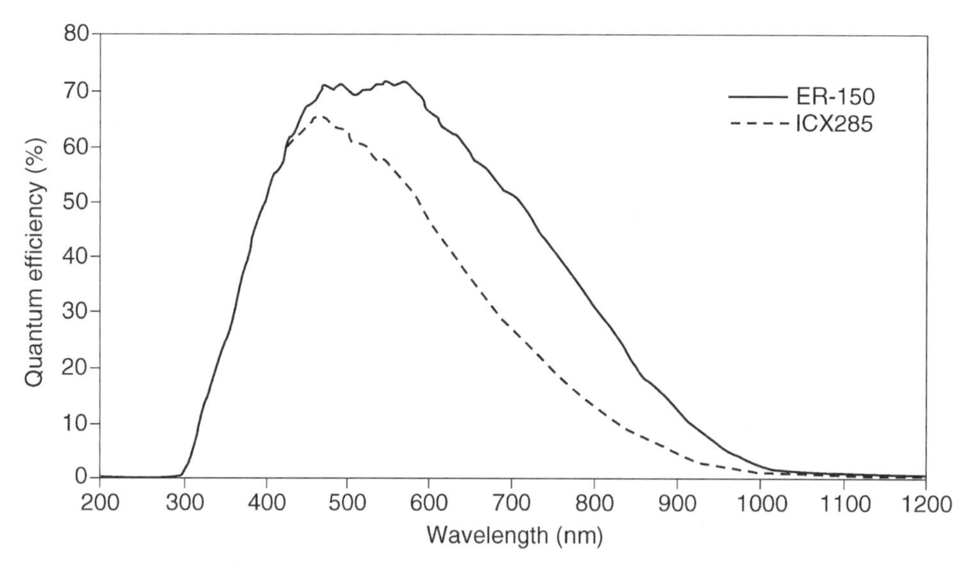

Figure 7.5.5 Quantum efficiency of an extended range (ER-150; solid line) and standard (ICX285; broken line) CCD. Note the extended range CCD has higher QE in the NIR. Originally published in *CP Molecular Biology Unit 10.5* (Medberry et al., 2004).

7

Why Use CCDs?

Quantitative digital imaging applications in the Life Sciences are demanding. CCD cameras used here require a unique combination of high sensitivity, a wide dynamic range, and low noise image capture at exposure times ranging from milliseconds to hours. Sensitivity in CCD technology is at the point where the conversion efficiency of photons to electron hole pairs can exceed 90%. This value is called the quantum efficiency of the CCD, and is the key factor in generating the signal in the detector (Fig. 7.5.5). On the opposite side of the signal level is how much noise is generated in the detector and the camera during the signal collection and transfer process. This relationship between the amount of signal collected and the amount of noise generated is called the signal-to-noise ratio, and is one of the key terms necessary to understand the choice of detector and choice of experimental design. In a CCD, this signal-to-noise ratio has reached the point where in most cases the limiting factor of the data accuracy is the signal itself. Few other detectors offer an equivalent signal-to-noise ratio at the data collection speeds of a CCD when a two-dimensional data array is required. CCD imagers range from relatively low cost for simple white light and fluorescent imaging to more sophisticated and costly cooled cameras for chemiluminescent and in vivo imaging (Table 7.5.3).

Two types of CCD sensors are typically used for scientific digital imaging. Full frame sensors, with a relatively simple architecture, have the full area of the CCD chip active when exposed to light (referred to as 100% fill factor). These show good sensitivity and operate like a film camera with an exposure period followed by a period of darkness while the readout occurs (Fig. 7.5.6). Interline transfer CCDs (Fig. 7.5.7), with every other column on the CCD masked, have lower fill factors and sensitivity, but allow the charge generated by the exposure to be shifted to the adjacent covered or masked column. This permits the new exposure to be captured while the previous exposure is being read off the CCD. The combination of relatively low cost, a wide range of CCD resolutions, and speed of preview have made the interline transfer CCD the most commonly used CCD for scientific imaging.

Table 7.5.3 CCD Camera Specifications

Price range	Low-end	Mid-range	High-end	Premium
Model/manufacturer	C9260-001, Hamamatsu	C8484-51-03G, Hamamatsu	ORCA AG, Hamamatsu	ORCA 2ER, Hamamatsu
Pixel array/size	1344 × 1024/ 6.45 × 6.45 μm	1344 × 1024/ 6.45 × 6.45 μm	1344 × 1024/ 6.45 × 6.45 μm	1344 × 1024/6.45 × 6.45 μm
CCD size	2/3 in.	2/3 in.	2/3 in.	2/3 in.
Readout noise/frame rate	20 electrons/4 Hz	6 electrons/8.9 Hz	6 electrons/8.9 Hz	6 electrons at 5.6 Hz; 3 electrons at 0.8 Hz
Dark current at indicated temperature[a]	8 e/p/s at 20°C	0.1 e/p/s at −10°C	0.03 e/p/s at 30°C	0.0045 e/p/s at −60°C
Camera noise	21.5 electrons	6 electrons	6 electrons	6 electrons/ 3 electrons
Full-well capacity (1 × 1)/(2 × 2)	18,000/23,000 electrons	15,000 electrons	18,000 electrons	18,000 electrons/ 40,500 electrons
Dynamic range (1 × 1)/(2 × 2)	832:1/1070:1	2500:01:00	3000:01:00	3000:1/13,500;1
Bit depth/gray levels	12 bit/4096	12 bit/4096	12 bit/4096	12 bit/4096 and 14 bit/16,384
Digitizer gain (1 × 1)/(2 × 2)	4.3 electrons/5.6 electrons	3.7 electrons	4.4 electrons	4.4 electrons and 2.5 electrons
Maximum quantum efficiency (QE) at indicated wavelength	50% at 525 nm	72% at 550 to 570 nm	72% at 550 to 570 nm	72% at 550 to 570 nm
Binning	1 × 1, 2 × 2, 4 × 4	1 × 1, 2 × 2, 4 × 4, 8 × 8	1 × 1, 2 × 2, 4 × 4, 8 × 8	1 × 1, 2 × 2, 4 × 4, 8 × 8

[a]Units of electrons per pixel per second, abbreviated e/p/s.

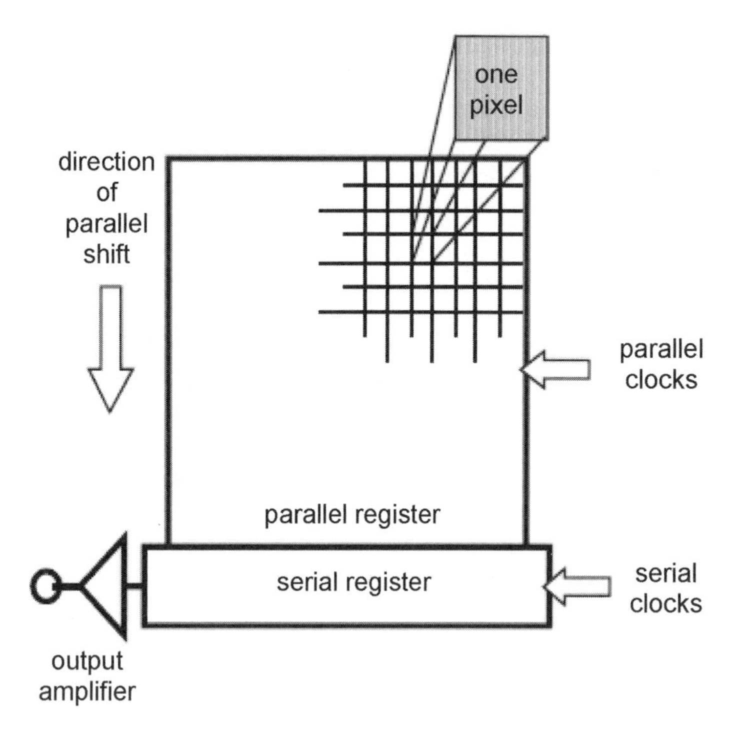

Figure 7.5.6 Full frame CCD Full frame CCDs are the highest resolution and most sensitive CCD architectures. However, mechanical shutters are needed.

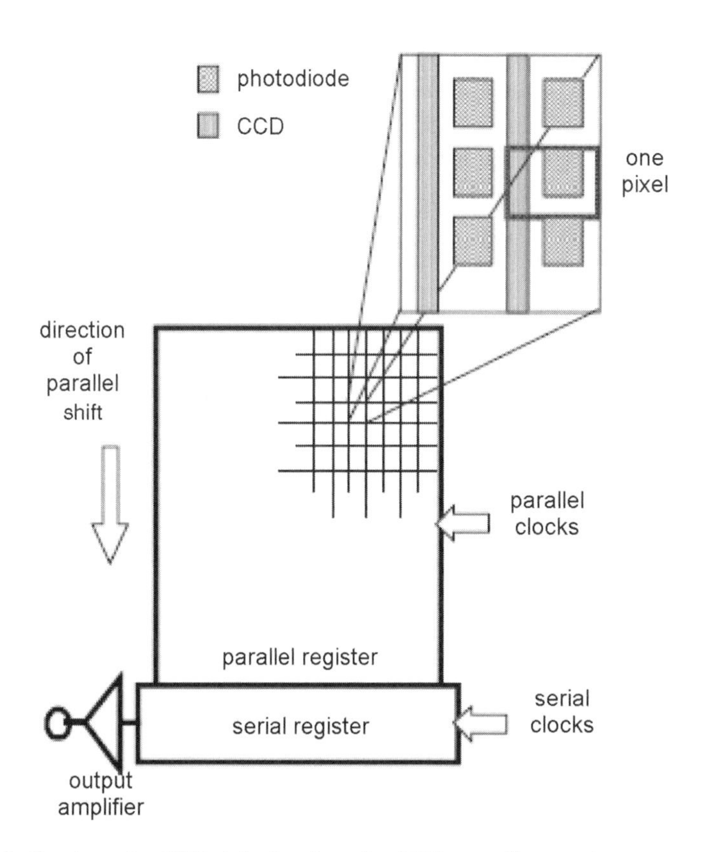

Figure 7.5.7 Interline transfer CCD. Interline transfer CCDs are the most common scientific CCDs in use, and are a good combination of high sensitivity, fast frame rate, and low noise.

Color imaging, while needed for some scientific applications, is not typically used for routine work. Color is derived from combining, e.g., red, green, and blue images into a single-color image (*APPENDIX 3B*). However, for scientific applications, quantitation of the amount of emitted light is the goal and generally a single wavelength of emitted light from a fluorescent sample is analyzed. The image is black and white, captured through a scientific-grade filter in front of the camera lens that selects for a particular wavelength of light emitted by the sample. In situations where color imaging is needed, such as teaching, illustration, or histology, there are both off-chip and on-chip methods used for color image capture (Figs. 7.5.8 to 7.5.10). The most quantitative approach uses a technique that combines separate captures off chip. Three separate monochrome (black and white) images can be captured each with a different color filter, for example red, green, and blue, and then recombined to create a color composite. Another approach uses a three-CCD camera that captures the three filtered (RGB) images at one time. The most common strategy with consumer cameras uses an integral color filter on the CCD, where each pixel has its own color filter patterned over it. Off-chip processing is required to recover a full-resolution color image. Effective resolution using an integral color filter on the CCD is lower since not all pixels are dedicated to a single color, and the missing color information needs to be synthesized in order to complete processing off chip.

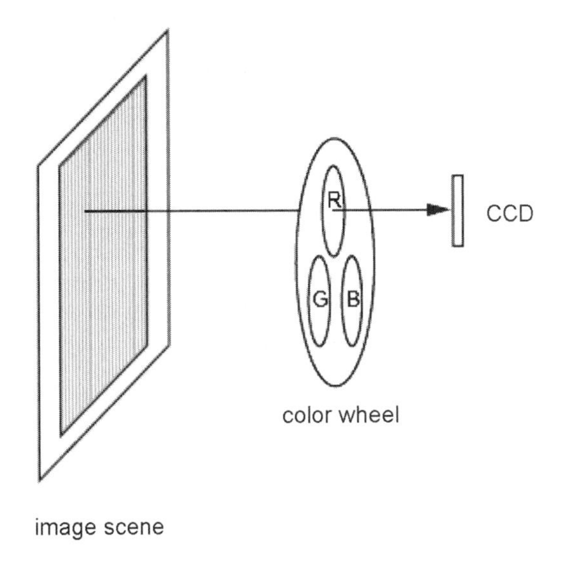

Figure 7.5.8 Color imaging with sequential capture. Automated capture of three separately filtered images enables full-resolution color capture.

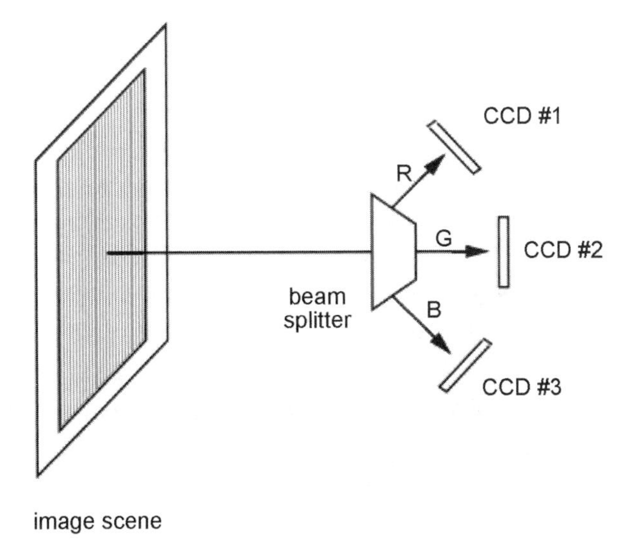

Figure 7.5.9 Color imaging with three CCD cameras. By combining 3 CCDs into a single camera, simultaneous full-resolution capture of a color image is possible.

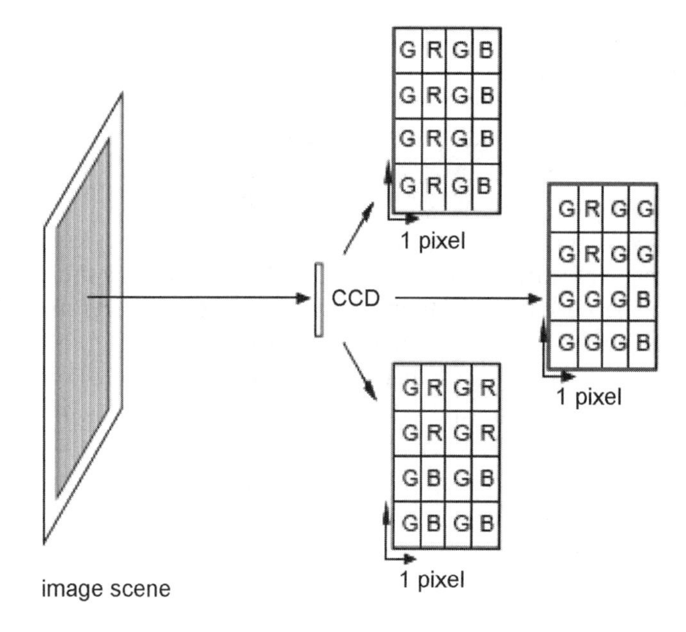

Figure 7.5.10 Examples of integral color filters used for color CCD. The most typical implementation of color imaging uses colored filters patterned onto the CCD, with a pixel permanently assigned a particular color. The color image is reconstructed off chip for display. Resolution, spectral selection and sensitivity are limited with integral color filters.

Glossary of CCD Terms

The following terms are used in the literature describing the range of capabilities, limitations, and applications appropriate for a particular digital imaging camera and system, and are important not only to understand an existing imaging system but also to provide background for purchasing a new system for the laboratory.

CCD. Charge-coupled device, the basis of operation for each detector. A CCD is a silicon-based detector in which an electronic charge (an electron and a hole pair) is produced in the silicon lattice by a photon.

Back-thinned CCD (Fig. 7.5.11). A CCD chip which has been reversed and has had a majority of the silicon substrate removed by either chemical or mechanical means is referred to as back-thinned. Incoming photons enter the silicon lattice from this back side, avoiding the layers of electronic circuits found on the front side of a typical CCD that reduce the spectral sensitivity and QE. This back side illumination of a back-thinned CCD results in a much broader spectral sensitivity range, a higher QE at all wavelengths, and a higher price.

Blooming. Overexposure is a special problem with digital imaging. Once a pixel becomes oversaturated, the accumulated charge will spill over and contaminate the adjacent pixels in a process called blooming. Anti-blooming strategies involve adding vertical or lateral overflow drains in the CCD, preventing the contamination from the adjacent pixels. The tradeoff is that adding anti-blooming drains lowers the overall full well capacity available to the pixel, thus lowering the dynamic range and sensitivity of the CCD.

Full frame CCD. CCDs with a relatively simple architecture have the full area of the CCD chip active when exposed to light (referred to as 100% fill factor). These show good sensitivity to low-light imaging. However, each exposure must be fully read off the chip before the next exposure is begun, requiring a mechanical shutter. Back-thinned versions are possible and offer high QE and wide spectral ranges.

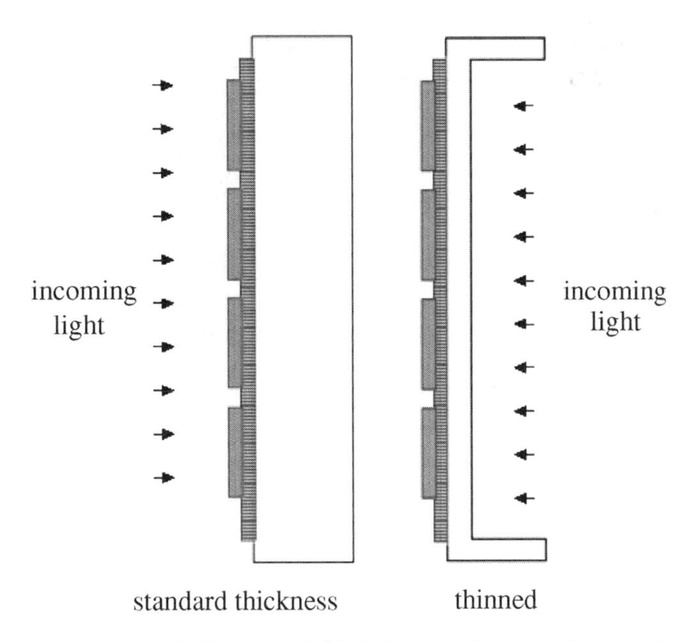

Figure 7.5.11 Back-thinned, back-illuminated CCD. Improved range of spectral response and QE are the hallmarks of back-thinned CCDs, which are used in the most demanding low-light applications.

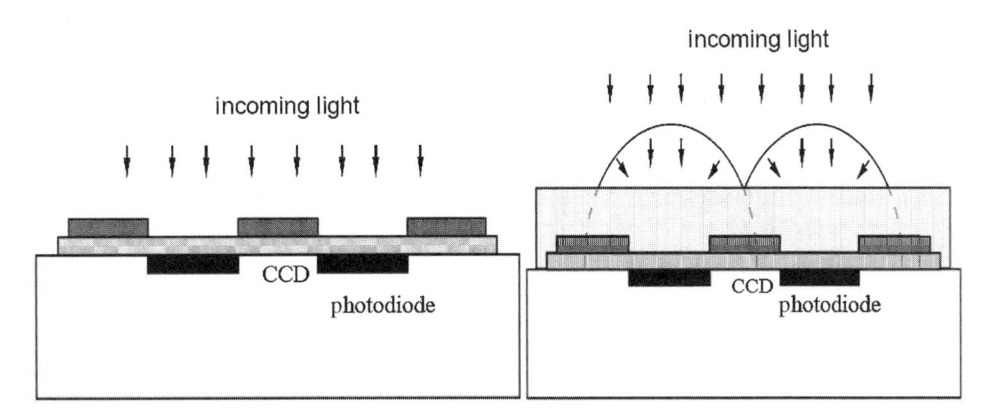

Figure 7.5.12 Microlenticular arrays on CCD. Through use of lenses over each light-sensitive element on the CCD, light-capturing capabilities are enhanced, and sensitivity can be improved by as much as 3-fold.

Interline transfer CCD (Fig. 7.5.7). A CCD chip with a masked area between each column of imaging pixels to which the image charges in every pixel are simultaneously transferred for readout. This rapid single step transfer of the image to the readout column means it is possible to acquire the next image while the previous image is read off the chip. The masked area results in a lower fill factor than a full frame CCD. Modern versions include microlenticular arrays (microlenses) to increase the effective fill factor to over 80%, making the devices useful for scientific imaging. No back-thinned versions of this device are possible at this time.

Electronic shutter. The period of time required to move the image charge in a CCD to the readout area.

Microlenticular arrays (Fig. 7.5.12; also called microlens arrays). Small lenses formed on the surface of each pixel to refract light onto the photosensitive area of a CCD, primarily used on interline CCDs to improve the effective fill factor.

Pixel. An electronically segregated area in the silicon lattice produced by microelectronic circuits on the surface of the silicon substrate, or chip.

Photon. The minimal unit of light, an electromagnetic energy with a particular wavelength (frequency) that interacts with the silicon lattice in a CCD to produce an electron and a hole pair.

Full-well capacity. The maximum number of electron hole pairs that can be contained in a pixel. Also defined as the upper limit of detection, the point at which additional signals can no longer be detected, and the saturation level of a detector.

Noise. A statistical description of the fluctuation in an otherwise stable current. Generally, noise is expressed as an RMS (root mean square) value. Noise terms are added or subtracted in quadrature; it is necessary to square each noise term, add or subtract the terms, then take the square root of the result.

Shot noise. See Signal noise.

Signal noise. The fluctuation in the number of photons in a given amount of time from a stabilized, uniform light source. Under these conditions, the signal noise should be equal to the square root of the signal level, expressed in electrons. Signal noise is not a property of a CCD or a camera, only the signal itself.

Dark current. The number of electrons that appear in a pixel when no light is falling on the detector. Dark current is caused by migration of electrons from areas outside the sensing area of the CCD into the sensitive area and is reduced by 50% with every 8°C reduction of the CCD temperature. Dark current is expressed as the number of electrons (*e*)/pixel (*p*)/sec (*s*). In modern cameras, this is generally so small that it can be disregarded at short exposure times.

Example: Specification for Dark Current = 25*e*/*p*/*s* = 25 electrons per pixel per sec.

Dark noise. The amount of fluctuation in the dark current. This should equal the square root of the dark current.

Example 1: dark current (D) = 25 *e*/*p*/*s*; exposure time 1 sec:

$$\text{dark noise}\,(N_D) = \sqrt{25}e = 5e$$

Example 2: dark current (D) = 25 *e*/*p*/*s*; exposure time 0.5 sec:

$$\text{dark noise}\,(N_D) = \sqrt{25 \times 0.5} = \sqrt{12.5} = 3.53e$$

Camera noise. The sum of the readout noise and the dark noise. This term refers only to a camera, not to an image.

Example: camera with read noise = 10*e* and dark noise = 5*e*:

$$\text{camera noise } = C_N = \sqrt{(N_R)^2 + (N_D)^2} = \sqrt{(10e)^2 + (5e)^2} = \sqrt{100e + 25e} = 11.18e$$

Readout noise. The fluctuation in a nominal zero level signal as it is transferred through the serial register and readout amplifier of a CCD. Often referred to as simply "read noise," the noise is related to the speed of the readout (pixel clock speed) and the related circuit design. In modern cameras, this is almost always the factor that limits the low-light-detection capability.

Total noise. The sum of the camera noise and the signal noise. This term only applies to an image.

$$\text{total noise } = \sqrt{(N_R)^2 + (N_D)^2 + (N_S)^2}$$

where N_R is read noise, N_D is dark noise, and N_S is signal noise.

Example: 1-sec exposure, camera with 10e read noise, 5 $e/p/s$ dark current, average gray value in one region of the image equals 3800, digitizer gain equals 4 electrons per count. To find the camera noise, the signal noise, and the total noise in the image:

1. read noise (N_R) = 10e (from camera data sheet or measured)
2. dark noise (N_D) (from camera data sheet or measured)

$$= \sqrt{D_s} = \sqrt{25e} = 5e$$

3. camera noise (N_C)

$$= \sqrt{(R_N)^2 + (D_N)^2} = \sqrt{(10)^2 + (5)^2} = \sqrt{100 + 25} = 11.18e$$

4. signal noise (N_S)

$$= \sqrt{S} = \sqrt{3800 \times 4} = \sqrt{15,200} = 123.2e$$

5. total noise (N_T)

$$= \sqrt{(N_R)^2 + (N_D)^2 + (N_S)^2} = \sqrt{(10)^2 + (5)^2 + (123)^2} = \sqrt{15,325} = 123.7e$$

Please note the difference the camera noise made in this image, only 0.5 electrons!

Dynamic range. A calculation of the biggest difference in brightness detectable by a device. To obtain this value, divide the Full-Well Capacity (FWC) of a detector by the camera noise to get this range as a ratio.

Example: Camera with 20,000 electrons FWC and 10 electrons camera noise:

20,000e/10e = 2000:1 dynamic range.

For dynamic range in decibels (dB), take the log of 2000 = 3.3 × 20 to get 66 dB.

Bit depth. The number of gray levels that the digitizer will create from the maximum signal level of a device. Bits refers to the exponent of 2 used to create the number of gray levels.

Examples:

10-bit = 2^{10} = 1024 gray levels,

12-bit = 2^{12} = 4096 gray levels.

A 10-bit digitzer will divide the maximum signal (full-well capacity of a CCD) by 1024 gray levels to produce 1024 gray levels in the image. Bit depth is determined by the digitizer in a digital camera; not the CCD. See example under "digitizer gain," below.

Digitizer gain/digitizer count. The number of electrons in the signal used to represent each gray level in an image. This is an essential number for making quantitative calculations and comparisons of digital camera data. A close approximation may be determined by dividing the full-well capacity by the number of gray levels in the digitizer. Camera data sheets should include the exact number.

Example: Camera with FWC = 20,000 electrons; what is the digitizer gain?

10-bit = 2^{10} = 1024 gray levels = 20,000 (electrons)/1024 (gray levels) = 19.5e

12-bit = 2^{12} = 4096 gray levels = 20,000 (electrons)/4096 (gray levels) = 4.9*e*

14-bit = 2^{14} = 16384 gray levels = 20,000 (electrons)/16384 (gray levels) = 1.2*e*

Always compare data and results by number of electrons in the signal, not by gray level, since different cameras have different digitizer gains. In many cases, larger gray values may actually represent smaller data values when making comparisons or calculating the signal noise in an image.

Digitizer offset. A value established in the analog-to-digital converter that represents the average zero value in an analog signal. Since a digitizer only has positive numbers but an analog signal will fluctuate both above and below an average value, a digitizer must have some positive values that can represent the negative portion of the analog signal to prevent clipping of the full data set. This range of positive values in the digitizer is called the digitizer offset. A digitizer with an offset of 200 means that the values from 1 to 199 are used to represent values less than zero in the analog signal. Gray level 200 represents the zero level signal and values above 200 represent signal levels above zero.

LIST OF TYPICAL CAMERA FEATURES AND BENEFITS

CCD camera features and their advantages and disadvantages that should be considered are listed below.

Number of Pixels

The number of pixels in a CCD of a given size will depend on the dimensions of the pixel. A typical 2/3-in. CCD might have pixels with dimensions of 6.45 × 6.45 μm in a CCD that has overall dimensions of 1344 (l) × 1024 (h) pixels or 1.376 million total pixels. While smaller pixels will increase the spatial resolution and the total number of pixels per chip, they do so at the expense of sensitivity and dynamic range because smaller pixels hold fewer electron-hole pairs in response to photon interactions. Larger pixels have larger full-well capacities, leading to more accurate and sensitive signal detection. In addition, larger pixels collect signal more quickly, leading to better sensitivity.

Size of Pixels

Smaller pixels provide higher spatial resolution. Cameras with larger pixels provide larger full-well capacity, bigger dynamic range, and better accuracy, and collect signal faster than small pixels.

Size of Detector

Larger detector dimensions mean a larger field of view, but require more expensive optics.

Cooling Temperature

Temperature alone is not meaningful. The dark current at that temperature must be determined. Cooling reduces the dark current. Cooling temperature may not be relevant for fast exposures

Readout Noise Specification

Lower values improve low-light sensitivity, but slower readout speeds may be required to achieve lower values. Always be sure the readout noise specification is valid at the desired frame rate.

Quantum Efficiency (QE) Value

Higher QE is better if it is in the same wavelength range as the signal, but is a problem if it is in a range other than the signal—i.e., near-infrared! QE value must be considered along with other factors, such as camera noise.

Major Considerations in Camera Selection

Detectability

Detectability is determined by the camera noise and QE. Once the camera noise is minimized, the largest gain in this area is obtained through greater QE.

Accuracy of a single measurement

Accuracy is determined by the noise of the signal itself in most applications. Since this noise is equal to the square root of the signal level, several points are worth noting. The signal level must always be calculated in electrons, using the digitizer gain value to multiply the gray levels in the image. The more signal collected, the better the accuracy. Since the square root of a larger signal is a larger number, this concept is not always obvious. The importance is found in the signal-to-noise calculation (S/N). As the signal becomes bigger, the noise becomes a smaller percentage of the signal. This percentage is a good approximation of the limit of accuracy.

Examples:

$$\sqrt{100} = 10e = 10\% \text{ accuracy}$$

$$\sqrt{500} = 22e = 4\% \text{ accuracy}$$

The accuracy of data in different areas of an image is dependent upon the signal level in each area. The full-well capacity is the limit to accuracy for the CCD. In absorption studies (measuring dark regions in a bright background), the need for accuracy means that devices with extremely large full-well capacities are needed. The dark regions need to have high signal and the background areas must still not be saturated.

It is not uncommon for a typical CCD camera to have 20,000 electrons full-well capacity and a 12-bit digitizer. Under these conditions, each of the 4096 gray values will represent $\sim$5 electrons of signal. With a digitizer offset of 200 gray levels, any signal from a sample will be superimposed on this 200 level.

In this example, we will assume that some region of the image has an average intensity of 500 gray levels and an adjacent region has an intensity of 600 gray levels. We want to find out how much confidence we can place in each of these measurements and how different they are.

We first must subtract the digitizer offset from the indicated signal which leaves gray levels of 300 and 400, respectively.

Then we multiply each data set by 5 electrons to get the actual signal values: 1500e in one region and 2000e in the adjacent region.

Now we take the square root of each to get 39e and 45e. These values represent the total noise in each region of the image. This is the amount of fluctuation we can expect in each signal. We can now make these numbers a percentage of the signal (2.6% and 2.3%) or convert them back to gray levels (8 gray levels and 9 gray levels) to establish the range of fluctuation, or accuracy, that we can expect from this data. If the expectation is accuracy of $\sim$2%, we have enough data to make this statement. If the need for accuracy is 1%, we will have to find some way to improve our data set. We would need about 10,000 electrons in one image to get close to 1% accuracy.

Improving Accuracy

Accuracy can be improved in two ways. First, one should collect more signals by increasing the QE, exposing longer, or increasing the signal intensity. These are the most common choices, but are not always possible. Second, measurements should be averaged. If it is possible to repeat the data collection process and mathematically average more than one image of the same data set, then the accuracy can be improved by the square root of the number of frames averaged. Even averaging two frames will increase the accuracy by 1.4 times.

ANALYSIS

Once the image has been captured, the data must be analyzed and distilled into information about the results of the electrophoresis experiment. Through the use of standards and experimenter input, this software-driven process can estimate mass and quantity of objects in an image and detect relationships between objects within one image and between similar images. The type of software used depends on the analysis to be performed. Images from single electrophoretic separations are examined by one-dimensional analysis software optimized for lane-based band detection. Images from two-dimensional electrophoresis are best handled by specific programs designed to detect spots and to assign two mobility values and a quantity value to the spot. After the initial characterization of bands and spots, comparisons are often made between bands or spots from different experiments through the use of database programs and matching algorithms.

Software for One-Dimensional Analysis

Table 7.5.4 lists popular sources of electrophoresis imaging software.

Lane positioning

For one-dimensional analysis, the first activity is to detect the lanes on the image. One of three different methods is commonly employed for this. For images with straight, well-defined lanes with a large number of bands, automatic lane-detection algorithms can quickly and accurately place the lanes. On images with very well-defined lanes, such as pseudoimages from finish-line type electrophoresis equipment, automated lane calling based on image position is possible. For images with "smiling," bent, or irregular lanes, manual positioning of the lanes is often the fastest and most accurate method of lane definition. Regardless of the method of identifying the lanes, the lane boundaries need to be carefully set for accurate quantitation and mass determinations. Lane widths should be wide enough so that the entire area of all bands in that lane are included, but they should not be so wide as to include bands from adjacent lanes. To accomplish this, curved or bent lanes might need to be used in order to follow the electrophoresis lane pattern. Lane length and position also must be adjusted as necessary so that all bands of interest are included. If mass determinations are necessary, the sample loading point should probably also be included in the lane or be the start

Table 7.5.4 Popular Sources of One- and Two-Dimensional Electrophoresis Analysis Software

Company	Web site	Software
1-D gels		
UVP, Inc.	*http://www.uvp.com*	VisionWorksLS, DocItLS
Nonlinear Dynamics	*http://www.nonlinear.com*	TotalLab series
MediaCybernetics	*http://www.mediacy.com*	Gel-Pro, ImagePro
NIH (free, with gel tutorial)	*http://rsb.info.nih.gov/nih-image/*	NIH Image
2-D gels		
ExPasy	*http://expasy.org/melanie/*	Melanie/ImageMaster
Nonlinear Dynamics	*http://www.nonlinear.com*	SameSpotseries
GE Healthcare	*http://www.ge.com*	ImageMaster
Bio-Rad	*http://www.biorad.com*	PDQuest

of the lane. At this point, lines of equal mobility (often called R_f or iso-molecular-weight lines) are added to the image as necessary. These lines allow for correction of lane-to-lane deviations in the mobility of reference bands and generate more accurate measurements of mass. A similar form of correction is also possible for within-lane correction of mobilities. This correction is important for accurate detection and quantitation of closely spaced bands.

Band detection

Once the lanes have been defined, the bands present in each lane need to be detected. There are many methods for detecting bands. One method is to systematically scan the lane profile from one end to the other, identifying regions of local maxima as bands. Another common method is to use first and second order derivatives of the lane image or lane profile in order to find inflection points in the change of slope in pixel intensity values (Patton, 1995). Regardless of the method used, it is often necessary to alter the search parameters so that they perform reliably under a given experimental condition. Typical search parameters include detection sensitivity, smoothing, minimum interband gaps, and minimum or maximum band peak size. Smoothing reduces the number of bands detected due to noise in the image. A minimum interband gap is often used to avoid detection of false secondary bands on the shoulders of primary bands. Limitations on peak sizes, especially for within-lane comparison to the largest band's peak, can be a useful way to allow sensitive detection of bands in underloaded lanes without detecting false bands in overloaded lanes.

Band edges are often detected in addition to band peaks in order to further define bands or to quantitate band amounts. This can be accomplished by using local minima, derivatives of the lane profile, or fixed parameters, such as image distances or a percentage of band peak height. The band edges can be applied as edges perpendicular to the long axis of the lane or as a contour of equal intensity circling the band. The perpendicular method is advantageous for bands with uneven distribution of material across the face of the band, while the latter method is better for "smiling" or misshapen bands.

Background subtraction

With nearly all electrophoresis procedures, the most informative images have a low level of signal intensity at each pixel that does not result from protein or nucleic acid. Instead, this background intensity is attributable to the gel medium, the visualization method, electronic noise, and other factors. Since this background tends to be nonuniformly distributed throughout the image, failure to subtract it can make band detection and quantitation less accurate. Many methods of background subtraction are possible. Sometimes it is possible to generate a second image under conditions that do not detect the protein or nucleic acid. The second image is then digitally subtracted from the data-containing image to remove the background. More often, background information must be obtained from a single image. If the background varies uniformly across the image, a line that crosses the variation can be defined at a point where no bands are present. The intensity values at each point on the line can be used as the background value for the pixels perpendicular to the line at that point. Commonly, background is also present as variations in intensity along the long axis within each lane. One simple method is to take the lowest point in the lane profile as the background. Another method is to use an average value of the edge of each band as the background for that band. More complicated methods such as valley-to-valley and rolling-disk use local minima points in the lane profile to define a variable background along the length of the lane. Because there can be many different causes and distributions of background, no single method of background determination can be recommended for all experiments.

Characterization

Once lanes and bands have been detected, it is possible to interpret the mobility of the nucleic acid or protein bands. Depending on the method of electrophoretic separation, information on mass (length or size), pI, or relative mobility (R_f) can be inferred from mobility information. The mobility is characterized using a standard curve with internal standards of known properties. The type of curve depends on several parameters. By definition, with R_f-based separation, a linear first-order curve is used since it represents the linear relationship between mobility and R_f. Similarly, pI and mobility are generally linear in isoelectric focusing separations. For separations based on size, a curve generated from mobility versus the log of the molecular weight provides a relatively good fit as measured by the correlation coefficient (R^2). Several other curve models have been suggested for size-based separations, including modified hyperbolic curves and curves of mobility versus (molecular weight)$^{2/3}$ that have good correlation coefficients (Plikaytis et al., 1986). In some cases, no single curve equation can adequately represent the data, and methods of fitting smooth contiguous curves using only neighboring points such as a Lagrange or spline fit (described in Hamming, 1973) are necessary. This is most common for size separations with a very large range of separation sizes and with nonlinear gradient gels. Care must be taken with multiple-curve techniques since they rely on only a few data points for any one part of the composite curve, and outlying data points can drastically affect the outcome.

For size and R_f determination, a uniform position must be found in each lane as a point from which to measure the mobility of each band. Many software analysis packages use the end of the lane as the measuring start point, so for them it is important to position each lane start point at an iso-molecular-weight or iso-R_f point. A convenient point is the well or sample-loading position since it is usually easily detectable and at an equal mobility position in each lane. A consistent point on each band must also be chosen to measure mobility. A band's peak is easily defined in digital image analysis and is commonly used. Since peak positions are harder to detect visually than edges on silver-halide images, the leading band edge is sometimes used when comparing digital results with silver-halide-based results.

Once lanes and bands have been detected, it is also possible to quantify the amount or at least relative amount of nucleic acid or protein present in each band. The amount in a band is related to the sum total of the intensity values of each pixel minus the background value for each pixel in a band. For absorptively detected bands, intensity values are converted to OD values. The total value calculated is equivalent to the volume of the band and can be directly compared to other bands that are within the linear range for the visualization method. If standards of known amounts are loaded onto the same gel, they can be used to generate a standard curve that converts band volume into standard units such as micrograms. For greatest accuracy, it is important to be able to generate multiple standard curves when using visualization methods, such as Coomassie blue staining, that are affected by band or spot composition.

Quantitation becomes more complicated when bands are not fully resolved. In this case, material from one band is contributing to the volume measurement of an adjacent band and vice versa. The simplest method for handling this is to partition each band, including only the volume within its edges. Alternatively, a Gaussian curve can be fitted to each band and the volume contained within the curve used to estimate the amount of the band. Since most electrophoresis bands have a pronounced skew towards the leading edge of the band, modified Gaussian curves have also been used (Smith and Thomas, 1990). In either case, the curve-fitting process is calculation intensive and can significantly increase analysis times for images with many bands.

Software for Two-Dimensional Analysis

In two-dimensional analysis, the first-dimension separation is performed in a single column or lane followed by a second separation performed perpendicular to the first. The result after visualization is a rectangular image of up to 10,000 spots. The most common two-dimensional gel type is one in which protein is separated first by apparent pI and second by molecular weight, although two-dimensional separation of nucleic acids is also possible. While many of the concepts and analysis techniques used with one-dimensional gels are applicable to two-dimensional gels, the complex nature of most two-dimensional gels requires somewhat different methodology. For example, spots are more difficult to detect since they are not conveniently arranged in lanes and can vary in shape and overlap. In addition, two-dimensional experiments usually require some method of comparing two images whereas one-dimensional images usually contain all of the information from an experiment.

See Table 7.5.4 for sources of two-dimensional electrophoresis imaging software.

Spot detection

Probably the most difficult aspect of two-dimensional analysis is efficient and accurate spot detection. If it is incorrectly done, it can lead to hours of manual editing. Due to the complexity and computational intensity of some algorithms, the detection process itself can last hours on relatively fast desktop computers. One theoretically effective but computationally intense method is to treat the image as essentially a three-dimensional image with spots treated as hills and background as valleys. A large number of Gaussian curves are then combined to describe the topology of the image. Many other methods make use of a digital-imaging technique known as filtering. In essence, filtering is a way to weigh the value of a pixel and its neighbors in order to generate a new value for a pixel. By passing a filter across an image pixel by pixel, a secondary image is generated. Filters can be designed for many tasks, including sharpening an image or removing high-frequency noise. Filters can also be generated to help detect spots by making images that are first and second derivatives of the original image. The derivative images indicate inflection points in the intensity pattern and can be used to detect spot centers and edges. In a different method, called thresholding, filters can be used to detect the edges of objects. Instead of looking for inflection points, threshold filters identify intensities above a set level or ratios between central and edge pixels above a set value. Since the edges on two-dimensional spots tend to be diffuse, sharpening filters are sometimes used prior to the thresholding filter. In some cases, multiple techniques are used to detect spots (Glasbey and Horgan, 1994).

Unlike one-dimensional detection, detection on images of two-dimensional experiments usually requires secondary processing to get acceptable performance. One example of a secondary process is to discard spots with sizes below a set minimum or above a set maximum. Another is to analyze spots that are oval for possible splitting into two spots. Even after secondary processing, it is likely that a small amount of manual editing will be necessary. When manually editing an image, care must be taken to use as objective criteria as possible, especially if two or more images are to be matched and spot volumes compared.

Characterization

In a two-dimensional system, determination of protein or nucleic acid mobility is complicated by the fact that there are two mobilities to account for, and that the second-dimension separation tends to make estimation of the separation which occurred in the first dimension more difficult. One method for dealing with this is to have a series of markers in the sample that, after both separations are completed, are evenly distributed within the gel and image. It is also possible to estimate separation characteristics from calibration points located at the periphery of the gel. For example, distance measurements can be used to pass calibration data from the first dimension separation, and standards can be separated at the ends of the gel to calibrate for mobility in the second dimension. Regardless of the method used, in many instances, a series of related images will be examined and similar spots in each image will be matched. When this occurs, it is possible to calibrate one image and then pass the calibration information via the matches to the related images.

Quantity determination is similar in many regards to that which occurs in one-dimensional analysis, but there are some differences. If spot edges are detected, a simple method of determining spot volume is to take the sum of the intensity value of each pixel in the spot minus a background intensity value. Multiple Gaussian curves can also be fitted to the spot to approximate the volume (Garrels, 1989). More difficult is attempting to compensate for a skewed distribution in a size-separating dimension while trying to use a regular Gaussian fit for a pI separation, such as is encountered with the most common form of two-dimensional protein separations. The distribution of background makes quantitation more difficult in two-dimensional gels. There is no lane-dependent component, so it is necessary to use other methods such as image stripes, finding local minima values, or using values derived from the spot edges to determine background values.

Matching

Matching is the process in which proteins or nucleic acids with similar separation properties are linked or clustered together. Matching can occur within one image or between multiple images as long as a frame of reference is established. Matching allows for comparisons between samples. It also makes annotation and data entry easier, since if one spot or band is matched to others and is characterized or annotated, this information is easily passed to all the other matches. An underlying assumption of matching is that objects with similar separation properties are actually similar. Care must be taken to confirm the identity of matched spots or bands by other methods on critical experiments.

A simple form of matching is to link bands or spots at similar positions on the gel images. This works well when separation and imaging conditions are uniform. This is very seldom the case, since slight differences in the electrophoresis, visualization, and imaging conditions across a gel and between gels generates incorrect matching with this method. Since bands on one-dimensional gels are relatively easy to calibrate for mobility, matching can occur along contours of equal mobility. This dramatically decreases, but does not eliminate the variability in detecting similar bands. Much of the remaining variability can be attributed to calibration errors. This error can often be compensated for by allowing a small tolerance in mobility values in determining whether a band is matched or not. Because of the difficulties in calibrating mobility in two-dimensional gels, it is often more practical to use matched spots for calibrating mobility than vice versa. Spot matching between two two-dimensional images starts by finding a small number of landmark spots that are used as anchors for subsequent matches (Appel et al., 1991; Monardo et al., 1994). There are many methods for finding the landmarks in both images, including finding the highest-intensity spots, spots in unique clusters, and manual positioning. The most common procedure from this point is to derive a vector that describes the direction and extent of the path from one matched spot to the other when the two images are superimposed. The vector is used as the basis for finding more matches near the landmark matches. To allow for error, the area within a small radius is searched extending from the end of the vector. Once another match is found, its vector is computed and used as the starting point for finding neighboring matches. From this progression, the entire gel is matched. If all vectors are displayed graphically when matching is complete, questionable matches can often be identified as vectors that are significantly different from neighboring vectors.

One specialized use of matching is as an estimator of the similarity and potential genetic relatedness of organisms. For example, on a one-dimensional gel image, a ratio of the matched to unmatched bands for each pairwise combination of lanes can be calculated. This ratio can be used as an indicator of similarity, with values near 1 indicating a pair of highly similar lanes, and values near zero indicating very dissimilar lanes. Assuming that the contents of the lanes are valid samples of the originating organism's genetic makeup, the information on lane similarity can be converted to estimates of genetic similarity. A convenient way to display this similarity data graphically is to generate a dendrogram with similar objects close to each other and less similar objects more distantly placed. An example of such a dendrogram is presented in Figure 7.5.13, where samples from *Listeria* isolates are arranged based on banding pattern.

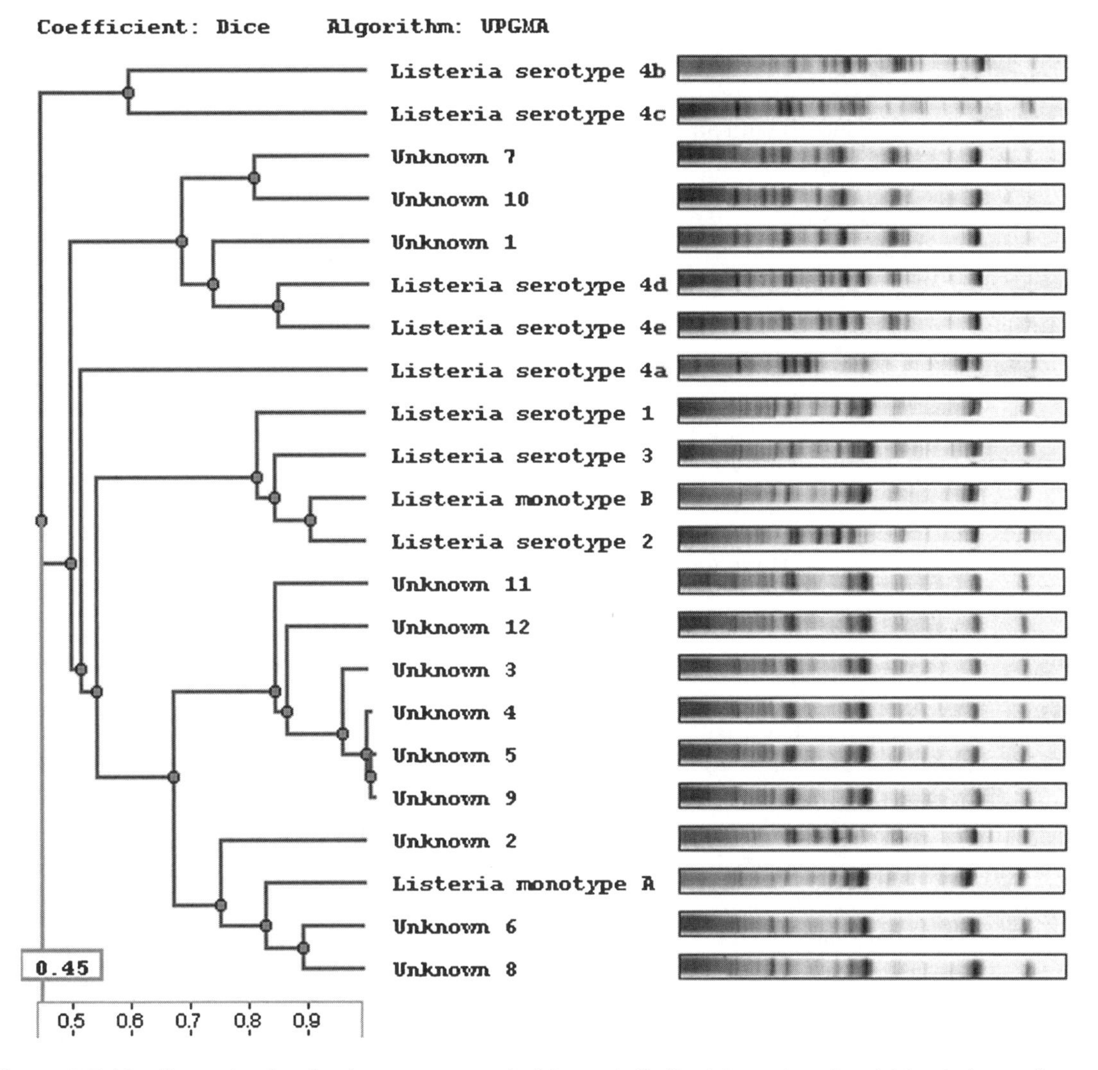

Figure 7.5.13 Example of a dendrogram generated from similarity data on band matching between lanes. DNA samples from 22 isolates of *Listeria* were subjected to Random Amplification of Polymorphic DNA (RAPD) analysis and the resulting electrophoresis was analyzed with ImageMaster software (Amersham Pharmacia Biotech). Clustering was performed using the Dice coefficient with a tree structure based on the Unweighted Pair-Group Method using Arithmetic Averages (UPGMA). Similarity values between isolates can be determined by locating the node that connects the isolates and reading the value from the scale on the lower left edge of the dendrogram. Originally published in *CP Molecular Biology Unit 10.5* (Medberry et al., 2004).

IMAGE DATABASES FOR IMAGE RETRIEVAL AND ANALYSIS

In many cases, image analysis is not the last step in the process. The image and analysis data need to be archived in a searchable format. There may be a need to analyze the data from multiple experiments conducted at different sites or in laboratories around the world. Bioinformatic links to diverse data sources might be desired to help develop a unified understanding of the biology behind particular phenomena. When these situations arise, database programs can be utilized to store, link, and search image analysis results.

As the number of images that are captured and analyzed grows, it becomes increasingly more difficult to find particular information from the large number of files that are stored. Relatively simple databases can be used if the major requirement is to find previously analyzed images and associated data. Such databases often display a miniaturized version of each image to aid in visual scanning for the file as well as simple searching for image-specific information, such as date of analysis, file name, or other information that was entered at the time of image capture. More powerful database products are also available that can perform complex searches on data generated during the analysis. For an example, a search on a two-dimensional database might include finding proteins exhibiting a specific expression profile and having a molecular weight >20 kDa with a pI between 3 and 5 or 8 and 10 with an amount <50 ng in a series of experiments conducted <1 year ago. Such searches can quickly target potentially interesting molecules for further analysis.

With the increasing ease of transferring data through the Internet as well as local- and wide-area networks, it has become practical to quickly find and examine data from distant locations. Of course, great care must be taken to ensure that similar experimental conditions are employed, otherwise the results will be difficult to compare. In this manner it is sometimes possible to dramatically increase the sample size and statistical accuracy as well as the probability of detecting rare events. In addition, if one data set is more completely characterized, this extra information can be extracted and applied to the other data set. For example if there is a band in common in two databases and there is sequence information for it in one database, that sequence information can be added to the other database. Currently most public electrophoresis database sites are two-dimensional protein databases. A list with links to many of these Internet database sites can be found at *http://www-lmmb.ncifcrf.gov/EP/table2Ddatabases.html.*

With biological questions becoming more complicated and the answers to the questions often requiring information from a variety of sources, it is becoming increasingly important to be able to move easily between information sources. A relational type of database can help achieve this. Unlike a conventional database with a fixed arrangement of data, relational databases have links between related files that allow for easy movement from one file to another. Another approach to interconnecting electrophoresis data with data from other sources is to generate a series of hypertext links between data sets, similar to what occurs on the Internet. Selecting a specific link moves the search to the related network site and the related information. Regardless of the method, the end goal is similar. An example of what is possible: a researcher selects a protein spot on a two-dimensional gel image which triggers accessing of related information on this protein. The protein sequence is accessed from mass spectroscopy analysis of the spot on a separate gel. The sequence of the gene and the cDNA that generated the protein is retrieved. The expression pattern of the gene in various tissues and conditions as well as information on similar genes in other organisms is incorporated. Citations and annotations to this are retrieved as well. All of this information is compiled automatically into an interactive report about the protein. From this report, the researcher can formulate a more refined hypothesis and plan the most appropriate experiments to test it.

LITERATURE CITED

Appel, R.D., Hochstrasser, D.F., Funk, M., Vargas, J.R., Muller, A.F., and Scherrer, J.R. 1991. The MELANIE project: From a biopsy to automatic protein map interpretation by computer. *Electrophoresis* 12:722-735.

Garrels, J.I. 1989. The QUEST system for quantitative analysis of two-dimensional gels. *J. Biol. Chem.* 264:5269-5282.

Glasbey, C.A. and Horgan, G.W. 1994. Image Analysis for the Biological Sciences. John Wiley & Sons, Chichester, England.

Hamming, R.W. 1973. Numerical methods for scientists and engineers, 2nd ed. Dover Publications, New York.

Huffman, D.A. 1952. A method for the construction of minimum-redundancy codes. *Proc. Inst. Elect. Radio Eng.* 40:9-12.

Janesick, J.R. 2001. Scientific charge-coupled devices. SPIE Publications, Bellingham, Wash.

Medberry, S., Gallagher, S., and Moomaw, B. 2004. Overview of digital electrophoresis analysis. *Curr. Protoc. Mol. Biol.* 66:10.5.1-10.5.25.

Monardo, P.J., Boutell, T., Garrels, J.I., and Latter, G.I. 1994. A distributed system for two-dimensional gel analysis. *Comput. Appl. Biosci.* 10:137-143.

Patton, W.F. 1995. Biologist's perspective on analytical imaging systems as applied to protein gel electrophoresis. *J. Chromatogr. A.* 698:55-87.

Plikaytis, B.D., Carlone, G.M., Edmonds, P., and Mayer, L.W. 1986. Robust estimation of standard curves for protein molecular weight and linear-duplex DNA base-pair number after gel electrophoresis. *Anal. Biochem.* 152:346-364.

Russ, J.C. 1995. The Image Processing Handbook. CRC Press, Boca Raton, Fla.

Smith, J.M. and Thomas, D.J. 1990. Quantitative analysis of one-dimensional gel electrophoresis profiles. *Comput. Appl. Biosci.* 6:93-99.

Sutherland, J.C., Lin, B., Monteleone, D.C., Mugavero, J., Sutherland, B.M., and Trunk, J. 1987. Electronic imaging system for direct and rapid quantitation of fluorescence from electrophoretic gels: Application to ethidium bromide-stained DNA. *Anal. Biochem.* 163:446-457.

Welch, T.A. 1984. A technique for high performance data compression. *IEEE Computer.* 17:21-32.

KEY REFERENCES

Glasbey and Horgan, 1994. See above.

Describes general image-processing techniques as they are applied to biological images.

Russ, 1995. See above.

A general reference book on digital image capture and analysis.

Sutherland, J.C. 1993. Electronic imaging of electrophoretic gels and blots. *In* Advances in Electrophoresis, Vol. 6. (A. Chrambach, M.J. Dunn, and B.J. Radola, eds.) pp. 1-41. VCH Verlagsgesellschaft mbH. Weinheim, Germany.

Provides an overview of image capture with particular emphasis on types of capture equipment.

INTERNET RESOURCES

http://rsb.info.nih.gov/nih-image

NIH Image is free software that provides basic image analysis tools for the Macintosh.

http://www.poynton.com/GammaFAQ.html

Contains an excellent description of gamma correction in the Gamma FAQ.

http://www-lmmb.ncifcrf.gov/EP/table2Ddatabases.html

A list of links to many two-dimensional electrophoresis gel databases that are available via the Internet.

Chapter 8

Blotting

UNIT 8.1

Overview of Blotting

Maria Cristina Negritto[1] **and Glenn M. Manthey**[2]
[1]Pomona College, Claremont, California
[2]City of Hope, Beckman Research Institute, Duarte, California

GENERAL OVERVIEW

Blotting is a technique by which a macromolecule such as DNA, RNA, or protein is resolved in a gel matrix, transferred to a solid support, and detected with a specific probe. These powerful techniques allow the researcher to identify and characterize specific molecules in a complex mixture of related molecules. Some of the more common techniques include Southern (DNA) blotting, northern (RNA) blotting, and immunoblotting (for protein; also known as western blotting). In this unit, a brief introduction to the main concepts used in blotting techniques will be discussed.

Blotting techniques such as Southern blotting or immunoblotting share some common steps that are described first in general terms and then more specifically. To access specific protocols on blotting methods please refer to *UNITS 8.2 & 8.3*.

The blotting procedures can be divided into six main steps, as illustrated in Figure 8.1.1.

Electrophoresis

Typically, the molecule of interest is present in a complex mixture of molecules. To detect the protein or nucleic acid sequence of interest, the mixture must be resolved, usually on the basis of size. This is achieved by separating the molecules by gel electrophoresis on either an agarose or polyacrylamide gel. See *UNIT 7.2* for DNA and RNA electrophoresis and *UNIT 7.3* for protein electrophoresis.

Transfer

Following separation, the molecules are transferred to a solid support such as a nylon, nitrocellulose, or polyvinylidene fluoride (PVDF) membrane. The transfer results in a replica of the molecules that were present in the gel and that are now immobilized on a membrane. The most common transfer techniques include capillary blotting, for use with Southerns or northerns, and electroblotting for immunoblots. These techniques are discussed in more detail in *UNITS 8.2 & 8.3*.

Blocking

Before detection of the target sample immobilized on the membrane, care needs to be taken to avoid nonspecific binding of the probe to the remaining binding sites on the membrane. Prior to the addition of the probe, the membrane is treated with general blocking agents such as proteins or detergent to reduce the nonspecific association of the probe molecule with the membrane: this is referred to as the blocking step.

Probing

Once the membrane has been blocked, it is incubated with a specific probe that binds to the protein or nucleic acid sequence of interest. In the case of a Southern or northern blot, the probe consists of complementary DNA or RNA sequences that will anneal to the target (see *UNIT 8.4*). The nucleic acid is labeled radioactively or enzymatically to allow for detection. By comparison, the probe used for an immunoblot is an antibody that recognizes a particular protein or epitope. A secondary antibody,

Current Protocols Essential Laboratory Techniques 8.1.1-8.1.19

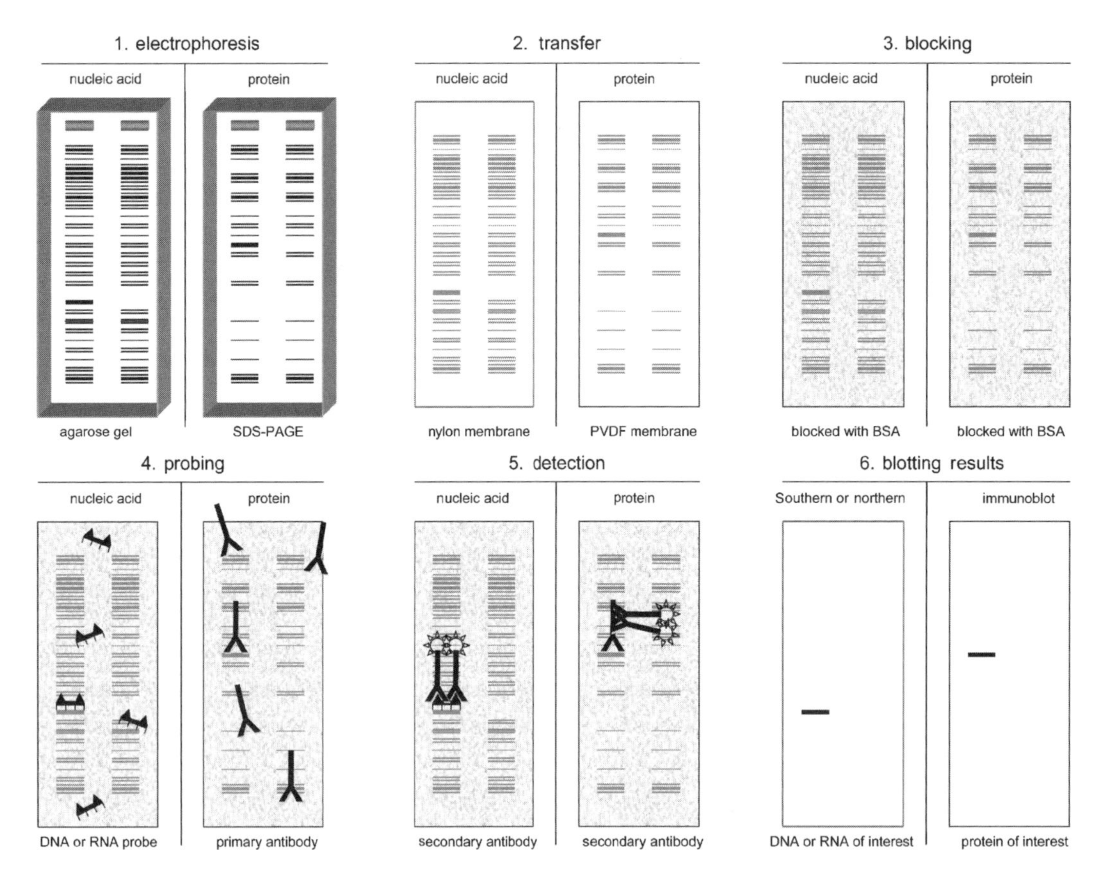

Figure 8.1.1 Blotting Procedure. General overview of the steps involved in a blotting procedure that are common to nucleic acid blotting (Southern and northern) and protein blotting (western or immunoblotting). (1) Separation of the molecules by gel electrophoresis on either an agarose (DNA or RNA) or a SDS-polyacrylamide gel (protein). (2) Resolved molecules are transferred to a membrane maintaining the same pattern of separation they had on the gel. (3) The blot is treated with blocking agents, such as proteins (BSA) or detergents that bind to unoccupied sites on the membrane. This is depicted as a gray background. (4) A specific probe that binds to the protein or nucleic acid sequence of interest is incubated with the blot. In the case of a Southern or northern blot the probe consists of a complimentary DNA or RNA sequence. For an immunoblot, the probe consists of a primary antibody that recognizes a particular protein or epitope. (5) Detection step. When using a radioactively labeled probe, the signal is detected by X-ray film or phosphorimager, resulting in the banding pattern depicted in step 6. Nonradioactive probes can utilize a reporter enzyme directly conjugated to the probe or a labeling moiety that is then detected by a specific antibody conjugated to a reporter enzyme. (6) The reporter enzymes are then presented with colorimetric, fluorogenic, or chemiluminescent substrates that produce signals, which can be detected as a colored product (analyzed visually), as a fluorescent precipitate (detected with a camera after excitation), or as a compound that emits light during its decomposition (detected with X-ray film or a cooled CCD camera).

conjugated to a reagent that allows for its detection, is incubated with the blot. A secondary antibody will bind to the primary antibody with high affinity to facilitate the generation of a specific signal. Following a period of incubation, the unbound probe or nonspecifically bound probe is removed by sequentially washing the membrane with increasingly stringent wash buffers.

Detection

The last step in a blotting experiment involves a detection step to visualize the bound probe. The method of detection will be determined by the nature of the probe. If a radioactive probe is used, exposure of the blot to X-ray film or a phosphorimaging device will allow for detection and quantitation of the bound probe. (Refer to *APPENDIX 1A* for information on the safe use of radiation.) If chemical- or enzyme-based detection systems are used, the appropriate substrates are added to the blot and the resulting signal is developed and can be documented by colorimetric, fluorescent, or chemiluminescent imaging (see *UNIT 7.5*).

Results and Analysis

Once the blot is developed, the resulting banding pattern can be analyzed. Analysis involves determining the amount and apparent molecular weight or size of the molecules on the blot and comparing the results to the predicted pattern. To determine the molecular weight of the molecules of interest, a standard curve of size versus migration distance is derived from the molecular weight markers (see *UNIT 7.2*). The standard curve is plotted on semilog paper or with a graphing and analysis program. The distance that each maker migrates from the origin is plotted on the *x*-axis and the corresponding size or weight of the marker is plotted on the *y*-axis. The resulting plot should produce a straight line that enables estimation of the size of the unknown. Automated analysis programs increase the accuracy and greatly simplify this process (see *UNIT 7.5*). In practice, the exact determinations of size are difficult with this type of approach because changes in a variety of parameters, including structure, salt concentration, and the speed at which the gel was run, can affect the resolution characteristics of a molecule. The resulting plot will allow for a reasonable determination of the molecule's size, allowing for a comparison of the predicted and observed outcomes.

Analysis of the blot can provide the researcher with a variety of details concerning the nature of the molecules being studied. In the case of a Southern blot, the structure of the gene of interest can be assessed. Initially, a restriction map of the gene was used to select restriction enzymes that would produce a distinct pattern of bands once the blot is developed. If the predicted banding pattern is observed, one can conclude that the structure of the gene of interest behaves as predicted. If large rearrangements such as insertions, deletions, or inversions have occurred, the banding pattern will deviate from the predicted pattern and the investigator can conclude that the gene of interest is altered in an unexpected fashion. Additionally, the relative dosage of a molecular species can be determined from a blotting experiment. The key to such an analysis is including an independent loading control to which the signal of the molecule of interest can be compared. If done carefully, one can differentiate the signal derived from a single copy of the gene as compared to two or more copies. A Southern blot may allow the researcher to detect related sequences that share homology to the molecule of interest but reside in different locations in the genome. Northern blots and immunoblots can be used to assess different levels of expression from a particular gene. Northern blots are also used to detect post-transcriptional modifications to the RNA, such as splicing. Immunoblots can be used in a similar fashion to detect post-translational modifications such as phosphorylation.

GENERAL CONSIDERATIONS

It is critical to remember that when running a blotting experiment only the molecules that your probe recognizes will be visualized. If a blotting experiment that is designed to detect a single species results in more than one band being detected, it may indicate one of several problems. The appearance of extra bands could be due to cross-reactivity of the probe with other molecules present in the original sample. This may occur if the antibody used in an immunoblot is polyclonal in nature and thus recognizes multiple proteins or antigens in the sample. Alternatively, if performing a Southern or northern blot, multiple bands may indicate that your probe sequence can associate with repetitive elements, that there is more than one copy of the sequence of interest in the genome, or that multiple isoforms of the RNA are expressed in your sample. In each case, selecting a different antibody or redesigning your nucleic acid probe may reduce this problem. Additionally, increasing the stringency of the hybridization or washing procedures may reduce the cross-reactivity of your probe with off-target species. On the other hand, the appearance of bands of lower molecular weight could be due to degradation of your sample of interest. Higher molecular weight bands could be the result of incomplete digestion of your DNA sequences (Southern) or incomplete denaturation or heat-induced aggregation of your protein sample (immunoblot). This is why it is very important to include controls so that one can distinguish among these possibilities. Such controls include samples that are identical to your experimental sample but are missing the target that the probe is supposed to recognize. This is called a negative control. Negative controls are very useful in determining the existence of any background that can be due to cross-reactivity between the probe and your sample.

Conversely, a positive control is a sample that contains the protein or nucleic acid of interest. For example, a plasmid containing a sequence of interest or the parental strain bearing the wild-type locus can be used as a positive control in a Southern blot. Similarly, a protein carrying the epitope of interest, such as FLAG or MYC, can be used as a positive control in an immunoblot. When included in the experiment the positive control allows the investigator to confirm that the experiment was successfully executed. Thus, if the samples being tested fail to produce a signal it may indicate that the problem lies with the experimental samples and not with the procedure.

SOUTHERN AND NORTHERN BLOTTING

The analysis of DNA and RNA through the use of Southern and northern blotting, respectively, allows the researcher to gain exquisite insight into basic biological processes. These techniques are among the most common applications in the molecular biology laboratory. Southern blotting is used to address a wide variety of basic biological problems, including defining the structure of a genomic locus, generating physical maps of a genome (Botstein et al., 1980), identifying genes involved in disease states (Gusella et al., 1983; Rommens et al., 1989), and analyzing replication and recombination intermediates. Similarly, northern blots have been instrumental in the elucidation of the transcriptional control paradigm, defining post-transcriptional modification such as splicing and poly(A) addition, and, more recently, defining processes involving siRNA and miRNA.

In 1975 Edwin M. Southern published a paper describing the technique, which became known as Southern blotting (Southern, 1975; Southern, 2000). Briefly, this technique allows the researcher to resolve DNA fragments of defined size on an agarose gel and transfer them to a membrane, which is subsequently hybridized with a probe generated from unique sequences. The result of this approach allows the researcher to visualize DNA sequences as discrete bands among a complex mixture of other DNA molecules (Fig. 8.1.1). Southern blotting was the result of a convergence of several independent technologies. The utilization of restriction enzymes to cleave the DNA into fragments of defined size allowed for the generation of discrete banding patterns for a particular digest (Kelly and Smith, 1970; Nathans and Smith, 1975). The development of slab gel electrophoresis systems to resolve the DNA fragments (McDonell et al., 1977) was a significant advance over the tube gels used previously. Prior to slab gel development, tube gels were the method of choice for resolving nucleic acids and proteins. Tube gels were difficult to prepare and process. Analysis of the resolved products was difficult, involving slicing the gel into discs, which were individually examined for the molecule of interest. By comparison, slab gels are easy to cast and multiple samples can be resolved simultaneously. Following resolution, the processing and subsequent analysis are much less laborious and more reproducible.

Finally, the development of membranes to serve as solid supports to which the resolved DNA fragments could be transferred and analyzed made it possible to carry out the Southern blotting technique. Shortly after Southern's ground breaking paper, the adaptation of a similar approach was used to analyze RNA resulting in the northern blot technique (Alwine et al., 1977; Alwine et al., 1979). The name northern was chosen to reflect Southern's contribution, and, henceforth, other blotting techniques such as western, southwestern, northwestern continued the tradition. Together, the Southern and northern blotting techniques opened avenues of investigation in the burgeoning field of molecular biology that were previously unassailable.

The following is a brief discussion of the issues that must be considered when designing a Southern or northern blot experiment. The specific details will be covered in greater depth elsewhere as noted. As an investigator designing a blotting experiment one must consider four distinct issues: (1) resolution and denaturation of the molecules, (2) membrane selection, (3) transfer methodology, and (4) hybridization methodology.

Resolution

The first issue to consider when designing a blotting experiment is how to resolve your molecules into a distinct pattern. When dealing with RNA, the molecules are of a defined size as determined by the transcriptional unit, i.e., mRNA, tRNA, etc. Because of its single-stranded nature, RNA molecules have the potential to form secondary structures by forming typical Watson-Crick base pairs between complementary bases that are part of the same strand, which may affect the resolution. To address this problem, the RNA should be treated with denaturants such as glyoxal or formamide prior to resolving the molecules in the gel (Thomas, 1980; also see UNIT 7.2). Alternatively, treatment of the RNA and resolution of the molecules in the presence of formaldehyde will also prevent the formation of secondary structures (Lehrach et al., 1977). As a result, the RNA will migrate as a function of its size rather than its structure.

In resolving DNA molecules, the size of a fragment will be determined by the particular restriction enzyme and the frequency of cutting within the interval of interest. Typically, the DNA is digested and then resolved on a gel. Many of the model systems used today have had their genomes sequenced; thus, it is possible to identify the interval of interest and select enzymes that will produce a defined pattern of fragments of discrete size to assay with the Southern blot.

Once the sizes of the molecules have been defined, the gel conditions to optimize resolution can be chosen as described in UNIT 7.2. In most cases an agarose slab gel will be sufficient to resolve the molecules. Because it is possible to adjust the concentration of agarose over a wide range, which will in turn affect the size of the pores in the gel, one can easily resolve fragments from very small (hundreds of base pairs) to very large sequences (thousands or tens of thousands of base pairs). It is even possible to resolve entire chromosomes for certain organisms using a CHEF (clamped homogeneous electrical field) apparatus (Chu et al., 1986; *CP Molecular Biology Unit 2.5B*, Finney, 2000). Once the size of the molecules of interest drop below 100 bp, it might be advisable to use polyacrylamide as the gel matrix instead of agarose. Because polyacrylamide can be used at much higher concentrations, the pore sizes associated with the gel matrix are much smaller. Thus, it is possible to achieve nucleotide resolution under the appropriate conditions. These gels may be prepared as denaturing or nondenaturing depending on the exact application, see *CP Molecular Biology Unit 2.12* (Ellington and Pollard, 1998) and Sambrook et al. (1989). Finally, optimizing the running conditions for the gel is primarily an empirical process. Using a standard laboratory electrophoresis unit (14 × 14–cm), a gel run at 50 to 100 V, or 1.9 V/cm^2, in TBE buffer for 4 to 7 hr will resolve the nucleic acids sufficiently for analysis. It is usually advisable to use a larger gel (14 × 14–cm) as compared to a mini-gel (7 × 10–cm), when running a blotting experiment. A larger gel allows for greater separation of the molecules of interest, as a result of the increase in distance migrated from the origin. This in turn will result in better resolution which will aid in the subsequent analysis of the blot. After running the gel, it is important to examine it by staining with ethidium bromide. This will allow the researcher to visualize the nucleic acids and the size markers on a UV transilluminator to assess the resolution of the gel and the appearance of the cut genomic DNA or RNA. It is important to note that your band of interest will not be able to be distinguished: instead a smear will be observed that corresponds to the entire genome cut into fragments of different sizes (Fig. 8.1.2A) or all the RNA molecules present in your experimental sample. At this point it is advisable to take a picture of the ethidium bromide stained gel aligned with a ruler (Fig. 8.1.2A). This will document the gel for later reference because the molecular weight standards will not show up in the blot since your probe will not anneal to them. This allows the investigator to compare the distances migrated by the molecular weight standards in the gel and the bands of interest on the blot during the analysis step. In order to determine the size of the observed bands on the blot, a standard curve can be generated by plotting the distance migrated by the standards versus their molecular weight (Fig. 8.1.2B). The investigator can measure how far the bands on the blot traveled from the origin and correlate this to the standard curve to determine the sizes of the bands. Alternatively, there are commercially available molecular weight standards that

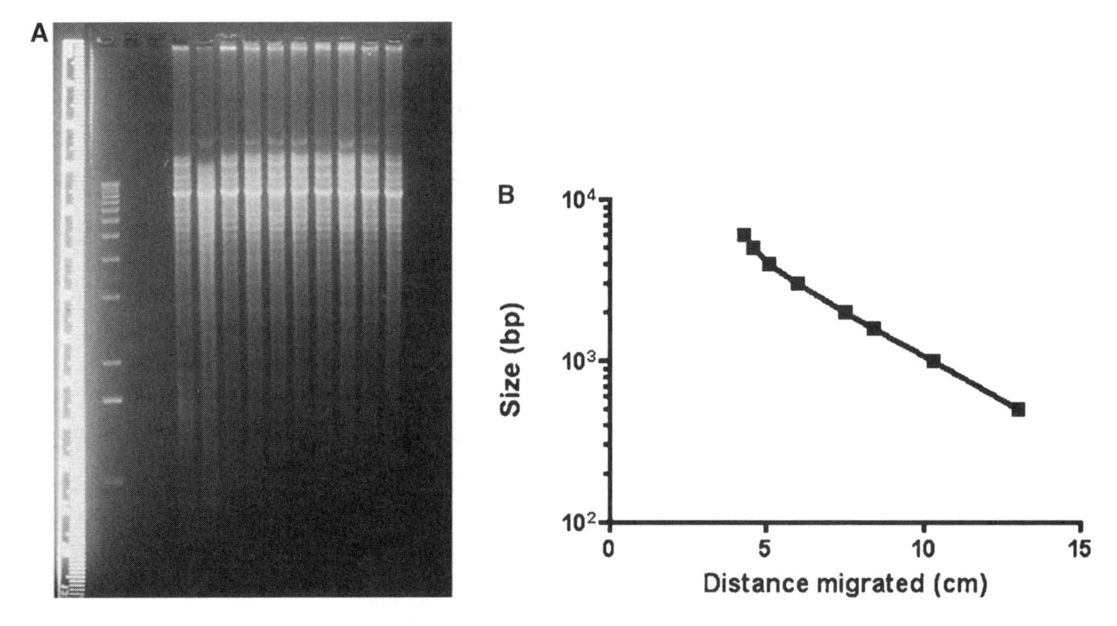

Figure 8.1.2 Gel documentation and standard curve. (**A**) Photograph of a 14 × 14–cm 0.8% agarose gel with samples of restricted yeast genomic DNA, stained with ethidium bromide and aligned with a fluorescent ruler. (**B**) Standard curve of the migration distances of the molecular weight standards resolved on the agarose gel and plotted as a function of size (bp) versus distance migrated from the origin (cm).

can be visualized with the detection system used, such that the molecular weight markers and the bands of interest will be simultaneously developed on your blot.

Denaturation

As discussed above, the RNA used for northern blot experiments is denatured with formaldehyde or glyoxal prior to running the gel to disrupt secondary structures that may affect the resolution of the molecules. In the case of a Southern blot, the DNA is resolved as double-stranded DNA fragments of defined size. If the expected fragments are >10 to 15 kb in size, an additional step involving depurination or UV radiation can be used to break the DNA into smaller, more readily transferable fragments. Depurination involves treating the gel, following resolution, with a solution of 0.1 to 0.25 M HCl for 15 min. This treatment results in the formation of abasic sites throughout the DNA. Subsequent incubation in a solution of 0.5 M NaOH leads to strand scission at the abasic sites. Excessive depurination can fragment the DNA into such small fragments that its ability to hybridize with a probe later in the procedure will be reduced. Treating the resolved DNA, which is intercalated with ethidium bromide, with short-wave, 240-nm UV will also generate strand breaks. As above, excessive irradiation can shear the DNA or lead to cross-linking, which will inhibit its ability to hybridize with the probe fragment. If done appropriately the resulting smaller DNA fragments should transfer efficiently. This step is discussed in *UNIT 8.2*. Once the fragments have been resolved and fragmented, the DNA needs to be denatured to allow for subsequent hybridization with the probe. There are two basic approaches to achieve this. The first involves incubating the gel in a denaturation buffer containing sodium hydroxide (NaOH). Alkaline denaturation results in the deprotonation of the atoms that form the hydrogen bonds between the base pairs of the duplex DNA. This reaction destabilizes the interaction between the DNA strands, generating single-stranded species and water by association of the extracted proton and the hydroxyl ion (Ageno et al., 1969). Following denaturation, the gel is incubated in a neutralization buffer containing 1 M Tris·Cl, pH 7.4. The DNA in the gel remains single stranded following the neutralization step. The single-stranded nature of the DNA is likely to be maintained for a variety of reasons. First, alternative interactions between the DNA bases (see Figure 8.1.2B) and the sugar moieties of the agarose, and later the amide moieties of the membrane, will limit reannealing. Whereas some limited reassociation may occur, it will probably result in mispaired sequences that will have significant amounts of single-stranded

DNA available to hybridize with the probe sequence later in the hybridization step experiment. Alternatively, one can use an alkaline transfer buffer, as discussed below, allowing for simultaneous denaturation, transfer, and cross-linking.

Membrane Selection

The membrane used in a blotting experiment serves as a solid support to transfer the nucleic acids and subsequently hybridize them with a probe derived from a unique sequence. When blotting technologies were first developed, there were several types of solid supports, including DBZ paper (Wahl, 1979) and nitrocellulose (Southern, 1975). Initially, nitrocellulose became the membrane of choice because of its relatively low cost and ease of handling. However, today a variety of membranes are available for blotting experiments (Table 8.1.1). Nylon membranes are the preferred substrate for most Southern and northern blotting experiments. Nylon has the advantages of being relatively cost effective, very stable, and having improved mechanical strength as compared to nitrocellulose. Thus, the blot can be stripped and rehybridized multiple times. Following hybridization, the blot can be either treated with an alkaline buffer or boiled in the presence of SDS to dissociate the bound probe from the blot. This process is termed "stripping" and allows the investigator to hybridize the same blot many times with different probes. The primary consideration in selecting a membrane that will be stripped and probed multiple times is mechanical strength. Nylon is preferable to nitrocellulose for this type of experiment. When using commercially available detection kits, make sure to follow the stripping protocol suggested in the user's manual.

Additionally, nylon membranes can be either charged or neutral, which facilitates the use of a variety of transfer protocols. The selection of a charged or neutral membrane may be determined by the

Table 8.1.1 General Properties of Blotting Membranes

	Type of membrane				
	Nitrocellulose	Supported nitrocellulose	Nylon	Charged nylon	PVDF
Application	Western	Western	Southern	Southern	Western
	Southern	Southern	Northern	Northern	—
	Northern	Northern	—	—	—
Binding capacity	80 to 150 μg/cm^2	75 to 90 μg/cm^2	>400 μg/cm^2	>600 μg/cm^2	>200 μg/cm^2
Transfer methods	Capillary blotting	Capillary blotting	Capillary blotting	Capillary blotting	Electroblotting
	Vacuum blotting	Vacuum blotting	Vacuum blotting	Vacuum blotting	—
	Electroblotting	Electroblotting	Electroblotting	Electroblotting	—
	—	—	Alkaline blotting	Alkaline blotting	—
Immobilization	UV-cross-linking	UV-cross-linking	UV-cross-linking	UV-cross-linking	Air drying
	Baking (80°C)	Baking (80°C)	—	—	—
Detection methods	Isotopic	Isotopic	Isotopic	Isotopic	Isotopic
	Chemi-luminescent	Chemi-luminescent	Chemi-luminescent	Chemi-luminescent	Chemi-luminescent
Reprobing	±	+	+	+	+

Table 8.1.2 Transfer Methods

	Capillary blotting	Electroblotting	Vacuum blotting
Technology requirements	Low tech	Requires a transfer chamber	Requires a vacuum blotting chamber
Speed	2 to 24 hr	1 to 2 hr	5 to 30 min
Applications	Nucleic acids	Protein	Nucleic acids
	—	Nucleic acids	—
Transfer efficiency	Good	Excellent	Excellent

nature of the fragments being transferred. If very small fragments (50 to 200 bp) are being blotted, a charged membrane that exhibits a higher degree of retention may be advantageous. Alternatively, a neutral membrane might be better for larger fragments due to a reduced background problem. Nylon membranes are commercially available from a number of companies including Hybond N and N^+ (GE healthcare), and Zetaprobe (Bio-Rad).

Transfer

Once the RNA or DNA is resolved and a membrane selected, the next step in the blotting process involves transferring the nucleic acid from the gel to the membrane. Over the years, numerous variations on the basic protocol have been developed. Most laboratories will have a basic protocol in place. It is well worth the investigators time to experiment with different conditions to arrive at an optimal protocol for their particular application.

There are three commonly used techniques to transfer nucleic acids from the gel to the membrane: capillary blotting, electroblotting, and vacuum blotting. The relative merits of each technique are outlined in Table 8.1.2.

Capillary blotting

Capillary blotting was used in the original Southern protocol (Southern, 1975). This approach involves the construction of a stack consisting of a wick, several pieces of blotting paper cut to the size of the gel, the gel, the membrane cut to the size of the gel, an additional piece of blotting paper, and finally a stack of paper towels. When assembled, the stack is placed on a platform such that the wick is in contact with a pool of the transfer buffer. As the name implies, the buffer wicks through the blotting paper and gel and into the paper towels on top of the stack. As the buffer moves upward thru the gel, the DNA migrates with it through the pores of the gel until it encounters the membrane and becomes trapped.

To carry out a capillary blotting experiment, there are several issues to be considered. First, one must select a transfer buffer. Traditionally, people have used ionic solutions such as 1× SSC (0.15 M NaCl/0.15 M NaCitrate)—e.g., Southern used this buffer in his original experiments with nitrocellulose. In doing so, he showed that increasing the ionic strength of the buffer improved the retention of the DNA to the membrane (Southern, 1975). In contrast, later studies using nylon membranes showed that it was possible to transfer the DNA in pure deionized water (Reed and Mann, 1985). Thus, the ionic nature of the transfer buffer is less important when using nylon membranes; however, most protocols still recommend the use of a 5× to 10× solution of SSC. Alternatively, as mentioned above, an alkaline transfer buffer can be used (Reed and Mann, 1985). This approach allows the investigator to skip the denaturation step described above. When used in concert with a charged membrane such as Hybond N^+ (GE Healthcare), the conditions lead to a cross-linking of the DNA when it encounters the membrane. The decision to use neutral or alkaline transfer protocols is dependent on the individual investigator.

Finally, there is the issue of how long to blot the DNA to the membrane. Most protocols suggest 16 hr or overnight. There are reports that 1 to 2 hr is sufficient for the transfer of most nucleic acids species (Reed and Mann, 1985). The caveat is that larger molecules tend to take longer to migrate out of the gel and onto the membrane surface.

Electroblotting

This approach is typically used to transfer small RNA or DNA molecules from polyacrylamide gels to a membrane. Electroblotting (Bittner et al., 1980) utilizes an apparatus identical to that used for immunoblotting. Once the nucleic acid has been resolved on the polyacrylamide gel, a sandwich of blotting paper, gel, membrane and blotting paper is built and placed in a cassette. The cassette is inserted into the buffer chamber filled with an ionic buffer such as TBE, so that the membrane is proximal to the positive electrode. When voltage is applied to the chamber, the negatively charged nucleic acid molecules will move from the gel to the membrane. Because of the small pore sizes associated with polyacrylamide gels, the capillary blotting approach is less efficient than electroblotting (Sambrook et al., 1989). As mentioned above, a charged membrane may be preferable for this approach. The primary advantage of this approach is the rapid transfer time, usually 1 to 2 hr, and the high recovery of nucleic acids. This is the method of choice for blotting nucleic acids when using a polyacrylamide gel to resolve the molecules.

If electroblotting is used with an agarose gel, it is important to monitor the current during the transfer. High current may lead to a significant increase in the buffer temperature, which could, in turn, melt the agarose gel. Adjusting the salt concentration of the transfer buffer will limit this problem. Polyacrylamide gels are much less sensitive to temperature fluctuations and are thus better suited for this approach. The only requirement for this technique is the availability of a transfer chamber large enough to accommodate the gel.

Vacuum blotting

The third approach that is commonly used to transfer nucleic acids to a membrane support is vacuum blotting (Olszewska and Jones, 1988). This approach takes advantage of the application of a vacuum to a transfer system, which requires the transfer buffer to pass through the gel and membrane to reach a solution trap. As with electroblotting, the advantages of this system are the quick transfer time, ranging from 5 to 30 min, the tight resolution of the resulting bands, and the improved recovery of the nucleic acids as compared to capillary blotting. This approach is applicable to both Southern and northern blots. These transfer systems are commercially available. With all approaches, monitoring transfer efficiency is important. One simple strategy is to stain the agarose or acrylamide gel post transfer to evaluate what is left behind. Typically, the higher molecular weight species will not transfer as well, with some remaining in the gel.

Cross-Linking

Once the nucleic acids have been transferred to a membrane, they have to be linked to the membrane. This process is called cross-linking. The approach to cross-linking varies depending on the membrane. As mentioned above, an alkaline transfer with a charged membrane often leads to spontaneous cross-links being formed between the membrane and nucleic acids, obviating the need for additional treatment. If a neutral membrane is used, the UV irradiation is required to covalently attach the nucleic acids to the membrane in preparation for hybridization. Application of UV radiation at a wavelength of 254 nm for 1 to 5 min results in the formation of a covalent bond between the amide groups on the nylon and the carbonyl groups found on the thymine and uracil bases. This is the same reaction that produces pyrimidine dimers in genomic DNA following exposure to UV radiation. The two most common sources of UV radiation are the UV transilluminator or a cross-linking instrument available from several suppliers (e.g., Stratagene, Bio-Rad, and

UVP). These instruments are built specifically for the purpose of applying a defined amount of UV radiation to a blot. If a transilluminator is used, the membrane needs to be exposed to 1.5 kJ/m^2 (Church and Gilbert, 1984).

Alternatively, the investigator can bake the membrane at 80°C for 2 hr. Baking leads to the dehydration of the nucleic acids on the blot, resulting in the generation of stable hydrophobic interactions between the nucleic acid and the membrane. Because water has been excluded during this process it is very difficult to disrupt these interactions, even when the membrane is rehydrated and processed. This approach leads to stable associations between the nucleic acids and membranes, allowing for multiple rounds of rehybridization (Nierzwicki-Bauer et al., 1990). Thus, baking is a very good method to immobilize nucleic acids to a solid support. If nitrocellulose is used as the solid support, the baking must occur under vacuum; otherwise the nitrocellulose will combust. Additionally, baking for longer than 2 hr can cause the nitrocellulose to become very brittle and hard to handle.

Hybridization

Once a blot has been generated, the next step involves applying a probe that will anneal to the sequence of interest and allow for the visualization and/or quantitation of the desired sequence. To accomplish this, the blot is processed in three successive steps: prehybridization, hybridization (probing), and washing.

Prehybridization

The prehybridization step is used to "block" the blot so as to limit nonspecific interactions of the probe fragment with the membrane. There are a wide variety of prehybridization solutions, of which the most popular are Denhardt's (Denhardt, 1966), Blotto (Johnson et al., 1984), and Church's buffer (Church and Gilbert, 1984). Although all three of these buffers are inexpensive and readily prepared with reagents commonly found in most molecular biology laboratories, Denhardt's will be used as an example for the remainder of the section. A 50× solution of Denhardt's consists of 1% BSA, 1% Ficoll, and 1% polyvinylpyrrolidone. The prehybridization solution consists of 5× Denhardt's, 6× SSC, 0.5% SDS, and 100 μg/ml single-stranded salmon sperm DNA. The components of a prehybridization solution typically consist of polar and nonpolar molecules that associate in a nonspecific manner with the polar and nonpolar moieties available on the membrane. The transferred nucleic acids only occupy a limited amount of the surface area of the membrane. The molecules in the prehybridization solution coat the rest of the membrane. In the absence of such a treatment, the probe would associate with the unoccupied sites on the membrane, resulting in very high background and a very low signal-to-noise ratio. To carry out the prehybridization step, the blot is placed in a container to which the prehybridization buffer is added. The blot is incubated at the desired temperature, typically between 50° to 65°C, for 3 to 24 hr. Once the membrane is blocked, the probe can be synthesized and added to the blot for hybridization.

Probing

To detect the desired sequence on the blot, a probe that anneals to the sequence must be synthesized, as discussed in UNIT 8.4. The probe will have two properties. First, it will anneal specifically with the sequence of interest. Second, it will be modified in such a way as to allow for the detection of the annealed sequences. To address the first condition, a unique sequence must be available to serve as a template for the probe. This sequence can be an oligonucleotide or a cloned DNA fragment. Depending on the nature of the template sequence, the hybridization and washing steps will be modified to accommodate the probe. If an oligonucleotide is used as the probe, it will be end labeled with ^{32}P-γ-ATP so that the 5′ phosphate of the molecule is now radioactive and ready to use as a probe (*CP Molecular Biology Units 3.4-3.15*, Ausubel et al., 2007; Sambrook et al. 1989). Alternatively, oligonucleotide probes come in a wide variety of modifications that enable nonradioactive (e.g., fluorescent or chemiluminescent) detection. If a cloned sequence is used as a template, the DNA fragment must first be denatured. The denatured DNA will be mixed with primers that will anneal to the template and prime DNA synthesis through the action of a DNA polymerase such as Klenow (Klenow and Henningsen, 1970). During this step, modified nucleotides are incorporated to generate the probe that can be detected after hybridizing with the target DNA. The nucleotides can be radioactive, such as ^{32}P-α-dCTP, and thus be readily detectable with a Geiger counter or as an autoradiogram. When synthesizing a radioactive probe, it is important to assess the quality of the resulting reaction. To assay the resulting probe for efficiency of incorporation of the radioactive nucleotides, a TCA (trichloroacetic acid) precipitation is utilized. This approach allows for the quantitation of incorporation of the labeled nucleotide into the newly synthesized probe. An incorporation of 40% to 70% of the total radioactive nucleotide is usually acceptable for generating a probe to use in a Southern or northern blot experiment. Alternatively, one can utilize nucleotides containing a moiety that is recognized by an antibody or protein, which is conjugated to an enzyme (see Table 8.1.3 for examples of detection systems used). The enzyme will react with a substrate reagent to generate a detectable signal. There are a variety of commercially available kits that utilize either the radioactive or nonradioactive detection systems providing the investigator with flexibility in deciding how to probe their blots. Traditionally, radioactive labeling methods were the method of choice because they were more sensitive, but nonradioactive methods of detection have gained great popularity as they are safer and have achieved the same level of detection. Additionally, nonradioactive methods are now of comparable cost relative to the radioactive methods. In the long term, nonradioactive methods may be more cost-effective because the waste products do not require special handling and disposal.

Once a probe is generated, it must be denatured and added to the blot for 1 to 24 hr. Determining how long to hybridize the blot depends on a variety of factors and must be determined empirically. It is often convenient to hybridize the blot overnight and thus be confident that the hybridization of the probe to the target molecule has been maximized. If an investigator is interested in accelerating the hybridization process, one can include reagents such as 10% dextran sulfate or polyethylene glycol molecular weight 8000 to the buffer (Wahl et al., 1979; Amasino, 1986). These reagents disrupt the water structure around the nucleic acids, thereby increasing the relative concentration of these species and improving the hybridization characteristics. Addition of these reagents can reduce the hybridization time to 1 to 2 hr. There are a variety of premade hybridization buffers commercially available for this purpose. Each probe will have specific characteristics that affect its ability to associate with the target. The most important characteristic for a hybridization experiment is the melting temperature of the probe sequence (T_m). The T_m is the temperature at which 50% of the base pairs in a duplex DNA have been denatured and will be affected by the base composition of the probe. GC base pairs have three hydrogen bonds as compared to two hydrogen bonds in an AT base pair; thus, GC base pairs are more stable. Probes with a high GC content have a higher T_m than probes with a high AT content. It is important to determine the temperature at which your probe will specifically interact with the target sequence, but at which other nonspecific interactions will be disrupted. To roughly calculate the T_m of an oligonucleotide 14 to 20 nucleotides in size, the Wallace rule (Wallace et al., 1979; Suggs et al., 1982) can be used. The Wallace rule uses the following equation, $2 \times (\#A + \#T) + 4 \times (\#G + \#C)$. Each AT base pair contributes 2°C to the melting temperature and each GC base pair contributes 4°C. For example, the T_m for the oligonucleotide AGTTGGCACTGGATTGCC can be expressed as $T_m = 2 \times (3+5) + 4 \times (6+4) = 56°C$.

Alternatively, the T_m of sequences longer than 50 base pairs can be determined using the %GC method (Meinkoth and Wahl, 1984): $T_m = 81.5 + 16.6[\log(Na^+)] + 0.41(\%G+C) - 500/N - 0.61(\%formamide)$; where N = length of the probe and the Na^+ refers to the concentration of monovalent cations in the buffer. Many buffers use sodium salts as one of the reagents. As an example, Denhardt's, described above, contains 6× SSC, which is made of a mixture of sodium chloride and sodium citrate. A 1× solution of SSC is 0.165 M Na^+. By changing the concentration of the SSC during the experiment, the T_m of the probe fragment can be altered. The presence of mismatches between the probe and the target also reduce the T_m by about 1°C/1% mismatching. Thus, probes associating with a sequence with which it shares 85% sequence homology would have

Table 8.1.3 Chromogenic and Luminescent Visualization Systems[a,b]

System	Reagent[c]	Reaction/detection	Comments[d]
Chemifluorescent			
Not enzyme-based, direct labeling of target protein	Antibody conjugated to fluorescent dye labels. Goat-IgG[a]-Cy3 (λ_{max} 570 nm) Goat-anti-rabbit IgG[a]-Cy5 (λ_{max} 670 nm)	Direct fluorescence detection using fluorescence scanner and CCD imagers. No need for enzyme-substrate amplification	Very sensitive (detects <1 pg protein). Reaction detected within a few seconds to 1 hr. Fluorescence can last for months. Two different proteins can be detected at the same time. Kits available from several different vendors. (e.g., GE Healthcare and Pierce)
Chemiluminescent			
HRPO[a]-based	Luminol/H_2O_2/*p*-iodophenol	Oxidized luminol substrate gives off blue light. There are a variety of systems that use similar substrates with slight variations. Detection with film, CCD cameras or laser scanners.	Very convenient, reaction detected within a few seconds to 1 hr. Very sensitive (detects < 1 pg to 50 pg of protein). Kits available from several different vendors.
AP-based	Substituted 1,2-dioxetane-phosphates (e.g., AMPPD[a], CSPD[a], Lumigen-PPD[a], and Lumi-Phos 530[g])	Dephosphorylated substrate gives off light. Detection with film, CCD cameras or laser scanners.	Reasonable sensitivity on all membrane types; reaction can last for >1 hr. Kits available from several different vendors. Consult reagent manufacturer for maximum sensitivity and minimum background.
Chromogenic			
AP[a]-based	BCIP[a]/NBT[a]	BCIP hydrolysis produces indigo precipitate after oxidation with NBT; reduced NBT precipitates; dark blue-gray stain results	More sensitive and reliable than other AP-precipitating substrates; note that phosphate inhibits AP activity
HRPO-based	*TMB*[a,f]	Forms dark purple stain	More stable, less toxic than DAB/$NiCl_2$; may be somewhat more sensitive[f]; can be used with all membrane types; kits available from several different vendors
	DAB[a]/$NiCl_2$[e]	Forms dark brown precipitate	More sensitive than 4CN[a] but potentially carcinogenic; resulting membrane easily scanned

continued

Table 8.1.3 Chromogenic and Luminescent Visualization Systems[a,b], *continued*

System	Reagent[c]	Reaction/Detection	Comments[d]
	4CN[a]	Oxidized products form purple precipitate	Not very sensitive (Tween 20 inhibits reaction); fades rapidly upon exposure to light

[a]Abbreviations: AMPPD or Lumigen-PPD, disodium 3-(4-methoxyspiro{1,2-dioxetane-3,2′-tricyclo[3.3.1.1^{3,7}] decan}-4-yl)phenyl phosphate; AP, alkaline phosphatase; BCIP, 5-bromo-4-chloro-3-indolyl phosphate; 4CN, 4-chloro-1-napthol; CSPD, AMPPD with substituted chlorine moiety on adamantine ring; DAB, 3,3′-diaminobenzidine; HRPO, horseradish peroxidase; NBT, nitroblue tetrazolium; TMB, 3,3′,5,5′-tetramethylbenzidine. IgG, Immunoglobulin G.

[b]Table adapted from *CP Molecular Biology Unit 10.8* (Gallagher et al., 2004).

[c]Recipes and suppliers are listed in *CP Molecular Biology Unit 10.8* except for TMP, for which use of a kit is recommended.

[d]See Commentary for further details.

[e]DAB/$NiCl_2$ can be used without the nickel enhancement, but it is much less sensitive.

[f]McKimm-Breschkin (1990) reported that if nitrocellulose filters are first treated with 1% dextran sulfate for 10 min in 10 mM citrate-EDTA (pH 5.0), TMB precipitates onto the membrane with a sensitivity much greater than 4CN or DAB, and equal to or better than that of BCIP/NBT.

[g]Lumi-Phos 530 contains dioxetane phosphate, $MgCl_2$, CTAB, and fluorescent enhancer in a pH 9.6 buffer.

a T_m 15°C lower than an association between a probe and target sharing 100% homology. These differences in T_m between mismatched and correctly paired probes allows the researcher to choose a washing temperature that will remove nonspecific binding while maintaining specific interactions. In addition to probe length and salt concentration, inclusion of reagents such as formamide can dramatically affect the T_m of the probe and thus affect the choice of hybridization conditions. Formamide is a denaturant, which tends to destabilize duplex nucleic acids by disrupting the hydration skeleton around the molecule. Additionally, formamide can provide alternative hydrogen bonding partners for the nucleotide bases in a single stranded molecule. As such, the predicted T_m of a probe is reduced substantially in the presence of formamide. This may be advantageous in situations where the investigator is trying to maintain stringent conditions at lower temperatures. This change in T_m is accounted for by including the %formamide as a factor in the %GC method.

Washing

Following hybridization, the blot must be washed to remove unassociated and nonspecifically annealed probe from the blot. By altering the stringency of the washing conditions, one can affect the degree of specificity obtained by the probe. The term "stringency" refers to conditions that affect the association of two single-stranded nucleic acids. In a standard protocol, there are two variables that can affect the stringency of hybridization, temperature and salt concentration. The association of two single-stranded nucleic acids is dependent on their relative melting temperatures. Thus, as the temperature is increased, the likelihood that the strands will separate also increases. Conversely, the presence of salt in the solution tends to increase the T_m for a given interaction. This effect can be explained by the interaction between cations such as sodium and the nucleic acids, which are negatively charged. When high concentrations of cations are present in the buffer there is a tendency to associate with the nucleic acids to neutralize their inherent negative charge. Thus, interactions between the probe fragment and the target are substantially stabilized. Therefore, as the salt concentration is decreased, the T_m decreases, leading to greater disassociation between the annealed strands. By increasing the temperature and decreasing the salt concentration in the wash buffers the stringency of the wash is also increasing. A typical wash protocol involves a series of sequential washes with decreasing salt concentration. For example, the first wash uses a solution of 2× SSC, 0.1% SDS and is carried out at room temperature for 15 min. This is followed by a wash in 1× SSC, 0.1% SDS at 65°C for 15 min. Finally, the last wash is the most stringent, utilizing a buffer of 0.1× SSC/0.1%SDS at 65°C for 15 min. Over the course of the washes from the original hybridization solution containing 6× SSC to the most stringent wash of 0.1× SSC, the salt concentration decreases from almost 1 M Na^+ to 0.0165 M Na^+. As mentioned above, interactions between sequences with <100% homology are more likely to be disrupted by the wash protocol because of their inherently lower T_m as compared to a perfectly base paired hybridization. By executing this wash protocol, one is selecting for highly specific associations between the probe fragment and the target sequences on the blot. It is possible to monitor the progress of the washing protocol, when using radioactive probes, with a Geiger counter (see *UNIT 2.3*).

Detection

Once the blot is washed, the next step is detection of the annealed probe as discussed in *UNIT 8.4*. If a nonradioactive probe kit is used, the detection protocol will be outlined by the manufacturer. If a radioactive probe is used, the blot is exposed to a sheet of film in a cassette. The film cassette will often have intensifying screens that, as the name implies, amplifies the signal generated by the probe. Following an exposure time ranging from hours to days, the film is developed and, hopefully, a pattern of bands is evident. Alternatively, an imaging device such as a phosphorimager can be used to visualize the banding pattern. This approach has the advantage of allowing the investigator to quantitate the intensity of the signal associated with a specific band. Such quantitation facilitates the determination of molecular half-life, relative abundance of a molecule compared to a loading control, etc. Such information can be invaluable to elucidation of the mechanisms involve in a particular biological process.

IMMUNOBLOTTING

Immunoblotting, also known as "western blotting" when following the same humor used to derive the term "northern" (a play on the name "Southern"), is basically a method used to study protein expression, purification, and modification, and has many applications of great relevance in both the research and clinical setting. The term western blotting was first used by Burnette (1981) when referring to the method used to assay proteins immobilized on a membrane originally developed by Towbin et al. (1979).

One of the main differences between immunoblotting and Southern or northern blotting is the nature of the target and probe. As discussed above, Southern and northern blots are used to assay nucleic acids and the probe of choice is a complementary nucleic acid that anneals to the target with high affinity through base pairing interactions. In the case of an immunoblot, the molecule of interest is a protein and the preferred probe an antibody. Antibodies (Fig. 8.1.3) recognize specific antigens associated with a particular protein. Because the proteins are denatured during the separation step, the particular antigen recognized by the antibody may be disrupted. Thus the effectiveness of a given antibody to recognize a specific protein in an immunoblot experiment must be determined empirically. Antibodies are often characterized by the manufacturer as being effective for a particular application including western blotting, immunoprecipitations, etc. Such information is invaluable in selecting an antibody.

Separation

The first step in an immunoblotting experiment requires the separation of a complex mixture of proteins into an ordered array based on the size of the proteins. This is accomplished with the separating power of a denaturing polyacrylamide gel electrophoresis (PAGE) system (Laemmli, 1970). As discussed in *UNIT 7.3*, this procedure requires the denaturation of the proteins in a mixture with sodium dodecylsulfate (SDS), also known as sodium lauryl sulfate. The denatured protein sample is loaded onto a discontinuous gel system, and resolved as described. The effective range of resolution on an SDS-PAGE depends on the concentration of acrylamide and bisacrylamide. As with agarose gels, the higher the concentration of the gel the higher the resolving capacity. *UNIT 7.3* provides the detailed information necessary for determining the optimal gel percentage to resolve

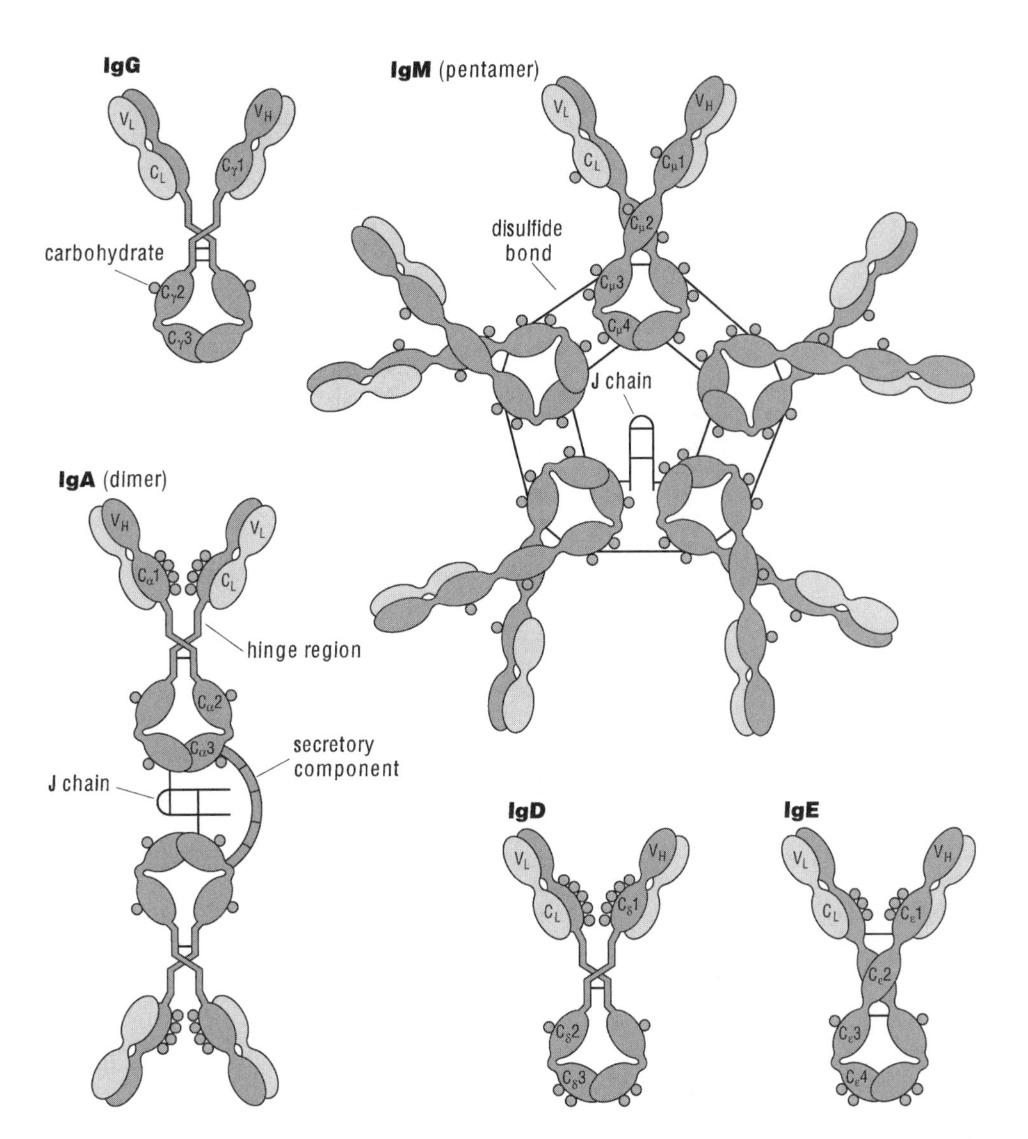

Figure 8.1.3 Structures of the five major classes of secreted antibody. Light chains are shown in light gray; heavy chains are shown in dark gray. Circles denote areas of glycosylation. The polymeric IgM and IgA molecules contain a polypeptide known as the J chain. The dimeric IgA molecule shown includes the secretory component. Reproduced with kind permission from Coico et al. (2003).

a protein of a particular molecular weight. Additionally, the thickness of the gel will affect the investigators ability to transfer the proteins from the gel to the membrane. Typically, gels between 1- and 0.4-mm are used for immunoblotting experiments. Thicker gels may impede the transfer step, while thinner gels are fragile and difficult to work with.

Membrane Selection

The choice of membrane (Table 8.1.1) used for an immunoblot is crucial to the success of the experiment (see *UNIT 8.3*; Sambrook and Rusell, 2001; Kurien and Scofield, 2006). The membrane serves as a solid support to which the proteins are transferred. The properties of the membrane will affect the retention of the protein, the ability to strip and subsequently the membrane, and the detection methodology. In the event the immunoblot experiment is not producing high-quality results, it is worthwhile to test different membranes to optimize the conditions. The most commonly used membranes for immunoblotting are nitrocellulose, polyvinylidene difluoride (PVDF), and nylon.

Some of the main characteristics of each type of membrane are as follow.

Nitrocellulose

Nitrocellulose is the standard membrane used for immunoblotting. It exhibits good binding capacity, between 80 μg/cm^2 and 250 μg/cm^2. Proteins interact with the membrane through hydrophobic interactions or possibly by hydrogen bonding between amino acid side chains and amino groups of the membrane. Nitrocellulose membranes tend to produce low background because they are readily blocked during the preincubation step. General blocking strategies are mentioned below and are explained in detail in UNIT 8.3. Blocking an immunoblot is analogous to the process of blocking a Southern or northern blot. By pretreating the blot with agents such as protein or detergent, the sites on the membrane that are not occupied with protein following the transfer step are coated or "blocked," thus preventing antibodies from binding to these sites nonspecifically. The limitations of nitrocellulose include a relatively weak association of the proteins with the membrane. As a result, during the subsequent processing steps, following transfer, some of the protein may be lost, leading to a reduction in signal intensity. Additionally, the membrane is brittle and can easily crack or break, so it is not recommended for subsequent probing, making it difficult to assay for multiple proteins on a single blot.

Nylon

Nylon membranes are very sturdy and can be probed multiple times with different antibodies. Additionally, the membrane has a high binding capacity, 150 to 200 μg/cm^2. Nylon membranes can be charged or display other chemical moieties that increase the association of the transferred proteins through electrostatic interactions in addition to hydrophobic interactions. However, one of the drawbacks of using a nylon membrane is that the unoccupied sites on the membrane are difficult to block. As a result, nylon membranes tend to have greater background problems. To address this issue, a more extensive blocking protocol or special blocking reagents can be considered.

Polyvinylidine difluoride (PVDF)

PVDF membranes are perhaps the most commonly used membrane for immunoblotting experiments. PVDF membranes have a high binding capacity (170 μg/cm^2) and exhibit strong hydrophobic interactions with proteins, resulting in tight binding. As a consequence, less protein is lost during processing. PVDF membranes have good mechanical strength and chemical stability, and are compatible with most detection procedures. In addition, proteins blotted to this membrane can be excised from the blot and processed for subsequent analysis, including mass spectrometry and sequencing. Having all these nice features in one membrane usually comes with a drawback, which in this case is its high cost; PVDF membranes tend to be the most expensive choice.

When preparing a membrane for the transfer step it is important to prewet the membrane. This will produce an even interface between the gel and the membrane to facilitate maximal retention of protein during the blotting procedure. With nitrocellulose and nylon membranes, this can be achieved by immersing the membranes slowly into a container with water and then into transfer buffer. With PVDF membranes, due to their hydrophobic nature, it is necessary to prewet the membrane in methanol first, followed by equilibration in transfer buffer. The membrane is now ready for the transfer protocol.

Transfer

Once your sample has been resolved by PAGE and a membrane has been chosen, the transfer can be carried out. This step can be achieved using different methods (Sambrook and Rusell, 2001; Kurien and Scofield, 2006). One approach utilizes simple diffusion, which involves the same basic principles explained for the transfer of DNA or RNA onto a membrane when discussing Southern and northern blots. A membrane is placed in contact with the gel and a stack of dry filter papers is placed on top. The advantage of this technique is that several blots can be obtained from the same gel. For example, bidirectional transfer can be achieved if the gel is sandwiched between two membranes with filter papers on both sides. Because transfer efficiency is only between 20% to 50%, more than one membrane can be blotted consecutively. As a result, several blots of the gel are obtained and each membrane can then be assayed with a different antibody. Sufficient protein may be left behind in the gel to allow for staining with Coomasie blue. This will allow the investigator to ascertain where the protein of interest resolved in the original gel. By superimposing the immunoblot with the stained gel, the band of interest can be identified, excised, eluted from the gel, and identified by mass spectrometry or sequencing. This transfer technique is not widely used when quantitative transfer of the protein is important for the analysis.

Electroblotting is the method of choice for the transfer of proteins from an acrylamide gel onto a membrane. It is much faster and far more efficient than capillary transfer, such that quantitative transfer of the proteins onto a membrane is possible. The method involves immersing the protein gel sandwiched by a membrane and filters into a chamber of transfer buffer. An electric current aligned perpendicular to the gel is applied so that the negatively charged proteins migrate towards the positive electrode and become trapped by the membrane that was placed on the side of the gel facing the anode. Alternatively, there is a method called eastern blotting which is modified so that the proteins are denatured with a detergent know as cetyltrimethylammonium bromide (CTAB), leading to a net positive charge compared to the negative charge achieved with SDS. As a result, proteins bound by CTAB will migrate to the cathode instead of the anode. This method is only used when working with highly charged proteins or glycoproteins that contain a high number of negatively charged carbohydrates that do not behave well under the standard electrophoresis and transfer conditions that make use of SDS (Buxbaum, 2003).

Blocking

Once the proteins have been transferred to a solid support, the transfer apparatus is disassembled and the membrane removed. Because the proteins are tightly associated with the membrane, a cross-linking step is not required. At this point, the membrane can be assayed immediately or stored. Prior to the addition of a primary antibody, the blot has to be blocked. Typically, this involves incubating the blot in a solution of protein, such as 5% low-fat dried milk or 5% BSA (bovine serum albumin), and detergents such as Tween 20. The most common blocking solution used is dried milk, which is the least expensive and compatible with most immunological detection systems used. As described in *UNIT 8.3*, the blocking solution uniformly coats the membrane and thereby reduces the level of nonspecific binding during the following step. If the blocking procedure is skipped or done inefficiently, the antibodies used in the detection step will associate with the unblocked sites on the membrane resulting in a high background. The blocking step is essential to reduce the level of background and to improve the signal to noise ratio.

Detection

Once the blocking step has been completed, the first step of immunodetection involves adding an antibody (monoclonal or polyclonal) that reacts with the epitope of interest. This reagent is referred to as the primary antibody. After incubating the blot with the primary antibody, the blot is thoroughly washed and a secondary antibody added. The secondary antibody recognizes epitopes associated with the primary antibody and not the proteins in the sample. If the primary antibody was raised against the target of interest in a mouse, the secondary antibody will likely be an anti-mouse antibody raised in another organism like goat, pig, or rabbit. Alternatively, one can use other factors such as Protein A, which will tightly associate with the primary antibody. Protein A is a cell wall component of *Staphylococcus aureus* that binds to the constant region of antibodies, mainly thru hydrophobic interactions. Because of the great affinity of Protein A for antibodies, it has been extensively used in a variety of immunological approaches. To facilitate detection of the protein of interest in an immunoblotting experiment, Protein A or the secondary antibody is conjugated to a compound that allows the investigator to visualize the location of the bound protein. The conjugated compound can include radioactive or enzyme moieties. An example of a radioactive conjugate is ^{131}I and ^{125}I. These compounds are readily detected by X-ray film or phosphorimaging. Examples of enzyme conjugates include horseradish peroxidase or alkaline phosphatase, which are less hazardous than radioactive conjugates and are very sensitive (Blake et al., 1984; Knecht and Dimond, 1984). Each of these enzymes, when presented with the necessary substrate, can facilitate a localized reaction to produce a visible band. Additionally, there are commercially available kits that utilize chemiluminescent reagents that luminesce when cleaved by the enzyme conjugates. This fluorescence luminesce can be detected with X-ray film, digital imaging, or phosphorimaging. Please refer to *UNITS 8.4 & 7.5* for specific information on detection systems.

LITERATURE CITED

Ageno, M., Dore, E., and Frontali, C. 1969. The alkaline denaturation of DNA. *Biophys. J.* 9:1281-1311.

Alwine, J.C., Kemp, D.J., and Stark, G.R. 1977. Method for detection of specific RNAs in agarose gels by transfer to diazobenzyloxymethyl-paper and hybridization with DNA probes. *Proc. Natl. Acad. Sci. U.S.A.* 74:5350-5354.

Alwine, J.C., Kemp, D.J., Parker, B.A., Reiser, J., Renart, J., Stark, G.R., and Wahl, G.M. 1979. Detection of specific RNAs or specific fragments of DNA by fractionation in gels and transfer to diazobenzyloxymethyl paper. *Methods Enzymol.* 68:220-242.

Amasino, R.M. 1986. Acceleration of nucleic acid hybridization rate by polyethylene glycol. *Anal. Biochem.* 152:304-307.

Ausubel, F.M., Brent, R., Kingston, R.E., Moore, D.D., Seidman, J.G., Smith, J.A., and Struhl, K. (eds.) 2007. Current Protocols in Molecular Biology. John Wiley & Sons, Hoboken, N.J.

Bittner, M., Kupferer, P., and Morris, C.F. 1980. Electrophoretic transfer of proteins and nucleic acids from slab gels to diazobenzyloxymethyl cellulose or nitrocellulose sheets. *Anal. Biochem.* 102:459-471.

Blake, M.S., Johnston, K.H., Rusell-Jones, G.J., and Gotschlich, E.C. 1984. A rapid, sensitive method for detection of alkaline phosphatase-conjugated anti-antibody on Western blots. *Anal. Biochem* 136:175-179.

Botstein, D., White, R.L., Skolnick, M., and Davis, R.W. 1980. Construction of a genetic linkage map in man using restriction fragment length polymorphisms. *Am. J. Hum Genet.* 32:314-331.

Burnette, W.N. 1981. "Western blotting": Electrophoretic transfer of proteins from sodium dodecyl sulfate - polyacrylamide gels to unmodified nitrocellulose and radiographic detection with antibody and radioiodinated protein A. *Anal. Biochem.* 112:195-203.

Buxbaum, E. 2003. Cationic electrophoresis and electrotransfer of membrane glycoproteins. *Anal. Biochem.* 314:70-76.

Chu, G., Vollrath, D., and Davis, R.W. 1986. Separation of large DNA molecules by contour-clamped homogeneous electric fields. *Science* 234:1582-1585.

Church, G.M. and Gilbert, W. 1984. Genomic sequencing. *Proc. Natl. Acad. Sci. U.S.A.* 81:1991-1995.

Coico, R., Sunshine, G., and Benjamini, I. 2003. Immunology: A short course. John Wiley & Sons, New York.

Denhardt, D.T. 1966. A membrane–filter technique for the detection of complementary DNA. *Biochem. Biophys. Res. Commun.* 23:641-646.

Ellington, A. and Pollard, J.D. 1998. Purification of oligonucleotides using denaturing polyacrylamide gel electrophoresis. *Curr. Protoc. Mol. Biol.* 42:2.12.1-2.12.7.

Finney, M. 2000. Pulsed-field gel electrophoresis. *Curr. Protoc. Mol. Biol.* 51:2.5B.1-2.5B.9.

Gallagher, S., Winston, S.E., Fuller, S.A., and Hurrell, G.R. 2004. Immunoblotting and Immunodetection. *Curr. Protoc. Mol. Biol.* 66:10.8.1-10.8.24.

Gusella, J.F., Wexler, N.S., Conneally, P.M., Naylor, S.L., Anderson, M.A., Tanzi, R.E., Watkins, P.C., Otinna, K., Wallace, M.R., Sakaguchi, A.Y., Young, A.B., Shoulson, I., Bonilla, E., and Martin, J.B. 1983. A polymorphic DNA marker genetically linked to Huntington's disease. *Nature* 306:234-238.

Johnson, D.A., Gautsch, J.W., Sportsman, J.R., and Elder, J.H. 1984. Improved technique utilizing nonfat dry milk for analysis of proteins and nucleic acids transferred to nitrocellulose. *Gene Anal. Tech.* 1:3-12

Kelly, T.J. Jr. and Smith, H.O. 1970. A restriction enzyme from *Hemophilus influenzae*. II. *J. Mol. Biol.* 51:393-409.

Klenow, H. and Henningsen, I. 1970. Selective elimination of the exonuclease activity of the deoxyribonucleic acid polymerase from *Escherichia coli* B by limited proteolysis. *Proc. Natl. Acad. Sci. U.S.A.* 65:168-175.

Knecht, D.A. and Dimond, R.L. 1984. Visualization of antigenic proteins on western blots. *Anal. Biochem.* 136:180-184.

Kurien, B.T. and Scofield, R.H. 2006. Western blotting. *Methods* 38:283-293.

Laemmli, U.K. 1970. Cleavage of structural proteins during the assembly of the head of bacteriophage T4. *Nature* 227:680-685.

Lehrach, H., Diamond, D., Wozney, J.M., and Boedtker, H. 1977. RNA molecular weight determinations by gel electrophoresis under denaturing conditions, a critical reexamination. *Biochemistry* 16:4743-4551.

McDonell, M.W., Simon, M.N., and Studier, F.W. 1977. Analysis of restriction fragments of T7 DNA and determination of molecular weights by electrophoresis in neutral and alkaline gels. *J. Mol. Biol.* 110:119-146.

McKimm-Breschkin, J.L. 1990. The use of tetramethylbenzidine for solid phase immunoassays. *J. Immunol. Methods* 135:277-280.

Meinkoth, J. and Wahl, G. 1984. Hybridization of nucleic acids immobilized on solid supports. *Anal. Biochem.* 138:267-284.

Nathans, D. and Smith H. 1975. Restriction endonucleases in the analysis and restructuring of DNA molecules. *Annu. Rev. Biochem.* 44:273-293.

Nierzwicki-Bauer, S.A., Gebhardt, J.S., Linkkila, L., and Walsh, K. 1990. A comparison of UV cross-linking and vacuum baking for nucleic acid immobilization and retention. *Biotech.* 9:472-478.

Olszewska, E. and Jones, K. 1988. Vacuum blotting enhances nucleic acid transfer. *Trends in Genetics* 4:92-94.

Reed, K.C. and Mann, D.A. 1985. Rapid transfer of DNA from agarose gels to nylon membranes. *Nucl. Acids Res.* 13:7207-7221.

Rommens, J.M., Iannuzzi, M.C., Kerem, B., Drumm, M.L., Melmer, G., Dean, M., Rozmahel, R., Cole, J.L., Kennedy, D., Hidaka, N., Zsiga, M., Buchwald, M., Riordan, R.R., Tsui, L., and Collins, F.S. 1989. Identification of the cystic fibrosis gene: Chromosome walking and jumping. *Science* 245:1059-1065.

Sambrook, J., Fritsch, E.F., and Maniatis, T. 1989. Molecular cloning: A Laboratory Manual, Second Edition. Cold Spring Harbor Press, Cold Spring Harbor, NY.

Sambrook, J. and Rusell, D.W. 2001. Molecular cloning: A Laboratory Manual, Third Edition. Cold Spring Harbor laboratory Press, Cold Spring Harbor, NY.

Southern, E.M. 1975. Detection of specific sequences among DNA fragments separated by gel electrophoresis. *J. Mol. Biol.* 98:503-517.

Southern, E.M. 2000. Blotting at 25. *Trends Biochem. Sci.* 25:585-588.

Suggs, S.V., Hirose, T., Miyake, T., Kawashima, E.H., Johnson, M.J., Itakura, K., and Wallace, R.B. 1982. *In* Developmental Biology Using Purified Genes (D. Brown, ed.) pp. 683-693. Academic: New York, N.Y.

Thomas, P.S. 1980. Hybridization of denatured RNA and small DNA fragments transferred to nitrocellulose. *Proc. Natl. Acad. Sci. U.S.A.* 77:5201-5205.

Towbin, H., Staehelin, T., and Gordon, J. 1979. Electrophoretic transfer of proteins from polyacrylamide gels to nitrocellulose sheets: Procedure and some applications. *Proc. Natl. Acad. Sci. U.S.A.* 76:4350-4354.

Wahl, G.M., Stern, M., and Stark, G.R. 1979. Efficient transfer of large DNA fragments from agarose gels to diazobenzyloxymethyl-paper and rapid hybridization by using dextran sulfate. *Proc. Natl. Acad. Sci. U.S.A.* 76:3683-3687.

Wallace, R.B., Shaffer, J., Murphy, R.F., Bonner, J., Hirose, T., and Itakura, K. 1979. Hybridization of synthetic oligodeoxyribonucleotides to $\Phi\ \chi$ 174 DNA the effect of single base pair mismatch. *Nucl. Acids Res.* 6:3543-3547.

UNIT 8.2

Nucleic Acid Blotting: Southern and Northern

Laura L. Mays Hoopes[1]
[1]Pomona College, Claremont, California

OVERVIEW AND PRINCIPLES

Southern versus Northern Blotting

E.M. Southern invented DNA blotting (Southern, 1975), and it spread throughout the molecular biology community like wildfire. His method enabled scientists to locate DNA sequences of interest on blots they had printed from agarose electrophoresis gels (see *UNIT 7.2*), where the DNA fragments were separated by size. Before this method became available, investigators had to cut up the gel into segments and test each nucleic acid band separately by liquid hybridization. The ability to detect DNA sequences by testing them simultaneously in Southern blotting was quite an improvement over the earlier methods. The innovation was that, as Southern said, "sequences can be detected as sharp bands."

Northern blotting was named to contrast with Southern blotting (molecular biology humor). In northern blotting, RNA rather than DNA is separated by size on electrophoresis gels (Alwine et al., 1977) before the blot is made. Major differences between northern and Southern blotting are summarized in Table 8.2.1. For Southern blotting, most applications involve cutting the DNA before electrophoresis, using restriction nucleases that recognize short specific DNA sequences. One popular application of Southern blotting is the identification of suspects from crime scene samples (see Internet Resources). In northern blotting, the RNA already occurs in short pieces, making cutting by nucleases before electrophoresis unnecessary. Northern blots can be used, for example, to study the synthesis of RNA in gene expression.

The primary purpose of nucleic acid blotting is to demonstrate the presence of a particular sequence of bases in the sample. The sequences are revealed by hybridizing the blot containing target DNA with a labeled nucleic acid probe (see *UNIT 8.4*) containing the base sequence of interest and then detecting the labeled sites. "Signal" is the indication that a particular probe is present. The signal is detected using fluorescence, chemiluminescence, or radioactivity detectors, depending on the type

Table 8.2.1 Major Differences Between Southern and Northern Blotting

	Southern	Northern
Nucleic acid separated by electrophoresis	Genomic DNA restriction digest	Total RNA
Time of target nucleic acid denaturation	In gel, after electrophoresis	In sample buffer, before electrophoresis
Molecular weight standards	Double-stranded DNA fragments of known sizes	Single-stranded RNA fragments of known sizes stabilized with RNase inhibitors
Problems encountered	Insufficient DNA concentration Incomplete digestion with restriction endonucleases	Degraded RNAs

Table 8.2.2 Methods for Detection of Labeled Probes for Southern and Northern Blots

Method	Molecular basis	Product	References
Colorimetry	Enzyme linked to affinity reagent	Colored compound	Boyle and Perry-O'Keefe (1992)
Chemiluminescence	Enzyme linked to affinity reagent	Chemiluminescent compound	Perry-O'Keefe and Kissinger (2000)
Autoradiography	Exposure of X-ray film	Spot on developed film	Sambrook et al. (1989)

label used in the probe. Methods used for detection of probes hybridized to the DNA on the blots are found in Table 8.2.2.

The second purpose of blotting is to show the sizes of the molecules that can base-pair with the labeled probe. That information comes from the log-inverse size dependency of the electrophoresis migration rate: the larger molecules move through the gel more slowly, while the smaller ones are faster. A standard curve is used to determine size (see *UNIT 7.2*).

Blotting techniques constitute one of the three most important methods in molecular biology (see *UNIT 8.1* for an overview), along with DNA sequencing (*UNIT 10.4*) and the polymerase chain reaction (*UNIT 10.2*). Together these three methods have had a revolutionary effect on the field.

Hybridization of Probe and Target

The results obtained in blotting depend on the recognition and pairing rate of the nucleic acid sequences (hybridization kinetics). In solution hybridizations, the typical unit of measure for hybridization rate is the initial concentration of the DNA (C_0) multiplied by the time of hybridization (t), or C_0t (pronounced "cot"). In blotting, the hybridization of probe and target is often observed after a single timed incubation (i.e., C_0t has t as a constant); therefore, the signal at the end of the experiment depends on the target nucleic acid sequence concentration (C_0). Other variables (known from liquid hybridization experiments) can also influence hybridization and affect the efficiency of probe-target base pairing, including the concentration of probe, length of matching and nonmatching sequences in the probe, secondary structure of the blot target and probe, temperature and time of hybridization, and ionic strength and hydrophobicity of the solvent used in the hybridization reaction (see Table 8.2.3). These factors are often standardized in a laboratory, but can be optimized for difficult targets on blots.

Repetitive Sequence Effects

A particularly important factor in the outcome of nucleic acid blotting is the presence and behavior of repeated sequences. Bacterial DNA is free of such problems, and simpler eukaryotes such as budding yeast have few repetitive sequences. However, in mammalian DNA, over half of the genomic DNA consists of repetitive sequences. A sequence with many copies per genome is at a higher concentration per unit of nucleic acid than a sequence with one copy per genome.

In mammalian genomic DNA that is cut into pieces 1000 bp in length, boiled to separate the strands, and allowed to reassociate in a solution hybridization experiment, three or more major rates of base pair formation typically are found (see Table 8.2.4). Highly repetitive DNA is first to base-pair (high C_0t or, for calf thymus, C_0t1), followed by middle repetitive sequences (intermediate C_0t), and finally the unique sequences (slowly hybridizing or low C_0t). In blotting experiments, high C_0t DNA (i.e., repeated sequences) will produce a strong band that may spread over the whole blot, making it impossible to see signal from rarer sequences. Low C_0t DNA (rarely occurring sequences often of most interest) will produce a weak band or no visible band. For hybridization

Table 8.2.3 Variables Affecting the Hybridization Reaction on Blots

Variable	Comment
Length of matching sequences	Longer is better due to more base pairing
Length of nonmatching sequences	Shorter is better due to less strain on paired bases
Secondary structure of target and/or probe	Less is better due to availability of bases to pair
Temperature of hybridization reaction	Optimum exists; lower is better for short, mismatched sequences
Time of hybridization	Optimum exists; 16 hr typical
Ionic strength	Optimum exists, but at typical concentrations, higher is better
Hydrophobicity of solvent (chaotropic effects of solvent mixture)	Higher chaotropic effects enable same hybridization with a lower temperature due to easier dissociation and reassociation

Table 8.2.4 Classification of Mammalian DNA According to Sequence Repetitions

Class	Example	Frequency
Highly repetitive	Centromeric repeats	>100,000 copies per genome
Middle repetitive	LINE elements, *Alu* sequences	Thousands per genome
Unique	Genes for globin, actin, or triose phosphate isomerase	One or a few per genome

of mammalian DNA, it is generally necessary to add unlabeled repetitive sequences (e.g., C_0t1 DNA, the most repetitive fraction of DNA from calf thymus or a similar organism, plus mixed tRNAs) to the reactions to avoid a high degree of smeary-looking hybridization from the repetitive DNA on the blots. This "cold" or unlabeled nucleic acid will hybridize with the target repetitive sequences, which will then not capture as much of the labeled probe. Repetitive sequences are not perfect repeats and are more variable in size after restriction nuclease cutting. If you have a lot of hybridization to them you will notice that the bands don't look sharp.

To the investigator performing blotting experiments, the presence of the target tends to be more important than its exact concentration, which is fortunate because nucleic acid blotting is not reliably quantitative in most cases. However, blotting can give good relative concentration estimates for targets on the same blot, depending on experimental conditions.

Stability and Denaturation of Target Sequences on Blots

DNA is double-stranded and quite stable (as it should be because it has to store the genetic information). For Southern blots, denaturation of the DNA (i.e., separation of the two strands) takes place late in the procedure (see Fig. 8.1.1 in *UNIT 8.1*) because the double-stranded DNA fragments need to be intact so that the fragment sizes will be correct for the electrophoresis step. The DNA samples loaded onto gels are usually first digested with a restriction endonuclease (*UNIT 10.1*). Each restriction enzyme cuts a specific double-stranded nucleotide sequence and the sizes of the DNA fragments that they produce are characteristic for a particular DNA sample. The double-stranded fragments are electrophoresed, but the two strands in each fragment must be separated (denatured) in order to stick very well to most blot membranes and hybridize with a probe.

RNA is already single stranded. It is much less stable than DNA for a number of reasons: It has a ribose ring with two adjacent hydroxyl groups. In chains of RNA, one of these ribose rings is linked with an adjacent ribose by a phosphodiester bond. Each sugar or ribose ring in the RNA has a free hydroxyl group unlike in DNA, which contains "deoxy" ribose. This hydroxyl group is a source of instability because it can attach the bond next to it and make a transient cyclic phosphodiester; when that happens, the RNA chain is broken (see Fig. 5.2.2 in *UNIT 5.2*). Metal ions or basic pH speed up the breakdown of RNA. The breakdown reaction is also catalyzed by RNase, the "enzyme from hell," which is very stable and can remain active even after boiling for 10 min. RNase is very available from fingerprints, bacteria in the air, dandruff, dust, paper, "pure" water, and many other sources. To accurately analyze the sizes of RNA samples by electrophoresis and northern blotting, it is essential to avoid breakage, so special methods must be employed, referred to here as "RNA paranoia" (see below).

Although RNA is single stranded, it has a tendency to bend back on itself and form base-paired loops, hairpins, cruciforms, and other secondary structures. These structures affect its migration in electrophoresis; denaturing agents (e.g., formamide) must be added to the electrophoresis buffer and sample loading buffer to prevent the formation of secondary structures (see discussion of denaturation in Strategic Planning). In this way, there will be accurate separation of the RNA by size. Furthermore, because RNA is so fragile, it is likely that some molecules that might have appeared to be "normal" in size without denaturing agents will fall apart due to broken backbone regions when denaturing agents are added. Therefore, electrophoresis of RNA without denaturation can give only a rough approximation, overestimating RNA's intactness, compared with denaturing electrophoresis. It is easy to detect the large and small ribosomal RNA bands in good preparations of RNA; their absence (or presence in smaller than expected size) is a good indication that the RNA has been extensively broken, and sizes estimated from the blot will not be those of intact RNA molecules.

STRATEGIC PLANNING

Southern Blotting

DNA samples

In planning an experiment using Basic Protocol 1, it is important to use the same concentration of DNA in lanes that are to be compared. It is even better to include different dilutions of DNA from each of the samples to enable to comparisons between lanes with very similar loads for each sample type. Another useful control is one to make sure the restriction enzyme is still active and is not inhibited by anything in the DNA sample. A subsample of the restriction enzyme reaction can be removed after it is set up, and an appropriate amount of control DNA, such as lambda virus, can be added. After the digestion, samples containing the viral DNA can be run on a minigel to see if the sample includes correct bands for digested viral DNA for the restriction enzyme being used (see *UNIT 10.1* for an overview), demonstrating that the enzyme was active in the DNA sample reaction tube. For example, one can predict the fragment sizes by downloading the sequence for lambda DNA from the NCBI nucleotide database and use software to locate all sites for the enzyme used or obtain this information from commercial catalogs. For genomic digestions, it is almost always safe to digest the DNA with enzyme overnight.

Agarose

The agarose selected for electrophoresis should be appropriate for the expected sizes of the DNA fragments or RNA molecules and the fate of the gel (also see *UNIT 7.2*). Some types of agarose used to separate small fragments do not allow effective capillary transfer of the nucleic acid to the membrane (read the fine print). If the fragments are too small to separate well on an agarose that allows capillary transfer, acrylamide gels can be used with electrophoretic transfer (see Alternate Protocol).

Molecular weight standards

Appropriate molecular weight standards are chosen so that well defined bands occur above and below the size expected for the fragments of interest. After the samples are electrophoresed, the gel is stained and photographed. A ruler is used in the photograph so that the final blot positions of bands can be compared with the known standard fragments, which often won't be detected by the probe used. Alternatively, commercially available standard fragments labeled so that the detection method will reveal them can be used. That method was used in the Southern blot shown in Figure 8.2.2A, where the standard is on the right side of the blot.

Membrane selection

The blot membrane can be nitrocellulose or one of a variety of charged and uncharged nylon membranes. Most people today select nylon membranes, but some still prefer nitrocellulose because of its lower background. The disadvantage of a nitrocellulose membrane is that it is much more fragile and easily broken (see *UNIT 8.1*).

Denaturation of double-stranded DNA

When the DNA has been electrophoretically separated in the double-stranded form, its strands are separated in the gel (denaturation) before transferring it to the blot. This step enables it to bind better to the blot medium, and it also leaves the DNA ready to recognize the probe via base-pairing. DNA is denatured in a basic solution since the alternative, heating to between 95°C and 100°C, would melt the gel and destroy the positional information it contains. After soaking in dilute base, the gel is neutralized to facilitate transfer, but the low temperature and dilute salt solution are not optimal for reforming base pairs, so the DNA remains single stranded.

Blot rinsing

When the blot apparatus is disassembled, the blot is briefly rinsed in 2× SSC to remove any flakes of agarose that would block hybridization. Because the sample is not yet fixed to the blot membrane, some DNA is lost in this step.

Cross-linking

The DNA is fixed to nylon by means of cross-linking the blot in a UV irradiator. If nitrocellulose membranes are used, the DNA is cross-linked to the blot by baking in a vacuum oven at 80°C for 2 hr. The vacuum is required because nitrocellulose explodes if heated in air. The DNA side should be up during the vacuum drying to avoid loss of DNA.

Hybridization conditions

The selection of the hybridization solution is based to some extent on trial and error. The solution described in Reagents and Solutions is a good starting point. If the hybridization is too weak, then the temperature can be lowered, the concentration of chaotropes (e.g., formamide) in the hybridization mixture can be lowered, or the ionic concentration (of cations) can be increased during hybridization. If the background (signal that is not on the bands but spread over other areas of the blot) is too high, then the temperature can be increased and/or more unlabeled repeated sequence DNAs added. Refer to Table 8.2.2 for conditions that can be manipulated and their effects on hybridization; also see *UNIT 8.1*.

Wash conditions

The signal from the probe seen on the hybridized gel will reflect both the blot hybridization conditions and the washing conditions. The stringency of the washes may need to be increased if there is a lot of background. Stringency is a measure of the difficulty of base pairing; high stringency (less salt and detergent and/or higher temperature) will disrupt base pairing except for long, perfectly matched nucleic acids. See *UNIT 8.1*.

Northern Blotting

Much of the technique for northern blotting is the same as that for DNA blotting. However, when using Basic Protocol 2, there are a few special considerations described below.

RNA paranoia

For northern blotting, RNA paranoia is very important from start to finish. The work area should be cleaned with RNase inhibitors. Several spray versions are commercially available (e.g., RNase Erase from QBiogene. Gloves should be changed if non-RNase-free items have been touched (e.g., your hair, your face, your arm, notebook paper). All equipment should have been ordered nuclease-free or baked (8 hr at 120°C) to get rid of RNase.

Water or dilute buffers can be treated with DEPC (see Reagents and Solutions) to eliminate RNase. DEPC combines with RNase and inactivates it irreversibly. It breaks down in dilute solutions, and solutions can be autoclaved to get rid of the DEPC after the treatment. Tris buffers cannot be freed of RNase in this way because Tris reacts with DEPC. It is recommended to use alternative buffer such as MOPS for RNA work. Water used for making solutions can be treated with DEPC, or nuclease-free water is commercially available at a reasonable price.

The electrophoresis apparatus used should be dedicated to RNA work, not used for DNA or protein electrophoresis experiments. It should be cleaned with RNase inhibitor (e.g., RNase Erase) and isopropanol and stored where dust will not accumulate on it. The area where northern electrophoresis is performed needs to be within a fume hood. The RNA is still unstable after electrophoresis, so all trays, buffers, transfer solutions, stain solutions, etc., need to be cleaned in the same manner as the electrophoresis apparatus and kept nuclease free. Do not let RNA analysis equipment sit around in the sink waiting to be cleaned up or it may become heavily contaminated with RNase.

It is not difficult to detect degradation in stained RNA. Take a look at the gel shown in Figure 8.2.3 to see what degraded RNA looks like. The rRNA is not usually of interest, but it is the major portion in total RNA and is similar in size to the mRNA. Thus, if the rRNA bands are too small, too light, or not distinguishable, it is a good indication that the mRNA is also damaged.

Denaturation

The northern blotting procedure calls for the RNA to be electrophoresed in its denatured, unwound state and compared to standards that are also single stranded and unwound. Thus, formamide (which destabilizes the foldback helices by making the base-pairing less energetically favorable; Sullivan and Lillie, 1988) is included in the sample buffer to denature the RNA. A gentle heating step is also used before loading, to ensure complete denaturation.

Hybridization of RNA with DNA probes

RNA-DNA base-paired double helices are more stable than DNA-DNA double helices, so there can be a more significant problem with background on northern versus Southern blots. Often, the conditions for washing a northern blot will appear harsh to people accustomed to Southern blots, but these stringent conditions as described in the protocols (e.g., note the higher wash temperatures for the northern blots) are required in order to avoid high nonspecific background.

SAFETY CONSIDERATIONS

Nucleic Acids with Dangerous Sequences

If analyzing human oncogenes or dangerous viruses use the precautions specified by governmental safety manuals (see Internet Resources). Such methods require avoiding the generation of aerosols in centrifugation and pipetting, collecting waste for safe disposal, and using self-protective clothing and devices.

Electrophoresis and Ethidium Bromide Staining

Refer to *UNIT 7.2* for precautions to avoid electrocution and prevent exposure to the potential carcinogen, ethidium bromide, which is often used to stain agarose gels. Also see *APPENDIX 1*.

Formamide and Formaldehyde

Formamide and formaldehyde are also compounds that require care in use and disposal. Formaldehyde must be used only in a fume hood; both compounds require proper storage of wastes and transfer to a hazardous waste manager. Refer to Internet Resources for obtaining material safety data sheets for these and any other potentially hazardous compounds that are used.

PROTOCOLS

Basic Protocol 1: Southern Blotting

This Southern blotting protocol is divided into four stages after running the gel (see *UNIT 7.2*):

1. treatment of the gel before it is blotted;
2. DNA transfer to a membrane;
3. DNA cross-linking;
4. hybridization and washing of the blot.

The starting material for this protocol is an agarose gel containing DNA fragments of interest. The fragments have been generated by restriction endonuclease digestion (*UNIT 10.1*), separated by agarose gel electrophoresis, and stained, photographed, and destained (see *UNIT 7.2*) before beginning this protocol. Procedures are given for using either a nylon or nitrocellulose membrane to create the blot.

Materials

Agarose gel containing electrophoretically separated DNA fragments that have been stained, photographed, and destained (see *UNIT 7.2*)
Depurinator: 0.25 M HCl (store up to 6 months at room temperature in an air-tight bottle)
Distilled water
Denaturant: 1.5 M NaCl/0.5 M NaOH (store up to 6 months at room temperature)
Neutralizer: 1.5 M NaCl/0.5 M Tris·Cl, pH 7.0 (store up to 6 months at room temperature)
2× and 20× SSC (see *UNIT 3.3*)
Prehybridization buffer with or without 50% formamide (see recipes), 42°C or 68°C
Double- or single-stranded probes (appropriate to the experiment; see *UNIT 8.4*)
Hybridization buffer with or without formamide (depending on hybridization temperature; see recipes)
2× SSC (see *UNIT 3.3*) containing 0.5% (w/v) and 0.1% (w/v) SDS
0.1× SSC (see *UNIT 3.3*)/0.5% SDS

Glass dishes (e.g., Pyrex baking pans) for washing gels
Rocking or gyrating platform with low speed setting
Gloves
10- or 25-ml glass pipet
Plastic wrap
Whatman 3MM filter paper
Paper towels
500-ml flask
UV cross-linker or UV transilluminator with 254-nm wavelength light (for nylon membranes), *or* vacuum oven set to 80°C (for nitrocellulose membranes)
Reclosable plastic bags (larger than the membranes)

Hybridization bottles (or sealable plastic bag)
Hybridization oven (or water bath), set to 42°C or 68°C
15-ml disposable, sterile test tube
Long forceps

Additional reagents and equipment for assembling the blotting transfer apparatus (Support Protocol)

CAUTION: Wear gloves to protect hands from denaturant and to protect the membrane from contamination.

Southern stage one: DNA fragmentation and denaturation

1. Place the gel in a clean glass dish.

2. *Optional, but recommended:* Recall the volume of agarose you used to prepare the gel, and put ~10× that volume of depurinator into the dish. Place on a gentle rocker for 30 min at room temperature.

   This depurination step will produce shorter DNA sequences, which will hybridize more successfully with the probe. This step may not be necessary if you took a prolonged sequence of pictures of the ethidium bromide–stained gel using UV illumination; the UV light will have caused breakage, producing short fragments that will hybridize efficiently. But if you don't fragment the DNA at all, the larger pieces may result in low signal in those bands at the end of the experiment. If your expected band is less than 4 kb in length, you probably don't need to fragment.

   This step is called depurination because the solution destabilizes the bond between purine bases and the sugars to which they are attached. Once the purine bases are released, the chain easily breaks at that location, resulting in shorter fragments that are easier to transfer.

3. Using a gloved hand to retain the gel, pour out the depurinator. Rinse the gel and pan gently with distilled water.

   The solutions in this protocol may be poured down the sink drain, assuming the solutions do not contain dangerous sequences or radiolabeled tags, or otherwise present a hazard (also see UNIT 2.3 and APPENDIX 1).

4. Add ~10 vol (same as for the depurinator) denaturant and shake or rock gently for 20 min.

5. Using a gloved hand, dump out the denaturant and replace with same amount of denaturant. Shake again for 20 min.

   The DNA is now single stranded.

6. Pour off the denaturant, rinse gel and pan gently with distilled water, and add ~10 vol neutralizer. Incubate 20 min at room temperature, with gentle shaking.

7. Pour off neutralizer and replace with a similar amount of neutralizer. Incubate 20 min at room temperature, with gentle shaking.

8. During this time, set up the transfer apparatus as described in the Support Protocol.

   By the end of the second neutralization, the gel will be at a pH <9, so the DNA fragments will bind to the membrane.

9. Pour off the second neutralizer solution and rinse with distilled water.

Southern stage two: Capillary transfer

10. Place the gel on the 3MM filter paper on the transfer apparatus assembled in step 8. Note the gel position (top up or bottom up).

    The transfer will work whether you place the gel with the top up or the bottom up; generally the DNA is nearer to the bottom of the gel. You should note the placement so that the position of each sample will be clear later.

11. To remove air bubbles under the gel, roll a clean glass pipet firmly over the gel, like a rolling pin used in baking. If a bubble resists removal, deliver more 20× SSC under the gel with a pipet and roll again.

12. With gloved hands, pick up the hydrated membrane (see Support Protocol), position it so that the cut corner is at the top right, and allow it to roll out onto the gel from one edge to the other. Roll with a pipet to remove air bubbles. Avoid moving the membrane around on the gel to avoid smears on the resulting image.

13. Tear off four strips of plastic wrap and drape them over the blot and gel just inside the edges to frame the gel and membrane. Drape the edges of each strip over the edge of the pan containing the 20× SSC to keep evaporation loss to a minimum.

    These plastic wrap strips are not shown in the figure, and the transfer will work without them, but they do have an important function. This setup forces any liquid coming up from the sponge to go through the gel and membrane to get to the absorbent padding that you are about to pile on top of your stack.

    The directions given in the Support Protocol are for a sponge transfer (see Fig. 8.2.1A for setup). If a platform with a 2-mm paper bridge is used instead (Fig. 8.2.1B), balance a Lucite sheet on the pan of transfer solution and use a 3-mm paper to provide wicking.

14. Stack five sheets of 3MM filter paper and dip them into the 20× SSC solution used to hydrate the blot membrane (in the Support Protocol).

15. Place them, dripping wet, on top of the blot and roll out air bubbles with a pipet.

16. On top of the filter papers, pile at least 2 in. of unfolded paper towels.

    You can be neat and cut them to fit, but as long as you've draped the plastic wrap carefully, they can be larger than the blot medium and not cause any problems.

17. Set a glass or plastic plate on top of the paper towels. Place a weight on top of the plate.

    For a 2% agarose gel, you can place a flask with about 1 liter of liquid on top, but the flask should contain only ~250 ml if using a 0.7% gel. (Lighter is better for lower percentage gels to prevent crushing.)

18. Allow transfer to occur overnight (~16 hr) at room temperature.

19. After the overnight transfer, remove and discard the paper towels and five sheets of 3MM filter paper from the top of the blot.

20. *Optional:* Before removing the blot from the gel, use a pencil or a nonbleeding pen to mark positions of the wells in the original gel.

    This can be quite useful later. Remember that the markings will be on the side of the membrane opposite the DNA.

21. Remove the membrane with gloved hands and rinse briefly in a glass pan filled with 2× SSC.

    This step removes tiny pieces of agarose that may block hybridization later, but you also lose DNA, so keep it fast.

22. Place membrane, DNA side up, on a sheet of 3MM filter paper to dry to an appropriate level.

    Make sure to consult your labeled probe detection method (see Table 8.2.2) to see how dry you want to make the membrane. Some methods require complete drying, others simply require removal of loose droplets, leaving the blot hydrated.

Southern stage three: DNA cross-linking

23a. *For cross-linking on nylon membranes using a UV cross-linker:* Put the nylon membrane with the DNA side up into a UV cross-linker for the amount of time determined by the manufacturer's instructions.

23b. *For cross-linking on nylon membranes using a transilluminator:* Wrap the membrane in UV-transparent plastic wrap and place directly onto a transilluminator that provides 254-nm wavelength light (short UV). Irradiate according to the nylon manufacturer's directions.

If no instructions exist, make a 5 inch long streak by micropipetting a sample of restriction-digested DNA onto the blot material ahead of time. Allow to dry completely. Cut into five pieces and irradiate on the transilluminator for 30 sec, 1 min, 1 min 30 sec, 2 min, and 5 min. Label each piece with its cross-linking time, hybridize the pieces with labeled probe, and rinse as you will do with the final blot. Develop and choose the cross-linking time that gave the best signal.

If cross-linking is not long enough, nucleic acid is lost from the blot during hybridization. If cross-linking is too long, the sample is tied to the membrane and can't hybridize.

23c. *For cross-linking on nitrocellulose membranes:* Place the blot between two pieces of 3 mM filter paper and bake in a vacuum oven at 80°C with the DNA side up for two hr.

Remember to use a vacuum oven for nitrocellulose membrane cross-linking: The vacuum is required because nitrocellulose explodes if heated in air.

24. Proceed directly to hybridization or store each cross-linked blot 1 year at room temperature between two sheets of supporting 3 mM filter paper in a reclosable plastic bag, keeping the blots dry.

Storage at 4°C may result in unwanted condensation on the blot.

Southern stage four: Blot hybridization and washing

25. Place the blot into a hybridization bottle with the DNA side facing away from the glass walls.

Take care not to bend the blot at sharp angles because this may cause cracks that later accumulate labeled probe.

26. To block all of the nonspecific nucleic acid binding sites on the membrane, add prehybridization buffer (with or without formamide, depending on the hybridization temperature to be used) to the bottle, using 1 to 2 ml per 10 cm^2 of membrane used. Incubate the membrane 1 to 2 hr at 42°C (or 68°C), with rolling, in the hybridization oven.

Choose the 42°C hybridization temperature or the 68°C hybridization temperature and use both prehybridization buffer and hybridization buffer appropriate for the hybridization temperature.

Being somewhat generous with the amount of prehybridization buffer in this step can be helpful in ensuring complete blocking.

The hybridization oven maintains constant temperature and rotates the bottles so that the small amount of liquid constantly washes over the blots.

27. Denature a double-stranded probe by boiling for 5 min before being added to the hybridization mixture and.

A double-stranded probe must be made single stranded so that it will base pair with the nucleic acids on the blot. Quickly cooling the denatured probe on ice prevents its reassociation.

If the probe is single-stranded DNA or an oligonucleotide, this is not necessary.

For a genomic DNA Southern blot using a probe labeled with radioactive phosphorus, using 10 μg of DNA with a specific activity of 10^9 cpm/μg is recommended.

28. Using a small disposable test tube, prewarm 1.3 ml per each 10 cm^2 of membrane of the hybridization solution to 42°C or 68°C (for solution with or without formamide, respectively).

29. Just before setting up the hybridization, add the denatured probe to the hybridization solution and mix thoroughly by inverting.

Do not vortex! Most hybridization solutions are so viscous that a vortex would introduce hundreds of unwanted bubbles that would interfere with hybridization.

30. Drain all of the prehybridization buffer from the hybridization bottle without losing the blot.

31. Transfer the hybridization solution with probe to the hybridization bottle. Incubate 6 to 16 hr at the appropriate temperature (42°C or 68°C for solution with or without formamide, respectively).

    The longer you incubate, usually the more signal you will get. However, sometimes getting the answer quickly is more important.

32. Remove the hybridization bottle from hybridization oven and unscrew the cap on one end. Carefully remove blot using long forceps restricted to touching the edges of the blot and place it in a glass dish with 200 ml of 2× SSC/0.5% SDS. Incubate 2 min at room temperature.

33. Drain the solution and replace with fresh 200 ml of 2× SSC/0.5% SDS. Incubate 15 min at room temperature, with gentle rocking.

    If results on similar blots have had high background signal, increase the stringency. One way to increase stringency is to heat this wash to 55°C before use.

34. Warm 2× SSC/0.5% SDS to 37°C and warm 0.1× SSC/0.5% SDS to 55°C (~200 ml for each blot) for use in steps 36 and 37, respectively.

35. With gloved hands, hold the blot while draining 2× SSC/0.5% SDS and replace with 200 ml of 2× SSC/0.1% SDS (room temperature). Rock gently 15 min at room temperature.

    Choosing the best wash protocol is a process of trial and error. The basic procedure works well for probes that are longer than ~40 bp and perfectly matched, but if the probe and the target are not well matched, it may be necessary to skip washes in SSC concentrations below 1× because higher stringencies will result in loss of the wanted base pairing. However, background signal all over the blot will be increased.

36. With gloved hands, hold the blot while draining the wash and replace with 2× SSC/0.5% SDS, prewarmed to 37°. Incubate 30 min at room temperature, with gentle rocking.

37. Drain and replace with 0.1× SSC/0.5% SDS at 55°C.

    It may be necessary to change the stringency of this step. If signal loss is a problem, it is possible to do this wash at room temperature. Alternatively, to reduce background, this wash can be heated to 68°C.

38. Briefly rinse the blot in 0.1× SSC and lay out to dry to the appropriate level on several sheets of 3MM filter paper at room temperature, with the DNA side up.

    Depending on the label detection method, the blot may need to be completely dry or may need to be kept moist (see Table 8.2.2).

39. Proceed to label detection (see Table 8.2.2).

Basic Protocol 2: Northern Blotting

For a northern blot, RNA is first separated on a denaturing agarose gel containing formaldehyde (to maintain the RNA in linear form). A detailed protocol for preparing and running this kind of gel can be found in (*UNIT 7.2*). The northern blotting protocol is divided into three stages after running the gel:

1. RNA transfer to a nylon membrane
2. RNA cross-linking, and hybridization
3. washing of the blot.

The first requisite for a successful northern blot is good quality RNA (see Fig. 8.2.3 for an example). The Troubleshooting section explains about RNA quality control. In performing the northern procedure RNA paranoia is essential throughout (see Strategic Planning): gloves must be worn, all equipment (dishes and pans, electrophoresis equipment, scissors) must be cleaned with RNase inhibitors and alcohol, and all solutions must be treated with DEPC and made in DEPC-treated water (or commercially available nuclease-free water). Equipment specifically dedicated to work with RNA is highly recommended, if financially feasible. Tables should be cleaned with RNase inhibitors, and micropipet tips and tubes must be certified nuclease-free. Paper is often a source of RNase so don't assume paper or wipes are safe.

Materials

DEPC-treated water (see *UNIT 3.3*)
Agarose gel containing electrophoretically separated RNA fragments that have been stained, photographed, and destained (see *UNIT 7.2*)
6× SSC treated with DEPC (see *UNIT 3.3*)
Prehybridization buffer (see recipe), 42°C
Double- or single-stranded probes (appropriate to the experiment; see *UNIT 8.4*)
Hybridization buffer with or without formamide (depending on hybridization temperature; see recipes)
1× SSC/0.1% SDS treated with DEPC (see *UNIT 3.3*)
0.2× SSC/0.1% SDS treated with DEPC (see *UNIT 3.3*), 68°C

Glass dishes (e.g., Pyrex baking pans), RNase free
Rocking or gyrating platform with low speed setting
10- to 25-ml glass pipet, RNase free
Gloves
Plastic wrap
Paper towels
500-ml flask
UV cross-linker or UV transilluminator with 254-nm wavelength light (for nylon membranes), *or* vacuum oven set to 80°C (for nitrocellulose membranes)
Reclosable plastic bags (larger than the membranes)
15-ml sterile disposable plastic test tube
Hybridization bottles
Hybridization oven, set to 42°C or 68°C
Long forceps

Additional reagents and equipment for assembling the blotting transfer apparatus (Support Protocol)

Northern stage one: RNA transfer to membrane

1. In order to remove the toxic formaldehyde from the gels, transfer the gel to a clean, RNase-free glass dish containing ~200 ml DEPC-treated water and incubate 5 min at room temperature, with gentle rocking.

   Formaldehyde gels are more fragile than plain agarose gels and break easily.

2. Repeat this wash three times. Discard the washings as formaldehyde waste (see the Safety Considerations section above).

3. While the gels are washing, set up the transfer apparatus as described in the Support Protocol.

4. Place the gel on the 3MM filter paper on the transfer apparatus (assembled in step 3). To avoid air bubbles under the gel, roll a clean RNase-free pipet firmly over the gel, like a rolling pin used in baking.

5. With gloved hands, pick up the hydrated membrane, position it so that the cut corner is at the top right, and allow it to roll out onto the gel from one edge to the other. Roll with a pipet to remove air bubbles. Avoid moving the membrane around on the gel to avoid smears on the resulting image.

6. Tear off four strips of plastic wrap and drape them over the blot and gel just inside the edges to frame the gel and membrane. Drape the edges of each strip over the edge of the pan (see Fig. 8.2.1A) to keep evaporation loss to a minimum.

   These plastic wrap strips are not shown in the figure, and the transfer will work without them, but they do have an important function. This setup forces any liquid coming up from the sponge will have to go through the gel and membrane to get to the absorbent padding that you are about to pile on top of your stack.

   The directions given in the Support Protocol are for a sponge transfer (see Fig. 8.2.1A for setup). If a platform with a 2-mm paper bridge is used instead (see Fig 8.2.1B), balance a Lucite sheet on the pan of transfer solution and use a 3-mm paper to provide wicking.

7. Stack five sheets of 3MM paper and dip them through the SSC solution used earlier to hydrate the blot medium (see Support Protocol).

8. Place them, dripping wet, on top of the blot and roll out air bubbles with pipet.

9. Pile on top at least 2 in. of unfolded paper towels.

   You can be neat and cut them to fit, but as long as you've draped the plastic wrap carefully, they can be larger than the blot medium and not cause any problems.

10. Set a glass or plastic plate on top of the paper towels. Place a weight on top of the plate.

   For a 1% agarose gel, you can place a flask with ~500 ml liquid on top, but the flask should contain only ~250 ml if using a 0.7% gel. (Lighter is better for lower percentage gels to prevent crushing.)

11. Allow transfer to occur up to overnight (6 to 18 hr) at room temperature.

   The longer transfer will result in a more complete movement of the RNA, but the shorter transfer is often sufficient for abundant RNA species.

12. After the transfer, remove and discard the paper towels and five sheets of 3MM filter paper from atop the blot. With a nonbleeding pen or pencil, mark positions of the wells in the original gel.

   The markings will be on the side opposite the RNA.

   Use RNA paranoia; if you get RNase on the blot now, all the targets can be chewed up.

13. Remove the membrane with gloved hands and rinse briefly in an RNase-free glass pan of 6× SSC.

   This step removes pieces of agarose that may block hybridization later, but you also lose RNA so make it fast.

14. Place membrane, RNA side up, on a sheet of 3MM filter paper to dry to an appropriate level.

   Consult your label detection protocol (see Table 8.2.2) to see how dry you want to make the membrane.

Northern stage two: RNA cross-linking

15a. *For cross-linking using a UV cross-linker:* Place the nylon membrane with the DNA-side up into a UV cross-linker for the amount of time determined by the manufacturer's instructions.

15b. *For cross-linking using a transilluminator:* Alternatively, wrap the membrane in UV-transparent plastic wrap, place directly onto a transilluminator that provides 254 nm wavelength light (short UV), and irradiate according to the nylon manufacturer's directions.

   Nitrocellulose membranes are not used for northern blots. They don't work very well for RNA binding.

16. Proceed directly to hybridization or store each cross-linked blot up to several months at room temperature between two sheets of supporting 3MM filter paper in a reclosable plastic bag, keeping the blots dry.

   Storage at 4°C may result in unwanted condensation on the blot.

Northern stage three: Blot hybridization and washing

17. Place the cross-linked blot into an RNase-free hybridization bottle with the RNA away from the glass walls.

    Take care not to bend the blot at sharp angles because this may cause cracks and eventually image artifacts due to labeled probe accumulation.

18. To block all of the nonspecific nucleic acid binding sites on the membrane, add prehybridization buffer (the same buffer as used for Southern blotting but treated with DEPC to destroy RNase; see *UNIT 3.3*) to the bottle, using 1 to 2 ml per 10 cm^2 of membrane used. Incubate the membrane 1 to 2 hr at 42°C (or 68°C), with rolling, in the hybridization oven.

    Being somewhat generous with the amount of prehybridization buffer in this step can be helpful in ensuring complete blocking.

19. Denature a double-stranded probe by, boiling for 5 min before adding to the hybridization mixture.

    If the probe is single-stranded DNA or an oligonucleotide, denaturation is not necessary.

    For a total RNA northern blot using a probe labeled with radioactive phosphorus, using 0.1 μg of probe with a specific activity of 2×10^8 cpm per μg is recommended.

    Note that since the RNA is single stranded, if you use a single-stranded probe, it must be complementary to the mRNA.

20. Using a small disposable test tube, prewarm 1.3 ml per 10 cm^2 of membrane of the hybridization solution to 42° or 68° (for solution with or without formamide, respectively).

21. Just before setting up the hybridization, add the denatured probe to the hybridization solution and mix thoroughly by inverting.

    Do not vortex! Most hybridization solutions are so viscous that a vortex would introduce hundreds of unwanted bubbles that would interfere with hybridization.

22. Drain out all of the prehybridization buffer from the hybridization bottle without losing the blot.

23. Transfer the hybridization solution with probe to the hybridization bottle. Incubate 6 to 16 hr at the appropriate temperature, (42°C or 68°C for solution with or without formamide, respectively).

    The stronger base pairing of the RNA/DNA hybrid molecules compared to a DNA double helix necessitates a higher temperature if a chaotropic molecule like formamide is not present in the hybridization solution.

24. Remove hybridization bottle from hybridization oven and unscrew cap on one end. Carefully remove blot using long forceps restricted to touching the edges of the blot and place in a glass dish with 200 ml of 1× SSC/0.1% SDS. Incubate 2 min at room temperature

25. Drain the solution and replace with 200 ml of fresh 1× SSC/0.1% SDS. Incubate 5 min at room temperature, with gentle rocking.

    Because RNA/DNA hybrids melt at higher temperatures than DNA/DNA hybrids, it may be necessary to use higher wash temperatures for northern blots.

26. Warm the 0.2× SSC/0.1% SDS to 68° (~200 ml for each blot) for use in steps 27 and 28.

27. With gloved hands, hold the blot while draining the wash and replace with 200 ml prewarmed 0.2× SSC/0.1% SDS. Incubate 20 min at 68°C, with gentle rocking.

    If it is not possible to keep the rocker at that temperature, at least ensure that the buffer is that hot when first added.

28. With gloved hands, hold the blot while draining the wash and replace with a second aliquot of 0.2× SSC/0.1% SDS, prewarmed to 68°. Incubate 20 min at 68° or room temperature, with gentle rocking. Repeat this step, making a total of three washes in hot 0.2× SSC/0.1% SDS.

29. Observing RNA paranoia, lay the blot out to dry on several sheets of 3MM filter paper at room temperature, with the RNA side up.

    Depending on the label detection method, the blot may need to be completely dry or may need to be kept moist (see Table 8.2.2).

30. Proceed to label detection (see Table 8.2.2).

Support Protocol: Assembling a Blotting Transfer Apparatus

The transfer apparatus should be assembled while the nucleic acid-containing gels are being treated in preparation for the transfer (see Basic Protocol 1, step 8, or Basic Protocol 2, step 3). The assembled transfer apparatus is shown in Figure 8.2.1.

NOTE: For northern blotting, all equipment must be washed with RNase inhibitor solution (e.g., RNase Erase from QBiogene). Also, all solutions must be treated with DEPC and then autoclaved to degrade the DEPC.

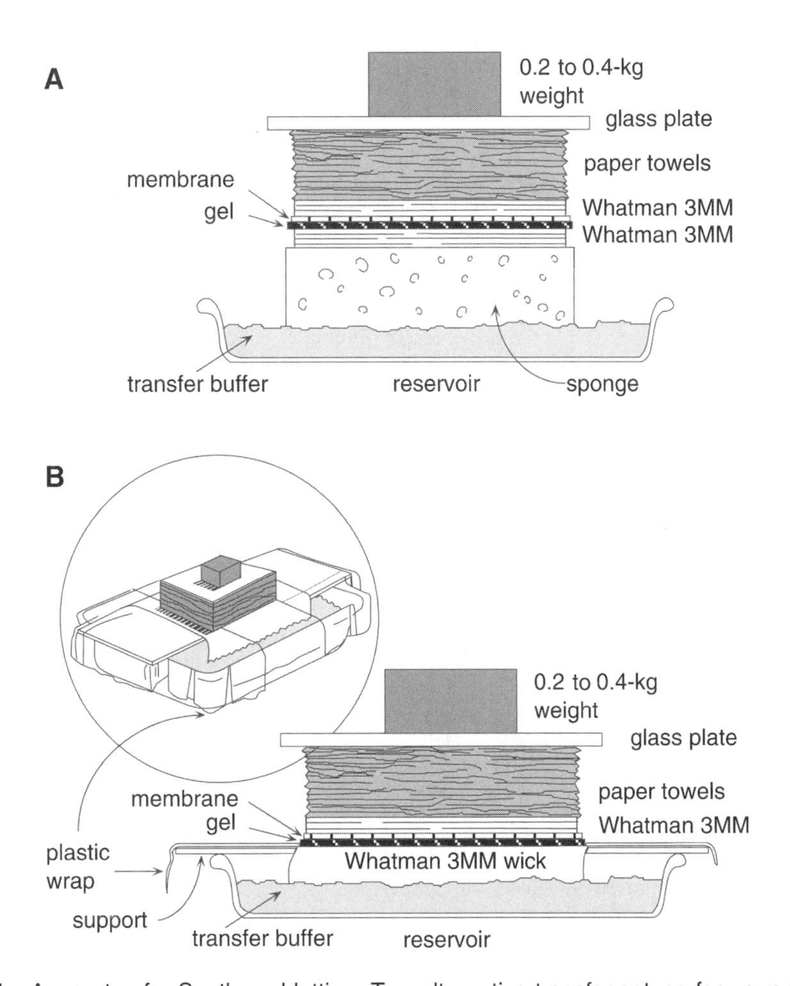

Figure 8.2.1 Apparatus for Southern blotting. Two alternative transfer setups for upward capillary transfer are shown: (**A**) sponge method and (**B**) Whatman 3MM filter paper wick method. Originally published in *CP Molecular Biology Unit 2.9* (Brown, 2004).

8

Materials

Absolute ethanol
Distilled water, sterile (treated with DEPC and then autoclaved for northern blots)
20× (for nylon membranes) or 2× (for nitrocellulose membranes) SSC (see *UNIT 3.3*)

Scissors
Gloves
Whatman 3MM filter paper sheets
Small glass or plastic dishes for hydrating membranes
Sponge, slightly larger than the gel being blotted
Uncharged nylon membrane (e.g., Hybond-N, Amersham; (Duralon-UV, Stratagene; or NEN GeneScreen, Dupont) *or* nitrocellulose membrane (for Southern blot only; e.g., BA 83 or BA 95, Schleicher and Schuell; Hybond-C, Amersham; or Biodyne A, Pall), cut to fit gel (see *UNIT 8.1* for overview of membrane types)
Large glass dish for transfer (to hold enough liquid to keep from going dry during an overnight capillary transfer period)

1. Clean the scissors with ethanol (or with RNase inhibitor followed by ethanol in the case of northern blots).

2. Using gloved hands, cut eight pieces of 3MM filter paper to the size of the sponge.

3. Cut a piece of the nylon or nitrocellulose membrane to the size of the gel. Cut off one corner of the blot to serve as a position marker for orienting the gel.

4a. *For nylon membranes:* In a small glass or plastic dish, immerse the membrane in distilled water so it can hydrate evenly. After 5 min, move membrane to 20× SSC for at least 10 min.

4b. *For nitrocellulose membranes:* Hydrate in 2× SSC for at least 10 min.

If white streaks and whorls appear on membrane and do not disappear during hydration, discard and replace with a new nitrocellulose rectangle.

Nitrocellulose membranes are not recommended for northern blots.

5. Place the sponge in a large clean glass dish.

If necessary you can use two sponges, but take care that the joint between them is nowhere near where you expect the fragments of interest.

6. Fill the dish with enough 20× SSC to soak the sponge and be about halfway up its side after the sponge has finished taking up fluid.

7. Place three of the 3MM filter paper rectangles on top of the sponge and wet them with 20× SSC. With a gloved hand, smooth out these sheets, avoiding wrinkles.

Alternate Protocol: Acrylamide Gel Vacuum Blotting

An exciting development recently has been the discovery of many small RNA species that are important in gene regulation (siRNA). Acrylamide gel electrophoresis is better that agarose gel electrophoresis for separating these small nucleic acids, as well as smaller fragments of DNA that are generated in procedures like PCR. However, acrylamide gels are not compatible with capillary blotting, and so electrophoretic blotting must be used. A procedure for this type of blotting is described below. If applying it to an RNA analysis, remember that RNA paranoia (e.g., use of DEPC water and RNase-free apparatus) will be required. The specific method below uses the Bio-Rad Trans-Blot apparatus, but it can be adapted to work with other manufacturers' units.

Materials

Acrylamide gel containing electrophoretically separated nucleic acids (see *CP Molecular Biology Unit 2.5A*; Voytas, 2000) that have been stained, photographed, and destained, if desired

0.5× TBE electrophoresis buffer: prepared in DEPC-treated water and sterilized by passing through a 0.22-µm filter (see UNIT 7.2), 4°C

Whatman 3MM filter paper

Uncharged nylon blotting membrane (e.g., Hybond-N, Amersham; (Duralon-UV, Stratagene; or NEN GeneScreen, Dupont)

Scotch-Brite pads (supplied with Trans-Blot)

10-ml pipet, sterile

Trans-Blot electrophoresis cell (Bio-Rad)

Blocks of ice frozen in plastic containers (containers supplied with Trans Blot)

Additional reagents and equipment for cross-linking by UV irradiation (see Basic Protocol 1 or 2)

1. Clean the scissors with ethanol (or with RNase inhibitor followed by ethanol in the case of northern blots).

2. Using gloved hands, cut a piece of nylon blotting membrane the size of the electrophoresis gel, and eight pieces of 3MM paper the same size.

3. Pour distilled water into a clean glass dish and put the nylon blot membrane into the water to hydrate it.

4. At the same time, put two Scotch Brite pads into a clean pan with fairly deep 0.5× TBE buffer and remove air by repeatedly bouncing and squeezing with gloved hands.

5. Remove one of the glass plates supporting the acrylamide gel containing the nucleic acids and lay a piece of 3MM paper onto the gel. Roll a pipet over it to remove bubbles between it and the gel; use the 3MM paper to lift the gel from the other glass plate.

   If the gel hasn't been stained and photographed, cut off the lane(s) of size standards that are intended for staining.

6. Open the gel holder of the Trans-Blot cell in a shallow tray, with the gray panel towards the bottom of the tray, and put a saturated Scotch Brite pad on the inside.

7. Dip three 3MM filter papers in the 0.5× TBE and roll each on onto the top of the Scotch Brite pad, then onto each other. As each paper is added, roll out air bubbles with a pipet.

8. Flood the 3MM paper stuck to the gel with 0.5× TBE from a pipet, and place it on top of the other papers.

9. Flood the gel top with 0.5× TBE from a pipet and then place the wetted nylon rectangle on the gel.

10. Flood the membrane with 0.5× TBE from a pipet, and then apply each of the other four 3MM papers, soaking each in TBE and removing bubbles.

11. Add the second Scotch Brite pad on top, and clamp the whole gel holder together with the attached clamps.

12. Half-fill the Trans-Blot cell with cold 0.5× TBE and place the gel holder in the cell with the gray panel facing toward the cathode.

13. Fill the cell with cold 0.5× TBE, add the frozen blocks, and electrophorese 4 hr at 30 V (~125 mA) or 2 hr at 40 V.

    If the Trans-Blot has a cooling coil, it's recommended that it be used for the 2-hr procedure.

14. Switch off the power and disassemble the apparatus.

15. Take out the gel holder and open it, cut off the upper right corner of the blot to mark the orientation, and/or mark the positions of the electrophoresis wells in pencil.

16a. *For nondenaturing DNA gels:* Place the membrane on three pieces of 3MM paper soaked in 0.4 M NaOH with the DNA side up. Allow to sit for 10 min.

16b. *For other types of gels:* Proceed immediately to step 17 without treatment with NaOH.

17. Rinse the membrane in 2× SSC and place on 3MM paper with nucleic acid up to dry sufficiently for the planned detection method (see Table 8.2.2).

18. Perform cross-linking by UV irradiation and storage as recommended for Southern (Basic Protocol 1) and northern (Basic Protocol 2) blots.

REAGENTS AND SOLUTIONS

Use commercial nuclease-free water or DEPC-treated water in all recipes and protocol steps.

Denhardt's solution, 50×

5 g Ficoll (type 400, Pharmacia)
5 g polyvinylpyrrolidone
5 g bovine serum albumin (Fraction V)

Adjust volume to 500 ml with DEPC-treated (see recipe) or nuclease-free (commercially available) water. Sterilize by passing through a 0.22-μm filter and store aliquots indefinitely at −20°C.

DEPC-treated water

Dissolve diethylpyrocarbonate (DEPC) in deionized, distilled water to a final concentration of 0.1%. Stir 12 hr at 37° (or if necessary, at room temperature) and then autoclave to break the DEPC down to CO_2 and ethanol. Store indefinitely at room temperature as long as not exposed to possible RNase contamination.

RNase in the water will be destroyed. DEPC cannot be used with Tris buffers; keep a special "for RNA" jar of Tris and protect it from contamination, or choose other buffers for RNA work. Refer to the Ambion Web site under Internet Resources for information about DEPC.

Hybridization buffer with 50% formamide

6× SSC (use 20× stock; see *UNIT 3.3*)
0.5% (w/v) SDS
100 μg/ml salmon sperm DNA (using 10 mg/ml stock; see recipe)

Measure all three stocks (to result in the indicated final concentrations) into the container where you're preparing the hybridization buffer, and then dissolve in 1:1 deionized water/formamide. Use DEPC-treated (see recipe) or nuclease-free (commercially available) water for northern blots. Sterilize by passing through a 0.22-μm filter and store up to 1 year at 4°C.

Use this hybridization buffer when hybridizing at 42°C.

Note that this is the same as prehybridization buffer without Denhardt's solution.

Hybridization buffer without formamide

6× SSC (use 20× stock; see *UNIT 3.3*)
0.5% (w/v) SDS
100 μg/ml salmon sperm DNA (using 10 mg/ml stock; see recipe)

continued

Measure all three stocks (to result in the indicated final concentrations) into the container where you're preparing the hybridization buffer, then dilute to final volume with deionized water. Use DEPC-treated (see recipe) or nuclease-free (commercially available) water for northern blots. Sterilize by passing through a 0.22-μm filter and store up to 1 year at 4°C.

Use this hybridization buffer when hybridizing at 68°C.

Note that this is the same as prehybridization buffer without Denhardt's solution.

Prehybridization buffer with 50% formamide

6× SSC (use 20× SSC; see *UNIT 3.3*)
5× Denhardt's solution (use 50× stock; see recipe)
0.5% (w/v) sodium dodecyl sulfate (SDS)
100 μg/ml salmon sperm DNA (use 10 mg/ml stock; see recipe)

Dissolve components to the indicated final concentrations in formamide (1/2 the final solution volume) Adjust to final volume with deionized water. Use DEPC-treated (see recipe) or nuclease-free (commercially available) water for northern blots. Sterilize by passing through a 0.22-μm filter and store up to 1 year at 4° C.

Use this prehybridization buffer when hybridizing at 42°C.

Prehybridization buffer without formamide

6× SSC (use 20× SSC; see *UNIT 3.3*)
5× Denhardt's solution (use 50× stock; see recipe)
0.5% (w/v) sodium dodecyl sulfate (SDS)
100 μg/ml salmon sperm DNA (use 10 mg/ml stock; see recipe)

Measure all three stocks (to result in the indicated final concentrations) into the container where you're preparing the prehybridization buffer, then dilute to final volume with DEPC-treated (see recipe) or nuclease-free (commercially available) water. Sterilize by passing through a 0.22-μm filter and store up to 1 year at 4°C.

SDS may precipitate if the solution gets cold, but it will go right back into solution as it warms to room temperature.

Use this prehybridization buffer when hybridizing at 68°C.

Salmon sperm DNA, 10 mg/ml

Dissolve Type III salmon sperm DNA (Sigma) in water at a final concentration 10 mg/ml by incubating overnight on a rotator at 4°C. (The resulting solution is very viscous.) To ensure purity, extract the DNA once with phenol and once with phenol/chloroform (see *UNIT 5.2*). Shear the DNA into random smaller fragments by passing rapidly passing the solution through a 17-G hypodermic needle twelve times or by sonicating ~1.5 min at a moderate setting. Precipitate the DNA by adding two volumes of ice-cold ethanol. Centrifuge 5 min at 2500 to 6000 × *g* (6000 to 10,000 rpm in a typical clinical centrifuge), 4°C. Remove the ethanol completely. Redissolve the DNA to a final concentration of 10 mg/ml in sterile, deionized water by rotating the tube overnight at room temperature. Determine the A_{260} and calculate the exact concentration of DNA (see *UNIT 2.5*). Boil the solution 10 min and quickly cool on ice (to make the DNA single stranded). Sterilize by passing through a 0.22-μm filter and store indefinitely in 200-μl aliquots at −20°C. After use, discard any remaining thawed material.

UNDERSTANDING RESULTS

One of the outcomes of blotting is to show the size of the nucleic acid fragment containing the sequence of interest, indicated by the location of the signal from the hybridized probe (reflecting the migration rate of the nucleic on the original electrophoresis gel). As noted earlier, you can use molecular weight standards during electrophoresis to bracket the position of the molecule containing the sequence of interest and estimate its size. The standards are often stained before blotting and care must be taken to make sure you can relate those sizes to the places where bands occur in the labeled blot by noting the locations of the stained bands prior to destaining and hybridizing the blots. Alternatively, you can use standards that are already labeled to provide the same signal that will be generated by the probes used for the unknown sequences (see Fig. 8.2.2A, where standards were loaded into the slot on the right side of the original gel). Autoradiography, fluorescence methods, or luminescence methods show where the probe has base paired with a target on the blot. It is always a good idea to have a good idea of the expected size of the DNA or RNA band containing the sequence of interest (where you expect to find it base paired to the probe being used). The ideal result would be one or more sharp bands of the expected size(s). See Troubleshooting (below) for possible causes when this result isn't achieved.

TROUBLESHOOTING

If unexpected bands on the blot are larger than the anticipated it's always a good idea to recheck the basis for your expectations to make sure you calculated correctly and took into account any restriction sites or alternative promoters where the RNA transcription could have begun at an unexpected spot. In addition, eukaryotic mRNAs can be spliced differently in different tissues or stages of development (Ferriera et al, 2007), resulting in RNA sizes different from that expected. In designing probes, it's important to make sure they cannot recognize more than one site in the genome (or in the RNA for northern blots). If the expectations still seem reasonable, then either the expected restriction site was mutated in your DNA, the restriction enzyme digestion was incomplete, or there was a transfer problem that distorted the gel.

If unexpected bands on the blot are smaller than anticipated, then a new restriction site may be present, the restriction enzyme may be contaminated or be exhibiting so-called star activity (see *UNIT 10.1*), or the gel may have been distorted.

If bands are seen that are fuzzy and indistinct, the electrophoresis may have been suboptimal, the gel may have been left too long before transfer so the bands diffused, or the molecules may have become degraded. Bands often look indistinct on blots that have low signal, and the bands may look sharper without changing anything else if you can achieve better labeling of the probe (see *UNIT 8.4*) before you hybridize.

If expected bands are absent or very weak, the blot often can be exposed longer to the detection mechanism to increase the signal. If still nothing is detected, then there are several possibilities. The stringency of hybridization or washing or both may also have been too high for the probe and target used. In this case you should repeat the experiment using more concentrated final wash solutions, e.g., try a calibration experiment where you cut the blot into equivalent series of lanes and treat with final washes at 0.2× SSC, 0.5× SSC, and 1× SSC. The temperature of the washes can also be decreased if needed to obtain a signal hybridizing to the band of interest. The target genomic DNA may have been of insufficient quantity to produce a detectable signal. The protocols recommend how much to use, but more may be required.

Southern Blotting

In Southern blotting, the stability of DNA is rarely a consideration. However, there can be incomplete digestion of the DNA by the restriction endonucleases (*UNIT 10.1*). Too much or too little DNA will obscure the results (Fig. 8.2.2B), and sometimes the denaturation and/or transfer from the gel can be incomplete. If the blot wasn't blocked successfully with nucleic acids that saturated all of the nonspecific binding sites, then a halo of dark signal will partially obscure the results (Fig. 8.2.2C). It's also possible to damage the blot, particularly while inserting it into a hybridization cylinder. This

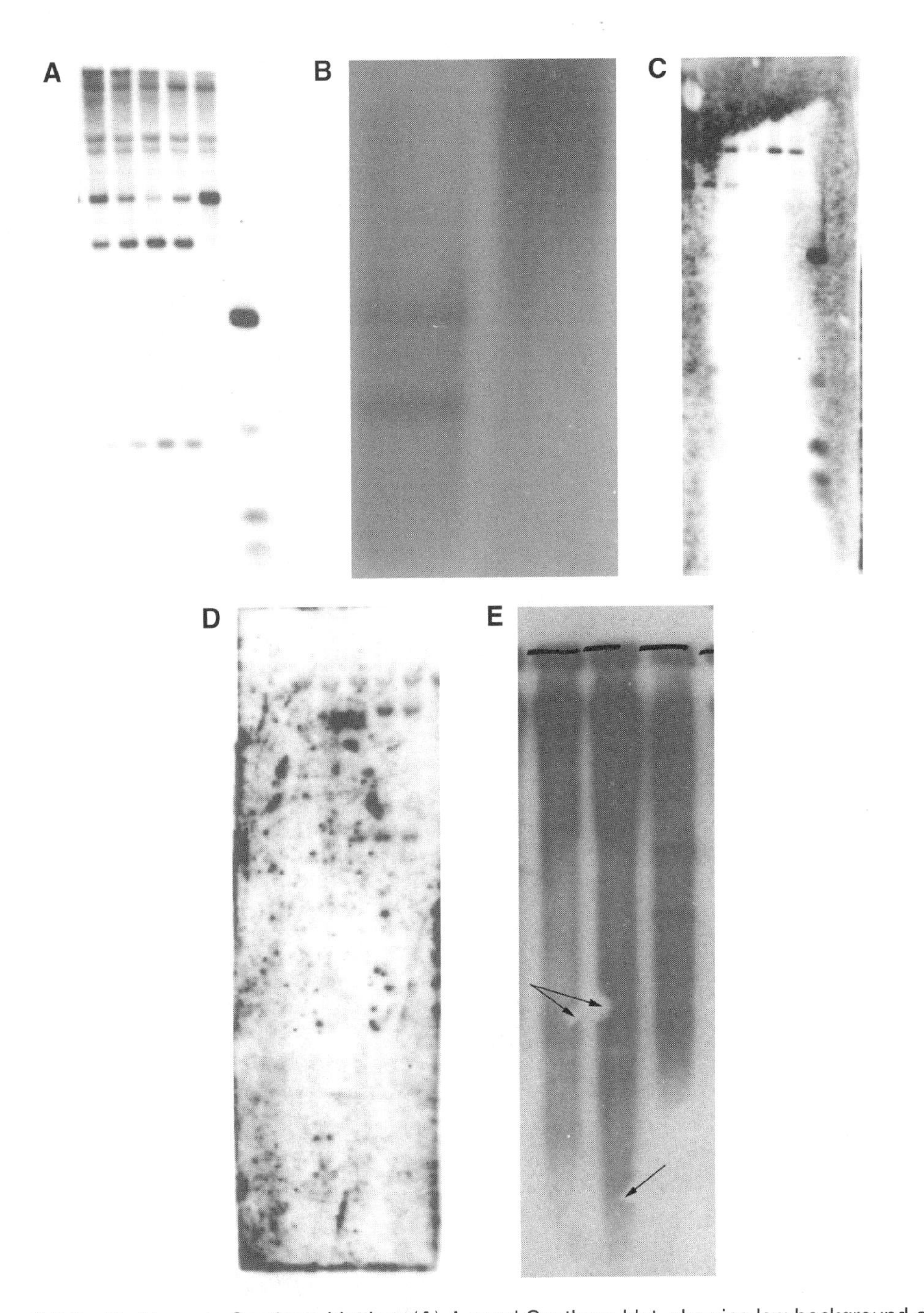

Figure 8.2.2 Problems in Southern blotting. (**A**) A good Southern blot, showing low background and sharp bands. (**B**) Excessive DNA (right lanes) and DNA incompletely digested by restriction enzymes (left lanes) can obscure Southern blotting results. Compare the excessive DNA at the top of the right lanes to the two bands of DNA in the left lanes, produced by completely digested DNA. (**C**) Effects of inadequate blocking on blots; note the edges of the blot have nonspecific label stuck on them. (**D**) Effects of damage to the blot medium on images. Note the cracks with label stuck in them, the results of damage to the blot membrane when inserting it into a hybridization tube. (**E**) Effect of a bubble between the gel and the blot medium causes a "pock" where no transfer occurs, making a white circular artifact on the blot.

damage will show up as label in strange lines and patterns on the blot (Fig. 8.2.2D). A particular artifact in blotting has been nicknamed "dreaded pox," which are regions where the transfer was blocked by an air bubble so that it appears blank on the final picture (Fig. 8.2.2E). Figure 8.2.2E also shows hybridization by repetitive sequences, hence the background "haze" all down the length of the DNA separation.

Northern Blotting

For northern blot bands that are the "wrong" size, see possible causes for these problems in Southern blots (above); RNA degradation, alternative promoters, or alternative splicing (described above; also see Ferriera et al., 2007) are also possibilities. When the signal from expected bands is weak or absent, the same possible problems described for Southern blots (see above) should be considered, as well as the quality of the RNA.

The integrity of RNA is a very important issue in northern blotting. It is necessary to perform quality control tests on the total RNA preparation. The extent of protein contamination of the RNA can be detected by the A_{260}/A_{280} ratio. This ratio should be high (1.8 to 2.0) for good RNA preparations because the bases of the nucleic acids absorb at 260 nm and the aromatic amino acids of proteins absorb at 280 nm. You need to dilute the RNA such that the Absorbance values are between 0.02 and 0.70 to get the most reliable spectrophotometric data, because below this range spectrophotometer accuracy and reproducibility are low, and above this range spectrophotometers can underestimate the concentration (see *UNIT 2.2*). However, the RNA can be nearly protein-free and still not be useful for a northern blot. The degradation of RNA, or its excessive contamination with

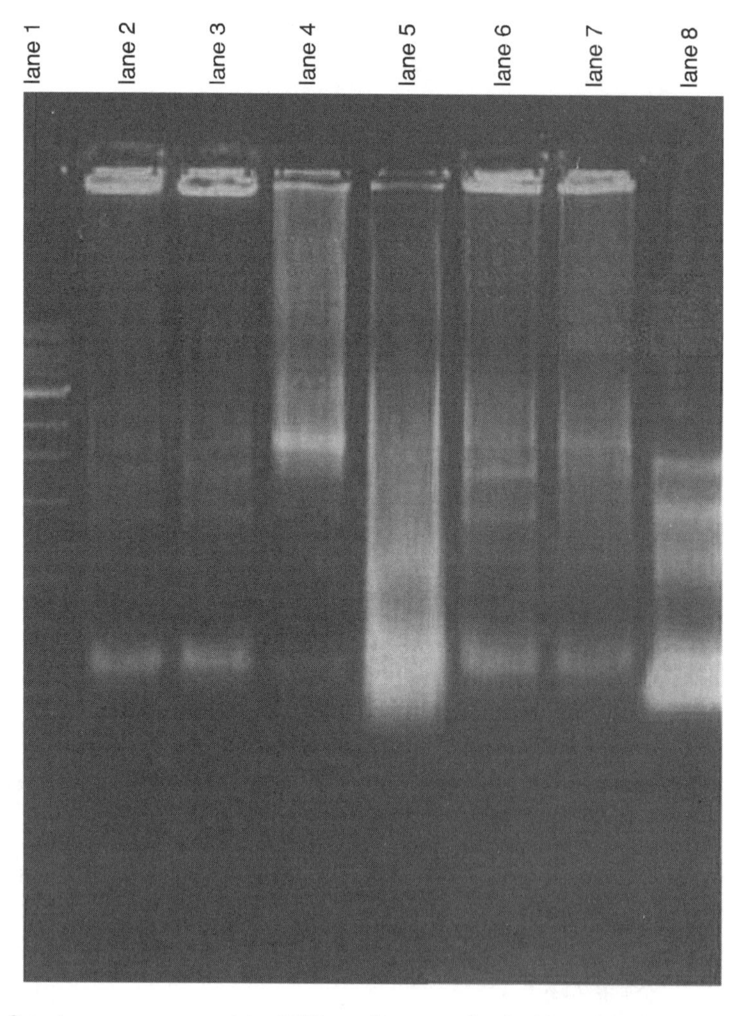

Figure 8.2.3 *Saccharomyces cerevisiae* RNA quality control gel with problems identified. Lane 1: Standards of double stranded DNA. Lane 2: RNA sample with mostly small RNA, possibly transfer RNA plus degraded fragments. Lane 3: RNA sample with some DNA (compare with lane 4), very light bands for the two large rRNAs (two bands running just ahead of the DNA). Lane 4: RNA sample that is almost all genomic DNA. Lane 5: Degraded RNA that runs as a smear and lacks distinct bands for the larger rRNAs. Lane 6: Good sample of RNA that has strong rRNA bands, not too much DNA or tRNA. Lane 7: Sample of RNA with excessive DNA so it may hybridize badly with probes. Lane 8: The rRNA bands are strong but the small RNA is excessive, suggesting degradation.

DNA, cause problems in blotting. These problems can be diagnosed by means of electrophoresis of the total RNA on a mini-denaturing gel (Fig. 8.2.3) where you would look for strong ribosomal RNA bands in a high-quality preparation. There are almost 100 times more of the rRNA molecules than total mRNA, let alone any individual mRNA molecule you might be trying to detect. However, the rRNA is similar in size to many mRNAs, so it's a sort of "canary in the mine" signal to the experimenter. If the rRNA is degraded, the mRNA is almost certainly also degraded. Often a strong band for small RNA (tRNA size, see Fig. 8.2.3) indicates that longer RNAs have been degraded. DNA contamination (also shown in Fig. 8.2.3) can compete with probe binding.

In some cases, the mRNA is very unstable. In such cases, the expected band will be observed, but there may be a smear of smaller bands that hybridize leading down from the expected band. If the expected band is seen, it is generally taken to mean that you've done the northern blot correctly. But if you want to quantitate the data, the existence of smaller molecules must be taken into account. One way to decide whether the degradation occurred naturally in the cell or during the isolation is to use an internal control. To do that, you can prepare an in vitro transcript of the gene of interest (or its cDNA if it has introns) and add a labeled sample of that in vitro transcript to a control sample at the start of the isolation. The amount recovered and its condition (one band or a smear) will indicate how much degradation has happened under your conditions during the isolation.

Unfortunately, even if the entire electrophoresis is ideal and your mRNA is good, that RNA band can still become degraded enough during the hybridization process not to hybridize with the probe unless strict "RNA paranoia" is observed (see Strategic Planning).

VARIATIONS

Dot or Slot Blotting

This method allows detection of sequences without providing information about size, since the electrophoresis step is omitted (see *CP Molecular Biology Unit 2.9*, Brown, 2004; *CP Molecular Biology Unit 4.9*, Brown et al., 2004). The protocol is very similar to those already described, except that no electrophoresis is performed. Using a filtration manifold, deposit samples of RNA (using RNA paranoia) or denatured and neutralized DNA onto nylon or nitrocellulose in a regular pattern, ideally spotting several concentrations of each nucleic acid. Then cross-link and hybridize according to Southern or northern blot protocols above.

LITERATURE CITED

Alwine, J.C., Kemp, D.J., and Stark, G.R. 1977. Method for detection of specific RNAs in agarose gels by transfer to diazobenzylmethoxymethyl-paper and hybridization with DNA probes. *Proc. Natl. Acad. Sci. U.S.A.* 74:5350-5354.

Boyle, A. and Perry-O'Keefe, H. 1992. Labeling and colorimetric detection of nonisotopic probes. *Curr. Protoc. Mol. Biol.* 20:3.18.1-3.18.9.

Brown, T. 2004. Analysis of DNA sequences by blotting and hybridization. *Curr. Protoc. Mol. Biol.* 68:2.9.1-2.8.20.

Brown, T., Mackey, K., and Du, T. 2004. Analysis of RNA by northern and slot blot hybridization. *Curr. Protoc. Mol. Biol.* 67:4.9.1-4.9.19.

Ferriera, E.N., Galante, P.A., Carraro, D.N., and deSouza, S.J. 2007. Alternative splicing: A bioinformatic perspective. *Mol. Biosyst.* 3:473-477.

Perry-O'Keefe, H. and Kissinger, C.M. 2000. Chemiluminescent detection of nonisotopic probes. *Curr. Protoc. Mol. Biol.* 26:3.19.1-3.19.8.

Sambrook, J., Fritsch, E.F., and Maniatis, T. 1989. Analysis of genomic DNA by Southern hybridization and northern hybridization *In* Molecular Cloning: A Laboratory Manual. 2nd ed. pp. E21-E25. Cold Spring Harbor Press, Cold Spring Harbor, N.Y.

Southern, E.M. 1975. Detection of specific sequences among DNA fragments separated by gel electrophoresis. *J. Mol. Biol.* 98:503-517.

Sullivan, K.M. and Lilley, D.M.J. 1988. Helix stability and the mechanism of cruciform extrusion in supercoiled DNA. *Nucl. Acids Res.* 16:1079-1093.

Voytas, D. 2000. Agarose gel electrophoresis. *Curr. Protoc. Mol. Biol.* 51:2.5A.1-2.5A.9.

KEY REFERENCES

Alwine et al., 1977. See above.

This paper is the classic description of northern blotting, and although it recommends a membrane that is no longer used, clearly describes why this technique was needed and how to perform the transfer.

Southern, 1975. See above.

This paper contains the original description of Southern blotting and describes the reasons for its superiority to the methods previously available.

INTERNET RESOURCES

http://www.dnalc.org/ddnalc/resources/shockwave/southan.html

Forensics Southern blot Internet animation requiring Shockwave software; leads user to identification of suspect.

http://www4.od.nih.gov/oba/rac/aboutrdagt.htm

Provides access to government safety regulations for recombinant DNA experiments, including those using dangerous viruses.

http://www.msdssearch.com

http://hazard.com/msds

Provide material safety data sheets for proper safe use and disposal of chemicals such as ethidium bromide, formaldehyde, and formamide that are used in these procedures.

http://www.ambion.com/techlib/tb/tb_178.html

Explains the use and misuse of DEPC in treating solutions to remove RNase activity.

http://www.accessexcellence.org/RC/VL/GG/ecb/southern_blotting.html

Southern blotting explained on Access Excellence, the National Health Museum Resource Center.

http://www.bio.davidson.edu/COURSES/GENOMICS/method/Southernblot.html

An explanation of Southern blotting by A. Malcolm Campbell (related to his genomics class) on the Davidson College Biology Department's home page.

Protein Blotting: Immunoblotting

Sean R. Gallagher[1]
[1]UVP, LLC, Upland, California

OVERVIEW AND PRINCIPLES

Immunoblotting (often referred to as "western blotting") is used to identify specific antigens recognized by polyclonal or monoclonal antibodies. With numerous applications from protein quantitation to diagnostics, western blotting remains a core technique for protein analysis (see Table 8.3.1). The typical steps for performing western blotting are described in Basic Protocols 1 and 2 and follow the sequence: SDS-PAGE, electroblotting, blocking, antibody probing, and visualization of the reaction.

While antibodies are exquisitely specific for the target antigen (sites on proteins in the case of western blotting), the small pores of the polyacrylamide gel used in SDS-PAGE (*UNIT 7.3*) prevent the antibody from binding to the electrophoretically separated proteins. Western blotting creates an exact replica of the protein gel, typically on nitrocellulose (NC) or polyvinylidene difluoride (PVDF) membrane, by electrophoretically removing the protein from the gel and, in the process, irreversibly binding the protein to the readily accessible membrane surface where reagents such as antibodies are free to interact with the bound protein (Towbin et al., 1979; Burnette, 1981; Kurien and Scofield, 2006). Membranes have an enormous capacity for protein binding, and before probing, any remaining binding capacity is blocked after transfer by soaking the membrane in a solution containing a nonspecific protein (e.g., purified casein) or, more typically, a detergent.

Table 8.3.1 Applications of Western Blotting

Application	Comment
Antibody development and characterization	Characterize antibody specificity
Subcellular localization of an expressed protein	Analyze isolated organelles (e.g., Golgi apparatus, plasma membrane, nucleus, etc.) via SDS-PAGE and immunoblotting for the presence of specific proteins; probe intact cells and tissue for subcellular localization using visible light, fluorescence, and electron microscopy for confirmation
Protein purification	Demonstrate enrichment of well characterized antibody by sampling, electrophoresing, and blotting at various stages of purification
Diagnostics	Separate and blot a viral lysate to test serum for antibodies that bind key viral proteins, indicating presence of antibodies against an infection (e.g., HIV testing)
Gene expression	Detect presence or absence and amount of a specific protein during gene expression; track markers and reporter genes (e.g., luciferase) in transgenic organisms to get a complete picture of transcription and translation
Post-translational modifications	Phosphoproteomics: determine phosphorylation status of the protein complement in the cell or tissue using antibodies specific for phospho amino acids
Protein sequencing by mass spectrometry	Characterize the readily available proteins blotted onto membranes using sensitive methods, e.g., peptide sequencing by MALDI (see *CP Protein Science Unit 16.1*; Carr and Annan, 1996)

Current Protocols Essential Laboratory Techniques 8.3.1-8.3.28

Nylon membranes can also be used for protein blotting, but are not as popular due to additional steps needed to minimize background signal.

Recent improvements in protein blotting have been driven by the need for faster and more convenient results. Precast small-format gels and blotting equipment are ideal (see Fig. 8.3.1 and Fig. 8.3.2). In the simplest and most common approach, the proteins are electrophoretically transferred in a semidry transfer apparatus to a nitrocellulose or PVDF membrane in a process that can be monitored for efficiency of transfer by staining both the membrane for both transferred proteins and the gel for any residual sample.

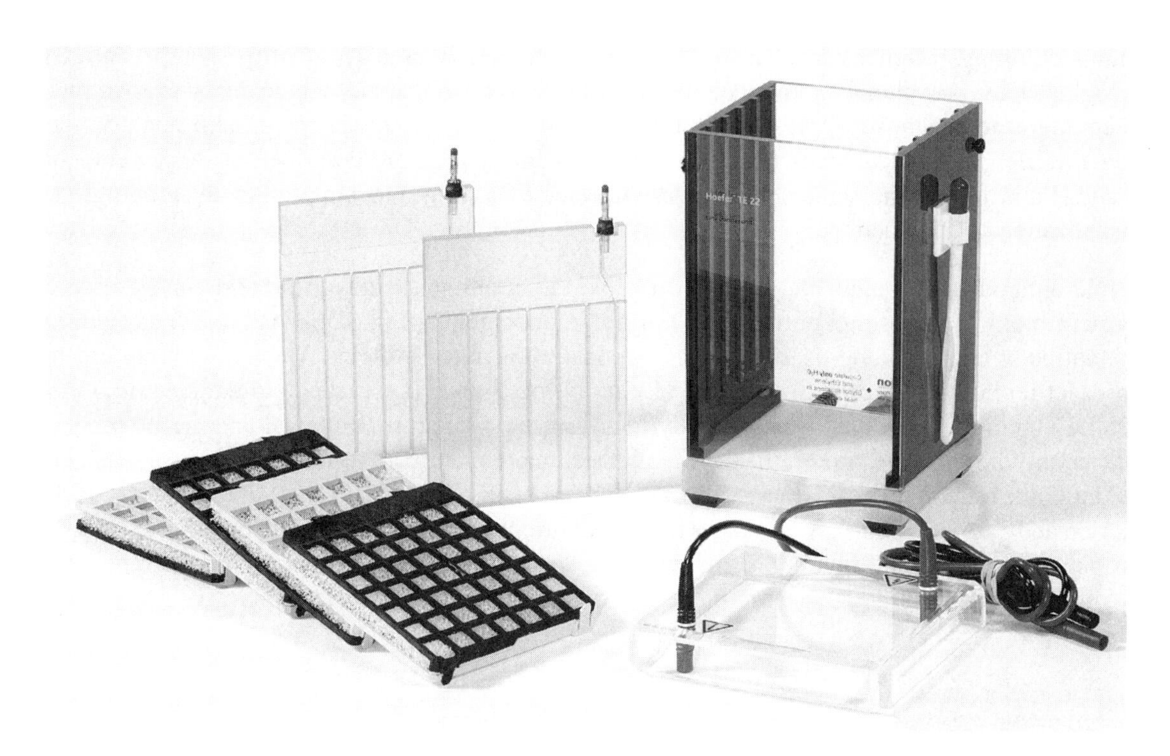

Figure 8.3.1 Minigel tank electrotransfer unit. Designed for smaller 8 × 10–cm gels, these units will process four gels at a time. Note that the two outside panels hold the electrode grid. Figure courtesy of Hoefer, Inc.

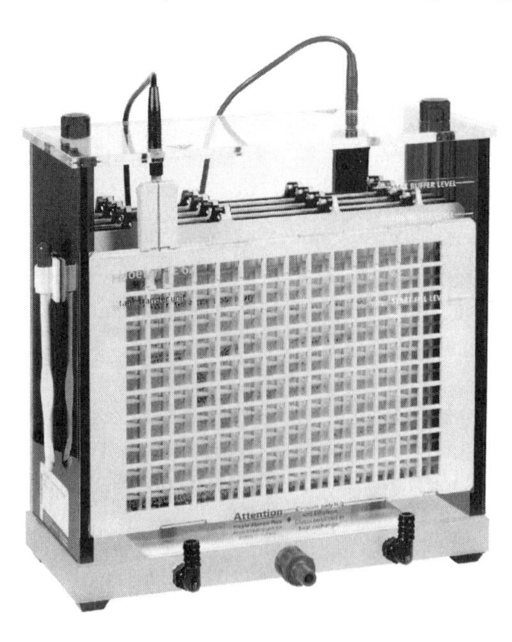

Figure 8.3.2 Large gel tank electrotransfer unit. Designed for larger 15 × 21–cm gels, these units will process four gels at a time. Note that the two outside panels hold the electrode grid. Figure courtesy of Hoefer, Inc.

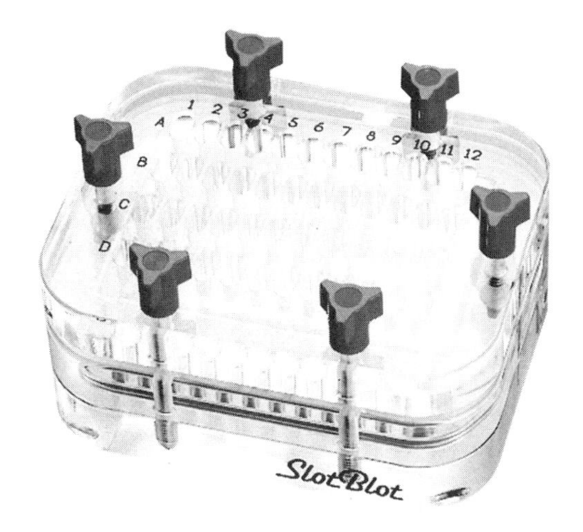

Figure 8.3.3 Slot-blot unit. Through use of a vacuum manifold or by simple hand spotting, up to 96 samples can be applied to a single NC or PVDF membrane for immunoblotting analysis. Although this approach cannot discriminate between the protein of interest and a cross reactive antigen, it is a quick way to perform preliminary characterization and high-volume, routine quantitation of a sample. Figure courtesy of Hoefer, Inc.

Dot and slot blots (Alternate Protocol 1) are alternatives useful for both preliminary and routine characterizations, providing a quick and simple way to determine the amount of an antigen in a sample without performing electrophoresis first. Briefly, the proteins are deposited onto the membrane either manually or through use of a vacuum manifold that pulls a solution containing the antigen through a nitrocellulose membrane (see Fig. 8.3.3), depositing the protein of interest onto the membrane. The membrane is probed with the same chromogenic or luminescence protocol as the western blot.

Visualizing Immunoblots

After probing with the primary antibody, the membrane is washed and the antibody-antigen complexes are identified with horseradish peroxidase (HRPO) or alkaline phosphatase (AP) enzymes coupled to the secondary anti-IgG antibody (e.g., goat anti–rabbit IgG). The enzymes are attached to the secondary antibody either directly (Basic Protocol 3) or via an avidin-biotin bridge (Alternate Protocol 2). Chromogenic or luminescent substrates are then used to visualize the activity (see Basic Protocol 4 and Alternate Protocol 3).

Total Protein Stains

Protocols for staining membranes to detect electroblotted proteins are straightforward and use either white light or fluorescence imaging (see *UNIT 7.4*). Detection limits for each staining method and the compatible blot transfer membranes and gels are presented in Table 8.3.2. Depending on the transfer membrane used and other considerations (e.g., the use of SDS-polyacrylamide gels or polyacrylamide gels without SDS and the level of sensitivity required), different total protein staining procedures must be selected.

Ponceau S staining

Ponceau S staining (Salinovich and Montelaro, 1986) is a simple method for visualizing proteins on nitrocellulose and PVDF membranes. Although relatively insensitive (see Table 8.3.2), Ponceau staining permits a quick visual inspection of the blot to verify transfer and to mark the positions of the molecular weight standards. The stain readily washes away and does not interfere with subsequent immunostaining.

Table 8.3.2 Properties and Compatibilities of Total Protein Stains for Detecting Membrane Transfers

Stain	Detection limit	Membrane types: Nitrocellulose	Membrane types: PVDF	Gel types: SDS-PAGE	Gel types: Native PAGE	Comments
Ponceau S	2 μg	+	+	+	+	Reversible, visible
India ink	50 ng	+	+	+	+	Permanent, visible
Gold	3 ng	+	+	+	+	Permanent, visible
SYPRO Ruby	2 ng	+	+	+	+	Reversible, fluorescent

Abbreviations: PVDF, polyvinylidene difluoride; SDS-PAGE, sodium dodecyl sulfate–polyacrylamide gel electrophoresis.

India ink staining

Probably the simplest approach is staining with India ink (Support Protocol 2), which is also compatible with chemiluminescent imaging (Eynard and Lauriere, 1998). The protocol for India ink staining of polypeptides is based on the procedure described by Hancock and Tsang (1983).

Gold staining

Gold staining (which stains the proteins pink) is very sensitive and easy to perform. Staining of membrane-bound polypeptides by gold sol is mediated by hydrophobic interactions and ionic interaction of negatively charged gold particles with positive groups on the proteins. Support Protocol 3 provides methods described by Moeremans et al. (1985).

Increasing staining by alkali treatment

Brief exposure to alkali as described in Support Protocol 4 significantly enhances staining of proteins by either India ink or colloidal gold. Sutherland and Skerritt (1986) suggest that the alkali treatment enhances retention of proteins on the nitrocellulose surface during extensive washing of the membrane; hence more protein is available for staining.

SYPRO Ruby staining

Fluorescent methods for total protein staining have gained popularity because of their high sensitivity and direct compatibility with colorimetric, fluorogenic, and chemiluminescent techniques. A method is described for staining blots with SYPRO Ruby (see Support Protocol 5), a sensitive fluorescent stain that works equally well on both nitrocellulose and PVDF blots. Support Protocol 6 provides detail for photographing fluorescently stained blots and recording the results.

Compatibility considerations

SYPRO Ruby staining, in contrast to India ink and colloidal gold, is compatible with colorimetric, fluorogenic, and chemiluminescent immunodetection techniques. Stains such as colloidal gold often block epitopes required for subsequent immunodetection. In addition, the dark color of the colloidal gold or India ink stains makes it difficult to visualize colorimetric or fluorogenic immunodetection reagents in subsequent immunodetection assays on the membrane.

Chromogenic versus Luminescent Assays

Immunoblotted proteins can be detected by chromogenic or luminescent assays; see Table 8.3.3 for a description of the reagents available for each system, their reactions, and a comparison of their advantages and disadvantages. Luminescence detection methods offer several advantages over traditional chromogenic procedures: (1) In general, luminescent substrates increase the sensitivity of both HRPO and AP systems without the need for radioisotopes. Substrates for AP systems

Table 8.3.3 Chromogenic and Luminescent Visualization Systems

System	Reagent[a]	Detection/reaction	Comments[b]
Chromogenic			
HRPO-based	4CN	Purple precipitate formed by oxidized products	Not very sensitive (Tween 20 inhibits the reaction); fades rapidly upon exposure to light
	$DAB/NiCl_2$[c]	Dark brown precipitate	More sensitive than 4CN but potentially carcinogenic; resulting membrane easily scanned
	TMB[d]	Dark purple stain	More stable, less toxic than $DAB/NiCl_2$; may be somewhat more sensitive[d]; can be used with all membrane types; kits available from KPL, Invitrogen, Sigma-Aldrich, and Vector Labs
AP-based	BCIP/NBT	Dark blue-gray stain: BCIP oxidized by NBP hydrolyzes to produces indigo precipitate; reduced NBT precipitates	More sensitive and reliable than other AP-precipitating substrates; note that phosphate inhibits AP activity
Luminescent			
HRPO-based	For example, luminol/H_2O_2/*p*-iodophenol	Blue light produced by oxidized luminol substrate; *p*-iodophenol increases light output	Very convenient, sensitive system; reaction detected within a few seconds to 1 hr; also see Haan and Behrmann (2007) for a lab-made reagent mix.
AP-based	For example, substituted 1,2-dioxetane-phosphates	Light produced by dephosphorylated substrate	Protocol described gives reasonable sensitivity on all membrane types; consult instructions of reagent manufacturer for maximum sensitivity and minimum background (see Troubleshooting); kits available from ABI, Pierce, Vector Labs, Beckman Coulter

[a]Recipes and suppliers are listed in Reagents and Solutions except for TMB for which use of a kit is recommended.
[b]See Commentary for further details.
[c]$DAB/NiCl_2$ can be used without the nickel enhancement, but it is much less sensitive.
[d]McKimm-Breschkin (1990) reported that if nitrocellulose filters are first treated with 1% dextran sulfate for 10 min in 10 mM citrate-EDTA (pH 5.0), TMB precipitates onto the membrane with a sensitivity much greater than 4CN or DAB, and equal to or better than that of BCIP/NBT.
Abbreviations: AP, alkaline phosphatase; BCIP, 5-bromo-4-chloro-3-indolyl phosphate; 4CN, 4-chloro-1-napthol; DAB, 3,3′-diaminobenzidine; HRPO, horseradish peroxidase; NBT, nitroblue tetrazolium; TMB, 3,3′,5,5′-tetramethylbenzidine.

are generally available as proprietary kits, although there is significant literature describing these reagents (see Gillespie and Hudspeth, 1991; Sandhu et al., 1991; Kricka et al., 2000). (2) Luminescence detection can be completed in as little as a few seconds; exposures rarely go more than 1 hr. (3) Depending on the system, the luminescence can last for 3 days, permitting multiple exposures of the same blot. (4) Furthermore, the signal is easily detected by digital charge-coupled device (CCD) systems, allowing simple digital overlay and comparison to blots stained for total protein. Compared to chromogenic development, the luminescent image recorded by CCD imaging is easier to photograph and quantitate because of the ability to record a wide dynamic range of signal over prolonged exposure. (5) Luminescent blots can be easily erased and reprobed because the reaction products are soluble and do not deposit on the membrane (see below). On the other hand,

AP-based luminescent protocols that achieve maximum sensitivity with minimum background can be complex, and the manufacturer's instructions should be consulted.

Where Immunoblotting is Used

Protein blotting has important clinical applications—it is the confirmatory test for human immunodeficiency virus Type 1 (HIV-1). SDS-PAGE-separated virus proteins are blotted onto nitrocellulose membranes and processed with patient sera. After washing and incubating (using a Tween 20/nonfat dry milk diluting and washing solution) with an anti-human IgG coupled to HRPO or alkaline phosphatase, the antigens are identified by chromogenic development. Typically, the prepared blots and all reagents needed for the test are purchased commercially (e.g., Immunetics).

Other areas of growing importance for immunoblotting are detection of post-translational modifications such as reversible protein phosphorylation, a widely used regulatory strategy in signal transduction. A challenge with phosphoprotein analysis is the sheer complexity of the number and type of phosphoproteins in a cell. Reducing the complexity requires some form of prefractionation and separation, including SDS-PAGE followed by immunoblotting and visualization with phospho-specific antibodies (Morandell et al., 2006). For a typical immunoblotting experiment to identify the phosphoprotein, a combination of cell fractionation, one-dimensional or two-dimensional electrophoresis, fluorescent phosphoprotein staining (ProQ Diamond), and protein immunoblotting with phosphospecifc antibodies is used.

STRATEGIC PLANNING

It is mandatory that all glassware and plasticware used with the gels and blot membranes be thoroughly cleaned in order to avoid staining artifacts. All blot membranes should be handled with forceps only. The total protein stains cannot be used for detecting polypeptides blotted onto nylon membranes, because these membranes are positively charged. Furthermore, due to its very high sensitivity, the gold stain is susceptible to impurities present in buffers and on the surface of staining boxes.

Choice of Materials

The following choices should be considered and made prior to beginning the immunoblotting experiment.

Primary antibody

First and foremost, the antibody being used should recognize denatured antigen. Nonspecific binding of antibodies can occur, so control antigens and antibodies should always be run in parallel. Dilution and incubation time for primary antibodies and conjugates should always be optimized.

Membrane blocking agent

A variety of agents are currently used to block binding sites on the membrane after blotting (Harlow and Lane, 1999). These include Tween 20, polyvinylpyrrolidone (PVP), nonfat dry milk, casein, bovine serum albumin (BSA), and serum. A 0.1% solution of Tween 20 in TBS (TTBS), a convenient alternative to protein-based blocking agents, is recommended for chromogenic development of nitrocellulose and PVDF membranes (Blake et al., 1984). In contrast to dry milk/TBS blocking solution (BLOTTO), TTBS is stable and has a long shelf life at 4°C. Furthermore, TTBS generally produces a clean background and permits subsequent staining with India ink.

Even with the application of such standard blocking procedures as 5% to 10% milk protein or 0.05% to 0.1% Tween 20, background can still be a significant problem. If this happens, using a blocking protein (e.g., goat, horse, or rabbit normal serum) from the same species as the primary antibody can reduce the background, presumably by reducing cross-reactivity between the primary antibodies and the blocking agent. If a secondary antibody detection is used, the blocking protein (i.e., normal serum) should be from the same species as the secondary antibody.

Combinations of blocking agents can also be effective. Thus, 0.1% human serum albumin (HSA) and 0.05% Tween 20 in TBS is recommended when probing PVDF membranes with human serum (Craig et al., 1993). However, this can also lead to overall loss of antigen signal, requiring a 10-fold increase in the primary antibody (serum) concentration to achieve an adequate background-free antigen signal.

Compared to dry milk, purified casein has minimal endogenous alkaline phosphatase activity (AP activity leads to high background) and is therefore recommended as a blocking agent for nitrocellulose and PVDF. Because nonfat dry milk and casein may contain biotin that will interfere with avidin-biotin reactions, subsequent steps are performed without protein-blocking agents when using these systems. If background is a problem, highly purified casein (0.2% to 6%) added to the antibody incubation buffers may help.

Type of membrane

Two membrane types, nitrocellulose (Burnette, 1981) and PVDF (Pluskal, 1986), are in wide use for protein blotting applications. Nitrocellulose is the historical choice, and it is still in wide use due to its lower cost and straightforward handling. Nitrocellulose is prepared for blotting by simply wetting it with water or aqueous buffer. In addition, it is tolerant of SDS in the transfer buffer. (SDS is added for more efficient transfer of proteins—particularly high-molecular-weight proteins—out of the gel and onto the membrane.) In contrast, PVDF membranes will not wet in aqueous buffers and must be equilibrated in 100% methanol prior to wetting in transfer buffer. However, PVDF has a much higher protein binding capacity (100 to 200 $\mu g/cm^2$ for PVDF) compared to nitrocellulose (80 to 100 $\mu g/cm^2$). PVDF is tough and will not crack or tear under normal use, unlike nitrocellulose. In addition, PVDF is popular for protein sequencing off of the membranes (e.g., see *CP Protein Science Unit 11.2*; Fernande and Mische, 1995).

General Considerations

When using chemiluminescent detection for immunoblotting, high background frequently occurs, particularly for strong signals (Pampori et al., 1995). Several methods are available for reducing the background from chemiluminescent reactions. These include changing the type and concentration of blocking agents (see above), optimizing antibody concentrations, letting the reaction proceed for several minutes before exposing to film, or simply limiting the exposure time of the film on the blot. These procedures are not always successful, however, and can lead to inconsistent results. An alternative approach is to reduce the concentration of reagents. A 10-fold reduction is a good place to start. This effectively removes the background and has a number of advantages which include lower cost, increased signal-to-noise ratio, and reduced detection of cross-reacting species.

If reprobing is desired, blots can be air dried and stored at 4°C for 3 months after chemiluminescence detection. After drying, store in a sealed freezer bag until use. (The probe is not typically removed before drying and storage.) Repeated probing will lead to a gradual loss of signal and increased background. However, this will depend in part on the properties of the sample.

Time Considerations

The entire immunoblotting procedure can be completed in 1 to 2 days, depending on transfer time and type of gel. Gel electrophoresis requires 4 to 6 hr on a regular gel and 1 hr on a minigel. Transfer time can be 1 hr (high-power transfer) to overnight (Ponceau staining takes <1 hr). The India ink staining procedure requires ~6 hr, while gold staining can be performed in ~3 hr, and an additional 30 min is required at the beginning of either protocol to perform the alkali pretreatment. SYPRO fluorescent staining is completed in <1 hr. Blocking, conjugate incubation, and washing each take

30 min to 1 hr. Finally, substrate incubation requires 10 to 30 min (chromogen) and a few seconds to several hours (luminescence).

PROTOCOLS

The typical steps for performing western blotting are described in Basic Protocols 1 (semidry systems) and 2 (tank transfer systems). Dot and slot blots (Alternate Protocol 1) are alternatives useful for preliminary or routine repeated characterizations. Antibody-antigen complexes on the blotting membranes are identified with horseradish peroxidase (HRPO) or alkaline phosphatase (AP) enzymes coupled to the secondary anti-IgG antibody (e.g., goat anti–rabbit IgG) either directly (Basic Protocol 3) or via an avidin-biotin bridge (Alternate Protocol 2). Chromogenic or luminescent substrates are then used to visualize the activity (see Basic Protocol 4 and Alternate Protocol 3). Support Protocols 1 to 6 are related to total protein staining used to visualize standards and to verify the transfer of proteins from the gels to the blotting membranes prior to immunostaining.

NOTE: All steps requiring membrane washing, staining, and equilibration should use continuous, gentle reciprocating or rotating shaking (e.g., on an orbital shaker at 50 rpm) for efficient and even coverage of the reagents across the membrane. For PVDF membranes, be sure to float the membrane face down on the solution. Staining in separate containers for each membrane is recommended to avoid contact transfer of the blotted proteins. Handle gels and membranes with forceps or powder-free gloves.

Basic Protocol 1: Protein Blotting with Semidry Systems

Even and efficient transfer of most proteins can be accomplished with semidry blotting. The gel is held horizontally between buffer-saturated blotting paper that is in contact with the electrodes (Fig. 8.3.4) and requires only a small amount of buffer. The electrodes are close together, giving high field strengths and rapid transfer with a standard electrophoresis power supply. Prolonged transfers (>1 hr) are not recommended; tank blotting (see Basic Protocol 2) should

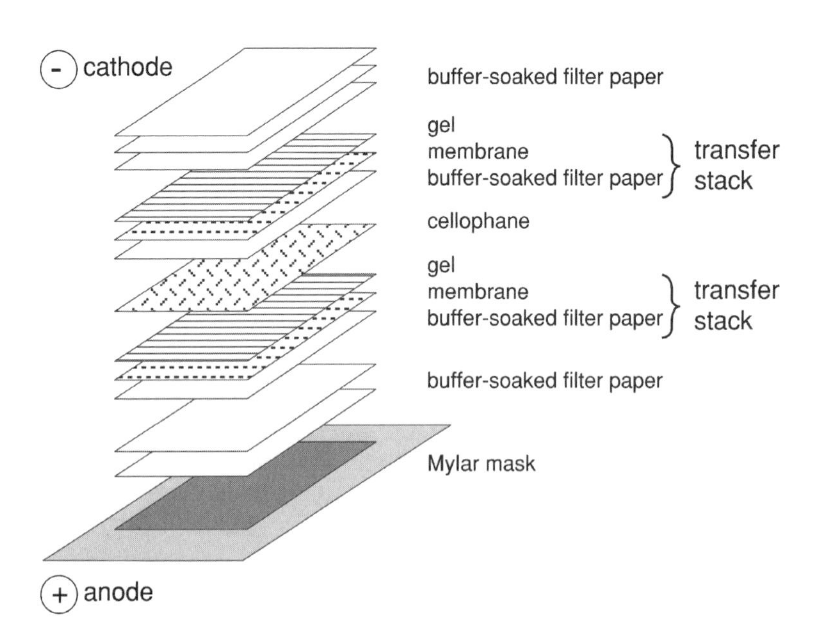

Figure 8.3.4 Immunoblotting with a semidry transfer unit. Generally, the lower electrode is the anode, and one gel is transferred at a time. A Mylar mask (optional in some units) is put in place on the anode. This is followed by three sheets of transfer buffer–soaked filter paper, the membrane, the gel, and finally, three more sheets of buffer-soaked filter paper. To transfer multiple gels, construct transfer stacks as illustrated, and separate each with a sheet of porous cellophane. For transfer of negatively charged protein, the membrane is positioned on the anode side of the gel. For transfer of positively charged protein, the membrane is placed on the cathode side of the gel. Transfer is achieved by applying a maximum current of 0.8 mA/cm^2 of gel area. For a typical minigel (8 × 10 cm) and standard-size gel (14 × 14 cm), this means 60 and 200 mA, respectively.

be used for proteins that require long blotting times for efficient transfer. Because transfer efficiency depends on many factors (e.g., gel concentration and thickness and protein size, shape, and net charge), results may vary. This protocol provides a guideline for 0.75-mm-thick SDS-PAGE gels transferred by semidry blotting.

Materials

Samples for analysis
Protein molecular weight standards (*UNIT 8.4*): prestained (Sigma or Bio-Rad), biotinylated (Vector Labs or Sigma), fluorescent (e.g., BenchMark fluorescent protein standards; Invitrogen), or compatible with other colorimetric and fluorescent detection method (e.g., MagicMark and MagicMark XP western protein standards; Invitrogen); see Table 8.3.4
Transfer buffer (see recipe)
100% methanol

UV transilluminator or overhead illuminator (e.g., UVP)
Transfer membrane: 0.45-µm nitrocellulose (Millipore or Schleicher & Schuell) or polyvinylidene difluoride (PVDF; Millipore Immobilon P)
Plastic trays (polypropylene for PVDF membranes), larger than the gels
Razor blade or spatula
Six sheets of Whatman 3MM filter paper or equivalent, cut to size of gel
Semidry transfer unit (Hoefer, Bio-Rad, or Sartorius)
Glass test tube (for removal of air bubbles by rolling over membrane)
Porous cellophane (Hoefer) or dialysis membrane (Bio-Rad or Sartorius), equilibrated with transfer buffer (see recipe), optional
Indelible pen (e.g., PaperMate) or soft lead pencil

Additional reagents and equipment for performing one-dimensional gel electrophoresis (*UNIT 7.3*) and staining proteins in gels (*UNIT 7.4*)

Electrophorese samples

1. Prepare samples and separate proteins using small or standard-size one-dimensional gels (*UNIT 7.3*) or gradient gels. Include protein markers when running the gels.

   The protein markers will be transferred to the membrane along with the proteins of interest and conveniently indicate membrane orientation and sizes of proteins after immunostaining.

   MagicMark western protein standards allow direct visualization of protein size standards on membrane blots without the need for protein modification or special detection reagents. The standard proteins are derived from E. coli cells containing a construct with repetitive units of a fusion protein forming the size variation and an IgG binding site. These protein standards do not have to be heated or reduced; they are in a ready-to-use format.

Table 8.3.4 Protein Standards for Western Blotting

Protein standard	Application
Unstained	Molecular weight calibration and transfer efficiency; can be visualized with total protein stains
Tagged	Molecular weight calibration and transfer efficiency; visualized during immunodetection steps; a variety of potential tags, including biotinylated and antibody-specific amino acid sequence engineered into standard proteins
Prestained	Excellent for checking transfer efficiency and visual inspection of the blot; typically do not produce as sharp a band as other standards, making precise molecular weight calculations difficult

Table 8.3.5 Recommended Acrylamide Percentages for Resolving Proteins

Percent acrylamide (resolving gel)	Size range transferred (~100% efficiency) in kDa
5-7	29-150
8-10	14-66
13-15	<36
18-20	<20

The proteins can be visualized with the colorimetric, chemiluminescent, or fluorescent detection system of choice simply by processing the membrane for the specific protein. The IgG binding site will allow all the standard proteins to react with the specific primary and secondary antibodies.

Alternatively, BenchMark Fluorescent Protein Standards are visualized directly via UV transillumination on a UV transilliminator (available from UVP) when wet, or via overhead UV illumination (apparatus also available from UVP) when dry.

A variety of gel sizes and percentages of acrylamide can be used (Table 8.3.5). Most routinely used are either 14 cm × 14 cm × 0.75–mm gels or 8 cm × 10 cm × 0.75–mm minigels. Acrylamide concentrations vary from 5% to 20%, but are usually in the 10% to 15% range.

Prepare transfer membrane

2. Cut the membrane to the same size as the gel plus 1 to 2 mm on each edge.

For nitrocellulose membranes

3a. Slowly place the membrane into distilled water in a plastic tray, holding one edge at a 45° angle.

The water will wick up into the membrane, wetting the entire surface. If it is inserted too quickly into the water, air gets trapped and will appear as white blotches in the membrane; protein will not transfer onto these areas.

Precut membranes, matched to precast minigels, are convenient and minimize handling that can damage the membrane and lead to artifacts.

4a. Decant the water and equilibrate 10 to 15 min in transfer buffer.

For PVDF membranes

3b. Immerse 1 to 2 sec in 100% methanol in a polypropylene plastic tray.

PVDF membranes are hydrophobic and will not wet simply from being placed into distilled water or transfer buffer.

4b. Decant the methanol and equilibrate 5 min with transfer buffer. Do not let membrane dry out at any time. If this occurs, wet the membrane once again with methanol and transfer buffer as described above.

Assemble transfer stack

5. Disassemble the gel sandwich after completion of SDS-PAGE separation. Excise the stacking gel (if present) with a razor blade or spatula and discard.

6. *Optional:* Equilibrate the separating gel 10 to 15 min in transfer buffer.

Equilibration of the separating gel with transfer buffer is not normally required for semidry blotting, but it may improve transfer in some cases.

7. Place three sheets of filter paper saturated with transfer buffer on the anode (Fig. 8.3.4) of the semidry transfer unit.

   Most transfer units are designed so that negatively charged proteins move downward toward either a platinum or graphite positive electrode (anode).

   The filter paper should be cut to the exact size of the gel. This forces the current to flow only through the gel and not through overlapping filter paper. Some manufacturers (e.g., Hoefer) recommend placing a Mylar mask on the lower platinum anode. With an opening that is slightly less than the size of the gel, the mask forces the current to flow through the gel and not the surrounding electrode area during transfer.

8. Place the equilibrated transfer membrane on top of the filter paper stack. Remove all bubbles between the membrane and filter paper by rolling a test tube over the surface of membrane.

   Any bubbles in the filter paper stack or between the filter paper, membrane, and gel will block current flow and prevent protein transfer. This problem is indicated on the membrane by sharply defined white areas devoid of transferred protein.

9. Place the gel on top of the membrane. Gently roll a test tube over the surface of gel to insure intimate contact between gel and membrane and to remove any interfering bubbles.

   Poor contact between the gel and membrane will cause a swirled pattern of transferred proteins on the membrane. Some proteins will transfer as soon as the gel is placed on the membrane; repositioning the gel or membrane can result in a smeared or double image on the developed blot.

10. Complete the transfer stack by putting the three remaining sheets of filter paper on top of the gel. Roll out bubbles as described above.

    Multiple gels can be transferred using semidry blotting. Simply put a sheet of porous cellophane (Hoefer) or dialysis membrane (Bio-Rad or Sartorius) equilibrated with transfer buffer between each transfer stack (Fig. 8.3.4). Transfer efficiency is dependent on the position of the transfer stack in the blotting unit, and for critical applications, transferring one gel at a time is recommended. The gel next to the anode tends to be more efficiently transferred when blotting more than one gel at a time.

Transfer proteins from gel to membrane

11. Place top electrode onto transfer stack.

    Most units have safety-interlock features and can only be assembled one way. Consult manufacturer's instructions for details.

    Once the transfer stack has been assembled with both electrodes, do not move the top electrode. This can shift the transfer stack and move the gel relative to the membrane. Some transfer will occur as soon as the gel contacts the membrane, and any shifting of the transfer stack after assembly will distort the transfer pattern.

12. Carefully connect high-voltage leads to the power supply (see *UNIT 7.1* for safety precautions). Apply constant current to initiate protein transfer.

    Transfers of 1 hr are generally sufficient.

    In general, do not exceed 0.8 mA/cm^2 of gel area, or overheating and drying of the gel will result in poor transfer efficiency and resolution. For a typical minigel (8 × 10 cm) and standard-size gel (14 × 14 cm) this means ~60 and 200 mA, respectively.

    Monitor the temperature of the transfer unit directly above the gel by touch. The unit should not exceed 45°C. If the outside of the unit is warm, too much current is being applied. Note that units with graphite electrodes are more prone to heating, because graphite has much more resistance to current flow than platinum or steel electrodes.

13. After transfer, turn off the power supply and disassemble the unit. Remove the membrane from the transfer stack and mark the side facing the gel and the top orientation with a soft lead pencil.

14. Proceed with staining for total protein (Support Protocols 1 to 6) and immunoprobing (see Basic Protocol 3 or Alternate Protocol 2).

Basic Protocol 2: Tank Transfer

In this method the gel and membrane are placed vertically into a tank filled with transfer buffer (Fig. 8.3.5). Tank blotting is the historical method of choice for immunoblotting. In contrast to the semidry technique (Basic Protocol 1), tank systems are capable of the high power and prolonged (overnight) transfers needed for difficult-to-transfer proteins (e.g., high-molecular-weight proteins from gradient gels). However, tank blotting is more complex to set up and uses large volumes of liquid. For simple routine work, the semidry technique is more convenient.

Materials

Samples for analysis
Protein molecular weight standards (*UNIT 8.4*): prestained (Sigma or Bio-Rad), biotinylated (Vector Labs or Sigma), fluorescent (e.g., Benchmark fluorescent protein standards; Invitrogen), or compatible with other colorimetric and fluorescent detection method (e.g., MagicMark and MagicMark XP western protein standards; Invitrogen); see Table 8.3.4
Transfer buffer (see recipe)
100% methanol

Razor blade or spatula
Plastic tray (polypropylene for PVDF membranes), larger than the gel
0.45-μm nitrocellulose (Millipore or Schleicher & Schuell) or polyvinylidene difluoride (PVDF; Millipore Immobilon P)
Transfer tank blotting apparatus and cassette with sponge (Hoefer, Bio-Rad, or Invitrogen; see Fig. 8.3.5)
Six sheets of Whatman 3MM filter paper or equivalent, cut to size of gel
Glass test tube (optional)
Heat exchanger and cooling recirculating water bath (optional)

Additional reagents and equipment for one-dimensional or gradient gel electrophoresis (*UNIT 7.3*) and staining proteins in gels (*UNIT 7.4*)

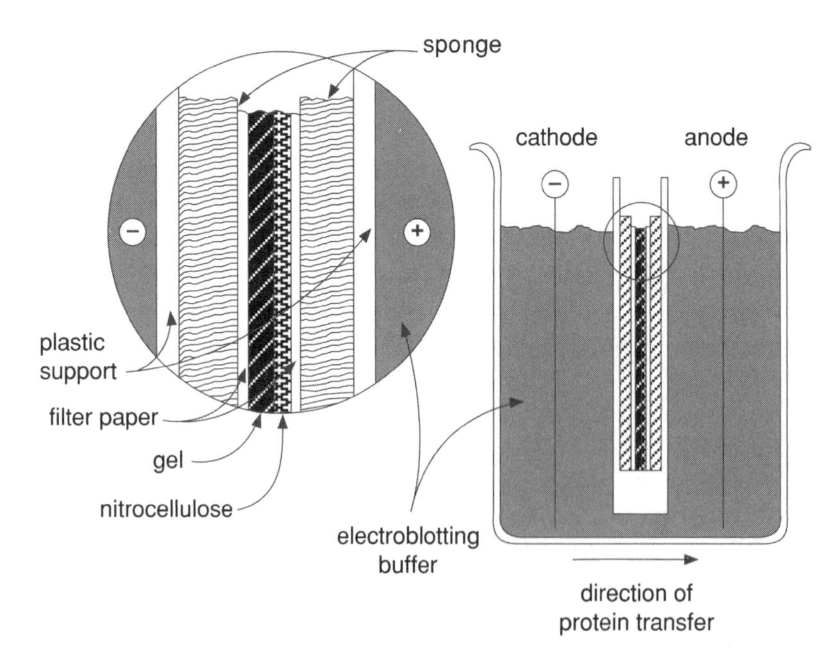

Figure 8.3.5 Immunoblotting with a tank transfer unit. The gel/membrane sandwich is held in a transfer cassette between two pads and assembled in the following order: three sheets of transfer buffer–soaked filter paper, transfer buffer–equilibrated gel, wet membrane, and three sheets of transfer buffer soaked filter paper. The sandwich is placed in the cassette and immersed in transfer buffer between the electrodes. For transfer of negatively charged protein, the membrane is positioned on the anode (+) side of the gel. For transfer of positively charged protein, the membrane is placed on the cathode side of the gel.

Prepare gel

1. Prepare samples and separate proteins using small or standard-size one-dimensional gels (*UNIT 7.3*) or gradient gels. Include protein markers when running the gels.

2. Remove the gel from the cassette or glass sandwich after completion of SDS-PAGE separation and excise the low percentage stacking gel (if present) with a razor blade or spatula.

   Precast gradient gels typically do not have stacking gels included.

3. Place the gel in a plastic tray with enough transfer buffer for the gel to freely float and equilibrate 5 to 15 min, with gentle shaking.

   Due to the differences between the electrophoresis gel buffer and the transfer buffer (pH, buffer strength, alcohol content) it is important to equilibrate the gel in the transfer buffer. This ensures equivalent pH and ionic strength between the gel and the transfer buffer, and that any shrinkage of the gel occurs before the actual transfer. Note the gel will also shrink in typical transfer buffer because of the 20% methanol, and shrinkage during the actual blotting will decrease transfer resolution. Low percentage or thin gels (<10%, 1 mm) should keep equilibration to a minimum (5 to 10 min) so no proteins are lost by diffusion out of the gel.

Prepare membrane

4. Cut the membrane to same size as gel plus 1 to 2 mm on each edge.

For nitrocellulose membranes

5a. Slowly place the membrane into distilled water in a plastic tray, holding one edge at a 45° angle.

   The water will wick up into the membrane, wetting the entire surface. If it is inserted too quickly into the water, air gets trapped and will appear as white blotches in the membrane; protein will not transfer onto these areas.

   Precut membranes, matched to precast minigels, are convenient and minimize handling that can damage the membrane and lead to artifacts.

6a. Decant the water and equilibrate 5 min in transfer buffer.

For PVDF membranes

5b. Immerse 1 to 2 sec in 100% methanol in a polypropylene plastic tray.

   PVDF membranes are hydrophobic and will not wet simply from being placed into distilled water or transfer buffer.

6b. Decant the methanol and equilibrate 5 min with transfer buffer. Do not let membrane dry out at any time. If this occurs, wet the membrane once again with methanol and transfer buffer as described above.

Assemble transfer sandwich

7. Immerse the sponge and filter paper in transfer buffer prior to assembly.

8. Start with one layer of sponge material and add two layers of buffer-saturated filter paper, equilibrated gel, membrane, and two sheets of buffer-saturated filter paper.

9. With a gloved hand push out any air trapped between the layers. Alternatively, roll a glass test tube over each layer during assembly to ensure good contact between the membrane and gel and the removal of trapped bubbles between the layers.

   The membrane side should face the positive electrode.

10. Fill the transfer tank with enough buffer to cover the top of the panel by at least 2 cm. Place the assembled cassette into the tank.

   This insures complete heat removal during transfer.

Table 8.3.6 Rapid High-Power Transfer Conditions for Tank Transfer

Gel/unit size	Number of gels	Voltage	Current	Transfer time at 15°C[a]
9 × 10 cm	1-4 gels	100 V	400 mA	1 hr
15 × 21 cm	1-4 gels	100 V	1 A	1-3 hr

[a]High-power transfer requires a transfer unit with a heat exchanger and a cooling recirculating water bath. For uncooled transfer, set the voltage to 10 V overnight constant voltage.

Transfer proteins from gel to membrane

11. Connect the power supply and set the cooling temperature (if available) to 15°C to 20°C.

    For high power transfers, cooling is required (see Table 8.3.6). For optimal reproducibility, use the same temperature for the transfer.

12. Carefully connect high-voltage leads to the power supply (see UNIT 7.1 for safety precautions). Electrophoretically transfer proteins from gel to membrane for 30 min to 1 hr at 100 V with cooling (see Table 8.3.6) or overnight at 14 V (constant voltage), in a cold room.

    Overnight transfers can be performed in a cold room at 10°C to 14°C without an external cooling unit. Lower percentage (protein of interest migrates to <0.5 R_f; see UNIT 7.3) and thinner gels (0.75 mm thick) are recommended for efficient transfer. Improved transfer efficiency may require SDS (up to 0.1%) in the transfer buffer and reduction or elimination of methanol. However, this may lead to reduced binding of the protein to the membrane, and conditions need to be optimized.

13. After transfer, turn the voltage on the power supply to 0 and turn off the power. Remove the safety lid and pull the transfer cassette out of the tank. Remove the membrane from the transfer stack and mark the side facing the gel and the top orientation with a soft lead pencil.

14. Stain and destain the gel (e.g., Coomassie blue; see UNIT 7.4).

    Any protein remaining in the gel indicates an incomplete transfer. Transfers are seldom 100%.

15. Stain the membrane with Ponceau S (Support Protocol 1) or SYPRO Ruby (Support Protocol 5) to confirm protein transfer from the gel to the membrane.

Alternate Protocol 1: Slot and Dot Blotting

Through use of a vacuum manifold or by simple hand spotting, up to 96 samples can be applied to a single nitrocellulose (NC) or polyvinylidene difluoride (PVDF) membrane for immunoblotting analysis. In slot and dot blotting, proteins are not separated by electrophoresis before immunoblot analysis. Instead, the entire sample is directly applied to (or spotted on) the membrane. Although this approach cannot discriminate between the protein of interest and a cross-reactive antigen, it is a quick way to perform preliminary characterization and high-volume routine quantitation of a sample.

Additional Materials (also see Basic Protocol 1)

<10 μg protein sample in <100 μl water or TBS (no detergent)
Tris-buffered saline (TBS; UNIT 3.3)

Slot and dot blotting apparatus (e.g., Hoefer, Bio-Rad, or Whatman)
Vacuum source

1. Prewet the membrane in distilled water (NC) or 100% methanol (PVDF) as described in Basic Protocol 1, steps 4 and 5.

2. Prepare the slot or dot blot manifold according to manufacturer's instructions.

3. Apply samples under a low house vacuum, followed by two equivalent volumes water or TBS, depending upon which was used to dissolved the sample, to rinse the wells of unbound protein.

   Typically, samples should contain <10 μg in under 100 μl. Overloading the wells will prevent the flow of the liquid through the well.

   Typical systems use a low house vacuum to pull liquid through each slot, binding the proteins to the membrane.

Support Protocol 1: Ponceau S Staining of Transferred Proteins

Nitrocellulose and PVDF membranes can be reversibly stained with Ponceau S to verify transfer efficiency and indicate the positions of the molecular weight markers. The stain is compatible with subsequent immunostaining procedures.

Materials

Membrane with transferred proteins (Basic Protocol 1 or 2 or Alternate Protocol 1)
Ponceau S solution (see recipe)

Plastic boxes
Pen with indelible ink

Additional reagents and equipment for photographing membranes (UNIT 7.5)

1. Stain the membrane in Ponceau S solution for 5 min at room temperature, with gentle agitation.

2. Destain 2 min in water.

3. Photograph membrane if required (UNIT 7.5) and mark the molecular-weight-standard band locations with indelible ink.

4. Completely destain membrane by soaking an additional 10 min in water and proceed to immunoprobing.

Support Protocol 2: India Ink Staining of Transferred Proteins

India ink is used to stain total protein on blot transfer membranes. The transferred proteins (~50 ng) appear as black bands on a gray background.

Materials

Membrane with transferred proteins (Basic Protocol 1 or 2 or Alternate Protocol 1)
Tween 20 solution (see recipe)
India ink solution (see recipe)
Plastic boxes

1. *Optional:* Pretreat proteins transferred onto nitrocellulose membrane with alkali (see Support Protocol 4).

2. Place blot transfer membrane(s) in a plastic box containing enough Tween to cover the membranes and wash in Tween 20 solution three times for 30 min each time at 37°C, with gentle shaking on an orbital shaker.

3. Wash the membrane in Tween 20 solution two times for 30 min each time at room temperature.

4. Stain the membrane in India ink solution 3 hr or overnight, room temperature.

5. Rinse the membrane twice in Tween 20 solution, destain in Tween 20 solution until an acceptable background is obtained, and then air dry the membrane for storage.

   Black bands appear against a gray background. While the membranes can be photographed dry or wet, the contrast of the bands will be greater in a wet membrane.

Support Protocol 3: Gold Staining of Transferred Proteins

A colloidal gold sol is used to stain proteins on blot transfer membranes. The transferred proteins (~3 ng) will appear as red bands on an almost white background.

NOTE: Do not attempt to stain nylon membranes using colloidal gold.

Materials

Nitrocellulose membrane with transferred proteins (Basic Protocol 1 or 2 or Alternate Protocol 1)
Tween 20 solution (see recipe)
Colloidal gold staining solution (Bio-Rad, Sigma, GE Healthcare)

Plastic boxes
Glass dish or heat-sealable plastic bags
Filter paper

1. *Optional:* Pretreat proteins on the nitrocellulose membrane with alkali (see Support Protocol 4).

2. Place blot transfer membrane(s) in a plastic box containing enough Tween 20 solution to cover the membranes. Wash in Tween 20 solution three times for 30 min each time at 37°C, with gentle shaking on an orbital shaker.

3. Continue to wash the membrane in Tween 20 solution three times for 5 min each time at room temperature.

4. Rinse well with water.

5. Place nitrocellulose blot or dot membrane with transferred proteins in a glass dish containing enough colloidal gold staining solution to cover the membrane. Stain the membrane 30 min to 1 hr at room temperature, with continuous shaking.

   A heat-sealed plastic bag is a convenient container for the staining. Extended staining periods do not impair visualization.

6. Rinse the membrane briefly in water and air dry on filter paper.

Support Protocol 4: Alkali Enhancement of Protein Staining

A brief pretreatment of nitrocellulose-bound protein with alkali enhances subsequent staining with either India ink or colloidal gold. This protocol is specifically for use with nitrocellulose membranes.

Materials

Nitrocellulose membrane with transferred proteins (Basic Protocol 1 or 2 or Alternate Protocol 1)
1% (w/v) KOH
Phosphate-buffered saline (PBS; UNIT 3.3)
Glass or Pyrex dish

1. Place nitrocellulose blot or dot membrane with transferred proteins in a glass dish containing enough 1% KOH to cover the membranes. Soak 5 min at room temperature, with gentle agitation.

2. Rinse twice in PBS for 10 min each time at room temperature.

Support Protocol 5: Fluorescent Protein Blot Staining of Transferred Proteins

The fluorescent SYPRO Ruby protein blot stain provides a rapid, simple, and highly sensitive method for detecting proteins on nitrocellulose or PVDF membranes (blots). Staining total protein before applying specific protein detection techniques provides an assessment of protein transfer efficiency, and makes it possible to detect contaminating proteins in the sample and to compare the sample with molecular weight standards. For blots of two-dimensional gels, total protein staining makes it easier to localize a protein to a particular spot in the complex protein pattern. The bright, orange-red fluorescent stain can be easily visualized using UV illumination or a laser scanner. The staining procedure is simple to perform and can be completed within 1 hr. SYPRO Ruby has a sensitivity limit of 2 to 8 ng/band, making it about 60 times more sensitive than reversible stains like Ponceau S and 20 to 30 times more sensitive than Amido Black or Coomassie Brilliant Blue stains. The SYPRO Ruby protein blot stain will not stain nucleic acids and is compatible with immunodetection and colorimetric, fluorogenic, and chemiluminescent detection techniques, as well as with Edman sequencing and mass spectrometry.

Materials

Nitrocellulose or PVDF membrane with transferred proteins (Basic Protocol 1 or 2 or Alternate Protocol 1)
7% (v/v) acetic acid/10% (v/v) methanol
SYPRO Ruby protein blot stain (purchase from Molecular Probes; also see recipe)
150 mM Tris·Cl, pH 8.8 (*UNIT 3.3*)/20% (v/v) methanol

Small polypropylene staining dish
Orbital shaker
Forceps

NOTE: Perform all washing, staining, and other incubation steps with continuous, gentle agitation (e.g., on an orbital shaker at 50 rpm). For PVDF membranes, be sure to float the membrane face down on the solution.

After electroblotting to nitrocellulose membranes

1a. Completely immerse the membrane in 7% acetic acid/10% methanol and incubate 15 min at room temperature in a small polypropylene staining dish.

2a. Wash the membrane in four changes of deionized water for 5 min each time.

3a. Completely immerse the membrane in SYPRO Ruby protein blot stain for 15 min.

4a. Wash the membrane in deionized water four to six times for 1 min each time to remove excess dye.

Membranes stained with SYPRO Ruby blot stain should be periodically monitored using UV overhead illumination to determine if background fluorescence has been washed away.

After electroblotting to PVDF membranes

1b. Allow the membrane to dry completely.

2b. Float the membrane face down in 7% acetic acid/10% methanol and incubate 15 min.

3b. Float the membrane for 5 min each time in four changes of deionized water, then float the membrane in SYPRO Ruby protein blot stain for 15 min.

4b. Wash the membrane two or three times in deionized water for 1 min each time to remove excess dye.

For subsequent Edman-based microsequencing

1c. Allow a PVDF electroblotted membrane to dry completely.

2c. Stain with SYPRO Ruby protein blot stain as described in steps 2b and 3b.

3c. Partially destain the blot by placing it face down in a solution of 150 mM Tris·Cl, pH 8.8 (*UNIT 3.3*)/20% methanol for 10 min, with gentle agitation.

4c. Rinse the blot four times in deionized water for 1 min each time.

5. Allow the membranes to air dry.

   After staining, wet membranes should not be touched because residue found on latex gloves may destroy the staining pattern. Use forceps to handle wet blots. Once dry, the membranes can be handled freely.

6. Proceed to Support Protocol 6 to view and photograph the protein blots.

Support Protocol 6: Viewing and Photographing SYPRO Ruby-Stained Protein Blots

SYPRO Ruby protein blot stain (Support Protocol 5) has two excitation maxima, one at ~280 nm and one at ~450 nm, and it has an emission maximum near 618 nm. Proteins stained with the dye can be visualized using a 300-nm UV overhead or transilluminator, a blue-light transilluminator, or a laser scanner. The stain is photostable, allowing long exposure times for maximum sensitivity. It is important to photograph or otherwise document the SYPRO Ruby stain before immunostaining because over 90% of the stain is washed off the blot during the blocking step. Detection limits and compatibilities of transfer membranes and gel types are summarized in Table 8.3.2.

UV Overhead Illuminator or Transilluminator

Proteins stained with SYPRO Ruby protein blot stain are readily visualized using UV illumination. The front face of the membrane can be illuminated using a hand-held UV-B (~300 nm) light source. Alternatively, a UV light box can be placed on its side to illuminate the blots, or a top-illuminating system such as the UVP Biochemi System or the Bio-Rad Fluor-S imager can be used to visualize the stain. Satisfactory results can also be obtained from direct UV transillumination through the blotting membrane. In either case, the use of a photographic or CCD camera and the appropriate filters is essential to obtain the greatest sensitivity. The camera's integrating capability can make bands visible that cannot be detected by eye. It is important to clean the surface of the transilluminator after each use with deionized water and a soft cloth (like cheesecloth); otherwise, fluorescent dyes (e.g., SYPRO stains, SYBER stains, and ethidium bromide) will accumulate on the glass surface and cause a high background fluorescence.

Digital documentation is the method of choice, and various types of cameras are available, ranging from simple digital compact consumer cameras to quantitative high–dynamic range cooled scientific imaging cameras and systems (*UNIT 7.5*). When using a CCD camera, the best images are obtained with high resolution (1.4 megapixel or greater) and 12- or 16-bit gray scale levels per pixel. The manufacturer of the particular imaging system should be contacted for recommendations on filter sets to use.

For those still in the analog world and using black-and-white print film, the highest sensitivity is achieved with a 490-nm long-pass filter (e.g., the SYPRO protein gel stain photographic filter; Molecular Probes S-6656). Blots are typically photographed using an *f*-stop of <4.5 for under 1 sec, using ASA 400 film.

Laser-Scanning Instruments

Blots stained with the SYPRO Ruby protein blot stain can be visualized using imaging systems equipped with lasers that emit at 450, 473, 488, or 532 nm.

Basic Protocol 3: Immunoprobing with Directly Conjugated Secondary Antibody

Immobilized proteins are probed with specific antibodies to identify and quantitate any antigens present. The membrane is immersed in blocking buffer to fill all protein-binding sites with a nonreactive protein or detergent. (Tween 20 is a common alternative to protein-blocking agents when using nitrocellulose or PVDF filters.) Next, it is placed in a solution containing the primary antibody (the antibody directed against the antigen). The blot is washed and exposed to the conjugated secondary antibody (an enzyme-antibody conjugate directed against the primary antibody; e.g., goat anti-rabbit IgG). Antigens are identified by chromogenic or luminescent visualization (see Basic Protocol 3 or Alternate Protocol 4, respectively) of the antigen/primary antibody/secondary antibody/enzyme complex bound to the membrane. The antibody probes are commercially available for a wide range of proteins and can also be prepared in the lab for a new protein so that simplified assays (e.g., for expression, subcellular location, or amount) can be developed. Custom antibodies based on a purified antigen can be prepared commercially (e.g., Covance Immunological Services). A quick way to purify a protein for antibody analysis is by SDS-PAGE. Although denatured, SDS-PAGE-isolated proteins typically generate antibodies that react well against separated and blotted proteins. Several hundred micrograms of highly purified protein can be obtained in this way, and equipment (e.g., Hoefer) is available for preparative electrophoresis. Once the protein is separated, it can be electroeluted from the gel (using, e.g., Bio-Rad PrepCell), concentrated, and used to generate polyclonal antibodies in the animal of choice (e.g., rabbit, goat, or chicken).

Materials

Membrane with transferred proteins (Basic Protocol 1 or 2 or Alternate Protocol 1)
Blocking buffer for colorimetric detection (see recipe) or blocking buffer for luminescence detection (see recipe)
Primary antibody specific for protein of interest (working concentration optimized; see Fig. 8.3.6)
TTBS (nitrocellulose or PVDF membranes) or TBS (nylon membranes; see *UNIT 3.3* for recipes)
Secondary antibody conjugate: horseradish peroxidase (HRPO)- or alkaline phosphatase (AP)-anti-Ig conjugate (MB Biomedical, Vector Labs, KPL, or Sigma-Aldrich; dilute as indicated by manufacturer

Heat-sealable plastic bag, plastic box, or slotted incubation tray
Orbital shaker or rocking platform

1. Place the membrane in a heat-sealable plastic bag with 5 ml blocking buffer and seal the bag. Incubate 30 min to 1 hr at room temperature, with agitation on an orbital shaker or rocking platform.

   Usually 5 ml buffer is sufficient for two to three membranes (14 × 14–cm size). If the membrane is to be stripped and reprobed (see CP Molecular Biology Unit 10.8; Gallagher et al., 2004), blocking buffer must contain casein (for AP systems) or nonfat dry milk.

   Plastic incubation trays are often used in place of heat-sealable bags, and can be especially useful when processing large numbers of strips in different primary antibody solutions.

2. Dilute primary antibody in blocking buffer.

   Primary antibody dilution is determined empirically but is typically 1/100 to 1/1000 for a polyclonal antibody (Fig. 8.3.6) or 1/10 to 1/100 for hybridoma supernatants and ~1/1000 for murine ascites fluid containing monoclonal antibodies. Ten to one hundred-fold higher dilutions can be used with alkaline phosphatase-based or luminescence-based detection systems. Both primary and secondary antibody solutions can be used at least twice, but long-term storage (i.e., >2 days at 4°C) is not recommended. Consult the antibody supplier's instructions for alternate dilution and blocking solution recommendations.

   To determine the appropriate concentration of the primary antibody, a dilution series is easily performed with membrane strips. Separate antigens on a preparative gel (i.e., a single large sample well) and immunoblot the entire gel. Cut 2- to 4-mm strips by hand or with a membrane cutter

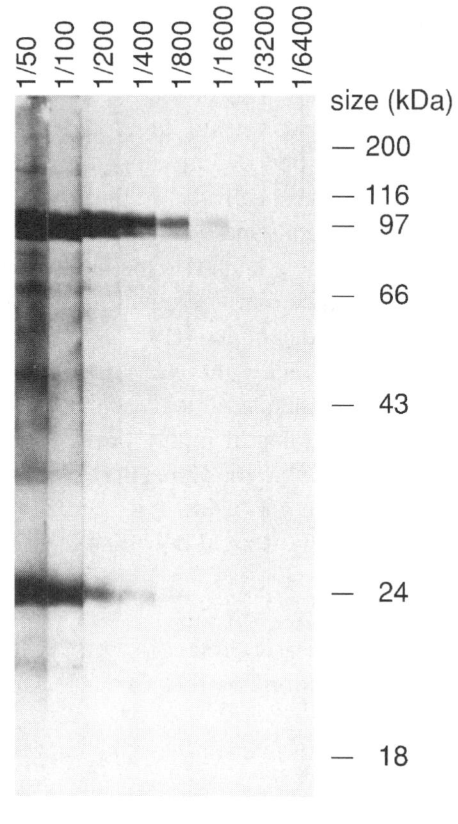

Figure 8.3.6 Serial dilution of primary antibody directed against the 97-kDa catalytic subunit of the plant plasma membrane ATPase. The blot was developed with HRPO-coupled avidin-biotin reagents according to the Alternate Protocol 2 and visualized with 4-chloro-1-naphthol (4CN; Basic Protocol 4). Note how background improves with dilution. In this case, a dilution of 1/400 to 1/800 is optimal, keeping the background to a minimum.

(Schleicher and Schuell; Inotech) and incubate individual strips in a set of serial dilutions of primary antibody. The correct dilution should give low background and high specificity (Fig. 8.3.6).

3. Open the bag and pour out the blocking buffer. Replace with diluted primary antibody and incubate 30 min to 1 hr at room temperature, with constant agitation.

   Usually 5 ml diluted primary antibody solution is sufficient for two to three membranes (14 × 14–cm size). Incubation time may vary depending on the conjugate used.

   When using plastic trays, the primary and secondary antibody solution volume should be increased to 25 to 50 ml. For membrane strips, incubation trays with individual slots are recommended. Typically, 0.5 to 1 ml solution/slot is needed.

4. Remove the membrane from the plastic bag with a gloved hand. Place in plastic box and wash four times with 200 ml TTBS (nitrocellulose or PVDF) or TBS (nylon) for 10 to 15 min each time, with agitation.

5. Dilute the secondary antibody (HRPO- or AP-anti-Ig conjugate) in blocking buffer.

   Commercially available enzyme–conjugated secondary antibody is usually diluted 1/200 to 1/2000 prior to use (Harlow and Lane, 1999).

6. Place the membrane in a new heat-sealable plastic bag, add diluted conjugated secondary antibody, seal, and incubate 30 min to 1 hr at room temperature, with constant agitation.

   When using plastic incubation trays, see step 3 annotation for proper antibody solution volumes.

7. Remove membrane from bag and wash as in step 4. Develop the color or luminescence according to appropriate visualization protocol (see Basic Protocol 4 or Alternate Protocol 3).

Alternate Protocol 2: Immunoprobing with Avidin-Biotin Coupling to Secondary Antibody

The following procedure is based on the Vectastain ABC kit from Vector Labs. It uses an avidin-biotin complex to attach horseradish peroxidase (HRPO) or alkaline phosphatase (AP) to the biotinylated secondary antibody. Avidin-biotin systems are capable of extremely high sensitivity due to the multiple reporter enzymes bound to each secondary antibody. In addition, the detergent Tween 20 is a popular alternative to protein-blocking agents when using nitrocellulose or PVDF filters.

Additional Materials (also see Basic Protocol 3)

Vectastain ABC (HRPO) or ABC-AP (AP) kit (Vector Labs) containing:
  Reagent A (avidin),
  Reagent B (biotinylated HRPO or AP)
  Biotinylated secondary antibody (request membrane immunodetection protocols when ordering)

1. Equilibrate the membrane in appropriate blocking buffer in heat-sealed plastic bag with constant agitation using an orbital shaker or rocking platform. For nitrocellulose and PVDF, incubate 30 to 60 min at room temperature.

   TTBS is well suited for avidin-biotin systems and is typically the only blocking agent needed. Because nonfat dry milk contains residual biotin, which will interfere with the immunoassay, it is typically not used in avidin-biotin systems.

   Plastic incubation trays are often used in place of heat-sealable bags, and can be especially useful when processing large numbers of strips in different primary antibody solutions.

2. Dilute the primary antibody solution in TTBS (nitrocellulose or PVDF) or TBS (nylon).

   Dilutions of sera containing primary antibody generally range from 1/100 to 1/100,000. This depends in large part on the sensitivity of the detection system. With high-sensitivity avidin-biotin systems, dilutions from 1/1000 to 1/100,000 are common. Higher dilutions can be used with AP- or luminescence-based detection systems.

   To determine the appropriate concentration of the primary antibody, a dilution series is easily performed with membrane strips. Separate antigens on a preparative gel (i.e., a single large sample well) and immunoblot the entire gel. Cut 2- to 4-mm strips by hand or with a membrane cutter (Schleicher and Schuell; Inotech) and incubate individual strips in a set of serial dilutions of primary antibody. The correct dilution should give low background and high specificity (Fig. 8.3.6).

3. Open bag, remove blocking buffer, and add enough primary antibody solution to cover membrane. Incubate 30 min at room temperature with gentle rocking.

   Usually 5 ml of diluted primary antibody solution is sufficient for two to three membranes (14 × 14–cm size). Incubation time may vary depending on conjugate used.

   When using plastic trays, the primary and secondary antibody solution volume should be increased to 25 to 50 ml. For membrane strips, incubation trays with individual slots are recommended. Typically, 0.5 to 1 ml solution/slot is needed.

4. Remove the membrane from the bag and place in the plastic box. Wash the membrane three times over a 15-min span in TTBS (nitrocellulose or PVDF). Add enough TTBS to fully cover the membrane (e.g., 5 to 10 ml/strip or 25 to 50 ml/whole membrane).

5. Prepare biotinylated secondary antibody solution by diluting two drops biotinylated antibody with 50 to 100 ml TTBS (nitrocellulose or PVDF).

   This dilution gives both high sensitivity and enough volume to easily cover a large 14 × 14–cm membrane.

6. Transfer the membrane to a fresh plastic bag containing the secondary antibody solution. Incubate 30 min at room temperature, with slow rocking, and then wash as in step 4.

   When using plastic incubation trays, see the step 3 annotation for proper antibody solution volumes.

7. While the membrane is being incubated with secondary antibody, prepare avidin-biotin-HRPO or -AP complex.

   a. Add two drops Vectastain reagent A and two drops reagent B to 10 ml TTBS (nitrocellulose or PVDF).

   b. Incubate 30 min at room temperature.

   c. Add 40 ml TTBS.

   Diluting the A and B reagents to 50 ml expands the amount of membrane that can be probed without greatly affecting sensitivity. Sodium azide is a peroxidase inhibitor and should not be used as a preservative. Casein, nonfat dry milk, serum, and some grades of BSA may interfere with the formation of the avidin-biotin complex and should not be used when employing avidin or biotin reagents (Vector Labs; Gillespie and Hudspeth, 1991).

8. Transfer the membrane to avidin-biotin-enzyme solution. Incubate 30 min at room temperature, with slow rocking, and then wash over a 30-min span as in step 4.

9. Develop the color or luminescence according to the appropriate visualization protocol (see Basic Protocol 4 or Alternate Protocol 3).

Basic Protocol 4: Visualization with Chromogenic Substrates

Bound antigens are typically visualized with chromogenic substrates. The substrates 4CN, DAB/$NiCl_2$, and TMB are commonly used with horseradish peroxidase (HRPO)–based immunodetection procedures, while BCIP/NBT is recommended for AP-based procedures (see Table 8.3.3). After incubation with primary and secondary antibodies, the membrane is placed in the appropriate substrate solution. Protein bands usually appear within a few minutes.

Materials

Membrane with transferred proteins, probed with antibody-enzyme complex (see Basic Protocol 3 or Alternate Protocol 2)
Tris-buffered saline (TBS; *UNIT 3.3*)
Chromogenic visualization solution (see Table 8.3.3)

Additional reagents and equipment for photographing gels (see *UNIT 7.5*)

1. If the final membrane wash (see Basic Protocol 2, step 7 or Alternate Protocol 2, step 8) was performed in TTBS, wash membrane in 50 ml TBS 15 min at room temperature.

   The Tween 20 in TTBS interferes with 4CN development (Bjerrum et al., 1988).

2. Place the membrane into the chosen chromogenic visualization solution.

   Bands should appear in 10 to 30 min.

3. Terminate the reaction by washing the membrane in distilled water. Air dry and photograph (*UNIT 7.5*) for a permanent record.

Alternate Protocol 3: Visualization with Luminescent Substrates

Antigens can also be visualized with luminescent substrates. Detection with light offers both speed and enhanced sensitivity over chromogenic and radioisotopic procedures. After the final wash, the blot is immersed in a substrate solution containing luminol for horseradish peroxidase (HRPO) systems or dioxetane phosphate for alkaline phosphatase (AP) systems, sealed in thin plastic wrap, and placed firmly against film. Exposures range from a few seconds to several minutes, although strong signals typically appear within 1 min.

Additional Materials (also see Basic Protocol 3)

Luminescent substrate buffer: 50 mM Tris·Cl, pH 7.5 (for HRPO; UNIT 3.3)
Luminescent visualization solution or kit for HRPO or AP (Table 8.3.3)
Clear plastic wrap
Film cassette

Additional reagents and equipment for autoradiography (*CP Molecular Biology Appendix 3A*; Voytas and Ke, 1999) or digital image capture (UNIT 7.5)

NOTE: See Troubleshooting section for suggestions concerning optimization of this protocol, particularly when employing AP-based systems.

1. Equilibrate the membrane by washing two times in substrate buffer for 15 min each time. For blots of whole gels, use 50 ml substrate buffer; for strips, use 5 to 10 ml/strip.

2a. Transfer the membrane to visualization solution. Soak 30 sec (HRPO reactions) to 5 min (AP reactions) in the same volumes as in step 1.

2b. Alternatively, lay out a square of plastic wrap and pipet 1 to 2 ml visualization solution into the middle. Place membrane on the plastic so that the visualization solution spreads out evenly from edge to edge. Fold wrap back onto membrane, seal, and proceed to step 4.

3. Remove the membrane, drain, and place face down on a sheet of clear plastic wrap. Fold wrap back onto membrane to form a liquid-tight enclosure.

   To ensure an optimal image, only one layer of plastic should be between the membrane and film. Sealable bags are an effective alternative. Moisture must not come in contact with the X-ray film.

4. In a darkroom, place membrane face down onto film.

   Do this quickly and do not reposition; a double image will be formed if the membrane is moved while in contact with the film. A blurred image is usually caused by poor contact between membrane and film; use a film cassette that insures a tight fit.

5. Expose the film for a few seconds to several hours.

   Typically, immunoblots produce very strong signals within a few seconds or minutes. However, weak signals may require several hours to an overnight exposure. If no image is detected, expose film 30 min to 1 hr, and if needed, overnight (see Troubleshooting).

6. *Optional*: Wash the membrane twice 50 ml TBS for 15 min each time and process for chromogenic development (see Basic Protocol 4).

   Chemiluminescent and chromogenic immunoblotting can be easily combined on a single blot to provide a permanent visual marker of a known protein. First probe membrane with the chemiluminescent reactions to record digitally or on film. For the last reaction, use chromogenic development to produce a permanent visual record of the blot. This results in a permanent reference stain on the blot for comparison to the digitally imaged chemiluminescent blot.

   Alternatively, staining with colorimetric stains such as India ink or fluorescent stains such as SYPRO Ruby will image all the proteins on the blot, permitting easy comparisons reference to the subsequent chemiluminescent blot (e.g., Eynard and Lauriere, 1998).

REAGENTS AND SOLUTIONS

Use deionized, distilled water in all recipes and protocol steps. For common stock solutions, see UNIT 3.3.

Alkaline phosphate substrate buffer

100 mM Tris·Cl, pH 9.5 (UNIT 3.3)
100 mM NaCl
5 mM $MgCl_2$

BCIP/NBT visualization solution

33 µl nitroblue tetrazolium (NBT) stock: 100 mg NBT in 2 ml of 70% dimethylformamide (DMF); store up to 1 year at 4°C
5 ml alkaline phosphate substrate buffer (see recipe)
17 µl 5-bromo-4-chloro-3-indolyl phosphate (BCIP) stock: 100 mg BCIP in 2 ml of 100% DMF; store up to 1 year at 4°C
Stable 1 hr at room temperature

Recipe is from Harlow and Lane (1999). Alternatively, BCIP/NBT substrates may be purchased from Sigma, Kirkegaard & Perry, Moss, and Vector Labs.

Blocking buffer for colorimetric detection

For nitrocellulose and PVDF: 0.1% (v/v) Tween 20 in TBS (TTBS; UNIT 3.3).

For neutral and positively charged nylon: Tris-buffered saline (TBS; UNIT 3.3) containing 10% (w/v) nonfat dry milk. Prepare just before use.

TTBS can be stored ~1 week at 4°C.

Blocking buffer for luminescence detection

0.2% (w/v) casein (e.g., Hammerstein grade or I-Block; Applied Biosystems) in TTBS (UNIT 3.3). Prepare just before use.

This blocking buffer can be used for nitrocellulose, PVDF, and neutral nylon (e.g., Pall Biodyne A) membranes.

4CN visualization solution

Mix 20 ml ice-cold methanol with 60 mg 4-chloro-l-napthol (4CN). Separately mix 60 µl of 30% (v/v) H_2O_2 with 100 ml TBS (UNIT 3.3) at room temperature. Rapidly mix the two solutions and use immediately.

DAB/NiCl$_2$ visualization solution

5 ml 100 mM Tris·Cl, pH 7.5 (UNIT 3.3)
100 µl 3, 3′-diaminobenzidine (DAB) stock (40 mg/ml in H_2O, stored in 100-µl aliquots at −20°C)
25 µl $NiCl_2$ stock (80 mg/ml in H_2O, stored in 100-µl aliquots at −20°C)
15 µl 3% (v/v) H_2O_2
Mix just before use

CAUTION: *Handle DAB carefully, wearing gloves and mask; it is a carcinogen.*

Suppliers of peroxidase substrates are Sigma, Kirkegaard & Perry, Moss, and Vector Labs.

Luminol visualization solution

0.5 ml 10× luminol stock [40 mg luminol (Sigma) in 10 ml DMSO]
0.5 ml 10× *p*-iodophenol stock: optional; 10 mg (Aldrich) in 10 ml DMSO
2.5 ml 100 mM Tris·Cl, pH 7.5 (UNIT 3.3)

continued

25 µl 3% (v/v) H_2O_2
H_2O to 5 ml
Prepare just before use

Recipe is from Schneppenheim et al. (1991). Premixed luminol substrate mix (e.g., Pierce; Amersham ECL; Perkin-Elmer Renaissance; Kirkegaard & Perry LumiGLO) may also be used. p-iodophenol is an optional enhancing agent that increases light output. Luminol and p-iodophenol stocks can be stored up to 6 months at −20°C

India ink solution

0.1% (v/v) India ink (e.g., Pelikan 17 black) in Tween 20 solution (see recipe)

SYPRO Ruby protein blot stain

SYPRO Ruby protein blot stain (Molecular Probes) is provided in a unit size of 200 ml. The 200-ml volume is sufficient for staining 10 to 40 minigel electroblots or four large-format electroblots (20 × 20 cm). SYPRO Ruby protein blot stain may be reused up to four times with little loss in sensitivity. The reagent is stable for at least 6 months to 1 year when stored at room temperature, protected from light.

Ponceau S solution

Dissolve 0.5 g Ponceau S in 1 ml glacial acetic acid. Bring to 100 ml with water. Prepare just before use.

Transfer buffer

Add 18.2 g Tris base and 86.5 g glycine to 4 liters of water. Add 1200 ml methanol and bring to 6 liters with water. For use with PVDF filters, decrease methanol concentration to 15% (v/v); for nylon filters, omit methanol altogether.

The pH of the solution is ∼8.3 to 8.4.

Tween 20 solution

0.3% (v/v) Tween 20 in phosphate-buffered saline (PBS; *UNIT 3.3*), pH 7.4

UNDERSTANDING RESULTS

Immunoblotting should result in the detection of one or more bands. Although antibodies directed against a single protein should produce a single band, degradation of the sample (e.g., via endogenous proteolytic activity) may cause visualization of multiple bands of slightly different size. Multimers will also form spontaneously, causing higher-molecular-weight bands on the blot. If simultaneously testing multiple antibodies directed against a complex protein mixture (e.g., using patient sera against SDS-PAGE-separated viral proteins in an AIDS western blot test), multiple bands will be visualized.

For immunoblot or protein dot blot chemiluminescent applications, the sensitivity using HRPO is ∼1 pg of target protein.

TROUBLESHOOTING

Immunoblotting Problems and Artifacts

There are several problems associated with immunoblotting. For instance, the antigen is solubilized and electrophoresed in the presence of denaturing agents (e.g., SDS or urea), and some antibodies may not recognize the denatured form of the antigen transferred to the membrane. The results observed may be entirely dependent on the denaturation and transfer system used. Gel electrophoresis under nondenaturing conditions can also be performed (see *CP Molecular Biology Unit 10.2B*; Gallagher, 1999).

Other potential problems include high background, nonspecific or weak cross-reactivity of antibodies, poor protein transfer or membrane binding efficiency, and insufficient sensitivity. For an extensive survey and discussion of immunoblotting problems and artifacts, see Bjerrum et al. (1988).

Transfer Efficiency

If no transfer of protein has occurred, check the power supply and electroblot apparatus to make sure that the proper electrical connections were made and that power was delivered during transfer. In addition, check that the correct orientation of filter and gel relative to the anode and cathode electrodes was used.

If the transfer efficiency using the tank system appears to be low, increase the transfer time or power. Cooling (using the unit's built-in cooling cores) is generally required for transfers >1 hr. At no time should the buffer temperature go above 45°C. Prolonged transfers (>1 hr) are not possible in semidry transfer units due to rapid buffer depletion.

If the protein bands are diffuse, check the transfer cassette. The gel must be held firmly against the membrane during transfer. If the transfer sandwich is loose in the cassette, add another thin sponge or more blotter paper to both sides.

Occasionally, a grid pattern will be apparent on the membrane after tank transfer. This is caused by having either the gel or the membrane too close to the sides of the cassette. Correct this by adding more layers of filter paper to diffuse the current flowing through the gel and membrane. Use a thinner sponge and more filter paper if necessary.

If air bubbles are trapped between the filter and the gel, they will appear as clear white areas on the filter after blotting and staining. Take extra care to make sure that all bubbles are removed.

Transfer Buffers

Alternatively, the transfer buffer can be modified to increase efficiency. Adding SDS at a concentration of 0.1% to the transfer buffer improves the transfer of all proteins out of the gel, particularly those above 60 to 90 kDa in size. Lowering the concentration of methanol will also improve the recovery of proteins from the gel. These procedures are tradeoffs. Methanol improves the binding of proteins to PVDF and nitrocellulose, but at the same time hinders transfer. With SDS present, transfer efficiency is improved, but the SDS can interfere with protein binding to the membrane. Nylon and PVDF membranes are particularly sensitive to SDS interference. If needed, 0.01% to 0.02% SDS may be used in PVDF membrane transfer buffers.

Gel cross-linking and thickness also have a profound effect on the transfer efficiency. In general, 0.5- to 0.75-mm-thick gels will transfer much more efficiently than thicker gels (e.g., 1.5 mm thick). Gels with a higher acrylamide percentage will also transfer less efficiently. Proteins can be particularly difficult to transfer from gradient gels, and a combination of longer transfer times, thin gels, and the addition of SDS to the transfer buffer may be needed.

Blocking

Insufficient blocking or nonspecific binding of the primary or secondary antibody will cause a high background stain. A control using preimmune serum or only the secondary antibody will determine if these problems are due to the primary antibody. Try switching to another blocking agent; protein blocking agents may weakly cross-react. Lowering the concentration of primary antibody should decrease background and improve specificity (Fig. 8.3.6).

When casein is used as a blocking agent, it can interfere with subsequent visualization procedures. When using AP staining, the casein must be heated to 65°C prior to use to reduce alkaline phosphatase activity inherent in the casein. In an avidin-biotin system, maximum sensitivity has been observed when free biotin or biotinylated proteins are removed by pretreating the casein with avidin-agarose (Sigma).

Luminescence Visualization

Due to the nature of light and the method of detection, certain precautions are warranted when using luminescent visualization (e.g., Harper and Murphy, 1991). Very strong signals can overshadow nearby weaker signals on the membrane. Because light will pipe through the membrane and the surrounding plastic wrap, overexposure will produce a broad diffuse image on the film. The signal can also saturate the film, exposing the film to a point whereby increased exposure will not cause a linear increase in the density of the image on the film.

Recording the image on the blot can be tailored by exposing bright signals for short exposure times (<5 sec) and weak signals for long exposure (up to 10 min). An advantage of digital imaging with CCD cameras is that the image capture can be programmed to automatically record a series of short to long exposures, ensuring that the full range of luminescence from the blot is represented. Light piping can be minimized through use of thin plastic overlays and minimal liquid.

The membrane can interact with the substrate as well. With the alkaline phosphatase substrate AMPPD, nitrocellulose and PVDF membranes require 2 and 4 hr, respectively, to reach maximum light emission. In addition, PVDF is reported to give a stronger signal than nitrocellulose (Applied Biosystems Western Light instructions).

VARIATIONS

While the most common procedure for immunoblotting is electrotransfer via a vertical tank or a horizontal "semi-dry" blotter, numerous variations in blotting exist. These include contact, pressure, and vacuum transfer (Kurien and Scofield, 2006).

Contact transfer is generally a problem to be avoided. During repositioning of the membrane while assembling the blotter paper, membrane, gel stack for tank or semi-dry transfer, protein immediately transfers and creates a new replicate of the gel every time the membrane is repositioned. However, if quantitation is not required, a quick press-and-lift contact transfer can be very useful for rapid identification.

Vacuum pressure uses a similar approach but places the gel under a vacuum, usually with a stack of buffer saturated blotting paper on top, to pull liquid and the protein from the gel and onto the membrane. The nonelectrical procedures, although very inefficient in transferring protein, are useful for charge-neutral transfers. No charge is required on the protein, allowing native, nondenatured proteins to be transferred regardless of the native charge.

LITERATURE CITED

Bjerrum, O.J., Larsen, K.P., and Heegaard, N.H.H. 1988. Nonspecific binding and artifacts-specificity problems and troubleshooting with an atlas of immunoblotting artifacts. *In* CRC Handbook of Immunoblotting of Proteins, Vol. I: Technical Descriptions (O.J. Bjerrum and N.H.H. Heegaard, eds.) pp. 227-254. CRC Press, Boca Raton, Fla.

Blake, M.S., Johnston, K.H., Russell-Jones, G.J., and Gotschlich, E.C. 1984. A rapid, sensitive method for detection of alkaline phosphatase-conjugated anti-antibody on western blots. *Anal. Biochem.* 136:175-179.

Burnette, W.N. 1981. Western blotting: Electrophoretic transfer of proteins from sodium dodecyl sulfate-polyacrylamide gels to unmodified nitrocellulose and radiographic detection with antibody and radioiodinated protein A. *Anal. Biochem.* 112:195-203.

Carr, S.A. and Annan, R.S. 1996. Overview of peptide and protein analysis by mass spectrometry. *Curr. Protoc. Protein Sci.* 4:16.1.1-16.1.27.

Craig, W.Y., Poulin, S.E., Collins, M.F., Ledue, T.B., and Ritchie, R.F. 1993. Background staining in immunoblot assays. Reduction of signal caused by cross-reactivity with blocking agents. *J. Immunol. Methods* 158:67-76.

Eynard, L. and Lauriere, M. 1998. The combination of Indian ink staining with immunochemiluminescence detection allows precise identification of antigens on blots: Application to the study of glycosylated barley storage proteins. *Electrophoresis* 19:1394-1396.

Fernande, J. and Mische, S.M. 1995. Enzymatic digestion of proteins on PVDF membranes. *Curr. Protoc. Protein Sci.* 0:11.2.1-11.2.10.

Gallagher, S.R. 1999. One-dimensional electrophoresis using nondenaturing conditions. *Curr. Protoc. Mol. Biol.* 47:10.2B.1-10.2B.11.

Gallagher, S.R., Winston, S.E., Fuller, S.A., and Hurrell, J.G.R. 2004. Immunoblotting and immunodetection. *Curr. Protoc. Mol. Biol.* 66:10.8.1-10.8.24.

Gillespie, P.G. and Hudspeth, A.J. 1991. Chemiluminescence detection of proteins from single cells. *Proc. Natl. Acad. Sci. U.S.A.* 88:2563-2567.

Haan, C. and Behrmann, I. 2007. A cost effective noncommercial ECL-solution for Western blot detections yielding strong signals and low background. *J. Immunol. Methods* 318:11-19.

Hancock, K. and Tsang, V.C. 1983. India ink staining of proteins on nitrocellulose paper. *Anal. Biochem.* 133:157-162.

Harlow, E. and Lane, D. 1999. Using Antibodies: A Laboratory Manual. Cold Spring Harbor Laboratory Press, Cold Spring Harbor, N.Y.

Harper, D.R. and Murphy, G. 1991. Nonuniform variation in band pattern with luminol/horseradish peroxidase western blotting. *Anal. Biochem.* 192:59-63.

Kricka, L.J., Voyta, J.C., and Bronstein, I. 2000. Chemiluminescent methods for detecting and quantitating enzyme activity. *Methods Enzymol.* 305:370-390.

Kurien, B.T. and Scofield, R.H. 2006. Western blotting. *Methods* 38:283-293.

McKimm-Breschkin., J.L. 1990. The use of tetramethylbenzidine for solid phase immunoassays. *J. Immunol.* 135:277-280.

Moeremans, M., Daneels, G., and De Mey, J. 1985. Sensitive colloidal metal (gold or silver) staining of protein blots on nitrocellulose membranes. *Anal. Biochem.* 145:315-321.

Morandell, S., Stasyk, T., Grosstessner-Hain, K., Roitinger, E., Mechtler, K., Bonn, G.K., and Huber, L.A. 2006. Phosphoproteomics strategies for the functional analysis of signal transduction. *Proteomics* 6:4047-4056.

Pampori, N.A., Pampori, M.K., and Shapiro, B.H. 1995. Dilution of the chemiluminescence reagents reduces the background noise on western blots. *BioTechniques* 18:588-590.

Pluskal, M.G., Przekop, M.B., Kavonian, M.R., Vecoli, C., and Hicks, D.A. 1986. A new membrane substrate for western blotting of proteins. *BioTechniques*. 4:272.

Salinovich, O. and Montelaro, R.C. 1986. Reversible staining and peptide mapping of proteins transferred to nitro cellulose after separation by sodium dodecyl sulfate-polyacrylamide gel electrophoresis. *Anal. Biochem.* 156:341-347.

Sandhu, G.S., Eckloff, B.W., and Kline, B.C. 1991. Chemiluminescent substrates increase sensitivity of antigen detection in western blots. *BioTechniques* 11:14-16.

Schneppenhein, R., Budde, U., Dahlmann, N., and Rautenberg, P. 1991. Luminography—a new, highly sensitive visualization method for electrophoresis. *Electrophoresis* 12:367-372.

Sutherland, M.W. and Skerritt, J.H. 1986. Alkali enhancement of protein staining on nitrocellulose. *Electrophoresis* 401-406.

Towbin, H., Staehelin, T., and Gordon, J. 1979. Electrophoretic transfer of proteins from polyacrylamide gels to nitrocellulose sheets: Procedure and some applications. *Proc. Natl. Acad. Sci. U.S.A.* 76:4350-4354.

Voytas, D. and Ke, N. 1999. Detection and quantitation of radiolabeled proteins and DNA in gels and blots. *Curr. Protoc. Mol. Biol.* 48:A.3A.1-A.3A.10.

UNIT 8.4

Labeling DNA and Preparing Probes

Karl A. Haushalter[1]

[1]Harvey Mudd College, Claremont, California

INTRODUCTION

It may seem so obvious that it does not need to be stated, but it is not possible to see DNA or RNA—at least not the microscopic quantities of DNA or RNA that are run on a gel or transferred to a blot. There are no special magnifying glasses that would allow us to visualize naked DNA or RNA when it is embedded in a gel matrix or affixed to a membrane. If we want to visualize DNA or RNA, we need the nucleic acid to be attached to *something*—something radioactive, something colored, something fluorescent, or something luminescent. For some applications, it is sufficient to treat the gel or the blot with a stain or dye that will bind to DNA or RNA regardless of the specific sequence. For example, the intercalating dye ethidium bromide is routinely used to visualize nucleic acids that have been separated by gel electrophoresis.

For some applications, however, dyes are not sufficient, due to the nonspecific nature of the staining process. In complex mixtures of nucleic acids, such as restriction digests of genomic DNA or total RNA preparations, you do not necessarily want to detect all of the DNA/RNA in a sample—you only want to visualize one (or a few) specific DNA/RNA molecule(s). Furthermore, many applications require a much lower limit of detection than conventional dyes can provide. This unit will discuss the preparation and purification of labeled DNA that can be used in applications requiring high specificity and high sensitivity.

Intended Application

The first step in designing a labeling strategy is to think about the intended application of the labeled nucleic acid. Table 8.4.1 summarizes some of the most common uses of labeled nucleic acids. Different downstream experiments place different constraints on what makes a good labeled molecule. You should consult the downstream protocol for the optimum size and concentration, as well as tolerance for chemical modifications to the labeled nucleic acid. In addition, some protocols work best if the labeled probe is labeled uniformly throughout the probe, while others work best if the label is restricted to one end or the other.

Uses of the labeled nucleic acids can be broadly broken down into two distinct and separate types. In the first (and more common) case, the labeled DNA itself is not the molecule of interest. Instead, the labeled DNA is complementary to the target nucleic acid. In this case, the labeled DNA is called a hybridization probe, or, more colloquially, just a probe. The target DNA in which you are ultimately interested can be detected because it will hybridize to the labeled probe. The second major class of experiments uses the labeled DNA molecule itself as the substrate for investigation. These experiments include biochemical or structural investigations of nucleic acids, or of proteins that interact with nucleic acids. For example, labeled probes can be used in mobility shift assays to measure the binding affinity of proteins for nucleic acids.

Type of DNA to Label

The next step in designing a labeling strategy is to identify the DNA that you want to label. The most common choices are oligonucleotides, restriction fragments (*UNIT 10.1*), PCR products

Table 8.4.1 Common Uses for Labeled Nucleic Acids

Experiment	References
Hybridization probe for Southern blotting	*UNIT 8.2*
Hybridization probe for northern blotting	*UNIT 8.2*
Monitoring real-time polymerase chain reaction	*UNIT 10.3*
S1 Analysis of Messenger RNA	*CP Mol. Biol. Unit 4.6* (Greene and Struhl, 1988)
Ribonuclease protection assay	*CP Mol. Biol. Unit 4.7* (Gilman, 1993)
Primer extension analysis of RNA	*CP Mol. Biol. Unit 4.8* (Triezenberg, 1992)
Screening recombinant DNA libraries	*CP Mol. Biol. Unit 6.3* (Strauss, 1993); *CP Mol. Biol. Unit 6.4* (Duby et al., 1993)
Mobility shift assay for protein-DNA interactions	*CP Mol. Biol. Unit 12.2* (Buratowski and Chodosh, 1996)
Interference assay for protein-DNA interactions	*CP Mol. Biol. Unit 12.3* (Baldwin et al., 1996)
DNase I footprinting analysis of DNA	*CP Mol. Biol. Unit 12.4* (Brenowitz et al., 1989)
In situ hybridization	*CP Mol. Biol. Units 14.1–14.16* (Ausubel et al., 2007)
Analysis of chromatin complexes	*CP Mol. Biol. Unit 21.4* (Vitolo et al., 1999)

(*UNIT 10.2*), or plasmids (*UNIT 4.2*). Each has distinct advantages and disadvantages that need to be considered, and you should consult the protocol of the downstream application for guiding information. Oligonucleotides are short DNA molecules, typically 10 to 60 base pairs long. You may hear someone in the lab refer to them as "oligos" or as primers. Oligonucleotides can be synthesized in-house (*CP Molecular Biology Unit 2.11*; Ellington and Pollard, 1998) or purchased relatively inexpensively from a third-party vendor. Due to the chemistry involved in DNA synthesis, the quality and yield of oligonucleotides go down dramatically with increasing length, while the price goes up. The longest oligonucleotides commonly synthesized using routine methods are 80 to 100 bp long. For longer DNA, it is typically necessary to use another method to prepare the substrate DNA. One option is to use the polymerase chain reaction (PCR) to copy a desired sequence of DNA from a template (see *UNIT 10.2*). By careful choice of the PCR primers and template, you can tailor the sequence and size of the probe for the application. It is also possible to isolate a useable probe from a restriction enzyme digest of a plasmid or other long template (see *UNIT 10.1*). By choosing the proper substrate and the right combination of restriction enzymes, it is usually possible to obtain the correct-size fragment. Finally, for some applications, it is possible to use very large DNA molecules, such as an intact plasmid, as the substrate for a labeling reaction.

Type of Label

The next step toward developing a labeling strategy is to determine what sort of label to incorporate. The three main choices are radioisotopes, fluorophores, and small-molecule binding partners (Table 8.4.2).

Radioisotopes

The most commonly used radioisotope for labeling nucleic acids is ^{32}P, which has a half-life of 14.3 days. When ^{32}P decays, it emits a high-energy β particle that can be easily detected by autoradiography (*UNIT 2.3*). A variety of nucleotides that incorporate ^{32}P into phosphate groups are

Table 8.4.2 Advantages and Disadvantages of Different Labels

Label	Advantages	Disadvantages
Radioisotope (commonly ^{32}P)	High sensitivity Minimal perturbation to structure of probe Does not require secondary reagents; can be directly visualized by autoradiography	Short-half life of ^{32}P means that labeled probe does not have long shelf-life Extra caution needed to handle radiochemicals safely Not all institutions licensed to use radioactivity
Fluorophore	Does not require secondary reagents; can be directly visualized by fluorescence scanner No special safety precautions necessary Long shelf life of probe	Sensitivity not as high as radioisotopes or small molecule binding partners
Small molecule binding partner (commonly digoxigenin or biotin)	Flexibility in detection mode: colorimetric, fluorescent, or luminescent No special safety precautions necessary Long shelf life of probe	Requires secondary reagents and extra steps for visualization Sensitivity not as high as radioisotopes (but higher than fluorophores)

commercially available. By using appropriate enzyme-catalyzed reactions, these phosphate groups can be readily incorporated into DNA. Other radioisotopes, e.g., ^{3}H, ^{33}P, and ^{35}S, are less commonly used to label DNA and RNA. While the protocols in this unit are written for ^{32}P incorporation, only minor modifications are necessary to adapt the protocols for other radioisotopes. The alternative radioisotopes emit β particles of different (much weaker) energy and have advantages for specialized applications, involving structural analysis of nucleic acids or ultrahigh-resolution gel electrophoresis.

Radiolabeled DNA can be characterized by its specific activity. Specific activity is defined as the ratio of the amount of radioactivity (in Ci or similar unit; see *UNIT 2.3* for radiation measurement) to the absolute amount of DNA (in moles or similar units; see *UNIT 2.2* for DNA calculation). All things being equal, probes with higher specific activity will give higher sensitivity for detection.

Fluorophores

Fluorophores are chemical groups that absorb light at one wavelength and then emit light at a different (longer) wavelength. The difference between the wavelength of the maxima of the absorbance and emission spectra is known as the Stokes shift. Importantly, the intensity of fluorescence of many fluorophores allows their detection at low abundance, although the sensitivity is reduced when compared to radioisotopes. A couple of representative fluorophores are shown in Figure 8.4.1, together with their absorbance and emission maxima. It is possible to incorporate fluorophores chemically during DNA synthesis, or enzymatically post-synthesis, using modified nucleotides conjugated to a fluorophore. To visualize a fluorescently labeled nucleic acid in a gel or on a blot, it is necessary to use imaging equipment to excite the fluorophore at an appropriate wavelength in its absorbance spectrum and to use the appropriate filters to detect the emitted light.

Small molecule binding partners (biotin and digoxigenin)

The last category of labeling uses small organic molecules that are recognized by antibodies or other protein binding partners. When using this method, label your nucleic acid with the molecule, perform the experiment (hybridization, etc), and then detect the small molecule–labeled nucleic

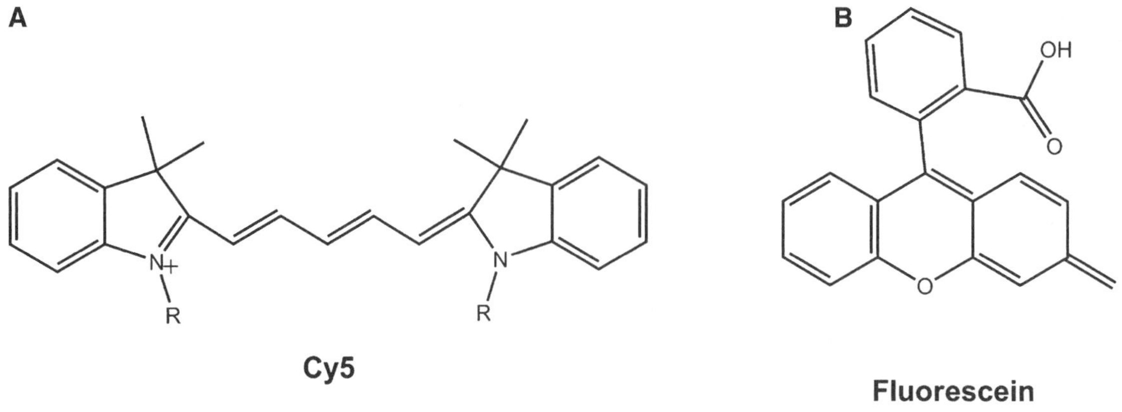

A

Cy5

B

Fluorescein

Figure 8.4.1 Representative fluorophores that can be used to label nucleic acids. (**A**) Cy5: maximum absorbance, 649 nm, maximum emission, 670 nm; (**B**) fluorescein: maximum absorbance, 494 nm, maximum emission, 518 nm.

acid by adding the appropriate binding partner, which is either directly labeled itself or can be used for detection without extrinsic labeling. While a variety of molecules can be used, the two most common are biotin and digoxigenin (Fig. 8.4.2).

Biotin, also known as vitamin H or vitamin B_7, is a water-soluble enzyme cofactor used broadly throughout metabolism. Biotin-labeled DNA can be detected on the basis of the strong affinity for biotin of the protein streptavidin. In fact, the biotin-streptavidin interaction is one of the strongest protein-ligand interactions known. With a K_D of $\sim 10^{-15}$ M, the binding between biotin and streptavidin is effectively irreversible under most experimental conditions. For detection, it is possible to use streptavidin that has been covalently conjugated with a detection moiety. For example, streptavidin derivatives can be purchased that are conjugated directly to a fluorophore or to enzymes such as horseradish peroxidase or alkaline phosphatase. Enzymes conjugated to streptavidin can be detected by their action on provided substrates that deposit products which are colored, luminescent, or fluorescent.

In a manner similar to the way in which biotin-labeled DNA can be detected through its association with streptavidin, DNA labeled with the small molecule digoxigenin can be detected by its association with antibodies that recognize digoxigenin. Digoxigenin is a steroid isolated from *Digitalis* plants. Antibodies are available that have high affinity and high specificity (low cross-reactivity)

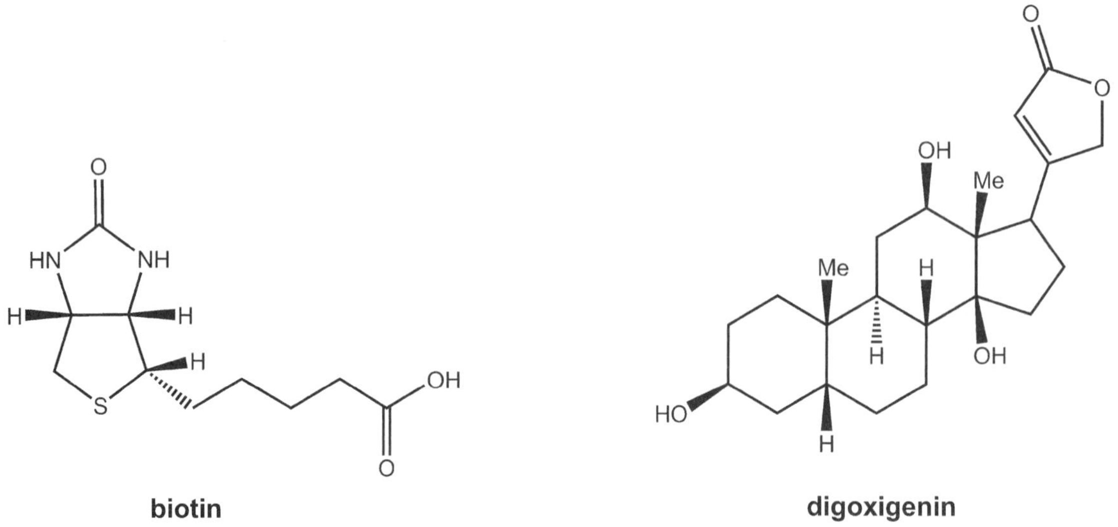

biotin

digoxigenin

Figure 8.4.2 Biotin and digoxigenin.

for digoxigenin and, as with streptavidin, the antibodies can be labeled with a variety of detection moieties including fluorophores, colloidal gold, and enzymes.

How to Incorporate the Label

Once you know why you want to label, what you want to label, and which label to use, the final step of planning is to pick a method for incorporating the label. This unit presents four common strategies for incorporating labels into nucleic acids. While these four methods will cover many applications, you may discover that an alternative labeling strategy would be more appropriate for your specific needs. At the end of this unit, the Variations section presents alternative approaches for labeling nucleic acids.

Incorporation of the label during DNA oligonucleotide synthesis

For oligonucleotides, the most straightforward method for incorporating a nonradioactive label (fluorophore or small molecule binding partner) is to insert the label directly during DNA synthesis. In addition to saving time, this method incorporates the label into essentially 100% of the synthesized oligonucleotides. Depending on the label, it may be possible to incorporate the label at the 5′ end, 3′ end, or at an internal site. For labels at a terminus, the chemical group is covalently bound to the 5′ or 3′ hydroxyl or phosphate. For internal labeling, the chemical group is covalently bound as an appendage to a DNA base in a modified nucleotide. The placement of the label depends on the intended purpose, and care should be taken to avoid having the label located at a site that would interfere with the desired properties of the oligonucleotide.

For reasons of time and money, most laboratories do not find it efficient to synthesize their own oligonucleotides in-house, and instead will outsource oligonucleotide synthesis to a commercial vendor or an institutional core facility (see *UNIT 10.4* for a discussion of ordering primers). Some commonly used vendors include Integrated DNA Technologies (Coralville, Iowa), Invitrogen (Carlsbad, California), Midland Certified Reagent Company (Midland, Texas), and Operon Biotechnologies (Huntsville, Alabama). Most vendors provide a convenient Web-based ordering form that allows you to type in your desired sequence and to specify the identity and location of desired modifications (labels) from a drop-down menu. For details on the chemistry involved in DNA oligonucleotide synthesis, see *CP Molecular Biology Unit 2.11* (Ellington and Pollard, 1998).

When ordering an oligonucleotide, you will be asked to specify the scale of the synthesis and the desired purification method. When thinking about the appropriate scale, look ahead to the downstream procedures for which you are ordering the oligonucleotide. Always order more than you think that you need, in case experiments need to be repeated unexpectedly. Note that you can only specify the scale of the synthesis (how much raw material is used in the synthesis)—not the final yield (how much product you will receive). For most applications involving labeled DNA, such as hybridization or gel shift assays, it is necessary to further purify the oligonucleotide, either by high-pressure liquid chromatography (HPLC) or polyacrylamide gel electrophoresis (PAGE). While it will cost more to have the oligonucleotide purified and the final yield will be decreased, the cost is well justified. Alternatively, you may elect to purify the oligonucleotide yourself (*CP Molecular Biology Unit 2.11*; Ellington and Pollard, 1998).

When the ordered DNA arrives, it will most likely arrive as a lyophilized (freeze-dried) powder. The package should also contain a detailed description of the synthesis, including the final yield. Some companies also provide quality-control testing data, such as MALDI-TOF mass spectrum or HPLC chromatogram, which can be used to check that the correct sequence has been synthesized and that the oligonucleotide is pure. While the labeled DNA can be stored in solution in an appropriate storage buffer (TE buffer or 10 mM Tris·Cl; *UNIT 3.3*) at −20°C, the best medium-term- and long-term-storage option is to resuspend the DNA in an appropriate storage buffer, separate the DNA solution into separate, single-use-aliquot tubes, and lyophilize (freeze-dry) the DNA. When needed, a single aliquot can be reconstituted with water to the original concentration and buffer composition.

5′ end-labeling of DNA with T4 polynucleotide kinase

Nucleic acids can be labeled with a radioactive phosphorus atom at their 5′ end in a reaction catalyzed by the enzyme T4 polynucleotide kinase (Tabor, 1987a; Sambrook et al., 1989). In this reaction, the terminal (γ) phosphate group of ATP is transferred to the free 5′-hydroxyl of the nucleic acid substrate to be labeled (Fig. 8.4.3). The products of the reaction are ADP and the substrate nucleic acid with a 5′ phosphate group. When using ATP in which ^{32}P has been incorporated into the γ phosphate group, i.e., [γ-^{32}P]ATP (Fig. 8.4.4), the product of the reaction catalyzed by T4 polynucleotide kinase will be radiolabeled at the 5′ end.

Polynucleotide kinase was first identified in extracts of *E. coli* infected with T2 and T4 bacteriophages (Richardson, 1965; Novogrodsky and Hurwitz, 1966; Novogrodsky et al., 1966). The bacteriophage enzyme has been proposed to function in vivo in the repair of tRNAs damaged by bacterial host defenses (Sirotkin et al., 1978; Amitsur et al., 1987), while the mammalian homolog has been shown to function in RNA and DNA repair (Jilani et al., 1999; Karimi-Busheri et al., 1999). By phosphorylating the 5′ end of DNA/RNA breaks, polynucleotide kinase enables ligases, which will only use 5′-phosphorylated DNA/RNA as substrates, to reseal the backbone. Stereochemical data indicates that the mechanism most likely proceeds through direct in-line attack of the 5′-hydroxyl on the γ-phosphate group of the donor nucleotide (Jarvest and Lowe, 1981). From the perspective of the molecular biologist at the lab bench, purified T4 polynucleotide kinase has become a workhorse enzyme used for radiolabeling, molecular cloning, and characterizing nucleic acids.

T4 polynucleotide kinase will phosphorylate the 5′ ends of either DNA or RNA, which may be single-stranded or double-stranded. However, as diagrammed in Figure 8.4.5, the nature of the termini will affect the efficiency of the reaction (Sambrook et al., 1989).

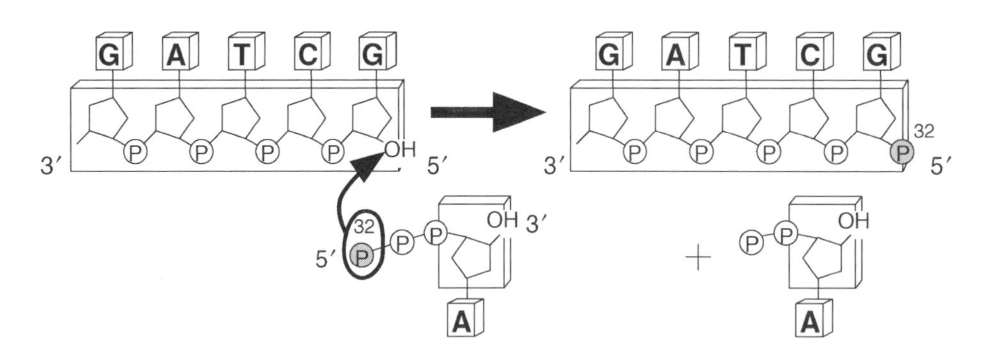

Figure 8.4.3 T4 polynucleotide kinase catalyzes the transfer of the γ phosphate from ATP to the 5′ hydroxyl of the DNA substrate.

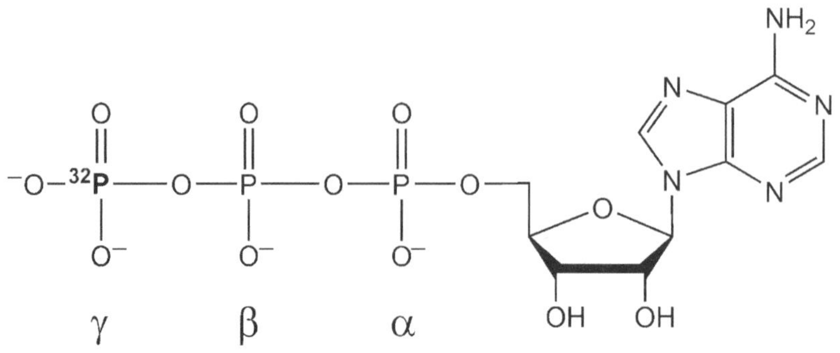

$^{-}$O–^{32}P–O–P–O–P–O– (adenosine); NH$_2$, N, OH OH

γ β α

Figure 8.4.4 Structure of [γ-^{32}P]ATP.

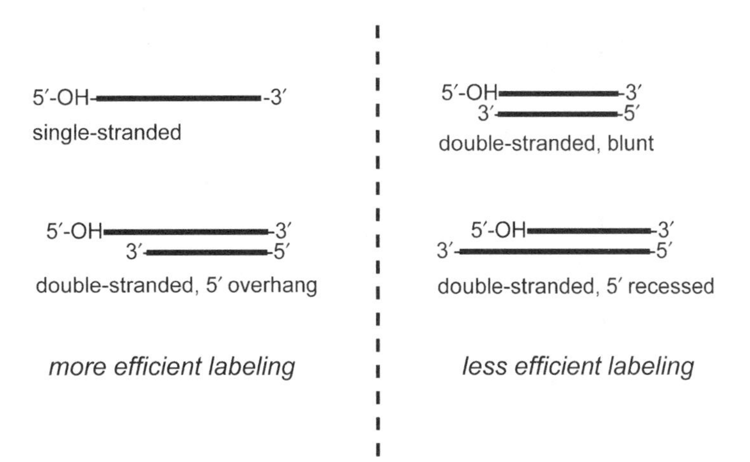

Figure 8.4.5 The nature of the ends of DNA affect the efficiency of labeling by T4 polynucleotide kinase.

The forward reaction of the enzyme, as drawn in Figure 8.4.3, requires that the substrate have a free 5′ hydroxyl, i.e., that no phosphate group be already present at the 5′ end. For synthetic oligonucleotides and PCR products, this is not generally a problem because oligonucleotide synthesis can be performed such that the final product has a free 5′ hydroxyl. However, DNA fragments produced by restriction digest from a larger piece of DNA will have 5′ phosphate groups. These DNA fragments can still be labeled with T4 polynucleotide kinase, but it is necessary to first pretreat the DNA with a phosphatase to remove the 5′ phosphate group (see *CP Molecular Biology Unit 3.10*; Tabor, 1987a). Alternatively, it is possible to take advantage of the reversible nature of the kinase reaction and use a variation of the basic protocol known as the exchange reaction (also described in Tabor, 1987a). In this variation, excess [γ-^{32}P]ATP is used in a buffer supplemented with ADP. The substrate first transfers the 5′ phosphate group to ADP and then is rephosphorylated using the phosphate from [γ-^{32}P]ATP.

Recombinant T4 polynucleotide kinase produced in *E. coli* is available from a wide variety of commercial suppliers. One unit of activity for the purified enzyme is defined as the amount of T4 polynucleotide kinase that will catalyze the incorporation of 1 nmol of ^{32}P into acid-insoluble polynucleotide in 30 min at 37°C (Richardson, 1981).

Labeling DNA by nick translation

In contrast to the 5′ end labeling reaction catalyzed by T4 polynucleotide kinase, described above, labeling DNA by nick translation will uniformly incorporate the label throughout the body of the DNA to be labeled (Rigby et al., 1977). In addition, nick translation can be used to incorporate both radioactive labels and nonradioactive labels, including a variety of fluorophores and small molecule binding partners. This method is appropriate for labeling double-stranded DNA substrates that are hundreds of base pairs to kilobase pairs long. Nick translation has become one of the standard methods for labeling DNA fragments to be used as hybridization probes (Sambrook et al., 1989; Tabor et al., 1997).

Nick translation is accomplished with the combined activities of two separate enzymes: bovine pancreatic deoxyribonuclease I (DNase I) and *E. coli* DNA polymerase I. Although the nick translation reaction occurs in a single reaction tube with all of the components and both enzymes mixed together, it is useful to describe the reactions catalyzed by the two enzymes in stages. In the first step, DNase I generates a nick in the substrate DNA by catalyzing the hydrolysis of a phosphodiester bond in the DNA backbone. The products of DNase I–catalyzed nicking are a free 3′ hydroxyl group and 5′ phosphate group (Fig. 8.4.6). Because DNase I is a relatively nonspecific endonuclease, the nick can occur anywhere along the length of the DNA substrate and on either strand.

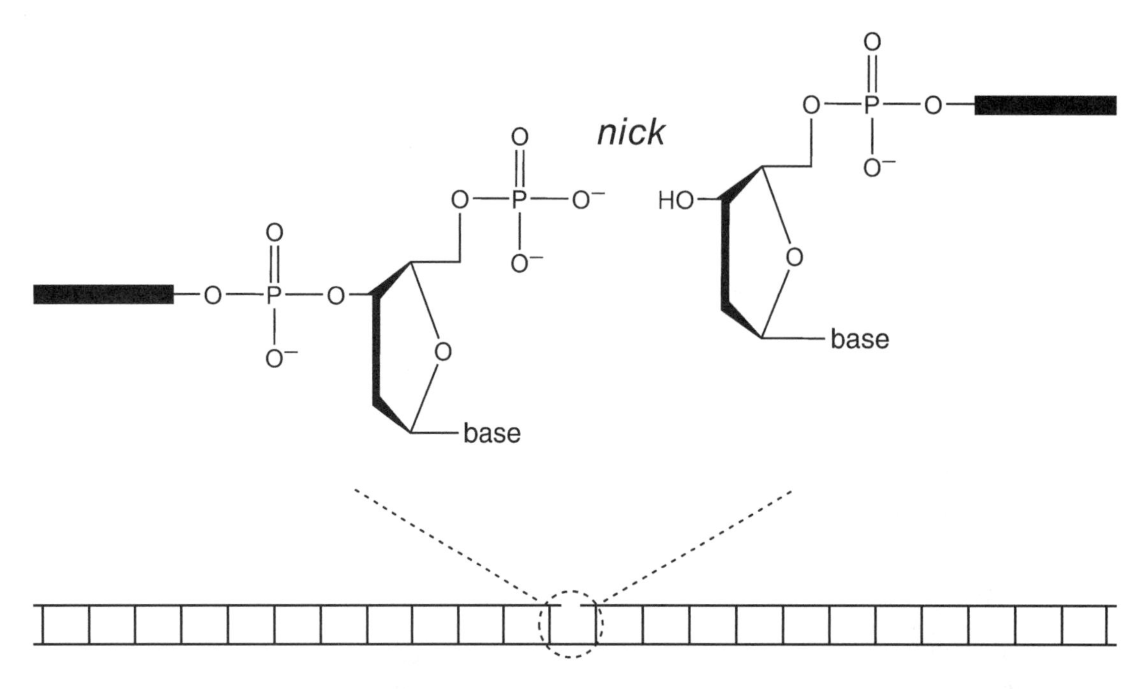

Figure 8.4.6 The nicking activity of deoxyribonuclease I (DNase I).

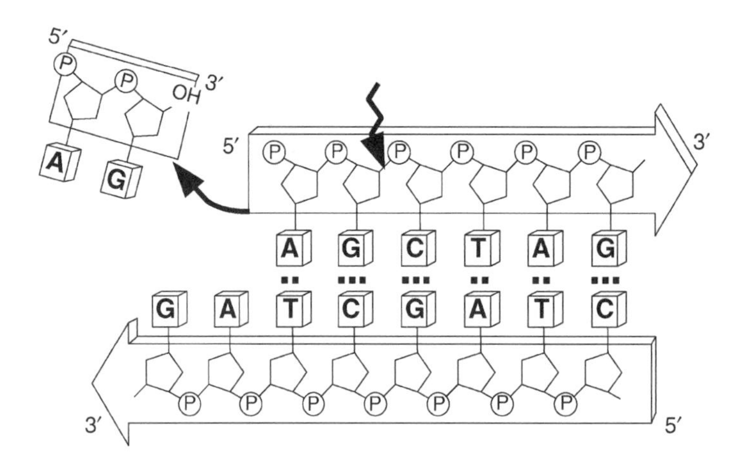

Figure 8.4.7 The 5′→3′ exonuclease activity of *E. coli* DNA polymerase I.

In the next step, the 5′→3′ exonuclease activity of *E. coli* DNA polymerase I removes one to several nucleotides downstream of the nick (Fig. 8.4.7). This will transform the nick created by DNase I into a gap of one to several nucleotides. While the 5′→3′ exonuclease activity is clearing a path ahead of the advancing enzyme, the 5′→3′ polymerase activity of *E. coli* DNA polymerase I will incorporate available deoxynucleotides into the extending chain by reaction at the exposed 3′ hydroxyl (Fig. 8.4.8). The combined effect of the two activities of the polymerase is that the nick is translated along the length of the substrate (Fig. 8.4.9). Nick translation will occur on both strands of the template DNA and when nicks on both strands of the DNA meet, a double-strand break is introduced. Depending on the reaction conditions, the typical product length is 200 to 500 base pairs of double-stranded DNA. The product length is tunable by varying the DNase I concentration. This control over product length is an important advantage when the probe will be used downstream for fluorescence in situ hybridization (FISH), which is very sensitive to probe length.

By using radiolabeled or chemically modified deoxynucleotides, it is possible to incorporate the desired label through DNA synthesis during nick translation. To radiolabel the DNA product, one of

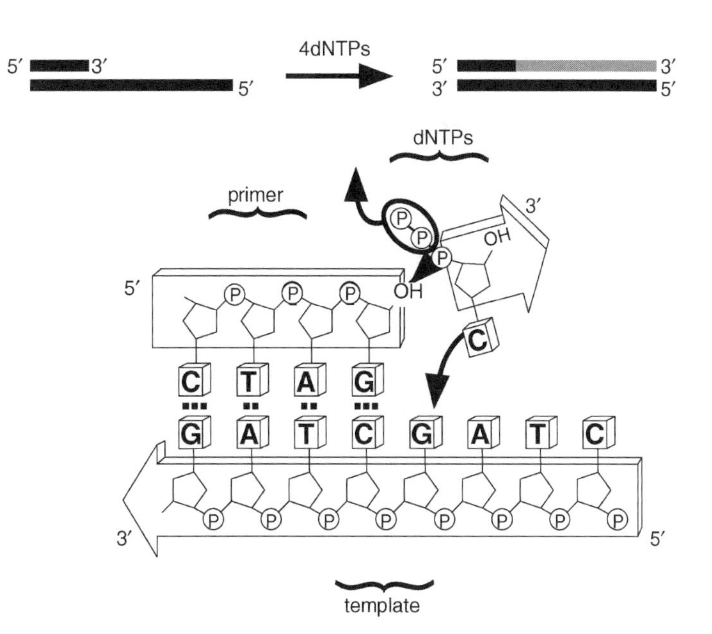
5′
3′
5′
4dNTPs
5′
3′
3′
5′
dNTPs
primer
3′
OH
5′
OH
C
C
T
A
G
G
A
T
C
G
A
T
C
3′
5′
template

Figure 8.4.8 The 5′→3′ polymerase activity of *E. coli* DNA polymerase I.

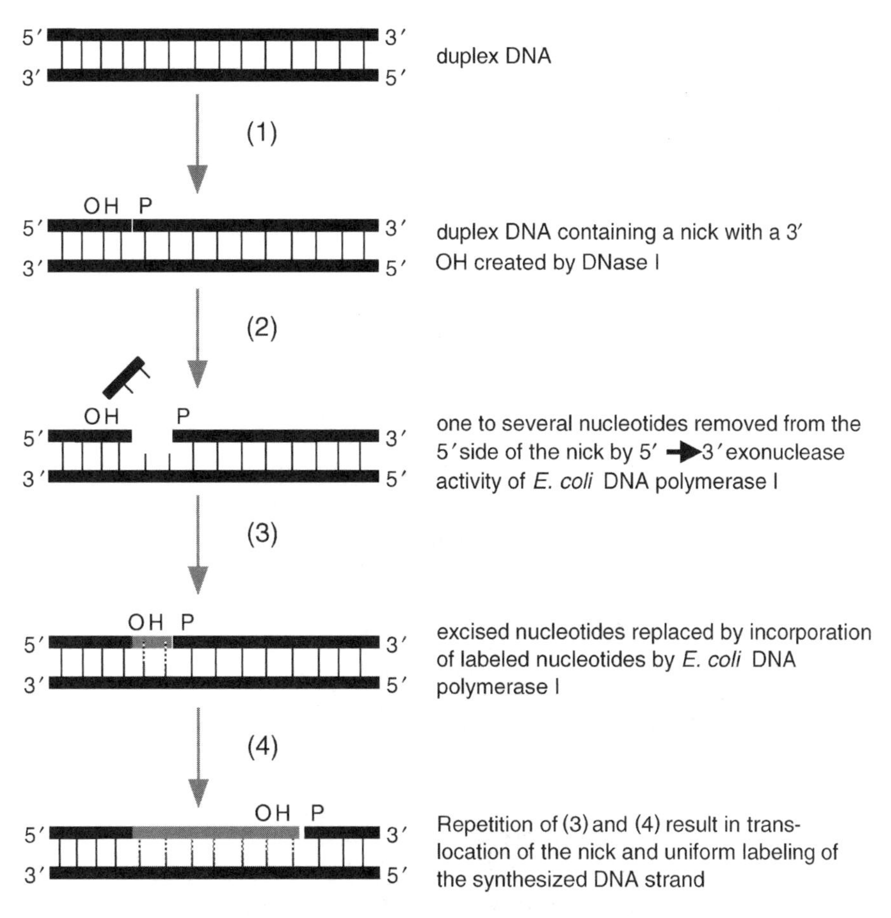
5′
3′
3′
5′
duplex DNA
(1)
OH P
duplex DNA containing a nick with a 3′
OH created by DNase I
(2)
OH
P
one to several nucleotides removed from the
5′side of the nick by 5′ →3′exonuclease
activity of E. coli DNA polymerase I
(3)
OH P
excised nucleotides replaced by incorporation
of labeled nucleotides by E. coli DNA
polymerase I
(4)
OH P
Repetition of (3) and (4) result in trans-
location of the nick and uniform labeling of
the synthesized DNA strand

Figure 8.4.9 DNA labeling by nick translation.

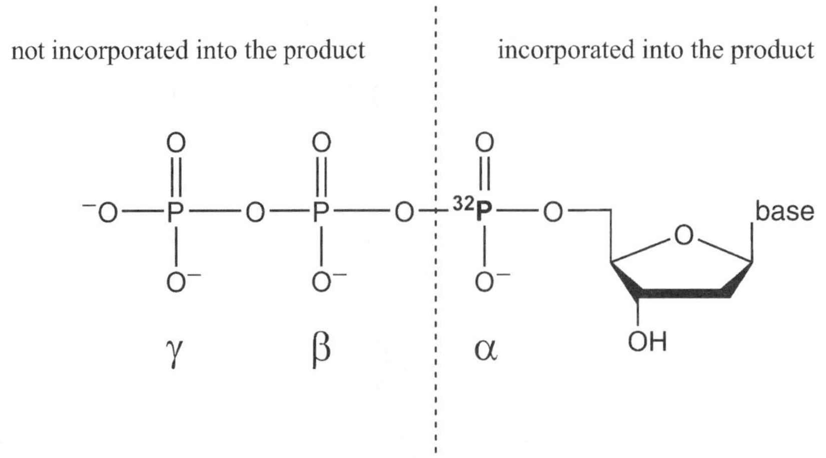

Figure 8.4.10 Structure of [α-^{32}P]deoxynucleotide used for radiolabeling DNA in nick translation or random primed synthesis.

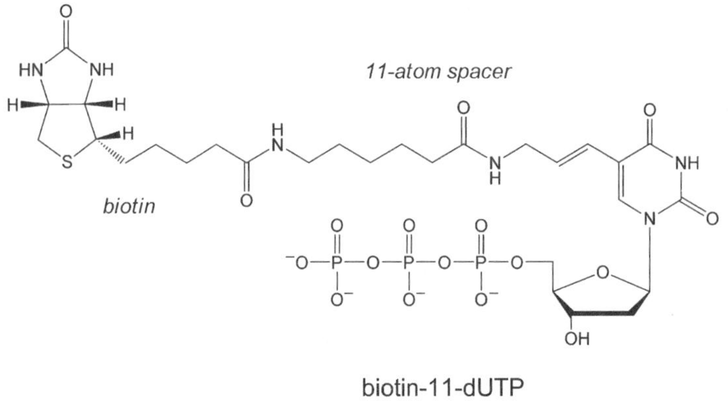

Figure 8.4.11 Biotin-11-dUTP: an example of a modified nucleotide than can be incorporated by *E. coli* DNA polymerase I.

the deoxynucleotides is partially or fully replaced with the corresponding radioactive [α-^{32}P]dNTP (Fig. 8.4.10). The choice of which deoxynucleotide is arbitrary. The specific activity of the product can be controlled by varying the concentration of the radiolabeled deoxynucleotide (relative to its nonradioactive counterpart) and varying the specific activity of the radiolabeled nucleotide precursor.

Alternatively, it is possible to supplement the nick translation reaction with deoxynucleotides conjugated to a fluorophore or small molecule binding partner. To varying degrees, *E. coli* DNA polymerase I will accept non-natural deoxynucleotides as substrates and incorporate the modified deoxynucleotide into the nick-translated product. Often, the label is appended to a long linker arm that is attached to an atom in the major groove of DNA. For example, biotin-11-dUTP, which can substitute for dTTP, contains an 11-atom spacer that links biotin to carbon 5 of dUTP (Fig. 8.4.11). The length and polarity of the linker will affect how easily the fluorophore or small molecule binding partner is recognized. A list of commercially available modified deoxynucleotides is tabulated in Table 8.4.3.

Labeling DNA by random primed synthesis

Random primed synthesis is an alternative method for achieving high-specific-activity labeling throughout the length of a probe. Like nick translation, random primed synthesis is appropriate for

Table 8.4.3 Some Commercially Available Modified Nucleotides

Modified nucleotide	Potential suppliers	Recommended activity/concentration
[α-^{32}P]dNTP (dATP, dCTP, dGTP, or dTTP)	GE Healthcare MP Biomedicals Perkin Elmer	10-20 μM dNTP (final conc. in reaction, 1-2 μM) *High-activity labeling:* Use a total of 100 μCi of dNTP labeled at 3000 Ci/mmol *Low-activity labeling:* Use a total of 25 μCi of dNTP labeled at 400-800 Ci/mmol
Biotin-dNTP (various linkers)	Invitrogen Perkin Elmer Roche Applied Science Sigma Aldrich	0.5 mM dNTP (final conc. in reaction ~50 μM) *High-activity labeling:* Use only labeled dNTP *Medium-activity labeling:* Mix labeled dNTP with 1 to 3 molar equivalents of the corresponding unlabeled dNTP *Low-activity labeling:* Mix labeled dNTP with 10-20 molar equivalents of the corresponding unlabeled dNTP
Digoxigenin (DIG)–dNTP (various linkers)	Roche Applied Science	
Cy3-dUTP Cy5-dUTP	GE Healthcare	
Alexa Fluor-dCTP Alexa Fluor-dUTP Rhodamine Green–dUTP (various linkers)	Invitrogen	
Fluorescein-dNTP Texas Red–dNTP (various linkers)	Invitrogen Perkin Elmer Roche Applied Science	

longer pieces of DNA, hundreds to thousands of base pairs. As shown in Figure 8.4.12, a collection of short, random primers are annealed to the template DNA. The large excess of primers used in the reaction and their short length make it likely that primers will be annealed at locations scattered throughout both strands of the template. Once annealed, the primers are extended by DNA synthesis with the large fragment (Klenow fragment) of *E. coli* DNA polymerase I (Fig. 8.4.8). The Klenow fragment lacks the 5′→3′ exonuclease activity of the full-length protein (Fig. 8.4.7). At the conclusion of the labeling, the DNA is denatured to produce a collection of newly single-stranded DNA molecules that vary in length, but together cover essentially the entire length of the template DNA.

Incorporation of the label during random primed synthesis occurs by adding modified deoxynucleotides containing the label to the reaction mixture, as described above for nick translation. An alternative approach for applications not requiring the highest amount of specific activity or requiring only end labels is to use normal nucleotides during the DNA synthesis reaction but use random primers that have previously been labeled (e.g., synthesized with a 5′ fluorophore or 5′ end-labeled with polynucleotide kinase).

For many applications, random primed synthesis and nick translation are nearly interchangeable as methods for labeling a desired probe. Compared with nick translation, some researchers find that random primed synthesis can give higher label incorporation and higher yield of labeled DNA, traits that are desirable for probing blots. On the other hand, random primed synthesis gives a slightly wider size range or products, which is undesirable for some applications, such as fluorescence in situ hybridization (FISH).

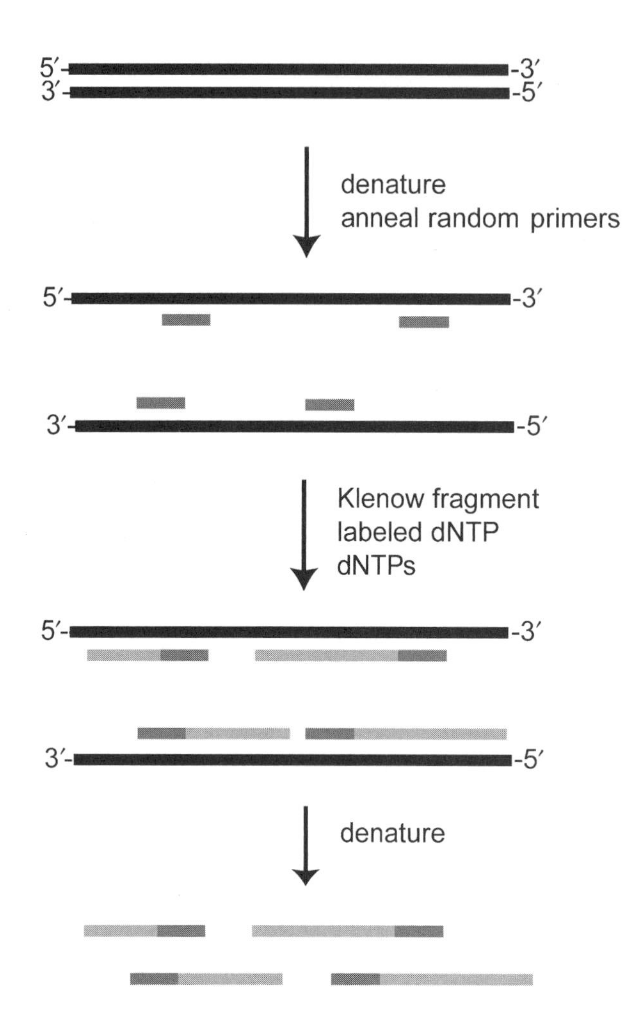

Figure 8.4.12 DNA labeling by random primed synthesis.

STRATEGIC PLANNING

General Experiment Guidelines

The protocols in this unit make extensive use of enzymes to modify or synthesize DNA. Enzymes should be handled with appropriate care, as detailed in *UNIT 10.1*. If the labeled DNA will be used to probe a northern blot, it is necessary to follow standard precautions to prevent RNase contamination (see *UNIT 8.2*).

Substrate DNA

The DNA that will be labeled should be high quality and purified prior to labeling. Oligonucleotides should be purified by HPLC or PAGE after synthesis and prior to labeling. For 5′ end labeling, it is very important to remove small nucleic acids, as these low-molecular-weight species can also be labeled by T4 polynucleotide kinase and can make up a substantial mole fraction of nucleic acid even if their weight fraction is low. T4 polynucleotide kinase is inhibited by ammonium ions and by inorganic phosphate, so these components should be avoided in buffers used to prepare DNA substrates prior to labeling by this method.

Similarly, for random primed DNA synthesis and nick translation, the purity of the input DNA is an important factor in the success of the labeling reaction. See *UNIT 5.2* for additional details about the purification of DNA fragments.

Radioisotopes

Follow your institution's guidelines for the procurement of radioisotopes and contact your radiation safety officer (RSO) if you have any questions. Manufacturers will only ship to institutions with a current license for the use of radioactive materials. Institutions vary in their radioisotope shipment receiving procedures. Do not assume that the radioisotope will be delivered you first thing the next morning—more commonly, the package will need to be logged and inspected by the radiation safety office and will be available later in the day. Find out the production schedule of your desired radionucleotide from the manufacturer and try to order material that is as fresh as possible. For example, if the manufacturer synthesizes a fresh lot of [γ-^{32}P]ATP every Tuesday, then you should avoid ordering this product on Monday and instead wait a day to get the product with the highest specific activity. Be sure to take the short half-life of ^{32}P into consideration when you plan your downstream experiments with the radiolabeled material. It is advisable to try to do all of your experiments with radiolabeled nucleic acids in a short time frame, so that a single (or small number) of radiolabeling reactions need to be performed. It is much less efficient (and much more costly) to wait several weeks between experiments, because the radioisotope will need to be reordered and the probe will need to be relabeled. For applications requiring the highest sensitivity, you will want to label a new batch of probe at least every 2 weeks, which corresponds to the half-life of ^{32}P. For other applications that are not as demanding, you can allow the probe to decay for 3 to 4 weeks or even slightly longer. For additional information, see *UNIT 2.3*.

Radiolabeled DNA can be characterized by its specific activity. Specific activity is defined as the ratio of the amount of radioactivity (in Ci or similar unit; see *UNIT 2.3* for radiation measurement) to the absolute amount of DNA (in moles or similar units; see *UNIT 2.2* for DNA calculation). All things being equal, probes with higher specific activity will give higher sensitivity for detection.

SAFETY CONSIDERATIONS

If radioisotopes are used to label nucleic acids, then the research must adhere to all institutional guidelines and follow the recommended safe practices for radioactive materials, as detailed in *UNIT 2.3*. Appropriate personal protection (gloves, safety glasses, lab coats, and personal monitoring badge) should be used at all times, appropriate shielding (acrylic shields) should be employed, and all materials should be disposed of properly. Before beginning work with radioisotopes, clean off a large work area on a lab bench and lay down two layers of disposable, absorbent bench paper. The bench paper is valuable in case of spills, to minimize contamination of surfaces—contaminated bench paper can simply be disposed of in the appropriate radioactive waste container. Throughout the experiment and again at the conclusion, monitor yourself, your pipets, and the working area for contamination with a hand-held Geiger counter.

PROTOCOLS

Basic Protocol 1: 5′ End-Labeling of DNA with T4 Polynucleotide Kinase

This protocol will transfer a radioactive phosphate from [γ-^{32}P]ATP to the free 5′ hydroxyl of a target DNA molecule. Single-stranded oligonucleotides will be labeled with a single phosphate at their 5′ end. Double-stranded DNA will be labeled at the 5′ ends of both strands, if both are available to be phosphorylated.

The labeling procedure should take ~90 min to complete. If the concentrated stock of radionucleotide is stored frozen, allow an extra 30 min for thawing. Most of the downstream experiments that you will perform will be done using dilutions from this labeling reaction, but in this protocol you will be working directly with the undiluted, concentrated stock of radionucleotide. Exercise extra caution and have an experienced lab member assist you if you do not have experience working with radioisotopes. Gloves, safety glasses, and a lab coat should be worn at all times during the procedure. Acrylic shielding should be employed to protect yourself and others from exposure to radiation. Also see *UNIT 2.3* for additional guidelines on using radioisotopes.

CAUTION: Radioactive materials require special handling. Follow the guidelines provided by your local radiation safety adviser. See *UNIT 2.3* for more information.

Materials

10 mCi/ml [γ-^{32}P]adenosine-5′-triphosphate ([γ-^{32}P]ATP; sp. act., 6000 Ci/mmol; lower-specific-activity ATP can be used, depending on the sensitivity needs)
10× PNK buffer (see recipe or use buffer supplied with the enzyme)
DNA substrate with free 5′-hydroxyl end(s)
10 U/μl T4 polynucleotide kinase
0.5 M EDTA, pH 8.0 (*UNIT 3.3*)

65° and 75°C water baths or heat blocks
Shielded radiation containers (*UNIT 2.3*)

Additional reagents and equipment for removing unincorporated nucleotides (Support Protocol) and radiation safety and measurement, including quantifying specific activity of radiolabeled DNA (*UNIT 2.3*)

Prepare end-labeling reaction mixtures

1. If necessary, remove the 10 mCi/ml [γ-^{32}P]ATP concentrated stock from the freezer and set it to thaw at room temperature behind appropriate acrylic shielding for at least 30 min.

   Some preparations of [γ-^{32}P]ATP contain a stabilizer which allows the nucleotide to be stored at 4°C. Other preparations of [γ-^{32}P]ATP, however, must be stored at −20°C and thawed before use.

2. Combine the following together in a microcentrifuge tube and mix well:

   4 μl 10× PNK buffer
   1 to 50 pmol DNA substrate to be labeled
   H_2O to 34 μl.

   For example, if the DNA substrate to be labeled is available in a concentration of 5 pmol/μl, then one might use 5 μl of this DNA substrate (for 25 pmol DNA total), and then add 5 μl of buffer and 24 μl of water.

3. *Optional:* Heat the DNA-buffer mixture at 75°C for 2 min, then immediately cool on ice for 2 min. Collect the condensate from the sides of the tube at the bottom of the tube by performing a brief "touch-down" microcentrifugation.

   The purpose of this step is to melt any secondary structure in the DNA that might interfere with the labeling reaction. This step has also been shown to improve phosphorylation efficiency for blunt and recessed 5′-end substrates (Sambrook et al., 1989).

4. Carefully add the following components to the DNA:

   4 μl 10 mCi/ml [γ-^{32}P]ATP (6000 Ci/mmol)
   2 μl 10 U/μl T4 DNA polynucleotide kinase.

   After each addition, use a hand-held Geiger counter to check your hands and the pipet for contamination.

Perform end labeling and purify product

5. Incubate the reaction mixture tube in a 37°C water bath for 60 min.

   During this 1-hr incubation, it is useful to prepare the spin column to be used for subsequent purification; see Support Protocol below.

6. After the incubation, perform a brief "touch-down" microcentrifugation to collect the condensate from the walls of the tube.

7. Add 1.5 μl of 0.5 M EDTA, pH 8.0, and incubate at 65°C for 20 min to inactivate the T4 polynucleotide kinase.

8. To remove incorporated nucleotides and purify the radiolabeled DNA, proceed to Support Protocol.

9. To quantify the specific activity of the radiolabeled DNA, refer to *UNIT 2.3*. To get a rough estimate of the labeling efficiency, compare the specific activity of small aliquots of the crude reaction mixture prior to removal of unincorporated nucleotides (before step 8) and the purified probe after purification (after step 8).

10. Store the final radiolabeled product in a shielded container at −20°C (*UNIT 2.3*).

11. Completely survey yourself and your working area with a Geiger counter when finished with the labeling procedure (*UNIT 2.3*).

Basic Protocol 2: Labeling DNA by Nick Translation

Nick translation is an easy and rapid method for generating medium-sized probes (200 to 500 bp), appropriate for most blotting and hybridization procedures (Rigby et al., 1977; Sambrook et al., 1989; Tabor et al., 1997). The method can be used to generate probes that are labeled with radioisotopes, fluorophores, or small molecule binding partners, depending on the identity of the deoxynucleotide used in the nick translation reaction. The DNA substrate to be labeled is typically a gel-purified restriction fragment (see *UNIT 7.2*).

The first time that a batch of DNase I is used, it is necessary to empirically determine the correct dilution of the enzyme and the reaction time to generate the highest-activity probe. Set up a few reactions in parallel with varying concentrations of DNase I and vary the reaction time. Once the correct concentration and reaction time have been determined for a given batch of DNase I, then these values can be used repeatedly without the need for recalibration. A typical starting concentration is a 10,000-fold dilution of a 1 mg/ml solution of DNase I (or ∼0.002 U/μl final concentration). The reaction time is generally between 15 and 45 min. The enzyme should be diluted using the standard enzyme diluent (see Reagents and Solutions).

The procedure will take ∼1 hr to complete. If using a frozen radionucleotide stock, allow an extra 30 min for thawing prior to use. Exercise extra caution and have an experienced lab member assist you if you do not have experience working with radioisotopes. Gloves, safety glasses, and a lab coat should be worn at all times during the procedure. Acrylic shielding should be employed to protect yourself and others from exposure to radiation.

CAUTION: Radioactive materials require special handling. Follow the guidelines provided by your local radiation safety adviser. See *UNIT 2.3* for more information.

Materials

dNTP with label (e.g., 10 μM [α-^{32}P]dNTP; see Table 8.4.3)
0.5 mM 3dNTP mix (omitting the labeled dNTP; see Table 8.4.4)
10× *E. coli* DNA polymerase I buffer (see recipe)
Deoxyribonuclease I (DNase I), diluted ∼10,000-fold from 1 mg/ml stock in enzyme diluent (see recipe)
5 to 15 U/μl *E. coli* DNA polymerase I
DNA to be labeled
0.5 M EDTA, pH 8.0 (*UNIT 3.3*)
TE buffer, pH 8.0 (*UNIT 3.3*)

15°C water bath
Shielded radiation containers (*UNIT 2.3*)

Additional reagents and equipment for removing unincorporated nucleotides (Support Protocol), radiation safety and measurement, including quantifying specific activity of radiolabeled DNA (*UNIT 2.3*), nucleic acid blotting (*UNIT 8.2*), agarose gel electrophoresis (*UNIT 7.2*), and digital image analysis (*UNIT 7.5*)

1. If performing radiolabeling, remove the [α-^{32}P]dNTP concentrated stock from the freezer and set to thaw at room temperature behind appropriate acrylic shielding (*UNIT 2.3*) for at least 30 min.

2. Mix the following components in a microcentrifuge tube on ice:

   10 µl solution of dNTP with label (e.g., 10 µM [α-^{32}P]dNTP; see Table 8.4.3)
   2.5 µl 0.5 mM 3dNTP mix (see Table 8.4.4)
   2.5 µl 10× *E. coli* DNA polymerase I buffer
   1 µl DNase I (diluted 10,000-fold from 1 mg/ml stock)
   1 µl 5 to 15 U/µl *E. coli* DNA polymerase I.

   Keep the mixture on ice until all additions are made. The reaction can be scaled up or down depending on the amount of probe that is desired—the protocol here is appropriate for 250 ng of DNA. If the scale of the reaction is changed, it is critical to maintain the above concentrations of enzymes, deoxynucleotides, and DNA. The proper dilution of DNase I is critical for optimal nick translation. The concentration of labeled dNTP and the specific activities listed in Table 8.4.3 are intended to serve as a starting point and can be adjusted according to the application. Choose only one labeled dNTP (e.g., biotin-11-dUTP) and use the appropriate 3dNTP mix corresponding to your choice of labeled dNTP (see Table 8.4.4).

3. Add 250 ng of the DNA to be labeled, in a total volume of 8 µl (add water if necessary).

   The total reaction volume should be 25 µl.

4. Immediately incubate the reaction mixture at 15°C for 15 to 45 min, depending on the DNase calibration curve (see above).

   During this incubation, it is useful to prepare the spin column to be used for subsequent purification; see Support Protocol below.

5. After the time determined for the particular batch of DNase I, stop the reaction by adding:

   1 µl 0.5 M EDTA, pH 8.0
   100 µl TE buffer, pH 8.0.

6. To remove incorporated nucleotides and purify the radiolabeled DNA, proceed to Support Protocol.

7a. *To quantify the specific activity of radiolabeled DNA:* Refer to *UNIT 2.3*. To get a rough estimate of the labeling efficiency, compare the specific activity of small aliquots of the crude reaction mixture prior to removal of unincorporated nucleotides (before step 6) and the purified probe after purification (after step 6).

7b. *To quantify the labeling efficiency for small molecule binding partners:* Perform a dot blot assay comparing serial dilutions of the labeled probe with small molecule–labeled standards of known concentration (*UNIT 8.2*).

7c. *To quantify the labeling efficiency for fluorophores:* Run a small aliquot of the labeling reaction on a 1% agarose gel (*UNIT 7.2*), then analyze with a fluorescence scanner or a Polaroid system with an appropriate filter for the fluorophore used (see *UNIT 7.5*).

8. Store radiolabeled probes in a shielded container at −20°C (*UNIT 2.3*).

9. If working with radioactivity, completely survey yourself and your working area with a Geiger counter when finished with the labeling procedure (*UNIT 2.3*).

Basic Protocol 3: Labeling DNA by Random Primed Synthesis

Random primed synthesis is an alternative to nick translation for uniform, high-specific-activity labeling of DNA probes with radioisotopes, fluorophores, or small molecule binding partners (Feinberg and Vogelstein, 1983; Boyle and Perry-O'Keefe, 1992; Tabor et al., 1997). Short (6-mer to 12-mer) oligonucleotides are annealed to template DNA and then extended by a DNA polymerase. For the DNA synthesis reaction, normal deoxynucleotides are mixed with radiolabeled nucleotides, in this case [α-^{32}P]dNTPs or nucleotides that are conjugated to a small molecule binding partner (biotin or digoxigenin) or a fluorophore. The extent of labeling can be adjusted by modifying the ratio of normal nucleotides to modified nucleotides. The labeling reaction should take ~30 min to set up, and then will proceed for 2 to 4 hr.

If using a frozen radionucleotide stock, allow an extra 30 min for thawing prior to use. Exercise extra caution and have an experienced lab member assist you if you do not have experience working with radioisotopes. Gloves, safety glasses, and a lab coat should be worn at all times during the procedure. Acrylic shielding should be employed to protect yourself and others from exposure to radiation.

CAUTION: Radioactive materials require special handling. Follow the guidelines provided by your local radiation safety adviser. See *UNIT 2.3* for more information.

Materials

dNTP with label (e.g., 10 µM [α-^{32}P]dNTP; see Table 8.4.3)
0.5 mM 3dNTP mix (omitting the labeled dNTP; see Table 8.4.4)
10× *E. coli* DNA polymerase I buffer
3 to 8 U/µl *E. coli* DNA polymerase I large fragment (Klenow fragment)
DNA to be labeled
Random hexanucleotide primers (Sigma cat no. H-0268)
0.5 M EDTA, pH 8.0 (*UNIT 3.3*)
TE buffer, pH 8.0 (*UNIT 3.3*)
Boiling water bath

Additional reagents and equipment for removing unincorporated nucleotides (Support Protocol), radiation safety and measurement, including quantifying specific activity of radiolabeled DNA (*UNIT 2.3*), nucleic acid blotting (*UNIT 8.2*), agarose gel electrophoresis (*UNIT 7.2*), and digital image analysis (*UNIT 7.5*)

1. If performing radiolabeling, remove the [α-^{32}P]dNTP concentrated stock from the freezer and set to thaw at room temperature behind appropriate acrylic shielding (*UNIT 2.3*) for at least 30 min.

Table 8.4.4 3dNTP Nucleotide Mixes for Nick Translation and Random Primed Synthesis

Modified nucleotide with label	Contents of 3dNTP mix (0.5 mM each)
dATP (e.g., biotin-11-dATP, [α-^{32}P]dATP)	dCTP, dGTP, dTTP
dCTP	dATP, dGTP, dTTP
dGTP	dATP, dCTP, dTTP
dTTP (dUTP)	dATP, dCTP, dGTP

2. Mix the following components in a microcentrifuge tube on ice:

   5 μl solution of dNTP with label (e.g., 10 μM [α-^{32}P]dNTP; see Table 8.4.3)
   2.5 μl 0.5 mM 3dNTP mix (see Table 8.4.4)
   2.5 μl 10× *E. coli* DNA polymerase I buffer
   1 μl 3 to 8 U/μl Klenow fragment.

   Keep the mixture on ice until all additions are made. The concentration of labeled dNTP and the specific activities listed in Table 8.4.3 are intended to serve as a starting point and can be adjusted according to the application. Choose only one labeled dNTP (e.g., biotin-11-dUTP) and use the appropriate 3dNTP mix corresponding to your choice of labeled dNTP (see Table 8.4.4).

3. In a separate microcentrifuge tube, combine the DNA to be labeled (30 to 100 ng) with the random hexanucleotide primers (1 to 5 μg) in a total volume of 14 μl. Heat the DNA mixture 2 to 3 min in a boiling water bath, then place on ice.

4. Add 11 μl of the reaction mixture from step 2 to the denatured DNA from step 3 (final volume, 25 μl). Incubate the mixture 2 to 4 hr at room temperature.

   During this incubation, it is useful to prepare the spin column to be used for subsequent purification—see Support Protocol below.

5. Stop the reaction by adding:

   1 μl 0.5 M EDTA, pH 8.0
   100 μl TE buffer, pH 8.0.

6. To remove incorporated nucleotides and purify the radiolabeled DNA, proceed to Support Protocol.

7a. *To quantify the specific activity of radiolabeled DNA:* To get a rough estimate of the labeling efficiency, compare the specific activity of small aliquots of the crude reaction mixture prior to removal of unincorporated nucleotides (before step 6) and the purified probe after purification (after step 6).

7b. *To quantify the labeling efficiency for small molecule binding partners:* Perform a dot blot assay comparing serial dilutions of the labeled probe with small molecule–labeled standards of known concentration (*UNIT 8.2*).

7c. *To quantify the labeling efficiency for fluorophores:* Run a small aliquot of the labeling reaction on a 1% agarose gel (*UNIT 7.2*), then analyze with a fluorescence scanner or a Polaroid system with an appropriate filter for the fluorophore used (see *UNIT 7.5*).

8. Store radiolabeled probes in a shielded container at −20°C (*UNIT 2.3*).

9. If working with radioactivity, completely survey yourself and your working area with a Geiger counter when finished with the labeling procedure.

Support Protocol: Purification of Labeled Probes Using Gel-Filtration Spin Columns

Following the labeling reaction, it is generally desirable to purify the labeled product and remove unincorporated nucleotides. One convenient method for this purification is to use spin columns loaded with a gel-filtration medium, such as Sephadex G-50 (GE Healthcare). The spin columns fit inside a standard microcentrifuge tube. The crude labeling reaction is loaded onto the spin column, and centrifugation is used to load and elute the sample. Small-molecule impurities, such as unincorporated nucleotides, are retained in small pores in the gel-filtration medium, while large biomolecules, such as the labeled DNA, will flow rapidly through the medium (see Chapter 6 and *UNIT 5.2*). An advantage of this approach is that it is possibly to simultaneously exchange the buffer of the radiolabeled DNA. Gel-filtration media are classified according to size resolution, so check the vendor's recommendations for the appropriate medium for the size of your probe. Some commonly used gel-filtration spin columns include CentriSpin 20 (Princeton Separations) and illustra Microspin Columns (GE Healthcare). Alternatively, you may perform gel filtration cleanup using gravity columns (e.g., illustra NAP columns, GE Healthcare) or push columns (e.g., NucTrap Probe Purification Columns, Stratagene). Finally, it is also possible to prepare your own spin or gravity columns (*CP Molecular Biology* *UNIT 3.4*; Struhl, 1993).

Follow the instructions provided by the manufacturer of the spin column. Typically, the procedure begins by hydrating/swelling the gel filtration media with the desired storage buffer (e.g., TE buffer, pH 8.0). Following preparation of the column, the labeling reaction is loaded carefully onto the spin column, which is inserted into a standard microcentrifuge tube. Microcentrifugation will elute the purified DNA into the microcentrifuge tube and the unincorporated nucleotide will be retained on the column. If purifying radiolabeled DNA, use extra caution when handling and disposing of the used spin column, as it will have a large amount of radioactivity.

REAGENTS AND SOLUTIONS

Use deionized, distilled water in all recipes and protocol steps. For common stock solutions, see **UNIT 3.3**.

E. coli DNA polymerase I buffer, 10×

500 mM Tris·Cl, pH 8.0 (*UNIT 3.3*)
100 mM $MgCl_2$
10 mM dithiothreitol (DTT)
Store up to 6 months at −20°C

Enzyme diluent

20 mM Tris·Cl, pH 7.5 (*UNIT 3.3*)
500 μg/ml bovine serum albumin (BSA)
10 mM 2-mercaptoethanol (2-ME)
Store up to 1 month at 4°C

PNK buffer, 10×

700 mM Tris·Cl, pH 7.6 (*UNIT 3.3*)
100 mM $MgCl_2$
50 mM dithiothreitol (DTT)
Store up to 6 months at −20°C

UNDERSTANDING RESULTS

It should be possible to synthesize high-specific-activity probes by any of the three protocols described above. The percentage incorporation of the radiolabel for probes generated by 5′ end labeling with T4 polynucleotide kinase is expected to be ∼30% to 50%. For DNA labeling by nick translation, radiolabeled probes with a specific activity of 10^8 cpm/μg should be obtained by this protocol. The product size will depend on the DNase I concentration and can be fine tuned depending on the intended application. The rate of incorporation of the labeled nucleotide should be ∼30% to 40%. No net synthesis of DNA occurs during nick translation, so the theoretical maximum amount of labeled probe is the complete replacement synthesis of the starting template. In contrast, for DNA labeling by random primed synthesis, each 1 μg of starting template should yield ∼250 ng of probe in a typical labeling reaction. The size of the random synthesis labeled product will be shorter than the template and will be in a fairly wide range. For example, a typical 2-kb template will give products of 300 to 1500 bp, with most at ∼800 bp.

TROUBLESHOOTING

See Table 8.4.5.

VARIATIONS

There are a large number of methods for labeling nucleic acids. The protocols presented in this unit are a subset of the available options, as listed in Table 8.4.6. The 3′ end-labeling reactions (first two entries in the table) are an alternative to the 5′ end-labeling procedure presented in this unit, and have the advantage of being able to incorporate nonisotopic labels as well as radioisotopic labels. The polymerase chain reaction (PCR; *UNIT 10.2*) can be a facile way to synthesize a labeled probe, and this method requires only minor modification of the standard PCR protocol. Finally, for RNA

Table 8.4.5 Troubleshooting Labeling Reactions

Problem	Possible causes	Recommendations
5′ end labeling with T4 polynucleotide kinase		
Low-efficiency labeling	10× PNK buffer not fresh; DTT is oxidized	Replace the buffer
	Inhibitory ions are present	Do not use ammonium ions or phosphate ions in the preparation of DNA substrate, as these ions can inhibit polynucleotide kinase
	DNA substrate not pure; low-molecular-weight nucleic acids are present	Purify substrate DNA prior to labeling (*UNIT 5.2*). Do not use tRNA during purification.
	5′ end is recessed	Change substrate or substrate preparation to give 5′ overhang Change labeling method
No (or very little) labeling	No free hydroxyl on 5′ end	Dephosphorylate 5′ end using a phosphatase[a] Use the exchange reaction of T4 PNK[a]
	Wrong nucleotide	Make sure that ^{32}P label is in the γ-phosphate—i.e., order [γ-^{32}P]ATP
Labeling DNA by nick translation		
Low-efficiency labeling	Labeling reaction not optimized	Adjust ratio of special nucleotide to normal nucleotide Vary concentration of DNase I
	Template not sufficiently pure	Purify template DNA (*UNIT 5.2*)
Labeling DNA by random primed synthesis		
Low-efficiency labeling	Template DNA not sufficiently pure	Purify template DNA (*UNIT 5.2*)
	Labeling reaction not optimized	Adjust ratio of modified nucleotide to normal nucleotide Let the reaction incubate for longer (to overnight) While leaving the amount of template constant, scale up the reaction volume and the other components
Probe too small (<200 bp)	Template too short	Switch to longer template Switch to 3′ end labeling

[a]*CP Molecular Biology Unit 3.10* (Tabor, 1987a).

Table 8.4.6 Alternative Approaches for Labeling Nucleic Acids

Method	Reference
3′ End labeling of DNA fragments with 5′ overhangs by fill-in with Klenow fragment	*CP Mol. Biol. Unit 3.5* (Tabor et al., 1997)
3′ End labeling by terminal transferase	*CP Mol. Biol. Unit 3.6* (Tabor, 1987b)
PCR with labeled primer or labeled dNTPs	*UNIT 10.2*
In vitro RNA synthesis	*CP Mol. Biol. Unit 3.8* (Tabor, 1987c)
Tailing RNA	*CP Mol. Biol. Unit 3.9* (Tabor, 1987d)

hybridization, such as in northern blotting, using RNA probes can give higher sensitivity, and the final two methods are suitable for labeling RNA instead of DNA.

LITERATURE CITED

Amitsur, M., Levitz, R., and Kaufmann, G. 1987. Bacteriophage T4 anticodon nuclease, polynucleotide kinase and RNA ligase reprocess the host lysine tRNA. *EMBO J.* 6:2499-2503.

Ausubel, F., Brent, R., Kingston, R., Moore, D.D., Seidman, J.G., Smith, J.A., and Struhl, K. (eds.) 2007. Current Protocols in Molecular Biology, Chapter 14. John Wiley & Sons, Hoboken, N.J.

Baldwin, A.S. Jr., Oettinger, M., and Struhl, K. 1996. *Curr. Protoc. Mol. Biol.* 36:12.3.1-12.3.7.

Boyle, A. and Perry-O'Keefe, H. 1992. Labeling and colorimetric detection of nonisotopic probes. *Curr. Protoc. Mol. Biol.* 20:3.18.1-3.18.9.

Brenowitz, M., Senear, D.F., and Kingston, R.E. 1989. DNase I footprint analysis of protein-DNA binding. *Curr. Protoc. Mol. Biol.* 7:12.4.1-12.4.16.

Buratowski, S. and Chodosh, L.A. 1996. Mobility shift DNA-binding assay using gel electrophoresis. *Curr. Protoc. Mol. Biol.* 36:12.2.1-12.2.11.

Cameron, V. and Uhlenbeck, O.C. 1977. 3′-Phosphatase activity in T4 polynucleotide kinase. *Biochemistry* 16:5120-5126.

Duby, A., Jacobs, K.A., and Celeste, A. 1993. Using synthetic oligonucleotides as probes. *Curr. Protoc. Mol. Biol.* 2:6.4.1-6.4.10.

Ellington, A. and Pollard, J.D. Jr. 1998. Synthesis and purification of oligonucleotides. *Curr. Protoc. Mol. Biol.* 42:2.11.1-2.11.25.

Feinberg, A.P. and Vogelstein, B. 1983. A technique for radiolabeling DNA restriction endonuclease fragments to high specific activity. *Anal. Biochem.* 132:6-13.

Gilman, M. 1993. Ribonuclease protection assay. *Curr. Protoc. Mol. Biol.* 24:4.7.1-4.7.8.

Greene, J.M. and Struhl, K. 1988. S1 analysis of messenger RNA using single-stranded DNA probes. *Curr. Protoc. Mol. Biol.* 1:4.6.1-4.6.13.

Jarvest, R.L. and Lowe, G. 1981. The stereochemical course of phosphoryl transfer catalysed by polynucleotide kinase (bacteriophage-T4-infected Escherichia coli B). *Biochem. J.* 199:273-276.

Jilani, A., Ramotar, D., Slack, C., Ong, C., Yang, X.M., Scherer, S.W., and Lasko, D.D. 1999. Molecular cloning of the human gene, PNKP, encoding a polynucleotide kinase 3′-phosphatase and evidence for its role in repair of DNA strand breaks caused by oxidative damage. *J. Biol. Chem.* 274:24176-24186.

Karimi-Busheri, F., Daly, G., Robins, P., Canas, B., Pappin, D.J., Sgouros, J., Miller, G.G., Fakhrai, H., Davis, E.M., Le Beau, M.M., and Weinfeld, M. 1999. Molecular characterization of a human DNA kinase. *J. Biol. Chem.* 274:24187-24194.

Novogrodsky, A. and Hurwitz, J. 1966. The enzymatic phosphorylation of ribonucleic acid and deoxyribonucleic acid. I. Phosphorylation at 5′-hydroxyl termini. *J. Biol. Chem.* 241:2923-2932.

Novogrodsky, A., Tal, M., Traub, A., and Hurwitz, J. 1966. The enzymatic phosphorylation of ribonucleic acid and deoxyribonucleic acid. II. Further properties of the 5′-hydroxyl polynucleotide kinase. *J. Biol. Chem.* 241:2933-2943.

Richardson, C.C. 1965. Phosphorylation of nucleic acid by an enzyme from T4 bacteriophage-infected *Escherichia coli*. *Proc. Natl. Acad. Sci. U.S.A.* 54:158-165.

Richardson, C.C. 1981. Bacteriophage T4 polynucleotide kinase. *In* The Enzymes (P.D. Boyer, ed.) pp. 299-314. Academic Press, San Diego.

Rigby, P.W., Dieckmann, M., Rhodes, C., and Berg, P. 1977. Labeling deoxyribonucleic acid to high specific activity in vitro by nick translation with DNA polymerase I. *J. Mol. Biol.* 113:237-251.

Sambrook, J., Fritsch, E.G., and Maniatis, T. 1989. Molecular Cloning: A Laboratory Manual, 2nd Ed. Cold Spring Harbor Laboratory Press, Cold Spring Harbor, N.Y.

Sirotkin, K., Cooley, W., Runnels, J., and Snyder, L.R. 1978. A role in true-late gene expression for the T4 bacteriophage 5′ polynucleotide kinase 3′ phosphatase. *J. Mol. Biol.* 123:221-233.

Strauss, W.M. 1993. Using DNA fragments as probes. *Curr. Protoc. Mol. Biol.* 24:6.3.1-6.3.6.

Struhl, K. 1993. Reagents and radioisotopes used to manipulate nucleic acids. *Curr. Protoc. Mol. Biol.* 9:3.4.1-3.4.11.

Tabor, S. 1987a. Phosphatases and kinases. *Curr. Protoc. Mol. Biol.*. 0:3.10.1-3.10.5.

Tabor, S. 1987b. Template-independent DNA polymerases. *Curr. Protoc. Mol. Biol.* 0:3.6.1-3.6.2.

Tabor, S. 1987c. DNA-dependent RNA polymerases. *Curr. Protoc. Mol. Biol.* 0:3.8.1-3.8.4.

Tabor, S. 1987d. DNA-independent RNA polymerases. *Curr. Protoc. Mol. Biol.* 0:3.9.1-3.9.2.

Tabor, S., Struhl, K., Scharf, S.J., and Gelfand, D.H. 1997. DNA-dependent DNA polymerases. *Curr. Protoc. Mol. Biol.* 37:3.5.1-3.5.15.

Triezenberg, S.J. 1992. Primer extension. *Curr. Protoc. Mol. Biol.* 20:4.8.1-4.8.5.

Vitolo, J.M., Thieret, C., and Hayes, J.J. 1999. DNase I and hydroxyl radical characterization of chromatin complexes. *Curr. Protoc. Mol. Biol.* 48:21.4.1-21.4.9.

Chapter 9

Microscopy

UNIT 9.1

Conventional Light Microscopy

Eric S. Cole[1]
[1]St. Olaf College, Northfield, Minnesota

PARTS OF THE LIGHT MICROSCOPE

Figures 9.1.1 and 9.1.2 show a conventional light microscope with all the commonly used parts labeled (also see Table 9.1.1).

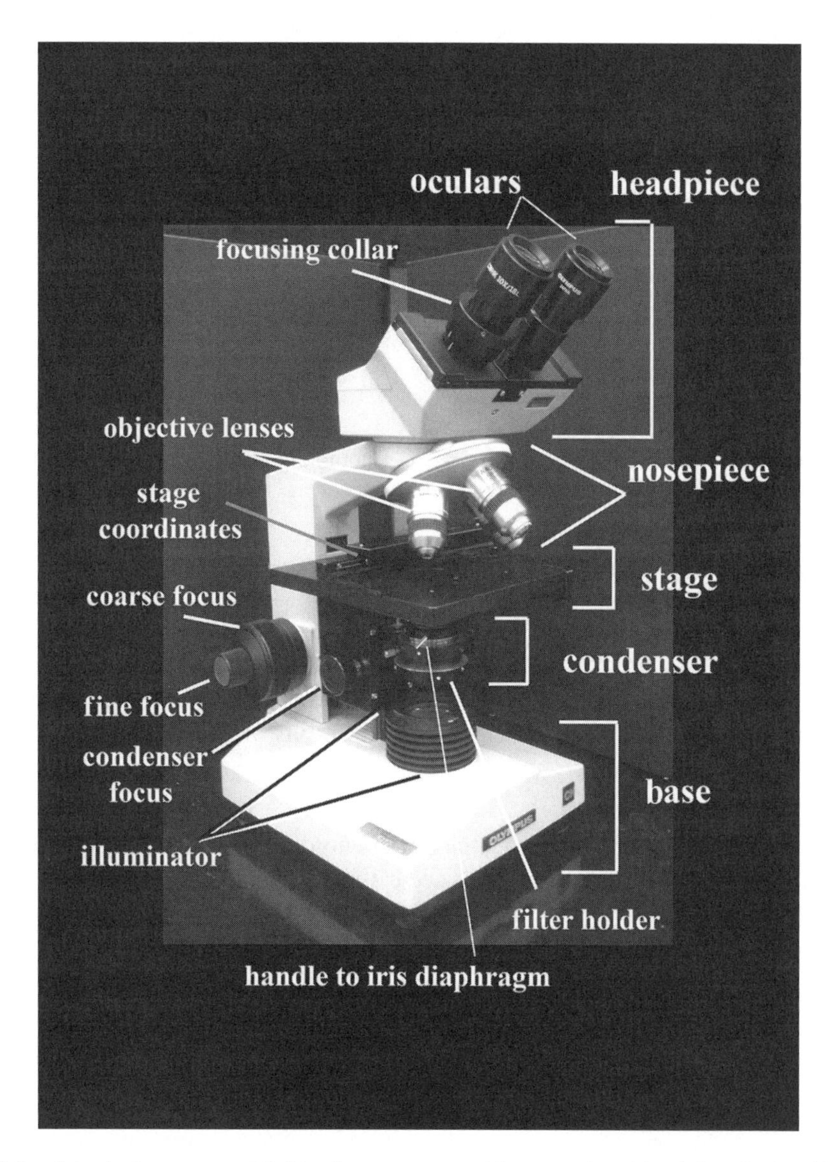

Figure 9.1.1 A typical compound light microscope used in many teaching laboratories. Not shown are the on/off toggle switch and the illuminator rheostat that controls the intensity of illumination.

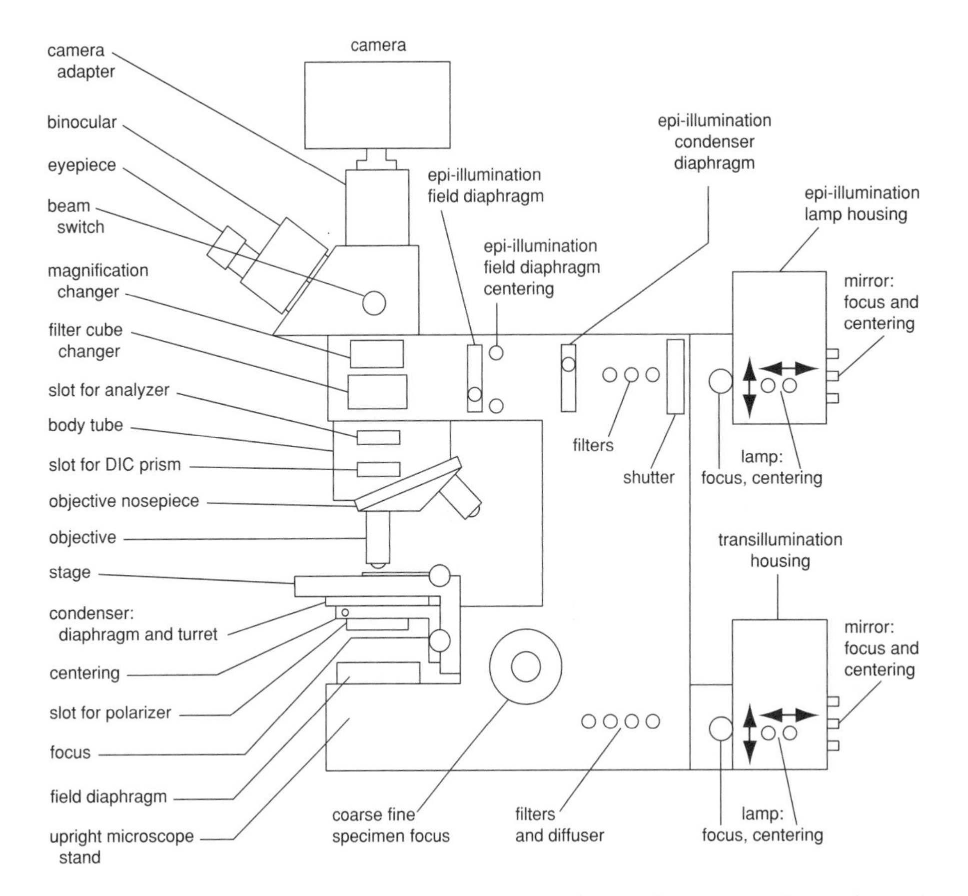

Figure 9.1.2 Diagram of the major component parts and centering screws for a research upright light microscope. Originally published in *Current Protocols in Microbiology Unit 2A.1* (Salmon et al., 2005).

CARE AND MAINTENANCE

Cords

Be sure that power cords, which have a tendency to catch on purses and key chains, do not drape over the edge of the table; slack cord should be wrapped around the microscope or the benchtop power outlet to minimize the possibility of accidentally upsetting and damaging the microscope.

Carrying

Use two hands, one supporting from underneath and one holding the neck. This prevents tipping, in which case ocular lenses can fall out, and adds security to prevent dropping.

Power

Begin with the rheostat turned low, then, after turning on the lamp, increase the light intensity until it is comfortable to view. To turn off, use the rheostat first to dim the light, then hit the "off" switch. Lamp filaments wear out when forced to go from cold to hot and vice versa. Gentle warm-up and cool-down ramps extend the life of the bulb.

Cleaning the Lenses

Use lens tissues. Other paper products (e.g., Kimwipes) may have abrasives. Do not use solvents; organic solvents will gradually dissolve cements that hold the lenses in place. Clean lenses after first moistening them with lens cleaner (either 50% ammonia/5% ethanol mix or a commercial glass cleaner).

Table 9.1.1 Components of the Conventional Light Microscope

Component	Function
Headpiece	Holds the oculars (or eyepieces)
Oculars	The ocular lens directs the final image into the observer's eye
Base	Holds the power supply, substage illuminator, and field condenser (if the microscope has one)
Nosepiece	Holds the objective lenses in a swiveling turret
Objective lenses	Gather light from the specimen and focus it into a magnified image
Stage	This platform holds the microscope slide and specimen
Slide calipers	These gently clasp the microscope slide for manipulation in *x*/*y* dimensions
x/*y* coordinates	These allow one to note/record specific locations on the microscope slide
Coarse focus	Allows rapid movement through the *z* (vertical) dimension, useful for finding the specimen in the first place (at low magnification)
Fine focus	Allows for more subtle, slower movement through the *z* dimension, making it possible to achieve crisp focus
Substage condenser and focusing knob	These allow one to focus a uniform light field (an image of the substage light diffuser) onto the specimen plane
Iris diaphragm with handle	Allows one to reduce or expand the diameter of the light cone that illuminates your specimen. Useful in establishing Köhler illumination.
Substage illuminator	Provides transmitted light to illuminate translucent specimens
Field diaphragm (sometimes absent)	Allows one to reduce or enlarge the column of light that fills the microscope light path. Useful in establishing Köhler illumination.
Filter holder	Allows one to do color corrections (e.g., approximate "daylight") or to slip a dark-field disk in for dark-field microscopy
On/off switch and adjustable light rheostat	Control power to the lamp

Quick clean

Wipe moistened lens surfaces with lens tissues. Never run tissues over a dry lens; dust and grit can serve as an abrasive without the lubricating action of cleaning fluid.

Careful clean

Try the "touch and draw" technique, in which the lenses are cleaned with a cotton-tipped applicator stick and plenty of lens tissue. First apply cleaner to lens, then place lens tissue into the puddle of cleaner on your lens. With a fresh (cotton-tipped) applicator stick, press gently down on the damp lens through the tissue. Draw the tissue away with one hand, while using gentle pressure on the applicator with the other to keep tissue in contact with the lens. Move to a clean place on the tissue and repeat. This insures that your finger oils do not contaminate the lens surface; repeated passes draw off cleaning fluid and dissolved oils and dirt. Inspect and repeat as necessary.

Work Station

Have a ready supply of cotton-tipped applicator sticks, lens cleaner, lens tissues, pipets and bulbs (or disposable pipets), slides, coverslips, clay (for clay-feet application; see below), and a sharps bucket for disposing of coverslips and pipets (no one in their right mind tries to wash coverslips).

BASIC PRINCIPLES AND DEFINITIONS

The Basics of Optical Magnification

Bright-field microscopy and its variations take advantage of the different kinds of interaction between matter and light. As light encounters matter, the light can be reflected, absorbed, or transmitted. Even when light is transmitted through a material, there are numerous ways in which it can be perturbed. Light can be polarized by selective absorption of light waves vibrating in particular orientations; it can be split into parallel (polarized) components as it passes through birefringent materials (i.e., those with the ability to rotate light); it can be diffracted as it interacts with edges and apertures whose dimensions are on the same order as the wavelength of the light itself; or it can simply bend by refraction when passing from one medium to another (such as from air to glass or from glass to air). The property of refraction is fundamental to all light microscopy, and deserves further explanation.

Refraction

Light moves at a constant velocity in a vacuum. Yet, in various (denser) materials light moves more slowly. The velocity of light in air is very nearly the same as that in a vacuum. The velocity in water, as well as in glass, is considerably slower. The ratio of the speed of light in vacuum to that in a second medium of greater density is known as the refractive index, or n. The refractive index of optical glass is $\sim$1.5. Light doesn't only slow down when moving from air to water or air to glass, it bends or refracts. This can be demonstrated by holding a stick in a clear container of water and observing it from several angles. The image of the stick above the water line does not match up with the image of the stick below water line. We can gain an intuitive appreciation of this phenomenon by looking at Figure 9.1.3. If we provide light with a vertical vibrational dimension and imagine it striking the water at an acute angle, the lower part of the wave will strike the interface first and will slow first; the top of the wave will strike the interface an increment later and be slowed an increment later. The result is refraction towards a line drawn perpendicular to the interface. To rephrase: when light passes from a medium of low refractive index (air) to one of higher refractive index (glass or water), the light will be refracted towards a line drawn perpendicular to the water's surface (Fig. 9.1.3A). Conversely, as light passes out of a high-refractive-index medium and into a lower-refractive-index material, (water to air), the "top" of the wave will accelerate first, before the "bottom" of the wave, and light will bend away from the perpendicular (Fig. 9.1.3B). A prism combines these refractive interactions to turn light toward the base of the prism (Fig. 9.1.4). By combining and stacking a series of such prisms (Fig. 9.1.5), each with a slightly more acute angle as one moves away from the middle, one gradually assembles a "lens" in which parallel light from some object will be refracted towards the lens' focal point. As an historical aside, the Fresnel lens invented for 19th-century lighthouses was in fact a series of cut glass blocks with increasingly acute angles that took light from a focal point, the lamp, and projected it as a bright, parallel beam (the converse of what a magnifying lens does).

In brief, a lens magnifies and inverts an image. Tracing ray diagrams can be helpful in understanding magnification and image inversion (Fig. 9.1.6). Three rays are particularly informative: (1) light from the object that strikes the lens perpendicular to the lens optical axis; (2) light that passes through the center of the lens; and (3) light that passes from the object through the focal point in front of the lens. These three rays will converge to form a real (inverted) image to the right of the lens in Figure 9.1.6A. For an excellent interactive Java tutorial based on ray diagrams, see *http://micro.magnet.fsu.edu/primer/java/lenses/converginglenses/index.html.*

If an object is placed at a distance equal to twice the focal distance of the lens, a "real image" will be formed with zero magnification (Fig. 9.1.6A). A real image is one that can be projected directly onto a screen or photographed. As one moves the object closer to the lens, this real image will be magnified (Fig. 9.1.6B). As the object approaches the focal point in front of the lens, the image will approach infinite magnification until, when placed at the focal distance of the lens, the light

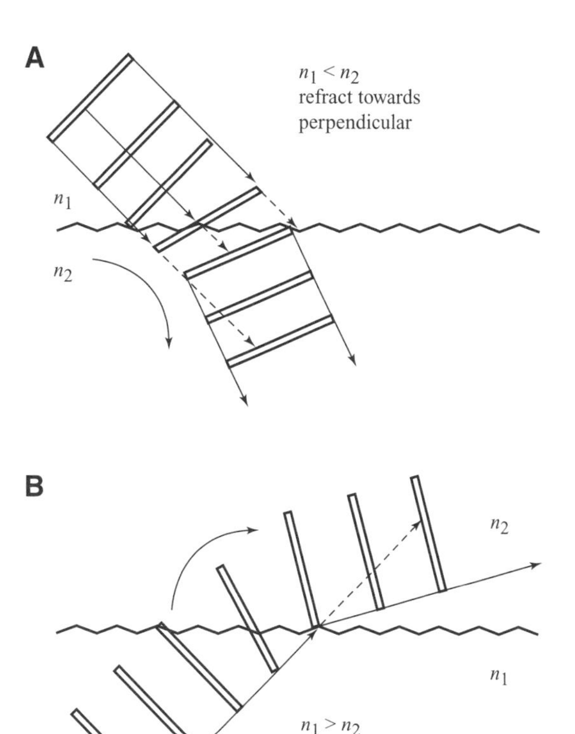

Figure 9.1.3 This figure illustrates the refraction of light as it passes from air to water (**A**) or water to air (**B**). Air is shown with refractive index n_1, and water with n_2. In general, light passing from a low- to high-density medium (A) bends towards an imaginary line drawn perpendicular to the interface. Conversely, light bends away from such a line as it passes from a high- to low-density medium (B).

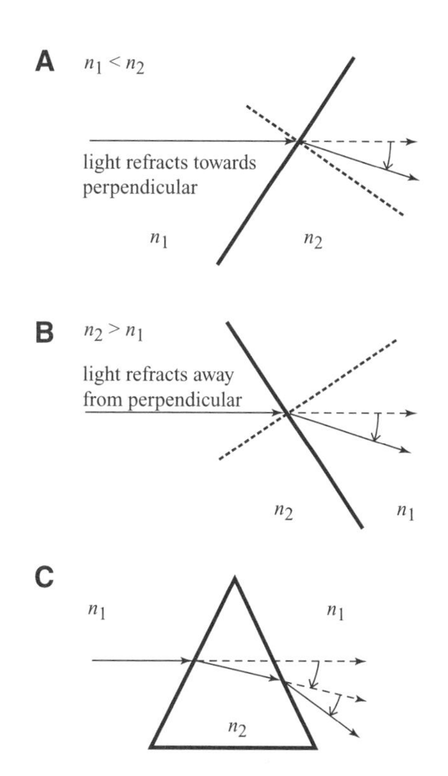

Figure 9.1.4 Building on Figure 9.1.3, this illustrates the principle of a prism in which light passes through two interfaces: air to glass (**A**), glass to air (**B**), and a combination found in a prism (**C**). Light passing through such an object is refracted towards the base of the prism.

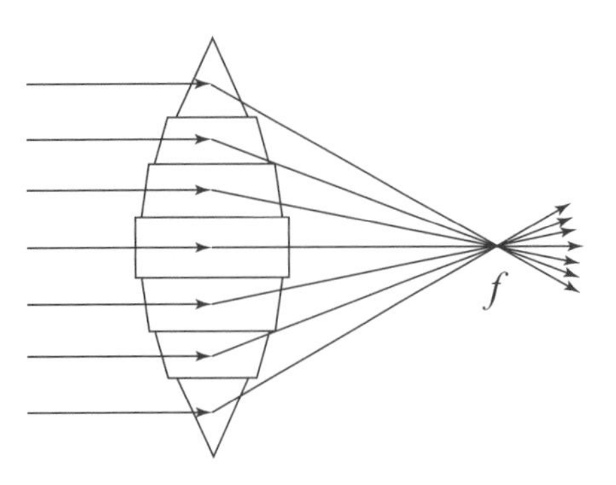

f
If the angle of the prism surfaces becomes more acute with distance from the center, light will be focused to a single point, the focal point "f".

Figure 9.1.5 When prisms are stacked as in this illustration, the compound effect is to refract light in such a way that it converges at a single point, the focal point (*f*). This is due to the increasingly acute angles of each prism as one moves away from the center of the lens. This is the principle behind a simple biconvex lens (a lens that curves "outward" on both surfaces).

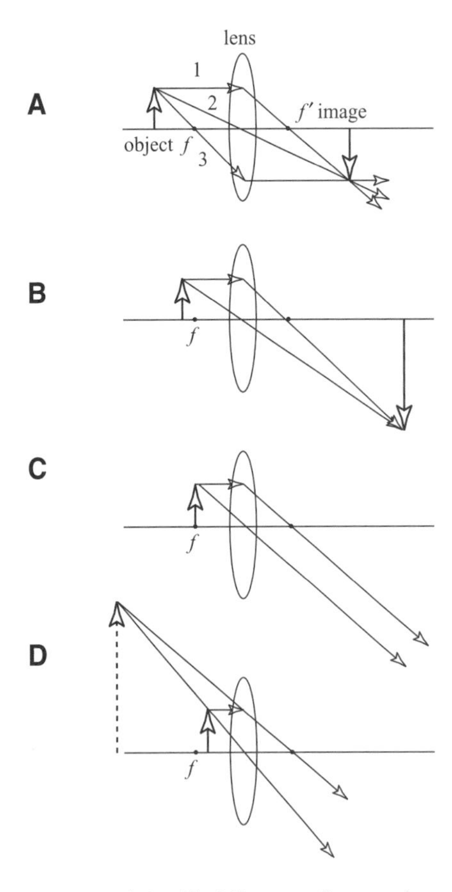

Figure 9.1.6 This figure illustrates two points. First the ray diagram in panel (**A**) traces the path of three particular rays of light. "1" represents the path of light from the object that strikes the lens at right angles to its optical axis. Such light will be refracted through the focal point (*f*). The light represented by "2" from the object that strikes the lens dead-center will not change trajectory, but pass straight through. The light represented by "3" from the object that passes through the "back focal plane" (*f*) will emerge from the lens perpendicular to the optical axis. All light emanating from *f* will be focused into a parallel beam as it emerges from the lens (the principle behind the Fresnel lens used in early lighthouse designs). Panels (**B**) to (**D**) illustrate the consequences of moving the object progressively closer to the lens.

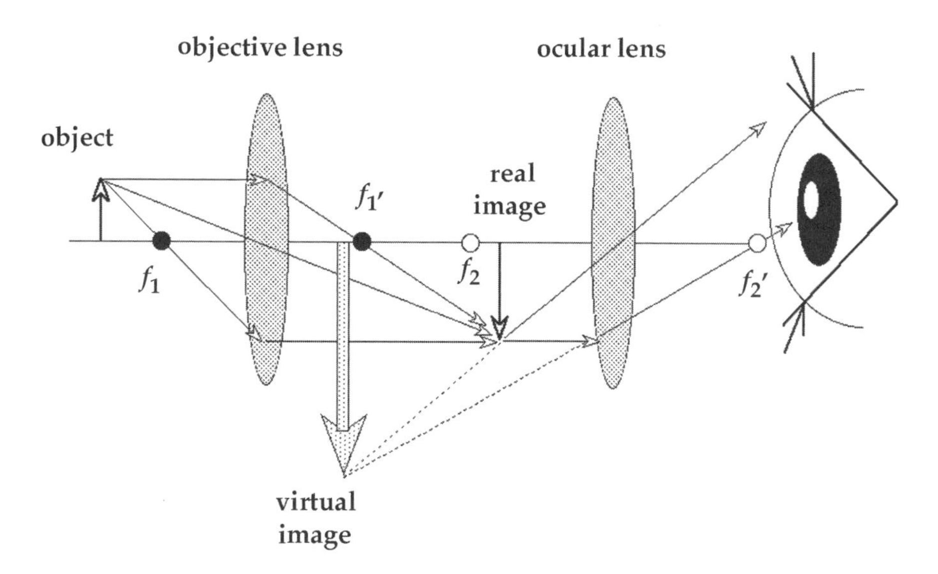

Figure 9.1.7 This line drawing illustrates the fundamental design of a compound microscope. The object is placed at less than $2 \times f_1$ in front of the objective lens, so that its image is magnified (as in Fig. 9.1.6B). This real image is projected onto a plane that lies inside the focal distance f_2 for the ocular lens. Consequently, as light from this real image passes through the ocular lens, its light diverges (as in Fig. 9.1.6D). Your eye collects this divergent light and creates a real image on the retina. This produces the effect of a virtual image.

from the object emerges from the lens as a series of parallel, nonconverging rays (Fig. 9.1.6C). This configuration is useful in lighthouses, but of no practical use for the microscope. It should be noted that, although infinite magnification can be achieved, such overmagnified images carry no useful information (the image becomes one of empty magnification; no details can be distinguished). This raises the important distinction between magnification and resolution (described below). When the object is moved even closer to the lens (in front of the focal point), the rays actually diverge as they exit the lens (Fig. 9.1.6D). This creates the rather interesting possibility of a "virtual image" constructed by the observer's eye. Such an image cannot be captured directly on film, but must be gathered and projected by yet another lens, in this case, the lens of the eye. The eye's lens gathers these diverging rays and projects a magnified, real image on the retina—note that this image is not inverted. The compound microscope makes use of both a real image (created by the objective lens) and a virtual image (created by the ocular, or eyepiece lens, and the viewer's eye; Fig. 9.1.7). The objective lens projects a real image onto a plane located between the ocular lens and its focal point (f_2). The ocular lens collects light from this real image and projects a diverging light path that the eye then collects as a virtual image, where it appears magnified on the retina.

MAGNIFICATION VERSUS RESOLUTION

As mentioned, magnification can be empty. Take a hand lens and magnify a letter on a page. Is there a limit to how big the image can become? Is there a limit to the image clarity, and *resolution*? Resolution is the ability to distinguish two parts of an object as visually discrete objects (also see *UNIT 7.5* and *APPENDIX 3B*). R is the distance between two spots that your optical system can resolve. In a world where bigger is often seen as better, it is counterintuitive to recognize that the best optical system is the one with the smallest R. If you can resolve two objects that are 1 mm apart, you have better resolution than an optical system that can only distinguish objects 1 cm apart.

More formally, $R = 0.61\lambda/\text{NA}$. λ is the wavelength of light being used (most conveniently measured in μm for this purpose). NA is the "numerical aperture" of the lens (stamped on the side of the objective lens). Numerical aperture is a lens characteristic that refers to how wide a cone of light can be admitted: the wider the cone, the better for resolution. If a lens is ground with a certain shape, one can move the specimen very close, and the lens captures light from a wide angle in the

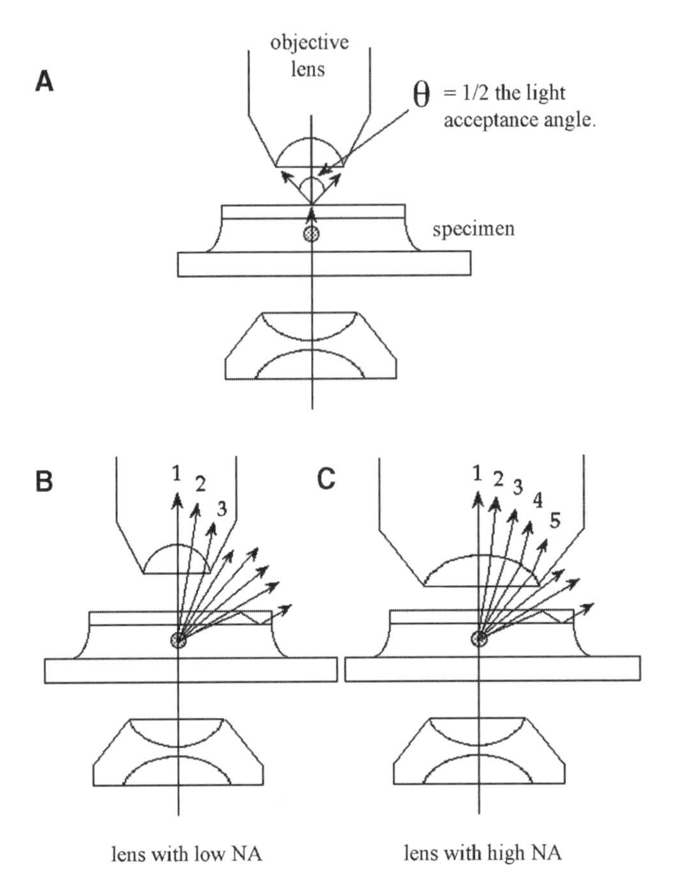

Figure 9.1.8 This illustration helps define the acceptance angle of an objective lens. (**A**) The numerical aperture of a lens is defined as NA = $n\sin\theta$, where θ is ½ the acceptance angle for a lens. Lenses with larger numerical apertures (NA) have better resolution. Panel (**B**) illustrates a lens with a lower acceptance angle (and hence lower NA) and panel (**C**) illustrates a lens with a higher acceptance angle (and hence higher NA). In panel B, only three light rays are admitted, whereas the lens shown in panel C can accept five of the rays indicated.

specimen plane (Fig. 9.1.8). This is desirable, but difficult to achieve without introducing optical aberrations. Formally, $\text{NA} = n\sin\theta$, where θ is half the angle of allowable light (Fig. 9.1.8A) and n is the refractive index of the medium between the specimen and the lens (for air, $n = 1.0$, for oil, $n = 1.5$). We can imagine that any lens accepts a cone of light. In our eyes, the edges of the cone define the limits of our peripheral vision. Lenses with small numerical apertures accept light from a very narrow field of view (Fig. 9.1.8B). Larger NA's allow light to be gathered from a wider field of view (Fig. 9.1.8C). Back to our formula: $R = 0.61\ \lambda/\text{NA}$, or $R = 0.61\lambda/n\sin\theta$. To "unpack" this formula, we can appreciate that R (resolution) is determined by three characteristics—the wavelength of light (λ); the refractive index of the medium that lies between specimen and lens (n); and the lens characteristic, i.e., the half-angle of the light cone accepted by our objective lens (θ). Since we want a small R (a small distance between closest discernable objects), we desire a small wavelength of light, a large refractive index, and a large θ (acceptance angle for our lens). Larger θ leads to larger sinθ. More sophisticated light microscopes (especially phase-contrast microscopes) are often outfitted with a green lens to select visible light with a smaller wavelength. Electron microscopy has taken this even further by going from 450 to 700 nm visible light to 0.005 nm electrons. Practically speaking, most of us have already been provided a microscope with fixed lens characteristics and a set light source. The one parameter that we have easy control over is the use of the oil-immersion lens. Going from air to oil can theoretically increase resolving power of the microscope by a factor of 1.5× (see below).

Limits to Resolution: Optical Aberrations

The use of light and glass as our medium for magnification introduces some physical limitations to improving resolution. Simple glass lenses, for example, bend blue light more severely than red light. Hence, light from a full-spectrum image will be focused over a range of focal distances (the blue-light image being focused well in front of the red-light image). This effectively means that one cannot achieve a fully focused, full spectrum image with a simple lens, and that there will always be color halos, or *chromatic aberration* (see *http://www.microscopyu.com/tutorials/java/aberrations/chromatic/index.html*).

This was first remedied in the latter part of the 18th century by the invention of *achromatic* lens combinations. These involved the use of two types of glass: flint glass (with a high lead content and high refractive index, $n = 1.75$) and crown glass (a typical soda-lime-silica formula, $n = 1.5$ to 1.6) within a compound lens. Combinations of two types of glass allow a compound lens to bring red and blue light from the object into very nearly the same focal plane (though green remains out of focus). Most teaching microscopes use achromatic lenses. All three colors can be made to focus in the same focal plane by using fluorspar in an appropriate lens combination. These *apochromatic* lenses are found in only the finest research microscopes.

Another form of optical aberration is *spherical aberration*. This occurs because lenses tend to be spherical in contour, and not parabolic (it is not practical to grind parabolic surfaces). The result is that one may focus the center of the image, but the edges will be slightly out of focus. Lens combinations can correct for spherical aberration as well as chromatic aberration. Achromatic lens combinations typically correct spherically only for green light, though they are corrected chromatically for both red and blue wavelengths. Apochromatic lenses are corrected spherically for two wavelengths, blue and green, and are corrected chromatically for three wavelengths (red, green, and blue). The highest-quality lenses are called *plan-apochromatic* lenses, which are spherically corrected for four wavelengths, and color corrected for three. Other types of lens aberrations are discussed in Murphy (2001).

GETTING COMFORTABLE

The Right Height

In a sitting posture, eyes should be comfortable looking through the oculars without bending over and without stretching. Looking through the microscope should promote good posture. Hands should rest on the benchtop comfortably; shoulders should not be hunched.

Binocular Customization: Secrets of the Headpiece

Most scopes these days are equipped with binocular headpieces. There are some individuals who simply cannot look through them. If you are having difficulty, simply close one eye and peer through a single eyepiece. Headpieces almost always offer two ways to customize viewing—adjustment of interpupillary distance and independent focusing of the two eyepieces. A properly customized optical pathway does not produce eye fatigue and can be very comfortable.

Interpupillary distance

Every individual has pupils set at a unique distance apart. This can be accommodated by widening or narrowing the distance between the oculars until they can comfortably fuse the binocular image.

Independent focus

In most modern binocular scopes, one can customize the focus for each eyepiece. Determine which ocular lens has an independent focusing collar on the eyepiece (this is often the leftmost ocular). With the other eyepiece, focus (using one eye only) on a slide specimen using the microscope coarse and fine focus. Next, look through the other eyepiece and adjust the ocular focus to bring that image into focus. You need to relax your eye for this, or it will struggle to compensate. For fine tuning, open both eyes and tweak the scope focus and the ocular focus to produce a comfortable, clear binocular image. Your microscope is now "customized." For those who really like detail and control, there are calibrations on the headpiece that you can record in your notebook to facilitate quick restoration of your own custom features.

Helpful hints

If fusing the binocular image proves to be a struggle, there are two tricks. First experiment with moving your head closer or further from the eyepieces. You will discover just the right distance for "catching the image." Second, while you peer through the microscope, have someone look at your face to see where the bright light discs are falling as the oculars project them into your eyes. If focused correctly, the light disks should be focused into the centers of both pupils.

FINDING THE OBJECT TO BE VIEWED

This sounds trivial, but it isn't.

Preparing Samples

Let us say that a pond sample is being examined. A couple of points may be helpful. Debris settles, and organisms like debris. To take a sample, first expel the air from the pipet (otherwise air bubbles will be blown into the sample jar and stir up the sediment, thereby diluting it). Next, while *watching the pipet tip*, chase the visible debris and suck it up.

Large Organisms

A nice trick for larger organisms is to produce "clay feet" between the coverslip and the slide. To do this, take a coverslip, and wipe each corner lightly across a piece of clay. When this coverslip is laid onto a drop of pond water, the clay feet will elevate the coverslip slightly and prevent crushing. Since the clay is soft, one can apply pressure to the coverslip and gently sandwich the organisms, reducing their mobility.

Moving Organisms

For slowing or immobilizing smaller specimens without propping the coverslip, there are several options. Some use methyl cellulose or cotton fibers to slow the faster protists long enough to observe. Most often, the author uses the "golden window." If 22-mm^2 coverslips are used, then 11 μl of sample will nicely flush the coverslip out when applied. As the light dries down the sample, most protists find themselves immobilized for about 10 min before lysing. This window of time is wonderful for close examination, as the specimen is immobilized yet alive. For research purposes, one can have a variety of roto-compressor devices manufactured by skilled mechanics (see Cole et al., 2002). These allow prolonged observation of specimens immobilized by pressure between slide and coverslip.

Center the Subject

Mount the slide onto the stage, clasping it gently with the stage calipers, and center the coverslip over the light source. Even better, try to center a visible piece of debris in the light path (look for it on the stage and watch it catch the light when centered).

CAUTION: Many high-school microscopes and most dissecting scopes have the kind of slide holders under which one jams the slide. Be aware that higher-quality microscopes are furnished delicate caliper-style slide holders that open laterally and clasp like a set of fingers from the side. Do not try to force the slide under the delicate calipers, mistaking them for the vertical-pressure variety. This can be an expensive mistake.

Find the Object

There are several quick fixes for the individual who cannot find an object.

1. Begin at low magnification.

   If you do not find it there, you will probably not find it at a higher magnification.

2. Find the edge of the coverslip and focus on it, then move into the center of the coverslip. Alternatively, if there is a bubble in the sample, use it to find the focal plane. This is especially useful in transparent samples.

3. Find the iris condenser, and shut it down completely (we'll optimize later, first let's find the "bug"). Dial up the illuminator to a comfortable level: not too bright and not too dim (a red or brownish background color usually indicates an illuminator that is too dim). Once the object has been found and focused, optimize the light path for "Köhler" illumination (or some variety of interference microscopy; see below).

   For fast-moving specimens, it can be useful to add impediments to the environment. A wisp of cotton fiber can provide a physical barrier to movement, or a drop of 1.5% methyl cellulose applied to a drop of pond debris can help retard mobility and aid observation.

   The substage condenser is a very useful object that seems to scare people. Imagine that you are sitting around the embers of a campfire at night. Suddenly an RV pulls up, its high beams on. Out of the corner of your eye, a moth flies across the visual field. As it passes in front of the headlights, what happens? This same phenomenon occurs when a common mistake is made, i.e., setting the substage illuminator on full with the iris diaphragm wide open. Any transparent protist passing in front of such a light-source will become completely invisible—a bug in the headlights.

MARKING THE LOCATION OF AN OBJECT: SECRETS OF THE MICROSCOPE STAGE

Once you have found an object, you might want to return to it one day. This is especially true of fixed histological preparations, or permanent slides. Most college-level microscopes have hidden virtuosities. One is the ability to read/record the *x*/*y* and *z* coordinates for a particular object. The *x* axis refers to movement left and right with respect to the viewer, and the *y* axis refers to movement toward and away with respect to the viewer. The *z* dimension refers to the vertical focal plane. There is an *x* and a *y* scale flanking the microscope stage (often referred to as the stage micrometer). These have a numbering system that one can read (Fig. 9.1.9). Decide on a particular place on a microscope slide to mark and return to. Focus on it, and center it in the visual field. For the *x* coordinate, look at the

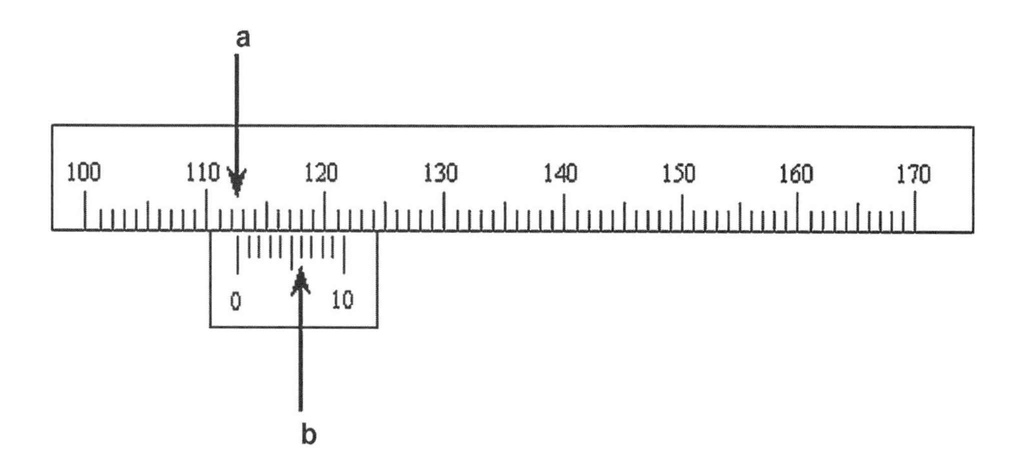

Figure 9.1.9 The microscope stage often has two sets of coordinates (*x* and *y*) that can be used to record the location of a specific object on a microscope slide. This figure illustrates the appearance of the *x* coordinate designating a location at 112.6. See text for details.

0 to 10 scale ("b" in Fig. 9.1.9) that slides back and forth across the higher-value 100 to 170 scale ("a" in the figure). Match the zero mark on the small sliding scale "a" to the closest integer on the large (fixed) scale ("b") and record that number. Next, look at which of the small-scale ("b") marks lines up most perfectly with some mark in the larger, fixed scale. Record that as the decimal. In the example in Figure 9.1.9, the *x* coordinate is 112.6, (112, read as the number directly across from the zero mark on the small scale, and 0.6 read as the best alignment between any mark on the large scale, "a," and any mark on the small scale, "b," which provides the decimal). The same operation can be performed with the *y* axis. One can record the *x* and *y* coordinates for a particularly choice location on ones' slide, returning to it years later with accuracy (provided that you use the same scope). Marks on the focus adjustment can also allow one to measure the optical depth of a specimen. One can focus on the top of the specimen (record the number) and then focus through to the bottom of the specimen and determine the depth measured in μm (microns), practically affording a microscopist's equivalent of GPS.

A QUICK GUIDE TO CHOOSING FROM VARIOUS OPTICAL TECHNIQUES

Different optical tricks are useful for different types of microscopic objects. Bright-field and dark-field are two easily developed techniques that can be applied on virtually any conventional compound microscope. Various forms of interference microscopy are valuable for highlighting transparent features inside cells or embryos. One should note that there are some very inexpensive tricks that can convert the image seen through a conventional compound microscope into something resembling the image obtained from a very expensive research microscope.

Bright-field microscopy (Köhler illumination). For use with translucent or transparent specimens (often used with stained histological preparations).

Oil immersion. For use with small objects, and in order to increase optical resolution.

Dark-field. This helps one to see tiny, highly refractile objects such as cilia and bacteria.

Rheinberg. To provide colorized images of highly refractile objects such as protists or microcrustaceans.

Polarized light. For specimens that are birefringent, or have optical crystals within them (pluteus larvae of echinoderms, protists with birefringent inclusions, plant stems, muscle tissue).

Phase-contrast. To distinguish various compartments within a completely transparent cell or organism at high resolution. A classic use is in examining the newt lampbrush chromosome.

DIC (Nomarski). To distinguish various compartments within a completely transparent cell or organism (embryos, invertebrates) by creating pseudo-landscapes based on edges of differing refractive index.

KÖHLER ILLUMINATION: SECRETS OF THE SUBSTAGE CONDENSER

Logistics of Köhler Illumination

The resolution of an image can be highly degraded by injudicious tinkering with the microscope. Conversely, one can take a miserable image (low resolution, glare-damaged) and work magic on it by "tuning" the microscope's light path. August Köhler redesigned the illumination logistics for light microscopy in 1893, and we honor his accomplishment by referring to this practice as establishing "Köhler illumination." Here is a practical walkthrough.

1. Starting at low magnification (10×), prepare a specimen for viewing and establish a comfortable focus—i.e., customize your microscope. Without changing focus, move to a clear area of the slide.

2. If your scope has a field diaphragm (Fig. 9.1.1), close it down so that you can see the iris' edges in the field of view. Without a field diaphragm, mark an "X" (1 to 2 cm in size) on a circular disk of transparent plastic with a felt-tip pen, and lay it onto the diffuser that covers the light source. Be sure it is flush with the surface.

3. Use the condenser-lens focusing knob to raise and lower the condenser until the image of the diaphragm (or your "X") is crisp.

   You have now focused the diffuse light that exits the illuminator onto the visual field that holds your object. This establishes a uniform field of light (rather than the annoying image of a lamp filament) covering the objects within your field of view.

4. You may notice that the diaphragm borders are off-center from the circle that forms the field of view. Find the centering screws that control the position of the substage condenser and adjust them to center the illuminator image in the field of view. You can widen the field diaphragm to nearly fill the field of view, to help with this. When centered, widen the field diaphragm to just clear it from view.

5. *Adjusting the condenser aperture:* Lift out one of the oculars (eyepiece) and look down the tube.

   You should see a bright disk of light.

6. Find the iris diaphragm control on the side of the substage condenser. Adjust the iris as you watch through the light tube (Fig. 9.1.10).

7. Open the iris until the light disk fills about 75% of the field of view.

   This cuts down on angular light that will reflect off of the light-tube walls as it travels to your eye, thereby reducing glare.

8. Replace the eyepiece.

9. If you switch to a different objective lens during use, repeat steps 1 to 8.

   Congratulations, you have just established Köhler illumination! (August would be proud). Reconfigure Köhler illumination for each different objective lens that is dialed in.

   To read more about Köhler's method see Köhler (1893), republished by the Royal Microscopical Society as a special centenary translation in 1994. Here, Köhler illumination is sometimes referred to as "bright-field" illumination, assuming that one has performed the Köhler procedure.

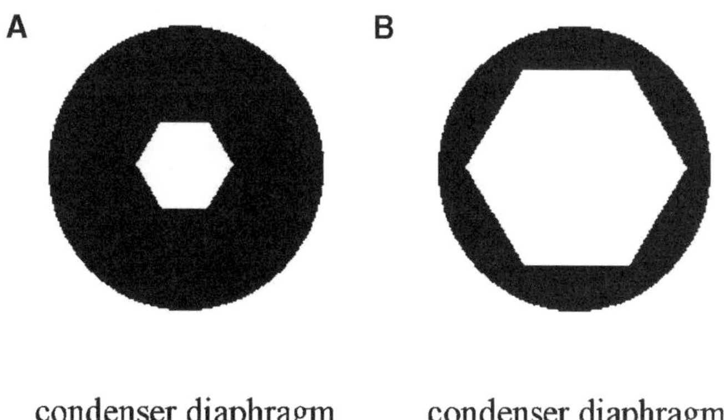

Figure 9.1.10 The condenser diaphragm is an essential piece with which to become familiar. (**A**) Condenser diaphragm closed down. The edges of the iris diaphragm appear as a polygon through the ocular light tube. (**B**) When adjusted for Köhler illumination (bright-field microscopy), the diaphragm should be open 75% as shown here. For dark-field, the diaphragm should be opened all the way to allow the greatest collection of tangential light (and the condenser should be racked all the way up, nearly touching the microscope slide).

Problems with Köhler Illumination

As mentioned earlier, those new to the microscope sometimes have a disproportionate fear of the substage condenser. A great strategy to get over this is to go back and forth between Köhler and dark-field illumination. This not only drives home the practical purpose of these devices, but also overcomes awe of those things lying beneath the stage. It also gives the young microscopist legitimate control over the apparatus.

Why Köhler Illumination Works

Establishing Köhler illumination optimizes image resolution, reduces glare, and creates a uniform field of illumination. Part of this is done for you in the design of the modern microscope (Fig. 9.1.11). When August Köhler was practicing microscopy, the light source was a gas lamp. The image of

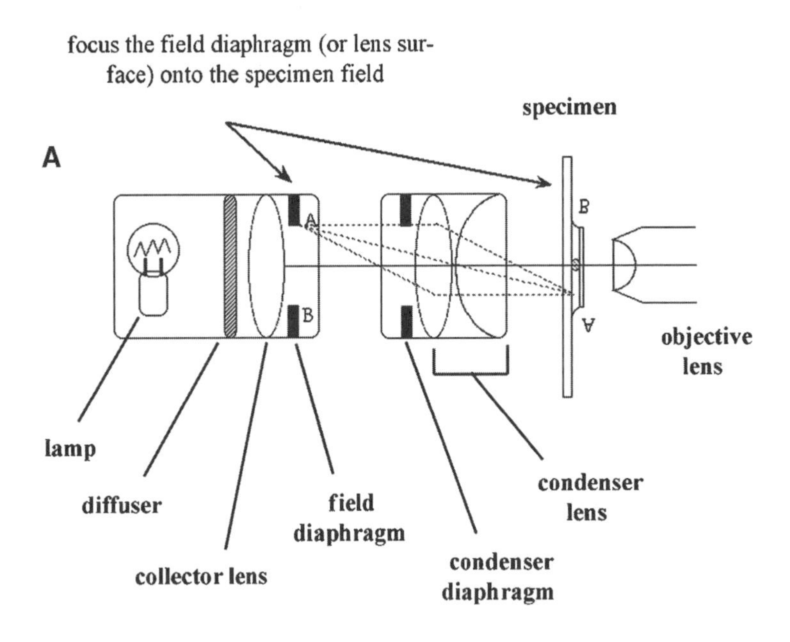

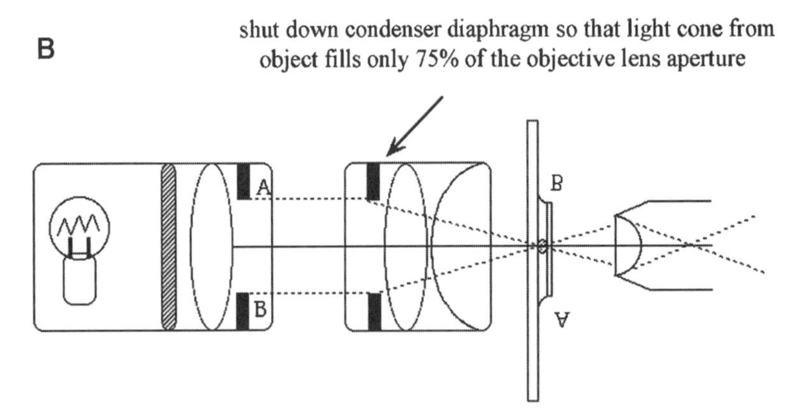

Figure 9.1.11 Establishing Köhler illumination. This illustrates the two major steps in establishing Köhler illumination. First, panel (**A**) shows how (after focusing the microscope on an object), one should focus the substage condenser using its own focusing knob in order to bring the image of the collector lens into focus. This can be done in one of three ways. (1) If the microscope has a field diaphragm, simply close the field diaphragm down and focus on the edges of the diaphragm polygon. Next, open the diaphragm back up until its image expands just beyond the field of view. (2) If the microscope doesn't have a field diaphragm, drop a clear plastic disk onto the illuminator marked with an "X," and focus on that, or (3) bring the ground glass of the diffuser into focus and then slightly defocus (the ground glass image can be too subtle for some to catch). Panel (**B**) shows the step (see step 7 under Logistics of Köhler Illumination) in which one adjusts the condenser diaphragm to the 75% image seen in Figure 9.1.10B.

the lamp was focused directly onto the specimen with the condenser lens, resulting in uneven illumination. Today, we typically use a halogen light bulb, but the problems are the same: to obtain the brightest illumination, one must focus the filament itself onto the specimen, creating uneven illumination. To create even illumination across the field of view, substage illuminators are outfitted with a diffuser (usually ground glass) that disperses the light, largely obscuring the image of the filament. Second, by focusing on the surface of the illuminator's collector lens (or effectively, the edges of the field diaphragm mounted on the collector lens), the light that one projects onto the specimen is not the image of the filament, but a disk of unfocused light emerging from the collector lens over the lamp (Fig. 9.1.11A). This bears repeating—every point on the lamp filament emits light in all directions. Light from a single point on the filament bathes the entire surface of the collector lens. Normally, we think that the job of a lens is to collect the light from an object and focus it into an image. The collector lens gathers the light from each point on the filament and refracts it. But the image of the filament does not form on the microscope slide and specimen. The image that the condenser projects onto the slide and specimen is the collected, unfocused light emerging from the surface of the collector lens itself. Consequently, light from each point on the filament contributes equally to illuminating every point on the specimen field.

Second, by restricting the field diaphragm to just fill the field of view, one excludes oblique light that would reflect off the walls of the light path, creating glare (Fig. 9.1.11B). Similarly, by reducing the condenser diaphragm to 75% of the ocular's field of view, one limits the internally reflected oblique light that creates glare. The result: uniform lighting, reduced glare, and high image resolution.

OIL IMMERSION

Logistics of Oil Immersion

As mentioned above, one tool available to most microscopists that can dramatically improve resolution is the refractive index (n) of the medium that lies between the specimen and the objective lens. The logistics are straightforward. For lenses that are designated "oil-immersion" lenses, proceed as follows.

1. Locate and focus the image of the object, moving from low to high magnification until you are at or just below the magnification of the oil-immersion lens.
2. Perform Köhler optimization.
3. Dial the objective half out of the way and place a drop of immersion oil onto the slide.
4. Dial the oil-immersion lens into place.

   The author sometimes rocks the lens back and forth to ensure good contact with the oil-drop.
5. After viewing, clean the lens and slide with lens tissues.

Problems with Oil Immersion

The following lists several possible reasons for blurred focus, as well as potential remedies. (1) The oil drop may be too small and may not be in contact with the lens. Add a larger drop and again rock the lens back and forth to insure a good seal. (2) There may be dried matter on the lens: clean as described above for careful cleaning. (3) You may have mixed oils from different companies (e.g., a new bottle but residue from the old on the lens): variation in oils can lead to mixing aberrations.

Why Oil Immersion Works

In brief, you are removing interfaces between media with different refractive indices. Recall that each time light passes from glass to air and air to glass, the light is refracted, or "bent." As light

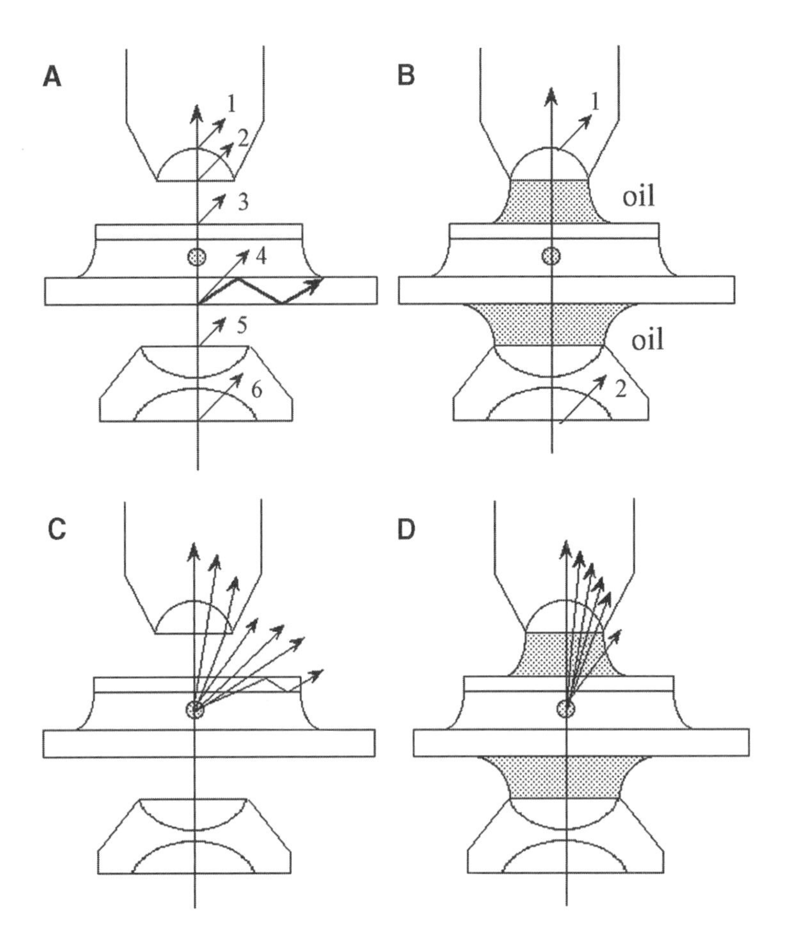

Figure 9.1.12 This illustrates the effect of oil immersion. Panel (**A**) indicates a total of six interfaces between glass and air, each of which is an opportunity for light to be refracted and information lost. Panel (**B**) shows that an appropriate oil ($n = 1.501$), establishes a light path with far fewer interfaces for light to be refracted through. The oil, glass, and mounting medium all share the same refractive index. Panels (**C**) and (**D**) illustrate how oil immersion helps a lens to capture more of the light from an illuminated object.

passes from air ($n = 1.0$) to glass ($n = 1.515$), it is bent towards the perpendicular. As light passes from glass to air, the light is bent away from the perpendicular. With air between the microscope slide and objective lens, there can be as many as six refractive interfaces resulting in lost light (numbered 1 to 6 in Fig. 9.1.12A). With oil ($n = 1.515 = n$ for glass and mounting medium), between slide and lens, one can reduce this to 4 (or, for the fanatic who oils the condenser lens as well as the objective lens, one can reduce the number of interfaces to as few as 2; Fig. 9.1.12B). This allows the objective lens to capture more of the light emanating from the object (Fig. 9.1.12C,D). From a naive perspective, the more light a lens can capture, the higher the resolution of the image.

A more sophisticated explanation requires an understanding of diffraction: the interaction of light with matter (or with the edges between transparent and opaque substances). As light waves pass through a pinhole (or around an "edge"), they emerge creating a diffraction pattern with concentric rings of light and dark emanating from a central illuminated disk (Fig. 9.1.13). The central disk of light (84% of the light from the point source pattern) is known as the Airy disk, after Sir George Biddell Airy (1801–1892), an extraordinary physical mathematician and astronomer who lists among his accomplishments details of planetary motion, establishment of the prime meridian, and an estimation of the mean density of the Earth. Each concentric ring of such a pattern is numbered, with zero-order light filling the central disk, first-order light just beyond, and so forth. Ernst Abbe first suggested that the more orders of diffracted light a lens can capture, the higher the resolution of the image achieved. In fact, if a lens captures only zero-order light, no image will be created. Apparently, image formation requires positive and negative interference that is created with the higher-order diffraction rings. Hence, the practical effect of oil immersion is to allow a lens to capture the higher-order diffracted light, and thereby generate a more highly resolved image. Microscopic objects can be viewed as assemblages of minute diffracting particles whose dimensions are on the order of the wavelength of light itself. Each such particle interacts with light by scattering it, creating an assemblage of Airy disks. "The diffraction of light by small structural elements in a specimen is the principal process governing image formation in the light microscope" (Murphy, 2001).

Figure 9.1.14, panels A and B, show the same object imaged at the same magnification with and without oil immersion, respectively. Arrows draw attention to fine cellular organelles that are resolved with oil and unresolved without.

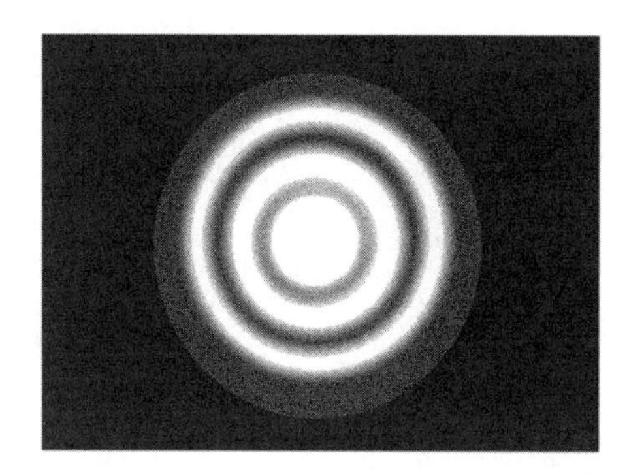

Figure 9.1.13 A cartoon of the diffraction pattern of light as it passes through a tiny aperture or is diffracted around some tiny microscopic object. The central light disk is referred to as the Airy disk and contains ~84% of the light. It is referred to as the zero-order light. The concentric rings of light represent the first-order, second-order diffracted light, etc., and are of decreasing intensity. All microscopic images can be modeled as a collection of such diffraction patterns and the composite interference pattern that occurs through their interaction.

DARK-FIELD, RHEINBERG, POLARIZED-LIGHT, PHASE-CONTRAST, AND DIC MICROSCOPY

Here we'll explore a series of variations on bright-field microscopy that (with the exception of phase contrast and DIC) require no specialized or expensive instrumentation. In each case one simply adapts a conventional (student-quality) compound microscope to specialized service.

Dark-Field Microscopy

Logistics of dark-field microscopy

It is possible that no single exercise packs more punch for less investment than the practice of dark-field microscopy. For pennies, a budding microscopist can transform an ordinary light microscope into an "interference microscope," and produce visually stunning images of highly refractile objects not available with conventional bright-field illumination (see Omoto and Folwell, 1999). Specimens that come to life using this technique include any living thing with cilia or flagella. To illustrate this technique, first the steps to establish Köhler illumination will be provided, and then those for dark-field.

Make a dark-field stop

1. Check your microscope condenser to determine if, like most, it has a filter holder that snaps in place to the underside of the condenser. Pop it out.

2. Cut a piece of clear plastic that will fit snugly into the filter holder on top of the color-correction filter that is already in place.

3. Cut out a circular disk of black electrician's tape, ~1 cm in diameter.

   One can experiment with the dimensions.

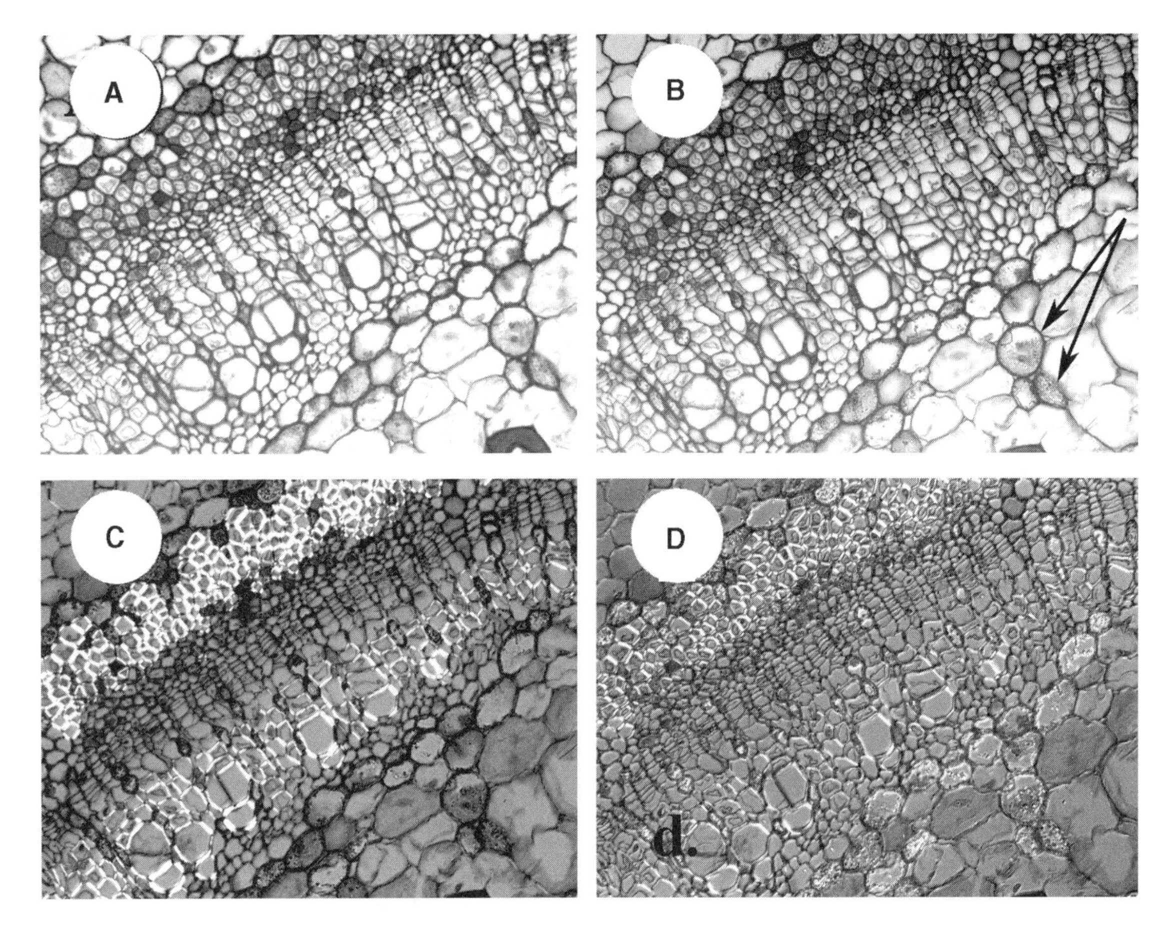

Figure 9.1.14 Images of a plant stem (*Tilia* sp.) at 400× magnification (a 40× objective lens and a 10× ocular). Panel (**A**) shows the image with a 40× dry lens. Panel (**B**) shows the same image with a 40× oil immersion lens. Cell walls and internal cytoplasmic inclusions appear crisper. Panel (**C**) shows the same image with polarized light. The polarizer and analyzer are not quite at right angles, so that background light is not fully extinguished. Polymers of lignin appear bright. Panel (**D**) shows a DIC (Nomarski) image of the same. A sense of topological relief is produced as steep gradients in refractive index are converted into changes in light intensity.

4. Center the tape disk onto the larger, clear plastic disk, and replace the filter holder on the condenser.

   If your microscope doesn't have a filter holder, simply tape the dark-field stop under the condenser, centering it as well as you can.

5. After checking the fit, remove the dark-field stop.

   Dark-field stops can also be purchased for most microscopes, including higher-NA lenses or even oil immersion. Commercial suppliers of dark-field stops include Olympus and Zeiss.

6. Without the dark-field stop, find and focus on your specimen.

   Ciliates are excellent subjects for this technique.

7. Establish Köhler illumination (see above) and observe. Focus particularly on the cilia or on tiny objects such as bacteria.

   Dark-field works best at lower magnifications (10× or 20× objectives).

8. Replace the dark-field stop on the substage condenser.

9. Raise the condenser all the way, until it contacts the underside of the slide.

   After contact is made, you may back off just a bit.

10. Open the iris diaphragm on the condenser all the way.

11. Observe.

    The background should be dark, with cells and dust appearing as self-luminous objects. Edges and tiny particles will diffract the most light and appear exceptionally bright. Cilia are often very prominent.

Problems with dark-field microscopy (the background does not appear uniformly dark)

There are four culprits for unsuccessful dark-field. (1) The "stop" is too large or too small, (2) the condenser lens is not up all the way, (3) the iris is not open all the way, or (4) the microscope is set up on the high-NA lenses (40× or higher magnification). These problems are easily remedied. Tinkering is useful: try tweaking the height of the condenser, the width of the diaphragm opening, and the size of the dark-field stop. This is, at least in part, an empirical process.

Why dark-field works

When a microscope is set up for bright-field work (Fig. 9.1.15A), the light interacting with the specimen passes through in two forms: it is either transmitted directly (undeviated light), or it is diffracted by the particulate nature of the specimen. Undeviated light makes up the majority of the image, creating low-contrast images against a bright background. Transparent objects appear...well...transparent! The dark-field stop blocks all the direct, undeviated light from entering

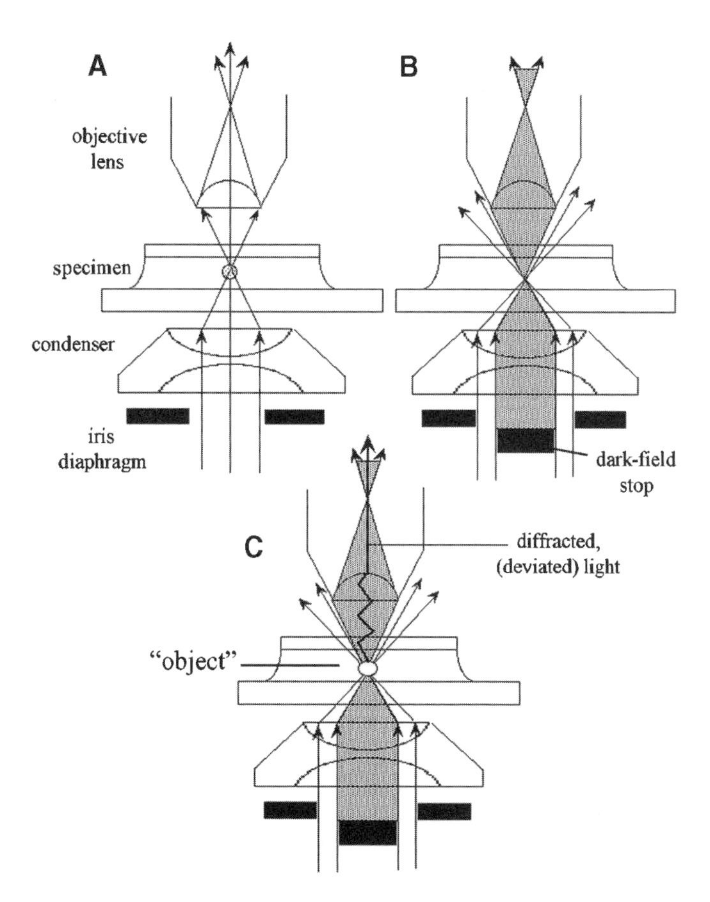

Figure 9.1.15 Dark-field microscopy. Panel (**A**) shows the light-path from a typical bright-field microscope. Most of the image is created by direct, undeviated light. Diffracted light is collected, but its signal is swamped by the undeviated light. Panel (**B**) shows the effect of placing a dark-field stop into the center of the light path between the illuminator and specimen. Only tangential light passes through the specimen pane. Having no specimen with which to interact, this tangential light misses the objective lens completely and the image is black. (**C**) When an object passes into the path of this hollow cone of light, light is diffracted, and that diffracted light will be collected by the objective lens. The result is a bright image against a dark background. (See Fig. 9.1.16B for an example).

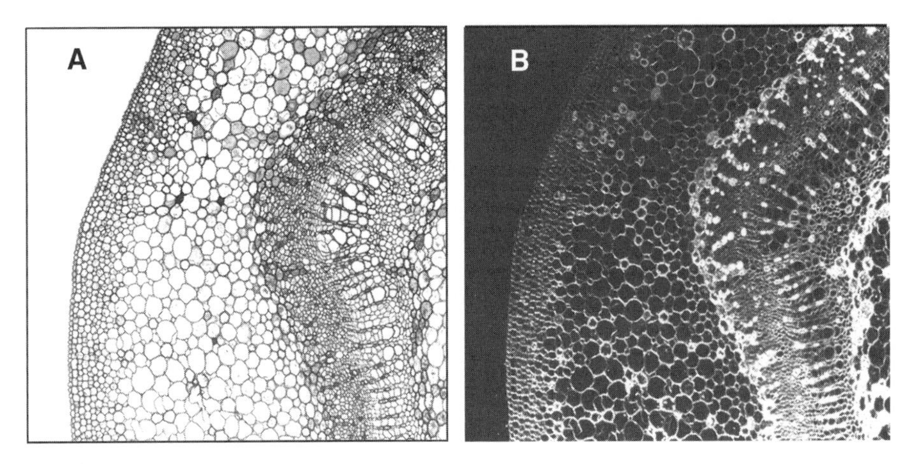

Figure 9.1.16 At 200× magnification, a plant stem in cross section is imaged with (**A**) bright-field and (**B**) dark-field illumination.

the objective lens (Fig. 9.1.15B). By raising the condenser and opening the diaphragm, one is illuminating the specimen only with oblique light coming from the edges of the condenser lens. A hollow cone of light passes through the specimen. If that light does not interact with the specimen, it emerges from the specimen plane undeviated, and misses the collecting surface of the objective lens: the image remains dark. Oblique light that does interact with the specimen will be diffracted, and emerges from each edge or particle as a new point source. This deviated light will be collected by the objective lens, and forms the image (Fig. 9.1.15C). Hence, the specimen appears as a luminous object against a dark background. Edges and particles (objects whose dimensions are close to the wavelength of light) interact most strongly, diffracting the most light, and hence appear accentuated in the image. The tiniest particles will, in fact, appear as Airy disks surrounded by diffraction rings. Figure 9.1.16 shows the same image with bright-field (A) and dark-field (B) illumination.

Rheinberg Microscopy

Logistics of Rheinberg microscopy

This is a particularly showy variation on dark-field microscopy resulting in specimens illuminated in one color against a contrasting color background. Many such images find their way onto promotional calendars, popular science magazines, and prize-winning photomicrographs (Abramowitz, 1985). This technique works especially well for protists and aquatic invertebrates.

To perform Rheinberg microscopy, one chooses contrasting colors for the annulus, which provides oblique illumination, and the central disk, which provides direct illumination (the latter is totally occluded in dark-field microscopy). The following steps assume a yellow annulus with a red inner spot.

1. Purchase colored sheets of clear plastic (or gels) and cut these to size to fit in the substage filter holder.
2. Cut a 3.5-cm (or whatever size fills the filter holder) disk of yellow plastic.
3. From the center, cut out a 1.5-cm hole, creating a yellow annulus with a clear center.
4. From white (opaque) tape, cut a 1.75-cm disk and cover the hole with it.

   This dampens the direct illumination, which can drown out the subtler oblique illumination, while providing an adhesive surface for mounting a 1.5-cm red disk.

5. Place your Rheinberg filter into the filter holder and adjust the condenser as for dark-field (condenser is racked up and diaphragm is opened wide).

Problems with Rheinberg microscopy

Sometimes the color differentiation of a Rheinberg image is poor. Experiment with the size of the inner spot, the condenser location, and the diaphragm setting. Best results come from color choices that are bold, complementary colors, e.g., red spot and yellow annulus, green spot and red annulus, blue spot and yellow annulus, red spot and green annulus. One can even experiment with partial occlusion using black electrician's tape. We have found that a black pie-shaped circle with a wedge removed, placed over the central, colored spot, produces an interesting hybrid of dark-field and Rheinberg (Leland Johnson, pers. comm.).

Why Rheinberg works

Essentially this follows the same basic principles as dark-field microscopy described above. In this case, the direct light is muted (not occluded) and tinted by the central disk. The opaque white tape reduces the amplitude of the direct light, while the colored central disk provides the background illumination. The annulus provides the oblique illumination that is diffracted within the specimen. This diffracted light from the specimen contributes to the image. Without some attenuation of the direct light, the effect is less striking, as the direct light can overwhelm the less intense oblique illumination. An excellent tutorial can be found at *http://micro.magnet.fsu.edu/primer/techniques/rheinberg.html.*

Polarized-Light Microscopy

Specimens that include birefringent materials or crystals often reveal lovely detail using polarized light to uncover their hidden structures. Again, this technique can be adapted to a conventional light microscope with minimal cost (though one can purchase more sophisticated microscopes dedicated for this purpose). All that is needed is a sheet of polarizing material that can be cut into disks. The uses for polarized-light microscopy are somewhat limited in biology, due to its reliance on birefringence; however, curiously, there are such materials in the life sciences. For instance, the pluteus larvae of most echinoderms produce a spicule skeleton made of crystalline calcium carbonate. These crystals are birefringent and form some of the most spectacular images to be found in nature. Mitotic spindles interact with polarized light, as do the actin and myosin polymers that form striated muscle, as well as the cellulose and lignins found in plant cell walls (Fig. 9.1.14C). Curiously, many protists (flagellates and ciliates) possess crystalline inclusions within their cytoplasm, and these are birefringent too. Polarized light applications for biology are discussed in Newton et al. (1998).

Logistics of polarized-light microscopy

1. Purchase a sheet of polarizing plastic.

   Unfortunately, there is now only one vendor, Edmund Scientific (http://www.edsci.com). One might need to make a special appeal to the manufacturer: Polarizing Optics, 320 Elm Street, Marlborough, Massachusetts 01752 (phone: 508-481-7495).

2. Cut out a disk that nicely covers the light source of the microscope.

3. Cut out a smaller disk that will cover the ocular lens. For practice, hold both pieces up to the light. While peering through them, rotate one. It should be possible to extinguish the light completely.

4. With an appropriate birefringent specimen in place, establish Köhler illumination (see above).

5. Attach one piece of polarizing material over the eyepiece using adhesive tape to secure it.

   This is the analyzer.

6. Drop the larger piece over the illuminator.

   This is the polarizer.

7. While peering through the eyepiece analyzer, rotate the polarizer until the background light is extinguished.

   Any subject that is birefringent will glow brilliantly against the darkened background.

Problems with polarized-light microscopy

Sometimes one cannot extinguish the background light, or geometric patterns of light and dark are seen. Some plastics have a randomizing effect on polarized light, thereby ruining the effect of the polarizer. It is unfortunate, but in today's competitive marketplace, microscope manufacturers have begun to use plastic instead of glass parts where they think they can get away with it. Sometimes you can circumvent this. Explore different placements of the polarizer. Try placing it above the condenser, but below the specimen; It must be below the specimen to produce the desired effect. If the condenser diaphragm is open too wide, one may see a "polarization cross." This can be minimized by shutting down the diaphragm until the background is more uniformly dark.

Why polarized light works

Photons have a plane of vibration or polarity. Most light sources emit photons vibrating in every possible plane. Certain materials or surfaces interact with light in an asymmetric way, reflecting or transmitting photons that vibrate in a particular plane. Highly reflective surfaces such as water or metal tend to reflect light preferentially in a given plane. Polarizing materials only transmit light that is vibrating in a highly selective plane, and absorb or reflect the rest. Randomized light passed through a polarizing filter emerges as polarized light, its photons vibrating in a particular plane. By orienting two pieces of polarizing material at right angles, one effectively blocks out all the light. Some substances possess a quality known as structural birefringence. These materials (crystals and oriented biological polymers) interact in a curious way with polarized light.

Simplified explanation of how a polarized-light filter works

Birefringent materials effectively twist the axis of polarization for light as it passes through, so that what would have been extinguished by the analyzer will now pass through and form a bright image against a dark background.

Less simplified explanation of how a polarized-light filter works

Polarized light entering a birefringent material splits into two waves oriented perpendicularly with respect to one another. The ordinary wave (O) emerges with a plane of polarization parallel to the optical axis of the crystal. The extraordinary wave (E) emerges with a plane of polarization at right angles to the O wave. These two component waves respond to the material they are traveling through as if they have different refractive indices. Apparently, light interacts more intensely with such materials depending on its orientation. Since the analyzer filter has an orientation perpendicular to the polarizer filter, light unaffected by birefringence will be extinguished, and the field will appear dark. If a birefringent object lies in the light path, the emerging O and E waves recombine and the resultant light has a component whose plane of polarization is now perpendicular to the initial plane of polarization, and will pass through the analyzer to form a bright image against this dark background. This can be a difficult concept to grasp. The Web sites *http://www.olympusmicro.com/primer/virtual/virtualpolarized.html* and *http://www.olympusmicro.com/primer/lightandcolor/birefringenceintro.html* can be helpful. A more thorough explanation can be found in Murphy (2001), Chapters 8 and 9.

Phase-Contrast Microscopy

Logistics of phase-contrast microscopy

A conventional light microscope exploits the fact that some objects (stained specimens) absorb light differentially, producing bright and dark regions, and even colors. Contrast is created by differential light absorption. Phase-contrast microscopy allows one to image structures within a completely transparent (unstained) object by taking advantage of subtle shifts in wave phase caused by diffraction. This technique requires use of a specialized microscope. Its design won the Nobel prize for Frits Zernike in 1953 (read his acceptance speech in Zernike, 1955). Details on operation depend upon the design of the microscope.

1. Establish Köhler illumination on a bright-field object.
2. Dial a specialized phase lens into place.

   This is outfitted with a phase plate, an optically clear disk in which an annulus has either been etched or raised to alter the effective length of the light path.

3. *Optional:* Put a low-wavelength green-light filter in place to increase the resolution.
4. Dial the substage phase annulus to match the objective lens chosen.

   The phase annulus dial is usually marked with individual settings for each phase objective.

5. Manually align the phase annulus with the phase plate.

   You are now ready for business.

Problems with phase-contrast microscopy

Most problems have to do with proper alignment of the phase plate and condenser annulus. One should also be sure that the proper phase annulus is matched to the proper phase-plate objective.

Why phase-contrast works

As mentioned earlier, a transparent microscopic object, such as a cell, can be viewed as an assemblage of tiny particles or edges (sizes on the order of the wavelength of light) suspended in a truly transparent background. One can imagine that membranes separating subcellular compartments, and regions of differing chemical composition (DNA-rich, protein-rich, carbohydrate-rich), all create optical edges varying in refractive index and degree of interaction with transmitted light. Light passing through a cell interacts with each particle or edge by diffraction (imagine waves striking a jetty or an aperture). Most of the light passing through an unstained cell fails to interact with the material of the cell, and passes through undeviated (this is the "S" wave, or surround wave). Some light does interact with the subcellular particles and edges by diffraction as it passes through. This is the "D" wave (or diffracted wave). There are two differences between the undeviated "surround" waves and the deviated or "diffracted" waves as they emerge from the object. First, the undeviated light is much brighter, that is, it has a higher amplitude since most the light does not interact with transparent objects. Second, the deviated light has become delayed or "phase shifted" by ¼ wavelength when compared to the undeviated light, as a result of its diffraction interactions. Let's imagine what happens to create an image of a diffraction object against a nondiffracting background. In a conventional light microscope, the light emerging from the transparent region will be bright. The light emerging from a region possessing a diffraction object will be subtly less bright (remember that most of the light will pass through such a region without interacting) and subtly phase delayed. Through a conventional light microscope, one will not detect the difference in intensities between two regions with such subtle differences in diffraction properties because the undeviated light swamps the deviated light, and the phase delay is subtle (i.e., interference will be negligible between deviated and undeviated light).

Zernike developed a method of equalizing the amplitude of the diffracted light and undiffracted light and exaggerating the phase shift between deviated and undeviated light. First, he put in place the condenser annulus. This created an effect similar to that described for dark-field microscopy. A hollow cone of light now illuminates the specimen, though this cone passes into the objective lens rather than around it (Fig. 9.1.17). The result is that undeviated light is projected as a hollow cone up through the specimen and into the objective lens. Deviated light that results from diffraction interactions with the tissue can pass up through the dark center of the cone. Deviated and undeviated light will be focused and recombined by the objective lens into an image. Positive and negative

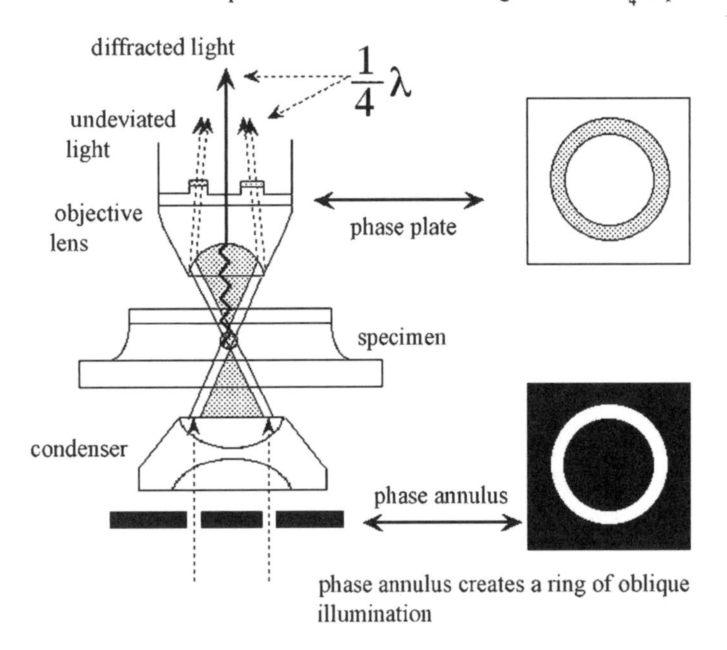
Elevated (or excavated) ring on phase plate is coated to reduce intensity of direct light. Differences in length of light path for direct and indirect light creates a $\frac{1}{4}\lambda$ phase shift.
diffracted light
$\frac{1}{4}\lambda$
undeviated light
objective lens
phase plate
specimen
condenser
phase annulus
phase annulus creates a ring of oblique illumination

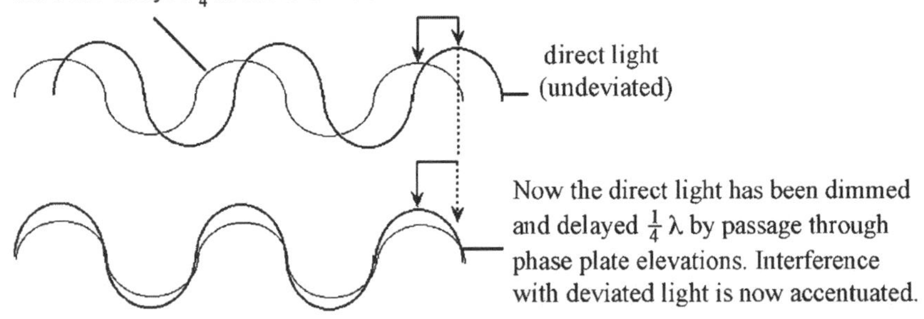
dim, indirect light deviated by passage through specimen has been delayed $\frac{1}{4}\lambda$ due to diffraction.
direct light (undeviated)
Now the direct light has been dimmed and delayed $\frac{1}{4}\lambda$ by passage through phase plate elevations. Interference with deviated light is now accentuated.

Figure 9.1.17 Phase-contrast microscopy. The upper figure illustrates the basic design of the phase-contrast microscope. A phase annulus lies just below the condenser lens, illuminating the specimen with a hollow cone of light, similar to the dark-field model, yet different. The hollow cone of direct (undeviated) light has a more acute angle, so that its light is collected by the objective lens and contributes to formation of the image. Within the objective lens, there is a phase plate through which the direct, undeviated cone of light is transmitted. In the model illustrated, this phase plate has a raised annulus of glass that is coated with a neutral-density filter to dim the undeviated light. The height of this raised ring is such that light passing through it is phase-delayed by $^1/_4\ \lambda$ (see lower diagram). Deviated light, i.e., light that has passed through a diffraction object in the specimen, passes through the unraised center of the phase plate. When light is diffracted, it is also phase delayed by $^1/_4\ \lambda$. Hence, in this design, the deviated light and the undeviated light are brought back in phase with one another, and will interfere positively to create regions of brightness in the image. Since diffraction is the result of light interfering with edges and tiny particles, these are the objects that become highlighted in the final image. Other models of phase microscopes etch the glass of the phase plate, reducing its effective light path so that the direct, undeviated light is advanced $^1/_4\ \lambda$ with respect to the deviated (diffracted) light. The net effect is that deviated light is now $^1/_2\ \lambda$ delayed with respect to undeviated light, and the deviated and undeviated light interfere negatively, producing areas of shadow around edges and particles.

interference will create subtle shifts in amplitude that are barely visible, since the undeviated light is so much more intense than the deviated light that it effectively swamps the signal.

Second, he placed a phase plate above the objective lens. This consists of a plate of optically clear glass with an annulus of coated material that is either raised higher or etched lower than the surrounding uncoated glass. Figure 9.1.17 depicts the "elevated" version. The annulus is designed to catch the undeviated light cone, which is projected onto this plate from the objective lens. The coating (a neutral-density filter) reduces the amplitude of the undeviated light, so that it is more commensurate with the diffracted light passing through the uncoated center. The fact that the annulus may be raised (or lowered) takes advantage of the fact that light travels more slowly in glass than in air. By increasing the distance light must travel through glass, one can slow this light and create a negative phase shift. The differences in light path are engineered so as to add or subtract a $^1/_4$-wavelength phase shift ($^1/_4$ λ). In the raised version, the undeviated light is retarded $^1/_4$ λ, so that it comes into phase with the deviated light (Fig. 9.1.17). Remember that deviated light has already been slowed by $^1/_4$ wavelength due to diffraction interactions with the specimen. If the undeviated light is retarded by $^1/_4$ λ (by passing the it light through a longer glass light path in the phase plate), the undeviated light will interfere positively with the deviated (diffracted) light, creating edges of brightness around regions of diffraction. If the undeviated light is accelerated ($^1/_4$ wavelength relative to the deviated light) by etching the annulus and reducing the distance that undeviated light must pass through glass, the deviated light will interfere negatively, creating dark shadows around phase objects. The combination of using a neutral-density filter to diminish the intensity of the undeviated light, combined with a $^1/_4$-wavelength phase shift to accentuate the positive or negative interference produced when combining the deviated and undeviated light, results in the production of a genuine contrast image from subtle differences in phase. A nice tutorial can be found at *http://www.microscopyu.com/articles/phasecontrast/phasemicroscopy.html*.

Specific example

One of my favorite subjects for phase-contrast microscopy works best with an inverted microscope. This is the preparation of lampbrush chromosomes from newt oocytes. Detailed protocols can be found at the following Web sites: *http://www.projects.ex.ac.uk/lampbrush/downloads/urodele.doc* and *http://www.projects.ex.ac.uk/lampbrush/downloads/starters.doc*.

Differential Interference Contrast (DIC) or Nomarski Optics

Logistics of DIC/Nomarski microscopy

Despite the remarkable power of phase-contrast microscopy to resolve details within transparent objects, another kind of "interference" microscopy has become more popular—Nomarski, or DIC microscopy. The DIC microscope was invented by the Polish-born, French-educated scientist, Georges Nomarski (Nomarski, 1955). This type of microscope utilizes both polarizing optics and birefringence to probe subtle differences in refractive index (n) within a specimen and convert these into a contrast image. The design is shown in Figure 9.1.18, with good reviews included in Allen et al. (1969) and Padawer (1968). Logistics typically involve establishing Köhler illumination on a DIC microscope and then inserting a polarizing filter below the stage, an analyzer above the stage, and a pair of Wollaston prisms both below and above (Wallaston I and II, respectively; the upper prism is often coupled with the analyzer in a slider). The upper Wollaston prism is then adjusted with a screw to achieve the desired optical effect.

Poor man's Nomarski

An interesting dark-field stop that produces a Nomarski-like image can be seen at *http://www.microscopy-uk.org.uk/mag/artnov02/diydic.html*. This can be a nice surrogate for those who cannot afford a DIC-Nomarski microscope.

Problems with DIC/Nomarski

Most common problems involve failure to insert all the components (e.g., leaving out the polarizer) or failing to match the substage components with the objective lens.

Why DIC/Nomarski works

The substage polarizer creates a beam of polarized light which subsequently passes through a Wollaston prism (the condenser prism). This prism is made of quartz or some other birefringent

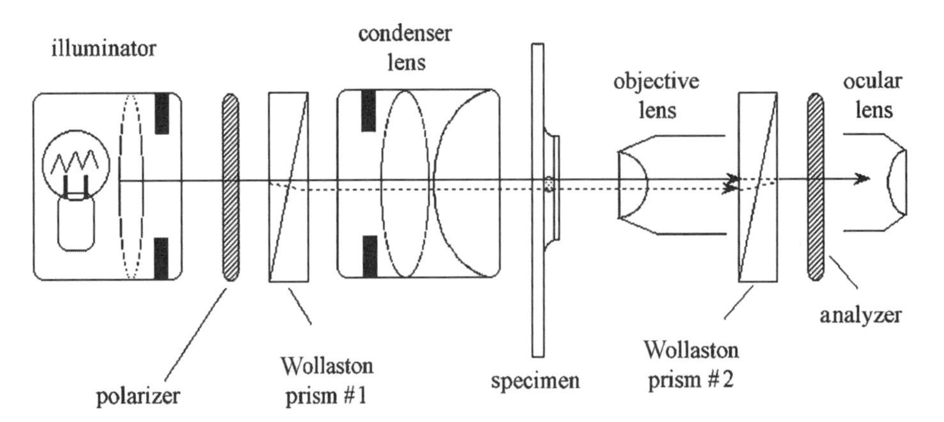

Figure 9.1.18 Nomarski or differential interference contrast (DIC) microscopy. This figure illustrates the basic design of the DIC microscope. Two features are familiar: the polarizer and analyzer, two pieces of cross-polarizing material that extinguish most light. The new features are the two Wollaston prisms, one located below the condenser lens and the other above the objective lens. Polarized light enters the first Wollaston prism and is split. Each ray is split into two rays that are cross-polarized and travel parallel (yet very close) paths. These split rays probe the specimen. When sharp gradients in refractive index are encountered (subcellular compartments with different chemical compositions), one of the twin rays passes through the lower-refractive-index medium, while the other passes through the higher-refractive-index medium. This results in a phase shift between the initially in-phase, cross-polarized beams. The second Wollaston prism recombines the "split rays." If a pair of split rays never encounter a difference in refractive index as they passed through the specimen, they are recombined exactly as they were, and their light is extinguished as it passes through the analyzer. If two rays do encounter differences in refractive index and become out-of-phase, when recombined by the second Wollaston prism, the resultant light has a component that vibrates at angle to the initial, polarized beam. This can then pass through the analyzer and form an image. This microscope turns steep gradients of refractive index (in an otherwise transparent object) into a landscape of contrast.

material; hence, the polarized light is split into the ordinary (O), and extraordinary (E) components prior to entering the specimen. The O and E rays are polarized at right angles relative to one another and separated by a shear distance (0.2 to 2.0 μm). This reflects the fact that the twin beams physically follow light paths through the tissue that are 0.2 to 2.0 μm apart. The result is that every component of the specimen is probed by a pair of cross-polarized beams. When a region of the specimen has a sharp gradient in refractive index (e.g., a membrane separating cellular compartments), the interface will result in differential retardation of the two beams: one beam will be slowed by passage through a material of higher refractive index with respect to the other. This will result in a slight phase shift between the two wave components as they emerge from the specimen. As the light is collected by the objective lens, it is passed through a second Wollaston prism (the objective prism) that recombines the twin beams, now phase-shifted. If no specimen is encountered, there is no resultant phase shift and the beams are recombined without modulation. The resultant beam is polarized in the same plane as the initial transmission axis of the polarizer. The analyzer (whose polarizing axis is oriented 90° to the polarizer) will block all such light. When a specimen is encountered, and the twin beams pass through material with a phase gradient, the objective prism recombines the light with an interesting "twist." It seems that when phase-shifted, perpendicularly oriented polarized light is recombined by a Wollaston prism. The resultant wave form has a component that vibrates in a plane different from the initial transmission axis of the polarizer. Hence, this light can pass through the analyzer. Actually, the recombination of perpendicular polarized light with varying degrees of phase shift results in circular or elliptically polarized light. This lies beyond the scope of this unit, and is explained in more depth at the following Web sites: *http://www.olympusmicro.com/primer/java/dic/wavefrontrelationships/* and *http://www.microscopy.fsu.edu/primer/java/polarizedlight/waveform3d/*. Regions of the specimen with sharp gradients in refractive index will create highly contrasting images. The result is an image with pronounced shadows and hot spots, and a false (but comforting) sense of three-dimensional relief (Fig. 9.1.14D). This is reminiscent of a favorite quote from University of Washington's microscopist, Professor Richard Cloney, *"There's nothing wrong with artifacts, if they help you to see the truth."*

9

ACKNOWLEDGEMENTS

There are four people I would like to acknowledge: Dr. John Wagner (Chicago Field Museum) who took me to buy my first microscope when I was 14, Dr. Frank Gwilliam (Reed College) who introduced me to the art of microscopy, Dr. John Giannini (St. Olaf College) who has been my partner in developing and teaching a course in Light Microscopy, and Dr. Leland Johnson (Augustana College) who has shared a wealth of cunning microscopy tricks with dozens of young faculty members at the Darling Marine Center workshop for Developmental Biology. I dedicate this unit to Leland. This work was also supported by a grant from the National Science Foundation, entitled *The Gene Stream: From sequence to cell function*; MCB: 0444700.

LITERATURE CITED

Abramowitz, M. 1985. Contrast Methods in Microscopy. Vol. 2. Olympus America Inc., Center Valley, Pa.

Allen, R.D., David, G.B., and Nomarski, G. 1969. The Zeiss-Nomarski differential interference equipment for transmitted-light microscopy. *Z. Wiss. Mikrosk.* 69:193-221.

Cole, E.S., Stuart, K.R., Marsh, T.C., Aufderheide, K., and Ringlien, W. 2002. Confocal fluorescence microscopy for *Tetrahymena thermophila*. *Meth. Cell Biol.* 70:337-359.

Köhler, A. 1893. A new system of illumination for photomicrographic purposes. *Z. Wiss. Mikroskopie* 10:433-440. Translated in *Royal Microscopical Society*-Köehler Illumination Centenary, 1994.

Murphy, D.B. 2001. Fundamentals of Light Microscopy and Electronic Imaging. Wiley-Liss, New York.

Newton, R.H., Haffagee, J.P., and Ho, M.W. 1998. Creating color contrast in light microscopy of living organisms. *J. Biol. Education* 32:29-33.

Nomarski, G. 1955. Microinterféromètre différentiel à ondes polarisées. *J. Phys. Radium* 16:9S-11S.

Omoto, C.K. and Folwell, J.A. 1999. Using darkfield microscopy to enhance contrast: An easy and inexpensive method. *American Biol. Teacher* 61:621-624.

Padawer, J. 1968. The Nomarski interference-contrast microscope. An experimental basis for image interpretation. *J. Royal Microscopical Society* 88:305-349.

Salmon, E.D., von Lackum, K., and Canman, J.C. 2005. Proper alignment and adjustment of the light microscope. *Curr. Protoc. Microbiol.* 0:2A.1.1-2A.1.31.

Zernicke, F. 1955. How I discovered phase contrast. *Science* 121:345-349.

KEY REFERENCES

Barker, K. 2005. At the Bench: A Laboratory Navigator. Updated Edition. Cold Spring Harbor Laboratory Press, Cold Spring Harbor, N.Y.

A useful (brief) practical guide.

Murphy, 2001. See above.

A recent book (the best in my opinion, which does a good job discussing practical microscopy and its theoretical underpinnings.

INTERNET RESOURCES

General Microscopy Resource Sites

http://micro.magnet.fsu.edu/primer/

http://micro.magnet.fsu.edu/primer/java/components/characteristicrays/index.html

http://www.olympusmicro.com/primer/index.html

http://web.uvic.ca/ail/techniques/scope_basics.html

http://www.ou.edu/research/electron/mirror/

Resolution and Numerical Aperture

http://www.microscopyu.com/articles/formulas/formulasresolution.html
http://micro.magnet.fsu.edu/primer/anatomy/numaperture.html

Köhler Illumination

http://micro.magnet.fsu.edu/primer/anatomy/kohler.html
http://www.aecom.yu.edu/aif/instructions/koehler/koehler.htm
http://www.emlab.ubc.ca/pKoehler.htm

Specialized Microscopy (Overviews)

http://micro.magnet.fsu.edu/primer/techniques/index.html
http://www.pirx.com/droplet/microscopes.html

Dark-Field Microscopy

http://www.ruf.rice.edu/~bioslabs/methods/microscopy/dfield.html
http://micro.magnet.fsu.edu/primer/techniques/dark-fieldindex.html
http://www.microscopyu.com/articles/stereomicroscopy/stereodark-field.html
http://www.olympusmicro.com/primer/techniques/dark-fieldindex.html
http://www.olympusmicro.com/primer/java/dark-field/cardioid/index.html

Rheinberg

http://micro.magnet.fsu.edu/primer/techniques/rheinberg.html
http://www.microscopy-uk.org.uk/mag/indexmag.html
http://www.microscopy-uk.org.uk/mag/artnov02/diydic.html

Polarized-Light Microscopy

http://www.microscopyu.com/articles/polarized/polarizedintro.html
http://www.microscopy.fsu.edu/primer/java/polarizedlight/waveform3d
http://www.microscopyu.com/articles/polarized/polarizedintro.html
http://www.olympusmicro.com/primer/techniques/polarized/polarizedhome.html
http://micro.magnet.fsu.edu/primer/techniques/polarized/polmicroalignment.html

Phase-Contrast Microscopy

http://www.microscopyu.com/articles/phasecontrast/phasemicroscopy.html
http://www.microscopyu.com/tutorials/java/phasecontrast/microscopealignment/index.html
http://micro.magnet.fsu.edu/primer/techniques/phasecontrast/phaseindex.html

DIC Microscopy

http://www.olympusmicro.com/primer/java/dic/wavefrontrelationships
http://micro.magnet.fsu.edu/primer/java/dic/wollastonwavefronts/index.html
http://www.microscopyu.com/articles/dic/dicindex.html

UNIT 9.2

Immunofluorescence Microscopy

David J. Asai[1]

[1]Harvey Mudd College, Claremont, California

OVERVIEW AND PRINCIPLES

Living systems are complex and dynamic. Each cell contains hundreds or thousands of different proteins and, in the case of eukaryotic cells, dozens of membranous organelles. In order to begin to understand the biology of the cell, it is important to visualize its macromolecules in four dimensions: the three spatial dimensions and the fourth dimension of time. Where is the molecule? How much of it is present in a particular location? With what other molecules does it interact? And how does this spatial pattern change with time?

Fluorescence is a useful method for visualizing specific molecules in the cell, and there are many different ways to use fluorescence to study cells. Expressing a gene of interest that has been fused to a fluorescent reporter gene (e.g., green fluorescent protein) can enable the investigator to watch the gene product function in a living cell. Fluorescence resonance energy transfer (FRET) can measure the interactions between two molecules. Ratiometric fluorescent dyes can be used to quantify the intracellular concentrations of specific ions (e.g., FURA-2 for calcium ion), and fluorescent lipid analogs can be used to visualize membrane dynamics.

A particularly powerful method for visualizing molecules in cells is immunofluorescence microscopy. While typically confined to examination of chemically fixed tissues and thus not generally used to visualize living cells, immunofluorescence microscopy presents the advantage of combining antibody specificity with fluorescence microscopy. The antibody acts as a "homing torpedo" that seeks out and binds its epitope with high affinity among all of the rest of the cellular components. Fluorescence enables the investigator to locate the bound antibody, even when the antibody is in a low concentration, because the fluorescence amplifies the signal. Today, there are hundreds of commercially available antibodies suitable for immunofluorescence microscopy; these antibodies are raised against a wide and diverse spectrum of molecules (e.g., see *http://www.antibodyresource.com*; US-Biological, *http://www.usbio.net*). The techniques of immunofluorescence microscopy are straightforward and widely accessible to investigators at all levels of experiences.

Principles of Fluorescence Microscopy

For additional details, the reader is referred to several excellent publications on microscopy, including chapters by Brian Herman (Herman, 2002) and Donald Coling and Bechara Kachar (Coling and Kachar, 1998) and a primer by Doug Murphy (2001).

Fluorescence is a property exhibited by some molecules in which energy is absorbed, electrons are excited within the absorbing molecule, and, as the electrons return to their ground energy state, energy is emitted as luminescence. Fluorescence requires the **excitation** of a molecule by **illumination** at a particular energy (measured in wavelength) and the **emission** of photons at a longer and thus lower energy wavelength. The separation of the wavelengths of the excitation and emission is referred to as the **Stokes shift** (Fig. 9.2.1).

Illumination

Bright, even illumination of the specimen is achieved using Köhler illumination, in which a condenser and collector lens focus a conjugate image of the light source on the back focal plane of

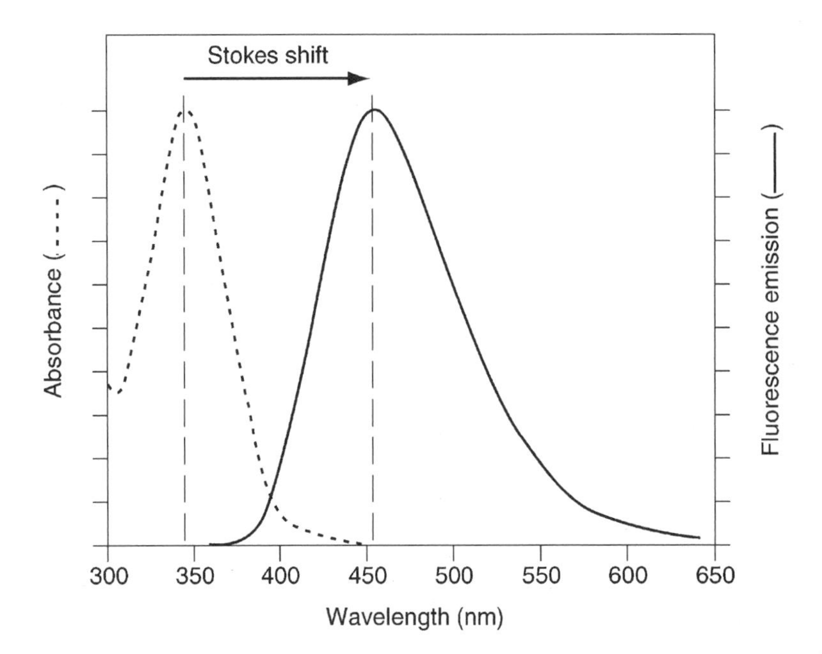

Figure 9.2.1 Stokes shift. Absorption and emission spectra of Hoechst 33342 (Molecular Probes), a dye that is used to stain DNA. The dye is excited with monochromatic light, typically near the λ_{max} of the absorption spectrum (in this case, ~350 nm), and the emission at 450 nm is measured with a spectrofluorometer. Spectra have been normalized. Spectra courtesy of Molecular Probes. Originally published in *CP Molecular Biology Unit 14.10* (Coling and Kachar, 1998).

the objective lens, which is then focused on the specimen (see Murphy, 2001). Most conventional fluorescence microscopes use epi-illumination, in which the illuminating light passes through the objective to the specimen, and the resulting emitted light returns back through the same objective. A common light source is a 50- or 100-W Hg lamp, which emits spectral maxima of 254, 365, 405, 436, 546, and 577 nm. In confocal laser fluorescence microscopy, the specimen is illuminated with lasers tuned to specific wavelengths.

Separating excitation and emission

In conventional epi-illumination microscopy, excitation light of a particular wavelength is reflected by a **dichroic beam-splitting mirror** onto the specimen. The dichroic mirror reflects light below a specified wavelength, and allows light of higher wavelengths to pass through the mirror. The objective focuses the light onto the specimen. The emitted fluorescence from the specimen then travels back through the objective, through the dichroic mirror, and to the observer (Fig. 9.2.2). Because it is often desirable to double- or triple-stain a specimen with different fluorescent molecules (e.g., DAPI, fluorescein, and rhodamine), it is important to selectively illuminate the specimen with a particular wavelength, and to capture a specific wavelength of emitted light. This specificity is accomplished by combinations of filters. A **short-pass filter** will pass light up to the limiting wavelength, but not light of higher wavelengths. A **long-pass filter** will pass wavelengths above the specified value, but not below it. **Band-pass filters** transmit light only in a specific range of wavelengths. Band-pass filters are named for the mean wavelength transmitted and the range or bandwidth. For example, a BP490/30 filter transmits light between 475 and 505 nm.

For example, consider the visualization of molecules using fluorescein and rhodamine, which are two commonly used fluorophores in immunofluorescence microscopy. Fluorescein absorbs maximally at 490 nm and emits maximally at 525 nm; rhodamine absorbs at 550 nm and emits at 580 nm. The microscope would be fitted with two separate filter sets, one for fluorescein and one for rhodamine. These two filter sets are typically on a slider or wheel so that it is easy to examine the very same portion of the specimen with each filter set. For fluorescein, the excitation light would be selected through a band-pass filter that transmits 450 to 490 nm (BP470/40) and then reflected by a

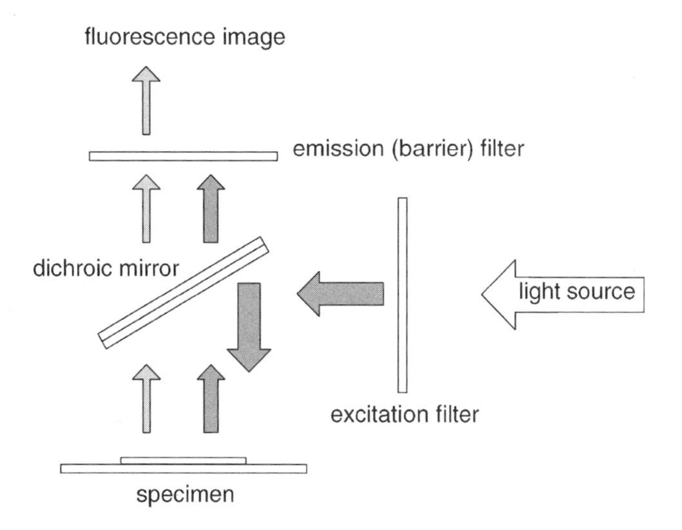

Figure 9.2.2 Arrangement of light source, filters, and dichroic mirror in a typical epi-illumination fluorescence microscope. The illuminating light passes through an excitation filter to select for the appropriate wavelength and then is selectively reflected by the dichroic mirror to illuminate the specimen. The emitted light (at a longer wavelength than the excitation light) passes through the dichroic mirror and an emission or barrier filter and is captured by the observer. Originally published in *CP Molecular Biology Unit 14.10* (Coling and Kachar, 1998).

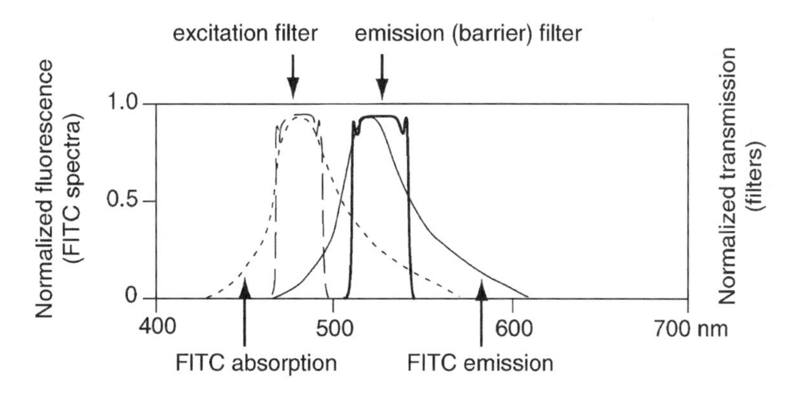

Figure 9.2.3 Fluorescein imaging using excitation and emission filters. In this example for fluorescein isothiocyanate (FITC), the excitation light passes through a band-pass filter (BP470/40), and the emission is filtered through another band-pass filter (BP540/50). In this figure, spectra are normalized for comparison. Originally published in *CP Molecular Biology Unit 14.10* (Coling and Kachar, 1998), courtesy of Omega Optical.

dichroic beam-splitting mirror that reflects below 510 nm (DM510). The fluorescein signal would be isolated through a long-pass filter that allows light above 520 nm to pass through it (LP520). The rhodamine filter set would comprise a BP546/12, DM580, and a LP590. Because the fluorescein excitation can also excite the rhodamine (bleed-through), the fluorescein signal can be isolated from the rhodamine with more selective emission filters (515 to 565 nm for fluorescein and LP610 for rhodamine). An example of the use of filters is shown in Figure 9.2.3.

Numerical aperture, resolution, and magnification

NOTE: For additional discussion of magnification versus resolution, see *UNIT 9.1*.

In fluorescence microscopy, we want to maximize the **numerical aperture** (NA) in order to collect the greatest amount of emitted fluorescence. NA is defined by the equation:

$$\mathrm{NA} = \eta \sin \alpha$$

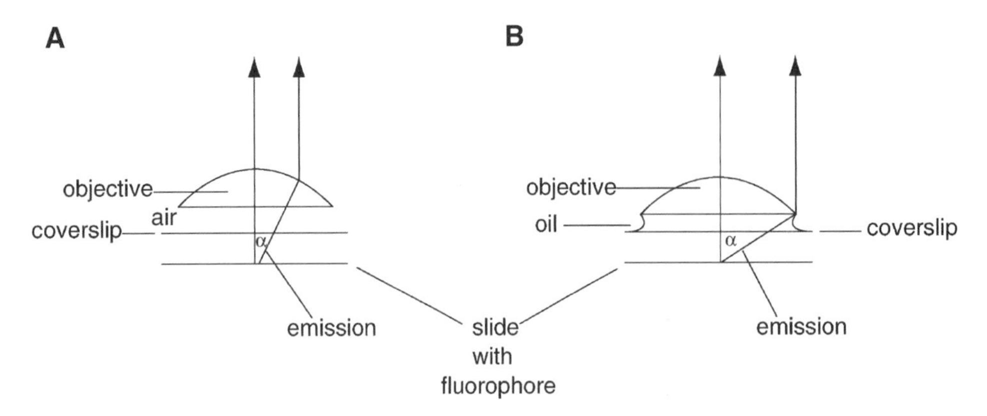

Figure 9.2.4 The numerical aperture (NA) is a measure of the light-collecting ability of an objective lens, with a larger NA corresponding to a greater quantity of light that can be collected. NA is proportional to the refractive index (η) of the medium between the lens and the specimen and the angle of the emission (α). Thus, a larger NA is achieved with a short working distance of the lens (distance between the front of the objective lens and the specimen) and by matching the refractive indices of the objective and the sample. (**A**) A "dry" lens. (**B**) An objective lens with oil immersion to match the refractive indices of the lens and sample. The shorter working distance and larger α in Figure 9.2.4B result in a larger NA. Originally published in *CP Immunology Unit 21.2* (Herman, 2002).

where η is the refractive index of the medium and α is the angle of the emitted light captured by the objective (see Fig. 9.2.4).

"Dry" lenses (i.e., objectives that do not use immersion oil) have a limited NA because of the mismatch between the refractive indices of the coverslip, air, and objective surface (Fig. 9.2.4A). To increase the NA, immersion oil is placed between the coverslip and the objective (Fig. 9.2.4B). The NA and the immersion medium are etched on the barrel of the lens.

Resolution (*R*) is the smallest distance between two objects so that they can be discerned as separate objects. *R* is defined by the equation:

$$R = (0.61\ \lambda)/\mathrm{NA}$$

where λ is the wavelength of the light and NA is the numerical aperture of the lens.

Magnification (*M*) is defined as the extent to which objects appear larger than they are. The useful range of magnification is 500 to 1000 times the NA. The brightness of an image decreases as the magnification increases. In fluorescence microscopy, the overall brightness is proportional to NA^4/M^2. Thus, the best fluorescence imaging will be accomplished by using an objective with the highest NA and the lowest magnification.

Antibodies

NOTE: For additional general information about antibodies, see Asai (1993).

Because it is possible to induce an immune response to virtually any biological molecule, antibodies are important tools in cell biology and are used to detect the presence (e.g., immunoblot, immunohistology) and quantity (e.g., ELISA) of an antigen. Indirect immunofluorescence microscopy depends on two antibodies, the primary antibody that binds the antigen of interest and the secondary antibody that binds the primary antibody and carries the fluorophore. Thus, the successful design of an immunofluorescence experiment requires an understanding of the structures and specificities of the antibodies.

Antibodies are proteins produced by the immune systems of all vertebrates (from sharks to humans). The hallmarks of antibodies are their **specificity** and **high affinity** for **antigen**. In cell biology, antibodies can be considered site-directed monkey wrenches that bind to a particular molecule and can inform the investigator about the location and quantity of the antigen in a preparation.

Some basic terminology:

Antigen. Anything that elicits an immune response. For our purposes, an antigen is a protein.

Immune response. Any way the vertebrate responds to an antigen. There are two kinds of immune response—a cellular response (e.g., the welt you see a couple days after a positive tuberculin skin test) and a humoral response (circulating antibodies). Another way the animal can respond to antigen is the failure to react against the antigen (tolerance). Cell biologists exploit the humoral response by raising specific antibodies to proteins of interest.

Antibody. An immunoglobulin. The basic unit of the immunoglobulin comprises two identical heavy chains and two identical light chains. There are five classes of immunoglobulins that are differentiated by the structure of their fraction crystallizable (Fc) domains: IgG is usually the major class in a hyperimmune serum (i.e., from an animal that has been repeatedly exposed to antigen); IgM is the major cell surface antigen receptor and the primary response in the serum; IgA is present in secretions and mucus membranes; IgE is the antibody that triggers mast cells to dump histamines in the allergic response; and IgD mediates the hyperimmune response. Cell biologists typically use IgGs and IgMs in immunofluorescence microscopy (see Fig. 9.2.5).

Epitope. The specific portion of the antigen to which the antibody binds. For a protein, the epitope is specified by a few amino acids or a post-translational modification, and very small differences in epitopes can be detected by different antibodies.

Monoclonal antibody. The product from a single clone of B cells (lymphocytes that secrete antibodies). All of the monoclonal antibodies from a single hybridoma (an immortalized proliferating B cell engineered to produce a desired antibody in large amounts) are structurally identical and therefore bind the same epitope with precisely the same degree of specificity and affinity.

Polyclonal antibody (or serum). The product from whole hyperimmune serum. Usually a polyclonal response involves antibodies produced by several different clones of B cells. Therefore, a polyclonal antiserum contains several different epitope specificities; ideally these different epitopes are on the same antigen. Generally, it is not possible to discriminate among epitopes with a polyclonal antiserum.

Antibody structure

IgG is a tetramer of two identical heavy chains with a relative molecular weight (M_r) of ~50 K and two identical light chains (M_r ~20 K). The tetramer is held together by disulfide bonds. There are two domains of each light chain (L): a variable region (V_L) and a constant (C_L) region. There are four domains of each heavy chain (H): a variable region (V_H) and three constant regions ($C_{H1\text{-}3}$). The variable regions fold to form the antigen-binding sites. Because there is mirror symmetry across the vertical plane of the Y-shaped antibody, it has two identical antigen-binding sites. The constant regions of the light chains determine the lambda and kappa allotypes (the allelic forms of the two light chain proteins). The structure of IgG is shown in Figure 9.2.5.

Antibody classes

The constant regions of the heavy chains determine the class of the antibody. There are five immunoglobulin classes: IgM, IgG, IgA, IgE, and IgD. IgG has two antigen-binding sites. IgM is pentameric and so has ten antigen-binding sites.

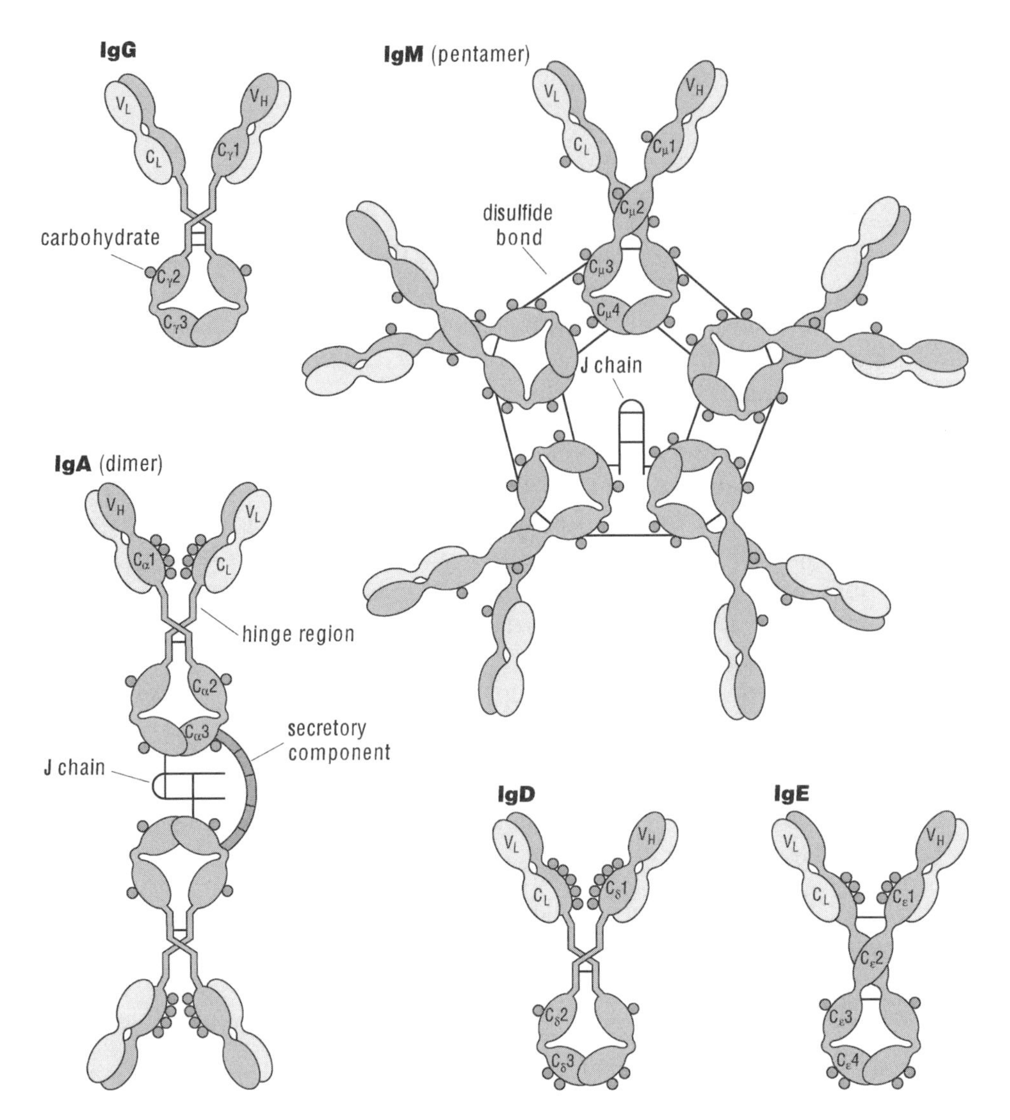

Figure 9.2.5 Structures of the five major classes of secreted antibodies. Light chains are lightly shaded; heavy chains are darkly shaded. Circles denote sites of glycosylation. The immunoglobulin G, D, and E molecules are Y-shaped, comprising two identical heavy chains and two identical light chains. The antigen-binding sites are near the tips of the two arms of the Y, formed from the variable regions of the heavy and light chains (V_H and V_L). The fraction crystallizable (Fc) portion of the antibody is composed of the CH domains (C2, C3, and C4 in this figure). IgA is a dimer, and IgM is typically a pentamer; the components are joined together by the J chain. IgG and IgM are typically used in cell biology, including Western blotting and immunofluorescence microscopy. Originally published in Coico et al. (2003).

Antibodies in cell biology

For cell biology, polyclonal antibodies are generally raised in rabbits, sheep, or goats; monoclonal antibodies are generally raised in mice or rats. The most useful antibody classes for immunofluorescence microscopy and immunoblotting (see *UNIT 8.3*) are IgG and IgM.

STRATEGIC PLANNING

Overview

The application of indirect immunofluorescence to visualize the cell cytoskeleton (Lazarides and Weber, 1974) revolutionized the way we think about the organization of a eukaryotic cell. However, fluorescent antibodies have been used in cell biology for many years, well before their application for the visualization of the cytoskeleton (e.g., see Frye and Edidin, 1970).

The term "indirect" refers to the fact that the method requires two antibody steps: (1) the primary antibody is unlabeled and is directed against a particular cellular component; (2) the secondary antibody is directed against the primary antibody and is covalently linked to a fluorescent molecule. The secondary antibody is raised against the C_H portion of the antibody that is species and class specific (e.g., rabbit anti-mouse-IgG for labeling of a primary IgG raised in mouse). Thus, a single secondary antibody can be used with a wide range of primary antibodies. The indirect method presents two advantages: (1) the signal is amplified because one antigen can be bound by several primary antibodies, and each primary antibody can be bound by several secondary antibodies; (2) it is easy to perform double or triple immunofluorescence with the appropriate combination of primary and secondary reagents.

There are several important considerations when developing an effective strategy for indirect immunofluorescence. These are discussed below.

Fixation and permeabilization

Preparing specimens for immunofluorescence microscopy involves a balance between two requirements—fixation and permeabilization. Many cellular structures interact noncovalently; therefore, it is essential to chemically fix the cell prior to dilution into an effectively infinite volume of antibody and buffers. Equally important is to remove the plasma membrane in order to allow the antibodies to interact with their antigens (if the antigens are inside the cell) or to **not** remove the membrane if the antigens are on the cell surface. A good chemical fixative is glutaraldehyde, which cross-links proteins through the primary amines on the protein surface. A workable and less expensive alternative is formaldehyde. Permeabilization of the cell is typically achieved with the nonionic detergent Triton X-100 and/or methanol. Finally, many cellular structures are labile (e.g., microtubules are sensitive to calcium ion and pH), and so it is important to identify conditions that preserve the integrity of the structures.

Exposing the epitopes

The reaction of antibodies with their antigens can be sensitive to the chemical condition of the antigen. For example, some antibodies react only with antigen that has been fixed with glutaraldehyde. Other antibodies react best when the antigen is denatured. Some antibodies—especially antibodies raised against a short peptide located in the middle of the protein sequence—react well on immunoblots of SDS-denatured proteins but do not bind their antigens in fixed cells because the epitope is hidden in the tertiary structure of the aldehyde-treated protein. Thus, when designing a peptide antigenic determinant, it is usually best to find a sequence near the N or C terminus of the protein. Sometimes a partial denaturation of the protein after fixation and before application of the primary antibody promotes binding of the antibody. In this case, the investigator can briefly treat the fixed cells with a cocktail of detergents (we use a "hardy wash" composed of 0.1% SDS, 0.5% NP-40, and 0.5% Tween 20; incubate cells for 5 min at room temperature).

Choice of antibody

The best antibody will bind with high affinity only the antigen of interest. Both monoclonal and polyclonal antibodies are used in cell biology. When performing double-staining (i.e., two different animals for two different antigens), it is convenient to select antibodies that are either from two different animals (e.g., mouse and rabbit) or antibodies of two different classes (e.g., IgG and IgM). In this way, indirect staining protocols can be employed, using commercially available fluorescent secondary antibodies that are specific for the species and class of each primary antibody.

A monoclonal antibody offers the potential advantage of a homogeneous preparation of antibodies, all of which bind exactly the same epitope with precisely the same affinity. For example, there are monoclonal antibodies that specifically recognize only a particular post-translational modification of a protein (e.g., acetylated tubulin). On the other hand, a potential disadvantage of using a monoclonal antibody is that it may significantly cross-react with another protein, and because the cross-reaction is likely due to the presence of a similar epitope on the second protein, it is difficult to selectively eliminate this unwanted reactivity.

The advantage of using a polyclonal antiserum (e.g., a hyperimmune rabbit serum) is that it usually comprises a rich mixture of antibodies reactive with several different epitopes all on the same antigen. Thus, even if one of the epitopes is also found on another protein, it is likely that the majority of the reactivities present in the polyclonal antiserum are not. By this "averaging" of reactivities, a polyclonal antiserum can significantly reduce the problem of a cross-reacting activity. Similarly a better signal can often be obtained by mixing several different monoclonal antibodies that together mimic a polyclonal serum by binding different epitopes on the same antigen.

Choice of fluorescent dyes

The choice of dyes depends on the microscope and the experiment: What are the available excitation wavelengths? What filters can be used? The fluorescent microscope should be tagged with information about the filter sets installed in the microscope, and the filters can usually be readily changed to customize the optics. (It is a good idea to consult the manufacturer before attempting to alter the optics on the microscope; there's a good reason there are screws on the access panel!) Does the investigator wish to double- or triple-stain the cells, in which case it is important to select dyes that are well-separated in excitation and emission wavelengths that can be distinguished with the available filters? Table 9.2.1 presents a summary of the absorption and emission maxima of some commonly used dyes for nucleic acids and secondary antibodies. An excellent resource for this type of information is Molecular Probes (*http://www.probes.invitrogen.com*).

Antibody specificity and important controls

The reliability of the information obtained by immunofluorescence microscopy is only as good as the specificity of the antibody. Even with (or, perhaps, **especially** with) "store-bought" antibodies, it is important to know the reactivity of the antibody. There are several ways to ensure antibody specificity; here are six:

Dilutions of antibodies. The appropriate dilutions of the primary and secondary antibodies must be empirically determined in preliminary experiments; it is typically easier to first test different

Table 9.2.1 Some Common Fluorescent Dyes Used for Nucleic Acid Visualization or Conjugated to Secondary Antibodies

Dye	Absorption maximum (nm)	Emission maximum (nm)
DNA		
Hoechst 33342	350	461
DAPI	358	461
BOBO-1	462	481
YOYO-1	491	509
SYTOX Green	504	523
TOTO-1	514	533
Secondary antibodies		
FITC	494	518
Rhodamine	555	580
Texas Red	595	615
Alexa Fluor 633	632	647

dilutions of the antibody on immunoblots before trying them in immunofluorescence microscopy. Generally, a hyperimmune polyclonal (e.g., rabbit) antiserum or a monoclonal ascites serum is used in immunofluorescence microscopy at a dilution of approximately 1/50 and 20- to 100-fold more concentrated than what is used in an immunoblot. Monoclonal antibodies in the spent hybridoma culture media are typically used with little or no further dilution. We follow the vendor's suggested dilutions for secondary antibodies, although it is often the case that the fluorescent secondary antibodies can be used at one-half the recommended concentration.

Immunoblot. If possible, the sample loaded onto the polyacrylamide gel should be a total extract of the cells to be stained. There should be an expectation of the apparent molecular weight of the antigen, and, ideally, this band and only this band should react with the antibody. (Refer to *UNIT 8.3* for more detailed information.)

Preimmune serum and "spontaneous" antibodies. Normal animal sera often contain antibodies that react with antigens that may be present in the cells being stained. This is especially true for plant and bacterial proteins because of the exposure to these antigens of the animal in which the antibody was raised. In addition, antibodies to common cellular proteins often appear in normal sera. For example, nearly all rabbit sera have significant titers of antibodies to cytoskeletal proteins (Karsenti et al., 1977; Asai and Brokow, 1980). Indeed, "spontaneous" antibodies to specific proteins have proven to be important reagents (H. Tjandra and D.J. Asai, unpub. observ.). One way to avoid these pre-existing antibodies is to prescreen the preimmune sera of several animals before initiating the immunization protocol. However, this requires the investigator to have direct control over the production of the antisera, which is often not practical. It is important to apply an equal quantity of preimmune serum (i.e., serum from the same animal that subsequently produced the primary antibodies) or nonimmune serum (i.e., serum from the same species of animal if the preimmune serum is not available) to the cells in order to determine the contributions of the pre-existing antibodies. Monoclonal antibodies are usually used as tissue culture supernatants or mouse ascites fluid. The appropriate nonimmune controls are culture media and normal mouse serum, respectively.

No primary antibody. Another important control is to apply only the fluorescently tagged secondary antibody to the fixed and permeabilized cells in order to determine the background staining due to the secondary antibody.

Affinity purification of the primary antibody. There are effective methods to affinity purify the desired antibodies from a serum (see Asai, 1993). Affinity chromatography over columns comprising beads conjugated with the antigen can be used to purify relatively large quantities of antibodies. On a much smaller scale, affinity purification of antibodies onto the antigen immobilized on nitrocellulose (e.g., immunoblot) can purify sufficient antibodies for staining. If possible, it is important to verify the purification of the antibodies by immunoblotting (see *UNIT 8.3*).

Absorption of the antibody reactivity. A complement to the affinity purification of the antibody is demonstration of antibody specificity by eliminating its reactivity with absorption by the appropriate antigen. This is especially important when using antibodies raised against short peptides because it is common for the short peptide sequence to occur in other proteins, increasing the likelihood of unwanted cross-reactivity. The strategy is to preincubate the working dilution of the primary antibody with a molar excess of the antigen/peptide prior to applying the antibody onto the fixed and permeabilized cells. Ideally, absorption with the appropriate antigen, and not with an unrelated protein (e.g., bovine serum albumin), should eliminate the staining.

SAFETY CONSIDERATIONS

The chemical fixatives used in these protocols (glutaraldehyde and formaldehyde) are extremely toxic and will damage exposed tissues, including eyes and airways. Always work with these chemicals in a properly exhausting fume hood and properly dispose of solutions. The high intensity of illuminating light (e.g., from a 100-W mercury lamp or a laser) will injure the retina if viewed directly.

CAUTION: Never look directly at the illuminating light source.

PROTOCOLS

The following protocols illustrate the methods of indirect immunofluorescence for staining two kinds of cells that have very different morphologies: mammalian fibroblasts (which are flat and grow on glass coverslips) and *Tetrahymena thermophila* cells (which are much thicker and grow in suspension). Our methods for staining microtubules with antibodies to tubulin are described. Microtubules are very labile, readily disassembled by cold temperature or calcium ions, and so the protocols illustrate the importance of controlling the conditions for fixation and permeabilization. Similar methods work for many other antigens, but the investigator will need to fine-tune the protocols for her/his particular antigen, cell type, and antibody.

Basic Protocol 1: Processing Fibroblasts

This protocol assumes the investigator has a source of fibroblasts plated onto sterilized glass coverslips. Cultured mammalian fibroblasts grown on glass coverslips are relatively flat. We use ceramic coverslip racks (12 slips per rack) to process the cells. Rinsing, permeabilization, and fixation are performed in 250-ml glass beakers, using 100 ml solution. This is sufficient volume to cover the coverslips in the ceramic racks. (See Thompson et al., 1984 for additional general information.) This protocol works well for staining the cytoskeleton, including microtubules, actin microfilaments, and intermediate filaments. It also can be used to stain organelles (e.g., the nucleus, Golgi apparatus, and mitochondria).

NOTE: The cold methanol step will prevent phalloidin from binding to actin filaments; methanol should be avoided if the investigator is using phalloidin as one of the cytoskeleton probes.

Materials

Mammalian fibroblasts: grown to ~50% confluence (typically 1 day, depending on initial cell density) on autoclaved no. 1, 22 mm × 22–mm glass coverslips (five to six) on the bottom of a sterile, plastic 150-mm petri dish using standard cell culture techniques (e.g., see *CP Cell Biology Unit 1.1*; Phelan, 1998)
Phosphate-buffered saline (PBS; see recipe), 37°C
0.1% (v/v) Triton X-100 in 1× microtubule stabilizing buffer (MTSB; see recipe)
3.7% (v/v) formaldehyde/1× MTSB (see recipe), 37°C
100% methanol, −20°C (optional)
Primary antibody, diluted (see Strategic Planning) to working concentration in 0.1% PBSA (see recipe)
Secondary antibody, diluted (see Strategic Planning) to working concentration in 0.1% PBSA (see recipe)
Other stains: e.g., 0.5 µg/ml 4′,6-diamidino-2-phenylindole (DAPI), 0.01 mM SYTOX (Molecular Probes), fluorescein isothiocyanate (FITC), or rhodamine
Mounting medium (see recipe)
Nail polish

Ceramic coverslip rack (Coors; Thomas Scientific)
Fine-tipped jeweler's forceps
250-ml beakers
Humidified chamber (e.g., Tupperware box with moistened paper towel) with grid (e.g., plastic gel spacers)
Microscope slides (Gold Seal; VWR)

1. Remove the coverslips from the medium and place in the ceramic coverslip rack. Keep track of which side of the coverslip has the cells!

   A curved fine-tipped pair of jeweler's forceps is useful for manipulating the coverslips.

2. Rinse the cells 3 to 5 min with 37°C PBS by placing the rack containing the coverslips into a 250-ml beaker containing 100 ml solution.

3. Demembranate the cells by incubating 5 min in 100 ml 0.1% Triton X-100 in 1× MTSB at 37°C.

4. Fix the cells 10 min in 100 ml of 3.7% formaldehyde/1× MTSB at 37°C.

   CAUTION: *Formaldehyde is extremely toxic and will damage exposed membranes (eyes, nose, throat). Always work in a properly exhausting fume hood and properly dispose of solutions.*

   Better preservation of cellular structures can be achieved by reversing steps 3 and 4, i.e., fixing the cells first in formaldehyde (step 4) followed by detergent extraction (step 3); however, the background can be higher in cells fixed before permeabilization. A good compromise may be achieved by reducing the time in the detergent. Figure 9.2.6 shows microtubule staining of fibroblasts fixed and permeabilized in different ways as described in Table 9.2.2.

5. *Optional:* Extract 10 min with 100 ml 100% methanol at −20°C.

   IMPORTANT NOTE: *If phalloidin will be used to stain the actin microfilaments, do* **not** *use methanol for the extraction step.*

   The cold methanol step serves to further extract the cells, usually resulting in a lower staining background. It is also possible to use only cold methanol or cold 100% acetone to permeabilize and fix the cells, i.e., skipping the detergent and aldehyde steps. In our hands, cold methanol alone tends to fragment the microtubules, presumably because the cells are being rapidly dehydrated.

6. Rinse coverslips briefly in 100 ml PBS.

7. Remove each coverslip from the rack, draw off excess fluid by touching the edge to a Kimwipe, and lay the coverslip cell-side up on a grid in a humidified chamber.

   We use a plastic box lined with a moistened paper towel.

   The coverslips are placed cell-side up on plastic strips (we use SDS-PAGE gel spacers).

   It is important to keep track of the arrangement of the coverslips.

8. Pipet 30 to 50 µl of the diluted primary antibody onto the coverslip, resulting in a drop of antibody solution on the coverslip.

   See Strategic Planning for a discussion of antibody dilutions.

   If the cells are to be double-stained (i.e., two different antibodies), it is convenient to mix the primary antibodies and to perform this step as a simultaneous co-incubation with both antibodies. 37°C for >1 hr.

   This is a good stopping point. The primary antibody incubation can extend for more than a day, as long as the coverslips are in a closed and moistened chamber. For incubations with primary antibody of overnight or longer, incubations at 4°C are recommended.

9. Replace the coverslips carefully into the ceramic rack. Rinse three times (5 min each) in 100 ml PBS at 37°C.

10. Remove each coverslip from the rack, draw off excess fluid with a Kimwipe, and lay the coverslip cell-side-up in the humidified chamber.

11. Pipet 30 to 50 µl of the diluted secondary antibody onto each coverslip.

    If double-staining, then both secondary antibodies can be simultaneously applied.

    Instead of or in addition to two different antibodies, the investigator may wish to include a stain for organelles (e.g., DAPI or SYTOX for nuclear DNA, wheat germ agglutinin for Golgi apparatus,

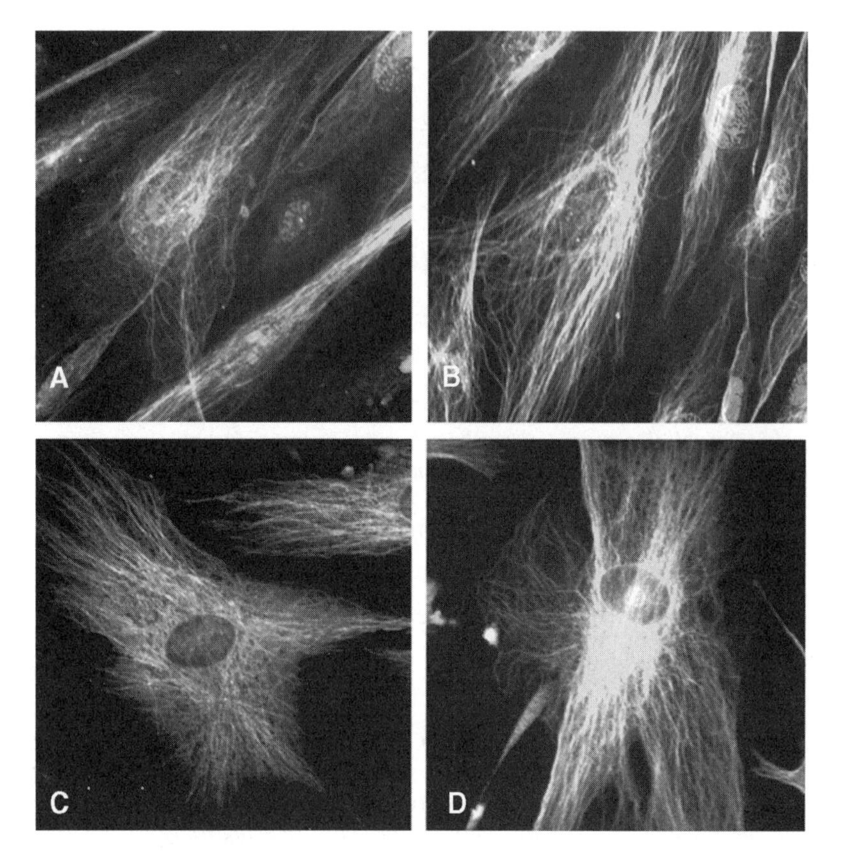

Figure 9.2.6 Microtubule arrays in cultured fibroblasts. Human corneal fibroblasts were grown on sterilized glass coverslips and processed for immunofluorescence microscopy. All slides were stained with a mouse monoclonal (IgM) antibody to β-tubulin (Asai et al., 1982) followed by a rhodamine-conjugated anti-mouse IgM antibody (Molecular Probes). Four different fixation and permeabilization protocols were used (see Table 9.2.2). Fewer microtubules were preserved when the first step was a prolonged treatment in detergent (panel **A**). The cultured fibroblasts were a gift from Dr. Elizabeth Orwin's laboratory.

Table 9.2.2 Treatments Used to Generate Panels Shown in Figure 9.2.6

Figure panel	Treatment
9.2.6A	(i) Triton X-100, 10 min; (ii) formaldehyde, 10 min; (iii) cold methanol, 10 min
9.2.6B	(i) Triton X-100, 2 min; (ii) formaldehyde, 10 min; (iii) cold methanol, 10 min
9.2.6C	(i) formaldehyde, 10 min ; (ii) cold methanol, 10 min (no detergent step)
9.2.6D	(i) formaldehyde, 10 min; (ii) Triton X-100, 5 min; (iii) cold methanol, 10 min

or phalloidin for actin). For example, if the microscope is fitted with the appropriate filters, the investigator may wish to stain the cells with (1) DAPI to visualize the nuclei, (2) anti-tubulin (e.g., FITC tagged) to visualize the microtubules, and (3) fluorescently tagged (e.g., rhodamine) phalloidin to visualize the actin stress fibers.

12. Incubate 1 hr at 37°C, in the dark.

    The secondary antibody staining step should not extend for more than a few hours.

13. Replace the coverslips carefully in the ceramic rack. Rinse the coverslips three times (5 min each) in PBS at 37°C.

14. After the last PBS wash, place the racked coverslips in a beaker of deionized water for a few minutes to remove salts.

15. Draw off the fluid from the coverslip with a Kimwipe and mount each coverslip on a cleaned glass microscope slide, inverting the coverslip (cell-side-down) onto a 10 μl drop of mounting medium.

    The mounting medium is a cryopreservative (glycerol) and inhibits fading of the fluorescence (because of a high pH and N-propylgallate).

16. Press gently (the eraser end of a pencil works well here) to squeeze out bubbles. Seal the coverslip to the slide with nail polish to prevent evaporation and the introduction of oxygen.

17. Store the well sealed slides up to 6 months in the refrigerator, in the dark.

Basic Protocol 2: Processing *Tetrahymena* Cells

Tetrahymena cells are much larger than cultured fibroblasts; thus, we use a different protocol to solubilize, fix, and stain *Tetrahymena* cells. The principle is the same, however: (1) remove the plasma membrane with nonionic detergent, (2) chemically fix the cells, (3) add primary antibodies, (4) add secondary antibodies, and (5) mount and visualize.

NOTE: For additional general information, refer to Stuart and Cole (2000).

Materials

Tetrahymena cells or other thick cells (e.g., sea urchin embryo cells)
PHEM buffer (see recipe)
10% (v/v) Triton X-100 in PHEM buffer
10% (w/v) paraformaldehyde in PHEM buffer: store up to 1 year at room temperature in a dark container in a fume hood
0.1% and 0.5% PBSA (see recipe)
Primary antibody, diluted (see Strategic Planning) to working concentration in 0.1% PBSA (see recipe)
Secondary antibody, diluted (see Strategic Planning) to working concentration in 0.1% PBSA (see recipe)
Nuclear stain: 0.5 μg/ml 4′,6-diamidino-2-phenylindole (DAPI) or 0.01 mM SYTOX (Molecular Probes), optional
Mounting medium (see recipe)
Nail polish

Tabletop centrifuge (e.g., VWR Galaxy Ministar personal centrifuge)
1.5-ml microcentrifuge tubes
Microscope slides

NOTE: All of the centrifugations in this protocol are performed for 2 min at 1000 to 2000 × *g*, room temperature.

1. Centrifuge cells in 1.5-ml microcentrifuge tubes.

   Throughout this protocol, cells are collected by centrifugation. We use a simple tabletop centrifuge (e.g., VWR Galaxy Ministar personal centrifuge).

2. Quickly but gently decant the supernatant before the cells swim out of the pellet.

3. Gently resuspend the cells in 950 μl PHEM buffer at room temperature.

4. Add 50 μl of 10% Triton X-100 in PHEM buffer (final concentration 0.5% Triton X-100). **Gently** mix by inverting the closed microcentrifuge tube several times. Incubate 5 min at room temperature.

5. Add 250 μl of 10% paraformaldehyde in PHEM buffer (2% final concentration) and mix gently. Incubate 10 min at room temperature.

   CAUTION: *Paraformaldehyde is toxic and will damage exposed tissues (including eyes, nose, throat). Handle in a fume hood.*

6. Centrifuge the cells and decant the supernatant.

7. Resuspend the cells in ~100 to 200 µl of 0.5% PBSA. Mix gently and incubate 2 min at room temperature.

   This step blocks any unoccupied reactive sites of the paraformaldehyde with bovine serum albumin, which should not react with the subsequent antibodies.

8. Repeat steps 6 and 7 and centrifuge the cells again.

   See the Alternate Protocol (below) for an alternative method for the following steps.

9. Decant as much fluid as possible and gently resuspend the cells in a ~100 µl diluted primary antibody.

   See Strategic Planning for a discussion of determining the appropriate working dilution of the primary antibody.

   If double staining, both primary antibodies can be mixed together and simultaneously incubated with the cells.

10. Incubate cells in the closed microcentrifuge tubes 1 hr at 37°C.

    This is a good stopping point. The primary antibody incubation can extend for more than a day, as long as the cells are in a closed tube. For overnight (or longer) incubations with primary antibody, incubation at 4°C is recommended.

11. Centrifuge the cells and decant the supernatant.

12. Wash the cells three times using 0.1% PBSA by incubating each wash 1 min at room temperature, with gentle mixing by inverting the tube. Centrifuge after each wash and discard the supernatant.

13. Resuspend the cells in ~100 µl diluted secondary antibody.

    If double-staining, then both secondary antibodies can be simultaneously applied.

    Instead of or in addition to two different antibodies, the investigator may wish to include a stain for organelles (e.g., DAPI or SYTOX for nuclear DNA).

14. Incubate the cells in closed tubes 1 hr at 37°C, in the dark.

    The secondary antibody staining step should not extend for more than a few hours.

15. Collect the cells by centrifugation and decant the supernatant.

16. Wash the cells three times with 0.1% PBSA by incubating each wash 1 min at room temperature, with gentle mixing by inverting the tube. Centrifuge after each wash and discard the supernatant. After the last wash, decant as much fluid as possible.

17. Resuspend the cells in ~20 µl mounting solution. Place 10 µl of the resuspended cells onto a clean glass microscope slide, drop a coverslip onto the drop, and press down gently (the eraser end of a pencil works well).

18. Seal the coverslip to the slide with nail polish. Store the well sealed slides up to several months at 4°C, in the dark.

Alternate Protocol: Staining Cells Adhered to Poly-L-Lysine-Coated Coverslips

After fixing the cells in paraformaldehyde and washing them (Basic Protocol 2, steps 1 to 8), the cells can be adhered to previously prepared coverslips coated with poly-L-lysine and stained. This is convenient if there is a very small quantity of cells. After adhering the cells to the coverslips, the coverslips are processed as described above for fibroblast staining (Basic Protocol 1). Washes are performed by placing the coverslips in the ceramic rack as described in that protocol.

Additional Materials (also see Basic Protocols 1 and 2)

Coverslips coated with poly-L-lysine (see recipe): prepared before beginning to work with the cells

1. Perform Basic Protocol 2, steps 1 to 8.
2. Decant as much fluid as possible.
3. Pipet the cells onto a coverslip coated with poly-L-lysine.
4. Allow the cells to settle onto the poly-L-lysine-coated coverslips for 30 to 60 min; keep covered to protect from dust and prevent evaporation.
5. Proceed as in Basic Protocol 1, steps 5 to 17.

Basic Protocol 3: Visualizing the Cells

Starting off

Whether using a conventional epi-illumination fluorescence microscope or a confocal laser scanning fluorescence microscope, the same rules apply for getting started. This is because it is always a good idea to first find the cell and focus on it using conventional optics before switching to the confocal system. Here are four suggestions:

Start small. Find the cells using a dry (nonimmersion) low magnification objective (e.g., 10×; see *UNIT 9.1* for details on conventional light microscopy). Staining the cells with a nuclear stain, e.g., 0.5 μg/ml 4′,6-diamidino-2-phenylindole (DAPI) or 0.01 mM SYTOX (Molecular Probes) makes it much easier to find the cells and focus the microscope on the cells.

Move up. After locating the cells, move to a higher-magnification objective if this is desirable. Most microscopes are parfocal so that all of the objectives will be focused on approximately the same focal plane. Note that as the magnification and numerical aperture (NA) increase, the depth of field decreases. Thus, the entire cell may appear to be in focus when viewed with a lower powered lens, and only part of it in focus at a higher magnification.

Oil. No going back and no mixing! Remember that once you go to an oil immersion objective, it is very difficult to go back to a dry lens. Further, different brands of oil are not necessarily compatible with one another, and mixing oils can produce distressing cloudiness. If it is necessary to remove the oil from a slide, try Windex first; if that doesn't work, then ether can be tried.

After using an oil immersion objective, remove all oil with lens paper, repeatedly blotting (NO WIPING!) the paper onto the lens. Never use anything except lens paper. Blotting the objective after each use should be sufficient. However, if excess oil accumulates, try Windex on a sheet of lens paper. As a last resort, we use lens paper soaked in ether. Always blot with lens paper; never wipe across the lens surface.

CAUTION: Ether is explosive and should be handled and stored in a fume hood.

Fading fluorophores. Prolonged excitation will bleach a fluorophore. FITC is more susceptible to bleaching than rhodamine because of the higher-energy illumination required for excitation. Although we include *N*-propylgallate as an anti-fade reagent in the mounting medium, several seconds of illumination will noticeably degrade the fluorescence signal, especially the fluorescein. Thus, it is important to find the cells and focus on the cells quickly. If possible, use rhodamine to find and focus to minimize FITC fading.

Confocal laser scanning fluorescence microscopy

For general references, refer to the operating manual for the LSM510 laser scanning microscope (Zeiss, release 3.2). Also see Hibbs (2004), *CP Cell Biology Unit 4.5* (Smith, 1999), and *CP Microbiology Unit 2C.1* (Smith, 2006).

To yield information on their inner structure using conventional transmitted-light microscopy, specimens must be very thin and translucent; otherwise, image definition will be poor. In many cases, cells and tissues do not conform to these requirements. For example, the *Tetrahymena* cell is ~30 μm thick. Confocal microscopy can circumvent this problem by collecting optical sections through a thick sample, and the equipment is increasingly available to investigators.

Unlike conventional microscopy, laser confocal imaging projects the light of a point light source (a laser) through a high numerical aperture (NA) objective onto a certain object plane of interest. In order to obtain an image of the selected object plane as a whole, the microscope scans the object plane in a point-by-point, line-by-line raster by means of an *x-y* light deflection system (a set of mirrors operated on galvos). The confocal microscope typically uses lasers to illuminate the specimen, thus providing specific wavelengths for excitation. For example, our Zeiss microscope is outfitted with three lasers: (1) an argon laser emitting at 458, 477, 488, and 514 nm; (2) a helium-neon mixed gas laser emitting at 543 nm; and (3) a helium-neon laser emitting at 633 nm.

The key feature of confocal microscopy is in the way the reflected fluorescence light is collected by the photomultiplier detectors. The stray light produced outside the object plane would degrade the in-focus image, resulting in a blurred image of poor contrast. The challenge is to capture **only** the light coming immediately from the object point in focus, while discarding the light coming from out-of-focus areas of the specimen.

The confocal microscope solves this problem simply. The reflected (fluorescence) light at the focus of the high-NA objective is projected onto a variable pinhole diaphragm by the same objective

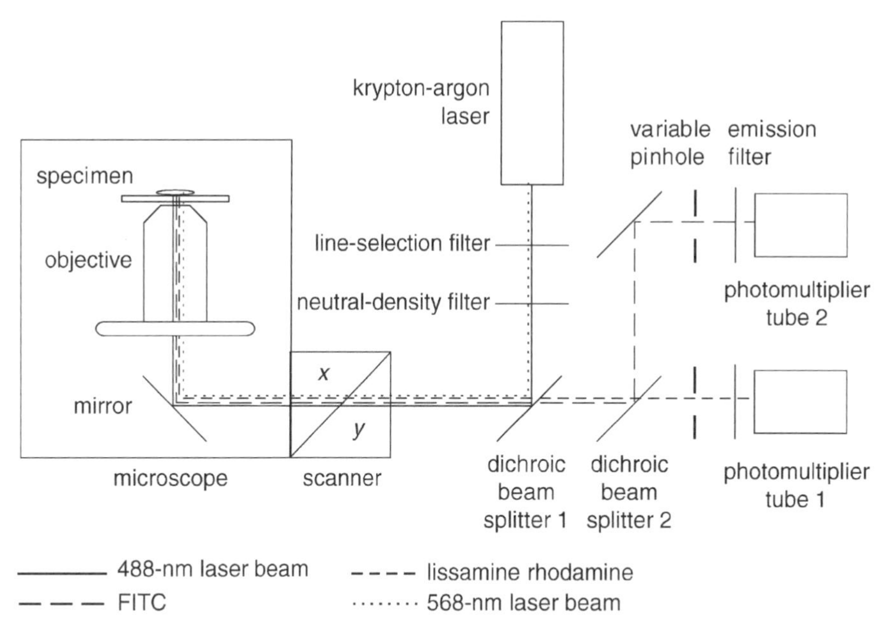

Figure 9.2.7 The light path of a confocal laser scanning microscope. The diagram illustrates the light path of a LSCM set up for simultaneous imaging of FITC and lissamine rhodamine. The 488- and 568-nm lines of a krypton-argon laser are reflected by dichroic beam splitter 1 into the optical axis of the microscope. The beam is reflected by a mirror into the microscope objective, which focuses the beam to a diffraction-limited spot in the specimen. The scanner consists of a pair of galvanometer mirrors that deflect the laser beams so as to scan the spot across the specimen in a raster pattern. Fluorescence emitted as each point is illuminated travels the reverse path through the scanning system. The FITC fluorescence (peak at 520 nm) and lissamine rhodamine fluorescence (peak at 590 nm) pass through dichroic beam splitter 1 to dichroic beam splitter 2, which transmits the lissamine rhodamine fluorescence to photomultiplier tube 1 and reflects the FITC fluorescence to photomultiplier tube 2. A variable pinhole in front of each photodetector blocks light from out-of-focus areas of the specimen while allowing light from the focal plane to reach the detector. Originally published in *CP Microbiology Unit 2C.1* (Smith, 2006).

and a tube lens. The focus inside the specimen and the pinhole are situated at optically conjugate points (confocal). The important advantage of this arrangement is that essentially no other light can pass the narrow pinhole and be registered by the detector. Unwanted light coming from other portions of the specimen is focused outside the pinhole, so it never finds the detector. The smaller the pinhole, the greater the confocality, at the expense of a reduced signal. The principle of confocal microscopy is diagramed in Figure 9.2.7. Because of the removal of out-of-focus signal by the

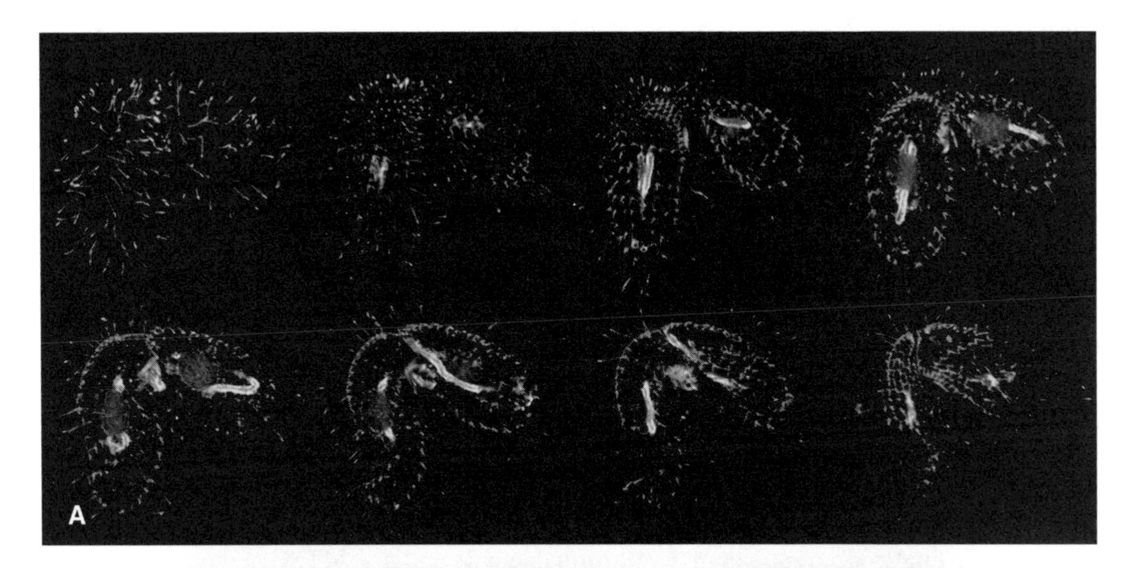

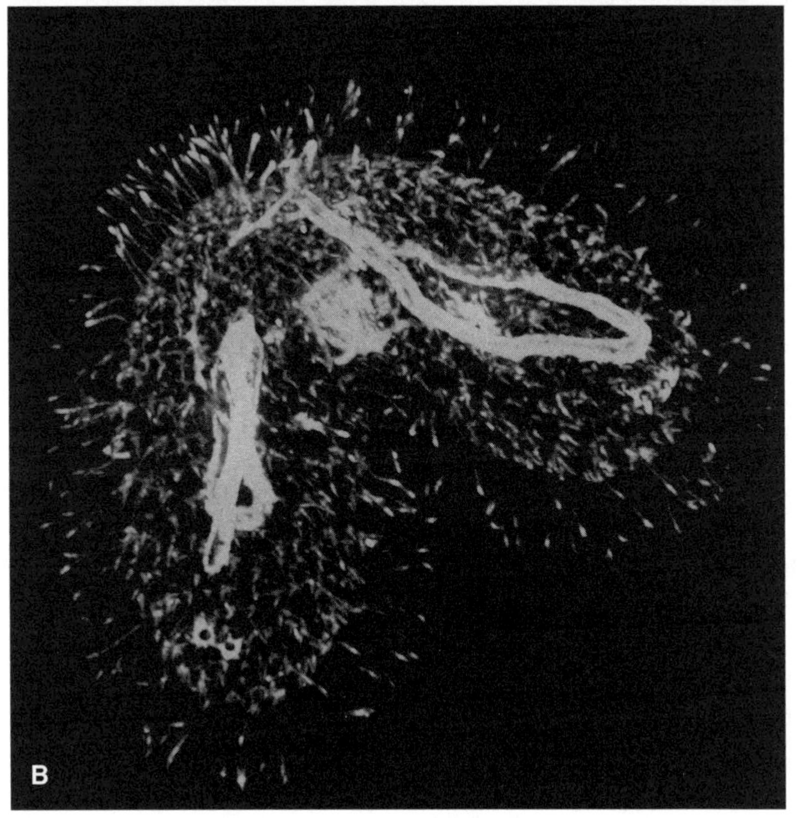

Figure 9.2.8 Example of optical sectioning by confocal microscopy. A conjugating pair of *Tetrahymena* cells, captured in prophase I (crescent stage), were fixed and stained with a cocktail of monoclonal antibodies to α-tubulin (Asai et al., 1982); the secondary antibody was rhodamine-conjugated anti-mouse IgG antibody (Molecular Probes). Mixed into the secondary antibody was SYTOX Green (Molecular Probes). The cells were then examined using confocal laser scanning fluorescence microscopy (Zeiss LSM510). (**A**) Selected optical sections through the pair of cells. (**B**) Projection of all thirty of the optical sections.

pinhole, a confocal microscope can optically section a cell or tissue in the vertical plane (*z*-section). Under optimal conditions, a confocal microscope should be able to achieve spatial resolutions of 0.2 μm in the *x*-*y* dimension and 0.1 μm in the *z* direction. An example of optical sections through a pair of *Tetrahymena* cells is shown in Figure 9.2.8.

REAGENTS AND SOLUTIONS

Use deionized, distilled water in all recipes and protocol steps. For common stock solutions, see **UNIT 3.3**.

Coverslips coated with poly-L-lysine

Holding the coverslip with a pair of sturdy forceps, pass the coverslip several times through the flame of a Bunsen burner.

This deposits a thin carbon coating on the glass.

Place the coverslips flat in a clean petri dish; pipet onto each coverslip a drop of 0.1% (w/w) poly-L-lysine.

Incubate coverslips under poly-L-lysine 30 to 60 min, at room temperature.

Remove coverslips and rinse in water.

Store the poly-L-lysine solution up to 3 months at 4°C (i.e., in a refrigerator).

The poly-L-lysine solution can be reused.

Formaldehyde, 3.7% (v/v)/1× MTSB

Dilute 37% formaldehyde stock (Sigma) 10-fold in microtubule stabilizing buffer (MTSB; see recipe) to give final concentrations of 3.7% formaldehyde and 1× MTSB. Prepare fresh before use.

Microtubule stabilizing buffer (MTSB), 2×

Combine the following in a final volume of 1 liter.

70 g disodium PIPES (0.2 M pipes)
3.04 g EGTA (2 mM)
80 g, polyethylene glycol, MW 6000 (8%)
0.5 g sodium azide (0.05%)
1.97 g $MgSO_4$, heptahydrate (8 mM)
Adjust to pH 6.8 using 1 M NaOH
Store up to 1 year at 4°C
Dilute as required with water to make 1× MTSB

CAUTION: *Azide is toxic!*

Final concentrations are given in parentheses.

Mounting medium

50% glycerol
50% 0.2 M borate, pH 9
0.7% *N*-propylgallate (antifade agent)
Store up to 6 months at 4°C, in a dark container

If glycerol/borate mounting medium is used, the coverslips must be sealed with fingernail polish.

PBSA

10% (w/v) bovine serum albumin (BSA) in 1× PBS (see recipe). Store up to 1 year at 4°C. Dilute as required with 1× PBS to make 0.1% and 0.5% PBSA.

PHEM buffer

Combine the following in a final volume of 500 ml.

9.1 g PIPES (60 mM)
2.98 g HEPES (25 mM)
1.9 g EGTA (10 mM)
0.2 g $MgCl_2$ (2 mM)
Adjust to pH 6.9 with 1 M NaOH
Store up to 6 months at 4°C

This buffer contains no azide.

Final concentrations are given in parentheses.

Phosphate-buffered saline (PBS), 10× stock solution

Combine the following in a final volume of 1 liter.

350.65 g NaCl (1.5 M)
2.64 g NaH_2PO_4, monohydrate (0.02 M)
11.5 g Na_2HPO_4, anhydrous (0.08 M)
5 g sodium azide (0.5%)
Adjust to pH 7.4 with phosphoric acid or 1 M NaOH, if necessary
Store up to 1 year at room temperature
Dilute to 1× with water before use

CAUTION: *Azide is toxic!*

We add sodium azide as a preservative. The azide does not affect these methods, although the PBS with azide is not appropriate for washing living cells.

Final concentrations are given in parentheses.

UNDERSTANDING RESULTS

"A picture is worth a thousand words." (Attributed to Fred Barnard in *Printer's Ink*, 8 December 1921).

The ability to gaze into the cell and determine the locations of specific molecules is the obvious strength of immunofluorescence microscopy. But it can also be a weakness: a seductive invitation to replace scientific rigor with pretty pictures. Investigators—especially students who may be visualizing their favorite molecules for the first time—should be reminded what an immunofluorescence image is **not**. It is not dynamic: it is important not to over-interpret the history or the future of a molecule based on a snapshot of its present location. It is not readily quantifiable: brighter does not necessarily mean "more" because the intensity of the signal depends on many things including the avidity of the antibodies, chemistry of the fluorophore, and optics of the microscope. It is not imaginary: what you see is what you get.

As summarized in this chapter, there are many factors the investigator should consider when interpreting his or her results. We offer five questions here to guide the analysis:

Is the signal specific? Is the investigator convinced about the specificities of the antibody reagents? What is the background fluorescence? Is there autofluorescence (i.e., the specimen fluoresces without any added fluorophore)?

Is the image real? The human mind tends to focus on structures and ignores the lack of organization. The interesting location of the antigen may, in fact, be diffused throughout the cell with no apparent organization.

Is the result reproducible? How is the staining pattern affected by the fixation and permeabilization conditions? The weather? The time of day? The person who does the staining?

What is the magnification of the image? It is always useful to determine the magnification as a check that the investigator understands her/his experimental system and the optics of the microscope.

How are the data recorded and reported? Whether the image was captured with film developed in a darkroom, a digitizing charge-coupled device (CCD), or photomultipliers communicating with a computer, it is affected by the conditions selected by the investigator. Further, after the image is captured, it often must be manipulated using software (*APPENDIX 3B*). Subjectivity is unavoidable; thus, the key is not to discourage manipulations, rather it is important to be consistent when capturing and reporting the images.

TROUBLESHOOTING

Immunofluorescence microscopy is technically challenging, with many steps that affect the final result. Thus, it is inevitable that something will go wrong, given enough opportunity, no matter how skilled and experienced the investigator. The first issue, of course, is to know if there is a problem to troubleshoot. It is helpful if the investigator has a good idea about what she/he expects to see. If the intracellular staining pattern is not known, then it is a good idea for the investigator to be certain of his or her techniques using reagents previously characterized with regard to a well characterized antigen, before embarking on an unknown.

Assuming that things aren't right, here are some possible problems and a few words about how the investigator might troubleshoot them. As is the case for all experimental techniques, the key to troubleshooting immunofluorescence microscopy is to approach the problem logically and, as much as possible, dissect it one variable at a time.

Cells are not stained at all. Can the cells be found using bright-field microscopy? If so, then the problem is with the staining (see below). If not, then the cells were either lost during processing (this includes insufficient fixation or upside-down coverslips) or not there to begin with. If the optics permit, counterstaining the cells with a nuclear stain, e.g., 0.5 μg/ml 4′,6-diamidino-2-phenylindole (DAPI) or 0.01 mM SYTOX (Molecular Probes), is a convenient way to find cells quickly.

Cells are there but not stained brightly. Is the antibody reactive? Is the antibody appropriately diluted? Check the reactivity and dilution of the antibody by immunoblotting (see *UNIT 8.3*) the proteins extracted from the same kind of cells that are being stained.

Cells are overstained. Is the antibody appropriately diluted? Perform a small experiment in which the amount of primary antibody is constant and the amount of fluorescent secondary antibody is varied. Reverse the constant and variable and repeat the experiment. Another possible reason is that the cells were not adequately blocked prior to antibody incubations to avoid the nonspecific "stickiness" of the primary and/or secondary antibodies. We dilute our antibodies in bovine serum albumin to reduce the background.

Cells are unevenly illuminated. For example, the cells in the upper left portion of the field may be stained brightly, but the cells in the center of the field are dim. The lamp is likely out of alignment. Consult the microscope manufacturer's instructions for aligning the optics.

Cells are stained but out of focus. Regardless of the focus adjustment, the images remain fuzzy. The optics (filters, objectives, and condenser) may be dirty. Consult the microscope manufacturer's protocol for cleaning the optical surfaces. Cleaning may be best done by a technician from the manufacturer. Avoid this problem by obtaining proper training prior to using the microscope.

Too much background staining. For example, the entire field is fluorescent whether or not there are cells present. The samples and coverslips may not have been adequately blocked with an inert protein (e.g., bovine serum albumin). Another possibility is that the glass slide requires better cleaning. We have observed lower backgrounds with microscope slides made by Gold Seal.

Nothing on the computer screen. When using a digital camera or a photomultiplier (confocal microscope), there is occasionally no signal on the computer screen. First, check to make sure that the cells can be readily seen by eye (i.e., looking through the eyepieces using epi-illumination). Assuming the cells are stained and the microscope is working, then the problem is in the light path to the camera or the camera or computer.

9

ACKNOWLEDGMENTS

Research in our laboratory is supported by a grant from the National Science Foundation. Our confocal fluorescence microscope was obtained through an MRI grant from the National Science Foundation.

LITERATURE CITED

Asai, D.J. (ed.) 1993. Antibodies in cell biology. *In* Methods in Cell Biology, Vol. 37 (L. Wilson and P.T. Matsudaira, eds.) Academic Press, San Diego.

Asai, D.J. and Brokaw, C.J. 1980. Effects of antibodies against tubulin on the movement of reactivated sea urchin sperm flagella. *J. Cell Biol.* 87:114-123.

Asai, D.J., Brokaw, C.J., Thompson, W.C., and Wilson, L. 1982. Two different monoclonal antibodies to alpha-tubulin inhibit the bending of reactivated sea urchin spermatozoa. *Cell Motility* 2:599-614.

Coico, R., Sunshine, G., and Benjamini, E. 2003. Immunology: A Short Course, 5th ed. John Wiley & Sons, Hoboken, N.J.

Coling, D. and Kachar, B. 1998. Principles and application of fluorescence microscopy. *Curr. Protoc. Mol. Biol.* 14:10.1-14.10.11.

Frye, L.D. and Edidin, M. 1970. The rapid intermixing of cell surface antigens after formation of mouse-human heterokaryons. *J. Cell Sci.* 7:319-335.

Herman, B. 2002. Fluorescence microscopy. *Curr. Protoc. Immun.* 21:2.1-21.2.10.

Hibbs, A.R. 2004. Confocal Microscopy for Biologists. Kluwer Academic/Plenum Press, New York.

Karsenti, E., Guilbert, B., Bornens, M., and Avrameas, S. 1977. Antibodies to tubulin in normal nonimmunized animals. *Proc. Natl. Acad. Sci. U.S.A.* 74:3997-4001.

Lazarides, E. and Weber, K. 1974. Actin antibody: The specific visualization of actin filaments in non-muscle cells. *Proc. Natl. Acad. Sci. U.S.A.* 71:2268-2272.

Murphy, D.B. 2001. Fundamentals of Light Microscopy and Electronic Imaging. John Wiley & Sons Hoboken, N.J.

Phelan, M.C. 1998. Basic techniques for mammalian cell tissue culture. *Curr. Protoc. Cell Biol.* 0:1.1.1-1.1.10.

Smith, C.L. 1999. Basic confocal microscopy. *Curr. Protoc. Cell Biol.* 1:4.5.1-4.5.12.

Smith, C.L. 2006. Basic confocal microscopy. *Curr. Protoc. Microbiol.* 0:2C.1.1-2C.1.19.

Stuart, K.R. and Cole, E.S. 2000. Nuclear and cytoskeletal fluorescence microscopy techniques. *Meth. Cell Biol.* 62:291-311.

Thompson, W.C., Asai, D.J., and Carney, D.H. 1984. Heterogeneity among microtubules of the cytoplasmic microtubule complex detected by a monoclonal antibody to alpha tubulin. *J. Cell Biol.* 98:1017-1025.

Chapter 10

Enzymatic Reactions

Working with Enzymes

David Skrincosky[1] and Cindy Santangelo[1]

[1]Worthington Biochemical, Lakewood, New Jersey

INTRODUCTION

The use of enzymes in life science research and diagnosis of disease are two of the important benefits derived from the intensive research in biochemistry since the 1940s. Enzymes have provided the basis for the field of clinical chemistry and are also useful in many pharmaceutical and therapeutic applications. Industrial grade enzymes are also widely used in the manufacturing processes of foods, beverages, and cleaning products. Other industrial applications include textile, paper, chemical manufacturing, and waste management. However, it is only within the past few decades that interest in diagnostic enzymology has intensified. Many methods currently on record in the literature are not in wide use, and there are still large areas of medical research in which the diagnostic potential of enzymatic reactions has not been explored at all. Enzymes are routinely used in laboratories for protein labeling, ELISA, immunoblots (*UNIT 8.2*; Harlow and Lane, 1998), DNA sequencing (*UNIT 10.4*; Sambrook and Russell, 2001), immunohistochemistry, protein sequencing, and DNA manipulation and analysis. This section summarizes in simple terms the basic principles and theories of enzymology and demonstrates how to properly apply these principles in setting up an enzymatic reaction.

OVERVIEW OF ENZYMES

The living cell is the site of tremendous biochemical activity called metabolism. This is the process of chemical and physical change that goes on continually in the living organism: buildup of new tissue, replacement of old tissue, conversion of food to energy, disposal of waste materials, reproduction—all the activities that we characterize as "life." This building up and tearing down takes place in the face of an apparent paradox. The greatest majority of these biochemical reactions do not take place spontaneously. The phenomenon of catalysis makes possible the biochemical reactions necessary for all life processes. Catalysis is defined as the acceleration of a chemical reaction by some substance which itself undergoes no permanent chemical change. The catalysts of biochemical reactions are enzymes, which are responsible for bringing about almost all of the chemical reactions in living organisms. Without enzymes, these reactions take place at a rate far too slow for the pace of metabolism. The oxidation of a fatty acid to carbon dioxide and water is not a gentle process in a test tube—extremes of pH, high temperatures, and corrosive chemicals are required. Yet in the body, such a reaction takes place smoothly and rapidly within a narrow physiological range of pH and temperature. In the laboratory, the average protein must be boiled for ~24 hr in a 20% HCl solution to achieve a complete breakdown. In the body, the breakdown takes place in 4 hr or less under conditions of mild physiological temperature and pH. It is through attempts to understand more about enzymes—what they are, what they do, and how they do it—that many advances in medicine and life sciences have been brought about.

Early Enzyme Discoveries

The existence of enzymes has been known for well over a century. Some of the earliest studies were performed in 1835 by the Swedish chemist Jons Jakob Berzelius who termed their chemical action "catalytic." It was not until 1926, however, that the first enzyme was obtained in pure form, a milestone accomplished by James B. Sumner of Cornell University. Sumner was able to isolate and crystallize the enzyme urease from the jack bean. His work earned him the 1947 Nobel Prize. John H.

Northrop and Wendell M. Stanley of the Rockefeller Institute for Medical Research shared the 1947 Nobel Prize with Sumner, for their discovery of a complex procedure for isolating pepsin. This precipitation technique devised by Northrop and Stanley has been used to crystallize several enzymes.

Chemical Nature of Enzymes

Almost all known enzymes are proteins, which are long chain polymers of amino acids. Another class of enzyme exists in nature that is composed of ribonucleic acid. Such "ribozymes" are often used to catalyze reactions involving deoxyribonucleic acid (DNA) or ribonucleic acid (RNA). Since the enzymes used in common laboratory practices are proteinaceous in nature, the remainder of this unit will pertain to this class of enzymes.

Enzymes can be denatured and precipitated with salts, solvents, and other reagents. They can vary significantly in size with molecular weights ranging from 10,000 to 2,000,000 Daltons. Many enzymes require the presence of other compounds (i.e., cofactors) before their catalytic activity can be exerted. This entire active complex is referred to as the holoenzyme, i.e., apoenzyme (protein portion) plus the cofactor:

$$\text{Apoenzyme} + \text{Cofactor} = \text{Holoenzyme.}$$

According to Holum (1968), the cofactor (see Fig. 10.1.1) may be a:

1. *Coenzyme.* A nonprotein organic substance that is dialyzable, thermostable, and loosely attached to the protein part.
2. *Prosthetic group.* An organic substance that is dialyzable and thermostable and is firmly attached to the protein or apoenzyme portion.
3. *Metal-ion-activator.* This includes but is not limited to K^{+}, Fe^{2+}, Fe^{3+}, Cu^{2+}, Co^{2+}, Zn^{2+}, Mn^{2+}, Mg^{2+}, Ca^{2+}, and Mo^{3+}.

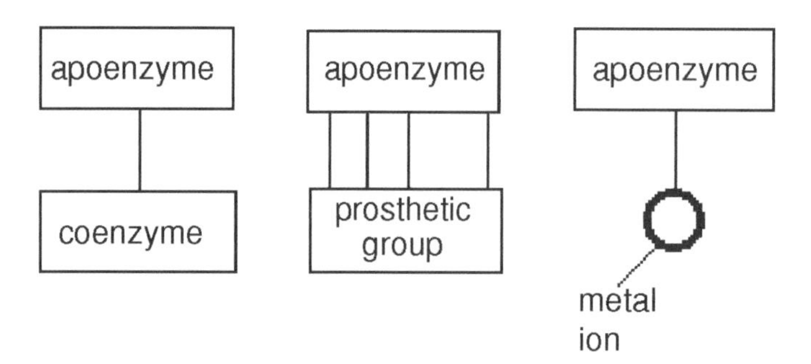

Figure 10.1.1 Holoenzymes: apoenzymes plus various types of cofactors. (Worthington, 1993; published with permission from Worthington Biochemical.)

Specificity of Enzymes

One of the properties of enzymes that makes them so important as diagnostic and research tools is the specificity they exhibit relative to the reactions they catalyze. A few enzymes exhibit absolute specificity, i.e., they will catalyze only one particular reaction. Other enzymes will be specific for a particular type of chemical bond or functional group. In general, there are four distinct types of enzyme specificity:

1. *Absolute specificity.* The enzyme will catalyze only one reaction (e.g., urease).
2. *Group specificity.* The enzyme will act only on molecules that have specific functional groups such as amino, phosphate, and methyl groups (e.g., hexokinase).
3. *Linkage specificity.* The enzyme will act on a particular type of chemical bond regardless of the rest of the molecular structure (e.g., trypsin and chymotrypsin).

4. *Stereochemical specificity.* The enzyme will act on a particular steric or optical isomer (e.g., L-amino acid oxidase).

Although enzymes exhibit great degrees of specificity, cofactors are nonspecific and may serve many apoenzymes. For example, nicotinamide adenine dinucleotide (NAD) is a coenzyme for a great number of dehydrogenase reactions in which it acts as a hydrogen acceptor. Enzymes for reactions that require NAD as a coenzyme include alcohol dehydrogenase, malate dehydrogenase, and lactate dehydrogenase.

Naming and Classification

The earliest studied enzymes—pepsin, trypsin, and rennin—were given names that did not relate to their functions. As more and more enzymes were discovered, the suffix "ase" was used to distinguish a protein as an enzyme, and the name often indicated the type of reaction catalyzed (e.g., oxidase or reductase) or the substrate that it acted upon (e.g., lipase, deoxyribonuclease). As the number of enzymes increased even further, the International Union of Biochemistry (IUB) initiated standards of enzyme nomenclature that recommend the enzyme names indicate both the substrate acted upon and the type of reaction catalyzed. Under this system, the enzyme uricase is called urate:oxygen oxidoreductase, while the enzyme glutamic oxaloacetic transaminase (GOT) is called L-aspartate:2-oxoglutarate aminotransferase.

Enzymes can be classified by the kind of chemical reaction catalyzed. The Enzyme Commission (EC) number is a numerical classification format based on the type of chemical reaction. The EC number has four elements separated by decimal points signifying the main class, subclass, sub-subclass, and serial number in sub-subclass. Reclassification is also common as technological improvements in enzyme characterization advance. The EC number for trypsin is 3.4.21.4, which identifies it as a hydrolase acting on peptide bonds, a serine endopeptidase serial number 4.

The six main classes of enzymes follow:

1. *Oxidoreductases.* Catalyze oxidation-reduction reactions (e.g., dehyrdogenases, oxidases, peroxidases).
2. *Transferases.* Transfer functional groups (e.g., kinases, polymerases).
3. *Hydrolases.* Catalyze the hydrolysis of a chemical bond (e.g., esterases, peptidases, ribonucleases).
4. *Lyases.* Remove or add groups from a double bond (e.g., decarboxylases, carbonic anhydrase).
5. *Isomerases.* Catalyze structural changes (e.g., phosphoglucomutase).
6. *Ligases.* Catalyze the joining of two molecules (e.g., T4 DNA ligase).

Refer to Internet Resources for Web sites relating to nomenclature.

Enzyme Units and Kinetics

Enzymes are catalysts that increase the speed of a chemical reaction without themselves undergoing any permanent chemical change. They are neither used up in the reaction nor do they appear as reaction products. The basic enzymatic reaction can be represented as:

$$E+S \rightarrow P+E$$

Equation 10.1.1

where E is the enzyme, S is the substrate, and P is the product. The activity of an enzyme is measured in terms of units. A unit refers to the amount of enzyme that will catalyze the transformation of a

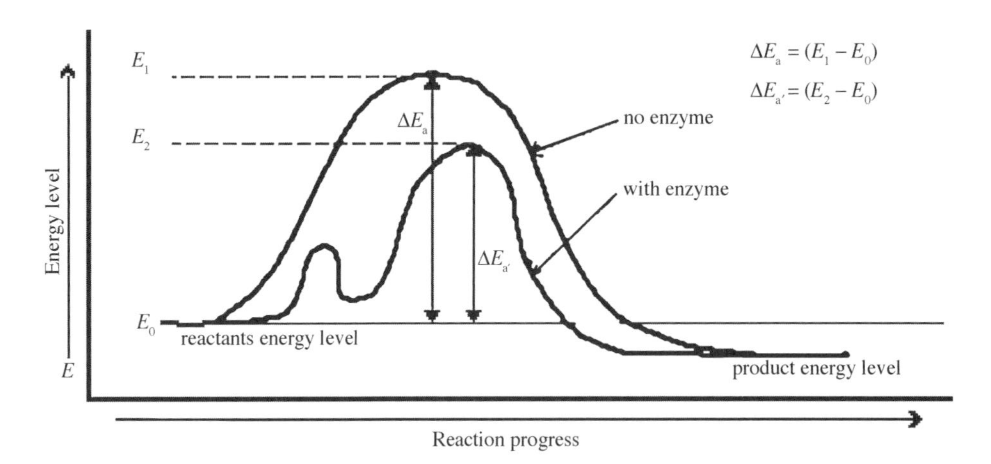

Figure 10.1.2 Change in energy of activation due to the effect of the enzyme on the substrate. The energy of activation ($\Delta E_{a'}$) is lower in the presence of enzyme than when the enzyme is absent (ΔE_a). The minor peak represents activation energy of the reaction intermediate. (Worthington, 1993; published with permission from Worthington Biochemical.)

given substrate into product(s) in a defined period of time under defined reaction conditions (e.g., pH, temperature, buffer composition).

Energy levels

Chemists have known for almost a century that for most chemical reactions to proceed, some form of energy is needed. They have termed this quantity of energy "the energy of activation." It is the magnitude of the activation energy that determines just how fast the reaction will proceed. It is believed that enzymes lower the activation energy for the reaction they are catalyzing. Figure 10.1.2 illustrates this concept and also shows a minor peak that represents activation energy for a reaction intermediate state (small energy barrier).

The enzyme is thought to reduce the "path" of the reaction (the time and energy required to convert a number of molecules of substrate into product). This shortened path would require less energy for each molecule of substrate converted to product. Given a total amount of available energy, more molecules of substrate would be converted when the enzyme is present than when it is absent. Hence, the reaction is said to go faster in a given period of time.

The enzyme-substrate complex

A theory to explain the catalytic action of enzymes was proposed by the Swedish chemist Svante Arrhenius in 1888. He proposed that the substrate and enzyme formed some intermediate substance, which is known as the enzyme-substrate complex. The reaction can be represented as:

$$E + S \rightarrow ES$$

Equation 10.1.2

where ES is the substrate enzyme complex.

If this reaction is combined with the original reaction shown in Equation 10.1.1, the following results:

$$E + S \rightarrow ES \rightarrow P + E$$

Equation 10.1.3

The existence of an intermediate enzyme-substrate complex has been demonstrated in the laboratory. For example, using catalase and a hydrogen peroxide derivative, Kurt G. Stern (1936) at Yale University observed spectral shifts in catalase as the reaction it catalyzed proceeded. This experimental evidence indicates that the enzyme first unites in some way with the substrate and then returns to its original form after the reaction is concluded.

Chemical equilibrium

The study of a large number of chemical reactions reveals that most do not go to true completion. This is likewise true of enzymatically catalyzed reactions. This is due to the reversibility of most reactions. In general this reversibility is demonstrated by:

10

$$A+B \xrightarrow{K_{+1}} C+D$$

Equation 10.1.4

$$C+D \xrightarrow{K_{-1}} A+B$$

Equation 10.1.5

where K_{+1} is the forward reaction rate constant and K_{-1} is the rate constant for the reverse reaction. Combining the two reactions gives the following equation.

$$A+B \underset{K_{-1}}{\overset{K_{+1}}{\rightleftarrows}} C+D$$

Equation 10.1.6

Applying this general relationship to enzymatic reactions allows the following equation.

$$E+S \underset{K_{-1}}{\overset{K_{+1}}{\rightleftarrows}} ES \underset{K_{-2}}{\overset{K_{+2}}{\rightleftarrows}} P+E$$

Equation 10.1.7

Equilibrium, a steady state condition, is reached when the forward reaction rates equal the backward rates. This is the basic equation upon which most enzyme activity studies are based.

HANDLING ENZYMES IN THE LABORATORY

Golden Rules for Handling Enzymes

1. Keep enzymes cold at all times.
2. Avoid repeated freeze-thaw cycles; store in small aliquots (that are used only once or very few times) or as 50% glycerol solutions (which prevents ice formation and loss of activity from denaturation).
3. Do not store enzymes in frost-free freezers.
4. Use proper techniques to avoid microbial or cross contamination.
5. Make stock solutions concentrated enough to avoid loss of activity due to protein binding to storage containers.
6. Provide optimal conditions (diluent, pH, inhibitors, and stabilizers) for storage, which vary for each enzyme.
7. Be aware of requirements for optimum activity (e.g., pH, temperature, ions or cofactors, and inhibitors).

Factors Affecting Enzyme Activity

Knowledge of basic enzyme kinetic theory is important in enzyme analysis, both to understand the basic enzymatic mechanism and to select a method for enzyme analysis. For example, the conditions selected for measuring the activity of an enzyme would not be the same as those selected for measuring the concentration of its substrate. Several factors affect the rate at which enzymatic reactions proceed: temperature, pH, enzyme concentration, substrate concentration, and the presence of any inhibitors or activators.

Enzyme concentration

In order to study the effect of increasing the enzyme concentration upon the reaction rate, the substrate must be present in an excess amount, i.e., the reaction must be independent of the substrate concentration. Any change in the amount of product formed over a specified period of time will be dependent upon the level of enzyme present. Graphically this is represented in Figure 10.1.3. These reactions are said to be "zero order" because the rates are independent of substrate concentration and are equal to some constant, K. The formation of product proceeds at a rate which is linear over time. The addition of more substrate does not serve to increase the rate. In zero order kinetics, allowing the assay to run for double time results in double the amount of product. See Table 10.1.1 for reaction orders and their meaning relative to reaction rate.

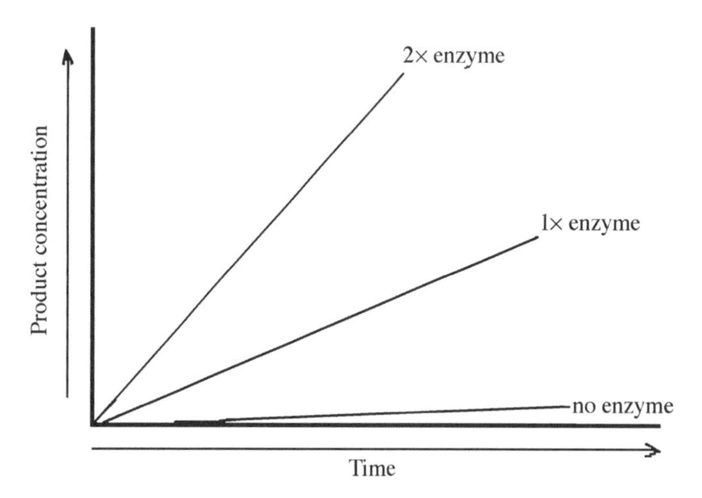

Figure 10.1.3 "Zero order" reaction rate independent of substrate concentration. The substrate is present in excess, and any change in the amount of product formed over a specified period of time will be dependent upon the level of enzyme present. (Worthington, 1993; published with permission from Worthington Biochemical.)

Table 10.1.1 Reaction Orders with Respect to Substrate Concentration

Order	Rate equation	Comments
Zero	rate = k	Rate is independent of substrate concentration
First	rate = k[S]	Rate is proportional to the first power of substrate concentration
Second	rate = $k[S][S] = k[S]^2$	Rate is proportional to the square of the substrate concentration
Second	rate = k[S1][S2]	Rate is proportional to the first power of each of two reactants

The amount of enzyme present in a reaction is measured by the activity it catalyzes. The relationship between activity and concentration is affected by many factors (e.g., temperature or pH). An enzyme assay must be designed so that the observed activity is proportional to the amount of enzyme present in order for the enzyme concentration to be the only limiting factor. It is satisfied only when the reaction is zero order. In Figure 10.1.4, activity is directly proportional to concentration in the area AB, but not in BC. Enzyme activity is generally greatest when the substrate concentration is in excess. When the concentration of the product of an enzymatic reaction is plotted against time, a similar curve results. In Figure 10.1.5, between A and B, the curve represents a zero order reaction, i.e., one in which the rate is constant with time. As substrate is used up, the enzyme's active sites are no longer saturated, substrate concentration becomes rate limiting, and the reaction becomes first order between B and C. To accurately measure enzyme activity, the measurements must ideally be made in that portion of the curve where the reaction is zero order. A reaction is most likely to be zero order at the initial time points because the substrate concentration is highest at that time.

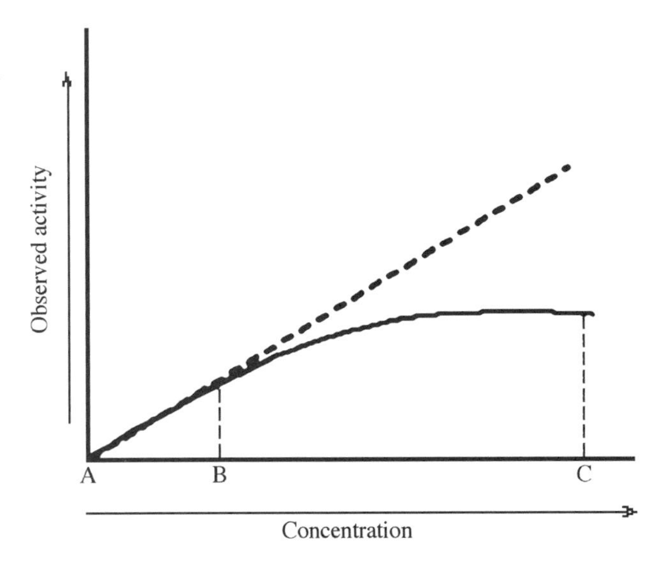

Figure 10.1.4 Activity vs. enzyme concentration. Enzyme activity is directly proportional to the enzyme concentration in area AB, but not in BC. (Worthington, 1993; published with permission from Worthington Biochemical.)

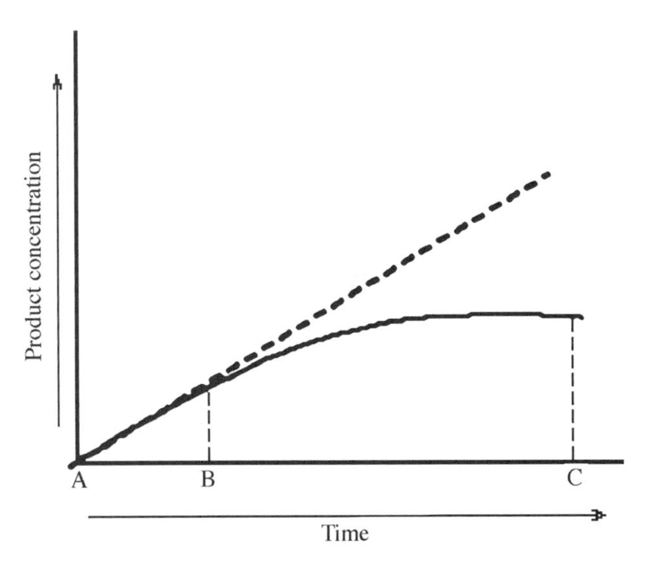

Figure 10.1.5 Reaction rate limited by substrate concentration. Between A and B, the curve represents a zero order reaction. As substrate is used up and its concentration becomes rate limiting, the reaction becomes first order between B and C. (Worthington, 1993; published with permission from Worthington Biochemical.)

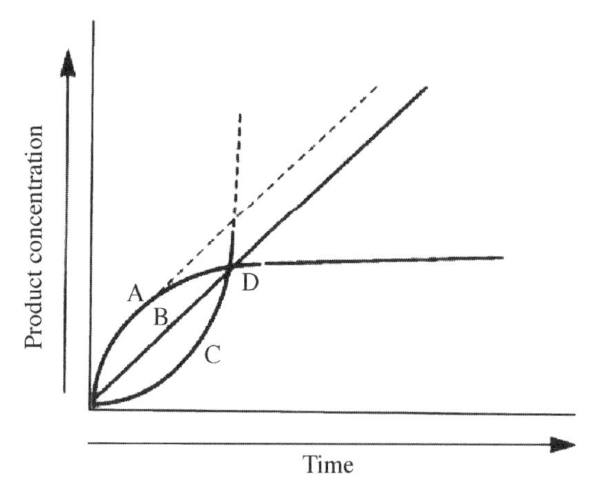

Figure 10.1.6 Leading, lagging, and linear reactions. Curve A represents a "leading" reaction where the reaction is initially zero order and then slows, presumably due to substrate exhaustion or product inhibition. Curve B is a straight line representing a zero order reaction. Curve C represents a reaction with an initial "lag" phase. The dotted lines represent the potentially measurable activity. A single end point determination at D would lead to the false conclusion that all three samples had identical enzyme concentration. (Worthington, 1993; published with permission from Worthington Biochemical.)

To be certain that a reaction is zero order, multiple measurements of product (or substrate) concentration must be made. Figure 10.1.6 illustrates three types of reactions that might be encountered in enzyme assays and shows the problems that might be encountered if only single measurements are made. Curve B is a straight line representing a zero order reaction that permits accurate determination of enzyme activity for part or all of the reaction time. Curve A represents the type of reaction shown in Figure 10.1.5. This reaction is zero order initially and then slows, presumably due to substrate exhaustion or product inhibition. This type of reaction is sometimes referred to as a "leading" reaction. True "potential" activity is represented by the dotted line. Curve C represents a reaction with an initial "lag" phase. Again the dotted line represents the potentially measurable activity. Multiple determinations of product concentration enable each curve to be plotted and true activity determined. Note that a single end point determination at D would lead to the false conclusion that all three samples had identical enzyme concentration.

Substrate concentration

It has been shown experimentally that if the amount of the enzyme is kept constant and the substrate concentration is then gradually increased, the reaction velocity will increase until it reaches a maximum. After this point, increases in substrate concentration will not increase the velocity ($\Delta A/\Delta T$). This is represented graphically in Figure 10.1.7.

It is theorized that when this maximum velocity has been reached, all of the available enzyme has been converted to enzyme-substrate complex (ES). This point on the graph is designated V_{max}. Using this maximum velocity and Equation 10.1.7, Michaelis developed a set of mathematical expressions to calculate enzyme activity in terms of reaction speed from measurable laboratory data. The Michaelis constant (K_M) is defined as the substrate concentration at 1/2 the maximum velocity. This is shown in Figure 10.1.7. Using this constant and the fact that K_M can also be defined as:

$$K_M = \frac{K_{+1} + K_{+2}}{K_{-1}} = [S]_{1/2V_{max}}$$

Equation 10.1.8

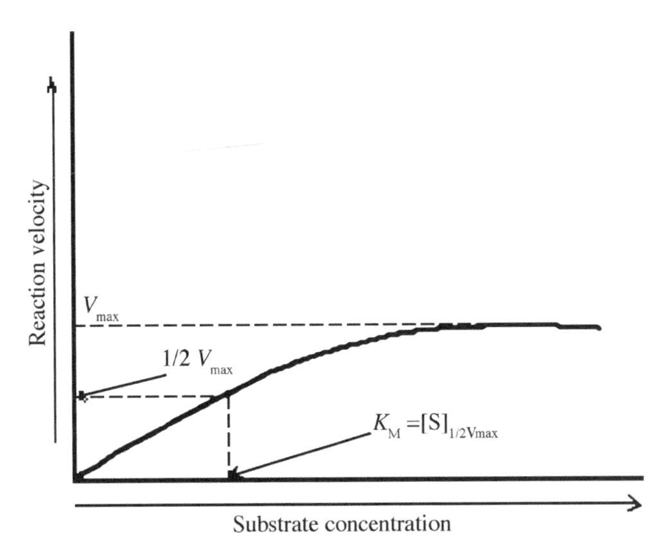

Figure 10.1.7 Effect of substrate concentration. The Michaelis constant (K_M) is defined as the substrate concentration at which half the maximum velocity (1/2 V_{max}) of the reaction occurs. (Worthington, 1993; published with permission from Worthington Biochemical.)

with K_{+1}, K_{-1}, and K_{+2} being the rate constants from Equation 10.1.7, Michaelis developed the following expression for the reaction velocity in terms of this constant and the substrate concentration:

$$V_1 = \frac{V_{max}[S]}{K_M + [S]}$$

Equation 10.1.9

where V_1 is the velocity at any time, [S] is the substrate concentration at this time, V_{max} is the highest velocity obtained under this set of experimental conditions (e.g., pH and temperature), and K_M is the Michaelis constant for the particular enzyme being investigated.

Michaelis constants have been determined for many of the commonly used enzymes. The size of K_M tells us several things about a particular enzyme.

1. A small K_M indicates that the enzyme requires only a small amount of substrate to become saturated. Hence, the maximum velocity is reached at relatively low substrate concentrations.
2. A large K_M indicates the need for high substrate concentrations to achieve maximum reaction velocity.
3. The substrate with the lowest K_M upon which the enzyme acts as a catalyst is frequently assumed to be enzyme's natural substrate, although this is not true for all enzymes.

Effects of inhibitors

Enzyme inhibitors are substances that alter the catalytic action of the enzyme and consequently slow down, or in some cases, stop catalysis. There are three common types of enzyme inhibition: competitive, noncompetitive, and substrate inhibition.

Most theories concerning inhibition mechanisms are based on the existence of the enzyme-substrate complex (ES). As mentioned earlier, the existence of temporary ES structures has been verified in the laboratory. Competitive inhibition occurs when the substrate and a substance resembling the substrate are both added to the enzyme. A theory called the "lock-key theory" of enzyme catalysts can be used to explain why inhibition occurs. The lock and key theory utilizes the concept of an

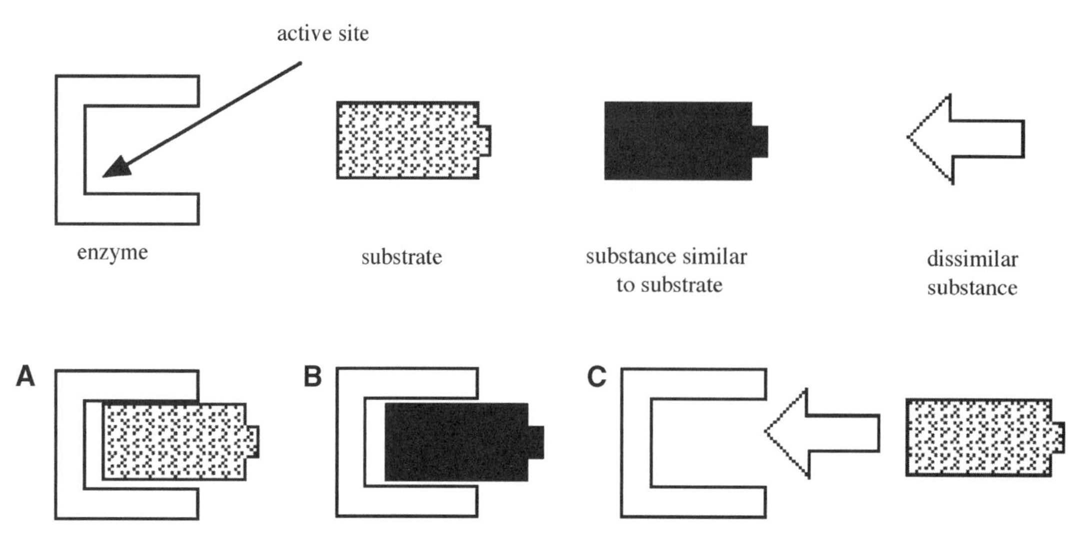

Figure 10.1.8 Lock and key theory of competitive inhibition. (**A**) The key fits the lock and turns it, thus opening the door for the reaction to proceed. (**B**) The key fits, but the lock will not turn. The reaction is slowed because the enzyme is occupied. (**C**) The enzyme rejects dissimilar substances and accepts the substrate. The reaction proceeds. (Worthington, 1993; published with permission from Worthington Biochemical.)

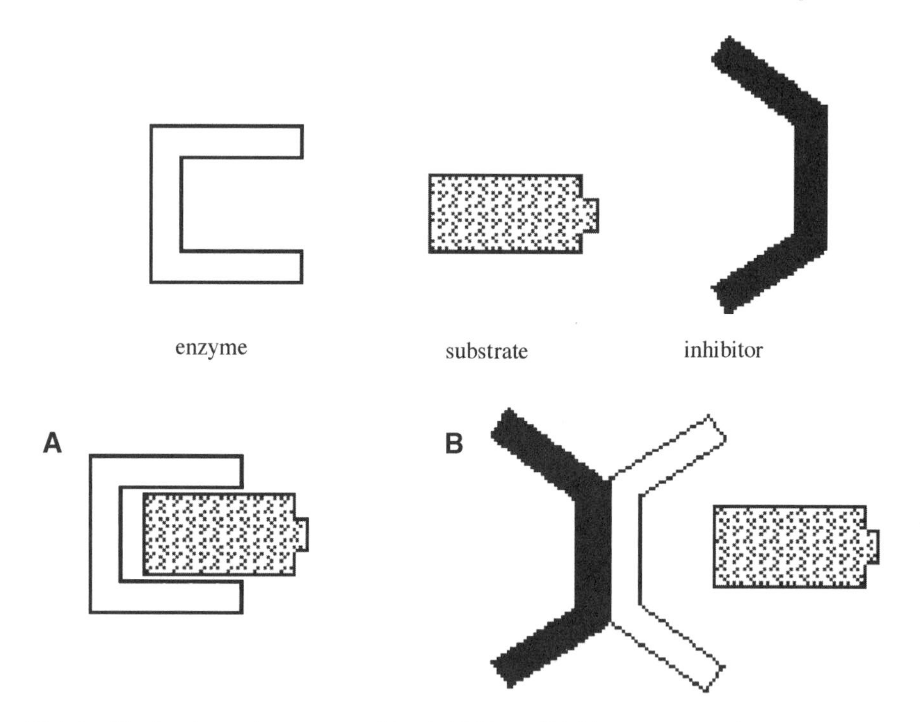

Figure 10.1.9 Mechanism of noncompetitive inhibition. (**A**) The substrate fits into the active site on the enzyme. (**B**) When noncompetitive inhibitors are added to the enzyme, they alter the enzyme in a way such that it cannot accept the substrate. (Worthington, 1993; published with permission from Worthington Biochemical.)

"active site." This concept holds that one particular portion of the enzyme surface has a strong affinity for the substrate. The substrate is held in such a way that its conversion to the reaction products is more favorable. If we consider the enzyme as the lock and the substrate the key (Fig. 10.1.8), the key is inserted into the lock, it is turned, the door opens, and the reaction proceeds. However, when an inhibitor resembling the substrate is present, it will compete with the substrate for the position in the enzyme lock. When the inhibitor wins, it gains the lock position but is unable to open the lock. Hence, the observed reaction is slowed down because some of the available enzyme sites are occupied by the inhibitor. If a dissimilar substance not fitting the active site is present, the enzyme rejects it, accepts the substrate, and the reaction proceeds normally.

Noncompetitive inhibitors are considered to be substances which when added to the enzyme alter it in a way such that it cannot accept the substrate (Fig. 10.1.9). In other cases, substrate inhibition will sometimes occur when excessive amounts of substrate are present. Figure 10.1.10 shows the reaction velocity decreasing after the maximum velocity has been reached. Additional amounts of substrate added to the reaction mixture after this point actually decrease the reaction rate. This is thought to be due to the fact that there are so many substrate molecules competing for the active sites on the enzyme surfaces that they block the sites (Fig. 10.1.11) and prevent any other substrate molecules from occupying them. This causes the reaction rate to decrease since not all of the enzyme present is accessible to the substrate.

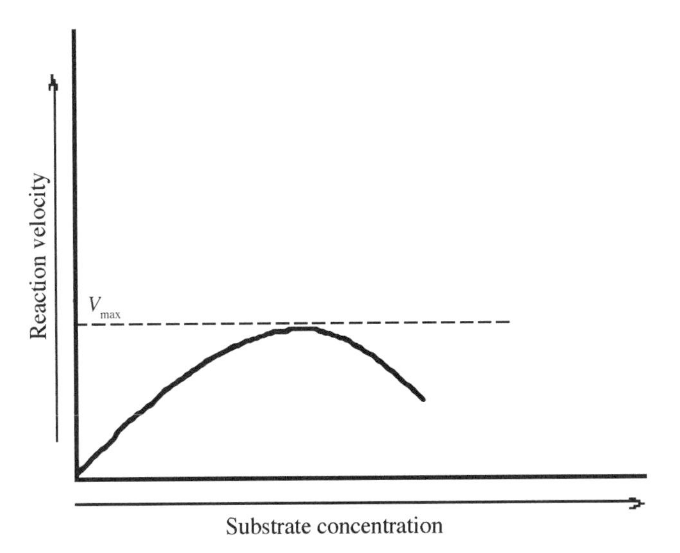

Figure 10.1.10 Inhibitory effect of substrate concentration. The reaction velocity decreases with increasing substrate concentration after the maximum velocity has been reached. (Worthington, 1993; published with permission from Worthington Biochemical.)

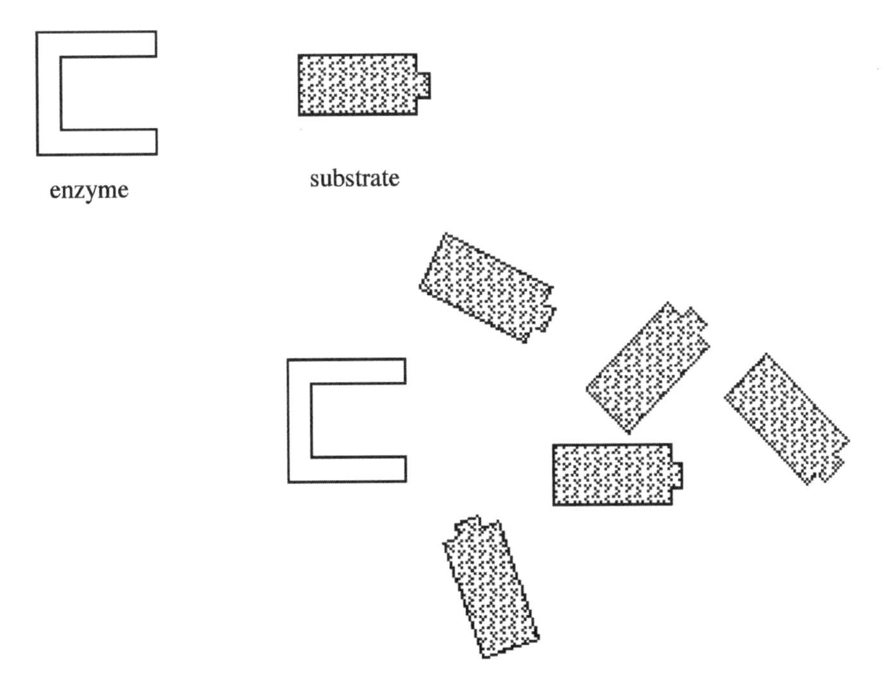

Figure 10.1.11 Mechanism of substrate inhibition. Excess substrate blocks site so that the enzyme goes unused and the reaction rate drops. (Worthington, 1993; published with permission from Worthington Biochemical.)

10

Temperature effects

Like most chemical reactions, the rate of an enzyme-catalyzed reaction increases as the temperature is raised to a maximum limit. A 10°C rise in temperature will increase the activity of most enzymes by 50% to 100%. Variations in reaction temperature as small as 1°C to 2°C may introduce changes of 10% to 20% in the results. In the case of enzymatic reactions, this is complicated by the fact that many enzymes are adversely affected by high temperatures. As shown in Figure 10.1.12, the reaction rate increases with temperature to a maximum level and then abruptly declines with further increase of temperature. Because most enzymes isolated from animal sources rapidly become denatured at temperatures above 40°C, enzyme activity determinations are usually carried out below that temperature. Over a period of time, enzymes will be deactivated at even moderate temperatures. Storage of enzymes at 5°C or below is generally the most suitable; however, some enzymes may lose their activity when frozen.

Effects of pH

Enzymes are affected by changes in pH. The most favorable pH (the point where the enzyme is most active) is known as the enzyme's optimum pH. This is graphically illustrated in Figure 10.1.13.

Extremely high or low pH values generally result in complete loss of activity for most enzymes. As with activity, for each enzyme there is also a region of pH optimal stability. As Table 10.1.2 demonstrates, the optimum pH value will vary greatly from one enzyme to another.

In addition to temperature and pH there are other factors, (e.g., ionic strength) that can affect the enzymatic reaction. Each of these physical and chemical parameters must be considered and optimized in order for an enzymatic reaction to be accurate and reproducible.

Storage and Handling

Enzymes are commonly stored at low temperatures to preserve their activity and integrity. Purified enzymes can quickly lose activity if stored under the wrong conditions. The shelf life or stability of an enzyme can vary considerably and is dependent on the nature of the protein, its concentration, and the diluent, pH, stabilizers, and storage temperature. Lyophilization (freeze-drying) of enzymes usually allows for long-term storage and shipment without special handling. Generalized guidelines

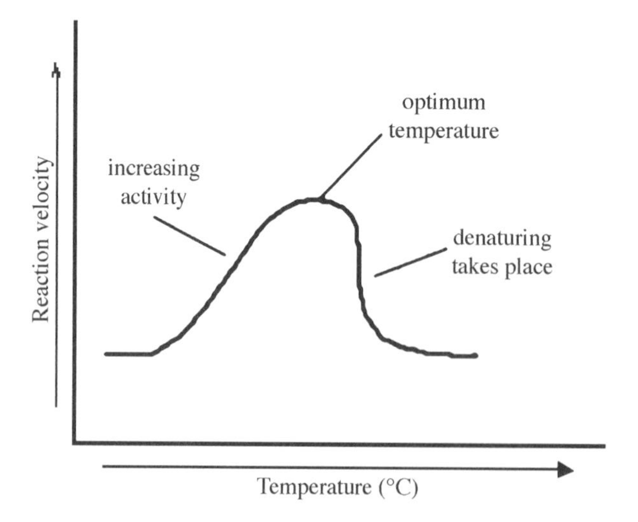

Figure 10.1.12 Effect of temperature on reaction rate. The reaction rate increases with temperature to a maximum level and then abruptly declines with further increase in temperature. (Worthington, 1993; published with permission from Worthington Biochemical.)

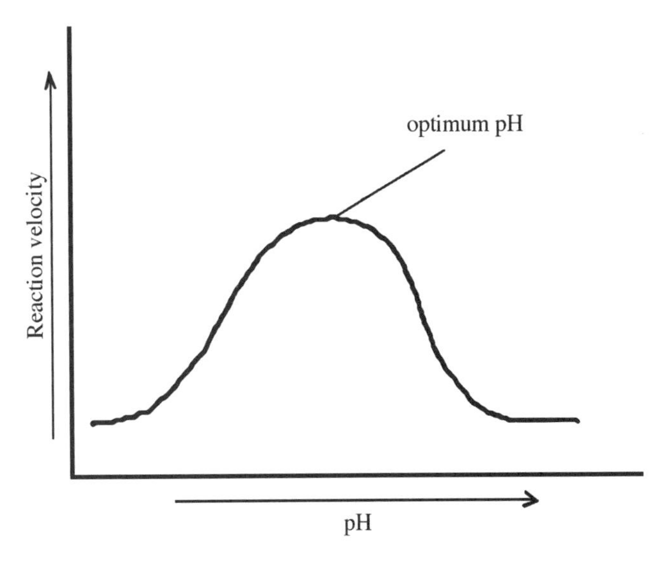

Figure 10.1.13 Effect of pH on reaction rate. The point where the enzyme is most active is known as the enzyme's optimum pH. (Worthington, 1993; published with permission from Worthington Biochemical.)

Table 10.1.2 pH for Optimum Activity

Enzyme	pH optimum
Lipase (pancreas)	8.0
Lipase (stomach)	4.0-5.0
Lipase (castor oil)	4.7
Pepsin	1.5-1.6
Trypsin	7.8-8.7
Urease	7.0
Invertase	4.5
Maltase	6.1-6.8
Amylase (pancreas)	6.7-7.0
Amylase (malt)	4.6-5.2
Catalase	7.0

for enzyme storage and handling follow, but each protein has distinctive characteristics that should be considered prior to dilution and storage.

All enzymes should be stored under cold conditions, because lower temperatures minimize loss of enzymatic activity. Frozen storage (−20°C or below) is usually recommended. Care must be exercised to avoid repeated freeze-thaw cycles that can lead to protein denaturation and loss of activity. When storing liquid enzyme preparations at −20°C, dispensing them into aliquots in single use vials is recommended. Alternatively enzymes can be stored at subzero temperatures in 50% glycerol, which prevents ice formation and subsequent loss of activity from denaturation caused by repeated freeze-thaw cycles. In fact, most enzymes used for molecular biology applications are supplied in 50% glycerol so they can be stored in the freezer and samples withdrawn without extended thawing. It is very important that only manual defrost freezers be used to store enzymes. The internal temperature of frost-free freezers fluctuates above and below −20°C in order to prevent the formation of frost. Elevated temperatures and continuous freeze-thaw cycles caused by the temperature fluctuations can result in loss of enzymatic activity.

Refrigerated storage at 2°C to 8°C is recommended only for the more stable lyophilized enzymes, enzymes supplied in ammonium sulfate, crystalline suspensions, or highly purified liquid preparations free of proteases and microbes.

If the sterility of any liquid preparation is in question, it should be passed through a low protein binding 0.22-μm membrane filter for prolonged storage at 2°C to 8°C (more than 3 days), in order to prevent growth of any microbial contamination.

It is best to prepare initial stock solutions as concentrated as possible because dilute solutions are more prone to lose activity due to protein binding to storage containers. Proteins can bind to glass, polystyrene, and quartz, so low–protein–binding vials of polyethylene or polypropylene are recommended when storing enzymes in small aliquots. Innocuous proteins like BSA and gelatin are sometimes added to dilute enzyme solutions so the vessel absorbs the added protein instead of the enzyme.

The optimum storage conditions for each enzyme need to be determined empirically because they vary widely based on the specific enzyme and its source. The main conditions to keep in mind when choosing the diluent are pH, inhibitors, and the presence of cofactors and/or stabilizers. Common stabilizers are reagents such as glycerol, which facilitates storage at freezing temperatures, and BSA or gelatin, which minimizes loss of enzymatic activity caused by adsorption to storage vessels or proteolytic degradation. Sometimes the optimum storage conditions are not the same as the optimum activity conditions. For example, trypsin has an optimum pH near 8.0 for activity yet is best stored at low pH where it is stable but inactive. Calcium ions retard trypsin autolysis, so calcium is usually added to buffers when using trypsin at neutral pH ranges. Most commercial suppliers provide product data sheets that list optimal storage and reaction conditions for individual enzymes. Additional information can also be obtained from a wide variety of reference materials, including Internet resources, product catalogs, and reference handbooks (e.g., Worthington, 1993).

Commonly Encountered Forms

Suspensions

Purification procedures for many enzymes include the use of salts at various ionic strengths that change the solubility of the protein (referred to as "salting-out" and "salting-in"). Ammonium sulfate is commonly used to precipitate proteins without irreversibly denaturing them. Suspensions of enzymes in ammonium sulfate are also generally stable. Since ammonium sulfate can interfere with certain procedures it should be removed by dialysis, centrifugation of enzyme crystals, or desalting by gel filtration chromatography. Enzymes in salt suspension form should generally not be frozen. Suspensions should be thoroughly mixed to ensure homogeneity prior to sampling from the container, and proper handling techniques should be used to avoid contamination.

Powders

Enzymes are also commonly supplied as lyophilized or "freeze-dried" powders. Lyophilization is the process of removing water through sublimation (freezing liquid to solid and removing water vapor under vacuum). Lyophilization causes greater stability at ambient temperatures, and many purified enzymes are offered as lyophilized powders to provide convenience, versatility, and ease in handling. Lyophilized proteins tend be very hygroscopic (water absorbing), so they should not be opened in humid areas while cold. Vials should be stored in desiccators and brought to room temperature before opening. Enzymes can also be immobilized, encapsulated, or bound to various support mediums.

Solutions

When working at the lab bench it is imperative to keep enzyme solutions cold for the entire time that they are removed from storage, until they are added to the reaction mixture. Most enzymes will lose some or all of their activity if left sitting on the bench top at room temperature even for a short time. The reasons for this vary from enzyme to enzyme. In the case of proteases (protein degrading enzymes, e.g., trypsin, pepsin) enzymatic activity is suppressed at lower temperatures. At elevated temperatures activity increases and protease molecules degrade themselves and neighboring protease molecules. Most proteolytic enzymes will lose their activity due to this autodegradation after a relatively short time at room temperature. As a general rule, all enzyme solutions should be kept on ice or in a portable cold box when working at the lab bench.

EXAMPLE OF SETTING UP AN ENZYMATIC REACTION: RESTRICTION ENZYMES

Up to this point, this unit has discussed principles applicable to enzymes in general. The purpose of the following discussion is to apply these principles to a specific class of enzymes in order to demonstrate how to correctly use these enzymes and to illustrate the factors that need to be considered in order for the enzymes to function properly. Restriction enzymes have been chosen for this purpose because they provide an excellent example of how reaction conditions can affect enzyme activity and substrate specificity.

Restriction enzymes are endodeoxyribonucleases that cleave both strands of double-stranded DNA within or near a specific recognition sequence. They are produced in prokaryotes and, to date, thousands of different restriction enzymes have been identified. Table 10.1.3, at the end of this unit, provides a list of common restriction enzymes as well as their recognition sequences. These enzymes are believed to serve a protective role in nature by attacking DNA from invading microorganisms, particularly bacteriophages. The DNA in the host cell producing the restriction enzyme is protected from digestion because the cell also contains modifying enzymes that methylate its own DNA, rendering it unrecognizable to the restriction enzyme. Each restriction enzyme has an associated modifying system. For some restriction enzymes, the restriction and modifying activities are contained on separate independent polypeptides, while for others the activities are contained in separate domains of a single polypeptide or in separate subunits of oligomeric proteins.

Restriction enzymes can be classified into one of three different types based on subunit composition, cofactor requirements, cleavage position in relation to the recognition sequence, and the relationship between restriction and modifying activities. It is not within the scope of this unit to provide an exhaustive review of all three types of restriction enzymes. Excellent reviews can be found in *Nucleases: Molecular Biology and Applications* (Mishra, 2002) and *Enzymes of Molecular Biology* (Burrell, 1993; also see Key References).

Type II endonucleases, in contrast to Types I and III enzymes, cleave DNA at defined locations within or very close to the recognition sequence and generate discrete fragments with defined ends. In other words, the sequence of the last few bases on the end of each fragment is known. This property makes these enzymes extremely useful for DNA analysis and gene cloning. For this reason the remainder of this discussion on restriction enzymes will pertain primarily to Type II enzymes.

Specificity of Restriction Endonucleases

Most restriction enzymes recognize a tetranucleotide or hexanucleotide palindromic sequence, meaning the 5′ to 3′ sequence of the sense strand is the same as the 5′ to 3′ sequence of the complimentary antisense strand; however, it should be noted that there are many exceptions to this rule. Some enzymes recognize sequences with one or more bases within the palindromic sequence, while others recognize nonpalindromic sequences altogether. Cleavage of DNA by restriction enzymes could occur at the same point in both strands and generate "flush" or "blunt" ends or can occur at staggered positions and generate 3′ or 5′ overhanging "sticky" or "cohesive" ends. Different restriction enzymes that recognize the same sequence are referred to as isoschizomers. Because recognition sites can differ from cutting sites, isoschizomers may or may not generate identical ends. In contrast, other restriction enzymes, regardless of whether they recognize identical or different sequences, may generate identical ends. Such endonucleases are said to produce "compatible" ends.

Reaction Conditions for Restriction Endonucleases

Optimal enzymatic activity and cleavage of the correct sequence by restriction enzymes are dependent upon proper reaction conditions with respect to pH, ionic strength, cofactors, stabilizers, and reaction time and temperature. Almost all restriction enzymes require the presence of Mg^{2+} ions and reducing reagents such as 2-mercaptoethanol or dithiothreitol (DTT). Restriction enzymes can become inactivated upon oxidation. DTT and mercaptoethanol help maintain restriction enzymes in a reduced, active state. Many restriction enzymes also require the presence of some type of stabilizer molecule. Commonly used stabilizers include innocuous proteins such as BSA and nonionic detergents such as Triton X-100. They mimic the actions of naturally occurring macromolecules by facilitating proper folding of an enzyme in a conformation favorable for optimum activity. Since the level of a stabilizer in a reaction exceeds the level of restriction enzyme, protein stabilizers also minimize the effect contaminant proteases can have on the restriction enzyme.

Most restriction enzymes work best in a pH range of 7 to 8 in a 10 to 100 mM Tris-based buffering system with an ionic strength ranging from 0 to 100 mM, provided by sodium chloride or potassium acetate. Most also work best at 37°C, although a significant number prefer lower (25°C) or higher (60°C to 75°C) temperatures.

Optimal reaction conditions can vary among different restriction enzymes. Fortunately, restriction enzymes used in the laboratory are almost always obtained from a commercial source. They are usually provided as a ready-to-use liquid preparation in optimal buffer containing 50% glycerol. Most suppliers also provide a 10× concentrated optimal reaction buffer for each enzyme; therefore, users should consult literature from the commercial source of the restriction enzyme to determine optimal reaction conditions.

Setting up a restriction enzyme reaction is relatively simple, especially when using commercially available enzymes. A 10× concentrated reaction buffer, DNA, and enzyme are added to an appropriately sized tube and incubated at the desired temperature for a period of time. Any size microcentrifuge tube will suffice. The incubation time and temperature will depend on the particular enzyme used and specific application.

The 10× concentrated reaction buffer provides an appropriate buffering agent to give the correct pH for optimal activity. It also provides DTT, Mg^{2+}, and enough salt to provide the proper ionic strength needed for optimal enzymatic activity. Concentrated reaction buffers for some enzymes that require stabilizers also contain BSA or Triton X-100. However, BSA may be lacking in certain reaction buffers of some enzymes that require it. In such cases a concentrated BSA solution is also provided and is added to the reaction separately.

The amount of DNA digested in a reaction could vary from $<$100 ng to $>$10 µg, depending on the specific application. Best results are obtained with relatively pure DNA preparations although preparations containing low levels of contaminants can successfully be digested. Common contaminants include protein, RNA, phenol, alcohol, and agarose. The level of contaminants in DNA preparations can vary and will depend on the source of DNA and method of purification.

The restriction enzyme should be the last component added to the reaction. Before it is added, the total volume of all the reaction components combined should be calculated. The balance of the desired reaction volume should be made up with deionized or distilled water.

Restriction enzymes are almost always measured in terms of units. One unit of restriction enzyme activity is the amount of enzyme needed to completely digest 1 µg of DNA in 1 hr in a 50-µl reaction. Using 3 to 10 units of restriction enzyme per µg of DNA is usually sufficient to completely digest DNA in most preparations.

After all the components have been added, it is very important to mix the reaction properly. Restriction enzymes are usually in 50% glycerol and will settle to the bottom of the tube. The reaction should be mixed by gently pipetting up and down several times. Vortexing or flicking the tube should be avoided. The tube should be centrifuged briefly if any liquid is on the walls of the microcentrifuge tube before incubation.

The description above provides general guidelines for setting up a restriction enzyme reaction. Several parameters such as the amount of enzyme used, incubation time, and reaction volume may be altered for specific applications without adversely affecting enzyme activity as long as optimal buffer conditions are maintained. The following components are typical for a restriction enzyme reaction:

Water
10× buffer (one-tenth volume of total reaction volume)
10× BSA (see discussion above)
DNA (0.1 to 10 μg)
Restriction enzyme (3 to 10 U/μg).

A typical reaction volume is 20 to 100 μl. Calculate the total volume of all reaction components and make up the balance of the volume with deionized water. The volume of the restriction enzyme should not be >5% of the total reaction volume.

Star Activity

Restriction enzymes, like all enzymes, work to varying degrees under suboptimal conditions, usually exhibiting lower activities. Using restriction enzymes under conditions dramatically different than optimal conditions can pose another problem in addition to decreased enzymatic activity. Under certain suboptimal conditions, restriction enzymes may cleave DNA at sites similar but not identical to their defined cleavage sequences. This phenomenon is referred to as "star" activity and can generate more fragments and fragments of different length than those obtained under optimal conditions. Obviously, this phenomenon would confound any cloning or DNA analysis experiment. Factors that can induce star activity include:

Glycerol concentration too high
Enzyme level too high
pH too high
Presence of other divalent cation in place of Mg^{2+}
Salt concentration (ionic strength) too low
Incubation time too long
Presence of organic solvents such as alcohol, DMSO, or ethylene glycol.

A common mistake made by novice users of restriction enzymes is believing that "more is better." Users may feel that using more enzyme or incubating the reaction for periods longer than the recommended time interval will give better results. In reality such approaches have a greater likelihood of causing cleavage at incorrect sequences and confusing results rather than improving them. This "more is better" approach is probably the most common cause of star activity. A general rule is to use 3 to 10 units of restriction enzyme per microgram of substrate DNA when digesting for one hr.

Using greater amounts of enzyme can also cause star activity by introducing too much glycerol into the reaction. Just about all restriction enzymes are supplied in a buffer containing 50% glycerol. Restriction enzymes are heat labile and need to be stored in subzero temperatures to preserve activity. The presence of glycerol prevents the solution from freezing and prevents denaturation and inactivation of the enzyme caused by repeated freezing and thawing. The presence of glycerol in a reaction does not present problems as long as its concentration does not exceed 5% (v/v) of the total reaction volume. Therefore, in a 50-μl reaction, no more than 5 μl of restriction enzyme should be added. Keep in mind that for double digests, the total amount of glycerol contributed by both enzymes has to be considered.

Star activity can also occur in double digests because of incorrect ionic strength. When low-salt- and high-salt-requiring enzymes are used together with a low-salt buffer, the high-salt-requiring enzyme can make incorrect cleavages, resulting in star activity. In such a case, a sequential digest is preferable to a simultaneous digest (see below).

Other factors listed above that cause star activity are usually introduced into the reaction by the DNA sample itself. In such cases it is desirable to purify the DNA prior to treatment with restriction enzymes.

Reactions Involving More Than One Restriction Enzyme

Certain applications require digestion of a DNA substrate with two different restriction enzymes. Unidirectional cloning of a cDNA insert into a vector is one such example. Digestion of the substrate could either be done simultaneously (both enzymes together in a single reaction) or sequentially (two separate reactions, one for each enzyme). Simultaneous digestion is often the method of choice since it is faster and easier than sequential digestions; however, this approach can only be used if reaction conditions can be achieved that are suitable for both enzymes. Simultaneous digestion should be reserved for enzymes with common optimal pH and temperature requirements and similar buffer requirements. Commercial suppliers provide information on conditions in which different restriction enzymes can be used in combination.

Digesting DNA with enzymes with dramatically different optimal reaction conditions should be done in a sequential fashion, which can be done in one of two ways: (1) The reaction involving the enzyme requiring less stringent conditions is done first. The reaction conditions are then adjusted to satisfy the more stringent conditions of the second enzyme. The second enzyme is then added in order to complete the digest. (2) An alternative sequential digestion approach involves digestion with the first enzyme, purification of the DNA by phenol/chloroform extraction and alcohol precipitation, suspension of the DNA in buffer compatible with the second enzyme, and finally, digestion with the second enzyme. Gel purification of the DNA after alcohol precipitation is sometimes included in this approach. This last approach is the most time consuming but offers the best chance of success. Often it is only used as a last resort after simultaneous digestions and the less detailed sequential digestions fail.

Reaction Termination

After the DNA is digested, the restriction enzyme or enzymes often must be inactivated or removed so as to not interfere with subsequent enzymatic manipulations of the DNA. Since restriction enzymes require the presence of Mg^{2+} for activity, the simplest method of inactivation is to add a ten-fold excess of the chelating agent, EDTA. However, this may inhibit subsequent enzymes acting on the DNA (e.g., polymerases and ligases) that require divalent cations for activity. Alternatively, heating to 65°C or 80°C for 20 min will inactivate most, but not all, restriction enzymes. Heat induces conformational changes in susceptible proteins leading to denaturation and inactivation. The commercial enzyme supplier usually provides heat inactivation information for a given restriction enzyme. Both inactivation methods are quick and easy; however, both also leave behind protein and other contaminants that can interfere with subsequent manipulations to the DNA such as sequence analysis. For these applications it may be necessary to remove the restriction enzyme by purifying the DNA. Again this can be done in a number of ways including gel purification, capture on spin columns, or phenol/chloroform extraction followed by alcohol precipitation (Sambrook and Russell, 2001).

In summary, DNA can be easily and successfully digested with restriction enzymes if two simple rules are followed: (1) Make sure the DNA is as clean as possible, and (2) use conditions as close to the defined optimal conditions as possible. Most failed reactions show no cleavage at all or show the presence of fragments of different number or size than expected. Problems can occur even if highly purified DNA is digested with a restriction enzyme under optimal conditions. Such failures are most attributable to improper storage and handling of restriction enzymes. Restriction enzymes are stored at subzero temperatures to maintain activity, and activity decreases with time and increased temperature. Often a failed restriction digest is due to an enzyme that has lost its enzymatic activity; therefore, the amount of time restriction enzymes are out of the freezer should be kept to a minimum. Remove the enzyme from storage immediately before it is to be used and keep it on ice at all times.

Failures showing an erroneous banding pattern can be attributed to contaminated enzymes as well as to the star activity discussed earlier. Commercially available restriction enzymes are usually so concentrated that 1 µl or less is sufficient for most applications. Using dirty pipet tips will cause cross contamination between different enzymes or different samples in an experiment. Change tips for every pipetting step when preparing samples. Never use a pipet tip that was previously used to dispense anything or to remove restriction enzymes from stock vials.

Hopefully this discussion will help even first-time users of restriction enzymes successfully digest DNA samples. Use of the information presented here should save time and money and help with the interpretation of restriction digest results.

LITERATURE CITED

Bloch, K.D. and Grossmann, B. 1995. Digestion of DNA with restriction endonucleases. *Curr. Protoc. Mol. Biol.* 31:3.1.1-3.1.21.

Burrell, M.M. (ed.) 1993. Enzymes of Molecular Biology. *In* Methods in Molecular Biology, Vol. 16. Humana Press, Totowa, N.J.

Harlow, E. and Lane, D. 1998. Antibodies: A Laboratory Manual. Cold Spring Harbor Laboratory Press, Cold Spring Harbor, N.Y.

Holum, J. 1968. Elements of General and Biological Chemistry, 2nd ed. pp. 377. John Wiley & Sons, New York.

Mishra, N.C. 2002. Nucleases: Molecular Biology and Applications. John Wiley & Sons, Hoboken, NJ.

Sambrook, J.and Russell, D. 2001. Molecular Cloning A Laboratory Manual. Cold Spring Harbor Laboratory Press, Cold Spring Harbor, N.Y.

Stern, K. 1936. On the mechanism of enzyme action. *J. Biol. Chem.* 114:473-494.

Worthington, V. (ed.) 1993. Worthington Enzyme Manual. Worthington Biochemical, Lakewood, N.J.

KEY REFERENCES

Barrett, A.J., Rawlings, N.D., and Woessner, J.F. 1998. Handbook of Proteolytic Enzymes. Academic Press, London.
Detailed information on proteases.

Burrell, M.M. (ed.) 1993. See above.
Provides a more extensive review of restriction enzymes than necessary for this chapter.

Coligan, J.E., Dunn, B.M., Speicher, D.W., and Wingfield, P.T. (eds.) 2003. Short Protocols in Protein Science. John Wiley & Sons, Hoboken, N.J.
Includes protocols for common enzyme techniques.

Copeland, R.A. 2000. Enzymes: A Practical Introduction to Structure Mechanism, and Data Analysis. Wiley-VCH, New York.
A comprehensive review of enzymology.

Mishra, N.C. 2000. See above.
Provides extensive reviews of restriction enzymes.

Whitford, D. 2005. Proteins Structure and Function. John Wiley & Sons, London.
An excellent reference for protein basics.

Methods In Enzymology series. Academic Press.
Highly respected series in life sciences publishing topics of current interest.

INTERNET RESOURCES

http://www.expasy.ch/enzyme

Web site containing information relative to nomenclature of enzymes and links to individual enzymes. A description of the Web site can be found in Bairoch, A. 2000. The ENZYME database in 2000. Nucleic Acids Res 28:304-305.

http://www.chem.qmul.ac.uk/iubmb/enzyme

Web site for enzyme nomenclature and links to details about specific enzymes. Contains information published in the book Enzyme Nomenclature (Academic Press), as well as in the European Journal of Biochemistry between 1994 and 1999 (details on the Web site).

http://www.worthington-biochem.com

Web site containing characteristics and references for numerous enzymes.

Table 10.1.3 starts on the next page.

Table 10.1.3 Recognition Sequences and Reaction Conditions of Commercially Available Restriction Endonucleases[a]

Name	Site[b]	Salt[c]	Rxn. temp.[d]	Inact. temp.[d]	Isoschizomers	Comments[e]
Restriction endonucleases						
*Aat*I	AGG↓CCT	M	37	75	*Eco*147I, *Stu*I	*dcm*
*Aat*II	GACGT↓C	M(K)	37	60		
*Acc*I	GT↓(A/C)(G/T)AC	—[f]	37	90		Increased activity at 55°C
*Acc*II	CG↓CG	L	37		*Bsh*1236I, *Bst*UI, *Mvn*I, *Tha*I	
*Acc*III	T↓CCGGA	H	60		*Bse*AI, *Bsi*MI, *Bsp*EI, *Kpn*2I, *Mro*I	
*Acc*65I	G↓GTACC	H(K)	37	65	*Asp*718I, *Kpn*I	
*Acc*B7I	CCAN$_4$↓NTGG	M	37		*Pfl*MI, *Van*91I	
*Aci*I	C↓CGC GGC↑G	H	37	65		
*Acs*I	Pu↓AATTPy	—[f]	50		*Apo*I	
*Acy*I	GPu↓CGPyC	H	50	65	*Bbi*II, *Bsa*HI, *Hin*1I, *Hsp*92I	
*Afa*I	GT↓AC	M	37	70	*Csp*6I, *Rsa*I	
*Afl*II	C↓TTAAG	M	37	65	*Bfr*I, *Bst*98I	
*Afl*III	A↓CPuPyGT	H	37	100		
*Age*I	A↓CCGGT	L	25		*Pin*AI	
*Ahd*I	GACN$_3$↓N$_2$GTC	—[f]	37	65	*Asp*EI, *Eam*1105I, *Ecl*HKI	
*Alu*I	AG↓CT	M	37	70		
*Alw*I	GGATCN$_4$↓ CCTAGN$_5$↑	—[f]	37	65		*dam*
*Alw*21I	G(A/T)GC(A/T)↓C	L	37	65	*Asp*HI, *Bsi*HKAI, *Hgi*AI	
*Alw*26I	GTCTCN↓ CAGAGN$_5$↑	L	37	65	*Bsm*AI	
*Alw*44I	G↓TGCAC	M	37	65	*Apa*LI	
*Alw*NI	CAGN$_3$↓CTG	—[f]	37	65		
*Aoc*I	CC↓TNAGG	L	37		*Bsu*36I, *Cvn*I, *Eco*81I, *Mst*II, *Sau*I	
*Aor*51HI	AGC↓GCT	M	37		*Eco*47III	
*Apa*I	GGGCC↓C	L	37	65	*Bsp*120I	*dcm*
*Apa*LI	G↓TGCAC	L	37		*Alw*44I	
*Apo*I	Pu↓AATTPy	H	50		*Acs*I	
*Asc*I	GG↓CGCGCC	—[f]	37	65		
*Ase*I	AT↓TAAT	H(K)	37	65	*Asn*I, *Vsp*I	Star activity
*Asn*I	AT↓TAAT	H	37		*Ase*I, *Vsp*I	
*Asp*I	GACN↓N$_2$GTC	H	37		*Tth*111I	

continued

Table 10.1.3 Recognition Sequences and Reaction Conditions of Commercially Available Restriction Endonucleases[a], *continued*

Name	Site[b]	Salt[c]	Rxn. temp.[d]	Inact. temp.[d]	Isoschizomers	Comments[e]
*Asp*700I	$GAAN_2{\downarrow}N_2TTC$	H	37		*Xmn*I	
*Asp*718I	G↓GTACC	H	37		*Acc*65I, *Kpn*I	
*Asp*EI	$GACN_3{\downarrow}N_2GTC$	L	37	65	*Ahd*I, *Eam*1105I, *Ecl*HKI	
*Asp*HI	G(A/T)GC(A/T)↓C	H	37		*Alw*21I, *Bsi*HKAI, *Hgi*AI	
*Asu*I	G↓GNCC	M	37		*Bsi*ZI, *Bsu*54I, *Cfr*13I, *Sau*96I	
*Ava*I	C↓PyCGPuG	M	37	100	*Bco*I, *Nsp*III	Increased activity at 45°C
*Ava*II	G↓G(A/T)CC	M	37	65	*Eco*47I, *Sin*I	*dcm*
*Avi*II	TGC↓GCA	H	37		*Fdi*II, *Fsp*I, *Mst*I	
*Avr*II	C↓CTAGG	M	37		*Bln*I	
*Bal*I	TGG↓CCA	L	25	65	*Mlu*NI, *Msc*I	*dcm* active >18 hr
*Bam*HI	G↓GATCC	H	37	60	*Bst*I	Star activity
*Ban*I	G↓GPyPuCC	L	50	70		
*Ban*II	GPuGCPy↓C	M	37	60		
*Ban*III	AT↓CGAT	M	37	70	*Bsi*XI, *Bsp*106I, *Bsp*DI, *Cla*I	
*Bbe*I	GGCGC↓C	L	37	65	*Ehe*I, *Kas*I, *Nar*I	
*Bbi*II	GPu↓CGPyC	L	37		*Acy*I, *Bsa*HI, *Hin*1I, *Hsp*92I	
*Bbr*PI	CAC↓GTG	H	37		*Eco*72I, *Pma*CI, *Pml*I	
*Bbs*I	$GAAGACN_2{\downarrow}$ $CTTCTGN_6{\uparrow}$	L	37	65	*Bpu*AI, *Bsc*91I	
*Bbu*I	GCATG↓C	L	37	65	*Sph*I	
*Bbv*I	$GCAGCN_8{\downarrow}$ $CGTCGN_{12}{\uparrow}$	L	37	65	*Bst*71I	
*Bca*77I	(A/T)↓CCGG(A/T)	—[f]	—[g]		*Bsa*WI	
*Bcg*I	${\downarrow}N_{10}GCAN_6TCGN_{12}{\downarrow}$ ${\uparrow}N_{12}GTN_6AGCN_{10}{\uparrow}$	H	37	65		Requires SAM
*Bcl*I	T↓GATCA	H(K)	50	100	*Bsi*QI	*dam*
*Bcn*I	CC↓(G/C)GG	H	37		*Nci*I	
*Bco*I	C↓PyCGPuG	H	65		*Ava*I, *Nsp*III	
*Bfa*I	C↓TAG	—[f]	37		*Mae*I	
*Bfr*I	C↓TTAAG	M	37		*Afl*II, *Bst*98I	
*Bgl*I	$GCCN_4{\downarrow}NGGC$	H	37	65		
*Bgl*II	A↓GATCT	H	37	100		
*Bln*I	C↓CTAGG	H	37		*Avr*II	

continued

Table 10.1.3 Recognition Sequences and Reaction Conditions of Commercially Available Restriction Endonucleases[a], *continued*

Name	Site[b]	Salt[c]	Rxn. temp.[d]	Inact. temp.[d]	Isoschizomers	Comments[e]
*Blp*I	GC↓TNAGC	—[f]	37		*Bpu*1102I, *Cel*II, *Esp*I	
*Bmy*I	G(A/G/T)GC(A/C/T)↓C	—[f]	37		*Bsp*1286I	
*Bpm*I	CTGGAGN$_{16}$↓ GACCTCN$_{14}$↑	H	37	65	*Gsu*I	
*Bpu*1102I	GC↓TNAGC	—[f]	37		*Blp*I, *Cel*II, *Esp*I	
*Bpu*AI	GAAGACN$_{2}$↓ CTTCTGN$_{6}$↑	H	37		*Bbs*I, *Bsc*91I	
*Bsa*I	GGTCTCN↓ CCAGAGN$_{5}$↑	—[f]	60	65		10% activity at 37°C
*Bsa*AI	PyAC↓GTPu	H	37			
*Bsa*BI	GATN$_{2}$↓N$_{2}$ATC	M	60		*Bsi*BI, *Bsr*BrI, *Mam*I	*dam* 20% activity at 37°C
*Bsa*HI	GPu↓CGPyC	—[f]	37	65	*Acy*I, *Bbi*II, *Hin*1I, *Hsp*92I	Increased activity at 37°C
*Bsa*JI	C↓CN$_{2}$GG	M	60			20% activity at 37°C
*Bsa*MI	GAATGCN↓ CTTAC↑GN	H	65		*Bsm* I	
*Bsa*OI	CGPuPy↓CG	M	50		*Bsi*EI, *Mcr*I	
*Bsa*WI	(A/T)↓CCGG(A/T)	M	60		*Bca*77I	
*Bsc*91I	GAAGACN$_{2}$↓ CTTCTGN$_{6}$↑	H	37		*Bbs*I, *Bpu*AI	
*BscB*I	GGN↓NCC	—[f]	55		*Nla*IV	
*Bse*AI	T↓CCGGA	—[f]	55		*Acc*III, *Bsi*MI, *Bsp*EI, *Kpn*2I, *Mro*I	
*Bse*RI	GAGGAGN$_{10}$↓ CTCCTCN$_{8}$↑	L	37	65		
*Bsg*I	GTGCAGN$_{16}$↓ CACGTCN$_{14}$↑	—[f]	37	65		
*Bsh*I	GG↓CC		37		*Bsp*KI, *Hae*III, *Pal*I	
*Bsh*1236I	CG↓CG	H	37	65	*Acc*II, *Bst*UI, *Mvn*I, *Tha*I	
*Bsi*BI	GATN$_{2}$↓N$_{2}$ATC		55		*Bsa*BI, *Bsr*BrI, *Mam*I	
*Bsi*CI	TT↓CGAA	L	65		*Bst*BI, *Csp*45I, *Nsp*V, *Nsp*7524V, *Sfu*I	
*Bsi*EI	CGPuPy↓CG	M	60		*Bsa*OI, *Mcr*I	
*Bsi*HKAI	G(A/T)GC(A/T)↓C	H	60		*Alw*21I, *Asp*HI, *Hgi*AI	
*Bsi*LI	CC↓(A/T)GG	—[f]	60		*Bst*GI, *Bst*NI, *Bst*OI, *Eco*RII, *Mva*I	
*Bsi*MI	T↓CCGGA	—[f]	60		*Acc*III, *Bse*AI, *Bsp*EI, *Kpn*2I, *Mro*I	

continued

Table 10.1.3 Recognition Sequences and Reaction Conditions of Commercially Available Restriction Endonucleases[a], *continued*

Name	Site[b]	Salt[c]	Rxn. temp.[d]	Inact. temp.[d]	Isoschizomers	Comments[e]
*Bsi*QI	T↓GATCA	—[f]	55		*Bcl*I	
*Bsi*WI	C↓GTACG	H	60		*Spl*I, *Sun*I	50% activity at 37°C
*Bsi*XI	AT↓CGAT	—[f]	65		*Ban*III, *Bsp*106I, *Bsp*DI, *Cla*I	
*Bsi*YI	$CCN_5{\downarrow}N_2GG$	M	55		*Bsl*I	
*Bsi*ZI	G↓GNCC	—[f]	65		*Asu*I, *Bsu*54I, *Cfr*13I, *Sau96*I	
*Bsl*I	$CCN_5{\downarrow}N_2GG$	H	55		*Bsi*YI	30% activity at 37°C
*Bsm*I	GAATGCN↓ CTTAC↑GN	M	65	90	*Bsa* MI	20% activity at 37°C
*Bsm*AI	GTCTCN↓ $CAGAGN_5{\uparrow}$	H	55		*Alw* 26I	10% activity at 37°C, 30% activity at 65°C
*Bsm*BI	CGTCTCN↓ $GCAGAGN_5{\uparrow}$	H	55			
*Bsm*FI	$GGGACN_{10}{\downarrow}$ $CCCTGN_{14}{\uparrow}$	L	37			
*Bso*FI	GC↓NGC	L	55	65	*Fnu*4HI	
*Bsp*106I	AT↓CGAT	—[f]	37		*Ban*III, *Bsi*XI, *Bsp*DI, *Cla*I	
*Bsp*120I	G↓GGCCC	L	37		*Apa*I	
*Bsp*1286I	G(A/G/T)GC(A/C/T)↓C	L	30		*Bmy*I	*dam*
*Bsp*1407I	T↓GTACA	—[f]	50	70	*Ssp*BI	
*Bsp*90I	GTA↓TAC	—[f]	45		*Bst*1107I	
*Bsp*CI	CGAT↓CG	—[f]	37		*Pvu*I, *Xor*II	
*Bsp*DI	AT↓CGAT	—[f]	37	65	*Ban*III, *Bsi*XI, *Bsp*106I, *Cla*I	*dam*
*Bsp*EI	T↓CCGGA	H	37		*Acc*III, *Bse*AI, *Bsi*MI, *Kpn*2I, *Mro*I	
*Bsp*HI	T↓CATGA	H(K)	37	65	*Rca*I	*dam*
*Bsp*KI	GG↓CC	—[f]	40		*Bsh*I, *Hae*III, *Pal*I	
*Bsp*MI	$ACCTGCN_4{\downarrow}$ $TGGACGN_8{\uparrow}$	M	37			
*Bsp*WI	$GCN_5{\downarrow}N_2GC$	—[f]	42		*Mwo*I	
*Bsr*I	ACTGGN↓ TGAC↑CN	H(K)	65			
*Bsr*BI	GAG↓CGG CTC↑GCC	M	37		BstD102 I	
*Bsr*BrI	$GATN_2{\downarrow}N_2ATC$	L	65		*Bsa*BI, *Bsi*BI, *Mam*I	

continued

Table 10.1.3 Recognition Sequences and Reaction Conditions of Commercially Available Restriction Endonucleases[a], *continued*

Name	Site[b]	Salt[c]	Rxn. temp.[d]	Inact. temp.[d]	Isoschizomers	Comments[e]
*Bsr*DI	GCAATGN_2↓ CGTTAC↑N_2	M	60			
*Bsr*BrI	GATN_2↓N_2ATC	L	65		*Bsa*BI, *Bsi*BI, *Mam*I	
*Bsr*FI	Pu↓CCGGPy	M	37		*Cfr*10I	
*Bsr*GI	T↓GTACA	M	37		*Ssp*BI	
*Bss*HII	G↓CGCGC	L[f]	50	90		
*Bss*SI	C↓TCGTG	H	37			
*Bst*I	G↓GATCC	H(K)	55	85	*Bam*HI	Star activity, 10% activity at 37°C
*Bst*1107I	GTA↓TAC	H(K)	37		*Bsp*90I	
*Bst*71I	GCAGCN_8↓ CGTCGN_{12}↑	H	50		*Bbv*I	
*Bst*98I	C↓TTAAG	M	37		*Afl*II, *Bfr*I	
*Bst*BI	TT↓CGAA	M	65		*Bsi*CI, *Csp*45I, *Nsp*7524V, *Nsp*V, *Sfu*I	10% activity at 37°C
*Bst*D102I	GAG↓CGG	M	37		*Bsr*BI	
*Bst*EII	G↓GTNACC	H	60	85	*Eco*O651	Star activity, 10% activity at 37°C
*Bst*GII	CC↓(A/T)GG	—[f]	—[g]		*Bsi*LI, *Bst*NI, *Bst*OI, *Eco*RII, *Mva*I	
*Bst*NI	CC↓(A/T)GG	H	60		*Bsi*LI, *Bst*GII, *Bst*OI, *Eco*RII, *Mva*I	Difficult ligation, 30% activity at 37°C
*Bst*OI	CC↓(A/T)GG	M	60		*Bsi*LI, *Bst*GII, *Bst*NI, *Eco*RII, *Mva*I	
*Bst*UI	CG↓CG	H	60		*Acc*II, *Bsh*236I, *Mvn*I, *Tha*I	20% activity at 37°C
*Bst*XI	CCAN_5↓NTGG	H	50	65		
*Bst*YI	Pu↓GATCPy	L	60		*Mfl*I, *Xho*II	30% activity at 37°C
*Bst*ZI	C↓GGCCG	H	50		*Eag*I, *Ecl*XI, *Eco*52I, *Xma*III	
*Bsu*23I	T↓CCGGA	—[f]	37	65	*Acc*III, *Bse*AI, *Bsi*MI, *Bsp*EI, *Kpn*2I, *Mro*I	
*Bsu*36I	CC↓TNAGG	H	37		*Aoc*I, *Cvn*I, *Eco*81I, *Mst*II, *Sau*I	
*Bsu*54I	G↓GNCC	—[f]	—[g]		*Asu*I, *Bsi*ZI, *Cfr*13I, *Sau*96I	
*Cac*8I	GCN↓NGC	H	37	65		
*Ccr*I	C↓TCGAG	H	37		*Pae*R7I, *Xho*I	

continued

Table 10.1.3 Recognition Sequences and Reaction Conditions of Commercially Available Restriction Endonucleases[a], *continued*

Name	Site[b]	Salt[c]	Rxn. temp.[d]	Inact. temp.[d]	Isoschizomers	Comments[e]
*Cel*II	GC↓TNAGC	H	37		*Blp*I, *Bpu*1102I, *Esp*I	
*Cfo*I	GCG↓C	L	37		*Hha*I, *Hin*P1I	
*Cfr*9I	C↓CCGGG	—[f]	37		*Psp*AI, *Sma*I, *Xma*I	
*Cfr*10I	Pu↓CCGGPy	L[f]	37	100	*Bsr*FI	
*Cfr*13I	G↓GNCC	M	37	100	*Asu*I, BsiZ I, Bsu54 I, Sau96 I	
*Cla*I	AT↓CGAT	M	37	65	*Ban*III, *Bsi*XI, *Bsp*106I, *Bsp*DI	*dam*
*Cpo*I	CG↓G(A/T)CCG	H	30	60	*Csp*I, *Rsr*II	
*Csp*6I	G↓TAC	L	37		*Afa*I, *Rsa*I	
*Csp*I	CG↓G(A/T)CCG	H(K)	30		*Cpo*I, *Rsr*II	
*Csp*45I	TT↓CGAA	M	37	65	*Bsi*CI, *Bst*BI, *Nsp*V, *Nsp*7524V, *Sfu*I	
*Cvi*JI	PuG↓CPy	—[f]	—[g]			
*Cvn*I	CC↓TNAGG	M	37	65	*Aoc*I, *Bsu*36I, *Eco*81I, *Mst*II, *Sau*I	
*Dde*I	C↓TNAG	H	37	70		
*Dpn*I	GA↓TC	H	37	65		Both strands methylated A^m
*Dpn*II	↓GATC	H	37	65	*Mbo*I, *Nde*II, *Sau*3A I	*dam*
*Dra*I	TTT↓AAA	M	37	65		
*Dra*II	PuG↓GNCCPy	H	37		*Eco*O109I	
*Dra*III	$CACN_3$↓GTG	H	37	65		Star activity
*Drd*I	$GACN_4$↓N_2GTC	M(K)	37			
*Dsa*I	C↓CPuPyGG	H	55			
*Dsa*V	↓CCNGG	H	60		*Scr*FI	
*Eae*I	Py↓GGCCPu	M(K)	37	65		*dcm*
*Eag*I	C↓GGCCG	H	37	65	*Bst*ZI, *Ecl*XI, *Eco*52I, *Xma*III	
*Eam*1104I	CTCTTCN↓ $GAGAAGN_4$↑	H(K)	37	65	*Ear* I, *Ksp*632I	
*Eam*1105I	$GACN_3$↓N_2GTC	—[f]	37	65	*Ahd*I, *Asp*EI, *Ecl*HKI	
*Ear*I	CTCTTCN↓ $GAGAAGN_4$↑	L	37	65	*Eam* 1104I, *Ksp*632I	
*Ecl*HKI	$GACN_3$↓N_2GTC	M	37	65	*Ahd*I, *Asp*EI, *Eam*1105I	
*Ecl*XI	C↓GGCCG	H	37		*Bst*ZI, *Eag*I, *Eco*52I, *Xma*III	
*Ecl*136II	GAG↓CTC	—[f]	37	65	*Eco*ICRI, *Sac*I, *Sst*I	

continued

Table 10.1.3 Recognition Sequences and Reaction Conditions of Commercially Available Restriction Endonucleases[a], *continued*

Name	Site[b]	Salt[c]	Rxn. temp.[d]	Inact. temp.[d]	Isoschizomers	Comments[e]
*Eco*105I	TAC↓GTA	L	37		*Sna*BI	
*Eco*130I	C↓C(A/T)(A/T)GG	H	37		*Eco*T14I, *Sty*I	
*Eco*47I	G↓G(A/T)CC	H	37	100	*Ava*II, *Sin*I	
*Eco*47III	AGC↓GCT	H	37	100	*Aor*51HI	
*Eco*52I	C↓GGCCG	H	37	80	*Bst*ZI, *Eag*I, *Ecl*XI, *Xma*III	
*Eco*57I	CTGAAGN_{16}↓ GACTTCN_{14}↑	L	37	65		
*Eco*O65I	G↓GTNACC	H	37	70	*Bst*EII	
*Eco*72I	CAC↓GTG	—[f]	37	65	*Bbr*PI, *Pma*CI, *Pml*I	
*Eco*81I	CC↓TNAGG	L	37	90	*Aoc*I, *Bsu*36I, *Cvn*I, *Mst*II, *Sau*I	
*Eco*ICRI	GAG↓CTC	M	37	65	*Ecl*136II, *Sac*I, *Sst*I	
*Eco*NI	CCTN_2↓N_3AGG	—[f]	37			
*Eco*O109I	(A/G)G↓GNCC(C/T)	L	37	65	*Dra*II	*dcm*
*Eco*RI	G↓AATTC	H	37	65		Star activity
*Eco*RII	↓CC(A/T)GG	M	37	60	*Bst*GII, *Bsi*LI, *Bst*NI, *Bst*OI, *Mva*I	*dcm*
*Eco*RV	GAT↓ATC	H	37	65		Star activity
*Eco*T14I	C↓C(A/T)(A/T)GG	H	37		*Eco*130I, *Sty*I	
*Eco*T22I	ATGCA↓T	H	37	100	*Nsi*I, *Ppu*10I	
*Ehe*I	GGC↓GCC	L	37	70	*Bbe*I, *Kas*I, *Nar*I	
*Esp*I	GC↓TNAGC	H	37	100	*Blp*I, *Bpu*1102I, *Cel*II	
*Esp*3I	CGTCTCN↓ GCAGAGN_5↑	—[f]	37	65		
*Fdi*II	TGC↓GCA	L	50	100	*Avi*II, *Fsp*I, *Mst*I	
*Fnu*4HI	GC↓NGC	L	37	65	*Bso*FI, *Ita*I	Difficult ligation
*Fok*I	GGATGN_9↓ CCTACN_{13}↑	L	37	65		
*Fse*I	GGCCGG↓CC	M	30			
*Fsp*I	TGC↓GCA	M	37	65	*Avi*II, *Fdi*II, *Mst*I	
*Gsu*I	CTGGAGN_{16}↓ GACCTCN_{14}↑	—[f]	37	65	*Bpm*I	
*Hae*II	PuGCGC↓Py	M	37	65		
*Hae*III	GG↓CC	M	37	90	*Bsh*I, *Bsp*KI, *Pal*I	Very stable active at 70°C
*Hap*II	C↓CGG	—[f]	37		*Hpa*II, *Msp*I	

continued

Table 10.1.3 Recognition Sequences and Reaction Conditions of Commercially Available Restriction Endonucleases[a], *continued*

Name	Site[b]	Salt[c]	Rxn. temp.[d]	Inact. temp.[d]	Isoschizomers	Comments[e]
*Hga*I	GACGCN$_5$↓ CTGCGN$_{10}$↑	M	37	65		
*Hgi*AI	G(A/T)GC(A/T)↓C	H	37	65	*Alw*21I, *Asp*HI, *Bsi*HKAI	
*Hgi*DI	GPu↓CGPyC				*Acy*I, *Bbi*II, *Bsa*HI, *Hin*1I	
*Hha*I	GCG↓C	M	37	90	*Cfo*I, *Hin*P1I	
*Hin*1I	GPu↓CGPyC	L	37	80	*Acy*I, *Bbi*II, *Bsa*HI, *Hsp*92I	
*Hinc*II	GTPy↓PuAC	H	37	70	*Hind*II	
*Hind*II	GTPy↓PuAC	M	37	65	*Hinc*II	
*Hind*III	A↓AGCTT	M	37	90		Star activity
*Hinf*I	G↓ANTC	H	37	80		Star activity
*Hin*P1I	G↓CGC	L	37		*Cfo*I, *Hha*I	
*Hpa*I	GTT↓AAC	M(K)	37	90		
*Hpa*II	C↓CGG	L(K)	37	90	*Hap*II, *Msp*I	No cleavage when internal C is methylated
*Hph*I	GGTGAN$_8$↓ CCACTN$_7$↑	L(K)[f]	37	65		*dam*
*Hsp*92I	GPu↓CGPyC	L	37		*Acy*I, *Bbi*II, *Bsa*HI, *Hin*1I	
*Hsp*92II	CATG↓	H	37		*Nla*III	
*Ita*I	GC↓NGC	H	37	65	*Bso*FI, *Fnu*4HI	
*Kas*I	G↓GCGCC	M	37	65	*Bbe*I, *Ehe*I, *Nar*I	Site-dependent activity
*Kpn*I	GGTAC↓C	L	37	60	*Acc*65I, *Asp*718I	Star activity
*Kpn*2I	T↓CCGGA	—[f]	55		*Acc*III, *Bse*AI, *Bsi*MI, *Bsp*EI, *Mro*I	
*Ksp*I	CCGC↓GG	H	37		*Sac*II, *Sst*II	
*Ksp*6321	CTCTTCN↓ GAGAAGN$_4$↑	H(K)	37		*Eam* 1104I, *Ear*I	
*Mae*I	C↓TAG	H	45		*Bfa*I	
*Mae*II	A↓CGT	H	50			
*Mae*III	↓GTNAC	H	45			
*Mam*I	GATN$_2$↓N$_2$ATC	H	37		*Bsa*BI, *Bsi*BI, *Bsr*BrI	
*Mbo*I	↓GATC	H	37	65	*Dpn*II, *Nde*II, *Sau*3AI	*dam*
*Mbo*II	GAAGAN$_8$↓ CTTCTN$_7$↑	L(K)	37	65		*dam*
*Mcr*I	CGPuPy↓CG	H	37		*Bsa*OI,*Bsi*EI	

continued

Table 10.1.3 Recognition Sequences and Reaction Conditions of Commercially Available Restriction Endonucleases[a], *continued*

Name	Site[b]	Salt[c]	Rxn. temp.[d]	Inact. temp.[d]	Isoschizomers	Comments[e]
*Mfe*I	C↓AATTG	—[f]	37	65	*Mun*I	
*Mfl*I	Pu↓GATCPy	L	37		*Bst*YI, *Xho*II	
*Mlu*I	A↓CGCGT	H	37	100		
*Mlu*NI	TGG↓CCA	—[f]	37	65	*Bal*I, *Msc*I	
*Mnl*I	CCTCN$_7$ GGAGN$_6$	M	37	65		
*Mro*I	T↓CCGGA	L	37	100	*Acc*III, *Bse*AI, *Bsi*MI, *Bsp*EI, *Kpn*2I	
*Msc*I	TGG↓CCA	M	37		*Bal*I, *Mlu*NI	Star activity
*Mse*I	T↓TAA	M	37	65	*Tru*9I	
*Msl*I	CAPyN$_2$↓N$_2$PuTG	M	37	65		
*Msp*I	C↓CGG	M	37	90	*Hap*II, *Hpa*II	No cleavage when 5′ C is methylated
*Msp*A1I	C(A/C)G↓C(G/T)G	H	37	65	*Nsp*BII	
*Mst*I	TGC↓GCA	H	37		*Avi*II, *Fdi*II, *Fsp*I	
*Mst*II	CC↓TNAGG	H	37	65	*Aoc*I, *Bsu*36I, *Cvn*I, *Eco*81I, *Sau*I	
*Mun*I	C↓AATTG	M	37	65	*Mfe*I	
*Mva*I	CC↓(A/T)GG	H	37	100	*Bsi*LI, *Bst*GII, *Bst*NI, *Bst*OI, *Eco*RII	
*Mvn*I	CG↓CG	M	37		*Acc*II, *Bsh*1236I, *Bst*UI, *Tha*I	
*Mwo*I	GCN$_5$↓N$_2$GC	—[f]	60		*Bsp*WI	
*Nae*I	GCC↓GGC	L	37	100	*Ngo*AIV, *Ngo*MI	Site-dependent activity
*Nar*I	GG↓CGCC	L	37	65	*Ehe*I, *Kas*I	Stable at 37°C for 24 hr site-dependent activity
*Nci*I	CC↓(G/C)GG	L	37	80	*Bcn*I	Difficult ligation
*Nco*I	C↓CATGG	H	37	80		
*Nde*I	CA↓TATG	—[f]	37	65		$t_{\frac{1}{2}}$ ~15 min at 37°C
*Nde*II	↓GATC	H	37	65	*Dpn*II, *Mbo*I, *Sau*3AI	
*Ngo*AIV	G↓CCGGC	—[f]	37		*Nae*I, *Ngo*MI	
*Ngo*MI	G↓CCGGC	—[f]	37		*Nae*I, *Ngo*AIV	
*Nhe*I	G↓CTAGC	M	37	65		
*Nla*III	CATG↓	L[f]	37	65	*Hsp*92II	
*Nla*IV	GGN↓NCC	L[f]	37	65	*Bsc*BI	

continued

Table 10.1.3 Recognition Sequences and Reaction Conditions of Commercially Available Restriction Endonucleases[a], *continued*

Name	Site[b]	Salt[c]	Rxn. temp.[d]	Inact. temp.[d]	Isoschizomers	Comments[e]
*Not*I	GC↓GGCCGC	H	37	100		
*Nru*I	TCG↓CGA	H	37	80	*Spo*I	*dam*
*Nsi*I	ATGCA↓T	H	37		*Eco*T22I, *Ppu*10I	
*Nsp*I	PuCATG↓Py	L	37	65	*Nsp*7524I	
*Nsp*III	C↓PyCGPuG	—[f]	37	85	*Ava*I, *Bco*I	
*Nsp*V	TT↓CGAA	L	50		*Bsi*CI, *Bst*BI, *Csp*45I, *Nsp*7524V, *Sfu*I	
*Nsp*7524I	PuCATG↓Py	—[f]	37		*Nsp*I	
*Nsp*7524V	TT↓CGAA	L	37	70	*Bsi*CI, *Bst*BI, *Csp*45I, *Nsp*V, *Sfu*I	
*Nsp*BII	C(A/C)G↓C(G/T)G	L	37		*Msp*A1I	
*Pac*I	TTAAT↓TAA	L	37			
*Pae*R7I	C↓TCGAG	—[f]	37		*Ccr*I, *Xho*I	Site-dependent activity
*Pal*I	GG↓CC	M	37	65	*Bsh*I, *Bsp*KI, *Hae*III	
*Pfl*MI	$CCAN_4$↓NTGG	H	37	65	*Acc*B7I, *Van*91I	
*Pin*AI	A↓CCGGT	M	37	65	*Age*I	
*Ple*I	$GAGTCN_4$↓ $CTCAGN_5$↑	—[f]	37	65		
*Pma*CI	CAC↓GTG	L	37	60	*Bbr*PI, *Eco*72I, *Pml*I	
*Pme*I	GTTT↓AAAC	—[f]	37	65		
*Pml*I	CAC↓GTG	L	37		*Bbr*PI, *Eco*72I, *Pma*CI	
*Ppu*10I	A↓TGCAT	—[f]	37	65	*Eco*T22I, *Nsi*I	
*Ppu*MI	PuG↓G(A/T)CCPy	—[f]	37		*Psp*5II	*dcm*
*Psh*AI	$GACN_2$↓N_2GTC	K	37			
*Psp*1406I	AA↓CGTT	—[f]	37	65		
*Psp*5II	PuG↓G(A/T)CCPu	H	37		*Ppu*MI	
*Psp*AI	C↓CCGGG	—[f]	37	65	*Cfr*9I, *Sma*I, *Xma*I	
*Pst*I	CTGCA↓G	H	37	70		Star activity
*Pvu*I	CGAT↓CG	H	37	100	*Bsp*CI, *Xor*II	
*Pvu*II	CAG↓CTG	M	37	95		Star activity
*Rca*I	T↓CATGA	M	37	65	*Bsp*HI	
*Rsa*I	GT↓AC	M	37	65	*Afa*I, *Csp*6I	
*Rsr*II	CG↓G(A/T)CCG	L	37	65	*Cpo*I, *Csp*I	
*Sac*I	GAGCT↓C	L	37	60	*Ecl*136II, *Eco*ICRI, *Sst*I	
*Sac*II	CCGC↓GG	L	37	80	*Ksp*I, *Sst*II	Site-dependent activity

continued

Table 10.1.3 Recognition Sequences and Reaction Conditions of Commercially Available Restriction Endonucleases[a], *continued*

Name	Site[b]	Salt[c]	Rxn. temp.[d]	Inact. temp.[d]	Isoschizomers	Comments[e]
*Sal*I	G↓TCGAC	H	37	80		Star activity
*Sap*I	GCTCTTCN↓ CGAGAAGN$_4$↑	—[f]	37	65		
*Sau*I	CC↓TNAGG	H	37		*Aoc*I, *Bsu*36I, *Cvn*I, *Eco*81I, *Mst*II	
*Sau*3AI	↓GATC	H	37	70	*Dpn*II, *Mbo*I, *Nde*II	
*Sau*96I	G↓GNCC	L	37		*Asu*I, *Bsi*ZI, *Bsu*54I, *Cfr*13I	*dcm*
*Sca*I	AGT↓ACT	H	37	100		Star activity
*Scr*FI	CC↓NGG	H	37	65	*Dsa*V	
*Sdy*I	GGNCC	—[f]	—[g]		*Asu*I, *Bsi*ZI, *Bsu*54I, *Cfr*13I, *Sau*96I	
*Sex*AI	A↓CC(A/T)GGT	M	37			
*Sfa*NI	GCATCN$_5$↓ CGTAGN$_9$↑	H	37	65		
*Sfc*I	C↓TPuPyAG	—[f]	37	65		
*Sfi*I	GGCCN$_4$↓NGGCC	M	50			
*Sfu*I	TT↓CGAA	H	37		*Bsi*CI, *Bst*BI, *Csp*45I, *Nsp*7524V, *Nsp*V	
*Sgf*I	GCGAT↓CGC	L	37			
*Sgr*AI	CPu↓CCGGPyG	—[f]	37			
*Sin*I	G↓G(A/T)CC	L	37	65	*Ava*II, *Eco*47I	
*Sma*I	CCC↓GGG	L(K)[f]	30	65	*Cfr*9I, *Psp*AI, *Xma*I	
*Sna*BI	TAC↓GTA	M	37		*Eco*105I	
*Spe*I	A↓CTAGT	M	37	65		
*Sph*I	GCATG↓C	H	37	100	*Bbu*I	
*Spl*I	C↓GTACG	H	55		*Bsi*WI, *Sun*I	
*Spo*I	TCG↓CGA	M(K)	37	65	*Nru*I	
*Srf*I	GCCC↓GGGC	M	37	65		
*Sse*8387I	CCTGCA↓GG	—[f]	37	60		
*Ssp*I	AAT↓ATT	—[f]	37	65		
*Ssp*BI	T↓GTACA	—[f]	50	70	*Bsr*GI	
*Sst*I	GAGCT↓C	M	37	65	*Ecl*136II, *Eco*ICRI, *Sac*I	
*Sst*II	CCGC↓GG	M	37		*Ksp*I, *Sac*II	
*Stu*I	AGG↓CCT	M	37	65	*Aat*I	*dcm*
*Sty*I	C↓C(A/T)(A/T)GG	H	37	65	*Eco*O130I, *Eco*T14I	

continued

10

Table 10.1.3 Recognition Sequences and Reaction Conditions of Commercially Available Restriction Endonucleases[a], *continued*

Name	Site[b]	Salt[c]	Rxn. temp.[d]	Inact. temp.[d]	Isoschizomers	Comments[e]
*Sun*I	C↓GTACG	M	55		*Bsi*WI, *Spl*I	
*Swa*I	ATTT↓AAAT	H	25			
*Taq*I	T↓CGA	H	65	90	*Tth*HB8I	*dam*
*Tfi*I	G↓A(A/T)TC	L	65			Star activity, 10% activity at 37°C
*Tha*I	CG↓CG	L	60		*Acc*II, *Bsh*1236I, *Bst*UI, *Mvn*I	
*Tru*9I	T↓TAA	M	65		*Mse*I	
*Tsp*45I	↓GT(G/C)AC	L	65			
*Tsp*509I	↓AATT	—[f]	65			
*Tsp*RI	N_2CAGTGN_2↓	—[f]	65			
*Tth*111I	GACN↓N_2GTC	—[f]	65		*Asp*I	Difficult ligation, 10% activity at 37°C
*Tth*HB8I	T↓CGA	H	65		*Taq*I	
*Van*91I	CCAN_4↓NTGG	H(K)	37	65	*Acc*B7I, *Pfl*MI	
*Vsp*I	AT↓TAAT	H	37		*Ase*I, *Asn*I	
*Xba*I	T↓CTAGA	M	37	70		*dam*
*Xcm*I	CCAN_5↓N_4TGG	M(K)	37	65		
*Xho*I	C↓TCGAG	H	37	80	*Ccr*I, *Pae*R7I	
*Xho*II	Pu↓GATCPy	L	37	65	*Bst*YI, *Mfl*I	
*Xma*I	C↓CCGGG	L	37	65	*Cfr*9I, *Psp*AI, *Sma*I	
*Xma*III	C↓GGCCG	L	25		*Bst*ZI, *Eag*I, *Ecl*XI, *Eco*52I	
*Xmn*I	GAAN_2↓N_2TTC	L	37	65	*Asp*700I	Star activity
*Xor*II	CGAT↓CG	L	37		*Bsp*CI, *Pvu*I	
Intron-encoded endonucleases						
I-*Ceu*I	TAACTATAACGGTCCTAA↓GGTAGCGA[f] ATTGATATTGCCAG↑GATTCCATCGCT					Rxn. temp. 37°C[f]
I-*Ppo*I	ATGACTCTCTTAA↓GGTAGCCAAA[f] TACTGAGAG↑AATTCCATCGGTTT					Rxn. temp. 37°C[f]
I-*Sce*I	TAGGGATAA↓CAGGGTAAT[f] ATCCC↑TATTGTCCCATTA					Rxn. temp. 37°C[f]
PI-*Psp*I	TGGCAAACAGCTATTAT↓GGGTATTATGGGT[f] ACCGTTTGTCGAT↑AATACCCATAATACCCA					Rxn. temp. 65°C[f]
PI-*Sce*I	ATCTATGTCGGGTGC↓GGAGAAAGAGGTAATGAAATGGCA[f] TAGATACAGCC↑CACGCCTCTTTCTCCATTACTTTACCGT					Rxn. temp. 37°C[f]

continued

Table 10.1.3 Recognition Sequences and Reaction Conditions of Commercially Available Restriction Endonucleases[a], *continued*

Name	Site[b]	Salt[c]	Rxn. temp.[d]	Inact. temp.[d]	Isoschizomers	Comments[e]
PI-*Tli*I	GGTTCTTTATGCGGACAC↓TGACGGCTTTATG[f] CCAAGAAATACGCC↑TGTGACTGCCGAAATAC					Rxn. temp. 50°C[f]

[a]Published previously in *CP Molecular Biology Unit 3.1* (Bloch and Grossmann, 1995).

[b]Abbreviations: N, any nucleotide (G, A, T, C); Pu, either purine (G or A); Py, either pyrimidine (C or T). Arrows indicate points of cleavage.

[c]Recommended concentrations of NaCl (or KCl, indicated by parenthetical K) where L is <50 mM, M is 50 to 100 mM, and H is >100 mM.

[d]Reaction temperature is the temperature (°C) at which the reaction should be performed; inactivation temperature indicates the temperature (°C) at which the enzyme is inactivated after 15 min of incubation.

[e]Abbreviations and other terminology: *dam*, activity blocked by *dam* or overlapping *dam* methylation (see Bloch and Grossmann 1995); *dcm*, activity blocked by *dcm* or overlapping *dcm* methylation (see Bloch and Grossmann 1995); difficult ligation, the enzyme produces single-bp 5′ overhanging ends that are difficult to ligate with T4 DNA ligase; SAM, *S*-adenosylmethionine; site-dependent activity refers to marked differences in rates of cleavage at various sites by a particular enzyme (probably determined by the surrounding sequence); star activity refers to altered specificity of a restriction enzyme that causes it to cleave sequences that are similar but not identical to its defined recognition sequence–conditions that may provoke this altered specificity include elevated pH, high glycerol concentration, low ionic strength, and high enzyme to DNA ratio; $t_{\frac{1}{2}}$, half-life.

[f]Refer to manufacturer's recommendations for information on salt requirements.

[g]Refer to manufacturer's recommendations for reaction temperature.

UNIT 10.2

Overview of PCR

Christine D. Kuslich,[1] Buena Chui,[2] and Carl T. Yamashiro[3]
[1]Molecular Profiling Institute, Phoenix, Arizona
[2]GE Healthcare, Piscataway, New Jersey
[3]Arizona State University, Tempe, Arizona

OVERVIEW AND PRINCIPLES

The polymerase chain reaction (PCR) was conceptualized by Kary Mullis in 1983 (Saiki et al., 1985; Mullis et al., 1986) and was rapidly put into practice. PCR has subsequently revolutionized molecular analysis in the life sciences to the point that Dr. Mullis received the 1993 Nobel Prize in Chemistry for his discovery. In fact, it has far exceeded the expectations of the inventor and the earliest developers. PCR has constantly evolved and amended itself to accommodate the specific needs of researchers. The simple elegance of PCR permitted the technique to be quite malleable and to be surrounded with supplemental techniques to enhance its capabilities.

The inherent power of PCR is its ability to exponentially amplify specific nucleic acid sequences in a short period of time. Deoxyribonucleic acid (DNA) amplification by PCR is achieved through multiple cycles of in vitro DNA replication. In its initial and simplest form, a DNA replication cycle for PCR proceeds as shown in Figure 10.2.1.

Reaction mix composition. The reaction solution within a tube contains the DNA template to be amplified (e.g., genomic DNA), two different species of short single-stranded oligonucleotides called primers, a DNA polymerase, the four deoxyribonucleotide triphosphates (dATP, dCTP, dGTP, and dTTP), a salt with a divalent cation (i.e. Mg^{2+}), and a buffer with a simple salt (i.e., KCl).

Denaturation. The solution is heated to a temperature for a period that promotes DNA strand dissociation from double-stranded to single-stranded form to prepare the DNA template for priming and subsequent DNA amplification (denaturation step).

Annealing. The temperature of the solution is brought down and held for a period sufficient to allow for hybridization of the primers to the single-stranded DNA template (annealing step).

Extension. The temperature is either kept constant (two-step PCR) or is raised for optimal DNA polymerase activity (three-step PCR) and is held for a period sufficient for all primed events to be fully extended, yielding the double-stranded products called amplicons (extension step).

Repetitive cycling. The denaturing, annealing, and extension steps are typically repeated 30 to 40 cycles.

In the original experiments to demonstrate proof-of-principle for PCR, the tube containing the reactants was manually transferred between water baths set and equilibrated to the desired temperatures. The added DNA polymerase was irreversibly inactivated during the denaturation step, making the process even more complicated; therefore, more enzyme had to be added manually just prior to each extension step. This was a very laborious and time-intensive methodology that was not amenable for routine use in the laboratory. Two key advances made PCR a technology that could be practiced in almost any laboratory. The first key improvement was the identification and implementation of thermostable DNA polymerases that eliminated the need for frequent manual addition of enzyme (Saiki et al., 1988). The second key advance was the development of an instrument called a thermal cycler that holds PCR reaction tubes and is programmed to adjust to specific temperatures for

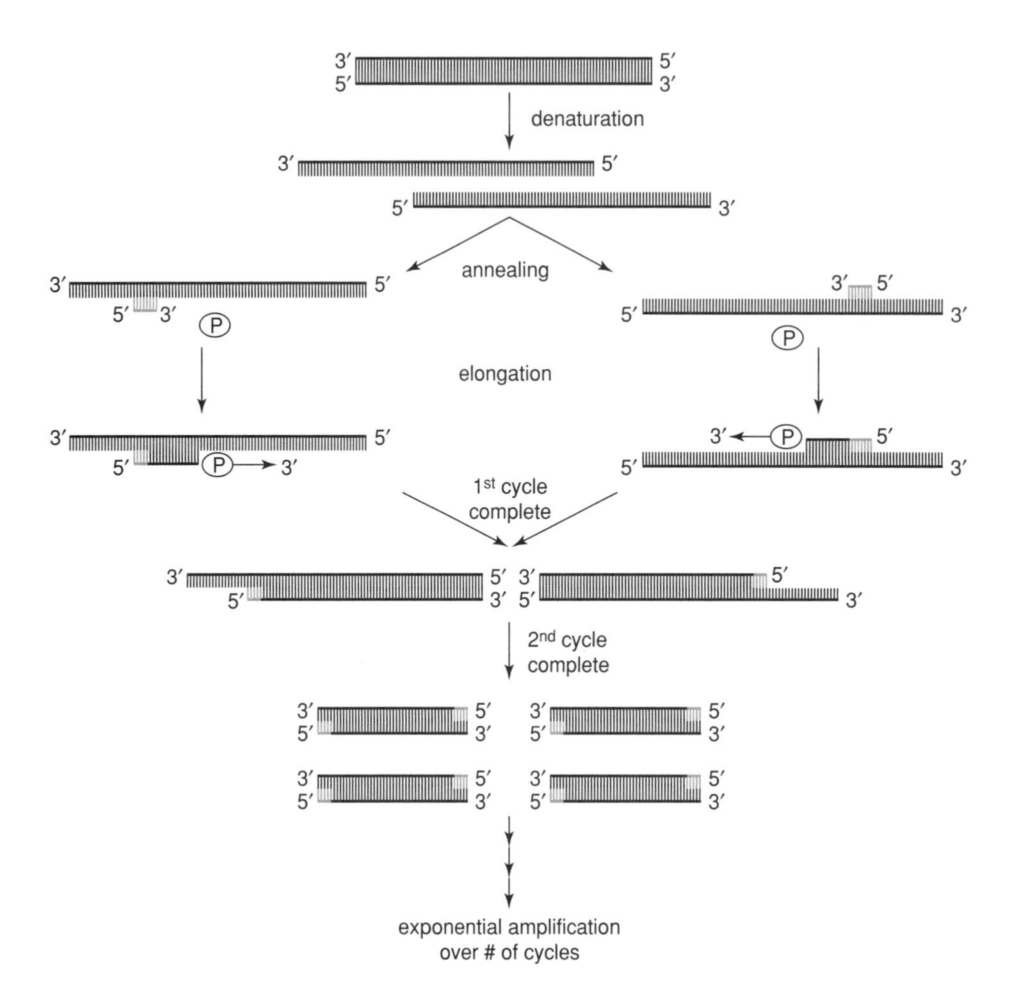

Figure 10.2.1 PCR process. Depiction of first cycle showing the denaturation of the target sequence using high temperature (92° to 96°C), primer (short gray segment) annealing (e.g. at 58°C), and elongation by DNA polymerase (denoted with P; e.g. at 72°C). Subsequent cycles will double the number of starting amplicons for that cycle.

discreet periods of time (Ehrlich et al., 1991). The thermal cycler was able to repeat this routine for a programmed number of cycles.

Once PCR became available as an accessible and common technique, many scientists began to imagine how to exploit the technique to address their particular research and development needs. Their creativity, now endowed with the overall simplicity of PCR, combined and spurred by the lack of effective alternative technologies, made the technique essentially ubiquitous. Its simplest utility is the immense amplification of a specific DNA sequence of defined length. Over the years there have been many modifications of the original PCR procedure to expand its capabilities and allow its use in many different applications (see Variations section at the end of this unit and *UNIT 10.3* on real-time PCR). This versatility of PCR has changed research in the life sciences in a most significant manner by reducing the time and expense of performing a multitude of genomic and genetic studies.

Polymerases

At the time of conceptualization of PCR, oligonucleotides could be chemically synthesized and DNA polymerase, from the common enteric bacterium *Escherichia coli*, could be purified for use in the first experiments testing PCR feasibility. However, as mentioned above, the polymerase had to be replenished during each cycle of the PCR because the high temperatures of the DNA denaturation step would inactivate the *E. coli* DNA polymerase. This was too impractical for its use to become commonplace in the laboratory. The key technical advance necessary for widespread use of PCR was the implementation of DNA polymerases that are relatively stable at high temperatures (~95°C) close to that of boiling water. It started with the thermostable *Taq* DNA polymerase from the thermophilic bacterium *Thermus aquaticus* (Saiki et al., 1988). Presently, there are dozens of commercial sources for PCR DNA polymerases which offer different versions of a variety of enzymes. The most common types of thermostable enzymes used for PCR are *Taq*, *Pfu*, and Vent DNA polymerases. *Taq* DNA polymerase, from *T. aquaticus* isolated at hot springs and other land-based thermal environments, is the most commonly used PCR enzyme and has the most variants commercially available. *Pfu* DNA polymerase comes from *Pyrococcus furiosus*, an Archaea hyperthermophile found in deep ocean vents where habitat temperatures can be >100°C (Lundberg et al., 1991). Vent polymerase, also known as *Tli* polymerase, is originally from *Thermococcus litoralis* that is also from deep ocean thermal vents (Mattila et al., 1991). All of these DNA polymerases function to replicate DNA but have functional qualities that differ between them. This includes whether the enzymes have other activities such as a 3′→5′ exonuclease that can improve the fidelity of DNA replication significantly. This activity allows for effective proofreading of the newly synthesized DNA strand by excising any mismatched nucleotides that were originally incorporated by the polymerase and resynthesizing the new strand correctly the second time. Thus, the error rate of incorrect incorporation of a nucleotide during DNA synthesis of PCR is significantly reduced, resulting in an improvement of DNA replication fidelity. A partial list of properties for the major PCR enzymes is shown in Table 10.2.1.

Even after the demonstration of the utility of thermophilic DNA polymerases, availability of these enzymes was very limited. This is primarily due to laboratories not being able to grow the required bacteria due to unusual growth conditions that standard laboratory equipment did not readily support. Therefore, life science companies quickly developed manufacturing and packaging processes for mass producing and selling these enzymes on a global basis. Initially the companies propagated the thermophilic organisms and purified the DNA polymerase. Yields were relatively poor and eventually the gene for the DNA polymerase from the thermophilic organism was cloned and then expressed in a host organism, such as *E. coli*, which can be rapidly propagated and induced to overexpress and synthesize the enzyme (Lawyer et al., 1989). Rapid and efficient purification processes were developed to maximize yields of functional enzyme that can be stabilized in a specific buffer and packaged in tubes and vials amenable for transportation and storage. The enzymes can be genetically modified to allow for performance of the polymerase to be altered to better serve specific applications. In addition, more stringent purification methods can be employed to greatly minimize the residual contaminating nucleic acid within the enzyme preparation which can be very

Table 10.2.1 Characteristics of Common DNA Polymerases for PCR

	DNA Polymerase	
	Taq	Vent/*Tli* (*Pfu* is similar)
Half-life @ 95°C	~90 min	~420 min
3′→5′ Exonuclease	No	Yes
5′→3′ Exonuclease	Yes	No
Extension rate (nt/sec)	75	>80
Error rate (errors/bp)	2×10^{-5}	4×10^{-6}
Resulting ends	3′ A	>95% blunt
Example application	Standard amplification of target sequences	Target amplification for cloning blunt-ended fragments requiring DNA sequence fidelity

important for highly sensitive work with bacterial sequences that could be adversely affected by host DNA being present. Today, isolation and purification of DNA polymerases for PCR has become relatively simple and straightforward for the individual investigator to perform. Typically, affinity chromatographic methods have been implemented for enzyme purification (Melissis et al., 2006, 2007). However, due to the continuous reduction of DNA polymerase prices from vendors through improved large-scale production capabilities, the cost of setting up a laboratory to propagate the bacteria and purify the polymerase may not justify this type of activity unless enzyme consumption is very high. An example of a useful genetically modified *Taq* DNA polymerase is a product from Applied Biosystems called AmpliTaq FS. The *Taq* DNA polymerase gene has an engineered mutation that results in a substitution of a phenylalanine for a tyrosine at position 667 that permits more efficient incorporation of dideoxyribonucleoside triphosphates (ddNTPs) widely used in chain-terminating DNA sequencing reactions (Tabor and Richardson, 1995). This modified enzyme allows for much more consistent peak profiles that are easier to read for making base calls and less prone to sequencing errors or nonreads due to erratic peak profiles coming from cycle sequencing (a variation of PCR; Parker et al., 1996).

Given the many choices of DNA polymerases available for PCR, it is important to match up the reported enzyme properties with the requirements of the application one wants to perform. One example is that for mutational analysis, it may be critical to use a DNA polymerase with the greatest fidelity of replication. This means that given a choice between a *Taq* and a *Pfu* DNA polymerase, the latter would be best given its significantly higher replication fidelity (see Table 10.2.1). However, *Pfu* enzymes are typically more expensive than *Taq* enzymes and this must be accounted for when budgeting for a given project.

Hot-start polymerases

Hot-Start PCR polymerases refer to the class of modified DNA polymerase systems that remain inactive until the highest temperature of PCR thermal cycling is initiated. Although thermostable DNA polymerases perform optimally at elevated temperatures, a significant amount of activity is still present at low and ambient temperatures, thus allowing unintended priming events. During PCR set up, the PCR reaction mixture is typically exposed to temperatures below the optimal primer annealing conditions that allow for specific hybridization to the intended target sequence. The lower temperatures create lower-stringency hybridization conditions. With lower stringency, primers tolerate mismatch pairing better and can bind to nonspecific sites, or the template may anneal onto itself long enough to be recognized as an initiation site for DNA elongation by the DNA polymerase. In either case, the initiation of mis-priming events results in spurious and unwanted products. Furthermore, primers may also form secondary structures or bind to each other, creating “primer-dimer” products. Active polymerases will extend annealed 3′ ends, and polymerases with 5′→3′ exonuclease activity will degrade any free 5′ end of annealed sequence, potentially destroying template and primers. Once unintended priming occurs, amplification of these undesired products continues throughout the remaining PCR cycles, consuming reagents and simultaneously inhibiting or significantly reducing the efficiency of the desired target amplification. This creates a combination of unwanted amplicons, a reduction in target product, and complications to downstream applications such as sequencing or cloning. The modified hot-start polymerase systems were designed to eliminate these events.

Early on, the hot start of a PCR was accomplished by manually adding a final component such as polymerase or Mg^{2+} to the reaction when the first annealing temperature was reached. However, manual methods are labor-intensive and prone to variability and contamination. As an alternative, some researchers create a physical barrier between components until higher temperatures are reached. This is typically accomplished with the use of paraffin wax or mineral oil (D’Aquila et al., 1991; Bassam and Caetano-Anolles, 1993; Wainwright and Seifert, 1993). For example, once the initial high temperature for DNA denaturation is reached, paraffin wax will melt and move to the surface of the solution since it is hydrophobic and more buoyant than the aqueous phase, effectively removing the physical barrier and allowing the once separated reagents to mix. Yet, this method often has the disadvantage of insufficient mixing of the enzyme within the reaction solution, thus resulting in suboptimal yields of amplicon.

The most prevalent method for hot-start PCR today is to use a DNA polymerase that is inactivated until the PCR reaction reaches the first denaturation step. The two principal means of DNA polymerase inactivation are ligand inhibition and chemical modification.

Ligand inhibition

Ligand inhibition involves the use of various ligands that specifically bind to the polymerase active site, effectively inhibiting enzyme activity in a temperature-dependent manner. For example, antibodies (Kellogg et al., 1994; Sharkey et al., 1994) and aptamers (Dang and Jayasena, 1996; Lin and Jayasena, 1997) have been demonstrated to achieve very specific, noncovalent binding to the polymerase substrate recognition site at low temperatures. Upon heating at 95°C for up to several minutes, these ligands dissociate from the polymerase and the substrate-binding site is now available for primer extension. The activation energy tends to be lower, and only short heat-activation periods are required. The ligands are still present during all PCR cycles and have the potential to refold into an active conformation at lower temperatures, potentially permitting substrate-binding competition at a low level, which may impact PCR efficiency. It should be noted that any additions to the PCR, such as ligands, should be controlled potential interference in downstream applications. One should be aware of differences in PCR buffer or DNA polymerase formulations in order to determine how the PCR may be affected by these additions. In most cases, the manufacturer of the hot-start enzyme provides recommendations for optimal reaction conditions.

Although not targeting the polymerase, ligands specific to single-stranded DNA have been recently shown to effectively achieve the same outcome (GE Healthcare, 2006). It has been demonstrated that proteins bound to the primers in a temperature-dependent manner block extension by DNA polymerase just as efficiently, and dissociate quickly, requiring a short heat activation period at elevated temperatures. The protein ligands are not thermostable and denature at high temperatures prohibiting any reassociation or low level of competition in later cycles. This approach is shown in Figure 10.2.2 as an example of the increased specificity that can be achieved by using a primer-inhibited hot-start PCR system (illustra; GE Health Care Life Sciences) compared to a standard non-hot-start *Taq* DNA polymerase.

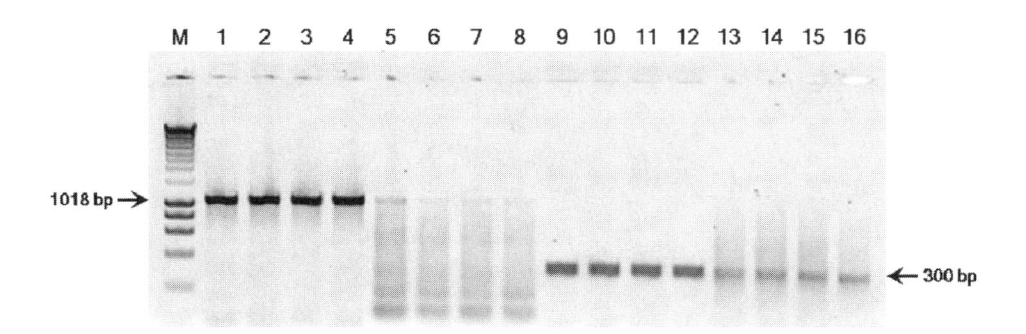

Figure 10.2.2 Inhibition of nonspecific product amplification using a hot-start PCR system. Amplification of a 1018 bp fragment (lanes 1 through 8) and a 300 bp fragment (lanes 9 through 16), randomly selected sequences within the human genome. PCR reactions with illustra Hot-Start Mix RTG are shown in lanes 1 through 4 and 9 through12, PCR reactions with standard *Taq* DNA polymerase, a non-hot-start system, are shown in lanes 5 through 8 and 13 through 16. In all reactions, 10 ng of human genomic DNA was used with 10 pmol of forward and reverse primers. The products and DNA size markers (M) were resolved on a 1.5% agarose TAE gel stained with ethidium bromide. In lanes 5 to 8, there are low molecular weight bands of which the lowest could be a primer dimer (unconfirmed); however, in the hot-start reactions, it is clear that nonspecific binding events were prevented, successfully leading to efficient, specific single product amplification of the expected size. Data and image used with permission from GE Healthcare.

Chemical modification

Inhibition of polymerase activity has also been achieved by covalent chemical modifications of specific amino acids in the DNA polymerase active site that are heat labile (Birch, 1996). Heating of the aqueous reaction mixture at low pH removes the modification to the amino acids and restores enzyme activity. Unlike the ligand inhibitors, the hydrolysis of the chemical modifiers involves the breaking of covalent bonds and is a permanent event as a result. This chemical activation process tends to require a longer heat incubation period, 10 to 15 min at 95°C. This longer incubation period at high temperature may increase the rate of random *N*-glycosylic bond breakage resulting in a higher rate of nicks present in the DNA template.

Finally, ligands and chemically modified hot-start enzymes typically cost more than unmodified enzymes due to the additional manipulation and components or the chemical modification of the enzymes followed by a second purification. However, the benefits of using a hot-start DNA polymerase, in terms of ease of use, and reduced opportunity for contamination, as well as increased specificity and reproducibility, are often considered to be well worth the additional cost.

Thermal Cyclers

Introduced in 1986, thermal cyclers, able to cycle and maintain precise temperatures for defined periods of time, have become standard equipment in most life science laboratories today. The first generation of thermal cyclers made use of water baths or heating blocks set to specific temperatures, requiring that the tubes be moved from one to another over multiple cycles. This mode of heating, being inefficient and laborious, has undergone significant innovations since its early inception. New instruments that carry out automated cycling have been engineered that make use of different technologies to regulate temperature change, dramatically reducing the reaction time. A couple of representative examples of commercially available thermal cycling instruments are shown in Figure 10.2.3.

The most common thermal cyclers used in laboratories today rely on Peltier technology. The Peltier effect first described by Jean Peltier in 1836 is used for the reliable cooling of temperature sensitive components, eliminating the need for refrigerants. Peltier devices are solid-state heat pumps that use a type of semiconductor called peltiers. When a low voltage direct current is applied to one side of this device, heat will travel in the direction of the current, thereby making one side of the block hot and the opposite end of the block cold. Thus, heat and cold can be directed from one side of the block to the other by simply changing the direction of the current.

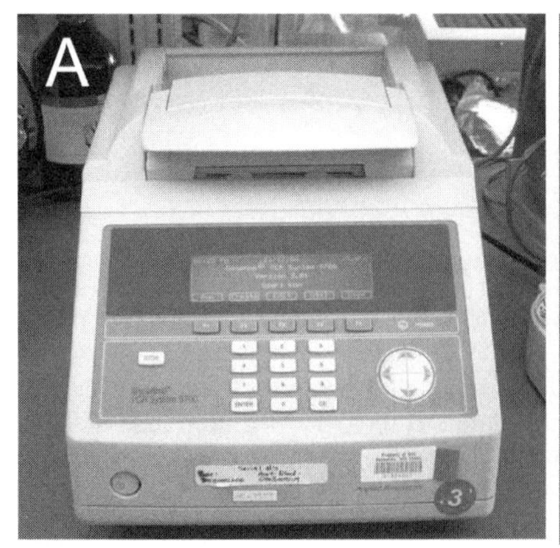

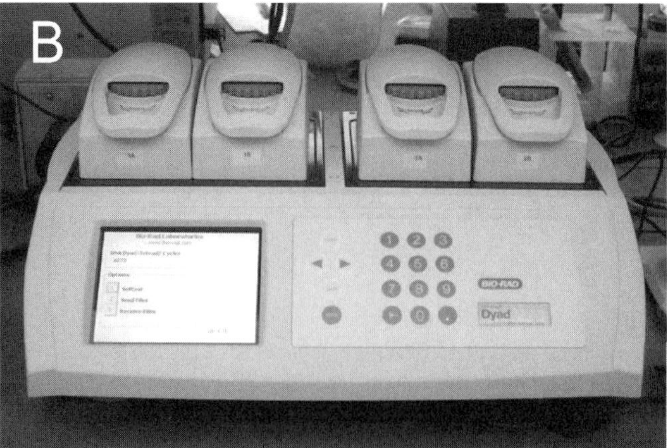

Figure 10.2.3 Thermal cyclers. Representative thermal cyclers within research laboratories: (**A**) thermal cycler (Applied Biosystems) with a single peltier block holding up to 96 tubes or a 96-well plate; and (**B**) thermal cycler (Bio-Rad Laboratories) with four independently controlled peltier blocks, each holding up to 96 tubes or a 96-well plate.

Resistive heating devices rely on a coiled heating element that contacts the wells of the thermal cycler directly to heat the reaction tubes. Passive air or water cyclers use heated air or water moved in and out of the instrument to heat and cool the reaction.

There are numerous additional features and functions added to the basic thermal cycling block to facilitate the reaction. Soon after PCR was adapted to the prototype thermal cycler, significant evaporation of the reaction solution was observed–the water condensed in the cap of the tube and could drip back down into the reactants. This would cause wild fluctuations in actual concentrations of reactants, thereby often rendering the PCR ineffective. The first solution was to overlay the reaction solution with buoyant mineral oil that provides a hydrophobic barrier to evaporation. However, this was often messy when extracting the finished reaction from the tube and often required extraction of the oil with a different organic solvent. The heated lid feature found on many thermal cyclers today is an effective and simple alternative to using mineral oil in the reaction vessel to prevent evaporation during heating cycles of PCR. The lid of these machines preheats before the reaction begins and stays at a constant temperature, usually >100°C, throughout the PCR profile.

A number of thermal cyclers can be purchased as multi-block units, which have two or more thermal cycling blocks that can be run independently of one another on the same machine. This configuration permits multiple PCR thermal cycling conditions to be run in parallel. Another feature many thermal cyclers have is the temperature control optimization to specific thin-walled tube/plate formats. These machines are designed to make very sensitive temperature estimations of the actual temperature in the sample tube as opposed to the temperature of the block. The time at a particular temperature does not begin until the liquid in the sample tubes reaches the target temperature.

Some thermal cyclers have a *gradient* function, which sets up a range of temperatures across the block for a given step of the PCR reaction. A gradient of temperatures in a specific range set by the user allows for multiple temperature profiles to be run concurrently. This function typically is used to quickly optimize thermal cycling temperature conditions for a given reaction (e.g., see Support Protocol 1). The use of this function precludes the need to run multiple experiments to optimize the cycling conditions (e.g., a gradient of different annealing temperatures) for a given set of primers. Table 10.2.2 provides a list of many of the major thermal cycler manufacturers.

Table 10.2.2 Thermal Cycler Suppliers

Applied Biosystems	*http://www.appliedbiosystems.com*
AlphaHelix AB	*http://www.alphahelix.com*
Barloworld Scientific Ltd	*http://www.barloworld-scientific.com*
Bioer Technology	*http://www.bioer.com.cn/en/index.htm*
Biometra	*http://www.biometra.de*
Bioneer	*http://www.bioneer.dk*
Bio-Rad	*http://www.bio-rad.com*
Corbett Life Science	*http://www.corbettresearch.com*
EDVOTEK, Inc.	*http://www.edvotek.com*
Eppendorf	*http://www.eppendorf.com*
Idaho Technology Inc.	*http://www.idahotech.com*
Labnet International, Inc.	*http://www.labnetlink.com*
MIDSCI	*http://www.midsci.com*
Stratagene	*http://www.stratagene.com*
Techne Inc.	*http://www.techneusa.com*
Thermo Electron Corporation	*http://www.thermo.com*

Contamination Control

PCR is a very powerful technology given its capability for amplifying specific DNA sequences easily. This power can also be a liability if proper precautions are not taken when setting up the laboratory and establishing procedures for carrying out PCR. The amplicons produced in previous reactions can be released into the local laboratory environment through various means, and such amplicons can become templates for future reactions that use the same or related primers for DNA amplification. This is especially true if the same primer sets are used repeatedly on a regular basis. When PCR cross-contamination occurs, one can have a false-positive result or may amplify a target different from what was intended. In addition to obtaining incorrect results due to contamination issues, this problem typically causes significant delays in conducting experiments, since considerable effort is required for finding the root cause of the contamination and then taking the proper corrective actions. The worst outcome is not to be aware of the contamination problem and obtaining incorrect results that can influence subsequent studies or be reported in a publication. To avoid this circumstance, implementation of proper controls should permit detection of a contamination problem (see Reaction Control section for more information).

There are several different levels of actions that can be taken to avoid or minimize the probability of experiencing contamination issues with PCR. The first level is to have good laboratory practices when performing PCR. The next level of prevention involves the use of alternative chemistries to negate the possible carryover of amplicons that may be present. Finally, specialized laboratory design can be a very effective means of preventing contamination; however, it is also the most expensive to implement.

Reaction controls

Including a positive and negative control in a PCR experiment is essential for confirming the validity of the experimental result. The positive control typically contains a template of known sequence, quality, and concentration which confirms the efficiency of the PCR master mix reagents and primers. Consider that the template selected for this control must contain the region of interest for proper hybridization of the PCR primers used in the reaction and subsequent amplification. These template controls can be purchased from numerous vendors or isolated and quantified in the laboratory then validated by another experiment prior to use. Typically, one positive control for each primer set used or specific set of reagents in a given PCR experiment is run.

A negative control contains all of the PCR components used in the reaction except a template. This control verifies that the reagents used in the experiment are not contaminated with template DNA. Typically a negative control reaction is run for every master mix used in a PCR experiment. There should be no amplification in this control reaction. If amplification is present in the negative control, the experiment should be considered invalid, as one cannot determine if amplification observed in the experimental samples was from contaminating template or the expected template. The entire PCR experiment will have to be repeated, carefully testing each reagent component to determine the source of contamination. This is a very time consuming and tedious process, which is why implementation of good PCR practices (see following section) can minimize the risk of having contamination problems. Proper use of positive and negative controls saves considerable time in the laboratory and helps to ensure reliable and meaningful results.

Good PCR practices

Common sources of contamination include stock solutions, pipettors, pipet tips, and PCR tubes. These can be contaminated through a number of different mechanisms, the most common of which is the improper handling of post-amplification products. Training those who perform PCR on good laboratory practices is often the most effective and least costly approach towards conducting high-quality, PCR-based experiments (Kwok and Higuchi, 1989). The following is a list of good practices:

1. Use a set of pipettors with pipet tips and tubes that are used only for setting up PCR reactions prior to amplification. It is best to use either positive-displacement pipettors and/or aerosol resistant tips.

2. All reagents used for setting up PCR should be dedicated for that purpose only and kept very close to where PCR set up occurs. This includes common laboratory reagents such as sterile distilled water, Tris buffer, and magnesium salt stocks.

3. Always wear gloves in the setup area, but never wear ones that have been outside of this area, especially where there may be released amplicons.

4. Clean up work area prior to PCR set up. For effective elimination of any nucleic acids present in the work area surfaces, wipe down with a 10% bleach solution. The same solution can be used to wipe down other items, such as pipettors and the outside of tip boxes, and to remove residual nucleic acids from surfaces that can be in contact with gloves.

5. Keep the analysis of PCR products in an area separate from where PCR is set up. This is especially crucial when PCR tubes with amplicon must be opened, such as for gel electrophoresis.

6. Acquisition of a PCR cabinet or workstation can be very useful, especially if laboratory redesign (described later) is not an option. PCR cabinets should be enclosures with a viewing window and a removable barrier for access with hands and arms of the user. It is also ideal to have a UV light source that can be turned on when not in use to destroy any nucleic acids that may be present (one can illuminate with UV light the entire laboratory or the part where PCR set up occurs, after hours, to provide additional protection from carryover contamination). Examples of such enclosures include those made by Labconco (*http://www.labconco.com*), UVP (*http://www.uvp.com*), CBS Scientific (*http://www.cbsscientific.com*), and Misonix (*http://www.misonix.com*).

Contamination control: Uracil-N-Glycosylase (UNG)

A different approach to controlling PCR contamination is to build in a special chemistry to support the specific removal of amplicons produced by PCR. An effective method involves the incorporation of 2′-deoxyuridine 5′-triphosphate (dUTP) in place of 2′-deoxythymidine 5′-triphosphate (dTTP) during the PCR and taking advantage of the enzyme uracil-*N*-glycosylase (UNG), which removes the uracil base from the amplicon (Thornton et al., 1992). Subsequent heating of the UNG-treated sample destabilizes the duplex DNA and is no longer a usable template for PCR amplification (Fig. 10.2.4).

To set this up, all reactions will have replaced all or a portion of dTTP with dUTP. The typical ratio of dUTP:dTTP is in the range of 1:8 to 1:10 but must be optimized depending on reaction conditions and DNA polymerase used. Within the same reaction mix, UNG is present. Normally the PCR is started with an incubation at a lower temperature (20° to 37°C for 5 to 20 min) to allow UNG to cleave any amplicon that may be present before the planned PCR has begun, and is followed by an incubation at 94° to 96°C to both inactivate the UNG and to denature the template DNA prior to the initiation of the actual PCR. These initial steps assure inactivation of all dUTP-containing amplicons, yet the high temperature inactivation step allows the PCR to generate more dUTP-containing amplicons. In addition, since there is an extended low temperature incubation for the UNG reaction, it is advantageous to use a hot-start method for the PCR to minimize the probability of producing unwanted PCR products.

Analysis of dUTP-containing amplicons should be done soon after completion of the thermal cycling as the UNG can renature to form active enzyme if left at 0° to 42°C for a few hours to days depending on the storage temperature. Freezing the sample at −20°C or below will allow for longer term storage of amplicons that are still amenable to analysis. One can also leave the PCR overnight at 72°C and the UNG will remain inactive; however, sometimes the PCR will yield a smear of products (higher background) when analyzed by gel electrophoresis due to initiation of nonspecific priming by the polymerase when held at an ideal extension temperature for an extended

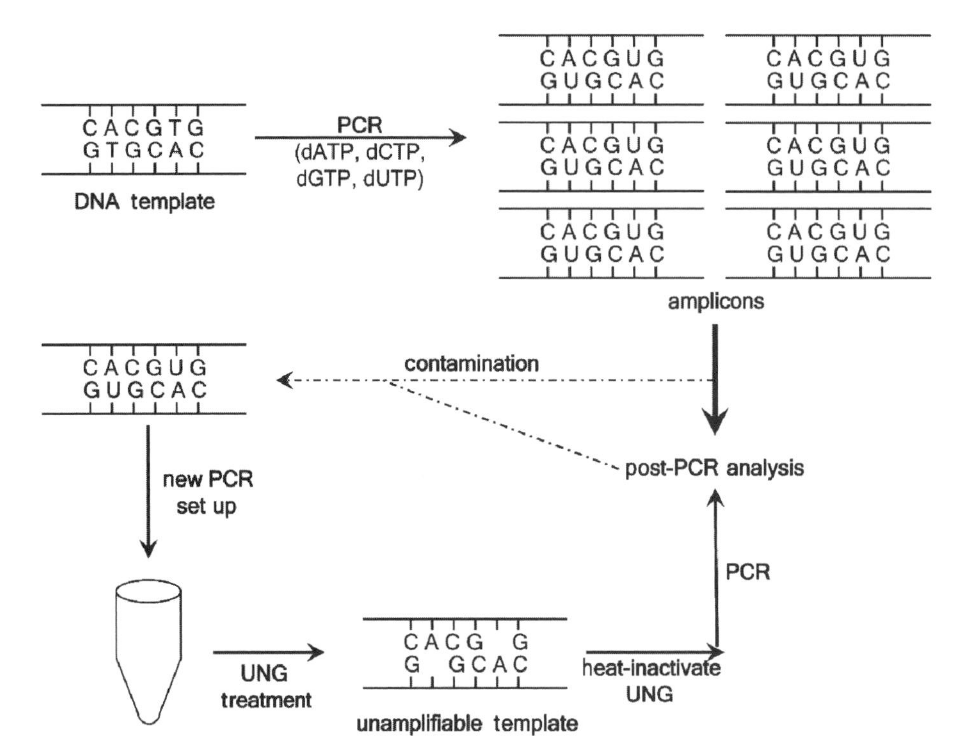

Figure 10.2.4 UNG contamination control process. PCR is performed normally except that dUTP replaces dTTP in the reaction, as well as inclusion of the UNG enzyme. The resulting amplicons will have incorporated dUTP. The dUTP will not affect downstream analysis of the amplicons (e.g. visualization of the amplicons by agarose electrophoresis). However, if the dU-containing amplicons contaminate a subsequent PCR that has UNG present, then during the UNG pre-treatment incubation any dU will be modified to render it unamplifiable in the PCR. The UNG enzyme is inactivated by high temperature treatment (usually coincides with the first cycle denaturation and Hot-Start enzyme activation if used) and therefore will not be able to degrade the newly created amplicons in the subsequent PCR reaction. PCR will only amplify natural DNA targets containing dT rather than dU.

period of time. This protocol does add significantly to the cost of the PCR since one has to include dUTP and UNG in the reaction mix.

Inactivation of DNA by ultraviolet (UV) light post-amplification

This is a simple and straightforward method for inactivating amplicons. After completion of thermal cycling, the unopened PCR tubes are exposed to UV light (254 to 300 nm wavelength) to induce the formation of thymine::thymine (TT) dimers which block the progression of most polymerases used for PCR (Mifflin, 1997; Fig. 10.2.5). The length of time needed for exposing DNA to UV light is dependent on the wavelength used since shorter wavelengths require less exposure time. For 254-nm UV light, typically no more than a 20 min exposure is required, but determination of the best exposure time should be empirically evaluated since light sources and plastics used for PCR tubes vary, which influences the effectiveness of TT dimer formation caused by UV light. This can be done by exposing amplicons in PCR tubes to UV light for varying amounts of time and then adding a small portion of the reaction to a new PCR and determining if new amplicon can be detected after thermal cycling. Note that this is the basis for UV treatments described in the good laboratory practices listed above.

A related method for inactivation of PCR amplicons is through the use of psoralen or isopsoralen with UV light to form adducts that inhibit subsequent amplification by PCR (Cimino et al., 1991; Fig. 10.2.6). The psoralens intercalate between stacked base pairs of double-stranded DNA and form adducts or inter-strand cross-links upon exposure to UV light. To implement this method, one typically includes isopsoralen within the PCR reaction mix and performs PCR in a normal

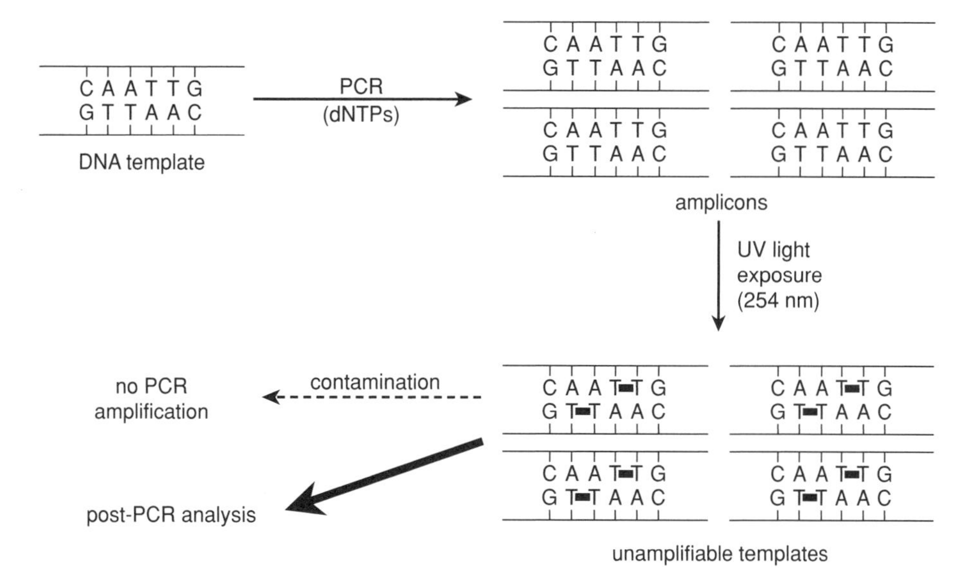

Figure 10.2.5 Thymidine dimer contamination. Standard PCR is performed. Prior to post-PCR analysis, the reaction mix is exposed to short-wave UV light to form thymidine (T-T) dimers which cannot participate in future PCRs should these amplicons contaminate a new PCR reaction. Black bars denote T-T dimer formation.

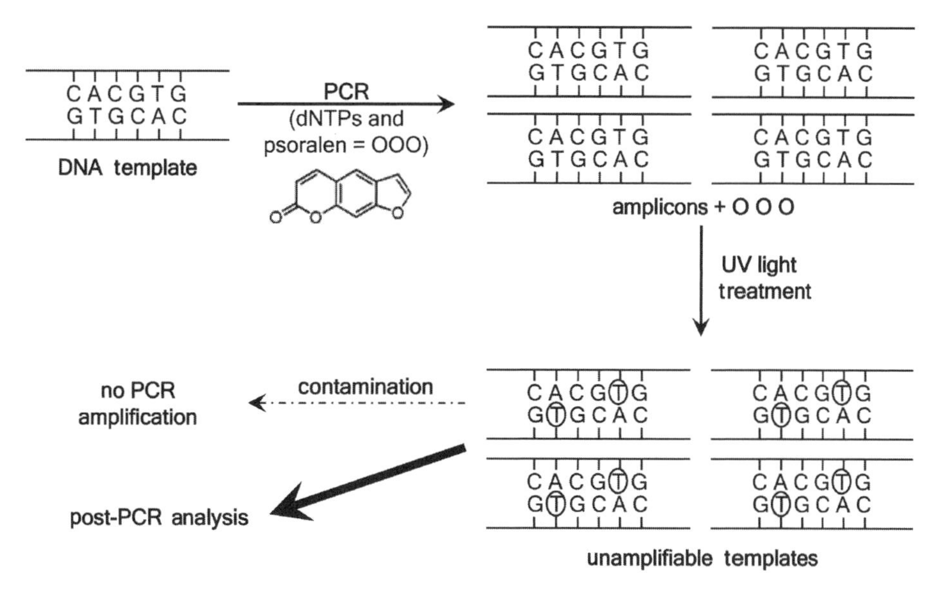

Figure 10.2.6 Psoralen contamination-containment procedure. Standard PCR with the inclusion of psoralen (or isopsoralen; shown as open circles) is run. Upon exposure of the completed reaction to UV light, psoralen will conjugate to dT bases (a T within a circle). The attached psoralen will inhibit any replication by DNA polymerases and thus is no longer a potential interfering contaminant for future PCRs.

fashion as the free isopsoralen neither is incorporated nor does it inhibit PCR. Immediately after the thermal cycling is complete, the unopened tubes are exposed to UV (300 to 400 nm wavelength) light. Upon adduct formation, during exposure to UV light, the tubes can be safely opened for subsequent analysis. Optimization of the psoralen or isopsoralen concentration and the length of UV light exposure (dependent on the wavelength used) is required using an empirical approach. This method also increases the reagent cost of each PCR to a similar degree as the UNG method.

Laboratory design

When the laboratory requires the same PCR reactions to be run repeatedly with the need to perform very sensitive analyses within a traditional laboratory environment, contamination becomes much more likely, with a greater overall impact. To mitigate this problem, more elaborate precautions than what was described above should be taken. The most effective actions taken to avoid future contamination problems will likely require significant redesign and renovations of existing laboratory space, or need to make special accommodations in the design of new laboratory space.

The key is both operational and physical separation of certain PCR processes, pre-amplification versus amplification and post-amplification (Mifflin, 1995). It is essential to prevent PCR amplicons from entering the area where pre-amplification set up of PCR occurs. To accomplish this, one must control the local environment of the laboratory areas, as well as the foot traffic through these areas. The most drastic and effective way to do so is to situate the set up area in a dedicated room that is located as far as possible from where thermal cycling and post-PCR work is performed, and controlling the ventilation of this laboratory area (Fig. 10.2.7A).

The physical layout for the laboratory set up for pre-PCR work ideally includes an anteroom where additional precautions can be taken in terms of controlling the foot traffic into and out of the setup laboratory. Outside the anteroom, there is a place to store the laboratory coats used in areas that may be exposed to amplicons. There is also a mat with a sticky or tacky coating on the floor in front of

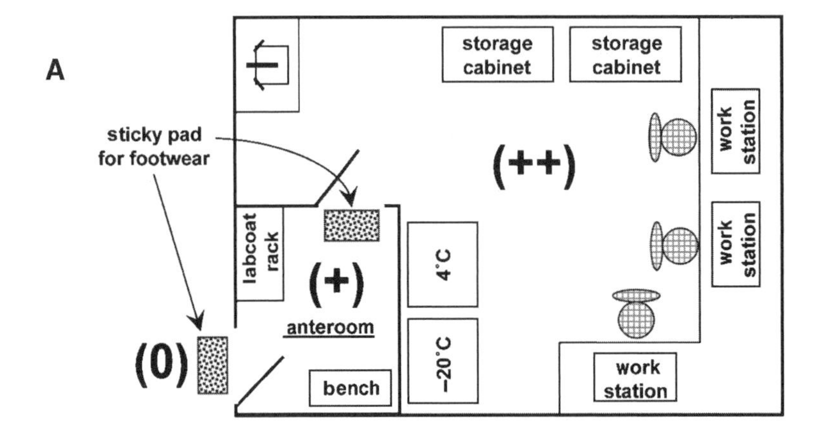

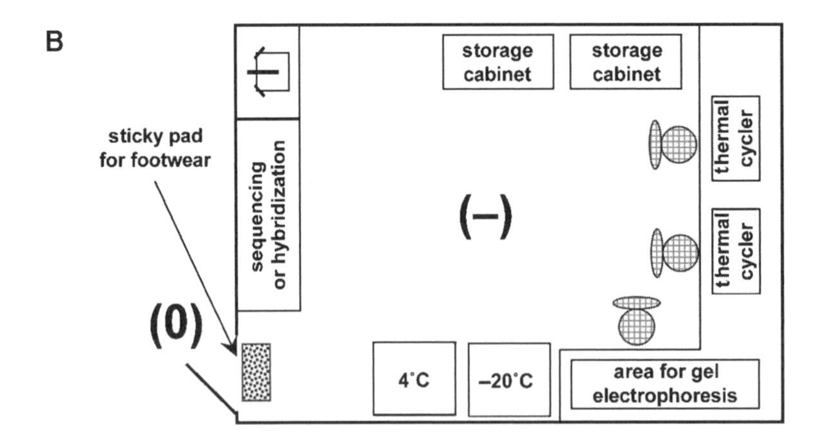

Figure 10.2.7 Laboratory design for contamination containment. (**A**) Pre-PCR set up laboratory with an anteroom; storage cabinets for materials and supplies; 4°C refrigerator and −20°C freezer for storage of enzymes, samples and other items requiring cold storage; workstations for setting up PCRs in a contained environment; and air pressure regulation for laboratory spaces: (++) denotes positive pressure, (+) denotes slightly less positive pressure, and (0) denotes normal ambient laboratory air pressure. (**B**) Thermal cycling and post-PCR analysis laboratory is typically not adjacent to the pre-PCR set up laboratory; contains thermal cycler(s), equipment for running and visualizing (including photo documentation) gels, cold storage for reagents, supplies and storing completed PCRs, and an area for other post-PCR analytical activities (i.e. sequencing or hybridization studies); and notation of laboratory air pressure, with the laboratory having negative pressure, (−) relative to the main laboratory area with standard ambient pressure.

the door to the anteroom so that prior to entering the anteroom one steps on the mat to remove loose material from the bottoms of shoes, including grime with amplicons picked up from the general laboratory and/or post-PCR areas. As an added precaution, disposable booties should be put on over the shoes prior to entry into the main pre-PCR setup area. In the anteroom, laboratory coats (these can be disposable) dedicated solely to the pre-PCR laboratory should be available for those entering that area to do work. After work is complete in that area and the individual has reentered the anteroom, the laboratory coat is put away or discarded, and the booties are discarded. At this time, the individual can leave the anteroom and return to the general laboratory or floor area.

In terms of ventilation, it is optimal to have a dedicated air-handling system for the pre-PCR laboratory and anteroom. The air pressure in these rooms needs to be adjusted such that the pre-PCR laboratory has positive pressure relative to the main building interior, and the anteroom has positive pressure that is between that of the pre-PCR and main interior areas. This positive pressure permits the flow of air out of the pre-PCR setup areas, thus preventing intake of particulates and aerosols that may contain amplicon contamination. Design of the air-handling system must take into account details such as positioning the air intake to not be in close proximity to the exhaust of the air-handling system for the area containing the PCR and post-PCR activities.

Other precautions mentioned earlier should be implemented even with inclusion of these more extreme laboratory design measures. For instance, all reagents used in the pre-PCR set up laboratory should only be prepared, stored, and used in that area. This is also true for all pieces of equipment and disposables. Some laboratories illuminate this laboratory area with UV light when not occupied, to inactivate any DNA present in exposed areas. The UV light should have a safety switch connected to a motion detector or to the visible light switch so that the UV light automatically shuts off if someone is in the room or turns on the regular light. Regarding housekeeping of this area, clean-up equipment and products must be kept in and used only for this area. Brooms, mops, and buckets used in other areas that are exposed to amplicons can become vehicles carrying contaminants into this very sensitive pre-PCR setup area.

The thermal cycling and post-PCR processing laboratory should be self-contained as well, when possible (Fig. 10.2.7B). The air-handling system should be set up for the laboratory air pressure to be slightly negative relative to the main building interior air pressure. This helps to keep the particulates within this area and not have them escape to the general area by possibly drifting or being carried to the pre-PCR set up area. In addition, the adhesive foot pad for removing loose material from the soles of footwear should be placed inside the post-PCR area next to the door so that those exiting can step on the pad to remove any contaminants they carry on the bottom of their shoes. Finally, all solid waste material from the post-PCR area should be bagged and secured using secondary containment, typically another waste bag that is tied after inserting the primary waste container.

STRATEGIC PLANNING

Careful planning and consideration of primer design, sample preparation, reagent and enzyme concentrations, as well as the environment in which the reaction will be set up, are all very important in PCR.

Primer Design

Effective primer design is critical to the specificity and sensitivity of a PCR reaction. When designing primers, the most important areas to consider are the relative sequence quality, length of the primer, melting temperature (T_m) of the primer, %G-C content, the 3′ end of the sequence, and potential secondary structures formed by either the template or primers.

The performance of a primer begins with the quality and specificity of the primer sequence. The sequence of a primer should be screened to be outside of low complexity regions (regions with short periodicity internal repeats, such as tri-nucleotide repeats), larger repeat sequence regions, and regions homologous to the target sequence desired (such as pseudogenes). If possible, it is optimal to be in the unique and invariate areas of a sequence, most usually in coding regions. Areas where the target sequences are unique and invariate will produce the most specific PCR products across multiple samples and reduce unintended priming. Overall, the best primer sequences will hybridize specifically with the target sequence without observable hybridization to any other sequences present in the template sample. The Basic Local Alignment Search Tool (BLAST; *http://www.ncbi.nlm.nih.gov/BLAST*) can be used to search for sequence similarities between the primer and other target template regions. Some amount of target sequence variability (e.g. single nucleotide polymorphisms; SNPs) is known. For instance, when working with human sequences from many different individuals, there are databases that list known SNPs that can be used when designing primers to determine potential variability of the target sequence. dbSNP is a commonly used database that is freely accessible for such comparisons (*http://www.ncbi.nlm.nih.gov/projects/SNP*).

The length of the PCR primer should be at least 17 nucleotides (nt) and typically no greater than 30 nt. At random, there is a 1 in 4 chance of a particular nucleotide (A, G, C, T or U) to be present at any given position. The chance of two specific nucleotide bases to occur together at random in a sequence string is $1:4^2$ or 1 in 16 chances. For a string of 16 nucleotides to occur together at random, there is a $1:4^{16}$ or 1 in 4,294,967,296 chances, a number which is a little larger than the size of the human genome. For this reason, primers that are a minimum of 17 nt in length are highly likely to be unique to the target region. It should also be noted that the shorter the sequence, the more efficient the annealing capability. Therefore, most primers are designed at a length between 18 to 25 nt.

The T_m of the primer is a principle element of specificity in a PCR reaction. The T_m is the temperature at which half of the primers are annealed to the template at any given time. The T_m of a primer is calculated from the number and sequence composition of the nucleotides comprising it–the longer the primer, the higher the T_m. Additionally, primers rich in G-C content will have higher melting temperatures than those with significant A-T composition. Both primers in a set should not have a T_m difference >2°C. The primers should have a T_m between 54° to 80°C and be ~5°C above the annealing temperature. The following equation works as a good rule of thumb for calculating the T_m of short sequences (<25 nt; Thein and Wallace, 1986):

$$T_m = 2C * (A+T) + 4C * (C+G)$$

The nearest neighbor calculation for short nucleotide sequences has been accepted as a more accurate measure for calculating T_m. The nearest neighbor equation shown below is based on the thermodynamic interactions between bases to account for the actual sequence of nucleotides. The equation also includes a correction for helix initiation and the salt concentration of the primers in solution. This calculation is widely used in many primer design software packages (Breslauer et al., 1986; Rychlik et al., 1990; Sugimoto et al., 1996).

$$\text{Nearest neighbor } T_m = (\Delta H - 3.4 \text{ kcal})/\{[A + \Delta S] + [R \ln(C/4)]\} - 273.15 + 16.6 \log_{10}[\text{salt}]$$

Where ΔH is the sum of nearest neighbor enthalpy changes, A is the helix initiation constant of 10.8 cal/K° mole for nonself complementary, −12.4 cal/°K mole for complementary sequences, ΔS is the sum of nearest neighbor entropy changes, R is the gas constant 1.987 cal/K° mole, and C is the concentration of primer.

The following is a brief list of characteristics of functional primers:

1. The percentage of G and C nucleotides in a primer sequence (%G-C content) is tightly related to the length and T_m of the primer. Good primers generally have between 40% to 60% G-C content and a random distribution of the bases.

2. The primers should not be biased to any stretches of purines or pyrimidines in the sequence.
3. The primer sequences should also be checked to ensure that palindromic sequences are not internal to the primer.
4. To prevent secondary structures such as hairpin loops, any self-complementarity within the primer sequence should be avoided. Available primer design software typically has the capability to either set design parameters to avoid sequences that would create secondary structures or can analyze individual designs and identify such structures.
5. The 3′ end of the forward and reverse primers should not be complementary to each other.
6. To promote specific binding between the primer and the template at the 3′ prime end of the primer, it is recommended to have Gs and Cs within the last five nucleotides of the primer but having no more than three Gs and Cs in that region. This is often referred to as a "GC clamp" due to the higher bond strength between G-C versus A-T. It has also been shown that primers that end with As at the 3′ end help reduce the presence of primer complementarity (Innis et al., 1999).

Table 10.2.3 provides some examples of software packages that are available to assist in the design of primer sequences and T_m calculations.

Sample Preparation

Theoretically, PCR only requires that a single intact DNA strand containing the target region be available. The very nature of the exponential amplification does not require a large amount of input DNA to acquire results. Typical amounts of template DNA range from 0.1 to 1 ng for plasmid or phage DNA, and between 0.1 to 1 µg for genomic DNA. However, the quality of the template and the presence of any inhibitory contaminants affect the PCR efficiency and overall yield of the PCR. For long-range PCR products, poor quality DNA (sheared or nicked), will greatly impact the efficiency of amplification. Therefore, isolation and purification of high quality nucleic acids for amplification is paramount for successful PCR.

The isolation of DNA from cells can be accomplished through a variety of routine methods. Selection of an appropriate method will depend on the sample source, (i.e., whole blood, tissue, bacteria, plant, or fungus), the amount of DNA required, and the resulting quality of DNA. Different methods may have different considerations in downstream PCR. Table 10.2.4 compares the benefits and disadvantages of some of the commonly used purification methods. Special considerations are needed for specific sample sources. For more details on this, please review *UNIT 5.2*.

There are classes of samples in which it is inherently difficult to prepare highly purified DNA. For instance, food and environmental samples are typically highly complex and contain many potential PCR inhibitors of which most have not been identified as such. One approach to counteract at least some of these inhibitors is to include additive(s) in the PCR buffer. Bovine serum albumin (BSA)

Table 10.2.3 Primer Design Software Packages

Software package	Provider	Web site
Oligo	Molecular Biology Insights	*http://www.oligo.net*
Primer 3	Whitehead Institute of Biomedical Research	*http://frodo.wi.mit.edu/cgi-bin/primer3/primer3_www.cgi*
Primer Express	Applied Biosystems	*http://www.appliedbiosystems.com*
PrimerSelect	DNAStar	*http://www.dnastar.com/products/primerselect.php*

Table 10.2.4 Comparison of DNA Purification Methods

Method	Features and benefits	Disadvantages
CsCl centrifugation	Highly pure DNA. Isolates large size of DNA. Proven very effective for plasmid purification.	Labor-intensive and expensive. This is a time-consuming process. Ethidium bromide (EtBr) used in protocol is carcinogenic.
Phenol/chloroform extraction	Scalable (small or large input amounts). Applicable for all sizes of DNA.	Organic solvents are toxic and not easy to dispose. The process is labor-intensive and cumbersome. Residual phenol or chloroform will inhibit polymerase activity.
Ethanol precipitation ("salting out")	Fast and easy	Precipitated pellet can be difficult to see and loss of sample could result. Over-drying of a sample sometimes results in insoluble DNA pellets. Carryover of ethanol will inhibit polymerase activity. The yields and purity are highly variable with this technique.
Silica membrane/ chaotropic salt	Rapid, easy, and elutes in ready-to-use low-salt environment	DNA tends to shear in the process resulting in smaller size DNA (affects long-PCR). Ethanol is used in the washes; thus care is required to ensure no ethanol carryover that could result in polymerase inhibition.
Anion exchange resin	Yields highly pure DNA. Isolates large DNA molecules.	Elutes in high-salt buffer. Desalting and/or concentration will be required downstream. Any salt carry-over may impact specific polymerase activity.

is a typical additive that acts as an inhibitor "sponge" by having 100 to 500 μg/ml in the reaction buffer (Kreader, 1996). T4 gene 32 protein is another additive but less commonly used (Kreader, 1996).

dNTP Concentration

The final concentrations of the deoxynucleotide bases range between 20 to 200 μM each in a conventional PCR reaction. It is critical to use equal concentrations of each dNTP to avoid the misincorporation of bases. If one nucleotide is present at a higher concentration, the polymerase will tend to misincorporate that particular nucleotide, and can slow down elongation, which can result in early truncation of products particularly in cases where the polymerase lacks $3' \rightarrow 5'$ exonuclease activity (proofreading). Lower concentrations of dNTPs increase fidelity and specificity of the reaction. In the case of long PCR, dNTP concentrations may need to be increased to account for the length of the product. Preblended mixtures of dNTPs for direct use in PCR reaction set up are available commercially for ease-of-use and to reduce manual dilution and solution transfer errors.

$MgCl_2$ Concentration

The optimal $MgCl_2$ concentration should be determined empirically for each PCR system. Most DNA polymerases require a divalent cation as a cofactor to properly function, in most cases Mg^{2+} is utilized, with Mn^{2+} being used in a few exceptions such as with *Tth* DNA polymerase. Cations will also form complexes with the negatively charged dNTPs, primers, and DNA template. As the concentration of dNTPs or primers is increased, the Mg^{2+} needs to be increased. Too many Mg^{2+} ions enhance the stability of mismatched primers and can result in an increase in nonspecific products. High Mg^{2+} ion concentrations also impact amplicon sequence accuracy by decreasing the enzyme fidelity of nucleotide incorporation. It is important to check, and sometimes optimize, the concentration of Mg^{2+} ions to ensure that specific amplification of the desired target sequence is reliably achieved. Ethylenediaminetetraacetic acid (EDTA) chelates Mg^{2+} ions and is present in many PCR buffers. If EDTA is present in the PCR buffers, the concentration of Mg^{2+} ion concentration must be increased proportionally (one molecule of EDTA binds one Mg^{2+}). The optimal Mg^{2+} concentration is dependent on the concentration of dNTPs, concentration of primers, amount of template, and the polymerase being used. Primer sequence can sometimes impact optimal Mg^{2+} concentration and requires testing of reaction conditions that vary Mg^{2+} and primer concentrations. In general, the Mg^{2+} concentration should be between 0.5 to 2.5 mM when used with *Taq* DNA polymerase. Note that Support Protocol 2 provides a method for optimizing $MgCl_2$ concentration.

10

Enzyme Concentration

The enzyme concentration for *Taq* DNA polymerase is between 1 to 2.5 U per 100 µl reaction. One unit of *Taq* DNA polymerase is defined as the amount of enzyme that will incorporate 10 nmol dNTP into acid-insoluble material in 30 min at 75°C (*http://www.neb.com/nebecomm/products_intl/productM0267.asp*). The unit definitions may vary based on the manufacturer and quality control of assays. When too much enzyme is added to the PCR reaction, high background amplification or increase in nonspecific products may result.

It should be noted that enzymes from different suppliers may have different specific activities due to different formulation processes and the effective unit definitions may be different.

PCR Additives

Enhancement of PCR performance is an ongoing endeavor for commercial sources of PCR-related reagents. A common way to improve PCR performance is to introduce chemical additives into the reaction buffers either supplied with the DNA polymerase or purchased separately and used with an enzyme that could be obtained from a different vendor. Typical additives found in reaction buffers include betaine and dimethylsulfoxide (DMSO) which provide more consistent amplification efficiencies by disruption of inhibitory secondary structures, especially for DNA templates that have high GC content of 60% to 70% (Baskaran et al., 1996; Henke et al., 1997). DMSO has also been found to improve yields of long PCR (Cheng et al., 1994), presumably due to the higher probability of encountering inhibitory secondary structures. Combinations of reaction enhancers have been examined and further improved formulations identified such as the addition of 7-deaza-dGTP to betaine and DMSO for amplification of high GC content templates (Musso et al., 2006). Typical working ranges of additive concentrations are 1% to 5% DMSO, 1.0 to 1.3 M betaine, and ~50 µM 7-deaza-dGTP. Additional systematic studies have characterized other general enhancers of PCR that include low-molecular-weight sulfones and amides (Chakrabarti and Schutt, 2001a,b). Implementation of additives requires empirical testing of the possible enhancers at different concentrations and possibly with different templates, as there may be specific interactions that could interfere with the overall reaction performance.

SAFETY CONSIDERATIONS

The most significant safety issues regarding PCR involve the front-end portion of the process, namely sample preparation. Any samples that potentially contain infectious disease pathogens must be handled in accordance with the requisite biosafety level guidelines pertaining to the most serious potential pathogen present in the samples being processed. Refer to APPENDIX 1 and other resources (e.g., *http://www.absa.org*) for more information on biosafety.

Another safety concern involves analysis of PCR products using gel electrophoresis. There is risk of electric shock or electrocution, as well as handling of gels that have been stained with potential carcinogens such as ethidium bromide and the use of UV light for visualization of stained DNA. These concerns are addressed in *UNITS 7.1 & 7.2*.

Finally, there is the concern of amplification of genetic material from infectious disease agents and mutated human genes that may have disease implications. In the great majority of cases, the amplicons produced by PCR will not replicate the entire genome of the pathogen or the entire mutated human gene, but safe disposal of the amplicons as biohazardous waste is still required so that one eliminates the chance that the DNA can transform cells of the user or other workers in the laboratory, as well as other organisms that may come in contact with the amplicon (e.g. bacteria in the laboratory or within people in the laboratory). Guidelines for safe handling of recombinant DNA molecules are available from the National Institutes of Health at *http://www4.od.nih.gov/oba/rac/guidelines_02/NIH_Gdlnes_lnk_2002z.pdf.*

10

PROTOCOLS

To carry out a PCR reaction will require the preparation of the template to be amplified, the design of primers specific to the region to be amplified on the template, the preparation of a master reaction mix, the design of a thermal cycling program specific to the primers and template, and analysis of the amplicons at the end of the PCR.

The protocols that follow provide the information necessary to conduct a routine PCR experiment. It is assumed that template DNA has been previously isolated and prepared (see *UNIT 5.2*). Reaction mix assembly, including a materials list, is covered in detail. Considerations and guidelines for amplification conditions as well as a basic PCR profile are provided. Additionally, amplicon analysis is covered. The preparation of an agarose gel (*UNIT 7.2*) or polyacrylamide gel (*UNIT 8.2*), as well as staining the gel (*UNIT 7.4*) for amplicon analysis, are briefly discussed; significantly more detail can be found in the corresponding units. Supporting protocols are provided to describe and implement the use of temperature gradients for PCR condition optimization, as well as an $MgCl_2$ titration protocol to determine the optimal $MgCl_2$ concentration for a given reaction.

Basic Protocol: Routine PCR

This protocol provides a generic starting point for conducting PCR using genomic DNA or cDNA as a template. The protocol, as stated, is designed for a 50-μl reaction volume; however, the reaction can be scaled up to 100 μl or down to 20 μl without changing reagent concentrations, to facilitate specific experimental needs. PCR is very sensitive to contamination from other DNA, particularly other amplicons. Thus, caution should be used to make sure the area where the PCR reaction is assembled, or any components to be used in the PCR reaction are made, is clean, reasonably free of dust (a PCR cabinet is highly recommended), and as far away as possible in the laboratory from where PCR products will later be analyzed. To reduce the chance for contamination, wear gloves, use only pipettors dedicated to making PCR reaction mixes and components (never used with amplified PCR product), and use aerosol tips (see Contamination Control section of this unit for further details).

There are numerous variables that can affect the outcome of a PCR reaction. The following protocol assumes the use of a hot-start *Taq* DNA polymerase (one does not have to add the enzyme at the beginning of the reaction but can add it in a master mix format), a PCR buffer that does not contain $MgCl_2$, and concentrations of reagents that represent a good starting point for primer-specific optimizations. Parameters that can be adjusted in a given reaction along with acceptable ranges are pointed out in the standard reaction table.

PCR thermal cycling profiles have three parameters that can be manipulated: time, temperature, and the rate at which the machine cycles between temperatures (ramp rate). The optimal conditions for a given reaction will be dependent on the type of thermal cycler in which the reaction will be run, the melting temperatures of the primers used, the size of the fragment to be amplified, the nature of the template (e.g., GC content, secondary structure), and the reaction vessel used. It is possible to use the ramp rate to further optimize the PCR reaction conditions. For most PCR reactions, ramp rates are set to manufacturer's default setting (often maximum settings) to rapidly cycle between denaturing, annealing, and elongation temperatures and to speed up the overall PCR. It is not commonly necessary to adjust this parameter. However, to compensate for a reduced rate of priming, for example, adjusting the ramp rate may increase the cycle range for successful amplification of the target sequence. In addition, when annealing temperatures are relatively high (e.g., ≥60°C), elongation can be efficient enough not to change the temperature for the elongation step, and this results in a two-step PCR which can reduce the overall cycling time since one ramping step is eliminated. This time savings is minimal since the overall ramping from annealment to elongation to denaturation covers effectively the same temperature difference as the two-step from annealing/elongation to denaturation.

Materials

10% (w/v) bleach solution
Master mix components (see Table 10.2.5):
  5 U/μl DNA polymerase (e.g., *Taq*)
  10× PCR buffer with $MgCl_2$ (e.g., Promega, no. M8295; also see recipe) or without $MgCl_2$ (e.g., Promega, no. M3005; also see recipe); typically optimized for DNA polymerase of choice and provided by the manufacturer of the enzyme
  25 mM $MgCl_2$ (if not already included in PCR buffer)
  40 mM dNTPs (dATP, dTTP, dCTP, and dGTP; e.g., Promega, no. C1141; also see recipe)
  10 μm forward (upstream) and 10 μM reverse (downstream) primer (see recipe for Primers; custom synthesis available from Invitrogen)
  Molecular-biology-grade, sterile, nuclease-free ddH_2O (e.g., Invitrogen, no. 10977-015)
DNA template (*UNIT 5.2*)
Sterile, nuclease-free mineral oil (e.g., Sigma, no. M5904; only necessary if thermal cycler does not have heated lid)
Gel loading buffer/dye
Agarose gel
100-bp PCR ladder (e.g., Sigma, no. D3687)
Ethidium bromide

0.2-ml (e.g., VWR cat. no. 10011-802) or 0.5-ml (e.g., VWR, no. 10011836) thin-walled reaction tubes (size dependent on thermal cycler block and manufacturer specifications)
Dedicated pipets used only for setting up PCR reactions (2 μl, 20 μl, 200 μl, and 1000 μl)
Sterile micropipettor tips (made for the pipets used for PCR reaction set up) with aerosol barrier (e.g., Rainin Instruments)
Vortex
Thermal cycler (see Table 10.2.2 for a list of manufacturers)
UV transilluminator

Additional reagents and equipment for preparing the DNA template (*UNIT 5.2*) and agarose gel electrophoresis including staining with ethidium bromide (*UNIT 7.2*)

NOTE: The time to complete reaction preparation will vary depending on the number of samples to be run. Typically setting up ten samples will take 30 to 45 min.

Prepare reaction mixture

1. Wipe down the work surface with 10% bleach solution.

2. Thaw PCR components on ice. Once all components have completely thawed, vortex each moderately before using.

   Note that some DNA polymerases can be sensitive to inactivation by vortexing too vigorously. It is a good rule to mix the DNA polymerase by flicking the tube to mix the contents and then returning it promptly back on ice.

Briefly centrifuging mixed tubes is often advisable to return disturbed contents to the bottom of the tube.

3. Prepare a master mix on ice according to the standard reaction table (Table 10.2.5). Add each reagent in the order listed in the table with the exception of the template. Briefly and moderately vortex the master mix.

It is common to assemble a master mix for a given PCR experiment, for each primer pair, that contains all components except the template DNA at n tubes + 2 (e.g., 10 samples will be run with a specific primer pair, prepare a master mix for 12 samples). This will allow for variations in pipetting; further, the master mix that is left over can be run, without template, in a separate tube as a negative control for the reaction. Any amplification seen in this tube could be an indication that the reaction mix was contaminated and the experiment would need to be repeated.

DNA template is left out at this step for the following reasons: (1) this offers the opportunity to provide consistency to all the reactions by using the same mix for multiple samples, and (2) a negative control (one that does not contain template and therefore should not amplify a product) for the overall reaction mix can be run to check for contamination.

4. Aliquot 49 μl of the master reaction mix into each thin-walled PCR tube and then add 1 μl template to each tube. Keep the tubes on ice.

Check the tubes for bubbles or reagents remaining on the sides of the tube. It is a good idea to quickly vortex and centrifuge to the tubes to ensure that all the reagents are mixed and located at the bottom of the tube.

5. If the thermal cycler does not have a heated lid, add 25 to 50 μl sterile, nuclease-free mineral oil to each tube. If the thermal cycler has a heated lid, make sure to adjust the lid so that it sits snugly against the tops of the reaction tubes.

As an alternative to mineral oil, paraffin beads may be used. Note these will solidify once the reaction has completed and the tubes are cooled to 4°C. Simply puncture the resulting wax seal with a pipet tip to access the PCR reaction mix for analysis.

Table 10.2.5 Standard PCR Reaction Mixture

Components	Final concentration	Per tube volume	Master mix for 10 tubes (prepare for 12 tubes)
10× PCR buffer $MgCl_2$-free	1×	5 μl	60 μl
25 mM $MgCl_2$[a]	1.5 mM	3 μl	36 μl
40 mM dNTP mix[b]	0.2 mM each dNTP	1 μl	12 μl
10 μM forward primer	1 μM[c]	5 μl	60 μl
10 μM reverse primer	1 μM[c]	5 μl	60 μl
Sterile, nuclease-free H_2O		30.75 μl (to a final volume of 50 μl)	369 μl
5 U/μl hot-start *Taq* DNA polymerase[d]	0.025 U/μl	0.25 μl	3 μl
>100 ng/μl template[e]	2 ng/μl	1 μl	—

[a] $MgCl_2$ concentration can be titrated to optimize the PCR reaction. $MgCl_2$ concentration ranges between 1.5 and 4.0 mM for most PCR.
[b] There are numerous commercially available dNTP mixes. Most are 40 mM (10 mM of each of the 4 dNTPs).
[c] Primer concentration can also be titrated to optimize the PCR reaction. Final primer concentrations generally can be adjusted within the range of 0.2 to 1 μM.
[d] Concentrations of *Taq* will vary. When using a non-hot-start polymerase, the enzyme should not be added to the reaction vessel until the tubes are at 95°C in the thermal cycler to avoid the formation of nonspecific products.
[e] Template DNA concentration can vary, but generally the final concentration of the template DNA will not exceed 10 ng/μl. The technique is sensitive enough to detect pmol quantities of template.

Amplification conditions

6. Program the thermal cycler with the desired PCR profile. The following is a generic profile that can be modified based on the specific characteristics of the primers, template, and size of the amplicon.

Initial step:	2 min	95°C	(denaturation)
35 cycles:	15 sec	95°C	(denaturation)
	30 sec	55°C	(annealing)
	60 sec	72°C	(extension)
Final elongation:	5 min	72°C	(extension)
Final step:	indefinite	4°C	(hold).

__Initial denaturation__: This step depends on the polymerase used (see Polymerase section) and the GC content and secondary structure of the target template. Completely denaturing the template DNA at the beginning of the PCR reaction ensures efficient use of the template in the first cycle of amplification. If a hot-start enzyme is used in the reaction, this initial step may also be important in the activation of the polymerase activity. If the GC content of the template is 50% or less, incubation at 95°C for 2 to 3 min is sufficient. The incubation can be increased up to 5 min for GC-rich templates.

__Annealing__: This step of the cycle is one of the most important to correctly adjust in the reaction. The temperature for this step can vary significantly based on the composition of the primer sequences. As a general rule, annealing temperatures are initially set at approximately 5°C below the melting temperatures (T_m) of the primers. The closer to the T_m of the primer, the more stringent the conditions for the reaction (i.e. there will be less nonspecific product). Annealing temperatures that are too high may result in no amplification due to inability of the primers to anneal to the target sequence. Quality primer design (see Primer Design in Strategic Planning) can reduce or eliminate many of these problems.

__Elongation/extension__: The length of time for the elongation step is directly proportional to the length of the amplicon. For example, 30 sec is sufficient for shorter amplicons (100 to 500 bp) whereas amplicons of >1 kb will require elongation times of 1 min or more. As a general rule of thumb, estimate 30 sec of elongation time per 500 bp of DNA in a standard PCR. The final elongation step is often 5 to 7 min to ensure that the PCR products amplified are full length.

__Cycle number__: Generally, 30 to 35 cycles of PCR are sufficient to amplify enough product to be visualized on a gel. Significantly increasing the number of cycles does not considerably increase the amount of amplicon. In fact, greater cycle numbers can increase the chance for nonspecific amplification. If there is no visible amplification of product after 35 cycles, simply increasing the cycle number will not improve the result.

Amplicon analysis

This portion of the protocol should be done in an area that is isolated from the PCR preparation area. Separate tips and pipettors should be used to aliquot PCR products. Agarose or nondenaturing polyacrylamide is typically used to make a gel for PCR analysis. See *UNIT 7.2* for agarose gel electrophoresis.

7. Mix 5 to 20 µl of amplified PCR product with a gel loading dye (typically 1:1 to 1:6 PCR product:loading dye).

   Do not overload the well with PCR product as this can interfere with visualization and size estimation of the amplicon. Generally, 5 to 10 µl of a robust reaction is sufficient to visualize the amplicon.

8. Load the PCR product/loading dye mix on a gel and electrophorese. Include a PCR sizing ladder composed of DNA fragments of known sizes in a well on the gel to help estimate the size of the PCR product.

   The type and concentration of the gel will depend on the expected PCR product size.

   0.5% to 2% agarose gels (UNIT 7.2) are sufficient for many PCR products.

9. Run the gel between 80 to 120 V for 20 to 45 min depending on the size of the amplicon, the concentration of the gel matrix and the size of the gel.

   Smaller amplicons will require a shorter run time than larger amplicons if run at the same gel matrix concentration.

   Many gel loading dyes have indicator bands that can be used to determine how far PCR products have run into the gel.

   Running the gel too quickly (i.e. at higher voltage) can result in bands that have a characteristic "smiling" appearance where the ends of the band appear to run slower than the middle of the band. It is also possible to run the gel at a high enough voltage to heat up the running buffer to the point that the agarose gel can melt and become useless. Running the gel too slowly can result in a band that looks diffuse, making it difficult to estimate amplicon size.

10. Stain the gel with ethidium bromide and visualize on a UV transilluminator (*UNIT 7.2*).

    CAUTION: *Ethidium bromide is a mutagen and potential carcinogen. Gloves should be worn and care should be taken when handling ethidium bromide solutions.*

    Gels can also be stained with SYBR Green or SYBR Gold (Invitrogen), which also intercalate double-stranded DNA and fluoresce under UV light (Schneeberger et al., 1995; Morin and Smith 1995; also see UNIT 7.2).

11. Analyze the gel and compare the position of the PCR product to the DNA size ladder loaded on the gel to enable estimation of amplicon size.

    Nonspecific products will often appear as smears on the gel. Primer dimers appear as hazy bands that run below the amplicon of interest (can be in the range of 25 to 50 bp in size–for an example, see the bottom band in lanes 5 through 8 in Fig. 10.2.2).

12. To confirm that the amplicon represents the product of interest, use a small amount of the amplification reaction for sequencing (Korn et al., 1993) or digest with a restriction enzyme (see *UNIT 10.1*) known to cut the product.

Support Protocol 1: Using Temperature Gradients for Rapid Optimization of PCR Cycling Conditions

Determining the proper annealing temperature for a new set of PCR primers is a key component to successful and selective amplification of the target DNA sequence. Many thermal cyclers available on the market today have a temperature gradient function that allows the researcher to designate a range of temperatures for the annealing step to be run in one experiment. Each column of wells on the block will have a different annealing temperature.

Additional Materials (also see Basic Protocol)

Thermal cycler with temperature gradient function

Amplification conditions

1. Program the thermal cycler according to manufacturer's specifications with the desired PCR profile, including a gradient function for the annealing step. The following is a generic profile. The annealing temperature gradient should be set up to flank the T_m of the primers (e.g. ± 5°C).

Initial step:	2 min	95°C	(denaturation)
35 cycles:	15 sec	95°C	(denaturation)
	30 sec	54° to 62°C	(annealing)
	60 sec	72°C	(extension)
Final elongation:	5 min	72°C	(extension)
Final step:	indefinite	4°C	(hold).

2. Prepare a master mix including the template for one well in each column on the thermal cycler block plus one additional well ($n + 1$) according to the standard reaction mixture table.

   Many thermal cycler blocks have a 96-well format. Assuming this format with twelve columns and eight rows, the gradient function will permit the analysis of up to twelve different annealing temperatures. Therefore, the master mix should be prepared for thirteen tubes to allow for pipettor inaccuracies.

3. Aliquot 50 µl master mix into each tube and place one tube in each column of the thermal cycler block.

4. Once the PCR profile is completed, analyze the samples according to the Amplicon analysis section of the Basic Protocol.

10

Support Protocol 2: Titration of $MgCl_2$ Concentration

It is well established that $MgCl_2$ concentration can have a significant effect on PCR outcome. Many commercially available PCR buffers on the market today contain $MgCl_2$ at a concentration that is generally successful for a majority of primer and template combinations. However, typically, to optimize PCR conditions it is necessary to titrate $MgCl_2$ concentration for best results.

For materials, see Basic Protocol.

Prepare reaction mixture

1. Wipe down the work surface with a 10% bleach solution.

2. Thaw PCR components on ice. Once all components have completely thawed mix well before using.

3. Create six tubes following the $MgCl_2$ titration mixture set up (Table 10.2.6) varying only the $MgCl_2$ and the water in each tube.

Table 10.2.6 $MgCl_2$ Titration Mixture Set Up

Components	Final concentration	Per tube volume
10× PCR buffer $MgCl_2$-free	1×	5 µl
Titration:	1.0 mM	1 µl
25 mM $MgCl_2$	1.5 mM	3 µl
	2 mM	4 µl
	3 mM	6 µl
	4 mM	8 µl
	5 mM	10 µl
40 mM dNTP mix	0.2 mM each dNTP	1 µl
10 µM forward primer	1 µM	5 µl
10 µM reverse primer	1 µM	5 µl
Sterile, nuclease-free H_2O		*X* µl (to a final volume of 50 µl)
5 U/µL hot-start *Taq* DNA polymerase	0.025 U/µl	0.25 µl
>100 ng/µl template	2 ng/µl	1 µl

4. Program the thermal cycler with the desired PCR profile. The following is a generic profile that can be modified based on the specific characteristics of the primers, template, and size of the amplicon.

Initial step:	2 min	95°C	(denaturation)
35 cycles:	15 sec	95°C	(denaturation)
	30 sec	55°C	(annealing)
	60 sec	72°C	(extension)
Final elongation:	5 min	72°C	(extension)
Final step:	indefinite	4°C	(hold).

5. Follow the amplicon analysis section of the Basic Protocol to visualize the reactions.
6. Choose the $MgCl_2$ concentration that gives the highest-intensity single band.

REAGENTS AND SOLUTIONS

Nuclease-free deionized, distilled water should be used in the preparation of all reagents to be used in the PCR reaction. For common stock solutions see UNIT 3.3.

dNTP stock (1 ml), 40 mM

100 µl 100 mM dATP
100 µl 100 mM dTTP
100 µl 100 mM dCTP
100 µl 100 mM dGTP
600 µl sterile, nuclease-free ddH_2O

Note each dNTP is 10 mM. The total concentration of the mix is then 40 mM.

Store up to 1 year at −20°C or as indicated on the product by the manufacturer.

PCR reaction buffer with $MgCl_2$, 10×

100 mM Tris·Cl, pH 9.0 (*UNIT 3.3*)
500 mM KCl
0.1% (v/v) Triton X-100
15 mM $MgCl_2$

Store in 1 ml aliquots at −20°C indefinitely.

PCR reaction buffer without $MgCl_2$, 10×

100 mM Tris·Cl, pH 8.3 (*UNIT 3.3*)
500 mM KCl
0.1% (v/v) Triton X-100

This is a generic PCR buffer that can be used with Taq polymerase. It is recommended to use the PCR reaction buffer supplied with the specific DNA polymerase being used in the reaction for optimal results. Store in 1 ml aliquots at −20°C indefinitely.

Primers

Primers (see Strategic Planning) are often shipped and received in a lyophilized state. First create a master 100× stock (see recipe) for each primer and then dilute it to a 10× working stock (see recipe). This reduces the number of freeze/thaw cycles that the master primer stock goes through and reduces the chances of contaminating the primary source for the primer.

Master stock, 100 µM
100 µM = *X* nmoles lyophilized primer + (*X* × 10 µl TE buffer, pH 7.5)

continued

The recipe for TE buffer can be found in UNIT 3.3. *Some researchers prefer to use 0.1 mM EDTA in the preparation of TE for primer stocks (TE buffer typically calls for 1.0 mM EDTA) to reduce the effect on the $MgCl_2$ concentration in the PCR reaction mix.*

To determine the amount of TE buffer to add to the lyophilized primer simply multiply the number of nmol of primer in the tube by 10 and that will be the amount of TE buffer to add to make a 100 μM primer stock. For example, if there are 38.2 nmol of primer then by adding 382 μl of TE buffer, a 100 μM primer stock is created.

Store up to 1 year at −20°C.

Working stock, 10 μM

Dilute the primer master stock (see recipe) in a new tube 1:10 with TE buffer, pH 7.5.

Store in 100 μl aliquots up to 1 year at −20°C.

UNDERSTANDING RESULTS

Following good primer design practices, condition optimizations, and laboratory techniques, PCR can be a powerful technique to amplify most target sequences. The target template should be visualized by gel electrophoresis as a crisp band without any smearing or other obvious signs of nonspecific amplification (e.g. additional unexpected products) during the optimization phase. The band should be of the size expected based on comparison to the DNA size standard run on the gel along with the PCR product and confirmed (initially) by sequencing the amplicons or restriction digestion. It is possible, depending on the application and reason for doing PCR analysis, that additional bands will be present after amplification. This may indicate the presence of an insertion (if the new band is larger than the anticipated product) or a deletion (if the new band is smaller) within the region flanked by the primers (Fig. 10.2.8). If the template population is heterozygous for the insertion or deletion then both the normal (the product designed for) and the variant band may be observed in a single reaction. If the template population is homozygous for the insertion or deletion then a band of unexpected size will be the only product visible on the gel. In this case, sequencing the PCR product can help to determine the nature of the variant observed. In some cases known, variants are being assayed for and restriction digests to differentiate the variants can be used to identify the new products.

TROUBLESHOOTING

Table 10.2.7 provides problems commonly encountered when performing PCR, as well as their possible causes and solutions.

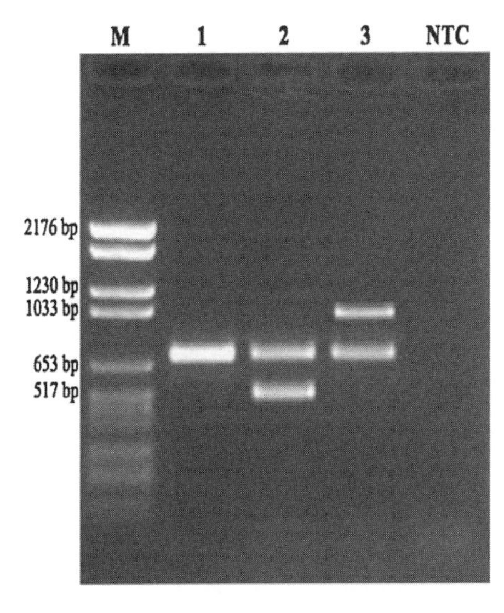

Figure 10.2.8 Understanding PCR results. An example of gel electrophoresis of a PCR for a given gene using a 1.5% agarose gel (*UNIT 7.2*). Primers were designed to flank a region known to have insertion and deletion mutations. Lane M has a DNA ladder loaded (various sized DNA bands are pointed out in the figure). Lane 1 is an example the normal band (680 bp). Lane 2 shows both the normal 680 bp product and the presence of a smaller 517 bp deletion product. Lane 3 shows both the 680 bp amplicon and the presence of a larger 1033 bp insertion product. Lane 4 (labeled NTC) is the no-template control reaction run with all of the reagents used in the PCR reaction (the remaining portion of the master mix) except the template. The absence of any amplification in the NTC demonstrates that there was no contaminating template in the reagents master mix that would account for the amplification seen in lanes 1 to 3.

Table 10.2.7 PCR Troubleshooting

Problem	Possible cause	Suggestions
Little or no product amplified	Thermal cycling conditions incorrect	Improper thermal cycling conditions may result in poor yield. Run diagnostic on thermal cycler according to recommendations of manufacturer. Consider using a prevalidated PCR reaction as a control to produce a specific known product.
	Thermal cycling conditions not optimal	Determine the optimal annealing temperature. Be sure that any initial hot-start activation steps are run according to recommendations of manufacturer
	Reagent was omitted	Consider running a known control to ensure that the master mix contains all the appropriate reagents
	Insufficient template	Increase the amount of input DNA not to exceed 1 μg/100 ml reaction. Long PCR will require more template.
	Quality of the DNA template was poor	Improperly stored template may lead to degraded material or inhibitors may be present from sample source. Create new template dilutions or isolate the DNA by another sample prep method.
	Mg^{2+} concentration not optimal	Perform a Mg^{2+} titration in 0.5 mM increments
	Not enough enzyme or loss of enzyme activity	Increase the amount of enzyme in 0.5 U intervals; enzyme reagent not stored properly, exposed to many freeze/thaw cycles, or reagent expired.
	Not enough cycles	Increase the number of cycles
	Primer sequence not optimal	Review considerations in primer design
Nonspecific product, background, or primer-dimer present (see Fig. 10.2.2)	Too much input DNA was added in the reaction	Reduce the amount of template DNA until background smearing is eliminated
	Too many cycles	Nonspecific background products may accumulate as the number of cycles increases. Reduce the total number of cycles until background is reduced.
	Primer sequence not optimal	Review considerations in primer design
	Too much enzyme was added to reaction	Reduce the amount of enzyme in the reaction
	Carryover contamination is present	Run a no-template control. If product or background is evident in the control, reagents are contaminated with PCR products or templates. Replace all reagents and review Contamination Issues section
	Annealing temperature is too low	Increase annealing temperature in 2°C degree increments.

continued

Table 10.2.7 PCR Troubleshooting, *continued*

Problem	Possible cause	Suggestions
	Too much primer was added to the reaction	Primers added in excess may create primer-dimer or other nonspecific products. Titrate the amount of primers until primer-dimer or other nonspecific products are greatly reduced.
	Low temperatures at beginning of PCR permit nonspecific priming and/or primer-dimer formation	Use a hot-start DNA polymerase system (see section of hot-start PCR; Fig. 10.2.2)

Table 10.2.8 PCR Variations and Methods with a Reference to One Example

Application	Description	Reference
Amplified fragment length polymorphism (AFLP)	Genetic fingerprinting technique requiring digestion of genome with restriction enzymes followed by ligation of special linkers with universal tails that eliminate the restriction site and then amplified by PCR with special sets of primers	Vos et al. (1995)
Alu-PCR	Differential comparison of Alu regions of the human genome by using primers against Alu repeats and analyzing the family of amplicons yielded from the PCRs.	Brooks-Wilson et al. (1990)
Asymmetric PCR	Used to produce an abundance of one amplicon strand by having a greater concentration of the primer for the desired strand.	Shyamala and Ames (1989)
Cloning PCR	Facilitated cloning of specific DNA regions by amplification using specific primers with a restriction enzyme recognition sequence added to their 5′ ends such that the amplicons are digested with that restriction enzyme and inserted into a cloning vector with that restriction site	Scharf et al. (1986)
Colony PCR	Screening of bacterial colonies for specific sequences by picking cells from a colony on a culture plate and placing directly into a PCR reaction tube	Zon et al. (1989); *UNIT 4.2*
Degenerate oligonucleotide-primed PCR (DOP-PCR)	Uses oligonucleotides against a partially degenerate sequence with lower annealing temperatures to allow for species-independent general DNA amplification	Telenius et al. (1992)
Differential display (DD) PCR	Identifies differentially expressed genes in different tissues by using short, nonspecific primers for reverse transcriptase (RT) PCR and analyzing products on a high resolution gel	Liang and Pardee (1992)
Inverse PCR	Amplification of DNA flanking a known sequence by restriction digesting DNA, ligating the fragments to form circles and using primers that extend outward from known sequence to amplify unknown region	Ochman et al. (1988)

continued

Table 10.2.8 PCR Variations and Methods with a Reference to One Example, *continued*

Application	Description	Reference
Long range PCR	Amplification of larger fragments of DNA which are typically longer than 1 kb using special DNA polymerases or combinations of enzymes with longer extension steps during the thermal cycling	Cheng et al. (1994)
Methylation-specific PCR	Identification of specific regions by using two primer sets, one recognizing cytosine where methylation is suspected and the other for thymine recognition to differentiate between methylation states by analyzing DNA treated with bisulfite to converts cytosine, not methylcytosine, to uracil	Herman et al. (1996)
Multiplex PCR	Performing PCR with multiple sets of specific primers within one tube that generate multiple amplicons for analysis, such as for mutations or pathogen identification	Ballabio et al. (1990)
Nested PCR	Uses two sets of primers in two sequential PCR runs which the second primer set amplifies a sequence internal to amplicon produced by the first set; a method to eliminate nonspecific products from the first primer set	Singh et al. (1996)
Quantitative real-time PCR	Utilizes either a double-stranded DNA-binding fluorescent dye or a fluorescent probe that interacts with the amplicons which are detected during each cycle; when the fluorescence intensity surpasses a threshold level, which is indicative of the starting amount of target as compared to a calibration curve with known quantities of control DNA (described in greater detail in *UNIT 10.3*)	Higuchi et al. (1993)
Random amplified polymorphic DNA (RAPD)/arbitrarily primed PCR (AP-PCR)	Amplification with random sequence primers permits molecular fingerprinting of genomes without requiring knowledge of genomic sequences, and often performed first at low and then at high stringency to provide more extensive fingerprinting for differentiating species	Williams et al. (1990)
Random mutation PCR	Introduction of random mutations within a defined DNA region using error-prone PCR, such as using Mn^{2+} instead of Mg^{2+} in the reaction buffer	Lin-Goerke et al. (1997)
Rapid amplification of cDNA ends (RACE)	Amplify either or both the 3′ and 5′ ends of cDNA using one known internal sequence and an arbitrary primer due to limited DNA sequence information available for that gene	Hebert et al. (1990)
Reverse transcriptase (RT) PCR	Amplification of RNA species through implementation of typically a two enzyme system with the first enzyme being reverse transcriptase to copy RNA to form a complementary cDNA strand and then followed by PCR with a DNA polymerase	Eldadah et al. (1991)

continued

Table 10.2.8 PCR Variations and Methods with a Reference to One Example, *continued*

Application	Description	Reference
Touchdown PCR	Commonly used to identify a DNA sequence from a protein sequence by having primers with limited mismatches due to alternative codon sequences and starting PCR at high annealing temperature followed by a gradual decrease in annealing temperature to promote amplification of a specific amplicon.	Don et al. (1991)

10

VARIATIONS

Table 10.2.8 provides an overview of some common variations to standard PCR. The reader is referred to *UNIT 10.3* for detailed methodology for real-time PCR.

LITERATURE CITED

Ballabio, A., Ranier, J.E., Chamberlain, J.S., Zollo, M., and Caskey, C.T. 1990. Screening for steroid sulfatase (STS) gene deletions by multiplex DNA amplification. *Hum. Genet.* 84:228-232.

Baskaran, N., Kandpal, R.P., Bhargava, A.K., Glynn, M.W., Bale, A., and Weissman, S.M. 1996. Uniform amplification of a mixture of deoxyribonucleic acids with varying GC content. *Genome Res.* 6:633-638.

Bassam, B.J. and Caetano-Anolles, G. 1993. Automated "hot start" PCR using mineral oil and paraffin wax. *BioTechniques* 14:30-34.

Birch, D.E. 1996. Simplified Hot Start PCR. *Nature* 381:445-446.

Breslauer, K.J., Frank, R., Blocker, H., and Marky, L.A. 1986. Predicting DNA duplex stability from the base sequence. *Proc. Nat. Acad. Sci. U.S.A.* 83:3746-3750.

Brooks-Wilson, A.R., Goodfellow, P.N., Povey, S., Nevanlinna, H.A., deJong, P.J., and Goodfellow, P.J. 1990. Rapid cloning and characterization of new chromosome 10 DNA markers by *Alu* element-mediated PCR. *Genomics* 7:614-620.

Chakrabarti, R. and Schutt, C.E. 2001a. The enhancement of PCR amplification by low molecular-weight sulfones. *Gene.* 274:293-298.

Chakrabarti, R. and Schutt, C.E. 2001b. The enhancement of PCR amplification by low molecular weight amides. *Nucl. Acids Res.* 29:2377-2381.

Cheng, S., Chang, S.Y., Gravitt, P., and Respess, R. 1994. Long PCR. *Nature* 369:684-685.

Cheng, S., Fockler, C., Barnes, W.M., and Higuchi, R. 1994. Effective amplification of long targets from cloned inserts and human genomic DNA. *Proc. Natl. Acad. Sci. U.S.A.*. 91:5695-5699.

Cimino, G.D., Metchette, J.W., Tessman, J.W., Hearst, J.E., and Isaacs, S.T. 1991. Post-PCR sterilization: A method to control carryover contamination for the polymerase chain reaction. *Nucl. Acids Res.* 19:99-107.

Dang, C. and Jayasena, S.D. 1996. Oligonucleotide inhibitors of *Taq* DNA polymerase facilitate detection of low copy number targets by PCR. *J. Mol. Biol.* 264:268-278.

D'Aquila, R.T., Bechtel, L.J., Videler, J.A., Eron, J.J., Gorczyca, P., and Kaplin, J.A. 1991. Maximizing sensitivity and specificity of PCR by pre-amplification heating. *Nucl. Acids Res.* 19:3749.

Don, R.H., Cox, P.T., Wainwright, B.J., Baker, K., and Mattick, J.S. 1991. 'Touchdown' PCR to circumvent spurious priming during gene amplification. *Nucl. Acids Res.* 19:4008.

Ehrlich, H.A., Gelfand, D.H., and Sninsky, J.J. 1991. Recent advances in the polymerase chain reaction. *Science* 252:1643-1651.

Eldadah, Z.A., Asher, D.M., Godec, M.S., Pomeroy, K.L., Goldfarb, L.G., Feinstone, S.M., Levitan, H., Gibbs, C.J. Jr., and Gajdusek, D.C. 1991. Detection of flaviviruses by reverse-transcriptase polymerase chain reaction. *J. Med. Virol.* 33:260-267.

GE Healthcare. 2006. Increased polymerase chain reaction (PCR) amplification specificity using illustra Hot Start Master Mix. *Application Note* 28-4073-38 AB.

Hebert, J.M., Basilico, C., Goldfarb, M., Haub, O., and Martin, G.R. 1990. Isolation of cDNAs encoding four mouse FGF family members and characterization of their expression patterns during embryogenesis. *Dev. Biol.* 138:454-463.

Henke, W., Herdel, K., Jung, K., Schnorr, D., and Loening, S.A. 1997. Betaine improves the PCR amplification of GC-rich DNA sequences. *Nucl. Acids Res.* 25:3957-3958.

Herman, J.G., Graff, J.R., Myohanen, S., Nelkin, B.D., and Baylin, S.B. 1996. Methylation-specific PCR: A novel PCR assay for methylation status of CpG islands. *Proc. Natl. Acad. Sci. U.S.A.* 93:9821-9826.

Higuchi, R., Fockler, C., Dollinger, G., and Watson, R. 1993. Kinetic PCR analysis: Real-time monitoring of DNA amplification reactions. *Biotechnology* 11:1026-1030.

Innis, M., Gelfand, D., and Sninsky, J. 1999. PCR Applications, Protocols for Functional Genomics. Academic Press, San Diego, California.

Kellogg, D.E., Rybalkin, I., Chen, S., Mukhamedova., N., Vlasik, T., Siebert, P.D., and Chenchik, A. 1994. TaqStart Antibody: "hot start" PCR facilitated by a neutralizing monoclonal antibody directed against *Taq* DNA polymerase. *BioTechniques* 16:1134-1137.

Korn, S.H., Moerkerk, P.T., and de Goeij, A.F. 1993. K-*ras* point mutations in routinely processed tissues: non-radioactive screening by single strand conformational polymorphism analysis. *J. Clin. Pathol.* 46:621-623.

Kreader, C.A. 1996. Relief of amplification inhibition in PCR with bovine serum albumin or T4 gene 32 protein. *Appl. Environ. Microbiol.* 62:1102-1106.

Kwok, S. and Higuchi, R. 1989. Avoiding false positives with PCR. *Nature* 339:237-238.

Lawyer, F.C., Stoffel, S., Saiki, R.K., Myambo, K., Drummond, R., and Gelfand, D.H. 1989. Isolation, characterization, and expression in *Escherichia coli* of the DNA polymerase gene from *Thermus aquaticus*. *J. Biol. Chem.* 264:6427-6437.

Liang, P. and Pardee, A.B. 1992. Differential display of eukaryotic messenger RNA by means of the polymerase chain reaction. *Science* 257:967-971.

Lin, Y. and Jayasena, S.D. 1997. Inhibition of multiple thermostable DNA polymerases by a heterodimeric aptamer. *J. Mol. Biol.* 271:100-111.

Lin-Goerke, J.L., Robbins, D.J., and Burczak, J.D. 1997. PCR-based random mutagenesis using manganese and reduced dNTP concentration. *Biotechniques* 23:409-412.

Lundberg, K.S., Shoemaker, D.D., Adams, M.W., Short, J.M., Sorge, J.A., and Mathur, E.J. 1991. High-fidelity amplification using a thermostable DNA polymerase isolated from *Pyrococcus furiosus*. *Gene* 108:1-6.

Mattila, P., Korpela, J., Tenkanen, T., and Pitkanen, K. 1991. Fidelity of DNA synthesis by the *Thermococcus litoralis* DNA polymerase–an extremely heat stable enzyme with proofreading activity. *Nucl. Acids Res.* 19:4967-4973.

Melissis, S., Labrou, N.E., and Clonis, Y.D. 2006. Nucleotide-mimetic synthetic ligands for DNA-recognizing enzymes One-step purification of *Pfu* DNA polymerase. *J. Chromatogr. A.* 1122:63-75.

Melissis, S., Labrou, N.E., and Clonis, Y.D. 2007. One-step purification of *Taq* DNA polymerase using nucleotide-mimetic affinity chromatography. *Biotechnol. J.* 2:121-132.

Mifflin, T.E. 1995. Setting up a PCR laboratory. *In* PCR Primer: A Laboratory Manual. (C. Dieffenbach and G. Dveksler, eds.) pp. 5-14. Cold Spring Harbor Laboratory Press, Cold Spring Harbor, N.Y.

Mifflin, T.E. 1997. Control of contamination associated with PCR and other amplification reactions. *http://www.mbpinc.com/html/pdf/techreport/MifflinReport.pdf.*

Morin, P.A. and Smith, D.G. 1995. Nonradioactive detection of hypervariable simple sequence repeats in short polyacrylamide gels. *Biotechniques* 19:223-228.

Mullis, K.B., Faloona, F.A., Scharf, S., Saiki, R.K., Horn, G., and Erlich, H.A. 1986. Specific enzymatic amplification of DNA in vitro: The polymerase chain reaction. *Cold Spring Harb. Symp. Quant. Biol.* 51 Pt 1:263-273.

Musso, M., Bocciardi, R., Parodi, S., Ravazzolo, R., and Ceccherini, I. 2006. Betaine, dimethyl sulfoxide, and 7-deaza-dGTP, a powerful mixture for amplification of GC-rich DNA sequences. *J. Mol. Diagn.* 8:544-550.

Ochman, H., Gerber, A.S., and Hartl, D.L. 1988. Genetic applications of an inverse polymerase chain reaction. *Genetics* 120:621-623.

Parker, L.T., Zakeri, H., Deng, Q., Spurgeon, S., Kwok, P.-Y., and Nickerson, D.A. 1996. AmpliTaq DNA Polymerase, FS dye-terminator sequencing: Analysis of peak height patterns. *BioTechniques* 21:694-699.

Rychlik, W., Spencer, W.J., and Rhoads, R.E. 1990. Optimization of the annealing temperature for DNA amplification in vitro. *Nucl. Acids Res.* 18:6409-6412.

Saiki, R.K., Scharf, S., Faloona, F., Mullis, K.B., Horn, G.T., Erlich, H.A., and Arnheim, N. 1985. Enzymatic amplification of beta-globin genomic sequences and restriction site analysis for diagnosis of sickle cell anemia. *Science* 230:1350-1354.

Saiki, R.K., Gelfand, D.H., Stoffel, S., Scharf, S.J., Higuchi, R., Horn, G.T., Mullis, K.B., and Erlich, H.A. 1988. Primer-directed enzymatic amplification of DNA with a thermostable DNA polymerase. *Science* 239:487-491.

Scharf, S.J., Horn, G.T., and Erlich, H.A. 1986. Direct cloning and sequence analysis of enzymatically amplified genomic sequences. *Science* 233:1076-1078.

Sharkey, D.J., Scalice, E.R., Christy, K.G. Jr., Atwood, S.M., and Daiss, J.L. 1994. Antibodies as thermolabile switches: High temperature triggering for the polymerase chain reaction. *Biotechnology* 12:506-509.

Shyamala, V. and Ames, G.F. 1989. Amplification of bacterial genomic DNA by the polymerase chain reaction and direct sequencing after asymmetric amplification: application to the study of periplasmic permeases. *J. Bacteriol.* 171:1602-1608.

Singh, B., Cox-Singh, J., Miller, A.O., Abdullah, M.S., Snounou, G., and Rahman, H.A. 1996. Detection of malaria in Malaysia by nested polymerase chain reaction amplification of dried blood spots on filter-paper. *Trans. R. Soc. Trop. Med. Hyg.* 90:519-521.

Schneeberger, C., Speiser, P., Kury, F., and Zeillinger, R. 1995. Quantitative detection of reverse transcriptase-PCR products by means of a novel and sensitive DNA stain. *P.C.R. Methods Appl.* 4:234-238.

Sugimoto, N., Nakano, S., Yoneyama, M., and Honda, K. 1996. Improved thermodynamic parameters and helix initiation factor to predict stability of DNA duplexes. *Nucl. Acids Res.* 24:4501-4505.

Tabor, S. and Richardson, C.C. 1995. A single residue in DNA polymerases of the *Escherichia coli* DNA polymerase I family is critical for distinguishing between deoxy- and dideoxyribonucleotides. *Proc. Natl. Acad. Sci. U.S.A.* 92:6339-6343.

Telenius, H., Carter, N.P., Bebb, C.E., Nordenskjold, M., Ponder, B.A., and Tunnacliffe, A. 1992. Degenerate oligonucleotide-primed PCR: General amplification of target DNA by a single degenerate primer. *Genomics* 13:718-725.

Thein, S.L. and Wallace, R.B. 1986. Human genetic diseases: A practical approach (K.E. Davis, ed.) pp. 33-50. IRL Press, Oxford, U.K.

Thornton, C.G., Hartley, J.L., and Rashtakian, A. 1992. Use of uracil DNA glycosylase to control carryover contamination in the polymerase chain reaction. *BioTechniques* 13:80-82.

Vos, P., Hogers, R., Bleeker, M., Reijans, M., van der Lee, T., Hornes, M., Frijters, A., Pot, J., Peleman, J., Kuiper, M., and Zabeau, M. 1995. AFLP: A new technique for DNA fingerprinting. *Nucl. Acids Res.* 23:4407-4414.

Wainwright, L.A. and Seifert, H.S. 1993. Paraffin beads can replace mineral oil as an evaporation barrier in PCR. *BioTechniques* 14:34-36.

Williams, J.G., Kubelik, A.R., Livak, K.J., Rafalski, J.A., and Tingey, S.V. 1990. DNA polymorphisms amplified by arbitrary primers are useful as genetic markers. *Nucl. Acids Res.* 18:6531-6535.

Zon, L.I., Dorfman, D.M., and Orkin, S.H. 1989. The polymerase chain reaction colony miniprep. *BioTechniques* 7:696-698.

KEY REFERENCES

Mullis et al., 1986. See above.

Saiki et al., 1985. See above.

Saiki et al., 1988. See above.

The three key references provided above are historical in nature and illustrate some of the noteworthy efforts made early on in the development of this prevalent technique.

UNIT 10.3

Real-Time PCR

Dean Fraga,[1] Tea Meulia,[2] and Steven Fenster[3]

[1]College of Wooster, Wooster, Ohio
[2]Ohio Agricultural Research and Development Center, Wooster, Ohio
[3]Ashland University, Ashland, Ohio

OVERVIEW AND PRINCIPLES

The real-time polymerase chain reaction (real-time PCR) is a recent modification to PCR (*UNIT 10.2*) that is rapidly changing the nature of how biomedical science research is conducted. It was first introduced in 1992 by Higuchi and coworkers and has seen a rapid increase in its use since (Higuchi et al., 1992, 1993). Real-time PCR allows precise quantification of specific nucleic acids in a complex mixture even if the starting amount of material is at a very low concentration. This is accomplished by monitoring the amplification of a target sequence in real-time using fluorescent technology. How quickly the amplified target reaches a threshold detection level correlates with the amount of starting material present.

Over the past decade, real-time PCR applications have become broadly used tools for the quantification of specific sequences in complex mixtures. For example, real-time PCR has been used for genotyping (Alker et al., 2004; Cheng et al., 2004; Gibson, 2006), quantifying viral load in patients (Ward et al., 2004), and assessing gene copy number in cancer tissue (Biéche et al., 1998; Königshoff et al., 2003; Kindich et al., 2005). However, the most common use for this technology has been to study gene expression levels by coupling it with a procedure called reverse transcription (Gibson et al., 1996; Biéche et al., 1999; Leutenegger et al., 1999; Livak and Schmittgen, 2001; Liss, 2002). By combining the two technologies (called real-time RT-PCR), one can measure RNA transcript levels in a quantitative fashion.

There are many different variations of real-time RT-PCR; the focus of this unit will be on describing a common approach that uses SYBR Green I. The use of SYBR Green for real-time PCR is relatively easy for the novice to set up and use, and has seen widespread adoption by many laboratories. Where appropriate, alternative technologies will be mentioned that may better suit particular experimental requirements. Readers desiring to learn more about advanced techniques or different chemistries are encouraged to explore the readings and Web sites listed at the end of this unit.

Real-Time RT-PCR is an Improved Method for Quantifying Gene Expression

Quantification of gene expression levels can yield valuable clues about gene function. For example, accurate measurements of gene expression can identify the type of cells or tissue where a gene is expressed, reveal individual gene expression levels in a defined biological state (e.g., disease, development, differentiation), and detect an alteration in gene expression levels in response to specific biological stimuli (e.g., growth factor or pharmacological agent). Other methods to study gene expression such as northern blot analysis (*UNIT 8.2*) and RNase protection assays are time consuming and sometimes prove difficult to reliably generate accurate measurements of gene expression levels. The use of real-time RT-PCR offers several advantages over these methods, including the relatively small amount of sample required for analysis, the ability to reproduce rapid and accurate data, and the capacity for analyzing more than one gene at a time.

Critical to the success of real-time RT-PCR for measuring gene expression levels was the development of procedures for converting an RNA population into a DNA copy (cDNA). Retroviral-derived, RNA-dependent, DNA polymerases (reverse transcriptases) were the first polymerases shown to be able to generate DNA copies of RNA templates (Temin, 1974, 1995). Over the last thirty years, the discovery and development of reverse transcriptase has led to a number of advances in biomedical research and have been instrumental in the development of the modern biotechnology industry.

PCR is the most powerful technique for detection of small amounts of nucleic acids (*UNIT 10.2*); it was recognized early in its development that it could be used to quantify gene expression (Carding et al., 1992; O'Garra and Vieira, 1992; Foley et al., 1993). Theoretically, during PCR each target sequence is amplified in proportion to the amount of target initially present in the sample. Appreciation of this led to the development of one common method for quantifying gene expression called quantitative endpoint RT-PCR (to learn more, see Kochanowski and Reischl, 1999). Quantitative endpoint RT-PCR analysis generally relies on measuring the end point of a PCR reaction by ethidium bromide visualization of the DNA product separated by agarose gel electrophoresis (Fig. 10.3.1). However, this approach presented many difficulties in accurately and reliably quantifying samples, due to the low sensitivity of the ethidium bromide–staining procedure and the variable kinetics of the PCR reaction. The variable kinetics of the PCR reaction derive from the variable efficiency of the amplification of a target sequence during a PCR reaction throughout the reaction time course (reviewed in Valasek and Repa, 2005). This variability can be readily seen in Figure 10.3.3, in which duplicate samples produce different amounts of product at the end of the PCR reaction. This results from PCR reactions not being 100% efficient and, as a consequence, amplification is not a deterministic process (Peccoud and Jacob, 1998). Real-time PCR improves upon quantitative endpoint PCR by measuring target amplification early in the reaction when amplification is proceeding most efficiently. To understand why real-time PCR is an improvement over quantitative endpoint PCR, it is important to understand in more detail how the PCR reaction proceeds.

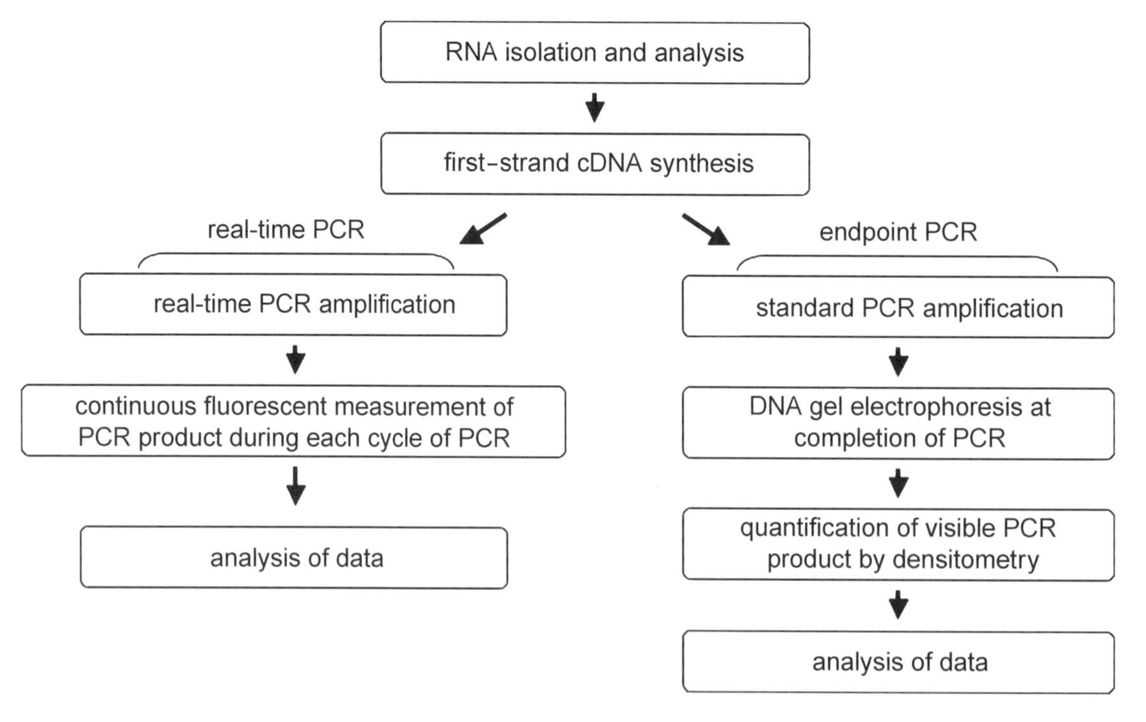

Figure 10.3.1 Comparison of endpoint RT-PCR and real-time RT-PCR. Both procedures begin with isolation of RNA followed by characterization for purity and integrity. Purified RNA is then used as template to generate first strand cDNA. During endpoint PCR, DNA is measured at the completion of PCR amplification. Quantification of DNA product is determined by gel electrophoresis, staining of separated DNA fragments with a fluorescent dye, and digital imaging densitometry to measure DNA band intensity. In real-time PCR, DNA is measured during the exponential phase of PCR amplification. Accumulating product is detected as it is being amplified using fluorescent DNA probes. Both endpoint PCR and real-time PCR data analysis require normalization of data to known standards to determine relative or absolute quantity of starting target gene expression.

A

Cycle nb	Theoretical doubling	Log10(amount of DNA)	Experimental	Log10(amount of DNA)
0	1	0.0000	1	0.0000
1	2	0.3010	2	0.3010
2	4	0.6021	4	0.6021
3	8	0.9031	8	0.9031
4	16	1.2041	16	1.2041
5	32	1.5051	32	1.5051
6	64	1.8062	64	1.8062
7	128	2.1072	128	2.1072
8	256	2.4082	256	2.4082
9	512	2.7093	322	2.5079
10	1,024	3.0103	1,004	3.0015
11	2,048	3.3113	1,987	3.2981
12	4,096	3.6124	3,932	3.5946
13	8,192	3.9134	7,782	3.8911
14	16,384	4.2144	15,401	4.1875
15	32,768	4.5154	30,474	4.4839
16	65,536	4.8165	60,293	4.7803
17	131,072	5.1175	115,343	5.0620
18	262,144	5.4185	225,444	5.3530
19	524,288	5.7196	440,402	5.6438
20	1,048,576	6.0206	859,832	5.9344
21	2,097,152	6.3216	1,677,722	6.2247
22	4,194,304	6.6227	3,187,671	6.5035
23	8,388,608	6.9237	6,039,798	6.7810
24	16,777,216	7.2247	11,408,507	7.0572
25	33,554,432	7.5257	21,474,836	7.3319
26	67,108,864	7.8268	40,265,318	7.6049
27	134,217,728	8.1278	75,161,928	7.8760
28	268,435,456	8.4288	139,586,437	8.1448
29	536,870,912	8.7299	257,698,038	8.4111
30	1,073,741,824	9.0309	472,446,403	8.6744
31	2,147,483,648	9.3319	858,993,459	8.9340
32	4,294,967,296	9.6330	1,460,288,881	9.1644
33	8,589,934,592	9.9340	2,405,181,686	9.3811
34	17,179,869,184	10.2350	2,646,522,913	9.4227
35	34,359,738,368	10.5360	2,876,350,229	9.4588

B

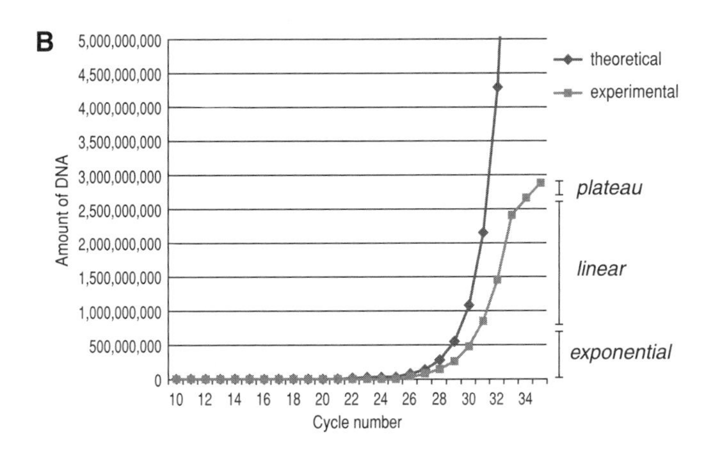

C

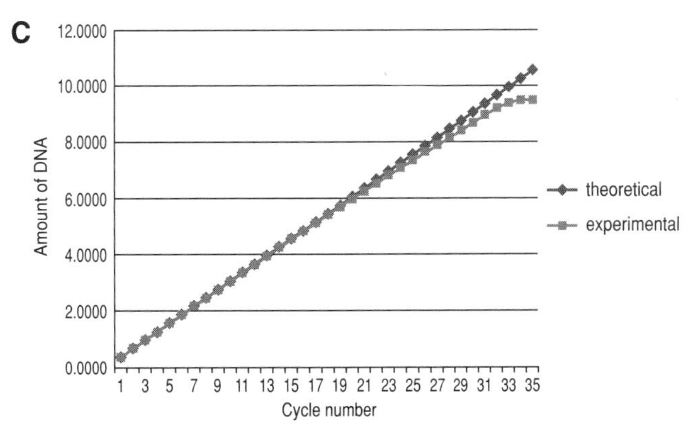

Figure 10.3.2 Theoretical doubling and experimental accumulation of target DNA during PCR. Starting with one molecule, the table in (**A**) shows DNA accumulation during each cycle under theoretical doubling conditions (diamonds), or as might be seen in an experimental assay (squares). The data in table (**A**) are plotted as a linear graph (**B**) and as a logarithmic graph (**C**). *x* axis: cycle number; *y* axis in (**B**): molecules of DNA; *y* axis in (**C**): $\log_{10}$[molecules of DNA].

What Happens During a PCR Reaction and How It Impacts PCR Quantification

During PCR amplification, short DNA sequences are copied at each cycle. Theoretically, the amount of DNA in the reaction should double at each cycle, resulting in an exponential amplification of the initial target DNA (Fig. 10.3.2). This is potentially true during the early cycles when the PCR components are in vast excess compared to the target sequence. However, as product accumulates, the substrates are depleted, resulting in the inhibition of the reaction. By examining how efficiently the amplicon is being produced, a PCR reaction can be broken into three distinct phases: *exponential*, *linear*, and *plateau* (Fig. 10.3.2B and Fig. 10.3.3).

The first phase in a PCR reaction is the *exponential phase* which, in a reaction proceeding with 100% efficiency, represents doubling of product at each cycle. Achieving 100% efficiency is not always possible and careful optimization of PCR conditions must be conducted to ensure that reactions are proceeding as efficiently as possible (ideally >90%; *UNIT 10.2*). As the amplicon exponentially accumulates in quantity, the PCR components are used up, the primer starts competing with amplicon reannealing to itself, and the reaction efficiency decreases. This generally becomes significant when the amplicon level reaches amounts on the order of several nanograms per microliter.

Over time, the reaction slows down and enters the next phase, referred to in this text as the *linear phase* but also known as the nonexponential phase. During this phase, there is no longer near doubling of the amplicon at the end of each cycle and the product formed is highly variable due to many factors, including differences in the rate at which specific components are depleted and the accumulation of products. Even replicate samples, which are identical during the exponential phase, may exhibit variability in this phase (Fig. 10.3.3).

Eventually the reaction will slow down and stop due to depletion of substrates and product inhibition and enter what is called the *plateau phase*. Compounded variation during the linear phase can lead to large differences in the final amount of product produced. As shown in Figure 10.3.3, each replicate reaction can plateau at different points due to different reaction kinetics unique to each sample. Typically, it is at this point or late in the linear phase, that traditional endpoint detection of PCR product is done. As can be seen, such variability complicates any subsequent quantitative analysis.

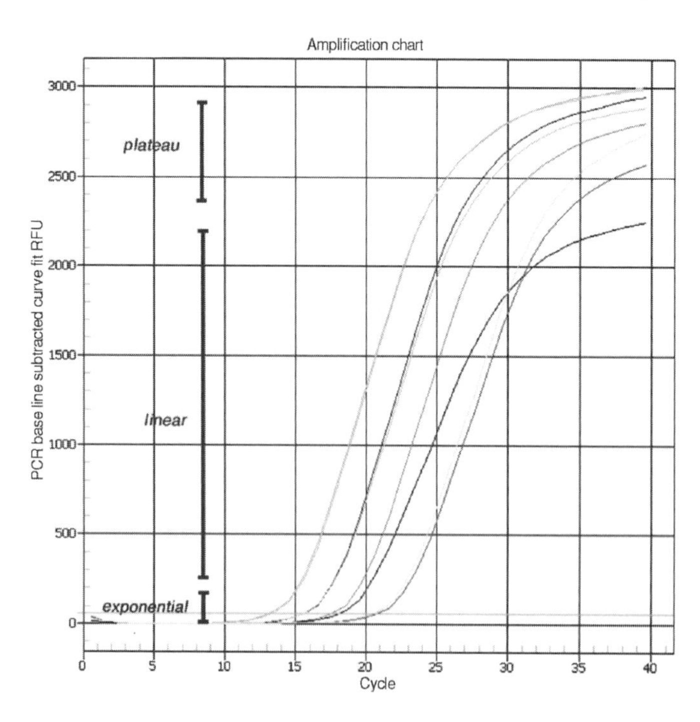

Figure 10.3.3 Plot of an experimental PCR reaction performed on a Bio-Rad iQ5 real-time PCR instrument with duplicates of each sample. *x* axis: cycle number; *y* axis: amount of DNA. Notice that the three PCR phases-exponential, linear, and plateau vary for the four samples.

Real-Time PCR Measures Amplicon Production During the Exponential Phase of PCR

Real-time PCR measures the amount of amplicon produced during each cycle of amplification using fluorescence-based technology. Real-time PCR can quantify amplicon production at the exponential phase of the PCR reaction in contrast to measuring the amount of product at the end-point of the reaction. The amplicon is monitored in "real-time," or as it is being produced, by labeling and detecting the accumulating product with a fluorescently tagged substrate during the amplification procedure. This method has many advantages over conventional PCR including increased speed due to reduced cycle number, lack of post-PCR gel electrophoresis detection of products, and higher sensitivity of the fluorescent dyes used for the detection of the amplicon. However, real-time PCR requires sophisticated equipment in comparison to conventional PCR and the procedure can be prohibitive for budget-oriented laboratories. Nonetheless, due to reductions in costs of equipment and reagents, real-time PCR is becoming more common in a wide-range of laboratories.

STRATEGIC PLANNING

Real-time RT-PCR is a multi-step protocol that requires: (1) high-quality RNA purification, (2) optimal conversion of RNA to cDNA, and (3) sensitive and accurate real-time detection of PCR products. To ensure purification of high-quality RNA template for cDNA synthesis, it is recommended that an RNA purification kit be purchased from one of the many commercial suppliers to eliminate the time required to prepare the reagents and then validate them. There are usually only minor additions to total costs and sometimes actual savings when working with inexperienced researchers (e.g., students). To perform cDNA synthesis, reverse-transcriptase enzyme complete with reaction buffer and other necessary reagents can be purchased from a number of manufacturers. Typically the manufacturer will provide a set of optimized conditions for efficient cDNA synthesis. Real-time RT-PCR analysis requires optimization of a number of reaction conditions and it is suggested that one use the recommended reagents and conditions of the manufacturer of your real-time thermal cycler to achieve reliable results. Each of these steps is dealt with in turn below.

Step 1: Converting an RNA Population into cDNA

RNA isolation

The most important consideration in generating useful data with real-time RT-PCR is the quality of the isolated RNA (Farrell, 1998). RNA is less stable than DNA due to the ubiquitous nature and stability of RNases compared with DNases. (See *UNIT 8.2* for a discussion of RNases.) Moreover, RNA is chemically less stable than DNA and, even when RNA is highly purified, it is more prone to spontaneous degradation in solution (*UNIT 5.2*). Thus, it is important to stabilize RNA preparations as quickly as possible. Typically RNA can be stabilized by the addition of RNase inhibitors (e.g., RNasin, diethylpyrocarbonate) and by storage at −70°C. In addition, DNA contamination can be a significant problem; thus, steps should be taken to reduce this possibility. The most straightforward approach to reduce DNA contamination is to treat RNA samples with RNase-free DNase, available from several manufacturers.

There are a variety of protocols for isolating total RNA from various tissues and protocols specific to the tissue under study should be used, if available (see *UNIT 5.2*). When selecting an RNA isolation protocol, one must decide whether to use messenger RNA (mRNA) or total RNA as the target for the RT-PCR assay. Isolating mRNA involves an additional step in which contamination or template loss can occur, but it is a better source if the target RNA template is present in very low amounts (Burchill et al., 1999). However, for most purposes, it is advisable to work with total RNA since the purification is simpler and generally provides RNA of sufficient amount and quality for subsequent real time RT-PCR. There are several manufacturers who supply easy-to-use kits for RNA isolation and these provide sound quality control and a standardization of isolation procedures for easy comparison between experiments (e.g., Ambion, Stratagene, Qiagen, and Invitrogen). Ambion provides a very useful Web site (*http://www.ambion.com*) that has an extensive background for those new to working with RNA. For a detailed discussion of RNA isolation procedures see Farrell (1998).

Generating cDNA from an RNA population

The first step in the RT-PCR reaction is to selectively convert only the RNA molecules that correspond to protein-encoding genes into cDNA. The RNA that encodes the protein sequence is called messenger RNA (mRNA) and it is purified and extracted as a fraction of total cellular RNA from a collection of cells or tissue. Experimentally, the process of reverse-transcription has many variations, but the essential step is the conversion of mRNA into a cDNA template in a reaction catalyzed by an RNA-dependent DNA polymerase enzyme called reverse-transcriptase (Fig. 10.3.4). A short DNA molecule termed an oligodeoxynucleotide primer is hybridized to complementary mRNA that allows the reverse-transcriptase enzyme (RT) to extend the primer and produce a complementary DNA strand. The sequence of the DNA primer can be designed to bind to a particular target gene or to all mRNA in a sample of purified mRNAs. A specific synthetic antisense oligonucleotide that hybridizes to the desired mRNA sequence is required for conversion of a specific gene sequence into cDNA. A universal primer capable of hybridizing to any mRNA's transcript in a defined sample is used when converting all mRNAs to cDNA. The two most common strategies include the use of an oligo (dT) primer that can hybridize to the endogenous poly (A)+ tails at the 3′ termini of nearly all eukaryotic mRNAs or random hexamer oligonucleotides which can serve as primers at many regions along all RNA templates. Random hexamers are the most nonspecific priming method, and are useful when attempting to amplify long mRNA templates or lowly expressed transcripts; however, they can hybridize to any RNA sequence and will also amplify non-mRNA templates such as ribosomal RNA (rRNA). Oligo (dT) is the preferred method of priming since it only hybridizes to the poly (A)+ tract of mRNA yielding cDNA that represents only transcribed gene sequences. Of note, bacteria do not have stable poly (A)+ tails associated with eukaryotic mRNA. Typically, random hexamers instead of oligo (dT) primers are used to convert bacterial mRNA into cDNA. This requires removal of rRNA from total RNA (e.g., through subtractive hybridization) prior to the RT reaction to prevent nonspecific binding of primers to rRNA.

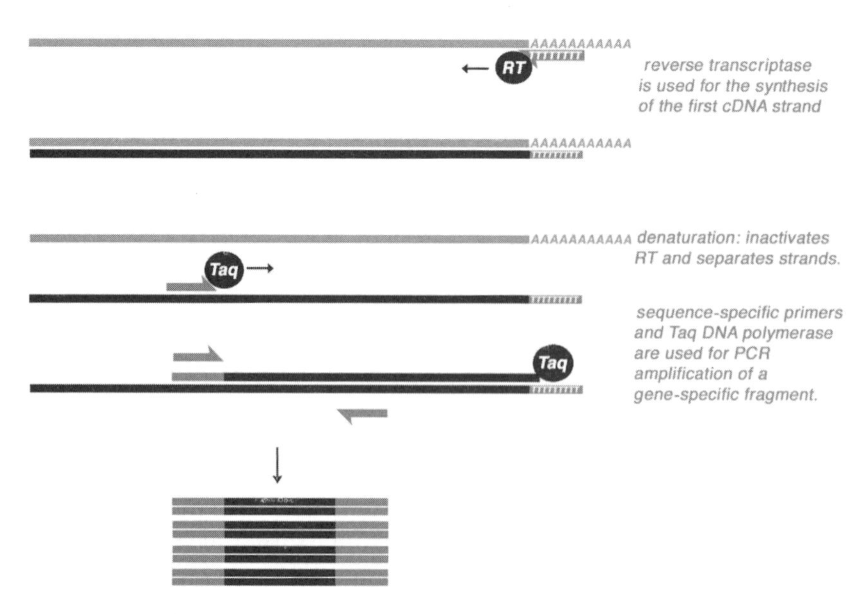

AAAAAAAAAAA
RT
reverse transcriptase is used for the synthesis of the first cDNA strand
AAAAAAAAAAA
AAAAAAAAAAA
denaturation: inactivates RT and separates strands.
Taq
sequence-specific primers and Taq DNA polymerase are used for PCR amplification of a gene-specific fragment.
Taq

Figure 10.3.4 Schematic drawing showing the steps of reverse transcription PCR reaction. mRNA is depicted as a gray line, while DNA is black. RT: reverse transcriptase; *Taq*: thermostable DNA polymerase. In the first step, RT transcribes a DNA copy of the message RNA using a poly-dT primer. Gene-specific primers are used to amplify specific targets using *Taq* polymerase as shown in the figure.

Several reverse-transcriptase enzymes derived from retroviruses can be used to generate DNA complementary to an RNA template. The most common RT enzymes used for cDNA synthesis are encoded by the avian myeloblastosis virus (AMV) and the Moloney strain of murine leukemia virus (MMLV). To initiate extension of complementary DNA, both enzymes require an RNA or DNA primer. Following extension of the DNA sequences, an intrinsic property of the RT enzymes termed RNase H activity degrades the RNA template by recognizing the RNA-DNA hybrid complex formed during the reverse transcription reaction. Commonly used in RT-PCR reactions are genetically engineered reverse transcriptases lacking RNase H activity. In these cases, the RNA/DNA duplex is simply denatured prior to the PCR reaction. This also inactivates the reverse transcriptase. Similarly, reverse transcriptases stable at higher temperature have been engineered. Performing the reverse transcription at a higher temperature may help denature RNA secondary structures and facilitate transcription.

Step 2: Optimizing the PCR Reaction for Real-Time Analysis

Following the generation of a single-stranded cDNA template, the next step in the RT-PCR reaction sequence is to amplify the template into double-stranded DNA (dsDNA) using the polymerase chain reaction with sense and antisense primers to prime new-strand synthesis.

Prior to performing real-time PCR experiments and analysis, one needs to consider the type of the real-time thermal cycler that will be employed, the method of amplicon detection, and how to optimize the PCR reaction. Factors that need to be considered for setting up a real-time PCR experiment are detailed below and summarized in the flowchart shown in Figure 10.3.5.

Optimization reactions can be run in a thermal cycler that does not have real-time analysis capabilities or on a real-time thermal cycler. When optimization reactions are performed in a regular thermal cycler, reaction amplification specificity needs to be verified by gel electrophoresis (see *UNIT 7.2*). On the other hand, if the optimization experiments are run in a real-time thermal cycler, a fluorescent dye may be included in the reaction mixture, so that amplification specificity can be also monitored by the "melting point analysis" (see below), besides verifying it by gel electrophoresis.

Primer design

Optimal primer design is critical for efficient amplification of desired sequences. This unit will focus on those aspects of the PCR reaction and primer design that are most pertinent for real-time RT-PCR. For more detailed information about PCR reactions and primer design, see *UNIT 10.2*.

Successful RT-PCR depends upon the ability to amplify a short product (<300 bp, ideally 100 to 200 bp) that is specific to the mRNA. Typically, this means that one or both primers straddle an intron by annealing to the exons at the 5′ end and the 3′ end of the intron (see Fig. 10.3.6). It is also possible to design effective primers that flank a large intron and are thus much more efficient at amplifying cDNA than any genomic DNA contamination. Effective design thus requires sequence information on the genomic DNA of a gene and direct alignment with the cDNA coding sequence. Sequence information can be downloaded from a number of publicly available Web sites including the NCBI (*http://www.ncbi.nlm.nih.gov/*) and EMBL sites (*http://www.ebi.ac.uk/embl/*).

To assist in the design of primers, a number of computer programs are available that can identify primers for the most efficient RT-PCR amplification of target sequence (for a list see Table 10.2.3 in *UNIT 10.2*, and also consider software packages such as Vector NTI, MacVector, or DNASTAR). These programs take into account several factors to optimize primer design, including: (1) elimination of primer dimers due to complementary sequence between primers, (2) matched annealing temperatures between the primers, and (3) suitable difference between the calculated annealing temperature of the primers and the annealing temperature of the PCR product. Once primers have been selected, a homology search should be performed against a genomic database to identify

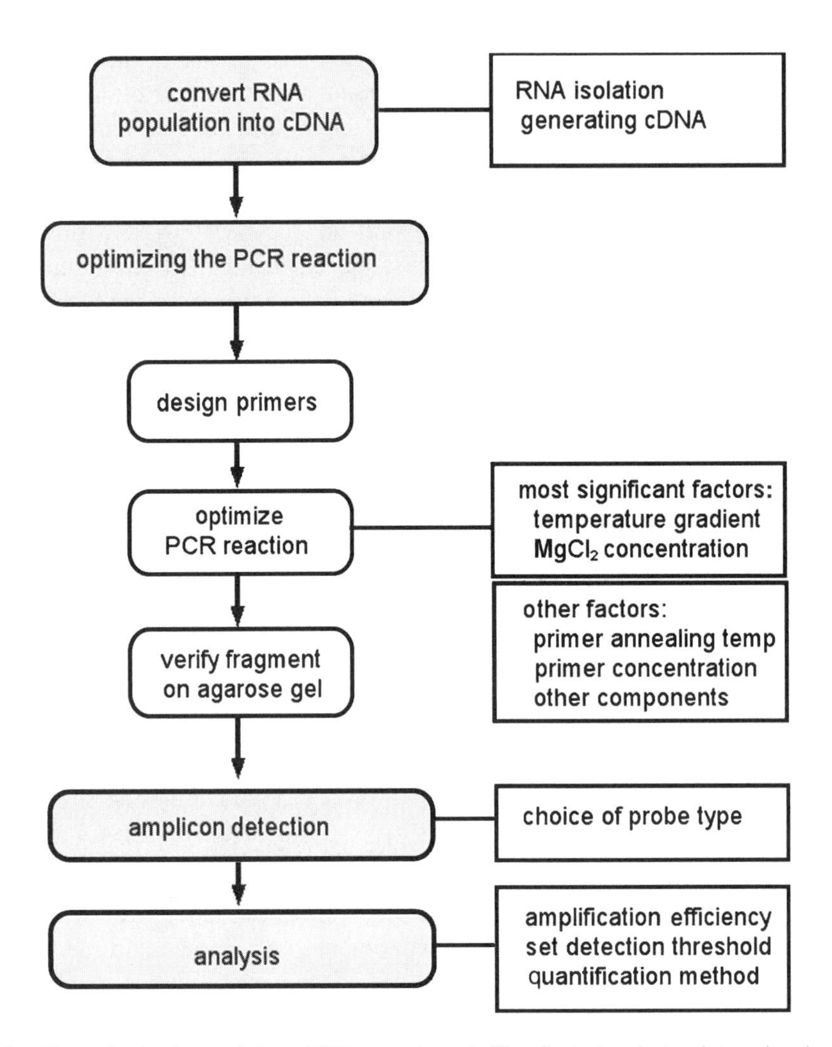

Figure 10.3.5 Flow chart of a real-time PCR experiment. The first step is to determine how to optimally isolate RNA and create the cDNA. Once a suitable cDNA template is created, it will be necessary to optimize the PCR reaction. The most critical aspects to consider when optimizing a PCR reaction are the annealing temperature and the $MgCl_2$ concentration. Other factors may be important if experiencing problems in generating a specific PCR product. After appropriate PCR conditions are determined, the amplification efficiency should be determined so that accurate comparisons can be made.

where the primers might possibly amplify homologous sequences. A few key features of effective real-time RT-PCR primer design are summarized below.

1. Primer pairs should amplify an amplicon of suitable size (<300 bp, ideally 100 to 200 bp).
2. Primers should straddle an intron or flank a large intron.
3. Primer binding sites should not have extensive secondary structure.
4. The 3′ ends of the primer should be free of secondary structure and repetitive sequences.
5. Primers pairs should not significantly complement each other or themselves.
6. Primer pairs should have approximately equal GC content (between 40% to 70%) and have similar annealing temperatures.

Annealing temperature

After primers have been synthesized, their annealing temperatures should be optimized. This can be done using a temperature gradient analysis to determine the optimal annealing temperature.

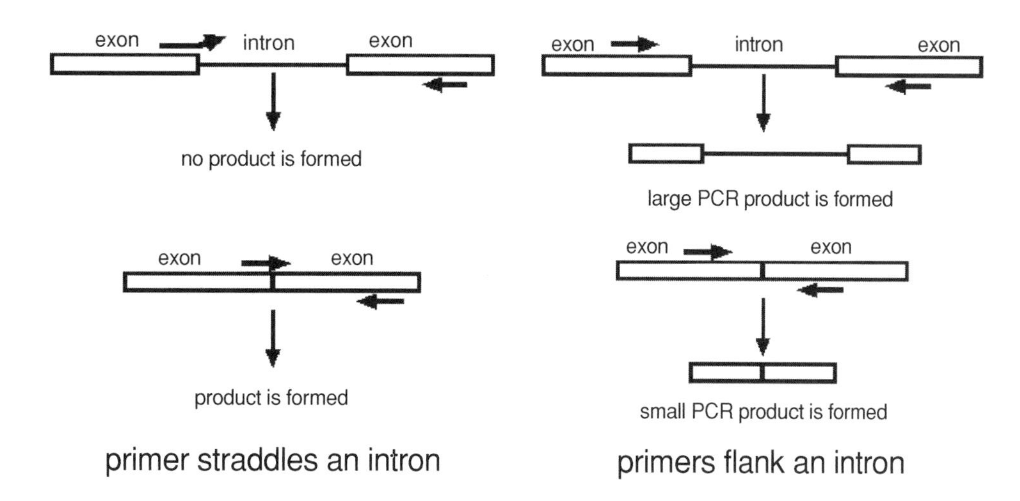

Figure 10.3.6 Designing primers for real-time RT-PCR. Ideally primers will be designed to straddle an intron such that it can anneal only to the cDNA synthesized from the spliced message RNA as shown on the left. Any contaminating genomic DNA would not be suitable for primer annealing. Alternatively, primers can be designed to flank a large intron and only cDNA synthesized from a spliced message RNA would efficiently produce a PCR product as shown on the right.

A temperature gradient analysis takes advantage of a feature found in most thermal cyclers that allows one to simultaneously conduct several PCR reactions at different temperatures. The different reactions can then be compared and a decision made about which annealing temperature will give the best product. To learn more about the details of conducting a temperature gradient PCR see *UNIT 10.2*, Support Protocol 1. Another optional method for determining the optimal annealing temperature is to take advantage of the melting point analysis feature of real-time PCR thermal cyclers (described above). To perform this analysis, complementary primers must be purchased for each primer and the two complementary pairs hybridized and a melting analysis performed to determine the actual T_m under experimental conditions. A melting point analysis will reveal the extent of annealing between two primers under incrementally increasing temperatures, and, from this, an ideal annealing temperature can be selected. This approach might be taken when troubleshooting a difficult PCR that is not producing an appreciable amount of product after attempts to optimize annealing temperature or $MgCl_2$ concentrations have failed. This approach may reveal that one primer is not annealing at the predicted temperature and may need to be redesigned.

Primer concentrations

Increasing primer concentration can enhance the detection of the PCR products but primer concentrations that are too high can lead to the production of nonspecific products such as primer-dimers. If the signal is weak or nonspecific products are too problematic, it is worthwhile to test varying amounts of each primer. Typically, optimal final concentrations will be between 0.05 and 0.9 μM, and occasionally optimal concentrations may not be the same for both primers. However, for most purposes, simply titrating equimolar amounts of the two primers will yield satisfactory results. The primer concentration that gives the lowest "cycle threshold" (C_T) value should be selected since lower C_T values correspond to a more efficient production of the product.

Other reaction components: Magnesium concentration and template

While it is clear there are many aspects to primer design and optimization that affect PCR efficiency, the other components included in a PCR reaction can also affect the efficiency of the reaction and, theoretically, could all be optimized. In practice, the most important factors are $MgCl_2$ concentration and template purity and concentration.

As described in *UNIT 10.2*, magnesium chloride ($MgCl_2$) concentration is a very important factor in PCR. Its impact ranges from no detectable product if too low, to a multitude of nonspecific products and PCR artifacts if too high. This impact can occur in a relatively narrow range of concentrations and typically an optimal concentration of $MgCl_2$ is between 1 to 5 mM. Most buffers supplied by the manufacturer will result in final $MgCl_2$ concentrations in this range. However, if one is having difficulty generating a detectable PCR product, it might be worthwhile to run a simple $MgCl_2$ titration to determine the optimal $MgCl_2$ concentration.

10

Template purity and concentration can also affect the efficiency of PCR product formation (also see *UNIT 10.2*). Typically this is due to the presence of PCR inhibitors in the template preparation, especially in blood and some environmental samples that contain humic acids or heavy metals. In some cases, using less starting template can result in better product production, as the inhibitors have been diluted. If this doesn't improve product production, then reprecipitation of the template in ethanol may remove the contaminant (see *UNIT 5.2*).

Verification of PCR specificity using an agarose gel

Before running the sample on the real-time thermal cycler, it is worthwhile to verify the specificity of the reaction early in the optimization process by running the PCR products on an agarose gel (see *UNIT 7.2 & 10.2*). Ideally, only one band will be visible at the expected size. This is particularly important if nonstrand-specific dyes will be used for detection (see below). If other nonspecific bands or primer dimers are detected on the agarose gel, the reaction needs to be optimized by changing the parameters mentioned above.

Step 3: Amplicon Detection in the Real-Time Thermal Cycler

Real-time PCR thermal cycler considerations

Real-time PCR thermal cyclers combine PCR product generation and recognition into one integrated format that allows for the subsequent analysis of the captured data (Fig. 10.3.7). To accomplish these two tasks, real-time PCR machines incorporate traditional PCR thermal cycler technology with integrated fluorimeters and detectors that provide the ability to both excite a fluorochrome and detect the emitted light (Fig. 10.3.7). There are several suppliers of such equipment and these

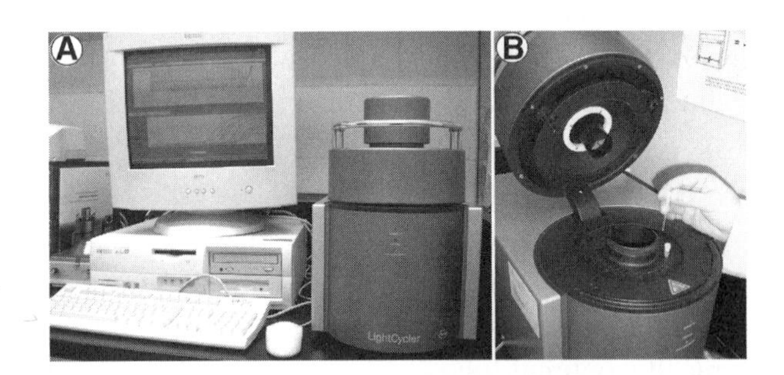

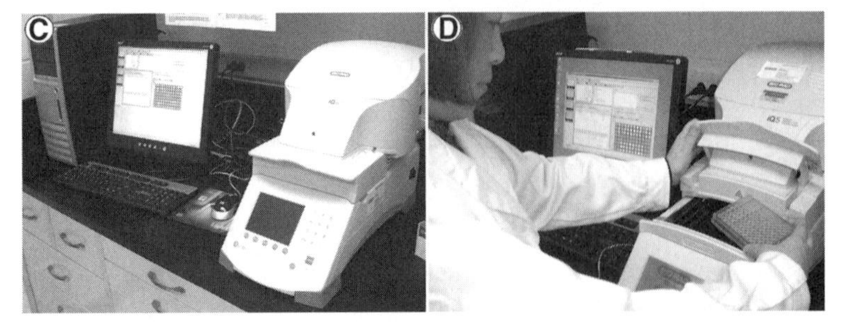

Figure 10.3.7 Real-time cyclers. Examples of two types of real-time thermal cyclers: (**A**) and (**B**), Roche LightCycler. In this system the reactions are set up in borosilicate 20 µl capillaries, and the circular "rotor" accommodates up to 32 reactions at one time. (**C**) and (**D**), Bio-Rad iQ5 which uses 96-well reaction format and up to five targets can be multiplexed in one well.

can range in cost depending upon the features. For many laboratories, this represents a substantial investment but can result in savings, as the analysis of PCR products can be conducted without running agarose gels, representing a savings in both time and money when processing a large number of samples. One fundamental difference between platforms is the ability to detect more than one fluorochrome with the addition of multiple excitation and detection channels. Many suppliers provide different models ranging from a relatively low-cost single-channel machine suitable for SYBR Green detection (e.g., Bio-Rad's DNA Engine Opticon) to a multichannel machine (e.g., the 7500 Real-Time PCR System from Applied Biosystems) that can be used for the detection of multiple fluorochromes at once (multiplexing). Many work best with specific chemistries and fluorochromes, so it is important to consider the cost and chemistry options available for each platform in addition to the cost and capabilities of the various thermal cyclers themselves. For a detailed assessment of the various platforms, see Logan and Edwards (2004).

Because real-time PCR reactions are quantified at each cycle by measuring fluorescence, most real-time thermal cyclers need to be calibrated for the particular tube or microtiter plate and microtiter plate seal used, and for the PCR reaction volume. Therefore, it is important to know which real-time thermal cycler will be used for the experiment before setting up the reactions.

Fluorescent dyes for monitoring real-time amplification

Detecting the PCR product in real-time involves the use of a fluorescent dye. These can be either nonspecific dyes, such as fluorescent DNA-binding dyes (e.g., SYBR Green I) or strand-specific probes (such as *Taq*man or Molecular Beacons), many of which use a phenomenon known as fluorescent resonance energy transfer (FRET) to distinguish between various products (to learn more about FRET, see Clegg, 1995).

For the initial optimization of the real-time PCR amplification, nonspecific fluorescent dyes from the SYBR Green family are most commonly used, because they are more economical compared to strand-specific probes, and easier to optimize. When using this method of detection, a single, specific DNA fragment has to be obtained during PCR amplification, because any additional nonspecific DNA fragment accumulation will contribute to the fluorescence measured.

Strand-specific probes need to be designed if nonspecific bands are amplified in the PCR reaction or if the amplification of more then one target sequence will be monitored in a single PCR reaction.

Nonspecific fluorescent probes

While there are other types of nonspecific probes available (e.g., Molecular Probes's YoPro and YoYo), use of the minor groove binding dyes, SYBR Green 1 and SYBR Gold, are particularly widespread. The emission maxima for SYBR Green 1 and SYBR Gold closely match that of fluorescein (around 520 nm) and most real-time PCR thermal cyclers have optics to detect at this wavelength. Their relative low-cost and generic nature make them ideal for optimizing a wide range of PCR reactions, and can provide quantitative data suitable for many applications (such as expression studies). However, for some experiments, such as detection of allelic variations, strand-specific approaches are recommended (e.g., *Taq*Man or Molecular Beacons).

The principle behind the SYBR Green family of dyes is that they undergo a 20- to 100-fold increase in their fluorescence upon binding dsDNA (Fig. 10.3.8) that is detected by the real-time PCR machine's detector (Fig. 10.3.9). Thus, as the amount of dsDNA increases in the reaction mix, there will be a corresponding increase in the fluorescent signal. However, this simplicity means that they do not distinguish between different dsDNA products, and it is important that PCR reactions be optimized so that only the target amplicon is present, or that other methods be employed to distinguish between different products (e.g., melting point analysis).

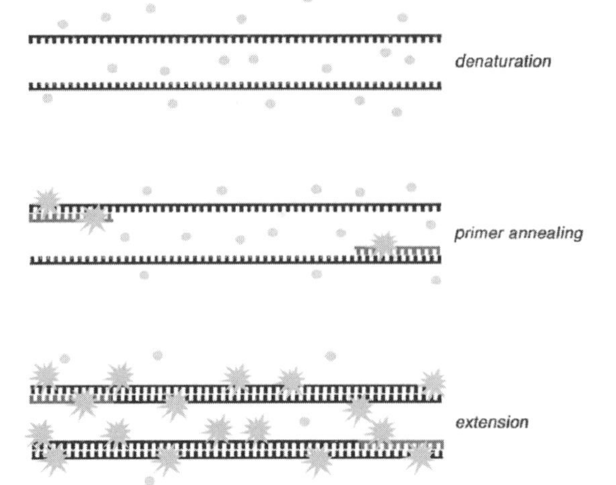

Figure 10.3.8 SYBR Green during PCR amplification: the fluorescence of this dye increases 100- to 200-fold when bound to the minor groove of double stranded DNA. This is used to measure the amount of DNA at the end of the elongation step of the PCR reaction.

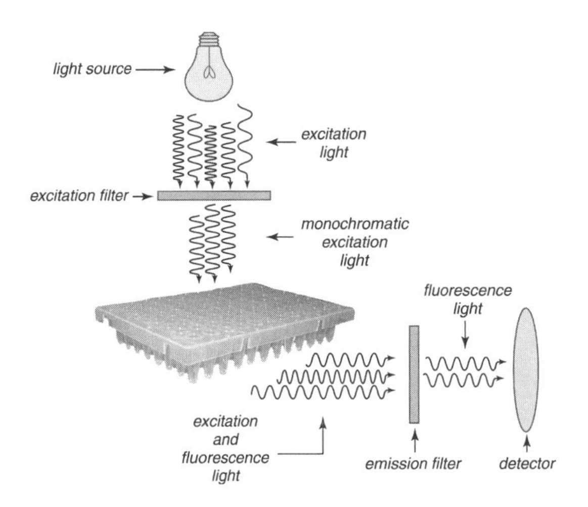

Figure 10.3.9 Optical design for detection of fluorescence in a real-time PCR thermal cycler. The detection system of the real-time thermal cycler is a excitation and emission filter-based system. Depending on the instrument manufacturer, the light source consists of a xenon, tungsten-halogen, or LED (light emitting diode) light, and the detector is a CCD (charge-coupled device) camera or PMT (photomultiplier tube). See *UNIT 9.2* for further discussion.

Strand-specific fluorescent probes

Fluorophore-coupled nucleic acid probes can be used as a detection system. These strand-specific probes interact with the PCR products in a sequence-specific manner and provide information about a specific PCR product as it accumulates. There are many variations on this general approach and it is beyond the scope of this short review to cover them all. However, one widely used example will be described to acquaint the reader with the basic principle behind strand-specific probes and to highlight some of the advantages of this approach over the use of nonspecific probes. Readers are referred to Lee et al. (2004) or various manufacturers' Web sites for a more complete description of the various systems available and how they work.

One of the most widely used strand-specific approaches involves hydrolysis probes, so named because their utility is based upon the 5′ nuclease activity of *Taq* polymerase. *Taq*Man probes are a well-known example and have been used extensively in a wide range of studies. *Taq*Man probes are sequence-specific oligonucleotides with two fluorophores labeled at either end (Fig. 10.3.10). One fluorophore is termed the quencher and the other the reporter. When the reporter fluorophore is

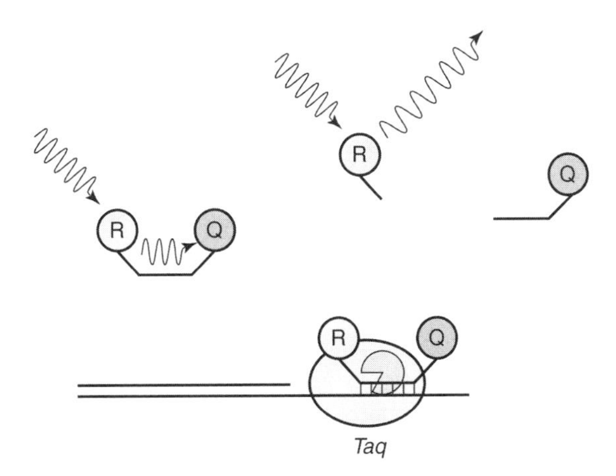

Figure 10.3.10 *Taq*Man probe used for sequence-specific amplification of DNA fragments. A laser excites the reporter fluorochrome and it then emits a light that is absorbed by the nearby quencher. After the 5′ nuclease activity of Taq polymerase, the quencher is no longer in close proximity to the reporter and thus the light emitted by the reporter can be detected. R: reporter; Q: quencher (adapted from Valasek and Repa, 2005).

excited by a laser at the appropriate wavelength, it absorbs that light and then emits a characteristic wavelength of light. As long as the quencher remains in close proximity it will absorb the emitted light by FRET. The quencher is selected based upon its ability to specifically absorb the emitted spectra of the reporter and should be spaced in the probe to optimize the capture of that light.

As shown in Figure 10.3.10, the probe is designed to anneal to the accumulating product at an internal site. Typically the 3′ end of the probe is blocked to prevent extension due to annealed probe. As product accumulates, the dual-labeled primer binds to the target sequence and gets degraded by the 5′ nuclease activity of the *Taq* polymerase. This is the result of the *Taq* enzyme encountering the probe during amplimer extension as shown in Figure 10.3.10. Degrading the probe separates the quencher from close proximity to the reporter, making it less able to absorb the light emitted by the reporter. Thus, there is a corresponding increase in fluorescence that is correlated with the specific amplification of the target sequence. The signal that is produced is cumulative, unlike what is seen with other strand-specific methods that rely on the measurement of the direct hybridization of probe to the target during each cycle. While the *Taq*Man approach is very sensitive, it is dependent upon a probe that can be efficiently hydrolyzed. To ensure efficient cleavage, the probe should have a T_m 5° to 10°C greater than the amplimer to facilitate its binding quantitatively to the template prior to extension. However, creating such a probe can be problematic with AT-rich targets.

One of the advantages of the strand-specific probe systems is that multiple probes can be combined, or multiplexed, and thus information about several target sequences can be obtained from one reaction. This can be very advantageous, as both controls and target sequences are amplified under identical conditions. A second advantage is when there are two potential PCR products that can be produced from the same primer set. Use of a strand-specific probe can help distinguish the two products. The hydrolysis probe system described above, as well as other sequence-specific probe systems, have also been widely used for genotyping and can identify point mutations, SNPs, and allelic variants which are not easily possible with SYBR Green I or SYBR Gold (Livak, 2003; Marras et al., 2003; Cheng et al., 2004; Eshel et al., 2006).

Melting point analysis to verify amplification specificity

Before proceeding with real-time PCR analysis, one should again verify that the PCR amplification is specific under the actual real-time PCR conditions. This can be done in the real-time thermal cycler using the melting curve analysis feature. It is good practice to perform melting point analysis after each real-time PCR run as a quality control step.

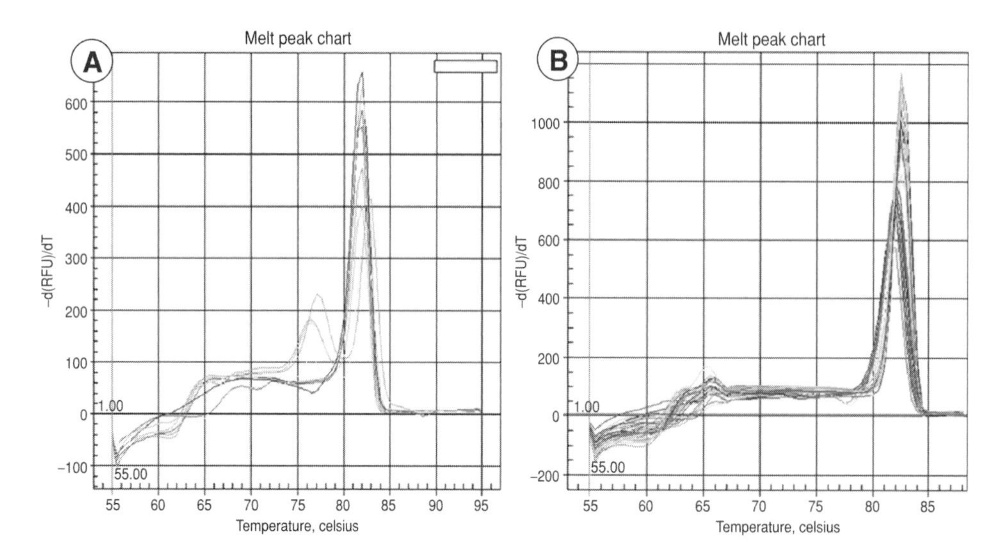

Figure 10.3.11 Melting curve analysis performed on the Bio-Rad iQ5 real-time PCXR instrument. Panel (**A**) shows double peaks. The lower peaks represent DNA fragment denaturing at lower temperature, most likely due to primer-dimers. Panel (**B**) shows a single peak melting curve, representative of a single species of DNA molecule in the reaction.

Melting point analysis is used to distinguish target amplicons from PCR artifacts such as primer-dimer or misprimed products. Melting point analysis is also useful for optimizing primer-annealing temperature if that becomes necessary.

The utility of melting point analysis derives from the observation that the temperature at which a DNA duplex will denature is dependent upon length and nucleotide composition. Fluorescence measurements are made while slowly increasing the temperature of the reaction products (for example, from 60° to 95°C). At the low temperature, the amplicons are all double stranded and thus bind the SYBR Green dye, producing a strong fluorescence signal. As the temperature increases, the PCR products are denatured, resulting in a decrease in fluorescence. The fluorescence is measured continuously and when the T_m of a particular dsDNA product is reached, there will be a rapid decrease in the fluorescence over a short temperature range. This is detected by the instrument and plotted as either total fluorescence or as the first negative differential of the fluorescence signal with respect to temperature. The advantage of the second method is that the plot will appear as one or more peaks centered on different temperatures (Fig. 10.3.11). These peaks represent the points at which the maximum rate of change in fluorescence is detected. Different amplicons can generate different peaks centered on different temperatures and, fortunately, PCR artifacts typically have lower melting temperatures than the target amplicon. By programming the instrument to detect fluorescence at the appropriate temperature, above the melting temperature of the misprimed or primer-dimer products but below the melting temperature of the target product, one can more accurately measure fluorescence of the target amplicon and not include the fluorescence from artifactual products. To conduct a melting point analysis, consult the manufacturer's manual of your real-time PCR thermal cycler for the details as to how to enter this feature into your PCR program. Generally it is a simple menu command.

Step 3: Real-Time RT-PCR Analysis and Quantification

Prior to setting up the actual experiments it is important to consider your experimental design and several quality control steps, including negative and positive controls. These will ensure that the real-time RT-PCR data being generated is of high quality.

In addition, for quality control reasons, each sample is run in triplicate. A dilution series is also run for each sample: as is discussed below, serial dilutions are needed to determine amplification efficiencies and to generate standard curves.

Experimental design and quality control

Real-time RT-PCR experiments should routinely include the following controls:

Negative controls (at least two):

1. No-template control.

2. No-reverse-transcriptase control.

Positive controls (help to ensure all reagents are working properly; chose one or both):

1. DNA (a source of quantified DNA containing the target sequence).

2. RNA (a sample of quantified RNA that contains the RNA target).

The no-template control confirms that there is no contamination in the PCR reagents. If a band is produced in the no-template control, this might indicate that one or more of the reagents is contaminated with template, or, more likely, previously amplified PCR product. The specific reagent that is contaminated will have to be determined by troubleshooting and it is often more cost effective to dispose of all the reagents except the *Taq* polymerase and retest with the negative control. Using barrier micropipet tips with all RT-PCR reagents can prevent most contamination problems. In some laboratories, separate and dedicated micropipets are used for PCR set up and analysis of PCR products. For more information about how to eliminate crossover contamination, see the discussion about contamination control in UNIT 10.2.

Running a no-reverse-transcriptase control helps detect the presence of contaminating DNA in the RNA. If the no-template control is negative and the no-reverse-transcriptase control produces a band, most likely there is contaminating DNA that is being recognized by the primers. Treatment of the RNA template with RNase-free DNase should alleviate the problem. In some cases, the primers may have to be redesigned so as to avoid amplifying potential DNA targets. Typically this may require straddling a different intron or adjusting the 3′ end of the primer slightly to have less potential to basepair at its 3′ end with a genomic DNA sequence.

When conducting real-time RT-PCR for the first time, it is helpful to also include a positive control. This is especially important if the absence of gene expression is a likely result. Positive controls can be either RNA, DNA, or both. Using an RNA control will provide valuable information about the RT-PCR steps but can be more problematic to prepare. A DNA positive control is often easier to prepare and provides information about the efficiency of the PCR reaction.

Finally, good quality control will insure that solid data is collected. Table 10.3.1 summarizes some additional tips or procedures that should be considered during the course of a real-time experiment to insure that the data collected are reliable.

Setting the threshold for detection of the amplicon

Setting the threshold determines the level of fluorescence signal that is sufficiently above background to be considered a reliable signal. The cycle at which the threshold is met or exceeded is called the cycle threshold (C_T), and is used for making comparisons between samples. Setting the threshold should take into consideration the detection limits of the equipment and background fluorescence due to the fluorescent chemistry used. Setting the threshold too low may result in unreliable data collection due to random fluctuations in the sample tube that result in premature ‘detection’ of product. Setting it too high may result in the detection of product after it has left the exponential phase and has entered the linear phase, again resulting in inaccurate data collection.

To set a threshold value, one must first identify the baseline value. Baseline values are generally determined from a plot of fluorescence versus cycle number. During the first few cycles, the amount

Table 10.3.1 Summary of Quality Control Measures

Protocol	Measure
RT-PCR quality control	The experiment should include no-template and no-reverse-transcriptase controls
	Use RNase-free tubes that are appropriate for the thermal cycler, barrier-tipped pipets, and gloves to avoid unwanted RNase contamination
Real-time PCR quality control	Set the threshold for detecting PCR product manually according to the standards set by the laboratory
	If setting the threshold manually, set it as low as possible to ensure that one is detecting product during the exponential phase of the amplification
	The melting curves of all samples should be checked to ensure that they are producing one peak
	The slopes of the samples when plotted in log view should be parallel
	Samples that can be detected at cycle 10 or earlier should be diluted
Standard curve quality control	The nucleic acid being used for the standard curve should be carefully quantified and serially diluted using a calibrated micropipette
	Include five or more points for standard curves used for absolute quantification
	Remove those dilutions whose fluorescence measurements cross the threshold line together
	The correlation coefficient for how well a series of diluted samples should fit a straight line in a log plot of the template should be equal to or higher than 0.99

of product formed is generally not detectable and the fluorescence will fluctuate around some value dependent upon the amount of fluorescent material present, the types of tubes used, and other factors. A set of these early data points (usually between cycles 3 to 15) is selected to serve as the basis for the baseline calculations. These early data points are averaged to form a mean baseline value, often shown as a straight line across the course of the reaction (Fig. 10.3.12). Typically, threshold values are set at three standard deviations above the mean baseline value. It is important that the threshold be set to allow detection of product while it is still in the exponential phase. The point at which the product's detected fluorescence crosses the threshold is called the C_T value.

The significance of threshold values is that they provide a useful measure for comparison between samples. In a 100% efficient PCR, the amplified product will double at each cycle. This means that differences in C_T between different samples correspond to differences in starting amount of the target sequence. For example, a C_T difference of 1 corresponds to a 2-fold difference in starting material and the sample with the lower C_T value has more starting material. Each difference in C_T value corresponds to differences in starting material of 2 raised to the power of the C_T difference.

$$2^{(C_{T1} - C_{T2})} = \text{fold difference in the amount of starting target}$$

Equation 10.3.1

In Equation 10.3.1, C_{T1} refers to the C_T value for sample one, and C_{T2} refers to the C_T value for sample two. Thus, a C_T difference of 4 equals 2^4 or a 16-fold difference in starting amount of target sequence. However, it is important that the two samples being compared have similar amplification

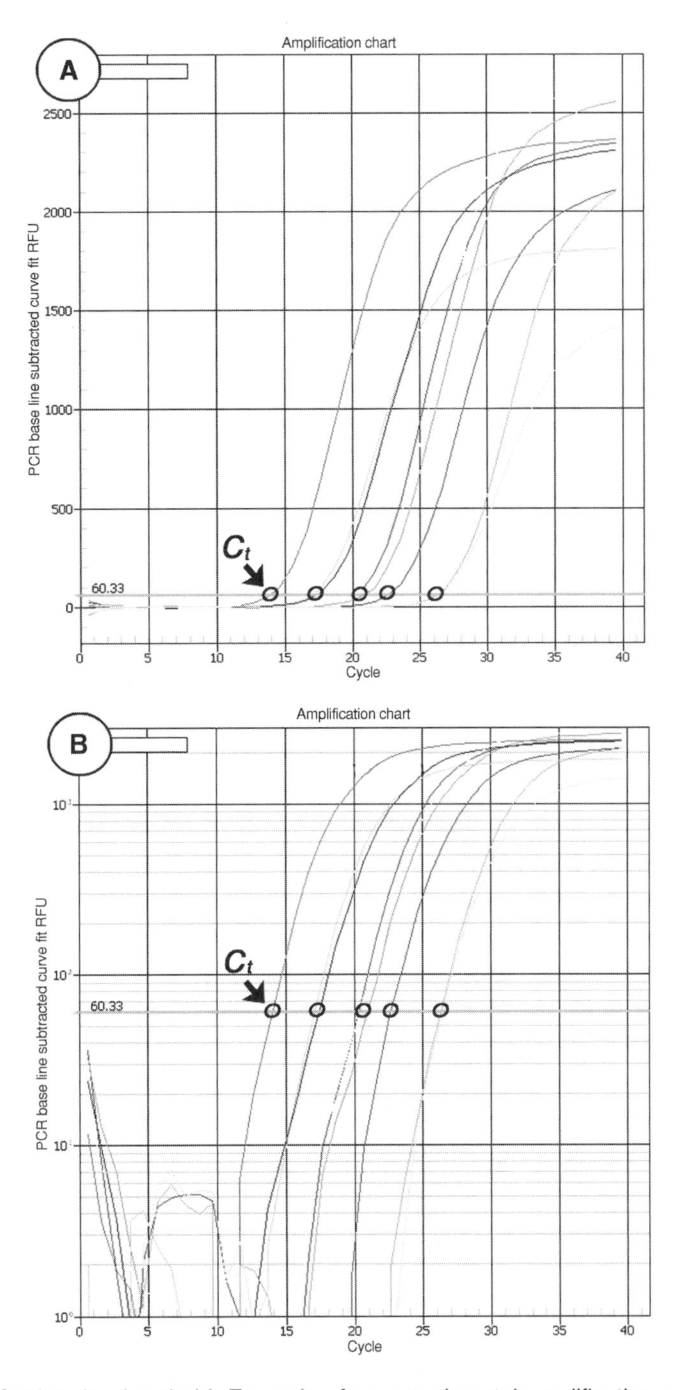

Figure 10.3.12 Setting the threshold. Example of an experimental amplification performed on the BioRad iQ5 real-time PCR instrument: (**A**) linear plot of the amount of DNA (*y* axis) and the cycle number (*x* axis); (**B**) logarithmic plot. The threshold line is used to quantify the amount of DNA and is set at the exponential phase of the amplification. C_T: cycle threshold.

efficiencies in order for the comparison to be valid. Amplification efficiency and how it impacts this calculation is discussed in more detail below.

PCR amplification efficiency

In order to make valid comparisons between different samples it is important to determine the amplification efficiency of your primers. This is especially true if using the recommended Pfaffl procedure for relative quantification of your samples (Pfaffl, 2001). Ideally, amplification efficiencies for each primer set should be roughly equal. However, the amplification efficiency for a pair of primers is affected by differences in primer binding sites, the sequence of the amplification product, and PCR product sizes, and thus should be determined experimentally.

The rationale behind conducting detailed assessment of amplification efficiency is that not every primer set and template are equally efficient in amplification. In an optimal PCR reaction, every amplicon will be replicated and the amount of product will double after each cycle. For example, if there were one copy of the target sequence present at the start of the first cycle, then there will be two at the end of cycle one, four at the end of cycle two, eight at the end of cycle three, and so forth. If the reaction is perfect (100% efficient) a plot of copy number versus cycle number produces a line as shown in Figure 10.3.13. However, if the reaction is only 90% efficient, the amount of product produced will not double after each cycle and the slope of the plot of amount of product versus cycle number will be less than the same plot assuming 100% efficiency. It is easy to see that if a difference in efficiency is not considered, there can be systematic errors in calculating the amount of target sequence present in a sample. This also underscores the importance of using as low a threshold value as possible, since errors due to differences in amplification efficiency compound.

To determine the amplification efficiency for a given primer set, it is necessary to perform PCR on a serial dilution series of the template. Theoretically, a PCR reaction proceeding at 100% efficiency would require 3.3 cycles to increase the amplicon concentration 10-fold. Plotting the C_T values for the log of each serial dilution generates a plot whose slope is related to the efficiency of the reaction. The slope of this line and amplification efficiency can be related using Equation 10.3.2:

$$E = 10^{(-1/\text{slope})} - 1$$

Equation 10.3.2

where E is the efficiency of the reaction and *slope* refers to the slope of the plot of C_T value versus the log of the input template amount. A slope between −3.6 and −3.1 corresponds to an efficiency between 90% to 110% (e. g., $E = 0.9 - 1.1$). Several authors have written about this topic at length and there are a variety of methods by which these measurements can be made (Pfaffl, 2001; Tichopad et al., 2003; Rasmussen, 2001).

Finally, one can determine *relative* primer efficiencies between different sets of primers by doing a simple dilution series of the template for each primer set. The resulting C_T values obtained are compared by subtracting the C_T values of primer set one from primer set two and plotting the difference against the logarithm of the template amount (serial dilution, copy number, or amount). If the slope of the resulting line is <0.1, the amplification efficiencies are comparable. If the slope is >0.1, consider redesigning the primers to improve the amplification efficiencies of one or both primers.

Amplicon quantification: Absolute versus relative comparisons

Before conducting a real-time RT-PCR experiment, a decision must be made as whether to quantify results using relative or absolute measures. Absolute quantification results in a measure of the amount of a target sequence in the sample expressed as either copy number or concentration. Relative quantification provides a relative value expressed as a ratio of the amount of initial target sequence between control and experimental treatments normalized relative to some internal standard such as the expression of a housekeeping gene. Both approaches are routinely seen but relative quantification requires less set up time and can be used for most experiments.

Absolute quantification

Absolute quantification requires generating a standard calibration curve using a standard that has been carefully quantified as to the amount or copy number (see *UNIT 2.2*, Quantification of Proteins and Nucleic acids). In addition, this method relies on identical amplification efficiencies for the

A

Cycle nb	Amount of DNA efficiency: 100%	Amount of DNA efficiency: 90%	Amount of DNA efficiency: 80%	Amount of DNA efficiency: 60%
0	1	1	1	1
1	2	2	2	2
2	4	4	3	3
3	8	7	6	4
4	16	13	10	7
5	32	25	19	10
6	64	47	34	17
7	128	89	61	27
8	256	170	110	43
9	512	323	198	69
10	1,024	613	357	110
11	2,048	1,165	643	176
12	4,096	2,213	1,157	281
13	8,192	4,205	2,082	450
14	16,384	7,990	3,748	721
15	32,768	15,181	6,747	1,153
16	65,536	28,844	12,144	1,845
17	131,072	54,804	21,859	2,951
18	262,144	104,127	39,346	4,722
19	524,288	197,842	70,824	7,556
20	1,048,576	375,900	127,482	12,089
21	2,097,152	714,209	229,468	19,343
22	4,194,304	1,356,998	413,043	30,949
23	8,388,608	2,578,296	743,477	49,518
24	16,777,216	4,898,763	1,338,259	79,228
25	33,554,432	9,307,650	2,408,866	126,765
26	67,108,864	17,684,534	4,335,959	202,824
27	134,217,728	33,600,615	7,804,726	324,519
28	268,435,456	63,841,168	14,048,506	519,230
29	536,870,912	121,298,220	25,287,311	830,767
30	1,073,741,824	230,466,618	45,517,160	1,329,228
31	2,147,483,648	437,886,574	81,930,887	2,126,765
32	4,294,967,296	831,984,491	147,475,597	3,402,824
33	8,589,934,592	1,580,770,532	265,456,075	5,444,518
34	17,179,869,184	3,003,464,011	477,820,935	8,711,229
35	34,359,738,368	5,706,581,621	860,077,682	13,937,966

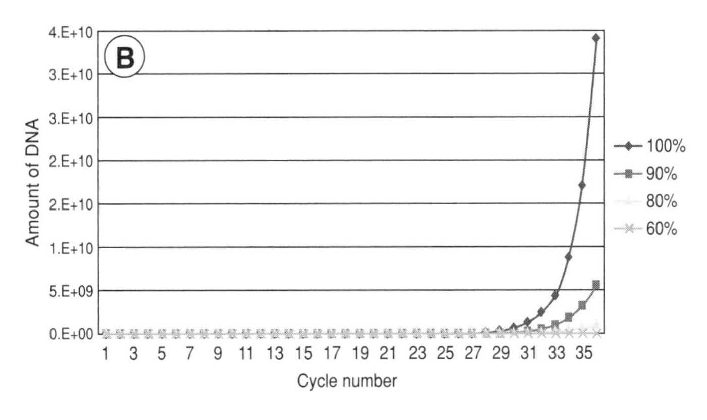

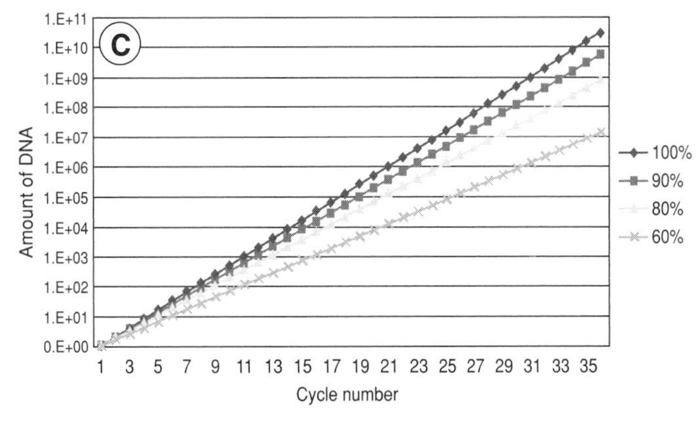

Figure 10.3.13 Effect of the PCR amplification efficiency on the accumulation of the DNA product. (**A**) The table shows values for 100%, 90%, 80%, and 60% efficient accumulation of product per cycle starting with 1 molecule. (**B**) and (**C**) are the linear and logarithmic plots respectively of values in the table.

control and target sequence. This approach is more accurate but often more labor-intensive (Bustin and Nolan, 2004). In addition, if quantifying mRNA levels, it is also necessary to control for the efficiency of the RT reaction. For these purposes, it is best to start with known amounts of target RNA that are converted to cDNA for subsequent real-time PCR. RNA can be prepared by cloning the DNA of the target of interest into a suitable vector that will allow the production of RNA in vitro. Typically these vectors use SP6, T3, or T7 phage RNA polymerase promoters. Examples include Promega's pGEM series, Stratagene's pBluescript, and Invitrogen's TOPO TA vector series. Several commercial kits are available that facilitate the production of RNA from these vectors, including Promega's Riboprobe system, Ambion's MAXIscript system, and Stratagene's RNAMaxx High Yield Transcription Kit.

If quantifying a DNA target, a suitable DNA starting template for the standard must be obtained. This can be genomic DNA (if gene copy number and genome size are known) or a plasmid containing the target gene. If using plasmid DNA, be sure that it is pure and does not contain contaminating RNA that can increase the A_{260} measurement and inflate copy number estimates.

The standard used for a standard calibration curve should have primer binding sites that are identical to those found in the target sequence and should produce a product of approximately the same size and sequence as the experimental target. Once the standard has been accurately quantified, it is serially diluted in increments of 3- to 10-fold and in a range that is sufficient to cover the likely amount of target expected to be present in the experimental samples. Each dilution should be run in triplicate. The average C_T values from each dilution are then plotted versus the absolute amount of standard present in the sample to generate a standard curve (see Fig. 10.3.14). Comparison of experimental C_T values to this standard curve produces an estimate of the amount of target present in the initial sample.

Relative quantification

Relative quantification measures the changes in a gene's expression in response to different treatments (e.g., control versus experimental) or state of the tissue (e.g., infected versus uninfected samples, different developmental states). It requires a suitable internal standard to control for variability between samples. Typically these are 'housekeeping' genes such as β-actin, cyclophilin, or glyceraldehyde-3-phosphate dehydrogenase (Bustin and Nolan, 2004; Valasek and Repa, 2005). The housekeeping gene serves as an internal reference between different samples and helps normalize for experimental error. Careful consideration should be given to the selection of the housekeeping gene to be used for your study and some effort should be made to confirm that the housekeeping gene is indeed invariant in its expression across your treatment regime.

The advantage of using a relative quantification approach is that there is no need for generating a standard calibration curve. If amplification efficiencies of the target and control sequences are identical, then one simply compares relative expression values for the target gene in different samples expressed as a ratio. The ratio is then normalized using the housekeeping gene's expression in the same samples. There are several mathematical models available for calculating relative expression (Meijerink et al., 2001; Pfaffl, 2001; Liu and Saint, 2002). Pfaffl (2001) developed a mathematical formula that takes into account the contribution of PCR amplification efficiencies and is widely used for the relative quantification of gene expression in real-time RT-PCR. This is discussed below in Support Protocol 2.

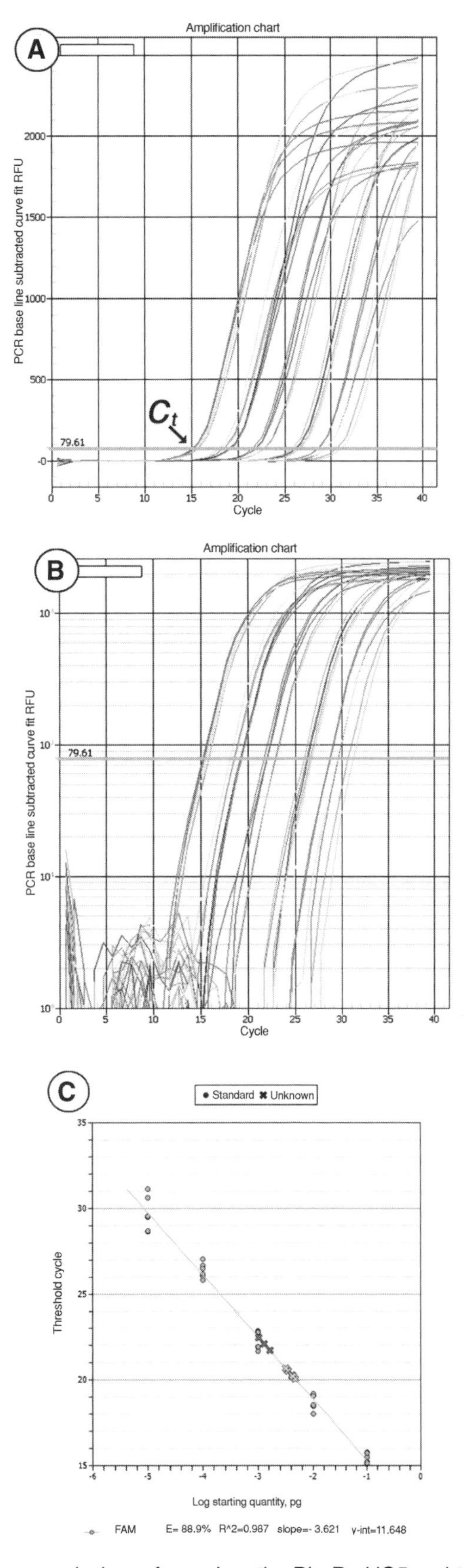

Figure 10.3.14 Standard curve analysis performed on the Bio-Rad iQ5 real-time PCR instrument. Panel A: linear plot; Panel B: logarithmic plot; Panel C: standard curve. Five dilutions of a sample were run in triplicate to determine the melting curve (circles). The concentration of DNA in the unknown samples, also run in triplicate, was determined by plotting the C_T values on the curve (crosses).

SAFETY CONSIDERATIONS

Anyone working in a laboratory should follow standard laboratory safety rules. These include how to handle hazardous chemicals properly, using electricity safely, and keeping the laboratory work area free of clutter to reduce accidents.

A number of chemicals used in the typical molecular biology can be hazardous and it is important that anyone working in a laboratory is aware of the information supplied in the form of Material Safety Data Sheets or MSDS. All manufacturers of hazardous materials are required by law to supply the user with MSDS sheets that describe all pertinent information about any hazards associated with their chemicals. In general, when working in a laboratory one should wear a laboratory coat, gloves, and protective eyewear. The most noteworthy chemical discussed in this unit is ethidium bromide. Ethidium bromide is a carcinogen and should be handled carefully and disposed of as described in the MSDS sheet or as described by the safety officer for the laboratory.

The other significant hazards mentioned in this unit are ultraviolet light (used with ethidium bromide–containing agarose gels) and electricity. Ultraviolet light can cause acute eye irritation. The human eye cannot detect overexposure and thus one should always wear appropriate eye protection (UV-blocking safety glasses). The voltages used for electrophoresis of agarose gels are sufficient to cause electrocution if not handled properly. Proper handling includes covering the buffer reservoirs during electrophoresis and turning off the power supply and unplugging the leads before removing a gel (*UNIT 7.2*).

NOTE: Refer to *APPENDIX 1* for information on general laboratory safety.

PROTOCOLS

Real-time RT-PCR requires two distinct procedures to be linked together. These are broken up here into two different basic protocols. The first (Basic Protocol 1) describes the conversion of isolated RNA into cDNA. The second (Basic Protocol 2) describes the actual real-time PCR reaction and its subsequent analysis. There are three support protocols that follow which are used in the analysis of the real-time PCR results.

The first and most critical step in the RT-PCR reaction is obtaining high-quality, intact, total RNA, free of contaminating RNases. Degraded or impure RNA can curtail the efficiency of the RT reaction and reduce the yield of cDNA. There is a large collection of methods that have been employed to purify RNA from either cells or tissue. Although a number of commercially available RNA purification kits allow for rapid isolation of RNA and eliminate the need for phenol in the purification step, the preferred method employs the single-step acid phenol–guanidium thiocyanate-chloroform extraction technique developed by Chomczynski and Sacchi (1987). Details about RNA isolation and purification are covered in *UNIT 5.2*, and readers are advised to review this information before beginning their RNA isolation procedure. In general, this step will require ~30 to 60 min, dependent upon the number of samples. The second step, synthesis of cDNA, will require about 90 to 120 min depending upon the number of samples. This step is described in Basic Protocol 1. Purification of mRNA (or Poly A+ RNA) requires additional steps and time and will not be discussed here (see Variations, below).

An additional and sometimes critical step prior to reverse transcription is elimination of contaminating genomic DNA. This requires incubating the purified RNA with RNase-free DNase prior to RT-PCR. The RNA must then be repurified to eliminate the contaminating DNase enzyme. This step can be eliminated by properly designing primers, as described above, to straddle an intron such that the primers' ability to prime synthesis of the target from its genomic DNA equivalent is disrupted. Once RNA has been successfully purified, it should be used immediately or stored at −80°C to reduce autocatalytic degradation.

The real-time RT-PCR protocols described below have been compiled from the instructions of several manufacturers of commercially available kits and modified based upon the authors' personal experiences. In general, any manufacturer of real-time reagents that are compatible with the thermal cycler being used are acceptable. In addition, it is important that the appropriate tubes be used in the thermal cycler. For example, the DNA Engine Opticon system (Bio-Rad) requires low-profile white tubes with clear caps for use in their machine. It is usually more convenient to use strips of eight tubes or plates for experiments involving a large number of samples.

Basic Protocol 1: Synthesis of cDNA by Reverse Transcription

The following protocol describes the synthesis of cDNA from total RNA purified from cells or tissues using the single-step acid phenol-guanidium thiocyanate-chloroform extraction technique (*UNIT 5.2*; Chomczynski and Sacchi, 1987). The following protocol is a typical reaction designed to convert 1 pg to 5 μg total RNA into cDNA.

Priming of the first strand cDNA synthesis reaction can be accomplished using random hexamers, oligo (dT), or gene-specific primers. Oligo (dT) is the preferred method for this protocol since it selectively hybridizes to the 3′ poly (A) tails found in nearly all eukaryotic mRNAs. Some protocols call for a mixture of random hexamers and Oligo (dT) although this protocol will use oligo (dT) in the first strand amplification method.

NOTE: This procedure was adapted from that of Invitrogen.

Materials

Purified total RNA dissolved in DEPC-treated water
First strand cDNA synthesis primers (50 μM)
10 mM dNTP (free nucleotides) mix (*UNIT 3.3*)
DEPC-treated H_2O (*UNIT 8.2*)
10× RT Buffer
25 mM $MgCl_2$
0.1 M DTT (dithiothreitol) used to reduce disulfide bonds
RNase inhibitor (e. g., RNasin Ribonuclease Inhibitor; often 20 to 40 U/μl)
M-MLV Reverse Transcriptase (RNase H^-)
RNase H

0.5-ml microcentrifuge tubes, RNase- and DNase-free
37° and 65°C incubators
85°C heating block

NOTE: To avoid contamination of samples with RNase, gloves should be worn at all times. Washing of gloved hands with a mild SDS solution (0.01%) can help remove contaminating RNases. Also, micropipettors equipped with barrier-filter tips should be used when pipetting all solutions. This will prevent the introduction of contaminants located in the chamber of the micropipettors. (Refer to *UNIT 8.2* for a discussion of RNases.)

NOTE: Quickly mix the sample by vortex and briefly centrifuge each component for 15 sec at full speed in a benchtop microcentrifuge to collect the samples at the bottom of the reaction tube. This can be done at room temperature.

1. Combine the following into an RNase and DNase-free 0.5-ml microcentrifuge tube

Component	Quantity
Total RNA	1 pg to 5 μg
Primer	1μl
50 μM oligo (dT)	1μl
10 mM dNTP mix	1μl
DEPC-treated water	to 10 μl final volume.

2. Incubate the reaction for 5 min at 65°C to denature RNA; briefly microcentrifuge to bring the solution to the bottom of the tube and place on ice.

3. Prepare the cDNA synthesis mix adding each component in the following order:

Component	Quantity
10× RT reaction buffer (as supplied by the manufacturer)	2 μl
25 mM MgCl	4 μl
0.1 M DTT	2 μl
RNase Inhibitor (e. g., RNasin 20 to 40 U/μl)	1 μl
Reverse transcriptase (M-MLV)	1 μl.

10

4. Add the cDNA synthesis mix to the RNA/primer mixture. Mix gently by pipetting and collect the liquid by brief microcentrifugation. Incubate the following reaction for 50 min at 50°C.

   The temperature can be varied between 42° and 50°C depending on the type of enzyme used in the reaction. Reactions performed at higher temperatures will resolve secondary RNA structure that can mask the RNA sequence and prevent the RT enzyme from copying the template strand in this region. However, many enzymes are not as active at temperatures higher than 42°C. Newly engineered RT enzymes that are optimized for temperatures higher than 42°C have overcome this obstacle. If one is having difficulties amplifying a specific target sequence, the use of more heat tolerant RT enzyme may be required.

5. Stop the reaction by heating the sample to 85°C for 5 min. Place the sample on ice for 5 min, spin down to collect the sample, and place the sample on ice again.

6. Add 1 μl of RNase H to the tube and incubate for 20 min at 37°C.

   cDNA can be stored at −20° or −80°C, or used immediately for PCR amplification.

Basic Protocol 2: Real-Time PCR Amplification and Analysis

The following protocol describes how to conduct a real-time PCR amplification using SYBR Green. There are three support protocols that are important for accurate analysis of the PCR results (Support Protocols 1 to 3). These include methodologies for the determination of amplification efficiency (Support Protocol 1), the Pfaffl method for relative quantification of the target sequence (Support Protocol 2), and construction of a standard calibration curve for absolute quantification (Support Protocol 3).

While it is possible to create reagents in house, many laboratories now use prepared reagent kits purchased from a manufacturer for real-time PCR. Typically, real-time PCR kits provide SYBR Green and polymerase at ideal concentrations for efficient amplification and detection of cDNA. A variety of manufacturers provide kits, some of which are tailored for specific chemistries or thermal cyclers. The Quantitect SYBR Green PCR kit sold by Qiagen is an example of one kit that has broad utility and can be used in a variety of real-time thermal cyclers. The protocol that is described below is adapted from the manufacturer's suggested protocol for the Opticon DNA Engine and can be used with the cDNA protocol described above. Each target sequence, including housekeeping genes, will have its own set of primers and can be assembled as described below.

This procedure will take ∼30 to 60 min to set up, about 3 hr to run the PCR program, and an additional 1 to 2 hr to analyze the results, depending upon the number of samples and the ease of use of the software.

Manufacturers generally do not reveal the details of what is in their reagent mixes. However, a few points can be made about the components to help explain why they are present. The protocol described below is based on the Qiagen QuantiTect SYBR Green kit. The reaction mix contains HotStar*Taq* DNA polymerase, QuantiTect SYBR Green PCR Buffer, dNTP mix, SYBR Green I, ROX passive dye, and 5 mM $MgCl_2$. The HotStar*Taq* DNA polymerase is a modified *Taq* DNA polymerase that is provided in an inactive state. Using a hot-start DNA polymerase lowers the chance of forming mis-primed products or primer-dimers (see UNIT 10.2, Hot-Start Polymerases). The polymerase typically becomes activated after a 10 to 15 min, 95°C incubation step. The QuantiTect SYBR Green PCR buffer is composed of a balance of KCl and $(NH_4)_2SO_4$ and, according to the manufacturer, is formulated to favor specific over nonspecific primer binding during the annealing step. The dNTPs that are included in a reaction mix are generally at a final concentration of 0.2 mM each. The SYBR Green is usually present in the final reaction as a 1:10,000 dilution of the SYBR Green stock. The ROX (carboxy-X-rhodamine) passive dye is used as reference dye for some machines (Applied Biosystems, Mx3000P, Mx3005P, or Mx4000) and does not participate in the reaction. The $MgCl_2$ is added at an optimal concentration for most reactions but additional $MgCl_2$ can be added if needed. The QuantiTect SYBR Green PCR kit can be obtained from Qiagen (no., 204143; *http://www1.qiagen.com/*).

10

Materials

PCR master mix:
- 2× reaction mix
- 50 μM forward primer
- 50 μM reverse primer
- RNase-free water

cDNA (Basic Protocol 1)
1.5-ml microcentrifuge tubes for preparing master mix
Thin-walled PCR tubes
Real-time thermal cycler

NOTE: This protocol describes how to perform a real-time PCR. When conducting an actual experiment comparing samples, it is important that the amplification efficiencies be known before analyzing the samples. Both housekeeping and target sequences can be prepared using separate master mixes.

1. Clean the work area.

   Make sure the work area is clean and free of distractions.

2. Assemble the components listed above on ice.

3. Prepare a PCR master mix in 1.5-ml microcentrifuge tubes, as shown in Table 10.3.2 to maximize uniformity and minimize labor when working with multiple tubes. Mix the master mix thoroughly and dispense equal aliquots into each thin-walled PCR tube.

   A 50× master mix is shown as an example. Typically, each sample is done in triplicate (target and housekeeping gene).

   Master mixes can be made in excess of what is actually needed to account for pipetting variations (see UNIT 10.1).

4. Add template (cDNA) to each reaction tube as needed (be sure to include a negative control that does not receive any template).

5. Set up and program the PCR machine according to the manufacturer's instructions.

   A typical program is shown in Table 10.3.3. In addition, a DNA melting curve analysis may be included (see above for an explanation).

6. Place the tubes in the thermal cycler and start the program.

   A melting curve analysis can be done to verify the specificity of the product formation. If primer-dimer or other nonspecific product is identified, the data acquisition step can be adjusted to a temperature that is above the T_m of the nonspecific product and 3°C below the T_m of the specific product. See above for an explanation of melting point analysis.

Table 10.3.2 Components of the Real-Time PCR Reaction Mix

Component	Volume/reaction	50× master mix	Final concentration
2× reaction mix	12.5 µl	625 µl	1×
Forward primer	0.5 µl	25 µl	1 µM
Reverse primer	0.5 µl	25 µl	1 µM
RNase-free water	Variable (bring to 25 µl per reaction)	—	—
Template cDNA (to be added at step 4 to each reaction tube)	Variable (1-5 µl if using preparation described above)	—	<500 ng/reaction
Total	**25 µl**	**1250 µl**	

Table 10.3.3 A Typical Real-Time PCR Program

	Time	Temperature	Comments
Step			
Initial PCR activation step	15 min	95°C	If the reaction mix does not have HotStar*Taq* or its equivalent, this can be reduced to 5 min
Cycle			
Denaturation	30 sec	95°C	Denatures dsDNA
Annealing	30 sec	45° to 65°C	Primer annealing step, typically 5° to 8°C below T_m of primers
Extension	30 sec	72°C	
Data acquisition		*x*°C	Follow manufacturer's protocol for programming the data acquisition of dsDNA product. See step below
Repeat cycles	Repeat cycle steps 35 to 45 times		Cycle number will depend upon the amount of input template.
Final extension	5 min	72°C	Final extension of PCR products
Melting curve analysis	Variable	45° to 95°C	Increase in 0.2°C increments with a hold time of 1 sec for each read (this may vary between thermal cyclers)

7. Run the PCR reaction to determine C_T values for each sample.

   It is important to have C_T values for the target sequence and the housekeeping gene for each sample.

8. If performing relative quantification see Support Protocol 2. If performing absolute quantification, compare C_T values to the standard curve as described in Support Protocol 3.

Support Protocol 1: Determination of Amplification Efficiency

In order to compare the results from different samples it is necessary to know the amplification efficiencies for each primer set. A PCR reaction proceeding at 100% efficiency will result in a doubling of product every cycle. However, several factors can interfere with PCR amplification efficiency and result in efficiencies that are <100%, even in the early cycles. Such factors include secondary structure in the template, contaminants in the template preparation, or poor optimization of PCR reaction conditions. For accurate assessment of differences between samples, it is advisable to determine the amplification efficiency of each primer set under the conditions they will be used. Once the amplification efficiencies are determined, they can be used to correct for this potential source of error before quantitative comparisons are made. The procedure outlined here will provide a measure of amplification efficiency that can be used in the relative quantification method described by Pfaffl (2001).

Materials

cDNA preparation (Basic Protocol 1) or other DNA preparation for amplification efficiency determination
PCR master mix:
- 2× reaction mix
- 50 μM forward primer
- 50 μM reverse primer
- RNase-free water

1.5-ml microcentrifuge tubes (for preparing master mix)
Thin-walled PCR tubes
Real-time thermal cycler

1. Assemble the components described above on ice.

2. Prepare a serial dilution of a DNA sample that contains the sequence to be amplified.

   This can be genomic DNA, cDNA, or a plasmid containing the sequence. There should be at least five data points representing at least five dilutions (typically ten or five fold dilutions). The actual copy number for the sequence is not important for this step.

3. Prepare a PCR master mix in 1.5-ml microcentrifuge tubes, as shown in Table 10.3.2, for the sample PCR to maximize uniformity and minimize labor when working with multiple tubes.

   Typically, each dilution point is done in triplicate.

   Master mixes can be made in excess of what is actually needed to account for pipetting variations.

4. Mix the master mix thoroughly, centrifuge briefly to collect disturbed contents, and dispense equal aliquots into each thin-walled PCR tube.

5. Add the serially diluted template to each reaction tube as needed (be sure to include a negative control that does not receive any template; see UNIT 10.1 for a discussion of controls).

6. Set up and program the PCR machine according to the manufacturer's instructions.

   A PCR program adapted from the DNA Engine Opticon system is shown in Table 10.3.3 as an example. In addition, a DNA melting curve analysis may be included after the final 72°C extension step to verify specificity of the PCR reaction.

7. Place the tubes in the thermal cycler and start the program.

8. Run the PCR reaction to determine C_T values for each dilution.

9. Plot the C_T values versus the logarithm of the concentration or the copy number of the template.

   These can be arbitrary values that reflect the serial dilution relationships (e.g., 1, 10, 100, 1000 etc.). The line drawn through these points should have a high R^2 value (>0.99), which is an indication of the goodness-of-fit of the line through the data points (see APPENDIX 4).

10. Use the slope of the line to determine the efficiency of the PCR reaction by using Equation 10.3.2.

11. Repeat for additional primer sets as needed.

12. Record the amplification efficiency values for each primer set.

    These will be used later when the differences between treatments are quantified using the relative quantification method described in Support Protocol 2.

Support Protocol 2: Analyzing Results Using the Pfaffl Method to Calculate Fold Induction

In 2001, Pfaffl described a method for expressing the ratio of a target gene transcript relative to a control calculated using solely PCR amplification efficiencies and crossing point (C_T) differences. This method is relatively easy to use and requires less set up than absolute quantification using a standard curve. However, it only provides relative changes and not absolute changes between treatments.

One significant problem with relative quantification is that it is sometimes difficult to find a suitable housekeeping gene whose expression is not affected by the treatment (Bustin and Nolan, 2004). This can result in erroneous conclusions about the extent of gene expression changes.

1. To calculate a relative quantification comparison, compare the C_T for the target sequence among the various treatments (e.g., $C_{T\ control} - C_{T\ experimental}$) to obtain the ΔC_T value. Do the same for the housekeeping gene.

2. Calculate the amplification efficiency values for each of your primer pairs (expressed as a value between 0 and 1) as described in Support Protocols 1 and 2. Convert each amplification efficiency measure into Pfaffl efficiency (E_P) measures by adding 1.0 (e.g., $0.9 + 1.0 = 1.9$).

    The reasoning behind this conversion is that an efficiency of 100% (or 1.0) is equal to a 2.0-fold increase per cycle. An efficiency of 90% (0.9) is equal to a 1.9-fold increase per cycle.

3. Calculate the fold change in expression between the two samples (e.g., control versus experimental) using Equation 10.3.3:

$$\text{Fold change between samples} = (E_P)^{\Delta C_T}$$

Equation 10.3.3

Do the same for the housekeeping gene.

For example, assume a difference in C_T values between a control and a treated sample of 2.7(ΔC_T =2.7). If the amplification efficiency of the primers set was 91% (E_P = 1.91), then the fold change for these two samples is $(1.91)^{2.7}$. This works out to a 5.65-fold change in gene expression. However, it still needs to be normalized with the housekeeping gene to correct for any random experimental error that might have occurred and differentially affected the two samples. This is described below.

4. To normalize the two samples using the housekeeping gene, simply divide the fold change for the target sequence by the fold change for the housekeeping gene (ideally 1.0 or very close to 1.0), expressed as Equation 10.3.4:

$$\text{Fold change (normalized)} = (E_{P\text{-target}})^{\Delta C_{T\text{-target}}} / (E_{P\text{-housekeeping}})^{\Delta C_{T\text{-housekeeping}}}$$

Equation 10.3.4

This will give the relative fold change in expression as a result of the treatment or condition of the target sequence. E_P refers to efficiency as defined by Pfaffl and is determined for the primer set that amplifies the target sequence ($E_{P\text{-target}}$) and the primer set that amplifies the housekeeping sequence ($E_{P\text{-housekeeping}}$). The term $\Delta C_{T\text{-target}}$ refers to the difference in C_T values between the control and treatment ($C_{T\text{-control}} - C_{T\text{-treatment}}$) for the target sequence and the term $\Delta C_{T\text{-housekeeping}}$ refers to the difference in C_T values between the control and treatment ($C_{T\ control} - C_{T\ treatment}$) for the housekeeping sequence.

For example, assume the following average C_T for your experimental and control treatments: experimental target $C_T = 27.1$; experimental housekeeping $C_T = 26.3$; control target $C_T = 31.3$; control housekeeping $C_T = 26.2$. In addition, assume that the efficiencies for both primer sets are the same at 90% ($E_P = 1.9$). Plugging these into the equations above:

$$\Delta C_{T\text{-}target} = (C_{T\,control} - C_{T\,treatment}) = (31.3 - 27.1) = 4.2$$

$$\Delta C_{T\text{-}housekeeping} = (C_{T\,control} - C_{T\,treatment}) = (26.3 - 26.2) = 0.1$$

$$\text{Fold change normalized} = (1.9)^{4.2}/(1.9)^{0.1} = 14.81/1.07 = 13.8.$$

This corresponds to a 13.8 fold increase in gene expression in the treated samples compared to the control samples.

Support Protocol 3: Serial Dilution for Standard Curve

Serial dilutions are done for two reasons: to determine amplification efficiencies (described above) and to generate a standard curve for absolute quantification of target sequences. If performing a standard curve for absolute quantification, it is vital that the actual amount of target or copy number be determined beforehand. The starting material should be free of contamination and carefully quantified; the copy number for the target sequence must also be determined. If using in vitro transcribed RNA it must first be converted to cDNA as described above (Basic Protocol 1). When comparing a sample's C_T value to the standard calibration curve, it is important to check that the amplification efficiencies are the same under both conditions (experimental and standard calibration curve).

Materials

DNA or RNA to be used for standard curve
In vitro transcribed RNA (if appropriate)
Real-time thermal cycler

1. Prepare a 5-fold serial dilution of the DNA or RNA to be used for the standard curve. Construct a standard curve for each target sequence including the housekeeping gene.

   There should be at least five data points in the standard curve.

2. If using RNA as the starting material, prepare a dilution of the transcribed RNA for use in the production of cDNA prior to running PCR.

   Below is a dilution series for use with an RNA molecule that is 1 kb in length. For the purposes of this calculation assume the stock is 1 μg/10 μl of the ssRNA. An average mol. wt. for a ssRNA molecule can be calculated using the following formula (the 159 g/RNA transcript accounts for the 5′ triphosphate cap):

   $$ssRNA = [(\text{no. of nucleotides}) \times 320.5\ g/\text{nucleotide}] + 159\ g/\text{RNA transcript}$$

   If determining the copy number value for a target contained in a plasmid or genomic DNA is desired, visit Applied Biosystem's technical support Web site (http://www.appliedbiosystems.com/support/apptech/) and click on the PDF entitled "Creating Standard Curves with Genomic DNA or Plasmid DNA Templates for Use in Quantitative PCR." This document provides a detailed step by step protocol for determining copy number for DNA samples.

   Ideally the dilutions should cover the range $1 \times 10^1 - 1 \times 10^6$ starting copy number (for absolute quantification). In the example given in Table 10.3.4, the dilutions numbered 3 to 8 would be used (after cDNA synthesis).

3. Run the PCR reaction to determine C_T values for each dilution.

4. Plot the C_T values versus the logarithm of the concentration or the copy number of the template.

   This line should have a high R^2 value (>0.99) which is an indication of the goodness-of-fit of the line through the data points (APPENDIX 4).

Table 10.3.4 Example Dilution Series for an RNA Standard[a]

Dilution	Source	Initial concentration	Volume of stock	Volume of diluent	Final volume	Final concentration	Copy no. present
1	Stock	100,000 pg/μl	10 μl	990 μl	1000 μl	1,000 pg/μl	1.88×10^9
2	Dil. 1	1,000 pg/μl	100 μl	900 μl	1000 μl	100 pg/μl	1.88×10^7
3	Dil. 2	100 pg/μl	100 μl	900 μl	1000 μl	10 pg/μl	1.88×10^6
4	Dil. 3	10 pg/μl	100 μl	900 μl	1000 μl	1 pg/μl	1.88×10^5
5	Dil. 4	1 pg/μl	100 μl	900 μl	1000 μl	0.1 pg/μl	1.88×10^4
6	Dil. 5	0.1 pg/μl	100 μl	900 μl	1000 μl	0.01 pg/μl	1.88×10^3
7	Dil. 6	0.01 pg/μl	100 μl	900 μl	1000 μl	0.001 pg/μl	1.88×10^2
8	Dil. 7	0.001 pg/μl	100 μl	900 μl	1000 μl	0.0001 pg/μl	1.88×10^1

[a]To convert this dilution series for use with a DNA standard use the following relationships; 1 μg of dsDNA that is 1 kbp in length is equivalent to 1.65 pmol or 9.91×10^{11} molecules.

5. Compare average C_T values for each sample to the standard curve to determine actual amount of starting target in the sample (see Fig. 10.3.14).

UNDERSTANDING RESULTS

The data obtained from real-time RT-PCR analysis are C_T values that are a measure of the amount of starting target material. The C_T values need to be converted using different procedures to make valid comparisons. Ultimately, what is obtained is a measure of differences in gene expression between two samples (e. g., treated and control).

When conducting a real-time PCR experiment that uses an absolute quantification procedure, it is necessary to compare the C_T value for each sample (determined in Basic Protocol 2) with the standard curve (Support Protocol 3). After locating the C_T value on the standard curve, one will have a measurement that corresponds to the copy number (or amount) of the target sequence in the starting template sample (Fig. 10.3.14). The absolute values of the treated and untreated target transcripts are then normalized using values obtained for the treated and untreated housekeeping transcripts. This provides a measure of the amount of message RNA in the treated and untreated samples.

When comparing samples using relative quantification, the C_T values from different treatments (determined in Basic Protocol 2) will need to be compared after correction for amplification efficiency and normalization with the housekeeping gene as described in Support Protocols 1 and 2. The normalized fold induction differences that are determined can be used for comparison purposes between treated and untreated samples. It is important to remember that these relative changes in gene expression are not a measure of how much actual message is present, and is simply a relative measure of how much that gene has increased (or decreased) in gene expression in relative terms. However, for many experimental purposes this is sufficient.

TROUBLESHOOTING

Low-Quantity or Poor-Quality PCR Product

Poor RNA quality may contribute to low yield in RT-PCR. Running the RNA preparation on a denaturing agarose gel containing formaldehyde can reveal the integrity of the RNA. (See *UNIT 5.2* for information on purifying RNA.)

No or low product yields in RT-PCR may be due to inefficient cDNA synthesis or low amplification of cDNA. This may require setting up different reactions where varying amounts of RNA are added to the cDNA synthesis reaction. Additionally, variable amounts of cDNA can be added to the PCR reaction.

Low yield can also be the result of poor primer design. If PCR product is low or nonexistent, design of new primer sets for amplification might be needed. Check the primer sequences to determine their propensity to form primer-dimers (see *UNIT 10.2*).

10

Negative RT-reaction control produces a band

If the negative RT-reaction control produces a band, the RNA may be contaminated. This may be due to amplicon contamination in the RT reagents or, in some cases, genomic DNA contamination in the RNA preparation. It is often best to simply repeat the RT reaction using new reagents (except RNA prep). If the problem disappears, one can proceed. If the problem persists, it will be necessary to either digest the RNA preparation with RNase-free DNase or re-isolate the RNA.

Negative PCR control produces band

If the negative PCR control produces a band, one may have amplicon contamination of the reagents. It is often best to simply repeat the PCR reaction using new reagents (except *Taq*). If the problem disappears, one can proceed. If the problem persists it will be necessary to test the *Taq* polymerase for contamination by securing a new tube of enzyme. The use of blocked pipet tips can help prevent amplicon contamination (see *UNIT 10.2*).

Excessive variability between duplicate or triplicate samples

If the samples display large variability between duplicates and triplicate there may be a problem in the preparation of the master mix or a pipetting error. Check that the master mix was prepared properly and well mixed, and that the micropipet is calibrated and is dispensing the proper amount of liquid reproducibly (*UNITS 1.1 & 3.1*).

Poor amplification efficiency

If the PCR amplification curve reaches an early plateau compared to other PCR reactions, the amplification efficiency may be poor (<90%). Check the amplification efficiency: if it is <90%, consider resynthesizing or redesigning the primers being used. Redesigning the primers may be done by moving them to a different location or extending their length. In addition, if not already done, consider performing a careful optimization of the primer concentrations using a primer matrix with varying concentrations for each primer and/or perform a careful optimization of the annealing temperature by temperature gradient analysis (*UNIT 10.2*).

Primer-dimer product seen in melting curve analysis and agarose gel

The presence of primer-dimers can be observed after running an agarose gel of the products or by a second peak in the melting curve analysis that melts at a lower temperature than the target amplicon. This may occur because the template concentration is too low. Increasing the template concentration may help.

Nonspecific amplification results in multiple peaks in melting curve analysis and multiple bands in agarose gel

Multiple peaks may be observed in the melting curve analysis that are confirmed by the observation of multiple bands in an agarose gel of the PCR product. To correct this problem, make sure a hot-start *Taq* polymerase is being used and that the annealing temperatures are sufficiently high for the primers (>60°C). It may be necessary to redesign the primers if the problem persists.

Multiple peaks (bimodal) in melting curve but no extra bands in agarose gel

It is possible to observe multiple bands or peaks in the melting curve analysis even though the agarose gel of the products shows only one discrete band. This can be due to internal sequence differences (high G:C content in one area versus another) that result in differences in melting characteristics. If the agarose gel analysis reveals only one discrete band, assume that the bimodal peaks in the melting curve analysis are specific for the target.

VARIATIONS

Working with mRNA rather than total RNA

If working with genes that have low expression levels, it may be useful to purify the mRNA from the total RNA. There are many procedures for doing this and various manufacturers offer slightly different variations. To learn more about mRNA considerations, visit the Ambion Web site (*http://www.ambion.com*) and navigate to the technical resources page. From this page one can search their extensive document collection for advice on a variety of RNA-related topics. In addition, *Current Protocols in Molecular Biology* (Ausubel, 2007) describes methods for purifying mRNA from total RNA. The book *RNA Methodologies: A Laboratory Guide for Isolation and Characterization* (Farrell, 1998) may also provide a useful reference as this approach is considered.

Multiplexing PCR

One of the advantages of using strand-specific detection technologies is the ability to perform multiplexing. Multiplexing refers to amplifying multiple targets in the same reaction tube. There are many factors that need to be considered when conducting a multiplex experiment, including manufacturer-specific issues. It is helpful when designing a multiplexing experiment to carefully review the manufacturers technical material, which can be quite helpful. In addition, Qiagen provides more generic advice (*http://www.qiagen.com*) that one may find helpful when considering which platform or chemistry to pursue. Finally, Logan and Edwards (2004) provide an excellent overview of different platforms and the different detection chemistries supported.

LITERATURE CITED

Alker, A.P., Mwapasa, V., and Meshnick, S.R. 2004. Rapid real-time PCR genotyping of mutations associated with sulfadoxine-pyrimethamine resistance in *Plasmodium falciparum*. *Antimicrob. Agents Chemother.* 48:2924-2929.

Ausubel, F.M., Brent, R., Kingston, R.E., Moore, D.D., Seidman, J.G., Smith, J.A., and Struhl, K. (eds.) 2007. Current Protocols in Molecular Biology. John Wiley and Sons, Hoboken, N.J.

Biéche, I., Olivi, M., Champéme, M.H., Vidaud, D., Lidereau, R., and Vidaud, M. 1998. Novel approach to quantitative polymerase chain reaction using real-time detection: Application to the detection of gene amplification in breast cancer. *Int. J. Cancer* 78:661-666.

Biéche, I., Laurendeau, I., Tozlu, S., Olivi, M., Vidaud, D., Lidereau, R., and Vidaud, M. 1999. Quantitation of *MYC* gene expression in sporadic breast tumors with a real-time reverse transcription-PCR assay. *Cancer Res.* 59:2759-2765.

Burchill, S.A., Lewis, I.J., and Selby, P. 1999. Improved methods using the reverse transcriptase polymerase chain reaction to detect tumor cells. *Br. J. Cancer* 79:971-977.

Bustin, S.A. and Nolan, T. 2004. Analysis of mRNA expression by real-time PCR. *In* Real-Time PCR: An Essential Guide (K. Edwards, J. Logan, and N. Saunders, eds.) pp. 125-184. Horizon Bioscience, Norfolk, U.K.

Carding, S.R., Lu, D., and Bottomly, K. 1992. A polymerase chain reaction assay for the detection and quantitation of cytokine gene expression in small numbers of cells. *J. Immunol. Methods* 151:277-287.

Cheng, J., Zhang, Y., and Li, Q. 2004. Real-time PCR genotyping using displacing probes. *Nucl. Acids Res.* 32:e61.

Chomczynski, P. and Sacchi, N. 1987. Single-step method of RNA isolation by acid guanidinium thiocyanate-phenol-chloroform extraction. *Anal. Biochem.* 162:156-159.

Clegg, R.M. 1995. Fluorescence resonance energy transfer. *Curr. Opin. Biotech.* 6:103-110.

Eshel, R., Vainas, O., Shpringer, M., and Naparstek, E. 2006. Highly sensitive patient specific real-time PCR SNP assay for chimerism monitoring after allogenic stem cell transplantation. *Lab. Hematol.* 12:39-46.

Farrell, R.E. 1998. RNA Methodologies: A Laboratory Guide for Isolation and Characterization, 2nd ed. Academic Press, San Diego, Calif.

Foley, K.P., Leonard, M.W., and Engel, J.D. 1993. Quantitation of RNA using the polymerase chain reaction. *Trends Genet.* 9:380-385.

Gibson, U.E., Heid, C.A., and Williams, P.M. 1996. A novel method for real time quantitative RT-PCR. *Genome Res.* 6:995-1001.

Gibson, N.J. 2006. The use of real-time PCR methods in DNA sequence variation analysis. *Clin. Chim. Acta* 363:32-47.

Higuchi, R., Dollinger, G., Walsh, P.S., and Griffith, R. 1992. Simultaneous amplification and detection of specific DNA sequences. *Biotechnology* 10:413-417.

Higuchi, R., Fockler, C., Dollinger, G., and Watson, R. 1993. Kinetic PCR: Real time monitoring of DNA amplification reactions. *Biotechnology* 11:1026-1030.

Kindich, R., Florl, A.R., Jung, V., Engers, R., Müller, M., Schulz, W.A., and Wullich, B. 2005. Application of a modified real-time PCR technique for relative gene copy number quantification to the determination of the relationship between NKX3.1 loss and MYC gain in prostate cancer. *Clin. Chem.* 51:649-652.

Kochanowski, B. and Reischl, U. 1999. Methods in Molecular Medicine: Quantitative PCR Protocols, 1st ed. Humana Press, Totowa, N.J.

Königshoff, M., Wilhelm, J., Bohle, R.M., Pingoud, A., and Hahn, M., 2003. *HER-2/neu* gene copy number quantified by real-time PCR: Comparison of gene amplification heterozygosity, and immunohistochemical status in breast cancer tissue. *Clin. Chem.* 49:219-229.

Lee, M.A., Squirrell, D.J., Leslie, D.L., and Brown, T. 2004. Homogenous fluorescent chemistries for real-time PCR. *In* Real-time PCR: An Essential Guide (K. Edwards, J. Logan, and N. Saunders, eds.) pp. 85-102. Horizon Bioscience, Norfolk, U.K.

Leutenegger, C.M., Mislin, C.N., Sigrist, B., Ehrengruber, M.U., Hofmann-Lehmann, R., and Lutz, H. 1999. Quantitative real-time PCR for the measurement of feline cytokine mRNA Vet. *Immunol. Immunopathol.* 71:291-230.

Liss, B. 2002. Improved quantitative real-time RT-PCR for expression profiling of individual cells. *Nucl. Acids. Res.* 30:e89.

Liu, W. and Saint, D.A. 2002. Validation of a quantitative method for real time PCR kinetics. *Biochem. Biophys. Res. Commun.* 294:347-353.

Livak, K.J. 2003. SNP genotyping by the 5′-nuclease reaction. *Methods Mol. Biol.* 212:129-147.

Livak, K.J. and Schmittgen, T.D. 2001. Analysis of relative gene expression data using real-time quantitative PCR and the $2^{-\Delta\Delta CT}$ method. *Methods* 25:402-408.

Logan, J.M.J. and Edwards, K.J. 2004. An overview of real-time PCR platforms. *In* Real-time PCR: An Essential Guide (K. Edwards, J. Logan, and N. Saunders, eds.) pp. 13-30. Horizon Bioscience, Norfolk, U.K.

Marras, S.A., Kramer, F.R., and Tyagi, S. 2003. Genotyping SNPs with molecular beacons. *Methods Mol. Biol.* 212:111-128.

Meijerink, J., Mandigers, C., van de Locht, L., Tonnissen, E., Goodsaid, F., and Raemaekers, J. 2001. A novel method to compensate for differential amplification efficiencies between patient DNA samples in quantitative real-time PCR. *J. Mol. Diagn.* 3:55-61.

O'Garra, A. and Vieira, P. 1992. Polymerase chain reaction for detection of cytokine gene expression. *Curr. Opin. Immunol.* 4:211-215.

Peccoud, J. and Jacob, C. 1998. Statistical Estimations of PCR Amplification Rates. *In* Gene Quantification (F. Ferre, ed.) pp. 111-128. Birkhuser New York.

Pfaffl, M.W. 2001. A new mathematical model for relative quantification in real-time RT-PCR. *Nucl. Acids Res.* 29:2002-2007.

Rasmussen, R. 2001. Quantification on the LightCycler instrument. *In* Rapid Cycle Real-time PCR: Methods and Applications (S. Meuer, C. Wittwer, and K. Nakagawara, eds.) pp. 21-34. Springer, Heidelberg, Germany.

Temin, H.M. 1974. On the origin of RNA tumor viruses. *Ann. Rev. Genet.* 8:155-177.

Temin, H.M. 1995. Genetics of retroviruses. *Ann. N.Y. Acad. Sci.* 758:161-165.

Tichopad, A., Dilger, M., Schwarz, G., and Pfaffl, M.W. 2003. Standardized determination of real-time PCR efficiency from a single reaction set-up. *Nucl. Acids Res.* 31:e122.

Valasek, M.A. and Repa, J.J. 2005. The power of real-time PCR. *Adv. Physiol. Educ.* 29:151-159.

Ward, C.L., Dempsey, M.H., Ring, C.J., Kempson, R.E., Zhang, L., Gor, D., Snowden, B.W., and Tisdale, M., 2004. Design and performance testing of quantitative real time PCR assays for influenza A and B viral load measurement. *J. Clin. Virol.* 29:179-188.

INTERNET RESOURCES

http://pathmicro.med.sc.edu/

University of South Carolina-School of Medicine's tutorial on real-time PCR. At the homepage click on the "Real-time PCR tutorial" link under the "Textbook" button.

http://dna-9.int-med.uiowa.edu/

University of Iowa DNA facility's on-line tutorial on real-time PCR. At the homepage, click on the "real-time PCR" button on the left.

http://www.ambion.com/

Ambion's Web site has many valuable technical reports. These can be searched for on the Technical Resources page.

http://www.gene-quantification.de/

A short primer on absolute quantification methods in real-time RT-PCR. From the homepage, click on the "RT.gene-quantification.info". From the subsequent page, click on the "REVIEW: Absolute quantification of mRNA using real-time reverse transcription PCR assays."

http://www.biocompare.com/

This tutorial presents researchers with an overview of real-time PCR, identifies the advantages and disadvantages of the various detection technologies, outlines the key issues for optimizing experimental design and offers a brief description of the various methods used for data analysis. Click on the "Video/Slide Show" button to be taken to a web seminar about real-time PCR.

UNIT 10.4

DNA Sequencing: An Outsourcing Guide

Jeffrey W. Touchman[1] and Stephen D. Mastrian[1]
[1]Translational Genomics Research Institute, Phoenix, Arizona

10

OVERVIEW AND PRINCIPLES

The discovery of the structure of deoxyribonucleic acid (DNA) is one of the most important scientific breakthroughs of the last century (Watson and Crick, 1953). Within DNA are instructions sufficient to make an organism and the means by which organisms pass information along to their offspring. Remarkably, this information is coded by only four nucleotides: adenosine (A), cytosine (C), guanine (G), and thymine (T). Understanding the order of these nucleotides in linear DNA molecules has been an active pursuit since the discovery of its structure. As a result, DNA sequencing has emerged as a fundamental tool in molecular biology research.

There are several ways to determine the sequence of nucleotides in a strand of DNA, but the most prevalent method is undoubtedly the dideoxyribonucleoside triphosphate (ddNTP)–mediated chain termination method of Sanger et al. (1977). In the basic Sanger sequencing reaction (Fig. 10.4.1), a synthetic oligonucleotide primer is annealed to one strand of a denatured DNA template. Once annealed to the template, the primer is extended in the presence of a DNA polymerase and the four deoxynucleoside triphosphates (dNTPs) A, C, G, and T. Importantly, the reaction also contains small amounts of four dideoxynucleoside triphosphates (ddNTPs) that terminate elongation when incorporated into the growing chain. Each ddNTP also has a unique fluorescent dye molecule attached to it that enables the detection of that ddNTP on an automated sequencing instrument.

After completion of the sequencing reaction, unincorporated nucleotides are removed from the reaction and the extension products are visualized by high-resolution electrophoresis on an automated sequencing apparatus like the one pictured in Figure 10.4.2. A generalized schematic of this process is shown in Figure 10.4.3. The sequencing reaction mixture is first loaded by the application of a brief electric current (called electrokinetic injection) into a glass capillary tube filled with a gel-like separation polymer. Then, a stronger current is applied to the capillary to draw the reaction products through the polymer by traditional electrophoresis (*CP Molecular Biology Unit 2.5A*; Voytas, 2000). Because DNA possesses a negative charge, it will migrate toward the (positive) cathode reservoir at the other end of the capillary tube. In this fashion, the labeled DNA fragments will separate by size as they move through the capillary, with the shorter fragments migrating faster than the longer ones. Fragments that differ in size by just one nucleotide will be discernable as they pass through a detection window, where a laser is focused on the capillary. The laser light excites the fluorescent dye molecules attached to the terminal ddNTPs of the sequencing fragments, which will emit a specific wavelength of light for each A, C, G, or T. The emission of light from the fluorescent dyes is captured by a detector and sent to a computer that constructs the order of nucleotides of the DNA fragment, based upon the order in which the labeled DNA strands passed the detector.

Improved DNA polymerases and instrumentation have ushered in an era of automated sequence analysis where thousands of sequencing reactions can be analyzed in one week by a single instrument. Dedicated core laboratories are frequently created to house and operate these instruments with the sole purpose of sequencing DNA in a rapid, accurate, and cost-effective manner. It is within the context of these service providers that approaches to sequencing are presented here. In this unit, methods and strategies are addressed to prepare DNA for automated sequencing. Particular focus is given to increasing the potential for positive results by paying careful attention to DNA template

Current Protocols Essential Laboratory Techniques 10.4.1-10.4.17

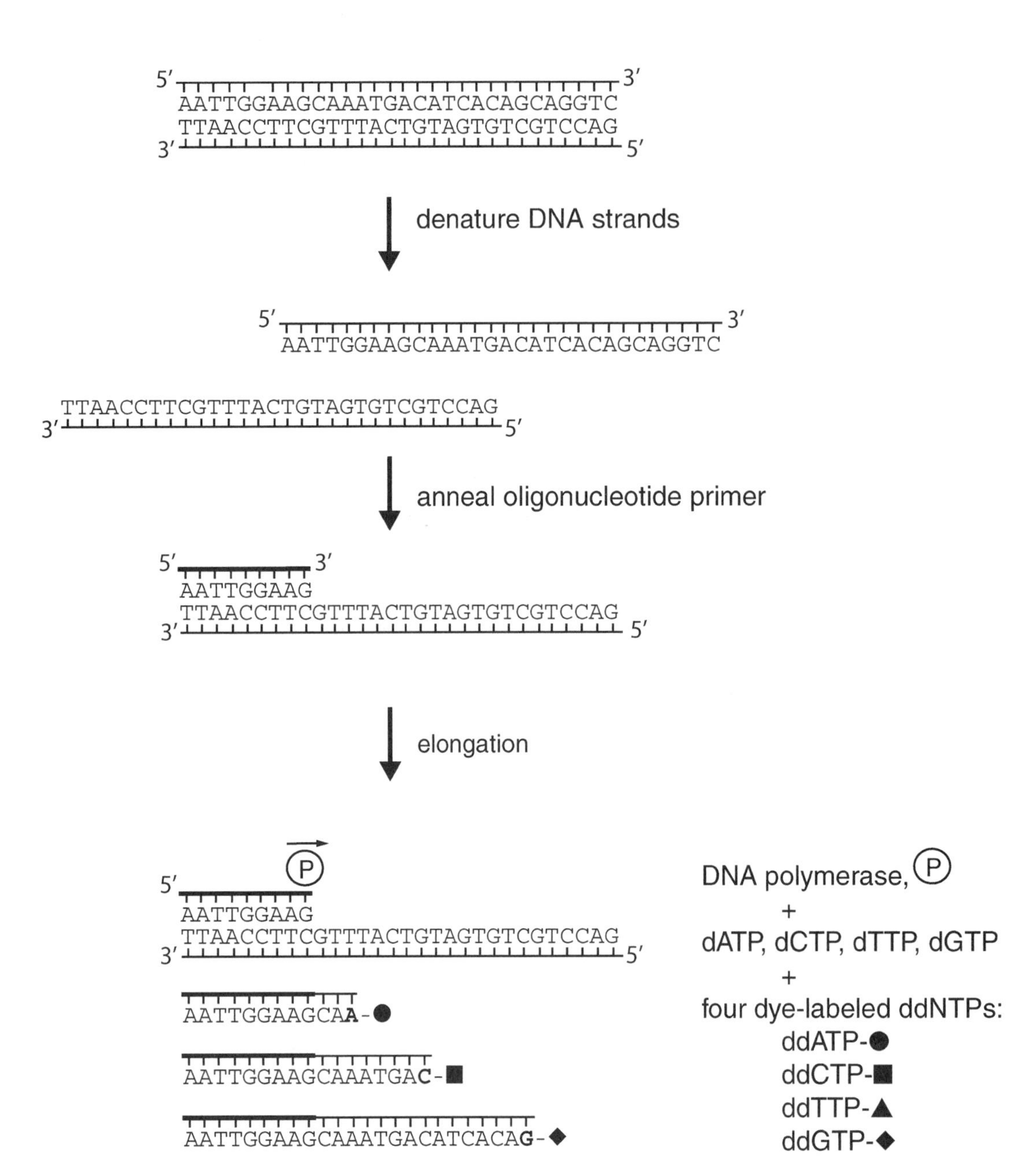

Figure 10.4.1 Steps used in Sanger DNA sequencing.

purification and primer design. Examples of poor quality data are given, as well as suggested solutions to common problems.

The basic strategy for outsourcing DNA sequencing consists of the following main steps (Fig. 10.4.4):

1. Purify template DNA to be sequenced.
2. Design oligonucleotide primers specific to your template.
3. Perform four-color fluorescent DNA sequencing reactions (service provider).
4. Conduct fragment analysis on a sequencing instrument (service provider).
5. Collect and troubleshoot results.

DNA sequence can be determined from DNA templates of varying size. Double-stranded DNA (e.g., plasmid clones and PCR products), single-stranded DNA (e.g., M13), large DNA clones (e.g., BACs, PACs, YACs, cosmids, and fosmids clones), and even bacterial genomic DNA can be used

Figure 10.4.2 The Applied Biosystems 3730×l DNA analyzer is a sequencing machine found in many modern facilities.

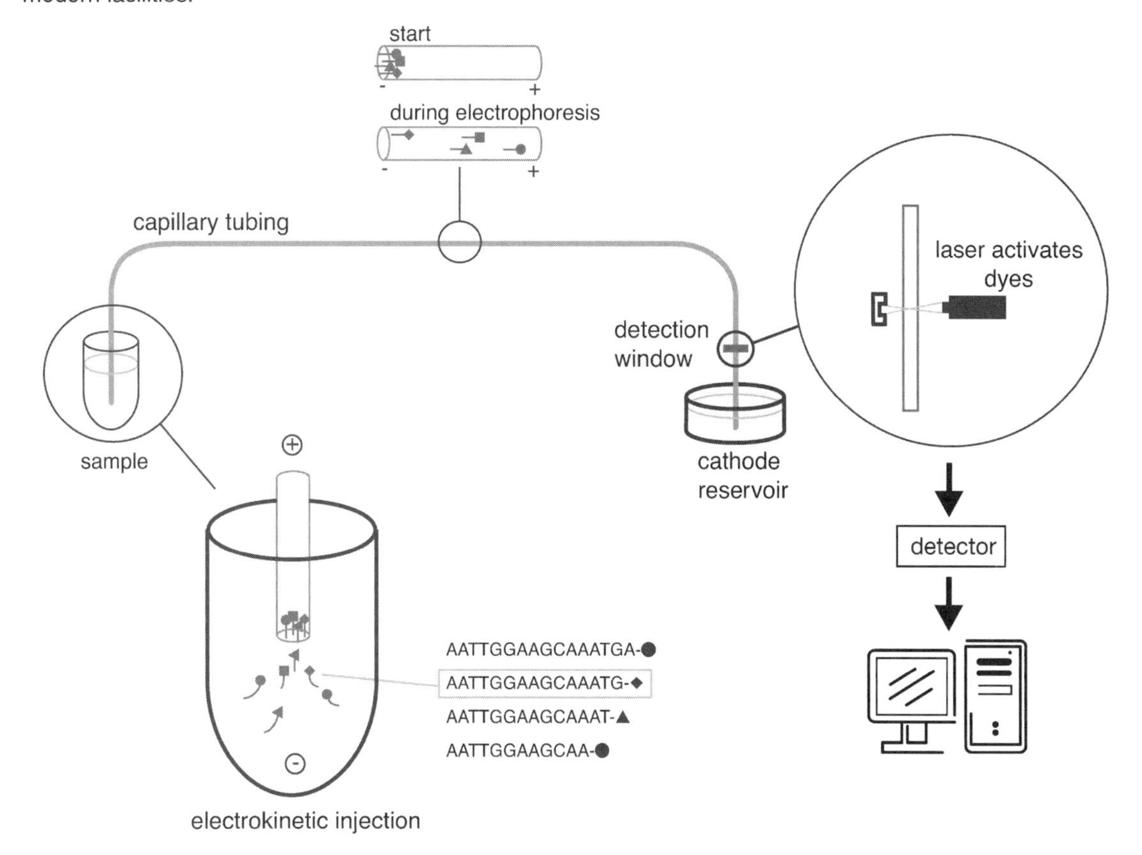

Figure 10.4.3 Automated capillary-based DNA sequencing. DNA fragments from a sequencing reaction are indicated by short gray lines. Fluorescent labels attached to each DNA fragment are indicated by a circle, diamond, triangle, or square. Movement of the fragments through the capillary is caused by an applied electrical current and proceeds toward the cathode. A laser excites the fluorescent labels and sends the signal to a computer attached to the sequencing instrument, which interprets the order of nucleotides in the DNA strand.

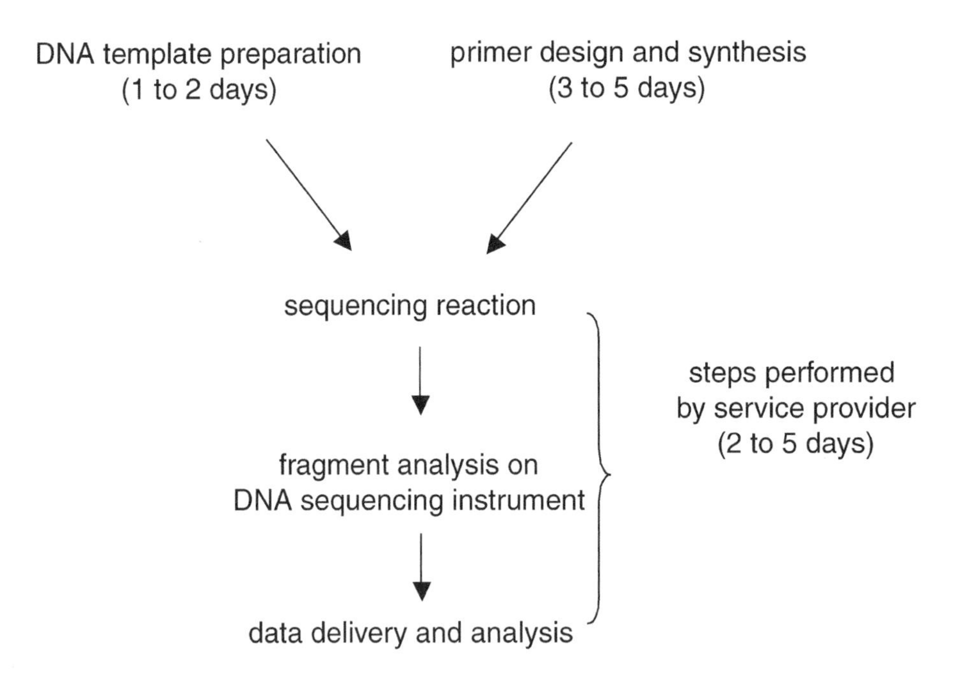

Figure 10.4.4 Key steps for DNA sequencing with a service provider. Steps performed by the service provider are indicated. Other steps are performed by the researcher.

as a template for DNA sequencing. But because of their ubiquity in the laboratory and ease of manipulation, plasmids and PCR products are by far the templates most frequently used for routine sequencing. In this chapter, the focus will be on preparing plasmid DNA and PCR products as templates for the sequencing reaction.

Finally, current automated sequencing technology is dominated by capillary instrumentation (e.g., 3730×l DNA Analyzer, Applied Biosystems; MegaBACE, GE Healthcare; and CEQ Genetic Analysis System, Beckman Coulter), and the use of four-color fluorescent dye-terminator cycle-sequencing reactions (e.g., PRISM BigDyeTerminator kit, ABI; DYEnamic ET Dye Terminator kit, GE Healthcare). The following protocols are presented with the assumption that this will be the DNA sequencing approach used by your service provider.

Template Quality

Automated sequencing is a relatively robust process, and today's success rates and read lengths (i.e., the number of consecutive base pairs that are determined in a single instrument run) are better than ever before. Successful read lengths as high as 800 bp can be routinely achieved. However, success depends on several factors. By far, the most important is the quality of template DNA. When preparing plasmid or PCR templates, it is critical to avoid the following:

1. Residual salts
2. Proteins
3. Residual detergents
4. Residual RNA.

Effect of salts

Capillary electrophoresis is very susceptible to salt in samples from template preparation, sequencing reactions, or precipitation methods using salts. The negative ions in salts can be preferentially injected into the capillary tube during electrokinetic injection, leading to lower signal. In addition, the negative ions compete and interfere with the injection of larger DNA extension fragments, leading to shortened read lengths. Both sodium acetate and EDTA are common sources of salts encountered during DNA sequencing.

Effect of proteins

Many DNA preparation methods for sequencing require the recovery of DNA from lysed bacterial cultures (Basic Protocol 1). Unless DNA is carefully purified, protein can remain in the DNA samples. Protein can be injected and will adhere to the walls of the capillary, adversely affecting data resolution and capillary array lifetime.

Effects of residual detergent

Detergents, e.g., sodium dodecyl sulfate (SDS), are used in plasmid purification protocols to lyse bacterial cells. Small, negatively charged detergents may be preferentially injected over DNA during electrokinetic injection. If present at high levels, such detergents will adversely affect the quality of the sequencing data and the life of the capillary.

Effect of residual RNA

Residual RNA that is present in DNA template preparations from bacterial clones competes with the DNA for injection into the capillary. Residual RNA has the same effect as excess salt, i.e., decreased signal and shortened read lengths.

Template Quantity

A sequencing service provider will usually provide guidelines regarding the quantity and concentration of DNA template required for the DNA sequencing reaction, but, generally, amounts will be in the ranges given in Table 10.4.1. Note that to maintain the same number of template molecules in a reaction, the quantity (measured in nanograms) of template required must increase proportionally as the template size increases.

Too little DNA template or primer in a sequencing reaction reduces the signal strength, and therefore the peak height of reaction products. In the worst case, the signal-to-noise ratio decreases such that bases cannot be called correctly.

Too much DNA template can affect both data quality from the sequencing reaction and sample loading onto the sequencing instrument. Excess template used in the sequencing reaction results in the biased generation of short extension fragments. During electrokinetic injection (Fig. 10.4.3), short fragments are injected more efficiently, resulting in "top-heavy" chromatogram characteristics and shortened reads (see Troubleshooting). Excess template also inhibits the injection of labeled

Table 10.4.1 Recommended Quantity of DNA for Cycle-Sequencing Reactions

Template	Acceptable quantity in sequencing reaction
PCR product	
100-200 bp	1-3 ng
200-500 bp	3-10 ng
500-1000 bp	5-20 ng
1000-2000 bp	10-40 ng
>2000 bp	40-100 ng
Double-stranded DNA (e.g., plasmids)	100-200 ng
Large DNA (e.g., BACs, PACs, YACs, cosmids, and fosmids)	0.5-1.0 µg
Bacterial genomic DNA	2-3 µg

extension fragments, thus affecting signals generated from the instrument. Excess template can behave similarly to proteins and accumulate in the capillary array, which adversely affects data resolution and capillary array lifetime.

Sequencing Primers

The Sanger sequencing reaction is essentially the synthesis of a DNA chain by a polymerase. Extension is initiated from a synthetic oligonucleotide primer, and careful selection of this primer will ensure high-quality results. The primers used in fluorescent sequencing with dye-labeled ddNTPs typically contain a custom sequence that is synthesized for a specific experiment or project. Several publicly available programs (Hillier and Green, 1991; Rozen and Skaletsky, 2000) can be used to design sequencing primers, and there are also several commercial software options (see UNIT 10.2). The parameters involved in manually designing successful sequencing primers are nearly identical to the parameters for designing primers for PCR. However, because sequencing requires one primer while PCR requires two, the potential interaction between two primers in the same reaction can be ignored. Thus, for DNA sequencing, the basic considerations for primer design should include the following:

Choose a primer sequence that is longer than 15 nucleotides.
Choose a primer sequence with a roughly equivalent number of A, C, G, and T bases and a G + C content of ~40% to 50%.
Avoid runs containing more than three of the same nucleotide.
Avoid regions with a potential for secondary structure formation or self-complementarity
Choose a primer sequence with a C or G at the 3′ end. This provides a higher degree of hydrogen bonding and thus ensures good annealing of the 3′ end.
Choose a primer sequence from a DNA segment for which high-quality sequence data are available.
Confirm that the primer binding site is unique in your template.
Choose a primer with a melting temperature (T_m) of 45°C to 55°C. If this in not possible because of the criteria described above, or if a longer primer with a higher T_m is to be used, the annealing temperature for thermal cycling should be adjusted to the primer T_m. The T_m of the primer can be estimated by using the equation

$$T_m = 69.3 + (0.4)(\%\,G + C) - (650/L)$$

where L is the length of the primer in nucleotides.

Purifying primers for DNA sequencing reactions is much more straightforward than in the past, primarily because of the high efficiency of nucleotide incorporation in automated synthesizers. In general, the primers should be very pure, with <5% failed products. Commercial suppliers of oligonucleotide primers normally meet or exceed this level of purity, although there is no direct way to detect the presence of poor-quality or failed products short of analyzing the primers on an acrylamide gel. Primers that are desalted (the most basic level of primer purification) are usually of sufficient purity for sequencing applications. Impure primers can cause a shadow sequence of $n-1$ to appear in the data. If this happens, request that the primers be purified by high-pressure liquid chromatography (HPLC), or polyacrylamide gel electrophoresis (PAGE). Both of these methods are exceptional at removing failed products from synthesized oligonucleotides.

Often, sequencing from a plasmid DNA template affords the advantage of using a universal primer that is complimentary to a sequence adjacent to the multiple-cloning site (Fig. 10.4.5). The most frequently employed universal primers are the M13 Forward and M13 Reverse primers, which flank the cloning site of pUC-based plasmids. These primers produce very high-quality data, and service providers will usually offer them for sequencing (when appropriate) at no additional fee.

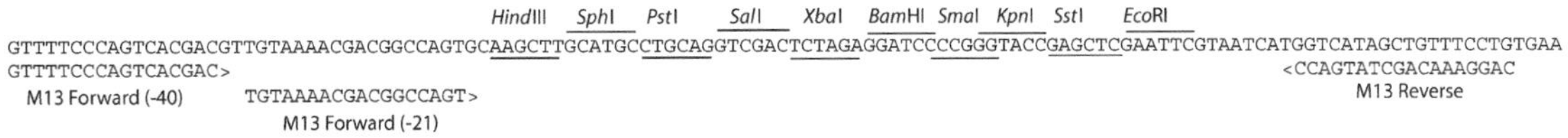

HindIII SphI PstI SalI XbaI BamHI SmaI KpnI SstI EcoRI
GTTTTCCCAGTCACGACGTTGTAAAACGACGGCCAGTGCAAGCTTGCATGCCTGCAGGTCGACTCTAGAGGATCCCCGGGTACCGAGCTCGAATTCGTAATCATGGTCATAGCTGTTTCCTGTGAA
GTTTTCCCAGTCACGAC> <CCAGTATCGACAAAGGAC
M13 Forward (-40) TGTAAAACGACGGCCAGT> M13 Reverse
M13 Forward (-21)

Figure 10.4.5 The sequence of the common plasmid vector pUC18 around the multiple-cloning site. Many commercial plasmid vectors are derivatives of pUC18. The sequences and positions of universal primers M13 Forward and M13 Reverse are shown. Restriction enzyme cutting sites are indicated by horizontal lines, with their corresponding names above the sites.

Table 10.4.2 Universal Primers

Primer name	Sequence (5′ to 3′)	Melting temperature (T_m)
M13 (-21) Forward	TGTAAAACGACGGCCAGT	54°C
M13 (-40) Forward	GTTTTCCCAGTCACGAC	52°C
M13 Reverse	CAGGAAACAGCTATGACC	54°C
SP6	ATTTAGGTGACACTATAG	47°C
T3	ATTAACCCTCACTAAAG	45°C
T7	TAATACGACTCACTATAGGG	54°C

These and other commonly used universal primers are listed in Table 10.4.2. Always check the sequence or map of your plasmid to confirm the presence or absence of universal primer binding sites. Plasmid sequences and labeled maps can usually be obtained from commercial suppliers of the plasmid. Additionally, many plasmid sequences are deposited in public repositories (e.g., GenBank; *http://ncbi.nlm.nih.gov/entrez*). If a labeled map is not available, a quick scan of the plasmid sequence in a word processing program using the sequence of your universal primer and the Find command should reveal the presence of a binding site. Remember to "reverse and complement" the sequence of your primer when searching the negative strand of the plasmid.

When sequencing from PCR products, it is typical to employ one of the primers used for the amplification of your product as the sequencing primer. Note however that you will use less primer in a sequencing reaction than in a PCR reaction. The service provider will guide you regarding the amount of primer they require from you.

STRATEGIC PLANNING

Finding a DNA Sequencing Service

Many institutions now have a dedicated laboratory or facility for DNA sequencing. If this is available, it is worthwhile to outsource DNA sequencing there because sample and data delivery are simplified and the cost is usually very competitive. Additionally, the personnel are often available to help troubleshoot poor quality results in person. If your institution does not have a sequencing laboratory, you may wish to explore nearby institutions. Barring that, there are a plethora of commercial DNA sequencing options available that will receive your DNA templates and primers by mail and return sequencing data electronically. Table 10.4.3 provides a list of some of these companies.

Communication of expectations with a DNA sequencing service provider is key to a positive outcome. Some important questions to ask prior to beginning work are:

What is the price per sequence read?
What is the turn-around time (i.e., how long will it take)?
What is the length of an average read?

Table 10.4.3 DNA Sequencing Service Providers

DNA sequencing company	Telephone number	Web site
Agencourt Bioscience	800-361-7780	*http://www.agencourt.com*
Cogenics	919-463-6738	*http://www.cogenics.com*
Eurogentec	+32-4-240-76-76	*http://www.eurogentec.be*
Lark Technologies	713-779-3663	*http://www.lark.com*
MACROGEN	+82-2-2113-7013	*http://www.macrogen.com*
MWG Biotech	877-694-2832	*http://www.mwgbiotech.com*
Polymorphic DNA Technologies	888-362-0888	*http://www.polymorphicdna.com*
Sequetech	800-697-8685	*http://www.sequetech.com/*
Seqwright	800-720-4363	*http://www.seqwright.com*
Qiagen	800-426-8157	*http://www.qiagen.com*

Who is responsible for "failed" reads?
Are positive controls run in parallel with the samples?
How is data delivered and in what format?

Compare the answers to these questions from several companies and select the one that meets your needs.

Preparing for DNA Template Isolation

Good laboratory practices should be followed when performing the DNA preparation protocols in this chapter. Clean the work area before beginning to minimize clutter, and wipe the surface down with a 10% bleach solution when working with bacterial cultures. Always wear latex gloves when working with chemicals and nucleic acids. Double check the materials list for each protocol and confirm that all necessary items are prepared and within reach. Fill a bucket of ice and have it available for use. Read each protocol thoroughly at least once before starting. Finally, anticipate the need for common instrumentation (e.g., centrifuges or agarose gel apparatuses) in a busy laboratory and reserve time on them, if necessary.

Purifying DNA Templates Using Commercial Kits

DNA templates can be purified for sequencing by using a wide variety of commercial kits and spin columns designed for this purpose. While generally more expensive, these purification solutions are perfectly acceptable alternatives to Basic Protocols 1 and 2, and the Alternate Protocol. The two most prevalent technologies are gel filtration and solid-phase extraction.

A gel filtration-based kit is usually composed of a column packed with a special gel matrix. The particles in this matrix consist of a porous organic polymer with a three-dimensional structure, resulting in a network of hydrophilic pores equilibrated with an aqueous buffer. Large molecules migrate faster through the column than small ones, since their size does not allow them to enter the pores of the matrix as freely. Small molecules are distributed in a larger solvent volume, and they penetrate the matrix more slowly. Thus, PCR products can be purified away from residual primers, enzymes, and unincorporated nucleotides left over from the PCR reaction using a gel filtration column.

Solid-phase extraction makes use of the reversible binding of nucleic acids to a physical matrix. The matrix is usually a special anion exchanger packed into polypropylene columns. DNA binding takes place under low salt concentration conditions. During the washing and elution steps, impurities are removed from the DNA sample, and the salt concentration is increased stepwise. Compared to the traditional liquid-liquid extraction, the major advantages are avoidance of dangerous organic solvents like chloroform and phenol, time savings, and the ultra-high purity of the nucleic acid product.

There is a veritable industry centered around the purification of nucleic acids, and such products are too numerous to list here. Consult your DNA sequencing service provider for recommendations, or visit the Web site *http://www.biocompare.com* for a description of some of the many available options.

SAFETY CONSIDERATIONS

Phenol is highly corrosive and can cause severe burns. Wear gloves, protective clothing, and safety glasses and always work in a chemical fume hood. Rinse any areas of skin that come in contact with phenol with a large volume of H_2O. Do not use ethanol!

Chloroform is irritating to the skin, eyes, mucous membranes, and respiratory tract. It is a carcinogen and may damage the liver and kidneys. Wear gloves and safety glasses and always work in a chemical fume hood.

Glacial acetic acid is volatile. Concentrated acids must be handled with great care. Wear gloves, safety glasses, and a face mask and work in a chemical fume hood.

Sodium hydroxide (solid NaOH) is caustic and should be handled with great care. Wear gloves and a face mask. Concentrated bases (e.g., 1 N NaOH) should be handled in a similar manner.

See *APPENDIX 1* for more information on general laboratory safety.

PROTOCOLS

Successful DNA sequencing is only achieved with a high-quality template DNA free of contaminants. The following protocols describe rapid and accessible methods for preparing DNA template free of residual salts, proteins, detergents, and RNA.

NOTE: Always use sterile, latex gloves and sterile pipet tips when handling DNA.

Basic Protocol 1: Isolating Plasmid DNA by Alkaline Lysis Miniprep for Use as Sequencing Template

Plasmid DNA can be isolated from small-scale (1- to 2-ml) bacterial cultures by treatment with alkali and SDS (Birnboim and Doly, 1979), providing plenty of pure material for DNA sequencing applications. Alkaline lysis in the presence of SDS is a flexible technique that works well with all strains of *E. coli*. This protocol can usually be completed in 2 days. It is essentially the same as one described in *UNIT 4.2* (Small-Scale Plasmid Isolation by Alkaline Lysis) except that a special formulation of TE buffer (sequencing TE buffer) is used here. Sequencing TE buffer has one-tenth the amount of EDTA as normal TE buffer because EDTA can negatively affect the outcome of a DNA sequencing reaction.

Materials

Plasmid-containing bacterial culture
YT medium (see recipe) with antibiotic appropriate for the plasmid used
GET buffer (see recipe)
Lysis solution (see recipe)
Potassium acetate solution (see recipe)
Absolute (100%) and 70% (v/v) ethanol
100 µg/ml DNase-free RNase A (Invitrogen) in TE buffer (10 mM Tris·Cl, pH 8.0/1 mM EDTA, pH 8.0)

Phenol/chloroform (1:1 v/v), pH 7.5 saturated with 10 mM Tris·Cl, pH 7.5 (*UNIT 3.3*), room temperature
3 M sodium acetate, pH 5.2 (*UNIT 3.3*)
Sequencing TE buffer (0.1 mM Tris·Cl, pH 7.5/0.1 mM EDTA, pH 8.0)

14-ml round-bottom polypropylene culture tubes
37°C shaking incubator
1.5-ml microcentrifuge tubes
Microcentrifuge, capable of at least 12,000 × *g*
Vortexer

Additional reagents and equipment for spectrophotometrically quantitating DNA (*UNIT 2.2*)

1. For each plasmid-containing culture, place 2 ml of 2× YT medium containing the appropriate antibiotic in a 14-ml culture tube and inoculate with a single colony. Incubate overnight at 37°C, with agitation at 300 rpm. (see *UNIT 4.2*).
2. Transfer 1.4 ml of the culture into a 1.5-ml microcentrifuge tube and microcentrifuge 10 to 15 sec at 12,000 × *g*, room temperature.
3. Aspirate and discard the supernatant and resuspend the cell pellet in 100 µl GET buffer with gentle vortexing.

   Make sure no clumps of cells remain before proceeding to the next step.
4. Add 200 µl freshly prepared lysis solution to the resuspended pellet and mix by gently inverting the tube several times. Incubate 10 min on ice.

   The mixture should become very viscous.

   Exposure of bacterial suspensions to an anionic detergent (e.g., SDS) at high pH opens the cell wall, denatures chromosomal DNA and proteins, and releases plasmid DNA into the supernatant.
5. Add 150 µl potassium acetate solution and mix by inverting the tube several times. Incubate 10 min on ice.
6. Microcentrifuge 5 min at maximum speed, room temperature.
7. Transfer 400 µl of the clear supernatant into a new 1.5-ml microcentrifuge tube. Be sure to avoid transferring any cellular debris. Use a 1000-µl pipet tip to remove any flocculent material from the supernatant. Discard the pellet and remaining supernatant.
8. Precipitate the DNA with 1 ml (2.5 volumes) absolute ethanol. Mix the solution by vortexing and incubate 30 min at −20°C.
9. Collect the precipitated nucleic acids by microcentrifuging 15 min at maximum speed.
10. Remove the supernatant by gentle aspiration and stand the tube in an inverted position on a paper towel to allow all of the fluid to drain away. Keep the open tube at room temperature until the ethanol has evaporated and no fluid is visible in the tube (2 to 5 min).

    Remove any droplets that adhere to the sides of the tube with a pipet.
11. Add 100 µl of 100 µg/ml DNase-free RNase A to the DNA pellet and dissolve by vortexing. Incubate 1 hr at 37°C.
12. Extract with an equal volume (100 µl) of Tris-saturated phenol/chloroform by mixing the organic and aqueous phases by vortexing and then microcentrifuging the emulsion 2 min at maximum speed.
13. Transfer the aqueous upper layer to a fresh tube. Precipitate the DNA in the aqueous phase with 10 µl (0.1 vol) 3 M sodium acetate (pH 5.2) and 250 µl (2.5 vol) absolute ethanol.
14. Mix the solution by vortexing and incubate 30 min at −20°C.

15. Collect the precipitated nucleic acids by microcentrifuging 15 min at maximum speed.

16. Remove the supernatant by gentle aspiration and stand the tube in an inverted position on a paper towel to allow all of the fluid to drain away (1 to 2 min).

    Again, remove any droplets that adhere to the sides of the tube with a pipet.

17. Add 1 ml of 70% ethanol to the pellet and recover the DNA by microcentrifuging 2 min at maximum speed, room temperature.

    Rinsing the plasmid DNA pellet with 70% ethanol is critical for removing residual salts from the sample.

18. Again remove all of the supernatant by gentle aspiration with a pipet and store the open tube at room temperature until the ethanol has evaporated and no fluid is visible in the tube (2 to 5 min).

    Take care with this step because the pellet sometimes does not adhere tightly to the tube.

19. Dissolve the plasmid DNA in 50 to 100 µl sequencing TE buffer.

    Sequencing TE buffer differs from conventional TE in that the concentration of EDTA is 10-fold less (0.1 mM instead of 1 mM). EDTA can inhibit polymerases by competing for magnesium ions in the sequencing reaction.

20. Quantitate the double-stranded DNA by measuring the absorbance at 260 nm (see UNIT 2.2).

21. Store the DNA solution up to several years at −20°C.

    This procedure can be modified for preparing large-insert DNA such as cosmids, BACs, and PACs (see Riethman et al., 1997).

Basic Protocol 2: Treating PCR Products with Exonuclease I and Shrimp Alkaline Phosphatase to Purify Sequencing Template

PCR reactions (see overview in UNIT 10.2) contain residual amplification primers and unincorporated dNTPs that interfere with sequencing reactions and must be removed. The combination of exonuclease I (Exo I) and shrimp alkaline phosphatase (SAP) accomplishes this efficiently in a single-tube reaction with no loss of PCR product. The entire procedure can be accomplished in less than 1 hr.

Materials

Molecular-weight markers (e.g., molecular-weight marker III; Roche)
10 U/µl exonuclease I (Exo I; GE Healthcare)
1 U/µl shrimp alkaline phosphatase (SAP; GE Healthcare)
Distilled, nuclease-free water (Invitrogen)

0.2-ml or 0.5-ml thin-walled PCR reaction tubes (appropriate for the thermal cycler block)
Thermal cycler (see UNIT 10.2)

Additional reagents and equipment for performing PCR (see UNIT 10.2) and agarose gel electrophoresis (UNIT 7.2)

1. Perform PCR as described in UNIT 10.2 to prepare a 50-µl PCR product.

2. Analyze a 2.5- to 5-µl aliquot of the PCR product and the appropriate molecular-weight markers on a 1% agarose gel to check the purity and the yield of DNA.

   Look for a single band on the gel. If multiple bands are evident, you will need to troubleshoot the PCR reaction (see UNIT 10.2).

   Agarose gel electrophoresis is described in UNIT 7.2.

3a. *If a single discrete PCR product is obtained:* Transfer a 5-µl aliquot of the PCR product into a new PCR tube.

3b. *If a single PCR product is not obtained:* Consult the PCR troubleshooting strategies described in UNIT 10.2.

4. Add 1 µl of 10 U/µl Exo I and 1 µl of 1 U/µl SAP to the tube.

   Exonuclease I acts specifically on single-stranded DNA (e.g., PCR primers), degrading it processively in the 3′ to 5′ direction producing 5′-mononucleotides. Shrimp alkaline phosphatase removes 5′ phosphates from dNTPs.

5. Incubate in a thermal cycler 15 min at 37°C and then 15 min at 85°C. Hold the sample at 4°C.

   Exonuclease I and shrimp alkaline phosphatase are heat-deactivated at 85°C

6. Add 3 µl nuclease-free water to the tube. Store the solution up to 1 year at −20°C.

   NOTE: *This procedure can be scaled up proportionally if desired.*

Alternate Protocol: PEG Purification of PCR Products for DNA Sequencing

A nonenzymatic alternative to the exonuclease I/shrimp alkaline phosphatase method for PCR clean-up is the PEG purification method. It is successful at removing residual PCR primers and unincorporated nucleotides from a PCR reaction and can be performed in less than 1 hr at room temperature.

Materials

5 M NaCl
TE buffer (10 mM Tris pH 8.0/1 mM EDTA pH 8.0)
40% (w/v) PEG 8000/10 mM $MgCl_2$
Absolute (100%) ethanol
Sequencing TE buffer (0.1 mM Tris, pH 7.5/0.1 mM EDTA, pH 8.0)

1.5-ml microcentrifuge tubes
Vortexer
Microcentrifuge capable of at least 12,000 × *g*
SpeedVac (ThermoSavant)

Additional reagents and equipment for performing PCR (see UNIT 10.2) and spectrophotometrically quantitating DNA (UNIT 2.2)

1. Perform PCR as described in UNIT 10.2 to prepare a 50-µl PCR product

2. To a 1.5-ml microcentrifuge tube, add:

   8 µl of 5 M NaCl
   8 µl of TE buffer
   14 µl of 40% PEG 8000/10 mM $MgCl_2$

   and vortex briefly to mix.

3. Transfer the 50 µl PCR product into the tube containing NaCl/TE/PEG/$MgCl_2$ and mix by vortexing. Incubate 15 min at room temperature.

4. Microcentrifuge 15 min at 12,000 × *g*, room temperature.

5. Carefully remove all of the supernatant by gentle aspiration.

6. Wash the pellet twice by adding 250 µl absolute ethanol, microcentrifuging 5 min at 12,000 × *g*, room temperature, and removing the supernatant by gentle aspiration.

7. Dry the DNA pellet under vacuum (e.g., a SpeedVac).

8. Dissolve each DNA pellet in 20 μl sequencing TE buffer.

   Sequencing TE buffer differs from conventional TE in that the concentration of EDTA is 10-fold less (0.1 mM instead of 1 mM). EDTA can inhibit polymerases by competing for magnesium ions in the sequencing reaction.

9. Quantitate the DNA by measuring the absorbance at 260 nm (*UNIT 2.2*). Store the DNA solution up to 1 year at −20°C.

REAGENTS AND SOLUTIONS

Use deionized, distilled water in all recipes and protocol steps. For common stock solutions, see **UNIT 3.3.**

GET buffer

0.9 g glucose (50 mM)
2 ml of 0.5 M EDTA, pH 8.0 (10 mM)
2.5 ml 1 M Tris·Cl, pH 8.0 at room temperature (25 mM)
H_2O to make 100 ml

Sterilize by passing it through a 0.22-μm filter. Store up to 2 months at room temperature.

Final concentrations of components are given in parentheses.

Lysis solution

1 g SDS (1%)
20 ml 1 N NaOH (0.2 N)
H_2O to 100 ml

Prepare just before use.

CAUTION: *See Safety Considerations for cautions related to sodium hydroxide.*

Final concentrations of components are given in parentheses.

Potassium acetate solution

60 ml 5 M potassium acetate (3 M potassium)
11.5 ml glacial acetic acid (5 M acetate)
H_2O to make 100 ml

Store up to 6 months at room temperature.

CAUTION: *See Safety Considerations for cautions related to glacial acetic acid.*

Final concentrations of components are given in parentheses.

YT Medium, 2×

To 900 ml H_2O, add:

16 g tryptone
10 g yeast extract
5 g NaCl

Shake or stir until the solutes dissolve. Adjust the pH to 7.0 with 5 N NaOH. Adjust the volume of the solution to 1 liter with H_2O. Sterilize by autoclaving 20 min at 15 psi (1.05 kg/cm^2) on liquid cycle. If required, add antibiotics after cooling to <50°C.

UNDERSTANDING RESULTS

Two types of data files are typically returned from a DNA sequencing service provider. The first is called a chromatogram (also called a trace), which is a graph of the separation of the sequencing reaction on a DNA sequencing instrument. The peaks on the chromatogram correspond to the different fluorescent ddNTPs at the end of a particular extension fragment as they pass the detector on the instrument. Shorter fragments pass the detector first, so the sequence is read from left-to-right on the chromatogram. The sequence generally begins 20 to 40 nucleotides from the primer binding site. An example of part of a successful chromatogram is shown in Figure 10.4.6. A good chromatogram is characterized by peaks with even spacing, consistent shoulder width, low background signal underneath the peak, and relatively even signal intensity throughout the run (some degeneration of signal is to be expected). Deviations from this may indicate an experimental problem (see Troubleshooting). Chromatogram files can be viewed with several free

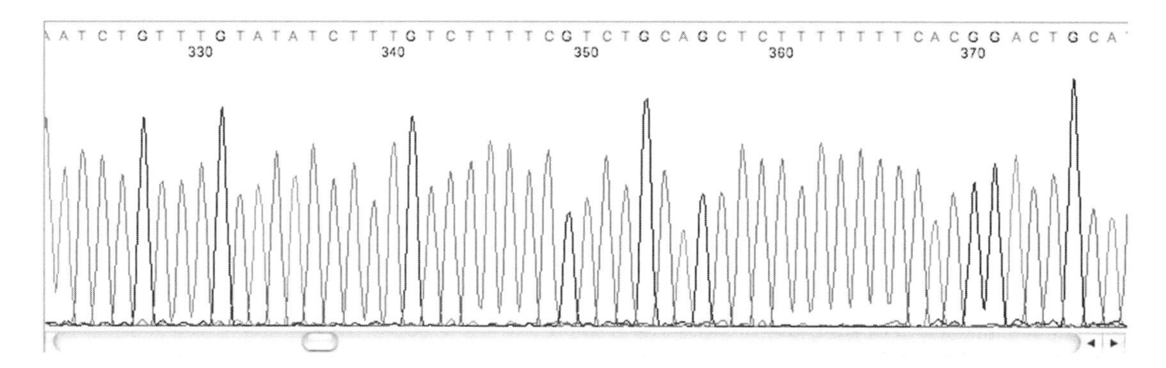

Figure 10.4.6 Typical chromatogram file (or "trace" file) from a successful DNA sequencing experiment using Applied Biosystems BigDye terminators. The peaks will usually be color coded to correspond to each base pair—A (green), C (blue), G (black), and T (red)—but are represented here in grayscale. The numbers above the peaks represent the nucleotide positions in the sequence, and above that are the nucleotide base calls for that position.

```
>Sequence_name
TGGATTGCATCCATCTCTCGCACTCAAGAGCAATACCAAAGGCGGCACTT
GACACTCTCTAAGCCCTATCAATTCTGAGCATATACAGTCGAACCATTAG
ATACTAAGGTGGATTCAATTCTAGCTAAATGCGGAATGATCTTTGATTTG
GCTTCAGGGTGGAGAGCGCAGAGAGTAATTCTAGGGAATTCATGACTTCC
CCTAAAAATGGCCAGAATACAGACATGCTCAGCGTGATGTATTGATTAGG
GGGTAACCCAGCCTTGAGCAAGAGATACGAATTTGAAGCCTGTTGCAGAA
AACGATGCAGTGCGATAGCGCAAAGACTGGATTCAATCTTGCTCTGTGGC
GGATGGTTAATGCTGCGATTGACGGTTTTATAACCATTGAGCAAGCCTGA
CTAATGACGATACAGCGCTTGTCAGATGATAGTTATCGCTCTTTCATCAC
TCTAGGCAGATAAAAACGGCGTTTTCACCCCAGGCGCAGCCCTGGGTATG
GCTCCATGGAGTTGATTCATCATCACAATCAATTGTGGAAATCCCAGTGA
TAATCTGGGAATGGAAAAACCTCCAACCAAGGTTGAATCTGATCGAATA
GAATCAACGATTGTATTATCACTAACCTCAAGTTGTTGAGCAGATGTTTC
CACCCTTTCCCTGTCTGCCAATATCGGTGCTGCGAAAACGTCTCTATCAC
AGGAGGGACTTGCTGTTTCATGGATAATCCTGTCGGTAGTTCAAACCTAT
CAGAGTGAAGACTCACCATCAGGTACGCACTCATGACGACCTGCCACCAT
CGTTCAATCTGGTCATAATCCGTGACCCGAAAATCGGCCCAGCCTAACTC
ATTCTTGCTTTGCTTGAGACCATATTCGACCCAGTTTCTCAGACCATAGA
AATTGCCTATTTGGTGATATTTGACGCCTTCTATTTCGCTCATTACATAC
CAGGTAGAGGCTTTAGGTAAGTTT
```

Figure 10.4.7 FASTA-formatted sequence file. This type of file is characterized by a short name or descriptor of the sequence preceded by a ">" character on the first line of the sequence.

viewing programs. Two popular options are 4peaks (*http://mekentosj.com/4peaks*) and FinchTV (*http://www.geospiza.com/finchtv*).

The second type of file, called the "flat" file or FASTA file, is simply a text file that contains the DNA sequence extracted from the matching chromatogram (Fig. 10.4.7). These files can be viewed with any text editor or word processing software.

TROUBLESHOOTING

There are several common problems in fluorescence-based DNA sequencing that can cause poor quality data. Examples of these problems and suggested solutions are described below. Note that while it is always tempting to blame sequencing failures on a service provider, keep in mind that such providers may process thousands of successful samples each week and often run experimental controls with your samples to monitor the performance of their instrumentation and sample handling. Thus, it is usually prudent to be aware of the many possible causes of poor quality data and to understand how the details of your experimental plan may influence such results.

Problems Caused by Templates

Low signal strength

Perhaps the most common cause of poor quality DNA sequence data is low signal strength. An example of this is shown in Figure 10.4.8. Note the increased level of background "noise" beneath the chromatogram peaks. Possible causes of weak signal and suggested solutions are the following:

Too little DNA in the sequencing reaction

Quantify the template sample again by measuring OD_{260} on a spectrophotometer (see *UNIT 2.2*) or by performing agarose gel electrophoresis (see *UNIT 7.2*) with a size standard of known quantity. Run the reaction again using more template DNA.

Poor annealing of the sequencing primer

Verify that the correct sequencing primer was matched to your DNA template. Next, check the concentration of the primer. Finally, verify that the correct T_m of your primer was used in the reaction conditions.

Use of a poor-quality DNA template

Run the sequencing reaction alongside a control template of known purity. If your service provider does not routinely use such controls, it is useful to include one with your sample.

Contaminated template DNA

The process of selecting bacterial colonies from an agar plate can sometimes result in poor-quality data if the template is prepared from a culture containing more than one subclone. An example of such data is shown in Figure 10.4.9. Here, the sequence quality is high until the cloning site of the plasmid is reached, after which sequence data from two different subclones are layered on top of each other. The template must be discarded and a new single colony selected for culturing.

Contamination can also occur when using PCR-based DNA templates if multiple products were produced in the PCR amplification reaction. Refer to the PCR troubleshooting section in *UNIT 10.2* to remedy this type of occurrence.

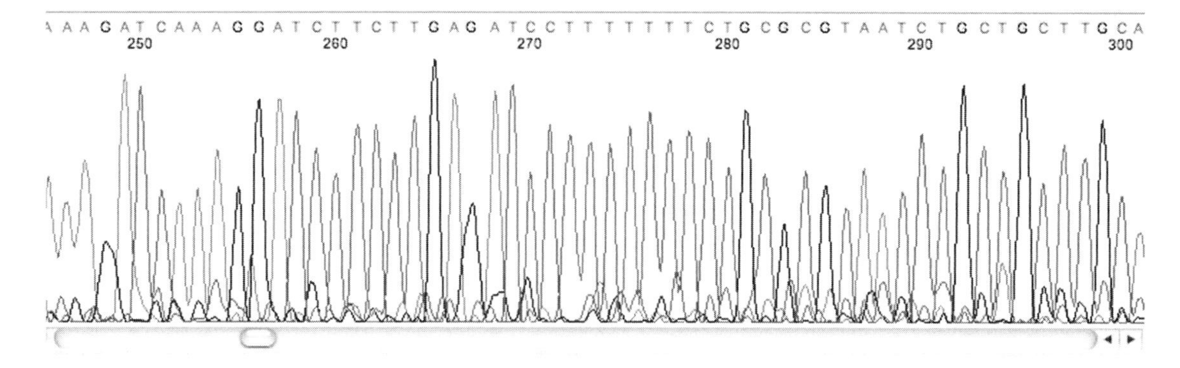

Figure 10.4.8 Chromatogram with low signal strength. The signal-to-noise ratio is increased in samples having a low signal strength as the sequencing software tries to normalize the peak height. Note the low, nonspecific background peaks that appear beneath the primary peaks in the data.

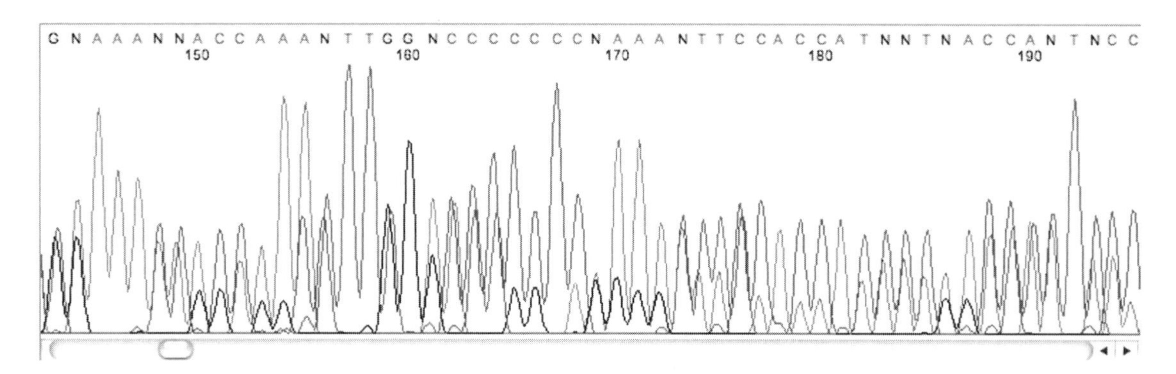

Figure 10.4.9 Chromatogram with contaminated template. This problem is typified by the appearance of two sequence traces overlaid on top of each other.

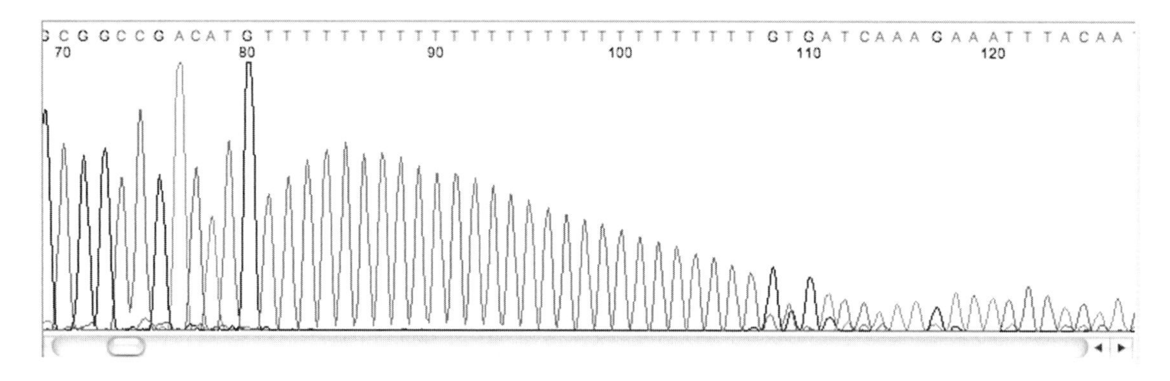

Figure 10.4.10 Chromatogram resulting from a homopolymeric run of A residues. Note the poor-quality data appearing after the run of A nucleotides, making it difficult to interpret the base calls.

Problems Caused by DNA Structure

Homopolymeric sequences and stops

Runs of mononucleotide or dinucleotide repeats can cause problems for fluorescence-based DNA sequencing. Runs of A or T residues are usually not a problem if they are short (<20 base pairs), but long runs can cause a "stuttering" effect after the run (Fig. 10.4.10). This is thought to result from offset annealing of the template and product strand during the sequencing reaction. The best way to resolve this type of region is to attempt to sequence into the run from both sides (or strands) of the template. This stuttering problem can be exacerbated if the DNA template is generated by PCR. In this instance, use a "proofreading" polymerase in the PCR reaction that has a 3′ to 5′ exonuclease activity (Table 10.4.2; also see *UNIT 10.2*).

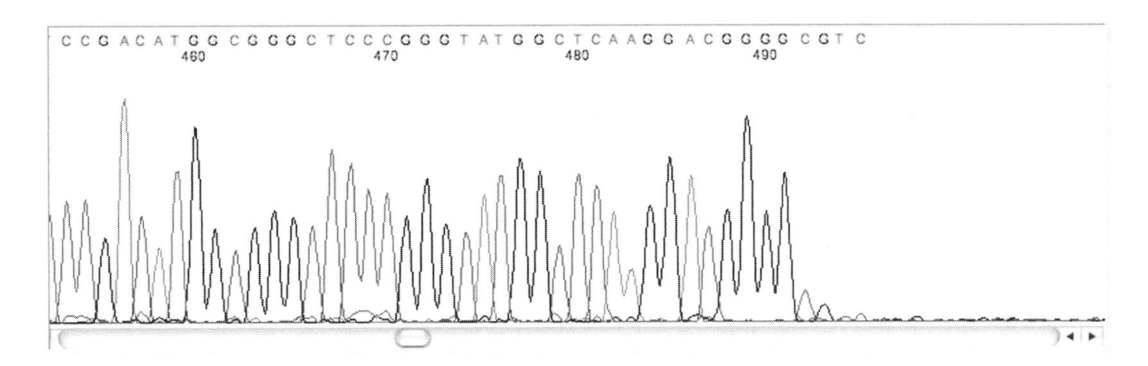

Figure 10.4.11 Chromatogram resulting from a run of C and G residues. Note the abrupt "stop" effect observed after the run of C and G nucleotides. Strong base-pairing of C and G nucleotides can sometimes block the activity of the DNA polymerase in a sequencing reaction.

10

Cytosine or Guanine Runs

Runs of C or G residues can also cause problems (Fig. 10.4.11). This type of artifact is commonly referred to as a "stop," and is thought to occur when a polymerase pauses at the strong base pairing of C and G residues between two stands of a double-stranded template. Very often, the weak signal beyond the stop is unintelligible. Problems with short runs of C or G nucleotides (<20 base pairs) can sometimes be solved by higher denaturation, annealing, and extension temperatures in the sequencing reaction. Also, the addition of 5% to 10% dimethylsulfoxide (DMSO) to the cycle sequencing reaction can help (Bergett and Rosteck, 1994). Longer C or G runs are more difficult (or impossible) to resolve.

VARIATIONS

An early approach to Sanger DNA sequencing employed radioisotopic labels (e.g., [α-^{35}S]dATP) and polyacrylamide gel electrophoresis, and indeed this strategy is still used by many research laboratories. It is a practical alternative to using a service facility if one is not readily available. The equipment used for this method is surprisingly accessible, requiring just a fraction of the capital investment compared to automated sequencing. Detailed protocols for this strategy can be found in other manuals (e.g., *CP Molecular Biology Unit 7.4A*, Slatko et al., 1999).

LITERATURE CITED

Bergett, S. and Rosteck, J.P.R. 1994. Use of dimethylsulfoxide to improve fluorescent *Taq* cycle sequencing. *In* Automated DNA Sequencing and Analysis Techniques. (M. Adams, ed.) Academic Press, New York, N.Y.

Birnboim, H.C. and Doly, J. 1979. A rapid alkaline extraction procedure for screening recombinant plasmid DNA. *Nucleic Acids Res.* 7:1513-1523.

Hillier, L. and Green, P. 1991. OSP: A computer program for choosing PCR and DNA sequencing primers. *PCR Methods Appl.* 1:124-128.

Riethman, H., Birren, B., and Gnirke, A. 1997. Preparing, manipulating, and mapping of HMW DNA. *In* Analyzing DNA: A Laboratory Manual. Genome Analysis. (B. Birren, E. Green, S. Klapholz, R. Myers, and J. Roskams, eds.) pp. 83-248. Cold Spring Harbor Laboratory Press, Cold Spring Harbor, N.Y.

Rozen, S. and Skaletsky, H. 2000. Primer3 on the WWW for general users and for biologist programmers. *Methods Mol. Biol.* 132:365-386.

Sanger, F., Nicklen, S., and Coulson, A.R. 1977. DNA sequencing with chain-terminating inhibitors. *Proc. Natl. Acad. Sci. U.S.A.* 74:5463-5467.

Slatko, B., Albright, L., Tabor, S., and Ju, J. 1999. DNA sequencing by the dideoxy method. *Curr. Protoc. Mol. Biol.* 47:7.4A.1-7.4A.39.

Voytas, D. 2000. Agarose gel electrophoresis. *Curr. Protoc. Mol. Biol.* 51:2.5A.1-2.5A.9.

Watson, J.D. and Crick, F.H. 1953. Molecular structure of nucleic acids: A structure for deoxyribose nucleic acid. *Nature* 171:737-738.

Appendices

APPENDIX 1

Laboratory Safety

George Lunn[1] and Warren Strober[2]

[1]Baltimore, Maryland
[2]National Institute of Allergy and Infectious Diseases, Bethesda, Maryland

Persons carrying out the protocols in this manual may encounter the following hazardous or potentially hazardous materials: (1) toxic chemicals and carcinogenic, mutagenic, or teratogenic reagents; (2) radioactive substances; and (3) pathogenic and infectious biological agents. Carcinogenic chemicals cause cancer in humans or animals; mutagenic chemicals have been shown to damage DNA in in vitro tests, such as the Ames test; and teratogenic chemicals cause birth defects in humans or animals. Most governments regulate the use of these materials: it is essential that they be used in strict accordance with local and national regulations. Cautionary notes are included in many instances throughout the manual, and some specific guidelines are provided below as well as in UNIT 2.3 (and references therein). However, the authors emphasize that users must proceed with the prudence and precautions associated with good laboratory practice, under the supervision of personnel responsible for implementing laboratory safety programs at their institutions, and in compliance with designated guidelines of federal, state, and local officials.

IMPORTANT NOTE: Please refer to UNIT 2.3 for information on the safe handling of radioisotopes.

HAZARDOUS CHEMICALS

In the space available it is not possible to list all the precautions to be taken when handling hazardous chemicals. Many texts have been written about laboratory safety: see Key References for a selected list of examples. Obviously, all national and local laws should be obeyed as well as all institutional regulations. Controlled substances are regulated by the Drug Enforcement Administration. By law, Material Safety Data Sheets must be readily available. All laboratories should have a Chemical Hygiene Plan [29CFR Part 1910.1450] and institutional safety officers should be consulted as to its implementation. Help is (or should be) available from your institutional Safety Office. Use it.

A listing of some commonly used hazardous chemicals and the specific hazards associated with each is provided below (Table A.1.1).

Store Chemicals Properly

Chemicals should be stored properly. For example, flammable chemicals (e.g., ethanol, methanol, acetone, methyl ethyl ketone, petroleum distillates, toluene, benzene, and other materials labeled flammable) should be stored in approved flammable storage cabinets, and flammable chemicals requiring refrigeration should be stored in explosion-proof refrigerators. Oxidizers should be segregated from other chemicals, and corrosive acids (e.g., sulfuric, hydrochloric, nitric, perchloric, and hydrofluoric acids) should also be stored in a separate cabinet, well removed from the flammable organics.

Use Appropriate Facilities and Equipment

Facilities should be appropriate for the handling of hazardous chemicals. In particular, volatile hazardous chemicals should only be handled in chemical fume hoods, not in laminar flow cabinets. Note that some finely divided solids may become electrostatically charged and react unpredictably during handling. These solids should also be handled in a fume hood. The functioning of these fume hoods should be periodically checked. Your institution may have particular regulations concerning the appropriate use of fume hoods.

Current Protocols Essential Laboratory Techniques A.1.1-A.1.12

Table A.1.1 Commonly Used Hazardous Chemicals[a]

Chemical	Hazards	Remarks[b]
Acetic acid, glacial	Corrosive, flammable liquid	
Acetonitrile	Flammable liquid, teratogenic	
Acridine orange	Carcinogenic, mutagenic	
Acrylamide	Carcinogenic	Use dust mask; polyacrylamide gels contain residual acrylamide monomer and should be handled with gloves; acrylamide may polymerize with violence on melting at 86°C
Brilliant blue R	Carcinogenic, mutagenic	
5-Bromodeoxyuridine (BrdU)	Mutagenic, teratogenic, photosensitizing	
Cetyltrimethylammonium bromide (CTAB)	Corrosive, teratogenic	
Chloroform	Carcinogenic, teratogenic	
Chromic/sulfuric acid cleaning solution	Carcinogenic, corrosive, oxidizer	Replace with suitable commercially available cleanser
Congo red	Mutagenic, teratogenic	
Coomassie brilliant blue G	Mutagenic	
Crystal violet	Carcinogenic, mutagenic	
Cresyl violet acetate	Mutagenic	
Cyanides (e.g., KCN, NaCN)	Severe poison	Contact with acid will liberate HCN gas
Cyanogen bromide (CNBr)	Severe poison, volatile at room temperature	
2′-Deoxycoformycin (dCF, pentostatin)	Teratogenic, mutagenic	
4′,6-Diamidino-2-phenylindole (DAPI)	Mutagenic	
Diaminobenzidine (DAB)	Carcinogenic	
1,4-Diazabicyclo[2,2,2]-octane (DABCO)	Skin and eye irritant	Forms an explosive complex with hydrogen peroxide
Dichloroacetic acid (DCA)	Carcinogenic, corrosive	
Dichloromethane (methylene chloride)	Carcinogenic, mutagenic, teratogenic	
Diethylamine (DEA)	Corrosive, flammable liquid	
Diethylpyrocarbonate (DEPC)	Carcinogenic	
Diethyl sulfate	Carcinogenic, teratogenic extremely poisonous by inhalation, skin absorption, and ingestion	
Diisopropyl fluorophosphate (DFP)	Highly toxic, cholinesterase inhibitor, neurotoxin	

continued

Table A.1.1 Commonly Used Hazardous Chemicals[a], *continued*

Chemical	Hazards	Remarks[b]
Dimethyl sulfate (DMS)	Carcinogenic extremely poisonous by inhalation, skin absorption, and ingestion	Inhalation causes severe lung injury
Dimethyl sulfoxide (DMSO)	Flammable liquid	Enhances absorption through skin
Diphenylamine (DPA)	Teratogenic	
Erythrosin B	Carcinogenic, mutagenic	
Ether	Highly flammable liquid	May form explosive peroxides on standing; do not dry with NaOH or KOH
Ethidium bromide (EB)	Mutagenic	
Ethyl methanesulfonate (EMS)	Carcinogenic	
Fluorescein and derivatives	Carcinogenic	
5-Fluoro-2′-deoxyuridine (FUdR)	Teratogenic	
Formaldehyde	Carcinogenic, flammable liquid, teratogenic	
Formamide	Teratogenic	
Formic acid	Corrosive	May explode when heated >180°C in a sealed tube
Glutaraldehyde	Corrosive, teratogenic	
Guanidinium thiocyanate	Toxic	Contact with acid may liberate hydrogen cyanide
Hoechst 33258 dye	Mutagenic	
Hydrochloric acid, concentrated	Corrosive, teratogenic, toxic	
Hydrogen peroxide (30%)	Carcinogenic, corrosive, mutagenic, oxidizer	Avoid bringing into contact with organic materials, which may form explosive peroxides; may decompose violently in contact with metals, salts, or oxidizable materials
Hydroxylamine	Corrosive, flammable, mutagenic	Explodes in air at >70°C
3-β-indoleacrylic acid (IAA)	Carcinogenic	
Iodine	Corrosive	
Iodoacetamide	Carcinogenic, mutagenic	
Janus green B	Carcinogenic, mutagenic	
Lead compounds	Carcinogenic	
2-Mercaptoethanol (2-ME)	Stench extremely pungent!	
Mercury compounds	Teratogenic, highly poisonous	
Methanol	Flammable liquid, teratogenic, toxic by ingestion, inhalation, and skin contact	
Methotrexate (amethopterin)	Carcinogenic, mutagenic, teratogenic	

continued

Table A.1.1 Commonly Used Hazardous Chemicals[a], *continued*

Chemical	Hazards	Remarks[b]
Methylene blue	Mutagenic	
Methyl methanesulfonate (MMS)	Carcinogenic	
Mycophenolic acid (MPA)	Teratogenic	
Neutral red	Mutagenic	
Nitric acid, concentrated	Corrosive, oxidizer, teratogenic	
Paraformaldehyde	Toxic	Gives off gaseous formaldehyde on heating
Phenol	Carcinogenic, corrosive, teratogenic	Readily absorbed through the skin
Phenylmethylsulfonyl fluoride (PMSF)	Enzyme inhibitor	
Phorbol 12-myristate 13-acetate (PMA)	Carcinogenic	
Piperidine	Flammable liquid, teratogenic	
Potassium hydroxide, concentrated	Corrosive	Produces a highly exothermic reaction when solid is added to water
Propidium iodide (PI)	Mutagenic	
Pyridine	Flammable liquid, unpleasant odor	
Rhodamine and derivatives	Carcinogenic, teratogenic	
Rose Bengal	Carcinogenic, teratogenic	
Safranine O	Mutagenic	
Sodium azide	Carcinogenic, severe poison	Adding acid liberates explosive volatile, toxic hydrazoic acid; can form explosive heavy metal azides, e.g., with plumbing fixtures–*do not* discharge down drain
Sodium deoxycholate (Na-DOC)	Carcinogenic, teratogenic	
Sodium dodecyl sulfate (sodium lauryl sulfate, SDS)	Sensitizing, skin and eye irritant	
Sodium hydroxide, concentrated	Corrosive	A highly exothermic reaction ensues when the solid is added to water
Sodium nitrite	Carcinogenic	
Sulfuric acid, concentrated	Corrosive, oxidizer, teratogenic	Reaction with water is very exothermic; always add concentrated sulfuric acid to water, *never* water to acid
Tetramethylammonium chloride (TMAC)	Irritating to eyes, respiratory system, and skin	

continued

Table A.1.1 Commonly Used Hazardous Chemicals[a], *continued*

Chemical	Hazards	Remarks[b]
N,N,N′,N′-Tetramethyl-ethylenediamine (TEMED)	Corrosive, flammable liquid, stench, extremely pungent!	
Texas Red (sulforhodamine 101, acid chloride)	Carcinogenic, toxic on prolonged exposure	
Toluene	Flammable liquid, teratogenic, toxic by inhalation	
Toluidine blue O	Mutagenic	
N′-p-Tosyl-L-lysine chloromethyl ketone (TLCK)	Toxic, enzyme inhibitor	
N-p-Tosyl-L-phenylalanine chloromethyl ketone (TPCK)	Toxic, mutagenic, enzyme inhibitor	
Trichloroacetic acid (TCA)	Carcinogenic, corrosive, teratogenic	
Triethanolamine acetate (TEA)	Carcinogenic	
Trifluoroacetic acid (TFA)	Corrosive	
Trimethyl phosphate (TMP)	Carcinogenic, mutagenic, teratogenic	May explode on distillation
Trypan blue	Carcinogenic, mutagenic, teratogenic	
Xylenes	Flammable liquid, teratogenic, toxic by inhalation	

[a]For extensive information on the hazards of these and other chemicals, as well as cautionary details, see Bretherick (1990), Lunn and Sansone (1994), Bretherick and Urben (1999), Furr (2000), Lewis (2004), and O'Neil (2006).
[b]*CAUTION:* These chemicals should be handled only in a chemical fume hood by knowledgeable workers equipped with eye protection, laboratory coat, and gloves. The laboratory should be equipped with a safety shower and eye wash. Additional protective equipment may be required.

Laboratories should also be equipped with safety showers and eye-washing facilities. Again, this equipment should be tested periodically to make sure that it functions correctly. Other safety equipment may be required depending on the nature of the materials being handled. In addition, researchers should be trained in the proper procedures for handling hazardous chemicals as well as other areas of laboratory operations, e.g., handling of compressed gases, use of cryogenic liquids, operation of high voltage power supplies, etc.

Plan for Accidents

Before starting work, have a plan for dealing with spills or accidents; coming up with a good plan on the spur of the moment is difficult. For example, have the appropriate decontaminating or neutralizing agents prepared and close at hand. Small spills can probably be cleaned up by the researcher. In the case of larger spills, the area should be evacuated and help sought from those experienced and equipped for dealing with spills, e.g., your institutional safety department.

Use Protective Equipment

Protective equipment should include, at a minimum, eye protection, a laboratory coat, and gloves. Sandals, open-toed shoes, shorts, and other clothing exposing the skin should not be worn. In certain circumstances other items of protective equipment may be necessary, e.g., a face shield.

A1

Different types of gloves exhibit different chemical resistance properties; listings of these properties are given in Table A.1.2. Gloves should, however, be regarded as the last line of defense and should be changed if they become contaminated, because many types of chemicals pass relatively freely through rubber. If possible, handling procedures should be designed so that gloves do not become contaminated. All common-sense precautions should be observed, e.g., do not pipet by mouth, keep unauthorized persons away from hazardous chemicals, prohibit eating and drinking in the lab, etc.

Table A.1.2 Chemical Resistance of Commonly Used Gloves[a,b]

Chemical	Neoprene gloves	Latex gloves	Butyl gloves	Nitrile gloves
*Acetaldehyde	VG	G	VG	G
Acetic acid	VG	VG	VG	VG
*Acetone	G	VG	VG	P
Ammonium hydroxide	VG	VG	VG	VG
*Amyl acetate	F	P	F	P
Aniline	G	F	F	P
*Benzaldehyde	F	F	G	G
*Benzene	P	P	P	F
Butyl acetate	G	F	F	P
Butyl alcohol	VG	VG	VG	VG
Carbon disulfide	F	F	F	F
*Carbon tetrachloride	F	P	P	G
*Chlorobenzene	F	P	F	P
*Chloroform	G	P	P	E
Chloronaphthalene	F	P	F	F
Chromic acid (50%)	F	P	F	F
Cyclohexanol	G	F	G	VG
*Dibutyl phthalate	G	P	G	G
Diisobutyl ketone	P	F	G	P
Dimethylformamide	F	F	G	G
Dioctyl phthalate	G	P	F	VG
Epoxy resins, dry	VG	VG	VG	VG
*Ethyl acetate	G	F	G	F
Ethyl alcohol	VG	VG	VG	VG
*Ethyl ether	VG	G	VG	G
*Ethylene dichloride	F	P	F	P
Ethylene glycol	VG	VG	VG	VG
Formaldehyde	VG	VG	VG	VG
Formic acid	VG	VG	VG	VG
Freon 11, 12, 21, 22	G	P	F	G
*Furfural	G	G	G	G
Glycerin	VG	VG	VG	VG

continued

Table A.1.2 Chemical Resistance of Commonly Used Gloves[a,b], *continued*

Chemical	Neoprene gloves	Latex gloves	Butyl gloves	Nitrile gloves
Hexane	F	P	P	G
Hydrochloric acid	VG	G	G	G
Hydrofluoric acid (48%)	VG	G	G	G
Hydrogen peroxide (30%)	G	G	G	G
Ketones	G	VG	VG	P
Lactic acid (85%)	VG	VG	VG	VG
Linseed oil	VG	P	F	VG
Methyl alcohol	VG	VG	VG	VG
Methylamine	F	F	G	G
Methyl bromide	G	F	G	F
*Methyl ethyl ketone	G	G	VG	P
*Methyl isobutylketone	F	F	VG	P
Methyl methacrylate	G	G	VG	F
Monoethanolamine	VG	G	VG	VG
Morpholine	VG	VG	VG	G
Naphthalene	G	F	F	G
Naphthas, aliphatic	VG	F	F	VG
Naphthas, aromatic	G	P	P	G
*Nitric acid	G	F	F	F
Nitric acid, red and white fuming	P	P	P	P
Nitropropane (95.5%)	F	P	F	F
Oleic acid	VG	F	G	VG
Oxalic acid	VG	VG	VG	VG
Palmitic acid	VG	VG	VG	VG
Perchloric acid (60%)	VG	F	G	G
Perchloroethylene	F	P	P	G
Phenol	VG	F	G	F
Phosphoric acid	VG	G	VG	VG
Potassium hydroxide	VG	VG	VG	VG
Propyl acetate	G	F	G	F
i-Propyl alcohol	VG	VG	VG	VG
n-Propyl alcohol	VG	VG	VG	VG
Sodium hydroxide	VG	VG	VG	VG
Styrene (100%)	P	P	P	F
Sulfuric acid	G	G	G	G
Tetrahydrofuran	P	F	F	F

continued

Table A.1.2 Chemical Resistance of Commonly Used Gloves[a,b], *continued*

Chemical	Neoprene gloves	Latex gloves	Butyl gloves	Nitrile gloves
*Toluene	F	P	P	F
Toluene diisocyanate	F	G	G	F
*Trichloroethylene	F	F	P	G
Triethanolamine	VG	G	G	VG
Tung oil	VG	P	F	VG
Turpentine	G	F	F	VG
*Xylene	P	P	P	F

[a]Performance varies with glove thickness and duration of contact. **An asterisk indicates limited use.** Abbreviations: VG, very good; G, good; F, fair; P, poor (do not use).
[b]Adapted from the July 8, 1998, version of the DOE OSH Technical Reference Chapter 5 (Appendix C at *http://www.eh.doe.gov/docs/osh_tr/ch5c.html*). For more information, also see Forsberg and Keith (1999) and Forsberg and Mansdort (2003).

Minimize Quantities of Hazardous Chemicals in Stock

Order hazardous chemicals only in quantities that are likely to be used in a reasonable time. Buying large quantities at a lower unit cost is no bargain if someone (perhaps the researcher) has to pay to dispose of surplus quantities.

Use Safer Substitutes

Look in the literature for techniques that use less hazardous chemicals. Techniques are available to reduce or eliminate the use of phenol and chloroform in DNA isolation and purification. It may be possible to substitute less toxic reagents for ethidium bromide. SYBR Safe (Invitrogen) or MegaFluor (*http://www.euroclone.net*) have been reported to be safer substitutes. Use biodegradable detergents in place of chromic acid cleaning solutions. Substitute alcohol-filled thermometers for mercury-filled thermometers. The latter are a hazardous chemical spill waiting to happen.

Pay Attention to Specific Hazards

Although any number of chemicals commonly used in laboratories are toxic if used improperly, the toxic properties of a number of reagents require special attention. Many chemicals are considered carcinogenic, corrosive, flammable, eye-irritating, mutagenic, oxidizing, teratogenic, or toxic. Chemicals labeled carcinogenic range from those accepted by expert review groups as causing cancer in humans to those for which only minimal evidence of carcinogenicity exists. Oxidizers may react violently with oxidizable material, e.g., hydrocarbons, wood, and cellulose. Before using any chemical, thoroughly investigate all of its characteristics. Material Safety Data Sheets are readily available; they list some hazards but vary widely in quality. A number of texts describing hazardous properties are listed in Further Reading. In particular, Sax's *Dangerous Properties of Industrial Materials*, 11th ed. (Lewis, 2004) and Bretherick's *Handbook of Reactive Chemical Hazards*, 6th ed. (Bretherick and Urben, 1999) give comprehensive listings of known hazardous properties. However, these texts list only the known properties. Many chemicals have been tested only partially or not at all. Prudence dictates, therefore, that unless there is good reason for believing otherwise, all chemicals should be regarded as volatile, highly toxic, flammable human carcinogens and should be handled with care.

Dispose of Waste Properly

Waste should always be disposed of in accordance with all applicable regulations. Waste should be segregated according to institutional requirements, for example, into solid, aqueous, nonchlorinated organic, and chlorinated organic material. A collection (Lunn and Sansone, 1994) of techniques for the disposal of chemicals in laboratories has been published. Incorporation of these procedures into laboratory protocols can help to minimize waste disposal problems.

BIOHAZARDS AND INFECTIOUS BIOLOGICAL AGENTS

The basic framework for safety in the research laboratory is encompassed in the concept of the biosafety level. This defines a set of procedures mandated by the type of microorganism being handled, or likely to be handled, as well as the number of such organisms involved and the nature of the manipulations to be carried out. Details concerning biosafety levels, data on the precautions necessary for individual infectious agents, and other pertinent information is provided in the Centers for Disease Control/National Institutes of Health handbook *Biosafety in Microbiological and Biomedical Laboratories* (U. S. Department of Health and Human Services, 1999), also available online (see Internet Resources). Every researcher engaged in the study of human tissue is advised to have a copy of this handbook in the laboratory. This appendix provides only a general discussion of biosafety levels and should not be considered comprehensive. Procedures for disposing of biohazardous waste vary at different institutions; consult the guidelines established at your institution. Biosafety level 1 (BSL-1; Table A.1.3), the lowest level, applies to ubiquitous microorganisms found in the general environment and requires that standard microbiological practices be followed, including the use of mechanical pipetting devices; daily decontamination of work surfaces; prohibition of eating, smoking, and application of cosmetics in the laboratory area; use of laboratory coats; and appropriate hand washing. Biosafety level 2 (BSL-2) applies to organisms capable of producing disease of moderate severity in normal healthy individuals. It assumes all the features of BSL-1 and, in addition, applies more stringent conditions for laboratory cleaning, use of containment equipment (e.g., biosafety cabinets) particularly if aerosols will be generated, and autoclaving laboratory wastes. Biosafety level 3 (BSL-3) applies to possible exposure to microorganisms capable of causing serious illness. This level differs from the BSL-2 level in that it provides for a double-door entry system into a sealed "inner" laboratory having a unidirectional airflow. At this biosafety level, tissue is manipulated only within biosafety cabinets, centrifugation of specimens is accomplished with sealed safety cups, and special clothing and gloves are routinely worn. Finally, BSL-4 applies to laboratory work with pathogens that pose a life-threatening risk to laboratory workers when even "casual" contact with tissue occurs. The safety practices employed at this level are an intensification of those at the BSL-3 level and include extensive clothing changes before entering the work area, decontamination of waste, strict training of all laboratory personnel, and the use of class II biosafety cabinets (which provide for physical isolation of specimens) or lower-class cabinets used in association with one-piece positive-pressure personnel suits ventilated by a life-support system.

Most laboratory work with materials containing blood-borne pathogens can be safely accomplished with BSL-2 level procedures, except when unusually large numbers of organisms must be handled or when aerosols will be generated. BSL-3, on the other hand, is reserved for the more dangerous infectious agents and is the preferred level for research involving, for example, HIV-1. Both the BSL-2 and BSL-3 levels require the presence of a laboratory leader who is fully versed in safety practices, and, in addition, require provision of safety training programs for workers and the establishment of laboratory-specific safety manuals. Particular attention must be given to the possibility of laboratory accidents that will result in infection of a laboratory worker (or waste handler) and the institution of strict procedures that prevent such accidents. It is essential that all persons working in a laboratory in which human tissues are being processed be fully aware of the *potential* for exposure to various pathogens present in the tissue. When a break in safety procedure

Table A.1.3 CDC Summary of Recommended Biosafety Levels for Infectious Agents[a,b]

Biosafety level	Agent characteristics	Practices	Safety equipment (primary barriers)[c]	Facilities (secondary barriers)
BSL-1	Not known to consistently cause disease in healthy adults	Standard microbiological practices	None required	Open bench-top sink
BSL-2	Associated with human disease, hazard from percutaneous injury, ingestion, mucous membrane exposure	Standard microbiological practices Limited access Biohazard warning signs "Sharps" precautions Biosafety manual defining any needed waste decontamination or medical surveillance policies	Class I or II biosafety cabinets (BSCs) or other physical containment devices used for all manipulations of agents that cause splashes or aerosols of infectious materials Laboratory coats and gloves Face protection as needed	Open bench-top sink Autoclave
BSL-3	Indigenous or exotic agents with potential for aerosol transmission; disease may have serious or lethal consequences	All BSL-2 practices Controlled access Decontamination of all waste Decontamination of laboratory clothing before laundering Baseline serum	Class I or II BSCs or other physical containment devices used for all open manipulations of agents Protective laboratory clothing and gloves Respiratory protection as needed	Open bench-top sink Autoclave Physical separation from access corridors Self-closing, double-door access Exhausted air not recirculated Negative airflow into laboratory
BSL-4	Dangerous/exotic agents which pose high risk of life-threatening disease, aerosol-transmitted laboratory infections; or related agents with unknown risk of transmission	All BSL-3 practices Clothing change before entering Shower on exit All material decontaminated on exit from facility	All procedures conducted in Class III BSCs, or Class I or II BSCs *in combination with* full-body, air-supplied, positive pressure personnel suit	All BSL-3 facilities plus: Separate building or isolated zone Dedicated supply and exhaust, vacuum, and decontamination systems Other requirements as outlined in the text

[a]Adapted from *Biosafety in Microbiological and Biomedical Laboratories, 4th Ed.* (GPO S/N 017-040-00547-4), available online at *http://www.cdc.gov/od/ohs/biosfty/bmbl4/bmbl4toc.htm* (also see Internet Resources).

[b]The practices, and primary and secondary barriers required for a given biosafety level include those of the all lower levels, as well as the additional required practices, equipment, and/or facilities described for the BSL in question.

[c]See *http://www.cdc.gov/od/ohs/biosfty/bsc/bsc.htm* for more information concerning biological safety cabinets (BSCs).

and undue exposure does occur, it is the responsibility of everyone concerned to investigate the incident and to bring the exposed individuals to immediate medical attention.

A1

ACKNOWLEDGEMENTS

The authors would like to thank Gretchen Lawler for contributing Table A.1.2, Chemical Resistance of Gloves.

LITERATURE CITED

Bretherick, L. 1990. Hazards in the Chemical Laboratory, 4th ed. Butterworth-Heinemann, London.

Bretherick, L. and Urben, P.G. 1999. Bretherick's Handbook of Reactive Chemical Hazards, 6th ed. Butterworth-Heinemann, London. [also available on CD-ROM]

Forsberg, K. and Keith, L.H. 1999. Chemical Protective Clothing Performance Index, 2nd ed. John Wiley & Sons, New York.

Forsberg, K. and Mansdorf, S.Z. 2003. Quick selection Guide to Chemical Protective Clothing, 4th ed. John Wiley & Sons, Hoboken, N.J.

Furr, A.K. 2000. CRC Handbook of Laboratory Safety, 5th ed. CRC Press, Boca Raton, Fla.

Lewis, R.J., Sr. 2004. Sax's Dangerous Properties of Industrial Materials, 11th ed. John Wiley & Sons, Hoboken, N.J.

Lunn, G. and Sansone, E.B. 1994. Destruction of Hazardous Chemicals in the Laboratory, 2nd ed. John Wiley & Sons, New York.

O'Neil, M.J. 2006. The Merck Index, 14th ed. Merck & Co., Whitehouse Station, N.J.

U.S. Department of Health and Human Services, Public Health Service, Centers for Disease Control and Prevention (CDC) and National Institutes of Health (NIH). 1999. Biosafety in Microbiological and Biomedical Laboratories (BMBL), 4th ed. (J.Y. Richmond and R.W. McKenney, eds.) U.S. Government Printing Office, Washington, D.C. (GPO S/N 017-040-00547-4).

KEY REFERENCES

General Safety

Furr, 2000. See above.

Picot, A. and Grenouillet, P. 1994. Safety in the Chemistry and Biochemistry Laboratory. John Wiley & Sons, New York.

Stricoff, R.S. and Walters, D.B. 1995. Handbook of Laboratory Health and Safety, 2nd ed. John Wiley & Sons, New York.

Chemical Safety

Sigma-Aldrich Co. 2005. Aldrich Handbook 2005-2006. Sigma-Aldrich Co., Saint Louis, Mo. (also available at http://*www.sigma-aldrich.com*)

Alaimo, R.J. 2001. Handbook of Chemical Health and Safety. American Chemical Society, Washington, D.C.

American Chemical Society, Committee on Chemical Safety. 1995. Safety in Academic Chemistry Laboratories, 6th ed. American Chemical Society, Washington, D.C.

Bretherick, 1990. See above.

Bretherick and Urben, 1999. See above.

Dawson, M.C., Elliott, D.C., Elliott, W.H., and Jones, K.M. 1986. Data for Biochemical Research. Alden Press, London.

Forsberg and Keith, 1999. See above.

Forsberg and Mansdorf, 2003. See above.

Lewis, 2004. See above.

Lunn and Sansone, 1994. See above.

Lunn, G. and Lawler, G. 2005. Safe use of hazardous chemicals. *Curr. Protoc. Microbiol.* 0:1A.3.1-1A.3.36.

National Research Council. 1995. Prudent Practices in the Laboratory: Handling and Disposal of Chemicals. National Academy Press, Washington, D.C.

O'Neil, 2006. See above.

Young, J.A. 1991. Improving Safety in the Chemical Laboratory: A Practical Guide, 2nd ed. John Wiley & Sons, New York.

Biological Safety

Coico, R. and Lunn, G. 2005. Biosafety: Guidelines for working with pathogenic and infectious microorganisms. *Curr. Protoc. Microbiol.* 0:1A.1.1-1A.1.8.

U.S. Department of Health and Human Services, Public Health Service, Centers for Disease Control and Prevention (CDC) and National Institutes of Health (NIH). 1999. See above.

INTERNET RESOURCES

http://www.ilpi.com/msds/index.html
Where to find MSDSs on the internet. Contains links to general sites, government and nonprofit sites, chemical manufacturers and suppliers, pesticides, and miscellaneous sites.

http://www.OSHA.gov
The OSHA Web site. Standards can be accessed by clicking on the link in the right sidebar labeled Standards.

http://www.osha.gov/pls/oshaweb/owadisp.show_document?p_table=STANDARDS&p_id=10106
Text of OSHA Standard 29 CFR 1910.1450: Occupational Exposure to Hazardous Chemicals in Laboratories.

http://www.osha.gov/pls/oshaweb/owadisp.show_document?p_table=STANDARDS&p_id=9992
Table Z-1 of OSHA Standard 29 CFR 1910.10000, which provides a list of permissible exposure limits (PELs) for air contaminants.

http://www.osha.gov/pls/oshaweb/owadisp.show_document?p_table=STANDARDS&p_id=9993
Table Z-2 of OSHA Standard 29 CFR 1910.1000, which provides a list of PELs for toxic and hazardous substances.

http://hazard.com/msds/index.php
Main site for Vermont SIRI. One of the best general sites to start a search. Browse manufacturers alphabetically (for sheets not in the SIRI collection) or do a keyword search in the SIRI MSDS database. Lots of additional safety links and information.

http://siri.uvm.edu/msds
Alternate site for Vermont SIRI.

http://www.eh.doe.gov/docs/osh_tr/ch5c.html
DOE OSH technical reference chapter on personal protective equipment.

http://www.cdc.gov/od/ohs/biosfty/bmbl4/bmbl4toc.htm
Online version of the Biosafety in Microbiological and Biomedical Laboratories (BMBL), 4th Ed.

http://www4.od.nih.gov/oba/
Homepage for the Office of Biotechnologies, NIH.

APPENDIX 2

Laboratory Notebooks and Data Storage

Michael Williams,[1] Donna Bozyczko-Coyne,[1] Bruce Dorsey,[1] and Scott Larsen[1]
[1]Cephalon, Inc., Frazer, Pennsylvania

A2

HARD COPY, PAPER NOTEBOOKS

Throughout this manuscript references are made to a hypothetical notebook prepared by John Hancock and verified by Alfred Nobel. Figures A.2.6 to A.2.15 at the end of this appendix show excerpts from the laboratory notebook, which has been maintained under best practices and provides examples of points made throughout the manuscript. The purpose of this illustration is to provide the reader with concrete examples of the proper maintenance of a laboratory notebook.

Notebook Format

A laboratory notebook should be hardbound, stitched but not glued, so that pages can neither be removed nor replaced (Fig. A.2.1). Because a notebook can frequently become unwieldy and cumbersome when excessive numbers of data records (e.g., worksheets, data printouts) are appended, supplemental notebooks are used. These supplemental notebooks should be (1) individually numbered and (2) specifically cross-referenced back to the primary notebook. They are preferably bound following the collection and cross-referencing of all information related to the experiments in the primary notebook. For electronic laboratory notebooks (ELNs), the experiments and data generated from them are entered directly into a database, but hard copies of notebooks can be generated to create a bound and signed copy of the electronic version.

For hardcopy laboratory notebooks, the paper used should be of archival quality, acid-free (neutral pH), lignin-free, white, nontransparent, and ~60 to 70 lbs. (~90 g/m^2), with a projected shelf life of ~50 years; a watermark is optional. Paper approaching newsprint quality has a limited life because it rapidly becomes brittle; therefore, it should not be used for recording experiments or archiving data. For patent purposes, a rule of thumb is that a laboratory notebook should be sufficiently durable to last for 6 years beyond the final expiration date of any patents resulting from the work archived in the notebook.

Notebook Numbering

Each notebook should have a distinct identification number (Fig. A.2.1) with each page (front and back if both are to be used) in the book being individually numbered (Fig. A.2.2). In a company, this might begin with the designator 0001 for the first notebook issued and be followed in numerical sequence as each notebook is signed out. Thus, the investigator may find that by the time notebook 0001 is complete, the next one available for issue may be 0123. In an academic laboratory, the notebook numbering system is the choice of the investigator and may use a numerical sequence based on the investigator's initials (e.g., JS1 and JS2 for John Smith's first and second notebooks), as long as it is systematic, easy to archive, and searchable. While notebooks with preprinted book and page numbers can be purchased (e.g., see supplies list in Table A.2.1), an investigator can number notebook pages by hand. The page shown in Figure A.2.2B is of this type, i.e., it does not have an assigned book number or page numbers.

The numbered notebook should have a title page and space for a table of contents. If it does not, the investigator should leave about five blank pages at the beginning of the notebook for these items. In a company notebook, each page should be stamped, back and front, with some version of the phraseology, "Confidential, Property of Company *X*."

Current Protocols Essential Laboratory Techniques A.2.1-A.2.28

A2

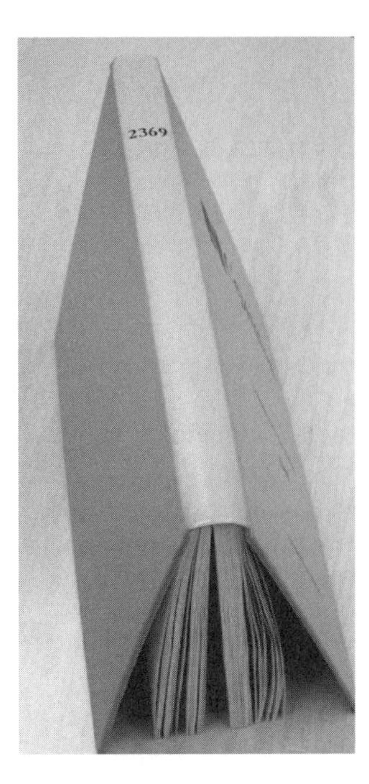

Figure A.2.1 Bound laboratory notebook photographed in open position.

Notebook Elements

Title page

The title page of the notebook should include the date the notebook was started and the date when it was completed, institutional/laboratory affiliation, and the investigator's name, e-mail address, and phone number to associate the book with the individual. One reason for this is that should the notebook be misplaced, it can be returned to the owner. A laboratory notebook should never leave the workplace, in industry for reasons of confidentiality, and in academia, so that it does not indeed get lost. A table of content*s* should document the list of experiments contained in the notebook in sequential order by start page and date. This provides a convenient list that can be used to archive and search the notebook in the future.

Entries

All notebook entries should be made in ink. Pencil must not be used because penciled entries can be readily changed (erased), thus defeating the function of the notebook, i.e., a true, chronological record of a scientist's bench work. A variety of nonerasable black ballpoint, roller ball, or gel pens are optimal because their ink does not run and can withstand water, buffer, and solvent spills. Blue and red inks can often fade and are not always reproduced clearly on a copy machine. Pens with waterproof, fade-resistant ink are readily available (e.g., see supplies list in Table A.2.1). Sharpies and their equivalent should not be used because they can bleed through the page. In some instances, computer-generated experimental protocols can be pasted into the laboratory notebook with an explanation along the edge of the pasted material noting that nothing is written underneath the pasted information. Such information is usually generic (e.g., protein determination and buffer solutions), and time is saved in not rewriting the details. Limited data sets (e.g., from a scintillation counter) can also be included in a bound notebook.

Experiment documentation

In addition to seminal texts (Kanare, 1985), there are a multitude of Internet Web sites from universities, government agencies, notebook manufacturers, and patent counsels across the world that

A

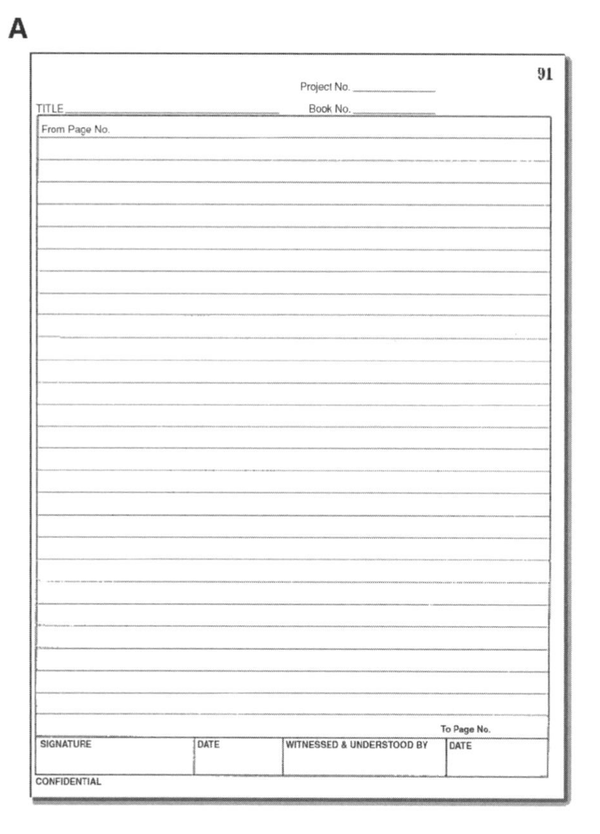

91

Project No. ____________

TITLE ______________________ Book No. ____________

From Page No.

To Page No.

SIGNATURE	DATE	WITNESSED & UNDERSTOOD BY	DATE

CONFIDENTIAL

B

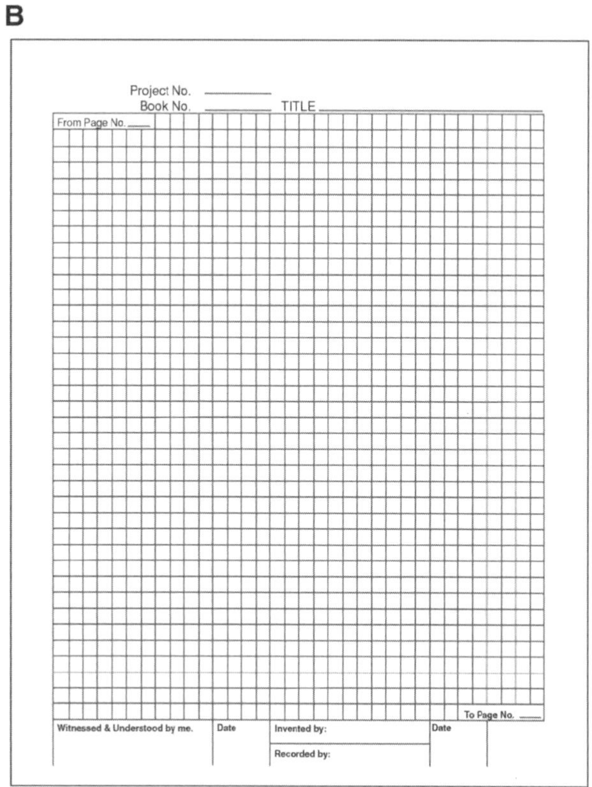

Project No. ________

Book No. ________ TITLE ______________________

From Page No. ____

To Page No. ____

Witnessed & Understood by me.	Date	Invented by:	Date
		Recorded by:	

Figure A.2.2 Two types of generic notebook pages with spaces to add notebook number and page number. Boxes for investigator signature and date and that of witness and date are also included. Reproduced with permission from (**A**) Eureka Laboratory Notebooks (*http://www.eurekalabbook.com*) and (**B**) Scientific Notebook Company (*http://www.snco.com*).

Table A.2.1 Supplies for Maintaining Laboratory Notebooks

Product	Suppliers	Remarks	URL
Hardcopy notebooks	Eureka Lab Book	See Figure A.2.2A	*http://eurekalabbook.com*
	Scientific Notebook Company	See Figure A.2.2B	*http://www.snco.com*
Electronic notebooks	Amphora Research Systems	OpenELN: uses Lotus 1-2-3 in conjunction with PatentSafe	*http://www.amphora-research.com*
	CambridgeSoft	E-Notebook Ultra 10: uses Microsoft Excel and Word documents together with ChemDraw drawings and reaction software and spectral data; can be searched by text, structure, or reaction; available only for Windows	*http://www.cambridgesoft.com/software/details/?ds=9&dsv=32*
	Gensys	Infotrieve ELN: an open, flexible discipline-neutral platform for laboratory data; endorsed by CENSA; compliant with 21 CFR 11	*http://www.gensys.com/home.html*
Pens	Pentel Hybrid gel roller, Zebra Sarasa	Non-erasable black ballpoint, roller ball or gel pens	—
Calculators	Hewlett Packard	Sole manufacturer of the only commercially available RPN-based handheld calculator (HP 33s)	*http://h20331.www2.hp.com*
	RPN Engineering Calculator 6.3.2.	Freeware/shareware	*http://www.cleansofts.com/get/203/10723/RPN_Engineering_Calculator.html*

Abbreviations: CENSA, Collaborative Notebook Systems Association; CFR, Code of Federal Regulations; ELN, electronic laboratory notebook; RPN, reverse Polish notation.

outline the basic requirements for entering information into a laboratory notebook. Not surprisingly, given the very basic and obvious nature of notebook record keeping, the majority are very consistent with one another. However, the challenge is for these guidelines to be consistently applied in daily activities by those conducting the experiment and using the notebooks. The following represents a general synopsis of the recommendations.

Language

All experimental procedures should be written in direct, colloquial language in both the present and past tense. The present tense is useful for documenting the actual experimental procedure, while the past tense is appropriate for the conclusions section of the experiment. Ideally, English should be used because this is the language used by the majority of scientific journals worldwide, and its use will facilitate the preparation of manuscripts and reports from notebook materials for publication and patent applications. The use of the first person, singular or plural (I or we) should be avoided, as recommended in the style guidelines for the majority of peer-reviewed journals.

Codification

In maintaining a notebook, an investigator has the obligation to document exactly what was done, when, by whom, and how, addressing this specifically to a hypothetical third party for reasons of clarity. This allows another scientist, "skilled in the art" (e.g., from the same or a related scientific discipline), to be able to independently repeat the recorded experiment in a similar manner at a future date, using the same procedures and experimental materials.

Documentation

The experimenter must sign and date the page(s) on which the particular experiment and its planning, preparation, execution, results, and conclusions are entered. Since the laboratory notebook serves as a chronological record of experimentation, signing and dating should take place upon completion of all entries on a page during the conduct of an experiment. The notebook entries should likewise be signed and dated by an independent witness (e.g., a scientist not involved in running the experiment) in a timely manner (within a week or so) with the statement, "Read, witnessed, and understood by (witness name)." See Figure A.2.2 and Figures A.2.6. to A.2.15.

Mapping discontinuous entries

A complexity of notebook recording can be encountered when a scientist runs different parts of multiple experiments on the same day. The best approach for recording in the notebook in this instance is to use a single different page for entering the information relevant to each different experiment. If the information recorded for each experiment on that day does not fill the entire page, the scientist draws a single line diagonally across the unused portion of the page. In the event that more space is needed for data recording and the scientist has already used the consecutive page to make entries for data collection for a different part of a different experiment conducted on the same day, the convention is to indicate on the bottom right hand side of the page the number of the page where continuation of data recording will be found for the particular experiment. The page on which the data is continued should have written on the top left hand side of the page "continued from page *x*." Consistent use of the "to and from" convention for data recording allows for reliable tracking of information recorded for any one particular experiment among several that may be ongoing simultaneously.

Good Laboratory Practices guidelines

Finally, basic biomedical (also called preclinical or nonclinical) and chemistry research does not have to be conducted under the Good Laboratory Practices (GLP) guidelines in accordance with U.S. Food and Drug Administration (FDA) regulations, 21 CFR Part 58. GLP refers to a system of management controls and a set of principles which provide a framework within which studies are planned, performed, monitored, recorded, reported, and archived. Conduct of studies under GLP principles and systems assures regulatory agencies (e.g., the FDA) that the data is a true reflection of the results obtained in a study. GLP studies are usually reserved for late stage safety testing of compounds in animals (see *http://www.fda.gov/ora/compliance_ref/bimo/glp/default.htm*).

Notebook Organization

The intent of the notebook is to allow others—e.g., a scientist from the same discipline, a patent examiner, or an informed legal authority—to clearly understand what was done and why, typically without any additional direct input from the investigator. Thus, in addition to being clearly written, the notebook should have a logical organization for each experiment, including its context, planning, execution, and results, as well as conclusions, which reflect the thought process of the investigator.

Any one laboratory notebook should be used by a single investigator to facilitate such organization and for ease of access to data. Recreating an experimental series across multiple notebooks with different investigators can be daunting enough. When these factors are mixed in the same notebook, it can be very easy to overlook data, especially when the table of contents is not properly maintained.

A2

Notebook Ownership

In industry, a central record of laboratory notebooks is typically maintained, indicating to whom and when a notebook was issued. An investigator will thus sign out a notebook for his or her use, signing a master notebook and using a page at the front of the newly acquired notebook (usually the inside front cover) to designate the name of the investigator and the date when it was signed out. For ELNs, this is less of an issue; in fact, one of the reasons for developing this format was to allow investigators to collaborate on experiments and share data from remote sites. An investigator may make exclusive use of a notebook, but when experimental work is performed under a grant or for a company, the notebook is not the property of the investigator. It belongs to the laboratory where the work was produced.

Steps in Recording the Experiment

Each experiment should be started at the top of a fresh page and work from the top to the bottom of the page. Some investigators prefer to use only the front of a notebook page (right-hand page in a bound notebook), leaving the back of the page (or left-hand page in a bound notebook) for additional experimental notations and to avoid overcrowding of information on the experimental pages. This preference should be immediately obvious in the layout of the notebook and its associated table of contents. A *single* line, in black ink, should be drawn across any open spaces on the page as well as any open space in between experiments. Figure A.2.3 provides a complete overview of the steps to be taken in recording experiments in a notebook.

Introduction

The first page of the experiment includes the date, title of the experiment, experiment number, if appropriate, and an introduction to the experiment (Fig. A.2.6) and should include the following:

1. *Title*. Make titles informative. In the example shown for illustration, the title is "Binding of new compounds to the adenosine A_1 receptor labeled by [^{3}H]JB007." Other useful titles would be

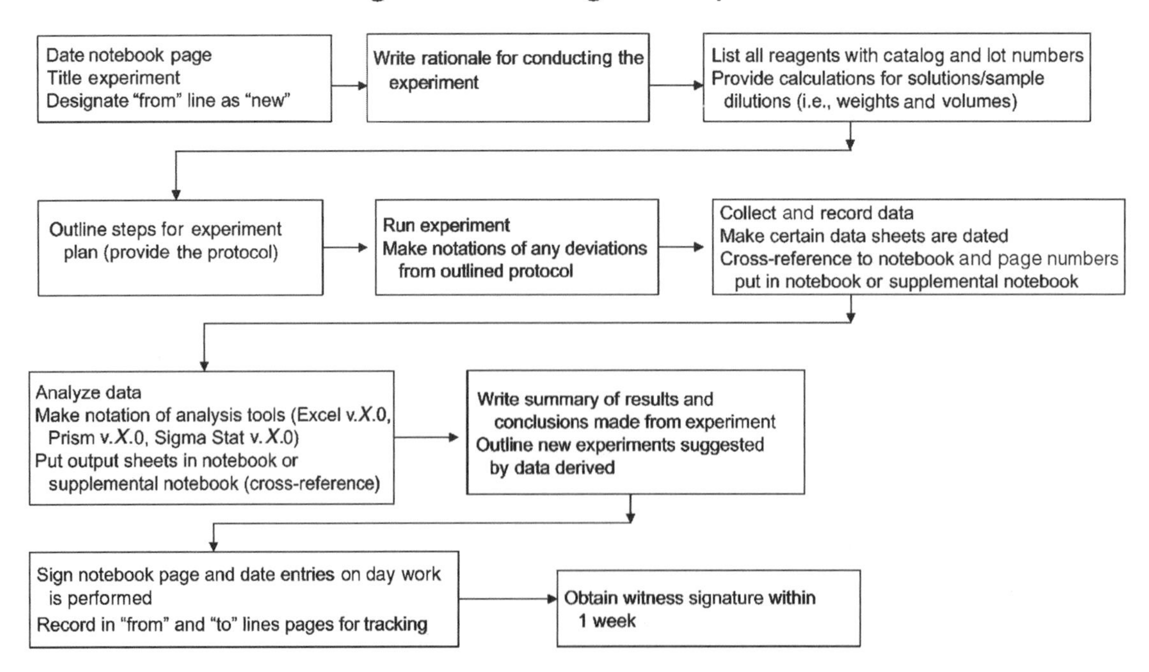

Figure A.2.3 Flow chart representing vital steps in recording and running the experiment.

"The effect of compound XYZ on urinary output" or "The synthesis of a series of heterocyclic molecules as part of project Y."

It is important (given that replicates of experiments are necessary to ensure the robustness of an observed outcome and to also derive statistical power) that a series of experiments be noted as such so that they can be easily identified, e.g., "The effect of compound XYZ on urinary output" or "The effect of compound XYZ on urinary output – experiment 2." In keeping with the table of contents, it can also be useful to number each experiment, e.g., "Experiment 1: The effect of compound XYZ on urinary output."

2. *Date.* Use a clear international convention that would be obvious in the country of origin (e.g., September 16th, 1996 would be 16/9/96,16/9/1996, or 1996/9/16 in the European convention; 9/16/96, 9/16/1996, or 1996/16/9 in the U.S.)

   However, it is critical to be consistent in use of this convention throughout the laboratory notebook. It is also advisable to designate the convention being used, e.g., month/day/year in any particular notebook.

3. *Rationale/purpose.* Provide an outline of the reasons for performing the experiment. Also include references to previous experiments or to work from other investigators.

   The rationale can be an expansion of the title outlining the context of the study, e.g., "Since compound ABC has been shown to be active in this model, and XYZ is an analog and has a similar mechanism of action, this experiment is being done to begin to derive a structure/activity relationship."

   References to previous experiments can take the form, "Based on the recent publication from Smith et al., (2007) it appeared worthwhile to examine . . ."

4. *Time conventions.* Where necessary, express time following either a 12-hr (a.m./p.m.) or 24-hr time convention.

   In the latter case, 9 a.m. would be noted as either 9:00 or 9h00, and 9 p.m. as either 21:00 or 21h00.

Experimental plan

The plan of the experiment should follow next in a sequential format. This part of the experimental detail should include:

1. *Materials.* List of all the materials (see Fig. A.2.6 to A.2.9) used in the experiment, including their source, batch number and expiration date, and the actual quantities weighed out for the particular experiment. Also indicate whether fresh solutions were made. Record all calculations used in weighing and preparing materials (including dilution steps) in the notebook. When new chemical entities are used, an indication of the procedures used to dissolve and dilute the compound should given (e.g., "100 μl DMSO + saline") to allow determination of the final concentration of solvent in the assay reaction. For shorter-lived radioisotopes (e.g., ^{125}I) include half-life calculations. If desired, use Web-based software programs for determining calculations (e.g., *http://www.filedudes.com/Molecular_Weight_Solver-download-32656.html*) or calculation programs (e.g., *http://casc-concentration-calculator.bpp-marcin-borkowski.qarchive.org*).

   IMPORTANT NOTE: *It is not sufficient to indicate that a 10 μM stock solution was made up without giving details of what quantity was weighed and into what volume of solvent it was dissolved. In more instances than the authors would care to remember, unexpected outcomes from an experiment have been traced to a mistake (they do happen) in the calculations and/or weighing of materials such that a 10-fold higher or 10-fold lower solution concentration was used. It is also important when water is used to indicate what type (deionized or distilled; double or triple distilled).*

   See Chapter 3 for information on preparing solutions.

2. *Special issues.* Indicate any comments or concerns regarding special handling conditions or safety issues.

   Are the compounds being used "high potency," e.g., cytotoxic agents that are extremely hazardous to handle and thus need to be weighed and put into solution in a hood? Are there material safety data sheets (MSDS) available? Has the supplier indicated special procedures for handling materials? If infectious biological materials are being used that are covered by the CDA/NIH biosafety levels guidelines (see APPENDIX 1*), are the appropriate laboratory conditions in place?*

3. *Procedures.* Provide a step-by-step outline of the procedures used in the experiment following, for example, the form shown in Figures A.2.9 to A.2.11), a protocol for running a hypothetical radioligand binding assay.

   Tactical planning for the experiment is critical to ensure that (a) all the materials are available in sufficient quantities (and a little more) to complete the experiment and (b) all the equipment is available and operational. An inoperable centrifuge, a water bath at too high a temperature at a critical point in the experiment, or an absence of dry ice at a critical point in a chemical reaction can all negate completion of an experiment.

4. *Controls.* Include appropriate controls and reference standards in the experiment.

   These provide an internal standard, indicating whether the experiment actually worked and are also a reference against which any new data can be compared.

   In the example shown in Figure A.2.7, the experimental compound A that had previously been examined was used as a reference for the new compound B to validate the experiment, thus establishing continuity between experiments. If compound A failed to replicate the activity previously recorded, then the experiment would need repeating. In other instances, a known commercially available reference compound could be used, where both the compound and the data are available in the public domain (Watling, 2006).

Running the experiment

1. While the experiment is ongoing, the investigator should actively note any deviations from the protocol, either designed or the result of a mistake, immediately in the notebook, e.g., "Tube 11 received twice the amount of tissue."

2. If any changes are made, this requires crossing out information already recorded. This should be done with a single line so as not to obliterate what was initially written (e.g., see Fig. A.2.5). When you cross out data, you must briefly state the reason for the cross-out and initial and date the change.

3. It is absolutely imperative that the notebook pages be used to record experimental observations, however messy this may make the notebook.

   Under no circumstances should experiments depend on notations on sticky notes, paper towels, lab coat sleeves, or similar places!

4. All information related to the experiment should be included either in the laboratory notebook or in a supplemental notebook with raw data.

   In the latter instance, the data contained in the supplemental notebook should not only be referenced in the primary notebook, e.g., "supplemental notebook 3, page 27," but there should be an indication as to what is contained on page 27, e.g., "NMR Spectra that show . . ." In that way, if the supplemental notebook is lost, there is an active record of what was contained there.

5. Specifically note if an expected result fails to occur, rather than leaving another reader to infer the event.

   A basic tenet in scientific exploration is that "an absence of evidence is not evidence of absence," leading to consideration of the differences between necessary and sufficient causality. For necessary causality, if x is a necessary cause of y, then the presence of y necessarily implies that x preceded it. The presence of x, however, does not imply that y will occur. For sufficient causality, if x is a sufficient cause of y, then the presence of x necessarily implies the presence of y. However, another different cause may also result in y. Thus the presence of y does not imply the presence of x.

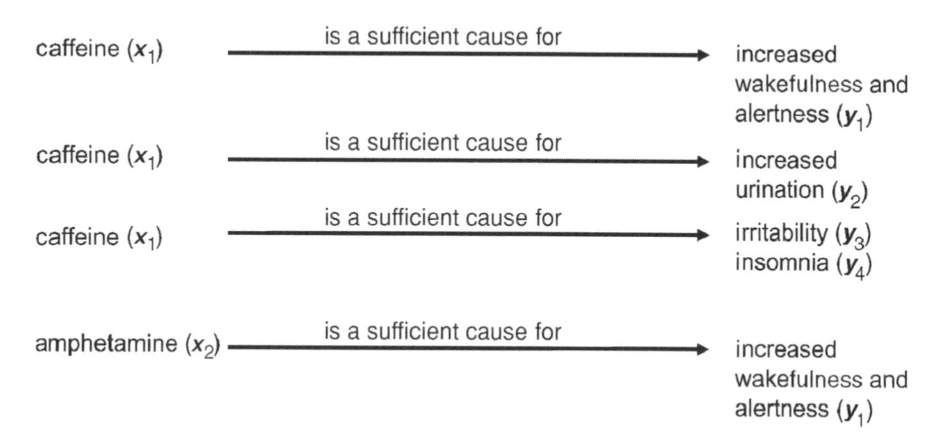

Figure A.2.4 Graphical representation of the difference between necessary and sufficient causality. Caffeine is a sufficient but not a necessary cause for increased wakefulness and alertness, increased urination, irritability, and insomnia because other substances can cause the same effects. For example, amphetamine is a sufficient cause for increased wakefulness and alertness.

This concept is illustrated in Figure A.2.4. Caffeine(x_1), the active ingredient in coffee and the most widely used psychoactive substance in the world, produces its pharmacological effects via antagonizing adenosine receptors in various tissues throughout the body. In the central nervous system, antagonism of adenosine A_1 receptors produces the effect of increasing alertness (y_1), a principal reason for coffee consumption. Additionally, caffeine antagonizes adenosine A_1 receptors in the kidney, producing a diuretic effect (i.e., an urge to urinate; y_2). Finally, in individuals who over-consume coffee, caffeine can cause irritability (y_3) and insomnia (y_4). Taking this example, caffeine (x_1) produces increased alertness and wakefulness (y_1), indicating (at face value) a linear causality. However, the psychostimulant, amphetamine(x_2) can also produce phenotypic effects similar to caffeine (i.e., increased wakefulness and alertness; y_1). Caffeine can increase urination, but this response can also occur from consuming too much water, from a urinary tract infection, or from taking an antihypertensive diuretic agent. Thus the presence a urination response (y_2), cannot be taken as evidence for the presence of caffeine in the system or of coffee consumption. Similarly, the irritability (y_3) and insomnia (y_4) resulting from excess coffee consumption can also be caused by anxiety or illness. Therefore, caffeine is a sufficient cause of alertness, irritability, wakefulness, and increased urination, but not a necessary cause because other causes can result in the same conditions.

A further elaboration on this key principle in the logic of scientific research is what is know as "argument from ignorance" (or argumentum ad ignorantiam), a process where an event is considered true only because it has not been proven false. The practical relevance of this concept is considered further in Altman and Bland (1995) and references therein.

In many instances, the astute recording of minor differences in the experimental protocol can make all the difference in understanding what precisely happened and what actually drove data generation. This can help in interpreting inconsistent data that may indeed be logical in cases where the protocol was changed.

Supporting information/supplemental notebooks

Once the experiment is completed and the data have been generated, the documentation needs to be either pasted in the notebook with one edge signed and dated by the investigator (Fig. A.2.13) or incorporated into and cross-referenced in a supplemental notebook. Notebook entries can be supplemented with other supporting material from the experiment (e.g., computer-generated Excel spread sheets, scintillation counter print outs, recordable compact discs of data, photomicrographs, gel films, and NMR or mass spectra printouts), all of which should be referenced to the actual experiment. A single page of material can be pasted in the notebook, but when the material is extensive, it can be stored in a marked loose-leaf binder with each piece of material properly noted, e.g., "supplementary table to experiment *x*, notebook *y*, page *z*." It is extremely important to reference and describe the content of supplemental material in the notebook properly, describing the attachment clearly and unambiguously (e.g., "HPLC trace from 11/7/04"), together with any conclusions that might be drawn from its content. It is preferable to bind the supplementary material once the primary notebook is completely filled up.

Data analysis

The data then need to be analyzed (Lutz and Kenakin, 1999). This usually involves a series of involved additional calculations, the nature of which should be clearly indicated in the primary notebook. For instance in Figures A.2.10 and A.2.11, the data presented were run in triplicate and the mean and standard deviation for these subsequently calculated from the scintillation counter print out (*APPENDIX*, page 137) as part of a data analysis program utilizing an Excel spreadsheet (Fig. A.2.13 and A.2.14). As can be seen in the data set, there is an outlier in tubes 10 to 12 (Fig. A.2.10 and A.2.12). Two of the numbers are close to one another while the third is twice that of the other two, reflecting the notebook notation that two aliquots of tissue preparation were added by mistake. Thus the latter number was discarded for the calculation, an average of two being calculated and a notation made.

In some instances, it is easier to use a hand-held calculator. The authors recommend one that is based on the reverse Polish notation (RPN) algorithm because this is more facile in handling large data sets and repetitive calculations.

Once the initial data are calculated, a data page should be used to store the scintillation counter data output sheet and associate it to the initial experimental design (Fig. A.2.13 and A.2.14). The data are then analyzed, in the present example using GraphPad Prism (Fig. A.2.13). Finally, the data are summarized and conclusions made (Fig. A.2.15). In the hypothetical example, the data are modest, reflecting the trivial nature of the experiment undertaken. In essence no hypothesis was being evaluated.

Conclusions

For more complicated experiments, the investigator should reflect on the data and then write a conclusion on the individual experiment. Was the hypothesis validated? Are the data consistent with previous experiments? How do they relate to other work in the same area? Were there errors in planning or execution that negated the value of the experiment such that the conclusion would be "this experiment did not work?" Did a water bath overheat? Did a centrifuge malfunction? Did a reaction vessel explode? Was there a fire alarm at a critical point? Was there a power outage that precluded the completion of the experiment? All such occurrences should be noted at the time they occur.

The major reason for conducting experiments is to see whether the initial hypothesis was correct. To ignore, without questioning, data that refute or confuse a hypothesis or fail to reproduce previous observations, directly negates the scientific approach based on the null hypothesis (see *APPENDIX 4*). Three examples worthy of comment are described in a section below (Discrepant Data Examples). However, in many instances experiments that failed to work have resulted in major breakthroughs in scientific knowledge because they challenged the investigator to seek other explanations for inconsistent data and avoided discounting data that did not fit into the initial hypothesis. Concerns related to the inability to publish negative data have led to two on-line journals: the *Journal of Negative Results in Biomedicine* and the *Journal of Articles in Support of the Null Hypothesis*.

The final caution about the conclusions for an experiment is that the outcome be clearly stated and quantified. For example, if compound x evoked a concentration- or dose-dependent effect, an IC_{50}, EC_{50}, pA_2 (see Neubig et al., 2003) should be determined and recorded as in the hypothetical example. So many biological experimental outcomes are still reported in purely qualitative terms (a cause of major concern to journal editorial boards) that provide no ability to compare data between experiments or laboratories (e.g., in experiments where the intensity of gel bands is the outcome and where data usually involve a single observation).

IMPORTANT NOTE: Web sites from law firms and university technology transfer functions caution against making categorical or colloquial statements about the worthlessness of an experiment that could be used in the future as evidence of the investigator's failure to appreciate the inventive significance of the recorded experiment, thus potentially compromising patentability of the invention. Such comments should not be confused with simple, accurate statements such as, "This experiment did not work."

Future ideas/concepts

Finally, the conclusion of an experiment should include the outline for any new experiments (if any) that are suggested by the derived data, as well as any notations on consistency with literature observations. This provides an opportunity to express a continuity of ideas (in much the same way as in a personal diary) and it can often be revisited by the investigator to check the logic behind a planned series of experiments in support of a publication. It is also possible that the witnessed notation of an idea at this stage (if followed up) could be used as the date of conception. For a complete overview of the steps to be taken in recording experiments in a notebook, refer to Figure A.2.3.

Witnessing

As already noted, it is an imperative for an experiment, once completed, to be witnessed by a competent, independent observer, skilled in the art, to verify the time at which the experiment was done. The absence of timely witnessing can confound patent cases and lead to questions about conception and reduction to practice—i.e., exemplifying that the concept works for its intended purpose (e.g., a compound interacts with the intended target). For patentability purposes, reduction to practice (see Patents and Intellectual Property) is considered to have occurred on the date that the witness signs the notebook, not the date on which the investigator's signature is appended. This reinforces the need to maintain a timely schedule of witnessing. The latter does not, however, provide any acknowledgement of the accuracy or quality of the experiment executed or the data derived from the experiment. It merely provides a record that the information on the pages signed was present on the date the signature was appended. In cases of scientific fraud, additional oversight is required to assess whether fraud has occurred. Of interest, an incident of a U.S. scientist sentenced to prison for fabricating or manipulating experimental data was reported recently in the *New York Times* (Interlandi, 2006).

In the hypothetical example shown in Figures A.2.6 to A.2.15, the experimental procedure took place between January 18th, 2006 and January 20th, 2006 (notebook pp. 131 to 140). The author, a certain John Hancock, dated his signature on the 19th and 20th of January. The witness, an Alfred Nobel, signed on January 25th, 2006. Were there a patent issue related to the biological evaluation of Compound B, the date of 1/25/06 rather than 1/18/06 would apply to due to the one week delay in witnessing.

In the hypothetical example shown in Figures A.2.6 to A.2.15, there are instances of crossing out information with a single line. Ideally, the crossed-out information should be initialed and dated by the notebook author; this is not exemplified. Also, in the example, there are several other instances where information should have been supplied. For example, each notebook page should be dated, and there should be an indication of the conclusion of recording for the particular experiment by designating "end" in the "continued to" line of the last page of the entry for this particular experiment. With respect to data recording there should have been included cross-references to the electronic data files. Thus, even in a good example of notebook recording, omissions were made attesting to the need for periodic review and reinforcement of the basic principles that should be maintained.

If an error is noticed after the notebook has been witnessed, the correction should be entered on a new page with cross-referencing back to the page where the data was in error, using the "from" convention (see Mapping discontinuous entries in the Experiment Documentation section). On the new page, any recalculation of the data should be included, and this page with the corrected data should be independently witnessed.

A2

Laboratory Notebook Storage

Once a notebook is completed, it typically remains on the bookshelf of the investigator indefinitely as an ongoing record and part of the historical record of the laboratory. In industry, notebooks and their supplemental data notebooks/binders are archived as soon as they are no longer needed as active reference works for ongoing experimental studies. The content of a completed notebook is usually duplicated or photographed and the notebook sent to a secure, fireproof, offsite storage facility (central repository) where it is kept at the correct temperature and humidity to maintain the physical integrity of the notebook.

Large companies and some universities have their own offsite storage facilities, while smaller organizations use commercial data security and management companies like Iron Mountain (*http://www.ironmountain.com*). Where patents are under application or issued, the notebooks supporting the intellectual property position should be cross-referenced to the patent. In industry, with investigators changing jobs and companies merging and becoming bankrupt, laboratory notebooks can readily be misplaced, and in such circumstances it is difficult to recreate an accurate record of an invention. A major loss of institutional knowledge and corporate value can occur when information is uniquely linked to individuals who change jobs or retire and leave no trace of what was in their notebooks or where the latter were archived. Retrieving such information can be extremely problematic.

Data Replication

Once an experiment is completed and its outcome determined to be of value, it is mandatory to replicate the experiment to an n value of at least three to allow statistical analysis of the data to derive a mean and standard deviation of the experimental outcome. Ideally, the experiment should be designed from a statistical perspective with the actual n value being delineated by power analysis (see *APPENDIX 4*). As noted above, a series of experiments could be run (e.g., 1, 2, or 3) with individual quantitative outputs (e.g., IC_{50} values) that could then be statistically analyzed and reported (e.g., an IC_{50} value of 27 ± 5 nM for compound x).

This final output from a laboratory notebook can then be included in a patent application or scientific manuscript. In industry, (e.g., pharmaceutical companies) interim internal reports also exist to support federal regulatory requirements for IND (Investigational New Drug; Ator, 2006) filing. In performing quality control on such data or researching them for a patent application, it is unfortunately not always obvious from where cited IC_{50} values are derived. In some instances, data from an experiment has been taken to no obvious conclusion, e.g., when raw dosage data is presented without a calculated IC_{50} value, requiring re-analysis to derive the IC_{50} value. The accuracy of this re-creation can be problematic, e.g., if a different software program or a handheld calculator were used, there were six IC_{50} values recorded (or findable) but only three appear to have been used, there were technical problems with three data points that were not recorded, or the reasons the best three values were selected may be unknown at this time, again confounding quality control and intellectual property issues. Ideally, after a series of experiments is completed, a summary table should be inserted on a new page in the laboratory notebook summarizing the data. This would include the individual values used to derive a mean and standard deviation as well as their source (i.e., page number) and any notation as to why any other numbers were excluded.

Discrepant Data Examples

The following relates a few real-world examples highlighting the importance of properly documenting all details of the experiment.

"Pouring" over your work

The impact of an overlooked minor point regarding solvent transfer in a chemical reaction that is critical to the experiment working is recorded by Kanare (1985). "How much detail should be recorded in your notes? Could another scientist who is competent in your field pick up your notebook and repeat your work solely from the written description without additional explanation? If the answer is yes, then you are doing a good job. Too many details are better than for you to assume that a future reader (perhaps you!) will know all about your work." The article goes on to illustrate the importance of documenting details by relating a story told by Nobel laureate Sir Geoffrey Wilkinson about A. J. Shortland's first synthesis of hexamethyltungsten, one of the biggest breakthroughs in transition metal organometallic chemistry. After publication of the work, others tried but were unable to reproduce it. After much investigation and a demonstration of the technique by Shortland, it was discovered that instead of transferring the petroleum to the reaction flask via a steel tube and serum cap technique, Shortland actually removed a stopper and poured the petroleum into the flask (even though the original papers on the subject stressed the necessity for rigorous oxygen-free conditions). This procedure resulted in immediate partial resaturation of the solvent with oxygen and a very different outcome. If it had not been for the thorough investigation and revelation of details, the work might have been abandoned as nonreproducible.

An incomplete washout

Approximately 30 years ago, research ongoing at a leading pharmaceutical company resulted in the question, "What is the molar descriptor below femtomolar (10^{-15} M)?" The answer, (after searching the library in the absence of immediate gratification from the Internet) was attomolar (10^{-18} M). The research prompting the question was the evaluation of the activity of a peptide neurotransmitter in the central nervous system (CNS). This peptide had equivalent efficacy in an *in vivo* mouse model at nanomolar (10^{-9} M), picomolar (10^{-12} M) and femtomolar (10^{-15} M) doses when given intracerebroventicularly. This information was both intriguing and puzzling, because the data being produced suggested that this was the most potent peptide ever evaluated for CNS activity. The peptide was then found to be fully efficacious at the attomolar (10^{-18} M) dose range. Another round of dilution then showed this peptide to be active at a concentration of 10^{-21} M (now termed zeptomolar). By this time, word had spread throughout the building, and the original investigator had now been joined by an audience of some 15 colleagues. While watching the final zeptomolar stages of the experiment, one observer noted that the same syringe was being used for the sequential dilutions without any intermediate washout with buffer. Carryover of the peptide due to faulty lab technique rather than an ultra-potent peptide turned out to be the reason for the observed effect.

One letter can make all the difference

In postdoctoral work, M.W. (one of the authors of this appendix) was researching changes in the phosphorylation of synaptic proteins as a molecular substrate of memory (Williams and Rodnight, 1977). After 10 months of intensive research, the initial increases in phosphorylation of the P2 synaptosomal fraction of rat brain isolated via a standard ultracentrifugation technique using molar sucrose concentrations could not be replicated.

The original investigator was called in to resolve the situation. He was immediately able to replicate the original findings much to the embarrassment of the author. A final experiment was planned to resolve the discrepancy. In preparing for this experiment, M.W. made up a solution of 0.32 M sucrose and added this to the remainder of an existing solution of sucrose used previously only to observe that the new solution was denser than the old. It turned out that the original investigator had made up a 0.32 M *molal* solution of sucrose rather than 0.32 M *molar* explaining the discrepant results and the inability to replicate because centrifugation with 0.32 molar sucrose removed the nuclear pellet from rat brain tissue while 0.32 molal did not.

ELECTRONIC NOTEBOOKS (ELNS)

With the advent of the personal computer and an exponential increase in data generation, the ability to use an electronic notebook (ELN) instead of a hard copy notebook provides a cost-effective solution when used for recording experiments with routine content and outcome. While the first ELNs became available in the late 1990s, these were custom in nature and better designed for use by chemists than biologists, given the routine procedures used in chemistry. There were also legal issues regarding the archival nature of ELNs for patent purposes and the issue of electronic signatures (Taylor, 2006).

An ELN can be viewed as the electronic equivalent of a hardcopy research notebook (Elliot, 2004). Instead of being recorded on paper, all the information outlined above for the hardbound notebook, including the supplementary materials and signatures, can be recorded on electronic notebook pages. Instead of writing with a pen and taping in images and graphs, a computer is used to add information from a keyboard, mouse, image files, and also directly from scientific instruments. Many different types of ELN software exist and vary to the extent they can recreate the "look and feel" of a paper notebook, but all recreate the basic functions of a paper notebook, and in using ELNs the investigator should follow the same guidelines discussed above for recording in hardcopy notebooks. The formal definition of an ELN has been developed by the Collaborative Electronic Notebook Systems Association (CENSA; Elliot, 2004).

A common ELN architecture was developed in 1997 by a consortium with Oak Ridge National Laboratories (ORNL), the Lawrence Berkeley National Laboratory, and the Pacific Northwest National Laboratory. A schematic for an ELN emerging from this consensus is shown in Figure A.2.5. The key features include a common notebook engine that can interface with secure data storage systems, data acquisition systems, and notebook client. The notebook interface, the actual notebook page, contains required input fields (e.g., name, time and date, equations or chemical structure, text, tables, and images). The notebook engine coordinates information and data transfer directly from laboratory instruments and data storage to the notebook interface. The data storage system

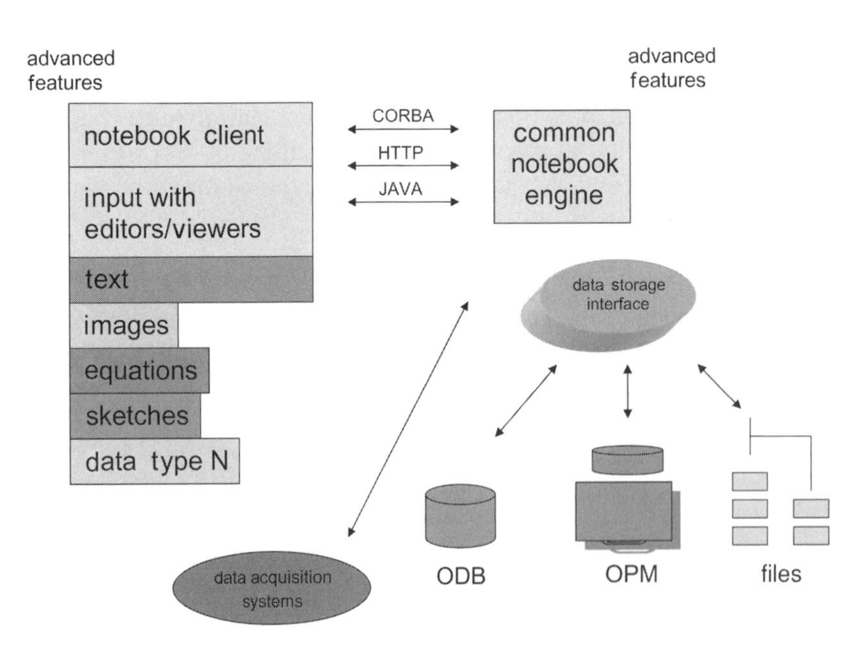

Figure A.2.5 Generic common framework of an ELN with cross-platform interoperating notebook implementations. Web-based access allows for facile interaction of the notebook client with a common notebook engine. The notebook client can be populated with text, images, equations, and chemical structures; they can be easily edited. Communication with data acquisition systems (laboratory instrumentation) linked through a data storage system and interfaced with optical databases (ODB) with output protection management (OPM) provides searchable and secure data retrieval.

allows for secure information retrieval, data sharing, and searching. In total, the system provides a platform to integrate research and development workflow. Another open source ELN is available from Electronic Laboratory Notebook (*http://collaboratory.emsl.pnl.gov*). While these open source solutions lack many of the database features of commercially supported products, they allow notebook sharing and data analysis at minimal cost suitable for academic and federal laboratories.

Interest in ELNs has increased significantly in the past few years, with many commercial products being available from scientific software, instrumentation, and database-management companies. Major vendors include Waters, CambridgeSoft, Elsevier MDL, and Symyx Technologies (see Ator et al., 2006). Each company approached the ELN from a unique perspective, in the case of Waters, interfacing with their HPLC and Laboratory Informatics Systems; and CambridgeSoft, their ChemDraw software platform. An overview of their particular strengths and weaknesses has been documented (Elliot, 2004).

ELNs can effectively provide benefit to the witnessing of workflows that are used to support due diligence (see below) during reduction to practice. Electronic data should be generated and archived in accord with the following guidelines:

1. Establish an official procedure for electronic record keeping and disseminate it to all ELN users.
2. Generate permanent electronic records with attention to the following:
    a. Back up and write-protect all electronic data sources.
    b. Reference the data in a signed and dated handwritten notebook.
    c. Store electronic records in a safe repository with custodians who can vouch for their authenticity.
    d. Use electronic or digital signature software to enhance the credibility of the electronic records.
3. Use hardware and/or software that prevent the ability to edit original research descriptions and data records.
4. Create records that accurately document the development of the research (e.g., conception, reduction to practice, and diligence related to an invention; see Patents and Intellectual Property below).
5. Use security measures to prevent unauthorized access to the system:
    a. Password protect the system and data.
    b. Limit system access only to authorized personnel.

Electronic Signatures and Records

Patent attorneys have been somewhat resistant to the concept of electronic signatures, and this has had a major impact on allowing ELN workflow to become fully electronic. This opposition is based on the lack of specific case law in the area of patent litigation; however, Federal Rules of Evidence do not currently exclude electronic documents. Such resistance is beginning to diminish, and vendors now provide electronic notary services, e.g., AbsoluteProof from Surety (*http://www.surety.com/absoluteservice.php*). In February 2005, eight major pharmaceutical companies formed the signatures and authentication for everyone (SAFE)-BioPharma Association (*http://www.safe-biopharma.org*) to address unique electronic identity credentials for legally enforceable regulatory compliant digital signatures across the global pharmaceutical environment. One alternative, PatentPad (Amphora Research Systems; *http://www.amphora-research.com*) is a documentation system comprising secure paper, where the authenticity of the paper and authenticity of the serial numbers are verified by a third party (SCRIP-SAFE; *http://www.scrip-safe.com*). Amphora controls the release of the paper and maintains a secure registry (SCRIP-SAFE). Many organizations view the added complexity and cost of this process worth the effort should they need to defend a patent. Those not yet willing to implement electronic signatures carry out a hybrid ELN approach. All data are entered electronically and are searchable, but notebook pages are printed, pasted into bound paper notebooks, signed, and witnessed. This remains the normal practice until electronic record systems become more widely accepted and increasingly cost-effective.

Interest in ELNs in biology, while behind that in chemistry, was driven by the exponential increase in data generation by high-throughput screening and genomics analysis in the majority of instances using Microsoft Excel or Lotus 1-2-3 as the data capture software. Early biology ELNs were generic blank-page or white-page notebooks that proved to be inconvenient to use (Taylor, 2006). ActivityBase (idbs solutions, *http://www.idbs.com*), its second-generation product (Deaffinity DAT-LA ELN), and Oracle (*http://www.oracle.com/index.html*) are three distinct but interrelated software programs that are routinely used as part of biological database platforms. There are many user-friendly software programs for data analysis that can integrate and manipulate newly generated data in the context of existing database information (e.g., genome maps, systems biology pathway databases) to analyze data via three-dimensional parameters (e.g., Spotfire; *http://www.spotfire.com/products/dxp.cfm*). Go to *http://www.knowledgestorm.com* to survey this and other such new software programs (see Internet Resources).

PATENTS AND INTELLECTUAL PROPERTY

The determination of invention priority in U.S. patent law is based on the premise that the *first* individual to make an invention that is deemed new, useful, and nonobvious is entitled to a patent on that invention. A constructive reduction to practice of an invention is the date that a regular, nonprovisional application disclosing the invention is filed in the U.S. Patent and Trademark Office (USPTO). An earlier presumed invention date may be obtained by claiming a priority date of:

1. an earlier filed U.S. provisional application;
2. an earlier filed foreign national patent application or;
3. an earlier filed Patent Cooperation Treaty (PCT) patent application, as long as the earlier filed application discloses the same invention and is within the time constraints as defined by U.S. law;
4. an earlier, properly recorded, and witnessed experimental notebook page.

Inventive activity as defined by U.S. patent law involves three distinct categories: *conception*, the formation in the mind of the inventor(s) of a specific invention; *reduction to practice* where the inventor(s) has either (1) developed a physical embodiment of the inventive concept that has been demonstrated to another and works for its intended purpose or (2) filed a patent application having a disclosure that teaches to "one of ordinary skill in the field/art" how to make and use the invention. The third aspect is *diligence*, which occurs between conception and reduction to practice.

An inventor is entitled to claim the date of conception as the invention date if reduction to practice is made within a reasonable period of time following conception and diligence can be proven using accurate, witnessed notebook records. The latter involves a continual chain of diligence activities from the filing date backwards and can obviously depend on the circumstances (e.g., illness or availability of equipment). However, a gap of more than 6 months in continuous reduction to practice can be viewed as questionable unless there is valid reason.

As U.S. patents are awarded on the basis of *first to invent*, not the first applicant to file, the ability to obtain a patent on an invention often depends upon how well research activity related to the invention is documented. Until January 1, 1996, inventive activity related to conception, actual reduction to practice, and diligence had to take place in the U.S. or one of its territories to claim the date of conception as the date of invention. After passage of the North American Free Trade Agreement (NAFTA) on December 8, 1993, and agreements with World Trade Organization (WTO), individuals or organizations seeking to obtain a U.S. patent had the same rights as a U.S. inventor. A major implication of this change in U.S. patent law is that investigators seeking to obtain a U.S. patent need to provide contemporary evidence of inventive activity (e.g., documentation generated at the time the inventive activity occurred). Thus, in the context of an invention, the relevant laboratory notebook should contain written records of the following:

Conception date. The recorded date when the invention was conceived.

Date of reduction to practice. The date on which "a working embodiment of the invention" was recorded.

Diligence in reducing the invention to practice. A record of the intention and efforts to make a working embodiment of the invention with dates and data that show what was done to reduce the invention to practice and when such activities were conducted. In periods when not working on reducing the invention to practice, diligence needs to be documented in the form of reasonable excuses for inactive periods by providing information as to why there was no activity during the period in question. (e.g., unavailability of test conditions or equipment).

How to make and use the invention. Provide documentation details sufficient to teach a colleague how to make and use your invention. This can also involve documenting the "best" mode of practicing the invention.

A2

COMMENTS

Despite the many guidelines and safeguards for transparent notebook keeping, the ultimate weakness in the chain is inevitably human. However many rules and regulations, effective notebook keeping can only occur where there is a clear understanding that the investigator is obligated to consistently implement the following:

1. Address the content of his or her notebook to a third party in much the same way as is done in publishing a scientific paper.

2. Write legibly and ensure that each and every experiment is written in the same generic format outlined above.

3. Ensure that the notebooks are witnessed in a timely manner.

4. Ensure that the whereabouts of all notebooks are known, and that they are available to appropriate parties.

LITERATURE CITED

Altman, D.G. and Bland, J.M. 1995. Absence of evidence is not evidence of absence. *B.M.J.* 311:485.

Ator, M.A., Mallamo, J.P., and Williams, M. 2006. Overview of drug discovery and development. *Curr. Protoc. Pharmacol.* 35:9.9.1-9.9.27. John Wiley & Sons, Hoboken, N.J.

Elliott, M. 2004. Electronic Laboratory Notebooks, A Foundation for Knowledge Management, 2nd ed. Attrium Research and Consulting, Wilton, Conn.

Interlandi, J. 2006. An unwelcome discovery. *The New York Times Magazine*. Oct. 22.

Kanare, M. 1985. Writing the Laboratory Notebook. American Chemical Society, Washington, D.C.

Lutz, M. and Kenakin, T. 1999. Quantitative molecular pharmacology and informatics. *In* Drug Discovery. John Wiley & Sons, Chichester, U.K.

Neubig, R.R., Spedding, M., Kenakin, T., and Christopoulos, A. 2003. International union of pharmacology committee on receptor nomenclature and drug classification. XXXVIII. Update on terms and symbols in quantitative pharmacology. *Pharmacol. Rev.* 55:597-606.

Taylor, K. 2006. The status of electronic laboratory notebooks for chemistry and biology. *Curr. Opin. Drug Disc. Develop.* 9:348–353.

Watling, K.J. 2006. The Sigma-RBI Handbook of Receptors and Signal Transduction, 5th ed. Sigma-RBI, Natick, Mass.

Williams, M. and Rodnight, R. 1977. Protein phosphorylation in nervous tissue: Possible involvement in nervous tissue function and relationship to cyclic nucleotide metabolism. *Prog. Neurobiol.* 8:183-250

INTERNET RESOURCES

A2

http://www.fda.gov/ora/compliance_ref/bimo/glp/default.htm
U.S. Food and Drug Administration. Code of Federal Regulations 21 Part 58, FDA Good Laboratory Practices.

http://www.censa.org
The Collaborative Electronic Notebook Systems Association (CENSA) Web site.

http://www.safe-biopharma.org
Signatures and authentication for everyone (SAFE). SAFE-BioPharma Association, New York.

http://www.scrip-safe.com
SCRIP-SAFE International: Securing documents for business and education.

http://www.idbs.com/products/abase

http://www.oracle.com/index.html
ActiveBase and Oracle software programs are routinely used as part of biological database platforms.

http://www.spotfire.com/products/dxp.cfm
A software program for data analysis that can integrate and manipulate newly generated data in the context of existing database information.

http://www.knowledgestorm.com/search/keyword/Biology
Use this Web site to survey new software programs useful for data analysis that can integrate and manipulate newly generated data in the context of existing database information.

Figures A.2.6 through A.2.15 appear starting on the following page.

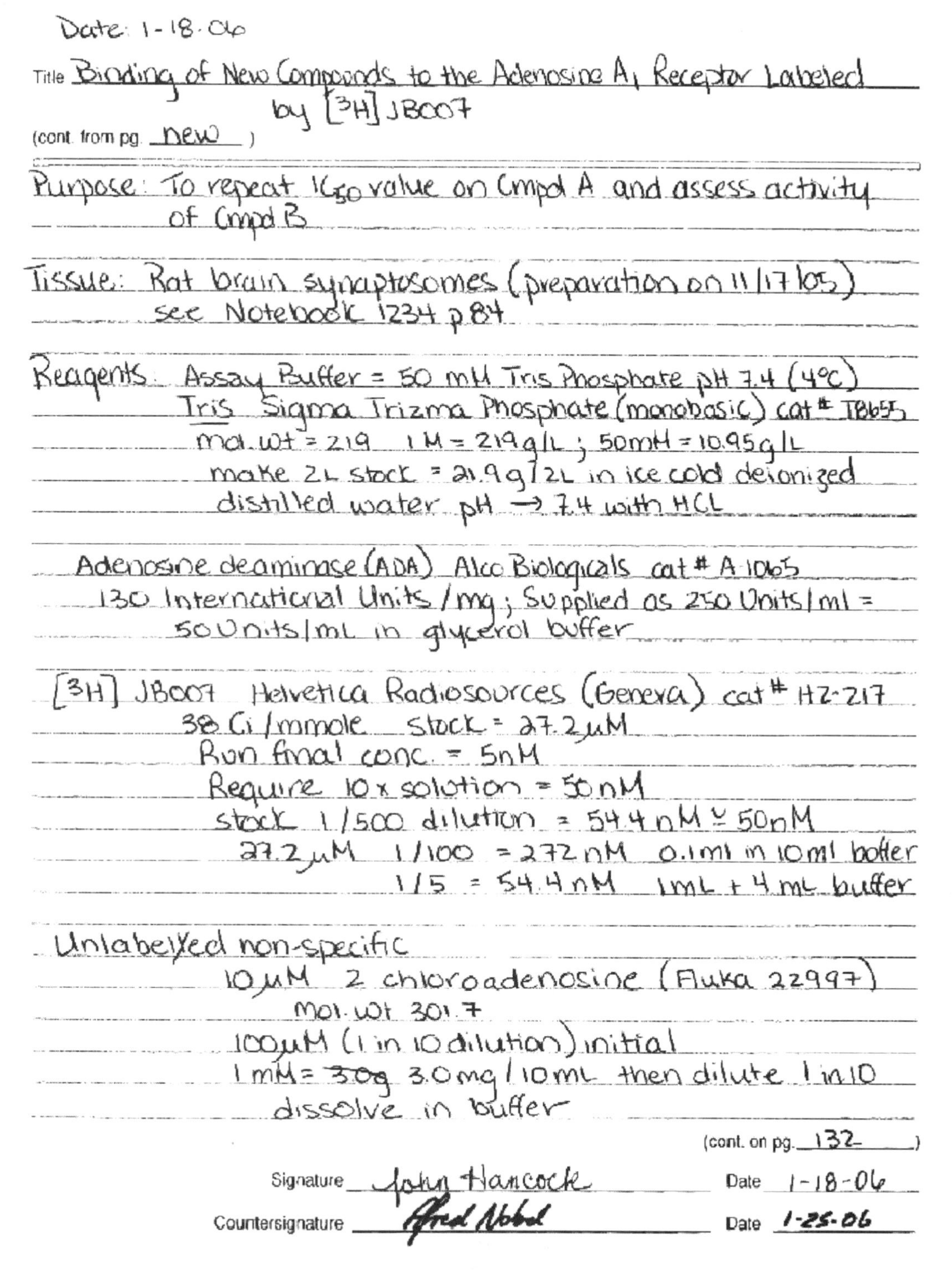

LABORATORY NOTEBOOK 1448 Page 131 of 200
Date: 1-18-06
Title Binding of New Compounds to the Adenosine A1 Receptor Labeled by [3H] JB007
(cont. from pg. new)
Purpose: To repeat IC50 value on Cmpd A and assess activity of Cmpd B
Tissue: Rat brain synaptosomes (preparation on 11/17/05) see Notebook 1234 p 84
Reagents. Assay Buffer = 50 mM Tris Phosphate pH 7.4 (4°C)
Tris Sigma Trizma Phosphate (monobasic) cat # T8655
mol. wt = 219 1 M = 219 g/L ; 50mM = 10.95 g/L
make 2L stock = 21.9 g/2L in ice cold deionized distilled water pH → 7.4 with HCL
Adenosine deaminase (ADA) Alco Biologicals cat # A 1065
130 International Units / mg ; Supplied as 250 Units/ml = 50 Units/mL in glycerol buffer
[3H] JB007 Helvetica Radiosources (Geneva) cat # HZ-217
38 Ci/mmole stock = 27.2 μM
Run final conc. = 5nM
Require 10x solution = 50nM
stock 1/500 dilution = 54.4 nM ≅ 50nM
27.2 μM 1/100 = 272 nM 0.1ml in 10ml buffer
1/5 = 54.4 nM 1mL + 4 mL buffer
Unlabeled non-specific
10 μM 2 chloroadenosine (Fluka 22997)
Mol. wt 301.7
100μM (1 in 10 dilution) initial
1 mM = 3.0g 3.0 mg / 10 mL then dilute 1 in 10
dissolve in buffer
(cont. on pg. 132)
Signature John Hancock Date 1-18-06
Countersignature Alfred Nobel Date 1-25-06

Figure A.2.6 A page in a hypothetical notebook prepared by John Hancock and verified by Alfred Nobel.

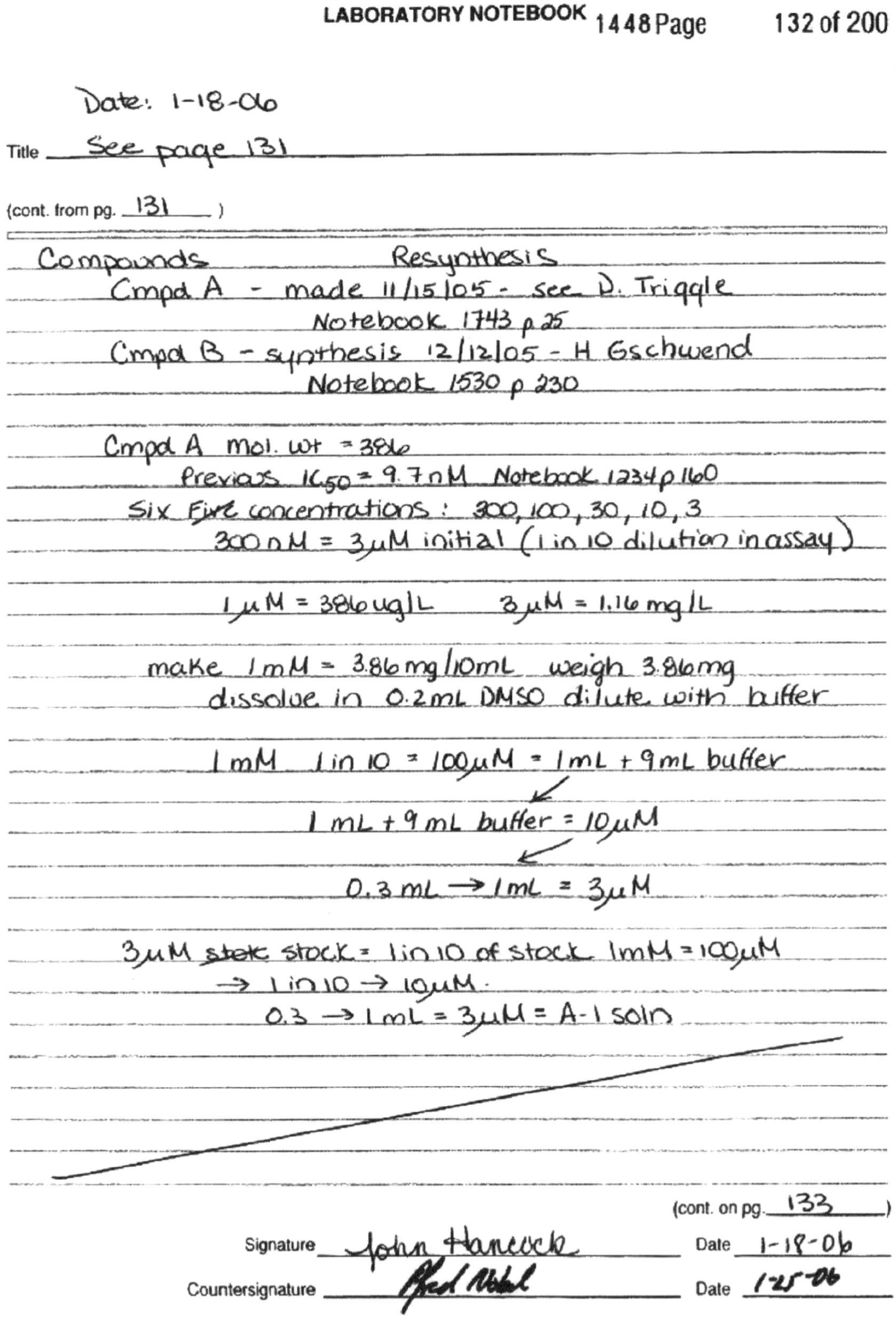

LABORATORY NOTEBOOK 1448 Page 132 of 200

Date: 1-18-06

Title See page 131

(cont. from pg. 131)

Compounds Resynthesis

Cmpd A - made 11/15/05 - see D. Triggle Notebook 1743 p 25

Cmpd B - synthesis 12/12/05 - H Gschwend Notebook 1530 p 230

Cmpd A mol. wt = 386

Previous IC_{50} = 9.7 nM Notebook 1234 p 160

Six ~~Five~~ concentrations: 300, 100, 30, 10, 3

300 nM = 3 μM initial (1 in 10 dilution in assay)

1 μM = 386 ug/L 3 μM = 1.16 mg/L

make 1 mM = 3.86 mg/10 mL weigh 3.86 mg

dissolve in 0.2 mL DMSO dilute with buffer

1 mM 1 in 10 = 100 μM = 1 mL + 9 mL buffer

1 mL + 9 mL buffer = 10 μM

0.3 mL → 1 mL = 3 μM

3 μM ~~state~~ stock = 1 in 10 of stock 1 mM = 100 μM

→ 1 in 10 → 10 μM.

0.3 → 1 mL = 3 μM = A-1 soln

(cont. on pg. 133)

Signature John Hancock Date 1-18-06

Countersignature Alfred Nobel Date 1-25-06

Figure A.2.7 A page in a hypothetical notebook prepared by John Hancock and verified by Alfred Nobel.

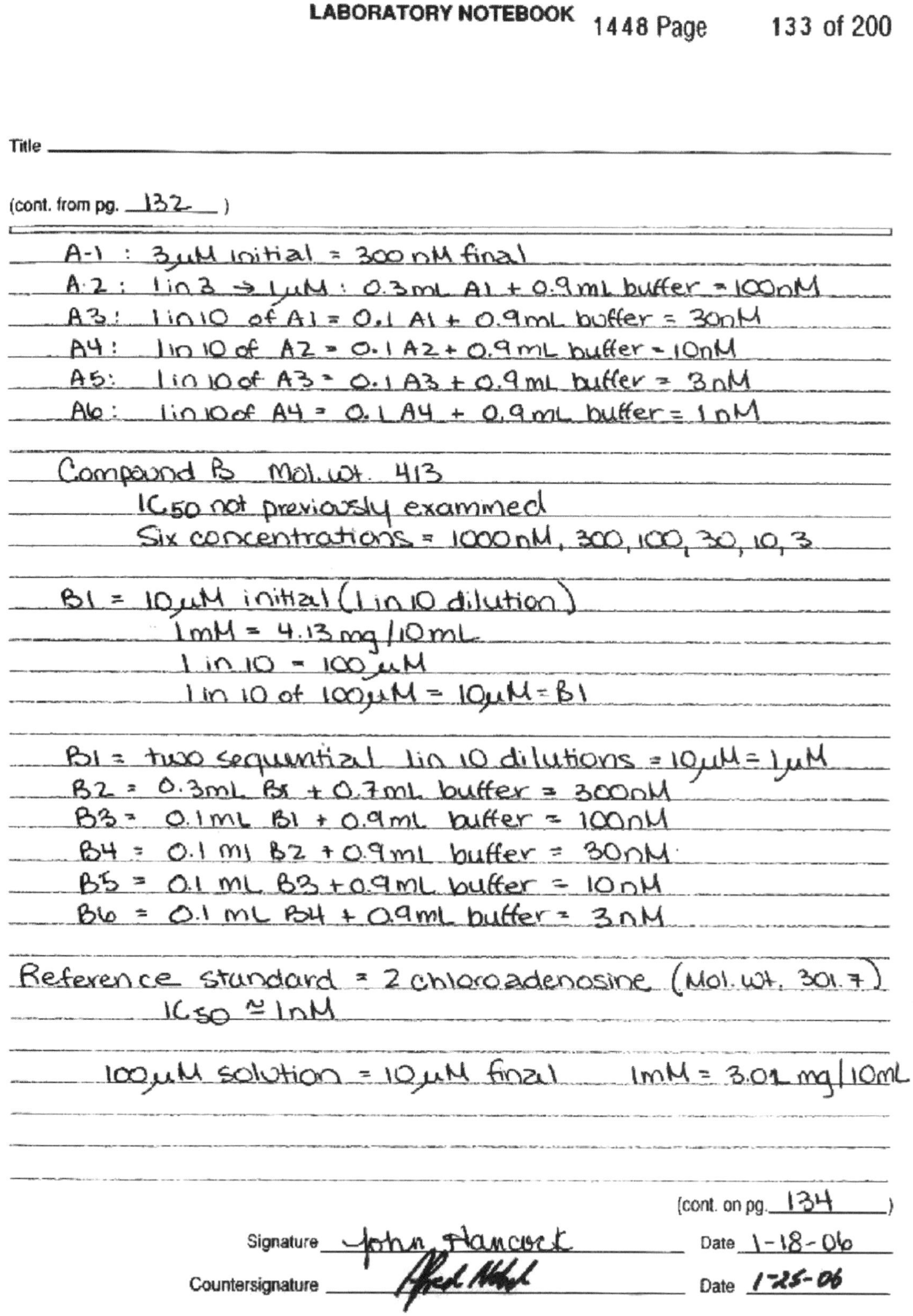

LABORATORY NOTEBOOK 1448 Page 133 of 200

Title

(cont. from pg. 132)

A-1 : 3 μM initial = 300 nM final

A-2 : 1 in 3 → 1 μM : 0.3 mL A1 + 0.9 mL buffer = 100 nM

A3 : 1 in 10 of A1 = 0.1 A1 + 0.9 mL buffer = 30 nM

A4 : 1 in 10 of A2 = 0.1 A2 + 0.9 mL buffer = 10 nM

A5 : 1 in 10 of A3 = 0.1 A3 + 0.9 mL buffer = 3 nM

A6 : 1 in 10 of A4 = 0.1 A4 + 0.9 mL buffer = 1 nM

Compound B Mol. wt. 413

IC50 not previously exammed

Six concentrations = 1000 nM, 300, 100, 30, 10, 3

B1 = 10 μM initial (1 in 10 dilution)

1 mM = 4.13 mg/10 mL

1 in 10 = 100 μM

1 in 10 of 100 μM = 10 μM = B1

B1 = two sequential 1 in 10 dilutions = 10 μM = 1 μM

B2 = 0.3 mL B1 + 0.7 mL buffer = 300 nM

B3 = 0.1 mL B1 + 0.9 mL buffer = 100 nM

B4 = 0.1 ml B2 + 0.9 mL buffer = 30 nM

B5 = 0.1 mL B3 + 0.9 mL buffer = 10 nM

B6 = 0.1 mL B4 + 0.9 mL buffer = 3 nM

Reference standard = 2 chloroadenosine (Mol. wt. 301.7)

IC50 ≃ 1 nM

100 μM solution = 10 μM final 1 mM = 3.01 mg/10 mL

(cont. on pg. 134)

Signature John Hancock Date 1-18-06

Countersignature Alfred Nobel Date 1-25-06

Figure A.2.8 A page in a hypothetical notebook prepared by John Hancock and verified by Alfred Nobel.

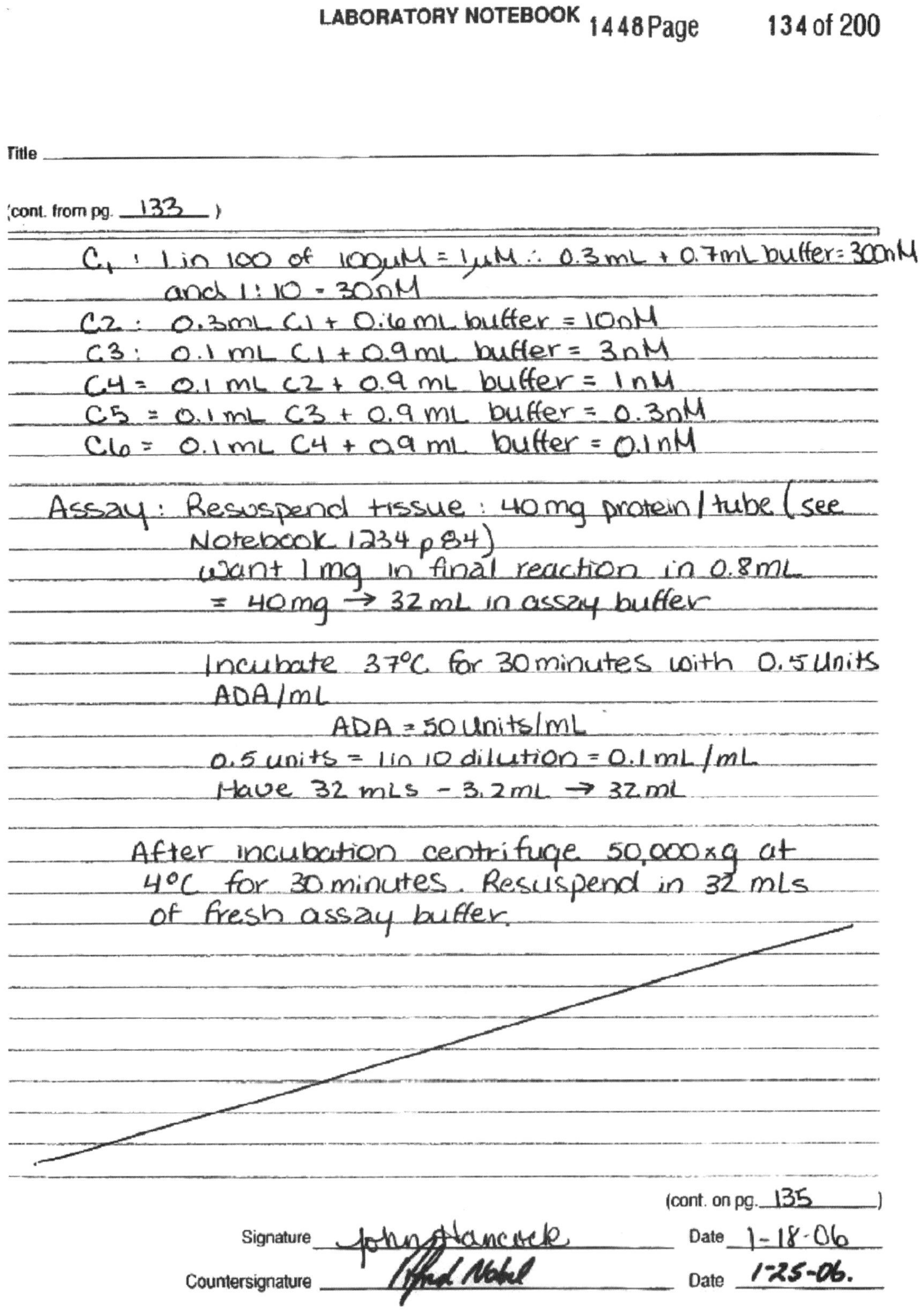

LABORATORY NOTEBOOK 1448 Page 134 of 200

Title

(cont. from pg. 133)

C1 : 1 in 100 of 100μM = 1μM ∴ 0.3mL + 0.7mL buffer = 300nM
and 1:10 = 30nM
C2 : 0.3mL C1 + 0.6mL buffer = 100nM
C3 : 0.1 mL C1 + 0.9mL buffer = 3nM
C4 = 0.1 mL C2 + 0.9 mL buffer = 1nM
C5 = 0.1 mL C3 + 0.9 mL buffer = 0.3nM
C6 = 0.1 mL C4 + 0.9 mL buffer = 0.1nM

Assay: Resuspend tissue: 40mg protein / tube (see
Notebook 1234 p 84)
want 1mg in final reaction in 0.8mL
= 40mg → 32 mL in assay buffer

Incubate 37°C for 30 minutes with 0.5 Units
ADA/mL
ADA = 50 Units/mL
0.5 units = 1 in 10 dilution = 0.1 mL/mL
Have 32 mLs - 3.2mL → 32mL

After incubation centrifuge 50,000xg at
4°C for 30 minutes. Resuspend in 32 mLs
of fresh assay buffer.

(cont. on pg. 135)

Signature John Hancock Date 1-18-06

Countersignature Alfred Nobel Date 1-25-06.

Figure A.2.9 A page in a hypothetical notebook prepared by John Hancock and verified by Alfred Nobel.

LABORATORY NOTEBOOK 1448 Page 135 of 200

Title

(cont. from pg. 134)

Set up tubes as follows:

Tubes	[3H] JB007	2-chloroadenosine	Buffer	Cmpd.	Tissue.
1-3	0.1	-	0.1	-	0.8
4-6	↓	0.1	-	-	↓
7-9		-	-	0.1 A1	
*10-12		-	-	0.1 A2	
13-15		-	-	0.1 A3	
16-18		-	-	0.1 A4	
19-21		-	-	0.1 A5	
22-24		-	-	0.1 A6	
25-27		-	-	0.1 B1	
28-30		-	-	0.1 B2	
31-33		-	-	0.1 B3	
34-36		-	-	0.1 B4	
37-39		-	-	0.1 B5	
40-42		-	-	0.1 B6	
43-45		-	-	0.1 C1	
46-48		-	-	0.1 C2	
49-51		-	-	0.1 C3	
52-54		-	-	0.1 C4	
55-57		-	-	0.1 C5	
58-60		-	-	0.1 C6	
61-63		-	0.1	-	
64-66	↓	0.1	-	-	↓

*note: tube # 11 received 2x amount of tissue

(cont. on pg. 136)

Signature John Hancock Date 1-18-06

Countersignature Alfred Nobel Date 1-25-06.

Figure A.2.10 A page in a hypothetical notebook prepared by John Hancock and verified by Alfred Nobel.

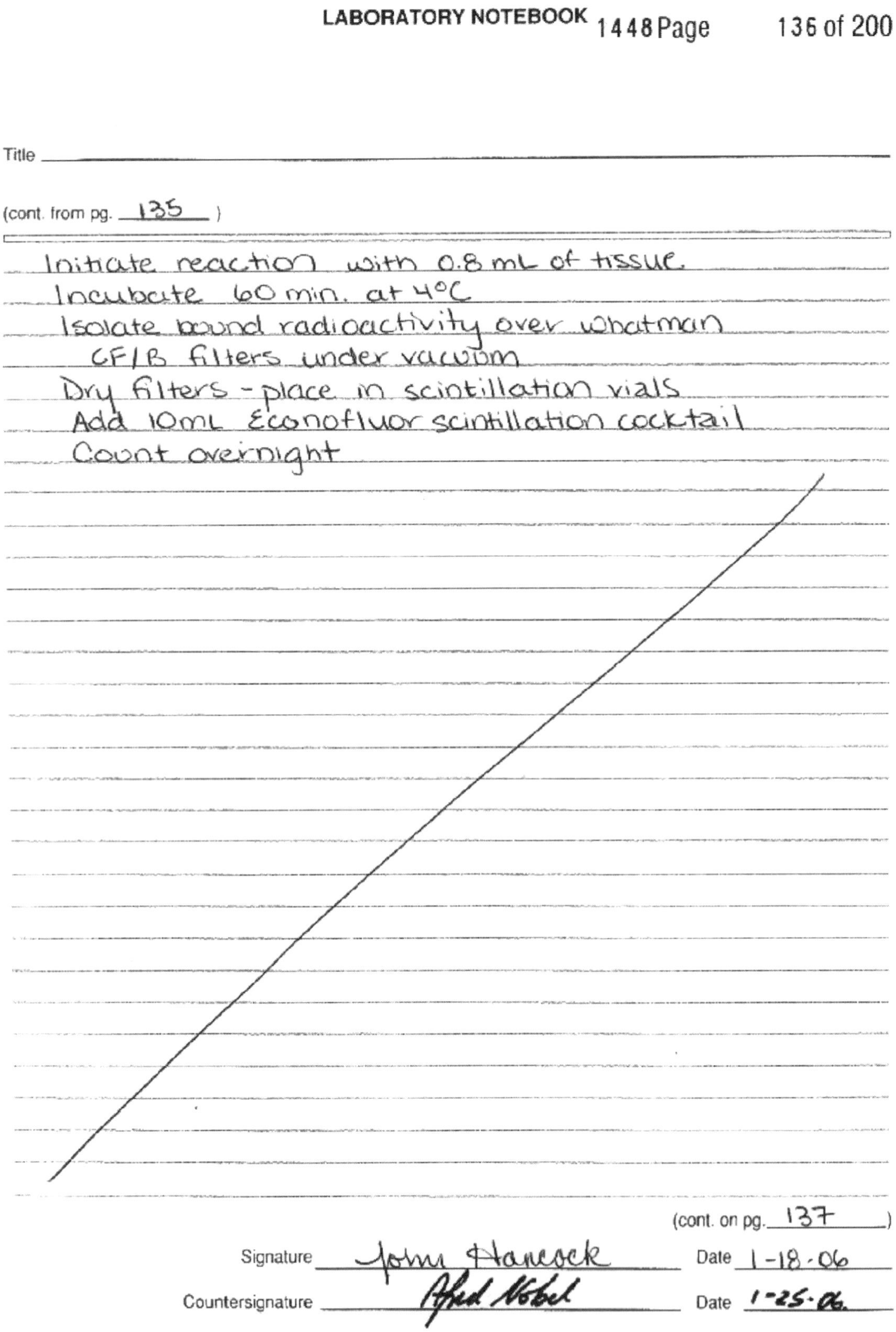

LABORATORY NOTEBOOK 1448 Page 136 of 200

Title

(cont. from pg. 135)

Initiate reaction with 0.8 mL of tissue
Incubate 60 min. at 4°C
Isolate bound radioactivity over Whatman
GF/B filters under vacuum
Dry filters - place in scintillation vials
Add 10mL Econofluor scintillation cocktail
Count overnight

(cont. on pg. 137)

Signature John Hancock Date 1-18-06

Countersignature Alfred Nobel Date 1-25-06

Figure A.2.11 A page in a hypothetical notebook prepared by John Hancock and verified by Alfred Nobel.

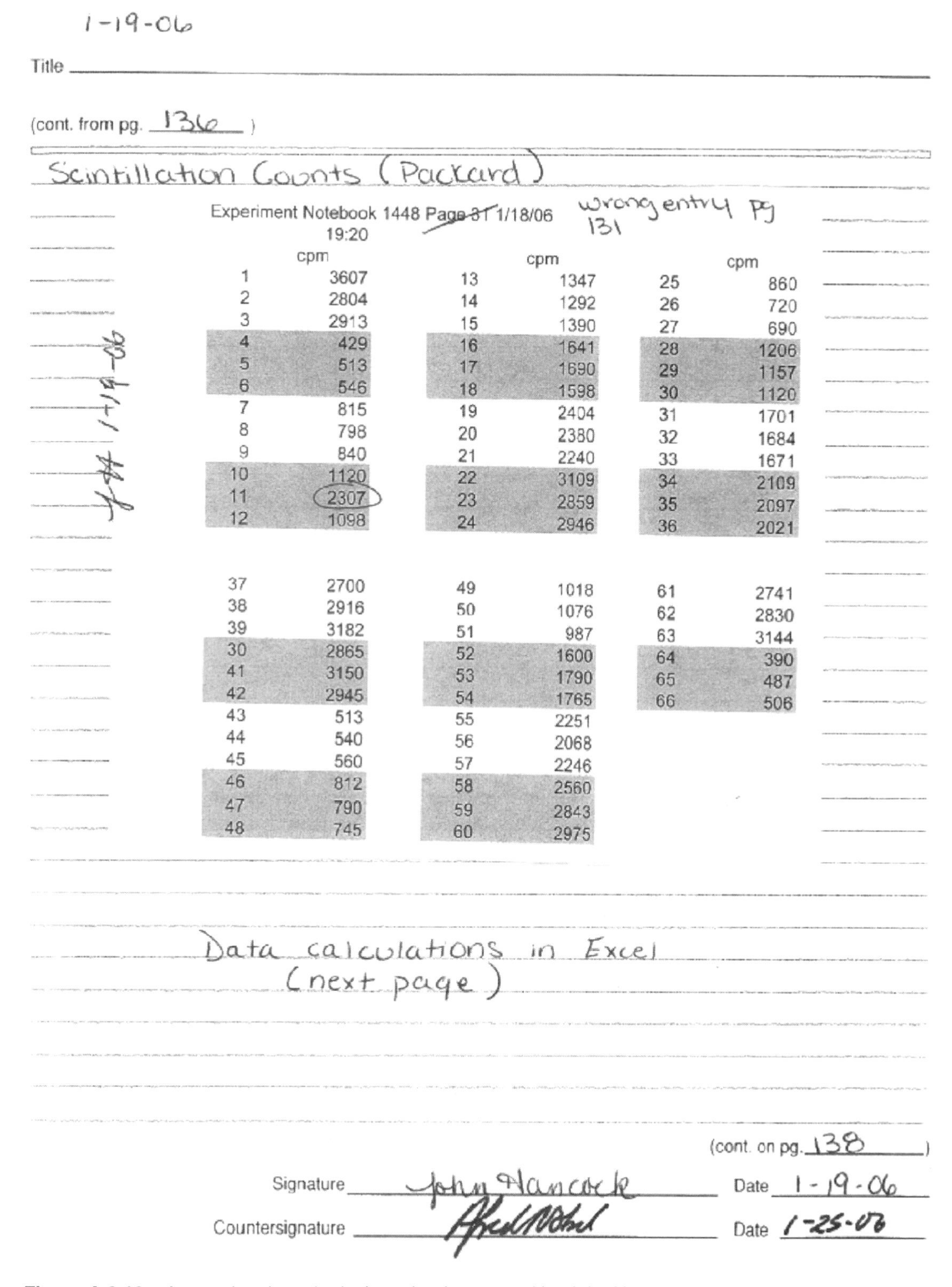

	cpm		cpm		cpm
1	3607	13	1347	25	860
2	2804	14	1292	26	720
3	2913	15	1390	27	690
4	429	16	1641	28	1206
5	513	17	1690	29	1157
6	546	18	1598	30	1120
7	815	19	2404	31	1701
8	798	20	2380	32	1684
9	840	21	2240	33	1671
10	1120	22	3109	34	2109
11	2307	23	2859	35	2097
12	1098	24	2946	36	2021
37	2700	49	1018	61	2741
38	2916	50	1076	62	2830
39	3182	51	987	63	3144
30	2865	52	1600	64	390
41	3150	53	1790	65	487
42	2945	54	1765	66	506
43	513	55	2251		
44	540	56	2068		
45	560	57	2246		
46	812	58	2560		
47	790	59	2843		
48	745	60	2975		

Figure A.2.12 A page in a hypothetical notebook prepared by John Hancock and verified by Alfred Nobel.

Title

(cont. from pg. 137)

Excel Calculations JH 1-19-06

		cpm/replicate			Mean	(-BKG)			Mean	STD	SEM	%CV	% total
total		3607	2804	2913	3006.5	3128.5	2325.5	2434.5	2528.0	325.9	133.0	12.9	100.0
		2741	2830	3144		2262.5	2351.5	2665.5					0.0
non-specific		429	513	546	478.5	-49.5	34.5	67.5	0.0	58.1	23.7		
		390	487	506		-88.5	8.5	27.5					
Cmpd A	300	815	798	840		336.5	319.5	361.5	684.5	21.1	12.2	3.1	27.1
	100	1120	2307	1098		641.5	1828.5	619.5	947.2	691.8	399.4	73.0	37.5
	30	1347	1292	1390		868.5	813.5	911.5	1014.5	49.1	28.4	4.8	40.1
	10	1641	1690	1598		1162.5	1211.5	1119.5	1513.7	46.0	26.6	3.0	59.9
	3	2404	2380	2240		1925.5	1901.5	1761.5	2177.8	88.6	51.1	4.1	86.1
	1	3109	2859	2946		2630.5	2380.5	2467.5	2492.8	126.9	73.3	5.1	98.6
Cmpd B	1000	860	720	690		381.5	241.5	211.5	480.3	90.7	52.4	18.9	19.0
	300	1206	1157	1120		727.5	678.5	641.5	944.7	43.1	24.9	4.6	37.4
	100	1701	1684	1671		1222.5	1205.5	1192.5	1401.8	15.0	8.7	1.1	55.5
	30	2109	2097	2020		1630.5	1618.5	1541.5	2025.5	48.3	27.9	2.4	80.1
	10	2700	2916	3182		2221.5	2437.5	2703.5	2481.2	241.4	139.4	9.7	98.1
	3	2865	3150	2945		2386.5	2671.5	2466.5	2508.2	147.0	84.9	5.9	99.2
+ control	30	513	540	560		34.5	61.5	81.5	181.5	23.6	13.6	13.0	7.2
2-chloroadenosine	10	812	790	745		333.5	311.5	266.5	426.2	34.2	19.7	8.0	16.9
	3	1018	1076	987		539.5	597.5	508.5	894.2	45.2	26.1	5.1	35.4
	1	1600	1790	1765		1121.5	1311.5	1286.5	1474.8	103.2	59.6	7.0	58.3
	0.3	2251	2068	2246		1772.5	1589.5	1767.5	2012.0	104.2	60.2	5.2	79.6
	0.1	2560	2643	2975		2081.5	2364.5	2496.5	2314.2	212.0	122.4	9.2	91.5

Entries in Prism Graph Pad JH 1-19-06

	X Values	A			B			C		
	Conc. (nM)	control			Cmpd A			Cmpd B		
	X	A:Y1	A:Y2	A:Y3	B:Y1	B:Y2	B:Y3	C:Y1	C:Y2	C:Y3
1	0.01									
2	0.03									
3	0.10	2081.5	2364.5	2496.5						
4	0.30	1772.5	1589.5	1767.5						
5	1.00	1121.5	1311.5	1286.5	2630.5	2380.5	2467.5			
6	3.00	539.5	597.5	508.5	1925.5	1901.5	1761.5	2386.5	2671.5	2466.5
7	10.00	333.5	311.5	266.5	1162.5	1211.5	1119.5	2221.5	2437.5	2703.5
8	30.00	34.5	61.5	81.5	868.5	813.5	911.5	1222.5	1205.5	1192.5
9	100.00				641.5		619.5	1630.5	1618.5	1541.5
10	300.00				336.5	319.5	361.5	727.5	678.5	641.5
11	1000.00							381.5	241.5	211.5

Data Transform X = Log X for non-linear regression analysis

(cont. on pg. 139)

Signature John Hancock Date 1-19-06

Countersignature Alfred Nobel Date 1-25-06

Figure A.2.13 A page in a hypothetical notebook prepared by John Hancock and verified by Alfred Nobel.

A2

Title

(cont. from pg. 138)

Linear Regression Analysis JH 1-19-06

		A	B	C
		control	Cmpd A	Cmpd B
		Y	Y	Y
1	Sigmoidal dose-response (variable slope)			
2	Best-fit values			
3	BOTTOM	0.0	0.0	0.0
4	TOP	2528	2528	2528
5	LOGEC50	-0.07207	1.107	1.972
6	HILLSLOPE	-0.9152	-0.7463	-0.7801
7	EC50	0.8471	12.78	93.78
8	Std. Error			
9	LOGEC50	0.03432	0.06063	0.1048
10	HILLSLOPE	0.06189	0.07838	0.1434
11	95% Confidence Intervals			
12	LOGEC50	-0.1448 to 0.0006991	0.9774 to 1.236	1.750 to 2.194
13	HILLSLOPE	-1.046 to -0.7840	-0.9133 to -0.5793	-1.084 to -0.4760
14	EC50	0.7164 to 1.002	9.494 to 17.21	56.21 to 156.5
15	Goodness of Fit			
16	Degrees of Freedom	16	15	16
17	R^2	0.9787	0.9406	0.8455
18	Absolute Sum of Squares	249484	566825	1.9619e+006
19	Sy.x	124.9	194.4	350.2
20	Runs test			
21	Points above curve	7	11	11
22	Points below curve	11	6	7
23	Number of runs	7	3	6
24	P value (runs test)	0.1448	0.001374	0.05995
25	Deviation from Model	Not Significant	Significant	Not Significant
26	Constraints			
27	BOTTOM	BOTTOM = 0.0	BOTTOM = 0.0	BOTTOM = 0.0
28	TOP	TOP = 2528	TOP = 2528	TOP = 2528
29	Data			
30	Number of X values	11	11	11
31	Number of Y replicates	3	3	3
32	Total number of values	18	17	18
33	Number of missing values	15	16	15

(cont. on pg. 140)

Signature John Hancock Date 1-19-06

Countersignature Alfred Nobel Date 1-25-06

Figure A.2.14 A page in a hypothetical notebook prepared by John Hancock and verified by Alfred Nobel.

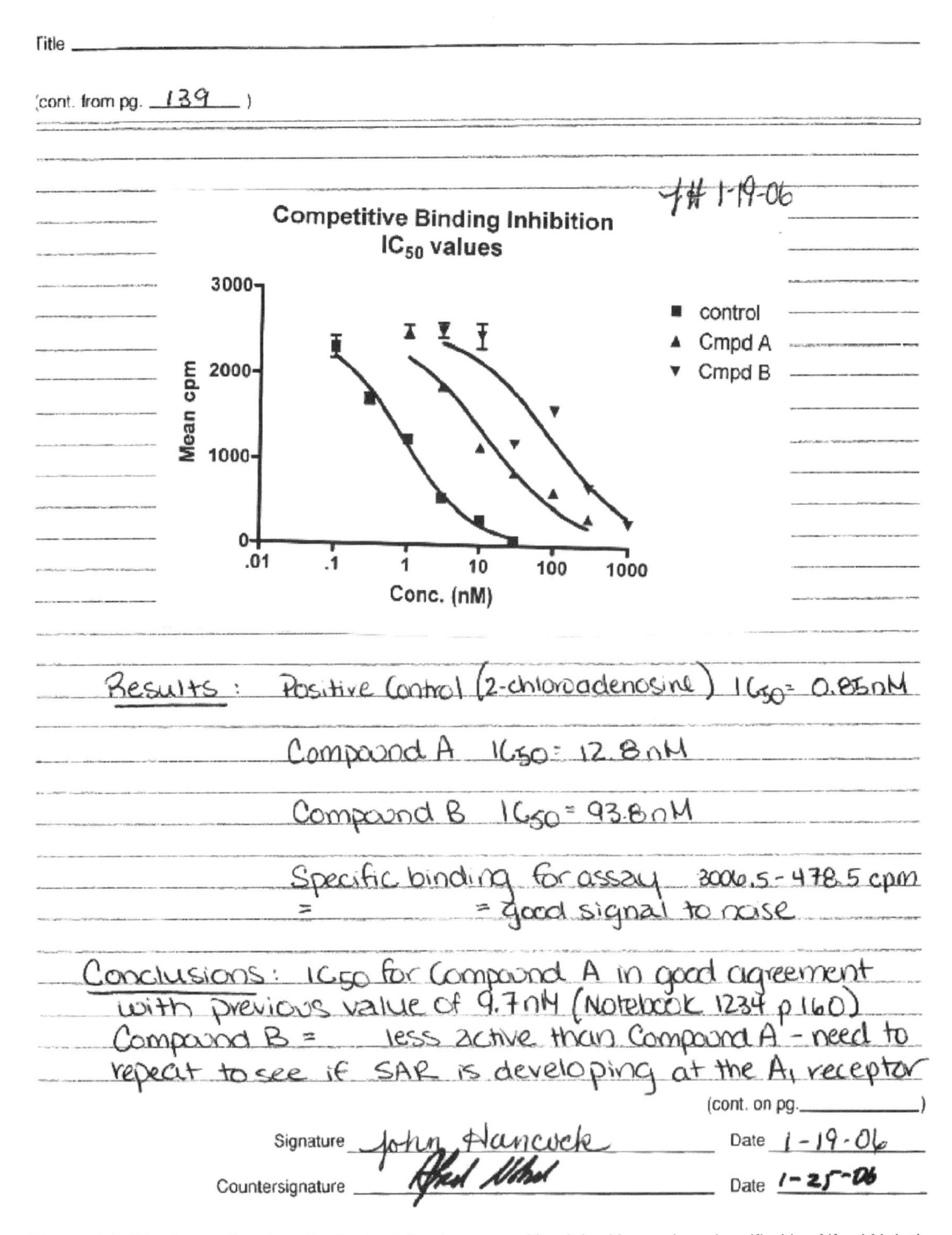

LABORATORY NOTEBOOK 1448 Page 140 of 200
Title
(cont. from pg. 139)
JH 1-19-06
Competitive Binding Inhibition
IC50 values
Mean cpm
3000
2000
1000
0
.01
.1
1
10
100
1000
Conc. (nM)
control
Cmpd A
Cmpd B
Results: Positive Control (2-chloroadenosine) IC50 = 0.85nM
Compound A IC50 = 12.8nM
Compound B IC50 = 93.8nM
Specific binding for assay 3006.5 - 478.5 cpm
= = good signal to noise
Conclusions: IC50 for Compound A in good agreement
with previous value of 9.7nM (Notebook 1234 p 160)
Compound B = less active than Compound A - need to
repeat to see if SAR is developing at the A1 receptor
(cont. on pg.)
Signature John Hancock
Date 1-19-06
Countersignature Alfred Nobel
Date 1-25-06

Figure A.2.15 A page in a hypothetical notebook prepared by John Hancock and verified by Alfred Nobel.

APPENDIX 3A

Ethical Considerations When Altering Digital Images

Sean R. Gallagher[1]
[1]UVP, LLC, Upland, California

INTRODUCTION

The growth of digital image acquisition and analysis in the life sciences continues unabated, and it is easy to see why. Digital imaging is faster, less expensive, and more sensitive than using film. The images are immediately available without developing film and printing, and where a film scan might offer a hundred-fold range of data, scientific digital cameras can far exceed this range, giving better data resolution. Quantitation should be a requirement for scientific publication and is easily handled by a number of image analysis programs (see *UNIT 7.5*), which enable accurate quantitation of image intensity. Furthermore, several different layout or photography-based programs such as Photoshop (*APPENDIX 3B*) can create the composite image by cropping and enhancing the images to improve the visual interpretation of an experiment. For example, if 15 lanes of a gel are used in an experiment with standards and samples, one may wish to show the relevant lanes along with the intensity, amount, and positional quantitation data as an inset table. Lanes can be cut, pasted, and aligned next to each other to show molecular-weight markers precisely next to a lane of interest. Specific regions of the lanes can be compared across the whole gel and enlarged to show detail. In addition, labels are easily added and the global image brightness and contrast can be improved to show more detail when published.

This very same flexibility in manipulating images presents serious ethical issues. Enhancing brightness can modify the original image data, making subsequent quantitative analysis flawed. While scientific image analysis programs have specific features to prevent modification of the original data, it is still possible to be careless. Cut-and-paste capabilities can create a virtual gel image that is really a composite of several experiments and that can be misleading or confusing without correct lane demarcations and annotation. There are two broad categories of imaging applications: non-quantitative (e.g., illustration and teaching) and quantitative (analysis and scientific publications). General and specific guidelines are given below with the intent of presenting an overview of what is and is not acceptable handling of image data. The underlying theme is to say what you do and do what you say, being careful not to modify your original data in the process. Lastly, whenever possible support the observations with quantitative intensity data.

GOLDEN RULES

For publications, confirm the anticipated publisher's digital imaging guidelines before starting experimentation. Although there are differences in requirements among the various journals, the general rules are similar. These guidelines are described below and are also summarized in Table A.3A.1.

Back Up the Data

Always work on a copy of the original image. The original image, typically acquired through digital scanning or charge-coupled device (CCD) image capture (*UNIT 7.5*), must be archived and stored with acquisition parameter details and referenced back to the experimental protocol (see *APPENDIX 2*). Remember, the time that went into preparing the experiment is significant, and the final result will depend on the acquired image. Ideally, data storage would be on a central server that is routinely

Current Protocols Essential Laboratory Techniques A.3A.1-A.3A.5

Table A.3A.1 General Guidelines for Digital Imaging[a]

Step	Application	
	Nonquantitative imaging for illustration or teaching	Quantitative imaging for scientific journal publication[b]
Film scanning		
Spot/scratch correction	Yes	No
Global changes (across entire image)		
Levels	Yes	Yes
Contrast	Yes	Yes
Brightness	Yes	Yes
Noise reduction	Yes	Yes
Color	Yes	Yes
Composites/photomontage[c]	Yes	Yes
Cut/paste multiple results[c]	Yes	Yes
Local (portion of image) changes		
Levels	Yes	No
Contrast	Yes	No
Brightness	Yes	No
Noise reduction	Yes	No
Color	Yes	No
Labels and artwork	Yes	Yes

[a]Refer to the particular journal for specific guidelines.

[b]Although several types of modifications are allowed, this assumes the quantitation of the image is performed using the unmodified data.

[c]The combination of specific elements needs to be clearly stated in figure legends and methods sections, with the paste/composite element bounded by a demarcation line.

backed up. Critical data should be backed up to least two different locations. (See *APPENDIX 3B* for additional details.)

Quantitate Only from the Original Image

Quantitation of the image should be performed only on the original, unmanipulated image data. Typically this involves defining two numbers: the intensity of the sample feature being analyzed (e.g., a DNA band on a gel) and a background intensity value. These numbers should be noted in the supplementary data or, ideally, in a table along with the size in pixels of the area being analyzed. The background value is subtracted to arrive at the intensity data specific to the feature being analyzed. The intensity data can be correlated to a known standard (e.g., *UNIT 2.2*) as long as the standard was internal to the experiment (e.g., on the same gel as the unknown) and treated the same as the unknown. In fact, if quantitative numbers for the unmanipulated image data are available, then much of the ambiguity associated with visual interpretation of a published image is minimized.

Disclose All Manipulations

Disclose in the materials and methods section or in figure legends all image acquisition parameters available, including the system, camera, bit depth, exposure time, filtration, lighting, and acquisition and analysis software with version number.

Disclose all global (whole image) and local (a portion of the image) changes made to the image, preferably using a history file associated with the image as the guide. Note that local changes are not typically allowed in scientific journals. The history file, useful as supplementary information, is available for many software packages (including Photoshop) and will list modifications made to the image.

Global Manipulations: Alright with Disclosure

Global changes include brightening or darkening the whole image using curves or levels commands. Use caution and carefully inspect the histogram before and after to be sure data are not being lost at the high or low end (see *APPENDIX 3B*).

Local Manipulation: Forbidden

Local (i.e., a portion of the image) changes in the image such as brightening (i.e., dodging) or darkening (i.e., burning) of a feature are not acceptable for scientific publications but may be needed for teaching and illustration purposes.

Cutting and Pasting: Sometimes Alright with Disclosure

Cut-and-paste and clone commands need to be used very carefully and in many cases are not permitted for scientific publications. However, composite images are permitted in most cases as long as there is a clear demarcation of the separate components in the composite and the data are legitimately comparable and collected under identical conditions. The caption must clearly explain that the composite image consists of portions of other images cut and pasted into the composite.

Avoid Lossy File Formats

Avoid using lossy file formats such as JPEG for quantitative imaging (see *UNIT 7.5* for discussion of file formats). The simplest implementation of JPEG can lead to a loss of information and resolution due to compression artifacts. More sophisticated implementations of JPEG have user selectable levels of compression. In situations where the image will be used in a qualitative fashion for PowerPoint illustration and teaching, JPEG may be an appropriate format.

Scanned Film

For the special case of scanning film, some spot and scratch correction on the copy of the scan image is usually appropriate for presentation and illustration, but not for scientific journal publication. Quantitative data analysis for publication should be performed on the uncorrected scan.

GUIDELINES FROM SPECIFIC JOURNALS

A general summary of the digital art guidelines from five respected journals is provided below with information about what needs to be disclosed and what manipulations are and are not appropriate. However, this list is not comprehensive, and prior to submitting a paper to any journal (or even before starting experimentation), it is the author's responsibility to check the detailed current guidelines.

NOTE: Every journal mentioned in this unit requires the author to provide original image files upon request. If they cannot be supplied, acceptance of the manuscript may be revoked (see Golden Rules).

Cell

In the case of image processing, alterations must be applied to the entire image (e.g., brightness, contrast, color balance). In rare instances where this is not possible (e.g., alterations to a single color channel on a microscopy image), any alterations must be clearly stated in the figure legend and in the methods section.

Groupings and consolidation of data (e.g., cropping of images or removal of lanes from gels and blots) must be made apparent and should be explicitly indicated in the appropriate figure legends.

Data comparisons should only be made from comparative experiments, and individual data should not be utilized across multiple figures.

In cases where data are used multiple times (e.g., multiple experiments were performed simultaneously with a single control experiment), this must be clearly stated within each figure legend.

Also see *http://www.cellpress.com/misc/page?page=dataprocessing*.

Journal of Cell Biology/Journal of Biological Chemistry

For microscopy, the journal must know the make and model of the microscope; type, magnification, and numerical aperture of the objective lenses; experimental conditions (e.g., temperature, imaging medium, and fluorochromes); camera make and model; and acquisition software. Additional software used for image processing, with details about types of operations involved (e.g., type of deconvolution, three-dimensional reconstructions, surface or volume rendering, and gamma adjustments) must also be disclosed.

No specific feature within an image may be enhanced, obscured, moved, removed, or introduced.

The grouping of images from different parts of the same gel, or from different gels, fields, or exposures must be made explicit by the arrangement of the figure (i.e., using dividing lines) and in the text of the figure legend.

Adjustments of brightness, contrast, or color balance are acceptable if they are applied to the whole image and as long as they do not obscure, eliminate, or misrepresent any information present in the original, including backgrounds. For example, without any background information, it is not possible to see exactly how much of an original gel is actually shown.

Nonlinear adjustments (e.g., changes to gamma settings) must be disclosed in the figure legend.

Also see *http://www.jcb.org/misc/ifora.shtml#Digital_images*.

Nature

Images should be minimally processed. The final image must correctly represent the original data and conform to community standards. Supplementary methods files listing experimental conditions, instrument settings, acquisition information and software used, and processing changes must be included.

Images gathered at different times or from different locations should not be combined into a single image, unless it is stated that the resultant image is a product of time-averaged data or a time-lapse sequence. Vertically sliced gels juxtaposing lanes not contiguous in the experiment must have a clear line for the boundary between the gels.

The use of touch-up tools or any feature that obscures manipulations is not allowed.

Processing must be applied equally across the entire image, including controls. High-contrast gels and blots are discouraged. Exposures with gray backgrounds are preferred. Multiple exposures should be presented in supplementary information if high contrast is unavoidable.

Positive and negative controls, as well as molecular size markers, should be included on each gel and blot.

Emphasis of one region at the expense of others must be avoided.

Cropped gels must retain important bands. Cropped blots should retain at least six band widths above and below the band. Immunoblots should be surrounded by a black line to indicate the borders of the blot, if the background (marking the edge of the original gel) is faint.

For microscopy: Threshold manipulation, expansion or contraction of signal ranges, and altering of high signals should be avoided. Use of "pseudo-coloring" and nonlinear adjustment must be disclosed. Adjustments of individual color channels should be noted in the figure legend.

Also see *http://www.nature.com/nature/authors/infosheets.html.*

Science

Figures assembled from multiple photographs or images must indicate the separate parts with lines between them.

Linear adjustment of contrast, brightness, or color must be applied to an entire image or plate equally.

Nonlinear adjustments must be specified in the figure legend.

Selective enhancement or alteration of one part of an image is not acceptable.

Also see *http://www.sciencemag.org/about/authors/prep/prep_subfigs.dtl.*

APPENDIX 3B

Practical Considerations When Altering Digital Images

Jerry Sedgewick[1]
[1]University of Minnesota, Minneapolis, Minnesota

INTRODUCTION

Central to any discussion of digital images are the concepts of resolution and resampling. Briefly, resolution refers to the number of pixels used to compose an image. This can be thought of in terms of a mosaic: the more tiles used to compose the mosaic, the greater the resolution. Resampling refers to the process of adding (upsampling) or removing (downsampling or subsampling) pixels from the image to change the resolution.

A3

In the most general terms, the initial resolution is first set when acquiring an image (see *UNIT 7.5*), while resampling occurs during downstream manipulation of the image, e.g., in programs such as Adobe Photoshop. These concepts are discussed in significantly more detail below.

IMPORTANT NOTE: Prior to performing any figure manipulation, review the information presented in *APPENDIX 3A*. Also consult the contributor's guide of the relevant publication. Note that many publishers now require that contributors be able to produce the original, unmodified image in order to publish.

SAMPLING RESOLUTION

In layman's terms, resolution is usually a reference to the ability of an imaging device to discriminate between closely spaced lines or points. The greater the ability to adequately separate two closely spaced lines, the better the resolution of the imaging device, and the greater its ability to reveal detail in fine structures. Much of that ability lies in the optics of the device (*UNIT 7.5*) and in the magnification used when taking a picture.

However, in many instances when the resolution of a device is noted, the reference is not at all to its ability to reveal fine detail. Rather, it is a reference to the device's ability to "preserve" the fine details projected by the optics. In a digital camera, that is done by projecting an image from the optics onto a detector comprised of densely packed light sensors in rows and columns. Increasing the number of these sensors (each measuring microns in diameter) and spacing them more closely progressively increases the likelihood of preserving fine details projected by optics. The most common reference to this kind of resolution is in advertisements of cameras describing the number of sensors resulting in so many "megapixels" (rows of sensors multiplied by columns).

How frequently the detector captures information from the projected image (based upon the number of sensors) is called the sampling rate. The higher the sampling rate, the greater the preservation of fine details, and the greater the preservation of optical resolution. Thus, the preservation of fine details is a result of the detector's "sampling resolution."

Pixels defined

A digital image is composed of picture elements called "pixels" (also see *UNIT 7.5*). These are much like small, individual tiles that make up a larger mosaic. In this example, if each tile could only be one color or one shade of gray (gray level) ranging from white to black, then greater numbers of these tiles would allow greater divisions of details. If the tiles were also smaller, in addition to being present in greater numbers, not only would there be greater divisions of details, but a point would be reached when it might be difficult to make out the tiles as individual entities. At that point, the mosaic would be a continuous-looking image and it would be difficult to recognize individual tiles.

But it is not quite that simple. If the viewer of a mosaic made of thousands of tiles were to move close, at some point the individual tiles could be seen. Conversely, if the viewer of a mosaic with only a dozen tiles moved far enough away, individual tiles would disappear into a larger mosaic. That distance between the mosaic and the viewer is called the viewing distance. The smaller the viewing distance, what is the same as zooming in on a computer, the more likely it will be to see the individual squares that make up the mosaic.

Range/bit depth

Pixels, like tiles, create a larger image from thousands of smaller rectangles arranged in rows and columns. Each pixel is dark to white with all gray tones in between if it is part of a grayscale image, or it is one of hundreds to millions of colors in a color image. Brightness levels at each pixel are predetermined by the digital imaging device (see *UNIT 7.5*). If the image is in grayscale, then each pixel is assigned a single, numeric value commonly in the range from 0 to 255, 0 to 4095, or 0 to 65,535. Camera systems capable of producing these ranges of values and consequent images are referred to as 8-, 12-, and 16-bit, respectively, which are known as the camera's "bit depth." Thus, an 8-bit (or 2^8) camera is capable of assigning a grayscale tone for each pixel from 0 for pure black to 255 for pure white, a 256-scale range (when zero is included).

Note the use of the word "assignment:" not all camera systems use an 8-, 12-, or 16-bit chip. Often the bit depth of the chip is lower than the resulting assignment of bit depth to pixels in an image. For example, a 10-bit chip can be used, but the assignment of grayscale values is in the 12-bit range. Note also that noise from heat and amplification of signal can result in a narrower range of values than what is assigned. A 16-bit camera, for example, may not always be capable of producing values that fill the entire 16-bit range from 0 to 65,535.

Color

Counter-intuitively, the assignment of grayscale values to each pixel is also how color images are produced. Cameras and scanners create color by collecting photons through red, green, and blue filters (see *UNIT 7.5*)—i.e., only photons that pass through the respective colored filters are collected. Thus, red-filtered light sensors collect photons emanating from red objects but not photons emanating from purely green or blue objects. These three colors make up the primary colors of light, as opposed to the primary colors of pigments (yellow, magenta, and cyan).

By combining these individual red, green, and blue images (called "channels"), an RGB Color image is created. Thus, at the positions in which pure red has been collected, there will be an absence of green and blue values.

These individual grayscale or colorized pixels then make up the elements of a larger image. Put another way, because the image is being broken up into more and more parts, the greater the number, the greater the sampling resolution.

Resolution guidelines

No particularly hard and fast limits have been defined for sampling resolution insofar as the number of pixels is concerned, but Table A.3B.1 presents a general guideline.

These guidelines assume that the resolution is relative to the viewing size. When the viewing size limit is exceeded, individual pixels can be seen and the image suffers from what is called "pixelation." Thus, a resolution that is defined as "ultra low" means that to avoid pixelation, the

Table A.3B.1 Image Resolution as a Function of Pixels

Relative resolution	General categories	Pixels in X dimension	Viewing size limit (in.)	File size (grayscale TIFF 8-bit)
Ultra low	Web-based images	<300 pixels in X	<2 (5 cm)	<0.25 Mb
Low	Video, laser/PMT	300–700 pixels in X	3–5	0.25–1 Mb
Mid	Computer screen	700–1200 pixels in X	5–12	1 MB–2 Mb
High	Digital cameras	>1200 pixels in X	>12	> 2 Mb

image must be reproduced at <2 in. In addition, fewer pixels result in fine detail being spread over a larger distance; thus details may not be resolved, e.g., two closely spaced lines can appear as one.

RESAMPLING

To view images at larger dimensions without seeing pixelation, it is necessary to add more pixels. This is equivalent to taking a mosaic made of a few tiles and arbitrarily adding more to make the mosaic larger without adding extra visual information. This process is called "resampling"—a certain number of pixels are resampled upon a larger (or smaller) number of pixels.

It is important to recognize that, at some point, the number of pixels will exceed the visual information available in the optical system. Thus, while two lines spaced closely together can occupy an infinite number of pixels, any more than what is needed for the final viewing distance only bloats the file size.

Finally, as long as the image is sampled upon enough pixels in the *x* and *y* dimensions ("enough" for adding pixels, not for matching optical resolution), it can be upsampled on more pixels ad infinitum to account for increasing viewing distance. The threshold at which "enough" pixels exist for upsampling lies at ~1200 or so pixels in the *x* dimension; with this pixel resolution as the starting point, an image can theoretically be resampled with enough pixels to fill a billboard and it would appear continuous from the highway. On the other hand, if the resolution of the image is less than ~500 pixels, it is difficult to upsample. This is because there are too few pixels at the borders of features to allow duplication. Thus, as these borders are expanded, the image appears blurred. "Sharpness" is the appearance of defined edges, especially at features where contrast is greatest.

Images that are resampled to lower pixel resolutions (called subsampling or downsampling) lose detail that is irretrievable; no amount of subsequent upsampling of pixels will remedy the problem.

Screen resolution

Digital cameras and scanners can image at pixel resolutions that easily exceed the resolution of a computer screen. As a result, the computer screen cannot adequately show detail until the image is seen at a closer viewing distance (zoomed in). This effect is most easily seen when thin, horizontal lines exist on the image. At certain viewing distances (zooms), horizontal lines will disappear as a result of poorer computer screen resolution; as the image is zoomed in, however, the lines can be distinguished.

Choices when resampling images

The whole idea of resampling is something of an anomaly to those accustomed to the Microsoft Office suite and graphing programs. These programs do not operate along the lines of sampling with so many pixels across and down. Instead, these are vector-based programs that operate by using lines and fonts to define how symbols are "traced" when printed. Vector programs produce what are called “scalable” graphics: symbols can arbitrarily be made larger or smaller without any loss of detail because these can be traced (and filled with color) regardless of dimension. Pixel-based programs, on the other hand, come with images and text defined by the number of pixels, and that number is fixed except when resampled.

Printers, e.g., deskjets, that use separated dots to make images (versus those that print by spraying dots, like inkjets, or that function by melting dots) are often made to interpret vector information. Pixels from nonvector programs like Photoshop are therefore poorly interpreted, and result in a loss of resolution. Thus, these images and associated lettering are best printed from Adobe Acrobat or MS Word.

PowerPoint is a mix of both vectors and pixels. The lettering and symbols remain vectors, but the images are defined by pixels. The density of pixels in the images does not create a consequent change in viewing distance, rather, more pixels at defined dimensions produce images at the same dimensions. Thus, when working with PowerPoint, it is best to insert images at defined heights and widths while leaving resolutions with original numbers of pixels. This will be made clearer later when describing methods for inserting images into PowerPoint.

When altering images, as a rule, it is best to keep the pixel resolution the same as that acquired by the imaging system. If the image has to be resampled to conform to the output, save a copy of the altered image at the original pixel resolution before proceeding. This will ensure that visual data is preserved in the event that mistakes are made when resampling.

The greatest mistake made when resampling is to subsample. Most subsampling decisions are made through ignorance of situations in which it is likely to occur. For example, when copying an image in Photoshop and pasting into PowerPoint, the resolution is likely to be reduced to computer screen resolution. Computer screen (or display) resolution is commonly referred to as 72 dpi (dots per inch), taken from the measurement of dots on an Apple 13-in. display. Thus, 72 dpi is often seen as a warning signal that images will be subsampled, and that pixelation will result. There are cases where subsampling is necessay, e.g., conforming the image to the pixel resolutions that work best on a website.

Table A.3B.2 outlines possibilities for subsampling when using different methods.

Note that in Table A.3B.2 the resolution of 72 dpi is most likely to be chosen. Many software writers believe that most users are interested in web placement of images and therefore provide web resolution as a default setting. Users must be educated for the correct choice of resolution, if that choice exists.

Table A.3B.2 Possibilities for Subsampling Graphics When Using Different Methods

Method	Consequent resolution	Subsampled in Photoshop?	Subsampled in PowerPoint?
Cut and paste	Computer display resolution	Yes, resolution of cut and pasted image	Yes, unless “Paste Special” options are used
Saved as TIFF from vector program	Most likely at 72 pixels per inch	Yes, if higher resolution settings were not used	Yes, if higher resolution settings were not used
Saved as JPEG from vector program	Most likely at 72 pixels per inch	Yes, if higher resolution settings were not used, or high compression	Yes, if higher resolution settings were not used, or high compression
Saved as Acrobat file when full version is used	If 300 to 2400 pixels per inch are chosen	No	No

Aliasing

An unfortunate consequence of using pixels to build images lies in the way in which symbols (such as lettering and numbering) appear. Often in low- to mid-resolution images, the rounded edges of symbols appear aliased like the sides of ziggurats, exhibiting pixelation (Fig. A.3B.1). Edges of image features may appear similarly, but this phenomena is overlooked because of both the complexity of edges of features in images, and the relative absence of high contrast between background and features of interest.

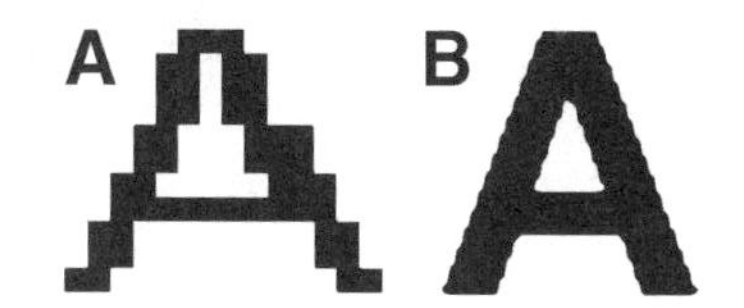

Figure A.3B.1 "Stairstepped" lettering (aliased; **A**) and anti-aliased lettering (**B**). Note ghosting of aliased sides of lettering.

The annoyance at the aliasing of symbols is so strong that its occurrence can lead the observer to conclude that images in the same figure are pixelated, simply by the power of association. In new versions of Photoshop, the problem of aliasing is overcome by keeping symbols vector based, or by anti-aliasing the edges of symbols. In the latter, offset aliased edges are "ghosted" alongside the defined edge to confuse the eye. These extra pixels help to give the appearance of roundness. Aliasing of symbols in Photoshop is overcome by resampling images before adding symbols. Especially when viewing distances are closer, pixel resolutions closer to those required for the output (e.g., publication) provide ample pixel density for anti-aliasing to be effective.

The other solution is to keep symbols as vectors. This will not, however, allow for output to programs other than those made by Adobe. Acrobat, for example, is a good choice for retaining vectors and images; however, these files will not import into PowerPoint nor are most publishers requesting files with images in the Acrobat format as of this writing.

Loss from Image Compression

Resolution has been referred to on the basis of the number of pixels. Resolution can also be considered along the basis of grayscale or color values. On that basis, the grayscale or color values of the original image must be retained to preserve detail in some instances and, in others, those values are altered to correct image artifacts or to reveal details.

The file format that is chosen when saving an image can contain all the visual information from the image, or it can group together grayscale or color values, without changing the total number of pixels, to save a subset of the visual information. Grouping is performed to reduce the file size of the image at the expense of "throwing away" some amount of image data and is referred to as file compression. The fact that a subset of visual information is lost gives it a second reference, "lossy compression." The most common example of a lossy compression file format is JPEG (Joint Photographers Expert Group; Fig. A.3B.2).

The JPEG format can often be saved at different compression strengths, from high to low. Any choice, however, will result in a loss of image data and will likely result in artifacts on the image

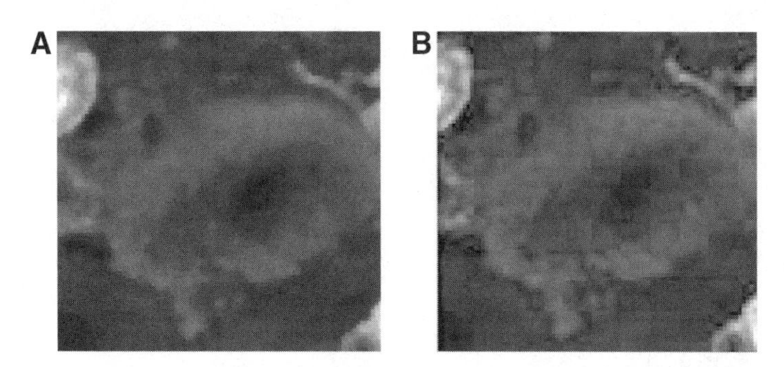

Figure A.3B.2 Noncompressed image (**A**) and JPEG compressed image (**B**). Compression was done in Photoshop at a low setting (for higher compression). Image in **B** was sharpened to aid in showing JPEG compression artifacts.

where pixel values have been grouped. Zooming in to view the resulting image will show these areas. These artifacts can be avoided by saving the original image only in non-lossy formats, such as TIFF (Tagged Information File Format). Duplicated images can the be saved as JPEG or PNG (Portable Network Graphics) files later on, especially when it is necessary to insert images into Microsoft programs (these programs function more reliably when using JPEG files).

NOTE: See the discussion under Digital Image Formats in *UNIT 7.5* for a thorough discussion of image formats.

ACQUIRING IMAGES

Clipping: Exceeding the Dynamic Range

Detail can be eliminated by choosing an exposure that exceeds the dynamic range (the thresholds of the brightest and darkest values that can be reproduced) of the imaging device that is used (see *UNIT 7.5*), by altering the image such that the dark-to-bright range is outside of the dynamic range of the output device, or both. Values brighter or darker than the maximum and minimum thresholds, respectively, are "clipped"—i.e., the value of the clipped pixel will be set at the appropriate threshold and thus will not reflect the true value (Fig. A.3B.3). Clipped sections are distinguished by untextured, flat expanses of pure white, black, or solid color. Like the untextured glare off of a car's bumper, all detail is missing.

The word "saturated" is also used as a reference to this phenomenon. However, while saturated also has a second meaning—i.e., colors that are pure, such as pure green or pure blue—in the context of dynamic range, clipping is a specific reference to those values that measure at the upper and lower limit—i.e., the values of 0 and 255 for an 8-bit image, 0 to 4095 on a 12-bit image, or 0 and 65,535 on a 16-bit image intended for a display on a computer screen. Again, these limits are not always hard and fast: some imaging devices may not operate within the entire dynamic range of their classification.

When acquiring images, care must be taken to avoid clipping of pixels, where detail in those areas will disappear. Often these are the pixels in the brightest parts of the image, and images are arbitrarily brightened to see detail in darker areas. As stated earlier, the human eye is good at dynamically lightening visual information hidden in the dark; the imaging device, on the other hand, shows a linear darkness-to-brightness range. If dark values are found to be above 0, then detail exists within the darker areas even if the human eye cannot readily make them out (especially if the computer screen is relatively dark). Many scientific imaging systems come with a graphic

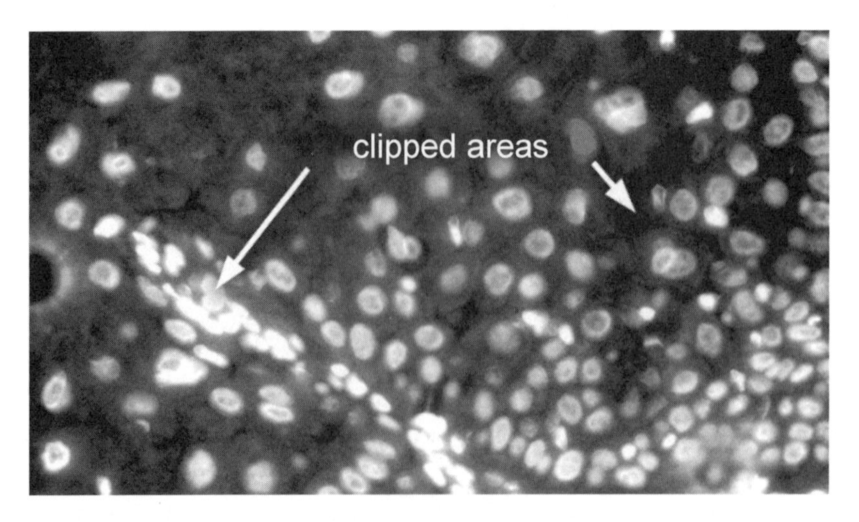

Figure A.3B.3 Arrows point to clipped or oversaturated parts of the image. Note loss of detail in white areas, and creeping of black ink into features in the clipped, black background portion of the image.

overlay to reveal pixels that are clipped during acquisition. This method is preferable to relying upon the computer screen, because although the display can be made lighter or darker by the user, doing so, more often than not, is misleading. When a graphic overlay is not available, and scientific imaging software is used, the darkest and brightest values in significant features can generally be determined by placing the cursor over these areas to see a live, numeric readout.

The same desire to brighten entire images with the result of clipping the brightest values also exists when altering images. Detail in the features that are clipped or saturated simply cannot be retrieved, and visual data and resolution of detail are lost.

Yet some altering of the image is necessary, not so much for reasons of brightening the image to reveal detail in the darker areas, but to fit the image within the limits of the dynamic range of the output device. Each output device has intrinsic limits on darkness and brightness, referred to as the black and white points, respectively. Output devices also have a specific range of colors that can be reproduced, called the gamut (Fig. A.3B.4). Output (and input) devices also have a way of interpreting the colors of the image based upon what is called the color table. Every model of every commercial imaging device comes with a color table, including models of computer monitors.

Because so many models of output devices exist, including a variety of computer screens, the task of matching color and contrast can easily become a full-time occupation. Thankfully, however, some target values can be set to accomplish reasonable reproduction with most output devices. This is done when grayscale and color numbers are relied upon for setting white points, black points, and colors, rather than evaluating the image by eye.

It is important to bear in mind, however, that especially with colors, the gamut of the device limits the colors that can both be displayed and reproduced. For instance, while computer monitors can reproduce many of the same colors that can be achieved on a printing press, some colors, which can be produced on the printing press, simply cannot be displayed accurately on a monitor (e.g., the saturated yellow range), and vice versa (e.g., saturated green and blue). Thus, the expectation of obtaining saturated greens at the brightness levels shown on a computer screen, for example, have to be reconsidered in light of the completely different hues reproduced by printing presses.

If the gamut of the output device is not considered, then certain colors will lose all textural information and become untextured and flat—e.g., while gradations of color can be seen on the computer screen, the output from a printer loses that kind of resolution. For instance, saturated

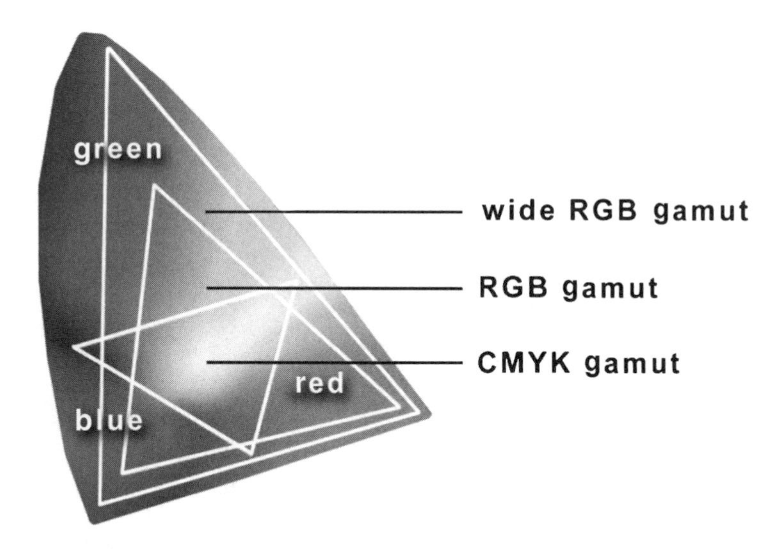

Figure A.3B.4 Areas of primary colors for light are marked red, green, and blue. Note the excessive lack of overlapped gamut in the green hues. Note, also, the variable overlap between what is displayed on a computer screen and what hues can be printed on a printing press in the blue and red colors.

(pure) colors often lead to saturated (untextured) values in the output. Again, ignoring what is seen on the screen and relying on numeric values yields a reasonable, and sometimes better, output.

Needless to say, because of these device-to-device differences, every image needs altering in order to output colors and grayscale values that match what was once seen by eye. Further, images must contain nonsaturated values to begin with to avoid the loss of resolution due to saturation.

Noise and Artifacts

Noise

Resolution can be lost due to the intrusion of noise, and artifacts can draw attention away from relevant details. "Noise" is a way of describing a phenomenon that occurs when images show graininess, or what is called "snow" on the TV (when the signal for a particular channel is weak). When this occurs, images are broken up into thousands of widely variable pixel values, such that a dark pixel can be next to a bright pixel on a feature that should contain constant values. These images are often acquired at longer exposures with higher amplification of signal, typical of confocal imaging devices when detecting weak, fluorescent signals.

Artifacts

"Artifacts" describe detritus in the image: unwanted and unintended particles such as dust, scratches, and fluorescent particulate. It can also be a general reference to shapes of darker or lighter background markings, indeterminate background fluorescence, and just about anything that cannot be explained or that might be introduced by the imaging device.

Color fringing is a specific artifact that occurs at the edges of features where an arbitrary color is deposited (Fig. A.3B.5). The alien color "fringes" the edge of features. This kind of artifact can be hard to see without zooming in, but the images that result from digital color cameras should be inspected for this artifact. Color fringing can be removed to regain original visual data by following the procedure described later in this appendix. The procedure can also be found in published Photoshop books or on the *http://quickphotoshop.com* website.

Removing noise

Removal of noise is a necessary step for revealing details. If possible, reduce noise when acquiring an image to obtain the best resolution (see *UNIT 7.5*). This can be done by acquiring several images and averaging them together, called frame averaging. As long as the noise is random from image

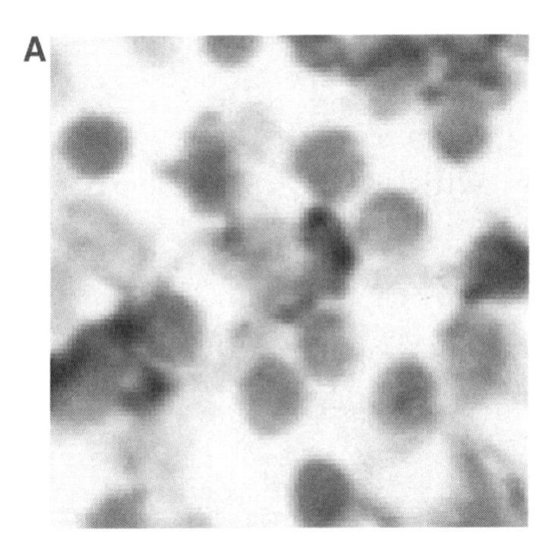

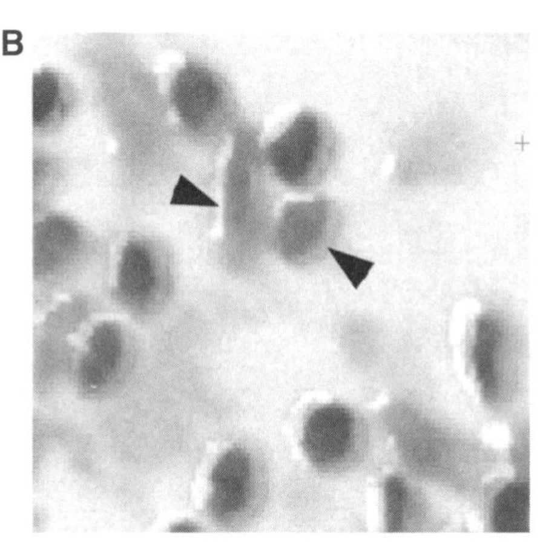

Figure A.3B.5 Image in **A** shows area of a single image in which no color fringing occurs. Image in **B** shows color fringing at edges of cells. These edges were colored cyan at the upper, left area of each cell, and a purplish hue at the lower, right of each cell. Color fringing artifacts disappeared when image was made from a RGB Color image to Grayscale. Fringed areas were amplified and made into distinct tones to show the effect.

to image, then an average pixel value can be obtained at each pixel position. The improvement of signal to noise increases exponentially with the number of images, or frames, that are averaged; thus, frame averaging four images reduces noise by one-half, by 16 images a fourth, and so on.

If noise cannot be removed by frame averaging at the time of acquisition, then smoothing, median, or blurring filters can be used to varying degrees of success after the image is acquired to improve resolution.

Removing artifacts

The issue surrounding the removal of artifacts is more troubling. Because artifacts can be introduced through poor laboratory techniques, bad reagents, and carelessness, it is more useful to control the introduction of artifacts via experimental procedure than image manipulation, for a long-term solution without uncertainties. Sometimes artifacts can be typical of the procedure itself, even if previously published literature and white papers may have had these removed digitally (one can only postulate). A change in the procedure to one that produces similar results, again, produces a longer term solution without uncertainties. Alternatively, the artifacts can be allowed to remain in the image, and can then be explained by honest scientists. If, however, artifacts cannot be controlled via the experimental process (e.g., the experiment would be difficult to repeat), the only solution may be to remove the artifacts. The principal investigator can make this decision; however, if doing so, ethical behavior dictates that the figure legends specifically state that the artifact has been removed and describe what the artifact is thought to be (see *APPENDIX 3A*).

PHOTOSHOP AND SCIENTIFIC IMAGE–ANALYSIS PROGRAMS

Who Photoshop was Made for

In pre-CS3 versions, Photoshop was intended for three major buyers: graphic designers, photographers, and art departments at printing presses. As a result, Photoshop contains features that allow photographers to perform on the computer what was once done in the darkroom, with more precise controls over altering specific parts of an image. It also contains features that allow the user to change black and white points, and colors, to fit the output, again allowing for the isolation of changes to specific parts of the image, if necessary.

Scientific imaging and analysis programs (e.g., Image J, Metamorph, Image ProPlus) on the other hand, often do not provide the ability to alter much of anything in the image except through filters that are applied to the entire image. The intent of these programs is to retain image information and pixel resolution, and change only visual information that improves signal to noise or separates relevant information from background. Scientific programs, in a sense, prevent the user from making the kind of changes that would affect pixel resolution, the relationship of grayscale and color values, and local versus global changes. These tools work very well for scientific images, but are, in effect, useless for conforming images to multiple output devices.

A local change is one that is applied to only one part of the image, such as what might be done if, on an image of an SDS-PAGE gel (*UNIT 7.3*), a single band is selected and darkened. A global change, on the other hand, is one that is done to the entire image, and then to every related image. Using the SDS-PAGE example, while the band might be darkened, it is darkened along with all the other bands, and then the same darkening procedure is applied in the same manner to other gels from the same set of experiments.

While the correct use of Photoshop can prevent the mistakes that are impossible to make in scientific programs, the plain fact that local (versus global) changes can be made makes the use of Photoshop troubling for altering scientific images. For instance, a researcher can cut and paste bands, or minimize undesired fluorescence in tissue. Because of the "local correction" ability, many publishers are using detection software to test images for indications that these have been locally manipulated in Photoshop. It has also resulted in some publishers requiring that gels, blots, and X-ray films cannot contain cut and pasted lanes, as well as any change, global or local, that alters the relationship of gray values (see *APPENDIX 3A*).

For researchers, the rule for use of Photoshop is clear: do not select local areas of the image to make changes. Instead, all changes should be made to the entire, global image, and then to all related images. Clearly, artifacts in the image cannot be isolated and removed when deeming local changes unethical: blurring techniques must be used instead. To be sure about alterations, check with publishers before making changes.

When the rule of avoiding local changes is broken, ethical behavior dictates that the change be clearly described in the publication. Always store the original, raw image in the event that the changes on the altered image are challenged.

NOTE: Refer to *APPENDIX 3A* for more information on ethical considerations when manipulating images as well as examples of specific publisher requirements.

Photoshop as an Image Analysis Program

Just as scientific programs are made for analysis and preservation of raw image information and not for use as an image-editing program, so, too, is Photoshop made for image editing but not for analysis. Plug-ins have been created for Photoshop to allow it to operate as an analysis program; however, on its own, it simply does not have features that make it consistent for measurement of multiple images. The exception is Photoshop CS3 Extended, with analysis features built in.

The most easily misused "analysis" feature in Photoshop is the measurement of surface areas by outlining selected feature(s) with the lasso tool or wand tool, or by using the Color Range, and then looking at the readout of the total number of pixels in the Histogram dialog box. Using this procedure, the measurement of the total number of pixels is achieved by including the squared sides of pixels on the perimeter of the selection when summing. A scientific program would not include the squared sides of pixels in the area: instead, the edges would be determined through a vector algorithm. In that way, the square edges would not be traced around, but, instead, a line would be drawn through the center of the square pixels to better mimic the shape. If a circle were drawn and the edges consisted of small squares, the perimeter that is drawn around the squares versus the one that mimics the shape and is drawn as a circle would then contain more pixels. If relative measurements are being done, that may not appear significant. However, it can matter when relative differences between experimental groups also involve variously sized shapes and lower pixel resolutions. Smaller shapes would yield greater differences than larger shapes, and small shapes with fewer pixels would create even larger differences than what would be obtained in scientific imaging programs.

Photoshop also creates what is called a cache. These are low-resolution images saved in computer memory for use as swap files—i.e., when changes are made to the image at different zooms, the low-resolution image is used in place of the original image to speed up previews of the changes. When the change is applied, it then occurs on the original image. To be safe, the cache should be set to 1 (Preferences>Memory and Image Cache) to avoid measuring low-resolution cached images.

Bit Depth and Photoshop: Not All TIFFs are Created Alike

Perhaps the most evident difference between Photoshop and scientific imaging programs lies in its lack of ability to open some files saved in scientific formats, at least at the present writing. For instance, for microscopists, the inability to open TIFF stacks (a series of images at evenly spaced depths, much like a series of CAT scans) in layers precludes its use for a substantial portion of imaging work (Photoshop CS3 does open image stacks, however).

One would expect that a 16-bit TIFF image would open as these appeared in scientific programs, but that does not always happen in science. Often, 16-bit images are truncated to a 12-bit range, and the display of the image does not resemble the display of the original image in a scientific imaging program. Measurement of 16-bit values in Photoshop CS2 confirms differences.

OPTIMIZING THE DISPLAY

Before altering images in Photoshop, the incorporation of some settings allows for a display that is closer to the actual output and a larger gamut. However, some frequently used saturated colors will never display to match the reproductions, nor will the other colors that should match (because both the computer display device and the output device have overlapping colors) be reproduced satisfactorily.

It is possible to calibrate the monitor to the output device that will be used, but that process involves contacting manufacturers and printing press companies to obtain data against which to calibrate. This involves more time and effort than many laboratories can invest. As mentioned above, many of the saturated colors used in science will not reproduce no matter how well calibrated the monitor. As an alternative, a device-independent approach can be taken. This approach, detailed in Margulis (2006), provides a method that does not take into account models and manufacturers of thousands of devices, but rather uses standard pixel values that will, on the average, work for most outputs. With this approach, the display is not meant to be set to a calibrated standard, but only within a range that is useful for evaluating prints on the basis of a generic output. When it comes to setting a display, the primary idea is to see images darker, so that the user is more apt to lighten shadow levels—the grayscale tones most likely to be lost when reproduced. It also provides the contrast (because contrast is often the addition of darker tones against lighter tones) that is needed when evaluating images by eye, a quality often sought by researchers. Lightened dark tones lead to a desire to darken for the effect of greater contrast, which in turn leads to the complete loss of detail in reproduction: a darker computer screen helps to prevent this tendency.

It is also useful to set every computer display within a workgroup to the same range so that images appear similar from one monitor to the next. Many experts in the color-correction field do not recommend changing the display of liquid crystal display (LCD) flat screens, however, unless these are made for professional use (e.g., Eizo LCD monitors). Consumer LCD screens can be adjusted by changing the backlight, whereas CRTs allow for adjustment of black level and brightness to provide greater contrast. To date, plasma screens for computer displays are not in high use because of cost, and the settings for these are thus outside the scope of this appendix. If only one CRT monitor is available in a workgroup, then that should be the screen upon which images can be evaluated so that consistency can be achieved.

Images are better evaluated by eye when the light level in the room is low. To be truly effective in evaluating color, walls should be colored gray, along with the sides of the monitors and the desk. Low light should be about as bright as darkroom lights so that the user is nearly in total darkness without a window, therefore, the contribution of subtle colors by ambient light sources do not affect judgment of images.

In practice, it is difficult to come by such an environment. It is more important to use a consistent lighting condition than a particular kind of light, though a lamp made specifically for color evaluation is useful for evaluating prints (D50 lights are recommended by ISO standards for color evaluation). A computer screen in a room with windows would be an example of an environment with inconsistent light. The coloring of light changes throughout the day—warmer colors (more red or yellow) in the morning and late afternoon and cooler colors (more blue) in the middle of the day—making room lighting extremely inconsistent.

USING IMAGES FROM VECTOR PROGRAMS AND POWERPOINT

Extra care must be taken when using images that originate from vector programs or from a vector/image program like PowerPoint. It is best to avoid the use of any images that originate from PowerPoint, and far better to use the image in a bitmap format (such as TIFF; see *UNIT 7.5*). The easiest method is to copy and paste the image into Photoshop, however, the copy-and-paste method is subject to subsample, and is not advised.

Instead, resolution of images can be maintained when images are either saved in a pixel-based format (ideally, the TIFF format) or if images are printed using the full version of Adobe Acrobat.

Saving images from PowerPoint and vector-based programs

Care must be taken when saving images in a pixel-based format from PowerPoint and vector-based programs when attention is not paid to correct settings, as incorrect settings may lead to subsamples and pixelated images. While an option is to save in a pixel-based format (as a JPEG file), the default settings will not produce an image with adequate resolution. If a button for settings or options is available when saving to a pixel-based format and the ability to set pixel resolution is available, then choose the correct resolution. Again, these are set at 300 or 400 dots or pixels per inch for images (check with the author's guidelines of the desired publication), and 1200 dots or pixels per inch for graphs and drawings. Note that these options are not available when saving in the native Acrobat format on Macintosh OSX. The dimensions can be set for the publication: an image that would take up a column in publication is ∼3–in. wide; a half page is 7.5-in. wide; and a full page is 7.5 × 9 in. (w × h); be certain to check the publication of interest for column and page widths. If the final dimensions are unknown, erring on the side of too much pixel resolution, the image can be saved at a nominal width of 7.5 in. at the largest axis.

More pixels may have been added to the originally composed image. Because these programs do not always provide information about pixels, it cannot be known that original pixel resolutions are maintained, but the greatest likelihood is that original pixel resolutions are not maintained.

With PowerPoint, for example, the default size of the template in a slide show is 7.5 × 10 in. The interpretation of how much space will be filled on the slide template does not depend upon pixels across and down, nor upon the interpretation of the image resolution in pixels per inch, but only upon the interpreted dimensions of the image. Both "pixels or dots per inch" and "height and width in inches or centimeters" is encoded in the image file. Except for images taken by a scanner in which there is a known dimension (the physical dimensions of the platen), images are, in effect, dimensionless. PowerPoint cannot interpret the dimensions or resolution (as specified by pixels or dots per inch) unless the user specifies it at acquisition (if available) or later on in the kind of software in which it can be specified (like Photoshop: see Image Size, below).

The net effect is that high-resolution images often overfill the PowerPoint slide template, and users scale the images to smaller dimensions and consequent resolutions; when such images are rescaled to fit publication requirements, subsampling and loss of detail results.

Printing images to Adobe Acrobat

The creation of TIFF files, even if done correctly by saving at 300 pixels per inch at specified dimensions, may yet suffer from aliasing of lettering and other symbols. Many vector-based programs do not include the option to anti-alias symbols. Furthermore, the option to save at 1200 dpi is included with the options, providing crisp lettering sans pixelation. For those reasons, it is preferable to "Print" to Acrobat. Acrobat provides anti-aliasing functionality as well as options to save at higher pixel resolutions.

It may seem odd to use the Print command to create an image file, but an Acrobat file is best thought of as what would be printed were it sent to a printer, thus explaining why so few editing features are available in Acrobat. It also explains why the file can be opened without loss of any visual information across computer platforms. Fonts that print to Acrobat are embedded in the program to make it "fontless," as though it had been printed on a sheet of paper.

The option for printing to an Acrobat file is available through the full (not free) version of the program. Freeware is available for accomplishing the same end, but the only guarantee that large file sizes can be maintained comes with purchasing the full version of Acrobat. It is well worth the investment.

When printing to an Acrobat file, be sure to choose Options from the print dialog box (Fig. A.3B.6). The default setting will be for low resolutions that conform to the web. In newer versions of Acrobat, choose High Quality for images, and Press Quality for graphs and drawings. The Acrobat file can then be saved as a TIFF file, or kept as an Acrobat file to be opened in Photoshop and saved in any number of formats.

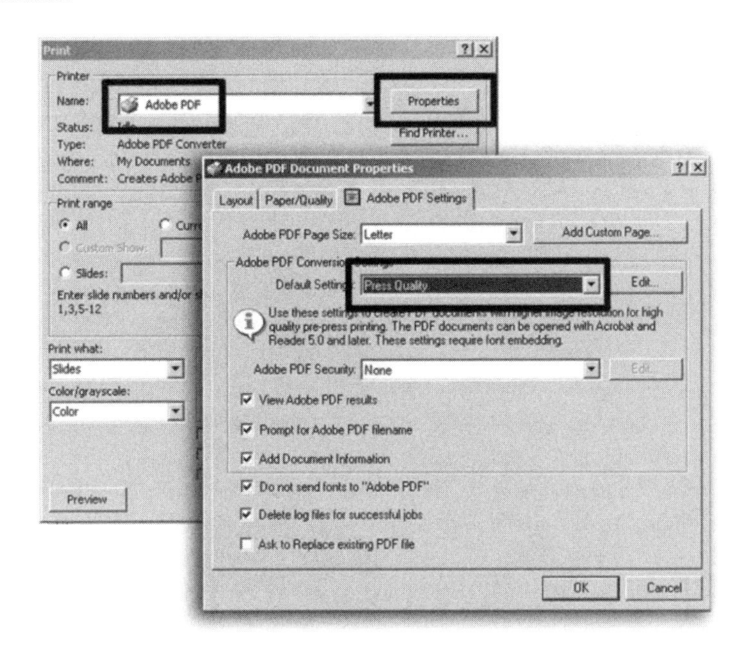

Figure A.3B.6 Images show two dialog boxes that appear when creating PDF documents on Windows with Adobe Acrobat 7 Professional. Note that the document is created by printing, not by saving or exporting. Click on "Properties" to open tab for "Adobe PDF Settings," and then choose "Press Quality" from the drop down list for best reproduction of both lettering and images.

ALTERING IMAGES USING PHOTOSHOP

Suggested Order of Operations

Once the Photoshop environment has been set up, then global image alterations can proceed. The recommended timeline for those changes is shown in Figure A.3B.7.

Note that setting pixel resolution is near the end of the alteration process so that pixel resolutions of the raw image can be maintained throughout the process. It is crucial to duplicate the raw image at the beginning so that it can be retained in its raw form. Raw images should be saved to at least two CDs or DVDs: one for the laboratory member and one for the group leader.

The order is set to make the process of alteration more efficient. Sharpening/blurring and Color Correction/Matching are set before Setting Black/White Limits and Midtones because the former alterations often brighten or darken pixel gray and color values. Other alterations can be out of order, depending on the circumstances.

Table A.3B.3 presents an overview of the features commonly used in Photoshop to alter scientific images.

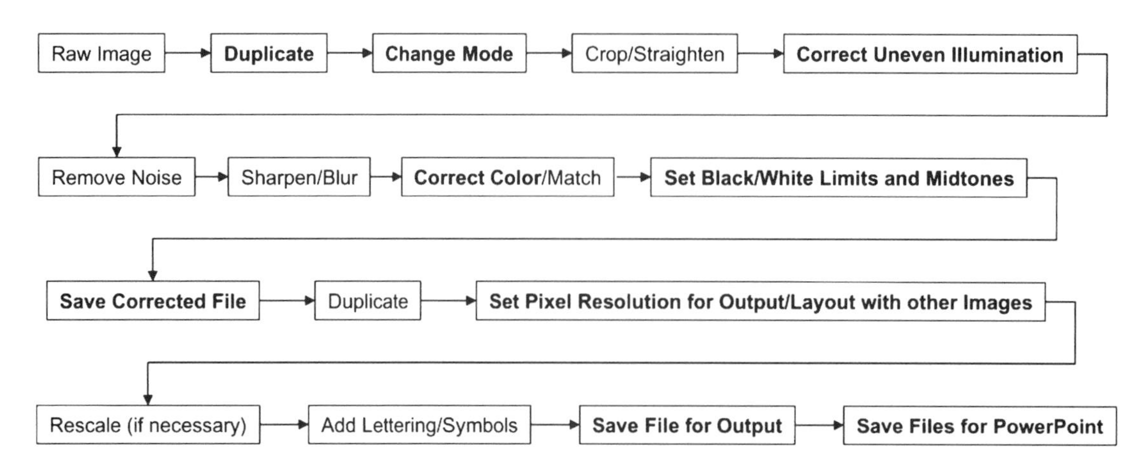

Figure A.3B.7 Recommended timeline for global image manipulations. Manipulations shown in bold are described in this appendix; other manipulations are beyond its scope.

Table A.3B.3 Overview of Features Commonly Used in Photoshop

Feature	What it does	Why it is useful	Potential problems with its use
Flatfield correction	Corrects uneven illumination	Essential for accurate optical intensity/density measurements and other quantization	Image degradation can occur with too much correction shown as a loss of contrast
Mode	Changes the components of images	Necessary for functions to work in Photoshop when Indexed Color	None
Bit depth	Changes the number of grayscale divisions in an image	Images need to be re-scaled to 8-bit for outputs: for 12-bit re-scaled to 16-bit	Misinterpretation of 12- and 16-bit images with certain manufacturers
Color correction	Removes overall shifts in hue toward a color	Returns image to its correct colors	When determining white background for auto correction, correct position must be found
Setting black and white limits	Conforms dynamic range of image to the output	Retains details in the brighter and darker regions of images	The setting is generic: setting may be too liberal or conservative
Brightness	Amplifies mostly lighter values in the image to brighten overall	Reveals details hidden in darkness	Images can be made too bright so that values clip
Contrast	Introduces deeper black values	Makes brighter features more apparent	Can hide significant features in shadows
Gamma (midtone correction in Photoshop)	Generally used to lighten darker values without affecting whites	Makes images appear in conformance to human vision	Changes the relationship of grayscale values: cannot be used with electrophoretic samples
Image size	Changes pixel resolution and image dimensions	Conforms images to outputs at optimal resolutions	Subsampling so that visual data is lost
Saving JPEGs	Reduces the file size of images	Works well in Microsoft products	Higher compression removes visual data

Cropping/straightening, noise removal, sharpening/blurring, color matching, rescaling, and lettering/symbols are outside the scope of this appendix.

Duplicating and the Image Window

Few comments are more frustrating than hearing that the raw image either cannot be found, or is irretrievably lost. Good laboratory practices include means to ensure that raw images are saved and not "saved over" by altered images. Writing raw images to a CD or DVD ensures an archived copy for at least 5 to 10 years.

In practice, the creation of a CD or DVD may not have been done because of deadlines, forgetfulness, and other factors. Thus, it is of crucial importance to duplicate the image file before doing any alterations. The added benefit is that the image is renamed by Photoshop with the word "copy" in the name of the image as a way to alert the user to the fact that it is not the raw image. To do this, under Image, select Duplicate and save in an appropriate location.

The image on the screen is surrounded by a frame called the "image window." At the top of the image window, the name of the image, the percentage of its ratio against the screen area of the monitor, and mode of the image/bit depth are shown (Fig. A.3B.8).

The percentage is misleading insofar as it is not a reference to how big the image is in inch or centimeter dimensions, but rather is a reference to its size on the screen in relation to the "screen area;" settings are found in the control panel. For example, if the screen area is set to 1024X728 pixels, a 1024X768 image will be at 100% when it fills or slightly overfills the screen (some pixels cannot be seen due to the physical space the Photoshop window takes).

Measurements in pixel units aid web designers because these percentages show the dimensions of the image for targeted screen areas. When the image is intended for the web, then these percentages are useful: for other outputs, percentages as a way of determining output dimensions are often incorrect.

Note that Photoshop opens images at the screen area resolution and above at <100% on the screen. This is done to allow the user to see the entire image from edge to edge. If image area is added using the Canvas command, or if the image is resampled to include more pixels, Photoshop does not automatically "fit" the image to the screen: the user must zoom out to see the entire image.

Mode and Bit Depth

The mode of the image is a reference to its composition. The mode for scientific images generally includes Grayscale, RGB Color, or Indexed Color. Grayscale is composed of a single channel and

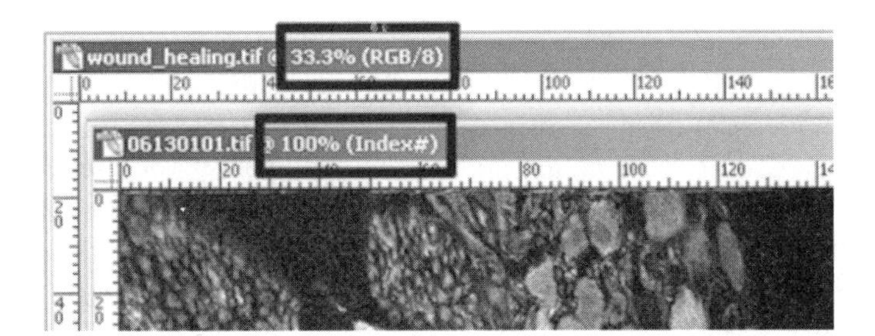

Figure A.3B.8 Two examples are shown of information presented in the image window in Photoshop. The top image shows that it is open at 33%, a reference to the size of the image in relation to a legacy Apple computer screen. It is not a reference to the dimensions of the image in inches or centimeters. The top image window bar also shows the Mode of the image, in this instance an RGB Color image at 8-bits per channel. The bottom bar shows a Mode of Index#, indicating that the image contains 256 levels of gray or of colors. That mode of image can be altered with full functionality of Photoshop after conversion to RGB Color or to Grayscale.

as many gray tones as the bit depth allows (see Bit Depth, above). RGB Color is composed of Red, Green, and Blue channels, as discussed previously (see *UNIT 7.5*). Indexed Color is either a colorized image, or a grayscale image, comprised of 256 color or grayscale levels. These images have been assigned gray or color levels based upon brightness to darkness levels in 256 “steps.”

Indexed Color mode is used fairly often in scientific imaging. When 256 levels are used, then known values can comprise an image, aiding in consistent visualization and measurement. In Photoshop, indexed color is primarily used for web-based images, often saved in the GIF format. Functionality is reduced in Photoshop for altering indexed color images, and so it is best to change the mode to either RGB Color (if it is a color image or intended as a color image) or Grayscale.

Bit depth has already been discussed, along with issues related to a truncation to 12-bit of pixel values when it is a 16-bit image. As far as 16-bit images are concerned, even when the image appears completely black (it could be the image of a control), it can be difficult to ascertain whether or not all the gray values have been included. There is a method, however, that will aid in this determination: looking at its histogram to view the range of gray values.

The histogram, located under Image in pre-CS versions of Photoshop, and under Window in CS versions, shows the range of pixels from 0 to 255 contained in the image on the x axis, with its summed occurrence on the y axis. Fluorescent images, as well as images with broad, white backgrounds generally take on a Poisson profile given the collective occurrence of large areas of black or white background. That appearance belies what is seen in many published histograms of everyday images out in the “real” world, in which the histogram is Gaussian. When the occurrence of pixels at any x position on the histogram is below a relative threshold, it does not appear at all, which can mislead the user into believing that those gray levels do not exist in the image, or that these values are not read by Photoshop. If the histogram could be enlarged, it would reveal that pixels exist at those gray levels.

Note that the histogram is a readout in 8-bit gray values (0 to 255) versus 16-bit values when the histogram is shown for a 16-bit image. Again, the histogram is misleading: it can be relied upon for its visual information insofar as the histogram will show the grayscale levels along a 16-bit x axis (0 to 65,535), but the numbers found along the histogram are 8-bit. The visual information does not match the numeric information. The eventual output for the image is inevitably 8-bit, so the readouts in 16-bit does not provide the information needed for the output.

The histogram, then, is best used for its visual information: if all the grayscale levels in x are shown at the extreme (the left end of the histogram where levels are near zero), then, very likely, the image has been truncated to 8-bit values (Fig. A.3B.9). As another more obvious clue, the image will appear completely black, even though white values are contained in it: the image is not a confirmation that the imaging system did not work.

Every Photoshop manual, including Photoshop’s help manual, will recommend that 16-bit images be changed to the 8-bit mode, but only after alterations on the image are made. These images, though opening as 16-bit, really are not 16-bit, but are interpreted that way by Photoshop. Therefore, these images must be fit into the 8-bit range of values first, and then the mode can be changed to 8-bit. If this is done in reverse, Photoshop re-scales the image incorrectly, according to the way the histogram looks. When the image is then seen in 8-bit, it is still dark and the range has to be expanded. Gray values were eliminated when the mode was changed from 16-bit to 8-bit, and a condition known as banding results visually. That is when gradients between gray levels disappear, replaced by distinct edges that look like outlines on a topographical map (Fig. A.3B.10).

The procedure, then, is as follows:

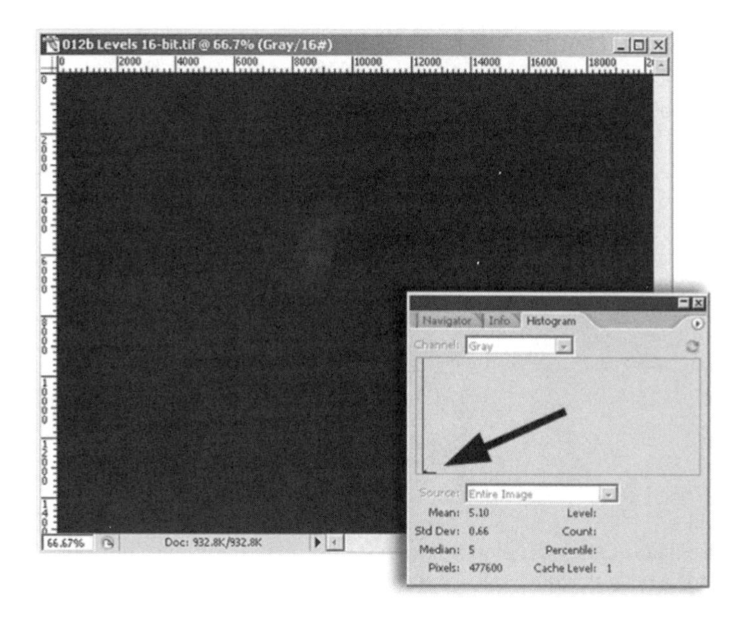

Figure A.3B.9 A completely black image contains detail, as shown by the histogram in the histogram dialog box (arrow). In this instance, it is an image converted from 12-bits to 16-bits in software that came with the camera. The 16-bit image was opened in Photoshop CS2 on Windows, only to contain an 8-bit readout of grayscale values, though the image Mode is yet considered to be 16-bits.

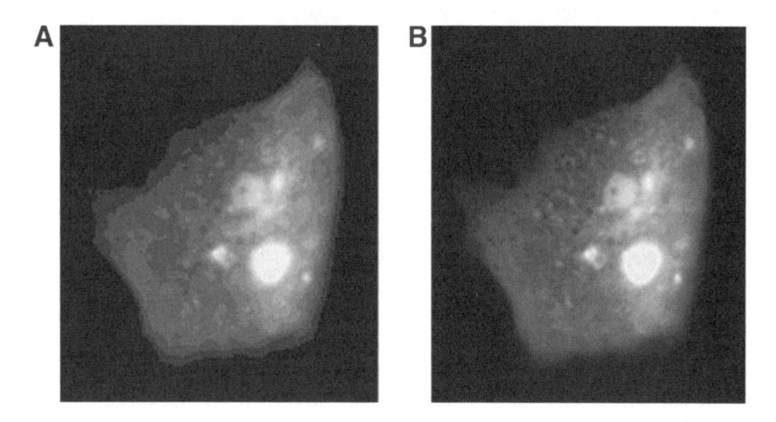

Figure A.3B.10 Image in **A** has been converted from 16-bits to 8-bits using the conversion option under Image>Mode>8-bits/channel. Then the image was brought back to values that existed in the original display using Levels. Note extreme loss of gradients in the gray values: instead, a topographical look containing limited and separate values results. Image in **B** was corrected in Levels first, and then converted from 16-bit/channel to 8-bit/channel using Mode. Note the absence of banding.

1. Duplicate the image.
2. Under Image>Adjust(ments), choose Levels. In the Levels dialog box, move the white slider until the brightest values readout at values no higher than 240.

   Generally, this occurs near the readout of 16, found in the rightmost box adjacent to Input Levels in the Levels dialog box. The readout of the whitest values can be found in the Info Box (Windows>Info) when moving the cursor over the brightest significant areas. If the image is in color, use the brightest of the red, green, and blue channels (the max value) for the correct setting. The brightest significant area can be found by eye: several areas may need to be examined before the brightest position is found. Finding that location on the image is not critical: later corrections will ensure that the brightest significant value in the image is at a 240 grayscale value.

   See Setting White/Black Limits in this appendix for explanation of the white limit.

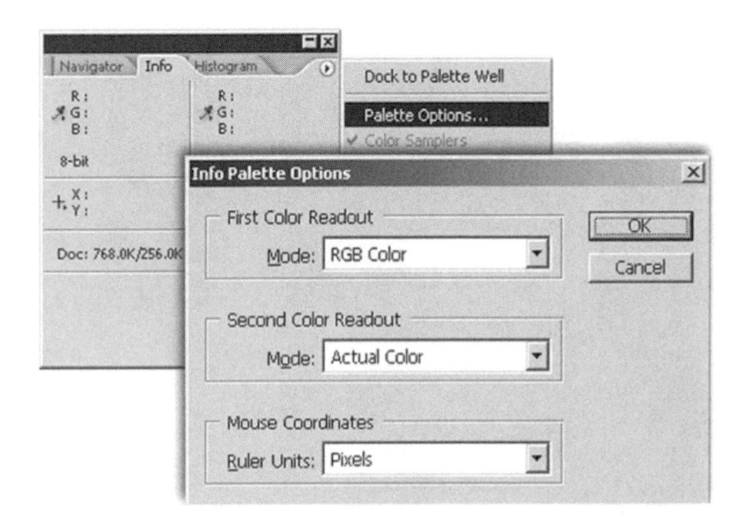

Figure A.3B.11 The Palette Options dialog box can be accessed from the Info Box by clicking on the top, right arrowhead to reveal a drop down box. To guarantee a display of RGB values (when the image is an RGB Color image), that is chosen as the first readout. The actual color is chosen for the second readout. Pixels are chosen for the unit. Note that Grayscale images will show only the percentages of black ink used when publishing to paper output. This is shown as the K: value.

3. If the Info Box is not visible, click cancel, and then open this dialog box. Be sure that the Palette Options in the Info Box are set to show RGB values.

   To show RGB values, click on the upper right-hand arrowhead to reveal the drop down box, select Palette Options, set First Color Readout Mode to RGB Color, and the Second Color Readout Mode to Actual Color (Fig. A.3B.11).

4. When reading out values for a grayscale image saved in RGB Color mode, use the RGB readouts versus the second readout: the K value.

   Readouts for grayscale images will show the same value for R, G, and B. The K value is a reference to the percentage of black ink that will be used at a printing press. A 0% K value indicates that no ink will be placed on the paper; a 100% K value means that ink will completely saturate the paper at that point in the image.

5. Once grayscale levels are expanded, change the bit depth by selecting Image->Mode, 8-bits per channel.

Uneven Illumination Correction

With the possible exception of scanners and laser scanning systems (with properly aligned mirrors), it is safe to say that nearly all scientific images are unevenly illuminated. Even when the imaging system itself provides even illumination, samples can contain backgrounds that lead to uneven illumination. For example uneven illumination is inevitable on the lanes of samples on nitrocellulose.

Uneven illumination problems can be eliminated in many digital imaging devices, most often in systems that use cameras. Depending on the manufacturer, within the software provided with the camera, the means exist for acquiring what is variously called a "flatfield," "shading," or "blank-field" image. Such an image is one that is acquired with the sample removed so that only the lighting is recorded. Because the background is dark when fluorescent samples are removed, a fluorescent reference slide replaces the sample to provide the flatfield image. This flatfield image, along with a background image taken with ambient light in the environment with the sample removed, is numerically divided into images from samples after the image is acquired. In that way, the illumination is made even.

When that correction is not available, or when it was not used, illumination correction can be done in Photoshop by either using the image itself as the flatfield image or by using a second, saved

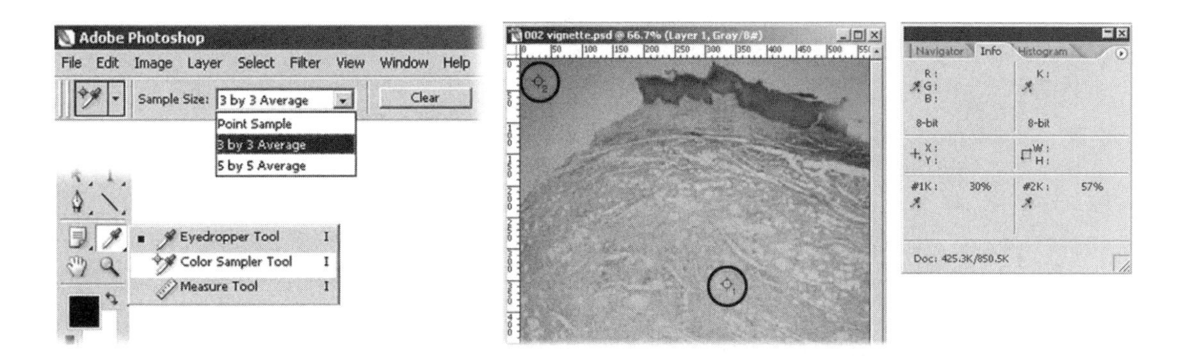

Figure A.3B.12 The Color Sampler tool is revealed from the toolbar by clicking and holding with the mouse on the eyedropper icon. Set the Sample Size to 3 × 3 or 5 × 5 from the submenu (in versions later than Photoshop 5.0) so that pixels surrounding areas subsequently chosen will be averaged together to obtain values. Once the brightest and darkest significant areas are chosen by clicking on the image with the Color Sampler tool, the numbered values appear at the bottom of the Info dialog box.

flatfield image. Flatfield correction can also be done by filtering, but only with samples with large expanses of constant gray values, such as those found in gels (e.g., *UNIT 7.2*), phase contrast, and differential interference contrast (DIC) images (*UNIT 9.1*). All three methods are outlined below.

Uneven illumination correction using the sample image (Leong et al., 2003)

1. Duplicate the image and close the original.

2. Make sure the Info Box is visible on the screen.

3. Place sampling markers on the image to mark the position of the dark part of the unevenness and the light part (Fig. A.3B.12).

   The darkest and brightest parts may be so far apart in values that some intermediate parts should be chosen for either the brightest or darkest value.

4. Click and hold the eyedropper icon to choose the color sampler tool.

   Pre-6.0 versions of Photoshop require double clicking on the tool to obtain more options: other versions show options in the submenu.

5. Choose a 3 × 3 or 5 × 5 sample size, depending upon the amount of noise of the image.

   Determine the amount of noise of an image visually: either choice is better than the default Point Sample.

   The positions marked will show up in the Info Box, marked by numbers from 1 to 4. There is a limit of four sampling positions.

6. Select the entire image by choosing Select All under the Select menu option. Under the Edit menu option, select Copy, then Paste.

   This will place the image on a layer above the existing "background" layer. Alternatively, the background layer can be dragged into the page icon in the Layers dialog box to be duplicated, or the choice can be made by clicking on the top, right arrowhead to reveal "Duplicate Layer" from the drop down list.

   The idea of "layering" images may be new. In Photoshop, layers can be created, one above another, much like sheets of paper (Fig. A.3B.13). Layers are not transparent until choices are made in the Layers dialog box, and so the creation of a layer is not apparent by eye. The layers can contain varying levels of transparency, depending on how the layer is blended with the layer below it.

Blending

Blending is achieved by changing the layer mode, or by adjusting the opacity slider. The "mode" refers to a list of layer blending options, each a mathematical formula that blends the layer of interest

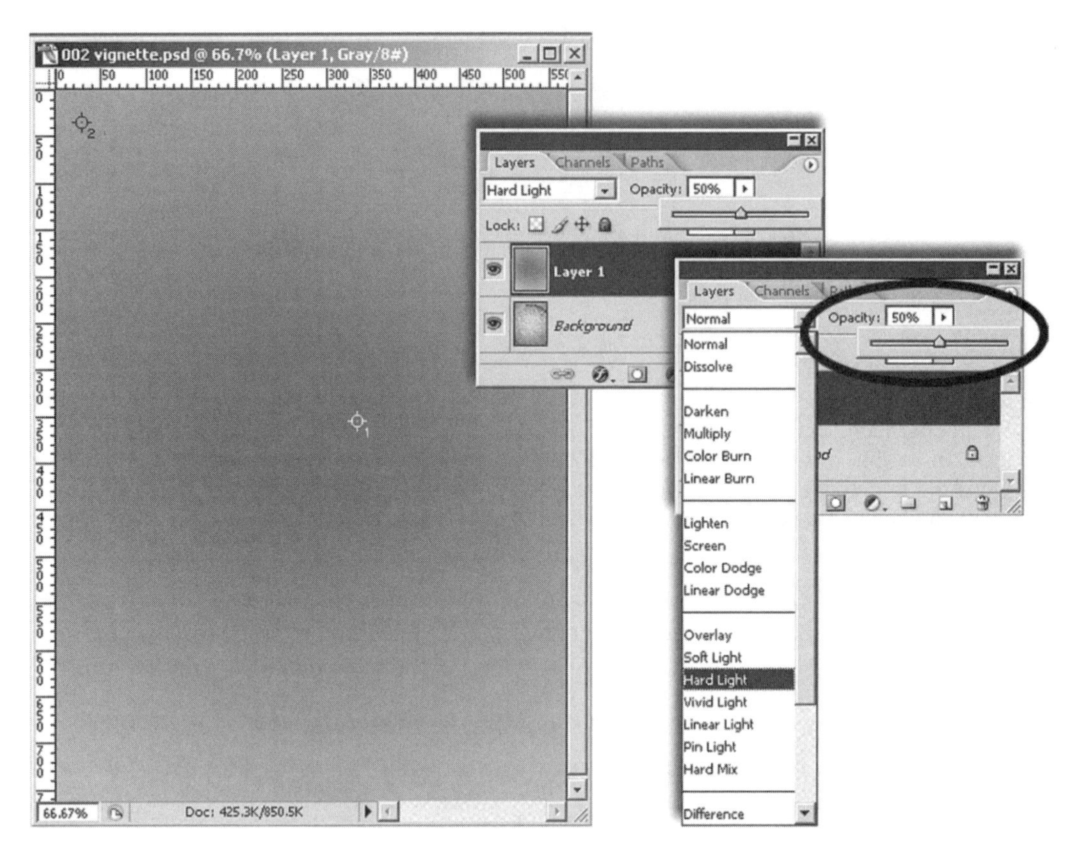

Figure A.3B.13 When layers are created on an image, the topmost layer "blocks" the view of other layers when the layer mode is set to Normal (the default setting). Varying levels of transparency can be obtained by clicking on the arrowhead adjacent to "Normal" and then choosing from the drop down list. In this instance, "Hard Light" is chosen. The opacity of the layer can be subsequently altered using the Opacity slider, in this instance to 50%.

with the layer below to achieve the desired results. The layer mode is physically located under the Layers tab at the top, left of the dialog box. The Layers dialog box is revealed by choosing Layers (or Show Layers) under Window in the menu. The raw image takes the bottommost, background layer, and in later versions of Photoshop, this layer is locked so that few alterations on the image can be made. This provides a way to preserve the original, but it is not foolproof: duplication, on the other hand, guarantees preservation of the raw image.

The background layer cannot be unlocked, but it can be duplicated by dragging the layer into the page icon. (Be sure to click on the area of the layer that does not contain text.) This will duplicate the background into an unlocked layer. Then the background layer can be dragged into the trash icon, if desired. Alternatively, the background layer can be duplicated and deleted by clicking on the top, right arrowhead and choosing either from the drop down list.

1. From Image>Adjust(ment), select Invert to invert the values of the image.

   This will reverse a brightfield image to make it dark and a darkfield image will be made bright.

2. Under Filter, select Blur and then Gaussian Blur. Set the slider in the Gaussian Blur dialog box to ~120.

   The degree of blurring can affect the correction of uneven illumination, but, in practice, it is not as important for minimally affected images as it is for images with greater degrees of uneven illumination. For highly affected images, an iterative approach will need to be taken in which varying degrees of blur are chosen and then evaluated in the next step.

3. From the Layer Mode, choose Hard Light. Then, using the opacity slider, set the opacity, at first, to 50%.

4. Look at the sampled points in the Info Box to determine if these are within 1 to 5 gray or color levels of each other. At the same time, evaluate the image visually to determine if illumination is evened to satisfaction.

   Generally, a 50% opacity corrects most images. If, however, the image both appears uneven and the numbers are spread too far apart, then adjust the opacity until the sampling numbers are closer together or the same. Depending on the computer and operating system, the numbers may not change as the opacity slider is moved without pausing to allow system to update the numbers. At other times, the degree of unevenness is so great that the image cannot be fully corrected. In the latter instance, consider cropping the image to select only the even portion.

Uneven correction using a saved flatfield image

To correct uneven illumination when a flatfield image has been saved at that imaging session, follow the "Uneven Illumination Correction Using the Sample Image," except that the flatfield image is substituted for the layer above the background layer. The flatfield image does not need the same high degree of blurring, however.

1. Duplicate the raw image and close the original.
2. Set sampling points as described above.
3. Open the Raw Image, and then the Flatfield image.
4. Select All (Select>All), Cut (Edit>Cut), then close the flatfield image. Paste (Edit>Paste) the flatfield image on the raw image.
5. Under Image>Adjust(ment)>Invert.
6. Under Filter>Blur, select Gaussian Blur. Set slider so that dust or other detritus on the flatfield image disappears, often set at <5.
7. In Layers box, set layer mode to Hard Light. Set opacity of the flatfield layer to 50% while looking at sampling points. Re-set, if necessary, depending both upon visual look of the image and readouts from the sampling points.

Uneven illumination correction using filtering

If the image contains broad areas of similar grayscale values, such as what would be found with images of gels, blots, phase contrast, and DIC (differential interference contrast), then a High Pass filter can be applied to correct uneven illumination.

1. Duplicate raw image and close the original.
2. Set sampling points.
3. Under Filter, choose Other > High Pass.
4. Carefully set the radius of the high-pass filter so that background is made even, but, at the same time, detail is not lost in significant features. Also look for banding artifacts by zooming in to examine carefully by eye, especially when the sample comprises molecular weight bands.

   If the image is a gel, blot, or X-ray film, and it is destined for publication, make certain that the journal to which it is sent allows manipulation in Photoshop: generally, it is not allowed. In any case, indicate in the caption that Photoshop was used to even out background values.

Color Correction

Because of color shifts with digital cameras, the raw image is rarely balanced for color. Even with careful white balancing, shifts can still occur. Furthermore, red values can saturate—i.e., become more pure in hue so that these look neon colored because of increased red sensitivity of the sensor.

The output image will create further color shifts, loss of color saturation, and gradations, depending on the output device. When outputting for publication, researchers must rely upon the art department at the publishing press to adjust colors. While some art departments are good at color adjustment of scientific images, others are not. In any case, rather than rely upon the color balancing expertise at presses, an image adjusted within a general target for the press will make their color adjustment less difficult.

Brightfield images

Brightfield images are less prone to contain colors outside the gamut of printing presses, and so the color correction is generally not as extensive or difficult as dark-field images. When the following steps are taken, the likelihood of gamut problems are less likely to occur, depending on the stain.

Correct the hue shift

1. Open image, duplicate, close raw image.
2. Under Image>Adjust(ments), select Levels.

Set limits (Fig. A.3B.14)

1. In the Levels dialog box, double click on the black eyedropper icon. If "Only Web Colors" is checked, uncheck. Set the Red, Green, and Blue values to 20.
2. Double click on the white eyedropper icon. Set Red, Green, and Blue to 240.

   You will likely get a prompt asking if these are the default values. Click okay so that these only need to be set once. If you are using a shared computer, occasionally check these values in the event that they have been changed.

3. Do one of two things to correct hue shift:

   a. Click the Auto button.

   b. Click on the white eyedropper tool, then click on a white, background area of the image (Fig. A.3B.15).

   You may have to click in more than one place until the image appears balanced. Do that by eye. When the ideal place is found, then click Okay. Depending on the image, the Auto button may or may not provide accurate correction. If Auto does not correct hue shift, use the white eyedropper method.

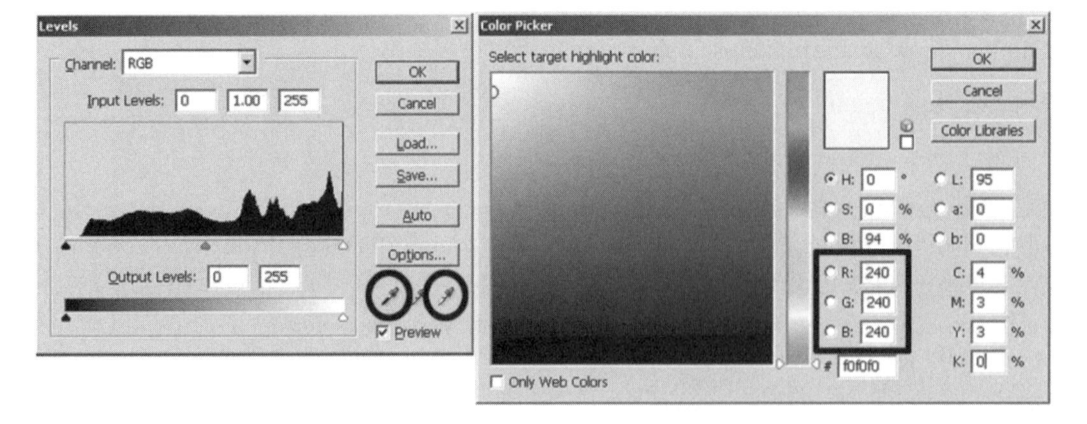

Figure A.3B.14 Limits to maximum white and minimum black values can be indicated in the Levels dialog box. In this instance, the white limit is being set by double clicking on the white eyedropper tool and setting the red, green and blue values to 240. That setting is generally within the range of most printing devices.

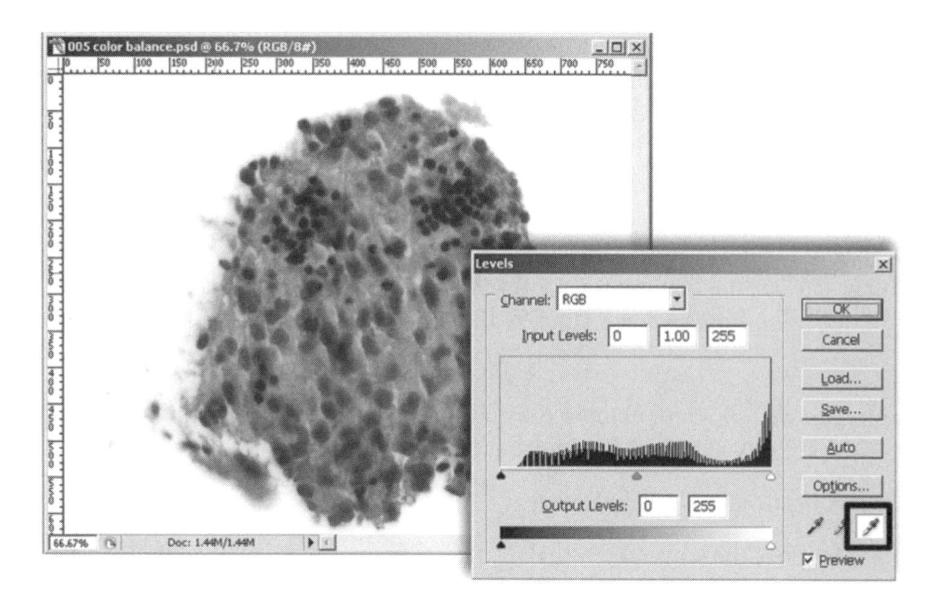

Figure A.3B.15 The white eyedropper tool can be chosen in the Levels dialog box, and then a white portion of the image can be clicked to auto-correct color. Several white areas may need to be clicked on until the image is fully corrected. Evaluation is done by eye.

Neon colors

Especially with eosin, the colors may appear to be incorrect. These colors are not necessarily incorrect but appear to be due to over-saturation. The hue is correct, but the intensity is too great. To correct for saturation, desaturate the offending color(s).

1. Under Image>Adjust(ment), select Hue & Saturation.
2. In the Hue & Saturation dialog box, choose the desired color.
3. Using the Saturation slider, desaturate by moving the slider to the left until the color appears correct by eye.

   The color can also be checked against whether or not it will reproduce by a printing press. By pressing the Shift + Control (or Apple Key) + Y keys, a gray color (if the default setting is used) will cover all areas that will not reproduce (Fig. A.3B.16). The saturation slider can be moved until the gray disappears (some gray areas may exist in darker areas, but if these are not significant values, then ignore). In that way, a more objective approach can be taken.

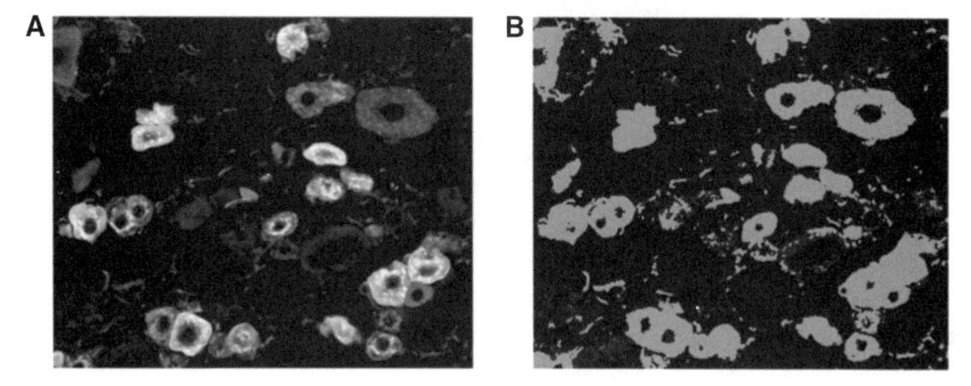

Figure A.3B.16 Image in **A** was blue colorized in confocal software with a standard blue LUT (Look Up Table) setting, then made into grayscale. Image in **B** shows solid gray toned areas that will not reproduce accurately. Note that all features have the overlay, indicating the lack of overlapped gamut values between a computer screen display and a printing press output.

Fluorescent images

When fluorescent dyes are used to label samples, resultant emissions wavelengths are saturated with pure color, much like the neon colors of some brightfield dyes. These dyes, especially when emitting in the green, blue, and violet wavelengths, are far more likely to fall outside the gamut of a printing press. While these colors can be corrected by desaturating—what was done above for eosin—the colors will become grayer in tone and dull in appearance. Some whiter values may also be diminished, thus reducing contrast. Because the colors of the dyes are secondary to the more important brightness levels, the method for conforming these to the output consists in changing the hue slightly by introducing other colors. Thus, in the following method, the hue is altered and not the saturation.

1. Use the eyedropper sampler tool to mark the darkest black in the background, and the brightest, significant white (Fig. A.3B.17). Mark more than one, if desired.
2. Under Image>Adjust(ments), select Levels.
3. In Output Levels, move the white slider to the left until significant white values read a maximum of 240 in the red, green, and blue values (K value of ~7%).
4. Move the black slider to the right until a limit of a minimum of 20 is reached (K value of ~90%).

   If black levels read at greater than a minimum of 20 in red, green, and blue, then increase levels of black by moving the black slider in Input to the right to increase blacks. If white levels read at less than 240, increase the brightness by moving the white slider to the left until a maximum of 240 reads out.

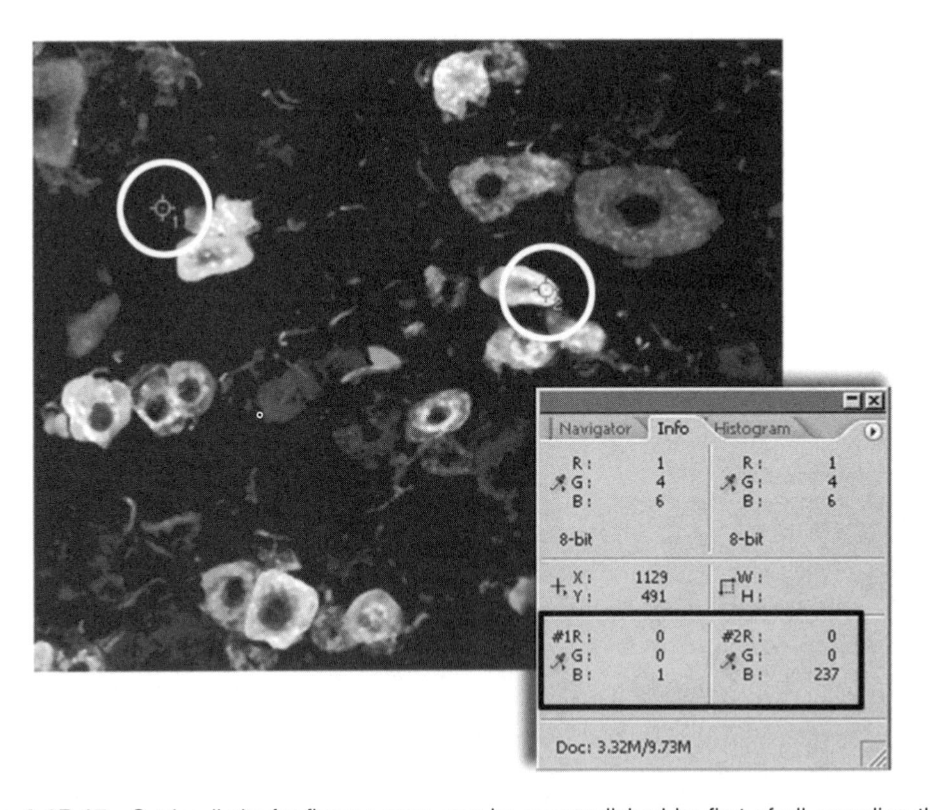

Figure A.3B.17 Setting limits for fluorescence can be accomplished by first of all sampling the brightest significant whites and blacks. These can be adjusted in Levels while viewing respective numeric readouts in the Info box. In this instance, a blue colorized image shows values only in the blue channel.

Changing Pure Colors

Nearly all fluorescent images are colorized according to what are called Look Up Tables (LUT). In this context, colorized, 8-bit images fall along 0 to 255 gradations of a single color. These colorized images fall along the popularly used pure colors of green, red, and blue. Less popular colors include yellow, magenta, cyan, and violet. Of these colors, yellow, magenta, cyan, and red have the best chance of reproducing as seen on the monitor. Green, blue, and violet are guaranteed to be reproduce poorly because of a gamut mismatch.

The pure colors must either be desaturated or replaced with gamut-matched colors. The following method will create colors that reproduce well.

1. Open image, duplicate, and close the raw image.
2. Change Mode (the type of image).

   In many instances, the color mode will need to be checked first. Under Image>Mode, look at the drop down list. If the image is RGB Color, then the mode will not need to be changed. If, however, the image is Indexed Color, then the mode will need to be changed to RGB Color. Many functions in Photoshop do not work with the Indexed Color Mode, the nomenclature for images colorized according to a LUT of 0 to 255 gradations.

3. Under Image>Adjust(ments), select Hue & Saturation. Press Shift + Control (or Apple Key) + Y to activate Gamut Warning (see above). This can be pressed again to eliminate overlay.
4. Choose Out-of-Gamut color from the Edit drop down list.
5. Move the Hue Slider to the left until the gray disappears in all but the darkest areas or at edges of features. Generally, for green and blue, the slider is moved to −32 (Fig. A.3B.18).

A3

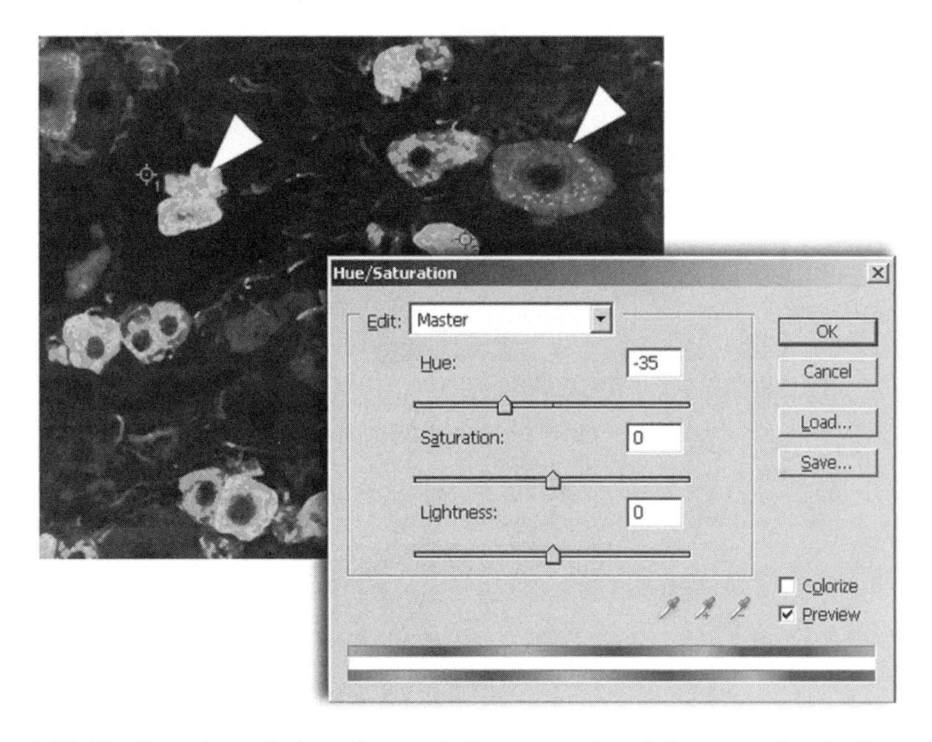

Figure A.3B.18 To conform displayed images to the gamut of a printing press, the Hue/Saturation tool can be used to desaturate offending colors. In this instance, a blue-colorized image is desaturated by sliding the Saturation slider to the left to a value of −35. Arrowheads show gamut overlay tones that still exist, showing likely areas of poor reproduction. These areas are ignored because of past experience when publishing, opting, instead, for a desaturation setting of approximately −32.

Setting Midtones

If the image is destined for publication, this correction may or may not be acceptable. Check with the publication to be certain, and indicate in the caption that the image has been improved in Photoshop. In any case, it should not be done on gels, blots, and X-ray films, as it changes the relationship of density values. Once black and white limits have been determined, then the midtone values can be altered to satisfactory results.

1. Under Image>Adjust(ments), choose Levels.
2. Use the middle slider to slightly improve appearance of contrast.

   If this slider is moved too far, then white and black limits will need to be set again.

Setting Limits, Midtones, and Color Corrections So That These Can Be Edited

Many color and contrast alterations can be applied to the image in such a way that these can be placed on a layer independent of the image. This method of applying corrections can be edited later on, and the change can easily be compared with the original, raw image.

1. To Make Editable Alterations (Fig. A.3B.19), under Window, select Layers.
2. At the bottom of the Layers box, click on the circle icon split into half black, half white. Choose desired tool from the drop down list.

   All alterations to the image will now occur on a layer above the image layer.

3. Turn the alteration on and off by clicking on the eye icon in the Layers dialog box.

Setting Resolution for Lettering, Symbols, and Output

Lettering and symbols require a high enough density of pixels upon which to be sampled so that the edges look defined. Because the required resolution for images is highest when publishing, it

Figure A.3B.19 Layers can contain the alterations that are the outcome of using Levels sliders. Rather than select Levels from the Image>Adjust(ment) drop down list, Levels can be found by choosing the circle icon in the Layers dialog box. When that selection is made, the alterations are applied to a layer that can be eliminated or edited later on without affecting visual data.

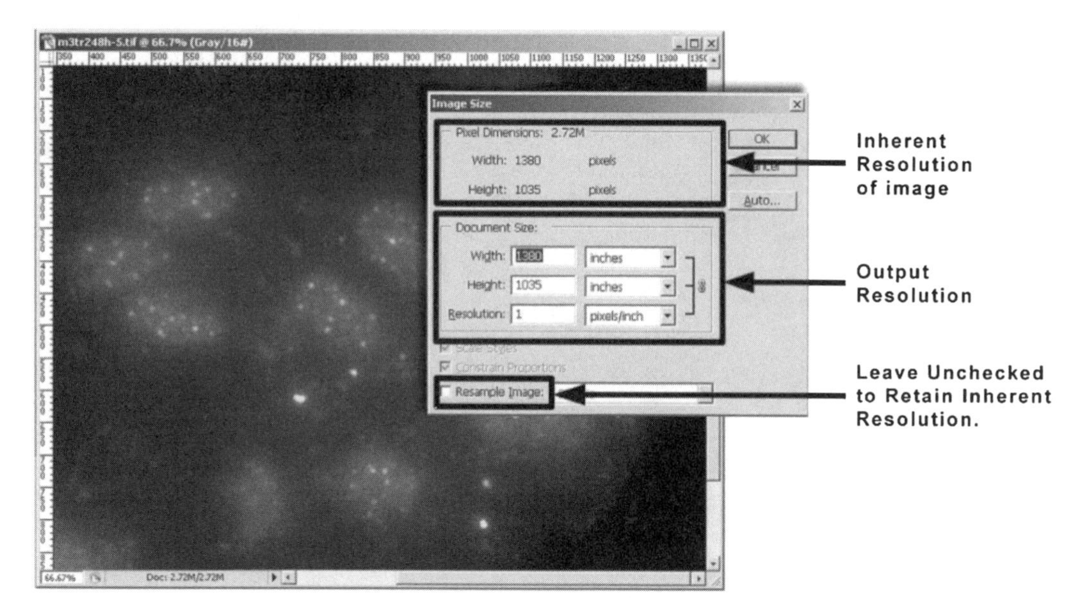

Figure A.3B.20 The Image Size dialog box can be divided into the inherent resolution readout (in pixels) and the resolution were it to be outputted. In this instance, the image is read in a curious way: the resolution is indicated as 1 with a width and height that matches pixels across and down. If this image is to be printed or prepared from PowerPoint with its inherent resolution maintained, Resample Image should be left unchecked. Simply set the width and height and ignore resolution. The output resolution only needs to be changed (with Resample Image checked) when outputting to specified resolutions, such as those given by publishers. Resampling may also need to be done when original resolution is low.

is recommended that this target resolution be set for all figures (plates). Publication resolutions guarantee that enough pixels exist for defined edges of lettering.

Generally, the resolution need only be changed when the image is being used with lettering or symbols, or for presentation. Resolution, more often than not, should not be changed for other purposes.

Image Size Box (Fig. A.3B.20). The inappropriate use of the Image Size box is, perhaps, the most common mistake in the use of Photoshop. The numbers in the more visible, center of the box are confusing because these represent the resolution and dimensions of the image were it to be outputted to a printer, therefore, this is the output resolution and not the intrinsic resolution of the image, which is shown in pixels at the top of the box (when Resample image is checked). Because many have learned that a resolution of 72 dots or pixels per inch is considered low resolution, many change the output resolution (in Photoshop CS2 & CS3 named "Document Size") in such a way that pixels are lost, not gained. Alternatively, the image is repeatedly resampled with a resultant loss of resolution.

Rather than use the Image Size Box at all, a method can be used to both provide a visual way to evaluate dimensions, and a means to guarantee that high enough resolutions are used:

1. Under File>New, create an image that is the size of a standard, published page.

   If the page is in the 8.5 × 11–in. dimensions, then a full-size, cover image would be slightly larger if it fills the entire page (what is called "bled to the edges" in the printing industry). If that same page size were to be printed on a desktop printer, then the largest size of reproduction would be less than a full page to make room for borders. Then, the page is closer to 8 × 10.5 in.

2. Set the page size in the New dialog box.

3. Set the resolution to 300 or 400 dots or pixels per inch for images, 1200 dpi for graphs and drawings.

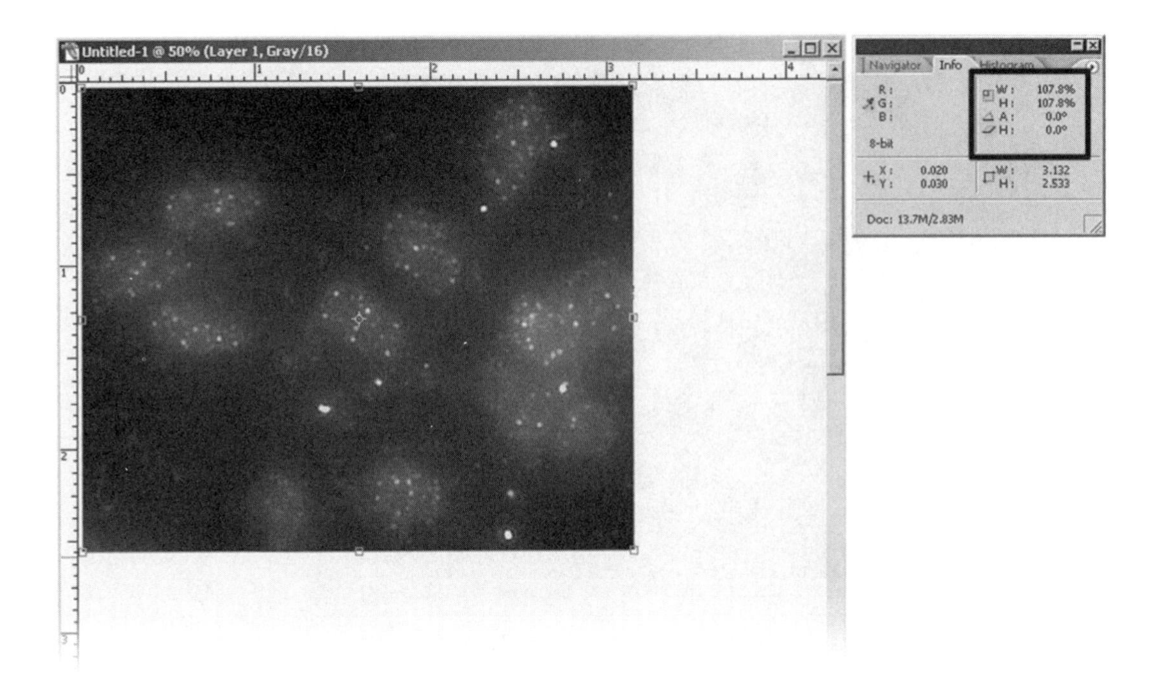

A3

Figure A.3B.21 Once an image is placed on a 300 pixel/dot per inch page for inclusion in a figure (or plate), then individual images can be rescaled in dimension to match sizes at desired output dimensions. Choose Scale under Edit>Transform and hold down the shift key when resizing to keep width to height ratios the same. If the Info box is open, then the percent scaling can be noted and repeated for subsequent images.

4. Set the color mode to RGB Color or to Grayscale, depending on nature of images.

   Many publications ask for 400 dpi for images. To save on disk space, 300 dpi can be used for general purposes. When 400 dpi is needed, simply resample the image. This can be done without a perceptible loss in resolution, though a slight sharpening filter may be necessary.

5. Once the page is created, place the images onto it and resize to fit the desired output dimensions.

   A grid can be shown to aid in lining up images (see below).

Resize the image (Fig. A.3.21)

6. Choose the layer of the image that has been placed on the page. Under Edit>Transform, choose Scale. While holding down the Shift key, push or pull on corners to enlarge or reduce the dimensions of the image.

To use grid (Fig. A.3B.22)

7. Under Edit>Preferences (Photoshop>Preferences on a Macintosh), choose Guides, Grid and Slices.

8. Set the grid to desired width in the box. Then, under View>Show, choose Grid.

   The grid overlay can be turned off by choosing View>Show>Grid again.

9. When all images have been placed on the page, use the crop tool to eliminate all but the desired parts.

10. Use the crop icon in the toolbar and pull or push on the corners.

    Excess areas will be eliminated without any change in resolution.

Saving Corrected Files

With the costs of hard disk memory and removable media becoming more and more affordable, storage of images is not a great hardship for most workgroups. Good laboratory practices dictate

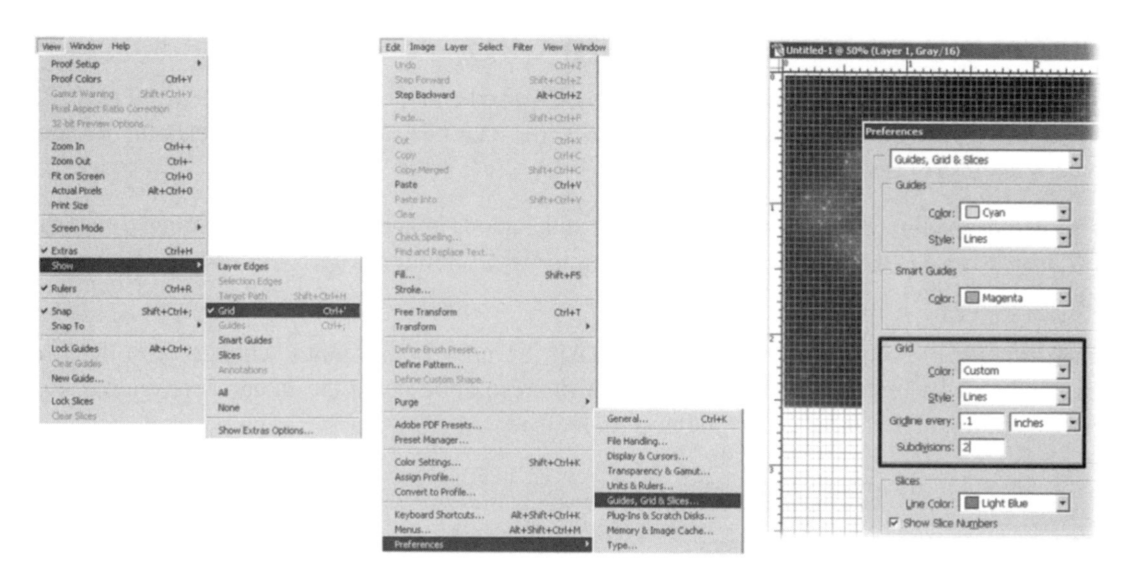

Figure A.3B.22 From left to right, the grid is revealed over the image when clicking on it from Window>View. The spacing of the grid can be changed by choosing Guides, Grid & Slices from Edit>Preferences (or Photoshop>Preferences on a Macintosh). The "Guides..." dialog box allows users to input values for grid sizes.

that images be stored in two places, both at work and at another location in the event of disaster. On top of that, two sets are kept for the laboratory worker and for the principal investigator (*APPENDIX 2*).

To ensure that images are saved in ways that provide output to various sources, three file formats are recommended: Photoshop files (.psd), TIFF (.tif), and JPEG (.jpg) or PNG (.png). After the discussion of lossy resolution, it may seem counter-intuitive to recommend JPEG files. Unfortunately, Microsoft products are heavily used in science and research, and these generally work poorly when non-lossy images are used. These products were not made with the idea that high resolution, large file size images are to be used, and so JPEG is often the only choice. Smaller file sizes can be made with non-lossy, compressed, PNG files, which also have the advantage of being displayed similarly on computer screens across software platforms.

Photoshop files are saved after alteration and addition of symbols and lettering to preserve layers so that images can be edited easily later on, in the event changes are needed.

TIFF files are saved for their universal application to other software programs, and for publication. These are saved without layers to make them more universal and appropriate for publication purposes.

TIFF files can be saved as both TIFF RGB Color files for applications other than publication, and as TIFF CMYK Color files for publication. TIFF files can be made into CMYK Color using Image>Mode. Some color shifts may occur.

JPEG files, as stated earlier, are saved mostly for Microsoft products. In Photoshop, very little visual information is lost when saving at the maximum compression. Photoshop is recommended for its superior performance in creating JPEGs. Except for purposes of inclusion in web pages, maximum compression retains virtually all the visual data.

PNG files were created to avoid lossy compression, as well as to contain parameters in the file to ensure that the image looks similar from computer to computer. These files are saved without interlacing, except when desired for web purposes.

NOTE: See the discussion of digital image formats in *UNIT 7.5* for further discussion.

INSERTING FILES INTO POWERPOINT

If the page method of enhancing resolution has been used and the resulting image saved as a JPEG at maximum resolution, for the most part these files can be inserted into PowerPoint without a perceptible loss of resolution.

When the page method is not used, PowerPoint understands images to be at the nominal dimensions of 7.5 × 10 in., in keeping with the aspect ratio for a monitor. The resolution of the image is effectively ignored by PowerPoint.

When images are brought in singularly, or when brought in as figures, change only the width and height of the image in Photoshop, and not the resolution using the Image Size dialog box of Photoshop, found under the Image menu (Fig. A.3B.23).

A3

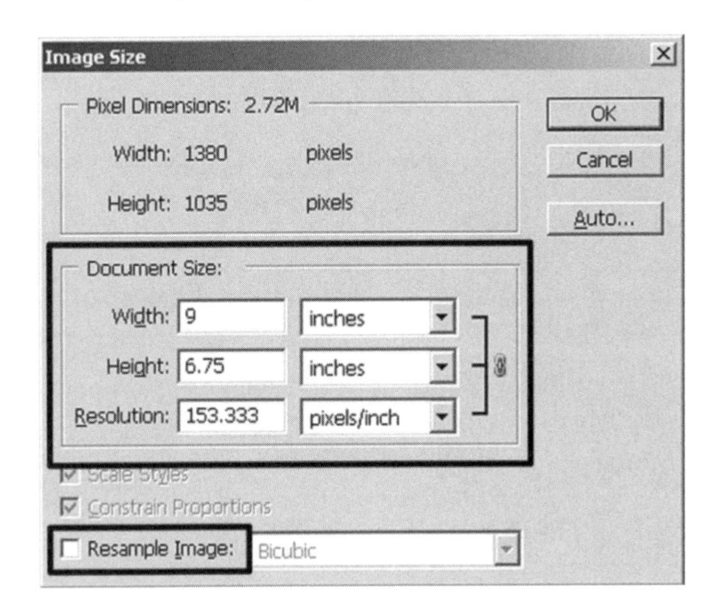

Figure A.3B.23 If images included in PowerPoint are derived from Photoshop, the document size is set with the default PowerPoint window size in mind: 10 in. in width, 7.5 in. in height. In this instance, the width is set to 9 in. to almost fill the PowerPoint slide window with the Resolution left at 153.333 because it is yet above poster printing resolutions of 150 dots/pixels per inch. Resample image is left unchecked so that the resolution at a particular output can be determined before making a decision to upsample, when necessary.

If images are already of high resolution, in the Image Size box, uncheck the Resample Image option in the Image Size box. Set the height or width to fit into 7.5 × 10 in.

If the images are low resolution (and have not been resampled at higher resolution by the Page Method), in the Image Size box, check Resample Image and set the resolution to at least 200 or 300 dpi for the additional benefit of preparing the image for the addition of lettering or symbols in Photoshop. Set the height or width to fit into 7.5 × 10 in.

Be sure to save as JPEG using the maximum setting before inserting into PowerPoint (see above).

Images that come into PowerPoint and then need to be scaled to a greater dimension are most likely to print at low resolution.

LITERATURE CITED

Leong, F.J., Brady, M., and McGee, J.O. 2003. Correction of uneven illumination (vignetting) in digital microscopy images. *J. Clin. Pathol.* 56:619-621.

Margulis, D. 2006. Professional Photoshop: The Classic Guide to Color Correction, 5th ed. Peachpit Press, Berkeley, CA.

APPENDIX 4

Statistical Analysis

John Kloke[1] and Johanna Hardin[1]
[1]Pomona College, Claremont, California

INTRODUCTION

This Appendix was written as a reference guide for data analysis of biological experiments. It should not replace a course in introductory statistics, but rather serve as a reminder of concepts learned in that course. For those interested in a deeper refresher or more derivations, we recommend the following textbooks (see Literature Cited for complete bibliographic information):

Introduction to the Practice of Statistics by Moore and McCabe (2006)
Mind on Statistics by Utts and Heckard (2004)
Primer of Biostatistics by Glantz (2005).

Notice that the approaches provided below, summarized in Table A.4.1, do not constitute a complete list of all possible statistics methods. It is possible that your data will need more sophisticated analyses, in which case you should consult a statistician or a post-introductory textbook.

We have tried to be clear about which method to use with which type of research question and data. However, if you are unsure, the references above can give a more complete picture of methods and their appropriate uses. In general, you should graph the data and check assumptions (e.g., independence) carefully.

The remainder of this Introduction section will give a sense of how to select an appropriate technique to analyze the research data of interest. We discuss different types of data, different methods for inferring statements about a population based on a sample, and different statistical concerns that arise when analyzing data. Additionally, Table A.4.1 gives a summary of the statistical analyses discussed in the text and Tables A.4.2 and A.4.3 define terms that are used throughout this Appendix. We recommend that you read the entire Introduction before analyzing any data.

Data

Before deciding what test or method to use, you have to know a little about the data you have collected and the questions you'd like to ask. Many methods require independent data. Two data points are independent if knowledge of the first point gives no information about the value of the second point. For example, pre-test and post-test scores on the same individual are not independent. Two scores on two different randomly selected individuals are independent.

We break data up three different ways: explanatory versus response, numeric versus categorical, and quantitative versus qualitative.

Explanatory versus response

Explanatory: An explanatory variable is used to explain or predict an outcome.

Response: A response variable measures the outcome of interest.

Example

If you are interested in measuring the difference in two diets, the explanatory variable would be the diets (diet 1 versus diet 2) and the response variable would be the number of pounds lost by each participant in the study.

Current Protocols Essential Laboratory Techniques A.4.1-A.4.30

Table A.4.1 Summary of the Statistical Analyses Discussed

Variables in dataset	Example	Method(s)
One continuous response variable (optional: a binary pairing variable)	Testing average blood pressure (against a known value)	One sample *t*-test Wilcoxon Rank Sum
	Estimating average blood pressure	CI for μ/θ
	Testing blood pressure before and after treatment to determine if there is a difference	Paired *t*-test Wilcoxon Rank Sum
	Estimating average change in blood pressure before and after treatment	CI for μ
One binary explanatory variable One continuous response variable	Testing the difference in average pounds lost for two different diets	Two sample *t*-test Mann-Whitney-Wilcoxon
	Estimating the difference in average pounds lost for two different diets	CI for difference in μ/θ
One binary response variable	Estimating the cure rate of some treatment or testing against a known value	Inference for a single p
One binary response variable One binary explanatory variable	Testing whether gender and pet ownership are independent	Tests of proportions Chi-square test
	Estimating the difference in proportion of men who are pet owners versus women who are pet owners	CI for difference in proportions
One multichotomous explanatory variable One continuous response variable	Testing the difference in average GPA for different grade levels	ANOVA Multiple comparisons
One continuous explanatory variable One continuous response variable	Predicting weight (response) from height (explanatory)	Simple linear regression
	Correlating weight and height	Correlation
Two binary or multichotomous variables	Testing whether race and political party are independent	Chi square test

Numeric versus categorical

Numeric: A numeric variable is one which takes numerical values. A numeric variable is either discrete or continuous as described below.

Continuous: A numeric variable which is measured on a continuous scale (e.g., height, weight, age, time).

Discrete: A numeric variable which may take on only a finite number of values, and which usually arises from counting situations (number of offspring).

Categorical variables: A categorical variable does not take on numeric values but rather can be placed into groups or categories such as those described below.

Binary or dichotomous: A categorical variable having only two levels, e.g., gender or on/off.

Multichotomous: A categorical variable having multiple levels, e.g., race or grade level.

Table A.4.2 Standard Symbols Used to Represent Parameters of Interest

Definition	Parameter
Population mean	μ
Population median	θ
Population variance	σ^2
Population standard deviation	σ
Population standard error of the mean	$\sigma\sqrt{n}$
Population proportion	p
Population regression slope	β_1
Population correlation	ρ

Table A.4.3 Standard Symbols Used to Represent Statistics of Interest

	Statistic	Definition
Sample mean	$\bar{x}$	$\frac{1}{n}\sum_{i=1}^{n} x_i$
Sample variance	s^2	$\frac{1}{n-1}\sum_{i=1}^{n}(x_i - \bar{x})^2$
Sample standard deviation	s	$\sqrt{s^2}$
Sample standard error of the mean	$s/\sqrt{n}$	$\sqrt{\frac{s^2}{n}}$
Sample proportion	$\hat{p}$	$\frac{\text{\# of successes}}{n}$
Sample regression slope	b_1	See Literature Cited
Sample correlation	R	See Literature Cited

Quantitative versus qualitative

Quantitative (numerical) variables are measured on a numeric scale. These can be continuous (e.g., height) or discrete (e.g., grade in school).

Qualitative (categorical) variables are measured as categories.

Parameters

Standard symbols will be used to represent parameters of interest, as presented in Table A.4.2.

Statistics

Standard symbols will be used to represent statistics of interest, as outlined in Table A.4.3.

INFERENCE

Typically, the goal of experiments and their statistical analysis is to infer something about a population based on a sample.

General Advice for Running Analyses

Typically there are two ways to analyze data: hypothesis testing and confidence intervals (CI). Both methods respond to questions about populations using sample data. Therefore, research questions should ask something related to a population of interest, and the analyses should be done using the sample data that were collected.

Hypothesis Testing

In general, there are two hypotheses. A null hypothesis is a statement about the population that nothing interesting is going on. An alternative hypothesis is a statement about the population which would be of interest to the research community. Generally, the alternative hypothesis corresponds to the research question being asked. If a null hypothesis is rejected in favor of the alternative hypothesis, this is an indication that the research question is true. The null hypothesis is rejected if the *p*-value of the test is small enough. *p*-values are discussed below.

In general, every test of hypothesis follows the same basic steps:

1. The null and alternative hypothesis (H_0, H_1, respectively) are formed based on some research question.

   For example, if testing a chemical's ability to act as an antibiotic (i.e., kill bacteria), the null hypothesis would be "The chemical does not kill more bacteria than a control compound," while the alternative hypothesis would be "The chemical kills more bacteria than a control compound."

2. A test statistic is calculated based on a sample of representative data.

   A number of plates would be grown with either the test compound or a control compound, and the number of colonies grown over a set period of time would be counted.

3. A decision to reject or to not reject the null hypothesis is made based on how likely the data are, given a true null hypothesis.

   If the p-value obtained by comparing the number of colonies appearing after incubation on plates with antibiotic versus those without is <0.05, the researcher can assume that the null hypothesis (i.e., "The chemical does not kill significantly more bacteria than a control compound.") is false.

A *p*-value is the probability of seeing the observed sample data, or data that are more extreme, if the null hypothesis is true. A very small *p*-value suggests that the original assumption (the null hypothesis being true) is false. A null hypothesis is rejected if the *p*-value is less than the *level* of the test, α, most often a level of $\alpha = 0.05$. Occasionally, levels of $\alpha = 0.10$ or $\alpha = 0.01$ are used. The α level represents how often you make a false-positive error. You are responsible for stating the false-positive error rate for your experiment. A rejected null hypothesis is often referred to as *statistical evidence* or *significant evidence* of the research question being true (i.e., H_1).

For example, suppose a coin is tossed 10 times, each time resulting in heads. You might reject the hypothesis that the coin is fair (H_0: $p = 0.5$, where p is the true probability of heads) in favor of the hypothesis that the coin is not fair (H_1: $p \neq 0.5$), since the probability of observing results as extreme when the coin is fair (i.e., that the null hypothesis is true) is small (*p*-value = 0.002).

If the *p*-value is greater than $\alpha = 0.05$, we are unable to say which of the hypotheses is true. A correct interpretation of a large *p*-value is: "The data do not provide evidence to reject the null hypothesis." A *p*-value of 0.05 is equivalent to saying that *if the null hypothesis is true*, data like those collected would happen 5% of the time.

Confidence Intervals (CI)

Sometimes an estimate of a population parameter is desired instead of a test regarding a particular claim about the population. For example, the interest might be in estimating the average blood pressure of women taking hormone replacement therapy (HRT). When estimating a population parameter, using a confidence interval is the appropriate method of analysis. A confidence interval is a set (an interval, in fact) of values which serves as an estimate of a population parameter.

Typically, 90%, 95%, or 99% confidence intervals are used. Consider a 95% confidence interval for some population characteristic, e.g., the population mean (μ). A mathematical derivation shows that out of all possible samples of size n, 95% of the intervals will contain the true population value.

The correct interpretation of a 95% confidence interval is "I am 95% **confident** that the true population parameter lies within the endpoints of the interval" or, "I am 95% **confident** that the true average blood pressure of women on HRT is between the bounds I have calculated." It would be incorrect to say "There is a 95% **probability** that the true blood pressure of women on HRT lies within the bounds I have calculated." Once the interval has been calculated, there is no more probability associated with the result.

Errors in Inference

As alluded to above, there is no way to know for sure which hypothesis is in fact true. One only has evidence to suggest which is true. At times, because of random variability, the data suggest a hypothesis which is in fact false. When this occurs, it said than an *error* has occurred. There are two possible types of errors which can be made when conducting statistical analysis.

The first type of error, a *type I error*, happens when the null hypothesis is incorrectly rejected. Type I errors happen, on average, at a rate of α, usually 0.05. Therefore, 5% of the time when the null hypothesis is in fact true, the data will indicate that the null hypothesis should be rejected.

A *type II error* happens when a false null hypothesis is not rejected. The rate of type II errors cannot be directly controlled and depends on what the actual (unknown) population values are. Because the rate of type II errors is unknown, the statement of "fail to reject the null hypothesis" is preferred to "accept the null hypothesis."

Note that the errors in inference are not the same as measurement variability. A larger sample size will reduce the type II error. Type I error is fixed (typically at 0.05).

Graphics

A graphical analysis of the data is a very important part of the data-analysis process. In addition to giving a pictorial representation of the data, graphics allow the analyst to check any assumptions for the desired statistical method/procedure. Several useful graphical tools exist for this particular purpose. We will primarily use boxplots and scatterplots, but histograms and quantile-quantile plots (not described here; see references in introduction for information on histograms and quantile-quantile plots) are also extremely useful tools.

A boxplot is a representation of a univariate (one-variable) set of data. The middle line represents the median (or 50% point), and the outer edges of the box represent the 25% point (lower quartile) and the 75% point (upper quartile). The whiskers extend to the minimum and maximum values within a certain threshold. If the minimum (or maximum) value is outside of a threshold, the minimum (or maximum) value will be represented by a point. See Figures A.4.1-A.4.6.

A scatterplot is a representation of a bivariate (two-variable) set of data. The x axis represents the value of the first variable and the y axis represents the value of the second variable. Each individual is given by a dot in the x-y plane. See Figures A.4.7-A.4.8.

Sample Sizes

For each of the methods described in this Appendix, we have given general sample size requirements. These should be used as a guide and not absolute law. In certain cases, larger sample sizes may be

Table A.4.4 Statistical Software

Name	URL	Interface[a]	Free or open source
Arc	*http://www.stat.umn.edu/arc/software.htmlwww.stat.umn.edu/arc/software.html*	GUI and command line	Yes
JMP	*http://www.jmp.com/www.jmp.com/*	GUI	No
Minitab	*http://www.minitab.com/www.minitab.com/*	GUI or command line	No
R	*http://www.r-project.org/www.r-project.org/*	Command line	Yes
SAS	*http://www.sas.com/www.sas.com/*	Command line	No
S-PLUS	*http://www.insightful.com/products/splus/default.aspwww.insightful.com/products/splus/default.asp*	GUI or command line	No
SPSS	*http://www.spss.com/*	GUI and command line	No

[a]GUI stands for graphical user interface; for instance the Windows and Macintosh operating systems use a GUI (arrows and icons). Command line indicates that commands are actually typed in at a prompt, more similar to DOS (or Unix).

A4

required for the methods to be valid. Also, in rare cases, smaller sample sizes may be sufficient. As a general rule, when it comes to sample sizes, bigger is better. As mentioned in the section on errors, a larger sample size will reduce the type II error. That is, if the true state of nature is "significant differences," the data are more likely to demonstrate significant differences with a larger sample size. The guidelines for sample sizes are given within each methodological section.

Note that n represents the number of independent measurements needed or taken. If there is measurement error, multiple measurements may need to be taken on each independent individual. Doing so will create two sources of variability: within variability and between variability. Analysis of such data is called repeated measures analysis and is outside the scope of this Appendix. However, taking the average (or median) of a few repeated measurements will usually provide reasonable data with which to work.

Parametric Versus Nonparametric

Parametric methods assume an underlying distribution (usually normality) of the data; nonparametric methods do not assume a known structure of the data. For the majority of topics which we discuss, we have included both parametric and nonparametric methods. In general the nonparametric approaches require fewer assumptions than their parametric counterparts. The parametric approaches are exact (has exact level α) when the underlying population is normal, otherwise the inference is approximate (has approximate level α). The approximation gets closer as the sample size increases. Although exact inference for the nonparametric methods exists regardless of sample size, we have chosen to discuss only the approximate versions. As with the parametric approaches, the approximations for nonparametric approaches decline as the sample size increases. Most software packages allow the user to chose between exact and approximate inference for the nonparametric methods. For a thorough treatment of nonparametric methods, see Hollander and Wolfe (1999).

Statistical Software

There are many statistical software packages; in Table A.4.4 we list only the ones which we are most familiar with. The analyses in this Appendix were done using the software package R (*http://www.r-project.org/*).

THE ONE SAMPLE LOCATION PROBLEM

The one sample location problem is concerned with inference on the center—mean (μ) or median (θ) of a single population. For example, one may want to estimate the average blood pressure of some population or to test if (on average) a certain drug contains the correct amount of active ingredient.

Parametric Procedures

Assumptions

1. The response variable is measured on a numeric scale.

2. For symmetric populations $n \geq 15$; otherwise $n \geq 30$. Symmetric populations are those that have the same type of distribution of values on the left and right side of the center. For example, income is typically not symmetric because of the extremely large values. Height is typically symmetric because there is the same spread of heights above the mean as below the mean.

Inference

Both the interval estimate and the hypothesis test are based on a *t*-test that gives a range of plausible values for the mean of the population (interval estimation) or tests a value of interest (μ_0) for the mean of the population (hypothesis testing).

Interval estimation

A $(1 - \alpha) \times 100\%$ confidence interval for μ is:

$$\bar{x} \pm t_{1-\alpha/2,n-1} \frac{s}{\sqrt{n}}$$

where $\bar{x}$ represents the sample average, $t_{1-\alpha/2,n-1}$ represents the $1 - \alpha/2$ cutoff of a t_{n-1} distribution, and s represents the standard deviation.

Hypothesis testing

The *t*-statistic is:

$$t^* = \frac{\bar{x} - \mu_0}{\frac{s}{\sqrt{n}}}$$

where t^* is defined as the number of standard deviations the sample mean is from the hypothesized mean.

Hypotheses are shown in Table A.4.5.

Table A.4.5 Decision Rules for the One Sample *t*-Test

Hypotheses	Decision rule	*p*-value
$H_0{:}\mu=\mu_0$ vs. $H_1{:}\mu\neq\mu_0$	Reject H_0 if $\lvert t^*\rvert > t_{1-\alpha/2,n-1}$	$2{*}p(t_{n-1} > \lvert t^*\rvert)$
$H_0{:}\mu\geq\mu_0$ vs. $H_1{:}\mu<\mu_0$	Reject H_0 if $t^* < -t_{1-\alpha,n-1}$	$p(t_{n-1} < t^*)$
$H_0{:}\mu\leq\mu_0$ vs. $H_1{:}\mu>\mu_0$	Reject H_0 if $t^* > t_{1-\alpha,n-1}$	$p(t_{n-1} > t^*)$

A

birth weight (n = 47):

74 81 87 88 88 89 89 92 98 98 99 100 100 100 102 104 104 104 106
106 107 107 108 109 109 111 111 113 113 113 115 117 117 119 120 121 121 121
123 123 124 125 130 132 141 142 147

B

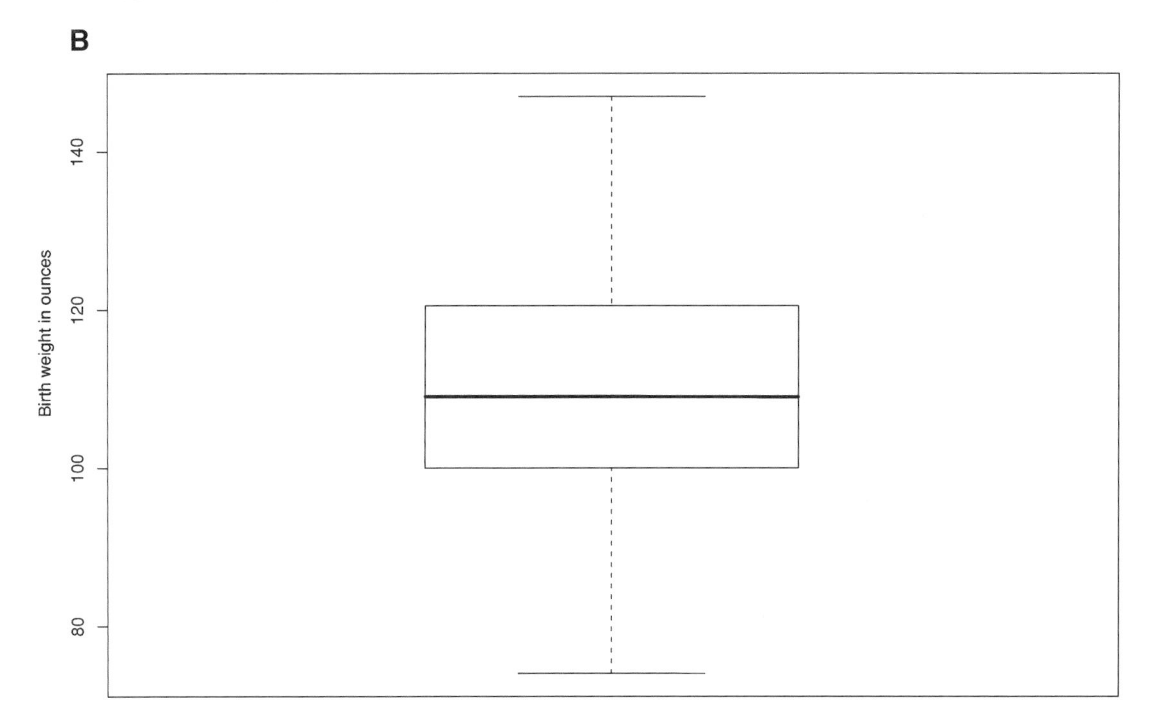

Figure A.4.1 (**A**) Sorted data of birth weights of 47 children whose mothers smoked. (**B**) Boxplot of these data. 50% of the mothers delivered babies who weighed less than 108.5 oz.; 25% of mothers delivered babies who weighed less than 99.5 oz.; 75% of mothers delivered babies who weighed less than 121 oz. The smallest baby in the dataset weighed 74 oz., and the largest weighed 147 oz.

Example

Suppose that we are interested in estimating the mean birth weight of children whose mother admitted to smoking during pregnancy. Suppose a sample of size $n = 47$ is taken, resulting in a sample mean of $\bar{x} = 109.5$ ounces and a sample standard deviation of $s = 15.8$ ounces. The actual data are given in Figure A.4.1, panel A; a plot of the data is shown in Figure A.4.1, panel B.

A birth weight of 116 ounces is considered typical. The research question that we wish to answer is: does smoking during pregnancy result in lower birth weights, on average?

That is, we wish to test the hypotheses:

$$H_0 : \mu \geq 116 \text{ versus } H_1 : \mu < 116$$

Using the equation for a t-test statistic from above, we see that the test statistic we observe for these data is:

$$t^* = \frac{109.5 - 116}{\frac{15.8}{\sqrt{47}}} = -2.82$$

Since $t^* = -2.82 < -1.68 = -t_{0.95,46}$, we conclude that there is significant evidence to reject the null hypothesis. There is significant evidence that smoking is linked with lower than typical birth weight. The t-value of -1.68 was computed from a t-table with 46 degrees of freedom and 0.95 probability to the left. Once you have a t-table, which can be obtained from software or a textbook, use degrees of freedom $= n - 1$. The degrees of freedom simply point to the correct table.

A 95% confidence interval for the true mean birth weight of children whose mothers admitted to smoking during pregnancy is $109.5 \pm 2.01 \times 15.8/\sqrt{47}$ (or 104.87 ounces, 114.13 ounces). Note that 2.01 is the t-value from a t-table with 46 degrees of freedom at 97.5% (which gives 2.5% error on each side of the confidence interval). We are 95% confident that the true mean birth weight for all children born to mothers who admitted to smoking during pregnancy is between about 104.87 and 114.13 ounces.

Nonparametric Procedures

There are several nonparametric approaches to the one sample location problem. The sign test is a very general test which requires almost no assumptions. We will take a large-sample Wilcoxon approach which assumes symmetry of the underlying population. See, for example, Hollander and Wolfe (1999) for Wilcoxon small-sample methods, as well as additional methods based on the sign test.

Assumptions

1. Data are measured on a numeric scale.

2. The data are fairly symmetric.

3. $n \geq 20$.

A4

Hypothesis testing

The Wilcoxon signed rank test statistic is defined as:

$$T^{+} = \sum_{i=1}^{n} R(|x_i - \mu_0|) I(x_i - \mu_0)$$

where $R(|x_i - \mu_0|)$ denotes the rank of $|x_i - \mu_0|$ among $|x_1 - \mu_0| \ldots |x_n - \mu_0|$ and $I(x_i - \mu_0)$ is the indicator function, which takes the value of 1 when $x_i - \mu > 0$, and 0 otherwise.

For large samples we may use the test statistic:

$$z^{*} = \frac{T^{+} - n(n+1)/4}{\sqrt{n(n+1)(2n+1)/24}}$$

with hypotheses as given in Table A.4.6.

Interval estimation

An interval estimate for the population median, based on the Wilcoxon signed rank, is the interval of values such that the null hypothesis is not rejected. See, for example, Hollander and Wolfe (1999).

Table A.4.6 Decision Rules for the Wilcoxon Signed Rank Test

Hypotheses	Decision rule	p-value
$H_0{:}\theta=\theta_0$ vs. $H_1{:}\theta\neq\theta_0$	Reject H_0 if $\lvert z^*\rvert > z_{1-\alpha/2}$	$2{*}P(Z>\lvert z^*\rvert)$
$H_0{:}\theta\geq\theta_0$ vs. $H_1{:}\theta<\theta_0$	Reject H_0 if $z^* < -z_{1-\alpha}$	$P(Z<z^*)$
$H_0{:}\theta\leq\theta_0$ vs. $H_1{:}\theta>\theta_0$	Reject H_0 if $z^* > z_{1-\alpha}$	$P(Z>z^*)$

Example

We use the same example as we did for the parametric analysis. The sorted data of sample birth weights are presented in Figure A.4.1, panel A.

Judging from the boxplot (see Figure A.4.1, panel B), the assumption of symmetry seems quite valid. That is, there are just as many points above as below the median; also, the distance from the 25% to the minimum value is about the same as the distance from the 75% to the maximum value. The plot tells us that the assumptions for the statistical method are not violated. The value of the Wilcoxon signed rank test statistic for these data is $T^+ = 303$. The large sample standardized test statistic is:

$$z^* = \frac{303 - 47 \times 48/4}{\sqrt{47 \times 48 \times 95/24}} = -2.76$$

Using $-z_{0.95} = -1.96$ as our critical value, we see that we can reject the null hypothesis and conclude that smoking is related to lower birth weight. Also, p-value $= 0.0029$.

The software package R reports (104.5 ounces, 114.0 ounces) as a interval estimate of the true median birth weight of children whose mothers admitted to smoking based on the Wilcoxon signed rank test.

A4

Paired Designs

A paired design, sometimes referred to as matched pairs, occurs when two repeated measurements are taken of the same individual or experimental unit or when measurements are taken on pairs of subjects or experimental units which are matched somehow. The parametric analysis of such data is often called a paired t-test. For other methods (including the nonparametric analysis), there is often a “paired” option in the statistical software. For two examples, measurements of siblings on two different treatments are matched; a baseline measurement on some individual and then a measurement on the same individual taken after some treatment is also matched.

A paired data analysis is a one sample analysis performed on the differences of the response variables. The null and alternative hypotheses almost always address the question of whether the pairs are different; that is, do they have a difference of zero?

Analysis

The data shown in Figure A.4.2, panel A, were taken from Hettmansperger and McKean (1998). The data consist of 15 pairs of heights in inches of cross-fertilized and self-fertilized plants, each pair grown in the same plot. A boxplot of these data is shown in Figure A.4.2, panel B.

Parametric analysis

Referring to the data in Figure A.4.2, the output from the software package R is as follows:

```
One Sample t-test
data: diff
t=2.1506, df=14, p-value=0.04946
alternative hypothesis: true mean is not equal to 0
95 percent confidence interval:
0.007114427 5.232885573
```

As we can see, the t-test is only marginally significant at the $\alpha = 0.05$ level (p-value $= 0.04946$).

A

cross (n = 15):
23.500 12.000 21.000 22.000 19.125 21.550 22.125 20.375 18.250 21.625
23.250 21.000 22.125 23.000 12.000

self (n = 15):
17.375 20.375 20.000 20.000 18.375 18.625 18.625 15.250 16.500 18.000
16.250 18.000 12.750 15.500 18.000

differences (n = 15):
6.125 -8.375 1.000 2.000 0.750 2.925 3.500 5.125 1.750 3.625
7.000 3.000 9.375 7.500 -6.000

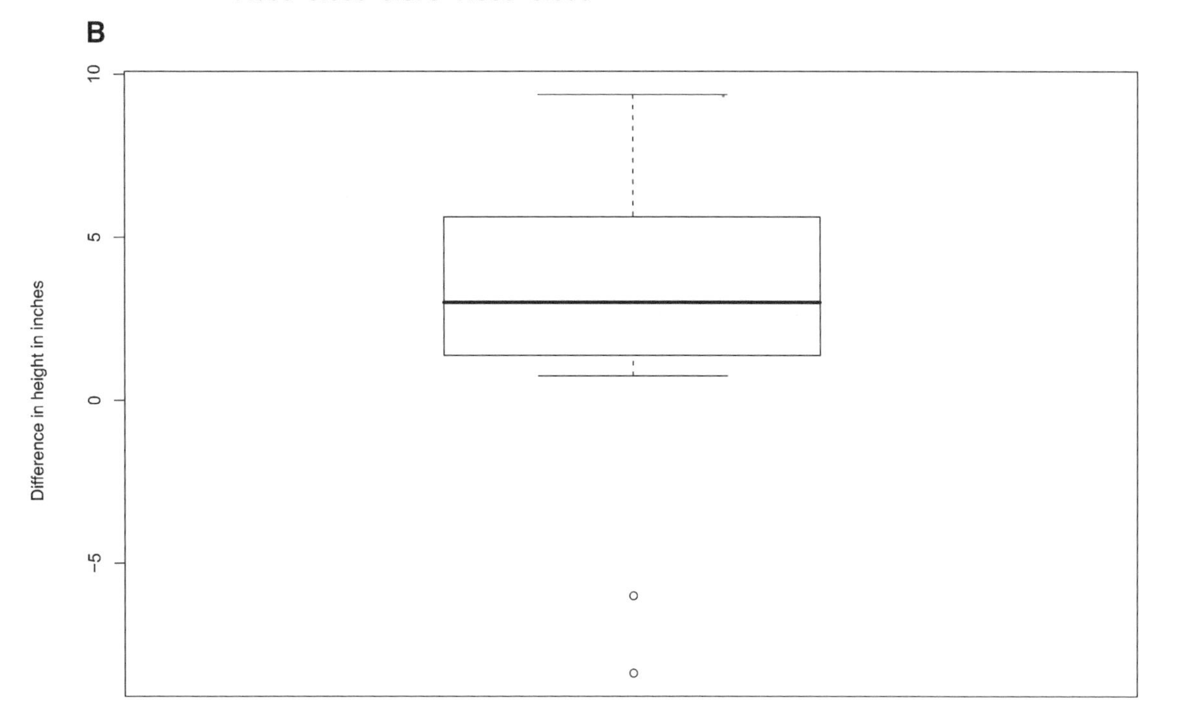

Figure A.4.2 Paired data analysis. (**A**) 15 pairs of heights in inches of cross-fertilized and self-fertilized plants, each pair grown in the same plot. (**B**) Boxplot of differences between cross and self.

Nonparametric analysis

Again referring to the data in Figure A.4.2, we include the output from the software package R.

```
Wilcoxon signed rank test
data: diff

V=96, p-value=0.04089
alternative hypothesis: true mu is not equal to 0
95 percent confidence interval:
0.4999872 5.2125081
```

The results for the Wilcoxon analysis are similarly borderline (p-value = 0.04089).

THE TWO INDEPENDENT SAMPLES LOCATION PROBLEM

The two sample location problem is concerned with inference on the change in means (parametric approach) or the change in medians (nonparametric approach) between two treatment or experimental groups. For example, one might be interested in testing if the LDL cholesterol levels of a group of treated quail are lower than a group of untreated quail. As another example, one might be interested in estimating how much taller, on average, adult males are than adult females. Note that if a population is symmetric, then the population mean is equal to the population median, which is the situation represented in the plots shown in Figure A.4.3 and discussed in the next paragraph.

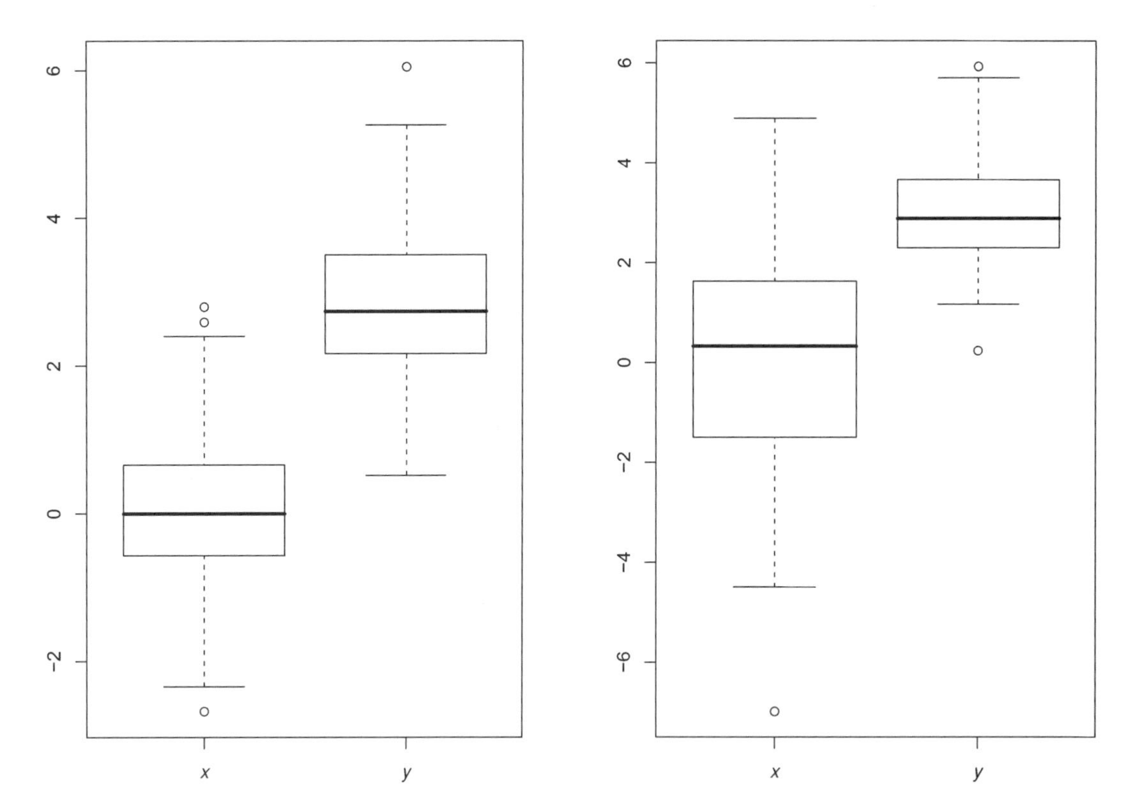

A4

Figure A.4.3 Comparison boxplots representing two typical situations for the two sample location problem. The plot on the left represents a case where either the MWW test for medians or the pooled *t*-test for means is appropriate. The plot on the right represents a case where the unpooled *t*-test is appropriate.

The boxplots in Figure A.4.3 represent two typical situations for the two sample location problem. The plot on the left represents a case where either the MWW test for medians or the pooled *t*-test for means is appropriate. The plot on the right represents a case where the unpooled *t*-test is appropriate.

Data

Suppose we are interested in comparing two populations, *X* and *Y*. We take a sample of observations from each of the two groups or under the two experimental conditions. Let $x_1,...x_m$ be a random sample of size *m* from population *X*. Let $y_1,...y_n$ be a random sample of size *n* from population *Y*.

Parametric Procedures

There are two types of *t*-tests for the two sample problem, the pooled and unpooled. The pooled requires the distributions differ only in location but that they be similar in terms of variability. The unpooled requires the shapes of the two populations to be the same, but does not require the variability of the two populations to be the same. When the assumptions of the pooled analysis are met, the pooled analysis results in a more efficient analysis (i.e., requiring fewer samples) than the unpooled counterpart.

In both cases, μ_X represents the true mean of population *X*, and μ_Y represents the true mean of population *Y*. Further $\bar{x}$ and s_X denote the sample mean and sample standard deviation of the *m* observations $x_1,...x_m$. Likewise $\bar{y}$ and s_y denote the sample mean and sample standard deviation of *n* observations $y_1,...y_n$.

Unpooled inference

Assumptions

1. The response variable is numeric.
2. The two distributions have the same shape.
3. Two independent samples are drawn from the two distributions.
4. If the distributions are symmetric, you should have at least $n > 15$ in each group. If the distributions are not symmetric, you should have at least $n > 30$ in each group.

Interval estimation

A $(1 - \alpha) \times 100\%$ confidence interval for the difference in the population means $\mu_X - \mu_Y$ is:

$$\bar{x} - \bar{y} \pm t_{1-\alpha/2,df}\sqrt{\frac{s_x^2}{m} + \frac{s_y^2}{n}}$$

A conservative estimate of degrees of freedom (*df*) is $df =$ minimum $(n - 1, m - 1)$. Some software programs will give a more precise estimate for *df*. The degrees of freedom are simply a mechanism for pointing us to the correct row of the appropriate table to use in constructing a confidence interval.

A4

Hypothesis testing

The unpooled *t*-statistic is:

$$t^* = \frac{\bar{x} - \bar{y}}{\sqrt{\frac{s_x^2}{m} + \frac{s_y^2}{n}}}$$

with hypotheses shown in Table A.4.7 and df as described above. Sometimes the difference in the two population means is called the *shift* and is denoted by $\Delta = \mu_X - \mu_Y$.

Example

Suppose we are interested in comparing the cholesterol levels of two groups of quail—one group that receives a treatment designed to lower cholesterol and one that does not. A random sample of $n = 30$ quail were selected for control, and a sample of $m = 20$ quail were selected for treatment. The data are presented in Fig. A.4.4, panel A, with comparison boxplots shown in Figure A.4.4, panel B. These indicate that the assumption of a common variance is not valid so we will perform an unpooled analysis.

A 95% confidence interval for the mean difference in the two populations $\mu_x - \mu_y$ (control – treatment) is (−2.14 mg/dl, 6.18 mg/dl). To test the hypothesis that the treatment is effective in

Table A.4.7 Decision Rules for the Two Independent Sample *t*-Test with Unpooled Variance

Hypotheses	Decision rule	*p*-value
$H_0{:}\mu_x{=}\mu_y$ vs. $H_1{:}\mu_x{\neq}\mu_y$	Reject H_0 if $\lvert t^*\rvert > t_{1-\alpha/2,df}$	$2*P(t_{df} > \lvert t^*\rvert)$
$H_0{:}\mu_x{\geq}\mu_y$ vs. $H_1{:}\mu_x{<}\mu_y$	Reject H_0 if $t^* < -t_{1-\alpha,df}$	$P(t_{df} < t^*)$
$H_0{:}\mu_x{\leq}\mu_y$ vs. $H_1{:}\mu_x{>}\mu_y$	Reject H_0 if $t^* > t_{1-\alpha,df}$	$P(t_{df} > t^*)$

A

control ($n = 30$)
44 51 50 52 41 69 56 67 45 37 40 44 46 55 50 60 65 53 46 38 58 58 48 29 65 62 59 45 61 56

treated ($m = 20$)
50 49 46 58 50 59 58 48 44 41 49 47 49 48 45 48 49 51 52 52

B

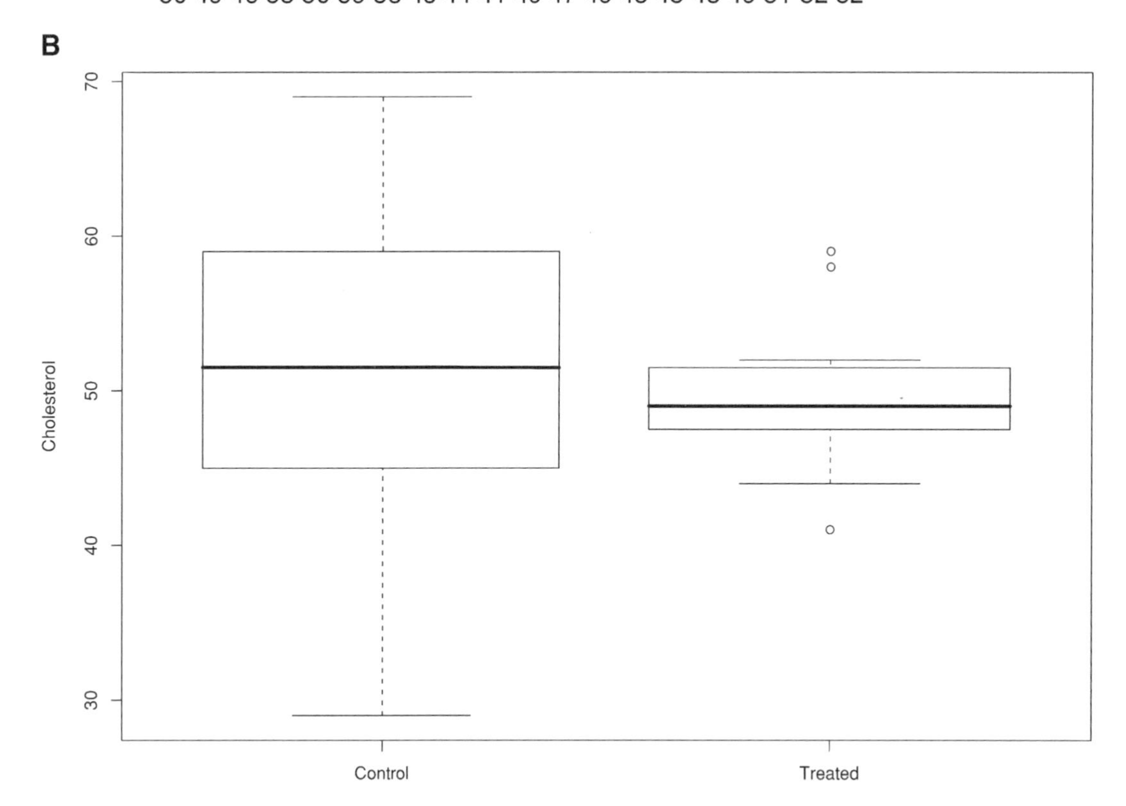

Figure A.4.4 (**A**) A random sample was taken of $m = 20$ quail receiving a treatment designed to lower cholesterol, to be compared with a random sample of $n = 30$ untreated quail selected for control. (**B**) Comparison boxplots of cholesterol levels of untreated and treated quail.

lowering cholesterol:

$$H_0 : \mu_x \leq \mu_y \text{ versus } H_1 : \mu_x > \mu_y$$

The t-test statistic is $t^* = 0.977$, which is not significant (p-value $= 0.1670$). So, based on these data we do not have significant evidence to conclude that the treatment is effective in lowering cholesterol.

Pooled inference

Assumptions

1. The response variable is numeric.
2. The two distributions have the same shape and variance.
3. Two independent samples are drawn from the two distributions.
4. If the distributions are symmetric, you should have at least $n > 15$ in each group. If the distributions are not symmetric, you should have at least $n > 30$ in each group.

Interval estimation

An interval estimate for the difference in the population means is:

$$\bar{x} - \bar{y} \pm t_{1-\alpha/2,m+n-2} S_p \sqrt{\frac{1}{m} + \frac{1}{n}}$$

where:

$$S_p = \sqrt{\frac{(m-1)s_x^2 + (n-1)s_y^2}{m+n-2}}$$

is a pooled estimate of the population standard deviation (σ) of the two groups.

Hypothesis testing

The pooled t-statistic is:

$$t^* = \frac{\bar{x} - \bar{y}}{s_p \sqrt{\frac{1}{m} + \frac{1}{n}}}$$

where s_p is as described above for the interval estimate. Hypotheses are shown in Table A.4.8.

Example

We wish to test the hypothesis that men are taller than women. Suppose that the heights, in inches, of a sample of $m = 20$ women and a sample of $n = 22$ men are as shown in Figure A.4.5, panel A. Let μ_x denote the true mean height for women and μ_y denote the true mean height for men. The comparison boxplots (Fig. A.4.5, panel B) for these data indicate that the assumption of equal variances is appropriate. The hypotheses are:

$$H_0 : \mu_x \geq \mu_y \text{ versus } H_1 : \mu_x < \mu_y$$

From software, we obtain the t-test statistic, $t^* = 6.8054$. Since the value observed is so extreme we may say that the p-value is approximately zero and conclude that indeed men are taller than women. The confidence interval for $\mu_x - \mu_y$ reported by the software is (−8.02 inches, −4.35 inches). This says that, on average, women are between about 4.5 and 8 inches shorter than men, with 95% confidence.

Nonparametric Procedures

Though the Mann-Whitney-Wilcoxon (MWW) test is general in that it tests if one population is larger than another, it is often used to test for differences in the medians of the two populations, in which case the populations are assumed to have the same shape. For example, one might want to know which bacterial strain, X or Y, has a longer lifespan. The distributions of the two random

Table A.4.8 Decision Rules for the Two Independent Sample t-Test With Pooled Variance

Hypotheses	Decision rule	p-value
$H_0{:}\mu_x = \mu_y$ vs. $H_1{:}\mu_x \neq \mu_y$	Reject H_0 if $\lvert t^* \rvert > t_{1-\alpha/2,m+n-2}$	$2*P(t_{m+n-2} > \lvert t^* \rvert)$
$H_0{:}\mu_y \geq \mu_z$ vs. $H_1{:}\mu_y <; \mu_z$	Reject H_0 if $t^* < -t_{1-\alpha,m+n-2}$	$P(t_{m+n-2} < \lvert t^* \rvert)$
$H_0{:}\mu_y \leq \mu_z$ vs. $H_1{:}\mu_y > \mu_z$	Reject H_0 if $t^* > t_{1-\alpha,m+n-2}$	$P(t_{m+n-2} > t^*)$

A

heights of women in inches ($m = 20$)
68 60 61 60 60 59 63 59 63 65 70 61 67 66 61 65 68 63 64 67

heights of men in inches ($n = 22$)
68 65 66 68 72 70 73 72 71 70 69 71 66 72 69 73 70 72 65 69 70 72

B

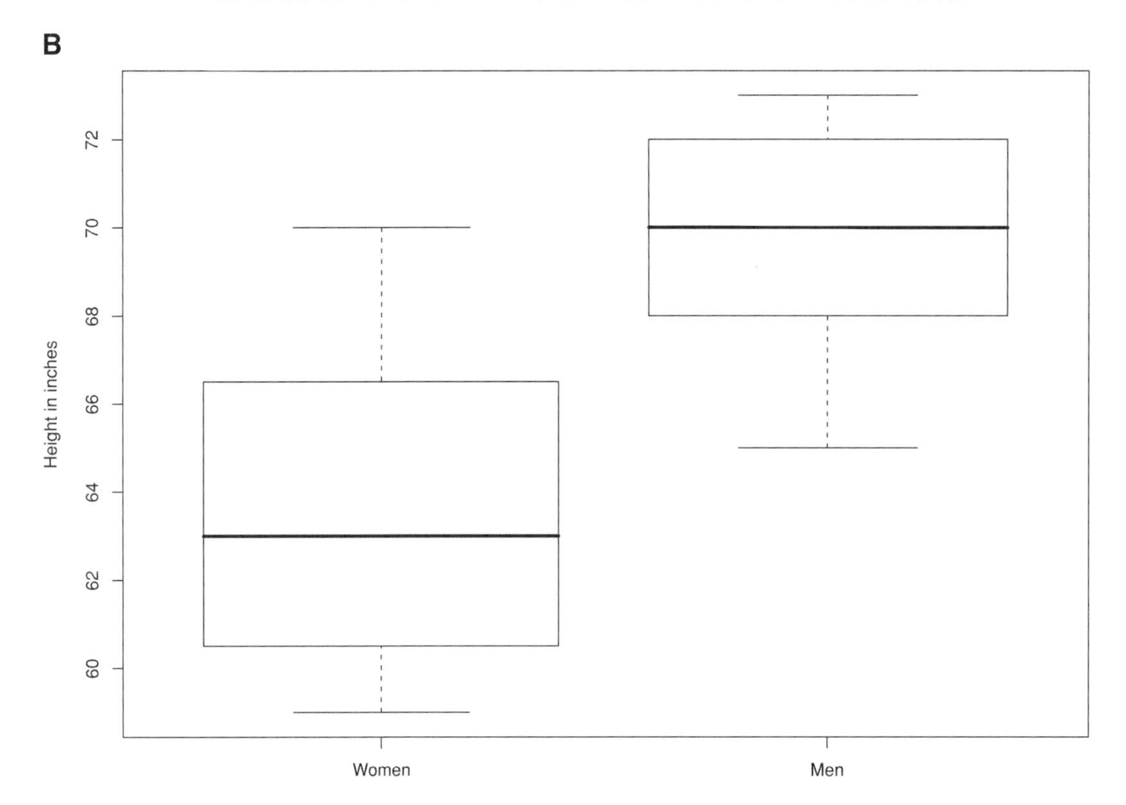

Figure A.4.5 (**A**) To test the hypothesis that men are taller than women, the heights in inches of $m = 20$ women and $n = 22$ men were measured. (**B**) Comparison boxplots of heights of women and men.

variables might be quite different in shape, etc. A rejected null hypothesis says that Y tends to outlive X. We will discuss only the inference on medians.

Inference

Assumptions

1. The two samples are independent.

2. The response variable is numeric.

3. The two population differ possibly only in location.

4. The sample sizes are sufficiently large: $m \geq 10$ and $n \geq 10$.

Test of hypothesis

The MWW test statistic is:

$$T = \sum_{t=1}^{n} R(x_i)$$

Table A.4.9 Decision Rules for the Mann-Whitney-Wilcoxon Test

Hypotheses	Decision rule	p-value
$H_0:\theta_x=\theta_y$ vs. $H_1:\theta_x\neq\theta_y$	Reject H_0 if $\lvert z^*\rvert>z_{1-\alpha/2}$	$2*P(Z>\lvert z^*\rvert)$
$H_0:\theta_x\geq\theta_y$ vs. $H_1:\theta_x<\theta_y$	Reject H_0 if $z^*<-z_{1-\alpha}$	$P(Z<z^*)$
$H_0:\theta_x\leq\theta_y$ vs. $H_1:\theta_x>\theta_y$	Reject H_0 if $z^*<z_{1-\alpha}$	$P(Z>z^*)$

where $R(x_i)$ denotes the rank of x_i among the combined sample $x_1,...,x_n y_1...,y_m$. The large sample standardized test statistic is:

$$z^* = \frac{T - m(n+m+1)/2}{\sqrt{nm(n+m+1)/12}}$$

Hypotheses are shown in Table A.4.9.

Interval estimation

Similar to the one sample location problem, an interval estimate based on the MWW test is the set of values for which the null hypothesis is not rejected.

Example

Using the height data from the pooled inference example (Fig. A.4.5, panel A), comparison boxplots (Fig. A.4.5, panel B) show that the assumption of a *shift model* is appropriate. The appropriate hypotheses to answer if men have a median height which is greater than the median height of all women are:

$$H_0 : \theta_x \geq \theta_y \text{ versus } H_1 : \theta_x < \theta_y$$

The value of the rank sum statistic, adjusted for ties, is $T = 244$, and the standardized version is $z^* = -4.684$. Since the p-value, equal to 1.406, is small (less than $\alpha = 0.05$), the null hypothesis is rejected. There is significant evidence that men are taller then women. A 95% confidence interval for the shift $\theta_x - \theta_y$ is (−9 inches, −4 inches).

THE ONE SAMPLE PROPORTION PROBLEM

The one sample proportion problem is concerned with inference on one population proportion or the probability of success of some process. For example, we might be interested in the success rate of a new treatment for some disease.

Data

Let x be the number of successes out of n. Then $\hat{p} = x/n$ is a point estimate of p.

Assumptions

1. The observations are independent.

2. The sample size n is large enough so that $np_0 > 5$ and $n(1 - p_0) > 5$ for testing.

3. The sample size n is large enough so that $n\hat{p} > 5$ and $n(1 - \hat{p}) > 5$ for interval estimation.

Interval Estimation

A $(1 - \alpha) \times 100\%$ confidence interval for p, the population proportion, is:

$$\hat{p} \pm z_{1-\alpha/2}\sqrt{\frac{\hat{p}(1-\hat{p})}{n}}$$

Hypothesis Testing

The test statistic is:

$$z^* = \frac{\hat{p} - p_0}{\sqrt{\frac{p_0(1-p_0)}{n}}}$$

Hypotheses are shown in Table A.4.10.

Example

It is well known that the cure rate for the standard treatment of a certain disease is 0.45. A new treatment has been developed, and we wish to test if the new treatment has a higher cure rate. The hypotheses of interest are:

$$H_0 : p \leq 0.45 \text{ versus } H_1 : p > 0.45$$

A4

Out of a sample of $n=99$ patients with the disease, $x=46$ of them were cured, so that $\hat{p}=46/99=0.4646$ is a point estimate of the true cure rate. An 95% confidence interval estimate is:

$$0.4646 \pm 1.96\sqrt{0.4646 \times 0.5354/99} = 0.4646 \pm 0.0982 \text{ (or } 0.3665, 0.5629)$$

That is, the true cure rate for the new drug is between about 37% and 56% with 95% confidence. The test statistic is:

$$z^* = \frac{0.4646 - 0.45}{\sqrt{\frac{0.4646 \times 0.5354}{99}}} = 0.2913$$

The p-value of the test is 0.3854, which indicates that there is not significant evidence to say that the new treatment is better at curing the disease than the current standard treatment.

The researcher may choose to do only an interval estimate or a hypothesis or both. The determination should be made based on the research question of interest. If interest is in getting an estimate of the rate, an interval estimate should be created. If interest is in testing a particular plausible value, a hypothesis test should be done. If the problem is new to the literature, both methods might be applied.

Table A.4.10 Decision Rules for the One Sample Test of a Proportion

Hypotheses	Decision rule	p-value
$H_0{:}p{=}p_0$ vs. $H_1{:}p{\neq}p_0$	Reject H_0 if $\lvert z^*\rvert > z_{1-\alpha/2}$	$2^*P(Z>\lvert z^*\rvert)$
$H_0{:}p{\geq}p_0$ vs. $H_1{:}p{<}p_0$	Reject H_0 if $z^* < -z_{1-\alpha}$	$P(Z<z^*)$
$H_0{:}p{\leq}p_0$ vs. $H_1{:}p{>}p_0$	Reject H_0 if $z^* > z_{1-\alpha}$	$P(Z>z^*)$

THE TWO INDEPENDENT SAMPLES PROPORTION PROBLEM

The two sample proportion problem is concerned with inference on the difference between two population proportions or the difference in the success rates of two treatments. For example if treatment 1 has success rate p_1 and treatment 2 has success rate p_2 (both unknown, both needing to be estimated from the data), we may want to test if treatment 2 has a higher success rate than treatment 1. We will provide information on the degree of difference of success rate.

Data

Let x_1 be the number of successes out of n_1 for treatment 1. Let x_2 be the number of successes out of n_2 for treatment 2. Then, $\hat{p}_1 = x_1/n_1$ and $\hat{p}_2 = x_2/n_2$ are the sample proportions.

Assumptions

1. The two samples are independent.

2. The sample sizes n_1 and n_2 are large enough so that $n_i \times \hat{p}_i > 5$ and $n_i(1 - \hat{p}_i) > 5$.

Interval Estimation

An interval estimate for the difference in the two populations' proportions $(p_1 - p_2)$ is:

$$\hat{p}_1 - \hat{p}_2 \pm z_{1-\alpha/2}\sqrt{\frac{\hat{p}_1(1-\hat{p}_1)}{n_1} + \frac{\hat{p}_2(1-\hat{p}_2)}{n_2}}$$

A4

Hypothesis Testing

The test statistic is:

$$z^* = \frac{\hat{p}_1 - \hat{p}_2}{\sqrt{\hat{p}(1-\hat{p})\left(\frac{1}{n_1} + \frac{1}{n_2}\right)}}$$

where:

$$\hat{p} = \frac{x_1 + x_2}{n_1 + n_2}$$

Hypotheses are shown in Table A.4.11.

Table A.4.11 Decision Rules for the Two Independent Sample Test of Proportions

Hypotheses	Decision rule	*p*-value
$H_0{:}p_1{=}p_2$ vs. $H_1{:}p_1{\neq}p_2$	Reject H_0 if $\lvert z^*\rvert > z_{1-\alpha/2}$	$2^*P(Z > \lvert z^*\rvert)$
$H_0{:}p_1{\geq}p_2$ vs. $H_1{:}p_1{<}p_2$	Reject H_0 if $z^* < -z_{1-\alpha}$	$P(Z < z^*)$
$H_0{:}p_1{\leq}p_2$ vs. $H_1{:}p_1{>}p_2$	Reject H_0 if $z^* > z_{1-\alpha}$	$P(Z > z^*)$

Examples

Suppose we are interested in knowing if more patients respond to a new treatment (treatment 1) than a standard treatment (treatment 2). That is, we are interested in comparing the response rates of the two treatments, $p_1$and p_2. In a sample of $n_1 = 100$ subjects on treatment 1, there were $x_1 = 33$ responders. In a sample of $n_2 = 110$ subjects on treatment 2, there were $x_1 = 20$ responders. The point estimates for the response rates are $\hat{p}_1 = 33/100 = 0.330$ and $\hat{p}_2 = 20/110 = 0.182$. A 95% confidence interval is:

$$0.330 - 0.182 \pm 1.96\sqrt{\frac{0.330 \times 0.670}{100} + \frac{0.182 \times 0.818}{110}}$$

or (0.031, 0.265). With 95% confidence, treatment 1 has a response rate that is between 3% and 26.5%, higher than the response rate of treatment 2.

Since we are interested in knowing if treatment 1 has a higher response rate than treatment 2 we perform a lower tail test (that is, we test whether treatment 1 has a higher response rate than treatment 2, instead of simply testing that the response rates are different):

$$H_0 : p_1 \leq p_2 \text{ versus } H_1 : p_1 > p_2$$

The test statistic we observe is:

$$z^* = \frac{0.330 - 0.182}{\sqrt{0.252 \times 0.748\left(\frac{1}{100} + \frac{1}{110}\right)}} = 2.477$$

since:

$$\hat{p} = \frac{33 + 20}{100 + 110} = 0.252$$

Because $z^* = 2.477 > 1.645 = z_{0.95}$, we would reject the null hypothesis and conclude that treatment 2 is probably doing better. Also p-value $= 0.0066 < 0.05 = \alpha$.

ANOVA

Analysis of Variance (ANOVA) is used to test for a difference in the means of three or more groups. For example, one might want to know if the mean blood pressure is the same or different across several ethnic groups. One could also use ANOVA to test for differences in blood pressure for three different treatment groups. The main assumption is that the variability across the groups is the same. If this assumption is not met, then pair-wise t-tests are appropriate, though a correction for multiple comparisons should be made (see below). For a thorough treatment of ANOVA and related topics, see Kutner et al. (2005).

Inference

Assumptions

1. The response variable is numeric.

2. The explanatory variable is categorical.

3. Variability across groups is the same ($\sigma_1 =...=\sigma_k = \sigma$).

4. If the distribution is symmetric, you should have at least $n > 15$ in each group. If the distribution is not symmetric, you should have at least $n > 30$ in each group.

Hypothesis testing

$$H_0 : \mu_1 = \mu_2 = \ldots = \mu_k$$
$$H_1 : \mu_i \neq \mu_j \text{ for some } i \neq j$$

The test statistic for the analysis of variance test of means is beyond the scope of this appendix. Computer software will provide you with a test statistic and a p-value. Reject H_0 if the p-value is less than the specified level of significance (usually $\alpha = 0.05$ or 0.01).

Confidence intervals

If, after running the hypothesis test, you want to find a confidence interval for a difference in means, refer to the section on Two Sample Location Problem (Parametric, Pooled Inference). However, instead of using the pooled estimate of variance (s_p^2), use the mean squared error, MSE, from the ANOVA output. Additionally, the degrees of freedom for the t critical value (see discussion of t^*, above) will be the degrees of freedom in the residual (or error) row from the ANOVA table.

Example

Suppose we want to compare the average life of three strains of yeast. The first strain (C) is a control strain, the second strain (D) has a transcription factor gene deleted, and the third strain (A) has an extra copy of the same gene added. We want to know whether modifying the transcription factor gene changes the average lifespan (in generations) of the strain of yeast. We collect 80 data points, summary statistics are shown in Table A.4.12.

As seen in the data table as well as the boxplots (Fig. A.4.6), the data are consistent with the assumptions (numeric, symmetric, constant variance).

Using statistical software, the ANOVA table shown in Table A.4.13 is obtained. The ANOVA table gives a small p-value, which leads to rejection of the null hypothesis that the average lifespans are the same across all three groups. However, it is not clear which of the three groups are different. Confidence intervals for each of the pair-wise differences of population means are calculated using the following formula, for an interval estimate for the difference in two population means ($\mu_1 - \mu_2$):

$$\bar{x}_1 - \bar{x}_2 \pm t_{\alpha^*/2,\mathrm{df}} \sqrt{\mathrm{MSE}}\sqrt{1/n_1 + 1/n_2}$$

Table A.4.12 Average Life Span of Three Strains of Yeast

	Control	Deletion	Addition
Sample size	30	27	23
Avg. lifespan	15.89	13.85	16.98
Std. dev.	2.84	2.85	2.65

Table A.4.13 ANOVA Table Obtained from Yeast Life Span Data

	Degrees of freedom (df)	Sum of squares	Mean square	F test statistic	p-value
Groups	2	128.7	64.35	8.259	0.0005631
Residual	77	599.95	7.79		

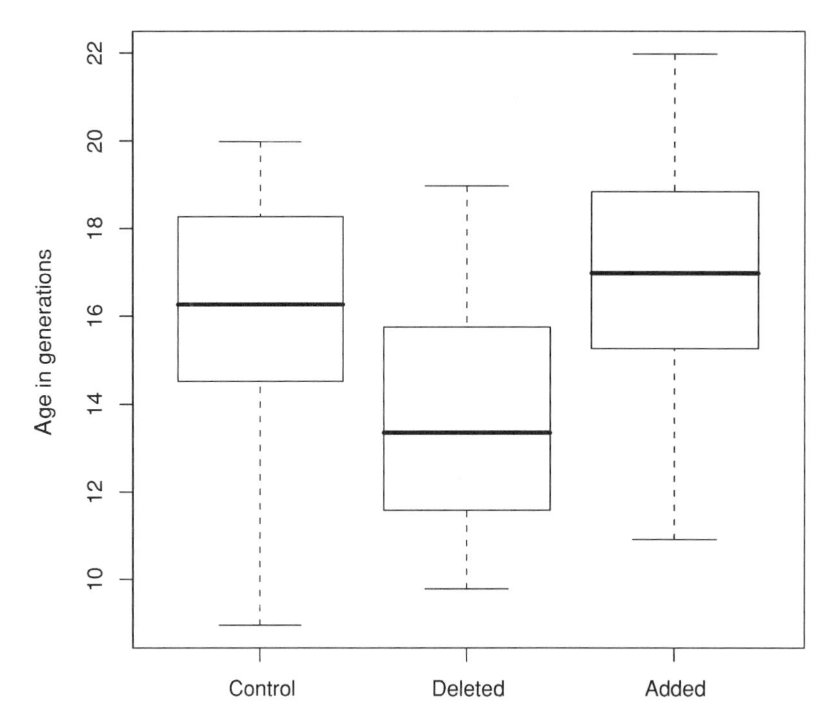

Figure A.4.6 Boxplot of the age in generations for three different strains of yeast. Corresponding data are tabulated in Table A.4.12.

Table A.4.14 Estimates of Pair-Wise Differences of Mean Lifespan for Different Strains of Yeast

Parameter	Estimate	Confidence interval
μ_C-μ_D	2.045	(0.233 generations, 3.857 generations)
μ_C-μ_A	−1.087	(−2.980 generations, 0.806 generations)
μ_A-μ_D	3.132	(1.194 generations, 5.070 generations)

Here, MSE = 7.79 and df = 77. Confidence intervals (Table A.4.14) for the three groups at the $\alpha^* = 0.05/3 = 0.017$ level (98.3% confidence) are created; see the Multiple Comparisons section below. The multiplier is $t_{.0083,77} = 2.45$.

Because neither of the confidence intervals for the deleted strain overlap zero, the average lifespan for the deleted strain is significantly different from either the wild-type or the addition strain, with a family-wise confidence rate of 95%. However the wild-type is not significantly different from the addition strain.

Multiple Comparisons

When computing a one-level α hypothesis test (in any setting, not just ANOVA), the probability of rejecting H_0 when H_0 is really true is set at α. Similarly, when finding a confidence interval, the probability of missing the true parameter is also α (or 1 minus the level of confidence). Notice that in 100 hypothesis tests where nothing interesting is going on (no alternative hypothesis is true), on average, five true null hypotheses will be rejected when using $\alpha = 0.05$. Similarly, out of 100 95% confidence intervals, on average, five of them will not overlap the true parameter of interest. This problem of multiple comparisons happens in any situation where there is more than one hypothesis test or confidence interval.

When finding confidence intervals for differences of means after running an ANOVA test, multiple comparisons should be considered. If we have three experimental treatments, we might want to find three confidence intervals for differences of means: 1 versus 2, 1 versus 3, and 2 versus 3. There are many ways to adjust for multiple comparisons; the Bonferroni correction is discussed here. Note that the Bonferroni adjustment is quite conservative.

Instead of controlling the type I error rate (rejecting a true null hypothesis) for each test, the Bonferroni adjustment controls the family-wise error rate. The family-wise error rate is the probability of rejecting one or more true hypotheses when doing numerous hypothesis tests. For confidence intervals, a family-wise error rate would be the probability of at least one confidence interval failing to capture the true parameter of interest.

When running k hypothesis tests, the level of significance should be adjusted to $\alpha^* = \alpha/k$. Then, α^* is used as the significance level for each individual test, and α is the conservative family-wise error rate. For example, if 12 hypothesis tests are to be performed, and a family-wise error rate of no more than 0.05 is desired, each test should be performed at level $\alpha = 0.05/12 = 0.00417$. That is, for each of the k hypothesis tests, reject the null hypothesis if the p-value is less than $0.05/12 = 0.00417$. In order to control the family-wise error rate for the 12 confidence intervals, create 99.583% confidence intervals.

Nonparametric Procedure: Kruskal-Wallis Test

An extension of the Mann-Whitney-Wilcoxon test (see above) is the Kruskal-Wallis test. Kruskal-Wallis is similar to the one-way ANOVA approach discussed previously. However, Kruskal-Wallis is a test for medians, not means.

A4

Assumptions

1. The k samples are independent.
2. The response variable is numeric.
3. The k populations differ only in location.

$$H_0 : \theta_1 = \theta_2 = \ldots = \theta_k \text{ versus } H_A : \theta_i \neq \theta_j \text{ for some } i \neq j$$

Example

Continuing the yeast example and using software to obtain the p-value $= 0.0011$, as with the parametric approach, the null hypothesis is rejected.

REGRESSION

Regression analysis is a statistical technique designed to fit a straight line through a cloud of points. Let the response variable be called y and the explanatory variable be called x; then the regression analysis will find b_0 (the y-intercept) and b_1 (the slope) such that:

$$y = b_0 + b_1 x$$

is the best-fit line between x and y. For data, we have a random sample of n pairs of observations: $(x_1,y_1),(x_2,y_2),\ldots,(x_n,y_n)$. For a thorough treatment of regression analysis, see Kutner et al. (2005).

Inference

Assumptions

1. y must be a continuous and numerical variable.

2. x must be a numerical variable.

3. y should be normally distributed around the regression line.

4. The variability around the regression line should be relatively constant for all values of x.

5. The y values should all be independent observations.

6. A sample of $n > 25$ pairs of observations is needed.

In most situations, the hypothesis of interest is whether or not y changes linearly with x (i.e., whether x and y are correlated). If the slope of the regression line is close to zero, we say that an increase in x is not statistically associated with any linear change in y, or that x and y are not significantly correlated.

Interval estimation

Let β_1 be the slope of the regression on the population of xs and ys. A $(1 - \alpha) \times 100\%$ confidence interval for β_1 is:

A4

$$b_1 \pm t_{(1-\alpha/2,n-2)} s(b_1)$$

where b_1 and $s(b_1)$ are computed using statistical software (formulae can also be found in the introductory statistics texts listed in Literature Cited).

Hypothesis testing

One-sided and two-sided tests concerning the populations slope, β_1, are constructed based on the following test statistic:

$$t^* = \frac{b_1}{s(b_1)}$$

Table A.4.15 contains the decision rules for the three possible cases, with the probability of making a type I error controlled at α.

Example

Consider a situation investigating the linear relationship between lizard speed and outside temperature. In particular, interest is in determining whether a change in lizard speed at 20°C (the explanatory variable, x) is linearly associated with lizard speed at 35°C (the response variable, y). A sample of 47 data points are collected (Fig. A.4.7); the first 12 points (in m/sec) are shown in Table A.4.16.

Table A.4.15 Decision Rules for Tests of the Slope for Linear Regression

Hypotheses	Decision rule	p-value
H_0:β_1=0 vs. H_1:$\beta_1 \neq 0$	Reject H_0 if $\lvert t^* \rvert > t_{1-\alpha/2,n-2}$	$2P(t_{n-2} > \lvert t^* \rvert)$
H_0:$\beta_1 \geq 0$ vs. H_1:$\beta_1 < 0$	Reject H_0 if $t^* < -t_{1-\alpha,n-2}$	$P(t_{n-2} < t^*)$
H_0:$\beta_1 \leq 0$ vs. H_1:$\beta_1 > 0$	Reject H_0 if $t^* > t_{1-\alpha,n-2}$	$P(t_{n-2} > t^*)$

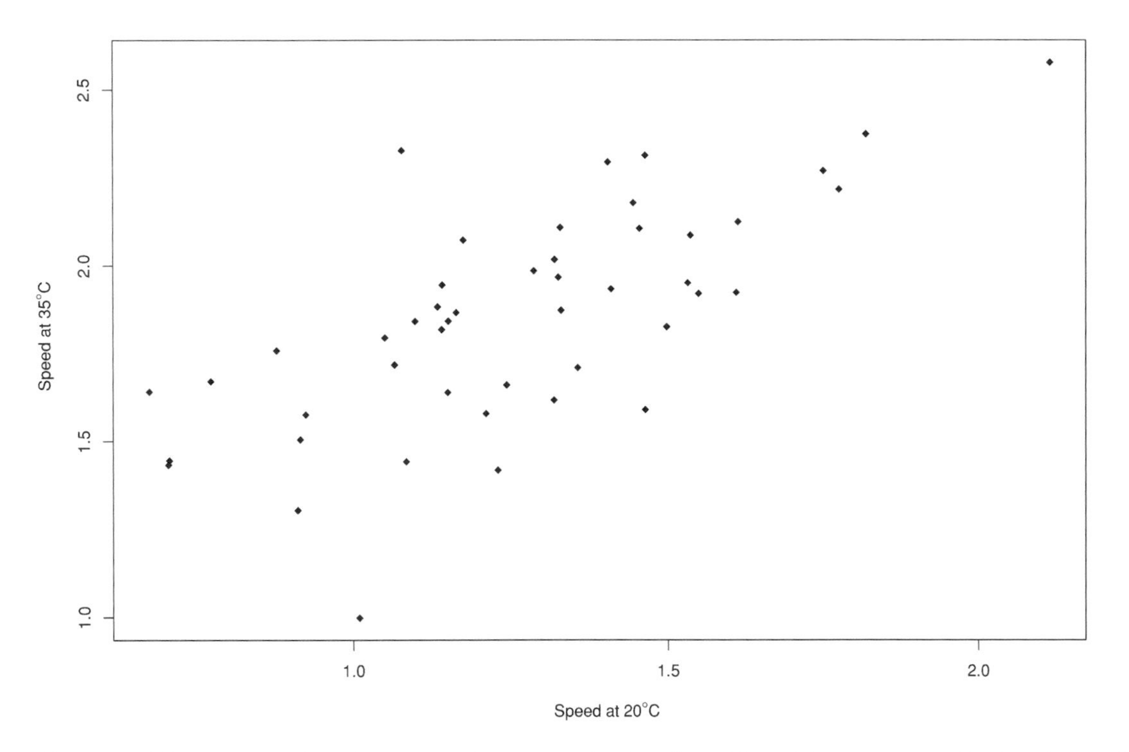

Figure A.4.7 Scatterplot of lizard speed on a racetrack at two different temperatures. Each point on the graph represents a particular lizard. The first 12 data points are tabulated in Table A.4.16. The speeds of 47 independent lizards were measured at 20°C and 35°C. We can see that there is a positive relationship between the two variables. However, because of the natural variability, we cannot predict perfectly the speed at one temperature from the speed at the other temperature.

A4

Table A.4.16 First Twelve Data Points Collected Correlating Lizard Speed and Outside Temperature

Lizard #	1	2	3	4	5	6	7	8	9	10	11	12
Speed at 20°C	0.96	1.05	1.18	1.32	1.37	1.59	1.39	1.23	1.74	1.13	1.06	1.30
Speed at 35°C	1.85	1.31	1.73	1.42	2.18	1.79	1.85	1.92	2.33	1.78	1.48	2.03

Analysis

Checking assumptions: the points are reasonably spread out around a line with constant variance across the explanatory variable. The relevant test is:

$$H_0 : \beta_1 \leq 0 \text{ no linear relationship or negative linear relationship}$$
$$H_1 : \beta_1 > 0 \text{ positive linear relationship}$$

From the software, $b_1 = 0.723$ and $s(b_1) = 0.109$, which give $t^* = 6.660$. The critical value is $t_{1\text{-}\alpha,n\text{-}2} = t_{.95,45} = 1.68$, so the null hypothesis is rejected; there is significant evidence to conclude a positive linear relationship between speed at 20°C and 35°C. Additionally, the associated *p*-value is 0.0000033, which says that if the two speeds were not linearly associated, we would only see data such as that obtained 0.00033% of the time, which is very unusual. The *p*-value is smaller than 0.05, again confirming our decision to reject H_0. Figure A.4.8 shows the resulting regression line.

$$b_1 \pm t_{0.975,10}\, s(b_1)\; 0.706 \pm 2.01 \times 0.109\; (0.504 \text{ m/sec}, 0.942 \text{ m/sec})$$

We are 95% confident that for every increase of 1 m/sec in speed at 20°C the speed at 35°C increases by between 0.504 m/sec and 0.942 m/sec.

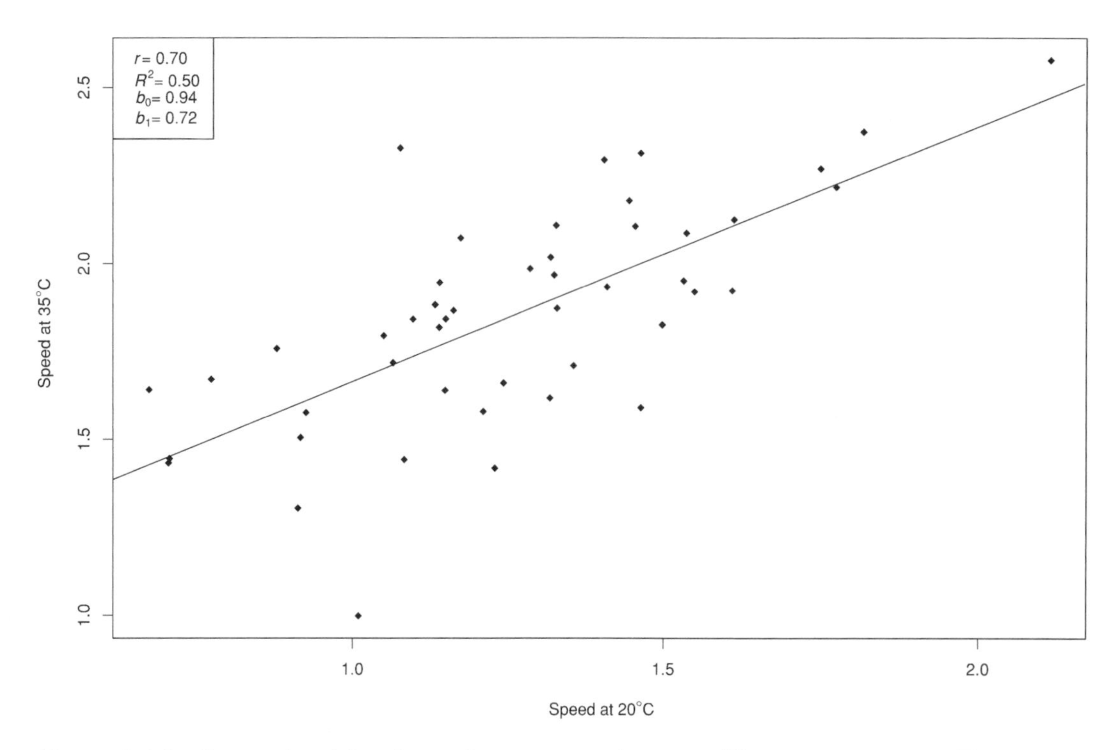

Figure A.4.8 Scatterplot of lizard speed on a racetrack at two different temperatures. Each point on the graph represents a particular lizard (also see Fig. A.4.7 and Table A.4.16). The line represents the regression line or least squares fit.

Note that regression analysis should only be done for *x*-axis values that are within the constraints of the data. It would not make sense to predict the running speed of a lizard who runs 10 m/sec or 0 m/sec, as neither of those explanatory values are realistic. Similarly, it is not wise to predict values outside the *x*-data, because one is never certain that the same linear model holds for all possible speeds.

CORRELATION

Related to linear regression are ideas of correlation. Correlation (typically denoted by *r*) measures the degree of linear association between two continuous variables. If two variables have a perfect positive linear relationship (e.g., miles and kilometers), they have a correlation of one (1). If two variables have a perfect negative linear relationship (e.g., number right and number wrong on an exam), they have a correlation of negative one (−1). If two variables are not correlated (e.g., age and beak length for adult chickens), they have a correlation of zero (0).

Inference

Although formulae exist for calculating the correlation between two continuous variables, we will use statistical software for calculations. The value of *r* gives an idea of how far the data points fall from a line. As mentioned above, the sign of *r* (either positive or negative) gives an indication of the relationship (either positive or negative) between the two variables. However, r^2 is often reported and interpreted as the proportion of variability explained by the regression model. Interest is usually in testing whether the population correlation (ρ) is significantly different from zero.

Hypothesis testing

The test for correlation at zero is the same test for the population slope at zero. Therefore, the summary of the correlation test is shown in Table A.4.17. An estimate of the population correlation using the sample correlation, *r*, is obtained using statistical software.

Table A.4.17 Decision Rules for Tests of the Population Correlation

Hypotheses	Decision rule	p-value
H_0:ρ=0 vs. H_1:ρ≠0	Reject H_0 if $\|t^*\|>t_{1-\alpha/2,n-2}$	$2P(t_{n-2}>\|t^*\|)$
H_0:ρ≥0 vs. H_1:ρ<0	Reject H_0 if $t^*<-t_{1-\alpha,n-2}$	$P(t_{n-2}<t^*)$
H_0:ρ≤0 vs. H_1:ρ>0	Reject H_0 if $t^*>t_{1-\alpha,n-2}$	$P(t_{n-2}>t^*)$

Example

Continuing with the lizard example above, correlation is tested using b_1, the estimated slope. Note that, because it was concluded that the population slope was significantly larger than zero we may conclude that the correlation between running speed at 20°C and 35°C is significantly positive. The estimated value is $r=0.704$. Note that here, $r^2=0.496$, which says that 49.6% of the variability in the running speeds at 35°C can be explained by the information given by the running speeds at 20°C.

Correlation does not imply causation

It is important to remember that when measuring the correlation between two variables, a significant correlation does not mean that one variable causes the other variable to increase or decrease. As an example, notice that ice cream sales and boating accidents tend to be strongly positively correlated. However, no one would believe that eating ice cream causes boating accidents. Both increase when the weather is warm. Be very careful with your conclusions when reporting correlations.

A4

CHI SQUARE (χ^2)

Chi square analysis is a technique used to compare two categorical variables. Usually, the question of interest is whether or not the two variables are independent—i.e., does information about one variable provide information about the second variable (and indicate a dependent relationship). The variables can be binary or have multiple levels.

Assumptions

1. Independent random samples from two or more populations, with each subject classified according to one categorical variable (the other categorical variable represents the population from which the subject came).

Or

2. A single simple random sample where each subject is classified into each of two categorical variables.

Additionally

3. All expected cell counts must be at least 1, and no more than 20% of the counts can be less than 5.

The data will be organized in a table such as that shown in Table A.4.18 (here we have M rows and N columns).

Inference

In general, the null hypothesis (H_0) is that there is no relationship between the two categorical variables. When the null hypothesis is true, the expected data values are easily calculated. That is, to investigate whether smoking and gender are related, start by figuring out what proportion of

Table A.4.18 Example of the Structure for a Data Set Which Has *N* Categories for the First Variables and *M* Categories for the Second Variable

	Variable 1			
Variable 2	Group 1	Group 2	...	Group N
Group A				
Group B				
...				
Group M				

smokers are female (and male) if the two variables are not related. That is, we would expect about half of the smokers to be male and half to be female if, in fact, the two variables were not related.

Expected counts

The expected count represents the number of subjects expected in each cell if in fact the null hypothesis is true (i.e., if there is no relationship between the two variables):

$$\text{expected counts} = \frac{\text{row total} \times \text{column total}}{\text{table total}}$$

A4

Hypothesis testing

With the chi square test, there is no obvious population parameter that is being estimated (unlike the one sample *t*-test of μ, where inference is about the population mean.) Here, the χ^2 test statistic combines the expected and observed counts. The test statistic will allow us to make conclusions about the null and alternative hypotheses:

$$\chi^{2*} = \sum \frac{(\text{observed count} - \text{expected count})^2}{\text{expected count}}$$

A chi square statistic is distributed according to the chi square distribution. There are chi square tables in the backs the introductory reference books, or such a table can be generated using statistical software. The chi square table is sorted by degrees of freedom. For the chi square test of independence, the degrees of freedom are based on the number of rows and columns you have in the data table:

$$\text{degrees of freedom} = \text{df} = (\text{number of rows} - 1)(\text{number of columns} - 1)$$

Note that the chi square test is always one-sided in the sense of rejecting the null hypothesis and calculating a *p*-value. Notice that if the null hypothesis is not true, the expected counts will be much different from the observed counts. It does not matter in which direction the null hypothesis is wrong; any direction will produce a large test statistic. Therefore, always reject the null hypothesis when the test statistic is big. The decision rule for a chi square test are given in Table A.4.19.

Example

Suppose that one is interested in determining whether or not there is a relationship between type of bird egg and ability of the egg to withstand an overnight freeze. Data on 100 eggs are collected and tabulated as shown in Table A.4.20.

Table A.4.19 Decision Rules for a Chi-Square Test

Hypotheses	Decision rule	p-value
H_0: there is no relationship between the two categorical variables	Reject H_0 if $\chi^{2*} > \chi^2_{1-\alpha,df}$	$P(\chi^2_{df} > \chi^{2*})$

Table A.4.20 Relationship Between Type of Bird Egg and Ability to Withstand Overnight Freezing

	Egg status		
Egg type	Okay	Cracked	Broken
Robin	9	8	4
Wren	11	12	7
Sparrow	5	10	9
Cuckoo	9	11	5

Table A.4.21 Expected Number of Eggs In Each Category Assuming No Relationship Between the Two Variables (i.e., Assuming H_0 is True)

	Egg status		
Egg type	Okay	Cracked	Broken
Robin	7.14	8.61	5.25
Wren	10.20	12.30	7.50
Sparrow	8.16	9.84	6.00
Cuckoo	8.50	10.25	6.25

Table A.4.21 shows the expected values computed assuming that the null hypothesis of no relationship across variables is true. Note that the expected values do not seem particularly different from the observed values.

The associated test statistic is $\chi^2 = 3.99$, with a p-value $= 0.678$. The p-value is large and says that if the null hypothesis is true, we would see data at least as extreme as ours about 67.8% of the time. The large p-value leads to failure to reject the null hypothesis. There is no significant evidence that egg type and egg status are related.

Only two categories per variable

A special case occurs when there are only two levels (or categories) for one of the variables. For example, suppose a cracked egg is considered to be broken (i.e., we have collapsed the cracked and broken data into one column). The data then become those shown in Table A.4.22.

The resulting test statistic is $\chi^2 = 2.73$ with a p-value $= 0.4355$. The null hypothesis of no relationship is still unable to be rejected. However, the hypotheses for this test are stated and interpreted slightly differently (though the computations are the same as when there are more than two rows or columns). Note that the hypotheses here can be stated as:

$$H_0: \quad p_R = p_W = p_S = p_C$$

H_1: at least one of the proportions is different

A4

Table A.4.22 Relationship Between Type of Bird Egg and Ability to Withstand Overnight Freezing with Cracked and Broken Collapsed

	Egg status	
Egg type	Okay	Broken
Robin	9	12
Wren	11	19
Sparrow	5	19
Cuckoo	9	16

where p_R is the true proportion of robin eggs, p_W is the true proportion of wren eggs, p_S is the true proportion of sparrow eggs; and p_C is the true proportion of cuckoo eggs broken after an overnight freeze. The chi square test of independence is actually testing whether or not the true proportion of broken eggs is the same across the four types of eggs.

A4

LITERATURE CITED

Glantz, S.A. 2005. Primer of Biostatistics, 6th ed. McGraw-Hill, New York.

Hettmansperger, T.P. and McKean, J.W. 1998. Robust Nonparametric Statistical Methods. Arnold Publishers, London.

Hollander, M. and Wolfe, D.A. 1999. Nonparametric Statistical Methods, 2nd ed. John Wiley & Sons, New York.

Kutner, M.H., Nachtsheim, C.J., Neter, J., and Li, W. 2005. Applied Linear Statistical Models, 5th ed. McGraw-Hill, New York.

Moore, D.S. and McCabe, G.P. 2006. Introduction to the Practice of Statistics, 5th ed. W. H. Freeman, New York.

Utts, J.M. and Heckard, R.F. 2004. Mind on Statistics, 2nd ed. Thomson Higher Education, Belmont, Calif.

APPENDIX 5A

Preparing and Presenting a Poster

Sharon Torigoe,[1] Nick Huang,[1] Matthew Hall,[1] and Benson Ngo[1]

[1]Joint Science Department, Claremont McKenna, Pitzer, and Scripps Colleges, Claremont, California

INTRODUCTION

A poster is an effective tool for summarizing research to present to a large group of people. Serving as an aid for an oral presentation (see *APPENDIX 5B*) or standing alone, a poster should tell the story of a research project. Both the content and formatting should contribute to the overall message and argument as in a term paper. For any situation, there are a number of qualities, both stylistic and substantive, which make posters more effective.

When starting a poster, first consider its purpose and the type of audience to which it is geared. The poster can be more specific and technical if the audience is familiar with the specific field, but it may need to be broader for a less knowledgeable audience. The approach will also be different, depending on whether the poster serves as a visual aid for an oral presentation or discussion, or will be displayed for an extended period of time. More figures can be included if the poster is going to be used only as a visual aid (Fig. A.5A.1); however, more text may be necessary if an individual will not always be present to offer explanations (Fig. A.5A.2).

A set of guidelines will often be provided for poster sessions. Regulations may specify the poster size, whether it should contain an abstract, what citation format should be used, or all of these parameters. Make sure you understand exactly what is required before you begin.

Reducing months of research into a single poster can be an intimidating proposition, but the guidance given below should enable anyone to synthesize a project into an effective poster. Make sure to start well before it is due, look carefully at the examples provided, and think about your project as you read this appendix.

CONTENT

When making a poster, first consider the content necessary to convey the findings of the research project. Aim to be brief and straightforward: each sentence should have a purpose, and readers should not be bogged down with unnecessary words. To express information effectively, a poster should be composed primarily of figures, augmented by small amounts of text to make the take-home message clear. A figure with a well written description will always be more effective on a poster than a paragraph of text. *The content should clearly address the following questions:*

What research question is being investigated?
How does the question relate to a broader scientific problem?
How was the question addressed?
What were the experimental results?
How were the results interpreted?

The most common components of a poster are listed below, with discussions about organizing their content given in the sections that follow:

Introduction/background: provides the background necessary to understand the goals and significance of the project
Methods and materials: indicates approaches taken and methods and materials used
Results: presents a summary of the most important data obtained
Conclusion: summarizes the take-home messages for the project.

Current Protocols Essential Laboratory Techniques A.5A.1-A.5A.14

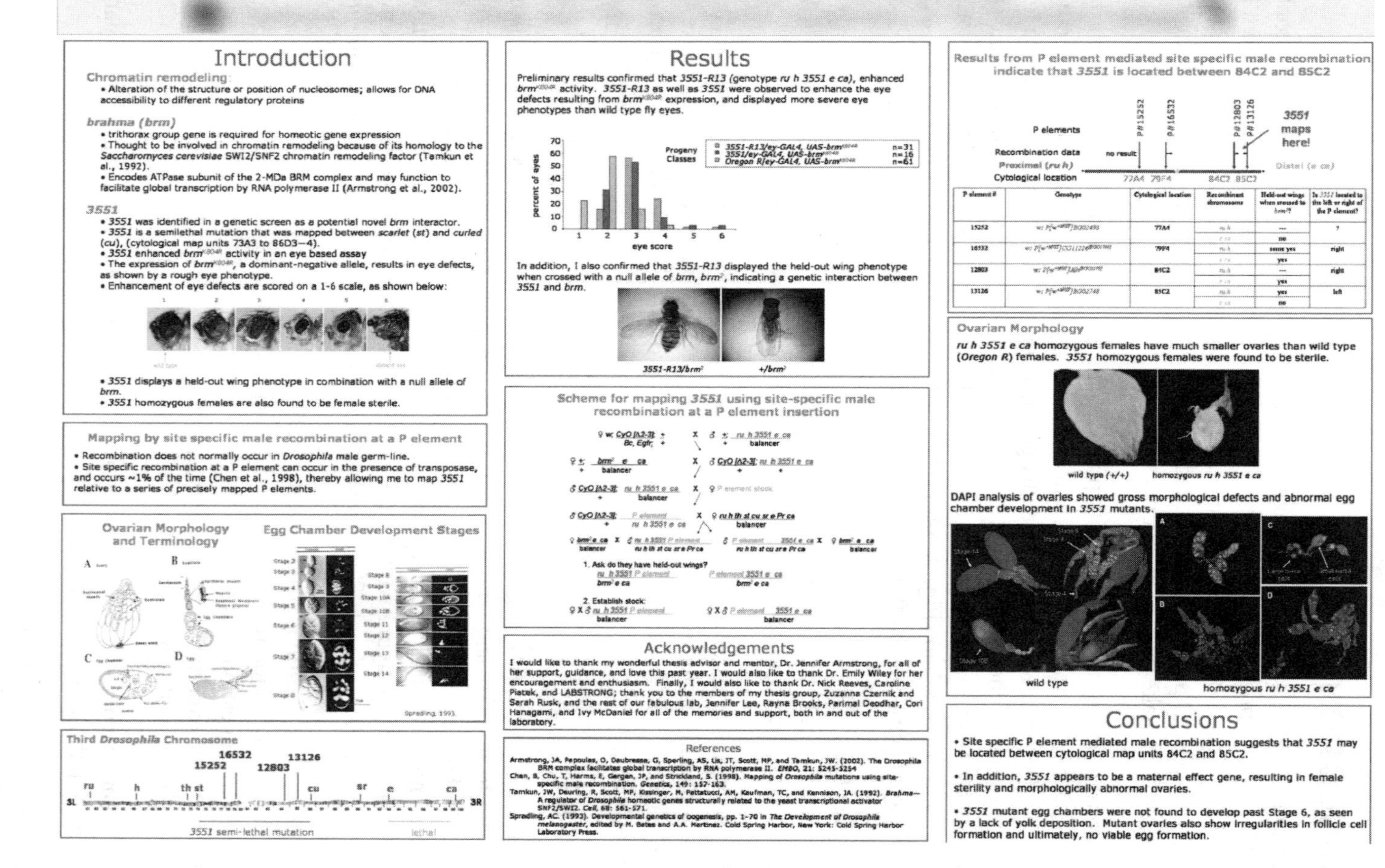

Figure A.5A.1 This poster would serve well as a visual aid in a poster presentation. The authors for this poster made use of more figures than text. The lack of written explanation for many figures suggests that assistance from an author is necessary to understand the poster.

CRE as a model system for the analysis of the structural effects of m^5C

To date, few studies have thoroughly investigated the structural and dynamical effect of DNA methylation in model systems aside from EcoRI. The cis-acting cAMP response element (CRE) has been of considerable interest as an evolutionarily conserved mechanism of transcriptional activation[1]. CRE is a promoter element that binds the transcriptional activator CRE binding protein (CREB) through a recognition site for the palindromic consensus sequence 5'-ACGT-3'.

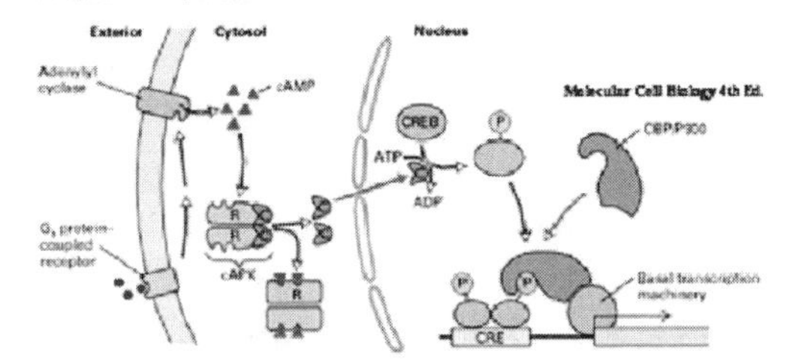

The CRE pathway is involved in numerous biological pathways, and is of biomedical interest in its role in opioid-dependence, endocrine tumorigenesis and the cellular stress response.

In this study we are using two dodecamers, CRE-1 and CRE-2, comparing native to methylated forms in DNA melting experiments and FTIR for structural analysis.

DNA sample	Sequence
ntCRE-1	d(ATGACGTCAT)
meCRE-1	d(ATGAC*GTCAT)
ntCRE-2	d(GAAAACGTTTTC)
meCRE-2	d(GAAAAC*GTTTTC)

Table 1. DNA samples and sequences in 5'-3' designation. *indicates methyl at C5. The CRE consensus sequence is in bold.

UV spectroscopy **yields quantification of the stability of the DNA samples based on the transition temperature (or melting temperature, Tm). UV absorption occurs through the nitrogenous bases, the hyperchromatic shift marking the transition from an ordered double-stranded conformation to single-stranded DNA upon disruption of the hydrogen bonds maintaining the double-helix. Thermal denaturation is based upon three main factors: GC/AT content, ionic strength and heat capacity (*Cp*).**

FTIR **has been shown to be a novel tool for analyzing the bending and flexibility effects of methylation on small oligomers[4,5,6]. Light within the infrared spectrum induces vibrational excitation of covalently bonded atoms and groups. Molecules experience a wide variety of vibrational motions characteristic of their component atoms, such that virtually all organic compounds will absorb infrared radiation that corresponds in energy to these vibrations. In this way, FTIR provides a unique reflection of the molecular structure of DNA in solution.**

Figure A.5A.2 This poster may be more suitable to display for a long period of time, or to post on the Web. While the poster appears heavy with text rather than images, the project has been concisely and thoroughly explained such that a reader does not need author assistance.

Introduction

The introduction (or background) must address two important questions about a project: (1) What is the research question being investigated? (2) How does the question relate to a broader scientific problem? Every research project has a question or goal, which must be clearly indicated and explained in this section of the poster.

When organizing the introduction, it is helpful to start with broader ideas and work towards a focused topic. It is easy to forget that not everyone will know the intricacies of a research project. Explaining how the project fits into the bigger picture of scientific research will make a project more accessible. This also allows a reader, who may be unfamiliar with the topic, to start with ideas he or she can easily understand. At the end of the introduction, clearly state the project goals, research questions, and/or hypotheses. This will set the reader up to understand the rest of the poster.

In this section, remember to cite references and resources for any background information. Refer to the section about references in this appendix for information on resource considerations and formatting.

Make use of figures to illustrate important concepts in this section. A figure will always be more effective than a block of text. Remember, a poster is a visual representation of a project, not a display of a written report.

Terms and abbreviations should also be defined in this section; commonly accepted abbreviations may not be known to all readers. Once they have been established in the poster (usually at their first occurrence), it is then acceptable to continue use of the abbreviation for the rest of the poster.

Methods and Materials

An overview of the experimental approach should be provided to answer the question, "How was the question addressed?" This section can have various forms, and it is wise to consider the best way to present methods on the poster. Often, it is written as a separate section on the poster, similar to journal articles. However, methods can also be incorporated with results, particularly for projects having many parts and steps.

This section should be limited to only key information. The purpose is to inform the reader of the experimental approach, but the reader does not need to be able to repeat the entire experiment. Protocols should be summarized in descriptive phrases, and specifics for measurements should be mentioned only when necessary. Consider the following two ways to describe a microscopy experiment to analyze nuclear DNA:

> Cells were fixed, stained by DAPI, and examined by fluorescence microscopy.
>
> Cells were fixed by 2% paraformaldehyde and dropped onto coverslips. Dry coverslips were stained by DAPI and then observed with a UV-microscope.

The first description briefly indicates how a cell sample was prepared for microscopy, with enough information so the reader will understand the general approach. The level of detail provided in the second description might be unnecessary for a poster presentation, depending on the experiment.

The appropriate level of detail for a methods section will depend on the audience and the purpose of the project. For example, a project aiming to develop a new procedure might require a more detailed description of the methods. However, if readers are likely to be familiar with the techniques used, fewer details are needed.

Diagrams and flowcharts are useful for indicating progression of a project, particularly when it includes many consecutive steps. Lengthy text describing a multistep experiment can be confusing to a reader. For example, if one were trying to clone a gene, a flowchart with diagrams of genes and vectors would be more understandable than a written explanation.

Results

Ultimately, after examining a poster, the reader should be able to answer the question, "What were the experimental results?" Results should be composed of experimental figures, each leading directly to a conclusion and together telling the research story. Raw data and results from troubleshooting and optimizations should not be included on the poster; they take up space and do not directly lead to the main conclusion. Only the most important pieces of information from the data (e.g., averages of numerical data from multiple experiments or a section of a gel with only the bands of interest) should be provided on a poster. If statistical analysis is performed on numerical data, remember to include relevant information (e.g., standard deviation, confidence level, and sample size).

When presenting results, conclusions based on the data should be obvious to the reader. Text describing the results should allow a reader to understand how to interpret the data and what conclusions to make. To most effectively communicate to the reader, use stylistic effects, e.g., bolding and underlining of conclusions.

Conclusions

At the end of the poster, it is important to relate the results back to the goals for the project and answer the question, "How were the results interpreted?" In this section, the goals stated in the introduction should be addressed, often as a summary of the conclusions from each individual figure. Suggestions for future work can also be included in this section. If the research goals have been met, questions for future projects can be suggested.

Other Important Sections

Besides the standard introduction, methods and materials, results, and conclusions sections, there are other important pieces of information that need to be included in a poster.

Heading

The heading provides information to identify a poster, e.g., a title and the authors and their affiliations.

Title

The title should be a phrase summarizing the main conclusion of the research project. It should inform readers of important aspects of the project (e.g., a specific protein, an experimental technique, and the model organism). It is recommended that the title be no more than two lines. Avoid using abbreviations because they make titles difficult to decipher for a viewer unfamiliar with the specific area of research. Consider the following two titles:

> Structure of a Chromodomain Bound to a Methylated Lysine on a Histone Tail in *Drosophila*
>
> Structure of the Chromodomain for HP1 Bound to Methylated Lys9 on the H3 Tail in *Drosophila*

Both examples summarize the project. The first example is more accessible to a reader who may be less familiar with the topic; thus, it is more appropriate for a general audience. The specifics and abbreviations used in the second example might be overwhelming or deterring to a reader; therefore, it should be used only for a more technically sophisticated audience.

Authors and affiliations

In addition to a title, the heading should include authors and their institutional affiliations. Check the requirements to determine who should be included. Others who have worked on the project may also need to be listed to give recognition for their contributions in the project. When multiple institutions are represented, the use of footnotes can help save space and prevent clutter in the heading.

Abstract

A good abstract explains an entire project in a single paragraph. An abstract is a concise summary of a project, with only a couple sentences each for introduction, methods, results, and conclusion. It can be useful for providing a good summary of the project for readers who do not choose to read the entire poster. However, it can also take up space that could be used for other sections of the poster. Whether an abstract is required depends on guidelines for the poster session. If an abstract is not required, consider whether it would be helpful for readers.

It should be noted that abstracts are sometimes published in conference proceedings. Particularly at large poster sessions, abstracts are important for attendees when planning which posters out of many that they will view.

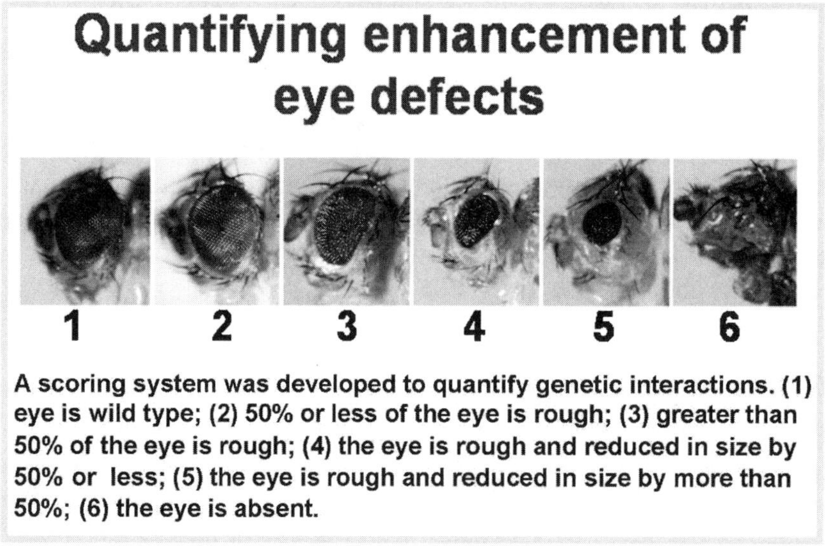

Quantifying enhancement of eye defects

1 2 3 4 5 6

A scoring system was developed to quantify genetic interactions. (1) eye is wild type; (2) 50% or less of the eye is rough; (3) greater than 50% of the eye is rough; (4) the eye is rough and reduced in size by 50% or less; (5) the eye is rough and reduced in size by more than 50%; (6) the eye is absent.

Figure A.5A.3 The above section of a poster demonstrates a good figure with a well-written caption. The writing is straightforward, and the figure is illustrative of the authors' ideas. It would be extremely difficult to present these ideas effectively with a paragraph of text.

Acknowledgements

Acknowledgements are an opportunity to recognize and thank those who have provided support for the project, e.g., primary investigators and advisors. It is also good practice to recognize granting agencies and other financial sponsors that provided support for the work presented. Many granting agencies (e.g., the National Science Foundation and the National Institutes for Health) require acknowledgements in publications and presentations, including poster presentations.

References

Just as it is important to recognize those who have contributed results to the poster, it is important to cite those who have provided you with background knowledge about the research topic. Provide only references for background information included on the poster. When selecting sources, it might be better to select reviews rather than individual papers, except for seminal papers in the field. For proper formatting of references, follow an established style used in relevant journals.

Writing Style

The style of writing directly impacts the effectiveness of a poster. In general, aim to be straightforward and brief. Each sentence should have purpose. Readers should not be bogged down with unnecessary words, or they may lose interest.

Figures are necessary tools for any poster. Figures with well written captions are always more illustrative of ideas than paragraphs of text (e.g., see Fig. A.5A.3). For presenting results, consider whether a graph or a table is most effective for organizing the data and making your point clear.

MECHANICS

This section addresses the mechanics of designing a poster, particularly in using a layout program and printing options. Professionalism is an important aspect of the poster presentation; a messy appearance will deter readers' attention. A layout program (see below) is useful for creating a professional look. It is very helpful to start with a template, and many can be easily found by searching online search for "science poster template" or "scientific poster template."

Microsoft PowerPoint

When asked which layout program to use, many will immediately recommend using Microsoft PowerPoint for designing a poster, and this program is used for illustrative purposes in the discussion below. There are many advantages to using PowerPoint: it is a common and readily available program which comes with Microsoft Office; it has many tools and options that are useful in producing a professional and finished poster; and it is compatible with other common Microsoft programs such as Word and Excel. Keep in mind that PowerPoint was not designed for printing purposes. As a result, PowerPoint may be incompatible with some printers.

There are many other layout programs that are very suitable for designing posters. Some of these include Adobe Illustrator, QuarkXpress, CorelDraw, InDesign, Pagemaker, Adobe Photoshop (see *APPENDIX 3B*), and Freehand. While Microsoft PowerPoint is used as an example here, use a program with which you are most familiar.

Size

When creating a new poster, begin by setting the appropriate size. Under the File menu, go to Page setup and select Custom from the drop-down menu under Slides sized for. You can now enter the width and height of the poster. Be aware that there are maximum dimensions for a slide in PowerPoint. If the dimensions of the poster will be greater than the maximums allowed in the program, you will need to scale appropriately (see *APPENDIX 3B*).

Figure File Guidelines

All graphics, diagrams, and figures ideally should be in lossless compression format (e.g., TIFF or RAW). Lossless compression format allows the images to be saved and edited without any loss in quality, and it is ideal for printing. For poster backgrounds, a high resolution image is necessary. In general, aim for 300 dots per inch (dpi) or more for acceptable print quality. File types with lower quality (e.g., JPEG and GIF) may appear to be acceptable on the computer screen, but they will often result in grainy figures when printed. Avoid these file types. Refer to *UNIT 7.5* and *APPENDIX 3B* for a more thorough analysis of figure file types and their appropriate uses.

Inserting Images and Text

There are a number ways of to add images and text to the poster. For images, one good method is to insert the image from a computer file. To insert an image, go to the Insert menu and, in the Picture option, choose Select From File. This will allow you to locate the image file and then insert it into the poster. You can also insert an image by cutting and pasting the image. However, this may compromise the quality of the image.

To insert text, simply cut and paste and resize text boxes as necessary. To add text manually, go to the Insert menu, and select Text box. This tool will allow you to create a new text box by clicking and dragging the mouse on the poster. Then type in the text.

If you are copying text from a separate file (e.g., a Word document) and would like to maintain its formatting, copy a selection of the text. In PowerPoint, go to the Edit menu and select Paste Special. In the resulting window, select Formatted Text to paste the copied text.

Alignment

Aligning images and text is important to a professional look. The best method for aligning groups of text and images is to use the alignment tool in PowerPoint. In the Drawing toolbar, select Draw. In that menu, go to Align or Distribute and select an option.

Printing Options

Universities often have poster printers available for student use at a reasonable price. However, there may be printing limitations, e.g., difficulty printing certain shades of color, poor paper quality, or inadequate size paper or printers for the poster. Therefore, you should inquire about such limitations before having the poster printed.

There are a number of alternatives to your institution's printing service. Copy and graphics vendors (e.g., Kinkos and AlphaGraphics) can prepare posters. There are also several commercial poster services, such as PosterSession (*http://www.postersession.com*), MakeSigns (*http://www.makesigns.com*), SciFor (*http://www.scifor.com*), and Posters1st (*http://www.posters1st.com*).

VISUAL STYLE

The visual presentation of a poster is very important because good layout and style are crucial to effectively communicating ideas. A good poster will attract attention, and readers will not struggle to decipher information because of any poor visual choices. Layout and style are meant to enhance the presentation of the content: they must never distract from it. Keep in mind that the poster is a visual presentation of a project and not a copy of the research paper. This section addresses a number of features in the layout and appearance of a poster.

Visual Consistency

It is extremely important to be consistent in the visual presentation of a poster. Features on a poster such as font sizes, colors, and spacing are visual cues about the material presented. Inconsistencies are either distracting or will lead readers astray from important messages. For example, using differing sizes of section headings for the Introduction and Results may unintentionally emphasize one section over another. The two sections from a poster shown in Figure A.5A.4 are an excellent demonstration of visual consistency. Choices in the text style for subheadings and body text are the same from one section to the next.

A5

Information Flow

Logical flow of information is one of the most important aspects of a poster. Poor organization will confuse readers. A poster should have the same organizational flow as in a journal publication, typically in the following order: Introduction, Methods and Materials, Results, and Conclusion. The abstract is often placed before the introduction, and acknowledgements and references are typically after the conclusions.

The visual organization on the poster must also lend itself to the above progression. When viewing a poster, readers generally begin at the upper left corner and will end in the lower right corner. Information should be organized with this in mind. Therefore, introductory material such as the abstract and introduction are placed in the upper left, while conclusions are towards the lower right of the poster. Most posters follow a multiple column format with the header at the top of the poster, as shown in the generalized sample in Figure A.5A.5.

Section headings are extremely useful for guiding the reader through information as they read through the poster. These section headings do not have to be labeled "Introduction" or "Conclusions." It can be more effective to have more descriptive headings, which indicate an important idea about that section. For example, rather than labeling an entire section as "Results," one could divide this up into smaller sections with headings more relevant to each experiment. The example in Figure A.5A.6 demonstrates an excellent use of more descriptive headings.

BRM Chromatin-Remodeling Complexes

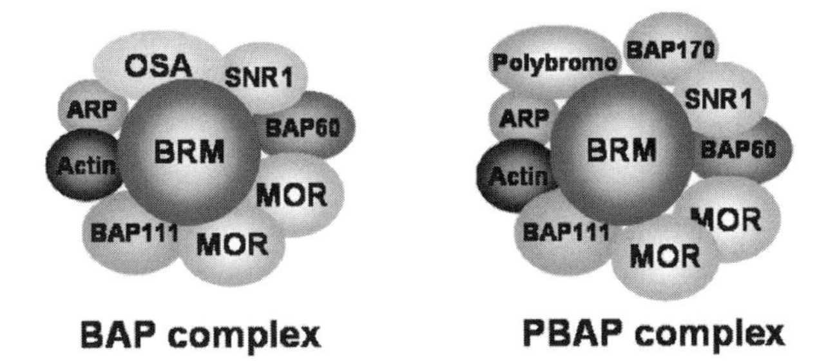

- The BAP complex contains OSA, while PBAP contains Polybromo and BAP170. BAP and PBAP are similar to yeast SWI/SNF and RSC, and human BAF and PBAF complexes (Mohrmann et al., 2004).
- The BRM ATPase is required for global transcription (Armstrong et al., 2002), the exact roles played by PBAP and BAP are unclear.

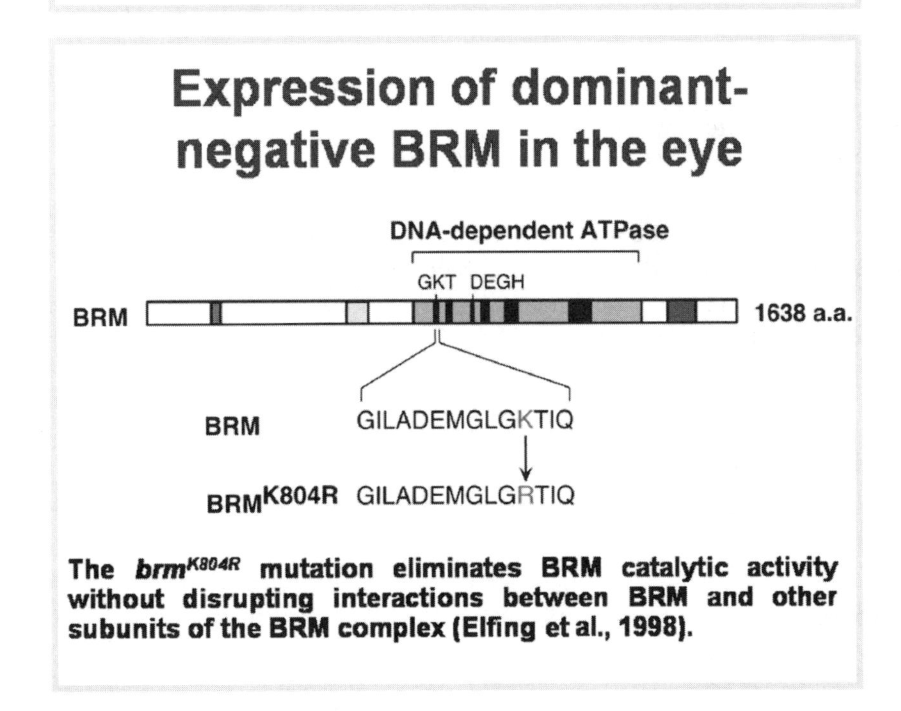

Figure A.5A.4 Visual consistency is important to maintain throughout a poster. Characteristics of the text for both the headings and the main body are consistent in all sections.

Borders and Spacing

Avoid filling the poster with excessive text and figures. Instead, space out the sections and subsections. The poster should allow the reader's eyes to stop and rest occasionally. Also, make sure that the poster has a good balance of text and space. If the text is heavy in any one section, try to reduce it by either omitting some or introducing empty space between crowded points.

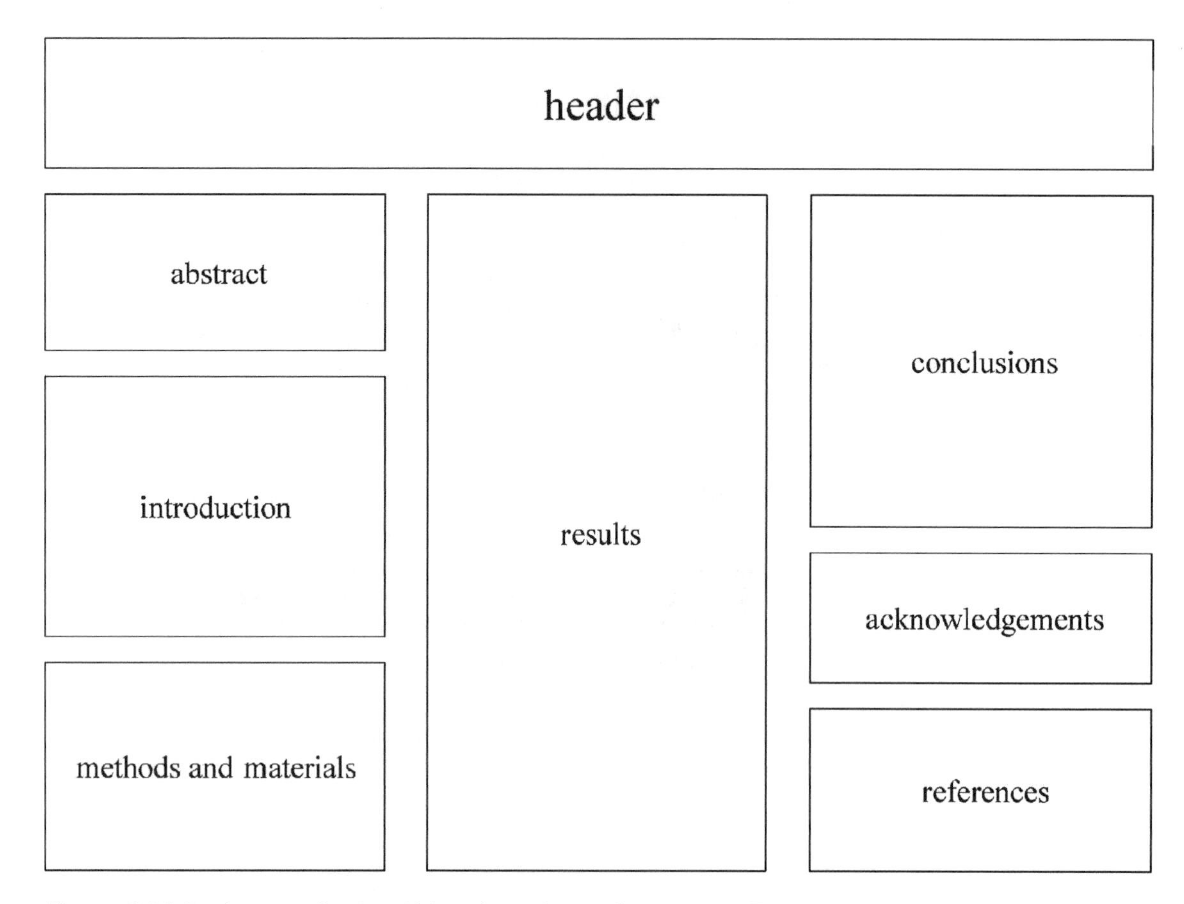

Figure A.5A.5 A generalized multiple-column format for a poster. The header is placed at the top of the poster, and the remaining content is divided among the columns below.

Another good way to ensure adequate spacing between sections is to use borders to define and organize the sections and subsections. Using borders can often make text much more readable and is a good tool for emphasizing certain important information. Another option is to use empty space as a way of defining a body of text (Fig. A.5A.7).

Text Style

Text style refers to the type of fonts and styles used, as well as the layout and presentation of the text on the poster. Regardless of how you ultimately choose to present text, you should aim for consistency and readability.

Font typefaces

Fonts come in two general types, sans serif and serif. Serif means "tails" in French and refers to the tails that are found on the tips and base of the letters. Serif fonts also have line widths that thin out on curves. Examples of serif fonts that are known to work well on posters include Times New Roman and Georgia. Sans serif fonts are fonts without serifs. Examples of sans serif fonts are Arial, Helvetica and Verdana. These fonts generally appear simpler than the serif fonts.

When choosing a font for the poster, aim for consistency and, most of all, readability. Many sources state that a serif font is best for blocks of text, particularly in books and newspapers, while a sans serif font is better for online publications. In the case of a poster presentation, it is appropriate to use either, provided that it is readable. There are some fonts, both serif and sans serif, which are often for decorative purposes and are difficult to read (e.g., French Script and Pristina).

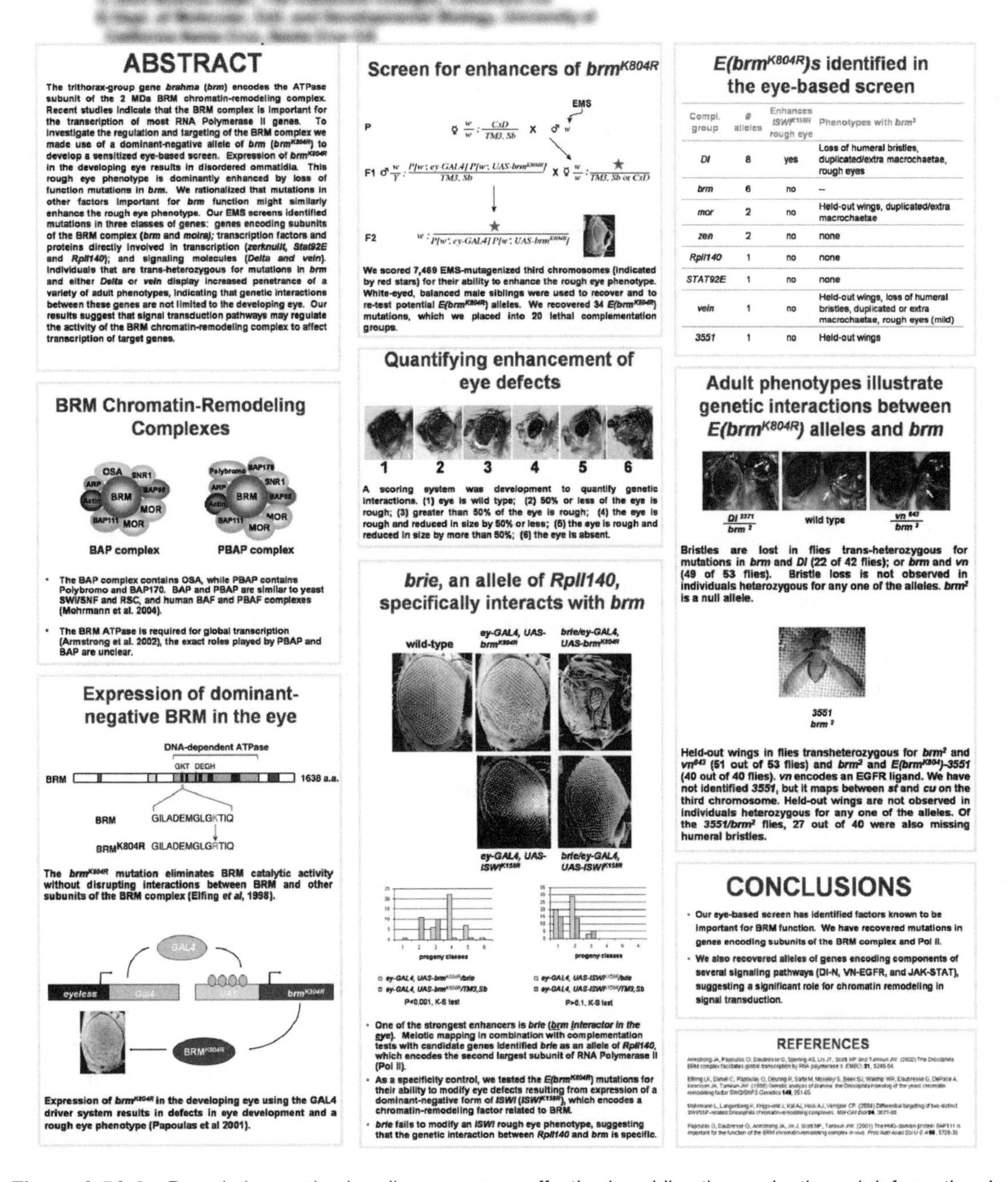

Figure A.5A.6 Descriptive section headings are more effective in guiding the reader through information. In this example, headings indicate the most important idea of each section.

Font effects

Using font effects (e.g., bold, italics, and underline) can be useful in highlighting important points. However, overuse of font effects is distracting and should be avoided. Be sure to be consistent when styling font. For example, you could always use bold for important conclusions in results.

Font size

The poster should be readable from several feet away. If you have not been given specific instructions on appropriate font sizes for the poster, there are a number of sizes that are typically suitable. Two

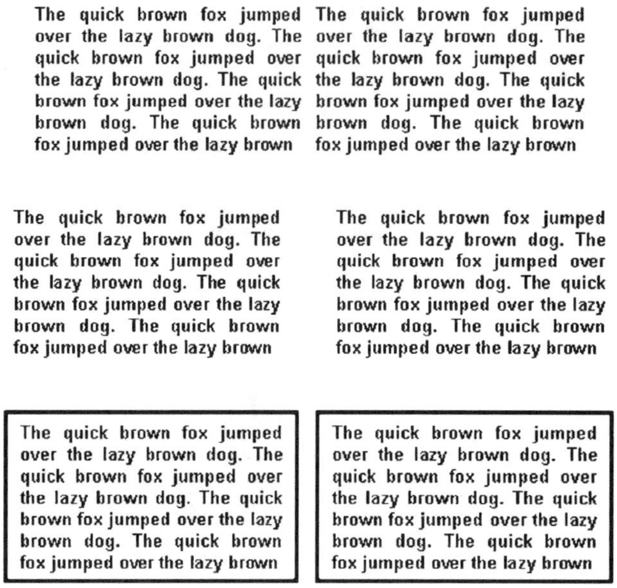

The quick brown fox jumped over the lazy brown dog. The quick brown fox jumped over the lazy brown dog. The quick brown fox jumped over the lazy brown dog. The quick brown fox jumped over the lazy brown

The quick brown fox jumped over the lazy brown dog. The quick brown fox jumped over the lazy brown dog. The quick brown fox jumped over the lazy brown dog. The quick brown fox jumped over the lazy brown

The quick brown fox jumped over the lazy brown dog. The quick brown fox jumped over the lazy brown dog. The quick brown fox jumped over the lazy brown dog. The quick brown fox jumped over the lazy brown

The quick brown fox jumped over the lazy brown dog. The quick brown fox jumped over the lazy brown dog. The quick brown fox jumped over the lazy brown dog. The quick brown fox jumped over the lazy brown

The quick brown fox jumped over the lazy brown dog. The quick brown fox jumped over the lazy brown dog. The quick brown fox jumped over the lazy brown dog. The quick brown fox jumped over the lazy brown

The quick brown fox jumped over the lazy brown dog. The quick brown fox jumped over the lazy brown dog. The quick brown fox jumped over the lazy brown dog. The quick brown fox jumped over the lazy brown

Figure A.5A.7 Boxes of text placed too closely together are difficult to distinguish (top). To define them, the two text boxes can be separated by empty space (middle) or by producing a border around each box of text (bottom).

common font typeface choices are Arial and Times New Roman, and a font size of 24 points works very well when using them for body text. Section headings range in size, but a good starting point is 36 points. The title in the header should be at least 84 points in size for Arial and Times New Roman. Ultimately, you may have to use your own judgment since the number of characters per pica (and therefore per line) differs between proportional typefaces (those where the character space is proportional to the size of the character, e.g., Arial and Times New Roman).

Paragraphs versus bullet points

Text can be organized by paragraph, bulleted points, or a combination of both. In general, the bullet format is easier to read and understand, but paragraphs may be more appropriate in some cases, depending on content matter. Paragraph style is generally used for the abstract and background sections.

Text spacing

Text should be double-spaced for maximum readability, if space permits. Where space conservation is important, single spacing should be reserved predominantly for content such as references. In Microsoft software, it is possible to select any spacing desired. In Word, go to the Format menu and select Paragraph. The resulting dialog box will have options for spacing under the Indents and Spacing tab. In PowerPoint, go to the Format menu and select Line Spacing to change the spacing of text.

Color Schemes

Color schemes can range from plain black text on white background to more elaborate combinations and shades of color. There are many color combinations that are acceptable, but there are several characteristics that should be avoided:

Low contrast color combinations for text (e.g., white text on a light blue background) are problematic for readability. Instead, aim for higher contrast combinations such as dark colored text on a light background and vice versa.

Jarring combinations (e.g., black with hot pink or red adjacent to blue) are distracting and hard on the eye. Choose colors that complement each other. For example, white with black text and accents in a couple shades of blue can work very well.

Very vibrant colors (e.g., some reds, oranges, and yellows) are also jarring. They can be difficult to read if used for text. Instead, select colors that are inherently not as bright, (e.g., blues and greens) or choose shades that are toned down or muted.

Keep in mind that the poster printing will likely not print the colors the same as they appear on a computer screen. The computer monitor settings can affect the appearance of colors while designing the poster. Shades of color can turn out differently than expected, depending on the ink used. In addition, choice of paper can also affect print quality. Not all white papers are the same shade or brightness, and they can alter the shades of the colors.

Background

The background of the poster should be carefully considered. A background that is too vivid and busy is distracting and will make the text unreadable. Aim towards a clean and simple background. Often a solid color that compliments and enhances the contrast of the poster is best. Gradients can be used, but they can make text difficult to read if not well selected.

An image can be used for the background if it relates to the experiments and (most importantly) is not distracting. Remember that the image should be of high quality and resolution to ensure good printing and prevent a grainy appearance. Effective use of an image background is demonstrated in the poster in Figure A.5A.8. The animals displayed are those which were studied in the experiment. Placement of the image also resulted in the faces of the animals appearing between boxes of text. To ensure that text was still readable, the authors filled in the textboxes with a background color.

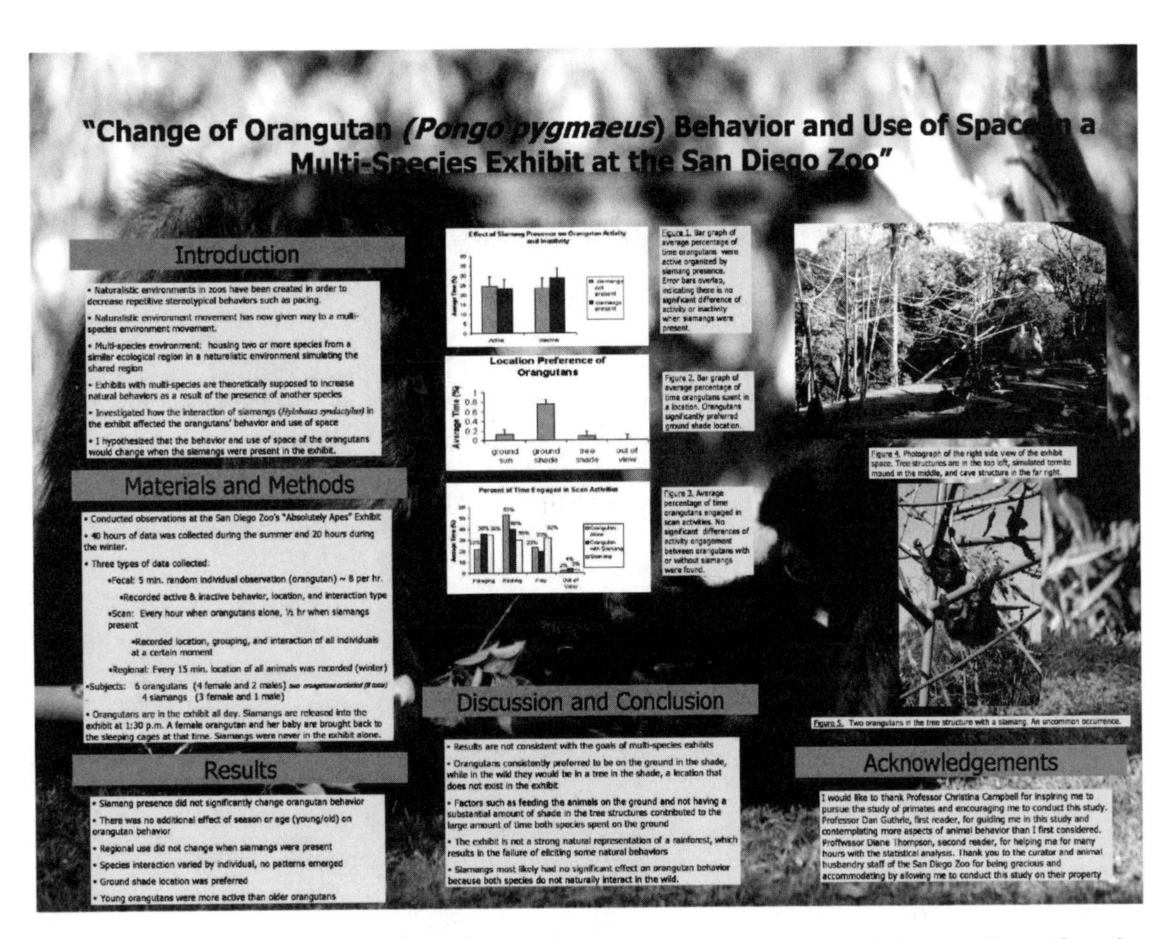

Figure A.5A.8 An appropriate usage of a background image. The image selected does not distract from the content of the poster, and features have been chosen to optimize readability of the text.

A5

Figure Usage

Use graphics, diagrams, and figures when possible to visually demonstrate complex experiments or concepts. Remember that a figure accompanied by a good explanation is more effective than blocks of text.

Avoid extraneous use of unnecessary results and pictures that have no relevance to the overall experiment. Graphics, diagrams, and figures should always be applicable to the project. Text should accompany such images if their significance is not clear.

Organize images in a logical progression. Important figures should be centered or displayed prominently for appropriate emphasis.

A CHECKLIST

The challenge of summarizing a research project into a poster is certainly daunting, and there are many aspects to consider, from the background information to the font choice. The poster should contain relevant information and results, and the visual presentation should facilitate effective communication of the content. Here is a checklist of important points to remember whenever designing a poster:

Don't wait until the last minute to start—give yourself a couple of weeks.
Know and follow all guidelines prescribed by the poster session.
Clearly state the research question.
Answer the research question using supporting data.
Include acknowledgements or references to the work and support of others.
Maintain consistent formatting.
Avoid dense text.
Make sure that the font is readable from a few feet away.
Label all figures.
Use figures in TIFF or RAW format (see *UNIT 7.5* and *APPENDIX 3B*).

ACKNOWLEDGEMENTS

For permission to use their posters as examples, we would like to thank Jennifer A. Armstrong, Mayra Garcia, Mary E. Hatcher-Skeers, and Caroline I. Piatek (Joint Science Department, The Claremont Colleges); Sabrina Dence and Patrick S. Michell (Pitzer College); Kristina Lee (Claremont McKenna College); and Adam S. Sperling and John W. Tamkun (Department of Molecular, Cell, and Developmental Biology, University of California Santa Cruz).

APPENDIX 5B

Preparing and Presenting a Talk

Christine Sjolander[1] and Stephen Chang[1]

[1]Keck Graduate Institute of Applied Life Sciences, Claremont, California

INTRODUCTION

The 300-seat lecture hall is filled with people you don't know, and they're waiting to hear about the research project that you've invested every spare minute of the last 9 months in completing. Your favorite professor is in the front row, and the one whose questions you dread sits next to her. Silence fills the cavernous room, you walk to the lectern, and begin to speak... Sound familiar?

From time you entered the scientific field, giving and hearing oral presentations has been a part of your day-to-day life. While these talks are accepted as a way to exchange information and ideas, the truth is that many of the talks you attend are flawed in their delivery and presentation. No matter how brilliant the research or fascinating the subject matter, a poor presentation can make the audience question the credibility of the scientist and lessen the impact of the discovery.

Studies show that we retain information best when we both see and hear it together (Pike, 1992). Remember show-and-tell back in kindergarten? Your scientific presentation is a more sophisticated opportunity to both show and tell the results of your hard work and for you to receive the recognition that it deserves.

The focus of this appendix is to provide you with tips to make your presentation stand out.

PREPARATION

Public speaking ranks as one of the most common fears for all humans—from students to seasoned professionals. Just like an athlete entering a big competition, the key to conquering that fear is preparation.

Familiarity with your subject and confidence in your research is not enough to give a good oral presentation. Those factors can, in fact, work against giving a good talk. They can produce a false sense of security, make you careless in your preparation, and let you take too much for granted with your audience. No matter how well you know your topic, taking each of the following steps is critical to preparing a good presentation.

Assessing the Audience

How often have you sat in a presentation where the speaker talked at a level far above the audience's knowledge base? How much of that information was retained? An extremely important part of effectively communicating what you know is tailoring your presentation to your audience's needs and level of understanding.

Before giving a presentation, make sure that you've addressed the following questions.

How large will the group be?

Are they experts in your field?

If the group is a combination of experts (colleagues in your specific field) and nonexperts (all others), what is the anticipated ratio?

If the group is composed of nonexperts, what is the age range and educational level?

It is easier to speak to either all experts or all nonexperts. If the entire audience is knowledgeable about the topic, you can speak to them as insiders and have an opportunity to dazzle and excite them with your knowledge and enthusiasm. If it is a group of nonexperts, you must adjust the scope and level of your material to engage and interest them.

The biggest challenge for scientific speakers arises when addressing a combined group of experts and nonexperts. As an expert, it is insulting and a waste of time to be talked down to; as a nonexpert, it is frustrating and a waste of time to be subjected to a talk that is beyond one's level of comprehension. Try to strike a happy medium in such a situation.

One useful suggestion is to devote half to two-thirds of your time to an introduction or overview of your subject and save the highly technical material for the remaining time.

Another useful approach is to give the more involved technical material, then to summarize with "in other words..." when you restate the information in simplified, plain English. These brief summaries should be done throughout the course of your talk and often enough to prevent those not following the technical information from drifting off.

Remember that your presentation is a dialog with the audience—not a monolog. Your goal should be to tell them the information you want to share in a way in which they are receptive to hearing it.

Structuring Your Material

After all of the research you've done and all of your hard work, you could probably speak for hours on your topic and not completely cover everything you've learned. However, the depth and scope of the scientific content you'll need to present are determined in large part by the audience profile and the time allotted, not by the topic itself.

Before you can begin drafting a talk, you must define the purpose, topic, and appropriate depth and scope of the information you will be presenting. The primary purpose of the scientific talk is to inform or instruct. You may also subtly try to persuade and even entertain your audience, but don't lose sight of the primary purpose.

In preparing your talk ask yourself a few questions.

Why does the audience care about my topic?

How can I generate excitement or interest in my subject even if the audience isn't familiar with it?

How might other research areas use this information?

Are there anecdotes I can include that will add interest, emphasis, or humor?

Drafting Your Talk

There are mixed views on the merits of writing out your entire presentation word-for-word. The pros are that, when written out, the speaker has the opportunity to choose the language carefully and organize thoughts effectively, and there are fewer opportunities to stray off topic. The cons are that writing it out word-for-word can inhibit the conversational style of the talk and alienate the audience. Whether you choose to write it out word-for-word or use an outline or index cards, it is not acceptable to read your presentation verbatim.

The talk must be well organized and logically structured. Be able to summarize the content of your presentation in two or three well constructed sentences. A great rule to follow is to tell the audience what you plan to tell in the whole talk as the introduction, use the body to present the information in the body, and summarize what you've already told them in the conclusion. A trap that inexperienced speakers often fall into is trying to cover far too much material in too short a talk. The language must be concise. Audiences become frustrated by talks they cannot understand and won't bother to struggle to follow along.

When drafting your talk, remember that the key to any scientific talk is clarity. As a rule of thumb, plan your presentation for 80% of the allotted time to allow time for questions from the audience. Preparing a short talk can be a very constructive exercise for the scientific speaker. The short time allocated will force you to assemble your talk very carefully, to be a severe editor of your words, and to be an exacting critic of your visual aids. Make every word count but remember you are writing for the ear and not the eye. People don't speak the way that they write. Your talk should flow like a story including:

Title slide. Include title, authors, and institutional affiliations. Titles should be attention getting and forceful and, if possible, include the key benefit the audience will receive by listening.

Introduction. Tell them what you're going to tell them, including the results of your research.

Methods. Clearly and simply state how and where the research was conducted

Results. This is the most important section of your talk. You can use charts and graphs to clearly illustrate results

Conclusions and discussion. Conclusions should be stated clearly with additional information on how your results fit into the big picture. Summarize what you've already told them.

Acknowledgement slide. It is important to recognize those who assisted you with any part of the research or presentation. Funding sources should also be stated at this time.

Know Your Stuff

When giving a scientific presentation, familiarity and comfort with the content of your presentation is critical. Whether your audience consists of experts or laypeople, you should anticipate questions and be able to answer them with authority. Accurate, complete, well-phrased descriptions of scientific information portray a speaker as a knowledgeable, reliable source of information. In contrast, glib, inaccurate statements that appear open to multiple interpretations will elicit skepticism and distrust.

Practice, Practice, Practice

Accomplished public speakers advise that rehearsals are almost as important to a good oral presentation as the actual text of the talk. It is not enough to read through your talk a couple of times. Things that read very well can sound very awkward. Speaking aloud while standing in front of some type of audience who will give you honest, constructive feedback will help you find the rough spots before your talk.

Another valuable rehearsal technique is to record your talk. Listen to the entire presentation without your notes. Do your thoughts flow logically? Are the transitions smooth? Do you vary your voice and your pace for emphasis, to avoid monotony, and as you transition to new thoughts? Do you hear any "ums?"

Videotaping a practice session is an extremely effective rehearsal technique. Many people loathe seeing themselves on tape, but it is a guaranteed way to improve your speaking skills. Run the tape first and just listen as though to a voice recording. Run it again and watch for these things: Do you make eye contact? What are your hands doing? Do you smile occasionally? How is your posture? Do you have distracting mannerisms? This method of critiquing your presentation, while somewhat painful, is brutally honest and, therefore, extremely valuable.

Rehearsals, including any visual aids, are also essential to timing your talk properly and achieving the comfortable, confident, conversational style considered good form in scientific circles. Practice with your visual aids. Many speakers undermine their own talks through clumsy handling of visual materials. Practice pointing to the image on the screen and turning back to the audience. If possible, practice in a room that is close to the size of the room in which you will deliver your talk.

Transforming a talk into a good or outstanding talk takes time. There is no way around it. A lack of practice will be very clear to your audience and will be interpreted as a lack of commitment, professionalism, and/or competence.

What to Wear

Like it or not, we all judged in part on our appearance. Extremes, either too casual or too overdone, are not good ideas if you want to be accepted by the audience. If you want credibility in the scientific community, dress the part for your presentation. In effect, you're playing a role, and it has nothing to do with who you are outside of the presentation or what you choose to wear on the weekends. Play it safe, use your common sense, and remember, neatness always counts. If you err, it's better to err on the side of being slightly overdressed and formal than not being dressed professionally enough.

Some very general guidelines for professional meetings and conferences

Men can't go wrong in a dark blue or black suit with a solid, light colored, well ironed shirt and a conservative tie that is not more than a year old. Tie styles change regularly, and it's worth the investment to pick up a new one at least once a year. Make sure your suit fits you well and that it is pressed. Remember the dark socks and nicely polished, black or brown leather shoes. Hair should be neat and clean. Facial hair needs to be well-kept and trimmed.

Suits for women are also appropriate. Generally, in the scientific community, suits for women can have either pants or skirts. Skirts should be no shorter than 2 in. above the knee. Choose a color that compliments your skin tone, keeping in mind that darker colors are more conservative that lighter ones. For the most professional appearance, wear hosiery and closed toes shoes with your suit. Blouses should coordinate with your suit and have a conservative neckline. Makeup and jewelry should be kept to a minimum, and hairstyles should be neat.

VISUAL AIDS

Many public speaking experts contend that visual aids ruin more speeches than they improve. Do you ever find yourself noticing the typos in a slide more than the material itself? How much do you absorb of a slide full of equations?

A visual aid becomes the focal point for the time it is in view. An audience's attention is quite naturally drawn from the speaker to anything put on a screen. Since they automatically assume center stage, it vitally important that all visual aids support your talk in an attractive, comprehensible manner, or they will detract and compete with it.

The three most important points for slides in any presentation are given below.

1. Each slide should be error free, consistent, and clean.
2. Each slide should be simple.
3. Each slide should be necessary to the story.

If you're also using a poster or handouts, your presentation should coordinate with these items. See the *APPENDIX 5A* for additional formatting suggestions.

Error-Free, Consistent, and Clean

At a minimum, run a spell-check and have at least one other person proofread your slides. Spell-check will only catch spelling errors and not word usage ones (e.g., "he" when you mean "the"). In addition, many scientific terms may not be included in the spell-check feature.

Carefully choose color schemes that convey professionalism to the audience. Limit colors to two or three, with a contrasting background color. A simple solution is to use one of the professional Microsoft PowerPoint styles already designed. Use the same one throughout your presentation.

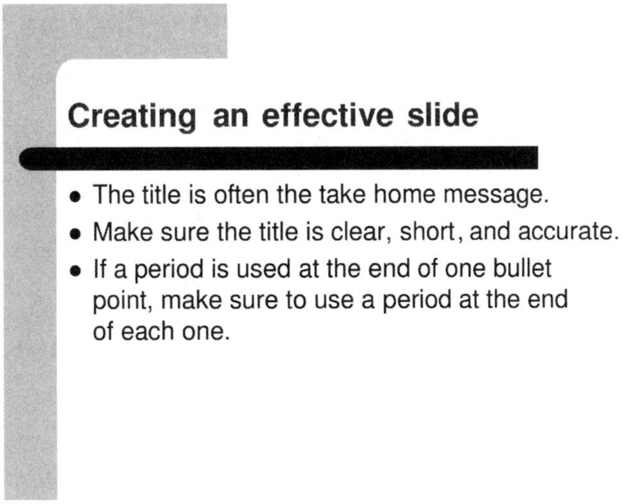

Creating an effective slide

- The title is often the take home message.
- Make sure the title is clear, short, and accurate.
- If a period is used at the end of one bullet point, make sure to use a period at the end of each one.

Figure A.5B.1 A sample slide made using Microsoft PowerPoint's Capsules template. Each bullet captures a point that supports the slide title. The presenter is expected to be able to elaborate with supplemental information concerning each bullet point made.

Simple

Slides should have the minimum number of words needed to convey the message. Data and information that is not referred to in the talk should not be included in presentation slides. If time restrictions limit the amount of information you can convey, backup slides can be included after the acknowledgements slide and used if an appropriate question is asked.

Since each bullet point is a brief summary of supporting information, more commentary must be verbally presented to expound on the contents of the slide. A good rule of thumb is to spend approximately 1 min in discussing each slide.

Type size should reflect the importance of the text. For example, main points should be larger than supporting points. However, never use a font less than 14 point. It needs to be visible from the back of the room—text can never be too big (see Fig. A.5B.1).

If you choose not to use a predesigned style, use one standard font throughout, e.g., Ariel, Times New Roman, or Verdana. Be consistent with your sizes so that all titles are the one larger font size and all supporting points are the same smaller font size.

Charts and graphs are great ways to convey data when properly used. Be sure all axes and components are clearly labeled on any slide. Avoid figure legends if the categories can be explicitly indicated on the chart. If a legend must be used, keep it short and easily distinguishable. Finally, keep charts and graphs simple—the less busy they appear, the more justice they do to the information you're attempting to communicate (e.g., see Fig. A.5B.2).

Necessary to the Story

Ask yourself if you need the slide to convey the story you're trying to tell. It may have great information but may not be relevant to the point you want to make. If it isn't needed, leave it out!

IMPORTANT NOTE: Do not read your visual aids to the audience instead of giving a talk.

First, the audience can read faster than they can hear you read. Second, if you are reading the screen, your back will be to the audience, and part of a speaker's job is to face the audience. Third, if your visual aids contain most or all of your talk, you should probably prepare a handout and relinquish your time at the lectern.

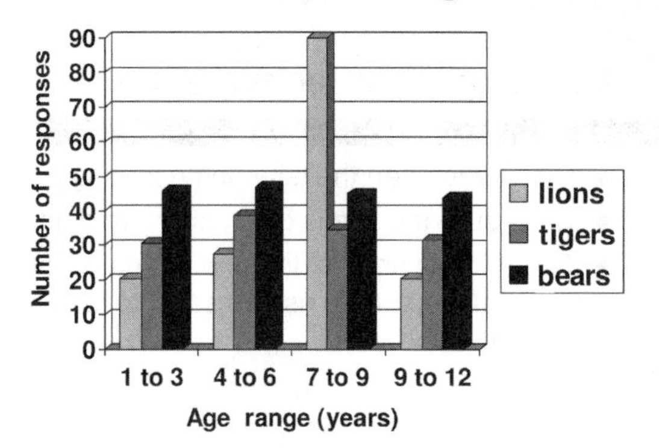

Figure A.5B.2 Charts and graphs can be easily generated using Microsoft PowerPoint or Excel. Slides containing charts and graphs should be simple, self-explanatory, and color coordinated to make your presentation more professional. The Pixel template was used to make this slide.

Handouts

Handouts give the audience something to take away from your presentation that they can refer to later. Handouts can take many forms:

Copies of your slides

Graphs or data charts that supplement your talk and may be too small to read on a slide

Supplemental reading, e.g., important reference articles or the complete research paper that you are presenting.

When using handouts, you can either pass them out at the beginning or during your presentation for the audience to refer to, or you can hold onto them until the end. A general rule of thumb is that if you will be referring to your handouts during your talk, for example, to highlight details, you should provide them at the beginning. This is a great tool to use if you're covering very technical material and you want to provide the data in detail but only have time to talk about the results and the conclusion. You can also hand them out at the beginning if you think that the audience may want to take notes on what you're presenting.

Supplemental reading material is usually best saved until the end of the presentation. You don't want your audience to be reading and not listening to your presentation. Make sure that any handouts complement your presentation and add value for the audience. If you provide copies of reference materials, make sure you have permission from the original author to distribute them.

DELIVERY

Studies of communication show that:

55% of interpersonal communication comes from facial expressions and other body language.

38% comes from vocal quality or tone of voice.

7% comes from the content, the actual meaning of the words.

In other words, how you look and how you sound while you present is key to whether or not your audience absorbs anything that you say. To be most effective, you'll need to develop a delivery style that incorporates good body language, pleasant facial expressions, and a confident, yet relaxed, tone of voice.

Body Language

Stand up straight. Don't be stationary and hide behind the lectern; use the whole stage, walking confidently from one side to the next, pausing en route. Avoid distracting mannerisms, e.g., rocking on your shoes or playing with your hair. Take your hands out of your pockets. It will feel odd, but either let them hang naturally at your side or use them to underscore your points.

Facial Expressions

Make the decision to smile. Even the most serious of topics will be better received when addressed with a smile. To effectively make eye contact, select an audience member and look directly at him or her for 2 to 3 sec. Select another audience member in another area of the room and do the same. Randomly select participants throughout the talk and each will feel like your speaking directly to him or her.

Tone and Quality of Voice

Slowing down is the remedy for 90% of most speakers' problems. Take the time to articulate each word clearly, speaking slightly louder than you think is necessary. Convey enthusiasm by varying the tones of your voice. To emphasize a point, speak slightly louder or deeper. Great public speakers almost seem "over-the-top" with their expressions and tones—push yourself just outside your comfort zone with your gestures and tones.

If you're a non-native English speaker, it is even more important to speak slowly and articulate each word. With a native speaker who will provide feedback, practice words that give you difficulty.

ANSWERING QUESTIONS

It's not enough that you made it through your talk. Now you must subject yourself to cross-examination and do so while thinking on your feet. Q & A sessions can definitely be tricky, but remember, while you are at the podium, you are in charge. You only have to take questions at your convenience.

Repeat all positive questions. This makes certain everyone heard the question and buys you a moment or two to compose your response. If necessary, rephrase the question for clarity. Paraphrase negative questions. By changing the slant or tone of the question, you can respond positively.

Respond simply and directly. Don't allow yourself to get sidetracked or to ramble. If you find yourself giving long answers or engaging in a debate with an audience member, tell the audience you will make yourself available after your talk to answer in more detail.

Don't bluff. If you don't have the answer, say so. Virtually no one has all of the answers all of the time. If you know where the questioner might find the answer, feel free to offer to put them in touch with the resource.

Don't lose your cool. Never respond defensively, with irritation, or with anger. Such responses show that you have lost control of yourself and your presentation.

If the question can be restated positively, do so, answer it, and move on. If not, firmly, yet diplomatically, state that this is not the time or place for that debate and move on.

If someone asks about something explicitly covered in your presentation, answer it anyway. Perhaps you did not make the point clearly enough. If someone repeats a question that's already been asked, don't answer it again. "I believe we've already covered that," usually works. On the other hand, if the second question indicates that your first response was inadequate or confused the audience, take another stab at it.

If someone asks a totally irrelevant question, respond by saying that really is not part of your topic. However, (if you're feeling generous) you could say, "It sounds like an interesting subject." It is always a good idea to be polite and gracious.

POSTER PRESENTATIONS

For many young scientists, the first opportunity for a scientific talk is a poster presentation. Consider the poster primarily as an opportunity to exchange ideas and dialog and not exclusively as a forum for presenting data.

Posters should be clean and aesthetically pleasing. Include only relevant information. The most important information should be in the upper left hand corner—ideally an introduction that indicates why the work presented is important within the context of a scientific principle. Describe the approach in an engaging, condensed style without excessive detail. In the lower, right-hand corner, state a number of well phrased conclusions and a summary statement.

Posters should be large enough to convey the necessary information, usually 2 × 3- or 3 × 4-ft. is sufficient. As with slide presentations, choosing a font style, color scheme, and overall look, and staying within in it, will give your poster a professional and coherent appearance. Use charts, graphs, and photos to make your poster more visually interesting. Make sure that anything attached to the poster is securely and neatly fastened.

When presenting at a poster session, the speaker should be the focal point with the poster available to provide clarity and reference. Acknowledge the visitor and introduce yourself. Engage in a dialog to find out what they're most interested in learning more about and tailor your 3- to 5-min talk to their interests.

See *APPENDIX 5A* for more detailed information about preparing posters.

A5

CONCLUSION

Common sense plays an important role in preparing and giving a good scientific talk. Scientists have ample opportunities for comparing good and bad talks. Learn to recognize what makes a talk good when you are in an audience, just as you recognize what makes a bad one, and try to incorporate the things you admire in your future presentations.

LITERATURE CITED

Pike, R.W. 1992. Creative Training Techniques Handbook. Lakewood Books, Minneapolis, Minn.

KEY REFERENCES

Anholt, R.H.R. 1994. Preparing a Scientific Presentation. W.H. Freeman and Company, New York.

Boothman, N. 2000. How to Make People Like You in 90 Seconds or Less. Workman Publishing, New York.

Peters, T.J. and Waterman, R.M. Jr. 1982. In Search of Excellence. Warner Books, New York.

Pike, 1992. See above.

Ron, H. 1992. I Can See You Naked. Andrews and McMeel, Kansas City, Mo.
These books are good references for presenting scientific talks and public speaking in general.

INTERNET RESOURCES

http://www.public-speaking.org
The Advanced Public Speaking Institute has more than 100 articles in 20 categories, ranging from using humor to stage fright, related to public speaking.

http://www.toastmasters.org
Toastmasters International has many on-line resources for public speaking. In addition, they have thousands of clubs in most geographic areas, where individuals can practice public speaking and leadership skills.

http://www.speechtips.com
In addition to details about how to write and a deliver a speech for all occasions, sample speech scripts are available for review and purchase.

Index

C

D

E

F

G

H

I

J

K

L

M

N

O

P

Q

R

S

T

U

V

W

X

CURRENT
CP
PROTOCOLS